Thin-Layer Chromatography

A Laboratory Handbook

Edited by
Egon Stahl

Second Edition, Fully Revised and Expanded

Translated by M. R. F. Ashworth

With 241 Figures and 3 Plates in Color

Springer-Verlag New York · Heidelberg · Berlin 1969

English translation by
Professor M. R. F. Ashworth, Institut für organische Chemie
Universität des Saarlandes, Analytische Abteilung,
D-6600 Saarbrücken 15

ISBN 0-387-04736-0 Springer-Verlag New York Heidelberg Berlin
ISBN 3-540-04736-0 Springer-Verlag Berlin Heidelberg New York

First Edition (1965)
ISBN 0-387-03416-1 Springer-Verlag New York Heidelberg Berlin
ISBN 3-540-03416-1 Springer-Verlag Berlin Heidelberg New York

Translation of Dünnschicht-Chromatographie, 2. Auflage (1967)
ISBN 0-387-03761-6 Springer-Verlag New York Heidelberg Berlin
ISBN 3-540-03761-6 Springer-Verlag Berlin Heidelberg New York

This work is subject to copyright. All rights are reserved, whether the whole or part of the material is concerned, specifically those of translation, reprinting, re-use of illustrations, broadcasting, reproduction by photocopying machine or similar means, and storage in data banks. Under § 54 of the German Copyright Law whehre copies are made for other than private use, a fee is payable to the publisher, the amount of the fee to be determined by agreement with the publisher. © by Springer-Verlag Berlin · Heidelberg 1965, 1969. Library of Congress Catalog Card Number 69-14538. Printed in Germany. The use of general descriptive names, trade names, trade marks, etc. in this publication, even if the former are not especially identified, is not to be taken as a sign that such names, as understood by the Trade Marks and Merchandise Marks Act, may accordingly be used freely by anyone.

Foreword to the 2nd Edition

The first edition, which came out four years ago, is recognised as a standard work on thin-layer chromatography. The subject has developed rapidly in the meantime and several thousand articles have since been published. With the exemplary cooperation of an appreciably widened group of contributors, the second edition has assimilated this mass of new data. The layout and presentation of the book, acclaimed as satisfactory, have been retained but new chapters and sections incorporated. Special attention has been devoted to the adsorbents, an integral part of the method. The new gradient-, transfer- and coupling procedures, preparative TLC, direct quantitative evaluation, reactions at the start point and new isotope techniques have likewise been included. More space has been allotted in the special section to application of TLC in clinical diagnosis, in investigations of foodstuffs and in the analysis of products of the organic chemical industry. The number of reagents has been increased to 264.

Almost all chapters have been rewritten and only a small number of the former figures and tables taken over. Despite strenuous contraction, the number of pages has almost doubled; a fourfold content of information has been provided, however. In other words, it is a new book which will be a reliable and indispensable laboratory aid for both beginner and specialist.

My thanks are due to all who have helped towards the reshaping of the book.

Saarbrücken, 18/10/1966   EGON STAHL

Translator's Note

As far as possible I have adopted English spelling usage, as befits my nationality. It has, however, often been necessary to consult American sources such as Chemical Abstracts or the Handbook of Chemistry and Physics (and, of course, American journals containing original work); some compounds, especially those which are less familiar, may thus turn out to be spelt in the American way. Moreover, I have left ChapterX, written by two American contributors, almost wholly unchanged. I hope that the resulting "mid-Atlantic" mixture will not succeed merely in distressing purists and devotees of consistency on both sides of that ocean.

To use or not to use capital letters: trade names force the writer into a Hamlet rôle. No decision has been necessary with many pharmaceuticals, antibiotics etc. which appear only in tables. I have used chiefly small letters, departing from this very rough rule with adsorbents, dyes, insecticides and other miscellaneous products which are newer and less "classical"; thus: silica gel, methyl red, aldrin but Sephadex, Palatine Fast Green, Rogor. Here, too, I hope to have offended neither firms nor individuals.

I should like to thank contributors in particular for most valuable help with terminology in their special subjects.

Saarbrücken, April 1969   M. R. F. Ashworth

Table of Contents

General Section

- **A. The Historical Development of the Method.** Egon Stahl 1
- **B. Adsorbents for TLC.** Egon Stahl 6
  - I. Silica Gel. H. W. Kohlschütter and K. Unger 7
    1. Formation of Silica Gel and Construction of its Framework . . . 8
    2. The Cavity System of Silica Gel 13
        - a) Capillary and Firmly Combined Water 20
        - b) Re-formation of Firmly Combined Water 21
  - II. Aluminas and Other Inorganic Adsorbents. H. Rössler 23
    1. Alumina 23
    2. Kieselguhr 27
    3. Silicates 29
        - a) Magnesium Silicate 29
        - b) Calcium Silicate 29
    4. Phosphates 30
    5. Calcium Sulphate 30
    6. Glass Powder 30
    7. Salts of Heteropoly-, Tungstic, Molybdic and Tetraboric Acids . 31
    8. Ferric and Chromic Oxides 31
    9. Zinc Carbonate and Zinc Ferrocyanide 31
    10. Active Carbon 31
    11. Zirconium Phosphate and Hydrous Zirconium Oxide 32
    12. Lanthanum Oxide 32
    13. Bentonites 32
    14. Combinations of Adsorbents 32
  - III. Organic Adsorbents. P. Wollenweber 32
    1. Cellulose and Derivatives 32
        - a) Normal Cellulose Powder 33
        - b) Acetylated Cellulose Powder 37
        - c) Cellulose Ion-Exchanger Powder 37
    2. Starch 39
    3. Sucrose 40
    4. Mannitol 40
    5. Dextran Gels 40
  - IV. Polyamides as Adsorbents. H. Endres 41
  - V. Ion Exchangers in TLC. K. Dorfner 44
  - VI. Modification of Adsorbents by Impregnation. Egon Stahl 48
    1. Impregnation before Coating 48
    2. Impregnation of the Ready, Dry Layer 48

C. Apparatus and General Techniques in TLC. EGON STAHL 52

I. Preparation of Thin, Uniform Layers 52
   1. Pouring Procedures 53
   2. Immersion Procedures 53
   3. Spreading Procedures 54
      a) Fixed Spreader (Kirchner-type) 54
      b) Movable Spreader (Stahl-type) 55
      c) Spreading Rods and Home-made Apparatus 56
   4. Spraying Procedures 57
   5. Automatic Coating 58
   6. Ready-coated Plates and Sheets 59

II. Preparation of TLC-Plates 60
   1. Drying, Storage and Handling 60
   2. Control and Marking of the Layer 61
   3. Application of the Substance Mixture for Separation 63
      a) Spot Application 63
      b) Band or Streak Application 64

III. Separation Chambers and Development 65
   1. Chamber Saturation and "Equilibration" of the Layer 66
   2. Setting up the Chambers 68
   3. Chambers for Ascending Development 68
      a) Rectangular Chambers 68
      b) S-Chambers 69
      c) Moisture Chambers 71
   4. Equipment for Descending TLC 72
   5. Devices for Horizontal Development 73
      a) Circular Technique 73
      b) Centrifugal Technique 75
      c) Horizontal Development, BN-Chamber and Automatisation . . 75

IV. Aids for Visualisation of Colourless Substances on Chromatograms . . 77
   1. UV-Lamps for Exciting Fluorescence 77
   2. Spray Apparatus and Fume Cupboards 79
   3. Apparatus for Heating the TLC-Plates 81
   4. Further Aids for Visualisation 81
      a) Direct Microsublimation 81
      b) Self-recording Measurement of UV-absorption 82
      c) Biological Visualisation Procedures 82

V. Laboratory and Basic Equipment for TLC 83
   1. Small Laboratories 83
   2. Medium-Sized Research and Control Laboratories 84
   3. Central TLC- and PC-Laboratories in Industry 84

VI. Standard Conditions in TLC 85

D. Special Techniques in TLC. EGON STAHL 86

I. Special Development Procedures 86
      a) Continuous Development 86
      b) Multiple Development 86
      c) Stepwise Development 87

     1. Two-Dimensional Separation and SRS-Technique 88
     2. Multidimensional Technique of Von Arx and Neher 88
     3. Wedged-tip Technique 89
  II. Gradient-Techniques in TLC 89
     1. Gradient-Elution 90
     2. TLC on Gradient Layers (Gradient-TLC) 91
  III. Temperature-TLC 94
     1. Apparatus for Temperature-TLC 95
     2. Possibilities of Application 96
  IV. Preparative TLC 97
     1. Improvement of Separation in Column Chromatography 97
     2. Adaptation of TLC to Larger Applied Amounts 98
        a) Increase of the Layer Thickness and Plate Size 98
        b) Band Application of Substance Solutions 98
        c) Types of Development 100
        d) Visualisation of the Separated Zones 100
        e) Collecting the Zones and Extraction 101
  V. Transfer Techniques 102
     1. Transfer: TLC → Solution 102
     2. Transfer: TLC → Reagent 104
     3. Transfer: TLC → Paper 104
     4. Transfer: Solution → Paper (PC) → TLC 104
     5. Transfer: Gas Chromatography → TLC 105

E. **Thin-Layer Electrophoresis.** K. Hannig and G. Pascher 105

  I. Principle of Thin-Layer Electrophoresis 105
  II. Theoretical Principles 106
     1. General 106
     2. Factors Influencing the Mobility of Particles in Electrophoresis
        with Supporting Medium (Thin-Layer Electrophoresis) 106
        a) Adsorption 107
        b) Diffusion 107
        c) Zone Flaws 107
        d) Zeta-potential and Electroosmosis 107
        e) "Suction Effects" 108
     3. Units of Reference 108
  III. Details of Apparatus and Methods 109
     1. Apparatus 109
     2. Methods 111
        a) Supporting Layers and Preparation of the Plates 111
        b) Buffer Solutions 111
        c) Current 111
        d) One-dimensional Thin-Layer Electrophoresis 113
        e) Two-dimensional Thin-Layer Electrophoresis 113
        f) Finger Print Technique 114

F. **Coupling of Gas- and Thin-Layer Chromatography.** R. Kaiser 114

  I. Principle of the GC-TLC-Coupling Analysis 115
  II. Equipment 118

III. Information from Coupling Analysis 120
IV. Details of Apparatus and Methods 121
    a) Deposition 121
    b) Example 122
V. Examples of Application 124

G. Documentation of Thin-Layer Chromatograms. H. GÄNSHIRT 125

I. Rf-values in TLC 127
II. Preservation of Thin-Layer Chromatograms 127
III. Graphical Copying 128
IV. Documentation with Light-Sensitive Paper 129
    1. Copies with Diazo-type Paper 129
    2. Copies with Blue-Print Paper 130
    3. Positive Copies and Photographs with Procedures Based on
       Diffusion of Silver Salts 130
    4. Photography 131
       a) General Points 132
       b) Black and White Photographs with UV-Source 366 or 254 nm . 132
       c) Colour Photographs with UV Light Sources 133
    5. Electrophotography 133

H. Quantitative Evaluation of Thin-Layer Chromatograms. H. GÄNSHIRT . . 133

I. Quantitative Evaluation on the TLC-Layers 135
    1. Visual Comparison 135
    2. Determination by Measurement of Spot Areas 135
    3. Transmission Measurements on Spots which are Coloured, Charred,
       or which absorb UV-Light 138
    4. Quantitative Determination of Fluorescing Spots 141
       a) Direct Evaluation 141
       b) Photographic Evaluation 142
    5. Evaluation of Reflectance (Remission) Spectra 142
II. Quantitative Determination after Extraction from the Adsorbent
    Layer 145
    1. Detection of the Separated Substances 145
       a) Use of Reference Chromatograms 145
       b) Use of Fluorescent Layers and Indicators 146
       c) Non-Destructive Detection by Spraying with Water 147
       d) Localisation of Separated Spots with Iodine Vapour or by
          Spraying with Iodine Solution 147
       e) Use of other Colour Reagents 148
    2. Removal of Spots from the Plate and Elution Technique . . . 148
    3. Eluents (Solvents for Extraction) 149
    4. Methods of Determination Used after Elution 151

I. Isotope Technique. HELMUT K. MANGOLD 155

I. Layers, Solvents and Chemical Methods of Detection 156
II. Procedures for Detection and for Measurement of Radioactivity . . 157
    1. Autoradiography of Thin-Layer Chromatograms 157
    2. Counting Tubes and Scintillation Counters 160

Table of Contents   XI

  III. Preparation of Radioactively Labelled Substances 167
  IV. Isolation of Radioactive Compounds by TLC 168
   V. Analysis by Means of Radioisotopes 170
       1. Indicator Analysis 171
       2. Isotope Dilution Method 171
       3. Activation Analysis 172
       4. Isotopic Derivative Method 172
          a) Fractionation before Radioactive Labelling 173
          b) Separation of Radioactive Derivatives 173
          c) Fractionation after Addition of a Radioactive Derivative to
             the Mixture of Non-labelled Derivatives 174
          d) Separation after Addition of an Inactive Derivative to the
             Mixture Containing the Radioactively Labelled Derivative of
             the Compound to be Determined 174
          e) Use of Two Radioactive Isotopes 174
  VI. Directions for Radioactive Labelling 175
       1. Esterification of Acids with Diazomethane ($^{14}CH_2N_2$) 175
       2. Acetylation of Alcohols with Acetic Anhydride ($^{14}CH_3CO)_2O$ or
          $(C^3H_3CO)_2O$ 175
 VII. Applications of TLC in Chemical and Biochemical Investigations
       with Radioisotopes 176

Theoretical Fundamentals of TLC 179
Bibliography of the General Section, Chapters A—I 180

Special Section

Introduction. EGON STAHL 201

J. Terpene Derivatives, Essential Oils, Balsams and Resins. EGON STAHL and
   H. JORK 206
   I. Separation of Lipophilic, Steam-volatile Mixtures 207
  II. Chromatographic Separation of Lipophilic, Steam-volatile Mixtures . 208
       1. Mono- and Sesquiterpene Hydrocarbons 210
       2. Oxides, Epoxides and Peroxides 212
       3. Steam-volatile Esters and Lactones 214
       4. Aldehydes and Ketones 217
          a) Separation of Free Carbonyl Compounds 217
          b) Separation of Derivatives of Aldehydes and Ketones 219
       5. Terpene and Sesquiterpene Alcohols 225
          a) TLC on Silica Gel Layers 225
          b) Paraffin-impregnated Silica Gel Layers 226
          c) Silver Nitrate-impregnated Silica Gel Layers 227
          d) Separation of the Dinitrobenzoate Esters (DNBs) 228
       6. Phenylpropane and Phenol Derivatives 229
          a) Silica Gel Layers 229
          b) Structure and hR_f-Value 231
          c) TLC of Phenol Esters, Coupling Derivatives and Other Con-
             densation Products 231
          d) Separation of the Methoxyallylbenzenes from their *Cis-Trans*
             Propenyl Isomers 233

III. Essential Oils 235
   1. Mixtures of Terpene and Sesquiterpene Derivatives 235
   2. Sulphur-Containing Oils 235
   3. Essential Oils with Polyacetylene Compounds 235
   4. Supplement 240
IV. TLC of Involatile Terpene Derivatives 240
   1. Diterpenes 240
   2. Triterpene Derivatives and their Glycosides 241
      a) Neutral Triterpenes 241
      b) Triterpene Acids 243
      c) Triterpene Glycosides 245
   3. Polyterpenes 247
V. Balsams and Resins 247
Bibliography for Chapter J. Terpene Derivatives 250

K. **Vitamins, Including Carotenoids, Chlorophylls and Biologically Active Quinones.** H. R. BOLLIGER and A. KÖNIG 259
   I. Method of Work and General Experience 259
   II. TLC of Fat-Soluble Vitamins, Carotenoids, Chlorophylls and Quinones 263
      1. Mixtures of Fat-Soluble Vitamins 263
         a) Separation 263
         b) Detection and Determination 265
      2. Carotenoids and Chlorophylls 266
         a) Separation 266
         b) Detection and Determination 272
      3. Vitamin A Group 273
         a) Separation 273
         b) Detection and Determination 275
      4. Vitamin D Group 275
         a) Separation 276
         b) Detection and Determination 278
         c) Tested Assay of Vitamin D 280
      5. Vitamin E Group 283
         a) Separation 284
         b) Detection and Determination 286
      6. Vitamin K Group and Related Quinones 288
         a) Separation 288
         b) Detection and Determination 291
   III. TLC of Water-Soluble Vitamins 292
      1. Mixtures of Water-Soluble Vitamins 292
         a) Separation 292
         b) Detection 294
      2. Vitamin B_1 Group 295
         a) Separation 295
         b) Detection and Determination 296
      3. Vitamin B_2 Group 296
         a) Separation 296
         b) Detection and Determination 297

4. Pantothenic Acid Group		297
a) Separation		298
b) Detection and Determination		298
5. Nicotinic Acid and Nicotinamide		299
a) Separation		299
b) Detection and Determination		299
6. Vitamin B_6 Group		300
a) Separation		300
b) Detection and Determination		301
7. Vitamin B_{12} Group		301
a) Separation		302
b) Detection		303
8. Folic Acid Group		303
a) Separation		303
b) Detection and Determination		304
9. Vitamin C		304
a) Separation		304
b) Detection and Determination		306
10. Biotin		306
a) Separation		306
b) Detection and Determination		307
11. Other Vitamins		307
Bibliography for Chapter K. Vitamins		308

L. TLC of Steroids and Related Compounds. R. Neher ... 311

   I. Nomenclature ... 312
   II. Range of Application of TLC of Steroids in Comparison with other Chromatographic Methods ... 313
   III. General Conditions ... 316
      1. Adsorbents ... 316
      2. Solvents ... 317
      3. Procedures ... 320
      4. Reactions for Detection ... 322
      5. Formation of Derivatives (Microreactions) ... 325
   IV. Structure and Chromatographic Behaviour ... 326
      1. Adsorption TLC ... 326
      2. Partition TLC ... 328
      3. Improvement of the Separation of Similar, Polyfunctional Steroids 328
   V. Sterols ... 329
   VI. Neutral C_{18}-C_{22}-Steroids ... 334
   VII. Cardiac Glycosides and Aglycones ... 341
   VIII. Saponins and Sapogenins ... 346
   IX. Aminosteroids, Steroid Alkaloids and Glycosides ... 348
   X. Phenolic Steroids (Oestrogens) ... 349
   XI. Bile Acids and Conjugates, Steroid Carboxylic Acids and Steroid Conjugates ... 351
      1. Bile Alcohols ... 355
      2. Steroid Carboxylic Acids ... 355
      3. Steroid Conjugates ... 355
Bibliography for Chapter L. Steroids ... 357

M. Aliphatic Lipids. HELMUT K. MANGOLD 363

    I. Introduction 363
        1. Neutral Lipids and their Hydrolysis Products 363
        2. Phospholipids, Sulpholipids and Glycolipids 363
        3. Older Methods of Lipid Analysis 366
        4. Newer Procedures for Separation of Lipids 367
        5. Processing of the Material for Analysis 368
            a) Homogenisation and Extraction 368
            b) Separation of Lipids from Non-Lipids 370
            c) Degradation of Lipids and Preparation of Derivatives 370
            d) Other Reactions 374
        6. Manufacturers and Suppliers of Pure Lipids 374

    II. Thin-Layer Chromatography of Lipids 374
        1. Separation of Lipids According to Compound Class 374
            a) Neutral Lipids and Their Hydrolysis Products 375
            b) Phospholipids, Sulpholipids and Glycolipids 388
        2. Fractionation of Pure Compound Classes 394
            a) Argentation Chromatography 396
            b) Chromatography of Mercuric Acetate Adducts 402
            c) Chromatography of Ozonides 406
            d) Reversed Phase Partition Chromatography 409
        3. Quantitative Evaluation of Thin-Layer Chromatograms 414
            a) Neutral Lipids and Their Hydrolysis Products 414
            b) Phospholipids, Sulpholipids and Glycolipids 415

Bibliography for Chapter M. Aliphatic Lipids 415

N. Alkaloids. F. ŠANTAVÝ 421

    I. Adsorbent Layers and Solvents for TLC 421
    II. Visualisation of the Alkaloids 423
    III. Separation Scheme for Alkaloids 424
    IV. Quantitative Determination of Alkaloids with TLC 425
    V. Special Section 425
        1. Colchicine Alkaloids 425
        2. Pyrrolidine, Pyridine and Piperidine Alkaloids 430
        3. Tropane Alkaloids 432
        4. Pyrrolizidine Alkaloids 435
        5. Quinolizidine Alkaloids 435
        6. Alkaloids of the Papaveraceae 436
        7. Bis(benzylisoquinoline) Alkaloids 444
        8. Ipecacuanha Alkaloids 444
        9. Amaryllidaceae Alkaloids 445
        10. Indole Alkaloids 446
            a) Indolyl Alkylamines 446
            b) Mavacurine, Fluorocurine, Ellipticine, Eburnamine, Aspidospermine and Strychnine Alkaloids 447
            c) Rauwolfia Alkaloids 448
            d) Vinca Alkaloids (Catharanthus Alkaloids) 449
            e) Ergot Alkaloids 450
            f) Oxindole Alkaloids 454

	11. Cinchona Alkaloids	454
	12. Furanoquinoline Alkaloids	456
	13. Purine Alkaloids	457
	14. Sterol Alkaloids	457
	a) Alkaloids of the Genera Solanum and Lycopersicum	458
	b) Alkaloids of the Genera Holarrhena and Funtumia	459
	c) Alkaloids of the Benzofluorene Type	461

General Bibliography on the Alkaloids 461

Special Bibliography for Chapter N. Alkaloids 462

O. "Simple" Indole Derivatives and Plant Growth Regulators. Urine Metabolites, Auxins, Gibberellins and Cytokinins. HARALD KALDEWEY 471

I. Introduction 471

II. Preparation of the Material for Analysis 473
  1. General Information 473
  2. Free Auxins 474
  3. Diffusible Auxins 474
  4. Indole Derivatives in Urine 474
  5. Gibberellins 475
  6. Cytokinins 475

III. Adsorbent and Solvents 475
  1. General Information 475
  2. Auxins and Urine Metabolites 476
  3. Gibberellins 479

IV. Multiple Development, Stepwise Development and Two-Dimensional Separation 481

V. Visualisation and Identification 483
  1. Chemical and Physical Methods of Detection 483
  2. Biological Methods of Detection 487

Bibliography for Chapter O. "Simple" Indole Derivatives 489

P. Amines and Tar Bases. EGON STAHL and P. J. SCHORN 494

I. Amines 494
  1. Aliphatic Amines 494
  2. Nitrosamines 497
  3. Aminoalcohols and Quaternary Ammonium Salts 498
  4. Catechol Amines (Phenylalkylamines) 499
  5. Aromatic and Heterocyclic Amines 500
  6. Thin-Layer Electrophoresis of Amines 502

II. Tar Bases 503

Bibliography for Chapter P. Amines and Tar Bases 505

Q. Synthetic Pharmaceutical Products. HERBERT GÄNSHIRT 506

  1. Antihistamines, Anti-Allergics and Structurally Related Compounds with Psychic Activity 507
    a) Phenothiazines and Diazepines 508
    b) Antihistamines 518

Table of Contents

    2. Analeptics, Psychotherapeutic Agents with Antidepressive Activity, Appetite Depressants and Some Carbamate Esters of Varied Activity 518
    3. Sympathomimetics of the Adrenaline Type 520
    4. Analgesics, Antipyretics and Antirheumatic Agents 524
        a) p-Aminophenol Derivatives, Pyrazolones etc. 524
        b) Analgesics with Narcotic Activity 527
    5. Anticoagulants of the 4-Hydroxycoumarin Group 532
    6. Hypnotics 533
        a) Barbiturates 533
        b) Hydantoins 537
        c) Bromoureides 538
        d) Other Hypnotics 538
    7. Bactericidal and Bacteriostatic Substances 540
        a) Phenols of Pharmaceutical Interest 540
        b) Sulphonamides 541
        c) Chemotherapeutic Agents of the Nitrofuran Series, etc. . . . 547
    8. Diuretics 547
    9. Purine Derivatives of Various Activities 548
    10. Oral Antidiabetic Agents 550
    11. Laxatives 551
    12. Local Anaesthetics 551
    13. Miscellaneous Other Active Substances 554

  I. Analysis of Various Drug Forms and of Commercial Preparations . . 554

  II. Stability Tests on Pharmaceutical Materials 558

Bibliography for Chapter Q. Synthetic Pharmaceutical Products 562

R. Antibiotics. K. H. WALLHÄUSSER 566

    1. Execution of the Microbiological Test 568
        a) Direct Method 570
        b) Contact Method (Reprint or Filter Paper Print Process) . . . 570
    2. Execution of the Chemical Test 570
    3. General Information about Layers and Solvents Used 570

  I. Polyenes and Polyacetylenes 571

  II. Macrolides 572

  III. Tetracyclines 573

  IV. Substances with Similar Structural Units 574

  V. Basic, Water-Soluble, Non-extractable Antibiotics 574

  VI. Nucleic Acid Derivatives 575

  VII. Acyclic Compounds 575

  VIII. Heterocyclic Compounds 575

  IX. Macrocyclic Peptides 576

  X. Other Peptides 576

  XI. Miscellaneous Antibiotics 576

Bibliography for Chapter R. Antibiotics 577

S. TLC in Clinical Diagnosis. NEPOMUK ZÖLLNER and GÜNTHER WOLFRAM . 578

Introduction 578

I. Investigation of Endogenous Substances 579
   1. Sugars, Their Derivatives and Metabolites 579
       a) Glycoproteins 580
       b) Ketone Bodies 581
   2. Amino Acids, Their Derivatives and Metabolites 581
       a) Amino Acids in Urine 582
       b) Amino Acids in Blood and Organs 585
       c) Amino Acids in Other Body Fluids 586
       d) Iodoamino Acids 586
       e) ε-Aminocaproic Acid 587
       f) Creatinine 588
       g) Amines and Metabolites 589
       h) Serum Proteins 591
   3. Lipids and Related Compounds 592
       a) Neutral Fat in Plasma 593
       b) Cholesteryl Esters in Plasma 593
       c) Phospholipids in Plasma 596
       d) Methods for Separating Other Medically Important Lipids . . 597
       e) Lipids in Secreted and Excreted Material 597
       f) Lipids in Tissue 598
   4. Steroids 599
       a) C_{21}-Steroids 600
       b) C_{19}-Steroids 602
       c) C_{18}-Steroids 602
       d) Bile Acids 603
       e) Sterols in Faeces 603
   5. Porphyrins and Metabolites 604
II. Investigations of Exogenous Substances 604
   1. Function Tests 605
   2. Poisoning 605
   3. Control of Therapy 606
Bibliography for Chapter S. TLC in Clinical Diagnosis 607

TF. Synthetic Colouring Materials. H. Schweppe 612
  I. Solvent Dyes 612
      Special Applications 615
       a) Dyes in Motor Fuels 615
       b) Colouring Matter in Naturally Occurring Fats and Oils 616
       c) Dyes in Polystyrene 616
 II. Disperse Dyes 616
III. Organic Pigments 617
IV. Basic Dyes 618
 V. Acid Dyes 620
VI. Direct Dyes 621
VII. Reactive Dyes 622
VIII. Metal Complex Dyes 623
IX. Synthetic Food Colorants 624

XVIII Table of Contents

    X. Dye Intermediates 626

Bibliography for Chapter TF. Synthetic Colouring Materials 628

TN. Foodstuffs and Their Additives. J. W. Copius-Peereboom 630

    I. General Applications 630
    II. Antioxidants 631
    III. Preservatives 636
    IV. Pesticides 638
        1. Phosphate Esters 639
        2. Chlorinated Hydrocarbons 642
        3. Pyrethrins and Synergists 645
        4. Herbicides 647
    V. Artificial Sweeteners 648
    VI. Emulsifiers and Swelling Agents 649
    VII. Alcohols and Glycols 650
    VIII. Organic Acids 650

Bibliography for Chapter TN. Foodstuffs and Their Additives 654

TS. Synthetic Organic Products. H.-J. Petrowitz 657

    I. Plastics and Plasticisers 657
        1. Polymers and Polymerisable Compounds 657
        1a. Urethanes 658
        2. Plasticisers 659
        3. Alcohols 660
            a) Simple Alcohols 660
            b) Polyalcohols 662
        4. Phenols 663
        5. Other Auxiliary Materials in the Plastics Industry 664
    II. Organo-Metallic Compounds 664
        1. Organo-Tin Compounds 664
        2. Ferrocenes 665
        3. Other Organo-Metallic Compounds 666
    III. Polyphenyls and Fused Polynuclear Aromatic Hydrocarbons 666
        1. Polyphenyls 666
        2. Polynuclear Aromatic Hydrocarbons 667
    IV. Explosives 669
    V. Industrial Additives 673
        1. Inhibitors and Antioxidants 673
        2. Detergents 674
        3. Optical Brighteners 675
        4. Wood Protective Agents 675
        5. Chemicals for Photography 676
        6. Special Mineral Oil Analyses 676
    VI. Intermediates in Organic Synthesis 677

Bibliography for Chapter TS. Synthetic Organic Products 684

U. Hydrophilic Plant Constituents and Their Derivatives 686

I. Plant Phenol Derivatives. KURT EGGER 687
1. Compound Classes and their Distribution 687
2. Concentration from Plant Material 690
   a) Acetone-extraction of Fresh Material 690
   b) Methanol-extraction of Dried Drugs 691
   c) Extraction in Stages 691
   d) Extraction with Water 691
3. Chromatographic Separation (without TLC) 691
4. Experimental Conditions for TLC 692
   a) Cellulose 692
   b) Silica Gel 693
   c) Polyamide and Other Polymers 699
   d) Polyacrylonitrile 704
   e) Ion Exchangers 704
5. Visualisation of Phenol Derivatives 705

II. TLC in the Characterisation of Animal and Plant Drugs. EGON STAHL and P. J. SCHORN 706
1. Anthraquinone Drugs 706
2. Lignan Drugs 709
3. Drugs with Phloroglucinol Derivatives 712
   a) Filix-Phloroglucinol Butanones 712
   b) Hop Bitter Principles 714
4. Drugs Containing Bitter Principles 715
5. Constituents of Hashish 715
6. Other Drugs and Mixtures of Natural Products 716

III. TLC as a Legally Binding Method for Characterisation of Drugs. EGON STAHL and P. J. SCHORN 720
1. General Information for Practical Directions 720
2. Two Examples of Special Procedures 722
   a) Rhizoma Filicis (Male Fern Rhizome) 722
   b) Radix Liquiritiae (Liquorice Root) 723

Bibliography for Chapter U. Hydrophilic Plant Constituents 724

V. Amino Acids and Derivatives. M. BRENNER, A. NIEDERWIESER and G. PATAKI 730

I. Introduction 730

II. General Technique 732
1. Preparation of the Layer 732
2. Chromatographic Techniques 734

III. Amino Acids 734
1. Preparation of the Solution for Investigation 734
2. Hydrolysis of Proteins and Peptides 735
3. Free Amino Acids in Biological Material 736
4. Solvents and Separation Efficiency 739
5. Detection of Amino Acids on the Chromatogram 746

IV. Peptides 751

V. N-(2,4-Dinitrophenyl)-amino Acids and 3-Phenyl-2-Thiohydantoins . 756
  A. Dinitrophenylamino Acids 756
    1. Dinitrophenylation 757
      a) Amino Acids........................ 757
      b) Peptides.......................... 758
      c) Polypeptides and Proteins 759
    2. Solvents and Separation Efficiency 760
      a) Solvents for Chromatography of Acid- and Water-soluble DNP-Amino Acids, not Extractable with Ether 762
      b) Solvents for Chromatography of Acid-insoluble DNP-Amino Acids, Extractable with Ether 763
    3. Documentation 771
  B. Phenylthiohydantoins 772
    1. Preparation of the Phenylthiocarbamyl-derivatives and their Conversion into PTH-Amino Acids 773
    2. Solvents and Separation Efficiency 774
    3. Detection of the Phenylthiohydantoins........... 777
  C. Other Amino Acid Derivatives 777
    1. Dinitropyridyl-Amino Acids 777
    2. 1-Dimethylamino-5-Naphthalenesulphonyl-Amino Acids (DANS-Amino Acids)....................... 778
    3. Carbobenzoxy Compounds 778
  D. Iodoamino Acids and Similar Compounds 779
Bibliography for Chapter V. Amino Acids and Derivatives 781

W. Nucleic Acids and Nucleotides. Helmut K. Mangold 786

  I. Introduction 786
    1. Nucleic Acids and their Hydrolysis Products 786
    2. Nucleotide-Coenzymes 788
    3. Older Methods of Nucleic Acid Analysis 788
    4. Newer Methods for Isolation of Nucleic Acids and Separation of Their Constituents 788
    5. Colour Reactions for Distinguishing Ribo- and Deoxyribonucleic Acids........................... 789
    6. Hydrolysis of Nucleic Acids 789
      a) Alkaline Hydrolysis 790
      b) Acid Hydrolysis..................... 790
      c) Enzymatic Hydrolysis 791
    7. UV-Spectra of Nucleic Acid Constituents 791
    8. Manufacturers and Suppliers of Pure Preparations 792
  II. Thin-Layer Chromatography of Nucleic Acids and their Constituents 792
    1. Purines, Pyrimidines and Nucleosides............ 792
    2. Nucleotides and Nucleotide-Coenzymes 794
    3. Oligonucleotides and Nucleic Acids 801
  III. Thin-Layer Electrophoresis of Hydrolysis Products of Nucleic Acids . 802
Bibliography for Chapter W. Nucleic Acids and Nucleotides 804

X. Sugars and Derivatives. B. A. Lewis and F. Smith 807

  I. Introduction 807

Table of Contents

II. Preparation of Plates 807
   1. Silica Gel G and Kieselguhr G Layers 808
   2. Kieselguhr G, Impregnated with Sodium Acetate 808
   3. Kieselguhr G, Impregnated with pH 5 Phosphate Buffer 808
   4. Impregnated Silica Gel G Layers 808
   5. Cellulose Layers (a) Cellulose MN 300 808
   5b. "Avirin" ("Avicel") 809
   6. ECTEOLA-Cellulose Layers for Sugar Phosphates 809
   7. Celite Layers (a) Filter-Cel and Hyflo Super-Cel 809
   7b. Celite 535-Starch 809

III. Visualization 810

IV. Chromatography of Sugars and Derivatives 811
   1. Sugars 812
      a) Separation of Sugars on Buffered Kieselguhr G Layers . . . 812
      b) Separation of Sugars on Buffered Silica Gel G 814
      c) Separation of Sugars on Cellulose 814
      d) Chromatography of Sugars on Miscellaneous Adsorbents . . . 815
   2. Oligosaccharides 816
   3. Amino Sugars 817
   4. Acids (Aldonic, Aldaric, Uronic and Saccharinic) 819
   5. Sugar Alcohols 820
   6. Methyl Glycosides 820
   7. Sugar Phosphates 821
   8. Acetates and Benzoates 822
   9. Hydrazones and Osazones 825
   10. Methyl Ethers 828
   11. Miscellaneous Derivatives 830

V. Monitoring Reactions by TLC 831

Bibliography for Chapter X. Sugars and Derivatives 834

Y. Inorganic Ions. H. SEILER 837

I. Preparation of the Solutions for Analysis 838

II. TLC of the Cations Separated in this Preliminary Stage 839
   1. Separation of the Cu-group (Solution I) 839
   2. Separation of the $(NH_4)_2$S-group (Solution II) 840
   3. Separation of the Ammonium Carbonate-group (Solution III) . . 841
   4. Separation of the Alkali-group (Solution IV) 842

III. Separation of Special Cation Mixtures 843
   1. UO_2^{2+} in a Mixture of Cations 844
   2. Ga^{3+} in Presence of Large Excess of Al^{3+} 844
   3. Sn, Cu, Hg, Pb, Bi, Cd and Zn as Dithizonates 844
   4. Ag, Pd, Au and Pt as Dithizonates 844
   5. Separation of Cations on Layers of Ion Exchangers 845
   6. Circular TLC of Cations 845
   7. Separation and Detection of Toxic Metals 846
      a) Qualitative Separation of Tl, Ni, Cu, Bi and Hg and of Ce,
         Ni, Cu, Be, Bi and Hg 846
      b) Determination of Hg 846

	Table of Contents

        8. Separation of *cis-trans* Co-Complex Isomers 846
        9. Separation of Radionuclides 847
   IV. Separation of Anions 847
        1. Separation of the Halides 847
        2. Separation of Phosphates 848
        3. Separation of Condensed Phosphates 848
        4. Separation of Sulphates and Polythionates 850
   V. Quantitative Determination 851

Bibliography for Chapter Y. Inorganic Ions 853

Z. Spray Reagents. K. G. KREBS, D. HEUSSER and H. WIMMER 854
   I. Preparation and Application of the Spray Reagents 855
   II. Compounds or Compound Classes and Reagents for their Detection . . 905
   III. Names and Abbreviations of Reagents 909

Conversion table for R_f into R_m and vice versa 910

Terms Frequently used in Thin-Layer Chromatography. HELMUT K. MANGOLD
  and M. BRENNER 912

List of Manufacturers and Suppliers 918

Author Index 924

Subject Index 1023

Contributing Authors*

BOLLIGER, H. R., Dr., Hoffmann-La Roche u. Co., AG., CH — 4000 Basel

BRENNER, M., Prof. Dr., Institut für Organische Chemie der Universität Basel, St. Johanns-Ring 19, CH — 4000 Basel

COPIUS-PEEREBOOM, J. W., Dr., Gouvernment Dairy Station, Vreewijkstraat 12 B Leiden/Niederlande

DORFNER, K., Dr., Badische Anilin- und Sodafabrik AG., 6700 Ludwigshafen

EGGER, K., Priv.-Doz. Dr., Botanisches Institut der Universität, 6900 Heidelberg, Hofmeisterweg 4

ENDRES, H., Priv.-Doz. Dr., Badische Anilin- und Sodafabrik AG, Abt. AWETA/Leder, 6700 Ludwigshafen

GÄNSHIRT, HERBERT, Dr., Farbenfabriken Bayer AG, 5090 Leverkusen

HANNIG, K., Priv.-Doz. Dr., Max-Planck-Institut für Eiweiß- und Lederforschung 8000 München 15, Schillerstraße 42—46

KAISER, R., Dr., Badische Anilin- und Sodafabrik AG, Analytisches Labor M 310, 6700 Ludwigshafen

KALDEWEY, H., Prof. Dr., Botanisches Institut der Universität des Saarlandes, 6600 Saarbrücken 15

KOHLSCHÜTTER, H. W., Prof. Dr., Eduard-Zintl-Institut für Anorganische und Physikalische Chemie der Technischen Hochschule Darmstadt, Lehrstuhl für Anorganische und Analytische Chemie, 6100 Darmstadt, Hochschulstr. 4

KREBS, K. G., Prof. Dr., Direktor des Kontroll-Laboratoriums der E. Merck AG, 6100 Darmstadt

LEWIS, B. A., Dr., New York State College of Home Economics, A Statutory College of the State University, Cornell University Ithaca, NY 14850/USA

MANGOLD, H. K., Prof. Dr., The Hormel Institute, University of Minnesota, 801 16th Avenue N. E., P. O. Box 367, Austin, Minn. 55912/USA

NEHER, R., Dr., Ciba Aktiengesellschaft, CH — 4000 Basel

PETROWITZ, H. J., Dr.-Ing., Bundesanstalt für Materialprüfung, Fachgruppe 2,4 „Biologische Materialprüfung, Holzschutz und Holztechnologie" 1000 Berlin-Dahlem, Unter den Eichen 87

RÖSSLER, H., Dr., E. Merck AG, Forschungsabteilungen, Hauptlaboratorium, 6100 Darmstadt

ŠANTAVÝ, F., Prof. Dr., Chemisches Institut der Medizinischen Fakultät, Palacký Universität, Olomouc/Tschechoslowakei, Hněvotinska 3

---

* The names of the co-workers who collaborated with the various authors are given in the individual chapter headings.

SEILER, H., Priv.-Doz. Dr., Institut für Anorganische Chemie an der Universität Basel, CH — 4000 Basel, Spitalstraße 51

SCHWEPPE, H., Dr., Badische Anilin- und Sodafabrik AG, Abt. AWETA I, 6700 Ludwigshafen

SMITH, F., Prof. Dr. †, University of Minnesota, Department of Biochemistry, Synder Hall, St. Paul, Minnesota/USA

STAHL, EGON, Prof. Dr., Direktor des Instituts für Pharmakognosie und Analytische Phytochemie der Universität des Saarlandes, 6600 Saarbrücken 15

WALLHÄUSSER, K. H., Dr., Farbwerke Hoechst AG, Mikrobiologisches Untersuchungslabor, 6230 Frankfurt-Hoechst

WOLLENWEBER, P., Dr., Machery, Nagel u. Co., 5160 Düren

ZÖLLNER, N., Prof. Dr., Medizinische Poliklinik der Universität München, 8000 München 15, Pettenkoferstraße 6a

General Section

A. The Historical Development of the Method

EGON STAHL

It may at first cause surprise that the principle of thin-layer chromatography was described as long as 25 years ago, yet has become of general value only in the last few years[1]. The first publication in this connection came out during the period of peak success of the Tswett column chromatography. Constant efforts were being made even at that time to achieve "micro-chromatography". ZECHMEISTER [781] summed up the situation in 1938 as follows: "The chief problem is not the development of suitable apparatus (referring to column chromatography) but the precise identification of the adsorbed substances". This problem was solved by the change-over from the "closed" to the "open" column, i. e., to the thin layer. This formulation, dating from 1958, demonstrated the general simplicity and wide applicability of thin-layer chromatography [661].

In 1938, N. A. ISMAILOV and M. S. SHRAIBER [313][2] described the basic principle of the procedure in an article entitled "Analysis by Drop-chromatography and its Application in Pharmacy". They applied the method to the separation and characterisation of extracts of medicinal plants (tinctures of the Soviet Pharmacopoeia VII).

They slurried the adsorbent (usually alumina) with water and spread this in a layer 2 mm thick on to glass microscope slides. Cracking was observed with thicker

---

[1] The development of other methods, even in wholly different domains, shows numerous parallels. The comment of a well known critic "only amateurs consider that an idea must be brand new in order to be good; important in reality is not he who first had the idea but he who expressed it better" is almost always valid.

[2] Prof. NICOLAI ARKADEVIC ISMAILOV (picture in Fig. 1 of [674]) was head of the physical chemistry laboratory in the Institute for Pharmaceutical Chemistry in Kharkov. He was born on 22. 6. 1907 and died on 2. 10. 1961 and was one of the best-known Soviet specialists in the field of solution electrochemistry, with about 240 publications to his name. He was a member of the Scientific Academy of the Ukrainian SSR and had received the Mendeleev Prize. The co-author, Dr. MARIA SEMENOVNA SHRAIBER, born on 11. 9. 1904, is still working today in the above named Institute. She has published over 50 papers on problems of pharmaceutical analysis (complexometry, paper chromatography and titrations in non-aqueous solutions, etc.) [511, 567, 633].

layers whereas they reported that thinner layers were technically impracticable. One drop of the alcoholic plant extract under examination was applied to the centre of the dried layer. Handsome "ultra-chromatograms", consisting of concentric rings, were frequently yielded; in other cases, alcohol was added until the zones were separated. They then compared their ring chromatograms with the corresponding column chromatograms (Fig. 1; cf. with Fig. 31) and pointed out the advantages of the new method.

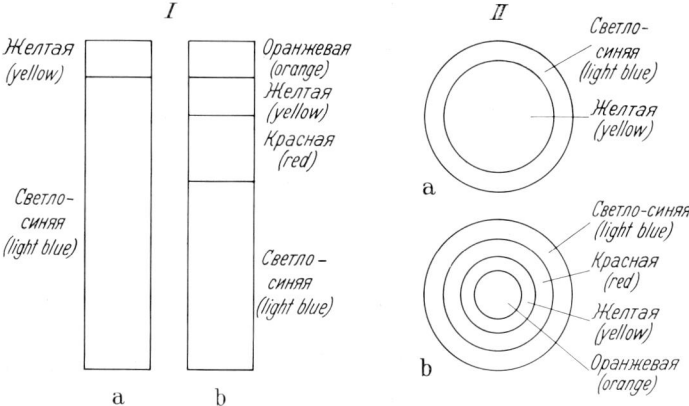

Fig. 1. Comparison of the fluorescence colours from a Belladonna tincture on an alumina column chromatogram (*I*) with those on a "drop chromatogram" (*II*) before (*a*) and after (*b*) development with alcohol. This diagram is taken from the work of ISMAILOV and SHRAIBER [313], the translation of the fluorescence colours having been added

The Russian article ends with the following summary:

"A method for chromatographic adsorption analysis is elaborated, based on the observation of the division of substances into zones on a thin layer of adsorbent, using one drop of the substance.

The results obtained by the method proposed are qualitatively the same as those obtained by the usual chromatographic adsorption method of analysis. The method enables to obtain satisfactory results using one drop of the substance under test, very small quantities of the adsorbent and minimal time.

The method may be used for the evaluation of galenical preparations and their identification as well as for preliminary test of the adsorbent and the kind of the developer. Sixteen galenical preparations are studied using the method proposed."

In 1941, CROWE [140], quoting this and other earlier publications, reported that his team had been using for some time thin layers of adsorbents in Petri dishes and, with this technique, had been able to ascertain rapidly the suitable solvent for elution in column chromatography.

Two procedures can be distinguished: one using firmly adhering layers, the other with loose, non-adhering layers. An improved form of the latter procedure was described in 1947 by WILLIAMS [759]. He protected the loose layer with a glass coverplate containing a central perforation through which the mixture to be separated and subsequently the eluent was added dropwise.

Several years earlier, MARTIN and SYNGE [433] had developed another procedure, namely partition chromatography, to separate amino acids and their derivatives. They fixed one phase (the stationary phase) on a support like powdered silica gel and introduced it into a column. As eluent, referred to as the "mobile phase", they used solvents like chloroform, for example.

In order to carry out partition chromatography on a micro scale, filter paper, i. e., an "open" column was then used (CONSDEN, GORDON and MARTIN [136] in 1944). Their results in the amino acid field found wholesale recognition and their method was generally adopted. The golden era of paper chromatography began and, by 1956, over ten thousand publications on the use of this "universal" method had come out [263]. Attempts were continually made, understandably under the influence of these impressive successes, to overcome the difficulties that were still arising, by change of degree of impregnation, by using new solvent combinations, by phase reversal and by chemical modification of the cellulose fibres. Hopes were high also that all adsorption effects would be eliminated by the use of glass fibre paper.

Attempts were made to separate lipophile mixtures by adsorption chromatography on filter paper and, later, on glass fibre paper, impregnated with silica or alumina. KIRCHNER [347] in 1950 was one of the first to do this. It is not clear why, in collaboration with MILLER, he took up the 1948 work of MEINHARD and HALL [439] in which the technique of Ismailov-Shraiber had been referred to as "surface chromatography". Perhaps the results on paper impregnated with silica gel, were unsatisfactory even at that time. It is however a fact that KIRCHNER and MILLER were the first who investigated more thoroughly the separation of terpene derivatives on thin layers of adsorbent and published their results in several papers [348, 349, 446—449]. They applied the method, termed "Chromatostrip-technique", to the separation of terpene derivatives and numerous publications appeared subsequently. One of these later authors (REITSEMA [571, 572]) in 1954 used broader carrier plates (12.5 × 17.8 cm) instead of the narrow glass strips and was able to separate several mixtures at a time, as in paper chromatography. It remains astonishing that practically all authors subsequently abandoned the method. Possibly this is related to the fact that the least satisfactory results were obtained in the separation of essential oils; ignorance of the controlling factors brought about considerable fluctuations in the "R_f"-values. It is also possible that at that time in the USA, gas chromatography was being regarded as the method to choose in this field. Anyhow, the universal applicability and the numerous further advantages remained unexplored. This is shown most clearly when the number of relevant articles published annually is considered.

Tswett's column chromatography underwent a similar development (cf. Fig. 2, 1st Edition). More than 20 years passed before the significance of this technique was recognised-by RICHARD KUHN's team. Paper chromatography required an even longer "incubation period". As long ago as 1906, GOPPELSROEDER [241] brought out a book entitled "Suggestions for the Study of Capillary Analysis, based on Capillarity and Adsorption Phenomena"; half a century elapsed before CONSDEN, GORDON, MARTIN and SYNGE developed a practicable technique.

A method is usually worked out for a particular research aim and this happened in my own work. Fifteen years ago we were wanting to separate the components of individual plant glandular fibres, after it became evident that histochemical methods would bring us no further; but no progress could be made with any of the chromatographic methods of that time. We devoted our study first to the structure of the layers. Filter papers, with their relatively coarse fibres and also the commercially available adsorbents, swamped the tiny amounts of compounds present in the glandular fibres, which were about 70 μ in size. We were forced to use thinner and thinner layers. Employing very fine grained silica gel layers of 20 μ thickness, we succeeded in 1955 in separating chromatographically the contents of some glandular fibres, practically invisible to the naked eye. Encouraged by this success and through contact with numerous other groups of natural products, we came to understand and appreciate the method more and more. Our first publication, in 1956, with the title "Thin-Layer Chromatography" was ignored, just like those of my predecessors. We began to consider why this useful method was not being used. We postponed the original research project and studied the thin-layer chromatography for about five years, trying to work out the best scheme. The following seemed important:

1. Rationalised preparation of uniformly thick layers.

2. Combination of the necessary apparatus into a basic kit, permitting immediate application of the method.

3. Testing the most universably applicable adsorbents.

4. Ascertaining the controlling factors and standardisation of the procedure.

5. Establishment of the scope of the method by means of examples with various compound classes.

These efforts prospered so that in 1958 reference could be made to the "new" method [661] and a basic equipment for thin-layer chromatography could be shown in the ACHEMA exhibition. The laboratories of the larger industrial concerns in South Germany and Switzerland were the first to take up and use this time-saving chromatographic method with success. The application extended rapidly over the whole western world. This is especially evident from the compilation of the annually published articles in which TLC was mentioned (Fig. 2).

After a hesitant start the method came into use in Eastern Europe and the Soviet Union, although still mostly with loose, more coarse-grained layers, like those used in 1951—1958 particularly by MOTTIER [461—463] in Switzerland. The more easily handled, adhesive layers are being increasingly used there also, however. In 1965 the first ready-made, coated plates and sheets became available commercially.

Up to the end of 1965 over 4500 publications had come out, as well as the following monographs on TLC:

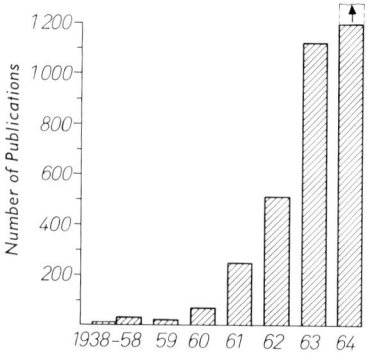

Fig. 2. Number of publications appearing annually in which TLC has been reported, at least incidentally

1. STAHL, E.: Dünnschicht-Chromatographie, ein Laboratoriums-Handbuch, 534 pages, published 1962 [673]. English edition, 553 pages, published 1965 [680]. Russian translation 1965. Peace press, Moscow [681].

2. RANDERATH, K.: Dünnschicht-Chromatographie, 242 pages, published 1962 [559]; 2nd Edition 1965, 291 pages, English edition 1963, 250 pages [564]. French edition 1964, 294 pages [565].

3. HASHIMOTO, Y.: Thin-layer Chromatography, 157 pages, published in 1962 in Japanese [277].

4. TRUTER, E. V.: Thin Film Chromatography, 205 pages, published in 1963 [720].

5. BOBBITT, J. M.: Thin-layer Chromatography, 208 pages, published in 1963 [76].

6. MARINI-BETTÒLO, G. B.: Thin-layer Chromatography (report of the 1st international TLC-Symposium in Rome, May 1963). 232 pages, published in 1964 [430].

7. ACHREM, A. A., and A. I. KUSNETSOVA: Thin-layer Chromatography, 175 pages, published in 1964 in Russian [8, 9].

8. LABLER, L., and V. SCHWARZ: Chromatografie na tenké vrstvě, 465 pages, published in 1965 in Czech [386].

9. MACEK, K., and I. M. HAIS: Stationary Phase in Paper and Thin-layer Chromatography; report of the 2nd international chromatography Symposium in Liblice (Prague), June 1964, 358 pages, published in 1965 [413].

Numerous review articles have been published also, including: In Spanish [22, 485]; Italian [125]; Polish [492, 588]; Greek [368]; Japanese [275, 778]; French [153, 221—222]; Russian [8]; Norwegian [327]; Dutch [158, 711]; English [51, 418, 421, 428, 590, 630, 775] and Serbo-Croat [657].

A few films on TLC have also been made (see Table 1).

Table 1. *Films about thin-layer chromatography*

Title	Author/address	Film type	Duration (min)	Availability for loan
"Unter Garantie"	E. Merck AG, 6100 Darmstadt, W. Germany	16 mm colour film with photographic sound recording	34	yes
"Dünnschicht-Chromatographie"	Camag AG, Muttenz, Switzerland	16 mm black and white film with magnetic tape recording [a]	14	yes
"Thin-Layer Chromatography"	E. H. AHRENS JR., The Rockefeller Institute, New York, USA	16 mm colour film with photographic sound recording	10	yes [b]
"Thin-Layer Chromatography"	O. S. PRIVETT, The Hormel Institute, Austin, Minnesota, USA	16 mm colour film with photographic sound recording	ca. 30	yes

[a] In German, French or English.
[b] On loan from Rothacker Inc., 241 West 175 St., New York, USA.

TLC has been officially recognised as a standard method by its inclusion in modern pharmacopoeias [142, 530, 531, 727].

B. Adsorbents for TLC

EGON STAHL

In all chromatographic procedures, the optimum conditions for separation are yielded through mutual harmonisation of the stationary and mobile phases. With TLC, the most suitable adsorbent (stationary phase) can be found easily and quickly. A whole spectrum of tested inorganic and organic adsorbents is available today, specifically for TLC. They differ from the usual materials for column chromatography in their fine-grained structure (Fig. 3). The thin layers prepared from these materials and their surprising separating ability, are a major feature of TLC.

When choosing the most suitable grain size, technical preparation and migration times on the chromatogram must be considered and a compromise arrived at. The grain size of most TLC adsorbents lies between 5 and 50 μm. Efforts must be made also to standardise the adsorbents so that their separation behaviour is as uniform as possible, showing no variations from manufacturer to manufacturer or from charge to charge.

Standard substance mixtures (test mixtures) from various compound classes have proved highly suitable for evaluating and comparing the separation properties of TLC adsorbents. This chromatographic test can be intensified by increasing the number of components in the mixture and including so-called "critical pairs". Porous adsorbents can be further specified through data like pore diameter (in Å); pore volume (cm^3/g); specific surface area (m^2/g) and specific gravity (g/cm^3).

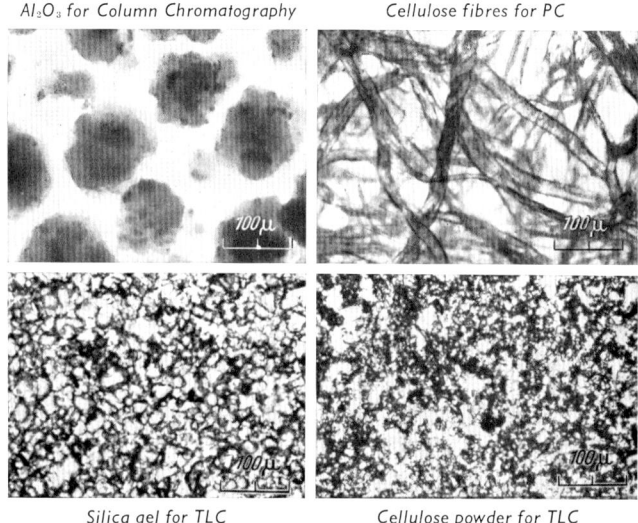

Fig. 3. Comparison of the structural differences between adsorbents for column and paper chromatography (above) and those for TLC (below). (STAHL [674])

The pH-value of a suspension in water (usually 10% adsorbent) serves as an additional characteristic. The purity of the adsorbent is often significant for the adsorption properties; occasionally inorganic or organic impurities are present which entered during the manufacturing procedure and were not properly removed.

The choice of adsorbent should present no difficulty even for the beginner if the following sections, the basic scheme (Fig. 99) and the examples quoted in the special sections are considered.

I. Silica Gel

H. W. KOHLSCHÜTTER and K. UNGER

Silica gel is an amorphous, porous substance. The properties important for chromatography are a combination of those of the solid skeletal, framework material and those of the cavity system. Accordingly some fundamental principles concerning the former are given in sub-section 1 and some concerning the latter in sub-section 2 [203, 237, 251, 310, 362].

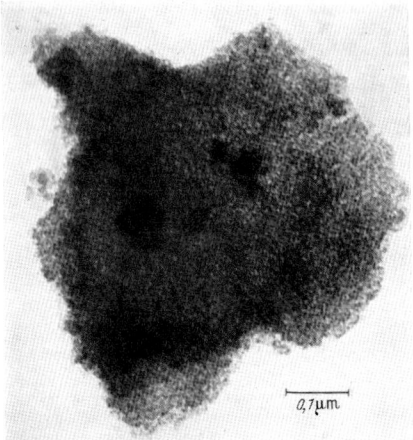

Fig. 4. Electron microscopic photograph of a very thin fragment from a grain of silica gel (G. KÄMPF)

1. Formation of Silica Gel and Construction of its Framework

The hydrolysis of silicon compounds constitutes the primary step in the formation of silica gel. Numerous reaction routes are possible.

Example of Silicon Tetrachloride

Stepwise hydrolysis of an ether solution of this compound leads via chlorosilanols to chlorosiloxanes or chlorosiloxanols of higher molecular weights [242]: $SiCl_3(OH)$, $Cl_3SiOSiCl_3$, $Cl_3SiOSiCl_2(OH)$ etc.

Complete hydrolysis yields polysilicic acids. Monosilicic acid is formed in dilute solution and when the hydrochloric acid set free according to the equation:

$$SiCl_4 + 4H_2O \rightarrow Si(OH)_4 + 4HCl$$

is taken up [760]. The molecular weight of the dissolved silicon-containing compound, which can be determined cryoscopically, is then about 60. This is interpreted as SiO_2 and 1 SiO_2 is taken as equivalent to 1 $Si(OH)_4$.

Example of Methyl Orthosilicate

By addition of water in the mole ratio $Si(OCH_3)_4:H_2O$ of $1:<1$, polysilicate esters of low to high molecular weight are formed as a result of hydrolysis and condensation [99, 369]. These contain Si—O—Si bridges as well as unchanged $-OCH_3$ groups [12]. The polysilicate esters give polysilicic acid sols or gels by total hydrolysis with excess water. Monosilicic acid can also be prepared by passing a diluted stream of $Si(OCH_3)_4$ vapour into 10^{-3} to 10^{-2} M hydrochloric acid at $0°$ C [90].

Compounds of the type $Si(OR)_4$ are termed tetraalkoxysilanes. Consequently information in the specialist literature about orthosilicate esters is to be found under this heading [90].

Example of Sodium Silicate

The customary preparation and industrial production of silica gel begins with the total hydrolysis of sodium silicate through treatment of water-glass solutions with acids. Water-glass solutions of various $Na_2O:SiO_2$ mole ratios and of various concentrations can be used. Conversely, water-glass solutions can be added to acids of various compositions, strengths and concentrations. All these parameters influence the course of the reaction and the properties of the resulting products. The form of the precipitates yielded by adding water-glass solutions to acids or vice versa, vary widely-an observation which can be made directly on the products; the precipitates may be:

granular, separating from the supernatant liquid and capable of being filtered; or plastic gel masses, uniformly filling the whole of the original solution and immobilising all the water.

Monosilicic acid can be prepared from the silicate-acid system, just as from the silicon tetrachloride-water or methyl orthosilicate-water systems [746]. Acetic acid or resin exchangers in the H-form can be added to dilute solutions of sodium silicate. Monosilicic acid is formed by reaction of solid calcium orthosilicate with hydrogen chloride in methanol, according to the equation:

$$Ca_2SiO_4 + 4HCl \rightarrow Si(OH)_4 + 2CaCl_2 \quad [209]$$

Table 2 contains selected references to simple procedures for preparing silicic acids.

Table 2. *Preparation of silicic acids* [92]

Products	Principle of the procedure
Solutions of monosilicic acid	Detachment of SiO_2 from silica gel with water. Ultrafiltration
Polysilicic acid sols	Mixing aqueous sodium silicate solution with 1:1 hydrochloric acid. Dialysis
Purest polysilicic acid gels	Hydrolysis of purified methyl orthosilicate with excess water at $40—50°$ C

Formal Derivation of the Silica Gel Framework

Monosilicic acid is not a starting material for preparation of silica gel. The silica gel framework may, however, be *derived formally* from it. Greatly simplified, there are five stages to this:

1. Monosilicic acid in solution.
2. Low molecular weight polysilicic acids in solution.
3. Formation of micelles through aggregation of low molecular weight polysilicic acids in sols and gels.

Formation of individual molecules of macromolecular polysilicic acids in sols.

4. Combination of freely mobile particles in sols (micelles and/or macromolecular atom associations) and in gels.
5. Dehydration of the plastic gels, forming harder and more porous xerogels.

Xerogels are the basis of the silica gel preparations which are subsequently worked up into TLC adsorbents. Cf. subsection 2.

Concerning 1. and 2. Some of the properties of soluble silicic acids of low molecular weight can be determined directly on the compounds themselves-e.g. molecular weight, acid strength and the rate of reaction with acidified ammonium molybdate solutions.

Monosilicic acid is termed "molybdate-active"; it reacts rapidly according to:

$$H_4SiO_4 + H_8Mo_{12}O_{40} \rightarrow H_4Si(Mo_{12}O_{40}) + 4H_2O .$$

The molybdenum content of the heteropoly acid formed can be reduced to molybdenum blue without reaction of the unused isopolyacid present. The colorimetric determination of soluble silicic acid depends on this principle. Polysilicic acids react more slowly in this analytical procedure because monosilicic acid must first be reformed from them [209, 746].

Information about how the molecules of soluble silicic acid are built up, can be derived from their way of formation. These compounds contain structural elements of crystallised silicates or of silicate esters, which can be distilled and from which they are formed by hydrolysis.

Monosilicic acid, H_4SiO_4, contains the tetrahedral SiO_4-group. Orthodisilicic acid. $H_6Si_2O_7$, contains in addition a Si—O—Si bridge, stereochemically two SiO_4-tetrahedra with a common corner.

Substantiated theoretical opinions are held on how monosilicic acid is converted to a di- or higher polysilicic acids. It is assumed that the coordination number of the silicon atom temporarily increases from four to six during polymerisation of the monosilicic acid molecules:

Water is eliminated in subsequent reactions, forming Si—O—Si bridges [749].

The chemical silicon-oxygen bond contains a polar contribution [508].

Concerning 3. and 4. Statements about the molecular structure of polysilicic acids in sols and gels are less definite than those concerning the structure of the lower molecular weight acids in solution. Both structural elements, SiO_4-tetrahedra and Si—O—Si-bridges retain their significance. A diversity of atom combinations, similar to those known for polysilicate esters, must however be reckoned with. Si—O—Si-bridges can be formed in two and three dimensions and association processes can occur also. Model conceptions of sols and gels are thus of two types [237, 310]:

a) Polysilicic acid molecules form *micelles* (definite particles of colloidal size with any arrangement form) primarily through association and condensation. Secondary linkages can occur between individual "adhesive" places on the surfaces of the micelles and also through condensation again. Solution → sol → plastic gel.

b) Polysilicic acid molecules grow immediately into atom combinations of high molecular weight. Condensation takes place between the individual atom groupings of these macromolecules. Solution → plastic gel.

If the processes assumed in a) and b) really occur, the ratio of a) to b) will depend on the procedure used for preparing the plastic gel.

Concerning 5. Dehydration of the polysilicic acid gels must involve elimination from the system of:

water occluded or immobilised by the polysilicic acids,
water held as complex or adsorbed on the hydrated polysilicic acids,
chemically bound water in the acids.

The transition from the plastic polysilicic acid gels to the hard xerogels takes place before the system is completely dehydrated. The plastic gels shrink considerably at first during dehydration. When hardening sets in, yet dehydration is still continuing, shrinking diminishes. Further marked shrinkage through sintering of the rigid framework sets in only when the last traces of water are eliminated at temperatures above 600° C.

Nomenclature

Suitable terms follow also from the formal derivation of the framework in silica gel preparations.

Polysilicic Acid Gel. Term for the (more or less) soft or plastic gel, saturated with liquid (e. g., water); that is, for a reaction product that can be derived from silicic acids of low molecular weight through processes of polymerisation, condensation and aggregation.

Silica Gel. Term for the hard, largely dehydrated xerogel that is formed from the polysilicic acid gel but which must be distinguished from it and which goes over into anhydrous silicon dioxide. In English the term has two words (silica gel) whereas in German, Spanish and Portuguese, usually one wort (Silicagel). Some commercial products are still called "Kieselgele" in German because the term "Silicagel" (in one word) was a protected trade term [582].

Structural Elements of the Silica Gel Framework

The concepts of molecule and micelle, used in describing the first stages of polysilicic acid formation, lose their significance in the description of the framework of silica gel. Infrared spectroscopy provides some information about the molecular structural elements of the framework [765].

Table 3. *Results of infra-red investigations of silica gel*

Absorption band (cm^{-1})	Allocation
7355 [765]	First overtone of the OH-group in SiOH
5265	Combination of stretching and bending (deformation) vibrations of H_2O
4545	Combination of stretching and bending (deformation) vibrations of SiOH
3400 [57, 77, 127, 235, 393, 412, 779]	OH-stretching vibration
1635	H_2O-bending (deformation) vibration
1200	Antisymmetric SiO-stretching vibration
1095	Antisymmetric SiO-stretching vibration
970	SiOH-bending (deformation) vibration
795	Antisymmetric SiO-stretching vibration
600	SiO-bending (deformation) vibration

The fundamental structural elements established during formation of polysilicic acid gel from mono- and polysilicic acids, persist right up to the formation of the related xerogel, i. e., silica gel: the tetrahedral distribution of oxygen atoms around each silicon atom; the siloxane groups or Si—O—Si-bridges; the silanol or SiOH-groups as end groups in highly condensed polysilicic acids. The properties of the framework can thus be discussed largely with the scheme:

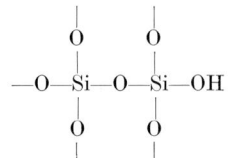

The SiOH-Groups

SiOH-groups which are accessible on the surface of the framework are capable of a variety of reactions.

Examples:

1. **SiOH** + ClSi(CH$_3$)$_3$ → **SiOSi(CH$_3$)$_3$** + HCl

The groupings in heavy type are part of the solid surface. The hydrophilic silica gel (SiOH) is transformed into hydrophobic silica gel [SiOSi(CH$_3$)$_3$]. Even when only part of the total SiOH-groups thus react, changes in the surface properties occur [360]

2. **SiOH** + CH$_2$N$_2$ → **SiOCH$_3$** + N$_2$

The Si-OH-groups reacting here can be determined [78]. Whether all the SiOH-groups react is uncertain.

The weakly acid reaction of silica gel suspensions in water is due to dissociation of SiOH-groups:

$$\text{SiOH} \rightleftharpoons \text{SiO}^- + \text{H}^+$$

The dissociation constants are of the order of 10^{-6} to 10^{-8} [173, 410].

Water is eliminated from SiOH-groups when silica gel is heated. Conversely, SiOH-groups can be reformed in liquid or with adsorbed water, according to:

$$\text{Si-O-Si} + \text{H}_2\text{O} \rightleftharpoons 2 \text{ SiOH}$$

Even on quartz surfaces this reaction leads to the formation of SiOH-groups. The functions of the SiOH-groups which are important for TLC are described in subsection 2.

The Si–O–Si-Bridges

Hydrolysis of Si–O–Si-bridges on the silica gel surface is possible. The socalled solubility of silica gel is attributed to such processes. A limiting concentration of about 0.01% SiO$_2$ is attained at 20° C in the system silica gel-water. It rises with increase in temperature and rises especially markedly above pH 9. It can be concluded from the solution properties that equilibrium states are reached and that supersaturation phenomena are possible [700].

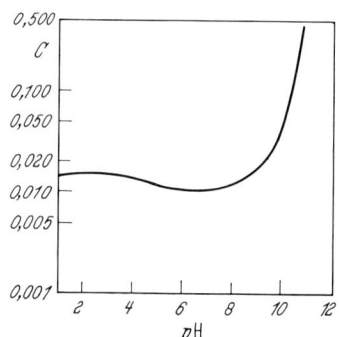

Fig. 5. Concentration of soluble silicic acid in equilibrium with silica gel [17, 49]

2. The Cavity System of Silica Gel

The cavity system of silica gel can be described by its structural elements just as was possible with the solid framework. The surface of the framework is one of these structural elements. It circumscribes the cavity system. Further structural elements are pore volume, pore radius and pore radius distribution.

Packing Density

Estimates of the ratio of the space occupancy of the solid framework to that of the cavity system, fluctuate widely. The socalled packing density, P can be calculated [422]:

$$\text{Packing density } P = \frac{v}{V + v}$$

where v = specific volume of the framework (reciprocal of the true specific gravity),
V = specific volume of the cavity system (pore volume).

If a value of about 2.2 is taken for the true specific gravity of the amorphous framework (quartz = 2.5), $v = 0.46$. A value of 0.5 found for V leads to P = 0.48, meaning that only about half of the total volume of a granule or fragment of silica gel is filled with solid framework.

Specific Surface Area

Silica gels of various average grain sizes are prepared by mechanical pulverisation and grading. Distinction must be made between the surface of the macro- or microscopic grains (= external surface) and that within the porous grains (= internal surface = sum of all the wall surfaces of all pores). The internal surface area is much larger than the external. Geometrical expansion of the total surface of a silica gel is described as the specific surface area (m^2/g).

Many of the data on specific surface areas depend on calculation from adsorption isotherms with nitrogen which covers the total accessible surface with a monomolecular layer. A surface equivalent of 16.2 Å per nitrogen molecule is taken. This is known as the BET-method [108].

Procedure

1. The adsorption isotherm for nitrogen on silica gel is recorded at 77.3° K in the low pressure region p/p_0 from 0 to about 0.25 (p_0 = saturation vapour pressure of pure liquid nitrogen at 77.3° K; p = vapour pressure of the nitrogen adsorbed on the silica gel).

2. Adsorption values are calculated with the help of the equation

$$\frac{p/p_0}{a\,(1-p/p_0)} = \frac{C-1}{v_m \cdot C} \cdot p/p_0 + \frac{1}{v_m \cdot C}$$

where p/p_0 = relative vapour pressure of nitrogen,
a = volume of gas adsorbed at pressure p,
v_m = volume of gas adsorbed when the entire adsorbent surface is covered with a monomolecular layer,
C = constant for the material.

Commercially available apparatus facilitates the adsorption measurements [515].

Approximate data, more quickly obtained and useful at least for *comparison* of various silica gels in routine tests, are based on titrations of a silica gel suspension, using dilute sodium hydroxide solution, from pH 4.0 to 9.0 at 25° C. G. W. SEARS [620] has established a linear relation between specific surface area (BET) of many silica gels and the consumption of 0.1 M sodium hydroxide solution.

Procedure

1. The silica gel is pulverised and a fraction of diameter 0.063—0.125 mm sieved out.
2. This is dried at 300° for 2 hours.
3. A sample of, e. g., 1.5 gram SiO_2 is taken, calculated from the gravimetrically determined loss in weight on igniting the silica gel.
4. Apparatus: ultra-thermostat; 250 ml 3-necked flask; stirrer; alkali-stable glass electrode with potentiometer; 25 ml-burette.
5. Titration: 30 gram reagent grade sodium chloride is added to the weighed silica gel sample in the 3-necked flask and the volume made up to 150 ml with distilled water. The pH of the suspension is brought to 4.00 with dilute alkali or hydrochloric acid. 0.1 M Sodium hydroxide solution is added slowly from the burette, stirring continuously, until pH 9.00 is reached. After each addition the attainment of a sensibly constant pH is awaited.
6. Result: The specific surface area in m^2/g is calculated with the help of the (empirical) equation:

specific surface area $O = 32 \times V - 25$ (m^2/g),

V = consumption of 0.1 M sodium hydroxide from pH 4.00 to 9.00, in ml.

The values are reproducible for different samples of one and the same silica gel. A special advantage of the method is that it can be applied also to plastic polysilicic acid gels, i. e., the *precursor* of silica gel. The values for the specific surface area of a plastic polysilicic acid gel and for the corresponding hard xerogel (silica gel) can be closely similar.

There is so far no final answer to the question how far agreement exists between the surface area values for a silica gel as obtained by the BET-method and by the titration method. The latter is nevertheless frequently used.

Specific Pore Volume

The pore volumes of solids can be determined through penetration of gases or liquids.

The adsorption and desorption isotherms, measured at 18° with water vapour in the pressure region p/p_0 from 0 to 1 are particularly informative about the hydrophilic silica gel (Fig. 6, p. 16)

p = vapour pressure of water adsorbed on silica gel at 18°,
p_0 = vapour pressure of pure water at 18°,
$\dfrac{x}{m}$ = adsorbed amount of (liquid) water per gram silica gel.

a) Commercial preparation, subsequently purified and dried at 100° before the measurements. The desorption isotherm returns to the starting point of the adsorption isotherm. Saturation volumes at $p/p_0 = 1$.

b) The same preparation as in a) but heated to 1000° before measurements. The desorption isotherm no longer attains the starting point since a small part of the adsorbed water has again been chemically bound by the silica gel surface.

The same procedure can be combined with determinations of thermodynamic and dielectric properties of the silica gel-water system.

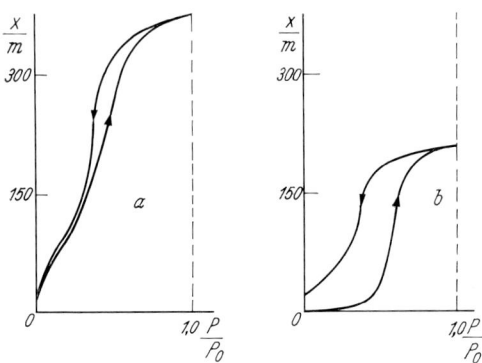

Fig. 6. Adsorption- and desorption isotherms of a silica gel preparation for water at 18° C

If, in addition, the specific surface area has been measured by the BET-method, the following picture is obtained [334]:

When silica gel takes up water, one may distinguish:

Capillary water, the water molecules of which are closely packed as in liquid water.

Water content of the first adsorption layer, the water molecules of which are less closely packed than in the capillary water because they are associated with the surface SiOH-groups and roughly according to the formula $SiOH \ldots OH_2$.

Water content of the SiOH-groups.

With silica gels of various surface areas and various pore volumes, the difference between the packing of the water molecules in the first adsorption layer and in the capillary water varies in significance [359]. Where this difference must be considered, the saturation volume derived from the end point of the adsorption isotherms cannot be taken as equal to the geometrical pore volume unless corrected.

N. E. FISHER and A. Y. MOTTLAU [193] have described a simple method for approximate determination of the specific pore volume. It is based on the observation that whereas dried, fine-grained silica gel is a mobile powder, the grains adhere to each other and cake when the pores have been filled with liquid (including water).

Procedure

1. The silica gel is powdered and a fraction of diameter 0.125 to 0.250 mm sieved out (about 60 mesh).

2. This is dried at 300° C for 2 hours and stored in a desiccator over phosphorus pentoxide.

3. A sample of 0.3—1.0 g. is taken, the smaller sized for silica gels of large specific pore volume and conversely.

4. Apparatus: 25 ml erlenmeyer flask with side tube; 5 ml microburette with 0.01 ml graduations, the fine tip of which extends into the flask through a rubber cork (pressure equilibration is ensured by connecting the burette and side tube); magnetic stirrer with teflon-coated stirring bar.

5. Titration is carried out with water (or other liquids) at one drop per minute, stirring continuously. After each addition, the mixture is left until the liquid is uniformly distributed and the silica gel powder appears dry again. The approach of the endpoint is heralded by parts of the silica gel powder adhering to the vessel wall. The end-point is a visible caking together of the powder.

6. Result: The titration in ml per gram silica gel sample corresponds to the ml pore volume per gram substance.

The technique requires practice but is suitable for rapid comparison of silica gel preparations.

Average Pore Radius

The classical method for calculating pore radii and therefore the distribution curve of the various pore sizes in the same silica gel preparation, is based on the Kelvin equation for the desorption isotherm (e. g., for water)

$$r = - \frac{2 \cdot M \cdot \sigma \cdot \cos \Theta}{d \cdot R \cdot T \cdot \ln p/p_0}$$

where $r =$ pore radius,
$M =$ molecular weight (of water),
$\sigma =$ surface tension of liquid water,
$\theta =$ contact angle,
$d =$ density of the water condensed in the pores,
$R =$ gas constant,
$T =$ adsorption temperature,
$p =$ vapour pressure of the adsorbed water,
$p_0 =$ vapour pressure of pure liquid water.

The vapour pressure of the capillary water above the meniscus which is formed in the pores, is a function of the pore radius. It is not necessarily the same as the geometrical radius of the pores at the point where the meniscus is. A layer of water adheres to the pore walls during dehydration of the pores [334]. C. PIERCE [532] has made a compilation of experience in this connection, with the help of which the radius calculated from the Kelvin equation can be converted into the effective pore radius. The sizes of larger pores can be determined with the electron microscope. The average silica gel preparations have a mean pore radius of the order of 25 Å.

The average pore radius can be calculated approximately by picturing the total number of pores as a single cylindrical pore [118] (Fig. 7); from the figure.

$$\text{average pore radius } r = \frac{2 \cdot \text{specific pore volume } V}{\text{specific surface area } O}.$$

r is a formal calculating entity in this expression.

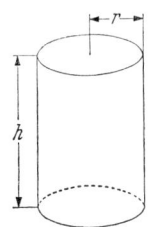

Fig. 7. Relation between radius, volume and surface area of a cylindrically shaped pore

Pore radii below 10 Å cannot be determined by using the Kelvin equation. There are however indications that the cavity system of silica gel contains, in addition to macropores, also micropores, the radii of which are of low molecular size.

Sintering

When silica gel is heated to higher temperatures (100 to 1000° C) the cavity system undergoes change. The porous material thickens. Silica gel sinters and the causes of this sintering are:

elimination of chemically linked ("structural") water from SiOH-groups and

the flexibility of the atoms or atom groups in the molecular associations of the framework.

Distinction must however be made between

a) heating silica gel in open vessels under normal pressure or in closed vessels under reduced pressure, using a pump system (Table 4).

b) heating with water in autoclaves under higher pressure, hydrothermal treatment.

When the "structural" water eliminated from the SiOH-groups is removed at once, as in a), secondary interactions between water and the framework can no longer occur. The solubility of silica gel also plays a part during heating with water as in b). Between 100 and 200° C, changes in the cavity system of silica gel are far more extensive during the hydrothermal treatment b) than on heating in vacuum a).

The results in Table 5 probably give clues concerning the real structure of the silica gel surface:

The silica gel *framework* is ruptured when water is split out. The same processes can lead to a *loosening* of the *surface*. Structural models, based on the crystal structure of the SiO_2-modification do not therefore fully depict the surface fine structure. These relationships are best expressed by using three concepts:

Geometrical expansion of the surface: m²/g, determined for example by the BET-method.

Ideal structure of the surface: derived for example from the crystal structure of the SiO_2 modification [334].

Real structure of the surface: defects, dislocations, loosening etc. of the crystallographically ideal surface.

The contrast between the concepts of ideal and real surface structure corresponds to the distinction between the notions of ideal and real crystals.

Table 4. *Sintering of a silica gel preparation[a] by heating in high vacuum[b]* [334]

Temp., °C	Specific surface area, m²/g	Specific pore volume as saturation volume, ml/g	Mean pore radius from the Kelvin equation [e] Å
100°	618	0,371	11.4
400°	609	0,373	11.1
700°	586	0,342 [c]	10.9
1000° [d]	341	0,216	11.1

[a] Commercial preparation, subsequently purified, grain size 0.2—0.5 mm.

[b] Each heated 1 h in high vacuum.

[c] Steep fall in the specific surface area above 600° C.

[d] Still amorphous, based on the X-ray picture.

[e] Sensibly constant. The method (Kelvin equation) does not respond to changes in the micropores.

Table 5. *Changes in grain size distribution through heating a silica gel preparation* [367] [a]

Sieve fractions; grain diameters in mm	Dried [b] at 100° C	Ignited [c] at 1000° C
0.25—0.5	100%	82.3%
0.2—0.25	0%	17.2%
0.125—0.2	0%	0.5%

[a] Commercial preparation, subsequently purified.

[b] Heated in an open vessel at ordinary pressure.

[c] Smaller grains, not larger, are formed through the ignition.

Quantitative Differentiation of the Water Content

The total amount of water which can be subsequently taken up by silica gel at room temperature, is altered by sintering. The proportions of chemically combined water (SiOH) and capillary water (H_2O) change also. A variety of methods is available for determination of these water amounts, e. g.:

Recording of complete adsorption and desorption isotherms (cf. Fig. 6);

Infra red spectroscopy (cf. Table 3 in subsection 1);

Titration using the Karl Fischer method [483];
Gravimetric determination.

Table 6 contains data obtained by gravimetric determination on an average silica gel [360].

Table 6. *Water content of the cavity system of a silica gel preparation* (commercial preparation, subsequently purified, grain size 0.2 to 0.5 mm)

	A	B	C
100° C	20 h	0.18	0.18
20° C		0.93	0.18
400° C	108 h	0.07	0.07
20° C		0.78	0.13
800° C	36 h	0.02	0.02
20° C		0.53	0.08
1000° C	50 min	0.00	0.00
20° C		0.25	0.02
1000° C	48 h	0.00	0.00
20° C		0.00	0.00

A: Heated in an open crucible (to 100° C, 400° C, 800° C and 1000° C); length of the heating given (50 min to 108 h). After heating, the samples were saturated with water at 20° C (column 2) and then again dehydrated in high vacuum (column 3).

B: Water content determined as loss in weight during ignition and expressed as analytical mole ratio H_2O/SiO_2.

Smaller value = water content after heating.

Larger value = water content after exposure to saturated water vapour at 20° C until attainment of constant weight.

C: Water content determined as loss in weight during ignition and expressed as analytical mole ratio H_2O/SiO_2.

Smaller (or first) value = water content attained on the preheated sample (column 1) at 20° C in high vacuum.

Larger value = water content attained again in high vacuum at 20° C.

Conclusions from Table 6:

a) Capillary and Firmly Combined Water

The silica gel, purified with acid and water and dried for a long period at 100° C retains a small water content (0.18 mole H_2O/SiO_2). This does not change in high vacuum at 20° C. In saturated water vapour at 20° C, it takes up water to 0.93 mole H_2O/SiO_2. This increased content is reduced to the original amount of 0.18 mole H_2O/SiO_2 in high vacuum at 20° C.

Accordingly the total water content of the preparation which has been saturated with water vapour at 20° C, can be divided into:

0.18 mole H_2O/SiO_2 firmly combined water
and 0.75 mole H_2O/SiO_2 capillary water
0.93 mole H_2/SiO_2 total water.

b) Re-formation of Firmly Combined Water

Through heating at 400° C, the silica gel loses some of its water (from 0.18 to 0.07 mole H_2O/SiO_2). The preparation is now also unable to take up as much capillary water; the total water content becomes smaller (0.78 mole H_2O/SiO_2). A small fraction of the capillary water is converted at 20° C into firmly combined water (0.13−0.07 = 0.06 mole H_2O/SiO_2). This re-formation of firmly combined water can be read from the data in column 3 for all sintering stages of the silica gel preparation:

Preheating	100° C	400° C	800° C	1000° C 50 min	48 hrs
Re-formation of firmly combined water in mole H_2O/SiO_2	0.0	0.06	0.06	0.02	0.0

Reproducibility of Silica Gel Preparations

The current status of silica gel research is such that not all properties of the silica gel substance can be quantitatively described. Some can be described only approximately and others are still largely unknown. The real surface structure is one of these last named properties. The more complete our knowledge of the formation, framework and cavity system becomes, the more easily possible it will be to reproduce accurately those properties which are important for the use of silica gel as an adsorbent. The following subjects can be regarded as fundamental for silica gel research:

Individual processes in the formation of the polysilicic acid gels.

Individual processes in the formation of xerogels (silica gels).

Molecular structure of the silica gel framework.

Structural elements of colloidal dimensions in the framework of the silica gels.

Formation in nature of macro- or microscopic grains in silica gel preparations (in contrast to grain formation through mechanical pulverisation).

Cavity system in the individual grain:

a) specific surface area; ideal structure of the surface; real structure of the surface;

b) specific pore volume;

c) pore radii and their distribution (macro- and micropores); degree of purity.

Grain size and purity requirements are generally modified to suit the particular application which is anticipated, after the hard xerogel has been formed. Theoretically they can however be fashioned by the method of formation.

Silica Gel Preparations for TLC

The final stages in the preparation of silica gels for TLC consist of:

mechanical pulverisation of pure silica gel preparations and grading of the grain sizes thus obtained;

preparation of aqueous suspensions of the pulverised material which can be spread uniformly on to supports such as glass plates;

drying the layers on these supports.

If the pulverisation procedure lowers the purity of the product so that it is inadequate for TLC purposes, subsequent purification is necessary; this may again bring about changes in the individual structural elements of its cavity system (specific surface area, real surface structure, specific pore volume or pore radius distribution). The general experience acquired concerning the silica gel/water system can be applied to the preparation of the aqueous suspensions (see a) below) and drying the layers (see b) except in special cases; these include: where the finely powdered silica gel contains small amounts of additives like inorganic or organic binders or fluorescent compounds; where suspensions are made in solutions of organic binders prior to preparing the layers; or where layers are spread on plastic sheets instead of glass plates.

a) Water is used in the suspensions in order to fill the cavity system of the grains. The ratio silica gel: water necessary to obtain a fluid system accordingly depends on the specific pore volume of the silica gel and on its initial water content (cf. Table 4).

b) In the drying process, water must be eliminated from the capillary system between the grains and from the cavity system of each individual grain. To avoid even the slightest sintering, the drying temperature must be between 100° and 200° C, as can be seen from the data on differentiation of water content (Table 6). This is the temperature range in which the entire capillary water is removed, yet no water split from the SiOH-groups.

The properties of the dried layers when chromatograms are made, are determined by the combined effects of the capillary system *between* the grains, the cavity system *within* the grains and the surface chemical groups in the cavity system. Since all the properties of neither a silica gel preparation, nor a silica gel layer can be quantitatively described or predetermined, the *chromatographic test* on the layer has a decisive significance. In this, the influence of silica gel on inorganic or organic substances in water or non-aqueous solvents is followed. These influences may be:

cation exchange with non-hydrolysing cations [366];
adsorption of the hydrolysis products of hydrolysing cations [363];
adsorption of colloidal hydrolysis products of cations [364];
adsorption of molecules of dissolved inorganic or organic compounds.

Differences in behaviour between various silica gel preparations are demonstrated especially sensitively through their action on hydrolysing salts in water.

In a communication "Dünnschicht-Chromatographie anorganischer Kationen an Silicagel" (Thin-layer chromatography of inorganic cations on silica gel), H. W. KOHLSCHÜTTER and L. SCHÄFER have described some simple basic experiments illustrating the manifold ways in which silica gel behaves towards dissolved materials.

II. Aluminas and other Inorganic Adsorbents

H. RÖSSLER

Some other inorganic substances besides silica gel have been used as adsorbents. Alumina and kieselguhr preparations are the most important of these. Their application to the study of particular compound classes is discussed in the special sections of this book; here, in this section, their preparation or isolation and properties are described. The comparatively few examples of application of other adsorbents to separations, are given below under the headings in this section (3—14).

1. Alumina

The hydroxides are the starting materials for preparing aluminas. The dehydration of these compounds has been studied in detail for many years, but the results have not always been reproducible. This is scarcely surprising since the reactions are influenced by many factors amongst which the most significant are: content of extraneous ions; particle size of the starting material; temperature; rate of heating up; duration of heating; and the atmospheric environment of the reaction. Only by combining the results from various methods like X-ray structure analysis, electron diffraction, calorimetric measurements, differential thermal analysis and infra-red determinations, has the present state of knowledge been reached. Some compilations summarising this have been made by the laboratories of the larger aluminium concerns [232, 474, 500] and by research and university institutes [592, 717].

Various aluminium hydroxides serve as starting materials for the preparation of oxides for chromatographic purposes. These are often denoted by a Greek letter prefix to the formula or chemical name. This characterisation is unfortunately not uniform in Europe and the USA and thus it is advisable to use the name of the mineral [232].

α) **Hydrargillite, Al(OH)$_3$.** This aluminium hydroxide is technically prepared on the large scale by the Bayer process, in which it is precipitated from sodium aluminate solutions at below 60° C by inoculation with a crystal and stirring. It forms the main constituent of tropical bauxite in nature.

β) **Bayerite, Al(OH)$_3$.** Another aluminium hydroxide was named in error after the Bayer process by FRICKE [202]. That hydrargillite is yielded was discovered only later. It is not certain if bayerite is a stable compound; it has so far not been found to occur naturally.

γ) **Böhmite, AlO(OH).** An aluminium oxide hydroxide, the principal constituent of European bauxite deposits, is obtained by hydrothermal modification of hydrargillite or bayerite in water or alkali solution. Hydrargillite prepared by the Bayer process always contains some böhmite.

Corundum is the final stage of dehydration of the hydroxide. The transition forms occurring during this have been denoted also with Greek letters by HABER. In contrast to that with the hydroxides, this use is uniform here because later authors have adhered to an initial scheme in which five other transition forms in addition to the γ-oxide were considered [704]. A further advance was the discovery that the dehydration of hydrargillite can occur via two different reaction routes [104]. The scheme[1] probably generally accepted nowadays is portrayed in Fig. 8.

Other transition forms obtained by dehydration in high vacuum (ϱ-Al$_2$O$_3$ [499]) or from especially pure starting material [234, 715, 716], have not been considered because they are not used as technical raw materials for preparing adsorbents. The structures of the transition forms in the scheme described have already been discussed in detail [474, 592]. The transition temperatures quoted are often markedly modified by impurities, e. g., alkalies; and also by particle size and the oven atmosphere.

An explanation of why both aluminium hydroxides follow two dehydration routes is that böhmite is formed under hydrothermal conditions [82, 84]. χ- and η-Aluminium oxides are the only products observed as first dehydration products of finely divided hydrargillite and bayerite respectively. Coarser starting material yields böhmite and its decomposition products in both cases. DEBOER and coworkers assume

---

[1] No structure suggestions have been made so far for those transition oxides which show only few and unclear X-ray diffraction patterns.

that water vapour supersaturation is established in larger crystals at the beginning of heating. The hydroxides thus yield the oxide hydroxide and then its transition forms. In the finely divided hydroxides there is negligible or no böhmite formation, so that only transition forms of the hydroxides are observed. The dotted rectangle in the diagram is intended to show that the two transition forms η- and γ-aluminium oxide are so similar that numerous authors have not distinguished them [592]!

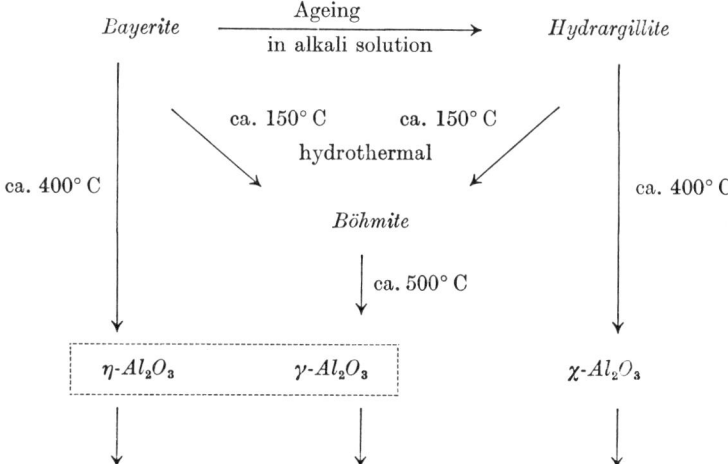

Fig. 8. Partial scheme of aluminium hydroxide decomposition (without the transition oxides formed at higher temperatures and the end product corundum)

Oxides used in chromatography contain either the γ- or the χ-forms or both. They are prepared mostly from hydrargillite. The dehydration products of gel-like böhmite [87, 538] are γ-oxides which are used principally for drying and as catalysts.

Oxide surface areas of 300–400 m²/g can be attained by dehydrating hydrargillite at temperatures of about 400° C [589]. Larger molecules are not able to reach all parts of such surfaces because the cross-section of some of the pores is too small [86, 87]. Crevice-like hollows of about 30 Å cross-section with a surface area of about 60 m²/g alongside capillaries of 10 Å cross-section forming a surface area of ca. 200–250 m²/g, have been described [82]. Heating to 600° C and above leads to further dehydration and sintering, accompanied by increase in size and uniformity of the pores. Larger molecules can then be accommodated, as seen from the results of DEBOER with lauric acid. The experiments were carried out on a technical hydrargillite and give an idea of the pore system. Since the dehydration processes are influenced by many factors, it cannot be expected that every batch of hydrargillite manufactured by the Bayer process will yield an oxide with the same surface area and pore structure. *Aluminas for chromatography have surface areas between 100 and 350 m^2/g*, measured by the BET-method, i. e., through covering with a layer of small molecules; and *average pore diameters between 20 and 80 Å*. *The water content*, measured by loss in weight on ignition at 1100° C, can still be as high as 8%, unless the activity has been reduced by subsequent addition of water which can raise the content to over 20%.

The oxide surface of the pore walls is covered with hydroxyl groups linked in various ways. Five different types of hydroxyl ion have been demonstrated by means of IR measurements and considerations of a statistical model [514].

Fluctuating results have been obtained when the transition oxides are rehydrated, depending on the experimental conditions. GLEMSER and RIECK have established that hydroxyl groups were built into the lattice [235] whereas DEBOER and coworkers considered the surface linkages in the first adsorption layer to be hydrogen bonds with the oxygen atoms in the oxide lattice [83]: they distinguished three types of adsorbed water in addition to capillary condensation.

All aluminas prepared from hydrargillite contain a few tenths of a percent of sodium oxide unless they have been subsequently treated with acid. This comes from the hydrargillite, the lattice of which is stabilised by alkali. An aqueous slurry of such an oxide reacts alkaline. The alkali part can be more or less fully removed from the surface by treatment with acid; the aqueous suspension then reacts neutral or weakly acid.

Aluminas for TLC. Suitable for TLC are χ- and γ-aluminium oxides, occasionally mixed and containing residual böhmite, and possessing a sufficiently large pore breadth; that is, they have been ignited at 500 to 800° C and have a surface area between 100 and 250 m^2/g. The finely divided oxide, the grain size of which should be less than 60 μm, is stirred into a suspension with water and formed into a layer by spreading. After air drying, the layer is activated at 110° C. Such a layer has an activity of ca. BROCKMANN II–III, in comparison with an alumina for column chromatography.

When *gypsum* is used *as binding material* with a basic alumina, the pH value of the aqueous suspension is reduced almost to neutral. This fact must be borne in mind when basic layers are necessary for a separation problem.

Since so many experimental factors influence the preparation of alumina for TLC, the same results cannot be expected with preparations from different manufacturers. For the same reason, varying amounts of water (1:1 to 1:3) must be used to prepare suspensions of the alumina.

Using the usual spreaders, 30–35 g adsorbent must be used to coat five 20 × 20 cm plates with layers of 250–300 μm thickness.

Preparations with and without binders are available; usually the manufacturer quotes any added adhesive agent. Aluminas containing finely divided inorganic or organic fluorescent additives are also available. Inorganic additives, principally activated silicates, are usually inert to the chemicals and solvents used in TLC. Excitation is carried out by short wave UV radiation (254 nm). Substances on the layer which absorb in this region then appear as dark spots on the uniformly fluorescing layer. Preparations of this sort are denoted by an extra suffix, e. g., F_{254} or UV_{254}. Pyrene derivatives have been recommended as organic fluorescent indicators, and also cyanine dyes, excited by long wave UV radiation (366 nm) have been used [510, 722, 267]. Substances which absorb only in the short wave UV region below 230 nm appear on such layers as light spots on a bluish fluorescing background. This abnormal behaviour has so far not been explained. Compounds absorbing at above 270 nm appear as dark spots on a light background when irradiated by longer wave UV light. Suffixes like F_{366} are used to characterise such products. The use of the indices 254 + 366 indicates that both fluorescent materials are present in the adsorbent. Only very small amounts of organic fluorescent indicators are needed on account of their high fluorescence intensities; TLC development is thus only seldom affected by their presence.

Alumina products are available also for separations on a preparative scale; these permit layers of up to 2 mm thickness to be made without cracks forming during drying. These adsorbents contain fluorescent indicators to facilitate the preparative work and are correspondingly labelled.

2. Kieselguhr

By the term "kieselguhr" is understood naturally occurring, amorphous silicic acid of fossil origin, also referred to as diatomaceous earth, bacillarieae earth or diatomite. The deposits, found in many parts of the globe, date from the later periods of the earth's history (quaternary and tertiary). They were formed from dead, single-celled plants, diatoms, whose shells consisting of silicic acid sank over very long periods of time to ocean or lake beds and formed deposits, sometimes very large [50, 103, 332, 335, 466]. These algae *(diatoms, bacillarieae, bacillariophytes)* are capable of life in marine, fresh or brackish water. Each cell has two siliceous valves fitting together like the halves of a pill-box. Each valve has a silicic acid coating of uniform pore size, the silica framework [285, 286, 310, 401]. The vast number of species (ca. 10 000) furnishes a multiplicity of structures [113]. There were also many early forms, varying according to the geographical environment. It is thus possible to establish the region of the deposit by microscopic examination of a kieselguhr.

The persistence of these delicately beautiful structures is amazing. During later geological periods the silica-coated valves became covered with sedimentary deposits, occasionally huge amounts, of higher specific gravity; the fine-pored structures survived nevertheless for 10^5 to 10^6 years [332]. They further withstood the heavy excavating machines for quarrying without loss of porous structure, even though many were broken by this treatment [308].

The kieselguhr excavated can seldom be used without treatment since it contains water, organic substances and other sediment material [308, 332].

The amounts of these impurities vary with the region of the deposit and the method of processing. The composition of two non-ignited, dried, light coloured kieselguhrs is given in the table:

Chemical Composition in %	Region and Age of Deposit	
	Central Europe, Lüneburger Heide (Diluvian)	North America California (Miocene)
SiO_2	89.2	89.7
Al_2O_3	1.9	3.7
Fe_2O_3	0.4	1.5
TiO_2	0.1	0.1
CaO	traces	0.4
MgO	0.2	0.7
Alkali (as Na_2O)	1.1	0.8
Loss on ignition (H_2O, CO_2 and organic substances)	3.6	3.7

For chromatography purposes especially, the earth must be carefully purified; this is usually done in several stages and may vary according to the region of the deposit (elutriation, drying, ignition, sometimes after addition of alkali chlorides or sodium carbonate). The porous structure must withstand these processing procedures, although some reduction of the surface area and widening of the pores cannot be avoided. A shrinkage from 12–40 m²/g to 1–5 m²/g has been reported [308, 710].

Some data for processed, ignited earths are in the following table [495, 577]. It is evident from the figures that the pore structure varies considerably, in contrast to silica gels[2].

---

[2] Celites (Firm 77) are products from Californian kieselguhrs.

Kieselguhr	Surface Area m²/g	Pore Volume cm³/g	Macro-pore volume %	Average Pore Diameter in Å
1	4.2	1.14	92	11 000
2	<1	1.7	88	> 70 000
3	1	2.78	—	110 000

The amorphous silicic acid of the crude earth becomes crystalline on ignition. X-ray crystal photographs show the interferences of the crystalline silica modifications [72]. Silanol and siloxan groups are to be found on the framework surface of the microcrystalline silica. Since there can be up to several percent of mineral impurities, the pore walls are not free from extraneous atoms.

A small, only slightly active surface and relatively large pore volume are available for chromatography. The use of kieselguhr in partition chromatography is thus understandable.

Finely divided fractions of grain size less than 60 μm are used in TLC. Gypsum is employed as binder. An adsorbent of this type is called Kieselguhr G for example (Firm 88).

For five 20 × 20 cm plates, 30 g kieselguhr are slurried with 60 ml water, spread immediately, dried at room temperature and activated by heating 30 min at 110° C.

3. Silicates

a) Magnesium Silicate

Precipitated magnesium silicates also have been used for separating sugars, sugar derivatives and other polyhydroxy compounds [244, 245, 767] and for circular chromatography of terpenes [109]. Aqueous suspensions of the preparations showed pH values between 8 and 10, depending on the manufacturer; this must be remembered when identical separation results are hoped for. Differing surface areas have also to be reckoned with so that the amount of water required to prepare the layers can vary from 1:2 to 1:3. The elution power of various solvents has been investigated on one preparation, *"Florisil"* (Firm 59), for which the manufacturer gave a specific surface area of ca. 300 m²/g [181].

Talc, a magnesium silicate mineral of composition $Mg_3[Si_4O_{10}](OH)_2$, has a layer lattice. The surface available for adsorption depends markedly on its state of division. Ordinary commercial preparations have been used to separate fatty acids [120] and lanatosides [787]. About 1.5 times its amount of propanol or ethanol is needed to prepare the slurry.

b) Calcium Silicate

A precipitated and hydrated product (about 15% water), termed Silene EF (Firm 40), with a particle size of about 0.03 μm and yielding an aqueous suspension of pH value 9.6, has been shown suitable for sugar separations. The slurry was prepared from 11 g silicate, 3 g kieselguhr, 700 mg sodium acetate and ca. 30 ml water [518].

4. Phosphates

Tricalcium phosphate decomposes slowly in water giving *hydroxylapatite* and cannot be obtained in a pure state from aqueous solutions. The effective surface in column chromatography of aqueous solutions on calcium phosphates has thus usually been an apatite surface.

There are several procedures for preparation of a suitable apatite [714, 20]. Proteins have been separated on a product prepared by the procedure of ANACKER and STOY [297]. An adhesive layer was made by stirring up 15 g adsorbent in 60 ml 70% ethanol, containing 40 mg of a polyamide (Zytel 61, Firm of DuPont). The layer was spread with an ordinary apparatus. The plates were ready for use after drying at room temperature. α- and β-Monoglycerides have been separated also on hydroxylapatite [298].

Magnesium hydrogen phosphate has been used amongst other adsorbents for separating carotenoids [663].

5. Calcium Sulphate

Gypsum layers on 20 × 20 cm plates of frosted glass to achieve better adhesion have been proposed for separating fatty acids and glycerides [199, 340]. The separation was influenced by the quality of the gypsum (alabaster gypsum). The layer was prepared by spreading a mixture of 50 g gypsum and 70 ml water with the usual apparatus, followed by 15 min drying at room temperature and 1 hr at 80–90° C. TLC of carbohydrates [783] and a biological test (wheat coleoptile test) [132] have been carried out likewise on gypsum layers. A successful electrophoretic separation of iodates and periodates has been achieved on thin gypsum layers, prepared by the pouring procedure from 25 g gypsum + 30 ml water [165].

6. Glass Powder

Glass is so hard that the particle surface becomes contaminated during mechanical pulverisation. Careful cleansing is thus necessary before use. The effect of the glass composition, layer thickness, temperature and solvent on the chromatographic development of various substances has been tested on pyrex glass powder and pulverised glass wool [464, 553, 554, 555]. There has also been a publication on the behaviour of crushed Jena glass which had been separated into two grain size classes by sedimentation [107].

The particles of preparations from thick glass possess no pores and a small surface area. In contrast, ground porous glass (Vycor-glass 7930, Firm of Corning-Glass) has a surface area of 150–200 m²/g and pores of 40 Å cross-section. A slurry with water of this powder, particle size 60–75 μm, together with 13% gypsum as adhesive, has been spread on plates in 0.3 mm layers. Three types of wax could be separated on it and the results compared with those of a separation on silica gel G and alumina G [380].

7. Salts of Heteropoly-, Tungstic, Molybdic and Tetraboric Acids

The alkali metal ions and also fission products of uranium have been separated on layers of the ammonium, 8-hydroxyquinolinium and pyridinium salts of dodecamolybdo-phosphoric -arsenic, -silicic and -germanic acids [110, 396, 397]. A paste of the salts with water was first made, suspended in acetone and layers of 250 μm thickness prepared, using 10–12 mg/cm² of plate surface. The preparation of the salts was described also. The use of kieselguhr, impregnated with sodium tungstate, molybdate or tetraborate as chelating agents has been published in two other articles (separation of adrenaline and catechol derivatives) [265].

8. Ferric and Chromic Oxides

The dark colour of these oxides is certainly one reason why they have been rarely used in chromatographic analysis. Their suitability for column chromatography has recently been pointed out [236]. Spraying with a silica gel G suspension has been recommended for identification of substances separated by TLC on these dark oxide layers [288]. Part of each adsorbed substance spot diffuses from the oxide into the silica gel layer and can be identified there in the usual way; the authors claim that this method is suitable also for use with active charcoal.

9. Zinc Carbonate and Zinc Ferrocyanide

A basic zinc carbonate containing 5% starch as adhesive has been used for the thin-layer chromatographic separation of aldehydes, ketones and other carbonyl compounds [302, 534, 32, 33, 34].

Fogg and Wood [196] have separated sulphonamides by TLC using 0.03–3.33 M acetic acid on layers of zinc ferrocyanide which they had prepared themselves.

10. Active Carbon

This adsorbent has only rarely found use. The difficulty of recognising the zones (cf. 8.) has been circumvented by using combined charcoal/silica gel layers (for ketone separation [350]); and by imprinting on an agar surface and developing according to the diffusion method with *Bacillus pumilus* (for separation of neomycin sulphate mixtures [102]).

11. Zirconium Phosphate and Hydrous Zirconium Oxide

The separation of inorganic cations on the preparations Bio-Rad ZP-1 (zirconium phosphate) and Bio-Rad HZO-1 (Firm 23), a hydrous zirconium oxide, has been the object of a study [780].

A mixture of 20 g with 20 ml water and 3% starch as binder was spread into 500 μm layers which were dried for 30 min at 40° C.

12. Lanthanum Oxide

Apart from its use with starch as binder in the TLC of phosphates and sulphates, employing radioactive isotopes [455], no other example has so far been reported.

13. Bentonites (Firm 91)

Lipophilic thickening agents are prepared by base exchange of the inorganic cations of bentonite or montmorillonite with organic quaternary ammonium ions. Polyphenyl isomers have been separated on layers of such material [579].

14. Combinations of Adsorbents

Mixtures of various substances have been used in addition to the inorganic adsorbents so far quoted: silica gel G and alumina G in 1:1 proportion for the separation of sugars [736, 752] and of phenols [622]; silica gel G and kieselguhr G in various proportions for the detection and determination of chlorine-containing organic herbicides [4] and for detection of antioxidants in fats [443], of steroids [138] and of carbohydrates [539]; zinc carbonate and silica gel, 1:1 for separating tocopherol derivatives. The employment and methods of testing adsorbent combinations are treated in the section "Gradient-TLC" on p. 91.

III. Organic Adsorbents

P. WOLLENWEBER

1. Cellulose and Derivatives

Whereas inorganic adsorbents — except kieselguhr — are used chiefly for separations of lipophilic compounds, organic adsorbents (this applies certainly to cellulose and its derivatives) are used almost exclusively in separations of hydrophilic compounds like amino acids, nucleic acid derivatives, sugars etc. The separation conditions on cellulose layers are analogous to those on paper. Consequently analogous modifications of cellulose powder exist and commercially available are: normal, naturally occurring material; chemically modified products, as strongly hydrophobic as desired, e. g. acetylated celluloses; or those with particularly marked hydrophilic character, e. g., ion exchange cellulose powders. The solvents customarily used in paper chromatography can be largely employed in chromatography on cellulose layers. The separation process on ordinary cellulose depends almost entirely on partition whereas with the cellulose ion exchangers, a part is played also by ion-exchanging forces. The special properties of starch, sucrose, dextran gels etc. are mentioned later since they have only a limited range of application at the moment.

a) Normal Cellulose Powder

Cellulose consists of a large number of cellobiose units linked via β-1,4-glycosidic bonds. It possesses hydrophilic properties due to the numerous hydroxyl groups of the units (Fig. 9). Normal cellulose powder is therefore suitable for separating hydrophilic compounds. More detailed information and examples of application are to be found in the review articles of P. WOLLENWEBER [770, 771, 772, 773].

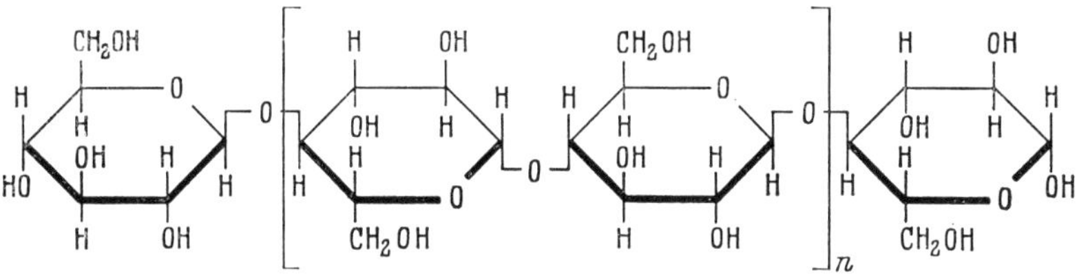

Fig. 9. Fraction of a cellulose molecule

Like many other high polymers, cellulose can form hydrogen bonds both within the cellulose structure and with liquids of low molecular weight such as alcohol and water. There is considerable evidence for the existence of the hydrogen bonds.

For example, a very slight degree of acetylation leads to increased hygroscopic character and further acetylation then to a diminution in these properties; this is due to destruction of the hydrogen bonds [14, 75]. Sheet formation in preparation of paper depends largely on formation of hydrogen bonds [719]. PIERCE [532] assumes that cellulose contains both inter- and intramolecular hydrogen bonds. The adsorption of water on cellulose — the basis of partition chromatographic processes — occurs through hydrogen bond formation. MUUS [468] considers that there are two types of hydrogen bond. Infra-red spectroscopy provides decisive evidence in the study of hydrogen bonds. The formation of such a bond is accompanied by a clear reduction in the frequency attributed to the OH-group; at the same time an ill-defined band structure is observed [105, 431].

Two types of normal cellulose powder are suitable for TLC:
1. native, fibrous cellulose,
2. "microcrystalline" cellulose.

Fibrous cellulose is high quality material prepared in the usual way in the cellulose or cotton industry and then reduced under mild conditions to a fibre grade suitable for TLC. The average degree of polymerisation of the native cellulose powder MN 300 (Firm 83) is of the order of 400–500. The fibre length of MN 300 ranges from 2 to 20 μm.

Preparation of the Layers. *An aqueous suspension containing about 15% cellulose powder, rendered homogeneous by 30–60 secs in an electric mixer, yields layers with the optimum uniformity, smoothness etc. The layers adhere excellently and can even be wiped in the dried state. Cellulose layers are best air dried.* It is unnecessary to activate them at higher temperatures. In fact, the separating properties of cellulose layers improve on long exposure to air.

"Microcrystalline" cellulose is prepared [44] through hydrolysis of extremely pure cellulose, such as regenerated cellulose, cotton linters and cotton of high purity, by 15 min treatment with boiling 2.5 N hydrochloric acid solution. Cellulose crystallites[3] with average degree of polymerisation from 40 to 200 are obtained, depending on the nature of the starting material and the subsequent processing. The microcrystalline structure of the cellulose thus prepared can be proved by X-ray studies. An amorphous structure is wholly excluded according to the present state of cellulose chemistry [481]. The expression "level-off degree of polymerisation cellulose products", used in the publication quoted above [44], is thus correct.

Preparation of Layers. Microcrystalline cellulose for TLC is marketed under the name "Avicel" (Firm 5). The grain size is 19 or 38 μm, depending on the modification. *TLC layers are made from aqueous suspensions containing 15–30% Avicel by weight. The manufacturers recommend 1 min vigorous stirring with an electric blender. Longer stirring leads to gel formation* [29].

The duration of homogenisation is thus of great importance. The data in the following Table 7 illustrate the influence of stirring time.

With the fibrous cellulose powder MN 300 (Firm 83), the time of homogenisation has an influence only when exceeding 10 min. On the other hand, in contrast to Avicel, a mixture satisfactory for layer preparation cannot be attained by merely mixing with a glass rod. The length of run on a MN 300 layer with the same solvents as that used

---

[3] JAYME and KNOLLE [325] have recently established through electron-microscopic studies that the crystallites have a marked tendency to form layers of secondary aggregates of a definite preferred form, mainly horizontal.

with the Avicel layers (Table 7), amounts to 7.5 cm/20 min. Avicel layers are treated subsequently in the same way as layers of native cellulose.

Magnified photographs of cellulose fibres from paper and cellulose powder for TLC are in Fig. 3. Rapid substance transportation along long fibres cannot occur on the short fibres of the cellulose powder; as a result, the spots from the same amounts of substance are more compact than on paper.

Table 7. *Influence of the duration of homogenisation of Avicel on the layer*

Amount of Avicel (g)	Duration in sec of homogenisation in electrical mixer	Necessary amount of water (ml)	Adhesion	Length of run of a NH_4OH-containing solvent
15	0 a	50	poor	10 cm in 15 min
15	60	100	good	6.5 cm in 30 min
15	150	120	good	5 cm in 30 min

a Stirred with a glass rod in a beaker.

The large specific surface area of TLC-cellulose powder (about 15000 cm²/g for MN 300 according to BLAINE's method) means that more substance can be taken up in a smaller space; this leads to compact spots.

Attention has been drawn to considerable variations in the separation properties of the two cellulose types, fibrous and microcrystalline, by e. g., WARING and ZIPORIN [740] in the TLC-separation of hexose- and triosephosphates and by BAUDLER and MENGEL [46] in the TLC-separation of phosphoric acids. WERNZE [747] has found that microcrystalline cellulose powder contained far more ninhydrin-positive contaminants than native, fibrous cellulose powder MN 300.

As a result of its preparation process, "microcrystalline" cellulose powder is purer than untreated, native, fibrous cellulose powder. All inorganic impurities have been removed since the drastic hydrolysis conditions completely break up the original fibrous structure and liberate the last traces of mineral matter. Native, fibrous cellulose of high purity is also commercially available (Firms 83, 121). Cellulose powder MN 300 HR (Firm 83) is prepared under mild conditions by acid washing, neutral washing and finally treatment with organic solvents to take out fats and resinous matter. Little hydrolysis thus takes place so that the adhesion of the starting material MN 300 is retained. A slightly yellowish solvent front is then no longer observed and the spots, even with Rf values of 1.0 remain compact and not branched. High purity cellulose powder is recommended especially for quantitative work, e. g., separation of carbohydrates where IR-spectroscopic measurements follow [257], separation of phosphoric acids [45, 46, 47] and phosphates [28, 583].

Neither the native, fibrous nor the „microcrystalline" cellulose requires **additives for improving adhesion.** The adhesion of the layers is many times greater than that of inorganic adsorbents; dried layers can be wiped. Addition of gypsum to cellulose powders may influence separations favourably or unfavourably. Thus it has interfered in amino acid separations [25, 769] and has improved the thin-layer chromatographic separation of nucleic acid products [130].

The **layer thickness** of the dry cellulose is never the same as that set on the spreader. A rule of thumb says that cellulose layers shrink to about half of the slit breadth set on the apparatus. Dry layers of about 0.12 mm are particularly good for most separations and reveal most convincingly the advantages of thin-layer chromatography. Thicker layers up to 0.5 mm of fibrous cellulose powder and up to 1 mm of microcrystalline powder, can be made without cracks.

Visualisation on cellulose thin-layer chromatograms can be carried out just as in paper chromatography.

In addition to the two types of cellulose mentioned, those with fluorescent additive are also commercially available. These additions are made to cellulose powders for separations of substances which can be detected in UV-light. An inorganic pigment which gives an intensely green fluorescence in UV-light of 254 nm is that almost always used. The separated substances become visible as dark spots on an intensely green fluorescent background, through quenching of the fluorescence. Not every solvent can be used with cellulose layers containing fluorescent additive. Some solvents quench the fluorescence of the whole layer.

Table 8. *Summary of the usual commercial types of cellulose powder for TLC*

Cellulose type	Manufacturer or source of supply [a]
Fibrous cellulose powder	33, 83
Fibrous cellulose powder of high purity	83
Cellulose powder, acid washed	121
Microcrystalline cellulose powder (Avicel)	5, 23, 83, 115, 117, 121, 127
Fibrous cellulose powder with fluorescent additive	33, 83
Microcrystalline cellulose powder with fluorescent additive	121
Acetylated cellulose powder	83, 121
DEAE-cellulose powder	23, 83, 115, 117, 127
ECTEOLA-cellulose powder	23, 83, 115, 117, 127
PEI-cellulose powder	23, 83, 127
CM-cellulose powder	23, 83, 115, 117, 127
P-cellulose powder	23, 83, 115, 117
Poly-P-cellulose powder	83
AE-cellulose powder	115

[a] See list of firms (p. 918) for addresses.

b) Acetylated Cellulose Powder

Acetylated cellulose powders, just like acetylated chromatography paper, are suitable for reversed phase chromatography. They are prepared by converting cellulose into its acetate esters. Up to three hydroxyl groups per cellulose structural unit can be esterified. The acetyl content may vary from a few % up to a maximum of 44.8, the latter value corresponding to the triacetate. Increase in the degree of esterification is accompanied by increase in the hydrophobic nature of the powder. A gradual and continuous change from the "aqueous" to the "organic-nonpolar" stationary phase can thus be effected through variation of the extent of acetylation. It must be borne in mind when choosing the solvent, that the various esters, mono-, di- and triacetates, are soluble in some organic solvents.

Preparation of the Layers. Acetylated cellulose powders (Ac-powder) of varying acetyl-content are available commercially (Firms 83, 121) and can be easily spread with the usual apparatus. The more or less hydrophobic Ac-powder (depending on the degree of acetylation) is best suspended in 95% ethanol (cf. instructions of Firm 83); mechanical stirring suffices. The layers are air dried and activation at higher temperature is unnecessary. The layers must however be pre-treated uniformly to ensure reproducible results. Although the adhesion of the Ac-layers is nowhere near as good as that of normal cellulose powder, it is nevertheless about the same as that of inorganic materials. The solvent used to dissolve the substance mixture under investigation has a by no means negligible influence on the separation.

Applications. Acetylated cellulose powder is in principle suitable for separation of all lipophilic substances. Separation alone is not enough however; the separated substances must be detectable. This fact has been responsible for the relatively modest extent of application of Ac-cellulose layers. They may be recommended for coloured substances and for those which are detectable in UV light. Some examples of use are: separation of polynuclear aromatic compounds [31, 309, 357, 599, 600, 755], anthraquinone dyes [768], antioxidants [596], cutin acids [98], sweetening materials [595], rhodanine derivatives of acetoacetic acid and acetone in urine [574] and ketocarboxylic acids [575].

c) Cellulose Ion-Exchanger Powder [4]

Cellulose ion-exchanger powders have been used for column-chromatographic separation of proteins, enzymes and numerous other biochemically interesting materials, since the fundamental work of PETERSON and SOBER [516] in 1956. Ion exchangers in a form suitable for TLC have been available for some years also. The types commercially available at present are listed in Table 9.

---

[4] Cf. Ion exchangers in TLC on p. 44.

Table 9. *Cellulose ion exchangers for TLC*

Abbreviation	Full name	Active groups	Capacity mval/g
DEAE	Diethylaminoethyl-cellulose	R—O—C_2H_4·$N(C_2H_5)_2$	0.7—1.0
ECTEOLA	Reaction product from epichlorohydrin, triethanolamine and alkali-cellulose	Structure not known	0.3—0.5
AE	Aminoethyl-cellulose	R—O—C_2H_4—NH_2	1.0
CM	Carboxymethyl-cellulose	R—O—CH_2COOH	0.7
P	Phosphorylated cellulose	R—O—PO_3H_2	0.7
PEI	Polyethyleneimine-impregnated cellulose	$(-CH_2-CH_2-NH-)_x$	1.0
Poly-P	Polyphosphate-impregnated cellulose	$(PO_3Na)_x$	1.0

With the exceptions of PEI- and Poly-P-cellulose, these are all native cellulose containing basic or acid groups attached via ether or ester links. By etherification of cellulose, products with various solubilities can be prepared, i. e., varying from water-insoluble to water-soluble. Not only the different substituents but also, and especially important, the number of groups introduced, plays a part. This is important particularly for those cellulose ethers destined to be used as ion-exchangers in chromatography. Since only few groups are introduced into cellulose for chromatographic purposes, it is better to speak of chemically modified cellulose rather than of cellulose ethers or esters.

The fibrous structure of the cellulose exchangers means that they possess a large surface area and this in turn means that most of the substituents lie compactly on the surface. Even large hydrophilic molecules like proteins can diffuse easily through the cellulose matrix, hydrophilic and capable of distension. Such penetration is not possible with exchange resins because the synthetic resin matrix is hydrophobic; such exchangers thus react only with the active groups on the rounded surface. The bulk of the active groups is however in the interior of the matrix of exchange resins. The active groups are about 50 Å apart in cellulose exchangers, i. e., large in comparison with resin exchangers (about 10 Å).

Cellulose exchangers, despite their far smaller exchange capacity, consequently possess a higher capacity for large molecules like proteins, than have synthetic resin exchangers. In addition, since the active groups are appreciably further apart, combination takes place on only one or very few positions; selective desorption is thus possible under very mild conditions, in contrast to resin exchangers. As a result of these properties, cellulose ion-exchangers have been most important for the separation, purification and isolation of sensitive substances in the biochemical field.

RANDERATH has introduced into TLC two new types of cellulose exchanger in addition to the above mentioned chemically modified forms. These were prepared by impregnation of native, fibrous cellulose (MN 300, Firm 83) with substances possessing ion-exchanging properties such as with polyethyleneimine; or by treating cellulose powder, which had been impregnated with polyethyleneimine, with polyphosphate (sodium metaphosphate). Cellulose impregnated with polyethyleneimine is termed PEI-cellulose; that impregnated with polyphosphate, PP- or Poly-P-cellulose. The former has anion-exchanging, the latter cation-exchanging properties.

The *manufacturers* of ion-exchangers for TLC are given in Table 10 and in the list of firms at the end of the book.

Layers of cellulose ion-exchangers are prepared with the usual commercial spreaders. A mixture is made of water or the solvent to be used, and about 10–20% cellulose powder. The amount of powder depends on the swelling tendency of the particular powder. The exchangers CM-, DEAE- and ECTEOLA-cellulose tend to swell considerably; PEI- and Poly-P-cellulose only slightly. This swelling is disadvantageous in that the completely dry layer shows a greater or lesser number of very fine cracks.

Note further: Cellulose exchange layers should not dry out completely before the chromatographic procedure is concluded. It is important that the active groups should be as far as possible in the most active exchange form in order to ensure reproducible separations. The following procedure is thus recommended:

Before the substances to be separated are applied, the layers are submitted to ascending chromatography with the proposed solvent (see p. 45). They are then air dried, the substance mixture is applied to the still moist layer and chromatographed without delay.

The rate of migration can be influenced or varied through addition of normal, non-ion exchanging cellulose powder.

Some examples of application of cellulose ion-exchangers to TLC are given in Table 10.

2. Starch

Starch is another polysaccharide which is suitable as an adsorbent for TLC. As a result of the linear combination of maltose units up to a molecular weight of at least a million, it contains many hydroxyl groups and is consequently strongly hydrophilic. Like cellulose, it is suitable for separating hydrophilic substances.

CANIĆ and PETROVIĆ [115] have used a Yugoslav maize starch for separating cations. After the starch had been purified by a special procedure, a suspension in water was applied, like other adsorbents, to glass plates, using an ordinary commercial spreader. Starch layers are ready for use after air drying.

DAVIDEK [149] has used starch layers for reversed phase chromatography, as in paper chromatography, by suitably impregnating them. He dissolved the substances used for impregnation (e. g., paraffin oil, vegetable oils, dimethylformamide) in a volatile solvent (about 20 + 80, v/v) and mixed this solution with soluble starch. The suspension thus obtained was spread on to glass plates. The plates are ready for use after evaporation of the solvent.

It may be mentioned here to complete the picture that RAMSEY [556] has employed the "hydrolysed starch", used by SMITHIES for starch gel block electrophoresis, for thin-layer chromatographic separation of proteins. The starch gel layer is poured, giving a layer about 1.5 mm thick. Further experimental details may be taken from the original article.

3. Sucrose

COLMAN and VISHNIAC [133] have used the disaccharide sucrose for separating chloroplast pigments. No water-containing solvent can be used with sucrose since it is easily soluble in water. The layer is *prepared* by mixing finely ground, commercial crystalline cane sugar, containing 3% starch, with the same amount of methanol (g/v) and spreading 0.25 mm thick with ordinary apparatus. The layer is dried for 2 hrs at 40° C. Longer drying yields a hard, glass-like layer without adsorbent properties. An ether or acetone solution of the sample is applied.

4. Mannitol

The hexahydric alcohol mannitol can be used like sucrose as an adsorbent for the thin-layer chromatographic separation of plant pigments [639]. Here too, water may not be used to prepare the layer, nor as solvent component on account of the considerable solubility of the adsorbent in water.

65 g mannitol and 100 ml acetone are mixed for 1 min with an electric mixer, 1 ml aqueous maize starch solution (5 g in 10 ml) is added to the paste thus formed and mixed 1 min more. The layers are prepared with the usual spreader. The plates can be used after 20 to 30 min air drying. Adhesion is good.

5. Dextran Gels[5]

Sephadex is a modified dextran of bacterial origin and is used in numerous modifications in gel filtration. This is mainly a column chromatographic procedure which permits separation of molecules on the basis of their differing sizes. This means that only those substances can be separated which differ sufficiently among themselves in molecular size, as for example proteins, peptides, enzymes, hormones, nucleic acids etc.

---

[5] See also Sephadex ion-exchangers (p. 45).

Special dextran gels, prepared similarly through cross-linking the linear dextran macromolecules, are available for TLC under the name of Sephadex Superfine (Firm 102). This Sephadex type contains many hydroxyl groups like all products of this sort and is thus strongly hydrophilic, swelling to a gel in water or electrolyte solutions. It is indifferent to cations and anions and possesses pores, the size of which depends on the degree of cross-linking.

To *prepare* a Sephadex layer, the material must be soaked previously for 5—72 hrs, depending on the type. According to the manufacturer [529], the soaked, swollen gel can be applied to carefully cleaned and dried glass plates, using any commercial spreader. An additive to improve adhesion is unnecessary. The layers are generally 0.2—0.5 mm thick. In contrast to other adsorbents, the layers are stored in a moist container; dried out layers can be regenerated by spraying with buffer solutions. The substance mixture to be separated is spotted on to the moist, soaked gel layer. The solvent used for development should be allowed to migrate through the gel layer for 10—15 hrs before spotting. Descending development is the usual procedure, achieved by slanting the plates. Solvent chamber and gel layer are connected via a filter paper wick. The speed of capillary movement of the developing solvent is varied or fixed by regulating the angle of slope to the horizontal. The best separations are obtained with a maximum rate of capillary movement of 1—2 cms/hr; separation times are thus 8—10 hrs.

Buffer and salt solutions are used as *solvents* with layers of cellulose ion-exchangers. The gel layers are very carefully dried at 50—60° C after chromatography and the spots visualised using the customary procedures.

IV. Polyamides as Adsorbents

H. ENDRES

It has been known for a long time that powerful hydrogen bonds are formed between phenolic hydroxyl groups and amide groups. The formation of definite addition compounds depends on these strong subsidiary valency forces, e. g., between phenol and urea in 2:1 ratio [177] or between phenol and diketopiperazine [527]. Further evidence that such bonds occur comes from IR spectra [116] and measurements of dielectric constants [231] of phenol/amide mixtures. FLETT [194] has determined by IR spectroscopy the energy of the hydrogen bond between phenol and dimethylformamide to be 6.4 kcal/mole. The affinity of substantive dyes for wool for example, has been accounted for by hydrogen bonding between phenolic groups and peptide protein links

[496, 696]. Further, these hydrogen bonds play an important part in tanning with vegetable tanning materials [249]. The strong affinity of phenol for peptides has been utilised for a long time for separating proteins from non-proteins by extraction with phenol [748]. Phenol-containing solvents are suitable for paper chromatographic separation of peptides; according to GRASSMANN and DEFFNER's experiments [246], the affinity of peptides for phenol increases with increase in chain length.

Based on these results, high molecular weight synthetic materials containing amide groups should be suitable for separating phenols. Commercially available polyamides of the Perlon-type (polycaprolactam, see Fig. 10) and Nylon-type (polyhexamethylenediamine adipate) come into consideration. As a result of cross-hydrogen bond linking, they are sufficiently poorly soluble in hydrophilic solvents like methanol, ethanol, acetone and dimethylformamide, yet still capable of swelling.

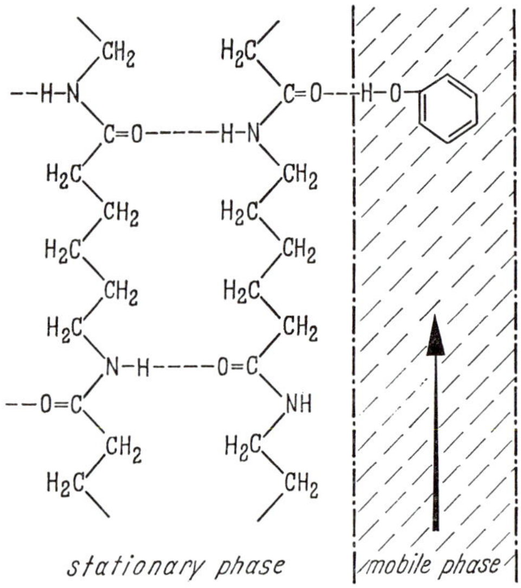

Fig. 10. Perlon as stationary phase in separation of phenols

The constancy of the partition coefficient up to relatively high phenol concentrations (the partition coefficient of phenol between polycaprolactam and water is linear up to the point where about one third of the polyamide peptide links is saturated) [247] creates ideal conditions for chromatographic separations. CARELLI et al. [117] and GRASSMANN et al. [247] described the first experiments with polyamide as filler in column chromatography.

The affinity for the polyamide increases in the order phenol, resorcinol, phloroglucinol and decreases in the series, phenol, catechol, pyrogallol [8]. As a rough rule it can be said that the strength of attachment of an aromatic compound to the polyamide is increased by a second or third hydroxyl group in m- or p-position but decreased when in the o-position. Evidently both hydroxyl groups in resorcinol and hydroquinone can interact simultaneously with different amide groups of the polyamide, so that the compounds are more strongly held back than phenol. The hydroxyl groups in catechol must, on the other hand, compete for the same amide group. The second hydroxyl group increases the affinity for the eluent (usually aqueous), thereby raising the migration rate further. A partial saturation through intramolecular hydrogen bonds must also be taken into consideration [8]. In addition to the influence of the number and position of the hydroxyl groups in an aromatic compound, the nature of the eluent also has an appreciable effect on the affinity of a phenol for the polyamide. Desorption takes place more or less rapidly, depending on the tendency of a solvent to compete with the particular phenol in forming hydrogen bonds with the polyamide, or even itself to show affinity for the material on the polyamide.

The desorption ability increases thus in the following order: *water < methanol < acetone < dilute sodium hydroxide solution < formamide < dimethylformamide.*

The high elution capacity of dimethylformamide depends evidently on its also containing the —CO—N= grouping, so that it can form hydrogen bonds with phenolic compounds in the same way as the polyamide.

The chromatography of phenolic compounds on polyamide was later extensively used for isolation and structure elucidation of various naturally occurring materials [229a]. As examples may be mentioned: the tannins of sprucewood bark [248] and Sumac [281]; isolation of two hydroxystilbenes from *Eucalyptus wandoo* [279]; resolution of several pharmacologically important plant extracts [294] and of the pigments of beets [613]; isolation of various ommochromes [112]; and the separation of a mixture of ε-rhodomycinone and ε-isorhodomycinone [100] (see the summarising work of H. ENDRES and H. HÖRMANN [186]).

Carboxylic acids are likewise bound to polyamides through hydrogen bonds [185]. The affinity of monocarboxylic acids for the polyamide is only moderate; dicarboxylic and aromatic carboxylic acids are more firmly held, especially when the aromatic part of the molecule is larger.

Strong attachment to polyamides is encountered with numerous nitro compounds. The addition of nitro compounds to a polyamide corresponds to the reaction of a Lewis acid with a base; separation of aromatic nitro compounds on polyamide can thus be regarded as ion exchange [293]. Elution with sharp zones is thus possible only with buffering solvents. A successful separation of dinitrophenylamino acids, arising from the determination of the amino end-groups of peptides and proteins, has been possible on polyamide columns [293, 697].

Quinones are irreversibly retained on polyamide, as a result of the free amino groups in the latter. A good separation of quinonoid and phenolic compounds is possible on acetylated polyamide [183].

Most of the work carried out on polyamide columns can be extended to polyamide-TLC; this has been shown by the separation of the contents of a *Solanum tuberosum* [292], a *Sumbuccus nigra* [148] and various tannin extracts [658]. The free amino groups must be blocked in TLC also if quinonoid substances are to be separated [184].

Polyamide powder for TLC is prepared by dissolving commercial Perlon powder, e. g., Ultramid BM 2 K 228 (Firm 16) in 35% hydrochloric acid and precipitating it with methanol/water, 1:1, stirring energetically. The precipitate is filtered off and dried. 10 g powder are suspended in 100 ml methanol for coating five 20 × 20 cm plates.

Polyamide layers are most simply prepared from powder, specially made commercially for TLC (Firms 83, 88, 153), following the instructions given.

V. Ion Exchangers in TLC

K. Dorfner

The presence of ion-exchanging groups is the dominating feature of an ion-exchanging adsorbent. Fine grain ion exchangers are used in TLC above all where ions or molecules with ionic or polar properties are to be separated and isolated on a preparative scale. All known effects in connection with ion exchange [169] can be utilised just as has already been done in paper and column ion exchange chromatography [168].

Ion exchangers are polyelectrolytes of high molecular weight, capable of exchanging their bound ions with ions of the same charge that are in the surrounding medium. One speaks of a cation or anion exchanger, depending on whether the framework of the exchanger is an acid or a base. The macromolecule of the ion exchanger is in general a three-dimensional, cross-linked structure, the matrix. A large number of ionisable groups is attached to it, usually referred to as "functional groups" ("anchor groups"). The exchangeable ions are called "counter ions". Ion exchangers can be polyfunctional, i. e., they contain various types of functional group; or monofunctional, with only one type. The spaces surrounding the framework of the ion exchanger are termed pores. The acidity or basicity of the exchanger is determined by the nature of the anchor or functional group. The strongly acidic synthetic resin exchangers may possess the $-SO_3^-$ functional group; the weakly acid, $-COO^-$; the strongly basic, $-N(CH_3)_2$; and the weakly basic, $-NH_2$.

The active groups of cellulose ion exchangers are given in Table 9.

DEAE-Sephadex ion exchangers contain the diethylaminoethyl group and SE-Sephadex, the sulphoethyl group.

The inorganic ion exchangers are aluminosilicates, the framework of which has a net charge. Synthetic inorganic exchangers are phosphates and tungstates which contain zirconium (IV). Mention may be made also of liquid ion exchangers; these may be considered for use as solvents.

Ion exchange materials occur in a definite form, termed for example, H^+-form, Na^+-form, Cl^--form or NO_3^--form. Their grain sizes are quoted in mm, or, in English-speaking areas, as mesh numbers (standard sieve sizes) (see [735]). Their volumes depend on the medium, meaning that change of medium is accompanied by a definite imbibition.

The most important property of an ion exchanger is its *exchange capacity*, since the number of counter ions which an ion exchanger can take up, may be calculated from it. It is usually expressed in the laboratory as milliequivalents/gram (mval/g). Ion exchangers can be selective and function specifically. Exchangers with chelate-forming groups are known as chelate ion exchangers.

Only a few of the vast number of available ion exchanger materials have been used so far in TLC. Those used or potentially useful are in Table 10.

Procedures for Preparing Layers

Preparation of suspensions and thin layers according to the original sources quoted in Table 10 is carried out as follows:

Dowex 50. The exchanger must be first converted into the H^+-form or Na^+-form. 5 g cellulose powder MN 300 is stirred with a few ml distilled water for some minutes. 20—30 ml water is then added while continuing stirring, followed by 30 g exchanger in small portions. Finally a further 25 ml water is added. The plates are coated in the usual way with 250 μm layers, air dried and stored in closed containers. They should be used within a week.

Dowex 1. The exchanger must be converted beforehand into the Cl^+-form or another desired form. The further preparation is as for Dowex 50.

BERGER et al. [63] have experimented also with the preparation and use of chromatographic films using Dowex 1 and Dowex 50 on a polyester base.

Wofatit CP 300. The exchanger is soaked in a 15% sodium acetate solution and buffered to pH 5.3 by stirring for several hours with acetic acid. It is then washed with water and dried in vacuo. The first coating on the plates is prepared from a suspension of the exchanger beads in 9 ml ether, 1 ml ethanol and 2 drops 4% collodion solution; or, for further layers, in ether alone. 150—200 μm layers are then prepared. The coated plates are stored in moisture-containing vessels.

Cellulose Ion Exchangers (see also p. 39)

DEAE-Sephadex. 25 g DEAE-Sephadex A 25 fine are treated with 0.5 N HCl, water, 0.5 N NaOH and water again, best in a centrifuge. The pH is then adjusted to the desired value (glass electrode) with the acid of the buffer mixture to be used subsequently. The exchanger is then carefully washed out with water and 0.2 mm layers prepared with the spreader. The plates are dried only up to the point where the gel grains are just visible.

Bio-Rad ZP-1. A mixture of 20 g of the H^+-form of the exchanger + 20 ml water + 3% starch as binder is heated until a thick gel is formed. It is then diluted with a little water, spread on plates in 500 μm layers and dried for 30 min at 40° C or overnight at room temperature.

Bio-Rad HZO-1. The exchanger, which can function also as an anion exchanger in acid solution, is best used in the NH_4^+-form or the H^+-form. The layers are prepared as under Bio-Rad ZP-1.

Table 10. *Ion Exchangers, Properties and Uses in TLC*

Type of Ion Exchanger	Commercial Products	Capacity mval/g	Grain size μm	Use and References
Resin Ion Exchangers				
Strongly acid cation exchangers	Dowex 50W (Firm 127); Lewatite S 100 (Firm 55); Amberlite IR 120 (Firm 127); Duolite C-20[a], Bio-Rad AG 50W-X8 (Firm 23)	4.5	40—80	Alkalies, alkaline earths [60]
Weakly acid cation exchangers	Wofatit CP 300 (Firm 146); Amberlite IRC-50 (Firm 127); Lewatite CNO (Firm 55)	10.0	64	Vitamin B [305]
Strongly basic anion exchangers	Dowex 1 (Firm 127); Permutite ES[b]; Amberlite IRA-400 (Firm 127); Bio-Rad AG 1-X 8 (Firm 23)	3.5	40—80	Halogens [59] tetrachlorofluorescein [59]
Weakly basic anion exchangers	Amberlite IR-45 (Firm 127); Merck Ion Exchanger II (Firm 88)	2.0	40—80	
Chelate-ion exchangers	Dowex A1 (Firm 127); Chelex 100 (Firm 30)		40—80	Alkalies, alkaline earths, heavy metals [62]
Cellulose Ion Exchangers	(see Table 9)			
SE-Cellulose	Serva SE-Cellulose (Firm 127)	0.6	10	
P-Cellulose	MN 300P (Firm 83); Cellex P (Firm 30)	0.7	2—20	
CM-Cellulose	MN 300 CM (Firm 83); Cellex CM (Firm 30)	0.7	2—20	
PP-Cellulose	MN 300 Poly-P (Firm 83)	0.7	2—20	Nucleobases, nucleosides [562]
DEAE-Cellulose	Serva DEAE-Cellulose ("TLC") (Firm 127); MN 300 DEAE (Firm 83); Cellex D (Firm 30)	0.7	2—20	Nucleotides [175, 558, 561], nucleosides, purines and pyrimidines [130], steroid sulphates and glucuronosides [490]
PEI-Cellulose	Cellex PEI (Firm 30); MN 300 PEI (Firm 83)	0.7	2—20	Nucleotide-coenzymes, mono- and oligo-nucleotides [560, 745]

[a] Firm Joh. A. Benckiser, GmbH., Ludwigshafen/Rhein, W. Germany.
[b] Firm Permutit AG., Berlin-Schmargendorf.

Table 10 (Continued)

Type of Ion Exchanger	Commercial Products	Capacity mval/g	Grain Size μm	Use and References
QA-Cellulose	Whatman thin-layer chromedia (Firm 115)			
ECTEOLA-Cellulose	Serva ECTEOLA-Cellulose "TLC" (Firm 127); MN 300 ECTEOLA (Firm 83); Cellex E (Firm 30)	0.35	2—20	Nucleic acid derivatives [557, 558], deoxyribo-nucleotides [48, 417] DNS steroid sulphates and glucurono-sides [490], sugar phosphates and nucleotides [162]
Sephadex Ion Exchangers	(Firm 102):			
SE-Sephadex	SE-Sephadex	2.3		
CM-Sephadex	CM-Sephadex	4.5		
DEAE-Sephadex	DEAE-Sephadex	3—4	40—80	Nucleic acid hydrogen-ases, adenine nucleotides [756]
Inorganic Ion Exchangers	(Firm 23)			
Zirconium phosphate	Bio-Rad ZP-1	1.9	2—44	Cation separations [780]
Zirconium tungstate	Bio-Rad ZT-1		2—44	
Zirconium molybdate	Bio-Rad ZM-1		2—44	
Molybdenum phosphate	Bio-Rad AMP-1		2—10	
Zirconium hydroxide	Bio-Rad ZHO-1		2—44	

Choice of Solvent in Ion-Exchange-TLC

It must be said in advance that in ion-exchange TLC, the regeneration or activation of the layer and the development of the chromatogram must be distinguished. Regeneration, i.e., conversion into the necessary ion-form, is carried out as far as possible before the material is spread on to the plates; it can also be performed on the plate with the solvent or buffer to be used subsequently. In cases of doubt, layers can always be activated with 0.1 N sodium chloride solution. In the examples so far known, chromatograms have been developed with: 0.01 to 1.2 N hydrochloric acid; acetate buffers; pure water or mixtures of alcohols with water and formic and acetic acids; 0.1–2.0 N sodium chloride solutions, with and without ammonia; citrate buffers; and other mixtures. Certain hints about the choice of solvent in a new separation problem can be obtained by consideration of experience gained from ion-exchange chromatography (see also [168, 169]). The techniques of using two solvents and of gradient development have been already described. Experience has shown in addition that a pre-treatment of the TLC-plate with distilled water often leads to better separations.

VI. Modification of Adsorbents by Impregnation

Egon Stahl

Successes in column chromatography using basic or acid impregnated aluminas, in paper chromatography using phase reversal and in GC using numerous organic phases, encouraged application to TLC.

The adsorbents can be impregnated in several ways:

1. Impregnation before Coating

The slurry is prepared with the aqueous solution of an acid, base or salt or of an organic, water-soluble compound, instead of with water.

2. Impregnation of the Ready, Dry Layer

a) By immersing the plates in a solution (usually 5—10%) of the involatile impregnation agent in a volatile solvent. The latter is then evaporated off.

b) By spraying the impregnation agent uniformly on the plate, removing solvent as in a).

c) By a type of ascending development with the impregnation solution, removing solvent as in a).

Information concerning the value of a given impregnation and the optimum degree of impregnation can be simply and rapidly obtained through gradient-TLC. For example, a 0–10% gradient is prepared and the mixture to be separated is chromatographed at right angles to the direction of the gradient.

Impregnation with Inorganic Compounds

As early as 1959, Stahl [665] pointed out the advantages of acidic or basic impregnation of silica gel layers; 0.1–0.5 N solutions of oxalic acid or potassium hydroxide were used.

"Acidic" silica gel layers can be profitably used to separate acid-reacting compounds (phenols, acids etc.); and "basic" layers for separation of alkaloids and amines etc. How an unknown compound reacts can be ascertained by TLC on layers of different basicities-best on pH-gradient layers. If the substances form salts with the impregnation agent, they usually do not migrate with weakly polar solvents, or at least migrate more slowly then the free bases and acids.

Buffered inorganic or even organic layers are comparatively often used, prepared with the usual buffer solutions.

A specific separation can be attained in some cases by impregnation with compounds which form coordination, chelate or inclusion complexes. Thus silica gel layers impregnated with silver nitrate have been used successfully for several years.

The separation depends on complex formation between Ag^+ ions and the π-electrons of one or several double or triple bonds in the substances to be separated; and on the strength of the complex-forming bond.

A marked improvement in separation is less frequently achieved with other complex-forming salts like sodium arsenite, tungstate, molybdate or metavanadate or basic lead acetate. Boric acid or borax are more often employed as impregnation agents for complex-formation. MORRIS [317a] and SCHORN [615a] have undertaken good compilations of the results so far obtained in this field.

Impregnation with Organic Compounds

The reversed phase technique in PC is generally known. It can be applied in TLC with at least as good if not better success. The layer is rendered water-repellent with on oil or fat and a hydrophilic mobile phase is chosen which is immiscible or only slightly miscible with it and which has been saturated with the impregnating agent. Reference is made to temporary or permanent impregnation, depending on the volatility of the impregnation agent.

The separated substances are often more easily identified when the impregnation liquid is removed; it is true that the degree of a temporary impregnation can less easily be reproduced than that of a permanent impregnation.

A selection of impregnation and separation conditions so far used is given in Table 11.

This table could be considerably extended since many "stationary phases" of GC could be in principle adapted. Firm 11 (catalogue No. 9, spring 1965) has made a compilation. Some ready impregnated adsorbents, such as silica gel, impregnated with silanes as described by KAUFMANN, are likewise commercially available (Firms 11, 88).

Application of Impregnated Layers

The useful effect of impregnation is clearly seen in the various chapters of the "special part" (of this book). A single example must suffice here: the terpene alcohols from C_5 to C_{20} appear more or less together on silica gel G layers. If this adsorbent is impregnated with paraffin, they

Table 11. *Compilation of Impregnation and Separation Conditions for TLC*

Impregnation agent	% in Solvent	Solvent	Layer	Development solvent	Examples of separation
Undecane	15	Petrol ether	Silica gel G	Acetic acid-acetonitrile (50 + 50) acetic acid-water (96 + 4) CHCl$_3$-methanol-water (25 + 75 + 5)	Fatty acids fatty acids diglycerides
Undecane	15	Petrol ether	Kieselguhr G	Acetic acid-water (80 + 20)	Lactones, keto-fatty acids
Tetradecane	5	Petrol ether	Kieselguhr G	Acetic acid-water (90 + 10)	Hydroxy-fatty acids
Decalin	5—10	Petrol ether	Silica gel G	Methanol-water (85 + 15)	2,4-Dinitrophenyl-hydrazones
Paraffinum subliquidum (= Nujol)	5—10 5—10 5—10 5—10 5—10	Petrol ether petrol ether petrol ether petrol ether petrol ether	Silica gel G kieselguhr G kieselguhr G cellulose silica gel-kieselguhr (1 + 1)	Butanone-acetonitrile (70 + 30) acetone-acetonitrile (80 + 20) 99—100% acetic acid acetone-methanol (66 + 33) methanol-isopropanol (90 + 10)	Cholesterol esters triglycerides fats + oils xanthophyll esters ubiquinones
Mineral oils (e. g., Shell Ondina 27)	10	Petrol ether	Silica gel G	Dioxan-water (60 + 40)	2,4-Dinitrophenyl-osazones
Squalane	5—10	Petrol ether	Silica gel G	Acetic acid-water (85 + 15) acetic acid-acetonitrile-water (10 + 70 + 25)	Fatty acids and esters
Silicone 1.5; 10 and 50 cSt.	5—7	Petrol ether	Silica gel G kieselguhr G	Acetic acid-formic acid-water (40 + 40 + 20) methanol-acetonitrile (50 + 40)	Fatty acids triglycerides
Vegetable oils and fats (Palmin/Livio)	7	Petrol ether	Kieselguhr G	Methanol-acetone-water (80 + 16 + 12)	Chloroplast pigments

Table 11 (Continued)

Impregnation agent	% in Solvent		Layer	Development solvent	Examples of separation
Polyethylene glycol 400	8 ml	In 48 ml 95% ethanol	14 g Seasorb + 7 g celite 545	n-Heptane	2,4-Dinitrophenyl-amines
Polyethylene glycol 1000	15 g	In 45 ml water	30 g kieselguhr G	Diisopropyl ether-formic acid-water (90 + 7 + 3)	Dicarboxylic acids
80—82% solution in methyl glycol of adipic acid triethylene glycol polyester	12 g	In 25 ml water + 25 ml ethanol	30 g silica gel G, and alumina G or kieselguhr G	m-Xylene-formic acid (98 + 2)	Benzophenones and other UV-adsorbers
	6.2 g	In 35 ml water + 35 ml ethanol	15 g cellulose		
	12 g	in 40 ml acetone + 20 ml water	30 g silica gel or kieselguhr G	Diisopropyl ether-petrol ether-CCl$_4$-formic acid-water (50 + 20 + 20 + 8 + 1)	Substituted amides of acetoacetic acid
2-Phenoxyethanol	10	Acetone	Kieselguhr G	Petrol ether (100—120° C)	2,4-Dinitrophenyl-hydrazones
Chlorobenzene	50	Ethanol	Silica gel G	Benzene-heptane (30 + 70)	Ketones or 2,4-dinitro phenylhydrazones
Nitromethane	50	Ethanol	Silica gel G	Petrol ether (60—70° C)	
Formamide	25	Acetone	Cellulose	Benzene-heptane-chloroform-diethylamine (60 + 50 + 10 + 0.2) etc.	Alkaloids
Dimethyl sulphoxide (DMSO)	50	Toluene	Silica gel	Ether; DMSO-tetrahydrofuran-diisopropyl ether (10 + 30 + 60)	Sugar acetates

can be separated according to the number of carbon atoms. Mixtures of alcohols with the same number of carbon atoms can be separated afterwards on a layer impregnated with silver nitrate (Fig. 105).

C. Apparatus and General Techniques in TLC

Egon Stahl

I. Preparation of Thin, Uniform Layers

A layer of finely powdered solids or their suspensions can be formed in several ways. The possible available procedures have been known in the film and varnish industries for a long time and they can be classified as follows:

1. pouring procedures (p. 53)   3. spreading procedures (p. 54)
2. immersion procedures (p. 53) 4. spraying procedures (p. 57)

There is thus a choice between these methods of application. A decision must be taken also about the sort of carrier material for the layer and which is the most advantageous form and size of this carrier.

Material, Form and Size of the Carrier

The carrier must be stable to all types of solvent and to reactive spray reagents, also to high temperatures. Mechanical strength and the possibility of repeated use are also desirable, combined with low cost.

Glass plates fulfil these varied requirements best and have been so far preferred. Machine drawn glass, that mostly used, is commercially available in the following thicknesses and shows departures from the norm as follows:

plate glass	thickness 1.8 mm + 0.2/—0.05
plate glass	thickness 2.8 mm + 0.2/—0.1
plate glass	thickness 3.8 mm + 0.2/—0.2
thick or	thickness 4.5 mm + 0.3/—0.2
building glass	thickness 5.5 mm + 0.3/—0.3

Poured and cut mirror glass is plane but difficult to obtain in sheets of precisely identical thickness. Borosilicate glass is used so that reactions may be carried out on the plate at temperatures of up to 250° C. Frosted glass has been suggested to improve adhesion of the layers [340]. Firmly adhering, dry adsorbent layers can be prepared on sand-blasted glass (verre sablé), ready for chromatography [766].

Advantages and disadvantages must be weighed against each other. Other material can be chosen also, e. g., plates and flexible sheets of stainless steel [134] or aluminium [378, 550, 641]; certain plastics, like frosted plexiglass, Vinylplaste VSA 3310 Cl (Firm 142) [398, 601, 656]; polyethylene terephthalate (Firms 52, 81); and even glass fibre carriers (Firm 63). The shape and size of the supports can likewise vary greatly. Rectangular, plane and more or less rigid plates or sheets are probably the easiest to handle. The concept of the "open column" has led to the use of commercial grooved glass as carrier plates [220]. Channels of this sort can also be etched into ordinary glass [270]. Further, a layer can be applied to the inside or outside of glass tubes or test tubes [403, 744] and to glass rods [191], all serving as rigid carriers.

It has been confirmed that a length of run exceeding 20 cm — in contrast to the so-called multiple development — yields no added advantage. The standard size of 20 × 20 cm has found general acceptance and accessories have been based on this size. The microscope slide size (26 × 76 mm) [296, 709] or cheap, washed photographic plates (5 × 5, 6.6 × 6.6 or 9 × 12 cm) [37, 469] doubtless suffice in many cases. Glass plates 20 × 40 cm and 20 × 100 cm are used as well as the standard size for preparative separations.

The following points should be considered when *choosing a procedure for coating:* number and nature of TLC plates needed daily; clean and rational work demands; layer evenness; reproducibility of the R_f-values; available laboratory apparatus; requirements of room space and personnel; the experience so far gained in any other laboratories of comparable size.

1. Pouring Procedures

Pouring procedures require no special apparatus. The necessary amount of adsorbent is weighed out and shaken to a homogeneous suspension with a suitable solvent. A previously determined volume of this is rapidly poured in one movement on to the middle of the plate to be coated. By gentle tilting and shaking, the suspension is distributed over the plate which is then dried in a horizontal position.

HÖRHAMMER and coworkers [295] have found in addition that a uniform layer over the whole plate is obtained only when the silica gel used (Firm 153) was suspended in pure ethyl acetate or acetone and not in water. To coat a 20 × 20 cm plate, 6.6 g silica gel were slurried with 16.5 to 23.1 ml ethyl acetate, depending on grain size. BHANDARI and coworkers [66] recommend a suspension of 6 g alumina or silica gel (Firm 153) + 96% ethanol/water (13.5 + 1.5 by volume = 86.4% alcohol). LEHMANN [392] shakes the adsorbents as a rule in 90% ethanol and pours mixtures of, e. g. 4.0 g silica gel G (Firm 88) and 12 ml ethanol or 3.0 g cellulose powder (Firm 83) and 14 ml ethanol.

PATAKI [505] has compared the fluctuations and standard deviations of the R_f-values on layers prepared in this way, with those prepared using spreaders and concludes that the reproducibility is better on the latter.

2. Immersion Procedures

Small glass plates can be rapidly coated by immersion in an appropriate suspension of the adsorbent. PEIFER [526] has carried out thorough investigations of this technique and quotes detailed working instructions and examples of use. The adsorbents must be evenly suspended in an organic solvent here, as in the pouring procedure, e. g., 35 g silica gel (Firm 88) in 100 ml chloroform or a 2:1 volume mixture of chloroform and methanol. Two equally sized, clean glass plates are placed face to face and then immersed in this suspension. Only one surface of each plate is thus coated through this "sandwich" technique. The solvent evaporates rapidly from the layer. The binder, in this case gypsum, is made effective by holding the strip briefly in steam. The layer is activated ready for use by drying for 1–3 min on a wire rack over a hot plate.

3. Spreading Procedures

Layers are prepared in most laboratories by spreading suspensions of adsorbents on glass plates. A series of devices has been developed for this technique and can be divided into two basic types. Common to both is that the suspension is run into a rectangular trough which has a suitable sized slit on one side. The Kirchner-type[1] of apparatus has a fixed trough. It is rigidly joined to the base at the side and the plates are pushed through in succession (Fig. 11). In the Stahl-type of apparatus, the trough is passed in one movement over a row of glass plates which are held firmly on an aligning tray (Fig. 13).

Fig. 11. Spreader for sliding through glass plates; with layer-thickness adjuster (photo, Firm 33)

a) Fixed Spreader (Kirchner-type)

In 1954, MILLER and KIRCHNER [448, 449] described an apparatus for coating narrow glass strips [23]. MUTTER and HOFSTETTER's apparatus, mentioned by WOLLISH [774] in 1961, also works on this principle. This apparatus has been obtainable in a variety of forms for a long time (Firm 33). JASPERSEN-SCHIB [323] regards it as dependable and easy

---

[1] BOBBITT [76] was the first to use this term and it is taken over here as practical. A number of "do it yourself" procedures have been described in addition to the commercial apparatus. These are discussed under 3c.

to handle. It is best operated by two persons, but one alone can manage. In this case, a suitable plate holder with two guide rails is made and the apparatus placed between them. Glass plates of 5, 10 and 20 cm width can be coated with various thicknesses using model B (Firm 33). The modifications of other manufacturers (Firms 106, 120, 137) are fairly similar to the original apparatus, at least externally.

b) Movable Spreader (Stahl-type)

If several carrier plates of the same width and thickness are laid together in a row on an aligning tray or template, they can be coated in a single passage over them of a spreader (Fig. 12 B). The procedure is simple, clean and rational and is thus used in many laboratories for coating (Fig. 12).

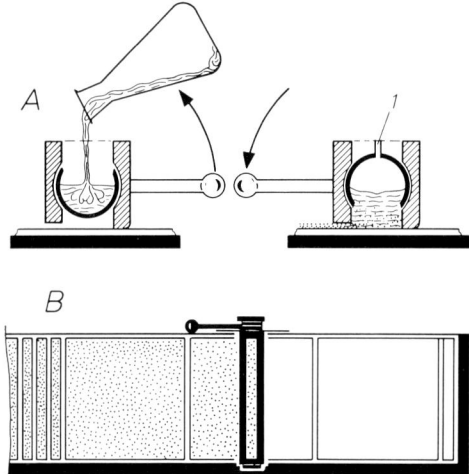

Fig. 12. Operation of the Stahl TLC-spreader. *A* Cross-section: filling with the slurry (left); position ready for spreading (right); note the opening for entry of air (*1*). *B* Aligning tray with row of glass plates and spreader viewed from above

The development of the apparatus began in 1956 with a small, cylindrical spreader possessing a device for adjusting layer thickness. It was however for glass strips of up to only 5 cm width [659], but permitted the study of the factors with an influence. It was the predecessor of later apparatus.

The 20 cm-spreader with rotatable jacket (Fig. 12 A) and with the possibility of simple adjustment of layer thickness between 0 and 2 mm, which was then developed [661, 685] was commercially available by 1958 (Firm 44) and is the most used TLC spreader[2]. The mass

---

[2] Working instructions for the use of spreaders can be omitted in this edition; they are provided with each commercial apparatus.

produced apparatus, denoted by "GM" and dating from 1964, can, with a few accessories, be rapidly and simply changed into a spreader for preparing gradient layers (p. 92). The same firm (44) manufactures

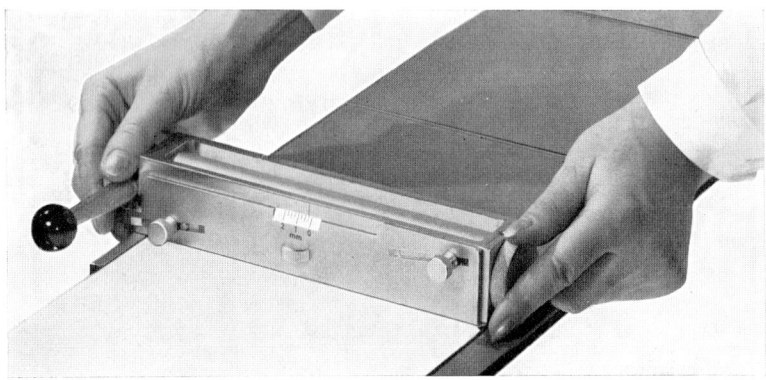

Fig. 13. Coating a row of plates using the spreader with layer-thickness adjuster, developed by STAHL (photo, Firm 44)

also the basic apparatus with larger capacity for preparative TLC and also a smaller and simplified form. Further, simple "spreading troughs" for 20 cm plates (Firms 111, 117, 129) and for smaller sizes (Firms 44, 111) in various materials are available commercially. Silver nitrate-silica gel layers are best prepared in V4A-steel (Firm 44), in silvered spreaders or in those made of a suitable plastic (Firm 11). Glass plates of equal thickness are used in the usual aligning tray for five 20 × 20 cm plates. Special holding trays have been developed (Fig. 14) for coating evenly plates of varying thickness. The plates are pressed from below against two guide brackets at the side so as to compensate for differences in thickness in the row to be coated (Firms 111, 129).

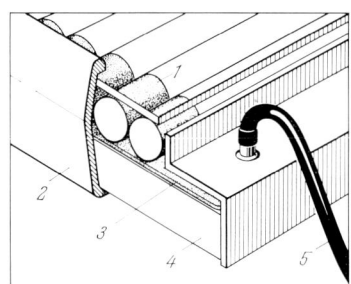

Fig. 14. Constructional details of a holding tray with compensation for different plate thicknesses. *1* rollers beneath the glass plates; *2* lateral guide brackets; *3* inflatable air cushion; *4* supporting frame; *5* rubber bulb for pumping up the air cushion *3* (according to Firm 129)

c) Spreading Rods and Home-made Apparatus

"Do it yourself" spreading troughs have been described [68, 80, 124, 144, 174, 407, 487, 742]. A "spreader" can be made out of the simplest laboratory material, e. g., from a glass rod [390] and two short pieces of rubber tubing (Fig. 15a). Next to it (Fig. 15b) is a TLC spreading bar (Firm 127) for layers 5 or 10 cm wide and 0.3, 0.5 or 1.0 mm thick. Figs. 15c—f illustrate further ways of fixing in advance the desired layer thickness by means of two lateral and supporting rails. After it has been poured, the slurry is then spread smoothly with a spatula [260] or metal rail etc.

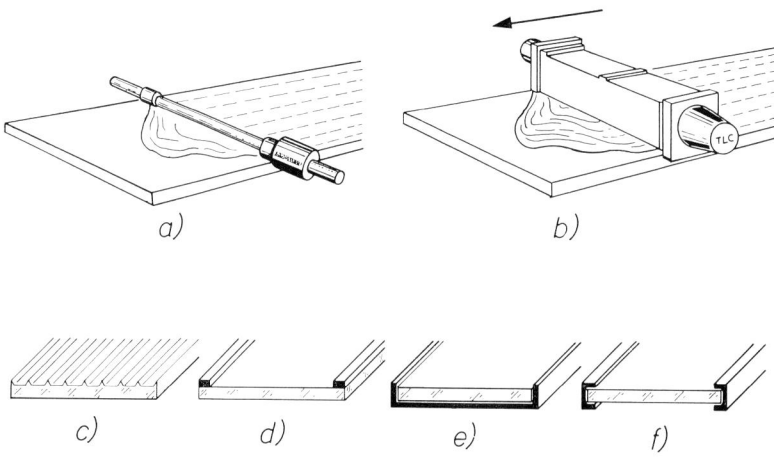

Fig. 15. Various ways of attaining a definite layer-thickness. *a*) glass rod as spreader; *b*) TLC spreading bar (Firm 127); *c*) grooved glass [220]; *d*) mounting of lateral glass strips; *e*) tray in which the glass plate lies; *f*) two attached U-type guide rails

4. Spraying Procedures

Suspensions of fine-grain adsorbents can be evenly spread on carriers using commercial sprayers. The uniformity of the layer depends however on the skill of the operator when the spraying apparatus is hand operated-more so than in the other procedures. SUTTER's proposed "TLC with the spraying procedure" [303] utilises a spray pistol in which the suspension is prepared in a container using an air current; a spray chamber, like a fume cupboard, containing a grating and a spirit level; and a compressor to develop the necessary pressure of 1—1.5 atmosphere (Firm 73).

MORITA and HARUTA [457] have used a simple glass wash-bottle to spray a suspension of 10 g silica gel G (Firm 88) in 30 ml water. They coated simultaneously 21 microscope slides (25 × 75 mm) which they had placed in three parallel rows of seven, a few millimetres apart and forming a rectangle. According to BEKERSKY [54], layers 0.25 to 1.1 mm thick can be prepared with the spraying technique. Thicknesses of ca. 1 mm are realised by spraying coat by coat at two minute intervals.

Suspensions of adsorbents, like many reagents, can be filled into Aerosol spray vessels, ready for use. A three-piece Aerosol sprayer of the "Spray-gun" type is very advantageous for spraying (Fig. 38).

The spraying procedure is that used in industrial preparation of ready coated plates and sheets on a conveyor band.

5. Automatic Coating

Quite early on, the idea arose of simplifying yet further the coating of plates. In this connection STAHL demonstrated in 1957 the first automatic thin layer spreader [660, 661]. The glass plates are pressed against the spreader part and glide through smoothly with the help of a small conveyor belt (Fig. 16). Apparatus of this sort has the advantage of giving uniform spreading and therefore coating evenly. Moreover, glass plates of varying thicknesses can be used without difficulty and considerable time is saved.

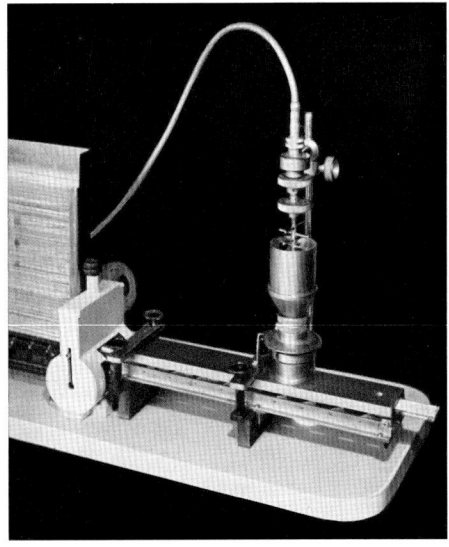

Fig. 16. Automatic coating device. On the left, the magazine with glass plates and beneath it the conveyor belt and motor drive. To the right, the aligning tray and layer-thickness adjuster. The slurry is fed into the funnel in which it is stirred [661]

TAKITANI and MATSUDA [706] have developed a transport system driven by an electric motor in order to draw the Stahl spreader regularly across the line of plates (100 cm in 30—40 sec). The device described by MARCUCCI and MUSSINI [429] is likewise equipped with an electric motor and the plates are drawn through beneath the trough for spreading.

There are two types of commercial motor-driven apparatus at present. In one, the spreader trough is carried over the row of plates by a motor drive (Firms 15, 35); in the other, the trough is fixed and the plates are passed beneath it using motor-driven rubber rollers (Firms 33, 54). A further development of the Stahl automatic apparatus permits the rational preparation of gradient layers [675] (Firm 44). This apparatus will have automatic devices for delivery, removal and stacking of plates so that they can be dried.

Such an apparatus is most time-saving when more than 100 TLC-plates are required daily. The experience gained in the film industry will have to be turned to account for mass preparation of TLC-plates or sheets, and preference given to a combination of conveyor belt and spraying procedure.

6. Ready-coated Plates and Sheets

A few years ago the coating of glass plates began industrially in the USA. An assortment of TLC-plates is offered, with names like "Uniplate" (Firm 8) or "Quick-Check" (Firm 86)[3]. The firm manufacturing the adsorbent can sometimes be recognised from the suffix. The layers can be impregnated also if desired. It is to be welcomed that the plates are provided in the sizes 20×5, 20×10 and 20×20 cm; these are now generally used and this means that all TLC apparatus may be standardised with these sizes. "Chromatogram Sheets" were then developed, profiting from the experience of the film coating industry [51, 163, 356, 398, 546]. Makrolon, a polyethyleneterephthalate plastic, serves as carrier sheet. A thin, porous, comparatively adhesive layer of only 100 μm thickness is spread on to it. Polyvinyl alcohol has been used as binder for silica gel (Firms 107, 148), as suggested earlier by ONOE [491]. The silica gel sheets manufactured in the USA are denoted with R (= Rochester, Firm 47) and have high migration speeds compared with the French sheets, denoted with V (= Vincennes, near Paris; Firm 81). Sheets coated with polyamide (K 541 V) and polycarbonate (K 511 V) are available in addition to the silica gel sheets (K 301). The sheets are light, flexible and unbreakable, in contrast to glass plates. Consideration must however be given to the varied stability towards chemicals and heat, both in the choice of solvent and in the use, often necessary, of aggressive spray reagents and subsequent heating. Attempts have been made again recently to introduce a flexible, glass-fibre material as support for adsorbents (Firm 63).

In the European industrial countries also, the manufacture of precoated plates and sheets has begun (Firms 83, 88).

---

[3] In 1965, a coated 20×20 cm plate cost $ 1.20.

II. Preparation of TLC-Plates
1. Drying, Storage and Handling

The adsorbent is usually spread as an aqueous slurry on the carrier plates. The water (6—8 ml per 20 × 20 cm plate) must be removed, a process termed drying. Water removal goes through the following stages:

The surface of the layer has initially a moist lustre.

After a few minutes this gives way to dullness.

When about half the water has evaporated off, the layer ceases to be transparent and becomes white.

Fig. 17 shows the removal of water during drying of a silica gel G layer.

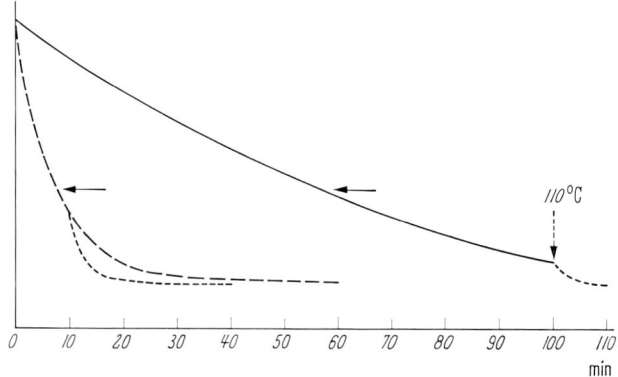

Fig. 17. Course of the release of water from freshly prepared silica gel G layers under various drying conditions. —— without drier at room temperature; - - - with drier; ...... drying oven at 110° C. The two arrows indicate the point at which the layer ceases to be transparent and turns white

It clearly saves time if the bulk of the water is first removed in a current of air (pre-drying) and the process then completed by drying the plates at higher temperature. The coated plates are left on the tray for pre-drying. The hot air drier, "Astron 2000" (Firm 132), has proved satisfactory for providing multiple stage, adjustable cold and warm air currents.

The drying is then completed at a higher temperature. Rectangular drying cabinets at least 42 cm wide (see preparative TLC) with an air blower for circulation, are the most suitable for this. Very active layers are obtained by heating silica gel and alumina plates for 3—4 hrs at 150° C.

High activation is worthwhile only if the substances are applied in a correspondingly dry atmosphere and if completely anhydrous solvents are used. This is necessary only occasionally, in the TLC of hydrocarbon mixtures. There is a grave danger that the substances will decompose on very active layers. The work of HESSE [287] on decompositions in chromatographic columns is worthy of notice here.

Opinions differ as to the most suitable duration and temperature of drying. In the TLC of polar compounds, e. g., amino acids, no pre- and hot air drying is carried out and the plates are allowed to dry overnight at room temperature. Layers of this type show good adhesion and yield more reproducible results than do activated plates.

Since 5 or 10 plates, 20 × 20 cm, are usually coated in one working operation, compact and handy drying racks are convenient (Fig. 18).

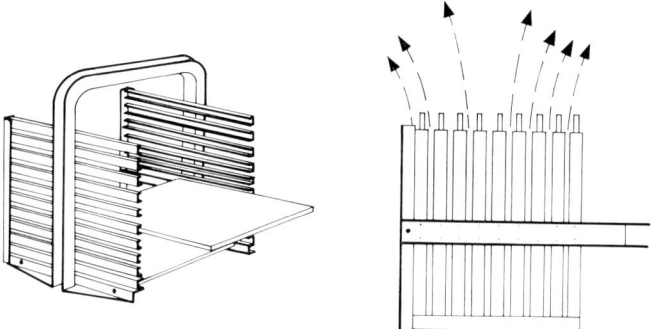

Fig. 18. Light alloy racks for drying and storing plates. Left: inserting the pre-dried plates; right: drying the plates in vertical position so that moisture (see arrows) can escape (manufacturer: Firm 44)

Method of Operation

The freshly coated plates are left on the tray until the transparence of the layer has disappeared. An air stream, hot or cold, gentle at first, accelerates this pre-drying. After about 10 min, the plates are stacked in a drying rack and heated in the vertical position for 30 min at 110° C. It is advantageous to open the door of the drying cabinet from time to time to allow the moist air to escape. The hot rack is taken out with a cloth or special carrying handle (Firm 44), tipped horizontally and placed in a dry container containing indicating silica gel.

As well as desiccators, other air-tight glass, plastic, wooden or metal containers may be used for dry storage and transport of the plates.

2. Control and Marking of the Layer

The uniformity of the layer must be tested before marking and spotting. It must appear homogeneous in transmitted light and show no unevenness or streaks. No larger grains should be recognisable in reflected light. The adhesive properties of the layer may be tested by gently rubbing a finger across the surface.

A TLC plate which is faulty only on part of the later length of run, may be used for preliminary experiments. It is best to strip off a 2–4 mm margin from the layer. This strip is generally slightly thinner from being under the supported surface of the spreader. The margin can be speedily and cleanly removed with the thumb, using the index finger as a guide. Alternatively, an easily made stripper, as shown in Fig. 9, 1st edition, or a similarly angled piece, may be used.

It is an advantage to mark on the layer the starting point, the line of the front and the nature of the applied substances. A sharp, hard pencil or a dissection pin from a biological or medical dissecting set can be used for this purpose. The layer is easily cut with these and the points or writing are engraved. A socalled template, which now has a wider sphere of application, is used to prevent damage to the rest of the layer. A centimetre scale is engraved along one edge of the transparent template, facilitating the marking of starting points at regular intervals (Fig. 19). The template can be used also to read off Rf-values and spot sizes.

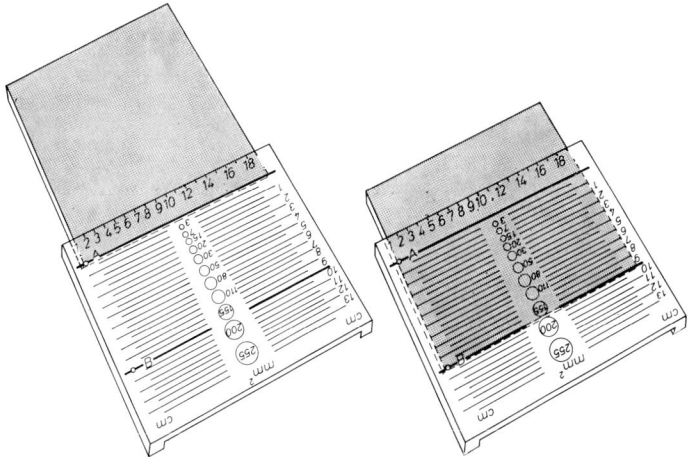

Fig. 19. Use of the multi-purpose template. Left: position for marking start points; right: marking the 10 cm run. Details in text (manufacturer, Firm 44)

Uses of the Multi-Purpose Template

The template (Firm 44) is placed astride the TLC plate so that the thick cross-line A coincides with the lower edge of the plate (Fig. 19, left). The starting points are marked by small dots or the sample can be applied directly at equal intervals with a micropipette. The points should lie 15—20 cm from each neighbour and the line of points 20 mm from the edge of the plate. In order to mark the subsequent "front line" as a dotted line, the template is moved forward until the lower edge of the plate coincides with the line B (Fig. 19, right). Each substance to be separated is noted above the "front line" or is numbered or lettered; solvent, other experimental details and the visualising agents are noted in addition.

To evaluate the chromatogram, the template is placed again in the second position and the Rf-values can be read off directly for a length of run of 10 cm. The approximate sizes of circular spots can be rapidly and semi-quantitatively estimated by comparison with the variously sized circles on the template.

3. Application of the Substance Mixture for Separation

The mixture is brought into solution and applied to the "start" (starting point or line) as a spot or band respectively.

Strongly polar or involatile solvents should not be used in TLC for dissolving solid samples or diluting liquid mixtures. These yield large starting spots and ring chromatograms. If really necessary, very small volumes must be applied successively and the solvent removed between additions with a current of warm air. Care must be taken to ensure that the substances do not crystallise out at the starting point. Should this happen, chromatograms with long streaks (tails, beards) are obtained, in the direction from start to front.

Most solutions for application are between 0.1 and 1%. From 1 to 20 mm^3 is normally applied[4].

The start points should be equally sized as far as possible and have a diameter between 2 and, at the most 5 mm.

The application of standard solutions and pre-determined volumes is strongly recommended, even in so-called qualitative chromatography. Various measuring devices can be used for this purpose, like those illustrated in Fig. 20.

a) Spot Application

α) Lüdy-Tenger[407] recommends using **platinum loops** (Emich[182]) for rapid application of usually 1% solutions. A loop for applying 10 µl water can be made from 0.4 mm thick platinum wire; the internal cross-section of the loop must thus be 1.5 mm (Fig. 20a).

β) **Micro bulb pipettes** exist in many forms. The "microcaps" (Firm 51) are intended for a single use. These are precision capillaries of capacity 1–100 µl. They may be used with or without a filling device (Fig. 20b). Accuracy of better than $\pm 1\%$ is claimed. The self filling and adjusting Lambda pipette (Fig. 20d) exists in the normal form with mark (Fig. 20e) and in several other forms from 1 to 1000 µl (Firm 117). The "automatic" micropipette is based on the same principle (Fig. 20c) (Firm 33).

γ) **Graduated micropipettes** with a capacity of 10 µl and bevelled tips are those mostly used for application in TLC (Fig. 20f). They are sometimes known as "blood sugar pipettes" from a former field of application. They often require standardisation.

δ) **Micro syringes** (Fig. 20g) are preferred in gas-liquid chromatography for injecting measured amounts of substances. They have been appreciably improved in recent years. They are frequently used in TLC with a micrometer screw and holding device (Fig. 20h) and are indispensable for preparing calibration curves by quantitative application of increasing amounts. Their manufacture demands the highest precision (Firms 29, 44, 69).

---

[4] 1 mm^3 = 1 λ (1 lambda) = 1 µl (1 microlitre).

Various syringe pipettes with standardised stroke, operated by pressing a knob, have become commercially available in connection with quantitative clinical chemical analysis in the µl domain (Firm 53). Continuous infusion apparatus with small syringe pipettes (Firm 27) can be used to apply aqueous solutions in the form of very small start points to several chromatograms simultaneously [441]. MORGAN [456] has described an easily constructable device (Firm 136) for simultaneous spot application of 19 different solutions or for band application of a single sample.

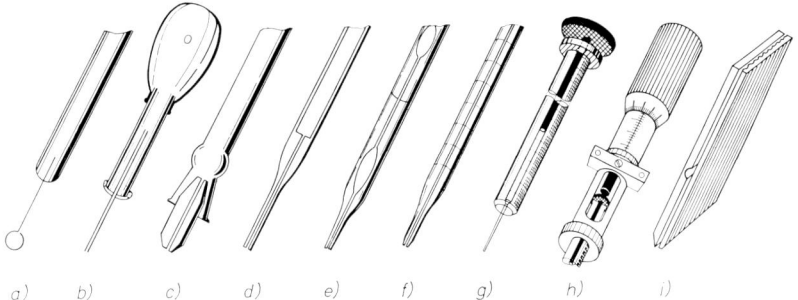

Fig. 20. Various types of equipment for applying substances in solution. Details in text

b) Band or Streak Application

The success of a band chromatogram depends largely on the uniformity of application. When the solution of the substance is applied spotwise to form a band, the zones are often blurred unless great care is taken (Fig. 21, right). A uniform spot application can be achieved with the Morgan-applicator mentioned above (Firm 136). Up to 37 glass capillaries are fixed alongside each other like the teeth of a comb. They can be tilted so as to dip into a narrow trough containing the solution to be applied. They fill by capillary attraction and are then touched on to the layer, emptying simultaneously. The Stahl broad band pipette (Fig. 20i) (Firm 44) can also be employed for band application, as described earlier. STOCKER [699] has proposed a 100 µl microcapillary pipette with the tip bent at a right angle and which is moved along a guide rail.

Since rather larger amounts are often needed in quantitative microdeterminations, band application of accurately measured volumes is important. The technique suggested by BACON [30] is impressive in its simplicity.

An Agla microsyringe is fixed horizontally on a small mobile platform travelling on four wheels and made from parts of toys. The screw-in type plunger is rotated by one set of wheels. As the platform is moved across the foot of the chromatogram, the movement of the wheels operates the plunger which ejects the solution at a uniform rate through a flexible tube passing over the layer.

Band application is important also in preparative TLC. Various devices have been developed for this purpose, with which millilitre amounts can be sprayed or squirted out in a fine jet (see p. 99, Figs. 50, 51).

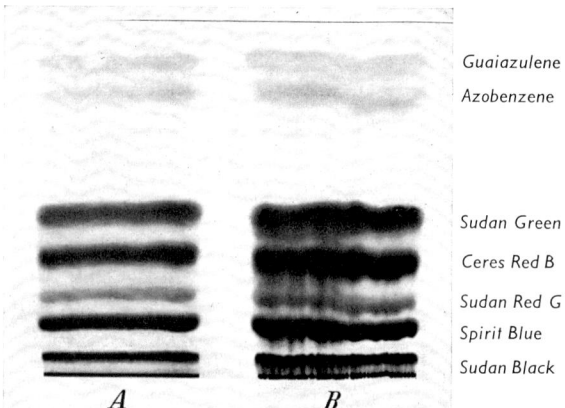

Fig. 21. TLC with band application of the substances. *A* Spraying or squirting the solution. *B* Spotting the solution. Compare the resolutions achieved

As WAGNER and POHL [734] have shown, solutions can be applied as a band using a soft brush. TAMURA [708] has employed for the same purpose, a capillary pipette in which a brush made of fine glass fibres had been inserted.

III. Separation Chambers and Development

The transition from the Tswett column to the "open" separation layer has led to the development of all types of chamber for separation. The early types resembled those used in PC and only their shape was modified. Newly acquired data then led to the use of special narrow chambers. A better overall picture is obtained when the types of chamber so far used are classified according to the nature of the development:

1. chambers for ascending development,
2. chambers for descending development,
3. chambers for horizontal development,
4. chambers for thin-layer electrophoresis (see p. 110).

Where possible, glass is preferable and only occasionally the dearer, stainless V4A-steel. Plastic chambers may be used with aqueous and less aggressive solvents.

Adsorption-TLC has created a series of special problems which have been only gradually recognised and solved. First, STAHL [665] drew attention to the importance of the degree of saturation of the atmosphere in the chamber and showed that it was necessary to adhere to certain conditions here also. He introduced the ratio *evaporation surface area: chamber volume* as a characteristic [673]. This ratio is 1:20 in the usual rectangular chamber but only 1:0.1 to 0.5 for narrow chambers. He attributed particular significance to chamber saturation after it had been possible to eliminate in this way the so-called edge effects.

1. Chamber Saturation and "Equilibration" of the Layer

In an ordinary chamber the observation is made that substances near the edge of a thin layer migrate further than the same substances in the centre (Fig. 22). This "edge effect", noted also by DEMOLE [152] occurs especially when mixtures of solvents are used which differ considerably in polarity, vapour density and density. The effect is due to inadequate chamber saturation. The plate divides the chamber in two. The developing solvent evaporates more rapidly at the edges of the side carrying the layer, in order to saturate the rear of the chamber as well. STAHL [665] therefore suggested chromatography in rectangular tanks with chamber saturation (CS).

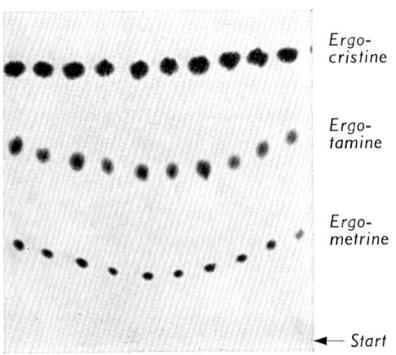

Fig. 22. "Edge effects" during TLC in a chamber with insufficiently saturated atmosphere (= NS). Solvent: chloroform-methanol (95+5) [665]

Establishment of Chamber Saturation (CS), see Fig. 23B

A smooth sheet of resin- and fat-free filter paper, about 15 × 40 cm, is laid in the form of a *U* in the glass trough and soaked in the solvent which has already been placed in the trough. After being thus moistened, the paper is then pressed against the sides of the chamber. It usually adheres but can be fixed more firmly with the help of a ring of thin steel wire which presses on the paper round the inside of the chamber. The paper is again soaked in the solvent by tilting the chamber, before the plate is introduced.

The technique just described is termed "chamber saturation" (CS; formerly known as chamber supersaturation) as distinct from the *normal* saturation (NS) (Fig. 23). GEISS and coworkers [224, 226] have studied in detail the processes occurring in various chambers for separation. They showed that in the saturated chamber (CS), the solvent evaporates slightly from the region of the front and that the layer ahead of the front ("fore-layer") takes up considerable amounts of solvent vapour (e. g., benzene). This "fore-layer" moreover gives up water-10 molecules benzene expel 1 molecule water. Chamber saturation

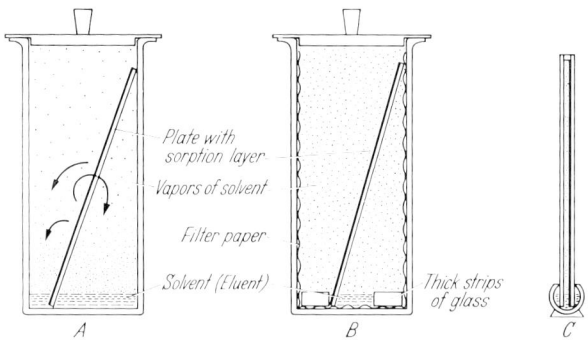

Fig. 23. Separation and saturation. *A* chamber with normal saturation (NS). The arrows indicate solvent evaporation from the layer and the dots symbolise the vapour density. *B* chamber saturated with solvent by means of filter paper lining, saturated with the solvent (CS). *C* reduction of the chamber volume using the S-chamber system

thus leads to advance saturation of the layer. This raises the rate of migration in the "fore-layer", whereby the activity is only slightly increased and the flow volume is less. The Rf-values are consequently lower than in a normal chamber without saturation (NS).

In a closed S-tank (without extra saturation) such equilibration does not take place or has as little an influence as the evaporation effect. The influence of the relative humidity first outside and then within the chamber must always be considered in adsorption-TLC. The TLC-experiments carried out in this connection, using various relative humidities, are impressive (Fig. 24, right) [226]. According to these, it appears necessary to develop thin-layer chromatograms in a moisture chamber in

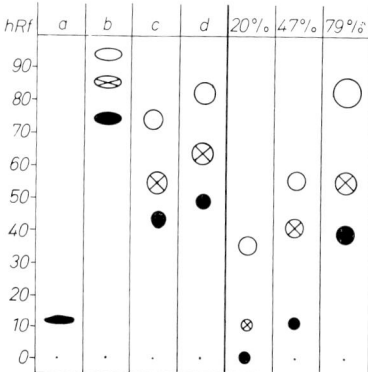

Fig. 24. The influence of various degrees of saturation (a—d) and relative humidities (20—79%). *a* test mixture, developed with methylene dichloride on a silica gel G layer without a chamber; *b* the same in a chamber at NS; *c* in the same chamber but at CS; *d* in the S-chamber. Alongside are developments in the GS chamber at different relative humidities. ○ = Butter Yellow; ⊕ = Sudan Red G; ● = Indophenol

which humidity is controlled and other conditions maintained constant, in order to obtain reproducible result (see p. 71).[5]

2. Setting up the Chambers
(Temperature, Light, Protection against Oxidation)

Separations are carried out generally at room temperature (about 20° C). One should however guard against heating or cooling on only one side when assembling chambers. Even small temperature differences within the chamber can bring about undesirable "oblique running" of the front and, in S-chambers, to unwanted condensation of the solvent on the cover glass.

On no account must the chambers be exposed to sunlight but should be placed in diffuse daylight. It is advisable in analytical work to cover the inside of the laboratory window with sheets opaque to UV-light (Firm 34) or an equivalent varnish (Firm 141). For the thin-layer chromatography of substances like carotenes which are sensitive to light, the outside of the chamber is covered with black foil or one works in a dark room in red or green light.

Oxidation of substances on the layer can be prevented by placing a so-called preparation box (Firm 44) over the plate during application and filling it with nitrogen or carbon dioxide. A fine spray of water often averts decomposition at the start point.

In order to develop chromatograms in absence of oxygen, carbon dioxide is passed into the chamber (as far as the substances allow it); or nitrogen is first swept through it. A cover equipped with tubulure and two-way tap is an advantage here.

3. Chambers for Ascending Development

a) Rectangular Chambers

Ascending development is that mostly used in TLC. The plate is placed in a chamber filled with solvent to a depth of about 0.5 cm. The lid of the chamber must fit tightly. Chambers specially made for TLC (21 × 21 × 9 cm) have a wide protruding edge and a plane cut knob cover glass. A cover fitted with standard joint tubulure is available (Firm 44) for passing through gases or liquids or for a thermometer. A Cryobox, shown on p. 95 (Firm 44) may be used for work at a particular temperature between -50 and $+50°$ C.

Note: the whole breadth of the TLC-plate must dip into the solvent to the same depth, i. e., ca. 0.5 cm, but not so as cover the spots at the start. Mixtures of

---

[5] It was repeatedly stated in the first edition that the Rf-values can serve only as guide values; they show the sequence and approximate position of substances on the chromatogram.

dissimilar solvents should be used only once or twice for development and from time to time the chamber should be opened for a moment. Other plates should not be placed in the chamber during the chromatographic procedure.

Two 20 × 20 cm plates can be placed like a V. Several plates may be developed simultaneously with the help of a holding frame made of stainless steel. The so-called "Chromatostack" technique [488] can be used also. In this, the plates are stacked, supported and separated at the corners by four plastic corks. The pile is held together with two rubber rings and put into the chamber.

A type of continuous chromatography can be realised even in the ordinary chambers. BENNETT and HEFTMANN [58] have attached a bent aluminium trough to the edge of the coated plate; loose adsorbent was filled into the trough and sucked up the solvent continuously which migrated through the layer (Fig. 25A). TRUTER [721] has described another scheme, whereby the end of the plate projects a few centimetres out of the top of the chamber, causing the solvent to evaporate. The chamber can be sealed with a cover made of two halves, enclosing the plate like a collar (Fig. 25B).

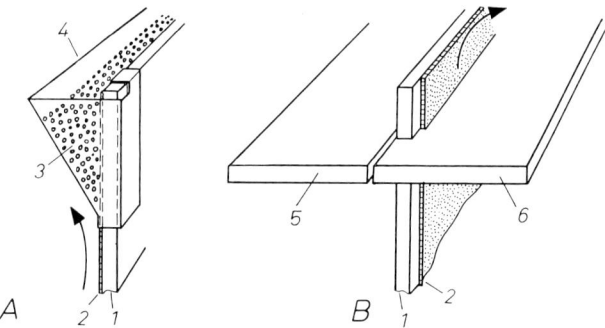

Fig. 25. Devices for continuous development. *A* angled trough containing adsorbent; *B* two-part special cover through which the plate projects. *1* carrier plate; *2* layer; *3* adsorbent for drawing up the solvent; *4* angled trough; *5* glass plate; *6* glass or metal plate with suitable slots

For preliminary tests, the 5 and 10 cm plates are more often employed. Development can be carried out in circular cylinders with a ground glass stopper; or in jamjars of suitable size. Further, as in PC, modified battery jars have been proposed as chromatography chambers [250].

b) S-chambers[6]

A new type of chamber, the S-chamber [673][7], developed out of the results of experiments on the influence of chamber size and of saturation.

---

[6] S stands for "schmal" (German for "narrow") or "sandwich".
[7] German patent [684].

The TLC-plate forms the rear wall of the S-chamber and a cover plate (frame plate) the front wall. The "side walls", only 2—3 mm thick, are narrow glass strips sealed on to the cover plate (Fig. 26). The necessary gap can be obtained also by inserting strips of cardboard [33], wire [793], glass [151] or plastic (Firm 44). The two plates are held together with strong clamps (see Fig. 26). One must take care that the S-chambers are air-tight on the two short sides-which is evidently not always the case with many simple "sandwich" chambers. The suggestion has been made of dipping the plate pair to a depth of about 0.5 cm into molten wax [593]. The "sandwich" is developed by standing in a suitably sized, gutter-shaped trough. A so-called S-chamber trough[8] has been made out of V4A-steel to ensure a tight fit. It consists of two jackets fitting in each other. The inner, fixed jacket forms the trough itself and the outer closes the system on turning.

Fig. 26. S-chamber system according to STAHL with a 40 cm wide thin-layer chromatogram (manufacturer Firm 44)

More and more interest is being shown in the S-chamber system. It was originally devised for developing TLC-plates 40 cm wide, using little solvent. Many separations are more successful in the S-chamber than in the normal types, e. g., of resins, as shown by JORK [333] (cf. Fig. 108). In order to be able to use flexible sheets also in an S-chamber, the frame plate has been covered with equally spaced glass blobs on the side nearest to the layer (Firm 52). Usually it is unnecessary to "saturate" these narrow chambers. There are cases, however, as cited by JÄNCHEN [314] where rapid conditioning of the layer with solvent vapour is an advantage. The relevant experiments of GEISS and co-workers [226]

---

[8] See footnote [7] on p. 69.

may be referred to in this connection. A fast "equilibration reversal" of the layer in the S-chamber can be achieved as follows; a smooth filter paper, soaked with solvent, is attached to the cover plate. Still better is to pour a thin adhesive layer of cellulose powder on to the cover plate (see pouring procedure, p. 53). After drying, the coated cover plate can be saturated with the particular solvent.

c) Moisture Chambers

It has been known for a long time that the chromatographic activity of "drying agents" like silica gel or alumina, depends on their water content. BADINGS [33] carried out TLC in a specially constructed rectangular chamber, in absence of atmospheric moisture. GEISS,

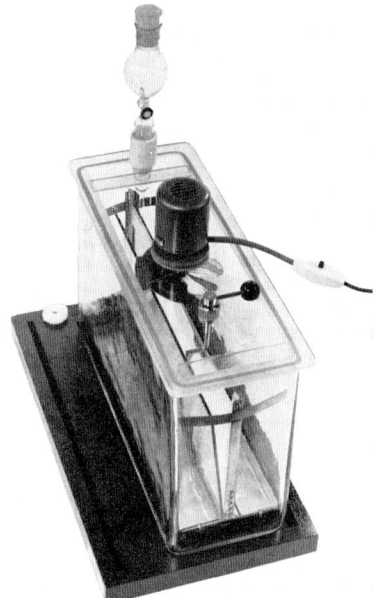

Fig. 27. GS moisture chamber for TLC under controlled moisture conditions (Firm 44) [227]

SCHLITT and KLOSE [226] then worked intensively on the problem of moisture control and TLC. They showed amongst other things that during chromatography, that part of the layer which has not yet been moistened, can take up solvent vapours and release water. The uptake of water and numerous other "equilibration reversals" can take place within a few minutes. They developed a so-called GS-moisture chamber (Fig. 27) in order to eliminate the shortcomings known up to that time. It consists of a large chamber in which a S-chamber system stands and two shallow troughs for the "equilibration solution".[9] A fan fitted in the cover, ensures rapid establishment of equilibrium. After equilibration, the solvent for development is introduced through a dropping funnel fitted into the cover. The same authors have described also a simple CS-chamber for obtaining orientation data about the sensitivity of a separation to moisture. In this, a TLC plate lies horizontally on a trough which contains the equilibration solution. Solvent is supplied from another trough through fritted glass.

General experience with moisture chambers is not yet available. TLC under moisture control in a special moisture chamber is indispensable for comparing Rf-values. Only in this way can adsorption chromatography be carried out in the tropics with their continually very high humidity.

4. Equipment for Descending TLC

Descending development is frequently used in paper chromatography, yet seldom in TLC. It neither shortens the migration time nor improves separation and continuous development is more easily achieved in another way (p. 69, 75).

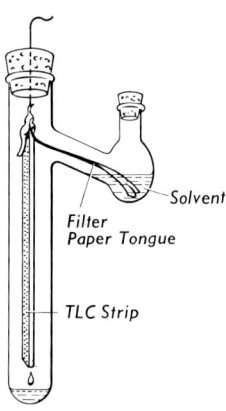

Fig. 28. Large test tube with side arm for descending development of a narrow TLC-strip (according to [688])

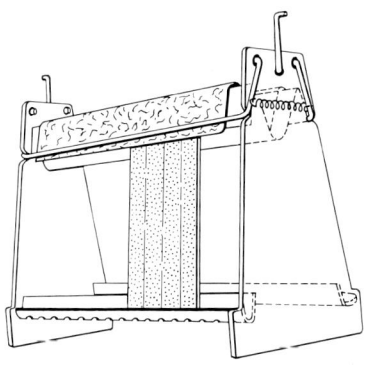

Fig. 29. Stand for prior washing of TLC strips and plates. The solvent is in the trough (above) and is fed to the thin layer via filter paper. The hoop presses the paper on to the layer (according to [689])

STANLEY and VANNIER [688, 689] were the first to describe a device for descending TLC (Fig. 28). It is commercially available (Firm 117). The stand recommended for prior washing of TLC strips (Fig. 29) can

---

[9] Saturated aqueous solutions of ammonium chloride, potassium thiocyanate or potassium acetate with relative humidities of 79.5, 47 and 20% respectively.

be used for the 10 or 20 cm wide plates which are preferred nowadays. Solvent evaporation may be prevented by standing it in a chamber with the appropriate saturation (CS). BIRKOFER and coworkers [71] have worked out another device for descending TLC. Here, the solvent is carried from a narrow trough to the layer via a roll of filter paper. Descending development can be carried out also in a BN-chamber, placed vertically, according to the same principle. A device has been manufactured for descending TLC and collection of eluent fractions, reminiscent of those for column chromatography (Firm 11).

5. Devices for Horizontal Development

A more or less horizontal plate is essential in the TLC on loose layers. A uniform supply of solvent must be ensured here also, as in descending development. The following procedures can be distinguished according to the nature of solvent supply and the technique of application:

a) Circular technique in open or closed systems. Solvent is supplied at a point, yielding ring-shaped zones. This technique is also termed radial or disc chromatography.

b) Centrifugal technique. The solvent is applied at a point and swept rapidly outwards through the centrifugal force of the rotating plate.

c) Horizontal technique with frontal supply of solvent.

a) Circular Technique

In its simplest form circular chromatography was used as early as 1938 by ISMAILOV and SHRAIBER [313] (Fig. 1). MEINHARD and HALL [439] worked with the same technique, likewise without chamber. The standing pipette for solvent supply, illustrated in their article, is most practical (Fig. 30). It is placed in the centre of the substance spot which has already been applied.

A little known publication of DATI and coworkers [146] from 1957 is interesting in this connection; they separated urinary steroids using circular chromatography on anode-oxidised aluminium plates in a desiccator. PEYRON [521–524] has worked extensively on the various techniques of "radial development" and made a compilation of the relevant literature.

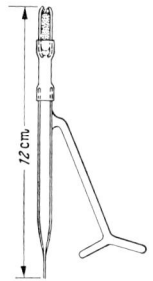

Fig. 30. Standing pipette for developing ring chromatograms (according to [439])

As STAHL [661, 662] showed in 1958, the microcircular technique is excellently suited for finding out quickly the requisite solvent (Fig. 31).

The chromatograms are only 1—2 cm in diameter, however, as a result of the rapid evaporation. Work must be carried out in closed chambers, saturated with the solvent, in the essential circular technique. There are two easily assembled chamber systems which have proved satisfactory for adhesive adsorbent layers (Fig. 32). Both of these (A and B) guarantee good chamber saturation.

In procedure A, the glass plate (1) with the layer undermost (2) rests tightly on a flat dish of suitable cross-section, containing the solvent (7). A wick of cotton wool (5) connects solvent and layer. The substance mixture to be separated can be applied as a spot within about 1 cm of the aperture (4). It is thus possible to compare a series of substances on a single layer.

Procedure B with its "pre-separation column" is preferable for larger amounts. In this case, the layer side is uppermost and the solvent (7) rises via the cotton wool pad (8) through the pre-separation column (3) to the thin layer (2). The substance is applied beforehand to the uncoated side of the pre-separation column (4). It is advisable to cover the layer to prevent too rapid evaporation of the solvent as it spreads out. A second but unperforated 20 × 20 cm plate can be used for this, on the edge of which a glass strip, 5 mm wide and 3 mm thick, has been fixed, giving a sort of picture frame (9).

Other simple devices for circular technique have been described by TUBARO and RUSTICI [723], STAMMBACH [687] and by HASHMI and coworkers for inorganic TLC [278].

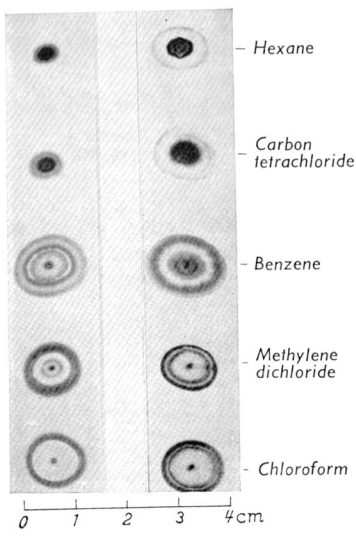

Fig. 31. Chromatograms obtained from two different dye mixtures, using the microcircular technique; preliminary test for rapidly ascertaining a suitable solvent [661]

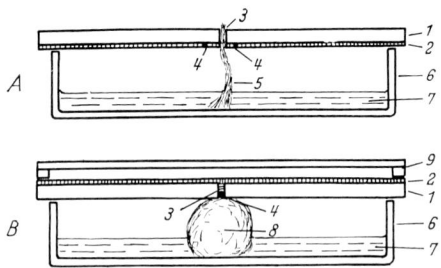

Fig. 32. Schematic representation of two devices for circular technique. A for separating several substance mixtures; B with a small preliminary separating column for a single substance mixture. 1 supporting 20 × 20 cm plate; 2 layer; 3 perforation of 2 mm diameter; 4 point of application; 5 cotton wool wick; 6 Petri dish cover; 7 solvent; 8 cotton wool pad; 9 glass strips [662]

b) Centrifugal Technique

The solvent migration rate can be accelerated through centrifugal force. The plate and layer are rotated in a horizontal plane and solvent introduced at the axis of rotation. ROSMUS, PAVLÍČEK and DEYL [585], who had previously studied centrifugal paper chromatography in detail, have tried out the technique with TLC. They showed that the Stahl test mixture and mixtures of 2,4-dinitrophenylhydrazones could be separated in 5—10 min on silica gel G layers of 20 cm diameter, using 500 revs/min.

The use of apparatus which has been developed for centrifugal paper chromatography (Firms 11, 58) can be extended to TLC also. KORZUN and BRADY [375] have thus separated in 10 min a mixture of dyes on circular glass or aluminium plates (20 cm diameter) at a speed of 500—700 revs/min. In the ordinary way, this separation requires 35 min.

c) Horizontal Development, BN-Chamber and Automatisation

Chambers for horizontal development can be constructed from simple materials. For example, as in Fig. 33, one can place the solvent in a shallow glass "developing" or "instrument" dish. A glass strip or rod (2 cm diameter), round which filter paper has been wrapped several times, serves to transport solvent to the layer. A glass rod or tube forms the second support. This and similar devices ensure good chamber saturation. LEES et al. [389] have accomplished continuous development

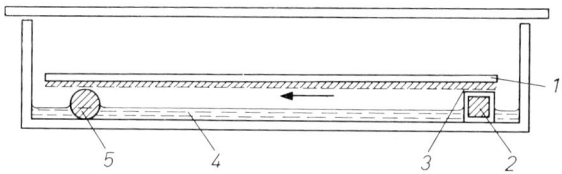

Fig. 33. Simple layout for horizontal technique in a glass dish with cover. *1* TLC plate, face down; *2* glass or metal rod, wrapped round with filter paper (*3*); *4* solvent; *5* glass rod or tube as support

using the same materials and a similar apparatus. Their plate, with the layer on the underside, was placed on a developing dish, effectively as a cover. The layer had been scraped off previously from the three supporting side strips. MISTRYUKOV [452, 453] has described corresponding devices for loose layers.

BRENNER and NIEDERWIESER [96] were the first who gave instructions for constructing and using a new type of chamber for continuous TLC with horizontal development (see first edition). This chamber has been improved in the meantime (Fig. 34) and the scope widened.

Solvent in the container (1) enters the trough (4) through a teflon tube (2) and then reaches the layer via a tongue of filter paper. The plate rests on a cooling block (7) and is fitted with a cover plate. The extremity of the TLC-plate lies on a narrow heating block (8). In this way, evaporation of high-boiling solvents at the plate end can be appreciably accelerated.

HÜTTENRAUCH and SCHULZE [306] have proposed thermostatisation of the cover to prevent condensation of solvent on it. RITSCHARD [576] has used a modified BN-chamber, easily convertible into an electrophoresis chamber for preparing "peptide maps" (cf WIELAND [757]).

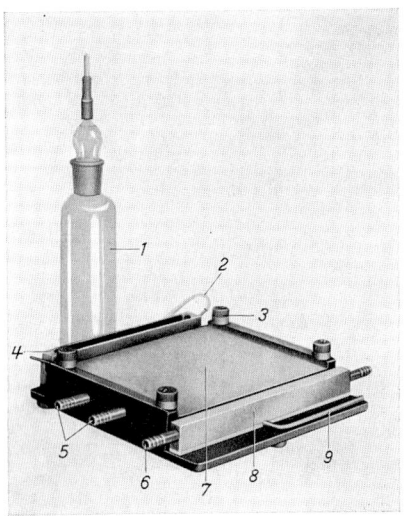

Fig. 34. BN-chamber. *1* solvent reservoir; *2* feed tube communicating with the trough (*4*); *3* holding or fixing device for the cover plate; *5* connections for water cooling; *6* connection for heating giving continuous flow; *7* cooling block on which the TLC-plate rests; *8* heating block; *9* foot-rests for holding the chamber (Firm 44's photograph)

Other development techniques also can be carried out with the BN-chamber, e. g., an "iterating", a "polyzonal" and the descending development already described (see p. 72).

The "iterating" technique is a variant of multiple development (see p. 86) in which the solvent trough can be moved back bit by bit.

In "polyzonal" development, the separation into phases of solvents during TLC, once regarded as undesirable, is utilised. Several "fronts" are then formed [476, 477, 673].

Finally, the apparatus of TAKITANI and MATSUDA [707] for "automatic development" may be mentioned. It is based essentially on the motor-controlled inclining or tipping of a shallow chamber for horizontal development (Fig. 33). The layer side is, however, uppermost and the plate is raised to a height of about 3—4 cm on one short side only.

By tipping the chamber, the solvent can be displaced so as to immerse and uncover the layer alternately. A photocell registers when the level of the front is reached and thus starts the tipping movement.

IV. Aids for Visualisation of Colourless Substances on Chromatograms

A chromatogram can be evaluated only when all the substances separated can be identified. Should the substances be colourless, the following can be utilised:

1. UV-lamps (emission maxima 254 and 365 nm) for identifying fluorescent compounds; the chromatograms are viewed in short and long wave UV light. Also for determining whether fluorescence-quenching spots are present on a fluorescing layer (e.g., silica gel GF_{254}).

2. Sprayers for atomising or spraying suitable reagents (see Special Section and Chapter Z) which give a coloured and/or fluorescent product.

3. Heaters for accelerating reactions and sometimes also to yield coloured or fluorescent pyrolysis products.

4. Other apparatus, e.g., for direct sublimation of the substance from the layer on to a cover plate.

Direct registration of the UV-absorption of the chromatogram (see Fig. 77). Application of sensitive biological tests.

A more detailed discussion of these various possibilities follows, based on available experience.

1. UV-Lamps for Exciting Fluorescence

Numerous substances absorb UV light of a particular wavelength. Some of them then emit it as visible light. This phenomenon, termed fluorescence or luminescence, is a first-rate, very sensitive means of identifying small amounts of substances on chromatograms. If the substances themselves are not fluorescent, they can often be converted into fluorescent derivatives or decomposition products. UDENFRIEND [726] and DANCKWORTT and EISENBRAND [145] give many suggestions for and examples of the successful application of "fluorescence analysis". FÖRSTER [195] has treated in detail the physico-chemical fundamentals of the fluorescence of organic compounds.

Even non-fluorescent compounds which however clearly absorb in the UV region, can be detected in UV light. Thus if a fluorescent layer such as silica gel GF_{254} is exposed to short wave UV light (254 nm), all substances which absorb in this region stand out distinctly as dark zones on the green fluorescing layer as background.

The possibilities of TLC today can be fully exploited only by using powerful emitters which render possible the evaluation of chromatograms in the short and long wave UV region. The current technical situation demands compromise solutions however.

Colourless compounds absorbing in long wave UV light can be similarly detected as dark zones on layers which have been impregnated with an indicator, fluorescing in this long wave UV light. The diazogram and electrovisualisation procedures proposed by SPRENGER [652, 653, 654, 655] can be used also (see the chapter "Quantitative Evaluation". pp. 133ff.).

In the long wave region around 365 nm[10], high pressure mercury lamps with dark glass bulbs as filters (Firms 96, 103), good value for money, yield a higher radiation intensity than the correspondingly sized dark glass fluorescence tubes (Firms 96, 103, 134, 145). The high pressure lamps however require 2–5 min to reach maximum brilliance, become relatively hot and, when turned off, must be cool before they can be put into action again. The weaker fluorescence tubes do not show these shortcomings. There is plenty of equipment for fluorescence excitation in the long wave region (Firms 44, 73, 106, 110, 129, 140 etc.).

For excitation in the short wave UV domain near 254 nm, low pressure mercury lamps, made from UV special glass (Firms 25, 110)

Fig. 35. Composite equipment for observing and photographing chromatograms in UV-light of 254 or 366 nm and in daylight (Firm 44's photograph)

with disc-shaped dark glass filters[11] (Firms 25, 123) are used. The handy "Germicidal-Lamp" (Firms 134, 145), 28.5 cm long and only 1.5 cm cross-section, has been built into some of the usual commercial apparatus.

Multiple-use, compact apparatus is preferable for evaluation; this permits illumination of the chromatogram with "daylight", long or

---

[10] 1 nm (nanometer) = 1 mµ = 10^{-7} cm = 10 Å.

[11] It must be noted that such filters age rapidly and can become almost opaque to short wave UV light.

short wave UV light. It should also guarantee uniform lighting for photographic purposes and have a holder for various types of camera. Commercially available are: Chromato-Vue (Firms 73, 140); Uvanalys (Firm 14); Fluotest (Firm 110); and UVIS (Firm 44). They are fitted with low pressure lamps; the Uvis contains altogether 7 interchangeable fluorescence tubes (Fig. 35). Others contain only one short and long wave source and are correspondingly cheaper (Firms 33, 44, 106, 129).

Note: In passing it may be mentioned that a small dark room-ca. 1×1.5 m-can easily be fitted up in a corner of the laboratory by means of black curtains.

2. Spray Apparatus and Fume Cupboards

Spraying with a reagent solution is the most widely used technique for visualisation of colourless substances on chromatograms. A uniform and finely atomised spray is necessary to colour evenly the mostly small and compact spots on the chromatogram. This is rarely achieved successfully with the "throat sprayers", customary in paper chromatography, or even with rubber bulb sprayers. It is essential to have fine spraying jets and compressed air or inert gas.

One- or two-part glass sprayers (Fig. 36) have proved satisfactory. The amount used for spraying is then easily controllable and they are

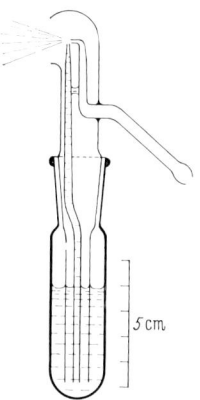

Fig. 36. Two-part glass sprayer with ground glass joint for atomising reagents with compressed air (Manufacturer, Firm 44)

Fig. 37. Diaphragm pump for producing oil-free compressed air, mounted on a multipurpose motor (Firm 126's photograph)

simple to clean. They can be connected to the compressed air source, if necessary via a T-piece. Compressed nitrogen in steel cylinders is also a convenient, inert propellent gas. *Oil-free* compressed air can be produced at any time by means of a small, efficient diaphragm pump (Fig. 37).

Sprayers operated with compressed propellent gas are being increasingly used for atomising reagent solutions. The three-part appliance of the "Spray-Gun" type (Firms 44, 117, 119, 129) is most practical. The propellent gas container and the spraying solution can be changed as desired (Fig. 38). The fine jet and operating valve are on the junction piece which is made of plastic. It may be mentioned in this connection that a number of spray reagent solutions are available commercially in ready-for-use aerosol-monobloc containers (Firms 86, 88, 117).

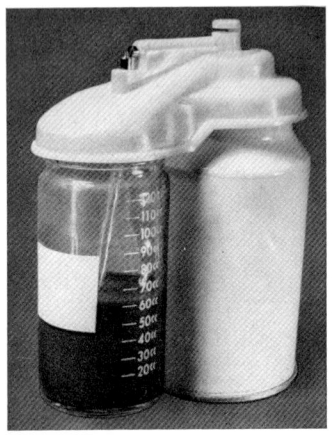

Fig. 38. Sprayer with exchangeable propellent gas container and reagent holder [736]

A fume cupboard with a good draught is essential when spraying with aggressive reagents. It it is not available, one can avail oneself of the following:

a) If a large but poorly drawing fume cupboard is available, all but one of its chimney exits are sealed off and a small spraying cabinet (Fig. 39) attached to the remaining one. The cabinet is made of acid resistant plastic (Trovidur) (Firm 44) and contains a removable frame of the same material, on which the plate is placed.

b) If no fume cupboard is there, a tube of Trovidur, diameter about 25—30 cm, is placed in communication with the open air and an acid-

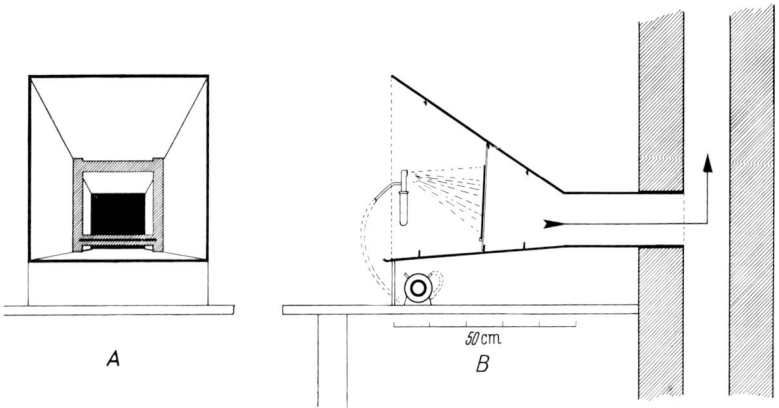

Fig. 39. Plastic spraying cabinet connected to the ventilation flue. *A* front view (shaded part is the frame for holding the chromatograms); *B* side view (section); the compressed air spraying equipment is also shown

resistant, explosion-proof fan built into it. The spraying cabinet is then attached to the tube.

A small fume cupboard, ca. 60 × 60 × 60 cm, with a built-in quiet-operating fan, is available commercially (Firm 44) and has proved its worth over a long period of use.

3. Apparatus for Heating the TLC-Plates

Optimum colour development after having sprayed with the reagent solution is often attained only after heating for a particular time at a particular temperature. A smaller, older drying oven with aluminium interior and temperature adjustable up to 250° C, is usually used for this purpose. These ovens are quite resistant to the acid vapours which are formed, even over a long period. Small drying ovens are now being manufactured, designed to take the 20 × 20 cm plates (Firms 32, 44, 70), Hot plates can be used also if an asbestos mat is placed on them; e. g., the flat laboratory hot plates, 20 × 40 cm, the temperature of which can be raised gradually to 300° C (Firm 70).

Better colour formation is sometimes achieved by preventing rapid evaporation of the reagent solution during the first stage of heating. To this end, the chromatogram is covered with a suitable plate. A cover of this sort can easily be made by sticking four glass strips on to a 20 × 20 cm plate [537].

Should high temperature be necessary for pyrolysis or to accelerate a reaction, the TLC plate is laid on an aluminium plate, itself lying on an asbestos mat. The layer is then heated with an infra red heater at a distance of 10–20 cm. The so-called quartz "surface evaporator" (30 cm diameter) has proved most useful in our work (Firm 70). The heating spirals reach temperatures up to 800° C and, like the quartz reflector, are not attacked by the usual acid vapours. HEIDBRINK [284] has recommended in connection with pyrolysis of the substances, that the chromatograms should be previously placed in a atmosphere of gas, e. g., Cl_2, Br_2 NO_2.

4. Further Aids for Visualisation

a) Direct Microsublimation

Many solids can be sublimed. This property can be made use of for separating such substances from mixtures and also for subsequent characterisation.

According to KOFLER [358], microsublimation is any sublimation procedure in which the sublimate is caught on a smooth plate-usually a small slide-for purposes of microscopic examination (melting point, refraction, crystal structure). Many substances can be directly sublimed in crystalline form from drugs containing them, e. g., caffeine; coumarin; umbelliferone; benzoic, ferulic, salicylic and cinnamic acids; emodin etc.

BAEHLER [35, 36] has employed microsublimation for direct detection of substances which he had previously separated on a thin-layer chromatogram. He demonstrated the advantages of this technique with a mixture of caffeine, theobromine and theophylline-not simple to identify through colour reactions. A very few micrograms of these substances can be recognised as white "sublimate spots". For observation, the plate is placed on a black background and illuminated from the side with light falling at a very small angle to the plate. A microscopic examination should be repeated after 6—24 hrs because the crystals can grow further after having been sublimed.

Sublimation-apparatus. The TLC-plate lies on a heated aluminium block, $3 \times 8 \times 25$ cm. It contains a lateral hole for inserting a thermometer. The developed plate (5×20 cm, with silica gel G coating) is placed in position and covered at the edges with asbestos strips, 1 mm thick. A 5×20 cm glass plate is laid on the strips to catch the sublimate. This plate is cooled by means of an aluminium block through which cold water is passed. A moist filter paper can sometimes be introduced between cooling block and plate. The lower block is then heated to 300—320° C with a burner.

Direct microsublimation is carried out in our laboratory using a rectangular, plane hot plate GP, 20×40 cm (Firm 70) and an aluminium cooling plate, 20×20 cm (Firm 44). V2A steel or aluminium plates are preferable to glass as supports since they appreciably reduce the undesirable drop in temperature between hot plate and layer.

This apparatus can be used for separating liquids like terpenes from sesquiterpenes, of different boiling points, as well as for sublimation. A cover plate with silica gel layer is then used to catch the substances "distilled over" (see also p. 226). It is important to seal the system all round to prevent any convection currents.

b) Self-recording Measurement of UV-absorption

A chromatogram can be automatically scanned with a suitable apparatus and the absorption at a previously set wavelength recorded. The necessary equipment is described in Chapter II, on quantitative evaluation (Fig. 77, p. 144).

c) Biological Visualisation Procedures

Biological procedures can be utilised in order to detect substances which have a particular physiological activity. Two types of procedure are basically possible, namely, detection directly on the layer and indirectly, after having scraped off the relevant chromatogram zones. So far, substances of only the following types have been detected by these two techniques: insecticides, haemolysing substances, bitter principles and antibiotics. The necessary apparatus and method of carrying out the test for the last named, are described in Chapter R.

To complete the picture it may be remarked that other procedures and apparatus can be used for visualisation. Combination of TLC with GC (p. 144) and with mass spectrometry (p. 204) are meant here in particular. Further procedures will be introduced and added to the list, extending the present range of possibility. This is only to be desired for really specific tests are rare. A section is devoted to suitable procedures for transferring very small amounts of substances (p. 102).

V. Laboratory and Basic Equipment for TLC

In any laboratory, TLC can be carried out in principle with a relatively small outlay. The size and equipment of a chromatography laboratory varies with the individual demands and problems, the number of chromatograms prepared daily and the number of the compound classes investigated (see Special Section).

1. Small Laboratories

The following equipment can be suggested for laboratories which pursue TLC only occasionally:

a) Small TLC equipment (Fig. 40) or other simple aids for preparing layers (p. 57) or pre-prepared plates or sheets (p. 59).

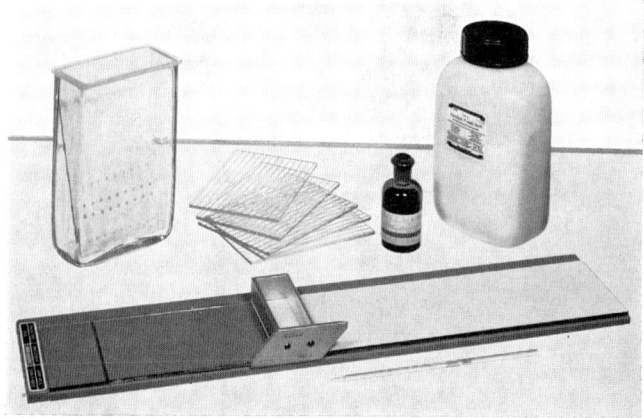

Fig. 40. Small basic equipment for TLC including a simple spreader for coating glass plates with a 250 μm layer and for grooved glass plates. A small development chamber for the 10 × 15 cm plates is in the left background (photograph from Firm 44)

b) Extra equipment: small oven for drying layers and carrying out visualisation reactions; short and long wave UV-lamp, e. g., Minivus (Firm 44); propellent gas sprayer (p. 80) for atomising reagent solutions; in certain cases, a small fume cupboard (p. 80).

2. Medium-Sized Research and Control Laboratories

TLC apparatus was first developed following the requirements of a medium-sized laboratory. The TLC apparatus, which at that time did not belong to the usual equipment of such a laboratory, was collectively termed "basic kit or equipment", in accordance with a suggestion of STAHL [661] (Firm 44). It thus became possible to apply the method without delay, using tested apparatus and accessories. The basic equipment contains the following items:

1. TLC-spreader "GM" with diagonal adjustment of layer thickness from 0 to 2 mm (Fig. 13); its use can be extended to gradient-TLC.
2. Aligning tray for coating five 20 × 20 cm plates.
3. Ten glass plates, 20 × 20 cm, of equal thickness.
4. Two end-plates (fore and aft), 20 × 5 cm.
5. A normal chamber with ground edges and plane ground knob cover.
6. A multiple purpose template, made of perspex, for marking, sample application and evaluation of the chromatograms (Fig. 19).
7. A drying rack of light alloy, taking ten 20 × 20 cm plates (Fig. 18).
8. Two special pipettes of 10 µl capacity (Fig. 20f).
9. A bottle containing a 3-colour standard solution.
10. 500 g silica gel for TLC and detailed instructions for preparing normal and gradient layers.

A number of other basic kits has become commercially available in the meantime (Firms 11, 15, 33, 106, 111, 117, 129 etc.). The following auxiliary apparatus is usually to be found in a modern TLC laboratory:

Several sprayers (Fig. 36) for atomising reagents; source of compressed air; oven with temperatures up to 250° for carrying out colour reactions; hot plate; infra red heater; circulation-drying oven, internal width 42 cm, for drying and activating the layers; high pressure UV-lamp; low pressure short and long wave UV-lamp; set of equipment for TLC-electrophoresis; various chambers for development, including an S-chamber. BN-chamber and Cryobox; extra accessories for gradient TLC; cupboards for storing variously coated TLC-plates; arrangements for documentation (photography, blue-print, photoprint); equipment for preparative TLC; apparatus for direct measurement and recording of absorption in the UV- and visible spectral regions; aids for TLC with radioactive compounds, including a scanner.

3. Central TLC- and PC-Laboratories in Industry

A few years ago, NEHER and V. ARX [24] described fitting-up such a laboratory. The change over from PC to the time-saving TLC has occurred in the meantime and the nature of the equipment in this type of laboratory shifted correspondingly in favour of TLC.

So-called automatic spreaders have been developed to rationalise the coating of hundreds of plates daily (p. 58). The requirements of apparatus in such a laboratory are otherwise the same as described under 2.; the apparatus is merely used more rationally. Centralisation, like with spectroscopy, is not however a desirable goal because TLC is a method carried out on the actual work bench of the research and works chemist. For him, it is the simplest, quickest and cheapest expedient for following chemical reactions and for testing purity and for identification.

VI. Standard Conditions in TLC*

Suggestions for unifying experimental conditions in TLC have had a universal response. These standards have created a good point of departure for the method. Only when really necessary and when experimentally justified, should one deviate from these conditions. The approved conditions are summarised as follows:

Dimensions of the layer: usually 200×200 mm for analytical purposes; 100×200 mm for preliminary experiments and 400×200 mm for serial analyses and preparative TLC.

Layer thickness: Normally ca. 150 μm for analytical separations; the moist layer, i. e., when just spread, is then ca. 250 μm thick (value set on the spreader). Layers up to 500 μm are used for certain quantitative determinations and up to 2 mm for preparative TLC.

Drying of the layer: a) Ca. 10 min preliminary drying on the aligning tray, if necessary in a current of warm air, followed by 30 min heating at 110° C in vertical position. A longer drying time is needed for layers thicker than 250 mμ. b) Air drying for about 12 hrs at room temperature in a dry atmosphere.

Storage: The plates must be protected from laboratory fumes (water, acids, bases) and are best stored over a drying agent like blue silica gel.

Starting points: The starting point should be 15 mm from the lower edge and the side of the plate; likewise 15 mm between neighbouring points.

Length of run: Normally 100 mm, i. e., from starting point to front.

Chambers: a) Chamber with plane, ground-glass cover and chamber saturation (CS); b) S-chamber system; c) BN-chamber.

Depth of immersion of the layer: The plate and layer should dip only about 5 mm into the solvent in the chamber and the S-chamber.

---

* See also "Suggestions for the Standardisation of Procedures and Terminology of Thin-Layer Chromatography. E. STAHL, J. Chromatog. **33**, 273 (1968).

Adsorbent: The use of established adsorbents, specially prepared for TLC, following the given instructions, is advantageous.

Standard substances: A so-called standard mixture is best used. It consists of 2—3 pure, commercially available, preferably coloured and/or fluorescent compounds, separable in the hRf-region between 20 and 80. Pure reference substances should be chromatographed simultaneously where possible.

D. Special Techniques in TLC

Egon Stahl

If the procedures discussed above fail to give a satisfactory separation or if a special problem is set, other techniques often effect progress.

I. Special Development Procedures

a) Continuous Development

When the substances to be separated are in the lowest third of the chromatogram after a single development, continuous or multiple development usually brings about a better separation. Reference has already been made under "Separation Chambers and Development" to various devices for continuous development (see p. 69, 72, 76); the BN-chamber was made specially for this purpose (Fig. 34). Since there is no front in this technique, a suitable reference substance is chosen which is chromatographed at the same time and R_{St}-values are quoted.

b) Multiple Development

As the name implies, the same solvent is used several times for development in this procedure. It must be evaporated from the layer between developments in order that it may migrate through again by capillary action. The method is well known from PC and its advantages have been studied several times [326, 394, 616]. ^nRf-values[1] were introduced in order to be able to quote the position of the separated substances. These can be calculated in advance from the relationship $^nRf = 1 - (1 - Rf)^n$. Further possibilities and variations of this procedure have been described along with instructive diagrams in the work of Schratz and Egels [616]. They showed also that, depending on the nature of the substance, the spots can shrink ("centralisation") or grow ("decentralisation") as the number of developments is increased. Multiple development can be employed also in TLC with similar success [612, 679]

---

[1] n = number of developments.

and is an integral part of Halpaap's method [266] for preparative TLC (p. 100).

c) Stepwise Development [665]

All the substances in a mixture cannot be satisfactorily separated if there are classes present which differ markedly from one another in their adsorption or partition behaviour. Separation is often achieved, however, if one develops successively to different heights with two contrasted solvent types. This has been demonstrated on a mixture of lignans and their glycosides (Fig. 41). In the first stage, a run of 6 cm, only the glycosides can be separated. The aglycones remain together in the region of the front but can be separated in a second stage with a more weakly polar solvent (12 cm length of run); the glycosides do not migrate in this solvent and thus remain in their original positions [670].

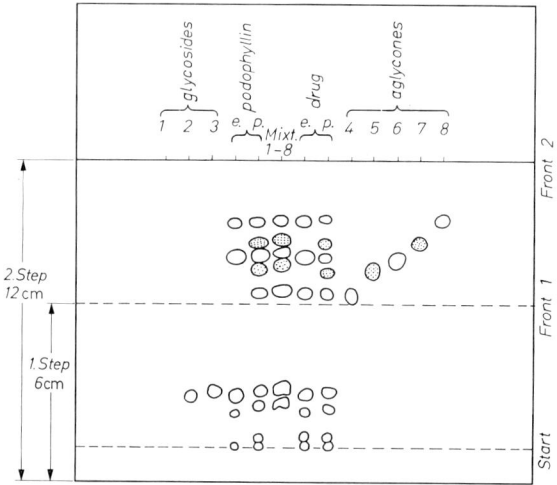

Fig. 41. Stepwise development for separating podophyllum constituents on silica gel G layers. Solvent: 1 step chloroform-methanol (90+10); 2 step chloroform-acetone (65+35). *1* α-peltatin-β-glucoside; *2* podophyllotoxin glucoside; *3* β-peltatin-β-D-glucoside; *4* 4′-demethylpodophyllotoxin; *5* α-peltatin; *6* podophyllotoxin; *7* β-peltatin; *8* 1-dehydroxypodophyllotoxin. 1.0 μg of each of the pure substances was applied and 5 μl of extracts of the podophyllins and of the drug in 10% alcohol (CS). Visualisation was with sulphuric acid-acetic anhydride (1+3), followed by 15 min/100° C. The zones marked with points reacted also with diazonium salts. (*e* = *emodi. p* = *peltatum*-drug) [670]

The technique can be used also for preliminary washing of chromatograms. In the first stage, which is then higher than the second, one carries out a preliminary wash with a more strongly polar solvent like chloroform-methanol, so as to be able to accomplish the desired separation in a second stage after drying (30 min at 110° C).

An interesting variation of gradual development for obtaining start lines is described on p. 100.

1. Two-dimensional Separation and SRS-Technique

Two-dimensional development is particularly valuable for mixtures of many components. The mixture is applied to the plate about 3–4 cm from one corner and is developed as usual. It is then turned through 90° and development carried out in the second direction. One must try to vary the solvent or method so that different separation effects are obtained in the second "dimension". Figs. 59, 143 etc., illustrate the advantages of a "two-dimensional separation" adequately. KAUFMANN and coworkers [339] have described combinations of adsorption TLC in the first and partition TLC in the second direction, constituting a useful methodic improvement. The combination of TLC-electrophoresis in one direction and adsorption or partition TLC in the other, is being increasingly employed [221, 299, 576 etc.].

In contrast to the above, the *SRS technique* (separation, reaction, separation) [668] employs the same conditions (e. g., the same solvent) in both directions. The compounds, assuming they have not decomposed during development, must then lie on a diagonal line between the two development directions. One can thus easily see whether any substance has undergone change in a solvent. If, after chromatographing in the first direction, one or other of the already separated components is submitted deliberately to chemical change, the second chromatographic stage with the same solvent in the other direction will show this. This simple technique permits rapid recognition of changes as a result of exposure to radiation (γ-, X-, UV-), gases, heat etc. Topical problems of great interest, like protection against radiation, photochemistry, stability testing, can be studied in this way. The first successful application of the SRS technique was an investigation of the inactivation of pyrethrins (Fig. 182) [668].

2. Multidimensional Technique of von ARX and NEHER [25]

This is also essentially a two-dimensional procedure. The mixture in question is first chromatographed on three TLC plates in one direction, using the same solvent. For the second direction, solvents are taken for each plate which yield separations differing as widely as possible from each other. Moreover, a different spray reagent is used for visualisation on each plate. Using this combined procedure, 52 amino acids could be classified and distinguished. It should be of advantage in the separation of other compound classes too.

JANÁK [320] used the same expression "multidimensional chromatography" for the combined GC-TLC procedure, recommended in 1961. The term "three-dimensional TLC" was first used for a two-dimensional gradient TLC [675]. It was then dropped since it did not fit in with the usual spatial conception of a third dimension [682].

3. Wedged-tip Technique

Separation can often be improved by using the horizontal circular technique (p. 73). By a suitable modification of shape, a similar separation effect can be obtained in ascending development also. This procedure is known in PC, especially in the work of MATTHIAS [438] and REINDEL and HOPPE [570]. It can be adapted to thin adsorbent layers without difficulty [662]. Dividing lines, 0.5–1 mm broad, are drawn in the layer with a narrow metal spatula or blunt pencil. The various possible shapes and distances from each other have been investigated. Good results are obtained with the shape shown in Fig. 42. The pentagons can be more easily scratched out with the help of a perspex or other stencil (Fig. 43). It is laid firmly on the lower edge of the TLC plate and the five edges marked round. Transverse lines are then drawn with the help of the template (p. 62). The substances are then applied to the lower part of the narrow wedge. The time of run of the chromatogram is increased as a result of the diminished supply of solvent.

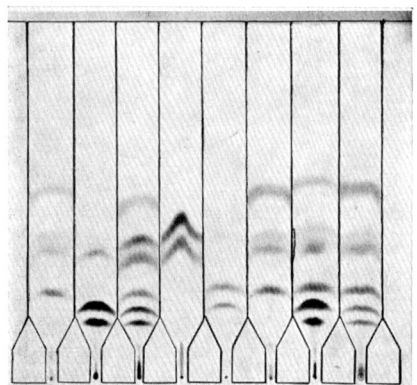

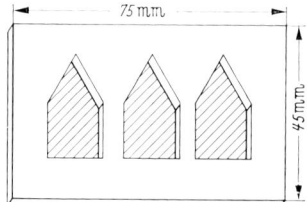

Fig. 42. Thin-layer chromatogram with wedged-tip division. The separation process assumes a band-like course. The ∧-shape (middle) occurs only with larger amounts

Fig. 43. Perspex stencil (2—3 mm thick) for drawing the dividing lines. The pentagons shaded in the diagram are cut out

II. Gradient[2]-Techniques in TLC

Many problems of separation can be solved rapidly and simply through the new gradient techniques. Two procedures of interest here may be distinguished. In *gradient TLC* a layer is used which, instead of being uniform as usual, shows a progressive change or gradient

---

[2] Continuous change from one property to another.

of separating ability. In *gradient elution* technique on the other hand, a uniform layer is employed but the composition of the solvent is changed continuously during development. The techniques can be combined, yielding further possibilities.

1. Gradient-Elution

TISELIUS et al. [714] and DONALDSON et al. [167] introduced this technique in column chromatography. Considerable gains have resulted from using it in the separation of mixtures of amino acids, peptides, proteins, nucleotides, nucleic acids and sugars.

Many simple devices are known by which two solvents may be mixed during chromatography in order to change continuously the elution power of the mobile phase [240, 283, 643, 650, 776]. The "Varigrad" chamber combination of PETERSON and SOBER [517] and the multiple chamber mixer described by WHITAKER and MITTELSTADT [750] enable the shape of the gradient to be pre-determined. WALLACH and NORDBY [738] have developed a solvent-resistant apparatus for linear and non-linear gradient elution at constant flow. KNEDEL et al. [353] have reviewed "gradient elution chromatography" (fundamentals, apparatus, applications).

RYBIČKA [591] described the first succesful attempts to introduce gradient elution into TLC. He separated a mixture of glycerides on a silica gel layer, increasing the % ether in his solvent from 10 to 60 during development. Ether was added to the solvent in the chamber (petroleum ether-ether-acetic acid $90 + 10 + 0.1$) from a burette, the tip of which passed through the cover and almost reached the bottom of the chamber. The solution was mixed with a magnetic stirrer. The start points were comparatively high since the level of the solvent rose as a result of the addition.

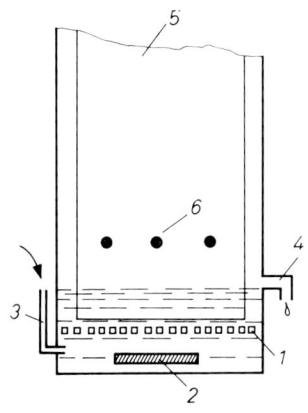

Fig. 44. Device for gradient elution according to [756]. *1* perforated plate holding the TLC-plate; *2* magnetic stirring bar for mixing; *3* supply tube for the second solvent; *4* overflow; *5* narrow TLC-plate; *6* start points

The device described by WIELAND and DETERMANN [756] for gradient elution TLC (Fig. 44) appears very convenient and capable of extension. The gradient is produced here by running the liquid to be added from a burette or measuring pump through a capillary (3) into the circular chamber (6 cm diameter) where it is then mixed with the existing solvent using a magnetic stirrer (2). The volume is kept constant by means of an overflow (4). The TLC plate, 5 × 20 cm, stands on a perforated plate (1). Care must be taken that the addition of solvent is completed soon enough before the final front line is reached. Using this gradient elution technique, mixtures of three adenine-nucleotides and two lactic acid dehydrogenases could be separated on DEAE-Sephadex layers.

In order to be abel to carry out gradient elution in the BN-chamber (p. 76) also, STICKLAND [698] divided the trough into two halves. The procedures mentioned above are in fact easy to carry out, yet, according to NIEDERWIESER and HONEGGER [478], they do not guarantee reproducibility of the gradient created. These authors have accordingly developed a new set-up, for which the following is needed:

a) a micro gradient-mixing battery with constant exit rate of flow. It consists of, e. g., 7 closed glass chambers, each with a capacity of 1.3 ml. The course of the concentration change is fixed by the fact that the same amount of liquid enters each chamber as leaves it. After leaving the battery, the solvent mixture is fed into a capillary tubing reservoir ("spiral").

b) the teflon tube functioning as solvent reservoir should have an internal diameter of 1—1.5 mm and can be up to 2.5 m long. It is wound spirally round a 13 cm thick cylinder and has to be attached to the BN-chamber at a definite height. It should on no account contain air bubbles.

c) a description is given of a device for spot supply like that used for circular chromatography and of a supply groove for band supply, connected to the "spiral".

Interesting examples of separation of synthetic dyestuffs and particularly of lipid mixtures are given in this and in another article [479].

2. TLC on Gradient Layers (Gradient-TLC)

Gradient layers provide three different types of running surface, in contrast to the *uniform* layers so far used (Plate Ia). In A (Plate Ia), chromatography is performed at right angles to the gradient and it can be rapidly decided which mixture of two adsorbents yields the best separation. It can further be ascertained whether, and at what concentration, a particular impregnation (silver nitrate [679], acids, bases, salts and organic stationary phases) furnishes any advantage. Substance-specific curves for the individual compounds in mixtures of acids and/or bases can be obtained on so-called pH-gradient layers [675].

The separation process runs in the direction of the gradient in the systems B and C (Plate Ia) and the property of the "open column" changes from start to front. The first impressive examples of application of this new technique were described in 1964 [675] and further examples came in 1965/66 [261, 615, 631, 682].

Gradient layers were first prepared using the GM-spreader [675] (Firm 44), which had been designed for this purpose. It is based on the established, adjustable spreader (Fig. 45), in general use for preparing uniform layers.

The so-called dividing hopper (Fig. 46/1) with the removable diagonal spacer, is placed on the filling opening. The mixing roller is placed in the jacket of the spreader. It can be operated by a gear wheel handle or, better, with a motor.

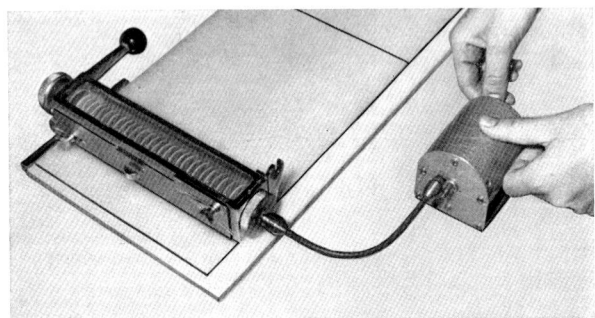

Fig. 45. GM-spreader with motor drive. The dividing hopper has been taken out and mixing is being carried out

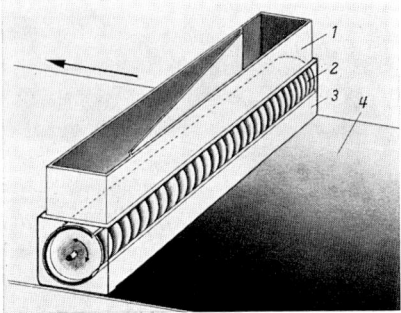

Fig. 46. Simplified, half-diagrammatic view of the GM-spreader during spreading a gradient layer. *1* dividing hopper with diagonal spacer in position; *2* mixing roller; *3* metal base of the spreader; *4* gradient layer

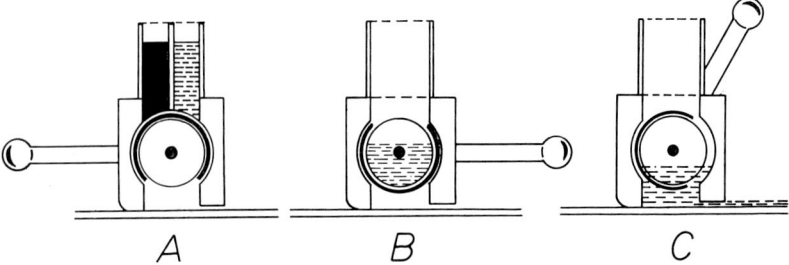

Fig. 47. Cross-section of the GM-spreader after filling the divider with two different adsorption suspensions (*A*), during mixing (*B*) and in position for spreading (*C*)

Principle of the method: The two different adsorbent suspensions are introduced into the dividing hopper (Fig. 47A). When the bottom is opened, both fall into a mixing chamber, divided up by numerous plates (Fig. 47B). After mixing, the suspension can be spread on glass plates in the usual way (Fig. 13). Preparation involves the following distinct stages:

1. Both parts of the dividing hopper must be equally full (Fig. 47A). The outlet slit of the spreader is closed.
2. The diagonal dividing spacer is carefully removed.
3. The bottom of the dividing hopper is opened by turning the jacket with the lever arm (Fig. 47B) or, with newer divider hoppers, by withdrawing the sliding bottom.
4. Homogeneous mixing is achieved by rotating the mixing roller for 1 min in both directions alternately (Fig. 47B).
5. While the mixing motor is running, the jacket is put first in position A, then C (Fig. 47); the latter serves to aerate the chambers.
6. The motor drive is stopped and the motor detached from the spreader. The outlet slit is now finally opened, about 0.3 mm, by sliding the frontal plate. The suspension is now spread as usual.
7. After drying the layer, the gradient must be controlled. This is easier if one adsorbent contains a fluorescent indicator. The intensity of fluorescence must then decrease uniformly from one side of the layer to the other.

Adherence to the established working instructions and a certain manual dexterity are required for preparing gradient layers (Fig. 44). Special attention must be paid to ensuring that each suspension has an optimum viscosity and that they dry out evenly and rapidly on the plate [682, 686]. Plate I and Figs. 105a, 152, 198, 204 etc. illustrate informative *examples* of use; others are to be found in the work of STAHL [675, 682], SCHORN [615] and SHELLARD et al. [631].

HONEGGER [301] has described a method of preparing layers with an activity gradient in which he deactivated active silica gel layers in stages by successive immersions in acetone-water mixtures.

WARREN [741] took over the essential construction principles of Stahl's GM method in 1965 and has attempted to reach a simpler solution.

A variation of the gradient layer technique is to use gradient layer thickness, e. g., from 2 down to 0.1 mm. Such layers can be easily made; one must simply ensure that the outlet slit has these dimensions. STAHL [678] has used such layers to demonstrate the influence of layer thickness on separation. ABBOTT et al. [1, 3] have chromatographed in the direction from thicker to thinner layer. They were thus better able to separate off polar compounds present in large excess; these undesirable compounds remained in the thick part of the layer.

These "wedge layers" can be regarded as a sort of preliminary stage of gradient TLC. They are somewhat analogous to the "stepwise development technique", previously described [665], which is nowadays better referred to as "stepwise elution technique". "Wedge layers" can be prepared with most spreaders. One or two transverse plates are inserted into the tank, the divisions thus formed are filled with different adsorbent suspensions and spreading is then carried out [1, 61].

Gradient columns have been rarely used up to now since their preparation by hand is fairly time consuming. A rational procedure for preparation has, however, been found. The GM-spreader can namely be used by taking out the mixing worm and attaching a funnel. A very slowly running motor, fixed to a retort stand, provides the drive. A special apparatus for filling gradient columns has been developed also [682].

The **gradient technique** will become more and more significant in all chromatographic domains. Apart from gradient layers and columns, these procedures have been developed most extensively in gas chromatography and have achieved considerable progress there — e. g., temperature programming, continuous change of carrier gas speed [619] and column gradient [405]. In 1966, STAHL reviewed chromatographic gradient procedures in detail [682].

III. Temperature-TLC

Few investigations have been made on the influence of temperature on separation in TLC. According to an older study it appears that in adsorption TLC, a lowering of temperature from 20° C to 4° C has a noticeable influence neither on the time of run nor on the Rf-values [659]. It is otherwise in partition chromatography. It has been observed here that some fatty acids could be much better separated on silica gel layers impregnated with silicones, at $4-6°$ C than at ordinary temperatures [420]. The maintenance of a definite temperature during development is of decisive importance in some separations [395, 465].

At about the same time, work was then published in which chromatography at temperatures down to $-20°$ C was described-e. g., a thesis of MATHIS [436]. STAHL [675] and ABBOTT et al. [5] have studied comprehensively the possibilities of low temperature TLC.

A type of high temperature TLC has also been carried out [172]. Development was performed at ca. 270° C with the $LiNO_3-KNO_3$ eutectic (43 mole-% $LiNO_3$) (= "molten salts mixture"). Silver, lead and mercury (I) ions were thus separated on a silica gel layer. A bath of molten salts was used also for the cylindrical development vessel, constructed of borosilicate glass.

1. Apparatus for Temperature-TLC

STAHL [675] has developed a chamber which can be thermostatically controlled between $+50°$ C and $-50°$ C. The usual glass chamber is surrounded with a plastic mantle in this so-called "Cryobox" (Fig. 48). The entry and exit connections for the thermostat fluid are fitted on the narrow ends of the chamber. For TLC down to $-20°$ C, it is enough to connect to a small table-cryostat capable of attaining $-35°$ C.

Fig. 48. Development chamber with mantle for temperature control. Right, below: entry connection with dial-type thermometer above it. The exit is on the upper part of the opposite side. The cover tubulure contains a thermometer and a device for holding both TLC-plates (photograph of Firm 44)

The Cryobox can be connected to an ultra-thermostat in order to carry out chromatography at constant room temperature or higher. The dial-type thermometer, attached to the end of the chamber, shows the temperature of the circulating fluid. A stem thermometer, fitted in the cover tubulure, enables the temperature within the chamber to be checked. A device for suspending the TLC plate is held in the second opening in the cover stopper. Temperature equilibration can thus be attained before the plate is dipped into the solvent.

It is advisable to surround the Cryobox (except the cover) with foam rubber sheets (e.g., styropor) about 2 cm thick, when working at below $-5°$ C. The chamber must be quite dry before use and the cover must be opened only briefly at lower temperature. A TLC plate at 20° C requires 30—45 min "acclimatisation time" to reach −15° C in the chamber.

For development at higher temperatures, i. e., over 25° C, the TLC plate must be pre-warmed [5]. Solvent vapours otherwise condense on it immediately.

2. Possibilities of Application

Chromatography at lower temperature is of practical interest because it has been found that Rf-values fall with decreasing temperature (Fig. 49). Since, however, not all compounds behave similarly, i. e., some show a more marked dependence on temperature, many separations are successful only at lower temperatures. Moreover, the spots are often more compact, i. e., smaller, than at ordinary temperature. Low temperature TLC is also called for in the separation of more volatile compounds like terpene hydrocarbons for example. A certain stabilisation of labile compounds has been observed also. It has been possible to employ low boiling solvents or liquefied gases as solvents in the Cryobox for the first time. STAHL and coworkers [675, 686] have achieved successful separations at − 10 to − 15° C with, for example, liquid butane and various liquefied propellent gases of the Freon type. The possibilities are by no means exhausted; in particular there have been no detailed studies of partition chromatography in this connection.

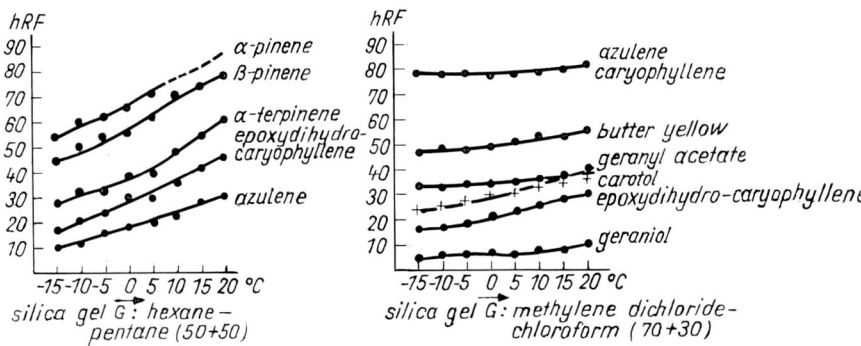

Fig. 49. Influence of temperature on the hRf-values

Development at different temperatures in the TLC of oligonucleotides on ion exchanger layers is important in view of the varying temperature-stability of complexes [563].

It must be clearly accepted that change in the temperature of development necessarily modifies the influence of all other factors. One is thinking here especially of the degree of chamber saturation and of layer changes through evaporation effects.

Plate I. Examples of Use of Gradient-TLC

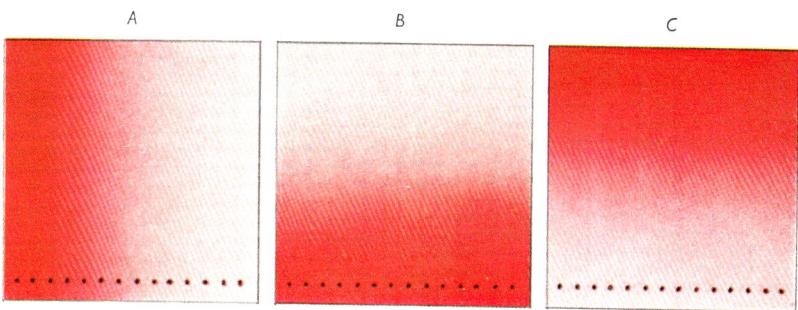

a) Three different separating surfaces in gradient TLC

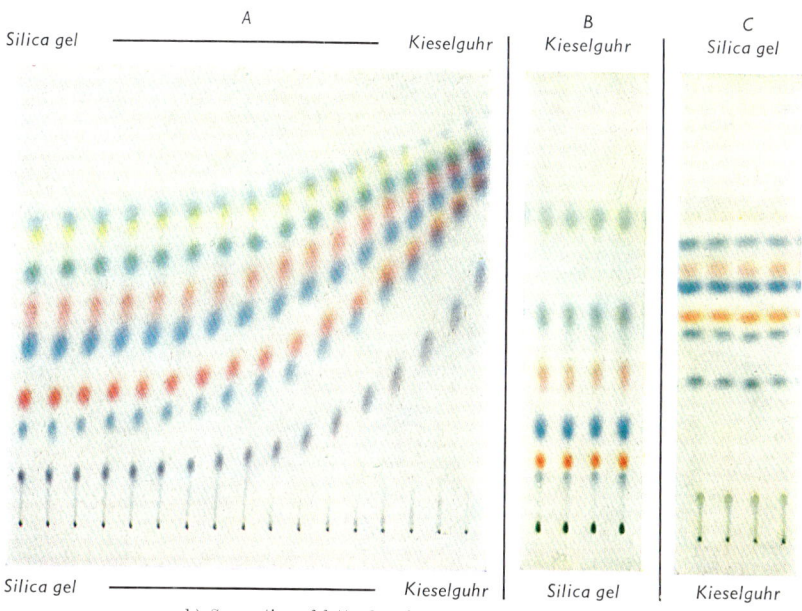

b) Separation of fatty dyes in the 3 running directions

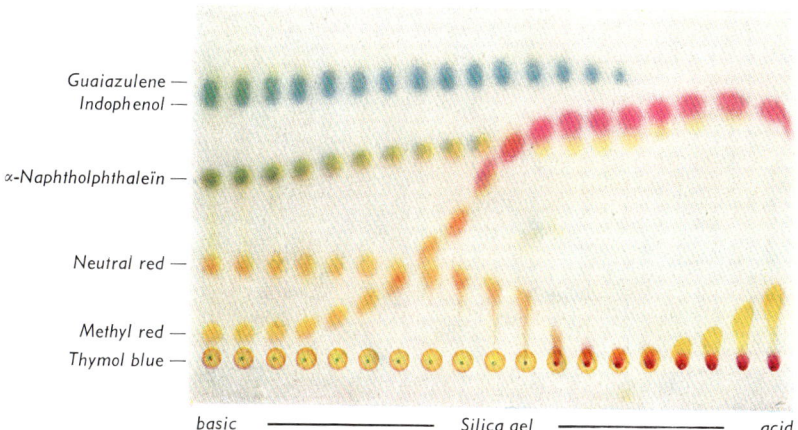

c) Curves for some dye indicators showing specific behaviour

Thin-Layer Chromatography, 2nd Edition

IV. Preparative TLC[3]

In general, microgram amounts of mixtures are separated in analytical TLC. The excellent results obtained here account for the wish to separate milligram or gram amounts in the same way.

It may be recalled that attempts have been made for a long time to carry out separations on a preparative scale in the other chromatographic methods (PC and GC). When all the results obtained there are regarded, they do not provide the same encouraging picture that might first be formed from considering the occasional individual work. Separation is often less good where the amounts were increased and the experimental conditions had been adapted to these increased amounts. A move in the other direction, i. e., to the submicro region, is more successful-for example in the GC-capillary columns or TLC itself.

As in distillation, it is advantageous to distinguish between "coarse" and "fine" separation problems. The former can mostly be adapted to the preparative scale. If the substances however lie close together on the TLC layer, adaptation to the preparative scale is unsatisfactory. Efforts are now being made to find suitable preparative procedures for these fine separations also.

1. Improvement of Separation in Column Chromatography

MILLER and KIRCHNER [446] (already in 1951) changed over from their chromatostrip technique to a rigid, self-contained, open column in order to separate volatile flavouring constituents in the 30–100 mg region; they termed the columns "chromatobars". Our earlier experiences with them were however not significantly better than those with the classical columns mostly used for this purpose. We therefore work even today according to the 1959 combination scheme [665] for separations of gram amounts:

Preliminary Trial	Main Experiment	Control
TLC $\longrightarrow$	Column $\longrightarrow$	TLC

DAHN and FUCHS [143] have developed a horizontal column chromatography in a cellophane tube, as analogy to TLC. These tubes, approximately 3.6 cm cross-section and 40–45 cm long, are filled with about 300 g especially fine-grained adsorbent. Several columns are joined in parallel in order to separate larger amounts. LÁBLER [385] and also LEFEMINE and HAUSMANN [391] have described devices for preparative

---

[3] the concept "preparative" is used in connection with gram amounts in organic synthesis. It is now being used in TLC, even though the "yield" per plate is usually of the order of a milligram. Many authors [672] used thus to refer to "micropreparative TLC". A recent comparison has been made between analytical and preparative TLC [Z. anal. Chem. **236**, 294 (1968)].

chromatography on a loose alumina layer, lying horizontally. Somewhat similar is the "dry column chromatography", claimed to have the resolution of TLC [405]. This technique has incidentally been described earlier in connection with the determination of activity grades [671].

The two-part "flat tank" of BEKESY in 1942 [55], subsequently modified by HALL [264], can be regarded as an interesting transition from the circular column to the preparative TLC-layer. It is essentially an S-chamber, filled with an adsorbent (p. 70).

2. Adaptation of TLC to Larger Applied Amounts

Many teams use rather thicker layers to separate milligram amounts. Usually 10 to 50 mg substance are applied as a band. Separations of amounts up to 2 g on 20×20 cm plates have only occasionally been reported.

A number of adaptations of the standard method are known and are treated here:

a) Increase of the Layer Thickness and Plate Size

Some investigations of the influence of layer thickness on separation have been made [300, 678]. Most authors have chosen layers between 0.5 and 1 mm thick; only occasionally have thicker layers been proposed. Slurries with lower water content and special drying procedures were then suggested to prevent formation of cracks [147, 300]. Special silica gels (P-series of Firm 88) and, recently, 20×20 cm pre-coated plates with 2 mm layers (Firm 88) have been developed also.

A strip of adhesive tape (Tesafilm, Scotch tape), 1—2 cm wide, is stuck all round the outside of the plate to be able to prepare thicker layers on glass plates, using the pouring procedure. The adsorbent suspension is poured into the shallow trough thus formed. The adhesive strip can be torn off after the layer has dried.

Plates of 20×20 cm or the equally handy 20×40 cm are mostly used as carriers for reasons of economy. HALPAAP [266], however, has preferred 1 m plates and constructed a correspondingly sized V2A development chamber and other accessories.

Various firms now offer collections of "basic equipment" for preparative TLC (Firms 33, 44, 129). "Do it yourself" spreaders for preparative TLC have also been described [56, 147, 374].

b) Band Application of Substance Solutions

Special techniques are necessary in order to apply larger volumes uniformly and as a narrow start band. A detailed description of the various methods is warranted since this is an important problem in preparative TLC.

Cumulative spot application, occasionally employed, is time consuming and generally leads to non-uniform zones (Fig. 21 B). A quite reasonable start band can be obtained using a suitably sized broad band pipette (Fig. 20i, Firm 44). Such two-part broad band pipettes can be prepared from narrow-grooved glass [684]. BENNETT and HEFTMANN [58] have also described a similar device. CONNOLLY et al. [135] have described the following technique for obtaining relatively slender start bands: two parallel cuts, ca. 1 mm broad, penetrating right through to the glass plate, are drawn on each side of the start band (ca. 0.5 mm broad). This prevents the applied solution from leaving the start band by capillary attraction. After the solution has been applied with a fine-tipped dropper, the cuts are refilled with loose adsorbent. A mask of aluminium foil with a 1 mm slit can be laid over each cut in turn, adsorbent shaken on to it and pushed through the slit with a spatula.

As has been already described [673], a solution can be evenly and rapidly sprayed as a band, using a micro spray gun (Firm 44). The layer on either side of the start band is then best covered (Fig. 21 A). In addition, devices have been developed which permit more uniform application. In this connection, the spreading apparatus for PC, constructed by MCKIBBINS et al. [414] first comes to mind. The newer commercial "Chromatocharger" (Fig. 51, Firm 33) functions according

Fig. 50. Motor-driven apparatus for automatic band application of solutions by injection (manufacturer, Firm 44)

Fig. 51. Hand operated apparatus for band application of a pre-determined amount of solution by injection (manufacturer, Firm 33)

to a similar principle. RITTER and MEYER [578] have worked out another idea; their more sophisticated, electrically driven apparatus for automatic band injection (not spraying) is commercially available under the name of Delfter-system (Fig. 50). A fully automatic apparatus for streak application is now commercially available under the name of "Autoliner" (Firm 44). See also footnote on p. 97. The apparatus described by COLEMAN [131] can be employed similarly for band application. Such spreaders can be recommended because they save time and spread more evenly. This latter is, in my opinion, a most important condition for preparative TLC. HALPAAP [266] has nevertheless shown that larger amounts can be spread as bands on 1 m plates by using a ruler and an ordinary blunt pipette.

Note: Usually 5–10% solutions are applied in preparative TLC. The solvent must be as volatile and non-polar as possible. If repeated application is necessary, one should wait after each addition until the solvent has evaporated. The formation of a chromatogram around the point or band of application, is undesirable. The band should terminate 1–2 cm from each side of the plate.

According to STAHL, stepwise development can be used for obtaining a very thin *start line* (p. 87). A solvent with strong elution properties is taken for the first stage, developing to a height of only 1–2 cm. This compresses the start band, which may be broad, to a single line. A shallow dish or suitable small vessel serves as development chamber for this. Development in the second stage is carried out in the normal way after having completely evaporated the solvent.

c) Types of Development (cf p. 68—77)

As a rule preference is given to ascending development, which requires less outlay. A form of "continuous development" is possible using a trough filled with adsorbent, as described by BENNETT and HEFTMANN [58]. SEIKEL et al. [625] have used a system for ascending continuous development, resembling Fig. 29 in principle. Successful separations with the circular technique (p. 73) have led v. SCHANTZ [602] to try this technique on the preparative scale. He prepared circular chromatograms on 40 × 40 cm plates, sucked off the zones with a micro "vacuum cleaner" and then extracted them.

Ascending development is that generally used. Multiple development, recommended for preparative TLC especially by HALPAAP [266], is very advantageous here. A marked improvement in separation is brought about when the solvent migrates 5–10 times through the layer. It must however be dried between each migration.

If 20 × 20 cm plates are used, several can be placed in the usual chamber, using a suitable holder. To develop 20 × 40 cm plates, either the S-chamber may be used or a suitably large chamber with a support and drying frame which can hold ten 40 cm plates (Firm 44).

The plates can also be packed together as "sandwiches", as described by NYBOM [488] (see p. 69) and placed in a chamber (Firm 33).

d) Visualisation of the Separated Zones

Procedures which do not harm the substances are used to identify the zones. The addition of suitable fluorescent indicators is probably the best method; detection is then in short or long wave UV light (see p. 77). TSCHESCHE and coworkers [722], for example, have used this. Another procedure frequently used is to cover the plate except for one or two strips; these exposed strips are then sprayed with a mild universal reagent like iodine vapour (Reagent No. 141) or potassium permanganate (Reagent No. 200) or others where heating is not necessary and also neither acid nor basic vapours are formed.

e) Collecting the Zones and Extraction (see also p. 148—151)

The substances are most simply obtained by scraping off the zones with a spatula or similar instrument and subsequently extracting or percolating. So-called "micro vacuum cleaners" like those described by RITTER and MEYER [578], GOLDRICK and HIRSCH [238] and STAHL [674] can be used for the scraping off and collecting. One must however realise that the substances on the adsorbent come into intimate contact with atmospheric oxygen and can easily undergo oxidation. Suitable apparatus for this transfer technique is seen in Fig. 52 B/C. To extract the substances with a volatile solvent, the percolation tube, previously used as micro vacuum cleaner, may be introduced into a glass apparatus which can be sealed [625]. Extraction in a Soxhlet apparatus is not advisable with substances sensitive to heat.

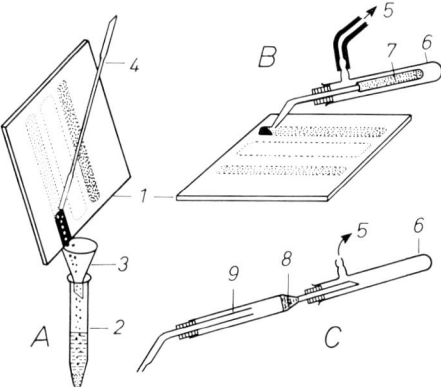

Fig. 52. Various ways of removing separated substance zones. *A* scraping off with a small spatula (*4*); *B* removal by suction with a small "vacuum cleaner"; *C* alternative device for a vacuum cleaner; *1* TLC plate with layer; *2* test tube containing solvent; *3* funnel; *4* spatula; *5* connection to reduced pressure; *6* test tube receiver; *7* soxhlet thimble; *8* cotton wool plug or sintered glass filter; *9* bent glass tube

Percolation is the least drastic and probably most suitable procedure for dissolving out substances from the adsorbent. The scraped-off adsorbent is placed in a conveniently sized glass tube, e. g., an Allihn tube or a tube with a sintered glass filter (G 3) and tap. An eluent is then added which dissolves out the substance. One must remember, however, that polar eluents, e. g., alcohol, can dissolve not inappreciable amounts of numerous inorganic adsorbents, such as colloidal silica from silica gel. The less polar the eluent, the less silica is found in the residue.

Further, it must not be forgotten when extracting the separated substances, that inorganic adsorbents may also contain small amounts of organic compounds which come from their method of preparation. GEISS and co-workers [226] have shown amongst other things that impurities can stem even from the PVC mouthpiece of the vacuum cleaner. Plasticisers can also pass from the plastic packing covers of the adsorbent into the adsorbent itself and are then found in the final extract.

In any case it is thus advisable to remove the solvent from the substances, to redissolve the latter and, after filtration through a very fine filter, to crystallise from an appropriate solvent.

V. Transfer Techniques [682]

Further study of the substance zones on a thin layer chromatogram is often desired. One may wish to confirm an identity by recording a spectrum, for example, or by carrying out some other physical, chemical or chromatographic analysis. Quantitative determination is frequently wanted and for this, the substance amount in the spot must be transferred.

The problem is thus the separation of very little substance from a relatively large amount of adsorbent in order to be able to pursue another method of investigation afterwards. It is, however, a matter not only of transferring micro amounts of the zones which have been separated by TLC, to the start of another analytical method, but also perhaps of the converse. Thus, for example, TLC control of a column, gas or paper chromatographic separation, is becoming increasingly important.

Procedures which enable such transport to be performed, are treated here under the heading of *"transfer"*-techniques.

1. Transfer: TLC → Solution

It is generally necessary to transfer the substance from the adsorbent (spot, zone) to a clear solution in order to be able to carry out subsequent spectroscopic investigation. A concentrated solution can also frequently serve as starting point for subsequent gas chromatographic study. There are no difficulties even with dilute solutions when using a gas chromatograph with a highly sensitive detector (e. g., flame ionisation detector). The investigation following initial TLC may also, however, be a further TLC procedure of the substance solution, using another developing solvent and/or on another layer. In this last named case, the special percolation tube (Fig. 53 b) can be conveniently used. It is a help here however, when the tip of the percolation tube is drawn out to a conical point so as to be able to employ it for direct application. The eluent can be largely expelled from the "micro-column" by blowing, if necessary with the help of a tube.

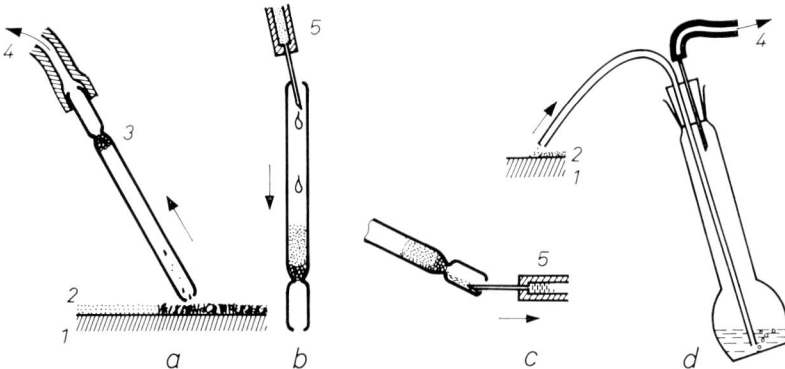

Fig. 53. Transfer of a substance from a TLC-spot. *a* suction of the scraped-off substance with a special percolation tube; *b* elution with a few drops of solvent; *c* withdrawal of the extract with a micro-syringe; *d* device for taking up the substance, together with adsorbent, in a definite amount of solvent. *1* glass plate; *2* adsorbent layer; *3* percolation tube with small plug of cotton wool; *4* communication with reduced pressure; *5* microlitre syringe

Apparatus; The special percolation tube [65], illustrated in Fig. 53 b, is made from glass tubing about 8–10 cm long and is of internal cross-section ca. 2 mm. A plug of fat-free cotton wool introduced into the tube serves as filter. The spot on the TLC plate is first dotted and then carefully scraped loose with a narrow metal spatula (2 mm broad). The powdered adsorbent can then be sucked directly into the percolation tube, as in Fig. 53 a. Depending on the filling of the column 20–100 µl solvent are then added with a microlitre syringe (Fig. 53 b). The eluent which has run through is collected in the short part of the tube and can in turn be taken up with a microsyringe (Fig. 53 c).

Fig. 53 d shows a device for direct transfer of the adsorbent of a zone into a 5 ml standard flask. It already contains 2 ml solvent for extraction. After sucking in the adsorbent and making up to the mark, rapid centrifuging is carried out. The supernatant liquid can be used for quantitative determinations [651].

Other devices, essentially similar, are described in the literature (see p. 148 and [70, 123, 318, 545]).

2. Transfer: TLC → Reagent

Direct transfer of the substance of the substance zone into a colour-yielding reagent solution is of interest in many quantitative determinations. The vacuum cleaner principle can be used here as well as the simple technique of scraping off (Fig. 52 A). Various other set-ups can be used besides that sketched in Fig. 53d [605] (Firm 119).

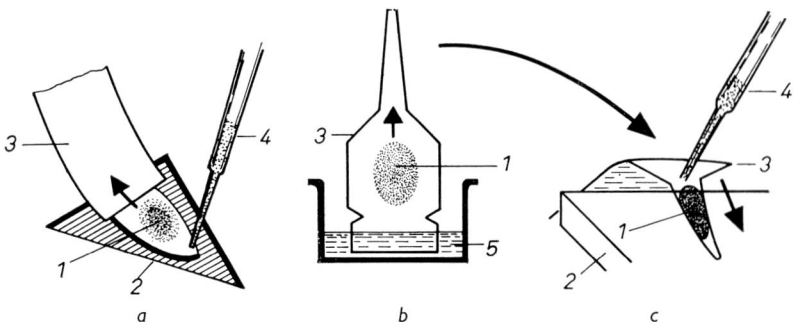

Fig. 54. Transfer TLG ↔ paper. *a* transferring a TLC-spot to a paper strip; *b* concentrating a PC-spot in the point of the paper; *c* delivering this concentrated substance spot to TLC. *1* substance spot; *2* scraped-off TLC-zone (shaded); *3* filter paper; *4* micropipette; *5* crystallising dish with solvent. For details, see text (according to [559])

3. Transfer: TLC → Paper

In many quantitative determinations, a preliminary transfer of the substance from the thin-layer chromatogram to filter paper is an advantage [566]. A triangular section of the layer is scratched off round the substance spot (Fig. 54a, shaded area). With the help of a micropipette (4), the substance is then washed into the filter paper strip (3) by capillary attraction. A solution of the substance, free from traces of colloidal adsorbent, is often more easily obtained by extraction from paper.

4. Transfer: Solution → Paper (PC) → TLC

Very dilute solutions which contain little substance, are best concentrated, i. e., enriched, in a preliminary procedure. For this, the solution can be spotted on to a tongue-shaped piece of filter paper (Fig. 54b). A stream of warm air accelerates evaporation of the solvent [440]. The ring-oven technique may also be employed [604]. In order to transfer the relatively large substance spot to TLC, the paper tongue is placed in a bowl filled with solvent (Fig. 54b). The substance is washed into the tip of the tongue. It can be transferred from here to the TLC start point in the simple way illustrated in Fig. 54c.

Note: Occasionally in trace analysis a particular substance must be concentrated on the TLC plate itself. Two-dimensional development of a substance applied in band form, is useful here. The substance of the particular band is displaced into the region of the front in the second dimension by using a strongly polar solvent [445].

5. Transfer: Gas Chromatography → TLC

Many liquids and solids, issuing in gaseous form, can be spotted directly on to an adsorbent layer. Prior condensation in a cooled tube and subsequent extraction are thus circumvented. Details of this useful combination can be found in Chapter F (p. 114).

E. Thin-Layer Electrophoresis

K. Hannig and G. Pascher

As long ago as 1946, Consden, Gordon and Martin [137] carried out electrophoretic separation of a mixture of amino acids and peptides on a thin layer of silica gel. Although this technique was accompanied by no particular difficulties, it could not at first keep up with the development at the same time of paper electrophoresis, which was so extremely simple to carry out [129, 268, 777]. Only in recent years, after Stahl in fundamental work had shown how to prepare and use thin layers of adsorbent, has the idea of thin-layer electrophoresis been taken up again. It has the advantage over paper electrophoresis, that interfering adsorption effects can often be avoided through appropriate choice of adsorbent; the scope of the method can thus be markedly broadened. In addition, the more uniform and finer structure of the adsorbent used for thin-layer electrophoresis, compared with filter paper, sometimes leads to better resolution.

All possible fields of application of thin-layer electrophoresis cannot be reviewed since the method is still too new. Only the essential aspects concerning theory, operation and application of the method are thus given below:

I. Principle of Thin-Layer Electrophoresis

An ordinary TLC plate is evenly sprayed with a suitable buffer solution. The substance mixture to be separated is applied to the moist layer as a spot or band and an electric field applied to the plate. The substances migrate at various speeds and in a direction depending on their electric charge. The individual zones are fixed at the conclusion on the carrier and can be detected with the customary methods.

A schematic representation of a separation is given in Fig. 55. The substances negatively charged at the pH used (1, 2, 3, 5) migrate towards the anode; the positively charged (4, 6) towards the cathode.

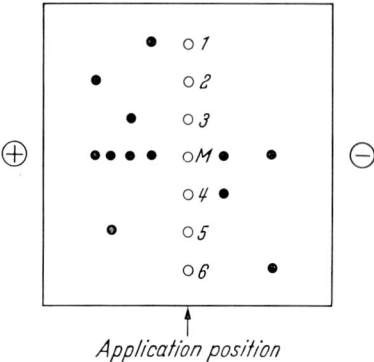

Fig. 55. Diagram of a thin-layer electropherogram. *1, 2, 3, 5*: negatively charged substances; *4, 6*: positively charged substances; *M*: mixture of *1—6*

II. Theoretical Principles [129, 268]

1. General

By the term "electrophoresis" is understood the migration of electrically charged particles in solution under the influence of an electric field. This migration depends on properties of the particle itself (like magnitude of its charge, hydration, tendency to dissociate, configuration etc.); on properties of the medium surrounding the particle (pH of the electrolyte solution, ionic strength, temperature, viscosity etc.); on the strength of the electric field applied; and on the time. In *carrier-free electrophoresis*, the electrophoretic mobility, u (in cm^2/$V \cdot s$) of a spherical particle is given by:

$$u = \frac{q \cdot s \cdot \varkappa}{i \cdot t}$$

q = cross-section in cm^2; s = path traversed in cm during time t; i = current in ampères; $\varkappa$ = specific conductivity in (ohm. cm)$^{-1}$.

2. Factors Influencing the Mobility of Particles in Electrophoresis with Supporting Medium (Thin-Layer Electrophoresis)

Thin-layer electrophoresis can in theory be used with all compounds which are soluble in the electrolyte medium and which are electrically charged, assuming that conditions can be found which yield adequate differences in the direction and speeds of migration of the compounds to be separated.

The charge (negative or positive net charge) of, e. g., a protein particle, is closely related to its electrophoretic mobility. Approximate information about the likely direction and magnitude of mobility of particles migrating in an electric field, can be obtained from titration results. There is, however, no quantitative agreement between the net charge as calculated from titration curves and from electrophoretic mobility.

Several quite general factors must be considered when working out mobility data in electrophoresis with supporting media; these may be mentioned here:

a) Adsorption

If a substance to be separated is adsorbed on the carrier material, only the non-adsorbed part of the substance migrates and a reduction of migration speed occurs. This adsorption effect influences not only the rate of migration but impairs the separation also. Adsorption brings about a non-uniform concentration distribution, so that the forward edge of the zones is more sharply defined than the rear edge which is more or less blurred.

b) Diffusion

Dissolved substances in a concentration gradient tend to diffuse into adjacent regions in accordance with the Brownian movement. The migration routes of particles in a carrier material are always tortuous, so that diffusion into neighbouring regions is increased through the varying concentration gradient. The influence of both factors is such that the definition of the zones in the supporting material can vary, depending on the migration distance which has been covered. Carrier materials of the highest possible grain uniformity are consequently preferable for obtaining the best resolution.

c) Zone Flaws

Somewhat irregular zone fronts occur always, since slight inhomogeneity of the support, non-uniform moisture content and electrolyte concentration and temperature differences in the electric field can never be avoided completely. This zone imperfection is of subsidiary importance in qualitative detection of individual fractions but interferes in a quantitative evaluation.

d) Zeta-potential and Electroosmosis

In all electrophoresis procedures in heterogeneous media, the solid phase is charged vis-à-vis the surrounding solution to an extent which depends on the pH. This is characterised as the zeta (ζ-) potential. Application of an electric field brings about a movement of the liquid with respect to the supporting medium which is fixed in space; this is termed electroosmosis. The electroosmotic flow can be so large under certain conditions that an apparent reversal of the migration direction of the studied particles occurs.

The electroosmotic effect increases with increase in the structural fineness of the supporting material and in the field strength. Increase in pH renders a negative zeta-potential smaller and a positive, larger.

e) "Suction Effects"

In addition to the usually unavoidable electroosmotic flow, movements in the electrolyte solution can also occur which are controllable. Should the supporting layer not have been completely saturated with liquid at the beginning of the experiment, an initial suction from both sides towards the centre occurs, even without application of an electric field. Evaporation takes place if the carrier layer becomes warmer than its surroundings, as a result of the heat generated by the current. The layer thus sucks buffer solution out of the electrode compartments. This evaporation suction causes electrolyte flow from the sides towards the centre of the coated plate.

Initial and evaporation suctions diminish as one moves towards the centre of the layer and become zero there. Substances which migrate rapidly are consequently relatively held up, which militates against the separation.

If the electrode compartments are not filled to equal heights or the apparatus is not horizontal, a flow of electrolyte occurs also towards the lower level (siphoning effect).

3. Units of Reference

It is evident from the foregoing discussion that an exact determination of the electrophoretic mobility, which can be utilised for identifying a substance, is accompanied in heterogeneous systems by many difficulties. In electrophoresis with supporting medium, one thus works in practice with the concept of apparent instead of real mobility.

Unfortunately there is no unambiguous definition of apparent mobility, similar to that of the exactly specified Rf-values in chromatography. Various terms have become accepted, causing confusion, such as M_{G}-, Ef-, R_{B}- and mr-values. The data of different authors are therefore not necessarily comparable.

FOSTER [198] has used 2,3,4,6-tetramethyl-D-glucose as non-migrating substance and glucose as reference substance for carbohydrate separations in borate buffer, introducing the term M_G[1]

$$M_G = \frac{\text{true distance of migration of the unknown substance}}{\text{true distance of migration of glucose}}.$$

A reference substance, electrically neutral under the experimental conditions, i. e., non-migrating, is applied together with the unknown substance on the start line.

Another method is to express the migration distance of the unknown substance in relation to that of a likewise migrating, but known substance. The migration is then calculated from the equation:

$$R_B = \frac{\text{distance of migration of the unknown substance}}{\text{distance migrated by the reference substance}} \quad [754].$$

HONEGGER [299] has called the same quotient the Ef-value.

Either value is adequate if one ensures that no suction effect occurs along with the unavoidable electroosmotic flow. Should this suction effect vary from separation to separation (see above), the following procedure is recommended:

The substance mixture to be investigated, an inert (i.e., nonmigrating under the conditions used) and a migrating reference substance are applied together. The substances are localised after separation and the migration of the unknown substance calculated according to Fig. 56.

Correction is made for both the suction and electroosmotic affects with the help of this method [758].

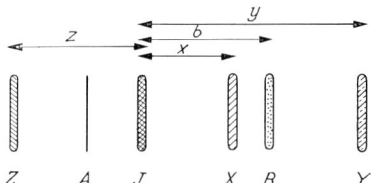

Fig. 56. Scheme of an electrophoretic separation for calculating the m_r-values. A point of application; B reference substance; J inert substance; X, Y, Z unknown substances after separation; x, y, z, b intervals between the inert substance and the unknown and reference substances;

substance X $m_r = \frac{x}{b}$; substance Y $m_r = \frac{y}{b}$; substance Z $m_r = \frac{z}{b}$

III. Details of Apparatus and Methods

1. Apparatus

One of the usual sets of equipment for paper electrophoresis can be used for thin-layer electrophoresis if necessary. Apparatus is, however, commercially available today, developed especially for the purpose (Firms 87, 129, 44).

---

[1] M = migration; G = glucose

The general view of an example of such apparatus is in Fig. 57 and, in Fig. 58, a cross-section through the electrophoresis chamber. The thin-layer plates (20 × 20 cm) rest on a water-cooled metal plate.

Fig. 57. General view of a thin-layer electrophoresis apparatus (photograph of Firm 44)

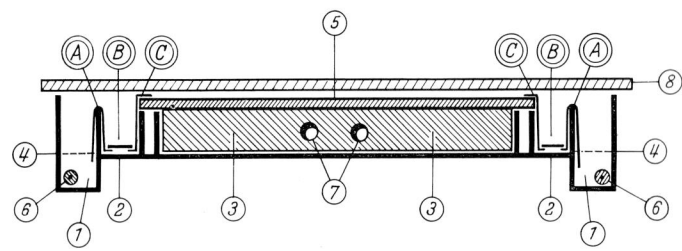

Fig. 58. Section through the electrophoresis chamber. *A* paper bridges; *B* filter cards; *C* filter paper tongues; *1* outer electrode compartment; *2* inner electrode compartment *3* cooling block; *4* level of buffer solution; *5* plate with thin layer; *6* platinum electrodes; *7* tubulures for cooling agent; *8* cover plate

Electrical contact is established between the electrode compartments, filled with buffer, and the adsorbent layer, by using a strip of filter paper saturated with buffer. Direct current is provided from a mains stabiliser with choice of 100, 200 or 400 V.

2. Methods

a) Supporting Layers and Preparation of the Plates

Some of the adsorbents being used most frequently at present may be mentioned here. WIELAND and PFLEIDERER [753] have separated protein hydrolysates, adenosine-mono-, di- and triphosphoric acids and also serum proteins, all on starch. On agar layers, MARTEN [432] has separated biological substances; and PFRUNDER and coworkers [528], inorganic ions. PASTUSKA and TRINKS [502–503] have reported separations on silica gel G and kieselguhr G and HONEGGER [299] has used these supporting materials and alumina G. Both JOHANSSON and RYMO [330, 331] and DOSE and KRAUSE [170] have separated proteins, peptides, amino acids and dyes on dextran gel (Sephadex G_{25}, G_{50} and G_{75}). RAMSEY [556] and BAUR [52] have reported separations on thin layers of starch gel. KECK and HAGEN [342] have employed cellulose layers to separate structural units of DNA. DOBICI and GRASSINI [165] have worked on the quantitative separation of iodate-periodate mixtures which was not possible on paper. They used gypsum as supporting material and applied the substances to the moist layer. The plates were then immediately placed in the apparatus and separation was begun. Many authors prepare a slurry of the adsorbent with the buffer ultimately to be used in the separation, instead of with water; they thus do not have to spray before separation. Such layers must then be prepared immediately before use.

b) Buffer Solutions

The substances to be separated must be soluble and stable in the chosen medium. The pH-value of the buffer solution must not change during the separation and the buffer concentration in the layer should remain constant.

Excessive heating of the layer occurs when the ionic strength is too high. The concentration of non-volatile buffers then rises, due to evaporation; this further increases the ionic strength and the heating in a vicious spiral. Unequal evaporation of cation and anion components of a volatile buffer can lead to changes in pH, e. g., with triethylamine (B. P. 89° C) and acetic acid (B. P. 118.1° C). Working with either lower current or more dilute buffer is then advisable. Table 12 contains a compilation of available buffer solutions for thin-layer electrophoresis.

c) Current

The current with which one can work depends on the degree of cooling. The heat generated must be removed or the layer will warm up continuously. The water cooling in the apparatus described above

Table 12. *Experimental Conditions for Separating Various Substance Groups*

Substance Group	Buffer Composition	pH	Voltage or Field Str.	Layer	Refer.
Amines; amino acids	2N Acetic acid-0.6 N formic acid (1 + 1)	2.0	460 V		[299]
	pyridine-acetic acid-water (1 + 10 + 90)	3.6		Silica gel kieselguhr	
	Na citrate buffer (0.1 M)	3.8	440 V	alumina G	
Amino acids; DNP-amino acids; peptides; serum proteins	0.02 M Phosphate buffer (containing 0.2 M NaCl)	7.0		Sephadex G_{25}, G_{50}, G_{75}, G_{100}	[330, 331]
Amino acids; trypt. degradation prods. of haemoglobin and ovalbumin	Pyridine-acetic acid-water (20 + 9.5 + 970)	5.2	60 V/cm	Silica gel H cellulose MN 300	[692]
Amino acids; amines; peptides; serum proteins; protein hydrolysates; biol. phosphate cpds.	0.075 M Veronal buffer	8.6	20 V/cm	Starch cellulose	[753]
Esterases	0.025 M Borate buffer	8.55	15 V/cm	Starch gel	[556]
Serum proteins; lactic acid dehydrogenase-isoenzymes	Tris-buffer: 9.3 g tris + 1.2 g Na-EDTA + 0.71 g boric acid/l	9.0	300 V	Acrylamide gel	[568]
Serum proteins; haemoglobin	0.1 M Tris, 0.0067 M citric acid, 0.04 M boric acid, 0.016 M NaOH	8.65	4—5 V/cm	Starch gel	[52]
Deoxyribonucleic acid units (bases, nucleotides, nucleosides)	Ammonium formate buffer (0.05 M)	3.4		Cellulose MN 300	[342]
Phenols, phenol-carboxylic acids, naphthols	80 ml Ethanol + 30 ml water + 4 g boric acid + 2 g Na acetate (with water of crystalln.)	4.5	20 V/cm	Silica gel	[502, 503]
	pH adjusted with acetic acid	5.5		Kieselguhr	
	pH adjusted with NaOH	7.8			
Dyes	pH adjusted with NaOH	12.0		Silica gel, G	
Inorganic cations and anions	0.05 M Lactic acid 0.1 M NaOH		13—46 V/cm 13—45 V/cm	Silica gel kieselguhr	[454]
Iodate-periodate	0.05 M ammonium carbonate		400 V	Gypsum	[165]
Tar dyes	0.05 M Borax	9.18	200 V	Kieselguhr silica gel alumina	[139]

is generally sufficient at 2—4 watts. If higher values are desired, cooling must be carried out with a cryostat [692, 754].

The strength of the electric field in the thin layer cannot necessarily be calculated from the voltage laid across the electrodes. The various bridges establishing contact between electrode compartments and thin-layer cause a more or less appreciable fall in voltage. The potential difference across the thin-layer plate would have to be measured with a voltmeter in order to obtain information about the field strength. It is thus customary to quote the voltage applied to the instrument (Table 12).

d) One-dimensional Thin-Layer Electrophoresis

The substance mixture to be separated and, if necessary, a reference substance, is applied with a micropipette as a spot or band to the start line on the plate which has been sprayed with buffer; the start line is at right angles to the applied electric field and is chosen according to the expected direction of migration of the substances. The start points can be marked with a metal or hard pencil. Neighbouring points should be 2—3 cm from each other and the diameter of the spots less than 1 cm.

e) Two-dimensional Thin-Layer Electrophoresis

The substance mixture is usually applied in one corner, about 4—5 cm from the plate edges. It is important to use in the first dimen-

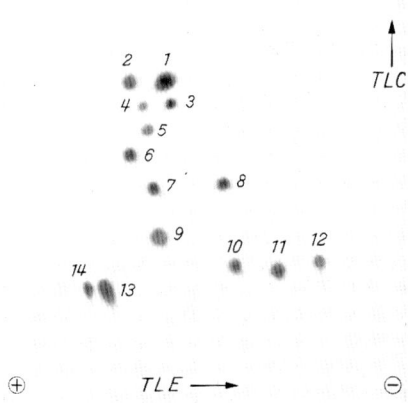

Fig. 59. Finger print of a casein hydrolysate on cellulose MN 300; plate size 20 × 20cm layer thickness 0.25 mm. 1 dimension: electrophoresis with buffer of 25 ml formic acid and 78 ml acetic acid per litre; 400 V and 75 min run. 2 dimension: TLC with chloroform-methanol-17.5% ammonium hydroxide (41+41+18). *1* leucine + isoleucine; *2* phenylalanine; *3* valine; *4* methionine; *5* proline; *6* tyrosine; *7* hydroxyproline; *8* alanine; *9* serine + threonine; *10* glycine; *11* histidine; *12* arginine + lysine *13* glutamic acid; *14* aspartic acid

sion a volatile buffer which must be completely removed before the second run.

f) Finger Print Technique

The combined electrophoresis-TLC procedure has proved very satisfactory [299, 342, 692]. The order in which the operations is carried out is immaterial in principle and will be dictated by the circumstances. It is important here also, however, that solvent and electrolyte be completely removed after the first run. If, therefore, salt-containing buffers are to be used for the electrophoresis, the chromatography stage must be the first. The plate for the subsequent electrophoretic separation must however be sprayed most cautiously so as not to wash away the substances already separated. The separation of a casein hydrolysate is shown in Fig. 59.

F. Coupling of Gas- and Thin-Layer Chromatography

R. Kaiser

Gas and thin-layer chromatography can be directly coupled. By "direct" one understands the immediate combination both in time and place.

If this combination be correctly used, i. e., limited to those analytical problems which can be attacked by either procedure, unexpectedly useful possibilities arise. The coupling can be regarded as a new analytical entity on account of the especially high provision of information.

There are three essential features:

1. GC-TLC-coupling permits a double chromatographic separation. The separation is two-dimensional. The final result is attained through comparison of a gas chromatogram with the GC-TLC-chromatogram.

2. The combination offers the possibility of independent and multiple qualitative identification as well as quantitative and qualitative determination of individual compounds.

3. The coupling method is a most critical control procedure. In particular, contradictions in qualitative results are disclosed and it therefore permits a retroactive check of quantitative GC-values. Since TLC results may readily be rendered quantitative with the help of GC, the accuracy of analytical work can be greatly enhanced through this type of control.

It has been stated above that correct application means working on those analytical problems alone, for which both methods are suitable. In amplification one may say: Coupling analysis often demonstrates whether the methods are suitable in a particular case. Moreover, experience has shown that many more problems can be successfully handled by GC or TLC than at first expected. Where the nature of the sample precludes the use of the individual procedures, partial problems can be solved in the majority of cases.

I. Principle of the GC-TLC-Coupling Analysis

The sample under investigation is subjected first to gas chromatography. As known, only compounds which volatilise without decomposition or which decompose reproducibly, or products of pyrolysis, are suited to this procedure (all gases, many liquids and solids with vapour pressures down to 10^{-2} mm at $200°$ C).

The sample is vaporised in an inert carrier gas at a temperature which can be selected between 20 and $500°$ C. If the whole sample cannot be vaporised, a measured amount of a known "internal standard" can be added beforehand.

The vaporised sample is now swept through a so-called separating column by means of a suitable carrier gas at a definite temperature. Separation into the individual components occurs in the column as a result of specific retention. Analogous to TLC, a qualitative characteristic of each substance is provided-here it is the retention time t-which can be used for identification or structure elucidation. It is unnecessary to discuss further details of GC here; see, for example, BAYER [53], KAISER [337] or KEULEMANS-CREMER [345].

The separated single compounds leave the column as "vapour clots" (peaks) in the carrier gas. The vapour concentration is usually ca. 10^{-5} to 10^{-10} g/ml and the carrier gas flow about 1—4 l/h. In direct GC → TLC coupling, the carrier gas leaving the apparatus is directed straight on to the thin layer placed some tenths of a millimetre from the column exit. This brings 10^{-5} to 10^{-10} g substance per sec on to the layer (Fig. 60). From 30—80% substance is retained, depending on temperature, gas flow, polarity-, concentration- and adsorption conditions. The "yield" can approach 100% in especially favourable instances.

The substances thus "transferred" from GC to a TLC start line can then be developed with a solvent in the usual way. Mixtures perhaps not separated in the GC stage may then be separated.

In any case, the characteristic hRf-values may be determined after development, so that each constituent of the original mixture can be characterised by two values, the retention index (GC) and the hRf-value

(TLC). Useful supplementary information may be obtained by skilful choice of retention in GC and TLC-for example, GC separation according to molecular weight, TLC separation according to polarity.

In GC-analysis, the chromatogram can be recorded only electronically or as a picture (pen recorder, magnetic recorder) and retained only as time values and intensities of a detector signal; when fed from the gas chromatogram on the TLC-plate, however, the materials are concretely

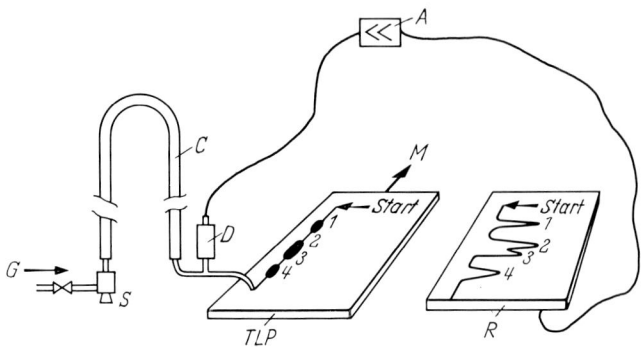

Fig. 60. Principle of GC/TLC direct coupling. G carrier gas; S sample introduction system; C column; D detector, to which only a fraction of the carrier gas containing the separated components is directed by the T-piece below it; TLP thin layer, moved in front of the depositing tube continuously or intermittently with the motor M; R recorder; A amplifier

present and can be submitted to the whole gamut of special TLC methods of chemical identification and treatment. The gas chromatographic identification from data derived from the response of various detectors and from differences in retention index is then amplified through visualisation with selective spray reagents as well as through the hRf-values. The final analytical result is thus doubly fortified qualitatively and finally quantitatively also.

A thin-layer chromatogram can be run also without GC, parallel to the coupling procedure (see Figs. 64, 65). This allows further information to be gained: completeness of substance take-up; changes of the compounds in the gas chromatograph or on the thin-layer chromatogram; interferences in the GC separation; falsification of the GC quantitative response, etc.

This check is important if one wishes to use GC as a means of rendering the TLC analysis quantitative. Should the sample components decompose or not be wholly taken up, large systematic errors easily arise.

NIGAM, SAHASRABUDHE and LEVI [480] were the first to describe the use of direct GC-TLC coupling and to point out its importance. They analysed the constituents of essential oils, depositing the individual GC peaks in the eluent on to the TLC layer.

At about the same time both KAISER [336] and JANÁK [319] independently described coupling techniques in which the plate moved along a so-called start line during deposition. JANÁK described the coupling procedure primarily as a deposition method for TLC and paper chromatography. KAISER pointed out that coupling served mainly to

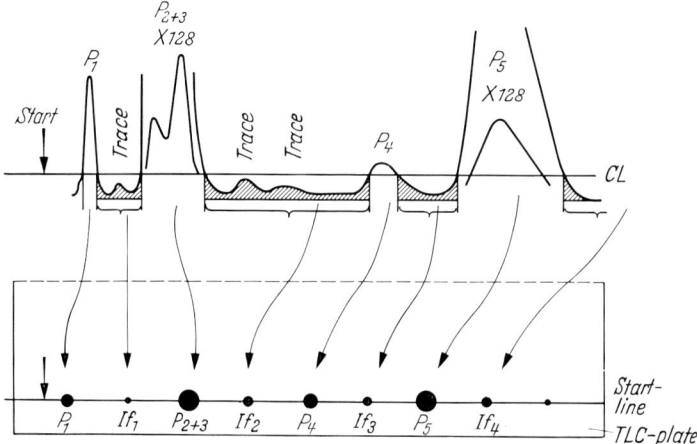

Fig. 61. Intermittent deposition. A contact at the level of the contact line CL is so adjusted on the recorder that each time the pen carriage traverses the electrical connection (Fig. 66), the motor M (Fig. 67) is switched on for rapidly transporting the plate 10—15 mm. In this way, each of the substances of peaks P_1, P_2 and P_3, P_4 and P_5 is automatically deposited as a single spot; further, the intermediate fractions If_1, If_2 etc are each deposited as spots on the start line

augment the analytical information provided by each method; he described a variation in which the plate movement was automatically controlled by the detector of the gas chromatograph so that each GC peak and each "intermediate fraction" were spotwise deposited (Fig. 61). This has the advantage that the "peaks", highly diluted with the carrier gas, are accumulated and concentrated by the layer, improving sensitivity enormously, especially in trace analysis. A linear movement of the plate is also an advantage, particularly when the column is temperature-programmed. Dilution of the deposited spot is then avoided; this is otherwise unavoidable as a result of the peak broadening that occurs under isothermal column conditions (Fig. 62).

The movement of the TLC plate also can be programmed. Isothermal technique in the GC can be retained if this plate movement is then logarithmically decelerated. For several reasons this type of coupling is nevertheless disadvantageous. JANÁK et al. [321] describe experimental details for it.

Indirect coupling of GC and TLC has been used for a long time, of course. The peaks from the gas chromatogram are isolated preparatively or the TLC spots are extracted and the materials thus isolated are subjected to the other chromatographic procedure; one cannot however profit in this way from the numerous and decisive advantages of the direct coupling and gains no new analytical weapon.

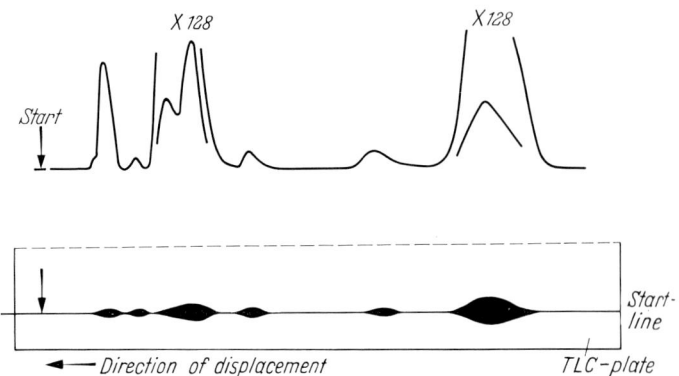

Fig. 62. Continuous deposition. The same mixture as in Fig. 61 is deposited on the continuously moving TLC-plate

Discontinuous coupling TLC → GC always occurs through a mechanical transfer step. This takes time, may be accompanied by loss of material and decisive information concerning the structure, is lost. This combination with time interval is not treated further here (see, e. g., [669] and Fig. 100). It is however not uninteresting to know that this discontinuous transfer with time lag is possible with the same high substance yield as in the direct coupling procedure. NEILL et al. [471] have stated that this transfer is still sufficiently accurate for biological analytical problems with 10^{-7} g substance.

The equipment necessary will be discussed below; this is followed by a description of some basic and special applications.

II. Equipment

The GC separating column terminates in a T-piece which divides the gas stream in two. The smaller part (5—20% of the total) flows over the detector; an ionisation detector or a detector combination are especially suitable. The major part (95—80%) flows out of a mouthpiece of ca. 0.8—1.5 mm diameter on to the thin layer about 1 mm vertically below it. The substances are directly transferred from the flowing carrier gas phase on to the thin layer. Up to the point when the adsorption capacity of the layer is exceeded, the area of application is thus scarcely larger than the cross-sectional area of the exit of the depositing tube (Fig. 63)[1].

When the thin layer plate is stepwise transported, each large or medium-sized peak is precipitated on to one point and all other small peaks, lying between the larger ones, are collected together on another

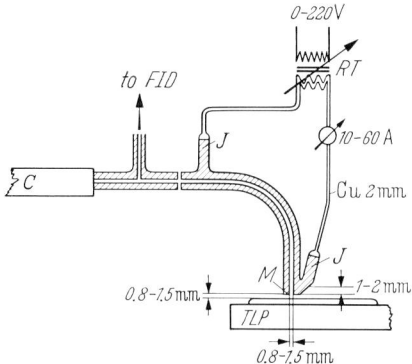

Fig. 63. Deposition tube with mouthpiece. Steel tube, about 2×0.5 mm, directly heated electrically with high current using a "jacket heater" J

point. To achieve this, a sliding contact on the GC-recorder, controlled by the pen carriage, despatches a current impulse to a mechanism for moving the plate step by step. Detection is highly sensitive when traces are collected in one spot. This stepwise spotting on the thin layer has advantages and disadvantages. A device with unbroken plate movement, in which the substance peaks are deposited rather as bands, has other pros and cons. Coupling with a temperature programmed GC apparatus is especially advantageous in this case. The speed of plate movement can be adapted to that of paper displacement on the recorder. When the available space on the start line (for spot or band application) of the TLC plate is filled, the customary development and visualisation converts the former GC peaks into visible TLC spots.

Mixtures which are only moderately or not at all separated and completely or partly overlapping peaks are thus subjected to a further chromatographic treatment. Any of the varied techniques of TLC separation can be applied-combinations of solvent, adsorbent and working procedure. A silica gel, suitable for TLC, will be used in the simplest case, separating according to polarity; and a normal solvent like chloroform for example. After separation, the spots are visualised as

---

[1] For description of a new device dividing the gas stream over the detector and on to the thin layer, see Chemistry in Britain (1968) (in the press).

usual (p. 77). The reaction used may yield a specific colour. Identification is sometimes possible already at this stage of the combined procedure, with the help of the TLC R_f-value, the GC retention index and the chemical test. All the vast spectrum of possible tests on TLC layers is available and a positive reaction provides more information than any specific GC-detector. The quantitative gas chromatogram must be critically compared with the qualitative thin-layer chromatogram. A suitable scheme is to apply the original sample to an extrapolation of the start line for deposited material from GC separation and to separate it on the same plate without previous GC treatment. The gas chromatogram, the "two-dimensional" GC/TLC separation and the chromatogram from TLC alone are thus available for comparison (see, e. g., Figs. 68 and 69).

III. Information from Coupling Analysis

The following questions can be answered:

1. Has the GC separation been complete or could any peaks, not separable in GC, be separated in the succeeding TLC (Fig. 64)?

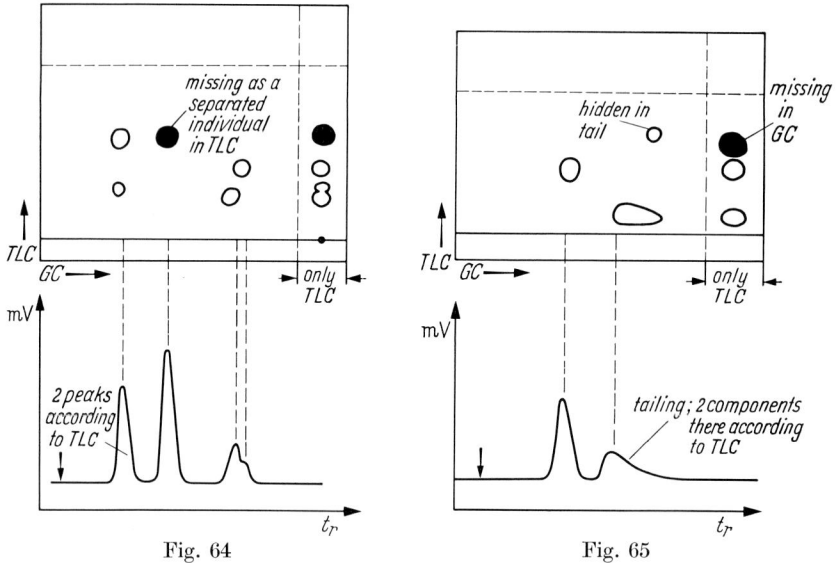

Fig. 64.   Fig. 65

Fig. 64. Separation on the TLC-plate of components not separated by GC; and GC separation of components not separable by TLC. Also good TLC-resolution of peaks only moderately separated by GC

Fig. 65. Control of the completeness of GC analysis. The spot in the thin-layer control chromatogram, drawn black, is missing in the gas chromatogram. A substance hidden in the tail of a drawn-out GC-peak is recognisable on the thin-layer chromatogram

2. Has it been possible to gas-chromatograph the entire sample or are there compounds in the sample which are separable by TLC but which do not appear in the gas chromatogram (comparison with the control TLC chromatogram); or which are hidden in the tail of a GC peak (Fig. 65)?

3. Have the components of the sample suffered change during the gas chromatographic process or in the deposition tube? Such components are missing in the control TL-chromatogram. Decomposition in the deposition tube can usually be modified by temperature reduction. An indirectly heated noble metal tube of internal diameter 0.5–1 mm is preferable in such cases to the directly heated steel capillary.

4. Is the separating column clean from a gas-chromatographic point of view and are fluctuations of the zero line due to electric interferences or impurities? The impregnation agent of the column (stationary phase) is then applied to the TLC plate. If the agent is not available, a portion of the column filling must be totally extracted with a suitable, very pure (TLC) solvent and the extract submitted to TLC control.

Answers to 3. and 4. are clearly most important for preparative gas chromatography. It is thus possible to control neatly numerous discontinuous coupling procedures in which gas chromatography is the first stage; interferences are disclosed.

Information concerning 1. and 2. is valuable especially for controlling quantitative gas chromatography. 1. and 3. provide important data for qualitative analysis. Direct coupling GC-TLC is primarily useful for checking an analytical procedure which has already been worked out. It can furnish necessary corrections to quantitative and qualitative GC values; it extends appreciably the scope of TLC.

IV. Details of Apparatus and Methods

a) Deposition

Between 30 and 80% of the substance which flows on to the TLC layer is retained when the deposition tube has the correct dimensions. Active layers take up the substance smoothly and the spots are hardly larger than the internal diameter of the outlet of the deposition tube. The spots grow only when the adsorption capacity of the layer is exceeded. They can become disturbingly large with large GC peaks; it is then better to deposit in several spots or to remove the plate from the proximity of the deposition tube during the time of exit of the principal peak (for this it is sufficient to increase the distance between plate and tube to 20 to 50 mm). How far the thin layer is capable of adsorbing various compounds, can be controlled by comparing the same amounts of compound directly and through deposition from GC.

b) Example

1 µl of a 10% solution and 1 µl of a 1% solution of a typical component of the mixture under investigation are introduced into the gas chromatograph. The ratio v, amount of gas directed on to the thin layer/total amount of gas (that directed on to the plate + that reaching the detector) is determined; let it be 0.8. The component separated in the GC is deposited on to the start line of the plate. Proportionate amounts of the same solutions are now applied directly to the continuation of the start line, alongside the spot formed from the GC effluent. This is done in the usual way, e.g., with a calibrated pipette. The amounts needed are, in the example chosen, 0.8 µl of the 10% solution and 0.8 µl of the 1% solution. The plate is developed and the spot rendered visible by a colour reaction or exposure to UV light and evaluated quantitatively or semi-quantitatively. The yield of deposited component is given by comparison. The size of the depositing tube, the shape of the mouthpiece, temperature and space between tube outlet and layer, have a decisive influence. Details are in Fig. 63.

Instructions for Preparation

The mouthpiece must be absolutely clean. The distance between plate and mouthpiece exit should not vary during transport of the former (changes up to ±0.1 mm have usually no effect). The deposition tube must be heated to a temperature a few degrees (ca 5) above that of the column; the mouthpiece in particular must be sufficiently warm. The temperature gradient can exceed $50°$ C/cm tube with steel tubes of suitable size and 0.5—1 mm wall thickness. It is consequently necessary, especially at high temperature, to work with an electrical "jacket heater". This is made of the same material as the deposition tube but of cross-section such that its conductivity is about one-half in comparison. The length of this jacket heater must be calculated so that it efficiently checks the flow of heat to the copper wire of the low voltage-high current heating. Normally 10—20 mm suffice, depending on the properties of the material.

Overheating of the deposition tube can be harmful to heat-sensitive compounds.

The length and diameter of the deposition tube should be as small as possible. An internal diameter of 1 mm is correct when the length is as small as possible. The walls should be 0.5—1 mm thick. Longer vertical sections of the tube should not be present outside the GC apparatus because these can attain a higher temperature than horizontal parts. The insulation of the sections of tube passing through the side of the apparatus into the column compartment, must be exactly right: adequately electrically, and thermally so that neither cooling nor overheating occurs at this point. This should be tested experimentally. The T-piece must be within the thermostatically controlled column compartment and must be without dead space. It should be fashioned so that the flow resistances yield an approximately constant partition of about 8:2 to 9.5:0.5, i.e., 80—95% of the total flows through the deposition tube on to the thin layer and 20—5% over the detector. The construction and calibration of the T-piece for this unequal partition requires some effort. It is however not satisfactory to do without it and to communicate the detector exit with the thin layer, as would be possible with apparatus equipped with a thermal conductivity detector, gas density balance, β-particle ionisation-, micro cross section- or electron capture detector. It is a fact that the interior of the detector becomes usually exceedingly contaminated; that both the detector volume and the cross section of the gas exit tubes are far too large in commercial apparatus; that their precise thermostatic control is difficult; and that sensitive substances may undergo considerable change (partial pyrolysis) in thermal conductivity cells, for example. The widely used flame ionisation detectors can still be used if the GC gas stream is diverted in front of the detector.

The electrical circuit for pen recorder control of the movement of the TLC plate is given in Fig. 66. The micro switch, type 02101 (Firm 152) is suitable as switch contact. Details of the device for a plate holder easily adjustable to any desired height and fitted with a motor for transport, are in Fig. 67. A purely electronic system can be used to control the plate, at rather greater expense, e.g., the control system of the automatic preparative GC apparatus APG 401 (Firm 74).

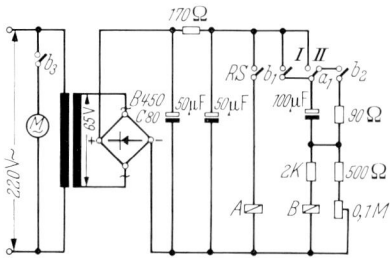

Fig. 66. Electric circuit for intermittent plate transport. Initiated by the pen carriage via a micro switch, with equal transport distances (pre-selected transport times). RS connection to the micro switch. Coil A of a Siemens relay type 65404/93cT.rls 154d is in series with the micro switch and contact a_1 switches from position II to I; this operates coil B, likewise of a relay of the same type, closing contacts b_1 and b_3 and opening b_2. This starts the cycle time, preselected with the potentiometer 0.1 M. At the end of the cycle, all contacts assume their original positions (b_1, b_3 open, b_2 closed) and a new cycle via micro switch RS can take place
(circuit constructed by K. H. Haas, BASF, Ludwigshafen, W. Germany)

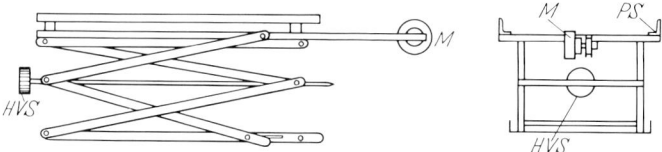

Fig. 67. Transportable plate holder. PS rails for holding plate; HSV screw for height adjustment; M motor which draws the plate with a pre-selected speed

Since the thin layer is exposed to the laboratory atmosphere and dust and there is a risk of touching or rubbing it during the deposition period, it should be protected with a glass cover plate (best attached at the sides with suitable adhesive tape or other means).

Close adherence to the optimum working conditions is necessary, both in the gas chromatographic and the thin-layer chromatographic parts of the coupling procedure. The temperature-programmed GC procedure is preferable because application then takes place with approximately linearly increasing molecular weight (if non-polar columns are used) and, further, the spot sizes are about the same for equal amounts of the individual components. Coupling with an isothermal GC-procedure causes increasing dilution with increase in the retention time. This can be combated with a programmed (progressively decelerated) plate movement. Coupling with capillary chromatography succeeds when the capillary extremity is directed on to the plate as a deposition tube; only the main components are detected in favourable cases.

Coupling with preparative GC can likewise be carried out with success, best with preparative TLC (see p. 97). The yield, which is only moderate, is a drawback here. Adequately deep penetration of a thicker layer is achieved when the carrier plate is relatively thin and is cooled from below with a current of cold gas (over 60 l/h nitrogen, passed through acetone/solid CO_2 or liquid nitrogen). Thin aluminium plates should be better than glass plates in such cases.

V. Examples of Application

Only two examples are cited here; others are to be found in [336, 337, 319–321].

Figs. 68 A–C contain the gas chromatogram (A), the thin-layer chromatogram (C) and the GC-TLC coupled chromatogram of a simple

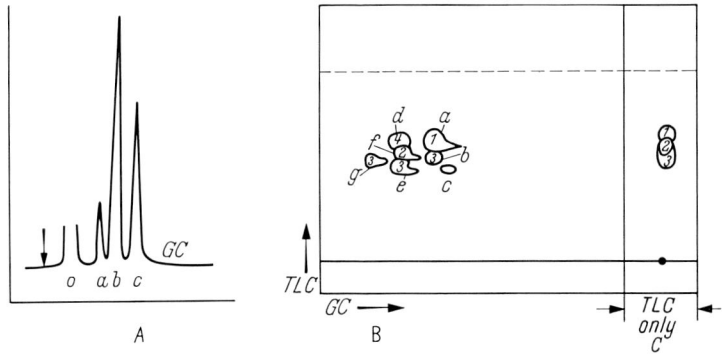

Fig. 68 A. Gas chromatogram (o: solvent; a: aniline; b: o-, p- and m-toluidines c: o-, p- and m-xylidines)

Fig. 68 B. Coupled GC/TLC: colours after visualisation with dibromoquinonechloroimide and exposure to HCl vapour: *1* green; *2* violet; *3* yellow ochre; *4* grey; *a* o- and p-xylidine; *b* m-xylidine; *c* unidentified; *d* o-toluidine; *e* p-toluidine; *f* m-toluidine; *g* aniline

Fig. 68 C. Thin-layer chromatogram of the mixture without coupling: *1* green; *2* violet; *3* yellow ochre; after visualisation as in Fig. 68 B

standard mixture. Both the GC and TLC data separately imply that the mixture consists of three components but seven individuals can be detected in the coupled chromatogram. In reality, eight compounds were present, one of which was an impurity of one of the components in the standard mixture.

Fig. 69 compares the coupling chromatogram with the one-dimensional thin-layer chromatogram of a xylenol distillation fraction, boiling within a narrow range. The non-phenolic compounds and sulphur compounds were required analytically. This problem can be solved neither with GC nor with TLC alone. It could however be solved qualitatively and quantitatively by coupling and using selective colour reagents on the TLC layer, together with measurements of the surface areas of the corresponding peaks in the gas chromatogram.

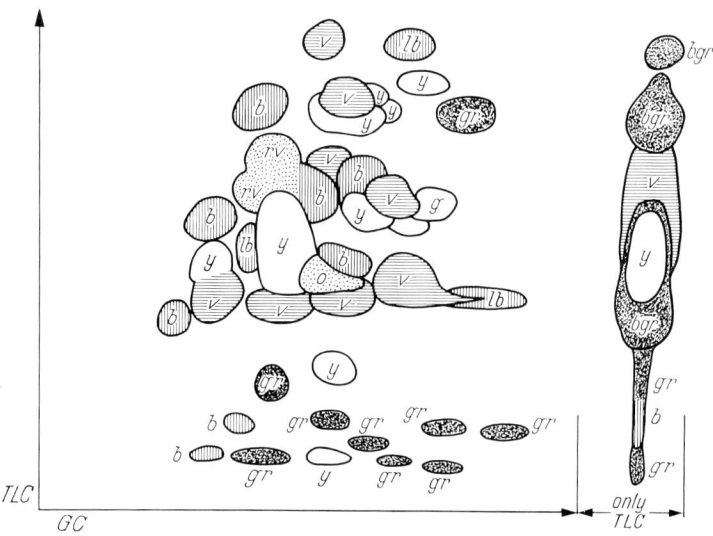

Fig. 69. Xylenol fraction. GC/TLC coupling and TL-control chromatogram. Over 42 components recognisable especially after colour development with dibromoquinonechloroimide and exposure to ammonia and then HCl vapour: v violet; rv red-violet; lb light blue; b blue; y yellow; gr grey; o orange. Chloroform solvent on silica gel GF_{254} layer

G. Documentation of Thin-Layer Chromatograms

H. GÄNSHIRT

All experimental conditions must be accurately noted in the documentation of thin-layer chromatograms. A printed form like in Fig. 70 should be used to facilitate precise repetition in the future [214, 344].

Thin-Layer Chromatogram No.

Compound class: ....
Literature: ....

Adsorbent layer: ....
Layer thickness: .... Size: ....
Preparation and Drying: ....

Chamber type: .... Separation technique: ....
Chamber saturation state: ....
Length of run: .... Time of run: ....
Solvent composition and total volume: ....

Purity data for solvent components: ....

Solvent and solution concentration used for application: ....
............Margin between start point and plate edge: ....
Amounts applied: ....
Special remarks: ....
....Date: ....

Substance	hRf	hRf	hRf	Detection and sensitivity
1.				
2.				
3.				
4.				
5.				
6.				
7.				
8.				
9.				
10.				

Fig. 70. Proposed scheme for a printed form giving the chromatographic experimental conditions

I. Rf-values in TLC

The migration distances of substances on thin layer chromatograms are generally fixed as Rf-values[1], like in paper chromatography. All Rf-values have been multiplied by 100 in this book and are therefore termed hRf-values. Whole numbers are then obtained, since Rf-values can usually be determined with an accuracy of only two places after the decimal point. If a length of run of 100 mm is used, the distance in mm from starting point to spot centre is identical with the hRf-value. Even though utilising the whole plate length would occasionally improve separation of mixtures of components with low hRf-values, it is nevertheless preferable to keep to a standard length of run of 100 mm (cf. p. 85). Experimental data can then be better compared.

The hRf-values are to be regarded only as guide values for the migration distances, even when all experimental data are accurately measured. Factors like layer thickness, chamber saturation, air humidity, separation effects of solvent mixtures etc., which are all difficult to reproduce, can exert a marked influence [614]. An investigation of the influence of layer thickness on Rf-values, using silica gel G layers, may be quoted as an example [504]. The age of solvent mixtures can also have a significant influence. Comparison of migration distances is especially difficult when the work has been carried out in different chambers and on layers which were hard to prepare. It is then useful not to measure hRf-values alone, but to compare the migration distances of the substances concerned with that of a simultaneously chromatographed reference substance; this last named should belong to the same or a similar compound class as far as possible. So-called hR_{St}-values[2] are thus obtained.

With especially important chromatograms, a purely numerical documentation does not suffice for comparing, if necessary, all details like spot shape, spot area, resolution, course of the solvent front, subsidiary spots etc. A number of possibilities has accordingly been described for preparing the chromatogram so that it can be kept or that copies as faithful as possible can be made.

II. Preservation of Thin-Layer Chromatograms

In this method, the chromatograms, after appropriate previous treatment, are peeled off the glass plate as films and stored. This preservation of adsorbent layers has so far been used for documentation mainly with silica gel and alumina layers.

---

[1] $Rf = \dfrac{\text{distance of spot centre from start point}}{\text{distance of solvent front from start point}}$

[2] $hR_{St} = \dfrac{\text{distance of sample spot from start point} \times 100}{\text{distance of reference material spot from start point}}$

Amino acid chromatograms have been preserved by impregnation with collodion to which a definite amount of glycerol had been added [42, 66]. Chromatograms have also been pasted over with self-adhesive plastic film and the layer subsequently peeled off the plate [354]. Spraying with polymer dispersions (polyacrylates, polyvinylidene chloride or polyvinyl propionate) is usually more suitable [402]. Such aerosols are commercially available (Firms 88, 153). Mention may be made also of impregnation of layers by immersion in solutions of polymers although this has a restricted application (Firm 60).

In spite of the advantage of being able to keep original chromatograms in this way, the method is limited to compound classes which are insoluble or sparingly soluble in the solvent used for the preserving suspensions or varnishes. Further, delicate features of the chromatograms are easily erased by the necessarily intensive spraying of the layer and the colours of the spots tend to fade rapidly when the plates are not kept in the dark. Handier copies which can be made with the help of procedures described below, are generally preferred in documentation.

Impregnation using the spray procedure: The layers are thoroughly moistened with, for example, "new Neatan" (Firm 88) or 5% permanent-adhesive Kiwotex D in ligroin [705] (Firm 79) and then air dried. The temperature must not exceed 50° C with "new Neatan" or the layer changes colour. In no case may further spraying be carried out after having dried. The layer is then scratched across its full breadth near the top edge, immersed in water and peeled off. A thin transparent adhesive sheet, specially prepared for the purpose (Firms 88, 125) is pressed on to layers which contain dividing lanes; this is done after spraying and drying but before immersion in water. The chromatograms are then dried completely in the air, stuck to cardboard and can be stored.

After spraying new Neatan, the jets of the sprayer are cleaned with dimethylformamide to prevent blockage.

If such lacquers are sprayed with ordinary atomisers, the lacquer particles may easily coagulate before reaching the layer surface, which is extremely unsatisfactory. A much more uniform impregnation is possible by spraying from "spray" bottles with propellent gas.

III. Graphical Copying

Tracing on transparent paper is the simplest technique for copying chromatograms (Fig. 71a). Spots which are visible only in UV light, are marked round with a pin and the outlines transferred to the tracing paper afterwards. The colour intensity of the spots can be indicated by shading. For this, it is better to use variously graded, transparent papers on which the spot outlines are drawn. These are then cut out and pasted on to writing paper in accordance with the migration distances [216]. Further, the chromatogram spots can be measured planimetrically and drawn half-quantitatively as equivalently sized rectangles with the same base line length (Fig. 71c) [519, 661, 667].

These simple copies are useful for internal use in serial investigations. This documentation method must likewise be used for chromatograms with poor contrast or which rapidly change. Since however such a subjective copy gives little information about the quality of the original chromatogram, one of the objective copying methods given below should be used.

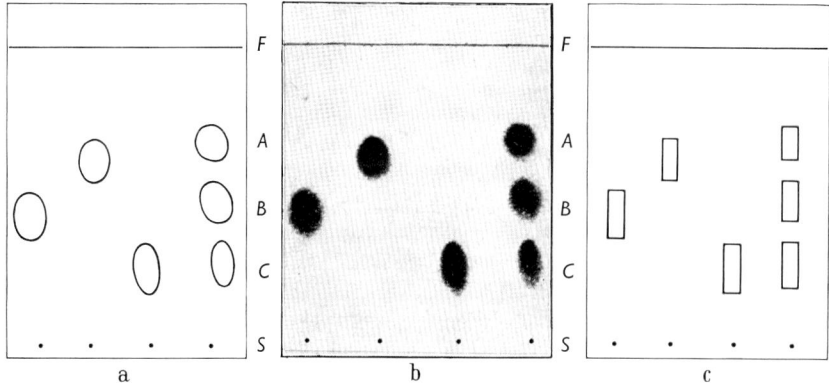

Fig. 71. Examples of ways of documenting thin-layer chromatograms. *a* chromatogram traced on transparent paper (transferred back to ordinary paper); *b* documentation by photostating; *c* as rectangles of the same area as the planimetrically measured spot areas. Experimental details: separation of bromoureides using cyclohexane-chloroform-pyridine (20 + 60 + 5) on silica gel G layers. *S* start; *F* solvent front; *A* carbromal; *B* acetylcarbromal; *C* bromisoval

IV. Documentation with Light-Sensitive Paper

The principle is to use on a paper base, special materials which are chemically modified by light or which change their net charge through illumination.

1. Copies with Diazo-type Paper

Paper is used in this procedure which has been prepared with light-sensitive diazonium compounds and which contains coupling components yielding azo dyes with the diazonium compounds in alkaline solution. The paper is maintained weakly acid to prevent premature coupling. The diazonium component decomposes under the influence of light. If a thin-layer chromatogram is now placed between light source and paper, any opaque spots prevent this decomposition. The paper is then developed in ammonia atmosphere and coupling to form a dye occurs in those places shielded by the light-absorbing spots on the chromatogram; these spots appear as positives on the diazo-type paper. Spots absorbing in long wave UV light can also be copied with this method if a suitable short UV-source is used.

Various papers [3-6] have been used [179, 552, 652, 654, 782]. Illumination with two 15-watt frosted glass bulbs serves for the copying; cellophane sheets are placed between layer and diazo-type paper to protect the former. The light is best directed through the uncoated side of the plate. The light source Ultra-Vitalux OUR 53, 300 watts (Firm 96), at a distance of 40 cm from the layer, has been used for copying spots which absorb in long wave UV light. To augment contrast, glass filters like GG 19 (1 mm)/BG 25 (3 mm) or UG 1 (7 mm) (Firm 123) can be interposed; or, best of all, glass carrier plates are used which absorb as little long wave UV light as possible, e.g., WG 7-Glas (Firm 123) [654].

2. Copies with Blue-Print Paper

This paper also can be used in a parallel way to diazo-type paper. A detailed report has been made on its use for making paper chromatogram copies in UV light [239].

The paper[7] is relatively insensitive to daylight. Long exposure times are needed even in UV light, so that a wide range of exposure is available. Development is carried out by moistening with water.

So little experience has been accumulated about the use of these two types of paper for documentation of chromatograms, that their value in this field cannot be estimated.

3. Positive Copies and Photographs with Procedures Based on Diffusion of Silver Salts

In this procedure, the silver halide of the negative emulsion which has not been modified by the action of light, is allowed to diffuse into the positive emulsion through a thin layer of the more or less viscous developer; its silver content is then precipitated to yield a positive picture [473]. A wet after-treatment is not necessary.

This method, not suitable for half tone reproduction in its original form, is used principally for preparing contact copies [304]. One can work with separate negative- and positive material or with a single paper on which negative and positive emulsion are layered, one on the other. The technique with separate papers is more adaptable and, moreover, several positives can be made from one negative if required. Although negative paper of various sensitivities is available, the exposure range for thin-layer chromatograms with poorly contrasted spots, is rather limited; the method is therefore of only restricted application.

According to SZÉKELY [705], better copies with varied grey tones are obtained by inserting transparent paper[8] between light-sensitive

---

[3] Ozalid 200 SS (Firm 97).
[4] Driprint HC 241 (F speed) (Firm 46).
[5] Safirpapier (Firm 116).
[6] Diazo 1200 SS (Firm 50).
[7] Dietzgen XL (Firm 46).
[8] Rolocor Screen tints 1204 (Firm Buma S. A., Basel, Switzerland)

layer and the thin adsorbent layer. The application of the method to copying thin-layer chromatograms has been described many times [230, 603, 739] (cf. Fig. 71 b).

The following equipment can be used for the Agfa photostat process [13]: *light source — glass plate of the copying apparatus — carrier glass plate — adsorbent thin layer — cellophane sheet — (transparent paper) — Agfa photostat negative paper.* Chromatograms which have been sprayed with aggressive reagents can also be copied if the cellophane sheet is inserted.

LAND [388, 409] has developed the diffusion transfer process a stage further, so that it can be used nowadays for normal photography. This involves a type of camera with which exposure and development can be carried out within seconds. Film material has also been introduced which, in speed and performance, can compete with the emulsion types in conventional photography. In this "polaroid photography", as it is now termed, roll film or film pack can be changed without trouble in daylight.

The directives concerning light sources for illuminating the TLC layers and concerning filters, given under "Photography" below, apply also to black and white polaroid photographs. HANSBURY [269] has already described the application to documentation of thin-layer chromatograms. If the polaroid camera type 110 A is used, which is suitable for photographing thin-layer chromatograms, the polaroid close-up lens attachment 1 + 2 is fitted, the bellows set to 15 m and the camera so fixed that the chromatogram is 34.5 cm from the front lens. An exposure time of less than one second is then required for UV-absorbing spots on fluorescing layers, using: polaroid roll film 47, of $36°$ Din = 3000 ASA; aperture f8; the Fluotest lamp (Firm 110) with the 254 nm emitter; and the liquid filter combination mentioned under "Photography" [218] (cf Fig. 73a, p. 136). Photographs taken with this type of film are then painted over with a varnish provided with the film pack and can be kept without undergoing change. The still existing disadvantage of the procedure is that only a single positive in one size can be made with the usual films. Colour photographs of thin-layer chromatograms, illuminated with UV-light, have not been obtainable free of flaws using the polaroid film material so far available.

4. Photography

Conventional photography is still the best process for obtaining the most faithful possible reproduction of a thin-layer chromatogram. This is ensured by the large choice of emulsion types for half tone reproduction and the fact that as many positives and in varied sizes as desired, can be made from the negative. Colour reproduction as picture or transparency, is also possible; daylight or UV-light can be used as source.

If chromatograms illuminated with UV light are to be photographed, this primary light must be removed as far as possible by placing a suitable filter in front of the camera objective. The choice of filter is not critical with black and white photographs. On the other hand, good colour reproductions, whether positives or transparencies, can be obtained only after careful choice of filter [178, 346]. Thin-layer chromatograms often give characteristically fluorescing spots if appropriate reagents have been used and the layers then exposed to light of principal wave-length 366 nm; these are valuable for documentation. Nevertheless, a black and white photograph usually suffices for reproduction of spots which absorb short wave UV light ("quenching spots") on a fluorescing background.

a) General Points

Reflex cameras are particularly well suited for photographing thin-layer chromatograms since the image appearing in the view finder, even where close-up lens attachments are used, outlines exactly the object. The small film (35 mm) size must always be enlarged; the square 6 × 6 cm shape, analogous to the square TLC plate, has the advantage that enlargement is not necessary.

The surface of the chromatogram must be well lit in order to obtain photographs without shortcomings. The angle of incidence should be 45° if possible for photographs taken in incident light. Chromatograms with poor contrast are best photographed in transmitted light. The usual photographic rules apply in choosing the black and white or colour material and for setting aperture and time of exposure.

b) Black and White Photographs with UV-Source 366 or 254 nm (see also pp. 77—78)

As sources of incident light of wave length 254 and 366 nm, the "Fluotest" apparatus (Firm 110), Uvis (Firm 44) or Uvanalys apparatus (Firm 14) for example can be employed. "G—E fluorescent lamps", ready equipped with the filters XX—15C, have proved satisfactory as sources for photography in transmitted light of 366 nm wave length (Firm 25). A 125 watt "Black light" H.P.V.-lamp with suitable power source (Firm 103) can also be used as primary 366 nm source. Sources of transmitted light are placed 10—15 cm below the glass side of the chromatogram. One must work rapidly when using high energy sources, in order to prevent changes in the chromatograms through the heat. Both solid filters and liquid absorbents can be used as UV filters in front of the objective. It is generally adequate to combine the UV filter commercially available for the camera chosen, with a medium yellow filter. A combination of the two liquid filters, 10% $NaNO_2$ and 10% K_2CrO_4 solutions, each 0.5 cm thick, has proved very good for both 254 and 366 nm. Jones [789] has described a 254 and 366 nm primary light source, consisting of a flat spiral, cold cathode mercury lamp "ozone-free" type 12555 (Firm 52a). It was used for working with transmitted light and the wave lengths for excitation were selected with the filters OX 7 and OX 9A. The light falls on the layer side of the plate, so that the glass plates themselves function as UV filters for the photographs. Writing scratched on the plates is recognisable on photographs of layers which fluoresce in 254 nm UV-light. If spots which fluoresce in 366 nm light are to be photographed on a non-fluorescing background, fluorescent material must be placed below the part of the plate on which writing has been scratched; a strip of filter paper which has been soaked in a 1% chloroform solution of quinine hydrochloride and exposed to hydrochloric acid vapours for activation, serves for this. Fine grained film material like Agfa Isopan 13 DIN is chosen and aperture f8 usually used if the reproduction of nuances in a chromatogram is desired; the exposure time must then be determined in a systematic preliminary experiment.

Direct copying on photographic paper with light in the visible or long wave UV region, is also possible [94, 739, 789].

c) Colour Photographs with UV Light Sources

The "G—E fluorescent lamp" with 366 nm light source (Firm 25), already mentioned, has been successfully applied to taking colour photographs with transmitted light of that wave length. The combined liquid filter, mentioned above, was used and Agfacolor negative film CN-17. About 1 min exposure time is then required at f5.6. Information can be found in the literature about the photographic technique with Kodak emulsions and the necessary filters [788]. Agfa reversal film CT 18 is also suitable for colour pictures [178]. Any desired number of positives can be made from the transparencies, using colour reversal paper (Agfa CT-copies). A mercury low pressure lamp (Firm 70) with principal wave length of 254 nm, and the filter UG 5 (Firm 123) is used as source. UV filters in front of the camera objective are: glass filter combinations WG 2 (2 mm), GG 13 (5 mm) and GG 3 (4 mm) (Firm 123).

5. Electrophotography

Reference has been made also to the possibility of documentation of thin-layer chromatograms through electrophotography [291, 653, 655]. The essential nature of electrophotography, which works according to two procedures, xerography (Firm 113) and the Electrofax process (Firms 10, 131), can be gathered from a review article of HAUFFE [280]. It is carried out with layers impregnated in the one case with selenium, in the other with zinc oxide. The method has so far found its most appropriate use in the reproduction of drawings; this is due to the sharp gradation yielded. The accent has been on mass production. At the moment, work is being carried out on electrophotographic half tone reproduction [315]. Aa a result of the sharp gradation mentioned and of the required equipment, the method has little point for thin-layer documentation, where usually only a limited number of copies is needed.

H. Quantitative Evaluation of Thin-Layer Chromatograms

H. GÄNSHIRT

Whether the evaluation is carried out on the layer itself or after extraction, depends on many factors: on the problem and associated demands of reproducibility; and on the properties and available amounts of the compounds to be separated.

In all procedures of quantitative evaluation it is imperative to expose the substances both at the start point and on the developed chromatogram as briefly as possible to the influence of light and air; oxidations and photochemical reactions (as has been described for example with polycyclic aromatic compounds [311]) are thereby minimised. Accurate application of the sample to the start point is also most vital. Suitable micropipettes, glass plunger pipettes or even glass capillaries with micrometer screw advance, can be used (p. 64, Fig. 20).

Micropipettes for applying microgram amounts are available commercially. Standard deviations of ± 0.2 to $\pm 0.4\%$ have been found in measurements of UV absorption, made on solutions prepared by pipetting p-hydroxybenzoic acid solutions (a solution of 5 µg acid per µl methanol, pipetted with lambda pipettes (Firms 67, 117) of 1.5 and 10 µl capacity and making up to 10 ml) [219]. The error in similar experiments was larger ($\pm 1\%$) when glass plunger pipettes with micrometer screw were used, e.g., the well known Agla-micro-syringe (Firm 29) [213]. These syringes have the advantage, however, that the volume applied can be varied over a wide range. Holders with exchangeable, self-filling capillaries of known volumes (Firm 33) have likewise found application. The capillary function depends however on the surface tension, viscosity and specific gravity of the solvent used. Such pipettes can really be recommended only for routine determinations on aqueous and alcoholic solutions. It is especially important to remove carefully all traces of surface-active cleansers. See p. 65 for information about band application.

PRIVETT et al. [544] have compiled a comprehensive review of the methods so far employed for quantitative evaluation of lipid chromatograms.

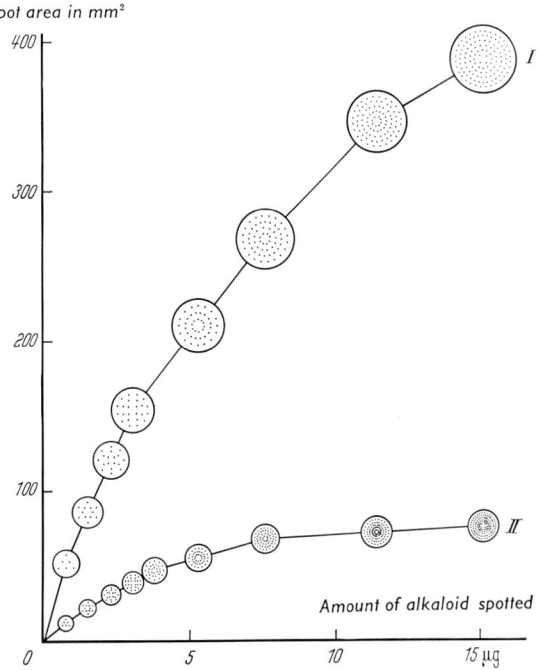

Fig. 72. Comparison of the spot areas of increasing amounts of alkaloids at equal length of run. Curve *I* on paper impregnated with formamide, curve *II* on a silica gel G layer. The stippling density is roughly equivalent to the number of molecules [664]

I. Quantitative Evaluation on the TLC-Layers

1. Visual Comparison

It was logical to attempt quantitative evaluation through visual comparison of spot areas. The relation between amount of substance applied and spot area is, for the same distance of migration, influenced by the adsorbent, solvent and chamber saturation. For this technique of evaluation, the chromatographic conditions must be directed not only towards achieving a good separation but also towards obtaining the largest possible changes in spot size per unit change in the amount applied. Visual comparison which can yield accurate results in paper chromatography [735] is more difficult in TLC because smaller areas are compared (Fig. 72).

Most determinations carried out using visual comparison are therefore semi-quantitative [322, 352, 763]. If the conditions are carefully standardised, quite acceptable data are obtained. Amounts of the solution under investigation and of a reference solution are applied alternately and alongside each other; the same amount always of the former and progressively increasing amounts of the latter are taken and then chromatographed. Examples demonstrating this are the determination of triacetylated adrenalin [737] and vitamin D [282].

2. Determination by Measurement of Spot Areas

Methods of area measurement, independent of judgment by the eye, more objective and without outlay of apparatus, have been studied intensively.

If the spots are outlined relatively distinctly, the areas can be measured in several ways: comparison with suitable standard sizes [459]; evaluation with a planimeter [493, 494]; copying on to writing paper and

Table 13. *Spot evaluation with a planimeter*

Spot area mm^2	Deviation with a single measurement (%)	Deviation with five measurements (%)
28	±10	±4
78	± 5	±2
113	± 3.5	±1.4

weighing [217]; copying photographically and weighing [608]; or copying on to squared paper and counting the number of square millimeters. Since the areas generally lie between 15 and 150 mm^2, considerable errors can arise in these measurements. Such errors have been determined in the planimetric[1] measurement of alkaloid spots; they could be reduced appreciably by making several measurements (cf. Table 13) [494].

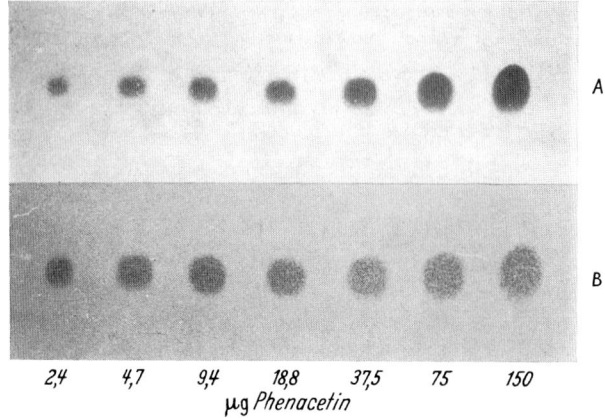

Fig. 73a. Comparison of surface areas. Increasing amounts of phenacetin applied and chromatographed according to PURDY and TRUTER's method [549]. A visualised by sparying with ferricyanide-ferric chloride reagent (No. 111); B inspected on a fluorescence layer in UV-light of 254 nm

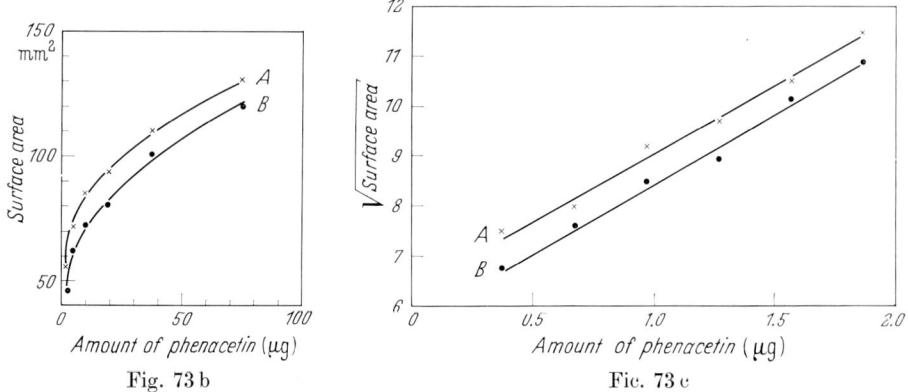

Fig. 73b. Dependence of surface area on applied substance amount in the chromatogram reproduced in Fig. 73a. A localised with a colour reagent; B in UV-light, 254 nm

Fig. 73c. Relation of spot area and applied substance amount in the chromatogram of Fig. 73a, according to the equation, log amount = $\sqrt{\text{area}}$

Should be spot edges be indistinct against the background, their more objective location may be attempted by copying on hard photographic paper [624].

---

[1] E.g., Kompensationsplanimeter III with magnifying glass (Firm 42).

In order not to have to prepare a calibration curve for each plate, using several standard solutions, as, for example, in Fig. 73b for the photographed chromatogram in Fig. 73a, a function has been sought, expressing a linear relation between applied substance amount and spot area. GIDDINGS and KELLER [114] treated theoretically the same problem for paper chromatograms and BRENNER et al. [94, 97] have discussed the application to thin-layer chromatographic data. The spot areas should be proportional to the logarithm of the applied substance amount, under definite conditions. PETROWITZ [520] has been able to determine constituents of tar oils, DDT and γ-BHC quantitatively within a narrow concentration range, using this relation.

The spot area however increases with increase in substance amount more rapidly than the theory predicts. The linear relation between the square root of the surface area and the logarithm of the applied substance amount, ascertained by PURDY and TRUTER [547, 548, 549, 720] is generally valid over a wider range of concentration (cf. Fig. 73c). With the help of a standard solution, the analysis solution and a diluted analysis solution, a quantitative determination can be carried out in accordance with the formula:

$$\log W = \log W_s + \left(\frac{\sqrt{A} - \sqrt{A_s}}{\sqrt{A_D} - \sqrt{A}} \right) \log D$$

where, D = dilution factor
 W = amount of the analysis material sought
 W_s = amount of standard applied
 A_s = spot area of the standard
 A = spot area of the material for analysis
 A_D = spot area of the diluted material for analysis.

The gradient of the calibration curve, established in this case with two spots, is a measure of the diffusion of the substance. The steeper the curve, the more accurate a determination is possible [94].

The technique of application is important for obtaining reproducible results. Equal volumes of the standards and the sample solutions must be applied as drops to the start line. An Agla micro-syringe, fixed so that the needle outlet is 2 mm above the adsorbent layer, can be used. The drop which emerges as a result of turning the micrometer screw can be brought on to the layer by cautiously tilting the plate towards the needle [548]. Methanol and ethanol are convenient solvents of intermediate volatility.

PURDY and TRUTER [549] have calculated average deviations based on their own results and on literature data; they found 3.1% for 600 separations depending on adsorption and 3.6% for 980 depending on partition. The same authors found an average deviation of 6.6% for 60 determinations of cholesterol; and MORRISON and CHATTEN [459], 7.7% for 580 determinations of antihistamines.

Reference may be made to further applications of this technique to determinations of: plasma lipids [608]; products from reaction of peptides with phenyl isothiocyanate [506]; tertiary amines [512]; and some amines in general [792]; and also to the investigations of AURENGE and coworkers [27].

The systematic error in this method is small because comparison is made always with values obtained by chromatographing standards simultaneously. The random errors are however generally high [217, 459, 493] as a result of the variations of spot shape in chromatography with reference substances and through difficulties of defining the spot outlines (cf. Fig. 73a). This fact must be taken into account in comparisons of error between this method of direct evaluation and methods after spot extraction, as long as the random error is not calculated in the same way [549]. If the substance amounts are small and can be easily visualised and yield sharply outlined circular to oval spots with the chosen solvent, one can nevertheless try quantitative evaluation with this simple technique without special aids. It has moreover the advantage over the densitometric determination described below, that the yield of coloured product in the colour reaction need be reproducible only on the edge of the spot where the reagent excess is largest [94].

3. Transmission Measurements on Spots which are Coloured, Charred or which Absorb UV-Light

In this technique, the light absorption of a section of the adsorbent layer in which the spot is situated, is compared with the light absorption of an identically sized section of layer without spot. Spots which absorb in the UV can be evaluated only when the carrier plate does not adsorb too strongly at the necessary wave length.

The spot can be measured as a whole or in segments when a suitably shaped light beam is used. Usually the spot area is scanned with a narrow beam and the diminution in light intensity summed for the spot area. One might expect that, under the conditions mentioned on p. 136, a linear relation would be found between the areas under the extinction curves thus obtained and the amounts of substance applied. As KLAUS [351] has shown, the absorption integrals are, however, independent of the spot shape only when the transmission is relatively high, the spot area not too small and the spot fairly symmetrical. Standard and sample must be chromatographed on the same layer in order to compensate for individual influences of the adsorption layer.

This working technique is more frequently used since, as with all evaluations directly on the layer, very small amounts of material can be determined and suitable measuring apparatus is commercially available. A" Chromoscan" densitometer (Firm 78) has been used for the evaluation of spots of various resins and balsams which had been visualised with ethanolic phosphoric acid [333]. The densitometric measurement of some ester spots, visualised with vanillin-sulphuric acid, has been possible within a narrow range of concentration by using a photovolt densitometer[2] [26]. Opium alkaloids have been analysed with a Zeiss-extinction recording apparatus[3] [536]. An Atago-densitometer[4] has been used for semi-quantitative determination of bile acids which had been sprayed with sulphuric acid and of amino acid spots, coloured with ninhydrin [274, 275]. Bile acids have been determined also with a photovolt densitometer after visualisation with 20% molybdophosphoric acid in ethanol [798]. Cholesterol ester spots have been determined with different sets of equipment [5, 6, 7] after having visualised with antimony trichloride [786]. The determination of fatty acids, visualised with cupric acetate, has been described also [256]. Another author [656] has quoted the use of clear plastic fragments[8] as carrier material. The silica gel layers were prepared on the matted side of the carrier. After development of the amino acids in the example cited, the layer was rendered transparent by spraying with a plastic spray[9]; the chromatogram strips could then be scanned with any suitable densitometer.

The area of charred lipid spots at constant substance amount has been found to have no influence on the light absorption in a range of R_f-values from 0.4–0.8 [73]; this contrasts with results on coloured spots. The areas obtained with densitometers[10] for the light absorption have been found to be proportional to the lipid amount applied, the gradient of the calibration curve depending on the particular lipid [540, 541].

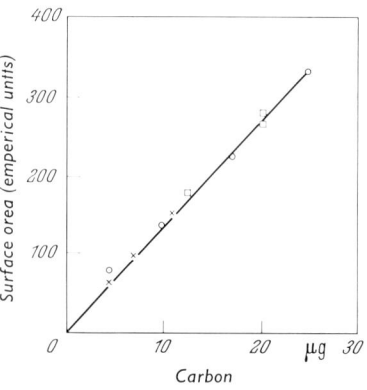

Fig. 74. Dependence of the densitometrically measured spot areas (transmitted light) on the applied lipid amount expressed as µg carbon, according to BLANK and co-workers [73]. ○ cholesterol palmitate; □ palmitic acid; × tripalmitin

BLANK et al. [73] have found that the calibration curves of various lipids when using another charring procedure and expressing the applied amount as the equivalent amount of carbon, all had the same gradient (Fig. 74). Charring was carried out with a fine spray of saturated potassium dichromate solution in 70% sulphuric acid and heating 25 min at 100° C. Spots from carbonisation of cholesterol with perchloric acid have been evaluated similarly [526]. The quantitative determination of charred spots has found frequent use so far, though principally in lipid analysis

---

[2] Photovolt densitometer, model 501 A (Firm 104).
[3] Extinction recording apparatus with integrator ERI 10 (Firm 156).
[4] Atago-densitometer, model AG-4 (Firm 12).
[5] Electrophoresis scanner (Firm 21).
[6] Chromatogram scanner for the Eppendorf photometer (Firm 53).
[7] Accessory for the Beckmann DU-Spectrophotometer (Firm 19).
[8] Product no. VCA 3310—C1 (Firm 143).
[9] Tuffilm-Spray No. 543 (Firm 66).
[10] See Photovolt-densitometers, footnote [11].

[73, 341, 427, 540, 541, 542, 543, 784, 785]; whether it can be applied successfully to the determination of other compound classes remains to be tested. Reproducible results are obtained when the oxidation is as fast as possible, thereby avoiding evaporation of the substances from the large surface. Uniform charring on layers impregnated with silver nitrate can be achieved only with difficulty [341]. Carbonisation by dry pyrolysis is likewise difficult since the heat radiation is reflected too strongly from the "white" adsorbent layer [284].

Carbonised spots can be directly evaluated with photometers, suitably equipped for scanning, or using so-called densitometers, which are built essentially for measurements of transmission and reflection on thin solid layers. Photovolt (Firm 104) and Chromoscan (Firm 78) densitometers of various types have been those so far principally used.

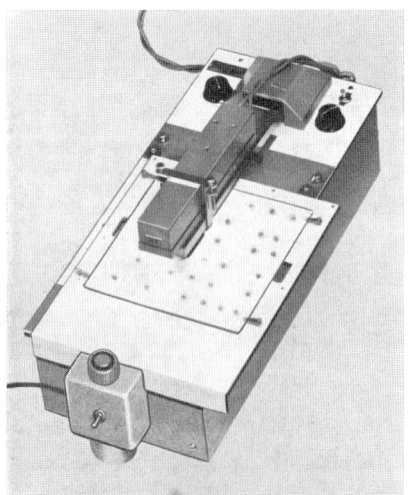 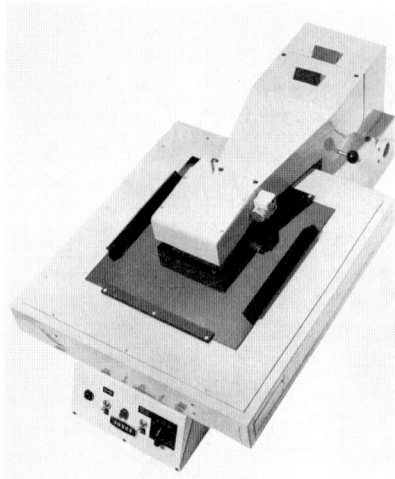

Fig. 75. Photovolt-densitometer; accessory for evaluation of thin-layer chromatograms[12] (Firm 104)

Fig. 76. Chromoscan-densitometer; accessory for evaluation of thin-layer chromatograms[13] (Firm 78)

The original work was carried out with Photovolt-densitometers[11] which had been constructed for evaluation of electrophoresis- and paper chromatogram strips [540, 541]. The TLC strips had therefore to be correspondingly shaped. The combined apparatus model 530[12] (Fig. 75) is obtainable today from the Photovolt firm; it offers semi-automatic or automatic evaluation of 20 × 20 cm plates. Nonlinear

---

[11] Photovolt-densitometer model 501A with light source model 52 and model 521A with light source model 52/C.

[12] Photovolt, multiplier-photometer model 520A, light source unit model 52C and accessory for evaluation of TLC plates with automatic drive and varicord-recorder model 42 = TLC densitometer model 530 (Photovolt information sheet 367/8—63) (Firm 104).

dependence of light absorbance on the amount of applied substance can be corrected electrically during recording with the help of the recorder belonging to the equipment [400]. A special TLC-accessory with quartz optical system and automatic recording of the light absorption can be obtained for the Chromoscan densitometer also[13]. During measurement, the spot is in a chamber shut off from outside light so as to reduce interference from stray radiation as far as possible. A mechanical correction can be applied in series analyses with this apparatus when light absorption does not depend linearly on the amount of substance applied; the profile of the comb which displaces the optical wedge in the reference beam is altered for this purpose [15]. The Photovolt and Chromoscan densitometers can also be used for measurements of reflection (cf p. 144)

4. Quantitative Determination of Fluorescing Spots
a) Direct Evaluation

Fluorescing spots can be evaluated directly on the adsorbent layer. The conditions for this are favourable: The relation $F \approx S \times f$ holds, where F = the fluorescence value, read from a suitable detector, S = radiation density of a fluorescent spot, regarded as a light source and f = spot area. The total fluorescence of a particular amount of substance is thus not significantly influenced by the spot shape.

A linear relation between the surface under the fluorescence intensity curve and the amount of substance chromatographed has been found within a useful measuring range, using the PMQ II spectrophotometer (Firm 155) [351, 626]; the calibration curves using a modified Amincomicrophotometer (Firm 4) were not linear [793].

Spots of fluorescing products of carbohydrate decomposition have been evaluated with a Turner fluorometer[14] [134]. Griseofulvin and griseofulvin-4′-alcohol have been quantitatively determined after TLC-separation by measuring the fluorescence intensities with a Photovolt 530 densitometer and corresponding accessories[12] and using the primary filter (catalogue No. 5267) and the Wratten filters 42A and 42 in the multiplier unit [791]. SAWICKI et al. [525, 598, 601, 690] have pointed out the possibility of measuring nanogram amounts, in particular of polycyclic compounds, directly on the plate after using appropriate reagents; the techniques were fluorometry [15a, b] and phosphorimetry [343]. These investigations are especially interesting because they extend down to sample magnitudes which are no longer detectable even with reflection spectra (cf. p. 142). Whether such spectra are quantitatively reproducible must await further tests.

---

[13] Chromoscan-densitometer with accessory for evaluation of thin layers (Firm 78).

[14] Turner fluorometer model 111 with adaptation for TLC (Firm 33); cf also Camag-information QTL-65 and QDC-6-65 (Firm 33).

[15a] Aminco-Bowman spectrofluorometer (Firm 4)

[15b] Fluorispec model SF/1, Fluorescence-spectrophotometer (Firm 15a).

b) Photographic Evaluation

If a film is exposed to a fluorescing spot as light source, so that the relation

$$Q = k \cdot \log I \qquad (1)$$

holds, where Q = substance amount, I = emitted light intensity and k = a property constant, the film blackens. If the total extinction (D) of such a spot is measured with a densitometer, then

$$D = \gamma \log I + B \qquad (2)$$

at constant time of exposure, where

$$B = \gamma (\log t - \log i) .$$

t = time of exposure of the film, i = film "inertia" and γ = film "gamma".

Equation (2) therefore includes the film factors i and γ. If the complete analysis is carried out on a layer and a piece of film is used for photography, the film factors are eliminated. One can then combine (1) and (2) to give:

$$D = K \cdot Q + B \quad \text{where} \quad K = \frac{\gamma}{k} .$$

There is thus a linear relation over a definite range between the total extinction measured by the densitometer and the amount of substance applied, provided the time of exposure is the same and all other photographic conditions are constant [316, 317].

The primary UV-light used for excitation, is removed with a filter attached to the objective (cf. p. 132).

The calibration curves plotted from certain steroids after having sprayed with sulphuric acid, do not pass through the origin; a small part of the film emulsion is always slightly altered for various reasons during development. The slopes of the calibration curves of the different individual compounds (steroids) differ as a result of differences in the intensity of the light emitted from the spots [316, 317]

5. Evaluation of Reflectance (Remission) Spectra

If light falls on to a white powder, a fraction of it is returned by regular reflection from the crystal surfaces, governed by the grain size distribution. The smaller the crystals are, the smaller is this regular reflection. The bulk of the incident radiation, however, penetrates into the layer of powder to a depth which depends on its wave length, returns to the surface after repeated scattering and is then emitted as a diffuse reflection (remission) in hemispherical form.

A "standard white", reflecting as uniformly as possible over the wave length range concerned, would naturally be best both as standard and as adsorbent. TLC is however restricted to particular adsorbents. Limits are thus set, resulting from the diminution in the degree of reflection as the wave length is reduced; this depends on the adsorbent chosen. The limits in each individual case can be accurately established by comparison with an aluminium standard which shows a sensibly constant degree of remission down to 215 nm.

If light-absorbing substances are mixed with the standard white or if the substances are present as absorbing spots on the surface of such a standard, the intensity of remission will be weakened; this weakening occurs in those wave length regions where absorption bands would occur in transmission measurements on solutions of the substances. The remission can be calculated quantitatively with the help of the Kubelka-Munk function.

The degree of reflection is proportional to the concentration of substances adsorbed on a standard white, under certain suppositions which have been studied in detail by KORTÜM and co-workers [371, 373] for powders. The relation is:

$$f(R) = \frac{(1-R)^2}{2R} = \frac{k}{s} = \frac{\varepsilon \cdot c \cdot (2.303)}{s}$$

where:

$R = \dfrac{\Phi \text{ sample}}{\Phi \text{ standard}}$, relative diffuse reflectance compared to a standard,

k = molar absorption coefficient defined in the Beer-Lambert law,
s = scattering coefficient,
ε = molar absorptivity (log to the base 10),
c = molar concentration.

The standard white can thus be used as a diluent for solids, analogous to the use of solvents when applying the Lambert-Beer law to determine concentrations in solution.

The distribution of adsorbent grain size must be standardised and air humidity and pH conditions reproducible if remission spectra are to be quantitatively evaluated. Water, for example, partly displaces the adsorbed substance molecules and thereby alters the form of the spectra [373]. FREI and co-workers [201] have investigated the reflectance spectra of some dyes adsorbed on "Merck" alumina, as a function of its thermal pre-treatment. They tested also the influence of neutral, acid or basic alumina on the form of the spectra. Modifications were observed due to interactions between adsorbed substance and adsorbent (chemisorption).

It is only partly known how far the mentioned laws for powders can be adapted to direct evaluation of the remission spectra of spots on adsorbent layers. It has been established from corresponding studies of paper chromatograms that the Kubelka-Munk function is valid only in particular regions of low concentration [372]. JORK [333a] has found linear dependence of degree of remission on applied substance amount over relatively wide ranges of concentration. He studied substances from various compound classes on silica gel and alumina layers; very small amounts of caffeine, for example (2.5 μg), could be determined quantitatively. A solution of 150 μg caffeine in 10 ml solvent would have been needed in order to obtain a spectrum in solution with the same absorbance. FRODYMA and co-workers [205] have compared quantitatively the diffuse reflectance spectra from spots on the layer and from the powder of the scraped-off spots. Better results were obtained from the powder method for eosin B and rhodamine B as test substances, adsorbed on alumina; this method was thus used for the determination of amino acids, visualised with ninhydrin [204]. A reproducible packing density must be ensured in the powder method [205]. A linear relation between concentration of substance and degree of remission according to the equation $\sqrt{c} = 2 - \log \% R$, was found for the dyes mentioned in evaluation directly on the layer within the measured range of concentration.

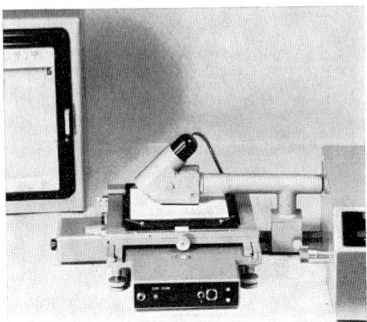

Fig. 77. Accessory for the PMQ II-spectralphotometer (Firm 155) for evaluating reflectance and fluorescence spectra of thin-layer chromatograms (according to STAHL)

Irradiation can be direct or diffuse with the help of a reflectance sphere. In the latter case, the conditions of irradiation must be maintained constant in the determination of sample and standard or else appreciable sphere errors can arise. Attachments for remission measurements with various angles of illumination and observation are obtainable for the spectrophotometer PMQ II (Firm 155). An attachment for this spectrophotometer has been recently developed by E. STAHL and coworkers, in collaboration with the firm of Zeiss (Firm 155); it permits quantitative evaluation of remission and fluorescence spectra of the individual spots on the chromatogram (cf Fig. 77). Work

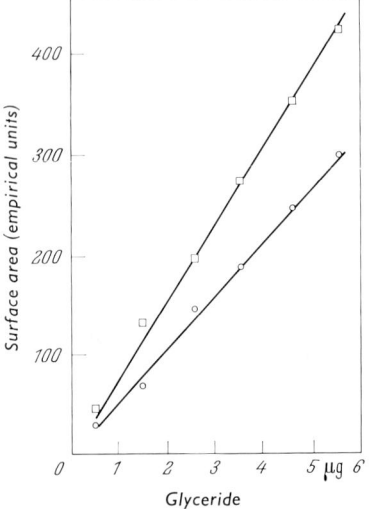

Fig. 78. Dependence of densitometer-spot area on the applied lipid amount, using the reflectance attachment of BARRET et al. [40, 41]. ○ triolein; □ tristearin

has likewise been carried out with reflection attachments for various densitometers[16], [17] [40, 41, 509]; carbonised lipid spots were evaluated (Fig. 78).

---

[16] Thin layer scanner attachment for Chromoscan (Firm 78), cf footnote 13
[17] Described in the publication [509].

II. Quantitative Determination after Extraction from the Adsorbent Layer

Procedures in which the substances are first separated with TLC, detected with a suitable method, eluted from the adsorbent and quantitatively determined, are in general use today. Disadvantages in comparison with direct evaluation on the layer are the greater time necessary and a certain unrecoverable amount of substance. These advantages are however often accepted since the random errors are generally lower than in the direct methods. Comparative studies have shown that quantitative TLC procedures using elution of the separated substances, have no greater range of error than other comparable analytical methods [259] (cf. Table 14). A clean, draught-free working place must be used if reproducible data are desired. Further, it must be ensured that adsorbent layers prepared for quantitative analysis are stored well away from laboratory fumes.

1. Detection of the Separated Substances[18]

Detection on the layer is easy if the substances are coloured or fluoresce in UV light.

In this way, azo dyes [638], dinitrophenylhydrazones [470], 2,4-dinitrophenylhydrazides of fatty acids [800], Rauwolfia alkaloids [607], aristolochia acids [507] and various polynuclear aromatic hydrocarbons [599] could be localised.

This is not possible with the vast majority of substances. They must be detected with the help of reference chromatograms or reagents since localisation through hRf-values is most unreliable [223].

a) Use of Reference Chromatograms

Reference chromatograms have to be used when the substances are colourless, cannot be detected in UV light, cannot be localised with iodine vapour and would be altered by the influence of light. In this method, already known from paper chromatography, at least two start spots are applied. After development, one chromatogram is sprayed with a colour-producing reagent, while the second chromatogram is covered. The second chromatogram is then marked at the positions corresponding to those of the coloured spots on the first. Identification with reference chromatograms is the least destructive but requires very homogeneous adsorbent layers so as to guarantee the most uniform possible solvent migration. Spots with only slightly differing migration distances cannot be localised with certainty in this way.

---

[18] see also pp. 77—82 and 141.

Examples of substances which have been thus detected before quantitative evaluation, are: vitamin B_6 factors [702]; vitamin D_2 [81]; opium alkaloids [434]; penicillin V [486]; emodi-podophyllin [695]; lanatosides [787]; hexadienolides [693]; ubiquinones [733]; homovanillic acid [597]; glucose and maltose [713]; sucrose esters and mixtures of raffinose and sucrose [223]; phenols [157, 623] and phenol aldehydes [790]. Small amounts of cobalamin were discovered with the help of bio-autographic reference chromatograms [128]. NISHIKAZE and STAUDINGER have employed reference substances on a two dimensional chromatogram in order to localise a particular substance (aldosterone) before extraction [482]. Penetration of the detection reagent into the chromatogram for analysis can be completely prevented by rolling a cylindrical wad of blotting paper, saturated with reagent solution, across the reference chromatogram instead of spraying [703].

b) Use of Fluorescent Layers and Indicators

The use of adsorbents containing fluorescent inorganic pigments is advisable for detection of substances which absorb in the UV region. In UV light, they appear as dark, quenching spots on a fluorescent background (cf. Fig. 73a/B, p. 136). The sensitivity of detection depends on the absorption maximum and the specific absorption of the chromatographed substances. The great advantage of the method is that the separated materials can be detected without being chemically modified in any reaction. This method has thus been used extensively. Numerous firms offer adsorbents containing mixed-in fluorescent indicators (e. g., Firm 88) (cf. Chapter B). If such adsorbents are not available, fluorescent indicators[19] can be introduced afterwards [67, 211, 213, 348, 349, 406, 437, 621]. From 0.2 to 5% is used, depending on the indicator itself and on the adsorbent; it is nevertheless difficult to attain the degree of homogeneity possessed by the commercially available, ready-mixed products.

The indicators mentioned do not usually interfere with the quantitative evaluation after extraction. They are however dissolved to some extent by acid extraction agents and may then interfere in certain analytical methods of determination, e. g., polarography [489].

Spraying with solutions of organic fluorescent indicators or mixing such indicators into the adsorbent, is often impracticable. These substances are soluble in solvents of polarity appropriate for development and elution. They have however been used with success in exceptional cases.

Thus lipid spots have been sprayed with rhodamine 6G solutions before gas chromatographic evaluation [475]. Steroids and insecticides [722] and oestrenols [217] could be localised before quantitative evaluation, on layers to which pyrene[20]

---

[19] E.g., zinc silicate and zinc cadmium sulphide (Firm 108), calcium halophosphate N 83 white, WQ, GQ or TQ (Firm 147); Leuchtstoff ZS-Super (Firm 118).

[20] Sodium salts of 3-hydroxypyrene-5,8,10-trisulphonic acid and of 3,5-dihydroxypyrene-8,10-disulphonic acid (Firm 55).

derivatives had been added as fluorescent indicators. Corticosteroids have been detected by first spraying with fluorescein solution and then observing in UV light; quantitative determination with blue tetrazolium followed [482]. Substances which absorb very strongly in the UV can be detected even on layers containing no fluorescent indicators if suitable radiation sources are used [70].

c) Non-Destructive Detection by Spraying with Water

Spots of lipophilic substances can be detected by spraying with water; this depends on the different wetting properties of adsorbent and substance. Detection is easiest in transmitted light. This method, which has been used occasionally to detect steroids [39, 70, 212, 794], is generally too insensitive.

d) Localisation of Separated Spots with Iodine Vapour or by Spraying with Iodine Solution

Substances which do not absorb in the UV can often be detected with iodine vapour [284, 636] (cf. Fig. 79). The iodine "coloration" may be due to the solubility of iodine in the substance concerned; on iodine adsorption or formation of adducts [91]; or on formation of definite addition compounds [475]. This in turn depends on the compound class. The influence of localisation with iodine on the quantitative method subsequently used, must therefore always be tested. After the spot has been detected, excess iodine is largely removed with a current of air or inert gas; this can easily be checked on the layer with starch solution.

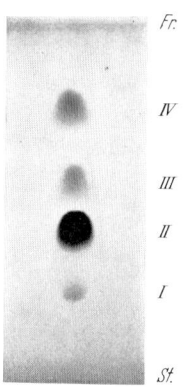

Detection with iodine vapour before quantitative determination has been used frequently, e.g.: analysis of lipid esters in conjunction with gas chromatographic determination [187] or with colorimetric determination after conversion to ferric hydroxamate complexes [731]; to detect analgesic drugs [215] (cf. Fig. 79) and steroids [217, 437] with subsequent spectrophotometric determination in the visible or UV regions; to detect phospholipids, where phosphorus determination followed [6, 79, 637].

Fig. 79. Detection with iodine, of a mixed pharmaceutical preparation after thin-layer chromatographic separation [215]. I caffeine; II aminopyrine; III phenacetin; IV benzyl mandelate. 15 cm run using cyclohexane-acetone (40 + 50) on silica gel G

Some individual substances may however be chemically modified during this test. NICHAMAN et al. [475] have reported the iodination of polyunsaturated fatty acids and the results of the subsequent gas-liquid chromatography which were thereby rendered inaccurate. Localisation with iodine, extraction and quantitative determination with dichromate solution has been described for lipids [19]; the interfering

10*

influence of adsorbents and solvent residues however militate against the reproducibility of this seemingly generally applicable quantitative method. DITTRICH [164] has evaluated compounds on paper chromatograms by determining the iodine in their iodine adducts; the attempt to adapt this to steroid determination on thin-layer chromatograms has also led to no reasonable results.

Detection has occasionally been carried out by spraying with iodine solutions [10], a method which easily leads to a considerable excess of iodine on the plate, then removable only with difficulty.

e) Use of Other Colour Reagents

Other colour tests which do not interfere in the subsequent quantitative determination, can be of only limited application. An example is with organic phosphorus compounds, where the substance can be ignited and determined through its phosphorus content [166, 258]. Glycolipoid fractions have been localised with ammoniacal bromothymol blue indicator and then, after extraction, evaluated with anthrone/sulphuric acid [324]. 17-Hydroxycorticosteroids have been sprayed with blue tetrazolium and then determined with phenylhydrazine/sulphuric acid [11]; the steroids partly reacted with the identification reagent which led to difficulties.

If the separated yet colourless substances can be converted with suitable spray reagents into coloured compounds, these may perhaps be determined directly after elution by a photometric method; the scope of this method is discussed on p. 154.

2. Removal of Spots from the Plate and Elution Technique

After the spots have been marked, they can be scraped off with a razor blade, spatula or similar tool [69, 122, 197, 213, 215, 636, 648]. The best way is first to remove the adsorbent from a zone around the spot and then to scrape off the spot itself on to a sheet of smooth cellophane. It is then extracted by shaking with the chosen solvent (cf. p. 153) and the extract freed from adsorbent by filtering or centrifuging.

Filter papers are unsuitable for filtration. The membrane filters so far commercially available, could not be used on account of their poor resistance to organic solvents. The successful use of methanol-insoluble membrane filters[21] in the extraction of TLC spots, has however been recently reported [799]. Glass or porcelain filters are generally used. The filtrate from a G4 sintered glass filter can be sucked directly into a small standard flask in an evacuated vessel [213, 215]. An extractor functioning under increased pressure has also been employed [695].

---

[21] Alpha-Metricel-Filter (Firm 63)

To prevent losses, air currents must be strictly avoided while scraping off the spot and transferring the powder to a suitable vessel. Various authors have accordingly used vacuum extractors, with which the localised spots could be sucked on to the filter surface or into the extraction flask and then eluted [65, 70, 123, 318, 349, 437, 451, 463, 578, 605, 651].

An extractor of this type is shown in Fig. 80 [437]. Continuous extraction according to the Soxhlet principle [329, 535] for example, is not recommended since even sparingly soluble components of the binder or impurities can be dissolved; these may interfere in the quantitative determination [338] (cf. p. 150). An automatic zone collector has been described for quantitative evaluation of radioactive spots [647].

After elution, adsorbents are most easily separated by centrifuging, provided that the solvents used are not too volatile. One then decants, repeats the procedure and makes up to a definite volume. A single extraction suffices with a solvent of high elution power. After centrifuging, an aliquot of the clear supernatant liquid can be taken for the quantitative determination. A centrifuge yielding a suitably high gravitational field (> 4000 g) must be used if the final stage is a spectrophotometric determination in the short wave UV region; this removes the fine adsorbent particles which otherwise give rise to light scattering. Through its simplicity, this method has found considerable application [69, 162, 206, 207, 208, 212, 217, 486, 594], the more so because the centrifuge tubes are easy to clean in contrast to glass and porcelain filters. RABENORT [550] has carried out chromatography on carrier sheets of aluminium, in order to do away with the scraping-off procedure. The sheets were stretched on frames and placed in the separation chamber. The spots on the chromatograms were cut out as a whole and eluted.

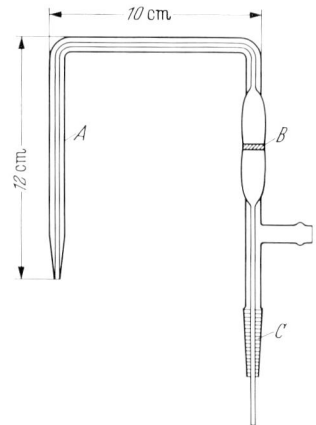

Fig. 80. Vacuum extractor of MATTHEWS et al [437], available also for removing spots from the plate by suction. A tube with which the spot is sucked off; B glass filter (G 4) on which the powder is collected; C ground glass joint fitting an appropriate standard flask

3. Eluents (Solvents for Extraction)

The choice of eluent is controlled primarily by the desorptive effect required. The elutive effect is tested by applying spots of the substances quantitatively to the adsorbent layer and trying to extract them without yet having carried out chromatography. Recovery must be at least 95%. Small losses must be reckoned with after the TLC procedure, depending on the distance of migration and contaminants.

Its potential interference in the subsequent quantitative stage must be considered when choosing the eluent. For example, if UV measurements are to follow, the solvent should absorb only weakly in this region; or one has to evaporate the eluate to dryness and take up the residue in a more suitable solvent.

Very fine particles of adsorbent which can scatter light, must be excluded by appropriate filtration or centrifuging at high speed [799] (cf. also p. 149).

The interference of impurities extracted from the adsorbent is appreciably harder to avoid. The usual silica gels and aluminas contain inorganic impurities, in particular iron and chloride; colour reactions influenced by iron show divergences when an eluent is used which dissolves iron [212]. Such divergences can be eliminated by using special preliminary purification procedures. The evaluation of bile acid chromatograms is an example illustrating this [797]; the silica gel G was treated previously with dilute sulphuric and hydrochloric acids. Preliminary washing with ethanolic hydrochloric acid has been used in the quantitative determination of phenol aldehydes [790]. The presence also of organic impurities in many cases can be deduced from UV-spectra (cf. Fig. 81) and ash analyses of extracts of adsorbents. For example, plasticisers or high-boiling hydrocarbons from the plastic packing appear to get into the adsorbent; this could be recognised in silica gel by gas chromatography of residues from chloroform extraction [225] and IR analysis of residues from acetone extracts [640]; and also from NMR spectra of chloroform extracts of adsorbents [795]. Some manufacturers quote the limits of known impurities [22] and have commercially introduced especially purified silica gels which are suited to quantitative analysis, e. g., silica gel HR (Firm 88). Preliminary purification of adsorbents by

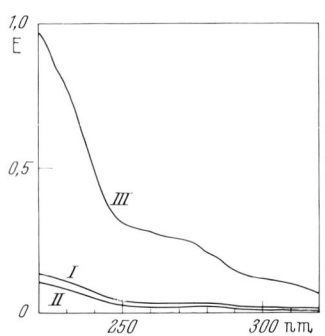

Fig. 81. Silica gel blanks in the UV region. Equal amounts of silica gel for TLC from different manufacturers I, II and III extracted with the same amount of methanol, centrifuged and light absorption compared with that of the methanol used

---

[22] "Merck" preparation for TLC according to E. STAHL, 1964 Issue, E. Merck, Darmstadt, W. Germany, p. 7.

extraction in the laboratory is difficult because unknown binding material is easily dissolved and fine adsorbent particles removed which are important for the adhesion of the layer; further, other new impurities may be introduced.

In practice, eluents of widely varying polarity have been used. Very weakly polar steroids (oestrenols) can be quantitatively eluted with hydrophobic solvents like methylene dichloride (cf Table 14) [217], whereas difficulties are reported in the elution of the more polar progesterone under the same conditions [691]. Chloroform has been used to elute dinitrophenylhydrazones from silica gel and opium alkaloids from alumina [470, 535]. Methanol and ethanol are often used for elution of substances of all types of compound class from silica gel [213, 215, 259, 434] or alumina [122, 437]. Butyl acetate has been chosen as the most suitable eluent for penicillin V [486]. Acetone has proved suitable for recovering ubiquinones [733]. Polar neutral, acid and basic aqueous eluents have also been employed, e.g., water for mucic acid derivatives [451] (cf Table 14), 1% Tween 80 solution for cobalamin [128], 0.2 N sulphuric acid for Vitamin B_6 factors [702], 34% ammonium persulphate solution for nicotinic acid [702] and ammonium hydroxide for azo dyes [638]. Nitro-4-acetaminophenetoles have been extracted from alumina with the highly polar dimethylformamide [489].

4. Methods of Determination Used after Elution

Evaluation by measuring UV absorption has been used in many cases for substances which absorb in this region because this is often possible directly after extraction with a suitable eluent, without any other stage. Limits are set only by the absorptivity of the compound to be determined.

Generally at least 5 ml solvent is used for extracting a spot, so as to be able to work with the normal 1 cm cuvettes. A 50 μg amount applied to the layer thus yields a concentration of 10 μg/ml after extraction. If such a procedure is used for example for the four pharmaceuticals aminopyrine, caffeine, benzyl mandelate and phenacetin, which absorb in the UV, the extinctions at the maxima are, respectively, 0.36, 0.46, 0.02 and 0.82 (cf. Fig. 82). It can be seen that the method is too insensitive for benzyl mandelate [215].

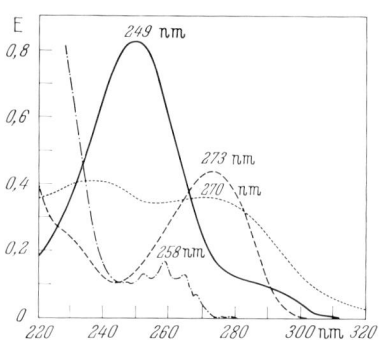

Fig. 82. UV spectra of the active substances in a mixed pharmaceutical preparation [215]. Solvent: methanol. anhydrous caffeine ------; aminopyrine ....; phenacetin ———; benzyl mandelate --------; 10 μg/ml of a, b and c; 100 μg/ml of d

Use of a micro-elution assembly has permitted determination of even small amounts of substances with low absorptivities, by extracting with 100 μl of solvent [451] (cf. Table 14). A microphotometer is then necessary, however.

Table 14. *Examples of quantitative*

Material for analysis	Adsorbent and solvent	Detection	Method of elution and evaluation
I. Methyl p-hydroxybenzoate II. Propyl p-hydroxybenzoate; 50 + 50% mixtures	Silica gel G + 2% fluorescent additive [a]; Pentane-acetic acid (88 + 12)	UV 254 nm	Micro-filtration with G 28 and G 4 glass filter rods [b]. Determination by UV
I. Mucic acid II. Hydroxymethylmucic acid; mixtures	Silica gel HF_{254}. $CHCl_3$-acetic acid (90 + 10)	UV 254 nm	Micro-extractor made by authors. Evaluation in UV with micro-photometer after extraction
I. Methylphenobarbital II. Phenobarbital mixture, 95% + 5% (only I detd.)	Silica gel HF_{254}. $CHCl_3$-acetone (90 + 10)	UV 254 nm	Centrifuging and UV measurement after addition of buffer
I. 6-Chloro-17α-hydroxypregna-4,6-diene-3,20-dione-acetate II. 17α-Ethynyl-oestradiol-3-methyl ether Mixture 85 + 15% (only I detd.)	Silica gel G. $CHCl_3$-ether (90 + 10)	Water	Constricted glass with cottonwool plug as filter; UV-measurement
17α-ethyl-17β-hydroxy-Δ^4-oestrene	Silica gel H. Heptane-acetone (60 + 30)	Iodine vapour	Centrifuged, solvent evapd., colour with H_2SO_4
Morphine in opium	Silica gel HF_{254}. $CHCl_3$-acetone-CH_3OH—$N(C_2H_5)_3$ (30 + 40 + 10 + 20)	UV 254 nm	Reagent for prepg. nitrosomorphine functioned as eluent; centrifuged and photometrically determined
Neomycin sulphate	Silica gel. 3% NH_4OH-acetone (80 + 20)	Reference chromatogram	Colour reagent served as eluent; photometry after elution

[a] Leuchtstoff ZS-Super (Firm 118).
[b] Firm 67.

Determination through measurements in the UV below 240 nm is generally possible only with a considerable error, as a result of the interference of co-extracted impurities and scattered light. The errors can be reduced by using blanks on extracted clean silica gel samples [70, 213, 451, 614] or by base line corrections if the absorption spectrum *determinations after extracting the spots*

Eluent	No. of detns.	Accuracy % recovered	Precison[c,d] $S_{rel.} \cdot \%$	µg used	Refs.
5 × 2 ml CH$_3$OH	14 sepns.	I. 96.0% II. 93.6%	±3.0 ±3.8	50 50	[213]
0.1 ml water	7 sepns. for each conc.	I. 100 % 100 % II. 97.2% 91.7%	±0.75 ±1.60 ±0.75 ±0.60	100 6 90 6	[451]
10 ml CH$_3$OH	12	97.0%	±2.4	450	[219]
5 ml abs. ethanol	32	97.6%	±3.5	60—80	[70]
3 ml CH$_2$Cl$_2$	30	95%	±2.1	60	[217]
—	15	—	±1.2	100—200	[289]
—	47	—	±1.9	50—100	[197]

[c] $S_{rel} = \dfrac{S \times 100}{\text{mean value}} \%.$

[d] Further details of errors when using analogous methods are to be found in the literature [122, 215, 415, 437, 535].

has a suitable form [213]. As far as possible, the blanks are taken from a point at the same migration distance from the start as the sample spot. Blanks from the region of the solvent front are useless.

IR-spectra of solutions have been little used up to now for quantitative evaluation on thin-layer chromatograms [383, 411]. The reason is that a high substance concentration is necessary for the measurements. Extraction has been performed in one case with a micro-extractor and measurement carried out in a micro-cuvette [411]. Successive drop application [640] or preparation of micro-discs with potassium bromide seems more promising.

Substances which absorb weakly or not all in the UV region have often been localised with iodine vapour, as already described, or with the help of a reference chromatogram, and then eluted with an appropriate solvent and determined quantitatively through a colour reaction [6, 79, 217, 637].

The reagent serves in many cases as eluent also and measurement is carried out at the suitable wave length after centrifuging. If strongly polar, acid or basic reagents are used, binder and impurities are easily dissolved out of the adsorbent; this can affect the colour reaction [212] (cf. p. 150).

This method has none the less been frequently used, since the working technique is simple. Examples of use are determinations of: bile acids [212, 376], pregnanediol [39], the dinitrophenylhydrazone of vitamin C [701], morphine [289, 290], neomycin sulphate [197] and pentaerythritol nitrate [162], with acid reagents; digitalis glycosides have been determined colorimetrically in a parallel fashion, using the xanthydrol reagent [289].

Evaluation by spraying the thin-layer chromatogram with a colour reagent, extracting the coloured product and determining it photometrically, is often beset with difficulties, as known already from experience in paper chromatography. The ratio of reagent concentration to amount of substance being determined in the middle of the spot differs from that at the edges; and on the upper surface of the layer from that on the underside towards the carrier plate. Reaction may thus occur to a varied extent. The coloured products are often harder to extract than the original substances. Moreover, the background is frequently unevenly coloured by the reagent, so that fluctuating blanks occur.

Difficulties of this sort have been reported, e.g., in the extraction of phenols which had been converted to azo dyes on the adsorbent layer [623]. The reaction of sulphonamides with a diazonium reagent has been taken to completion by adding excess reagent again after having extracted the coloured product [69]. The determination of monosaccharides with benzidine-glacial acetic acid has been carried out similarly [38]. The method has been used also for the quantitative analysis of: amino acids, after colour reaction with ninhydrin [42, 95]; corticosteroids, after reaction yielding formazans [64]; and of cannabinols after conversion into azo compounds [370].

Quantitative determination by fluorescence analysis is possible for very small amounts of substances under certain conditions; the substance to be determined may itself fluoresce or it may be transformed with a suitable reagent into a fluorescent derivative. This method should be used only if absolutely necessary, since the evaluation of fluorescence spectra is rather susceptible to interferences.

In this way, generally after localisation through the fluorescence of the substances themselves, determinations have been carried out of: coumarin derivatives, following chromatographic separation from lemon and grapefruit oils [729]; quinoxaline derivatives of 2-ketoacids [651]; benzo-(α)-pyrene [599]; and vitamin B_6 factors [702]. Cortisol [228] and aldosterone [229] which had been extracted from urine and purified by TLC, have been localised with the help of a reference chromatogram and then determined fluorometrically after treatment with sulphuric acid. After extraction from the thin layer, nicotine, nornicotine and anabasine could be determined phosphorimetrically (see p. 141) with a relative standard deviation of 6% [761].

Some substances have been determined polarographically following extraction. 2- and 3-Nitro-4-acetaminophenetoles have been determined with an error of $\pm 3\%$ [489] and nitrosomorphine with a relative standard deviation of $\pm 2.9\%$ [289, 290]. Tocopheronolactone, extracted from a thin-layer chromatogram, has likewise been determined polarographically [610]. An eluent should be chosen that cannot dissolve cations out of the adsorbent which could function as depolarisers (cf. p. 149).

Under suitable conditions, micro amounts could be determined coulometrically; there is no example so far of this use. Very small amounts of vitamin B_{12} have been determined with a microbiological turbidity test after they had been localised bioautographically on a reference chromatogram (cf. p. 82) and extracted [128]. The quantitative determination of radioactive substances is treated in the following chapter.

I. Isotope Technique

Helmut K. Mangold

The distinctive advantage of using radioactive isotopes in chemical investigations is well known: radioactive elements can be detected with high sensitivity, independently of their chemical binding.

Chromatographic procedures have been used in recent years for separating and isolating radioactive elements which are chemically very similar [192]. The same methods have often been applied also to the fractionation of radioactively labelled organic substances. Chromatographic separation methods have furthered considerably the use of isotopes, especially in biochemical research. Applications of radiochromatographic procedures are described in several reviews and handbooks [101, 387, 484, 581].

Within a few years, TLC has largely ousted other radiochromatographic techniques. Its advantages over column and paper chromatography are:

As a micropreparative method, the high capacity of the thin layer; it is easily possible to separate several milligrams of a mixture within an hour on a single 20 × 20 cm plate and to isolate radioactive compounds in amounts which are adequate for most purposes.

As an analytical method, TLC possesses particular advantages:

The entire path of migration is freely visible, as in paper chromatography but in contrast to column chromatography.

Almost every substance can be detected, very volatile compounds being exceptions.

Resolutions are usually sharper than those obtained in paper and column chromatography.

The sensitivity of detection is even higher than in paper chromatography because the separated substances are concentrated in much smaller spot areas. Substances of lower activity and isotopes with low radiation energy can therefore be detected on a thin-layer chromatogram. Quenching by the adsorbent layer is generally negligible.

Manipulation is simple and most separations can be performed in less than an hour.

These advantages are especially valuable when working with "hot" radioisotopes and with those of short-half-life.

Almost only fairly long-lived radioisotopes, mostly weak β-emitters, have been used in chemical and biological investigations. Radiocarbon, ^{14}C(half-life 5568 years, maximum energy 0.155 mev) and radioactive hydrogen, ^{3}H(tritium) (half-life 12.26 years, maximum energy 0.018 mev) are particularly important. Radioactive phosphorus, ^{32}P(14.2 days, 1.71 mev), sulphur, ^{35}S(87.1 days, 0.167 mev) and iodine, ^{131}I(8.04 days, 0.608 mev(β) and several β- and γ-radiations) have also often been used.

The physical unit of activity of a radioactive product is the curie (Ci); this is the amount of substance in which the same number of disintegrations occur per second as in 1 g radium. 3.7×10^{10} atoms disintegrate per second in 1 g radium, i.e., 2.2×10^{12} disintegrations per minute (dpm).

Calculations in chemical studies are usually in millicuries (mCi) (2.2×10^9 dpm), microcuries (μCi) (2.2×10^6 dpm) or nanocuries (nCi) (2.2×10^3 dpm).

I. Layers, Solvents and Chemical Methods of Detection

The experimental separation conditions quoted in the different chapters of this book apply also when the various compound classes are present as radioactively labelled substances. The fractionation may be effected by adsorption and/or partition on adsorbents; by partition in reversed phase chromatography on hydrophobic layers;

by ion exchange; or by TL-electrophoresis or TL-electrophoresis-TLC. The same solvents are used also.

Chemical test reagents can be applied also to thin-layer chromatograms which contain radioactive substances. In addition, various procedures for detection and quantitative determination of labelled compounds can be employed. Both chemical and radiometric methods of detection should always be used on the same thin-layer chromatogram; this yields information about both the chemical and radiochemical purity of the sample studied.

II. Procedures for Detection and for Measurement of Radioactivity

Autoradiography is particularly suited for detecting radioactive materials on coated plates. Detection by means of a Geiger-Müller counter is possible but more tedious; this is true also for proportional counters.

The sensitivity and accuracy of measurement of radioactivity with liquid scintillators is far superior to that of all other counting methods in quantitative analysis.

1. Autoradiography of Thin-Layer Chromatograms

Photographic emulsions are darkened by α-, β- and γ-rays. Radioactive substances are photographically detected on a coated plate by carefully pressing the dry chromatogram against an X-ray film, wrapping this "sandwich" in black cloth and keeping it in the dark. A plate holder for making autoradiographs from thin-layer chromatograms is available commercially (Fig. 83). Simpler apparatus has been described in the literature [573, 645]. It must be borne in mind that small amounts of many chemicals used in the solvents may adhere to the layer and lead to formation of artifacts on the film. The solvent should thus always be completely removed before the chromatogram is brought into contact with the film. Even components of the material for analysis can cause artifacts, e. g., inactive oestradiol rapidly darkens the X-ray film [573].

If several thin-layer chromatograms are to be evaluated, it is a good idea to mark the individual plates with fluorescent material which is clearly registered on the X-ray film (Firm 24).

The film most commonly used is "No-Screen Medical X-Ray Safety Film" (Firm 52). The emulsion of this material is quite sensitive and the resolution [611] suffices for TLC requirements. Suitable films are manufactured by other firms also (Firms 1, 9, 75, 107). "Supermix developer" or some other X-ray film developer (Firms 1, 9, 75, 107) is suitable for processing the films. This is carried out in the dark or in red safety light. The development time depends on temperature and the age of the developer, ranging from 3 to 6 min at 20° C. X-ray fixing salt is obtainable from several photographic firms (Firms 1, 9, 64, 75, 107). Fixing times of 10–30 min are used and the films then rinsed for 30 to 60 min in running tap water.

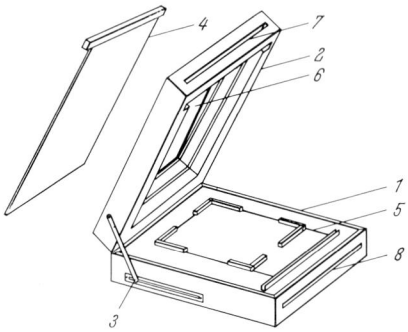

Fig. 83. Container for autoradiography of thin-layer chromatograms [73] (Firm 7). *1* lower compartment, containing a coated 20 × 20 cm plate on a platform (*5*); *2* upper compartment, functioning as film cassette; *3* sliding hinge; *4* slide; *5* platform with ridges, serving as support for the chromatogram; *6* X-ray film (20.3 × 25.4 cm) held firmly by guide rails; *7* slot for closing the film cassette with the slide (*4*); *8* slot for raising the platform (*5*) with the slide (*4*)

The contact time is governed by the activity of the separated substances, the nature and energy of the radiation of the radioisotopes used for labelling and by the effect desired. The film is adequately blackened when one to ten million β-particles per cm^2, depending on the isotope, fall on the film during the contact time. A rough rule with radiocarbon is that substance amounts which show in the Geiger-Müller counter an activity double that of the background activity, produce a distinct darkening of the X-ray film with a contact time exceeding 2 days. Detection of radioactive carbon compounds is ten to a hundred times more sensitive than on paper chromatograms.

Generally, identical chromatograms of ^{14}C-labelled compounds are left 1, 2, 4 and 8 days in contact with X-ray film. Autoradiographs are obtained in this way, in which the ratios of the amounts of the separated components are approximately correctly reproduced; and also autoradiographs in which the main components yield photographically over-dense spots and minor contaminants also show up. A complete picture of the qualitative composition of the separated mixture is thus obtained.

If exposures of several weeks are needed, the thin-layer chromatogram and film are kept in a refrigerator to prevent secondary photographic effects.

Thin-layer chromatograms of tritiated substances usually need more than a week's exposure, whereas compounds containing ^{32}P and ^{131}I often yield good autoradiographs within 30 min and generally in less than 6 hours [89, 426].

It is often necessary to detect on the same chromatogram, substances which contain different radioisotopes or which are doubly labelled. This is easily possible through autoradiography when the half-life and/or the hardness of the β-radiation of the two isotopes differ widely.

A chromatogram containing ^{14}C- and ^{35}S-compounds can be brought into contact with X-ray film immediately and again after several months. The first autoradiograph shows dark spots produced by the radiation of both isotopes; the second, spots only from the radiocarbon, since the sulphur isotope has by then virtually fully disintegrated. The second autoradiograph should be taken when five to ten times the half-life period of the short-lived isotope has elapsed.

An example of differentiation through differing energies of radiation may be given here: *Two* X-ray films are placed, one above the other on the chromatogram containing ^{14}C- and ^{32}P-labelled substances. After development, the film next to the chromatogram shows black areas coming from both radioelements. The β-radiation of radiocarbon is so weak however, that it is unable to pass through the first film; the second film thus shows spots which are produced only by the appreciably harder β-rays from the radioactive phosphorus.

^{131}I can be detected alongside ^{35}S or ^{14}C in a similar way: A sheet of aluminium foil, less than 0.1 mm thick, suffices to screen off the soft radiation from the radioactive sulphur and carbon, whereas the harder radiation from the radioactive iodine can pass through it. Cellophane screens off the softer radiation from tritium and permits distinction of spots containing ^{14}C and ^{3}H of comparable activity [573].

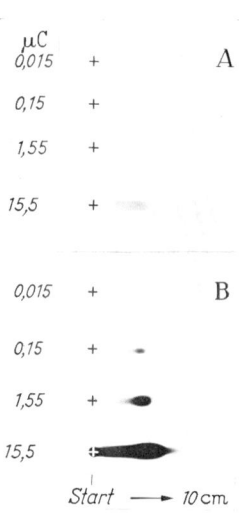

Fig. 84. Autoradiograph (*A*) and "fluorogram" (*B*) of two thin-layer chromatograms of ^{3}H-labelled diallylnortoxiferine (^{3}H-alloferine) [408]. Layers: *A*. silica gel G; *B*. silica gel G and anthracene (50 + 50) (g/g). Solvent: acetonitrile-hexane-diethylamine (75.5 + 17.5 + 5). Film: Eastman Kodak "Kodirex"; contact time: 18 h at —70° C

Photographic detection of ^{3}H-labelled substances on silica gel, is time-consuming and insensitive [262, 426]. "Nuclear-Track Emulsion, Kodak NTB" (Firm 52) is more sensitive for tritium than is X-ray film [632]. The sensitivity of autoradiography of tritium-labelled substances can be raised also by impregnating the thin-layer with a photographic emulsion like, e. g., "Ilford XK" (Firm 75) [126]. This is however a comparatively complicated procedure. The autoradiographic method can be appreciably improved by inducing fluorescence with tritium at low temperature [408]. This technique is described below:

Directions

Silica gel G and anthracene are mixed in the ratio 1:1 and pulverised in a ball mill to a grain size of 1—5 µm (about one third of the range of ^{3}H β-radiation!). 30 g of this powder are slurried with 80 cm^3 of 96% ethanol and spread on to 5 20 × 20 cm glass plates as layers 0.25 mm thick, using the standard method (see p. 85). The coated plates are air dried. The customary solvents used for silica gel G layers can be used with this mixed layer too. The anthracene has little influence on the separating power of the adsorbent.

Chromatograms of tritiated compounds are pressed on to X-ray film and exposed at —40° C.

These "Fluorograms" of tritium-labelled compounds on a silica gel G-anthracene layer are substantially more sensitive than the autoradiographs on a pure adsorbent layer (see Fig. 84). Sensitivity is higher at low temperature than at room temperature but lowering the temperature beyond — 70° C effects no further improvement. "Fluorography" offers no advantages over autoradiography for ^{14}C-labelled compounds [408].

Autoradiographs of paper and thin-layer chromatograms can be quantitatively evaluated by photodensitometry or according to the spot sizes. These possibilities are further discussed below.

It may be pointed out here that non-radioactive compounds can be detected autoradiographically after neutron activation or in the form of labelled derivatives. These procedures are discussed in more detail in section V, "Analysis by Means of Radioisotopes".

2. Counting Tubes and Scintillation Counters

The quantitative evaluation of paper chromatograms of radioactive substances by means of a Geiger-Müller counter, a gas-flow proportional counter or a scintillation counter, has often been described. Detectors for radioactive radiation, with devices for automatic transport of a paper strip or of a two-dimensional chromatogram and for recording quantitative data, are commercially available. Such apparatus can be used for thin-layer chromatograms also if the layer is impregnated with a plastic emulsion and peeled off as a sheet [141, 618]. "Strip-scanners" can be modified for use with thin-layer chromatograms [484, 584].

Instruments are obtainable for *direct* quantitative evaluation of radioactive thin-layer chromatograms (Firms 13, 22, 44, 61, 99, 131a, 135). The "Thin-layer scanner", constructed by Schulze and Wenzel [617] is illustrated in Fig. 85; two other counters are seen in Figs. 86 and 87.

The sensitivity of these instruments for a particular radioisotope depends on the nature of the detector. Open methane gas-flow proportional counters give the highest count yields and have the additional

Fig. 85

Fig. 86

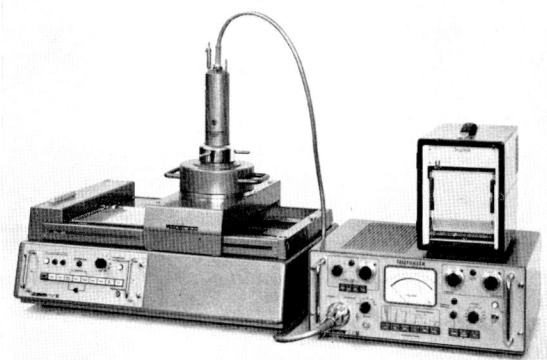

Fig. 87

Figs. 85—87. Apparatus for recording the distribution of activity on thin-layer chromatograms. Fig. 85. "Thin-Layer Scanner" (Firm 44); Fig. 86. "Model" RSC-363, Deluxe Scanner" (Firm 13); Fig. 87. "Radio-Chromatograph" (Firm 135)

advantage that they can be used to determine tritium-containing compounds [74, 88, 606, 617].

The influence of the adsorbent and of layer thickness on the count yield of three different isotopes is seen from Table 15.

Table 15. *Count yields of ^{14}C, ^{35}S and ^{3}H, measured with a methane gas-flow counter* [88]

Layer	Layer thickness	Surface density	Count rate/decays per min (cpm/dpm)	
			^{14}C, ^{35}S (with window, 0.7 mg/cm²)	^{3}H (without window)
	μm	mg/cm²		
Silica gel C	1000	34	1.8%	0.15%
Silica gel G	250	8.5	4.5%	0.4%
Alumina G	250	8.5	4.9%	0.4%
Kieselguhr G	250	7.5	5.5%	1.4%
Cellulose MN 300 G	250	3.0	5.7%	1.3%

The accuracy of measurement and the resolution of an instrument are fixed by the plate movement and the width of the detector. With some instruments, e. g., the "Radio-Chromatograph" (Firm 135), the speed of the plate movement and also the slit width of the detector can be chosen according to the requirements of the determination. The activity on the thin-layer chromatogram is first measured in an orientation experiment. With the help of a nomogram, the registration time can be worked out which is necessary not to exceed a chosen statistical error.

A mean statistical error of 3—5% should generally be chosen [88]. This is a compromise between the crudeness of the recorded curves and the necessary duration of measurement.

A slit width of one order of magnitude smaller than the spot diameter is recommended in order to obtain satisfactory resolution [88].

The duration of measurement can be appreciably reduced by regulating the plate movement. The "Radio-Chromatograph" (Fig. 87) permits scanning of radioactive spots at the optimum speed and of inactive zones at higher speed.

If thin-layer chromatograms are sprayed with scintillation solution, ^{3}H-labelled substances of relatively low activity can be registered and measured with photo-multipliers [586].

The curves recorded using the instruments described above can be quantitatively evaluated by graphical integration or with digital counters. Elution of the radioactive substance from the adsorbent, followed by measurement in a conventional counter has however usually been preferred [238, 426]. A useful glass apparatus for eluting radioactive substances can be obtained commercially (Firm 82). Should the different fractions contain various amounts of "cold" carrier substances — this is likely to be the case in biological studies — the measurement must be carried out in a liquid scintillation counter in order to eliminate self-absorption [106]. Liquid scintillation counters are more suitable than any other method for determining ^{14}C- or ^{3}H-labelled compounds. Even very low activities of these isotopes, or of other weak β-emitters, can be measured in the presence of one another with high accuracy.

Numerous scintillation solutions are suitable as solvents [649]. Most substances can be eluted with the aqueous scintillation solutions 3 and 4 quoted in Table 16. Complete mixing of sample and scintillation solution ensures greater dependability and sensitivity and thereby facilitates quantitative evaluation of radioactive thin-layer chromatograms. The composition of commonly used "cocktails" is given in Table 16:

Table 16. *Scintillation solutions ("cocktails")*

1. *Toluene* — ("Cab · O · Sil")[a] [649]
    PPO[b] . . . . 5.0 g
    Dimethyl-POPOP[b] . . . . 0.3 g
    Toluene, made up to . . . . 1000.0 cm³
    (Cab · O · Sil . . . . 40.0 g)
2. *Dioxan — Naphthalene — Methanol* — ("Cab · O · Sil[a]) [642]
    PPO[b] . . . . 6.5 g
    POPOP[b] . . . . 0.13 g
    Naphthalene . . . . 104.0 g
    Toluene . . . . 5.0 cm³
    Dioxan . . . . 500.0 cm³
    Methanol . . . . 300.0 cm³
    (Cab · O · Sil . . . . 32.0 g)
3. *Dioxan — Naphthalene — Water* — ("Cab · O · Sil")[a] [642]
    PPO[b] . . . . 10.5 g
    POPOP[b] . . . . 0.45 g
    Naphthalene . . . . 150.0 g
    Dioxan, made up to . . . . 1500.0 cm³
    Water, made up to . . . . 1800.0 cm³
    (Cab · O · Sil . . . . 72.0 g)
4. *Toluene — Methyl cellosolve — Naphthalene — Water*
    BBOT[b] . . . . 4.0 g
    Naphthalene . . . . 80.0 g
    Methyl cellosolve[c] . . . . 400.0 cm³
    Toluene . . . . 600.0 cm³
    Water . . . . 35.0 cm³

[a] Cab · O · Sil, a highly dispersed silica gel which forms a thixotropic gel when shaken with scintillation solutions.
[b] PPO (2,5-diphenyloxazole) serves a primary scintillator.
POPOP (1,4-di[2-(5-phenyloxazolyl)]-benzene) and
Dimethyl-POPOP serve as secondary scintillators.
BBOT (2,5-bis-2-[5-*tert.*-butylbenzoxazolyl]-thiophene) requires no secondary scintillator.
Cab · O · Sil and scintillator substances are commercially available (Firm 99).
[c] Methyl cellosolve is 2-methoxyethanol.

Scintillation solutions 1 and 2 are suitable for determining radioactive lipids in particular. The third cocktail resembles the well known Bray's solution and, like this, is used chiefly for measurements on aqueous solutions. Scintillation solution 4 contains methyl cellosolve instead of dioxan as it is generally obtainable in a purer state.

Measurements are carried out at $+10°$ C to prevent formation of precipitates in the scintillation solution.

The activities of labelled compounds can be measured on silica gel G without prior elution [642, 649]. The adsorbed radioactive substance + adsorbent is scraped off the plate and the activity measured in suspension. The adsorbent is vigorously shaken with a gel prepared by homogenising 4% "Cab · O · Sil" with one of the four solutions in Table 16. The activity of the gelatinous mass can then be determined in the scintillation counter with a high counting efficiency. No quenching effect of ^{14}C-labelled substances is caused by 200 mg silica gel (corresponding to about 10 cm^2 layer) per 15 ml scintillation gel (see Table 17). Activity in suspension can be measured with silicone-impregnated adsorbents also [718].

Table 17. *Influence of the adsorbent on the count yield* [642]
[200 mg silica gel was added to each scintillation solution (15 cm^3)]

"Cocktail"	Oleic acid 1-^{14}C	Palmitic acid-9,10-^{3}H	Tripalmitin ^{14}C (impure)	Total lipids from	
				Liver[a]	Marrow[b]
Dioxan-naphthalene-water-Cab · O · Sil	97	101	100	98	98
Dioxan-naphthalene-water	99	100	100	99	98
Toluene-Cab · O · Sil	94	78	97	98	97

[a] 100 = no quenching effect.
[b] After feeding a rat with palmitic acid-1-^{14}C.

Fluorescence quenching by aldehydes, mercaptans, nitro compounds and other substances interferes with the liquid scintillation counting ("chemical quenchers") [387]. Most coloured compounds, such as spray reagents on the layer, act likewise ("optical quenchers") [387]. The extent of quenching by a given substance depends also on the composition of the scintillation solution [106]. Quenching effects can as a rule be recognised and correction made for them by using an internal standard [387]. If the quenching is too powerful, it is advantageous to ignite the sample and to determine the activity of the $^{14}CO_2$ [171]. The quenching effects of some indicators used in the TLC of lipids, with three different scintillation solutions, are quoted in Table 18.

Table 18. *Quenching effects of various adsorbents and indicators* [642][a]

Adsorbent	Indicator	"Cocktail" (cf. Table 16, p. 163)		
		Toluene-Cab·O·Sil	Dioxan-naphthalene-water	Dioxan-naphthalene-water-Cab·O·Sil
1. —		100	100	100
2. Adsorbosil-1		100	99	99
3. Adsorbosil-2		100	100	100
4. Silica gel G		100	100	99
5. Silica gel H		100	100	101
6. Adsorbosil-1 + 1% $AgNO_3$		99	98	100
7. Adsorbosil-1	Iodine	99	97	99
8. Adsorbosil-1 + 1% $AgNO_3$	vapour	95	98	99
9. Adsorbosil-1	Charring	67	82	31
10. Adsorbosil-1 + 1% $AgNO_3$	with H_2SO_4	52	70	64
11. Adsorbosil-1	Rgt. No. 63	100	100	99
12. Silica gel D5	Rgt. No. 63	100	101	100
13. Adsorbosil-1	Rgt. No. 221	100	100	99

[a] 100 = no quenching effect.

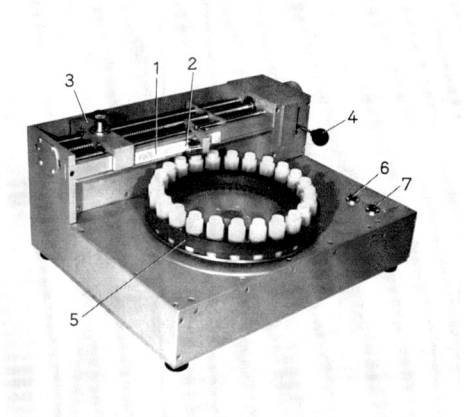

Fig. 88. Automatic apparatus for scraping off narrow zones from a thin-layer chromatogram (2 × 20 cm) [647]. *1* thin-layer chromatogram (2 cm wide); *2* scraping knife (razor blade); *3* slide; *4* lever for setting the advance; *5* set of 24 counting vials (Firm 99); *6* starting knob; *7* stop-knob. The chromatogram is attached to a strip mounted on the slide. The slide is transported stepwise from left to right by a worm; the advance can be set so that the scraper takes off zones 1,2 or 5 mm wide from the layer. After each operation the strip carrying the chromatogram is vibrated so that loosely adhering adsorbent is completely shaken off. The counting vials then advance one further and the knife scrapes off another zone

SNYDER [642, 647] has developed instruments ("zonal scrapers") with which the layer of a chromatogram (2 × 20 cm) can be scraped off rapidly and cleanly in zones 1, 2 and 5 mm broad. The activity of each zone is then determined by liquid scintillation counting in suspension (see Table 16, p. 163). An electrically driven model of the "zonal scraper" is illustrated in Fig. 88. Plans for constructing this apparatus and also a hand operated model, can be obtained from ORINS, Oak Ridge, Tennessee, USA.

"Zonal scanning" is a most sensitive detection procedure, also for ^{3}H-labelled substances. The resolution of the method for scraping off and measuring 1 mm zones, is substantially better than directly recording the activity in a G-M-counter or a methane gas flow proportional counter and nearly as good as with autoradiography. The first two methods mentioned are compared in Fig. 89.

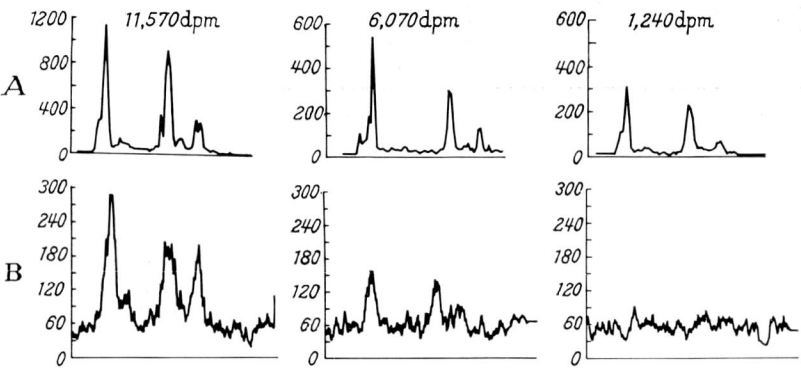

Fig. 89. Comparison of the sensitivity and resolution in direct activity determination on the layer using a methane gas flow proportional counter (*B*) and a liquid scintillation counter (*A*); measurements on zones scraped from the same thin-layer chromatogram [646]

The reliability and accuracy of liquid scintillation counting in suspension is superior to that of all other radiometric methods for quantitative analysis of whole spots or of narrow zones. This is for the following reasons:

Direct measurement of radioactivity on adsorbent layers is insensitive because of the high self-absorption of the radiation and the "softness" of the (glass) background; it is unreliable as a quantitative procedure.

The yield from elution of the substance to be determined, is uncertain.

Quantitative evaluation of autoradiographs with a photodensitometer is tedious. Substances occurring in small amounts are hard to determine on autoradiographs in the presence of the principal components of the mixture for analysis. The radiation from these components photographically supersaturates the film material before the radiation from the minor components has had a detectable influence. Photodensitometric or planimetric evaluation of autoradiographs fails completely if the different radioactive fractions are diluted with varying amounts of "cold" carrier material.

III. Preparation of Radioactively Labelled Substances

Compounds, radioactively labelled in one or more positions, can be synthesised chemically. Procedures for preparation of specifically tagged organic substances are to be found in a very comprehensive monograph [467]. EURATOM [189] publishes short summaries of new work in the realm of the chemistry of labelled compounds.

Non-specifically or randomly tagged compounds can be obtained by biosynthesis [101, 121, 635]. For example, ^{14}C-labelled carbohydrates can be obtained by allowing leaves of the plant *Canna indica* to assimilate radioactive carbon dioxide for 10—20 h. "Hot" starch can be similarly prepared photosynthetically from tobacco leaves. Radioactive peptides and amino acids are obtained from the yeast *Torula utilis* or, more usually, from algae like *Chlorella vulgaris* or *Chlorella pyrenoidosa*, which are grown in an atmosphere of radioactive carbon dioxide for several days. The latter alga yields chiefly lipids in a nutrient medium deficient in nitrogen; it is consequently excellently suited for preparing radioactive fats and fatty acids. The fungus *Phycomyces blakesleeanus* can likewise be used to prepare ^{14}C-tagged fatty acids. "Hot" nucleic acids, nucleotides and nucleosides are obtained from *Torula utilis*. Soya can be employed for biosynthesis of labelled natural products too.

Many specifically labelled compounds can be obtained commercially. Radioactive yeasts or algae as well as protein hydrolysates, saccharide- and lipid extracts of these organisms are also marketed.

A brochure published by the International Atomic Energy Agency lists the commercially available preparations, their specific activities and the manufacturers' addresses [312]. The commercially available preparations are often very impure and can be used for chemical and biochemical studies only after careful purification. The classical methods for processing crude synthetic products are generally unsatisfactory when applied to small amounts of radioactive compounds. TLC is particularly convenient for purifying radioactive compounds. A detailed account of such preparative applications of the method is given in a special section of this chapter.

The Wilzbach procedure is a simple method of labelling organic compounds with tritium: The substance is exposed to tritium gas in a sealed glass vessel for several days or weeks. During this period, some of the hydrogen atoms of the molecule to be labelled are substituted by ^3H atoms. Chemical side reactions may take place however at the same time, e.g., unsaturated compounds may be partly or completely hydrogenated. Several firms carry out this type of labelling on request [312].

The difficulties of the Wilzbach procedure arise in the working up of the tritiated substances and the purification of the compounds which suffer radiolysis during prolonged storage. TLC can be used successfully for this purification purpose also (see Fig. 93).

IV. Isolation of Radioactive Compounds by TLC

Usually only a few mg of labelled compound are necessary for radiochemical investigations. Such a small amount can be isolated in a particularly elegant fashion through TLC. Fig. 90 shows the autoradiograph of a thin-layer chromatogram of various commercial fatty acids with tagged carboxyl group. Each preparation contains several impurities. It is striking that many of these substances (intermediate products in the synthesis) are highly polar. This facilitates the isolation of the pure acids by TLC.

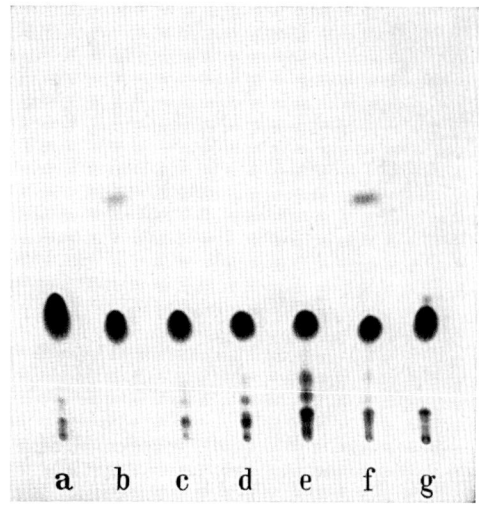

Fig. 90. Autoradiograph of a thin-layer chromatogram of radioactively labelled fatty acids [424]. Layer: silica gel G; solvent: petrol ether (BP 60—70° C)-diethyl ether-acetic acid (90 + 10 + 1); time of run: 40 min; film: Eastman Kodak "No-screen medical X-ray safety film". a lauric acid-1-^{14}C; b myristic acid-1-^{14}C; c palmitic acid-1-^{14}C; d stearic acid-1-^{14}C; e oleic acid-1-^{14}C; f linoleic acid-1-^{14}C; g linolenic acid-1-^{14}C

Solvents suitable for elution of chromatographically separated radioactive compounds can be ascertained by chromatographing the substances in solvents of various polarities; those which carry a substance with the solvent front, are appropriate for elution. The adsorbent is scraped off the plate and digested with several portions of the solvent. The clear solution is decanted and filtered through a small sintered glass funnel. Losses generally ensue from the use of filter paper.

When the compound to be purified is contaminated with substances of similar polarity from which it can be separated only with difficulty, TLC of derivatives can often achieve success. For instance, acids can be purified as methyl esters; alcohols and amines as acetyl compounds. Derivatives of this sort are more easily eluted than the polar starting materials.

Fig. 91 shows photodensitometric curves of the autoradiograph of a thin-layer chromatogram made of ^{14}C-tagged tripalmitin, before and after purification.

These experiments have shown the advantage to be gained by de-activating layers through leaving them at least one day in the air, in the chromatography of substances containing ester groups. The esters are slightly hydrolysed during chromatography on freshly prepared, highly active silica gel.

Amounts of several grams are best purified by column chromato-

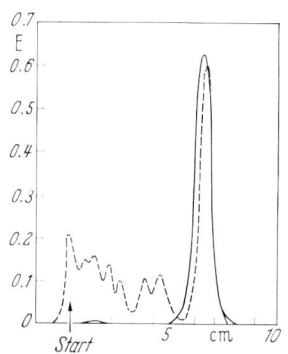

Fig. 91. Purification of radioactive tripalmitin through TLC [426]. Densitometer curves of the autoradiograph of a chromatogram from purchased (- - - -) and purified (———) tripalmitin

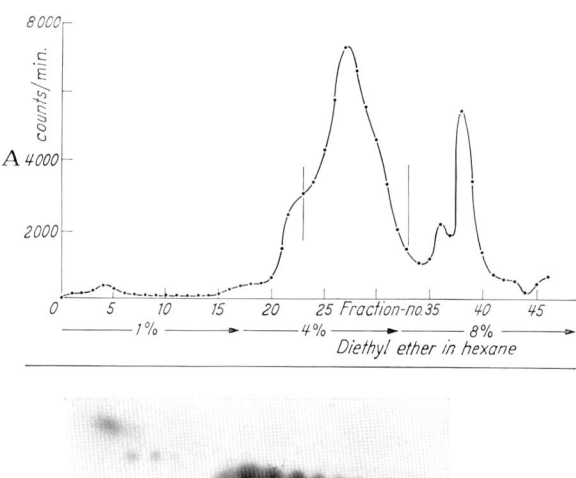

Fig. 92. Purification of triolein containing ^{131}I by means of column chromatography [724]. Aliquot portions of eluate fractions were counted overnight (A) and analysed within 2 h through autoradiography of a thin-layer chromatogram (B)

graphy. The separation effect is measured by carrying out a count or, more dependably and often more rapidly, by TLC. An example is seen in Fig. 92, of column chromatographic fractionation of a sample of triolein containing ^{131}I. The commercial product contains iodinated mono- and diolein, the corresponding acetyl compounds, iodinated oleic acid and a little methyl oleate [724].

Fig. 93 shows the separation of tritiated steroids on silica gel G. The distribution of activity was measured on the plate with a special counting tube, the active zones scraped off and the individual substances eluted [617].

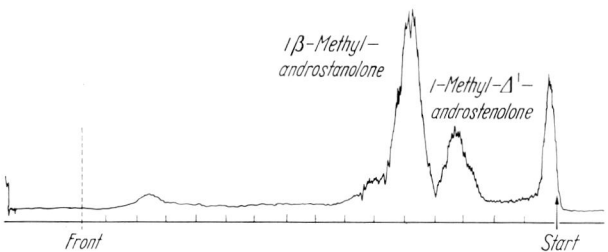

Fig. 93. Isolation by TLC of two steroids, tritiated by the WILZBACH method [617]. Layer: silica gel G (0.9 mm thick); solvent: cyclohexane-ethyl acetate (60 + 40); time of run: 40 min; amount: 8 mg of the reaction mixture. The distribution of activity was measured directly on the layer with a counting tube. Zones showing activity were scraped off and eluted

Radioactively tagged compounds of value, e. g., steroids, can be recovered from the scintillation solutions with the help of TLC [580].

The homogeneity of the substances purified by TLC should be tested with the help of other methods. Compounds isolated through adsorption-TLC ought always to be analysed with partition TLC, paper or gas chromatography.

V. Analysis by Means of Radioisotopes

The following radiochemical analytical procedures can be distinguished according to the way in which the radioactive material is used [101, 387, 460, 611]:

Indicator analysis.
Isotope dilution method.
Activation analysis.
Isotopic derivative method.

These techniques have been often used in combination with PC; there have been few instances to date of the use in TLC.

1. Indicator Analysis

This procedure is suitable for testing the resolution of chromatographic methods.

One can demonstrate, for example, that the fractions of a naturally occurring lipid mixture, separated by TLC, are not contaminated by one another. This is shown in Fig. 94. A small amount of "hot" tripalmitin was mixed with shark liver oil and separation then carried out with adsorption TLC. An autoradiograph of the chromatogram showed that the total radioactivity was in the triglyceride fraction. The glyceryl ether diesters were not contaminated with the triglycerides despite their very similar structure.

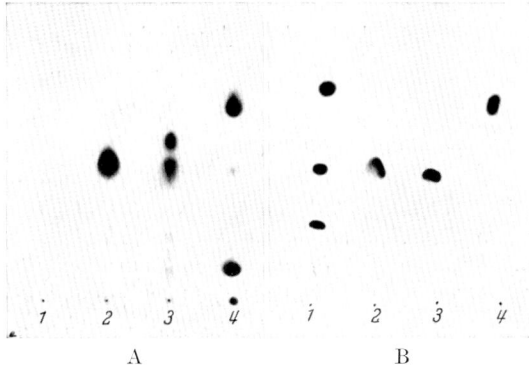

Fig. 94. Test of the resolution of thin-layer chromatographic separations [725]. Layer: silica gel G; solvent: petrol ether (BP 60—70° C)-diethyl ether-acetic acid (80 + 20 + 1); time of run 1 h; indicator: A iodine vapour; B autoradiography; amounts: about 200 µg of each of the naturally occurring lipids; *1* cholesteryl palmitate-1-^{14}C, tripalmitin-1-^{14}C and palmitic acid-1-^{14}C; *2* human depot fat and tripalmitin-1-^{14}C; *3* dogfish shark liver oil and tripalmitin-1-^{14}C; *4* lipid extract of the "calcification" of a human aorta and cholesteryl palmitate-1-^{14}C

2. Isotope Dilution Method

This procedure is used especially in biochemistry for the quantitative determination of small amounts of compounds which cannot be performed accurately by any other method. The substance to be determined must be available in radioactive form and its activity must be known. A weighed amount of the radioactive compound is mixed with the sample and some of the compound to be determined is then isolated in a pure state. The amount of compound originally present (M_1) in the mixture for analysis can be calculated from the formula:

$$M_1 = M_2 \left(\frac{A_2}{A_3} - 1 \right)$$

where M_2 = amount of radioactive substance added, A_2 = its activity and A_3 = specific activity of the isolated pure compound. The amount of radioactive material added is usually so small that it can be neglected in relation to the amount of "cold" substance. The ratio of the activities of the added and the isolated substance thus gives the yield of the separation; the amount of the compound to be determined is therefore easy to work out.

The isotope dilution method would appear suitable for quantitative determination of substances which can be isolated through TLC even when in only small amounts.

In the reverse isotope dilution method, radioactive material for analysis is mixed with a known amount of the "cold" component to be determined.

3. Activation Analysis

Non-radioactive elements can be converted into radioactive isotopes by neutron bombardment in a reactor, cyclotron or van de Graaff generator; they can then be detected by radiometric methods. Phosphorus-, sulphur- and chlorine-containing compounds for example, have been detected on paper chromatograms in this way. In order to determine traces of a compound quantitatively, known amounts of it are irradiated on the same chromatogram.

Applications of activation analysis in TLC are not so far known. The multitude of elements in adsorbent, binder and glass plate would demand an extensive modification of the chromatographic procedure in order to prevent formation of long lived, active isotopes in the carrier material. The advantages of TLC, like high layer capacity and speed would be virtually forfeited in relation to the high sensitivity of the measuring process and the appreciable expense associated with the activation.

4. Isotopic Derivative Method

The isotopic derivative method may help where other methods fail because too little material is available for analysis or the concentration of the component to be determined is too low in the sample. The procedure is useful even where the isotope dilution technique cannot be employed because the substance to be determined is not available in radioactive form.

The method is based on the reaction of an inactive element, radical or molecule with a radioactive reagent, yielding a "hot" derivative, which can be detected and determined with radiometric measuring methods.

The analysis can be carried out according to one of the following techniques:

a) Fractionation before Radioactive Labelling

The inactive sample is separated and the fractions obtained are treated with a radioactive reagent, yielding products which are susceptible to radiometric measurement.

Heavy metal ions, for example, can be separated, treated on the layer with $H_2{}^{35}S$ and identified and quantitatively determined as radioactive sulphides (Fig. 95).

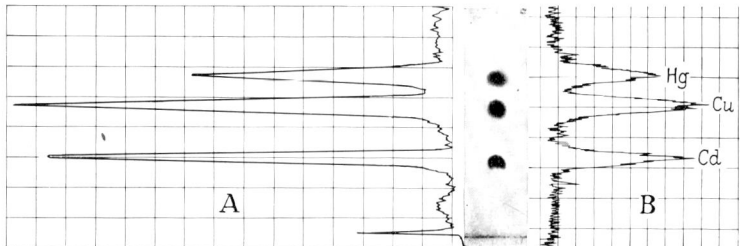

Fig. 95. Separation of cadmium, copper and mercury salts on a coated microscope slide [379]. Layer: silica gel G; solvent: n-butanol-1.5N HCl-acetonylacetone (100 + 20 + 0.5); time of run 2 h; detection with $H_2{}^{35}S$. The chromatogram was evaluated by means of a Geiger-Müller counting tube (B) and the autoradiograph photometrically (A)

b) Separation of Radioactive Derivatives

One milligram or less of a mixture of compounds is first treated with a radioactive reagent and the "hot" derivatives then fractionated. 3-Chloroanisoyl-3-^{36}Cl, 4-iodobenzoylchloride-^{131}I and, in particular, acetic anhydride-1-^{14}C or -2-^{3}H, have been recommended as radioactive reagents for labelling inactive hydroxy- and amino-compounds. Procedures for acetylating sterols and aliphatic lipids are given on p. 175. Acids can be reacted with 4-chloroaniline-^{36}Cl or with ^{14}C-labelled diazomethane, $^{14}CH_2N_2$ (tritiated diazomethane cannot be used (see however [355])). Directions for esterifying higher fatty acids with ^{14}C-tagged diazomethane are on p. 175.

The author [423] has described a scheme for analysing radioactively labelled lipid derivatives by means of TLC or column chromatography and PC: mixtures of fatty alcohols, mono- and diglycerides and other lipids capable of acetylation are reacted with radioactive acetic anhydride and the acetyl derivatives fractionated according to compound classes by adsorption chromatography on silica gel layers or silica columns. The relative amounts of the classes of acetylated lipids are determined by measuring the eluate activity. Each of these compound classes can be further separated by chromatography on siliconised paper. The quantitative evaluation of paper chromatograms of radioactive lipid derivatives is carried out with the help of gas flow proportional counters, equipped for automatic transport of the paper strip and for recording the results.

Fatty acids in castor oil have been determined in a like fashion [426]. They were reacted with radioactive diazomethane. The methyl esters of the ordinary fatty acids were separated from those of the monohydroxy- and dihydroxy acids, using adsorption TLC. All three classes of methyl ester were eluted and the ratio of the amounts determined by measuring their radiation intensities. The mixture of esters of non-hydroxy-acids was then fractionated further by reversed phase PC and its composition quantitatively determined in the gas flow proportional counter. The ratio of monohydroxy- to dihydroxy- fatty acids may be more easily determined through the "hot" acetylated acids or esters. The small amount of dihydroxy fatty acids could then be labelled with two radioactive acetyl groups per molecule, which would double the specific activity and increase the accuracy of the result still further.

The isotopic derivative method is suitable for analysis even of complex naturally occurring mixtures. Thus, classes of hydroxy compounds in vegetable oils have been fractionated through TLC after acetylation with tagged acetic anhydride, identified and quantitatively determined by counting in suspension [419].

c) Fractionation after Addition of a Radioactive Derivative to the Mixture of Non-Labelled Derivatives

The sample for analysis is treated with "cold" reagent, yielding a mixture of derivatives; a small amount of the radioactively tagged derivative of the component to be determined is added to this mixture. The derivative (hot + cold) is isolated and its activity determined; the amount of original "cold" derivative can then be calculated with the help of the formula for the isotope dilution method (p. 171) and hence the amount of the original substance in the sample.

d) Separation after Addition of an Inactive Derivative to the Mixture Containing the Radioactively Labelled Derivative of the Compound to be Determined

This procedure reverses the isotope dilution method. A mixture is treated with a radioactive reagent yielding quantitatively a stable derivative of the desired constituent. Excess of the unlabelled derivative is added and the derivative isolated in a pure state. From the amount of unlabelled derivative added and the activities of the pure isotopic derivative and of the mixture with inert derivative, the amount of the original radioactive derivative and hence of the original constituent, can be calculated. This procedure was developed more than 20 years ago and has proved of especial value in analyses of protein hydrolysates.

e) Use of Two Radioactive Isotopes

The material for analysis is treated with a radioactive reagent giving a reaction mixture which contains a "hot" derivative of the constituent to be determined. Some of this same derivative, but labelled with another isotope, is then added to the reaction mixture. After fractionating the mixture, the radiation intensities of each radioactive isotope in a fraction and the total activity of the same fraction, are measured. From the results, the efficiency of the fractionation and the amount of the original constituent present can be calculated. The procedure was developed for analysing protein hydrolysates. As radioactive reagents, ^{131}I- and ^{35}S-tagged p-iodobenzenesulphonyl chloride ("pipsyl chloride") and ^{3}H- and ^{14}C-labelled acetic anhydride are used.

VI. Directions for Radioactive Labelling

The procedures given here are suitable for labelling small amounts of inactive acids and alcohols by "hot" esterification. Acids are reacted with radioactive diazomethane, alcohols with radioactive acetic anhydride.

1. Esterification of Acids with Diazomethane (^{14}CH$_2$N$_2$)

Radioactive diazomethane can be obtained from nitrosomethyl-^{14}C-urea. The more stable N-methyl-^{14}C-N-nitroso-p-toluenesulphonamide (radioactive "diazald") is however preferable. Nitrosomethyl-^{14}C-urea and radioactive diazald are commercially available (Firms 76, 92, respectively).

A test tube with fused-on side arm, or a micro gas generator serve for preparing and distilling radioactive diazomethane.

Procedure

A solution of 10 mg (0.05 mmole) N-methyl-^{14}C-N-nitroso-p-toluenesulphonamide (specific activity about 0.6 mCi/mmole) in 2 ml of diethyl ether, is mixed with 2 ml of an ice-cooled solution of 0.1 g sodium hydroxide in ethanol-water (10 + 1) and heated in a stream of nitrogen to 60—70° C on a water bath. The diazomethane which distils over is collected in two test tubes joined in series, each containing 1—2 ml ice-cooled diethyl ether. The two ethereal diazomethane solutions are combined and used *immediately* for esterifying the fatty acids in a mixture of diethyl ether-methanol (9 + 1).

Diazomethane forms polymethylenes easily and other substances of as yet unknown structures. Pure methyl esters can be separated quickly from these side products by means of adsorption TLC.

A procedure for preparing ^{3}H-labelled methyl esters by reaction with tritium-containing water and inactive diazomethane, has been published recently [355].

It is frequently preferable to reduce acids or esters quantitatively to alcohols using lithium aluminium hydride and then to label these with radioactive acetic anhydride.

2. Acetylation of Alcohols with Acetic Anhydride (^{14}CH$_3$CO)$_2$O or (C^{3}H$_3$CO)$_2$O

Acetylation is particularly easy to carry out. ^{14}C- or ^{3}H-labelled anhydride can be obtained from several firms [312].

Procedure

A mixture of 10—20 mg of a lipid sample containing compounds with free alcohol or amine groups and a 20% excess of a solution of acetic anhydride-1-^{14}C (specific activity 0.6 mCurie/mmole) in pyridine (1 + 10) is sealed into a glass ampoule, 5 × 150 mm. The ampoule is heated 30—60 min at 100° C. After cooling and opening, the reaction mixture is diluted with 10 ml N sulphuric acid. The acetylated lipids are extracted with several portions of ether, the combined extract washed neutral with water and dried over anhydrous sodium sulphate.

Compounds containing ester groups, e. g., mono- and diglycerides, undergo slight acetolysis; the acetylation of these compounds should thus be used with great care. Purer acetylation products can be obtained by reaction with radioactive ketene [16].

A special apparatus has been described for acetylating amino acids under mild conditions [751].

VII. Applications of TLC in Chemical and Biochemical Investigations with Radioisotopes

TLC is ideally suited for analysing the products of radiochemical syntheses and isolation of labelled compounds from complex reaction mixtures. The method has proved its value, for example, in the preparation of ^{3}H- or ^{14}C-tagged fatty acids [628], cholesterol esters [416], glyceryl ethers [425], phospholipids [569] and steroids [328].

The autoradiograph in Fig. 96 shows the analysis of the intermediate products formed in a synthesis of several stages. Radioactively labelled chimyl alcohol was prepared in accordance with the following scheme ($R=C_{15}H_{31}-$):

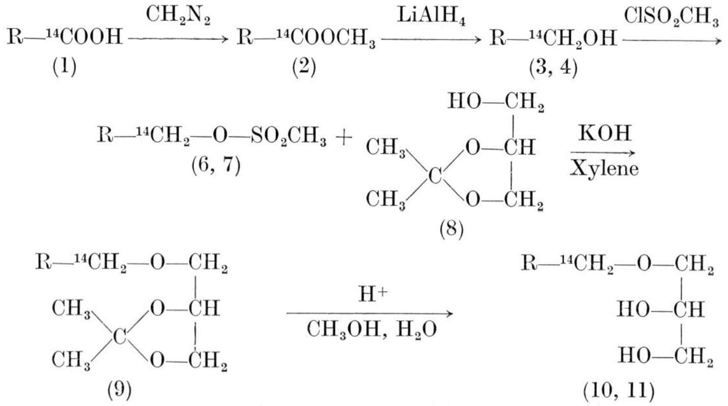

$$R-^{14}COOH \xrightarrow{CH_2N_2} R-^{14}COOCH_3 \xrightarrow{LiAlH_4} R-^{14}CH_2OH \xrightarrow{ClSO_2CH_3}$$
$$(1) \qquad\qquad (2) \qquad\qquad (3, 4)$$

$$R-^{14}CH_2-O-SO_2CH_3 \;+\; \begin{array}{c}CH_3\\ \\ CH_3\end{array}\!\!>\!\!C\!\!<\!\!\begin{array}{c}O-CH_2\\ |\\ HO-CH\\ |\\ O-CH_2\end{array} \xrightarrow{KOH}_{Xylene}$$
$$(6, 7) \qquad\qquad\qquad (8)$$

$$\begin{array}{c}R-^{14}CH_2-O-CH_2\\ |\\ CH_3\!\!>\!\!C\!\!<\!\!O-CH\\ CH_3 \quad\;\;\;\; O-CH_2\end{array} \xrightarrow[CH_3OH,\,H_2O]{H^+} \begin{array}{c}R-^{14}CH_2-O-CH_2\\ |\\ HO-CH\\ |\\ HO-CH_2\end{array}$$
$$(9) \qquad\qquad\qquad\qquad\qquad (10, 11)$$

The composition of all the intermediates, including recrystallised compounds and their filtrates, was determined [425] by scintillation counting of scraped-off zones [647]. The amounts needed for the analysis did not impair the yield (40%).

It has likewise been possible to separate and analyse mixtures of water-soluble labelled substances, using TLC. An example is the ^{131}I-labelled S-sulphonated A- and B-chains of insulin, using TLC on silica gel G or Amberlite IR-120 [435].

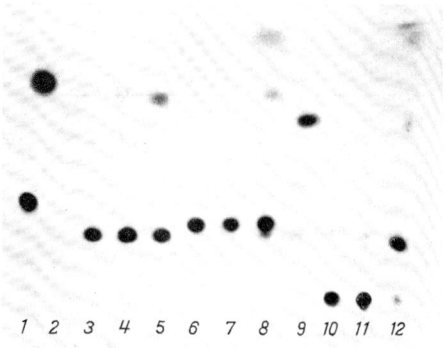

Fig. 96. Application of TLC in the control of the synthesis of radioactively labelled chimyl alcohol (α-hexadecyl glycerol ether) from palmitic acid-1-^{14}C [425]. (see reaction scheme above) Layer: silica gel G; solvent: petrol ether (BP 60—70° C)-diethyl ether-acetic acid (80 + 20 + 1); time of run 1 h; autoradiograph. *1* palmitic acid; *2* methyl palmitate; *3* crude palmityl alcohol; *4* recrystallised palmityl alcohol; *5* filtrate from (*3*); *6* crude palmityl mesylate; *7* recrystallised palmityl mesylate; *8* filtrate from (*6*); *9* crude acetone ketal of chimyl alcohol; *10* crude chimyl alcohol; *11* recrystallised chimyl alcohol; *12* filtrate from (*10*)

Several authors have reported the analysis and purification by TLC, of substances which had been tritiated according to the Wilzbach procedure. Thus glycerol ethers [262], the insect hormone ecdysone [338], vitamin D_3 [501] and other steroids [501, 513, 617] (see Fig. 93, p. 170) and cardiac glycosides [551] have been purified on silica gel G layers. Unsaturated fatty acids have been isolated by TLC on silver nitrate-impregnated silica gel layers [425, 628].

In recent years, TLC has been used most successfully as an analytical tool in research on the biosynthesis of long-chain fatty acids in microorganisms [188, 307, 442], and in plant [497] and animal tissues [150, 160, 210, 276]. The method has become very nearly indispensable in investigations of the metabolism of neutral lipids and phospholipids [271, 694, 712]. Thus it has been shown for example, that the methyl esters of higher fatty acids can be taken up as such by the animal body, i. e., without having to be hydrolysed [159]: Guinea pigs were fed on radioactive methyl elaidate-1-^{14}C. After four hours, the animals were killed and the lipids extracted from the blood, liver, kidneys, lungs, heart, spleen and depot fat. The cholesterol esters, triglycerides, sterols and phospholipids were isolated by chromatography on a silica gel column and their respective radioactivities measured. Almost the entire activity was found in the cholesterol ester fraction. It could however be shown by TLC that the "cholesterol esters" contained radioactive methyl elaidate and that this alone was "hot". These results can be seen clearly in Table 19:

Table 19. *Adsorption of elaidic acid in the animal body; testing the eluate from a silica gel column by means of TLC* [159]

Solvent for elution from the silica gel column	Identified through TLC	Specific activity (dps/mg)
2% Diethyl ether in pentane	Cholesterol esters	0.5
	methyl esters of higher fatty acids	708
5% Diethyl ether in pentane	Triglycerides	4.1
20% Diethyl ether in pentane	Sterols	0.4
Methanol	Phospholipids	3

Cholesterol esters and the methyl esters were separated by TLC on silica gel G using hexane-diethyl ether (98.5 + 1.5).

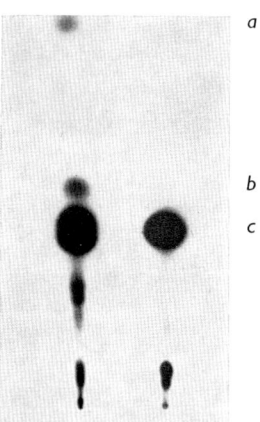

Fig. 97. Use of TLC in following the biosynthesis of terpenes in *Mentha piperita* [43]. Peppermint plants were grown in an atmosphere of $^{14}CO_2$ and radioactive pulegone isolated from the essential oil of these plants, using preparative TLC. This compound was incubated with sections of young leaves from peppermint plant shoots. The autoradiograph of a chromatogram demonstrates formation of *menthone* (b) and *menthofuran* (a) from *pulegone* (c) in this leaf tissue

TLC has proved a useful aid also in biological work with labelled terpenes [43, 58]. An example is shown in Fig. 97.

WINTERSTEIN and coworkers [764] have isolated a series of carotenoid aldehydes from plant and animal tissues. This was rendered possible in some cases only by a markedly improved technique, especially by using TLC and radioactive derivatives [762].

TLC has played a significant part in research on the biosynthesis and metabolism of steroids and bile acids [255, 377, 444], gibberellins [243], macrolides [7] and aromatic carboxylic acids [253, 254]. Austrian research workers reported its use in studies of the biogenesis of lignin [381, 382] already in 1960. They injected coniferin-3-^{14}C into the branches of a birch tree and, with the help of TLC, isolated two radioactive products which had been formed in the plant by methylation [381]. TLC has been used also in a study of the biogenesis of isoflavones [252].

Recent publications have reported the application of TLC in research on biosynthesis of inositol [180] and in an investigation of the metabolism of glucose [384]. The autoradiograph of a thin-layer chromatogram of a ^{32}P-labelled ribo- and deoxyribonucleoside triphosphate from *E. coli* is shown in Fig. 98.

Metabolites of labelled pharmaceuticals have been detected by TLC in experiments on animals [272, 732]. The method has been used also in the investigation of the choleretic activity of a preparation, using "hot" bile acids [376].

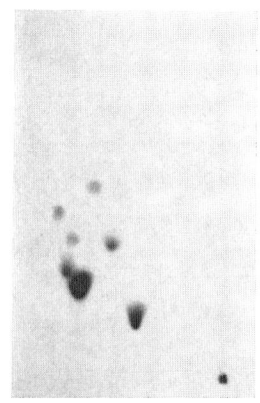

Fig. 98. Separation through ion exchange-TLC, of ^{32}P-labelled ribo- and deoxyribonucleoside triphosphate from *E. coli* (details in Chapter W)

In 1961/62, the considerable impurity of many of the radioactively labelled compounds commercially available at that time, was demonstrated with the help of TLC [424, 426]. The first explanation was a radiolytic(self-) decomposition [644]. Two to three years ago, the manufacturers of radioactive substances began to analyse their products by TLC. It soon became evident that the commercial preparations were contaminated largely with intermediates from the syntheses and not with products formed by radiolysis.

Although pure labelled products are obtainable nowadays, one should make sure before using them that they are chemically *and* radiochemically uniform.

Several authors have reported the application of radioactive substances and TLC in clinical analysis and diagnosis [190, 724, 730]. Others have tested and compared the accuracy of routine chemical methods with isotope techniques [399, 419, 634, 728].

TLC has been used in inorganic chemistry for separating radioactive cations [93, 379, 455, 627] (see Fig. 95, p. 173) and anions [111, 455].

Theoretical Fundamentals of TLC*

About 60 pages of the general section of the 1st edition [675] were devoted to the theoretical fundamentals of TLC. The state of knowledge

---

* Editor's comment.

concerning the chromatographic behaviour of a substance in relation to its chemical structure, was discussed also. The significance and application of the Martin-relation for partition chromatographic separations was discussed in this chapter, written by BRENNER and co-workers, and a table for converting Rf-into Rm-values was given there also. For two reasons, "theoretical fundamentals" has not been reprinted in this edition. First, shortening did not seem advisable on the grounds of comprehension; secondly, there has been such an enormous growth of material from other TLC domains. The more frequently used Rf-Rm conversion table has been retained and is to be found at the end of the book.

Bibliography of the General Section. Chapters A—I

1. ABBOTT, D. C., and J. THOMSON: Analyst **89**, 613 (1964).
2. — — Chem. & Ind. (London)**1965**, 310.
3. — — Chem. & Ind. (London) **1964**, 481.
4. — H. EGAN, E. W. HAMMOND, and J. THOMSON: Analyst **89**, 480 (1964).
5. — —, and J. THOMSON: J. Chromatog. **16**, 481 (1964).
6. ABRAMSON, D., and M. BLECHER: J. Lipid Res. **5**, 628 (1964).
7. ACHENBACH, M., u. H. GRIESEBACH: Z. Naturforsch. **19**b, 561 (1964).
8. ACHREM, A. A., and A. I. KUZNETSOVA: Russ. Chem. Rev. **32**, 366 (1963).
9. — — Thin-layer chromatography (Russian). Moscow: Isdatelsvo Nauka 1964.
10. ADAM, G., u. K. SCHREIBER: Z. Chemie **3**, 100 (1963).
11. ADAMEC, O., J. MATIS, and M. GALVANEK: Steroids **1**, 495 (1963).
12. AELION, R., A. LOEBEL et F. EIRICH, Rec. trav. chim. **69**, 61 (1950).
13. Agfa company: Manual for fast copying with Copyrapid paper, Agfa A.G., Leverkusen-Bayerwerk, W. Germany.
14. AIKEN, W. H.: Ind. Eng. Chem. **35**, 1206 (1943).
15. ALBERT-RECHT, F., and J. A. OWEN: Clin. Chim. Acta **10**, 577 (1964).
16. ALDERHOUT, J. J. H., G. K. KOCH, and A. H. W. ATEN JR.: Rec. trav. chim. **76**, 712 (1957).
17. ALEXANDER, G. B., W. M. HESTON, and R. K. ILER: J. Phys. Chem. **58**, 453 (1954).
18. ALM, R. S., R. J. P. WILLIAMS, and A. TISELIUS: Acta Chem. Scand. **6**, 826 (1952).
19. AMENTA, J. S.: J. Lipid Res. **5**, 270 (1964).
20. ANACKER, W. F., u. V. STOY: Biochem. Z. **330**, 141 (1958).
21. ANDREWS, P.: Biochem. J. **91**, 222 (1964).
22. ANGUERA, P., y L. CODERN: Medicamenta (Madrid) **25**, 203 (1961).
23. APPLEWHITE, T. H., M. J. DIAMOND, and L. A. GOLDBLATT: J. Am. Oil Chemists' Soc. **38**, 609 (1961).
24. ARX, E. v., u. R. NEHER: J. Chromatog. 8, 145 (1962).
25. — — J. Chromatog. **12**, 329 (1963).
26. ATTAWAY, J. A., R. W. WOLFORD and G. J. EDWARDS: Anal. Chem. **37** 74 (1965).
27. AURENGE, J., M. DEGEORGES et J. NORMAND: Bull. soc. chim. France **1963**, 1732.
28. — — — Bull. soc. chim. France **1964**, 508.

29. *AVICEL Applications Bulletin*, FMC Corporation, American Viscose Division, AVICEL Sales, Marcus Hook, Pa., USA.
30. BACON, M. F.: J. Chromatog. **16**, 552 (1964).
31. BADGER, G. M., J. K. DONNELLY, and T. M. SPOTSWOOD: J. Chromatog. **10**, 397 (1963).
32. BADINGS, H. T.: J. Amer. Oil Chemists' Soc. **36**, 648 (1959).
33. — J. Chromatogr. **14**, 265 (1964).
34. — en J. G. WASSINK: Ned. Melk Zuiveltijdschr. **17**, 132 (1963).
35. BAEHLER, B.: Helv. Chim. Acta **45**, 309 (1962).
36. — Pharm. Acta Helv. **39**, 457 (1964).
37. BANCHER, E., H. SCHERZ u. V. PREY: Mikrochim. Acta **1963**, 712.
38. — — u. K. KAINDL: Mikrochim. Acta **1964/65**, 652.
39. BANG, H. O.: J. Chromatogr. **14**, 520 (1964).
40. BARRETT, C. B., M. S. J. DALLAS, and F. B. PADLEY: Joyce-Loebl Rev. **1**, 8 (1963).
41. — — — J. Am. Oil Chemists' Soc. **40**, 580 (1963).
42. BARROLLIER, J.: Naturwissenschaften **48**, 404 (1961).
43. BATTAILE, J., and W. D. LOOMIS: Biochim. et Biophys. Acta **51**, 545 (1961).
44. BATTISTA, O. A.: Ind. Eng. Chem. **42**, 502 (1950); US-Patent 2.978.446.
45. BAUDLER, M., u. M. MENGEL: Z. anal. Chem. **206**, 8 (1964).
46. — — Z. anal. Chem. **211**, 42 (1965).
47. BAUDLER, M., u. F. STUHLMANN: Naturwissenschaften **51**, 57 (1964).
48. BAUER, R. D., and K. D. MARTIN: J. Chromatog. **16**, 519 (1964).
49. BAUMANN, H.: Beitr. Silikoseforsch. Heft 37, 47 (Bochum 1955).
50. BAUMANN, J., u. G. FORTHMANN: Kieselgur. In: W. FOERST: Ullmanns Encyklopädie der technischen Chemie, Bd. 15, S. 727. 3. Aufl. München-Berlin: Urban & Schwarzenberg 1964.
51. BAUMANN, W. J., and H. K. MANGOLD: In: R. PAOLOTTI, and D. KITCHEVSKY: Advances in Lipid Research. New York-London: Academic Press (in the press).
52. BAUR, E. W.: J. Lab. Clin. Med. **40**, 166 (1963).
53. BAYER, E.: Gas-Chromatographie, 2. Aufl. Berlin-Göttingen-Heidelberg: Springer 1962.
54. BEKERSKY, I.: Anal. Chem. **35**, 261 (1963).
55. BEKESY, N. VON: Biochem. Z. **312**, 100 (1942).
56. BELL, C. E.: Chem. & Ind. (London) **1965**, 1025.
57. BENESI, H. A., and A. C. JONES: J. Phys. Chem. **63**, 179 (1959).
58. BENNETT, R. D., and E. HEFTMANN: J. Chromatog. **12**, 245 (1963).
59. BERGER, J.-A., G. MEYNIEL et J. PETIT: Compt. rend. **255**, 1116 (1962); **257**, 1534 (1963).
60. — — — et P. BLANQUET: Bull. soc. chim. France **1963**, 2662; ibidem **1964**, 3179.
61. — — P. BLANQUET et J. PETIT: Compt. rend. **257**, 1534 (1963).
62. — — et J. PETIT: Bull. soc. chim. France **1964**, 3176.
63. — — — Compt. rend. **259**, 2231 (1964).
64. BERNAUER, W.: Klin. Wschr. **41**, 883 (1963).
65. BEROZA, M., and T. P. MCGOVERN: Chemist Analyst **25**, 82 (1963).
66. BHANDARI, P. R., B. LERCH u. G. WOHLLEBEN: Pharm. Ztg. **107**, 1618 (1962).
67. — Pharm. Ztg. **110**, 687 (1965).
68. BHATNAGAR, J. K., K. K. KAPUR, and C. K. ATAL: Indian J. Pharm. **26**, 103 (1964).
69. BIĆAN-FIŠTER, T., and V. KAJGANOVIĆ: J. Chromatog. **16**, 503 (1964).

70. BIRD, H. L. JR., H. F. BRICKLEY, J. P. COMER, P. E. HARTSAW, and M. L. JOHNSON: Anal. Chem. **35**, 346 (1963).
71. BIRKOFER, L., CH. KAISER, H.-A. MEYER-STOLL u. F. SUPPAN: Z. Naturforsch. **17b**, 352 (1962).
72. BLANDENET, G., and J. P. ROBIN: J. Gas Chromatog. **2**, 225 (1964).
73. BLANK, M. L., J. A. SCHMIT, and O. S. PRIVETT: J. Am. Oil Chemists' Soc. **41**, 371 (1964).
74. BLEECKEN, S., G. KAUFMANN u. K. KUMMER: J. Chromatog. **19**, 105 (1965).
75. BLETZINGER, J. C.: Ind. Eng. Chem. **35**, 474 (1943).
76. BOBBITT, J. M.: Thin-Layer Chromatography. New York: Reinhold Publ. Co. 1953.
77. BOEHM, H.-P., u. G. KÄMPF: Z. physik. Chem. (N.F.) **23**, 257 (1960).
78. — u. M. SCHNEIDER: Z. anorg. u. allgem. Chem. **301**, 326 (1959).
79. BOHNER, DE, L. S., E. F. SOTO, and T. DE COHAN: J. Chromatog. **17**, 513 (1965).
80. BOLL, P. M.: Chemist Analyst **51**, 52 (1962).
81. BOLLIGER, H.-R.: In: E. STAHL: Thin-Layer Chromatography, I. Edition. p. 228. Berlin-Göttingen-Heidelberg: Springer 1965.
82. DE BOER, J. H.: Angew. Chem. **70**, 383 (1958).
82. — J. M. H. FORTUIN, B. C. LIPPENS, and W. H. MEIJS: J. Catalysis **2**, 1 (1963).
84. — — en J. J. STEGGERDA: Proc. Koninkl. Ned. Akad. Wetenschap., Ser. B. **57**, 170 u. 434 (1954).
85. — G. M. M. HOUBEN, B. C. LIPPENS, W. H. MEIJS, and W. K. A. WALRAVE: J. Catalysis **1**, 1 (1962).
86. — J. J. STEGGERDA en P. ZWIETERING: Proc. Koninkl. Ned. Akad. Wetenschap. B **59**, 435 (1956).
87. — — J. M. H. FORTUIN, and P. ZWIETERING: Proc. 2nd Int. Congr. of Surface Activity II, 93 (1957).
88. BOUCKE, G.: Atompraxis **11**, 263 (1965).
89. BOVÉ, J. N.: Bull. soc. chim. biol. **45**, 421 (1965).
90. BRADLEY, D. C.: Polymeric Metal Alkoxides, Organometalloxanes and Organometallanosiloxanes. In: F. G. A. STONE and A. W. G. GRAHAM: Inorganic Polymers, New York: Academic Press 1962.
90a. BRAND, J. M.: J. Chromatogr. **21**, 424 (1966).
91. BRANTE, G.: Nature **163**, 651 (1949).
92. BRAUER, G.: Handbuch der präparativen anorganischen Chemie, 2. Auflage. Bd. 1, S. 619ff. Stuttgart: Ferdinand Enke Verlag 1960.
93. BRECCIA, A., and F. SPALLETTI: Nature **198**, 756 (1963).
94. BRENNER, M., A. NIEDERWIESER u. G. PATAKI: In: E. STAHL: Thin-Layer Chromatography, I. Edition, p. 391. Berlin-Göttingen-Heidelberg: Springer 1965.
95. — — Experientia **16**, 378 (1960).
96. — — Experientia **17**, 237 (1961).
97. — — and G. PATAKI: In: A. T. JAMES, and L. J. MORRIS: New Biochemical Separations, p. 123. London: D. van Nostrand Company Ltd. 1964.
97a. — — — R. WEBER: In: E. STAHL: Thin-Layer Chromatography, I. Edition, p. 123—125. Berlin-Göttingen-Heidelberg: Springer 1965.
98. BRIESKORN, C. H., u. J. BÖSS: Fette, Seifen, Anstrichmittel **66**, 925 (1964).
99. BRINTZINGER, H., u. B. TROEMER: Z. anorg. u. allgem. Chem. **181**, 237 (1929).
100. BROCKMANN, H., u. H. BROCKMANN jr.: Chem. Ber. **94**, 2681 (1961).

101. BRODA, E., u. T. SCHÖNFELD: Die technischen Anwendungen der Radioaktivität. Band 1 u. 2. Leipzig: VEB Deutscher Verlag für Grundstoffindustrie 1962.
102. BRODASKY, T. F.: Anal. Chem. **35**, 343 (1963).
103. VAN DEN BROECK, J.: La Diatomite, 3. Aufl. Paris: Astorg 1960.
104. BROWN, J. F., D. CLARK, and W. W. ELLIOTT: J. Chem. Soc. **1953**, 84.
105. BROWN, L., P. HOLLIDAY, and I. F. TROTTER: J. Chem. Soc. **1951**, 1532.
106. BROWN, J. L., and J. M. JOHNSTON: J. Lipid Res. **3**, 480 (1962).
107. BRUD, W. S.: J. Chromatogr. **18**, 591 (1965).
108. BRUNAUER, ST., P. H. EMMETT, and E. TELLER: J. Am. Chem. Soc. **60**, 309 (1938).
109. BRYANT, L. H.: Nature **175**, 556 (1955).
110. BUCHTELA, K., u. M. LESIGANG: Mikrochim. Acta **1965**, 67.
111. BURIÁNEK, J., u. J. CÍFKA: Z. anal. Chem. **213**, 1 (1965).
112. BUTENANDT, A., E. BIEKERT, H. KÜBLER u. B. LINZEN: Hoppe-Seyler's Z. physiol. Chem. **319**, 238 (1960).
113. CALVERT, R.: Diatomaceous Earth, American Chemical Society Monograph. New York: The Chemical Catalog Co., Inc. 1930.
114. GIDDINGS, J. C., and R. A. KELLER: J. Chromatog. **2**, 626 (1959).
115. CANIĆ, V. D., u. S. M. PETROVIĆ: Z. anal. Chem. **211**, 321 (1965); — — and A. K. BEM: **213**, 251 (1965).
116. CANNON, C. G.: Mikrochim. Acta **1955**, 555.
117. CARELLI, V., A. M. LIQUORI, and A. MELE: Nature **176**, 70 (1955).
118. CARMAN, P. C.: J. Phys. Chem. **57**, 56 (1953).
119. CARNEGIE, P. R., and G. PACHECO: Proc. Soc. Exptl. Biol. Med. **117**, 137 (1964).
120. CARREAU, J.-P., et J. RAULIN: J. Chromatog. **15**, 186 (1964).
121. CATCH, J. R.: Carbon-14 Compounds. Washington: Butterworth & Co. Publishers, Ltd. 1961.
122. CAVINA, G., and C. VICARI: In: G. B. MARINI-BETTÒLO: Thin-Layer Chromatography, p. 180. Amsterdam: Elsevier Publ. Co. 1964.
123. ČERNÝ, V., J. JOSKA, and L. LÁBLER: Collection Czech. Chem. Commun. **26**, 1658 (1961).
124. ČERNY, J.: Českoslov. farm. **13**, 266 (1964).
125. CERRI, O., e G. MAFFI: Boll. chim. farm. **100**, 940 (1961).
126. CHAMBERLAIN, J., A. HUGHES, A. W. ROGERS, and G. H. THOMAS: Nature **201**, 774 (1964).
127. CHEVET, A.: J. phys. radium **14**, 493 (1953).
128. CIMA, L., e R. MANTOVAN: Farmaco (Pavia), Ed pract. **17**, 473 (1962).
129. CLOTTEN, R., u. A. CLOTTEN: Hochspannungselektrophorese. Stuttgart: Georg Thieme 1962.
130. COFFEY, R. G., and R. W. NEWBURGH: J. Chromatog. **11**, 376 (1963).
131. COLEMAN, M. H.: In: Thin Layer Chromatography, a series of articles reprinted from Laboratory Practice, London: United Trade Press Ltd. 1964.
132. COLLET, G.: Compt. rend. **259**, 871 (1964).
133. COLMAN, B., u. W. VISHNIAC: Biochim. et Biophys. Acta **82**, 616 (1964).
134. CONNORS, W. M., and W. K. BOAK: J. Chromatog. **16**, 243 (1964).
135. CONNOLLY, J. P., P. J. FLANAGAN, R. Ó. DORCHAÍ, and J. B. THOMSON: J. Chromatog. **15**, 105 (1964).
136. CONSDEN, R., A. H. GORDON, and A. J. P. MARTIN: Biochem. J. **38**, 224 (1944).
137. — — — Biochem. J. **40**, 33 (1946).
138. CRÉPY, O., O. JUDAS et B. LACHESE: J. Chromatog. **16**, 340 (1964).

139. CRIDDLE, W. J., G. J. MOODY, and I. D. R. THOMAS: Nature **202**, 1327 (1964).
140. CROWE, M. O'L.: Ind. Eng. Chem., Anal. Ed. **13**, 845 (1941).
141. CSALLANY, A. S., and H. H. DRAPER: Anal. Biochem. **4**, 418 (1962).
142. DAB 7 (Deutsches Arzneibuch = German Pharmacopoeia, GDR.) 7. Edition. Berlin: Akademie-Verlag 1964
143. DAHN, H., u. H. FUCHS: Helv. Chim. Acta **45**, 261 (1962).
144. VAN DAM, M. J. D., and S. P. J. MAAS: Chem. & Ind. (London) **1964**, 1192.
145. DANCKWORTT, P. W., u. J. EISENBRAND: Lumineszenz-Analyse in filtriertem ultraviolettem Licht. 7. Auflage. Leipzig: Akademische Verlagsgesellschaft Geest u. Portig 1964.
146. DATI, T., G. DE ANGELIS, P. IPPOLITI e C. LULY: Ricerca sci. **27**, 2988 (1957).
147. DAUVILLIER, P.: J. Chromatog. **11**, 405 (1963).
148. DAVÍDEK, J.: Nature **189**, 487 (1961).
149. — In: G. MARINI-BETTÒLO: Thin-Layer Chromatography. p. 117. Amsterdam-London-New York: Elsevier Publishing Company 1964.
150. DAVIDOFF, F., and E. D. KORN: J. Biol. Chem. **240**, 1549 (1965).
151. DAVIES, B. H.: J. Chromatog. **10**, 518 (1963).
152. DEMOLE, E.: J. Chromatog. **1**, 24 (1958).
153. — J. Chromatog. **6**, 2 (1961).
154. Desaga, C. Nachfolger E. Fecht, 6100 Heidelberg, Hauptstr. 60. W. Germany.
155. DETERMANN, H.: Experientia **18**, 430 (1962).
156. — u. W. MICHEL: Z. anal. Chem. **212**, 211 (1965).
157. DETERS, R.: In: Institut für Baustoffkunde und Stahlbetonbau der Technischen Hochschule Braunschweig, Braunschweig **1962**, Heft 1.
158. DHONT, J. H.: Chem. Techniek (Dordrecht) **15**, 340 (1960).
159. DHOPESHWARKAR, G. A., and J. F. MEAD: J. Lipid Res. **3**, 238 (1962).
160. — — Proc. Soc. Exptl. Biol. Med. **109**, 425 (1962).
161. DICARLO, F. J., J. M. HARTIGAN JR., and G. E. PHILLIPS: Anal. Chem. **36**, 2301 (1964).
162. DIETRICH, C. P., S. M. C. DIETRICH, and H. G. PONTIS: J. Chromatog. **15**, 277 (1964).
163. Distillation Products (Firm 47).
164. DITTRICH, S.: J. Chromatog. **12**, 47 (1963).
165. DOBICI, F., and G. GRASSINI: J. Chromatog. **10**, 98 (1963).
166. DOIZAKI, W. M., and L. ZIEVE: Proc. Soc. Exptl. Biol. Med. **113**, 91 (1963).
167. DONALDSON, K. O., V. J. TULANE, and L. M. MARSHALL: Anal. Chem. **24**, 185 (1952).
168. DORFNER, K.: Ionenaustausch-Chromatographie. Berlin: Akademie-Verlag 1963.
169. — Ionenaustauscher, 2. Aufl. Berlin: Walter de Gruyter & Co. 1964.
170. DOSE, K., u. G. KRAUSE: Naturwissenschaften **49**, 349 (1962).
171. DRAWERT, F., O. BACHMANN, and K.-H. REUTHER: J. Chromatog. **9**, 376 (1962).
172. DRUDING, L. F.: Anal. Chem. **35**, 1744 (1963).
173. DUGGER, D. L., J. H. STANTON, B. N. IRBY, B. L. MCCONNELL, W. W. CUMMINGS, and R. W. MAATMAN: J. Phys. Chem. **68**, 757 (1964).
174. DUNCAN, G. R.: J. Chromatog. **8**, 37 (1962).
175. DYER, T. A.: J. Chromatog. **11**, 414 (1963).
177. ECKENROTH, H.: Arch. Pharm. **224/24**, 623 (1886); see also A. BAEYER u. V. VILLIGER: Ber. deut. chem. Ges. **35**, 1201 (1902); K. FREUDENBERG, Collegium **616**, 353 (1921).

178. EGGERS, J.: Phot. u. Wiss. **10**, 40 (1961).
179. EISENBERG, F. jr.: J. Chromatog. **9**, 390 (1962).
180. — and A. H. BOLDEN: Biochem. Biophys. Res. Communs. **12**, 72 (1963).
181. ELBERT, W. C.: Chemist. Analyst **54**, Nr. 3, 68 (1965).
182. EMICH, F.: Lehrbuch der Mikrochemie. München: Bergmann 1926.
183. ENDRES, H.: Z. anal. Chem. **181**, 331 (1961).
184. — In: K. MACEK u. I. M. HAIS: Stationary Phase in Paper and Thin-Layer Chromatography. Amsterdam: Elsevier Publ. Co. 1965; W. GRAU u. H. ENDRES: J. Chromatogr. **17**, 585 (1965).
185. — W. GRASSMANN u. M. OPPELT: Hoppe-Seiler's Z. physiol. Chem. **317**, 21 (1959).
186. — u. H. HÖRMANN: Angew. Chem. **75**, 288 (1963).
187. ENG, L. F., Y. L. LEE, R. B. HAYMAN, and B. GERSTL: J. Lipid. Res. **5**, 128 (1964).
188. ERWIN, J., and K. BLOCH: J. Biol. Chem. **238**, 1618 (1963).
189. *Euratom Reports EUR 2212e, EUR 2212e* Suppl.
190. EVANS, J. R., R. W. GUNTON, R. G. BAKER, D. S. BEANLANDS, and J. C. SPEARS: Circulation Research **16**, 1 (1965).
191. FELTKAMP, H.: Deut. Apotheker-Ztg. **102**, 1269 (1962); FELTKAMP, H., u. F. KOCH: J. Chromatog.**15**, 314 (1964).
192. FINSTON, H. L., and J. MISKEL: Ann. Rev. Nuclear Sci. **5**, 269 (1955).
192a. FISCHER, L. J., and S. RIEGELMAN: J. Chromatog. **21**, 268 (1966).
193. FISHER, N. E., and A. Y. MOTTLAU: Anal. Chem. **34**, 714 (1962).
194. FLETT, M. ST. C.: J. Soc. Dyers Colourists **68**, 59 (1952).
195. FÖRSTER, T.: Fluoreszenz organischer Verbindungen. Göttingen: Van den Hoeck u. Ruprecht 1951.
196. FOGG, A. G., and R. WOOD: J. Chromatog. **20**, 613 (1965).
197. FOPPIANO, R., and B. B. BROWN: J. Pharm. Sci. **54**, 206 (1965).
198. FOSTER, A. B.: J. Chem. Soc. **1953**, 982.
199. FRANZKE, CL., u. A. JANTZ: Nahrung **8**, 637 (1964).
200. FRAY, G., et J. FREY: Bull. soc. chim. biol. **45**, 1201 (1963).
201. FREI, R. W., and H. ZEITLIN: Anal. Chim. Acta **32**, 32 (1965).
202. FRICKE, R.: Z. anorg. u. allgem. Chem. **175**, 249 (1928); **179**, 287 (1929).
203. — u. G. F. HÜTTIG: Hydroxyde und Oxydhydrate. Leipzig: Akademische Verlagsgesellschaft mbH. 1937.
204. FRODYMA, M. M., and R. W. FREI: J. Chromatog. **15**, 501 (1964).
205. — —, and D. J. WILLIAMS: J. Chromatog. **13**, 61 (1964).
206. FROSCH, B.: Arzneimittel-Forsch. **15**, 178 (1965).
207. — u. H. WAGENER: Z. klin. Chem. **2**, 7 (1964).
208. — — Klin. Wschr. **42**, 901 (1964).
209. FRYDRYCH, R.: Chem. Ber. **97**, 151 (1964).
210. FULCO, A. J., and J. F. MEAD: J. Biol. Chem. **236**, 2416 (1961).
211. GÄNSHIRT, H.: Dissertation, p. 9. Karlsruhe 1953.
212. — F. W. KOSS u. K. MORIANZ: Arzneimittel-Forsch. **10**, 943 (1960).
213. — u. K. MORIANZ: Arch. Pharm. **293/65**, 1065 (1960).
214. — In: E. STAHL: Thin-Layer Chromatography, I. Edition, p. 41. Berlin-Göttingen-Heidelberg: Springer 1965.
215. — Arch. Pharm. **296**, 129 (1963).
216. — Arch. Pharm. **296**, 132 (1963).
217. — and J. POLDERMAN: J. Chromatog. **16**, 510 (1964).
218. — Unpublished.
219. — Lecture: Univ. Saarbrücken 1964.

220. GAMP, A., P. STUDER, H. LINDE u. K. MEYER: Experientia 18, 292 (1962).
221. GAREL, J.-P.: Bull. soc. chim. France 1964, 653.
222. — Bull. soc. chim. France 1965, 1899.
223. GEE, M.: J. Chromatog. 9, 278 (1962).
224. GEISS, F., u. H. SCHLITT: Naturwissenschaften 50, 350 (1963).
225. — A. KLOSE u. A. COPET: Z. anal. Chem. 211, 37 (1965).
226. — H. SCHLITT u. A. KLOSE: Z. anal. Chem. 213, 321 (1965).
227. — — — Z. anal. Chem. 213, 331 (1965).
228. GERDES, H., u. W. STAIB: Klin. Wschr. 43, 744 (1965).
229. — — Klin. Wschr. 43, 789 (1965).
229a. German Patent No. 106383, Max-Planck-Institut für Eiweiß und Lederforschung.
230. GETZ, H. R., and D. D. LAWSON: J. Chromatog. 7, 266 (1962).
231. GILES, C. H., T. J. ROSE, and D. G. M. VALLANCE: J. Chem. Soc. 1952, 3799; F. M. ARSHID, C. H. GILES, S. K. JAIN, and A. S. A. HASSAN: J. Chem. Soc. 1956, 72.
232. GINSBERG, H., W. HÜTTIG u. G. STRUNK-LICHTENBERG: Z. anorg. u. allgem. Chem. 293, 33 u. 204 (1958).
233. GLASSTONE, S.: Sourcebook on Atomic Energy. 2nd ed. Princeton, Toronto, London, New York: D. van Nostrand Company, Inc. 1958.
234. GLEMSER, O., u. G. RIECK: Angew. Chem. 67, 652 (1955).
235. — — Z. anorg. u. allgem. Chem. 297, 175 (1958).
236. — — u. H. LACKNER: Chem. Ber. 92, 662 (1959).
237. Gmelins Handbuch der anorganischen Chemie, 8. Auflage, Silicium. Teil B. Weinheim/Bergstraße: Verlag Chemie GmbH. 1959.
237a. GNEHM, R., H. U. REICH u. P. GUYER: Chimica 19, 585 (1965).
238. GOLDRICK, B., and J. HIRSCH: J. Lipid Res. 4, 482 (1963).
239. GORDON, H. T.: Science 128, 414 (1958).
239a. — J. Chromatog. 22, 60 (1966).
240. GORDON, A. H., and J. E. EASTOE: Practical Chromatographic Techniques. Princeton-Toronto-New York-London: D. van Nostrand Company, Inc. 1964.
241. GOPPELSROEDER, F.: Anregungen zum Studium der auf Capillaritäts- und Adsorptionserscheinungen beruhenden Capillaranalyse. Basel: Helbing und Lichtenhahn 1906.
242. GOUBEAU, J., u. R. WARNCKE: Z. anorg. u. allgem. Chem. 259, 109 (1949).
243. GRAEBE, J. E., D. T. DENNIS, CH. D. UPPER, and CH. A. WEST: J. Biol. Chem. 240, 1847 (1965).
244. GRASSHOF, H.: Deut. Apotheker Ztg. 103, 1396 (1963).
245. — J. Chromatog. 14, 513 (1964).
246. GRASSMANN, W., u. G. DEFFNER: Hoppe-Seyler's Z. physiol. Chem. 293, 89 (1953).
247. — H. HÖRMANN u. A. HARTL: Makromol. Chem. 21, 37 (1956).
248. — H. ENDRES, W. PAUCKNER u. H. MATHES: Chem. Ber. 90, 1125 (1957).
249. — — M. OPPELT u. H. EL SISSI: Leder 10, 149 (1959); other literature quoted there.
250. GREF, C.-G., and J. J. SAUKKONEN: Anal. Biochem. 8, 132 (1964).
251. GRIESSBACH, R.: Chemiker-Ztg. 57, 253 (1933).
252. GRISEBACH, H., u. G. BRANDNER: Z. Naturforsch. 16b, 2 (1961).
253. — u. K.-O. VOLLMER: Z. Naturforsch. 18b, 753 (1963).
254. — — Z. Naturforsch. 19b, 781 (1964).

255. GRUNDY, S. M., E. H. AHRENS JR., and T. A. MIETTINEN: J. Lipid Res. **6**, 397 (1965).
256. GRYNBERG, H., i M. BELDOWICZ: Tluszcze; Srodki Piorace **7**, 188 (1963), C. A. **61**, 8899 G (1964).
257. GÜNTHER, H., and A. SCHWEIGER: J. Food Sci. (in preparation).
258. HABERMANN, E., G. BANDTLOW u. B. KRUSCHE: Klin. Wochschr. **39**, 816 (1961).
259. HAEFELFINGER, P., B. SCHMIDLI u. H. RITTER: Arch. Pharm. **297**, 641 (1964).
260. HÄUSSER, H.: Mitteilungsbl. Ges. deut. Chem., Lebensmittelchem., gerichtl. Chem. **13**, 194 (1959).
261. HAEUSSLER, H.: Lecture Nordwestdeut. Chemiedozenten-Tagung, June 1965.
262. HAIGH, W. G., and D. J. HANAHAN: Biochim. et Biophys. Acta **98**, 640 (1965).
263. HAIS, I. M., u. K. MACEK: Handbuch der Papierchromatographie, Bd. 1 und 2. Jena: VEB-Fischer-Verlag 1958 u. 1960.
264. HALL, R. J.: J. Chromatog. **5**, 93 (1961).
265. HALMEKOSKI, J.: Suomen Kemistilehti B **35**, 39 (1962) u. B **36**, 58 (1963).
266. HALPAAP, H.: Chem.-Ing.-Tech. **35**, 488 (1963).
267. — Chemiker-Ztg. **89**, 835 (1965).
268. HANNIG, K.: In: HOPPE-SEYLER-THIERFELDER: Handbuch der physiologisch- und pathologisch-chemischen Analyse II/1, S. 143. Berlin-Göttingen-Heidelberg: Springer-Verlag 1960.
269. HANSBURY, E., J. LANGHAM, and D. G. OTT: J. Chromatogr. **9**, 393 (1962).
270. HANSBURY, E., D. G. OTT, and J. D. PERRINGS: J. Chem. Educ. **40**, 31 (1963).
271. HANSEN, I. A.: Arch. Biochem. Biophys. **110**, 485 (1965).
272. HANSSON, E., P. HOFFMANN, and L. KRISTERSON: Acta Pharmacol. Toxicol. **22**, 231 (1965).
273. HARA, S.: Japan Analyst **12**, 199 (1963).
274. — M. TAKEUCHI, M. TASCHIBANA, and G. CHIHARA: Chem. Pharm. Bull. (Japan) **12**, 483 (1963).
275. — H. TANAKA, and M. TAKEUCHI: Chem. Pharm. Bull. (Japan) **12**, 626 (1964).
276. HARLAN, W. R. JR., and S. J. WAKIL: J. Biol. Chem. **238**, 3216 (1963).
277. HASHIMOTO, Y.: Thin-Layer Chromatography (Japanese). Tokyo: Hirokawa-Shotten Ltd. 1962.
278. HASHMI, M. H., M. A. SHAHID, and A. A. AYAZ: Talanta **12**, 713 (1965).
279. HATHWAY, D. E., and J. W. T. SEAKINS: Biochem. J. **72**, 369 (1959).
280. HAUFFE, K.: Angew. Chem. **72**, 730 (1960).
281. HAWORTH, R. D.: Lecture during the Symposium "Current Chemical Research on Plant Phenolics", Egham, (England) 1960
282. HEAYSMAN, L. T., and E. R. SAWYER: Analyst **89**, 529 (1964).
283. HEFTMANN, E.: Chromatography. New York: Reinhold Publishing Company 1961; see also 2. Edition, 1966.
284. HEIDBRINK, W.: Fette, Seifen, Anstrichmittel **66**, 569 (1964).
285. HELMCKE, J.-G.: Naturwissenschaften **41**, 254 (1954).
286. — u. W. KRIEGER: Atlas der Diatomeenschalen im elektronenmikroskopischen Bild. Teil I, 1953, Teil II, 1954. Berlin-Wilmersdorf: Transmare-Photo G.m.b.H.
287. HESSE, G.: Z. anal. Chem. **211**, 5 (1965).
288. — et M. ALEXANDER: Journées Intern. Etude Methodes Séparation Immediate Chromatogr. Paris 1961, 229 (pub. 1962).
289. HEUSSER, D.: Planta med. **12**, 237 (1964).
290. — u. E. JACKWERTH: Deut. Apotheker-Ztg. **105**, 107 (1965).

291. HILTON, J., and W. B. HALL: J. Chromatog. **7**, 266 (1962).
292. HÖRHAMMER, L.: Lecture during the II. Chromatographie-Symposium, Brussels 1962.
293. HÖRHAMMER, L., u. H. WAGNER: Pharmaz. Ztg. **104**, 783 (1959).
294. — — u. G. BITTNER: Deut. Apotheker **14**, 148 (1962).
295. HÖRMANN, H., u. H. v. PORTATIUS: Hoppe-Seyler's Z. physiol. Chem. **315**, 141 (1959); see also **321**, 120 (1960).
296. HOFMANN, A. F.: Anal. Biochem. **3**, 145 (1962).
297. — Biochim. et Biophys. Acta **60**, 458 (1962).
298. — J. Lipid Res. **3**, 391 (1962).
299. HONEGGER, C. G.: Helv. Chim. Acta **44**, 173 (1961).
300. — Helv. Chim. Acta **46**, 1772 (1963).
301. — Helv. Chim. Acta **47**, 2384 (1964).
302. TEN HOOPEN, H. J. G.: Z. Lebensm.-Unters. u. -Forsch. **119**, 478 (1963).
303. Hormuth (Firm 73).
304. HORNUNG, W.: Handbuch der Agfa-Photopapiere. Düsseldorf: Karl Knapp-Verlag 1955.
304a. HORVATH, C.: J. Chromatog. **22**, 52 (1966).
305. HÜTTENRAUCH, R., L. KLOTZ u. W. MÜLLER: Z. Chem. **3**, 193 (1963).
306. — J. SCHULZE: Pharmazie **19**, 334 (1964).
307. HULANICKA, D., J. ERWIN, and K. BLOCH: J. Biol. Chem. **239**, 2778 (1964).
308. HULL, W. Q., H. KEEL, J. KENNEY, and B. W. GAMSON: Ind. Eng. Chem. **45**, 256 (1953).
309. IKAN, R., I. KIRSON, and E. D. BERGMANN: J. Chromatog. **18**, 526 (1965).
310. ILER, R. K.: The Colloid Chemistry of Silica and Silicates, p. 157/158 und 280/281. Jthaka, N. Y.: Cornell University Press 1955.
311. INSCOE, M. N., Anal. Chem. **36**, 2505 (1964).
312. *International Directory of Radioisotopes*, Vol. I and II. Vienna: The International Atomic Energy Agency 1962.
313. ISMAILOV, N. A., i M. S. SHRAIBER: Farmatzija (Moscow) **3**, 1 (1938).
314. JÄNCHEN, D.: J. Chromatog. **14**, 261 (1964).
314a. JACKSON, R.: J. Chromatog. **20**, 410 (1965).
315. JAENICKE, W., u. B. LORENZ: Z. Elektrochem. **65**, 493 (1961).
316. JACOBSOHN, G. M.: Anal. Chem. **36**, 275 (1964).
317. — Anal. Chem. **36**, 2030 (1964)
317a. JAMES, A. T., and L. J. MORRIS: New Biochemical Separations. London: van Nostrand Comp. 1964.
318. JANÁK, J.: Nature **195**, 696 (1962).
319. — J. Gas Chromatog. **1**, No. 10, 20 (1963).
320. — J. Chromatog. **15**, 15 (1964); see also [321].
321. — I. KLIMEŠ, and K. HANÁ: J. Chromatog. **18**, 270 (1965).
322. JANECKE, H., u. L. MAASS-GOEBELS: Z. anal. Chem. **178**, 161 (1960).
323. JASPERSEN-SCHIB, R., u. H. P. JASPERSEN-SCHIB: Schweiz. Apotheker-Ztg. **102**, 339 (1964).
324. JATZKEWITZ, H.: Hoppe-Seyler's Z. physiol. Chem. **326**, 61 (1961).
325. JAYME, G., u. H. KNOLLE: Makromol. Chem. **82**, 190 (1965).
326. JEANES, A., C. S. WISE, and R. J. DIMLER: Anal. Chem. **23**, 415 (1951).
327. JENSEN, A.: Tideskr. Kemi, Bergvesen Met. **1**, 14 (1961).
328. JERCHEL, D., S. HENKE u. KL. THOMAS: In: Sitzungsberichte der Konferenz über Verfahren zur Herstellung und Aufbewahrung markierter Moleküle, Brüssel, 1963. Brüssel: Euratom, EUR 1625e, 1964.
329. JOHANNESEN, B., og A. SANDEL: Medd. Norsk. Farm. Selskap. **23**, 205 (1961).

330. JOHANSSON, B. G., and L. RYMO: Acta chem. scand. **16**, 2067 (1962).
331. — — Acta chem. scand. **18**, 217 (1964).
332. JOHNS-MANVILLE: The Story of Diatomite. New York: Johns Manville 1953.
332a. JONES, C. R.: Chem. and Ind. (London) **1965**, 1999.
333. JORK, H.: Deut. Apotheker-Ztg. **102**, 1263 (1962).
333a. — Z. analyt. Chemie **221**, 17 (1966).
334. KÄMPF, G., u. H. W. KOHLSCHÜTTER: Z. Elektrochem. **62**, 958 (1958).
335. KAINER, F.: Kieselgur, 2. Aufl. Stuttgart: Ferd. Enke Verlag 1951.
336. KAISER, R.: Z. analyt. Chem. **205**, 284 (1964).
337. — Chromatographie in der Gasphase. I. Gas-Chromatographie, 2. Auflage, 1965; II. Kapillar-Chromatographie, 2. Auflage, 1966; III. Tabellen, 1963; IV. Quantitative Auswertung, 1965. Mannheim: Bibliographisches Institut.
338. KARLSON, P., R. MAURER u. M. WENZEL: Z. Naturforsch. **18b**, 219 (1963).
339. KAUFMANN, H. P., u. Z. MAKUS: Fette, Seifen, Anstrichmittel **62**, 1014 (1960).
340. — u. T. H. KHOE: Fette, Seifen, Anstrichmittel **64**, 81 (1962).
341. — u. K. D. MUKHERJEE: Fette, Seifen, Anstrichmittel **67**, 183 (1965).
342. KECK, K., and U. HAGEN: Biochim. et Biophys. Acta **87**, 685 (1964).
343. KEIRS, R. J., R. D. BRITT jr., and W. E. WENTWORTH: Anal. Chem. **29**, 202 (1957).
344. KELEMEN, J., u. G. PATAKI: Z. anal. Chem. **195**, 81 (1963).
345. KEULEMANS, A. I. M.: Gas-Chromatographie; translated and revised by E. CREMER. Weinheim/Bergstr.: Verlag Chemie 1959.
346. KINGDON, F., and R. E. SCHRANZ: Hercules Chemist **1963**, 1 (Fa. 71).
347. KIRCHNER, J. G., and G. J. KELLER: J. Am. Chem. Soc. **72**, 1867 (1950).
348. — — J. M. MILLER, and G. J. KELLER: Anal. Chem. **23**, 420 (1951).
349. — —, and R. G. RICE: J. Agr. Food Chem. **2**, 1031 (1954).
350. — Abstr. of papers, 147. Meet. Am. Chem. Soc. 1964, 2 B.
351. KLAUS, R.: J. Chromatog. **16**, 311 (1964).
352. KLAVEHN, M., u. H. ROCHELMEYER: Deut. Apotheker-Ztg. **101**, 477 (1961).
353. KNEDEL, M., u. A. FATEH-MOGHDAM: Glas-Instrum.-Tech. (GIT) **9**, 675 (1965).
354. KNIGHT, C. S.: Nature **199**, 1288 (1963).
355. KOCH, G. K., and G. JURRIENS: Nature **208**, 1312 (1965).
356. KODAK-PATHÉ: French Patent 1.370.780; ref. C. A. **62**, 2243h (1965).
357. KÖHLER, M., H. GOLDER u. R. SCHIESSER: Z. anal. Chem. **206**, 430 (1964).
358. KOFLER, L.: In: R. WASICKY: Leitfaden für die Pharmakognostischen Untersuchungen im Unterricht und in der Praxis, I. Teil. Leipzig und Wien: Deuticke 1936; s. auch Hdb. d. Mikrochem. Methoden Bd. I/Teil 1. Wien: Springer 1954.
359. KOHLHEPP, E.: Dissertation, Darmstadt 1963.
360. KOHLSCHÜTTER, H. W., P. BEST u. G. WIRZING: Z. anorg. u. allg. Chem. **285**, 236 (1956).
361. — u. G. KÄMPF: Z. anorg. u. allgem. Chem. **292**, 298 (1957).
362. — Chimia (Switz.) **14**, 285 (1960).
363. — H. GETROST u. S. MIEDTANK: Z. anorg. u. allgem. Chem. **308**, 190 (1961).
364. — u. G. HOFMANN: Z. anorg. u. allgem. Chem. **327**, 51 (1964).
365. — u. W. KATZENMAYER: Z. anorg. u. allgem. Chem. **329**, 163 (1964).
366. — A. RISCH, K. UNGER u. K. VOGEL: Z. Elektrochem. **69**, 849 (1965).
367. — u. M. DAUM: Unpublished.
368. KOKOTI-KOTAKIS, E.: Chim. Chronika (Athens, Greece) **27A**, 59 (1962).
369. KONRAD, E., O. BÄCHLE u. R. SIGNER: Liebigs Ann. Chem. **474**, 276 (1929).
370. KORTE, F., u. H. SIEPER: J. Chromatog. **14**, 178 (1964).

371. Kortüm G. u. J. Vogel: Z. physik. Chem. N. F. **18,** 110 (1958).
372. Kortüm, G., u. J. Vogel: Angew. Chem. **71,** 451 (1959).
373. — W. Braun u. G. Herzog: Angew. Chem. **75,** 653 (1963).
374. Korzun, B. P., L. Dorfman, and S. M. Brody: Anal. Chem. **35,** 950 (1963).
375. — and S. Brody: J. Pharm. Sci. **53,** 454 (1964).
376. Koss, F. W., G. Beisenherz, R. Engelhorn u. U. Chuchra: Arzneimittel-Forsch. **12,** 1026 (1962).
377. — — U. Chuchra u. I. Huber: Arzneimittel-Forsch. **14,** 191 (1964).
378. — u. D. Jerchel: Naturwissenschaften **51,** 382 (1964).
379. — — Radiochim. Acta **3,** 220 (1964).
379a. Kottke, B. A., J. Wollenweber, and C. A. Owen: J. Chromatog. **21,** 439 (1966).
380. Kramer, J. K. G., E. O. Schiller, H. D. Gesser, and A. D. Robinson: Anal. Chem. **36,** 2379 (1964).
381. Kratzl, K., u. G. Puschmann: Holzforschung **14,** 1 (1960).
382. — Holz Roh- u. Werkstoff **19,** 219 (1961).
383. Krell, K., and S. A. Hashim: J. Lipid Res. **4,** 407 (1963).
384. Kuzuya, T., E. Samols, and R. H. Williams: J. Biol. Chem. **240,** 2277 (1965).
385. Lábler, L.: In: G. B. Marini-Bettòlo: Thin-Layer Chromatography. Amsterdam-London-New York: Elsevier Publ. Co. 1964.
386. — a Vl. Schwarz: Chromatografie na tenké vrstvé. Praha: Nakladatelstvi Československé Akademie VĚD 1965.
387. Lambie, D. A.: Techniques for the Use of Radioisotopes in Analysis. Princeton, Toronto, New York, London: D. van Nostrand Company, Inc. 1964.
388. Land, E. H.: J. Opt. Soc. Am. **37,** 61 (1947).
389. Lees, T. M., M. J. Lynch, and F. R. Mosher: J. Chromatog. **18,** 595 (1965).
390. — and P. J. deMuria: J. Chromatog. **8,** 108 (1962).
391. Lefemine, D. V., and W. K. Hausmann: Antimicrob. Agents Chemotherapy **1963,** 134.
392. Lehmann, G.: private communication.
393. Lehmann, H., u. H. Dutz: Tonind.-Ztg. **83,** 219 (1959).
394. Lenk, H. P.: Z. anal. Chem. **184,** 107 (1961).
395. Lennart-Harthon, J. G.: Acta Chem. Scand. **15,** 1401 (1961).
396. Lesigang, M.: Mikrochim. Acta **1964,** 34.
397. — u. F. Hecht: Mikrochim. Acta **1964,** 508.
398. Lestienne, A., E. P. Przybylowicz, W. J. Staudenmayer, E. S. Perry, A. D. Baitsholts et T. N. Tischer: In: Société belge des sciences pharmaceutiques: Chromatographie Symposium III. Bruxelles 1964.
399. Levin, E., and C. Head: Anal. Biochem. **10,** 23 (1965).
400. Levy, G. B.: Medical Electronics News March 1963. Reprint from Photovolt Corp. (Fa. 104).
401. Lewin, J. C.: In: R. A. Lewin: Physiology and Biochemistry of Algae. Chapt. 27, p. 445. New York and London: Academic Press 1962.
402. Lichtenberger, W.: Z. anal. Chem. **185,** 111 (1962).
403. Lie Kian Bo, and J. F. Nyc: J. Chromatog. **8,** 75 (1962).
404. Locke, D. C., and C. E. Meloan: Anal. Chem. **36,** 2234 (1964).
405. Loev, B., and K. M. Snader: Chem. & Ind. (London) **1965,** 15.
406. Ludwig, E.: Z. Chemie **5,** 186 (1965).
407. Lüdy-Tenger, F.: Pharm. Acta Helv. **37,** 770 (1962).
408. Lüthi, U., and P. G. Waser: Nature **205,** 1190 (1965).

409. MAAS, H.: Praxis der Polaroid Land Photographie. Seebruck (Chiemsee): Hering-Verlag 1965.
410. MAATMAN, R. W.: see [173].
411. MCCOY, R. N., and E. C. FIEBIG: Anal. Chem. **37**, 593 (1965).
412. MCDONALD, R. S.: J. Phys. Chem. **62**, 1168 (1958); J. Am. Chem. Soc. **79**, 850 (1957).
413. MACEK, K., and I. M. HAIS: Stationary Phase in Paper and Thin-Layer Chromatography. Proceedings of the 2nd Symposium held at Liblice. Prague: Publishing House of the Czechoslovak Academy of Sciences; Amsterdam: Elsevier Publishing Company 1965.
414. MCKIBBINS, S. W., J. F. HARRIS, and J. F. SAEMAN: J. Chromatog. **5**, 207 (1961).
415. MCLAUGHLIN, J. L., J. E. GOYAN, and A. G. PAUL: J. Pharm. Sci. **53**, 306 (1964).
416. MAHADEVAN, V., and W. O. LUNDBERG: J. Lipid Res. **3**, 106 (1962).
417. MAHAPATRA, G. N., and O. M. FRIEDMAN: J. Chromatog. **11**, 265 (1963).
418. MAIER, R., and H. K. MANGOLD: In: REILLEY, C. N.: Advances in Anal. Chem. and Instrum. **3**, 369 (1964).
419. — — Chem. Phys. Lipids (in the press).
420. MALINS, D. C., and H. K. MANGOLD: J. Am. Oil Chemists' Soc. **37**, 576 (1960).
421. — — In: F. J. WELCHER: Standard Methods of Chemical Analysis, Vol. 3, p. 738. Princeton-Toronto-New York-London: van Nostrand and Company 1966.
422. MANEGOLD, E.: Kolloid-Z. **96**, 186 (1941).
423. MANGOLD, H. K.: Fette, Seifen, Anstrichmittel **61**, 877 (1959).
424. — J. Am. Oil Chemists' Soc. **38**, 708 (1961).
425. — W. J. BAUMANN, and C. R. HOULE: Microchem. J. (in the press).
426. — R. KAMMERECK, and D. C. MALINS: Microchem. J., Sympos. Vol. II, 697 (1962).
427. — — J. Am. Oil Chemist's Soc. **39**, 201 (1962).
428. — H. H. O. SCHMID, and E. STAHL: In: GLICK, D.: Methods of Biochemical Analysis, Vol. **12**, p. 393. New York: Interscience Publ. 1964.
429. MARCUCCI, F., and E. MUSSINI: J. Chromatog. **11**, 270 (1963).
430. MARINI-BETTÒLO, G. B.: Thin-Layer Chromatography, a Scientific Report of the Istituto Superiore di Sanitá, Roma. Amsterdam-London-New York: Elsevier Publ. Co. 1964.
431. MARRINAN, H. J., and J. MANN: J. Appl. Chem. (London) **4**, 204 (1954); J. Polymer Sci. **21**, 301 (1956).
432. MARTEN, G.: Pharmazie **10**, 602 (1955).
433. MARTIN, A. J. P., and R. L. M. SYNGE: Biochem. J. **35**, 91 and 1358 (1941).
434. MARY, N. Y., and E. BROCHMANN-HANSSEN: Lloydia **26**, 223 (1963).
435. MASSAGLIA, A., and U. ROSA: J. Labelled Comp. **1**, 141 (1965).
436. MATHIS, C.: Dissertation, Strasbourg 1963.
437. MATTHEWS, J. S., A. L. PEREDA, and A. AGUILERA: J. Chromatog. **9**, 331 (1962).
438. MATTHIAS, W.: Naturwissenschaften **41**, 17 (1954).
439. MEINHARD, J. E., and N. F. HALL: Anal. Chem. **21**, 185 (1949).
440. METZ, H.: Naturwissenschaften **48**, 569 (1961).
441. METZNER, H., u. H. VOLCSIK: Experientia **20**, 104 (1964).
442. MEYER, F., and K. BLOCH: J. Biol. Chem. **238**, 2654 (1963).
443. MEYER, H.: Deut. Lebensm.-Rundschau **57**, 174 (1961).

444. Miettinen, T. A., E. H. Ahrens jr., and S. M. Grundy: J. Lipid Res. **6**, 411 (1965).
445. Milborrow, B. V.: J. Chromatog. **19**, 194 (1965).
446. Miller, J. M., and J. G. Kirchner: Anal. Chem. **23**, 428 (1951).
447. — — Anal. Chem. **24**, 1480 (1952).
448. — — Anal. Chem. **25**, 1107 (1953).
449. — — Anal. Chem. **26**, 2002 (1954).
450. Mima, H.: Kagaku-no-Ryoiki (Tokyo) **17**, 189 (1963).
451. Millett, M. A., W. E. Moore, and J. F. Saeman: Anal. Chem. **36**, 491 (1964).
452. Mistryukov, E. A.: Collection Czech. Chem. Commun. **26**, 2071 (1961).
453. — J. Chromatog. **9**, 311 (1962).
454. Moghissi, A.: Anal. Chim. Acta **30**, 91 (1964).
455. — J. Chromatog. **13**, 542 (1964).
456. Morgan, M. E.: J. Chromatog. **9**, 379 (1962).
457. Morita, K., and F. Haruta: J. Chromatog. **12**, 412 (1963).
458. Morris, C. J. O. R.: J. Chromatog. **16**, 167 (1964).
459. Morrison, J. C., and L. G. Chatten: J. Pharm. Sci. **53**, 1205 (1964).
460. Moses, A. J.: Nuclear Technology in Analytical Chemistry. (International Series of Monographs on Analytical Chemistry, Vol. 2). Oxford-London-New York: Pergamon Press 1964.
461. Mottier, M.: Mitt. Gebiete Lebensm. u. Hyg. **47**, 372 (1956).
462. — Mitt. Gebiete Lebensm. u. Hyg. **49**, 454 (1958).
463. — et W. Potterat: Anal. Chim. Acta **13**, 46 (1955).
464. Mouton, M., S. Jaquard et M. Sagot-Masson: Ann. pharm. franç. **21**, 233 (1963).
465. Müller, K. H., u. H. Honerlagen: Mitt. deut. Pharmaz. Ges. **30**, 202 (1960).
466. Mulryan, H.: Diatomite. In: Encyclopedia of chemical Technology, Vol. 5, p. 33. 1. Ed. New York: The Interscience Encyclopedia, Inc. 1950.
467. Murray, A., III., and D. L. Williams: Organic Syntheses with Isotopes. Vol. 1 and II. New York and London: Interscience Publishers, Inc. 1958.
468. Muus, L. T.: Tidsskr. Textiltek. **13**, 139 (1955).
469. Naff, M. B., and A. S. Naff: J. Chem. Educ. **40**, 534 (1963).
470. Nano, G. M.: In: G. B. Marini-Bettòlo: Thin-Layer Chromatography, p. 138. Amsterdam-London-New York: Elsevier Publishing Company 1964.
471. Neill, J. D., B. N. Day, and G. W. Duncan: Steroids **4**, 699 (1964).
472. Neuhard, J., E. Randerath, and K. Randerath: Anal. Biochem. **13**, 211 (1965).
473. Newmann, A. A.: Brit. J. Phot. **108**, 54 (1961).
474. Newsome, J. W., H. W. Heiser, A. S. Russell, and H. C. Stumpf: Alumina Properties, Technical Paper No. 10, second revision. Pittsburgh Pennsylvania: Aluminium Company of America 1960.
475. Nichaman, M. Z., C. C. Sweeley, N. M. Oldham, and R. E. Olson: J. Lipid Res. **4**, 484 (1963).
476. Niederwieser, A., u. M. Brenner: Experientia **21**, 50 (1965).
477. — — Experientia **21**, 105 (1965).
478. — u. C. G. Honegger: Helv. Chim. Acta **48**, 893 (1965).
479. — J. Chromatog. **21**, 326 (1966).
480. Nigam, I. C., M. Sahasrabudhe, and L. Levi: Can. J. Chem. **41**, 1535 (1963).
481. Nikitin, N. I.: Die Chemie des Holzes. Berlin: Akademie-Verlag 1955.
482. Nishikaze, O., u. Hj. Staudinger: Klin. Wschr. **40**, 1014 (1962).
483. Noll, W., K. Damm u. R. Fauss: Kolloid-Z. **169**, 18 (1960).

484. *Nuclear Chicago Technical Bulletin*, No. 16: How to Use Radioisotopes with Thin-Layer Chromatography. Nuclear Chicago Corp., Des Plaines, Ill., U.S.A.
485. NÜRNBERG, E.: Rev. univ. ind. Santander **4**, 259 (1962).
486. NUSSBAUMER, P. A.: Pharm. Acta Helv. **38**, 758 (1963).
487. NYBOM, N.: Nature **198**, 1229 (1963).
488. — J. Chromatog. **14**, 118 (1964).
489. OELSCHLÄGER, H., J. VOLKE u. G. T. LIM: Arch. Pharm. **298**, 213 (1965).
490. OERTEL, G. W., M. C. TORNERO, and K. GROOT: J. Chromatog. **14**, 509 (1964).
491. ONOE, K.: J. Chem. Soc. Japan, Pure Chem. Sect. **73**, 337 (1952); ref.: Chem. Zentr. **127**, 3958 (1956).
492. OPIENSKA-BLAUTH, J., H. KRACZKOWSKI i H. BRZUSZKIEWICZ: Postepy Biochem. **11**, 211 (1965).
493. OSWALD, N., u. H. FLÜCK: Pharm. Acta Helv. **39**, 293 (1964).
494. — — Sci. Pharm. **32**, 136 (1964).
495. OTTENSTEIN, D. M.: J. Gas Chromatog. **1**, (4) 11 (1963).
496. OTTO, G.: Leder **4**, 1 (1953).
497. OVERATH, P., and P. K. STUMPF: J. Biol. Chem. **239**, 4103 (1964).
498. OVERMAN, R. T., and H. M. CLARK: Radioisotope Techniques. New York, Toronto, London: McGraw-Hill Book Company, Inc. 1960.
499. PAPÉE, D., et R. TERTIAN: Bull. soc. chim. France **1955**, 983.
500. — — In: H. F. MARK, J. J. MCKETTA, and D. F. OTHMER: Encyclopedia of Chemical Technology. 2nd Edit., Vol. 2, p. 41. New York-London-Sydney: Interscience Publishers 1963.
501. PAREKH, C. K., and R. H. WASSERMAN: J. Chromatog. **17**, 261 (1965).
502. PASTUSKA, G., u. H. TRINKS: Chemiker-Ztg. **85**, 535 (1961).
503. — — Chemiker-Ztg. **86**, 135 (1962).
504. PATAKI, G., u. M. KELLER: Helv. Chim. Acta **46**, 1054 (1963).
505. — Helv. Chim. Acta **47**, 784 (1964).
506. — Helv. Chim. Acta **47**, 1763 (1964).
507. PATT, P.: Arzneimittel-Forsch. **15**, 90 (1965).
508. PAULING, L.: The Nature of the Chemical Bond and the Structure of Molecules and Crystals. Third Edition, p. 102. Ithaca, N. Y.: Cornell University Press 1960.
509. PAYNE, S. N.: J. Chromatog. **15**, 173 (1964).
510. PEEREBOOM, J. W. C.: J. Chromatog. **4**, 323 (1960).
511. PELICK, N., H. R. BOLLIGER, and H. K. MANGOLD: In: GIDDINGS, J. C., and R. A. KELLER: Advances in Chromatography. Vol. 3, p. 85. New York: M. Dekker, Inc. 1966.
512. PELKA, J. R., and L. D. METCALFE: Anal. Chem. **37**, 603 (1965).
513. PENG, C. T.: J. Pharm. Sci. **52**, 861 (1963).
514. PERI, J. B.: J. Phys. Chem. **69**, 211 u. 220 (1965).
515. Perkin-Elmer Sorptometer.
516. PETERSON, E. A., and H. A. SOBER: J. Am. Chem. Soc. **78**, 751 (1956).
517. — — Anal. Chem. **31**, 857 (1959).
518. TORE, J. P.: J. Chromatog. **12**, 413 (1963).
519. PETROWITZ, H. J.: Materialprüfung **2**, 309 (1960).
520. — Mitt. deut. Ges. Holzforsch. **48**, 57 (1962).
521. PEYRON, L.: Bull. soc. chim. France **1958**, 889.
522. — Chim. anal. **43**, 364 (1961).
523. — Bull. soc. chim. France **1962**, 891.
524. — Chim. anal. **45**, 186 (1963).

525. PFAFF, J. D., and E. SAWICKI: Chemist Analyst. **54**, 30 (1965).
526. PEIFER, J. J.: Mikrochim. Acta **1962/3**, 529.
527. PFEIFFER, P.: Organische Molekülverbindungen. Stuttgart: Thieme 1927; P. PFEIFFER, O. ANGERN, L. WANG, R. SEYDEL u. K. QUEHL: J. prakt. Chem. **126**, 97 (1930).
528. PFRUNDER, B., R. ZURFLÜH, H. SEILER u. H. ERLENMEYER: Helv. Chim. Acta **45**, 1153 (1962).
529. Pharmacia AB, Uppsala, Sweden, Firm pamphlet: SEPHADEX Thin-Layer Gel-Filtration.
530. Pharmacopoeia Helv. VII, in preparation.
531. Pharmacopoeia Nordica, Editio Danica, Addendum 1965.
532. PIERCE, C.: J. Phys. Chem. **57**, 149 (1953).
533. PEIRCE, F. T.: Trans. Faraday Soc. **42**, 545 (1946).
534. POEL, G. H. VAN DER: Ned. Melk. Zuiveltijdschr. **15**, 98 (1961).
535. POETHKE, W., u. W. KINZE: Arch. Pharm. **297**, 593 (1964).
536. — — Pharmaz. Zentralhalle **103**, 577 (1964).
537. PORGESOVÁ, L., u. E. PORGES: J. Chromatog. **14**, 286 (1964).
538. PRETTRE, M., B. IMELIK, L. BLANCHIN u. M. PETITJEAN: Angew. Chem. **65**, 549 (1953).
539. PREY, V., H. SCHERZ u. E. BANCHER: Mikrochim. Acta **1963**, 567.
540. PRIVETT, O. S., and M. L. BLANK: J. Lipid Res. **2**, 37 (1961).
541. — — and W. O. LUNDBERG: J. Am. Oil Chemists' Soc. **38**, 312 (1961).
542. — — J. Am. Oil Chemists' Soc. **39**, 520 (1962).
543. — — J. Am. Oil Chemists' Soc. **40**, 70 (1963).
544. — — D. W. CODDING, and E. C. NICKELL: J. Am. Oil Chemists' Soc. **42**, 381 (1965).
545. PROCHÁZKA, Z.: Chem. listy **55**, 974 (1961).
546. PRZYBYLOWICZ, E. P., W. I. STAUDENMAYER, E. S. PERRY, A. D. BAITSHOLTS, and T. N. TISCHER: Pittsburgh Conference on Anal. Chem. and Appl. Spectroscopy, 1965.
547. PURDY, S. J., and E. V. TRUTER: Chem. & Ind. (London) **1962**, 506.
548. — — Analyst **87**, 802 (1962).
549. — — Lab. Practice, **1964**, 500.
550. RABENORT, B.: J. Chromatog. **17**, 594 (1965).
551. RABITZSCH, G., u. H. HERZMANN: Liebigs Ann. Chem. **685**, 261 (1965).
552. RADIN, N. S.: J. Lipid Res. **6**, 442 (1965).
553. RAHANDRAHA, TH.: In: Chromatogr. Symp. 2ième Brussels **1962**, 261.
554. — Deut. Apotheker-Ztg. **102**, 1500 (1962).
555. — M. CHANEZ, P. BOITEAU et S. JAQUARD: Ann. pharm. franç. **21**, 561 (1963).
556. RAMSEY, H. A.: Anal. Biochem. **5**, 83 (1963).
557. RANDERATH, K.: Angew. Chem. **73**, 436 u. 674 (1961).
558. — Angew. Chem. **74**, 484 (1962).
559. — Dünnschicht-Chromatographie, 1. Aufl. Weinheim/Bergstr.: Verlag Chemie 1962; 2. Aufl. 1965.
560. — Angew. Chem. **74**, 780 (1962); Biochim. et Biophys. Acta **61**, 852 (1962).
561. — Nature **194**, 768 (1962).
562. RANDERATH, E., and K. RANDERATH: J. Chromatog. **10**, 509 (1963).
563. RANDERATH, K., u. G. WEIMANN: Biochim. et Biophys. Acta **76**, 129 (1963).
564. — Thin-Layer Chromatography, 1st Edit. Weinheim/Bergstr.-New York: Verlag Chemie/Academic Press 1964.
565. — Chromatographie sur couches minces. Paris: Gauthier-Villars éditeurs 1964.

566. RANDERATH, E., and K. RANDERATH: Anal. Biochem. **12**, 83 (1965).
567. RATSHINSKY (Prof.) (Moscow) and Dr. SHRAIBER (Kharkov): Private communications.
568. RAYMOND, S., and B. AURELL: Science **138**, 152 (1962).
569. REHBINDER, D., u. D. M. GREENBERG: Liebigs Ann. Chem. **681**, 182 (1965).
570. REINDEL, F., u. W. HOPPE: Naturwissenschaften **40**, 245 (1953).
571. REITSEMA, R. H.: J. Am. Pharm. Assoc. Sci. Ed. **43**, 414 (1954).
572. — Anal. Chem. **26**, 960 (1954).
573. RICHARDSON, G. S., I. WELIKY, W. BATCHELDER, M. GRIFFITH, and L. L. ENGEL: J. Chromatog. **12**, 115 (1963).
574. RINK, M., and S. HERRMANN: J. Chromatog. **12**, 249 (1963).
575. — — J. Chromatog. **14**, 523 (1964).
576. RITSCHARD, W. J.: J. Chromatog. **16**, 327 (1964).
577. RITTER, H. L., and L. C. DRAKE: Ind. Eng. Chem. Anal. Ed. **17**, 782 (1945).
578. RITTER, F. J., and G. M. MEYER: Nature **193**, 941 (1962).
579. — —, and F. GEISS: J. Chromatog. **19**, 304 (1965).
580. RIVLIN, R. S., and H. WILSON: Anal. Biochem. **5**, 267 (1963).
581. ROCHE, J., S. LISSITZKY et R. MICHEL: In: E. LEDERER: Chromatographie en chimie organique et biologique, Vol. I, p. 321. Paris: Masson et Cie. 1959.
582. RÖMPP, H.: Chemie-Lexikon, 4. Auflage, S. 4047. Stuttgart: Franckhsche Verlagsbuchhandlung 1958.
583. RÖSSEL, T.: Z. anal. Chem. **197**, 333 (1963).
584. ROSENBERG, J., and M. BOLGAR: Anal. Chem. **35**, 1559 (1963).
585. ROSMUS, J., M. PAVLÍČEK, and Z. DEYL: In: G. B. MARINI-BETTÒLO: Thin-Layer Chromatography, p. 119. Amsterdam-London-New York: Elsevier Publishing Co. 1964.
586. ROUCAYROL, J. C., et P. TAILLANDIER: Compt. rend. **256**, 4653 (1962).
587. RUDDAT, M., E. HEFTMANN, and A. LANG: Arch. Biochem. Biophys. **110**, 496 (1965).
588. RUSIECKI, W., i M. HENNEBERG: Farm. polska **18**, 203 (1962).
589. RUSSELL, A. S., and C. N. COCHRAN: Ind. Eng. Chem. **42**, 1336 (1950).
590. RUSSEL, J. H.: Rev. Pure Appl. Chem. **13**, 15 (1963).
591. RYBICKA, S. M.: Chem. & Ind. (Lond.) **1962**, 308.
592. SAALFELD, H.: N. Jahrb. mineral. Abhandl. **95**, 1 (1960).
593. SACHS, L., u. Z. SZEREDAY: J. Chromatog. **18**, 170 (1965).
594. SAHLI, M., u. M. OESCH: Pharm. Acta Helv. **40**, 25 (1965).
595. SALO, T., E. AIRO u. K. SALMINEN: Z. Lebensmittel-Untersuch. u. -Forsch. **125**, 20 (1964).
596. — u. M. SALMINEN: Suomen Kemistilehti A **37**, 161 (1964).
597. SANKOFF, I., and T. L. SOURKES: Can. J. Biochem. Physiol. **41**, 1381 (1963).
598. SAWICKI, E., T. W. STANLEY, and W. C. ELBERT: Occupational Health Rev. **16**, 8 (1964).
599. SAWICKI, E., T. W. STANLEY, W. C. ELBERT and J. D. PFAFF: Anal. Chem. **36**, 497 (1964).
600. — — J. D. PFAFF, and W. C. ELBERT: Chemist Analyst **53**, 497 (1964).
601. — — and H. JOHNSON: Microchem. J. **8**, 257 (1964).
602. SCHANTZ, M. VON: In: [*430*].
603. SEHER, A.: Fette, Seifen, Anstrichmittel **61**, 345 (1959).
604. SCHERZ, H., E. BANCHER u. K. KAINDL: Mikrochim. Acta **1965**, 255.
605. SCHILCHER, H.: Z. anal. Chem. **199**, 335 (1964).
606. SCHILDKNECHT, H., u. O. VOLKERT: Naturwissenschaften **50**, 442 (1963).
607. SCHLEMMER, F., u. E. LINK: Pharm. Ztg. **104**, 1349 (1959).

608. SCHLIERF, G. and P. WOOD: J. Lipid Res. **6**, 317 (1965).
609. SCHMANDKE, H.: J. Chromatog. **14**, 123 (1964).
610. — u. H. GOHLKE: Clin. Chim. Acta **11**, 491 (1965).
611. SCHMEISER, K.: Radionucleide. Berlin-Göttingen-Heidelberg: Springer 1963.
612. SCHMIALEK, P.: In: LENK, H. P.: Private communication.
613. SCHMIDT, O. TH., u. W. SCHÖNLEBEN: Z. Naturforsch. **12 b**, 262 (1957); O. TH. SCHMIDT, P. BECHER u. M. HÜBNER. Chem. Ber. **93**, 1296 (1960).
614. SCHORN, P.-J.: Dissertation, Saarbrücken 1963.
615. — Lecture, III. Internat. Symposium der Chromatographie. Brussels: Sept. 1964.
615a. — Z. analyt. Chem. **205**, 298 (1964).
616. SCHRATZ, E., u. W. EGELS: Planta med. **6**, 148 (1958).
617. SCHULZE, P.-E., u. M. WENZEL: Angew. Chem. **74**, 777 (1962); Internat. Edit. **1**, 580 (1962).
618. SCHWANE, R. A., and R. S. NAKON: Anal. Chem. **37**, 315 (1965).
619. SCOTT, R. P. W., and A. T. JAMES: Lecture, Unilever Res. Symposium, Welwyn/Herts. 1964.
620. SEARS, G. W. JR.: Anal. Chem. **28**, 1981 (1956).
621. SEASE, J. W.: J. Am. Chem. Soc. **69**, 2242 (1947).
622. SEEBOTH, H.: Chem. Tech. (Berlin) **15**, 34 (1963).
623. — u. H. GÖRSCH: Chem. Tech. (Berlin) **15**, 294 (1963).
624. SEHER, A.: Mikrochim. Acta **1961/62**, 310.
625. SEIKEL, M. K., M. A. MILLETT, and J. F. SAEMAN: J. Chromatog. **15**, 115 (1964).
626. SEILER, N., G. WERNER u. M. WIECHMANN: Naturwissenschaften **50**, 643 (1963).
627. SEILER, H., u. M. SEILER: Helv. Chim. Acta **48**, 117 (1965).
627a. SEMENUK, G., and W. T. BEHER: J. Chromatog. **21**, 27 (1966).
628. SGOUTAS, D. S., and F. A. KUMMEROW: Biochemistry **3**, 406 (1964).
629. — M. J. KIM, and F. A. KUMMEROW: J. Lipid Res. **6**, 383 (1965).
630. SHELLARD, E. J.: Research and Develop. No. **21**, 30 (1963).
631. — M. Z. ALAM u. J. ARMAH: In the press.
632. SHEPPARD, H., and W. H. TSIEN: Anal. Chem. **35**, 1992 (1963).
633. SHOSTENKO, U. V., and V. V. RATSHINSKY: J. phys. Chem. (Moscow) **39** (7), 1802 (1965).
634. SIEGEL, E. T., and R. I. DORFMAN: Steroids **1**, 409 (1963).
635. SIMON, H., u. G. MÜLLHOFER: In: H. M. RAUEN: Biochemisches Taschenbuch, 2. Auflage, I. Teil, S. 913. Berlin-Göttingen-Heidelberg: Springer 1964.
636. SIMS, R. P. A., and J. A. G. LAROSE: J. Am. Oil Chemists' Soc. **39**, 232 (1962).
637. SKIPSKI, V. P., R. F. PETERSON, and M. BARCLAY: Biochem. J. **90**, 374 (1964).
638. SMITH, G. A. L., and P. J. SULLIVAN: Analyst **89**, 1058 (1964).
639. SMITH, L. W., R. W. BREIDENBACH, and D. RUBENSTEIN: Science **148**, 508 (1965).
640. SNAVELY, M. S., and J. G. GRASSELLI: Develop. applied spect. **3**, 119 (1963).
641. SNYDER, F.: Anal. Chem. **35**, 599 (1963).
642. — Anal. Biochem. **9**, 183 (1964).
643. SNYDER, L. R.: J. Chromatog. **13**, 415 (1964).
644. SNYDER, F.: In: Symposium on Advances in Tracer Methodology. Vol. **2**, 107. New York: Plenum Press 1965.
645. SNYDER, F.: In: Symposium of Radioisotope Sample Measuring Techniques in Medicine and Biology. Vienna: The International Energy Agency, p. 521.

646. SNYDER, F.: Separation Science **1**, 655 (1966).
647. — and H. KIMBLE: Anal. Biochem. **11**, 510 (1965).
648. — and F. STEPHENS: Anal. Biochem. **1**, 427 (1961).
649. — — Anal. Biochem. **4**, 128 (1962).
650. SNYDER, L. R., and H. D. WARREN: J. Chromatog. **15**, 344 (1964).
650a. SPENCER, R. D., and B. H. BEGGS: J. Chromatog. **21**, 52 (1966).
651. SPIKNER, J. E., and J. C. TOWNE: Chemist Analyst **52**, 50 (1963).
652. SPRENGER, H.-E.: Z. analyt. Chem. **199**, 241 (1964).
653. — Z. anal. Chem. **199**, 338 (1964).
654. — Z. anal. Chem. **204**, 241 (1964).
655. — Z. anal. Chem. **207**, 90 (1965).
656. SQUIBB, R. L.: Nature **198**, 317 (1963).
657. SREPEL, B.: Farma. Glasnik **18**, 64 (1962).
658. STADLER, P., u. H. ENDRES: J. Chromatog. **17**, 587 (1965).
659. STAHL, E.: Pharmazie **11**, 633 (1956).
660. — Lecture, Hauptversammlung Deutsch. Pharmaz. Ges. Freiburg/Br. 1957; Arch. Pharm. **290/27**, 121 (1957).
661. — Chemiker-Ztg. **82**, 323 (1958).
662. — Parfümerie u. Kosm. **39**, 564 (1958).
663. — H. R. BOLLIGER u. L. LEHNERT: Wissenschaftl. Veröffentl. d. deut. Ges. Ernährung **9**, 129 (1963).
664. — Z. analyt. Chem. **181**, 303 (1961).
665. — Arch. Pharm. **292**, 411 (1959).
666. — Pharm. Rdsch. **1**, No. 2 (1959).
667. — Naturwissenschaften **47**, 114 (1960).
668. — Arch. Pharm. **293**, 531 (1960).
669. — u. L. TRENNHEUSER: Arch. Pharm. **293**, 826 (1960).
670. — u. U. KALTENBACH: J. Chromatog. **5**, 351 u. 458 (1961).
671. — Chemiker-Ztg. **85**, 371 (1961).
672. — Angew. Chem. **73**, 646 (1961).
673. — Dünnschicht-Chromatographie, ein Laboratoriumshandbuch, 1. Aufl. Berlin-Göttingen-Heidelberg: Springer 1962.
674. — In: MARINI-BETTÒLO, G. B.: Thin-Layer Chromatography. Amsterdam: Elsevier Publ. Co. 1964.
675. — Chem.-Ing.-Techn. **36**, 941 (1964); (in English): Angew. Chem. Intern. Ed. **3**, 784 (1964).
676. — u. H. KALDEWEY: Planta **62**, 22 (1964).
677. — Arch. Pharm. **297**, 500 (1964).
678. — Lab. Practice **13**, 496 (1964).
679. — u. H. VOLLMANN: Talanta **12**, 525 (1965).
680. — Thin-Layer Chromatography, a Laboratory Handbook, 1st. Edit. Berlin-Göttingen-Heidelberg/New York: Springer-Verlag/Academic Press 1965.
681. — Russian translation, Peace publishers Moscow 1965, quoted from Pharmaz. Zentralhalle **104**, 459 (1965).
682. — Z. analyt. Chemie **221**, 3 (1966).
683. — H. K. MANGOLD, and H. H. O. SCHMID: In: D. GLICK: Methods of Biochemical Analysis, vol. **12**, 393. New York-London-Sydney: Intersciences Publ. 1964.
684. — Patents: DBP 1.180.166 (Broad band pipette) and DBP 1.175.013 (S-chamber).
685. — Patents: DBP 1.129.730 and DBP 1.175.912 (Spreaders for normal and gradient-TLC).

686. STAHL, E.: Unpublished; see also DUMONT, E., Dissertation Saarbrücken 1968.
687. STAMMBACH, K.: Private communication, quoted from [559].
688. STANLEY, W. L., and S. H. VANNIER: J. Assoc. Offic. Agr. Chemists **40**, 582 (1957).
689. — — and B. GENTILI: J. Assoc. Offic. Agr. Chemists **40**, 282 (1957).
690. STANLEY, TH. W., and E. SAWICKI: Anal. Chem. **37**, 938 (1965).
691. STANSFIELD, D. A.: Biochem. Biophys. Research Communs **16**, 398 (1964).
692. STEGEMANN, H., and B. LERCH: Anal. Biochem. **9**, 417 (1964).
693. STEIDLE, W.: Liebigs Ann. Chem. **662**, 126 (1963).
694. STEIN, Y., and O. STEIN: Biochim. et Biophys. Acta **54**, 555 (1962).
695. STEINEGGER, E., u. J. GEBISTORF: Pharm. Acta Helv. **38**, 840 (1963).
696. STEINHARDT, J., C. H. FUGITT, and M. HARRIS: J. Research Nat. Bur. Standards **25**, 519 (1940); **26**, 293 (1941); **30**, 123 (1943); J. B. SPEAKMAN, and E. STOTT, Trans. Faraday Soc. **31**, 1425 (1935).
697. STEUERLE, H., u. E. HILLE: Biochem. Z. **331**, 220 (1959).
698. STICKLAND, R. G.: Anal. Biochem. **10**, 108 (1965).
699. STOCKER, H. R.: Helv. Chim. Acta **46**, 2050 (1963).
700. STÖBER, W.: Kolloid.-Z. **147**, 131 (1956).
701. STROHECKER, R. jr., u. H. PIES: Z. Lebensm.-Untersuch. u. -Forsch. **118**, 394 (1962).
702. STROHECKER, R., u. H. HENNING: Vitamin-Bestimmungen. Weinheim/Bergstraße: Verlag Chemie 1963.
703. STRUCK, H.: Mikrochim. Acta (Vienna) **1961**, 634.
704. STUMPF, H. C., A. S. RUSSELL, J. W. NEWSOME, and J. W. TUCKER: Ind. Eng. Chem. **42**, 1398 (1950).
705. SZÉKELY, G.: Lecture, Chromatographie Symposium III, Brussels 1964.
706. TAKITANI, S., u. K. MATSUDA: Japan Analyst **13**, 562 (1964).
707. — — Japan Analyst **14**, 479 (1965).
708. TAMURA, Z.: J. Chromatog. **19**, 429 (1965).
709. TAYLOR, E. H.: Am. J. Pharm. Educ. **28**, 205 (1964).
710. TEICHNER, S., and E. PERNOUX: Clay Minerals Bull. **1**, 145 (1951).
711. TEIJGELER, C. A.: Pharm. Weekblad **97**, 43 u. 401 (1962); **99**, 101 (1964).
712. TERNER, CH., and F. R. DAREY: Biochim. et Biophys. Acta **98**, 194 (1965).
713. THOLEY, G., et B. WURTZ: Bull soc. chim. biol. **46**, 769 (1964).
713a. THOMPSON, A. C., and P. A. HEDIN: J. Chromatog. **21**, 13 (1966).
714. TISELIUS, A., S. HJERTÉN, and Ö. LEVIN: Arch. Biochem. Biophys. **65**, 132 (1956).
715. TORKAR, K., u. Mitarb.: Monatsh. Chem. **91**, 400, 450, 653, 658 (1960).
716. — — Monatsh. Chem. **92**, 512 (1961).
717. — — Monatsh. Chem. **94**, 110 (1963).
718. TOVE, B. S.: Private communication.
719. TREIBER, E.: Die Chemie der Pflanzenzellwand, S. 154. Berlin-Göttingen-Heidelberg: Springer 1957.
720. TRUTER, E. V.: Thin Film Chromatography, p. 112. London: Cleaver-Hume Press Ltd. 1963.
721. — J. Chromatog. **14**, 57 (1964).
722. TSCHESCHE, R., G. BIERNOTH, u. G. WULFF: J. Chromatog. **12**, 342 (1963).
723. TUBARO, E., e L. RUSTICI: Boll. chim. farm. **103**, 205 (1964).
724. TUNA, N., H. K. MANGOLD, and D. G. MOSSER: J. Lab. Clin. Med. **61**, 620 (1963).
725. — R. KAMMERECK, and H. K. MANGOLD: Unpublished.

726. UDENFRIEND, S.: Fluorescence Assay in Biology and Medicine. New York-London: Academic Press 1962.
727. United States Pharmacopeia, U.S.P. XVII, p. 442. New York 1965.
728. VAHOUNTY, G. V., C. R. BORJA, and S. WEERSING: Anal. Biochem. **6**, 555 (1963).
729. VANNIER, S. H., and W. L. STANLEY: J. Assoc. Off. Agr. Chem. **41**, 432 (1958).
730. VERMEULEN, A., and J. C. M. VERPLANCKE: Steroids **2**, 453 (1963).
731. VIOQUE, E., and R. T. HOLMAN: J. Am. Oil Chemists' Soc. **39**, 63 (1962).
732. WAGNER, H.: Mitt. Gebiete Lebensm. u. Hygiene **51**, 416 (1960).
733. — u. B. DENGLER: Biochem. Z. **336**, 380 (1962).
734. — u. P. POHL: Biochem. Z. **340**, 337 (1964).
735. WALDI, D.: In: E. Merck AG. Chromatographie. S. 67. Darmstadt 1959.
736. — In: E. STAHL: Thin-Layer Chromatography, 1. Edition, p. 29. Berlin-Göttingen-Heidelberg: Springer 1965.
737. — Mitt. deut. pharm. Ges. **32**, 125 (1962).
738. WALLACH, D. F. H., and G. L. NORDBY: Biochim. et Biophys. Acta **70**, 188 (1963).
739. WALZ, D., A. R. FAHMY, G. PATAKI, A. NIEDERWIESER u. M. BRENNER: Experientia **19**, 213 (1963).
740. WARING, P. P., and Z. Z. ZIPORIN: J. Chromatog. **15**, 168 (1964).
741. WARREN, B.: J. Chromatog. **20**, 603 (1965).
742. WASICKY, R.: Anal. Chem. **34**, 1346 (1962).
743. — Naturwissenschaften **50**, 569 (1963).
744. — Rev. Fac. farm. bioquim. (Sao Paulo) **1**, 135 (1963).
745. WEIMANN, G., and K. RANDERATH: Experientia **19**, 49 (1963); see also [563] and Biochim. et Biophys. Acta **76**, 622 (1963).
746. WEITZ, E., H. FRANCK u. M. SCHUCHARD: Chemiker-Ztg. **74**, 256 (1950).
747. WERNZE, H.: Private communication, May 1965.
748. WESTPHAL, O., O. LÜDERITZ u. F. BISTER: Z. Naturforsch. **7b**, 148 (1952).
749. WEYL, W. A.: A new Approach to Surface Chemistry and to heterogeneous Catalysis. Mineral Industries Exp. Sta. Bull. No. **57** 46 (1951).
750. WHITAKER, D. R., and K. A. MITTELSTADT: Can. J. Biochem. **42**, 149 (1964).
751. WHITEHEAD, J. K.: Biochem. J. **68**, 662 (1958).
752. WIEDENHOF, N.: J. Chromatog. **15**, 100 (1964).
753. WIELAND, TH., u. G. PFLEIDERER: Angew. Chemie **67**, 257 (1955).
754. — — Angew. Chemie **69**, 199 (1957).
755. — G. LÜBEN u. H. DETERMANN: Experientia **18**, 430 (1962).
756. — u. H. DETERMANN: Experientia **18**, 431 (1962).
757. — u. D. GEORGOPOULUS: Biochem. Z. **340**, 476 (1964).
758. WIEME, R. J.: In: H. RAUEN: Biochemisches Taschenbuch, 2. Aufl. II. Teil, S. 947. Berlin-Göttingen-Heidelberg: Springer 1964.
759. WILLIAMS, T. I.: Introduction to Chromatography. Glasgow: Blackie & Son 1947.
760. WILLSTÄTTER, R., H. KRAUT u. K. LOBINGER: Ber. deut. chem. Ges. **58**, 2462 (1925).
761. WINEFORDNER, J. D., and H. A. MOYE: Anal. Chim. Acta **32**, 278 (1965).
762. WINTERSTEIN, A.: Angew. Chem. **72**, 902 (1960).
763. — u. B. HEGEDÜS: Hoppe Seyler's Z. physiol. Chem. **321**, 97 (1960).
764. — A. STUDER u. R. RÜEGG: Chem. Ber. **93**, 2951 (1960).
765. WIRZING, G.: Naturwissenschaften **50**, 466 (1963).
766. WOHNLICH, J. J.: J. pharm. Belg. **18**, 255 (1963); **19**, 53 (1964).

767. WOLFROM, M. L., R. M. DE LEDERKREMER, and L. E. ANDERSON: Anal. Chem. **35**, 1357 (1963).
768. WOLLENWEBER, P.: J. Chromatog. **7**, 557 (1962).
769. — J. Chromatog. **9**, 369 (1962).
770. — Lab. Pract. **13**, 1194 (1964); see also Thin-Layer Chromatography (A series of articles reprinted from Lab. Pract.), p. 76—83. London: United Trade Preess Ltd. 1964.
771. — In: G. MARINI-BETTÒLO: Thin-Layer Chromatography, p. 14. Amsterdam-London-New York: Elsevier Publishing Company 1964.
772. — Österr. Chemiker-Ztg. **66**, 207 (1965).
773. — In: K. MACEK, and I. M. HAIS: Stationary Phase in Paper and Thin-Layer Chromatography, p. 98. Amsterdam: Elsevier Publishing Company 1965.
774. WOLLISH, E. G., M. SCHMALL, and M. HAWRYLYSHYN: Anal. Chem. **33**, 1138 (1961).
775. — Microchem. J., Symposium Vol. II, 687 (1962).
776. WREN, J. J.: J. Chromatog. **12**, 32 (1963).
777. WUNDERLY, C.: Die Papierelektrophorese, II. Aufl. Aarau-Frankfurt a. M. 1959.
778. YAMAGUCHI, M.: Hakkô Kyôkaishi **21**, 361 (1963).
779. YOUNG, G. J.: J. Colloid Sci. **13**, 67 (1958).
780. ZABIN, B. A., and CH. B. ROLLINS: J. Chromatog. **14**, 534 (1964).
781. ZECHMEISTER, L., u. L. V. CHOLNOKY: Die chromatographische Adsorptionsanalyse. 2. Aufl. Wien: Springer 1938.
782. ZEITMANN, B. B.: J. Lipid Res. **5**, 628 (1964).
783. ZHDANOV, Y. A., et al.: Doklady Akad. Nauk. S.S.S.R. **149**, 355 (English translation).
784. ZÖLLNER, N., u. G. WOLFRAM: Klin. Wschr. **40**, 1098 (1962).
785. — — Klin. Wschr. **40**, 1101 (1962).
786. — — and G. AMIN: Klin. Wschr. **40**, 273 (1962).
787. ŽURKOWSKA, J., u. A. OZAROWSKI: Planta med. **12**, 222 (1964).
788. JACKSON, R.: J. Chromatog. **20**, 410 (1965).
789. JONES, C. R.: Chem. and Ind. **1965**, 1999.
790. BRAND, J. M.: J. Chromatog. **21**, 424 (1966).
791. FISCHER, L. J., and S. RIEGELMAN: J. Chromatog. **21**, 268 (1966).
792. GNEHM, R., H. U. REICH u. P. GUYER: Chimia **19**, 585 (1965).
793. GORDON, H. T.: J. Chromatog. **22**, 60 (1966).
794. HORVATH, C.: J. Chromatog. **22**, 52 (1966).
795. JAMES, C. N. MA: J. Chromatog. **21**, 151 (1966).
796. JORK, H.: Chromatography Symposium III, 1964 Brussels, Report p. 295.
797. KOTTKE, B. A., J. WOLLENWEBER, and C. A. OWEN: J. Chromatog. **21**, 439 (1966).
798. SEMENUK, G., and W. T. BEHER: J. Chromatog. **21**, 27 (1966).
799. SPENCER, R. D., and B. H. BEGGS: J. Chromatog. **21**, 52 (1966).
800. THOMPSON, A. C., and P. A. HEDIN: J. Chromatog. **21**, 13 (1966).

Special Section

Introduction

EGON STAHL

Suitable methods of separation and detection are part of the equipment for working with mixtures of chemical compounds. There is no universal method. A complex mixture is submitted first to preliminary separation into, e. g., a lipophilic and a hydrophilic fraction. Each fraction is then subdivided further until mixtures of substances with fairly similar properties are obtained, for the separation of which, chromatography is then used with benefit.

Numerous separation problems can be easily and quickly solved by TLC. However, one should always remember the importance of attaining optimum separation conditions by the right mutual harmonisation of stationary and mobile phases. Fig. 99 shows the correlations between the three main chromatographic elements which must be considered. This scheme is based primarily on the wealth of experience gained from adsorption chromatography; this experience is summarised as follows:

Mixture to be Separated

a) *Saturated hydrocarbons are adsorbed only slightly or not at all and thus migrate fastest. Unsaturated hydrocarbons are more strongly adsorbed, the more double bonds they contain and the more these double bonds are conjugated with one another.*

To separate hydrocarbons, an active adsorbent and an only slightly polar solvent must be used. Alternatively, "phase reversal" may be employed, in which the stationary phase is impregnated with a lipophilic material and a hydrophilic solvent as used a mobile phase.

b) *If functional groups are introduced into a hydrocarbon, the adsorption affinity is increased in the following sequence;* $-CH_3 < -OAlkyl < > C=O < -NH_2 < -OH < -COOH$.

When benzene for example is used as solvent on silica gel or alumina layers, ethers and esters are found in the front part of the chromatogram, ketones and aldehydes approximately in the middle, alcohols behind them and the acids still at the starting point. The sequence of separation thus follows the polarity of the compounds. Organic acids and strong bases can be chromatographed either as their less polar derivatives or by using acid or basic solvents.

Solvents

The solvents used in chromatography can be arranged according to their eluting effect in a so-called "eluotropic series". TRAPPE had suggested this for column chromatography as long ago as 1940. One notes from Table 20 that the elution power increases with increase in polarity of the solvent. The dielectric constant can be used as an indicator of polarity. An even better relation is seen by comparing the interfacial tension values with water.

Table 20. *Eluotropic series of solvents*
(The elution effect increases from top to bottom)

Solvent[a]	Boiling Point (BP 760 mm) in °C	Dielectric Constant ε 20° C	Interfacial tension with water in dynes/cm (20° C)	Viscosity in cp (20° C)
n-Hexane	68.7	1.890	51.1	0.326
Heptane	98.4	1.924	50.4[b]	0.409
Cyclohexane	81.4	2.023	—	1.02
Carbon tetrachloride	76.8	2.238	45.1	0.969
Benzene	80.1	2.284	35.0	0.652
Chloroform	61.3	4.806	27.7	0.580
Diethyl ether	34.6	4.34	9.7	0.233
Ethyl acetate	77.1	6.02[b]	6.3[c]	0.455
Pyridine	115.3	12.3[b]	miscible	0.974
Acetone	56.5	20.7[b]		0.316[b]
Ethanol	78.5	24.30[b]	with	1.2
Methanol	64.6	33.62		0.597
Water	100	80.37	water	1.005

[a] Additional solvents and their properties can be found in Biochemical Handbook, Springer-Verlag, Berlin-Göttingen-Heidelberg 1964 and in Handbook of Chemistry and Physics, Chemical Rubber Publishing Co., Cleveland, Ohio, USA, 1959.
[b] At 25° C.   [c] At 30° C.

The speed of migration of a solvent depends on, amongst other things, its viscosity. The migration time for a 10 cm length of run increases linearly with the viscosity within a homologous series, e. g., aliphatic hydrocarbons (C_5-C_{11}) and the normal alcohols (C_1-C_8).

Note: Only the purest solvents should be used in TLC. It must be borne in mind that mixtures should be used for development only two or three times; the initial composition suffers change through adsorption or evaporation of a component; further, that individual components may react with each other; and that ether and chloroform contain 0.5—1% ethanol as stabiliser.

Adsorbents

Adsorbents can be classified similarly and one distinguishes those with high adsorption activity and those showing only a meagre activity.

BROCKMANN introduced so-called "activity grades" as a relative measure for the adsorption power of alumina. The highest activity was denoted by "I" and the lowest by "V". Silica gels and aluminas usually used in TLC have activities between II and III at a relative humidity of 40 to 65%.

The grain sizes and packing density of the layer also influence the rate and therefore the time of migration. The run durations are appreciably larger on layers prepared from very fine-grained silica gel (0.1 to 10 μm diameter) than on those of grain sizes between 10 and 40 μm.

Correlation of the 3 Main Elements

The fundamental rules of adsorption chromatography and the correlations can be summarised most instructively in a triangular diagram (Fig. 99). Imagine that the shaded triangle can be rotated. If, e. g., a lipid mixture is to be separated, one apex of the triangle is pointed towards the word "lipophilic" in "mixture to be separated" (indicated with the dotted triangle in the diagram); the other two apices of the dotted triangle now show that a non-polar mobile phase (solvent) and an active stationary phase (adsorbent) are required. Conversely, for separation of a polar mixture, the other two apices show that a hydrophilic solvent and a not too active adsorbent are needed.

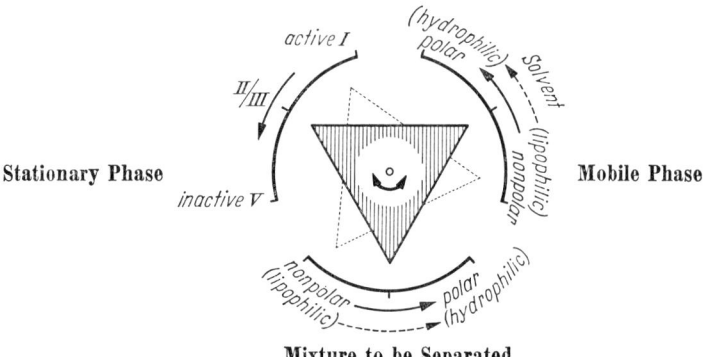

Fig. 99. Diagram showing the close relationship between the three main variable elements in chromatography as illustrated from adsorption chromatography. See text for details

The scheme can be applied to other chromatographic procedures: For example, it is valid for phase reversal procedures in PC and TLC. Then, under "Stationary phase", one replaces "active I" with *"lipophilic"* or *"non-polar"* and "inactive V" by *"hydrophilic"* or *"polar"*; and, under "Mobile phase", one replaces "polar" by *"non-polar"* and *"hydrophilic"* by *"lipophilic"*.

Combination of Several Procedures

Chromatography is essentially a separation proceure and the Rf-values can generally be used only as guides (cf. p. 127). They are totally inadequate for identifying a compound. Only the combination of chromatographic procedures with one another and subsequent coupling with appropriate methods of identification, provide sufficient information about a substance and guarantee the reliability of statements about it. Tested combinations are therefore given here:

1. Combination of Chromatographic Procedures with one Another

a) Adsorption with b) partition,
a) or b) with c) ion exchange (see p. 44),
a) or b) with d) electrophoresis (see p. 105),
b) with e) "molecular sieve separation" (see p. 40).

Procedures a) to e) can be carried out with various stationary phases (see also p. 48 and Table 11) and various types of solvent.

2. Coupling of Chromatography with Identification Procedures

1. With chemical treatment (colour reactions, derivatives of various types, pyrolysis products).

2. With spectrographic procedures (spectra: UV to IR, mass spectra, NMR).

3. With other physical micromethods (sublimation, polarography etc.).

4. With biological and/or physiological procedures (testing an antibiotic, insecticidal or haemolysing effect; or taste or activity as a growth or insect attraction agent).

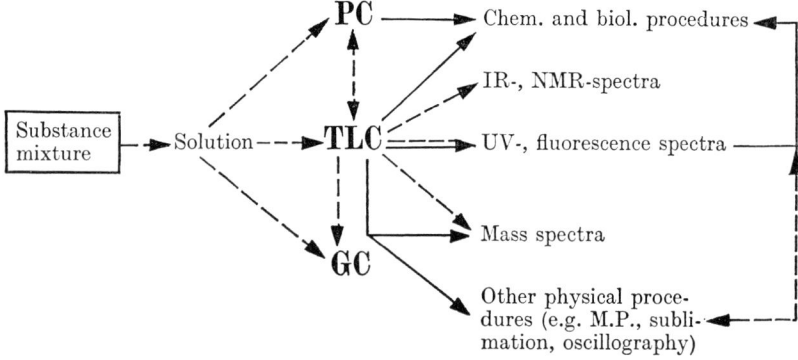

Direct coupling = ⟶ ; Mechanical transfer = --▶

Fig. 100. An example of the combination of procedures in parallel and/or in series

Introduction 205

Chapter D, section V (p. 102) has been devoted to the transfer techniques which are often required in combinations of procedures; and chapter F (p. 114) to TLC-GC coupling. References are made to the best ways of identification in many chapters of the special section of the book.

By coupling separation and identification procedures in parallel and/or in series, one can nowadays provide reliable information even in the microgram region. An example of such types of combination is given in Fig. 100. This scheme can be used in an abbreviated or an expanded form.

Table 21. *Some possible reactions*[a] *at the start point or in application capillaries*

Reaction	Reagent	Conditions
Oxidation	a) chromic (VI) oxide 20% in acetic acid	at start point: 5—30 min at 20° C
	b) p-nitroperbenzoic acid 2% in ether	in capillary: 60 min at 100° C
	c) heating in the air, if necessary + 1 drop 1% $KMnO_4$ solution	at start point: 60 min at 120° C
	d) photooxidation in UV light, e. g., 254 nm	at start point: 1—60 min lamp 5—10 cm distant
Epoxidation	p-nitroperbenzoic acid 1% in ether	in capillary: 3—6 h at 20° C; or at start point
Hydrogenation	a) sodium borohydride 10% in absolute ethanol	in capillary: 2—3 h at 20° C
	b) sodium borohydride 5% in absolute ethanol; 3 drops acetyl chloride after drying	at start point; a few min at 20° C
	c) lithium aluminium hydride 10% in absolute ether	at start point; a few min at 20° C
	d) H_2; colloidal 1% palladium solution (Firm 123a); dried and impregnated with undecane	at start point: 60 min hydrogenation (desiccator filled with H_2)
Dehydration	conc. sulphuric acid	at start point: a few min at 20° C
Saponification	1N alcoholic potassium hydroxide	in capillary: 1—2 h at 50—100° C
Coupling	Diazonium salt from 2,5-dimethoxyaniline	at start point: 2—5 min at 20° C
Acetylation	a) acetic anhydride/pyridine 1:2	in capillary: 3—6 h at 50° C
	b) acetyl chloride	at start point: 2—3 min at 20° C; then 10 min at 120° C to evaporate excess reagent
	c) acetyl chloride	in capillary: 60 min at 100° C

[a] Preparation of other derivatives, e. g., with: benzoyl chloride, 3,5-dinitrobenzoyl chloride, 4'-nitroazobenzene(4)-carboxylic acid chloride (Firm 88), 2,4-dinitrophenylhydrazine etc., is possible.

Reactions at the Start Point or in the Capillary

Within the realm of combinations of methods, chemical reactions at the start point are simple to carry out and require a minimum of expense. MILLER and KIRCHNER[1] drew attention to the advantages of this technique and gave numerous examples. MATHIS and OURISSON[2,3] have recently studied this again in detailed fashion, widening the scope by carrying out the reaction in the capillary used for application. This "chromatographie fonctionelle sur couche mince" (CFCM) has a broad range of application, capable of considerable extension. A compilation of some established reactions together with the conditions is in Table 21.

α) *at the start point*, the substance to be treated is applied as a spot (3—5 mm diameter) in the usual way and excess of reagent solution is then added to it. Care must be taken that no "circular chromatograms" are formed. Suitable solvents for substance and reagent must be chosen. Both the starting material to be treated and the reagent should be applied separately close to the start point for control comparison.

β) *in the capillary*, such reactions can be carried out more systematically. Melting point capillaries (internal diameter 1—1.5 mm) are used, drawn out in the form of pipettes, and the desired amount of substance solution and of reagent solution are allowed to rise into two separate capillaries. The contents of one capillary are then transferred to the other by bringing the two pipette tips into suitable contact. Mixing is performed by tilting the capillary. Both ends of the capillary can now be sealed off and the contents warmed for a while if necessary. It is then cautiously opened at room temperature and the contents are directly applied to the layer.

This procedure is suitable for better characterisation of functional groups and also for following the course of reactions carried out on the ultramicro scale in capillaries. It is also valuable for identification of single substances. With mixtures, the SRS-technique is used (p. 88).

J. Terpene Derivatives, Essential Oils, Balsams and Resins

EGON STAHL and H. JORK

This chapter contains a discussion of possible ways of separating lipophilic mixtures of natural products, the components of which are built up from isoprene units in accordance with Ruzicka's isoprene rule [219]. WALLACH [321] has divided the hydrocarbons and their derivatives into the following groups:

---

[1] MILLER, J. M., and J. G. KIRCHNER: Anal. Chem. **25**, 1107 (1953).
[2] MATHIS, C., et G. OURISSON: J. Chromatog. **12**, 94 (1963).
[3] MATHIS, C.: Ann. pharm. franç. **23**, 331 (1965).

Hemiterpenes	C_5H_8	and derivatives	volatile in steam;
Monoterpenes BP 140–180°	$C_{10}H_{16}$	and derivatives	principal components of essential oils;
Sesquiterpenes BP over 200°	$C_{15}H_{24}$	and derivatives	including phenylpropane compounds
Diterpenes BP about 300°	$C_{20}H_{32}$	and derivatives	involatile or scarcely volatile in steam;
Triterpenes	$C_{30}H_{48}$	and derivatives	components of balsams, resins,
Polyterpenes[4]	$(C_{10}H_{16})_x$	and derivatives	waxes, rubber

The phthalides and the important group of naturally occurring C_9-compounds, which can formally be considered as phenylpropane derivatives, are included after the hemiterpenes.

The many compounds of each group are differentiated by the type of cyclisation (open-chain, monocyclic, bicyclic etc.), the number and position of double bonds, the centres of asymmetry and the nature and number of functional groups. The individual terpene derivatives are most conveniently treated in the order of increasing polarity; hydrocarbons first, then esters and lactones, carbonyl compounds, alcohols and finally phenols and acids.

The differences in structure and the physical and chemical properties of the terpene- and phenylpropane derivatives encountered in essential oils, have been described in the standard works of GILDEMEISTER-HOFFMANN [72], GUENTHER [82], SIMONSEN [240], DE MAYO [160], W. KARRER [117] and MORITZ [173]. HAAGEN-SMIT[86] has dealt with sesqui- and diterpenes. Both STEINER and HOLTZEM [721] and JONES and HALSALL [109] have compiled comprehensive articles on triterpenes.

I. Separation of Lipophilic, Steam-volatile Mixtures

It is frequently necessary to concentrate the terpenes and their derivatives in a reaction charge or in plant or animal material before separating them. As mentioned already in the introduction, the C_5- to C_{15}-compounds which come into consideration are volatile in steam and can thus be isolated, using appropriate apparatus [81, 174]. The "Karlsruhe" apparatus (Firm 21) has proved the best for quantitatively isolating small amounts (1–1000 mg) of such mixtures (Fig. 101). The lipophilic fraction which has distilled over can be transferred, free of water, to a small microflask for weighing; it can then be investigated according to the scheme in Table 22.

About 300–400 mg oil are needed for determination of density according to FURTER [67a], refractive index with the Abbé-refractometer

---

[4] No attempt is made here to subdivide further into tetra-, penta- and the real polyterpenes.

and optical rotation in precision polarimeter tubes, 10 cm long and 1.6 mm internal diameter. The C_{10}- and C_{15}-hydrocarbons can then be determined quantitatively in duplicate by adsorption chromatography (cooling column, neutral alumina activity grade III, pentane).

Possibilities for identifying the separated substances are:
a) melting or boiling point, density, refraction, rotation,
b) preparation of derivatives,
c) spectra (UV-, IR-, NMR-, Raman-, mass-).

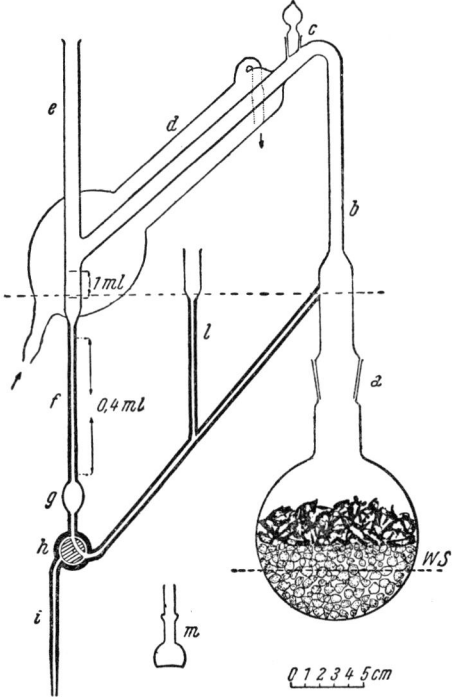

Fig. 101. "Karlsruhe" apparatus for isolation of small amounts of steam-volatile lipids. *a* standard ground glass joint B 29; *b* ascending tube; *c* inlet for washing; *d* condenser; *e* pressure compensator with 1 ml graduations; *f* graduated capillary with divisions of 1/200; *g* bulb; *h* three-way tap; *i* outlet; *l* filling tube; *m* microflask for gravimetric estimation of collected lipids; *WS* water level in flask [245]

II. Chromatographic Separation of Lipophilic, Steam-volatile Mixtures

In any further examination of the essential oils separated by steam distillation from the remaining components of the material to be investigated, the working procedure adopted depends on the amount of oil available. Fractional distillation and crystallisation are preferable if larger amounts are available. On the micro scale, the scheme reproduced in Table 22 has proved helpful. Column chromatography is combined with chemical separation procedures.

Table 22. *Separation and identification of essential oils on the micro-scale*

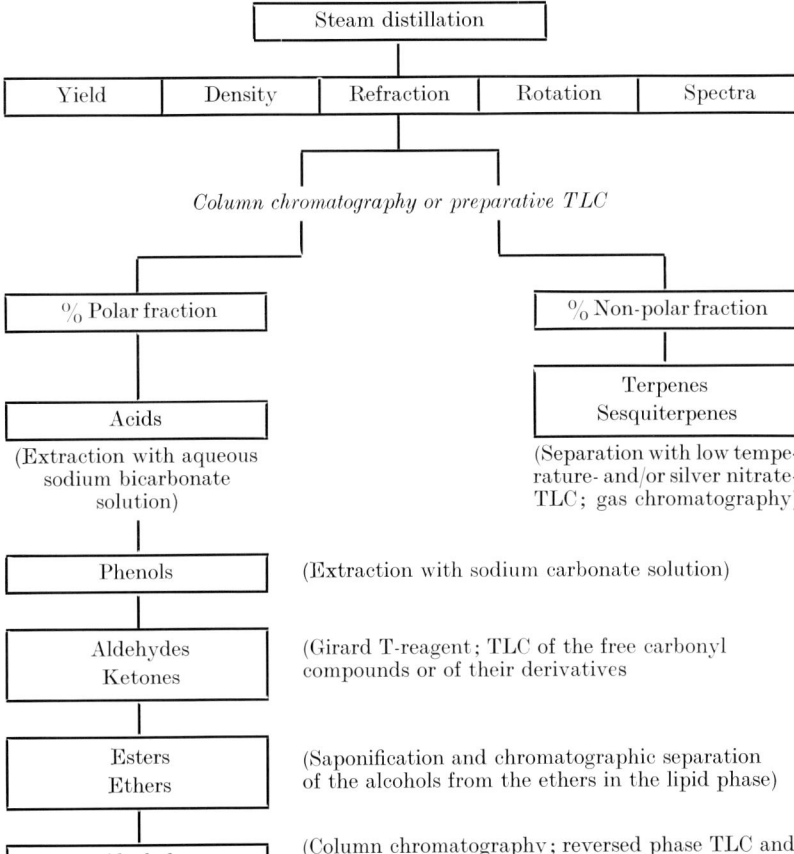

A preliminary separation of this type is advantageous for both TLC and gas chromatography. If these two techniques are combined [266], the results yield more information (p. 114). Colour reactions, based on genuine reference material[5], can be used together with the hRf-values for identification of the separated individual compounds.

---

[5] Remission spectra can also be recorded and used for detection without loss of material, especially in the UV region [114].

Specific reactions, carried out on the layer as has been often suggested [29, 158, 168], likewise permit conclusions to be drawn. The definitive identification of compounds ought always to be attempted with the help of established classical micromethods (see Table 22).

1. Mono- and Sesquiterpene Hydrocarbons

As already mentioned, dielectric constant values give an idea of the adsorption affinity of the compounds to be separated. There are no significant variations within the terpene group (ε between 2.24 and 2.76) so that the usual adsorption chromatography leads merely to a subfractionation of the hydrocarbons. The terpenes and sesquiterpenes are found in the upper part of the chromatogram ($hRf = 80-100$) when benzene ($\varepsilon = 2.284$) or chloroform ($\varepsilon = 4.806$) are used as solvents. This applies also to descending technique [268, 331]. The monoterpenes migrate in fact further than the sesquiterpenes. Chamazulene, as a polar C_{15}-hydrocarbon, has a medium Rf-value with n-hexane as solvent [224, 246]. Humulene, caryophyllene and isocaryophyllene are also separable with success in this region, using petroleum ether-carbon tetrachloride (75 + 25) [14].

The investigations of ATTAWAY and co-workers [6, 6a] are noteworthy in this connection and with regard to the combination of TLC with GC (see p. 114). They were able to achieve a group separation of the C_{10}- and C_{15}-hydrocarbons on alumina G layers, using perfluorinated alkanes (BP 70–80° C) (Firms 41, 100) as solvent. With a length of run of 7.5 cm, the 19 monoterpenes remained at the start and the 11 sesquiterpene hydrocarbons were separated in the hRf range of 10–71.

In order to obtain a better separation of monoterpenes and to eliminate the interference of evaporation during chromatography, ADHIKARI [1] and MATHIS [159] have developed at $-15°$ C, using silica gel G layers and n-hexane (chamber saturation, hRf 25–85). The time of run was shorter and the hRf-values lower than under normal conditions, as STAHL [254] has been able to show (cf. Fig. 49). The mono- and sesquiterpene fractions of *Daucus* oils have been chromatographed also at $-20°$ C, on silica gel H layers [255]. A separation in accordance with the influence of the C=C double bonds was accomplished with n-hexane-pentane (50 + 50) as solvent.

MILLER and KIRCHNER [168] had also suggested activation of the adsorbent layer back in 1953. They achieved this by drying the chromatogram strips over phosphorus pentoxide in a vacuum desiccator at 3 mm mercury pressure. Since the danger of inactivation by air humidity is so great with the standard sizes used today and the standard 250 μ thicknesses, the procedure is of no more than historical interest.

If mono- and sesquiterpene hydrocarbons are to be distinguished after chromatography, their differing vapour pressures may be used to advantage, according do DEMOLE [42]:

A second TLC plate with silica gel layer is placed on the chromatogram which has been freed from solvent. The layers face each other and are 1 mm apart. By cautious, even heating, the lower boiling monoterpenes are the first to evaporate from the lower chromatogram. They are readsorbed on the upper layer. Reduced pressure is an advantage in that lower temperatures then suffice. Both chromatograms are subsequently sprayed for visualisation. The upper layer should show chiefly the monoterpene hydrocarbon zones; the lower, original one, the sesquiterpenes.

The ability of mono- and sesquiterpenes to form complexes or adducts is utilised in order to achieve separation on silica gel G-silver nitrate layers; the separation is governed by the number and position of the C=C double bonds [223, 326]. Adsorption properties and adduct-forming properties of the layer are superimposed. The best degree of impregnation can be ascertained without particular difficulty, with the help of the gradient technique (p. 91; cf. Fig. 105a). GUPTA and DEV [83] have employed silica gel layers containing 15% silver nitrate to separate 5 sesquiterpenes, whereas MATHIS [157] has chromatographed on layers with 5%.

PREY et al. [205] have pointed out another way of separating olefines of low molecular weight. They chromatographed stable mercuric acetate adducts on ordinary silica gel G layers. Since these adducts are more polar, the authors were able to use more polar solvent mixtures; this ought to be possible in terpene separation also. The following hRf-values were obtained, using n-propanol-triethylamine-water (50 + 25 + 25): ethylene 7; propylene 13; 1-butene 17; 1-pentene 29; 1-hexene 31; the time of run was 90 min but, according to BRAUN and VORENDOHRE [21], could be shortened by using a mixture of butanone-n-propanol-ethanol-25% ammonium hydroxide (50 + 5 + 20 + 35). GAREL [70] maintains that a better separation can be obtained with this solvent.

Preparation [205].

The gaseous olefines, if necessary after chromatographic separation, are passed into a 1% solution of mercuric acetate in methanol; liquid olefines are added directly to it. The reaction takes place quantitatively and almost momentarily. This methanol solution may be used directly for chromatography. 5—10 µg of adduct are applied.

According to the experience so far gained, gas chromatographic separation of mixtures of this compound class is better, despite all the TLC possibilities [115, 228, 251].

Visualisation of the substance zones can be carried out on normal silica gel layers by using conc. sulphuric acid, if necessary with addition of aldehyde. The anisaldehyde-sulphuric acid reagent (No. 11) is preferable. The fluorescein-bromine test (No. 118) and the antimony(V)-

chloride reagent (No. 18) are also used. Grey-blue zones on a yellow background are given with the molybdophosphoric acid reagent (No. 168) after heating. BERGSTRÖM and LAGERCRANTZ [11] have recommended the diphenylpicrylhydrazyl reagent (No. 88). It possesses no advantage over the anisaldehyde reagent mentioned. For the detection of mercuric acetate adducts, the chromatogram is sprayed with 2% diphenylcarbazide solution in alcohol and heated briefly to 80° C. The olefine adducts turn violet blue.

2. Oxides, Epoxides and Peroxides

Terpene oxides, epoxides or peroxides are encountered more often than is usually believed during TLC of essential oils [248]. Eucalyptol (1,8-cineole) and ascaridole, which is easily accessible by photochemical treatment of α-terpinene, are the best known. These compound types play a special part in the determination of the value of individual oils.

GYANCHANDANI [85] has thus chromatographed caraway seed oil and dill oil under standard conditions and determined the position of the limonene epoxide (1,2) which had been formed by autoxidation of limonene. NIGAM, SAHASRABUDHE and LEVI [185] have detected piperitone oxide in peppermint oils in a similar way. Attention has been drawn repeatedly to the bisabolene oxides I, II and III along with chamazulene in the assessment of camomile oils [78, 224, 305, 306].

The position of these substances in the chromatogram is determined by the polarity of the parent compound. Menthofuran migrates just behind guaiazulene [182a, 247]. The terpene and sesquiterpene epoxides follow, lying in the upper part of the ester zone on silica gel G layers (Table 23) (cf. also [155]). in agreement with EL-DEEB [55]. The carbonyl and alcohol oxides follow with lower hRf-values. This sequence holds even at $-9°$ C using Freon (Frigen 21) as solvent [255].

Terpene derivatives are often characterised specifically through epoxide formation. They can be microanalytically prepared directly on the TLC plate. OURISSON and co-workers [157, 158, 159], like KAUFMANN's team, apply the reagent solutions (e. g., an ether solution of p-nitroperbenzoic acid) and the compound to be treated on to the starting point together, chromatographing then in the usual way. PESNELLE et al. [195] have likewise used the method in order to characterise more definitively the two sesquiterpene alcohols guaiol and bulnesol. KLEIN, ROJAHN and HENNEBERG [129] have prepared epoxides of geraniol, nerol and linalool and separated them from the starting materials using benzine (BP 40—50° C)-ethyl acetate (85 + 15). Linalool oxide was detected and identified in this way in geranium, lavender and rosewood oils. Gas chromatographic differentiation of the various epoxides is not easy on account of their thermal instability. According to our experience, this can be achieved only with difficulty on the usual stationary phases.

Table 23. *hRf-values and colour reactions of some terpene- and sesquiterpene oxides, epoxides and peroxides on silica gel layers* [a]

Substance	Solvent			Colour reactions with various reagents		
	I	II	III	Anisaldehyde/ H_2SO_4 (Rgt. No. 11)	Antimony(III)-chloride (Rgt. No. 15)	
					Daylight	UV-light (365 nm)
Menthofuran	71	82	85	light brown	grey-brown	ochre
Bisabolene oxide I	40	67	35	brown-grey	brown-grey	brown
Bisabolene oxide II	25	—	24	brown-grey	brown-grey	brown
Limonene epoxide-(1,2)	26	66	63	red-blue	light brown	dark
Rose oxides, L. and R.	19	64	60	violet-blue	grey	light brown
Eucalyptol (1,8-cineole)	19	58	65	grey-blue	grey-blue	yellow-brown
α-Pinene epoxide	19	60	69	brown grey	grey	dark
Epoxidihydro-caryophyllene	18	63	52	red-violet	violet	red-blue
Myrcene oxide	18	62	49	brown-green	grey-blue	ochre
Ascaridole	18	66	54	grey-brown	grey-brown	brown
Limonene hydroperoxide	10	32	—	brown-red	—	—
Linalool oxide	5	25	24	orange-brown	ochre	yellow-brown
Daucol	4	18	17	blue violet	brown-grey	brown

[a] Chamber saturation (CS) (the solvent was introduced into the chamber 30 min before chromatography and shaken energetically in order to improve the reproducibility of the *hRf*-values); solvent I = benzene; II: chloroform; III: n-hexane-diethyl ether (80 + 20), double development, 10 cm.

Furan derivatives occur alongside terpene- and sesquiterpene epoxides in essential oils. The 5-ring parent structure may be part of a polycyclic molecule as, e. g., in daucol; or may have unsaturated side chains, occasionally with acetylenic links (see p. 235). Ipomeamarone contains a furan ring, connected with a 6-membered side chain via methyltetrahydrofuran. Myoporone and ipomeanine may be mentioned as further examples of furanoterpene derivatives; they can be chromatographed with n-hexane-ethyl acetate (90 + 10) [2]. The separation of 6 sesquiterpene furans (linderene and lindestrene derivatives), quoted by TAKEDA and co-workers [280, 281], can be used for obtaining these substances on a preparative scale. They carried out chromatography on silica gel G layers, 500 μm thick, using benzene-ethyl acetate (90 + 10). Menthofuran, a known constituent of peppermint oils, would be found in the front zone using this solvent system (cf. also NIGAM and LEVI [182]).

Terpene peroxides may be discussed in connection with the epoxides and oxides; the example of ascaridole has already been mentioned.

It is the chief constituent of the poisonous American wormseed oil and is found in amounts of up to 40% in oil of *Chenopodium ambrosioides*, although not found in other oils of the same type [52]. Ascaridole, as an inner peroxide, is found on chromatograms at a higher position than the hydroperoxides which have been repeatedly detected as intermediates during epoxide formation. Using n-hexane-diethyl ether (87 + 13), the three "limonene peroxides" can also be separated under standard conditions (hRf 22, 27 and 33) [85].

Oxides and peroxides can be **visualised** with the usual reagents (Nos. 11, 15, 18) or with an acidic p-dimethylaminobenzaldehyde solution (No. 70). The iodide-acetic acid-starch test (Rgt. No. 139) has usually been better than the ferrous thiocyanate reagent (No. 115) as a more specific test for peroxides (cf. also KNAPPE and PETERI [133]).

3. Steam-volatile Esters and Lactones

Terpene esters constitute an important compound class in the essential oils. Acetate esters are those most frequently encountered; more rarely, formates, propionates, butyrates, valerates and caprylates. Experience accumulated with over 30 different esters of lower fatty acids, shows that the zones on silica gel G layers, when developing with benzene or chloroform, are about 2.5 times higher than those of the corresponding alcohols (chamber saturation). BRUD and DANIEWSKI [28] have confirmed our finding [260] for loose silica gel layers also. The largest differences in hRf-values have been observed when using normal saturation (NS).

Table 24. *Separation of terpene esters on silica gel G layers*

Esters	hRf-values, solvent and layer thickness			
	I 60 μm [a]	I 250 μm [b]	II 60 μm [a]	III 250 μm [b]
Formates	62—75	42—47 Linalyl formate 37	38—44	71—74
Acetates	50—62	25—33	31—35	63—65
Propionates	62—73	42—47	38—41	71—74
Butyrates	57—71	42—47	27—43	71—74
Caprylates	63—85	—	40—57	—
Alcohols (see Table 28)	—	5—8	—	26—33 (Linalool 38)
Standards				
Butter yellow	—	40	—	73
Sudan Red G	—	15	—	63
Indophenol	—	5	—	58

Solvents I = benzene, II = trifluorotrichloroethane-methylene dichloride (60 + 40); III = chloroform.

[a] Compare ATTAWAY et al. [7], also ATTAWAY [5].
[b] Chamber saturation.

Comparison of the hRf-values of terpene esters of homologous lower fatty acids shows that the values increase with increasing number of carbon atoms [120]. They approach a limiting value asymptotically [7, 251]. Formates are exceptional since they appear at the level of the propionates (Table 24) and can be separated only from acetates when using benzene or chloroform as solvent.

The intramolecular esters, the *lactones*, may be mentioned here. The antihelminthic agent santonin, for example, belongs to this class. A well known lactone is coumarin, the odour of which resembles dried woodruff and which is found in numerous essential oils, also as derivatives[6]. SUND and SACCARDI [273] have succeeded in separating 5 coumarins on silica gel G layers, using petroleum ether (50—75° C)-ethyl acetate (67 + 33) or n-hexane-ethyl acetate (72 + 29). The sequence found was: coumarin, 6-methylcoumarin, dihydrocoumarin, 3-methylcoumarin, 3-ethylcoumarin.

KORTE and VOGEL [140] have chromatographed lactones, lactams and thiolactones of organic chemical interest. They used standard conditions on silica gel G layers with diisopropyl ether, ethyl acetate and isooctane alone and in mixtures.

Another group of lactones is that of the phthalides which occur in oil of lovage and celery oil for example and are responsible for the odour. MITSUHASHI and co-workers [170] carried out orientation tests of purity

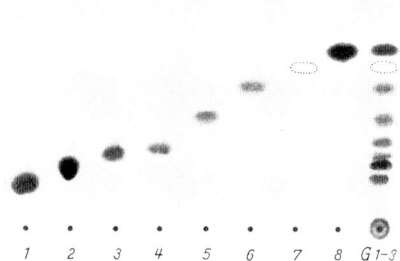

Fig. 102. Separation of some phthalides from coumarin. Layer: silica gel GF_{254}; detection: fluorescence quenching in short wave UV light [258]. *1* phthalide; *2* coumarin; *3* isobutylidene-3a, 4-dihydrophthalide; *4* phenylphthalide; *5* butylphthalide; *6* benzalphthalide; *7* ligustilide; blue fluorescence in 365 nm light; *8* butylidene phthalide. *G* mixture *1—8*

of some phthalides using TLC. STAHL and BOHRMANN [258] then succeeded in separating 7 phthalides on silica gel GF_{254} layers using two runs with n-hexane-diethyl ether (85 + 15) (Fig. 102). Chromatography on 500 μm thick silica gel GF_{254} layers at chamber saturation using benzene and also with a double run, serves for obtaining milligram amounts for preparative purposes. The zones on the chromatogram can

---

[6] See Chapter U, I for other coumarin derivatives which are involatile in steam.

be easily detected through fluorescence quenching or sometimes through fluorescence of the phthalides themselves. Scraping off and elution follows. HERZ and INAYAMA [89] have carried out similar separations on normal silica gel layers in order to purify the sesquiterpene lactones pulchellin A, B and C.

The ketolactone isolated from *Jasminum grandiflorum* by DEMOLE and co-workers [41, 44] can be satisfactorily separated from the other compounds on silica gel G layers, using petroleum ether (80—100° C)-ethyl acetate (75 + 25) (hRf 31). Investigations so far carried out indicate that it contains a 10-membered lactone ring. Higher membered lactone rings are found in the naturally occurring musk odorous principles like civetone, muscone etc.

Detection

a) *Transformations.* If the presence of an ester is established, the following can be made use of, in addition to what has been given above:

1. Saponification with alcoholic potassium hydroxide directly on the layer. Coumarin, for example, yields the potassium salt of coumaric acid which fluoresces yellow-green.

2. Reduction with lithium aluminium hydride [168]. The terpene alcohol and the alcohol from reduction of the acid component are formed. Lactones yield the corresponding diols.

The ester is spotted on to the start point and a drop of a 10% solution of lithium aluminium hydride in absolute ether is added. Chromatography is then carried out. If reduction is complete, a zone is found in the alcohol region only.

3. The acid components can be detected as hydroxamic acids by reacting esters and lactones with hydroxylamine. It must be tested whether these hydroxamic acids can be separated by TLC.

b) *Colour reactions.* DEMOLE [40] describes the application of the ester test mentioned under 3., as follows:

After development, the layer is freed from solvent and sprayed with water until it is transparent. A still moist strip of filter paper, freshly impregnated with a hydroxylamine reagent, is then pressed on to the layer with a glass plate. The impregnation solution is 100 ml of a mixture of equal parts of 7% aqueous hydroxylamine hydrochloride solution and 32% potassium hydroxide in methanol. The carrier plate is placed on a hot plate at 35—45° C. The esters evaporate and are converted in the paper to potassium hydroxamates. After 15—30 min, these can be visualised as red zones by spraying with a 5% solution of ferric chloride in 0.5 N hydrochloric acid.

All the colour reagents described for the terpene alcohols (p. 229) can be used successfully for the normal esters. The molybdophosphoric acid reagent [No. 168] can also be employed. The sensitivity of detection and the colour intensity are about the same. The coloration with the ester does not seem to be related to the acid component.

The potassium permanganate reagent (No. 200) is suitable for visualisation of the phthalides and coumarins. It gives yellow or white zones on a reddish background. Pyrolysis can also be utilised for localising spots (Rgt. No. 241). Mention has already been made of the test for phthalides through fluorescence quenching (silica gel GF_{254} layers) or through their own fluorescence. Anthranilate esters can be easily identified in the same way through their own fluorescence ($\lambda = 366$ nm) [45].

4. Aldehydes and Ketones

Numerous aldehydes and ketones, like the terpene alcohols, possess a characteristic odour. Some are easily accessible synthetically and are often used in the perfume industry. Commonly known examples are cinnamaldehyde with the typical smell of cinnamon; vanillin with the vanilla odour; and carvone smelling of caraway seed.

Both the aldehydes and ketones directly and also their crystalline derivatives may be investigated chromatographically.

a) Separation of Free Carbonyl Compounds

It is evident from published work that silica gel G layers are the best also for separating *aldehydes* and ketones [136, 176, 183, 273]. Only KOTAKIS and co-workers [278] have used layers of silica gel-celite-magnesium oxide (63 + 32 + 5) for chromatographing vanillin, ethylvanillin and other aromatic aldehydes. KLOUWEN and co-workers [132] have separated 24 benzaldehyde derivatives using the standard method and 3 different solvent systems (Table 25/I, II and III). They discussed in detail the dependence of the hRf-values on the position of the hydroxyl or methoxyl group in the benzaldehyde molecule. PETROWITZ [197] has carried out similar investigations on the same hydroxyaldehydes, using benzene-methanol (95 + 5) as solvent.

Ketones are closely related to aldehydes. Consideration of the polarity of the functional group leads us to expect that their hRf-values will lie in the same region as those of the aldehydes and consequently corresponding to the terpene esters (cf. Table 24). This is apparent from the published data of NIGAM and co-workers [183], SEVERIN [237a] and SCHRATZ and QEDAN [228].

DHONT and DIJKMAN [47] have separated the isomeric ionones and methylionones on air-dried silica gel G layers. developing 6 times with benzene. Whereas the lower acyclic ketones (up to about C_{10}) occur in essential oils, the homologous products of breakdown of keto-fatty acids are no longer volatile in steam. Separation can be achieved on silica gel G layers with benzene [152, 153] or n-hexane-diethyl ether (50 + 50) [190].

Table 25. hRf-values and colour reactions of some aldehydes and their 2,4-DNP derivatives [a]

Substance	hRf-values			Fluorescence 365 nm	Hydrazinium sulphate (No. 130)		hRf DNP IV
	I	II	III		Daylight	UV-light 365 nm	
Citral	13	42	51	0	0	dark	—
Citronellal	31	65	64	0	0	0	—
Hydroxycitronellal	1	—	11	0	0	0	—
Cyclamenaldehyde	37	63	65	0	weakly yellow	ochre	—
Safranal	10	—	—	0	—	—	—
Furfuraldehyde	13	—	50	0	—	—	—
Benzaldehyde	32	56	59	0	0	0	60
p-Tolualdehyde	27	50	63	0	0	blue	—
Cuminaldehyde	33	54	65	0	0	blue	—
α-Phenylpropionaldehyde	35	—	—	0	0	0	—
β-Phenylpropionaldehyde	30	52	55	0	0	0	—
Cinnamaldehyde	19	45	52	dark	dark yellow	orange	—
α-Amylcinnamaldehyde	33	63	66	dark	yellow	ochre	—
α-Hexylcinnamaldehyde	36	64	65	dark	yellow	ochre	—
Salicylaldehyde	—	48	57	light yellow	light yellow	orange	48
m-Hydroxybenzaldehyde	3	21	9	0	0	dark brown	32
p-Hydroxybenzaldehyde	2	16	6	0	yellow	yellow-green	30
Syringaldehyde	—	18	19	dark	brownish	0	5
Protocatechualdehyde	0	5	1	dark	dark yellow	weakly yellow	2
β-Resorcylaldehyde	2	21	8	weakly yellow	yellow	yellow	—
Gentisaldehyde	—	19	8	dark yellow	0	orange	—
3,4-Dihydroxy-5-methoxybenzaldehyde	—	7	4	0	yellow	yellow-brown	—
o-Vanillin	7	43	47	yellow	0	dark brown	32
o-Ethylvanillin	—	49	56	yellow	0	dark brown	—
Isovanillin	3	20	21	bluish	yellow	brown-yellow	23
Ethylisovanillin	—	25	29	grey-yellow	yellow	brown-yellow	—
p-Vanillin	5	27	27	dark	dark yellow	orange	23
p-Ethylvanillin	8	32	38	dark	yellow	orange	42
o-Anisaldehyde	21	51	57	blue	yellow	yellow-green	49
m-Methoxybenzaldehyde	—	55	60	0	0	brownish	50
p-Anisaldehyde	18	45	56	0	yellow	yellow	48
2,3-Dimethoxybenzaldehyde	11	48	57	yellow	weakly yellow	ochre	47
2,4-Dimethoxybenzaldehyde	9	38	50	dark blue	yellow	yellow-brown	41
2,5-Dimethoxybenzaldehyde	12	49	58	blue-green	yellow-orange	yellow-orange	45
Veratraldehyde	6	8	—	blue	yellow	yellow	34
3,5-Dimethoxybenzaldehyde	—	55	64	light brown	0	brown	47
Piperonal	18	45	57	blue	yellow	yellow-orange	—
3,4,5-Trimethoxybenzaldehyde	5	34	41	blue	yellow	ochre	—
2,4,5-Trimethoxybenzaldehyde (Asaraldehyde)	3	21	40	blue	weakly yellow	yellow	—

[a] Chamber saturation; silica gel G; solvent I: benzene; II: benzene-ethyl acetate-glacial acetic acid (90 + 5 + 5); III: chloroform; IV: ethyl acetate-ligroin (75—120° C) (33 + 67) [218]. Length of run: 10 cm, for IV, 14 cm.

Table 26. hRf-values of some ketones in various solvents
(Chromatographic details, see Table 25)

Substance	hRf-values			Substance	hRf-values		
	I	II	III		I	II	III
Camphor	6	33	27	Acorone	3	35	47
Carvone	13	48	63	Isoacorone	1	37	48
Dihydrocarvone	14	48	65	Frambinone	1	21	9
Zingerone	3	26	30	Phyllantone	18	60	—
Fenchone	—	—	33	Methyl naphthyl ketone	18	54	59
Menthone	24	62	—	Acetophenone	17	50	62
Piperitone	4	—	45	Methoxyacetophenone	10	38	55
α-Thujone	29	62	63	Benzophenone	31	—	—
β-Thujone	25	62	64				

Colour reactions have not been quoted. Yellow to orange zones are formed in some cases by spraying with 2,4-DNP (Rgt. No. 82); only a few ketones ($< 50 \mu g$) react with hydrazinium sulphate (Rgt. No. 130); methyl naphthyl ketone alone fluoresces itself (blue).

b) Separation of Derivatives of Aldehydes and Ketones

Oximes or 2,4-dinitrophenylhydrazones (DNPs), which crystallise well, are the principal derivatives used for concentrating and isolating carbonyl compounds from substance mixtures.

Thus it has been possible to separate a mixture of isomeric benzaldoximes, benzoin oximes and anisoin oximes on 200 μm silica gel layers, using benzene-ethyl acetate (83 + 17) chamber saturation [192]. o-, m- and p-Nitrobenzaldoxime isomers have been separated in the same way. The α-isomers always migrate further than the β-isomers.

The results of investigations on separation of DNPs of carbonyl compounds have been compiled in Table 27. It can be seen here also, from the very first publications in 1952 (ONOE [187]), that silica gel G is a most suitable layer material. Both ascending development and two-dimensional chromatography can be carried out, as performed by PAILER et al. [188] for low molecular weight carbonyl compounds. Better separations are accomplished in some special cases by impregnating with mineral oil [149], 2-phenoxyethanol [310], polyethylene glycols [9, 61] or undecane [153, 154]. These types of partition chromatographic separation systems, using multiple development, are especially good for differentiating homologous carbonyl compounds [310].

Separations into groups according to the number of C=C double bonds can be achieved on layers impregnated with $AgNO_3$. Appropriately treated alumina or kieselguhr layers are evidently better suited to chromatography here than are impregnated silica gel layers.

Inspection of the contents of Table 27 shows that the investigations of URBACH [310] (Fig. 103) and of BADINGS and WASSINK [9] are

Table 27. *Survey of the possibilities for chromatographic separation of dinitrophenylhydrazones of carbonyl compounds (2,4-DNP)*

Layer/Substance	Solvent	Remarks	Reference
a) Silica gel G layer			
19 aliphatic carbonyl compounds	CCl$_4$-n-hexane-ethyl acetate (77 + 15 + 8)	layer thickness 500 µm, double development	[162]
10 cyclic carbonyl compounds	petroleum ether (40—60° C)-diisopropyl ether (88 + 12)	layer thickness 500 µm, double development	[162]
4 aromatic carbonyl compounds	benzene-n-hexane (40 + 60)	layer thickness 500 µm, triple development	[162]
41 carbonyl compounds (mono-, bis-, tris-)	petroleum ether (80—100° C)-ether (70 + 30)	horizontal, continuous, SRS- and other techniques	[29]
5 lower fatty aldehydes, furfuraldehyde, acetone	I: toluene; II: CCl$_4$-diethyl ether (66 + 33)	layer thickness 270 µm, two-dimensional technique, run direction 1 = I, 2 = II	[188]
6 volatile aldehydes, glyoxal from alcoholic drinks	benzene-petroleum ether (60—80° C)-ethyl acetate (85 + 13 + 2)	methylglyoxal detected; time of run 1—2 h	[213] [275]
4 aliphatic methyl ketones from alcoholic drinks	benzene-petroleum ether (50—75° C) (75 + 25)	adsorbent slurried with ethyl acetate	[143]
volatile carbonyl compounds from *Streptomyces odorifer*	petrol ether-diethyl ether-ethyl acetate (90 + 5 + 5)	carbonyl compounds not identified, except acetaldehyde	[69]
9 ketones (menthone, pulegone, fenchone, etc.)	benzene-cyclohexane (50 + 50)	α-, β-thujone separated	[179]
α-, β-ionone, methylionones, pseudoionone	benzene	3 h activation in the air, six times developed	[47]
6 aliphatic and 7 aromatic aldehydes, α- and β-ionone	benzene-petrol ether (60—80° C) (75 + 25)	good separation of the aliphatic aldehydes with the first solvent	[48]
citral, citronellal from *Thymus spec.*	benzene-ethyl acetate (95 + 5) benzene-ethyl acetate (80 + 20)	layer thickness 250 µm, standard conditions	[228]
9 aliphatic and 3 cyclic carbonyl compounds	benzene-ethyl acetate (95 + 5)	detected with 10% potassium hydroxide; standard conditions	[172]
11 aliphatic aldehydes, 7 ketones	petrol ether (60—80° C)-ether (70 + 30)	reference standard: HCHO; 4 further solvents	[332]
glyoxal hydrazone	benzene-petroleum ether-ethyl acetate (85 + 12 + 3)	colour distinction between glyoxal mono- and di-DNP	[212]

Table 27 (Continued)

Layer/Substance	Solvent	Remarks	Reference
8 aldehydes	benzene-methanol-petrol ether-ethyl acetate (57 + 33 + 8 + 2)	investigations of wine distillates	[276]
15 oxoterpenes	I: benzene-petrol ether (70 + 30) II: $CHCl_3$–CCl_4	I was best; II in ratios 10 + 90, 15 + 85 and 5 + 95 = hRf values < 50	[315]
3 aliphatic methylketones	I: n-hexane-ether-ethanol (88 + 10 + 2); II: n-hexane-benzene-ether (48 + 48 + 4)	separation according to increasing C-number	[18]
27 aromatic aldehydes and 5 ketones	ethyl acetate-ligroin (75–120° C) (33 + 66)	layer: 500 μm; 14 cm run; colour intensified with NH_3	[218]
4 low molecular weight aldehydes + acetone	petroleum ether-dioxan (90 + 10)	solvent free of carbonyl groups	[34]
5 saturated and unsaturated cyclohexanones	petrol ether (60–80° C)-ethyl acetate (50 + 50)	cyclohexanone hRf 90; cyclohexan-1,2-dione hRf 50 etc.	[151]
b) Impregnated silica gel layers			
10 aliphatic aldehydes and 7 ketones	benzene-petrol ether (40–60° C) (60 + 40)	impregnation: 25% $AgNO_3$; further adsorbents and solvents	[20]
7 hexenal and 9 heptenal derivatives	benzene	impregnation: 30% $AgNO_3$; layer; 500 μm	[164]
11 aldehydes, 9 ketones	diisopropyl ether-formic acid-water (90 + 7 + 3)	impregnation: carbowax 600, 1500, 4000; separation according to increasing no. of C atoms	[61]
C_1–C_{14} n-aldehydes	dioxan-water (65 + 35)	impregnation: mineral oil "ondina"; 6 h time of run; reversed phase	[149]
homologous osazones	dioxan-water (60 + 40)	impregnation: mineral oil "ondina"; 7 h time of run	[32]
C_8- and longer-chained aldehydes and ketones	1. benzene-diethyl ether mixture 2. methanol-water mixtures or acetonitrile-acetic acid (75 + 25)	1. direction on silica gel 2. direction on silica gel with undecane impregnation	[153] [154]
isomeric methylcyclohexanones	benzene-heptane (30 + 70)	impregnated with chlorobenzene or nitromethane	[28a]
c) Alumina layer			
6 aldehydes and 6 ketones	I: benzene-n-hexane (50 + 50); II: ether	loose layer (800 μm), horizontal technique	[214]

Table 27 (Continued)

Layer/Substance	Solvent	Remarks	Reference
7 aliphatic carbonyl compounds (C_1—C_3)	cyclohexane-nitrobenzene (66 + 33) or hexane-chloroform-nitrobenzene (72 + 18 + 9)	silica gel, magnesol etc. not so suitable	[178]
21 isomeric ketones	cyclohexane-ether (80 + 20)	two-dimensional separation, small plate size	[239]
homologous carbonyl compounds	petrol ether-diethyl ether (96 + 4)	two-dimensional technique; multiple development, sometimes on impregnated adsorbents	[310]
d) Impregnated alumina layer			
saturated and unsaturated carbonyl compounds	petrol ether (30—40° C)-ether (84 + 16)	impregnation: 20% $AgNO_3$	[310]
unsaturated carbonyl compounds	I: $CHCl_3$-petrol ether (40—60° C) (40 + 60) II: ether-petrol ether mixtures of various compositions	impregnation: 31% $AgNO_3$; layer 200 or 800 μm; separation into groups according to no. of C=C bonds and configuration	[110]
e) Miscellaneous layers			
23 saturated and unsaturated aliphatic aldehydes and ketones	petrol ether (60—70° C)-benzene-pyridine (70 + 10 + 20)	layer: zinc carbonate + 6% amylopectin; other impregnated adsorbents	[9]
6 volatile carbonyl compounds	petrol ether-pyridine-benzene (80 + 8 + 8)	layer: zinc carbonate; colour differentiation between sat. and unsat.	[96]
homologous carbonyl compounds	petroleum ether (100—120° C)	layer: kieselguhr, impregnated with 2-phenoxyethanol	[310]
23 saturated and unsaturated alkanals and alkanones	petrol ether (60—70° C)	layer: kieselguhr, impregnated with 20% silver nitrate	[8]
glyoxal, methylglyoxal, diacetyl, C_5- and C_7-2,3-diketones	benzene-methanol (92 + 8), saturated with ethylamine	layer: sea-sorb 43-silica gel G (50 + 50)	[31]
glyoxal, methylglyoxal, diacetyl	$CHCl_3$-tetrahydrofuran-methanol (75 + 20 + 5)	layer: sea-sorb 43-celite-$CaSO_4$ (50 + 42 + 8)	[31]
13 saturated and unsaturated aldehydes, 3 ketones	$CHCl_3$-n-hexane (85 + 15)	layer: sea-sorb 43-celite (72 + 28), slurried with ethanol	[233]
7 dicarbonyl compounds	acetone-benzene-methanol (75 + 23 + 2) and (75 + 15 + 10)	layer: sea-sorb 43-celite (72 + 28); separation into 3 groups	[232]

especially worthy of emphasis, since they have dealt systematically with the separation of the DNPs of carbonyl compounds. RUFFINI [218] chromatographed 32 DNPs of aromatic aldehydes and ketones. His studies correspond to those of KLOUWEN and co-workers [132] on the free carbonyl compounds and are included along with the latters' results in Table 25.

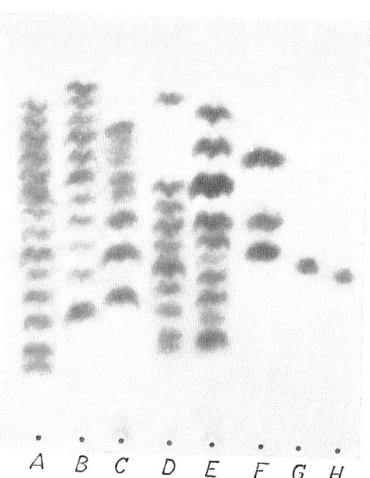

Fig. 103. Multiple development of homologous 2,4-dinitrophenylhydrazones on kieselguhr G-layers, impregnated with 10% phenoxyethanol-acetone solution [310]. Solvent: petrol ether (BP 100 to 120°C). 1—3. 9 cm developments; 4. 11 cm development

A Alkanal-DNP: C_{1-11}
B Alkan-2-one-DNP: C_{3-13}
C Alk-1-en-3-one-DNP: C_{4-10}
D Alk-2-enal-DNP: $C_{3-11, 16}$
E Alk-2,4-dienal-DNP: $C_{5-12, 14, 16, 18}$
F Alk-3-en-2-one-DNP: $C_{6, 7, 10}$
G Nona-all *trans*-dienal-(2,6)
H Nona-*trans*-2, *cis*-6-dienal

In his work on chromatography of 41 mono-, bis- and tris-DNPs, BYRNE [29] has particularly stressed separation conditions which are generally applicable over a wide range. Hydroxy-compounds are acetylated directly on the alumina or silica gel G layer; various TLC techniques (SRS-, multiple development, horizontal continuous development etc.) may follow. Like DHONT and DIJKMAN [47], MEHLITZ and co-workers [162] have used multiple development to separate 33 aldehydes and ketones on 500 μm-thick silica gel G layers, with carbon tetrachloride-n-hexane-ethyl acetate (81 + 16 + 8). Particularly interesting is how they distinguished between the yellow or orange zones by means of potassium ferricyanide-hydrochloric acid reagent (cf. test, p. 224). The low molecular weight carbonyl compounds can be differentiated in this way also. Both LIBBEY and DAY [149] and DHONT and DEROOY [48] have studied these compounds (Fig. 104).

After it had been shown that free aldehydes and ketones and also their derivatives could be separated with a one- or two-dimensional procedure, the problem remained of how to distinguish remaining critical pairs. The SRS-technique (p. 88) is often successful in this.

The mixture under investigation is chromatographed in direction 1 using chloroform; after covering up the remainder of the chromatogram, the shaded part (Fig. 182) is sprayed with a 0.5% solution of 2,4-dinitrophenylhydrazine in methanol, containing 1% added hydrochloric acid (25%), or with alkaline hydroxylamine hydrochloride solution. The aldehydes and many ketones react in the former case on the layer, yielding an orange-yellow colour. After 10—15 min, chromatography is carried out in direction 2 with, e.g., the solvent carbon tetrachloride-n-hexane-ethyl acetate (81 + 16 + 8), proposed by MEHLITZ [162]. At chamber saturation, anisaldehyde and cinnamaldehyde for example, can be thus separated from each other and from citral and veratraldehyde which migrate further or less far, respectively.

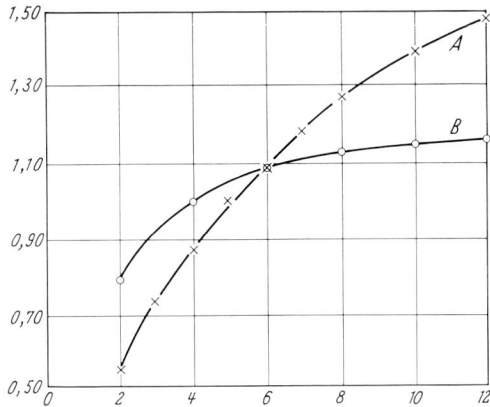

Fig. 104. R_B-values of 2,4-dinitrophenylhydrazones of aliphatic aldehydes, showing dependence on chain length. Solvent A: benzene-petrol ether (75 + 25); B: benzene-ethyl acetate (95 + 5)

Special separation problems may be solved by trying out "reaction layers", easily prepared by adding sodium bisulphite to the slurry.

Detection. Aldehydes and ketones are visualised most often with an acid 2,4-dinitrophenylhydrazine reagent (Rgt. No. 82). The corresponding hydrazones appear as coloured spots after brief reaction at room temperature. Non-cyclic aldehydes and ketones give mostly a yellow colour; and cyclic, orange to orange-brown colours. The sensitivity of detection varies. Aldehydes in amounts as small as 1—5 µg usually yield very distinct spots. Larger amounts (about 80 µg) are necessary with numerous ketones like menthone, fenchone, piperitone and camphor. According to GÄNSHIRT, [68a], camphor itself can be visualised more sensitively with the Dragendorff reagent (No. 96). The lower limit of detection is about 20 µg [264]. Unfortunately, compounds without an aldehyde or ketone group may give yellow or orange colorations with the DNP-reagent (No. 82); e.g., the hydroxyphenylpropane derivatives trans-isoasarone, apiole and myristicin etc.

The molybdophosphoric acid reagent (No. 168) does not react well with aldehydes and ketones (5 µg). Antimony (III)- and (V)-chloride mixture can, however, be sprayed on after having used the DNP reagent

(No. 82). A saturated solution of o-dianisidine in glacial acetic acid, as described by WASICKY and FREHDEN [323], is satisfactory; aldehydes, especially furfuraldehyde, yield yellow to brown zones even in the cold. The colour intensifies on heating and the layer darkens. Larger amounts are required of some ketones like carvone, menthone and piperitone (20—50 µg). This reagent is not specific; colorations are given by, e. g., ascaridole, 1,8-cineole, hydroxyphenylpropane compounds and some terpenes.

A 0.5% solution of cupric chloride (Rgt. No. 54) can be used for visualising *oximes*. α-Isomers become coloured (usually weakly green-brown) only after heating to 110° C; the β-oxime complexes show as green spots immediately after spraying.

According to MEHLITZ and co-workers [162], *2,4-dinitrophenyl-hydrazones* can be distinguished with a 0.2% solution of potassium ferricyanide in 2N hydrochloric acid, which serves for characterising the DNPs. The DNPs of saturated ketones give a blue colour very soon after spraying. Derivatives of saturated aldehydes react appreciably more slowly, turning olive green. Derivatives of unsaturated carbonyl compounds react either not all (up to ca. C_{10}) or colour feebly only after a long interval.

5. Terpene and Sesquiterpene Alcohols
a) TLC on Silica Gel Layers

The hRf-values of many alcohols are about the same when chromatographed with benzene-ethyl acetate (80 + 20) [151a] or chloroform on silica gel layers [183, 251]. Subdivision only into C_{10} and C_{15} alcohols can be recognised (Table 28). The alcohols which are not found in the "normal" region, are mentioned specially.

Table 28. *Separation of terpene and sesquiterpene alcohols by adsorption chromatography on a silica gel G layer using chloroform*

Compound	hRf-values[a]		
	NS[b]	CS[b]	S-chamber
13 Monoterpene alcohols	30—40	25—35	18—24
Thujyl alcohol	47—55	26—33	20—24
Menthols	48—62	32—44	22—33
Terpineol-(4)	68	45	36
17 Sesquiterpene alcohols	35—70	30—40	18—30
Daucol	16	14	14
α-Caryophyllene alcohol	77	41	30
Junenol	88	57	46
Carotol	93	56	45

[a] Only to be regarded as guide values.
[b] Chromatography chamber: NS = normal saturation at 24° C; CS = chamber saturation at 24° C.

It is surprising that many stereoisomers can be comparatively well separated, such as the menthols quoted in Table 29 [79, 105, 196, 260].

Table 29. *Separation of the menthol isomers*

Compound	hRf-value in solvent[a]			
	I	II	III	IV
Menthol	32	47	67	36
Isomenthol	29	41	62	37
Neomenthol	40	64	73	51
Neosiomenthol	36	68	76	55

[a] Solvent I = Chloroform-analytical grade, standard conditions, CS; II = n-hexane-ethyl acetate (90 + 10), standard conditions, NS; III = benzene-methanol (75 + 25), layer prepared according to [196]; IV = benzene-methanol (95 + 5), layer prepared according to [196]; hRf-values to be regarded only as guide values.

The 4 isomeric thujyl alcohols also have different hRf-values [177]. The sequence found, using chamber saturation and triple development with benzene-chloroform (50 + 50), was: isothujyl alcohol (hRf 47), thujyl alcohol (50), neothujyl alcohol (57) and neoisothujyl alcohol (67) [264].

The stereoisomeric farnesols have been similarly investigated [300]. These possess juvenile hormone activity [225, 226] and could be detected thin-layer chromatographically as the sexual attractant of the male bumble bee [269, 270]. The chromatography was carried out with benzene-ethyl acetate (95 + 5) on silica gel V- and szial gel 47-layers (Firm 114) [303, 309]. The *trans-trans*-farnesol (hRf 27) then lay below the *cis-trans*-farnesol (hRf 36). This sequence remains also when pure benzene is used for development [304] and is observed with the farnesol esters too [307–309]. SEIKEL and ROWE [237] have studied the TLC of the eudesmols. They separated milligram amounts of α- and β-eudesmol with benzene-petroleum ether (50 + 50) on alumina layers, 500 μm thick. The use of ascending technique is said to augment the differences in the hRf-values.

GUTZWILLER and co-workers [84] have purified and checked two furanoid sesquiterpene alcohols by chromatography also on silica gel layers, using mixtures of chloroform and methanol.

The "diffusion separation technique" of DEMOLE [42] can be applied to special problems of separation. As described on p. 82, the less polar, less strongly adsorbed compound can be concentrated and detected on the cover plate

b) Paraffin-impregnated Silica Gel Layers

The alcohols considered here have been succesfully fractionated into groups with the same number of C atoms by impregnating silica gel layers with paraffin or silicone oil and then using suitably hydrophilic solvents (Table 30 [112]). It is now possible to say with certainty whether a hemi-, mono-, sesqui- or diterpene alcohol is present; this is normally very difficult to decide in the microgram region when using non-impregnated layers [113].

Table 30. *Separation of alcohols on a paraffin-impregnated silica gel G layer, using 70% methanol as solvent*

No. of C atoms	Alcohols	hR_f-values[a]	
20	Phytol, isophytol	2—5	
15	Farnesol, nerolidol, cedrol, guaiol, carotol, junenol, elemol (hR_f 25)	17—20	
10	Geraniol, nerol, isoborneol, borneol, linalool, menthol, terpineol etc.; cumic alcohol (hR_f 50)	42—48	
9	Cinnamyl alcohol	55	separated
7	Benzyl alcohol	59	in a
5	Furfuryl alcohol	63	mixture

[a] The hR_f-values decrease with increasing number of carbon atoms. They are to be regarded as guide values only. They were obtained in the S-chamber system, applying 5 µg of each.

Method. The dried silica gel G layer, after cooling to room temperature, is immersed for one minute in a 5% solution of paraffin or of silicone oil DC 550, in petroleum ether (see p. 48). The petrol ether (BP 40 to 60° C) is allowed to evaporate for 15 min with the plate horizontal and the layer then used in the normal way. Impregnation by ascending treatment with the paraffin-petrol ether solution has no advantage in this case.

A mixture of 70 ml of analytical grade methanol and 30 ml distilled water, saturated with liquid paraffin DAB 6 or with silicone oil DC 550, is the best solvent.

MCSWEENEY [151a] has chromatographed 6 mono-, sesqui- and diterpene alcohols in like fashion on paraffin-impregnated kieselguhr layers, using acetone-water (65 + 35). GAREL [70] has summarised briefly the conditions for separating alcohols of chemical interest.

c) Silver Nitrate-impregnated Silica Gel Layers

Once the alcohols have been subdivided into groups containing the same number of C atoms, they can be further separated on the basis of the number of C=C bonds each contains. Using gradient-TLC, STAHL and VOLLMANN [267] have been able to show that 2.5% was the optimum and most economical content of silver nitrate in the layer (Fig. 105a). A mixture of methylene dichloride-chloroform-ethyl acetate-n-propanol (45 + 45 + 4.5 + 4.5) was used to separate 9 mono- and sesquiterpene alcohols. A good separation was accomplished with a run of 15 cm but it could be improved further by following up with double development using methylene dichloride-chloroform (40 + 60) (Fig. 105b II, III).

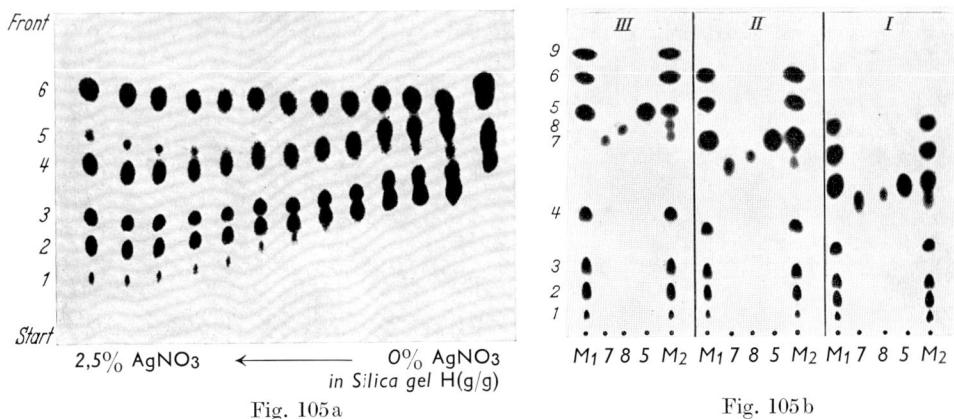

Fig. 105a.   Fig. 105b

Fig. 105a. Establishment of the optimum AgNO$_3$-concentration on silica gel H layers for separation of terpene alcohols (1—6, see below under Fig. 105b)

Fig. 105b. Multiple development of a mixture of 4 mono- and 5 sesquiterpene alcohols on silver nitrate-silica gel layers [267]. I: single development with methylene dichloride-chloroform-ethyl acetate-n-propanol (45 + 45 + 4.5 + 4.5); II and III: second and third developments, with methylene dichloride-chloroform (40 + 60)

1 nerolidol	5 borneol	9 carotol
2 geraniol	6 cedrol	M_1 mixture 1—6,9
3 nerol	7 cumic alcoho	M_2 mixture 1—9
4 guaiol	8 daucol	

PESNELLE et al. [195] have used 10% silver nitrate-silica gel G layers and cyclohexane-ethyl acetate (60 + 40) in order to separate galbanol [327], formerly believed to be a single compound, into bulnesol, guaiol and β-eudesmol.

d) Separation of the Dinitrobenzoate Esters (DNBs)

Another way of characterising alcohols is through TLC after having treated them with 3,5-dinitrobenzoyl chloride (preparation, see Chapter TS, 3a). DHONT and DEROOY [49] were the first to show that the dinitrobenzoate esters could be separated by using benzene-petrol ether (50 + 50) on silica gel G layers. Butter yellow (p-dimethylamino-azobenzene) was chosen as reference substance to determine the positions in the chromatogram. The following R_B-values,

$$R_B = \frac{\text{distance of migration of the substance}}{\text{distance of migration of the butter yellow}}$$

were found for the dinitrobenzoates of: furfuryl alcohol 0.71; benzyl alcohol 0.76; maltol 0.92; geraniol 1.26; citronellol 1.69.

BRAUN [20], GRAF and HOPPE [79] and MEHLITZ and co-workers [163] have used similarly non-polar solvent combinations. Menthol and isomenthol dinitrobenzoates can be separated with ligroin (105–120° C)-isopropyl ether (95 + 5) on silica gel G layers under normal conditions [79]. The dinitrobenzoates of other terpene and sesquiterpene alcohols could be clearly differentiated by double development on 500 µm-silica gel G layers using petrol ether-isopropyl ether (94 + 6). Solvent mixtures containing twice as much ether are more suitable for TLC of the dinitrobenzoates of lower alcohols.

Visualisation. The molybdophosphoric acid reagent (No. 168) serves for sensitive, though non-specific, detection of alcohols. Amounts of 0.5–1.0 µg can be detected on silica gel layers as blue-grey spots on a yellow background, after spraying and heating. The antimony(III)- and (V)-chloride reagents (Nos. 15 and 18) are less sensitive. When they are used, the chromatogram should be examined in daylight and in long wave UV light (365 nm) both before and after heating. Amongst the alcohols studied (5 µg amounts), geraniol, nerol, linalool, terpineol, nerolidol, farnesol, guaiol and phytol react without heating to give a grey to violet colour. After heating, these and the remaining alcohols of Table 30 mostly turn brown. The majority of the alcohols and their esters fluoresce brown-red in long wave UV, in contrast to the phenol ethers (see there).

The sensitivity of detection with the anisaldehyde-sulphuric acid reagent (No. 11) is greater than that with the antimony chloride reagents; the former yields a certain colour differentiation, ranging from grey-blue through violet, rose-red to green. It is noticeable that the esters give colours similar to those from the corresponding alcohols. The lower limit of detection of esters is of the same order as that of the free alcohols (cf. also [312]). Rhodamine B solutions (Rgt. No. 220) and the fluorescein-eosin test (Rgt. No. 118) are other detection agents which have been used with success.

6. Phenylpropane and Phenol Derivatives
a) Silica Gel Layers

Among the substances which are treated in this section, the methoxyphenylpropane derivatives are of more interest in medicine and in the perfume industry than are the steam-volatile phenols. Compounds of this type occur in a number of plants or in the essential oils obtained from them; cloves, pimento, anise, fennel, parsley, dill, calamus and sassafras may be mentioned.

Numerous compounds can be separated successfully on 250 µm-silica gel GF_{254} or HF_{254} layers, prepared in the usual way and with benzene as solvent (Fig. 106) (cf. also [282]). PASTUSKA and PETROWITZ [191a], FISHBEIN and co-workers [63] and FIORI and MARIGO [62] have recommended equipolar solvents like petrol ether (BP 50–75° C)-diethyl ether (85 + 15) or n-hexane-ethyl acetate (95 + 5). Catechol and the other "di- and triphenols" then remain at the start point (Table 31). They can, however, be separated with more polar solvents [70].

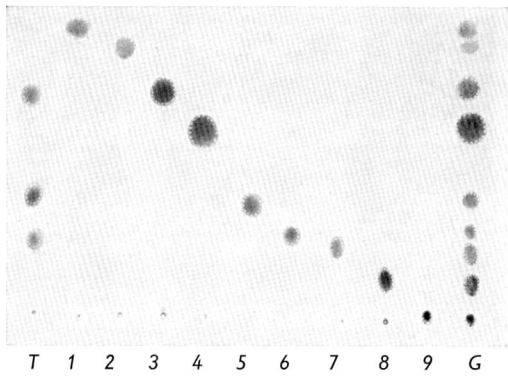

Fig. 106. Separation of methoxyphenylpropane derivatives (4 µg of each) on a silica gel G layer, using benzene. S-chamber; 10 cm run, lasting 35 min; spray reagent: molybdophosphoric acid solution (No. 168). *T* "Desaga" test mixture; *1* safrole; *2* methylchavicol; *3* myristicin; *4* apiole; *5* eugenol methyl ether; *6* asarone; *7* tetramethoxyallylbenzene; *8* elemol; *9* catechol; *G* mixture (2 µg of each)

LAPINA [146] and KLOUWEN and TER HEIDE [131] have suggested two-dimensional TLC. If benzene or chloroform is used in the first direction, acidic or basic conditions are used in the second. KRATZL and MIKSCHE [141] have employed an acetic acid-containing solvent with water-saturated benzene for separating the phenolic alcohols 1-guaiacyl-propane-1,3-diol, coniferyl alcohol and dihydroconiferyl alcohol. SEEBOTH [234–236] and PASTUSKA and co-workers [191a] likewise quote acid mobile phases. The former then concluded nevertheless that the diphenols are best separated with chloroform-acetone-diethylamine (64 + 32 + 3) on "acidic" alumina-silica gel[7]-gypsum layers (42 + 42 + 17). GABEL and collaborators [68] have distinguished the isomeric phenols, thymol and carvacrol, on ordinary silica gel G layers, using chloroform-pyridine-benzene (65 + 5 + 30) (cf. also [228]). If alkalised silica gel G or alumina layers [126] are used instead of a basic solvent, the phenols migrate less than the neutral phenol ethers as a result

---

[7] Supergel (Firm 146).

of phenolate formation. Conversely, larger hRf-values are observed on acidic reaction-layers in comparison with neutral stationary phases [70]. Acetylated polyamide layers [80a] or kieselguhr, impregnated with formamide, can be used successfully for differentiation [242]. Since the degree of saturation and the nature of the chamber have a not negligible influence on the hRf-values in adsorption chromatographic separation (cf. p. 66), these guide values are quoted in Table 31 for S-chamber and for chambers with and without chamber saturation. Standard conditions were otherwise used.

b) Structure and h*Rf*-Value

The following *rules of thumb* apply to the separation and identification of phenylpropane and phenol derivatives when standard conditions are used; they are the result of systematic investigations and have been confirmed by KLOUWEN and TER HEIDE [131].

1. Free phenol groups have a powerful influence on the adsorption affinity. Monophenols migrate but diphenols remain at the start point in benzene or chloroform solvents.

2. Phenol ethers show distinctly lower adsorption affinity than the corresponding phenols (phenol-anisole; resorcinol-resorcinol dimethyl ether).

3. The hRf-value falls as the content of methoxyl groups increases (anethole = 1 —OCH_3 group: hRf 83; isoeugenol methyl ether = 2 —OCH_3 groups: hRf 22; isoasarone = 3 —OCH_3 groups hRf 13). The position of the methoxyl groups influences the position of the compound in the chromatogram. Elimicin with 3 vicinal —OCH_3 groups is more strongly adsorbed than isoasarone which can be regarded as a hydroxy-hydroquinone derivative. At least a partial separation of the 6 isomeric trimethoxyallylbenzenes should be possible.

4. The more identical substituents are present in the ring, the less marked is the influence of the individual group on the adsorption. The adsorption affinity often appears to fall as, e.g., with tetramethoxyallylbenzene which has the same hRf-value as elemicin (trimethoxyallylbenzene).

5. The polarity of the compound is reduced if two neighbouring methoxyl groups in an aromatic ring are replaced by a methylenedioxy group (eugenol methyl ether hRf 22-safrole hRf 89; elemicin hRf 12-myristicin hRf 61; tetramethoxyallylbenzene hRf 12-parsley apiole hRf 45).

6. Introduction of an aliphatic side chain (methyl-, ethyl-, allyl-, propenyl-,) into the nucleus influences only slightly the adsorption affinity and hence the hRf-values on normal silica gel G layers.

c) TLC of Phenol Esters, Coupling Derivatives and Other Condensation Products

Phenols may be better characterised by conversion into *3,5-dinitrobenzoates*, as with terpene alcohols (p. 228). In TLC on silica gel G layers under standard conditions using butter yellow as reference substance, the following R_B-values have been found for dinitrobenzoates of: eugenol 0.55; isoeugenol 0.56; phenol 0.76; α-naphthol 0.89; β-naphthol 0.93; cresols 1.00–1.05; thymol 1.42 [49].

Table 31. *Phenol and hydroxyphenylpropane derivatives, hRf-values and colour reactions*
(Silica gel G layer; benzene solvent)

Substance	hRf-values Chamber CS	NS	S-chamber	Colour reactions Antimony(III)- and (V)-chloride (1 + 1) at room temp.	after heating[a]	(Nos. 15 and 18) UV-fluorescence after 24 h	Anisaldehyde-H$_2$SO$_4$ reagent (No. 11) after heating[b]
Safrole	57	90	89	—	grey-violet	brown-red, light edge	blue-green-grey
Isosafrole	57	90	89	grey-violet	blue-violet	violet-grey, light edge	grey-violet
Anethole	56	87	83	—	violet-grey	violet-pink	grey-red (bluish edge)
Methylchavicol	55	85	80	—	olive-green-grey	pink	violet-grey (after 15′)
Myristicin	46	70	61	light brown	grey-brown	dark, reddish edge	brown-green-grey
Isomyristicin (*trans*)	46	70	61	—	brown-violet	brown violet	grey-brown
Resorcinol dimethyl ether	40	72	56	grey-brown	brown-green	dark	red
Apiole	35	58	45	light brown	olive-grey	dark	violet-grey
Isoapiole (*trans*)	35	58	45	violet	brown-violet	dark	blue-violet
Hydroquinone dimethyl ether	31	54	45	—	yellow-green to brown	dark	weakly violet-grey
Carvacrol	25	47	33	—	red to brown	red	light red-red-brown
Thymol	23	47	33	—	blue-red	red violet	red
Catechol diethyl ether	22	47	32	—	blue-grey	dark	red
Guaiacol	18	40	27	—	grey-black	dark	brown-grey
Eugenol	18	40	27	grey	brown-grey	violet-grey	dirty green
Eugenol methyl ether	15	31	22	violet-brown	brown-violet	beige, greenish edge	brown, blue edge
Isoeugenol methyl ether	15	31	22	violet-green	violet	reddish-brown, light edge	dark violet
Veratrol	13	29	17	—	blue-black	dark	weak reddish-lilac
Tetramethoxyallyl-benzene	11	20	12	—	olive green	dark	beige-brown (after 15′)
Phenol	10	19	13	—	grey-brown	dark	red-orange
Asarone (*trans*-Iso)	10	20	13	grey-yellow	grey-green	dark	red-lilac
Elemicin	8	12	8	weakly grey	grey-brown	dark reddish	brown-violet
Catechol	0	0	0	—	light pink-lilac	dark	red
Homocatechol	0	0	2	brown-grey	brown	dark	red
Resorcinol	0	0	0	—	brown	dark violet	red-orange
Hydroquinone	0	0	0	—	yellow-brown	dark	brown-grey
Butter yellow	32	60	53				
Sudan red	10	24	17				
Indophenol	3	10	8				

CS = chamber saturation; NS = without saturation (see p. 66).
[a] 10′ at 100—105° C.   [b] No colour reaction before heating.
Times for the 10 cm run: NS = 35′; CS = 50′; S-chamber = 40′.

Their *condensation products* with 4-aminoantipyrine can be used also to separate phenols, possibly critical pairs. GABEL and co-workers [68] quote chloroform-pyridine-benzene (65 + 5 + 30) as solvent.

Should this method also fail to provide information about the phenols or phenol isomers present, TLC of suitable *coupling products* may be attempted. KNAPPE and ROHDEWALD [133a] have separated phenol, the 3 cresol and 6 xylenol isomers in this way. They chromatographed the coupling products with Fast Red AL Salt (Firm 127) on silica gel G-potassium carbonate layers in combination with oxalic acid-silica gel G layers; solvents were methylene dichloride-ethyl acetate-diethylamine (92 + 5 + 3) and chloroform-ethyl acetate-ethanol (93 + 5 + 2).

d) Separation of the Methoxyallylbenzenes from Their Cis-Trans Propenyl Isomers

Chromatography on silver nitrate-impregnated silica gel G layers has proved a good method for separation and characterisation of isomeric phenylpropane derivatives [177a, 259]. The degree of impregnation was determined in advance with the help of the gradient technique (p. 89). Petrol ether (50—75° C)-diethyl ether (85 + 15) can be used as solvent. The n-compounds, which form more stable complexes, thus migrate less far then the corresponding iso-forms. This was observed by PEYRON [199] for anethole and methylchavicol even on ordinary silica gel G layers, using benzene-pyridine as solvent.

Detection of the substances listed in Table 31 is possible through fluorescence quenching in short wave UV light, on silica gel GF_{254} or HF_{254} layers. Since the compounds are aromatic, they absorb strongly in the 250—280 nm region, in contrast to most terpene- and sesquiterpene hydrocarbons or alcohols. The intensity of fluorescence of the adsorbent layers is thus reduced or completely eliminated and the substances, mostly without suffering decomposition, appear as dark zones on the yellow-green fluorescing background.

The spots can be visualised with molybdophosphoric acid (Rgt. No. 168). After heating they show as blue-grey zones on a yellow background. The best colour distinctions are obtained with the anisaldehyde-sulphuric acid reagent (No. 11) or with a 1:1 mixture of antimony (III)- and (V)-chlorides (Rgts. 15 and 18). The addition of antimony(V)-chloride intensifies the colour but has the disadvantage that the coloured spot shows no fluorescence at first in long wave UV light. This fluorescence does appear a day later but with the pure antimony(III)-chloride solution, fluorescence appears directly after heating, permitting certain isomers to be distinguished through this fluorescence. Thus eugenol methyl ether fluoresces yellow green, clearly distinguishable from the reddish brown of isoeugenol methyl ether (*cis-trans* mixture) [248].

Gabel and co-workers [68] have described a colour reaction between phenols and 4-aminoantipyrine in alkaline solution in the presence of potassium ferricyanide (Rgt. No. 3). Salmon pink (carvacrol) to red-orange (thymol) zones are formed on the thin-layer chromatogram.

The colour reactions generally depend on the substance and reagent amounts and also on temperature and duration of heating. Insufficient reagent is often used. A 20×20 cm layer should be sprayed with 10 to 15 ml solution. An alkaline diazonium salt solution (Rgt. Nos. 100, 238) can be used as spray reagent in orientation tests whether a compound containing a free phenol group is present. Phenols couple giving coloured products; some other substances like azulenes and amines also react.

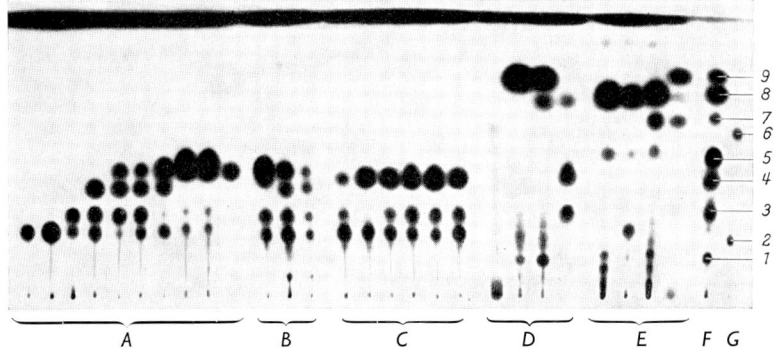

Fig. 107. TLC-comparison of the essential oils of various Daucus species using hexane-ether $(87 + 13)$ on a 40 cm wide silica gel GF_{254} layer at chamber saturation. Visualisation with molybdophosphoric acid (Rgt. No. 168)

A *Daucus carota subsp. maximus*
B *D. carota*, mixed forms
C *D. carota subsp. hispanicus*
D *D. carota subsp. carota*
E *D. carota subsp. sativus*

F mixture of
  1 geraniol
  4 eugenol methyl ether
  5 unidentified alcohol
  8 carotol
G test mixture of
  2 Sudan red
  3 elemicin with asarone
  7 epoxydihydro-caryophyllene
  9 geranyl acetate
  6 butter yellow

Application: TLC has been, generally speaking, the prerequisite for recognition of "chemical races" in medicinal plants. Differences of this sort, fixed by heredity, have been found in species which contain hydroxyphenylpropane compounds, like, for example, calmus [248–250], parsley [261], asarum [113, 262], carrots [255] and dill [1, 257]. The same plant organs must be investigated each time so that natural differences in the composition of the oil do not lead to false conclusions (cf. Reitsema, Cramer and Fass [210]).

As Gabel and co-workers [68] and Frömming [67] have been able to show, synthesis and control of quality are other spheres of application of TLC.

III. Essential Oils

1. Mixtures of Terpene and Sesquiterpene Derivatives (Chapter J II, 1—6)

Essential oils are steam-volatile, lipophilic mixtures of naturally occurring substances from plants; they usually possess a characteristic odour. In addition to the one or several chief constituents which are specific for the oil concerned, numerous (up to ca. 50) other compounds of the types discussed in the preceding paragraphs may be present.

2. Sulphur-Containing Oils

The sulphur-containing essential oils occupy a special place. After Curtis et al. [3, 4, 48] had separated many thiophene derivatives on silica gel G or alumina G layers and had pointed out [230] that thiophene alkynes frequently occur together with polyacetylenes in plants, Schultz and co-workers [231] and Wagner et al. [320] carried out TLC of the various allyl isothiocyanates. Schultz's team has separated the components of garlic oil in the S-chamber, using carbon tetrachloride-methanol-water (60 + 30 + 3) or (60 + 30 + 6) on silica gel G layers. Wagner and co-workers used the same adsorbent. The latter, however, prepared the corresponding thiourea derivatives of the isothiocyanates and chromatographed these with the upper phase from the mixture ethyl acetate-chloroform-water (30 + 30 + 40). A good separation of aliphatic and aromatic isothiocyanate homologues was achieved.

Preparation of the thiourea derivatives: 0.3 g essential oil is dissolved in 1 ml 95% ethanol and treated with the same volume of 25% ammonium hydroxide solution. After warming this mixture on the water bath until the exothermic reaction sets in, it is left to cool until the thiourea derivative crystallises out. The products are recrystallised from ethanol-water mixtures.

3. Essential Oils with Polyacetylene Compounds

Polyacetylene derivatives form a special group of compounds which are more and more often encountered. They are isolated preferably by extraction, however, since they are partly destroyed by steam distillation [134].

These compounds have been investigated thoroughly by Bohlmann's and Schulte's teams. Up to now, about 150 alkynes have been isolated [217, 333]. Schulte et al. [230] recommend silica gel G layers for separating naturally occurring polyacetylene compounds which are conspicuously often found along with thiophene derivatives in plants (p. 235). Mixtures of petrol ether (BP ca. 60° C) and diethyl ether in various proportions have proved to be good solvents. KNÜTTER and POHLOUDEK-FABINI [134] have suggested chloroform-benzene-cyclohexane (33 + 33 + 33) for TLC of the matricaria ester in oils from *Solidago virgaurea*. STAHL and SCHEU [264] have accomplished a good separation of such types of compounds in oils from the roots of *Chrysanthemum vulgare* (tansy), using two runs with petrol ether (50—75° C)-methylene dichloride (70 + 30).

Preparative TLC on silica gel HF_{254} layers can be employed for isolation of milligram amounts of the alkynes [15, 16]. Most polyacetylenes quench fluorescence and are, like the phenylpropane derivatives, easily detectable as dark zones.

The following must be borne in mind in the microanalytical characterisation of essential oils with TLC and/or GC:

a) commercial essential oils have been nearly always subjected to rectification so as to conform to the demands concerning odour and/or taste. Fractionation in this way often removes low- and high-boiling parts. Essential oils which have been largely freed from terpenes and sesquiterpene hydrocarbons are regarded as especially high grade. Oils of different qualities are often mixed.

b) the composition of the essential oils depends on numerous factors:

α) on the botanical uniformity of the plant material used ("chemical races") (cf p. 234).

β) on the plant organ used (flower, leaf, bark or root).

γ) on the place of origin, climate, harvest and storage.

δ) on the processing procedures and the chemical changes which occur during isolation.

c) only through combination of several analytical procedures can identification and quantitative determination of the individual components be conclusive [260, 266] (cf p. 114, 203).

A discussion of each of the newer articles[8] summarised in Table 32 would be outside the scope of this chapter. Its general value would be small in any case, since most of the articles are concerned with special problems. The publications are classified alphabetically according to species.

Table 32. *Separation of essential oils and terpene-derivatives*[a]

Plant	Layer	Solvent	Reference
Achillea fragrant.	KG	Benzene-ethyl acetate (95 + 5)	[238]
Achillea millefolium	KG	Benzene-ethyl acetate (95 + 5)	[221]
Acorus calamus	KG	n-Hexane-ethyl acetate (90 + 10)	[155]
	KG	Benzene	[222, 314, 315]

[a] KG = Silica gel G; HF_{254} = Silica gel HF_{254}; Al = Alumina.

[8] Work up to 1961 was compiled in Table 24 of the 1st Edition [251].

Table 32 (Continued)

Plant	Layer	Solvent	References
Amomum subulatum	KG	no details given	[184]
Amyris balsamifera	KG	Benzene-chloroform (50 + 50)	[325]
Anethum graveolens	KG	Benzene-chloroform (50 + 50)	[12]
and sova	KG, HF$_{254}$	Benzene-methylene dichloride (50 + 50)	[1]
	KG	Benzene-ethyl acetate (95 + 5)	[93]
Angelica archangelica	KG	Benzene-chloroform (50 + 50)	[12]
Araucaria imbricata	KG	no details given	[30]
Artemisia absinthium	KG	Benzene	[302]
		no details available	[316]
Asarum europaeum	KG	Trichloroethylene-chloroform (75 + 25)	[77, 113, 262]
Cannabis sativum	KG	no details given	[241]
Capsicum annuum	KG	Benzene; benzene-ethyl acetate (98 + 2)	[283]
Carica papaya	KG	Benzene-petrol ether (50 + 50)	[120]
Carum carvi	KG	Benzene-chloroform (50 + 50)	[12]
	KG	various solvents	[56]
	KG	n-Hexane-diethyl ether (87 + 13)	[203]
	KG	Benzene-ethyl acetate (95 + 5]	[93]
Cedarwood oil	KG	Benzene-ethyl acetate (95 + 5)	[108]
Centaurea species	KG	Benzene-chloroform (90 + 9), also on Al	[3]
Ceratocystis species	KG	impregnated with undecane or silicone oil	[244]
Chenopodium ambrosioides	Al	Benzene-ethyl acetate (95 + 5)	[52]
Chrysanthemum vulgare	KG	Benzene	[222]
	KG	no details given	[223a, 263, 265]
Cinnamomum species	KG	Benzene-chloroform (50 + 50)	[325]
	KG	Benzene	[14, 221]
	KG	Methylene dichloride-isopropyl ether (97 + 3)	[189]
	KG	Cyclohexane-ethyl acetate (95 + 5)	[75, 155]
Citrus bergamia	KG	n-Hexane-ethyl acetate (70 + 28)	[274]
Citrus decumana	KG + starch	Benzene-ethyl acetate (85 + 15)	[181]
Citrus media and aurantium	KG	Methylene dichloride-isopropyl ether (93 + 7)	[189]
Citrus limetta var.	KG	Methylene dichloride-isopropyl ether (90 + 10)	[189]
Citrus oils	KG	Benzene	[116b]
	KG	Benzene-chloroform (50 + 50)	[221]
	Al	Hexane or methylene dichloride	[6a]
	KG	various solvents	[56, 315a]
	KG	Benzene-chloroform (50 + 50)	[325]
Grapefruit oil	KG	Petrol ether (50—70° C)-ethyl acetate (75 + 25)	[211]
Coriandrum sativum		no details available	[194]
	KG	Benzene-chloroform (50 + 50)	[325]
	KG	Benzene-ethyl acetate (95 + 5)	[93]
Cuminum cyminum	KG	Benzene-chloroform (50 + 50)	[12]
	KG	Benzene-ethyl acetate (95 + 5)	[58]

Table 32 (Continued)

Plant	Layer	Solvent	Reference
Cymbopogon species	KG	Benzene-chloroform (50 + 50)	[325]
	KG	n-Hexane-diethyl ether (90 + 9)	[183]
	KG	Cyclohexane-methyl ethyl ketone (95 + 5)	[277]
Daucus carota	KG + starch	Benzene-ethyl acetate (85 + 15)	[181]
Daucus carota subsp.	KG, HF$_{254}$	Frigene (Freon) 21 at —9° C	[255]
Elettaria cardamomum	KG	Benzene	[221]
Eucalyptus globulus	KG	Benzene	[221]
	KG	Benzene-chloroform (50 + 50)	[325]
	KG	n-Hexane-ethyl acetate (95 + 5)	[155]
Eugenia caryophyllata	KG	Benzene	[221]
	KG + starch	Benzene-ethyl acetate (85 + 15)	[181]
	KG	Benzene-isopropyl ether (80 + 20)	[189]
Foeniculum vulgare	KG	n-Hexane or benzene	[93, 221]
	KG	Benzene-chloroform (50 + 50)	[325]
Gaultheria procumbens	KG	n-Hexane-ethyl acetate (95 + 5)	[155]
Geranium oil	KG	n-Heptane-benzene (50 + 50)	[198]
	KG	Benzene-chloroform (50 + 50)	[325]
Hamamelis virginiana	KG	Benzene	[107]
Humulus lupulus	KG	Petrol ether; chloroform; stepwise development and SRS-technique	[220]
Hypericum perforatum	KG	various solvents at —15° C	[157—159]
Illicium verum	KG	Benzene-chloroform (50 + 50)	[325]
		Benzene-ethyl acetate (95 + 5)	[93]
Juniperus sabinae	KG	Benzene	[221]
Lavandula species	KG	Benzene-ethyl acetate (95 + 5)	[92, 108]
	KG	no details given	[194]
	KG	Benzene	[222]
	KG	Benzene	[161b]
	KG	Benzene-chloroform	[325]
	KG	Methylene dichloride-isopropyl ether (90 + 10)	[189]
	Al	impregnated with silver nitrate	[329]
Loroglossum hircinum	KG	Chloroform-methanol (95 + 5)	[311]
Lycopus europaeus	KG	Benzene	[94]
Majorana hortensis	KG	Benzene	[221]
Matricaria chamomilla	Szial gel	Benzene	[305, 306]
	KG + Al	Benzene	[156]
Mentha species	KG	Benzene	[215a, 221]
		no details available	[193]
	KG	Benzene-isopropyl ether (80 + 20)	[189]
	KG	Benzene-chloroform (50 + 50)	[325]
	Szial gel V	Benzene	[284]
	KG	n-Hexane-diethyl ether (90 + 9)	[183]
Myristica fragrans	KG	no details given	[194]
	KG	Benzene	[221]
Nigella sativa	KG	Petrol ether; petrol ether-benzene mixtures	[54]
Ocimum gratissimum	KG + starch	Benzene-ethyl acetate (85 + 5)	[181]
Ocimum species	KG	Benzene-ethyl acetate (95 + 5)	[90]
Origanum vulgare	KG	Benzene	[221]

Table 32 (Continued)

Plant	Layer	Solvent	Reference
Pastinaca sativum		no details given	[13]
Petroselinum hortense	KG	Trichloroethylene-chloroform (75 + 25)	[113, 261]
	KG	Benzene-ethyl acetate (95 + 5)	[92]
	KG	Benzene-chloroform (50 + 50)	[12]
Peumus boldus	KG	Benzene	[322]
Pimpinella anisum	KG	Benzene	[93, 221]
	KG	Benzene-chloroform (50 + 50)	[12, 279, 325]
		no details given	[64]
		Methylene dichloride-isopropyl ether (93 + 7)	[189]
	KG	Benzene-ethyl acetate (85 + 15)	[116a]
Pinus silvestris	KG	no details given	[116]
Rosa species	KG	Benzene-ethyl acetate (95 + 5)	[108]
	KG	Benzene-chloroform (50 + 50)	[325]
	KG	n-Hexane; benzene-ethyl acetate (75 + 25)	[115]
	KG	impregnated with silver nitrate	[329, 330]
Rosmarinus species	KG	Benzene-ethyl acetate (95 + 5)	[108]
	KG	Benzene-chloroform (50 + 50)	[325]
	KG + starch	Benzene-ethyl acetate (85 + 15)	[181]
	KG	Benzene	[221]
Rudbeckia species	KG	Benzene-chloroform (90 + 9)	[3]
Salvia species	KG	Benzene-ethyl acetate (95 + 5)	[224a, b]
	KG or magnesium silicate	Benzene	[22, 23, 39]
	KG	Benzene	[221]
Solanum laciniatum	Szial gel-47	Benzene	[301]
Tagetes minuta	Al	Petrol ether (40—60° C) or	[4]
	KG	Benzene-chloroform (90 + 9)	
Turpentine oil	KG	Benzene; benzene-ethyl acetate (95 + 5)	[221]
	KG	Benzene-chloroform (50 + 50)	[325]
		no details available	[145]
Thymus species	KG + starch	Benzene-ethyl acetate (95 + 5)	[181]
	KG	Benzene	[68, 221]
	KG	Benzene-chloroform (50 + 50)	[325]
	KG	no details given	[166]
	KG	n-Hexane-diethyl ether (60 + 40), two-dimensionally	[167]
	KG	Benzene-ethyl acetate (90 + 10)	[228]
Trachyspermum ammi	KG	Benzene-chloroform (50 + 50)	[12]
Valeriana procurrens	KG	Chloroform-benzene (80 + 20)	[317]
	KG	Benzene, normal saturation	[253]
	Al	Benzene	[142]

Summarising from all the articles cited, it may be said that as a rule, detailed fundamental work still has to be carried out on every essential oil. A beginning can be made with the separation on silica gel GF_{254} layers, using benzene or chloroform as solvent.

4. Supplement

Brief mention may be made of the separation of some "wax-like" compounds which have been detected in essential oils although, in accordance with general ideas, they should not be found in the steam-distillate. Such compounds are, firstly, the higher fatty and wax acids like palmitic and myristic acids [256], the chromatography of which has been studied by KARTNIG [119] in connection with the waxes of the Umbelliferae; and, secondly, plant waxes, which includes the "stearoptenes" of rose oils, investigated by ŠORM [243] and determined quantitatively by WOLLRAB and co-workers [329, 330]; the C_{13}-C_{34} alkanes and alkenes are considered here.

Visualisation of the terpenoid substances is best with the anisaldehyde-sulphuric acid reagent (No. 11), antimony chloride reagents (Nos. 15, 18) or molybdophosphoric acid (No. 168) (cf. also [312]).

The polyacetylene compounds yield colours only poorly with these reagents, however. The dicobalt octacarbonyl reagent (No. 68 [230]) or a permanganate-sulphuric acid reagent (No. 200) have thus been recommended for these compounds [51], if the method of fluorescence quenching on silica gel F layers cannot be used.

The isothiocyanates can be detected with ammoniacal silver nitrate solution (No. 225) or with potassium ferricyanide reagent (No. 111).

Waxes and the hydrocarbons in them can be detected by ignition or by spraying with a 1% sodium fluorescein solution, as suggested by HYYRYLAINEN [100].

IV. TLC of Involatile Terpene Derivatives

1. Diterpenes

The diterpenes are discussed in connection with the mono- and sesquiterpene derivatives with which they are closely related according to modern ideas of biosynthesis. BRIESKORN's team [24, 24a] has studied the bitter tasting picrosalvin in *Salvia* species. It is a diterpene-o-diphenol lactone [24a] which occurs also in other Labiatae. NICHOLAS [180] has worked likewise on *Salvia sclarea* and utilised successfully two-dimensional TLC on silica gel G layers. He used ethyl acetate as solvent and was able to justify the hypothesis about the incorporation of $^{14}CO_2$ into sclareol.

KAUFMANN and co-workers [121, 122, 123] have also used silica gel G layers for separating kahweol and cafestol. They chromatographed with benzene-cyclohexane (50 + 50) for 5–6 h after having impregnated 15 cm of the run with propylene glycol-methanol (33 + 66) and dried in air. FASSINA and co-workers [59, 60] have used n-propanol-xylene-water (70 + 20 + 10) as solvent and acid-impregnated silica gel G layers (N/15 KH_2SO_4) for separating aglycones from the corresponding glycosides.

Back in 1958, DEMOLE and LEDERER [43] reported the separation of 4 diterpene derivatives. They chromatographed with n-hexane-ethyl acetate (85 + 15) on silica gel layers which had been prepared according to REITSEMA [209]. The following hRf-values serve as rough guides: phytol 35, isophytol 50, geranyl-linalool 44, phytyl acetate 66.

Visual detection can be achieved by spraying with an aqueous 0.2—0.5% potassium permanganate solution. The reducing substances stand out as yellow spots against the moist, red-violet background. The zones become brown after excess reagent has been carefully washed out [40a]. These compounds can be visualised better with the anisaldehyde-sulphuric acid reagent (No. 11) or the antimony chloride reagents (Nos. 15, 18), either as individuals or in the mixture. Kahweol turns blue-violet and cafestol reddish yellow with an acetic acid-containing antimony chloride reagent (No. 16). Preliminary detection with iodine vapour does not interfere with the identification of the zones using the antimony chloride reagents [151a].

2. Triterpene Derivatives and Their Glycosides
(Saponins and bitter principles)

Compounds of the triterpene series may be regarded as $C_{30}H_{48}$-derivatives. They are cyclic compounds with 4 or 5 ring systems, apart from a few exceptions such as squalene. Both classes can be termed formally phenanthrene derivatives. They are thus related to the steroids, which partly expresses itself in their similar chromatographic behaviour [229, 298] (p. 347).

There seems little point in subdividing the triterpenes into classes according to polarity, as done with the C_{10}-derivatives. With a view to preliminary separation of the substances during isolation, a division into neutral and acid triterpene derivatives is to be recommended. The hydrophilic triterpene glycosides can then profitably be separated from these ("bitter principles" and "saponins").

a) Neutral Triterpenes
(Hydrocarbons, esters, alcohols, etc.)

One recognises among the variety of solvent systems which have so far been used for chromatographic separation of the neutral C_{30}-derivatives, that here too, benzene, chloroform or diisopropyl ether have had preference as major solvent components (Table 34).

Only non-polar hydrocarbons like squalene [124] or β-amyrene migrate on silica gel layers when n-hexane is used. Separation of compounds of equal polarity is still difficult. Whereas esters can sometimes be distinguished by adsorption chromatography (Table 33) [297], this usually fails with C_{30}-alcohols having the same number of hydroxyl groups. This applies to both alumina G and silica gel G layers [101, 103]. Individual compounds can, however, always be characterised [27, 71] especially when they belong to different triterpene types [299, 318] or differ in the number of hydroxyl groups. In this connection, it is noteworthy that IKAN and co-workers [102] have been able to separate epi-β-amyrin and β-amyrin as well as epi-lupeol and lupeol. The epimeric compounds migrate further than the n-compounds when n-heptane-benzene-ethanol (50 + 50 + 0.5) is used. A total of 35 tetra- and pentacyclic neutral triterpenes has been investigated.

Table 33. *hRf-values of neutral triterpenoids on silica gel G layers* [293]

Neutral triterpenoid	hRf-values[a]		
	I	II	III
α-Amyrin		38	12
β-Amyrin acetate			45
α-Amyrenone			31
Lanosterol	75	40	14
Dihydrolanosterol acetate			43
Methyl acetylursolate		77	26
Methyl acetyloleanolate		77	24
Methyl ursonate	85	51	IV
Methyl monoacetylcrataegolate	40	13	80
Methyl diacetylcrataegolate	72	37	92
Methyl diacetyldehydrocrataegolate	69	38	
Methyl monoacetyl-11-ketocrataegolate			77
Methyl monoacetylacantholate			73
Methyl echinocystate	59	15	
Dimethyl emmolate	73		

[a] Solvent I: diisopropyl ether; II: methylene dichloride; III: benzene; IV: diisopropyl ether-acetone (75 + 30); hRf-values are to be taken only as rough guides.

Special separation problems can occasionally be solved by changing the separation technique or adsorbent. Thus TSCHESCHE and co-workers [288, 290] have distinguished α- and β-amyrin by partition chromatography using n-butanol-2N ammonium hydroxide solution; and HUNECK [97] has separated the *Sorbus* triterpenes on fibrous alumina. PONSINET and OURISSON [202] have achieved a better characterisation of some triterpene alcohols by preparing their mono- or diepoxides directly on the silica gel layer (see p. 206 für directions) and then chromatographing twice with cyclohexane-ethyl acetate (85 + 15).

If the number and position of the double bonds in the substances studied, differ, their silver nitrate-complexes can often be separated on the basis of their stability [101]. Chloroform or carbon tetrachloride is recommended as solvent [137].

Table 34. *Survey of chromatographic investigations of neutral tripertenes on silica gel layers*

Substances	Solvent	References
Cycloartenediol derivatives	Petrol ether-ethyl acetate (95 + 5)	[50]
Germanicol	Petrol ether-ether (90 + 10)	[27]
Methyl oleanolate and ursolate	Petrol ether-ether (50 + 50)	[25]
Methyl esters of triterpene acids from *Liquidamber orientalis*	Diethyl ether	[98, 99]
Arburinol	Benzene	[318]
Aescin, aescinidin etc.	Benzene-ether (60 + 40)	[144]
Gypsogenin (hRf 25) Gypsogenin lactone (hRf 85)	Ether-benzene (80 + 20)	[127]
Triterpenes from *Lycopodium*	Benzene-chloroform (90 + 10)	[297]
Triterpene lactones	Benzene-CHCl$_3$-methanol (43 + 43 + 13)	[106]
Cucurbitacins *(Ecballium)*	Benzene-ethyl acetate (70 + 30)	[147]
4 isomeric 2,3-diols of methyl $\Delta^{2,12}$-oleanadienoate	Diisopropyl ether	[293]
6 Methyl esters of triterpene acids	Diisopropyl ether	[293]
6 Cincholic acid derivatives	Diisopropyl ether	[288]
Dimethyl esters of bremedolic acid and senegenin	Diisopropyl ether-acetone (95 + 5)	[291]
5 Sapogenins from *Quillaja* (hRf 20, 40, 60, 80, 93)	Diisopropyl ether-acetone (75 + 25)	[227]
Methyl betulinate (hRf 30)	Chloroform	[165]
Esters of triterpene acids from *Commiphora glandulosa*	Chloroform-ethyl acetate (80 + 20)	[285]
Methylated aralosides	Chloroform-ethyl acetate (80 + 20)	[135]
Aescigenin (hRf 24)	Chloroform-methanol (90 + 9)	[287]
Cycloartenol, parkeol (comparison with steroids)	Chloroform-methanol (60 + 40) and other mixtures	[229]
Gratiogenins from *Gratiola*	Chloroform-acetone (84 + 14)	[289]
4 *Primula* sapogenins	Chloroform-acetone (84 + 14)	[296]
Cucurbitacins from *Iberis* (Cruciferae) and *Bryonia*	Chloroform-ethanol (95 + 5)	[73, 74]
Sapogenins from *Styphnodendrum coriaceum*	various solvents	[299]
Baccatin	n-Hexane-acetone (78 + 13) or CH$_2$Cl$_2$-acetone (75 + 25)	[204]

b) Triterpene Acids

Data on triterpene acids are obtained by chromatography of the free acids (Table 35) or of their esters, using the solvent systems mentioned above (cf. Table 34). Weakly acid silica gels have proved suitable for separating the acids. The presence of other polar functional groups in addition to the carboxyl group, influences choice of the solvent polarity. Sometimes acid [130, 324] or basic [293] components are added to prevent tailing [290]. Multiple development can help also in such cases. With dichloroethylene-petrol ether-acetic acid (45 + 30 + 9) for example, the acids are separated into fractions with the same number of hydroxyl groups [57]. Those acids with only one hydroxyl group

Table 35. hRf-values of triterpene acids on silica gel G layers [290, 293]

Triterpene acid	hRf[a]	Triterpene acid	hRf[a]
Acetylacacic acid	83	Emmolic acid	59
Rehmannic acid	82	Chinovasic acid	55
Acetylursolic acid	76	Cincholic acid	53[b]
Pyroquinovic acid	71[b]	Masticadienonic acid	47
Pyrocincholic acid	70[b]	Isomasticadienonic acid	47
Ursolic acid	68	Polyporenic acid	43[b]
Oleanolic acid	68	Guaiiavolic acid	35
Betulinic acid	68	Acantholic acid	29
Oleanonic acid	68	Medicagenic acid	29[b]
Acetylacantholic acid	66	Machaerinic acid	23[b]
Siaresinolic acid	66	Bayogenin	21
Acacic acid	63	Cochalic acid	18[b]
Morolic acid	59	Proceric acid	4
Ceanothic acid	59	Tumulosic acid	2
Bremedolic acid	59[b]		

[a] Solvent: diisopropyl ether-acetone (75 + 30); hRf-values are to be regarded as only rough guides.
[b] Sign indicating tailing.

migrate to the upper part of the chromatogram (hRf 60—80). They are clearly separated from those with two (hRf 30—50) or three (hRf < 30) hydroxyl groups. Isomers are, however, fractionated only as a group in this system, as TSCHESCHE and co-workers [293] have found. ELGAMAL and FAYEZ [57] have accordingly recommended kieselguhr layers for separation of the frequently occurring isomeric acids like betulinic, oleanolic and ursolic acids (Table 36).

Table 36. Separation of isomeric triterpene acids on silica gel G layers [293]

Compound	hRf-value in solvent [a]			
	I	II	III	IV
Betulinic acid	75	84	80	85
Oleanolic acid	50	65	40	70
Ursolic acid	20	42	15	20

[a] Solvents: I: petrolether (100—200° C)-dichloroethylene-acetic acid (50 + 50 + 0.7); II: toluene-acetone-acetic acid (100 + 3 + 0.07); III: carbon tetrachloride-petrol ether (70—80° C)-acetic acid (66 + 33 + 0.07); IV: petrol ether (100—120° C)-ethyl formate-formic acid (93 + 7 + 0.7). The hRf-values must be regarded only as guide values.

A further and specific example of the application of TLC is the determination of glycyrrhizin derivatives and glycerrhetic acid in biological material [53, 87, 88]. BONATI [17] has differentiated chromatographically 18-α- and 18-β-glycerrhetic acids by using ethyl acetate-methanol-diethylamine (70 + 20 + 15). TSCHESCHE and STRIEGLER [294] have demonstrated that tenuifolic acid and senegenin are identical, using the solvent diisopropyl ether-acetone (70 + 28), previously employed for neutral triterpenes. A similar solvent combination has been used to detect ursolic acid in apple peel [26] and rosemary wine [80].

c) Triterpene Glycosides
(Saponins and bitter principles)

Neutral and acid triterpenes have been treated in the two preceding sections. These compounds can also be combined in glycosides. One may therefore expect that neutral and acid glycosides will occur in nature, all the more so because uronic acids are not excluded as sugar components. Chromatographic investigations should be aimed in this direction.

α) Saponins (cf. also the steroid saponins, p. 346)

Saponins are naturally occurring materials, the aqueous solutions of which foam like soaps (sapones). They have a more or less strong haemolysing effect on blood erythrocytes in vitro (295, 316a]. They can be detected and quantitatively determined with the help of this property [118].

The various teams [19, 125, 128, 135, 191, 245, 256, 290, 319] have used mostly aqueous-alcoholic mobile phases like, for example, chloroform-methanol-water (65 + 35 + 10) for separation of mixtures of *neutral* triterpene glycosides. The ethyl acetate-carbon tetrachloride mixtures which LINDE [150] has employed for separation of *Cimicifuga* glycosides, are less polar (cf. also [35, 36]).

Mobile phases containing acetic acid have been only rarely used [324] but ammonia-containing systems have found more frequent application [288]. Even *basic* saponins can be chromatographed in this way [295]. Ammoniacal n-butanol mixtures have been proposed for differentiating the acid *Gypsophila-*, *Aralia-* or *Clematis*-saponins [135, 288, 295].

TSCHESCHE and co-workers recommend the "wedged-tip" technique (p. 89) for special separation problems. Should this prove inadequate, descending chromatography can be tried. GÖLDEL and co-workers [76] have applied this method in cases where hR_f-values were too similar; KHORLIN and co-workers [127, 128] and COLEMAN and PARKE [33] have preferred two-dimensional TLC on alumina layers. The last named developed with acetone-chloroform (50 + 50) in the first direction and methanol-33% ammonium hydroxide (80 + 20) in the second.

If the different separation techniques do not succeed as desired, the adsorbent should be varied; lower activity (air drying) can operate favourably [290]. Spraying the silica gel layer slightly with 2N ammonium hydroxide solution has also been recommended. Polyamide would appear to have no advantage as adsorbent [33]. Layers of powdered glass are of specific interest only, although RAHANDRAHA et al. [206, 207] have used them for separating asiatic acid and the asiaticosides. These materials have too low a capacity for normal separations.

β) Bitter principles. The numerous bitter-tasting compounds of the plant world may belong to one of various chemical classes. The tetracyclic *cucurbitacins* [73, 74, 147, 185a] are outstanding members of the triterpenoid bitter principles which are of current interest. Their aglycones differ from the other tetracyclic triterpenes in having a methyl or hydroxymethyl group in the C_9-position instead of in the customary C_{10}-position. In the TLC procedure, fat-free acetone or chloroform extracts of seeds are applied to silica gel G layers and development carried out with chloroform-ethanol (95 + 5) under standard conditions [21a, 73]. The aglycones are thus separated in a run of 15 cm; the glycosides can be distinguished only with more polar solvents.

Tests

Specific Physiological Detection. Both saponins and bitter principles possess physiological activities which can be utilised for their detection [81].

Saponins. These haemolysing substances are detected by contact with a blood-gelatine solution on the plate [89a, 251, 324, 328]; "haemolytical halos" are observed. These zones, where the saponins have diffused out of the chromatogram are, transparent and almost colourless in contrast to the normally cloudy and red blood-gelatine layer.

Preparation and application of the blood-gelatine solution

α) 100 ml 0.9% sodium chloride solution is added to 4.5 g gelatine powder; this is left standing 30 min and the gelatinous mass heated to 80° C on the water bath, with stirring.

β) after cooling solution α) to 40° C, 6 ml defibrinated cow blood (see below) is added with stirring; 50 ml of this blood-gelatine suspension is poured immediately on to the thin-layer chromatogram to be tested, forming a thin film. To prevent the suspension from running away, adhesive tape (1 cm wide) is stuck all round the edge of the plate to form a trough. The plate is then left in an accurately horizontal position, best on a cooling block, until the film has set.

After 1 h at the most, the red blood-gelatine film has become transparent where the saponins were present on the chromatogram; the remainder of the film remains opaque and red.

Note: Acid or basic solvents, if used, must be completely removed from the adsorbent layer before the blood-gelatine is poured on.

Defibrinated blood: About 200 ml blood of a freshly slaughtered cow is collected in a 1-litre wide-necked flask and immediately stirred vigorously with a wooden stick until the fibrin agglutinates. The gelatinous residue is removed by "filtration" through several layers of muslin. The defibrinated blood thus obtained can be kept 1—2 days at 3—4° C. As soon as the cloudy "solution" becomes transparent, it can no longer be used (haemolysis).

Bitter principles. To determine the position of these substances on the chromatogram, the layer above the start region is scraped off in successive millimetre zones and each suspended in 0.5 ml pure ethanol. Each "stock solution" is progressively diluted by water, permitting testing according to the Wasicky determination of bitterness factor [81]. Distinct differences in the composition of bitter principles between individual extracts can be established in this way. The position of the chromatographically separable hop bitter principles [145a] and bitter-tasting substances from extracts of bitter wood have been determined likewise (see Chapter U, II.4 and Fig. 200).

Spray Reagents. The chlorosulphonic acid reagent (No. 45) has been often mentioned as especially suitable for detection of triterpenes. As little as 0.2 μg of oleanolic acid can be detected with it. The separated zones usually colour violet to brown; betulinic acid appears pale blue, oleanolic and ursolic acids form reddish zones. They can be distinguished in long wave UV light (365 nm) [57, 293].

According to [95], unsaturated hydroxytriterpene carboxylic acids react well with the Liebermann-Burchard reagent (No. 1). The antimony chloride reagents (Nos. 15, 18) can also be used with profit [89a, 91]. Zones of different colours are yielded with concentrated acids (sulphuric or phosphoric), with or without added aldehyde (vanillin, anisaldehyde, furfuraldehyde) after heating. This applies also to the cucurbitacins even if they have previously been sprayed with the ferric chloride reagent (No. 102) in order to visualise selectively the diosphenol type of bitter principle as brown zones. KLINGMÜLLER [130] recommends a vanillin-containing antimony trichloride solution for detection of *Primula* saponins with which a red colour is yielded.

Stannic chloride (Rgt. No. 236) is now only rarely used for detection. If no success is obtained with fluorescence layers, iodine vapour (Rgt. No. 141) or rhodamine B reagent (No. 220) can be used for non-destructive detection.

3. Polyterpenes

The TLC of polyterpene mixtures, particularly of carotenes, carotenoids and ubiquinones, is described in the next chapter (K).

V. Balsams and Resins

Resins and balsams are plant excreta like essential oils. In contrast to these last named, they are only slightly or not at all volatile in steam. Resins are yellow to brown, solid, brittle, amorphous masses which soften only at higher temperature and have no sharp melting point. They are insoluble in water and only moderately so in alcohol; chloroform or ethyl acetate are the best solvents for them. Whereas resins contain only little essential oil, balsams contain larger amounts and are thus generally viscous liquids.

There has been no lack of attempts to characterise resins and balsams chromatographically. ROTHENHEIMER [215] and STOCK [272] have used capillary analysis; VALENTIN [313], chromatography on alumina colums; MILLS and WERNER [169] and RAWLINGS and WERNER [208], reversed phase procedures on paper. Successes have been only partial and none of these procedures has attracted any particular interest in practice.

The desired success came first through TLC, demonstrated early on by STAHL's team [247, 249, 253]. JORK [111] obtained particularly good separations using the S-chamber (Fig. 108). The usual silica gel G layer was used, developing twice with benzene-methanol (95 + 5). Comparison experiments were carried out by applying 1 μl amounts of 3% solutions of the resins and balsams in ethyl acetate or chloroform. The individual constituents seen on the chromatogram have for the most part not yet been identified (cf. also [65]). Only FRAUENDORF and AUTERHOFF [66] have carried out detailed investigations on Peru balsam from this point of view; MOREIRA and CECY [171] and MASSE and PARIS [155] have studied gum (resin) benzoins.

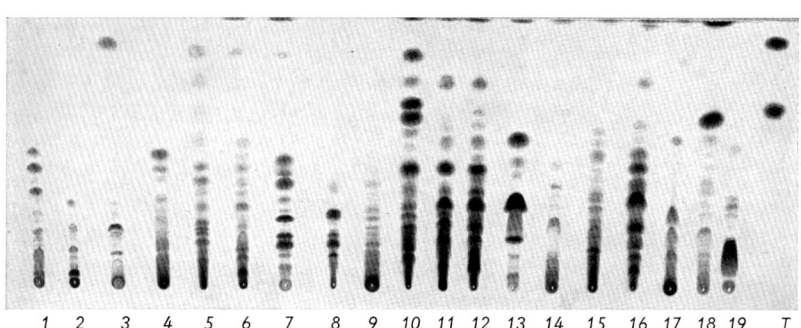

Fig. 108. Comparison of the most important resins and balsams on a 40 cm TLC plate. Detection with antimony(III)- and (V) chlorides (Rgt. Nos. 15 and 18). See text for details

1 Siam benzoin
2 Sumatra benzoin
3 Styrax
4 Peru balsam
5 Perugen
6 Tolu balsam
7 Asa foetida
8 Galbanum
9 Myrrh
10 Olibanum
11 Mastic
12 Dammar
13 Terebinthina
14 Rosin
15 Canada balsam (genuine)
16 Synthetic Canada balsam
17 Sandarac
18 Copaiba balsam
19 Gamboge
T Desaga test mixture

The TLC of glycoside resins and resin acids has been described in other work [148a]. The last named can be separated on silver nitrate-impregnated layers after methylation. NORIN and WESTFELT [186] impregnated silica gel layers using the method of BARRETT and co-workers [10]; ZINKEL and ROWE [334] preferred correspondingly impregnated alumina G layers. These allegedly do not peel off so easily as AgNO₃-impregnated silica gel G layers. As a result of their using another layer, ZINKEL and ROWE were not able to use benzene like

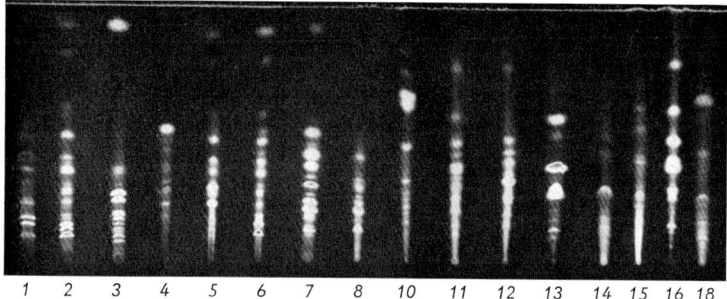

Fig. 109. Thin-layer chromatogram of resins and balsams, photographed in UV light. See Fig. 108 for separation conditions and legend. Spray reagent: antimony(III)-chloride (Rgt. No. 15)

NORIN and WESTFELT; they had to take a mixture of petrol ether (30—60° C) and anhydrous diethyl ether (75 + 25). The zone sequence on the chromatogram was consequently different. This change can be utilised for further characterisation. The results are compared in Table 37. The hR_f-values in column II were taken from a schematic diagram and are only approximate.

Table 37. *Separation of some resin acids on $AgNO_3$-impregnated layers*

Basic structure of the acids	Methyl esters of the acids	hR_f-values I[a]	II[a]	Visualisation with SbCl₅ (No. 18)
	Pimaric acid	40	44	blue
	Sandaracopimaric acid	27	—	violet
	Isopimaric acid	32	12	brown
	Levopimaric acid	50	21	grey
	Palustric acid	60	26	yellow
	Dehydroabietic acid	83	82	yellow-brown
	Abietic acid	75	58	blue-grey
	Neoabietic acid	73	65	grey
	Di- and tetrahydro-resin acids	—	94	

[a] Solvent I: benzene; layer: silica gel-silver nitrate (75 + 25) [186]. Solvent II: petrol ether (30—60° C)-diethyl ether (anhydrous) (75 + 25); layer: alumina G-silver nitrate (75 + 30) [334].

In conclusion, reference may be made to a publication of MECKEL and co-workers [161] who have separated hydrolysates of synthetic resins on silica gel G layers. Methanol-acetic acid (90 + 10) was the solvent.

Detection. The antimony(III)-chloride reagent (No. 15) is suitable for visualisation of resin and balsam components. Although the addition of antimony(V)-chloride reagent (No. 18) intensifies the colour and renders it greyer, an undesirable weakening in the fluorescence is brought about. It is best to examine the chromatogram in daylight and in long wave UV light, both before and after spraying. Only after this should it be heated about 10 min at 110° C. In Fig. 108 is seen a thin-layer chromatogram of some resins and balsams, photographed in visible light after having been heated. The zones are mostly brown, violet or grey. The yellow spots from artificial Peru balsam (Perugen) are especially striking. The majority of resins and balsams can be clearly distinguished from one another through the number, size and colour of the zones (finger print technique).

Inspection in long wave UV light (Fig. 109) provides further characterisation information. Siam or Sumatra benzoins or their tinctures may in practice be clearly differentiated in this way [66, 155].

Numerous other reagents may be used for visualisation but they are less sensitive.

Note: The resins and balsams in the chromatograms illustrated here are a selection of products supplied by the German drug trade between 1957 and 1962. Usually 5—10 different samples were available. Only when the route between plant source and commercial product is known unequivocally, is it possible to decide whether a material is really genuine. When comparing materials one must remember that the methods of processing and grading, the age of the sample and many other factors, can influence the composition. The same problems arise when judgment must be passed on resins or balsams which are found in ancient tombs or vessels at the chemists'. Nevertheless it has often been possible recently by means of the fingerprint technique, to give information about the nature of the resin in question.

Bibliography for Chapter J. Terpene Derivatives

1. ADHIKARI, V.: Dissertation Saarbrücken 1965.
2. AKAZAWA, T., J. URITANI, and Y. AKAZAWA: Arch. Biochem. **99**, 52 (1962).
3. ATKINSON, R. E., and R. F. CURTIS: Tetrahedron letters, No. 5, 297 (1965).
4. — —, and G. T. PHILLIPS: Tetrahedron letters No. 43, 3159 (1964).
5. ATTAWAY, J. A.: Analyt. Chem. **36**, 2224 (1964).
6. — L. J. BARABAS, and R. W. WOLFORD: Analyt. Chem. **37**, 1289 (1965).
6a. — A. P. PIERINGER, and L. J. BARABAS: Phytochemistry **5**, 141 (1966).
7. — R. W. WOLFORD, and G. J. EDWARDS: Analyt. Chem. **37**, 74 (1965).
8. BADINGS, H. T.: J. Chromat. **14**, 265 (1964).
9. — and J. G. WASSINK: Neth. Milk and Dairy J. **17**, 132 (1963).
10. BARRETT, C. B., M. S. J. DALLAS, and F. B. PADLEY: Chem. Ind. **1962**, 1050.
11. BERGSTRÖM, G., u. C. LAGERCRANTZ: Acta Chem. scand. **18**, 560 (1964).

12. BETTS, T. J.: J. Pharm. Pharmacol. Suppl. **16**, 131 (1964).
13. BEYRICH, T., u. R. POHLOUDEK-FABINI: Pharmazie **16**, 360 (1962).
14. BHRAMARAMBA, A., u. G. S. SIDHU: Perf. Essent. Oil Rec. **54**, 732 (1963).
15. BOHLMANN, F., C. ARNDT, K.-M. KLEINE u. H. BORNOWSKI: Chem. Ber. **98**, 155 (1965).
16. — K.-M. KLEINE u. H. BORNOWSKI: Chem. Ber. **98**, 369 (1965).
17. BONATI, A.: Fitoterapia **34**, 19 (1963); ref. C. A. **59**, 10136d (1963).
18. BORDET, C., et G. MICHEL: C. R. hebd. Séances Acad. Sci. **256**, 3482 (1963).
19. BORKOWSKI, B., u. B. PASICH: Farm. Polska **19**, 435 (1963); ref. C. A. **61**, 4493a (1964).
20. BRAUN, D.: Chimia (Zürich) **19**, 82 (1965).
21. — u. G. VORENDOHRE: Z. analyt. Chem. **199**, 37 (1964).
21a. BREDENBERG, J. B., u. R. GMELIN: Acta Chem. scand. **16**, 1802 (1962).
22. BRIESKORN, C. H., u. S. DALFERTH: Deut. Apoth.-Ztg **104**, 1388 (1964).
23. — — Lieb. Ann. Chem. **676**, 171 (1964).
24. — u. A. FUCHS: Dtsch. Apoth.-Ztg. **102**, 1268 (1962).
24a. — — Chem. Ber. **95**, 3034 (1962).
25. — u. H. KLINGER: Z. Lebensmitt.-Untersuch. **120**, 269 (1963).
26. — H. KLINGER u. W. POLONIUS: Arch. Pharm. **294**, 389 (1961).
27. — u. W. POLONIUS: Pharmazie **17**, 705 (1962).
28. BRUD, W., i. W. DANIEWSKI: Chem. analit. (Warszawa) 8, 753 (1963).
28a. BUSLANOVA, M. M., i. W. F. STEPANOVSKAYA: Zhur. analyt. Chem. **20**, 859 (1965); from Pharm. Zentralhalle **104**, 725 (1965).
29. BYRNE, G. A.: J. Chromatog. **20**, 528 (1965).
30. CHANDRA, G., J. CLARK et al.: J. chem. Soc. **1964**, 3648.
31. COBB, W. Y.: J. Chromatog. **14**, 512 (1964).
32. — L. M. LIBBEY, and E. A. DAY: J. Chromatog. **17**, 606 (1965).
33. COLEMAN, T. J., and D. V. PARKE: J. Pharm. Pharmacol. **15**, 841 (1963).
34. COLLINS, R. P., and K. KALNINS: Lloydia **28**, 48 (1965).
35. CORSANO, S., e. L. PINIZZI: Atti Accad. Italia, Rend. Cl. Sci. fisiche, mat. natur. **32**, 601 (1962).
36. — e. G. SPANO: Atti Accad. Italia, Rend. Cl. Sci. fisiche, mat. natur. **32**, 674 (1962); ref. C. A. **58**, 11408 (1963).
37. COUCHMAN, F. M.: Tetrahedron **20**, 2037 (1964).
38. CURTIS, R. F., and G. T. PHILLIPS: J. Chromatog. **9**, 366 (1962).
39. DALFERTH, S.: Dissertation Würzburg 1963.
40. DEMOLE, E.: Thèses de Doctorat, Paris Ser. A. n° 844, N° d'Ordre 870 (1958).
40a. — Chromat. Rev. 1, 8 (1959); **4**, 31 (1961).
41. — Helv. Chim. Acta **45**, 1951 (1962).
42. — In G. B. MARINI-BETTÒLO: Thin-Layer Chromatography. Amsterdam-London-New York: Elsevier 1964.
43. — et E. LEDERER: Bull. Soc. Chim. France **1958**, 1128.
44. — B. WILLHALM u. M. STOLL: Helv. Chim. Acta **47**, 1152 (1964).
45. DESHUSSES, J., u. A. GABBAI: Mitt. Lebensmitt.-Hyg. **53**, 408 (1962).
46. DENTI, E., and M. P. LUBOZ: J. Chromatog. **18**, 325 (1965).
47. DHONT, J. H., and G. J. C. DIJKMAN: Analyst **89**, 681 (1964).
48. DHONT, J. H., and C. DE ROOY: Analyst **86**, 74 (1961).
49. — — Analyst **86**, 527 (1961).
50. DJERASSI, C., and R. MCCRINDLE: J. Chem. Soc. **1962**, 4034.
51. DRESSEN, F.-P.: Dissertation Münster 1965.
52. DUŠINSKÝ, G., a M. TYLLOVÁ: Chem. Zvesti **16**, 701 (1962).
53. VAN DUUREN, A. J.: J. Amer. Soc. Sugar Beet Technol. **12**, 57 (1962).

54. EL-DAKHAKHNY, M.: Planta medica 11, 465 (1963).
55. EL-DEEB, S. R.: J. Pharm. Sci. U.A.R. 3, 41 (1962); ref.: C. A. 61, 6853f (1964).
56. — M. S. KARAWYA, and S. K. WAHBA: J. Pharm. Sci. U.A.R. 3, 81 (1962).
57. ELGAMAL, M. H. A., and M. B. E. FAYEZ: Z. analyt. Chem. 211, 190 (1965).
58. EL-HAMIDI, A., and G. RICHTER: Lloydia 28, 252 (1965).
59. FASSINA, G.: Boll. Soc. ital. Biol. sper. 36, 1417 (1960).
60. — A. R. CONTESSA e C. E. TÓTH: Boll. Soc. ital. Biol. sper. 38, 260 (1962).
61. FEDELI, E., P. CAPELLA e L. TADINI: Riv. Ital. Sost. Grasse 40, 669 (1963).
62. FIORI, A., u. M. MARIGO: Minerva med. leg. 82, 350 (1962).
63. FISHBEIN, L., and J. FAWKES: J. Chromatog. 20, 521 (1965).
64. FLÜCK, H., u. K. BICHSEL: Lecture, Internat. Kongr. Pharmaz. Wissenschaften, Münster 1963.
65. — u. C. WINDECK-LUTZ: Lecture in Pharmac. Acta Helv. 40, 637 (1965).
66. FRAUENDORF, H., u. H. AUTERHOFF: Dtsch. Apoth.-Ztg. 103, 1299 (1963).
67. FRÖMMING, K.-H.: Arch. Pharmaz. 297, 172 (1964).
67a. FURTER, M.: Helv. chim. Acta 21, 1666 (1938).
68. GABEL. E., K. H. MÜLLER u. J. SCHOKNECHT: Dtsch. Apoth.-Ztg. 102, 293 (1962).
68a. GÄNSHIRT, H.: Unpublished.
69. GAINES, H. D., and R. P. COLLINS: Lloydia 26, 247 (1963).
70. GAREL, J.-P.: Bull. Soc. chim. Fr. 1965, 1563.
71. GIACOBAZZI, C., e G. GIBERTINI: Bull. chim. farm. 101, 490 (1962).
72. GILDEMEISTER, E., u. FR. HOFFMANN: Die ätherischen Öle. Berlin: Akademie-Verlag 1956—1962.
73. GMELIN, R.: Arzneimittel-Forsch. 13, 771 (1963).
74. — Arzneimittel-Forsch. 14, 1021 (1964).
75. GODON, M.: Dissertation Paris 1963.
76. GÖLDEL, L., W. ZIMMERMANN u. D. LOMMER: Hoppe-Seylers Z. physiol. Chem. 333, 35 (1963).
77. GRACZA, L., u. A. ZARÁNDI: Pharmazie 19, 228 (1964).
78. GRÄB, R.: Dtsch. Apoth.-Ztg. 103, 1424 (1963).
79. GRAF, E., u. W. HOPPE: Dtsch. Apoth.-Ztg. 102, 393 (1962).
80. — — Dtsch. Apoth.-Ztg. 104, 287 (1964).
80a. GRAN, W., u. H. ENDRES: J. Chromatog. 17, 585 (1965).
81. GSTIRNER, F.: Prüfung und Verarbeitung von Arzneidrogen. Bd. I. Berlin-Göttingen-Heidelberg: Springer 1955.
82. GUENTHER, E.: The Essential Oils. New York: D. van Nostrand Company 1952—1955.
83. GUPTA, A. S., and S. DEV: J. Chromatog. 12, 189 (1963).
84. GUTZWILLER, J., R. MAULI, H. P. SIGG u. CH. TAMM: Helv. chim. Acta 47, 2234 (1964).
85. GYANCHANDANI, N.: Dissertation Freiburg 1964.
86. HAAGEN-SMIT, A. G.: In L. ZECHMEISTER: Fortschritte der Chemie organischer Naturstoffe, Bd. XII. Vienna: Springer 1955.
87. HELBING, A. R.: Pharm. Weekbl. 99, 1116 (1964).
88. — Clin. chim. Acta 8, 756 (1963).
89. HERZ, W., and S. INAYAMA: Tetrahedron 20, 341 (1964).
89a. HILLER, K., B. LINZER u. S. PFEIFER: Pharmazie 21, 182 (1966).
90. HÖRHAMMER, L., A. EL-HAMIDI, and G. RICHTER: J. pharm. Sci. 53, 1033 (1964).
91. — H. WAGNER u. B. LAY: Pharm. Ztg. (Frankfurt) 106, 1307 (1961).
92. — — u. G. RICHTER: Dtsch. Apoth.-Ztg. 103, 1737 (1963).

93. HÖRHAMMER, L., H. WAGNER, G. RICHTER, H. W. KÖNIG u. J. HENG: Dtsch. Apoth.-Ztg. **104**, 1398 (1964).
94. — — u. H. SCHILCHER: Arzneimittel-Forsch. **12**, 1 (1962).
95. HOFMANN, H.: Dissertation Würzburg 1962.
96. TEN HOOPEN, H. J. G.: Z. Lebensmitt.-Untersuch. **119**, 478 (1963).
97. HUNECK, S.: J. Chromatog. **7**, 561 (1962).
98. — J. org. Chem. **28**, 2390 (1963).
99. — Tetrahedron **19**, 479 (1963).
100. HYYRYLAINEN, M.: Farm. Aikakauslehti **72**, 161 (1963).
101. IKAN, R.: J. Chromatog. **17**, 591 (1965).
102. — J. KASHMAN, S. HAREL, and E. D. BERGMANN: Israel J. Chem. **1**, 248 (1963).
103. — — u. E. D. BERGMANN: J. Chromatog. **14**, 275 (1964).
104. IKEDA, R. M., W. L. STANLEY, S. H. VANNIER, and L. A. ROLLE: Food Technol. **15**, 379 (1961).
105. ITO, M.: J. Chem. Soc. Japan, Pure Chem. Sect. **78**, 172 (1957); ref. Chem. Zbl. **42**, 11595 (1957).
106. JACOB, J.: Dissertation Bonn 1963.
107. JANISTYN, H.: Parfümerie u. Kosm. **45**, 335 (1964).
108. JASPERSEN-SCHIB, R., e H. FLÜCK: Boll. chim. farm. **101**, 512 (1962).
109. JONES, E. R. H., u. T. G. HALSALL: In: L. ZECHMEISTER: Fortschritte der Chemie organischer Naturstoffe, Bd. XII. Vienna: Springer 1955.
110. DEJONG, K., K. MOSTERT et D. SLOOT: Rec. Trav. chim. Pays-Bas **82**, 837 (1963).
111. JORK, H.: Dtsch. Apoth.-Ztg. **102**, 1263 (1962).
112. — J. Pharm. Belg. **1963**, 213.
113. — Dissertation Saarbrücken 1963.
114. — J. Pharm. Belg. (Symposium volume) **1966**, 295.
115. JUVONEN, S.: Planta med. (Stuttg.) **12**, 488 (1964).
116. — Lecture Internat. Kongr. Pharmaz. Wissenschaften, Münster 1963.
116a. KARAWYA, M. S., and S. K. WAHBA: Egypt. pharm. Bull. **44**, 23 (1964); C. A. **62**, 10288e (1965).
116b. — — Egypt. pharm. Bull **44**, 31 (1964); C. A. **62**, 10288f (1965).
117. KARRER, W.: Konstitution und Vorkommen der organischen Pflanzenstoffe. Basel-Stuttgart: Birkhäuser 1958.
118. KARTNIG, TH., F. J. GRAUNE, and R. HERBST: Planta med. **12**, 428 (1964).
119. — Pharm.Ztg **110**, 1051 (1965).
120. KATAGUE, D. B., and E. R. KIRCH: J. pharm. Sci. **54**, 891 (1965).
121. KAUFMANN, H. P., u. A. K. SEN GUPTA: Chem. Ber. **96**, 2489 (1963).
122. — — Chem. Ber. **97**, 2652 (1964).
123. KAUFMANN, H. P., u. A. K. SEN GUPTA: Fette, Seifen, Anstrichmittel **65**, 529 (1963).
124. — — Fette-Seifen-Anstrichmittel **66**, 461 (1964).
125. KAWASAKI, T., and K. MIYAHARA: Chem. pharm. Bull. **11**, 1546 (1963).
126. KHEIFITS, L. A., G. J. MOLDOVANSKAYA i L. M. SHULOV: Akad. Nauk S.S.S.R. **28**, 267 (1963).
127. KHORLIN, A. Y., L. V. BAKINOVSKIĬ, V. E. VAS'KOVSKIĬ, A. G. VENYAMINOVA i Y. S. OVODOV: Isvest. Akad. Nauk S.S.S.R. **1398**, 2008 (1963).
128. KHORLIN, A. Y., Y. S. OVODOV i. N. K. KOCHETKOV: Zhur. obshchei Khim. **32**, 782 (1962).
129. KLEIN, E., W. ROJAHN u. D. HENNEBERG: Tetrahedron **20**, 2025 (1964).
130. KLINGMÜLLER, L.: Dissertation Hamburg 1961; abstract in: Dtsch. Apoth.-Ztg. **104**, 927 (1964).

131. KLOUWEN, M. H., u. R. TER HEIDE: Parfümerie u. Kosm. **43**, 195 (1962).
132. — R. TER HEIDE u. J.G.J. KOK: Fette, Seifen, Anstrichmittel **65**, 414 (1963).
133. KNAPPE, E., u. D. PETERI: Z. analyt. Chem. **190**, 386 (1962).
133a. — u. J. ROHDEWALD: Z. analyt. Chem. **200**, 9 (1964).
134. KNÜTTER, S., u. R. POHLOUDEK-FABINI: Pharmazie **17**, 456 (1962).
135. KOCHETKOV, N. K., A. Y. KHORLIN i V. E. VAS'KOVSKIĬ: Isvest. Akad. Nauk S.S.S.R. **1398**, 1409 (1963).
136. KOHAN, S., and J. FITELSON: J. Ass. off. Agric. Chem. **47**, 551 (1964).
137. KOHEN, F., B. K. PATNAIK, and R. STEVENSON: J. org. Chem. **29**, 2710 (1964).
138. KORTE, F.: Arch. Pharm. **286**, 295 (1953).
139. — H. BARKEMEYER u. J. KORTE: In: L. ZECHMEISTER: Fortschritte der Chemie organischer Naturstoffe, Bd. XVII. Vienna: Springer 1959.
140. — u. J. VOGEL: J. Chromatog. **9**, 381 (1962).
141. KRATZL, K., u. G. E. MIKSCHE: Mh. Chem. **94**, 530 (1963).
142. KŘEPINSKÝ, J., M. ROMAŃUK, V. HEROUT a F. ŠORM: Collect. Čs. Chem. Commun. **27**, 2638 (1962).
143. KUBECZKA, K.-H.: Dtsch. Apoth.-Ztg **104**, 369 (1964).
144. KUHN, R., u. I. LÖW: Tetrahedron Letters **1964**, 891.
145. KUNOVITS, G.: Seifen, Öle, Fette, Wachse **90**, 895 (1964).
145a. KUROIWA, Y., and H. HASHIMOTO: J. Inst. Brew. **67**, 347, 352 (1961).
146. LAPINA, T. G.: Trudy Khim. i Khim. Tekhnol. **1962**, No. 2, 424.
147. LAVIE, D., and B. S. BENJAMINOV: Tetrahedron **20**, 2665 (1964).
148. LEA, C. H.: Chem. and Ind. **1963**, 1406.
148a. LEGLER, G.: Phytochemistry **4**, 29 (1965).
149. LIBBEY, L. M., and E. A. DAY: J. Chromatog. **14**, 273 (1964).
150. LINDE, H.: Arzneimittel-Forsch. **14**, 1037 (1964).
151. LITTLER, J. S., and I. G. SAYCE: J. chem. Soc. **1964**, 2545.
151a. MCSWEENEY, G. P.: J. Chromatog. **17**, 183 (1965).
152. MARCUSE, R.: Fette, Seifen, Anstrichmittel **66**, 192 (1964).
153. — J. Chromatog. **7**, 407 (1962).
154. — U. MOBECH-HANSSEN u. P. O. GÖTHER: Fette, Seifen, Anstrichmittel **66**, 192 (1964).
155. MASSE, J., et R. PARIS: Ann. pharm. franc. **22**, 349 (1964).
156. MÁTHÉ, I., és E. TYIHAK: Herba hung. **1**, 31 (1962).
157. MATHIS, C.: Dissertation Strasbourg 1963.
158. — et G. OURISSON: J. Chromatog. **12**, 94 (1963).
159. — and G. OURISSON: Phytochemistry **3**, 115, 133 (1964).
160. DE MAYO, P.: Mono- and Sesquiterpenoids. New York-London: Interscience Publishers 1959.
161. MECKEL, L., H. MILSTER u. U. KRAUSE: Textil-Prax. **16**, 1032 (1961).
162. MEHLITZ, A., K. GIERSCHNER u. T. MINAS: Chem. Ztg. **87**, 573 (1963).
163. — — — Chemiker-Ztg. **89**, 175 (1965).
164. MEIJBOOM, P. W., u. G. JURRIENS: J. Chromatog. **18**, 424 (1965).
165. MENARD, E. L., J. M. MÜLLER, A. F. THOMAS, S. S. BHATNAGAR u. N. J. DASTOOR: Helv. chim. Acta **46**, 1801 (1963).
166. MESSERSCHMIDT, W.: Planta med. **12**, 501 (1964).
167. — Planta Med. **13**, 56 (1965).
168. MILLER, J. M., and J. G. KIRCHNER: Analyt. Chem. **25**, 1107 (1953).
169. MILLS, J. S., and A. E. WERNER: Nature (Lond.) **169**, 1064 (1952).
170. MITSUHASHI, H., U. NAGAI, T. MURAMATSU, and H. TASHIRO: Chem. Pharm. Bull. **8**, 243 (1960).
171. MOREIRA, E. A., y C. CECY: Trib. Farm. **32**, 55 (1964).
172. MORGAN, M. E., and R. L. PEREIRA: J. Dairy Sci. **45**, 457 (1962).

173. Moritz, O.: In: W. Ruhland: Handbuch der Pflanzenphysiologie, Bd. X. Berlin-Göttingen-Heidelberg: Springer 1958.
174. — In: K. Peach u. M. V. Tracey: Moderne Methoden der Pflanzenanalyse, Bd. III. Berlin-Göttingen-Heidelberg: Springer 1955.
175. Mouton, M., S. Jaquard et M. Sagot-Masson: Ann. pharm. franç. **21**, 233 (1963).
176. Nadal, N. G. M., C. M. C. Chapel, and C. Lecumberry: Amer. Perfumer, Cosmet. **79**, 43 (1964).
177. Nano, G. M., e A. Martelli: Gazz. chim. Ital. **94**, 816 (1964).
177a. — — J. Chromat. **21**, 349 (1966).
178. — u. P. Sancin: Experientia (Basel) **19**, 329 (1963).
179. — u. P. Sancin: Ann. Chim. Roma **53**, 677 (1963).
180. Nicholas, H. J.: Biochim. biophys. Acta (Amst.) **84**, 80 (1964).
181. Nigam, S. S., and G. L. Kumari: Perfum. Essent. Oil Rec. **53**, 529 (1962).
182. Nigam, I. C., and L. Levi: J. Pharm. Sci. **53**, 1008 (1964).
182a. — — Parf. Cosm. Savons **8**, 423 (1965).
183. Nigam, M. C., I. C. Nigam, and L. Levi: J. Soc. Cosm. Chem. **16**, 155 (1965).
184. — and R. M. Purohit: Indian Perfumer **5**, 3 (1961); ref. Parf. u. Kosm. **45**, 283 (1964).
185. Nigam, I. C., M. Sahasrabudhe and L. Levi: Canad. J. Chem. **41**, 1535 (1963).
185a. Noller, C. R., A. Melera and M. Gut: Tetrahedron Letters No. 15 (1960).
186. Norin, T., and L. Westfelt: Acta chem. scand. **17**, 1828 (1963).
187. Onoe, K.: J. Chem. Soc. Japan, Pure Chem. Sect. **73**, 337 (1952); from: Analytical Abstr. **1953**, 3757.
188. Pailer, M., H. Kuhn u. I. Grünberger: Facgl. Mitt. Österr. Tabakregie.
189. Paris, R., et M. Godon: Recherches (Paris) **13**, 48 (1963).
190. Parks, O. W.: J. Lipid Res. **5**, 232 (1964).
191. Pasich, B.: Planta med. **11**, 16 (1963).
191a. Pastuska, G., u. H.-J. Petrowitz: Chemiker-Ztg. **86**, 311 (1962).
192. Pejković-Tadić, I., M. Hranisavliević-Jakovljević u. S. Nesić: In: G. B. Marini-Bettolo: Thin-Layer Chromatography. Amsterdam-London-New York: Elsevier 1964; cf. also J. Chromatog. **21**, 247 (1966).
193. Pertsev, I. M., i G. P. Pivnenko: Pharmaz. J. (Kiev) **16**, 28 (1961).
194. — — Farmatsev, Zh. (Kiev) **17**, 35 (1962).
195. Pesnelle, P., P. Teisseire u. M. Wichtl: Planta med. **12**, 403 (1964).
196. Petrowitz, H.-J.: Angew. Chemie **72**, 921 (1960).
197. — Z. analyt. Chem. **183**, 432 (1961).
198. Peyron, L.: Perfum. Cosm. Savons **5**, 1 (1962).
199. — Chim. analyt. **45**, 186 (1963).
200. Pisters, H.: Dissertation Cologne 1960.
201. van der Poel, G. H.: Nederl. Melk- en Zuiveltijdschr. **15**, 98 (1961).
202. Ponsinet, G., and G. Ourisson: Phytochemistry **4**, 799 (1965).
203. Preuss, R.: Dtsch. Apoth.-Ztg. **104**, 1797 (1964).
204. — u. H. Orth: Pharmazie **20**, 698 (1965).
205. Prey, V., A. Berger u. H. Berbalk: Z. analyt. Chem. **185**, 113 (1962).
206. Rahandraha, T., M. Chanez et P. Boiteau: Ann. pharmac. franç. **21**, 313 (1963).
207. — — — et S. Jaquard: Ann. pharmac. franç. **21**, 561 (1963).
208. Rawlings, F. I. G., and A. E. Werner: Endeavour **13**, 140 (1954).
209. Reitsema, R. H.: Analyt. Chem. **26**, 960 (1954).
210. — F. J. Cramer, and W. E. Fass: Agr. and Food Chem. **5**, 779 (1957).

211. Rispoli, G., A. di Giacomo e M. E. Tracuzzi: Riv. Ital. Essenze-Profumi, Piante Offic.-Aromi-Saponi Cosmet. **45**, 62 (1963).
212. Ronkainen, P., D. Kaempgen u. H. Suomalainen: Z. analyt. Chem. **201**, 14 (1964).
213. — T. Salo u. H. Suomalainen: Z. Lebensmittel.-Untersuch. **117**, 281 (1962).
214. Rosmus, J., u. Z. Deyl: J. Chromatog. **6**, 187 (1961).
215. Rothenheimer, C.: Pharm. Ztg. **74**, 712 (1929).
215a. Rotbacher, H., C. Crisan i E. Bedo: Farmacia (Bucharest) **12**, 733 (1964); C. A. **62**, 10288c (1965).
216. Rowe, J. W.: Tetrahedron Letters **1964**, 2347.
217. Rücker, G.: Pharm. Ztg. **108**, 1169 (1963).
218. Ruffini, G.: J. Chromatog. **17**, 483 (1965).
219. Ruzicka, L.: Experientia (Basel) **9**, 357 (1953).
220. Schäfer, F.: Dissertation Saarbrücken 1964.
221. von Schantz, M.: Farmaseutt. Aikakauslehti **71**, 52 (1962).
222. — Farmaseutt. Aikakauslehti **72**, 95 (1963).
223. — S. Juvonen, and R. Hemming: J. Chromatog. **20**, 618 (1965).
223a. — Lecture, Berlin 1965.
224. Schilcher, H.: Dtsch. Apoth.-Ztg. **104**, 1019 (1964).
224a. — Dtsch. Apoth.-Ztg. **105**, 1067 (1965).
224b. — Dtsch. Apoth.-Ztg. **106**, 231 (1966).
225. Schmialek, P.: Z. Naturforsch. **16**b, 462 (1961).
226. — Z. Naturforsch. **18**b, 462 (1963).
227. Schmittmann, B.: Diplomarbeit, Bonn 1962; quoted by: A. T. James and L. J. Morris: New Biochemical Separations. London: van Nostrand Company 1964.
228. Schratz, E., u. S. Qedan: Pharmazie **20**, 710 (1965).
229. Schreiber, K., O. Aurich, and G. Osske: J. Chromatog. **12**, 63 (1963).
230. Schulte, K. E., F. Ahrens u. E. Sprenger: Pharm. Ztg. **108**, 1165 (1963).
231. Schultz, O. E., u. H. L. Mohrmann: Pharmazie **20**, 379 (1965).
232. Schwartz, D. P., M. Keeney, and O. W. Parks: Microchem. J. **8**, 176 (1964).
233. — and O. W. Parks: Microchem. J. **7**, 403 (1963).
234. Seeboth, H.: Chem. Techn. **15**, 34 (1963).
235. — Mber. deutsch. Akad. Wiss. Berlin **5**, 693 (1963).
236. — u. H. Görsch: Chem. Techn. **15**, 294 (1963).
237. Seikel, M. K., and J. W. Rowe: Phytochemistry **3**, 27 (1964).
237a. Severin, M.: Bull. Inst. Agron. Sta. Rech Gembloux **32**, 122 (1964); C. A. **62**, 9755e (1965).
238. Shalaby, A. F., u. G. Richter: J. Pharm. Sci. **53**, 1502 (1964).
239. Shevchenko, Z. A., i J. A. Favorskaya: Vestn. Leningr. Univ. **19**, 107 (1964); ref. C. A. **61**, 8874a (1964).
240. Simonsen, J.: The Terpenes. Cambridge: University Press 1952.
241. Smith, M. D., and C. G. Farmilo: Communication Commonwealth Lab., Melbourne.
242. Smith, G. A. L., and P. J. Sullivan: Analyst **89**, 312 (1964).
243. Šorm, F.: Chem. and Ind. (Lond.) **1964**, 1833.
244. Sprecher, E.: Planta med. **11**, 119 (1963).
245. Stahl, E.: Mikrochim. Acta **40**, 367 (1953).
246. — Pharmazie **11**, 633 (1956).
247. — Parfümerie u. Kosm. **39**, 564 (1958).
248. — Chemiker-Ztg. **82**, 323 (1958); Il Laborat. Scientifico No. 6, 171 (1959).
249. — Pharm. Rdsch. **1**, No. 2, 1 (1959).

250. STAHL, E.: Arch. Pharm. **292**, 531 (1960).
251. — Thin-Layer Chromatography, 1. Edition. Berlin-Göttingen-Heidelberg: Springer 1965.
252. — Farmaseutt. Aikakauslehti **72**, 213 (1963).
253. — In: H. F. LINSKENS u. M. V. TRACEY: Moderne Methoden der Pflanzenanalyse, 5. Band. Berlin-Göttingen-Heidelberg: Springer 1962.
254. — Angew. Chemie, International edition **3**, 784 (1964).
255. — Arch. Pharm. **297**, 500 (1964).
256. — and co-workers: unpublished.
257. — u. V. ADHIKARI: Arch. Pharm. (in the press).
258. — u. H. BOHRMANN: Naturwissenschaften **54** (in the press).
259. — u. J. FUCHS: Unpublished.
260. — u. H. JORK: In [*251*].
261. — — Arch. Pharm. **297**, 273 (1964).
262. — — Arch. Pharm. **299**, 670 (1966).
263. — u. D. SCHEU: Naturwissenschaften **52**, 394 (1965).
264. — — Unpublished.
265. — u. G. SCHMITT: Arch. Pharm. **297**, 385 (1964).
266. — u. L. TRENNHEUSER: Arch. Pharm. **293**, 826 (1960).
267. — u. H. VOLLMANN: Talanta **12**, 525 (1965).
268. STANLEY, W. L., R. M. IKEDA u. S. COOK: Food Technol. **15**, 381 (1961).
269. STEIN, G.: Biol. Zbl. **82**, 343 (1963).
270. — Naturwissenschaften **50**, 305 (1963).
271. STEINER, M., u. H. HOLTZEM: In: K. PAECH u. M. V. TRACEY: Moderne Methoden der Pflanzenanalyse, Bd. III. Berlin-Göttingen-Heidelberg: Springer 1955.
272. STOCK, E.: In: K. DIETERICH u. E. STOCK: Analyse der Harze, Balsame und Gummiharze, 2. Aufl. Berlin: Springer 1930.
273. SUNDT, E., and A. SACCARDI: Food Technol. **16**, 89 (1962).
274. — B. WILLHALM u. M. STOLL: Helv. chim. Acta **47**, 408 (1964).
275. SUOMALAINEN, H.: Branntweinwirtschaft **105**, 1 (1965).
276. — u. P. RONKAINEN: Teknillisen Kem. Aikakauslehti **20**, 413 (1963).
277. SWALEH, M., B. BHUSMAN, and G. S. SIDHU: Perf. Essent. Oil Rec. **54**, 295 (1963).
278. SYNODINOS, E., E. KOKOTI-KOTAKIS u. G. KOTAKIS: in the press.
279. SZASZ, Z. M., u. G. SZASZ: Fette, Seifen, Anstrichmittel **67**, 332 (1965).
280. TAKEDA, K., H. MINATO, M. ISHIKAWA u. M. MIYAWAKI: Tetrahedron **20**, 2655 (1964).
281. — M. IKUTA u. M. MIYAWAKI: Tetrahedron **20**, 2991 (1964).
282. TANKER, M.: Ist. Tip Fak. Mec. **26**, 26 (1963).
283. TATAR, J.: Herba hung. **3**, 457 (1964).
284. TÉTÉNYI, P., u. D. VÁGUJFALVI: Herba hung. **2**, 185 (1963).
285. THOMAS, A. F., u. J. M. MÜLLER: Experientia (Basel) **16**, 62 (1960).
286. TSCHESCHE, R., F. LAMPERT, and G. SNATZKE: J. Chromatog. **5**, 217 (1961).
287. — u. U. AXEN: Ann. Chem. **669**, 171 (1963).
288. — J. DUPHORN, and G. SNATZKE: Ann. Chem. **667**, 151 (1963).
289. — G. BIERNOTH, and G. SNATZKE: Ann. Chem. **674**, 196 (1964).
290. — J. DUPHORN, and G. SNATZKE: In: A. T. JAMES, and L. J. MORRIS: New Biochemical Separations. London: Van Nostrand Company 1964.
291. — u. A. K. SEN GUPTA: Chem. Ber. **93**, 1903 (1960).
292. — E. HENCKEL and G. SNATZKE: Ann. Chem. **676**, 175 (1964).
293. — F. LAMPERT, and G. SNATZKE: J. Chromatog. **5**, 217 (1961).

294. TSCHESCHE, R., u. H. STRIEGLER: Naturwissenschaften **52**, 303 (1965).
295. — u. G. WULFF: Planta med. **12**, 272 (1964).
296. — u. N. ZIEGLER: Ann. Chem. **674**, 185 (1964).
297. TSUDA, Y., T. SANO, K. KAWAGUCHI, and Y. INUBUSHI: Tetrahedron Letters **1964**, 1279.
298. TURSCH, B., u. E. TURSCH: Bull. Soc. chim. belg. **70**, 585 (1961).
299. — — and I. T. HARRISON: J. org. Chem. **28**, 2390 (1963).
300. TYIHÁK, E.: Sci. Pharm. **30**, 185 (1962).
301. — u. D. FÖLDESI: Naturwissenschaften **49**, 469 (1962).
302. — u. M. IMRE: Herba hung. **2**, 157 (1963).
303. — u. G. MOLNAR: Z. allg. Mikrobiol. **4**, 161 (1964).
304. — — Z. Lebensmitt.-Untersuch. **123**, 362 (1963).
305. — J. SÁRKÁNY-KISS u. I. MÁTHÉ: Pharm. Zentralhalle **102**, 128 (1963).
306. — — — Herba hung. **2**, 173 (1963).
307. — u. D. VÁGUJFALVI: Herba hung. **1**, 97 (1962).
308. — — u. P. L. HÁGONY: J. Chromatog. **11**, 45 (1963).
309. — — — Act.: Pharm. hung. **196**, 176 (1963).
310. URBACH, G: J. Chromatog. **12**, 196 (1963).
311. URECH, J., B. FECHTIG, J. NÜESCH u. E. VISCHER: Helv. chim. Acta **46**, 2758 (1963).
312. VÁGUJFALVI, D., és E. TYIHÁK: Herba hung. **2**, 363 (1963).
313. VALENTIN, H.: Pharm. Ztg. **80**, 469 (1935).
314. VASHIST, V. N., and K. L. HANDA: Soap, Perfum. Cosm. **37**, 135 (1964).
315. — — J. Chromatog. **18**, 412 (1965).
315a. VERDERIO, E., e D. VENTURINI: Boll. chim. farm. **104**, 170 (1965).
316. VIOQUE, E., and R. T. HOLMAN: J. Amer. oil Chem. Soc. **39**, 63 (1962).
316a. VOGEL, G.: Planta Med. **11**, 362 (1963).
317. VOLK, O. H., u. R. SCHUNK: Dtsch. Apoth.-Ztg. **104**, 187 (1964).
318. VORBRÜGGEN, H.: Ann. Chem. **668**, 57 (1963).
319. WAGNER-JAUREGG, TH., u. M. ROTH: Pharmac. Acta Helv. **37**, 352 (1962).
320. WAGNER, H., L. HÖRHAMMER u. H. NUFER: Arzneimittel-Forsch. **15**, 453 (1965).
321. WALLACH, O.: Terpene und Campher. 2. Aufl. Leipzig: 1914.
322. WASICKY, R.: Rev. Fac. Bioquim. Sao Paulo **1**, 69 (1963); ref. Anal. Abstr. **11**, No. 3772 (1964).
323. WASICKY, R., and O. FREHDEN: Mikrochem. Acta **1**, 55 (1937).
324. WEIGERT, E., y P. J. SCHORN: Trib. Farmaceut. **30**, 48 (1962).
325. WELLENDORF, M.: Dansk Tidskr. Farm. **37**, 145 (1963).
326. WESTFELT, L.: Acta Chem. scand. **18**, 572 (1964).
327. WICHTL, M.: Planta Med. **11**, 53 (1963).
328. WINKLER, W.: Kolloid. Z. **177**, 63 (1961).
329. WOLLRAB, V.: Riechst., Aromen, Körperpflegemittel Nr. **10**, 321 (1964).
330. — M. STREIBL a F. ŠORM: Collect. Čs. Chem. Commun. **30**, 1654 (1965).
331. WROLSTAD, R. E., and W. G. JENNINGS: J. Chromatog. **18**, 318 (1965).
332. ZAMOJSKI, A., i F. ZAMOJSKA: Chem. analit. (Warszawa) **9**, 589 (1964).
333. ZECHMEISTER, L.: Fortschritte der Chemie organischer Naturstoffe, Bd. 14. Berlin-Göttingen-Heidelberg: Springer 1957.
334. ZINKEL, D. F., u. J. W. ROWE: J. Chromatog. **13**, 74 (1964).

K. Vitamins, Including Carotenoids, Chlorophylls and Biologically Active Quinones[1]

H. R. BOLLIGER and A. KÖNIG

TLC has become an indispensable technique in almost all scientific and practical fields connected with the research and analysis of these compound classes. The methods of investigation and separation hitherto used, have been supplemented and refined by this technique with its striking advantages. It may be used for checking these methods and simplifies in part working up and purification procedures, isolation, identification, control of purity, determination and control of syntheses of the active materials. Fractionation on the layer has made use of the following principles in practice, often in combination: primarily adsorption and partition; also complex formation on silver nitrate-impregnated carrier material; polyamide- and ion exchange chromatography. The dissolved substance mixtures, sometimes radioactively labelled, have been applied as spots or bands to thin and thick layers. Separation, using various solvents, has been mostly by ascending chromatography but horizontal, descending, circular, two-dimensional and other variations have been used. Visualisation, evaluation or elution has then followed.

Up to the end of 1965, over 500 publications had appeared on the TLC of vitamins and related compounds. About 80% of these concerned the fat-soluble class. Five surveys have been made [8, 70, 77, 110, 130]. Methods of proved value and also hitherto unpublished experience is described here, based on this mass of work. Many additional systems and reactions for detection will certainly be still found, which will help in solving specific problems

The carotenoids have been treated in this chapter because they show pro-vitamin A activity to some extent; likewise chlorophylls which occur along with the carotenoids in the chloroplasts. Further, active quinones similar to vitamins E and K have been included; in nature they play a part as catalysts and electron carriers.

I. Method of Work and General Experience

Reference should be made to the original articles and relevant handbooks [1, 46, 54, 57, 78, 90, 130, 142] for information about occurrence,

---

[1] The IUPAC suggestions have been mostly adopted for the nomenclature of these compounds which differ so much amongst themselves in their chemical structure. (Inform. Bull. No. 25, 1966).

biological function and activity, properties, structural chemical formulae and methods hitherto used for isolation and determination of the active compounds. This knowledge, along with experience, skill and fantasy, are the success requirements for the worker on TLC alone or in combination with other procedures.

Testing vitamins through TLC combines extraction, separation, identification and determination. The choice of processing procedure for obtaining an extract for TLC is determined by the nature of the starting material, the form and amount of the vitamins present and the other components of the sample. In naturally occurring products, fat-soluble vitamins, for example, are rarely in a free state but are linked chemically to lipids and proteins or occur even in a water-soluble form. Artificial preparations are often stabilised by coating with gelatine. The water-soluble vitamins occur also as esters of fatty acids or in combined form as coenzymes. Vitamins are extracted, using known methods, from a representative, homogeneous sample (solution, powder) either directly or after hydrolysis or enzymatic liberation. The extracts thus obtained are concentrated, or, if necessary, purified further or treated chemically; they are then applied to the adsorbent layer and chromatographed with suitable solvents. Detection on the layer can be carried out for analytical and preparative purposes by examination in light of various wave lengths or by treatment with reagents[2], often with heating. Since the hR_f-values are only guide values, the substances must be characterised unambiguously by measurements of spectra after elution. Quantitative evaluation can be performed directly on the layer or after elution.

Sources of error which must be borne in mind during the determinations, are sensitivity towards atmospheric oxygen, light, heat, solvents, alkalies, acids and active adsorbents; these may bring about changes in sensitive vitamins, carotenoids and lipoquinones. These changes can be simply and well recognised through TLC (cf. Fig. 112, degradation of vitamin A on the dry layer) and may be reduced or eliminated by appropriate protective measures and care. Interferences can be caused also through impure solvents (Fig. 110), damage to the adsorbent layer during application (Fig. 113), contaminants which have not been separated (e. g., oils, Fig. 111) or overloading of the layer material by using too concentrated solutions.

Our own experience has shown that it is advisable to carry out vitamin analyses in laboratories facing north (northern hemisphere!) maintained at temperatures as constant as possible (about 22° C) and in subdued light. Colourless foil[3] or varnishes[4] which contain UV-

---

[2] A list of the reagents (alphabetically arranged) is at the end of the book (p. 918).
[3] e.g. special foil UVEX (Firm 34).
[4] e.g. commercial nitrolack with 1—2% UV-absorber (Firm 88).

absorbing materials, covering over or painted on the windows, respectively, have proved to be good protective devices. Blinds need then be drawn only in sunshine. All glassware must be scrupulously clean. Ground glass joints must be lubricated with water and not with grease for work on lipophilic compounds. Solutions containing unstable vitamins and carotenoids can be kept only a short time. During all stages of processing they must be maintained in a nitrogen atmosphere and should be chromatographed immediately after application to the plate. The layer should be scraped off for elution while it is still moist from solvent.

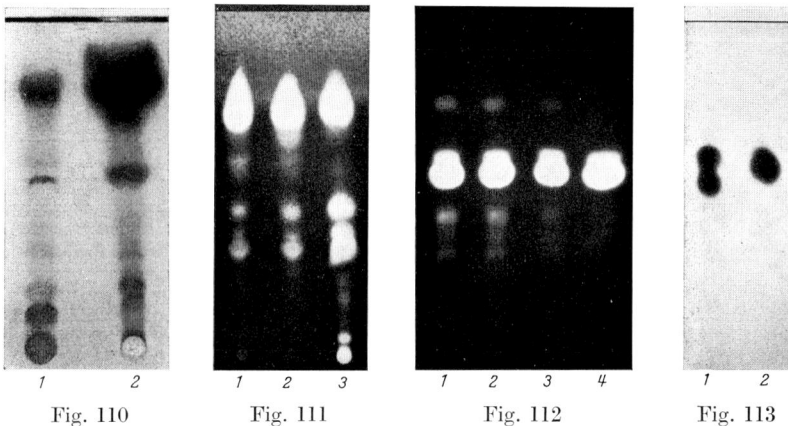

Fig. 110   Fig. 111   Fig. 112   Fig. 113

Interferences in TLC of vitamins (Layer: silica gel GF_{254}, activated and about 0.25 mm thick; CS)

Fig. 110. Solvent purity. Residue from evaporation of 400 ml applied and chromatographed with cyclohexane-ether (80 + 20). Chromatogram photographed after spraying with molybdophosphoric acid reagent and heating. *1* peroxide-free ether; *2* absolute alcohol

Fig. 111. Oils in which fat-soluble vitamins are dissolved. Solvent: cyclohexane-ether (50 + 50). Chromatogram photographed in UV light (365 nm) after spraying with conc. sulphuric acid and heating. 500 µg of each applied: *1* peanut oil; *2* cottonseed oil; *3* wheat-germ oil

Fig. 112. Stability of vitamin A-acetate on silica gel. Solvent: cyclohexane-ether-pyridine (79 + 20 + 1). Chromatogram photographed in UV light (365 nm). 30 µg amounts applied at intervals and chromatographed after 5 min (*1*), 3 min (*2*), 1 min (*3*) and 0 min (*4*)

Fig. 113. Duplicate spot formation through damage to the layer while applying certain vitamins. Chromatogram photographed in UV light (254 nm). Application: *1* layer damaged with the pipette tip; *2* layer undamaged

Quantitative isolation and purification without losses of the active components, which are mostly present only in trace amounts, is beset with difficulties due to the many interfering contaminants; it must be carried out under the mildest possible conditions. Consequently every step in the analysis should be checked, whether simple or complex material is being investigated. Pure standard preparations are indispensable for this purpose. One must remember that the carotenoids, certain vitamins and quinones occur in isomeric forms of differing biological activity. These may be easily interconvertible. It must also be established whether decomposition products were present already in the material to be tested or have been formed during treatment.

We use the standard TLC method (p. 85). STAHL's gradient layers [126] have been of value for finding out a suitable layer composition for compounds difficult to separate. Where possible, adsorbents with a fluorescent additive are used, so that the chromatograms can always be tested for light absorption and fluorescence in short- and long-wave UV light, both before and after spraying. The layer thicknesses quoted by us are those given on the scale of the spreader. Our limits of detection have been determined on 0.25 mm layers; they depend also, however, on the spot area, the intensity of spraying and how long before inspection is carried out. Before application, the extracts and pure substances are dissolved in the prospective solvent itself or in one of its components, in order to avoid artifacts as a result of using other solvents. Fresh solvent is used in the chamber for each new chromatogram and it is shaken briefly to achieve atmospheric saturation when CS or NS are used.

Band application of relatively large amounts of extract (0.1–2.0 ml) on thicker (0.4–2.0 mm) and broader (up to 40 cm) [131] layers[5] has proved valuable for preparative separation and quantitative determination of vitamins. Useful also have been reference chromatograms (cf. Fig. e in Plate II, Figs. 119 and 121). These furnish immediate proof of identity and any possible displacement through overloading of the layer can be recognised; the active substance, often not directly detectable, can be localised with the help of the reference strip and is then scraped off into an elution column fitted with a sintered glass filter (cf. Fig. 120). Separation of silica gel by filtration is a practical procedure. The eluate, after concentration, can then be characterised and evaluated by the most suitable method. The level of interference of the adsorbent must always be established and taken into consideration. This effect can be eliminated by preliminary washing of the coated plate or by employing adsorbents of high purity [131]. Instruments are available nowadays for direct evaluation of the spots on chromatograms (remission, transmission, fluorometry); these are useful for many active compounds.

There is unfortunately no general analytical procedure for vitamins which is suitable for all samples. The analyst must try out and combine

---

[5] Still thicker layers and broader plates are used in preparative work [60].

the individual stages for the particular sample to hand. To this end, TLC must be properly applied along with other methods. The factors of time, instrumentation and expense play a part here.

II. TLC of Fat-Soluble Vitamins, Carotenoids, Chlorophylls and Quinones

The lipophilic compounds are in general less stable than the water-soluble vitamins and in TLC also, they must be handled with the care mentioned above. Crude extracts of plant and animal tissues, or of artificial preparations, often contain larger amounts of fat and protective materials with similar properties; this can interfere with analytical and preparative work. In this case, they must be adequately purified beforehand.

1. Mixtures of Fat-Soluble Vitamins

a) Separation

There are many examples of separation of these compounds. Numerous authors have used adsorption TLC for this purpose [e. g., 8, 25, 70, 75, 130]. After concentration, the solution studied is generally applied to activated silica gel or alumina or, also, to air dried layers [83] and chromatographed without delay, using various solvents. Runs last 40—60 min, a migration distance of 12—18 cm. Good solvents are chloroform, benzene, cyclohexane-ether, cyclohexane-ethyl acetate etc. New separation effects can be obtained, depending on the problem, by varying the composition of the mixtures mentioned. Thus the spots of vitamin A-alcohol and vitamin D, which lie very close together, can be made to migrate away from each other; or substances with higher Rf-values can be made to merge together in the front, or vice versa. These resolutions succeed just as well on thicker layers and with larger amounts of vitamin and/or of extract. DAVÍDEK and BLATTNÁ [25] have tested 14 solvents, almost all of which gave the same substance sequences as chloroform, petrol ether or carbon tetrachloride. Their technique using loose layers has, due to the lesser handiness, scarcely succeeded in competing with the use of fixed layers, although equivalent results have been obtained with it.

Success has been achieved also with reversed phase partition chromatography on impregnated layers (e. g., activated silica gel GF_{254}, 0.25 mm thick, immersed in a 5% solution of paraffin oil in petrol ether and dried) [8]. As in adsorption chromatography, the separation sequence is modified by varying the composition of solvent mixtures. The substances are stabilised efficiently by being dissolved in the paraffin oil. Tailing is encountered when larger amounts of certain vitamins are used, due to the lower adsorption capacity of such layers.

Table 38. hR_f-values of fat-soluble vitamins, using various separation systems (12—16 cm lengths of run in 40—60 min)

Layer (ca. 0.25 mm):	S_1	S_1	S_1	S_1	S_1	S_2	S_2	S_2	S_3	S_4	S_4
Solvent:	F_1	F_2	F_3	F_4	F_5	F_4	F_4	F_5	F_6	F_7	F_8
References:	[8]	[10]	[10]	[70,75]	[70,75]	[8]	[70,75]	[70,75]	[25]	—	—
β-Carotene[a]	84	90	87	100	100	89	78	86	90	2	17
Vit. A-alcohol[a]	10	35	35	8	22	16	8	28	—	70	87
Vit. A-acetate[a]	45	74	65	41	69	72	62	78	63	62	86
Vit. A-palmitate[a]	72	87	84	75	94	84	74	82	—	3	16
Vit. D_2 or D_3	15	40	45	9	14	22	17	51	9	54/50	85/84
α-Tocopherol	32	68	66	35	56	53	37	62	—	30	83
α-Toc.-acetate	40	77	68	40	76	71	60	80	54	—	—
Vit. K_1[a]	61	82	75	67	81	79	73	80	74	8	40
Vit. K_3[a]	38	—	—	29	49	—	63	75	49	—	—

Layer:

S_1 = silica gel G or GF_{254} (Merck), activated, CS
S_2 = alumina G (Merck), activated, CS
S_3 = alumina (Lachema Inc., Activity III—IV), loose
S_4 = silica gel GF_{254} (Merck), activated and impregnated with paraffin oil

Solvent:

F_1 = cyclohexane-ether (80 + 20)
F_2 = cyclohexane-ether (50 + 50)
F_3 = cyclohexane-ethyl acetate (75 + 25)
F_4 = benzene
F_5 = chloroform
F_6 = carbon tetrachloride
F_7 = acetone-water (80 + 20)
F_8 = acetone-water (90 + 10)

[a] All-*trans* compounds.

Table 38 gives information about the hR_f-values of fat-soluble vitamins in mixtures, using the solvent systems quoted above (cf. also Fig. a in Plate II). According to STROHECKER [130, 131], silica gel H and HF_{254} yield the same chromatographic behaviour. It is evident that

Table 39. *Detection of fat-soluble vitamins directly* [8] *and* (0.25 mm thickness set on

Vitamins	Appearance in		
	UV (254)	UV (365)	Daylight
β-Carotene	dark[1]	dark[1]	orange
A-alcohol, -acetate, -palmimate	dark[1]	yellow-green[2]	—
D_2 or D_3	dark[1]	—	—
α-Tocopherol	dark[1]	—	—
α-Toc.-acetate	dark[1]	—	—
K_1	dark[1]	dark[1]	yellow

1 = absorption; 2 = fluorescence; 3 = unstable colours which change more or less still distinguishable).

Plate II. *TLC of vitamins and carotenoids on silica gel GF_{254}* (ca. 0.25 mm thick, activated; CS)

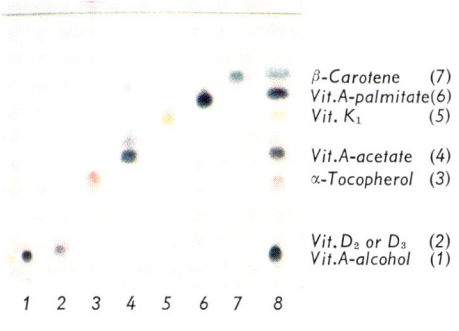

Fig. a. Separation of fat-soluble vitamins [8]. Solvent: cyclohexane-ether (80 + 20). Colours 30 sec after spraying with $SbCl_5$-reagent. *1, 4, 5, 6* and *7* all-trans compounds; *8* mixture *1—7*; 20—40 µg of each

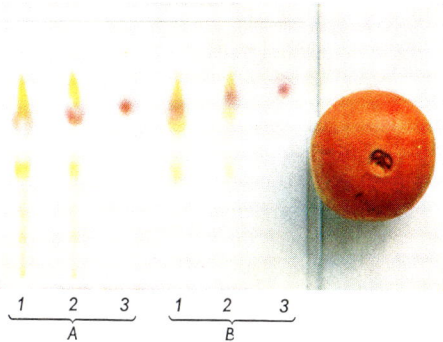

Fig. b. Detection of 8′-apo-β-carotinal in oranges [132]. Solvent: benzene-methanol-ether (85 + 5 + 10). *1* orange extract; *2* 1 + 8′-apo-β-carotinal; *3* 8′-apo-β-carotinal; *B* indandione derivatives of *A*

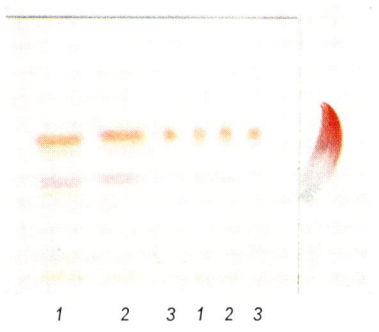

Fig. c. Detection of canthaxanthin in flamingo feathers [134]. Solvent: methylene dichloride-ether (90 + 10). *1* extract of the feathers; *2* 1 + all-trans-canthaxanthin; *3* all-trans-canthaxanthin

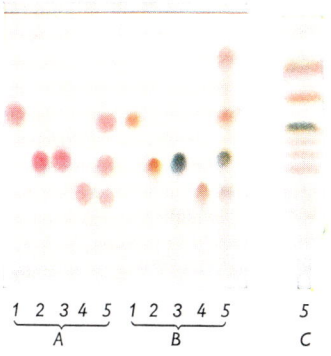

Fig. d. Separation and detection of various tocopherols [79]. Solvent: chloroform. *1* α-; *2* β-; *3* γ-; *4* δ-T (30 µg of each); *5* concentrate of naturally occurring vitamin E (100 µg). Spray-detection: *A* Dipyridyl-iron reagent; *B + C* $SbCl_5$ reagent. Technique: *A + B* one-dimensional; *C* wedge-tip

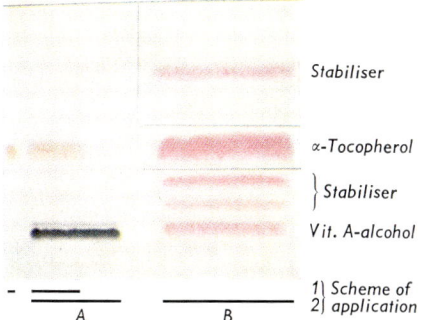

Fig. e. Quantitative determination of α-tocopherol in water-soluble vitamin A + E concentrate, after hydrolysis [8, 130]. Solvent: cyclohexane-ether (80 + 20). α-T zone located in UV (254 nm) in part B, scraped off, eluted and submitted to colour reaction. *1* α-T standard; *2* extract of the hydrolysed sample solution; *A* Reference chromatogram, sprayed with; *A* $SbCl_5$-reagent; *B* dipyridyl-iron reagent

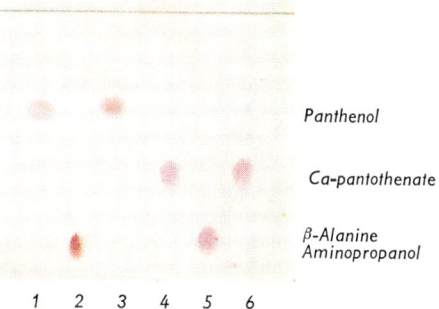

Fig. f. Detection of panthenol and Ca pantothenate. Solvent: ethanol-water (80 + 20). Plate heated, sprayed with ninhydrin reagent and again heated to 160° C. 5—10 µg of each active material: *1* panthenol; *2* aminopropanol; *3* extract from a cosmetic preparation; *4* Ca pantothenate; *5* β-alanine; *6* extract from multivitamin tablets

there are differences between the migration distances on alumina (S_2) using benzene (F_4), found by KATSUI et al. [70, 75] and by ourselves. Such differences are due to different activities of the layers. This can occur particularly when they are stored in containers which are not airtight. We have unfortunately been unable to confirm the hRf-values for vitamin A-alcohol and vitamin D_2 [70, 76], obtained with the system S_2/F_5; if this had been so, the system could have been considered for determination of vitamin D. Comparison of pure chloroform with chloroform stabilised with 1% alcohol, shows a marked influence of the alcohol on the distance of migration.

EGGER et al. have demonstrated the usefulness of polyamide layers in the separation of lipoquinones and carotenoids [41, 42]. This methiod which depends on partition and adsorption processes, offers separaton, possibilities for the fat-soluble vitamins also.

b) Detection and Determination

Small amounts of the active substances can be identified directly in light of various wave lengths through their own colour (carotenoids), absorption and fluorescence. This is useful as regards subsequent quantitative elution. Exposures to UV light must, however, be brief to prevent photochemical decomposition. A universal and differentiating method of visualisation is spraying with concentrated sulphuric or nitric acid, 70% perchloric acid [25] or antimony(III)- or (V)-chloride (Rgt. Nos. 15 and 18). The colours are, however, rather unstable and change more or less rapidly (cf. Table 39 and Fig. a in Plate II). The tints depend on the humidity of the layer, the concentration of substance, the intensity of spraying and the interval before assessment. Traces of vitamins can be rendered visible using molybdo- and tungstophosphoric acids (Rgt. Nos. 168 and 256); grey blue spots are yielded [8]. Further specific detection methods are referred to in the following paragraphs.

immediately after moderate spraying, on silica gel GF_{254} spreader; activated layer)

Limit of detection in μg			Colours[a] with		Limit of Detection in μg	
			Conc. H_2SO_4	$SbCl_5$-Reagent		
0.1	0.1	0.05	blue	blue	0.1	0.1
2	0.05	—	blue	blue	0.1	0.1
0.5	—	—	brown[4]	orange	1	0.5
20	—	—	yellow-green	orange-red	5	2
20	—	—	—	yellowish	—	50
0.5	5	50	yellow-brown	yellow	1	2

quickly; 4 = after about 3 min, D_2 is brown, D_3 is dirty green (with at least 30 μg

Lipophilic active substances can be determined in the presence of one another by using the procedure described; this may be done in commercial preparations of the pharmaceutical, food and feeding-stuff industries, qualitatively and partly even quantitatively [8, 10, 13, 25, 70, 130]. Numerous experiments have been described also, in which these compounds were detected alongside one another in plant and biological extracts [e. g., 5, 30, 45, 141].

2. Carotenoids and Chlorophylls

One cannot stress enough the value of TLC in biochemical studies of these naturally occurring pigments and in their synthesis. It is certain that many important experiments in the carotenoid field would have been far more difficult, perhaps even impossible, without this simple and elegant technique. Harmful influences and artifacts can be eliminated as a result of the present day knowledge that the colouring materials may decompose easily in any stage of working up. Many of these substances suffer relatively rapid change in solution, even at room temperature. Only freshly prepared solutions should thus be investigated and chromatographic reference substances used which are vital for the identification. The carotenoids and chlorophylls remain mostly unchanged during the brief time of run. The use of alkalised layers (prepared with 3% aqueous potassium hydroxide) prevents possible isomerisation of certain carotenoids during chromatography on silica gel G. Carotenoids begin to decompose on the dry, solvent-free layer. As a safety measure, one can work in an inert gas. The colours can be protected against oxidation and at the same time be intensified by spraying with 5% paraffin oil in petrol ether, immediately after development.

a) Separation

Almost all adsorbents known from column chromatography are suitable for adsorption TLC of carotenoids, the procedure most used hitherto; they may also be combined. The solvent chosen is always determined by the components to be separated. The chromatographic behaviour on magnesium oxide layers of the six carotenes of type $C_{40}H_{56}$, depends on the position of the double bonds in the ring systems or on the ring opening; the structure without closed rings is the one most strongly adsorbed [9]. As expected, the carotenoid hydrocarbons are retained less strongly than the oxygen-containing compounds. In C_{40}-carotenoids, for example, polar substituents act in the sequence $-OR < =O < -OH$ in diminishing the Rf-value; if, however, a second substituent is introduced into an already modified ionone ring, these rules do not apply any longer for the new group [39]. The adsorption affinity of carotenoid aldehydes (apo-β-carotenals) is not a function of the number of double bonds (as, for example, in the homologous polyene series) but of the constitution. The C_{40}-, C_{35}-, C_{30}- and C_{25}-aldehydes are more easily eluted than the C_{37}-, C_{32}- and C_{27}-aldehydes; in each group, the distance of migration increases with fall in molecular weight. Decisive here is therefore the presence in α-position to the aldehyde group, of a methyl group or a hydrogen atom. Another example is retinal (vitamin A_1-aldehyde) with five double bonds which is distinctly more strongly adsorbed on silica gel than is 8'-apo-β-carotenal with its nine [148, 149].

Examples of adsorptive separation systems are in Table 40, some in Table 41. The separation into groups due to STAHL et al. [8, 125] has been modified by using magnesium oxide layers for the carotenes [9]. Plant extracts have been chromatographed according to this analytical scheme and then quantitatively evaluated, directly or indirectly. Better results were obtained in all experiments using the S-chamber than with the usual glass chambers. Stepwise development by using solvents of varying elution properties on silica gel G, has been employed in the fractionation of such colouring materials [124]. Adsorption procedures have been used successfully by many research teams in studies of biogenesis, distribution, function and transformation of carotenoids in plants, algae, bacteria, fungi etc. and in human and animal tissue (WINTERSTEIN et al. [148—150], GOODWIN et al. [26, 27, 55, 85, 147], GROB et al. [e. g., 43, 44, 56], THOMMEN et al. [133 to 135 etc.] and others). DAVIES et al. [27] have isolated some important carotenoid precursors like phytoene, phytofluene and related compounds, using petrol ether (BP 40—60° C) on silica gel G layers. WINTERSTEIN et al. [148, 149] have used calcium hydroxide-silica gel layers (80 + 20) and usually petrol ether-benzene (50 + 50) as solvent for separation and determination of carotenoid aldehydes in biological material. The solvent composition was varied in certain cases and the elution effect augmented by adding 1% methanol. THOMMEN et al. have demonstrated the versatile applicability of silica gel-TLC by detecting and identifying the following carotenoids in the presence of others: 8'-apo-β-carotenal in oranges [132] (Fig. b, Plate II) and in alfalfa meal [133]; canthaxanthin in the lesser flamingo [134] (Fig. c in Plate II) and in hen egg yolks [135]; canthaxanthin, β-carotene and astaxanthin in trout [137]; and canthaxanthin, echinenone and β-carotene in crustaceans [136]. Labelled pigments have likewise been chromatographed since the high sensitivity and specificity of the isotope technique enable special information to be acquired [e. g. 55, 135,]. Mixtures of isomers are best separated through adsorption chromatography; the optimum separation conditions must, however, be sought for each case individually because no generally applicable rules can be established [8]. Chlorophylls and

Table 40. hRf-values of carotenes and carotenoids in adsorption-TLC
(12—16 cm run in 40—60 min)

Layer (ca. 0.25 mm):	S_1	S_2	S_3	S_4	S_4	S_5^a	S_5^a	S_6^a	S_6^a	S_2^a
Solvent:	F_1	F_2	F_3	F_4	F_5	F_6	F_7	F_8	F_1	F_9
References:	[54]	[8]	[28]	[8]	[8]	[9]	[9]		[8, 125]	
ε-Carotene	—	—	—	—	—	70	84	↑	↑	↑
α-Carotene	—	—	—	—	—	66	80			
β-Carotene	81	82	96	—	—	49	74			
δ-Carotene	—	—	—	—	—	20	55	97	100	100
γ-Carotene	—	—	—	—	—	11	41	↓	↓	
Lycopene	—	74	—	—	—	1	13	↓		
Torulin	—	—	—	—	—	—	—	—	98	
β-Zeacarotene	75	—	—	—	—	—	—	—	—	
ζ-Carotene	62	—	—	—	—	—	—	—	—	↓
Torularhodin methyl ester	—	—	—	—	—	↑	↑	83	94	100
12′-Apo-β-carotenal	—	—	—	—	—			70	—	—
8′-Apo-β-carotenal	—	15	—	—	—			64	—	—
10′-Apo-β-carotenal	—	—	—	—	—			53	—	—
Lycopinal	—	—	—	—	—			43	—	—
Methylbixin	—	—	—	—	—			13	81	97
Canthaxanthin	—	0	38	43	—			↑	63	90
Cryptoxanthin	—	—	—	34	74				54	75
Lutein	—	—	—	—	55				10	35
Antheraxanthin	—	—	—	—	—				10	32
Zeaxanthin	—	—	17	—	57				↑	24
Violaxanthin	—	—	—	—	—	0	0	0	5	21
Capsanthin	—	—	—	—	—					16
Capsorubin	—	—	—	—	—				↓	13
Bixin	—	—	51	—	—				0	5
Azafrin	—	—	—	—	—	↓	↓	↓	0	2
Echinenone	—	10	—	82	—	—	—	—	—	—
Rhodoxanthin	—	—	—	16	94	—	—	—	—	—
Isozeaxanthin	—	—	63	—	—	—	—	—	—	—
Methyl 8′-apo-β-carotenoate	—	26	—	—	—	—	—	—	—	—
Butter yellow	—	—	—	—	—	25	51	94	95	100
Indophenol	—	—	—	—	—	17	45	58	86	100
Sudan red	—	—	—	—	—	4	11	68	90	100

Layer:
S_1 = Silica gel G-MgO (1 + 1), activated, CS
S_2 = Silica gel G, activated, CS
S_3 = Silica gel-rice starch (98 + 2), activated, NS
S_4 = Ca(OH)$_2$-silica gel G (6 + 1), activated, CS
S_5 = MgO ("Darlington" light B. P.), activated, CS
S_6 = sec. Magnesium phosphate, activated, CS

Solvent:
F_1 = Petrol ether (40—60° C)-benzene (90 + 10)
F_2 = Petrol ether (90—110° C)-benzene (50 + 50)
F_3 = n-Hexane-ether (30 + 70)
F_4 = Benzene
F_5 = Benzene-methanol (98 + 2)
F_6 = Petrol ether (90—110° C)-benzene (50 + 50)
F_7 = Petrol ether (90—110° C)-benzene (10 + 90)
F_8 = Carbon tetrachloride
F_9 = Methylene dichloride-ethyl acetate (80 + 20)

[a] Modified separation into groups according to STAHL et al. [8, 125].

related compounds also may be isolated on activated adsorbents, as seen in Table 41 [12, 113]; silver nitrate-silica gel layers can also be used for this [20]. Ways of applying silica gel-TLC to determination of carotenoids in food and drink have been described many times [4, 89, 114 etc.].

Both non-impregnated and impregnated layers have been proposed for partition chromatography of naturally occurring colouring materials. Amongst the former have been: cellulose [2, 39], sugars [18, 122], mannitol [122] and a mixture of kieselguhr, silica gel, calcium carbonate

Table 41. hRf-values of carotenoids and chlorophylls in partition- and adsorption-TLC (12—16 cm lengths of run in 30—60 min)

Layer (ca. 0.25 mm):	S_1 K'guhr	S_1 K'guhr	S_2 K'guhr	S_3 Cell.	S_3 Cell.	S_4 Mann.	S_5 K'guhr	S_6 S. Gel
(Impregnation)	Par.	Par.	Trigl.	—	—	—	—	—
Solvent:	F_1	F_2	F_3	F_4	F_5	F_6	F_7	F_8
References:	[37]	[37]	[36]	[39]	[2]	[122]	[12]	[113]
Neoxanthin	—	—	95	—	—	—	0	23
Violaxanthin	—	—	84	—	—	—	65	37
Lutein epoxide	—	—	72	—	—	—	—	—
Lutein	100	97	56	—	—	55	83	42
Canthaxanthin	94	86	—	35	—	—	—	—
Cryptoxanthin	91	80	7	58	—	—	—	—
Zeaxanthin	—	—	55	10	—	—	—	—
Rhodoxanthin	—	—	26	33	—	—	—	—
8'-Apo-β-carotenal	80	65	—	—	—	—	—	—
Echinenone	69	42	—	90	—	—	—	—
Torularhodin methyl ester	57	25	—	—	—	—	—	—
β-Carotene	22	3	0	97	98	99	99	87
Lutein dipalmitate	11	0	0	—	—	—	—	—
Chlorophyll a	—	—	25	—	60	73	92	77
Chlorophyll b	—	—	13	—	35	40	84	60
Phaeophytin a	—	—	1	—	93	—	—	82
Phaeophytin b	—	—	7	—	80	—	—	—

Layer and Solvent:

S_1 = Kieselguhr G (Merck), mixed with dioxan-water (3 + 1), dried 4 h at 100° C, impregnated to within 3—4 cm of the upper edge by ascending treatment using paraffin oil-petrol ether (100—140° C) (8 + 92) and then heated 1 h at 70° C to remove the petrol ether.
S_2 = Prepared like S_1 and partly impregnated with an 8% solution of a triglyceride (as far as possible a vegetable oil, free of acid, like olive oil or livio etc.).
S_3 = Cellulose MN 300 (standard method), CS
S_4 = Mannitol (65 g in 100 ml acetone + 1 ml 50% aqueous starch solution), air dried for 30 min
S_5 = Kieselguhr G (Merck), activated
S_6 = Silica gel G, activated, NS
F_1 = Acetone-methanol-water (50 + 47 + 3)
F_2 = Acetone-methanol-water (20 + 76 + 4)
F_3 = Acetone-methanol-water (15 + 75 + 10)
F_4 = Petrol ether-carbon tetrachloride (60 + 40)
F_5 = Petrol ether (60—80° C)-acetone-n-propanol (90 + 10 + 0.45)
F_6 = Petrol ether (30—60° C)-n-propanol (99.5 + 0.5)
F_7 = Petrol ether (65—90° C)-n-propanol (99 + 1)
F_8 = Petrol ether (60—80° C)-ethyl acetate-diethylamine (58 + 30 + 12)

and calcium hydroxide containing ascorbic acid as antioxidant, dried at 50° C [59]; fractionation of the substances on these layers is very similar to that in adsorption (cf. Table 41). Amongst the latter, impregnated layers, paraffin oil or triglycerides have proved especially suitable agents.

In reversed phase partition, lengthening of a chain, also in the alkyl group of an ester, reduces the Rf-values; these values are generally increased by polar substituents. Impregnated kieselguhr and cellulose layers yield similar results. On the other hand, differences in polarity of the stationary phase are significant. Impregnation with paraffin oil permits separation of the less polar carotenoids and the polar compounds migrate with the front. These latter can be fractionated on layers

Table 42. hRf-values of carotenoids on polyamide [42]
(15—18 cm runs in 2—2.5 h)

Compound	Substituents etc.	unpolar ← PE:M/M			Solvent → polar H$_2$O:M/M			
		4:1	2:1a	1:1	M/M	1:10	1:6	1:3
Physalien	2—COOR	100	100	95	30	0	0	0
β-Carotene	—	100	100	100	80	25	10	0
Lycopin	2 R!	95	90	75	60	5	0	0
Cryptoxanthin	1—OH	62	70	76	74	39	21	4
Rubixanthin	1—OH + 1 R!	45	60	64	45	15	4	0
Lycoxanthin	1—OH + 2 R!	29	37	40	32	8	0	0
Lutein	2—OH	35	57	93	95	68	45	14
Zeaxanthin	2—OH	30	54	78	82	55	35	10
Isozeaxanthin	2—OH	34	56	92	91	57	36	10
Eschscholtzxanthin	2—OH + I! + II!	12	22	25	22	8	1	0
Lycophyll	2—OH + 2 R!	8	20	22	20	7	0	0
Violaxanthin	2—OH + 2 > 0	30	52	76	93	88	73	35
Neoxanthin	3—OH + 1 > 0	22	42	82	96	90	78	40
Echinenone	1 C=O	91	92	90	72	35	18	2
Euglenanone	2 C=O	62	68	81	80	54	34	9
Canthaxanthin	2 C=O	58	65	79	80	55	37	20
Rhodoxanthin	2 C=O + I! + II!	28	42	43	40	14	7	1
Astacin	4 C=O	34	50	69	72	42	25	7
Capsanthin	1 C=O + 2—OH	24	42	79	81	62	42	15
Capsorubin	2=O + 2—OH	19	37	74	76	60	42	15
Fucoxanthin	2=O + 3—OH + ?	27	44	84	98	95	85	55
8'-Apo-β-carotenoic acid	1—COOH	28	38	30	15	5	0	0
Torularhodin	1—COOH	6	10	9	2	1	0	0

Layer (0.25 mm): Polyamide (Merck)-Cellulose MN 300 (85 + 15) slurried in methanol, then air dried 30 min.

Solvent: PE = petrol ether (100—120° C); M/M = methanol-methyl ethyl ketone (50 + 50).

Substituents etc.: > O = epoxide; R! = ring opening between C_1 and C_6 or between C_1' and C_6'; I! = conjugated double bonds —C_6=C_7— and —C_6'=C_7'—; II! = additional —C_4'=C_5'— and —C_4=C_5—.

a Formation of a second front here.

impregnated with triglycerides, however [36—38]. Table 41 and Figs. 114 and 115 give information about these results.

WINTERSTEIN et al. [148] have separated the carotenoid aldehydes on silica gel layers, impregnated with paraffin oil, using methanol,

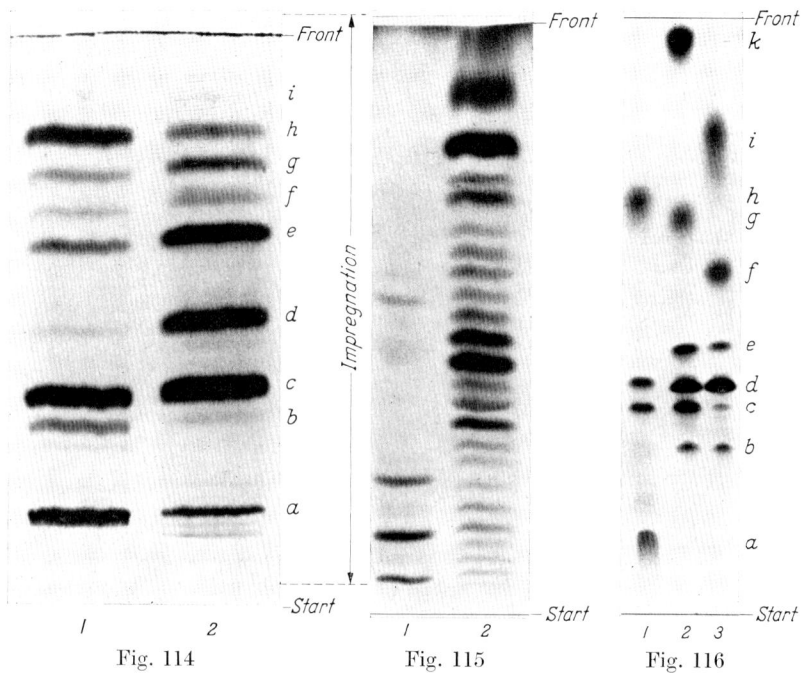

Fig. 114   Fig. 115   Fig. 116

Separation of carotenoids and chlorophylls on various adsorbents (see text):

Fig. 114. Layer: cellulose, impregnated with triglyceride; solvent: methanol-acetone-water (74 + 20 + 6), saturated with triglyceride. Applied in non-impregnated part: *1* extract of *Fucus serratus* (brown seaweed); *2* extract of *Chara fragilis* (stonewort). Substances: a β-carotene; b isomer of chlorophyll a; c chlorophyll a; d chlorophyll b; e lutein; f and g (in *1*) unknown xanthophylls; f (in *2*) lutein epoxide; g (in *2*) violaxanthin; h fucoxanthin; i neoxanthin [36, 37]

Fig. 115. Layer: cellulose, impregnated with paraffin oil; solvent: acetone-methanol (66 + 34), saturated with paraffin oil. Application in non-impregnated part: *1* lutein diester from the autumn leaves of *Aesculus hippocastanum* (horse-chestnut), containing in ascending order: lutein dipalmitate, lutein palmitate-linoleate and lutein dilinoleate; *2* total extract from the peripheral flower parts of *Helianthus annuus* (sunflower), containing numerous lutein esters and presumably taraxanthin [38]

Fig. 116. Layer: polyamide (perlon)-cellulose (80 + 20); solvent: petrol ether (BP 100—120° C)-butanone-methanol (78 + 11 + 11). Substances: a eschscholtzxanthin; b capsorubin; c capsanthin; d (in *1*) zeaxanthin; d (in *2* and *3*) fucoxanthin; e lutein; f fucoxanthin acetate; g lutein monoacetate; h cryptoxanthin; i by-product from acetylation of f; k lutein diacetate. Applied: *1* $a + c + d + h$; *2* $b + c + d + e + g + k$; *3* $b + c + d + e + f + i$ [42]

saturated with the paraffin oil, as solvent; they found the following hRf values, which correspond to theoretical expectations:

C_{40}	C_{37}	C_{35}	C_{32}	C_{30}	C_{27}	C_{25}	C_{20}
28	34	38	47	49	63	69	85

EGGER [36, 38, 39] has investigated and developed reversed phase partition chromatography further. He introduced partial impregnation. When application is carried out on the untreated part, the substances reach the stationary phase as narrow zones at the solvent front; from this point, particularly sharp resolution is accomplished. Triglyceride-containing layers also yield excellent separation of chlorophylls [36]. Further, zones of carotenoid esters, seemingly homogeneous after adsorption chromatography, have been separated into a number of components by partition-TLC [38, 39].

EGGER and VOIGT have recently described and discussed the separation of 31 carotenoids on polyamide layers, using 9 different solvent mixtures [39, 42] (cf. Table 42 and Fig. 116). The polyamide swells in alcohols. These then function as a stationary "liquid" phase, into which molecules of the compounds to be separated can penetrate deeply. The polarity of this phase has a favourable intermediate value, so that both very polar and relatively nonpolar solvents can be chosen. Phase reversal may thus be achieved without impregnation. In the authors' opinion, new fractionations which are related to the oxygen-containing functional groups and to the position of double bonds in the molecule, are brought about here as a result of interaction of partition processes and special adsorption phenomena (hydrogen bond formation), peculiar to the polyamide.

b) Detection and Determination

Very small amounts of the intensely coloured pigments can be detected in daylight and also in UV light (cf. Table 39). 0.01 µg of the deep red carotenoids may sometimes still be detected on the layer. Certain decomposition products fluoresce in the light of the quartz lamp. Spray reagents of general application, like, for example, concentrated sulphuric acid, are less sensitive but they compensate for this by yielding differentiating colours.

Carotenoid aldehydes form intensely coloured condensation products with the rhodanine reagent (No. 222) after drying, which serve for identification. As little as 0.02—0.03 µg retinal can be seen as an orange-red spot [150]. Blue to deep violet complexes, depending on the compound, are yielded with the indandione reagent (No. 138). The sensitivity is similar and they also fluoresce orange in UV light [133] (Fig. b, Plate II).

Carotenoids and chlorophylls can be quantitatively determined either on the layer by visual comparison with a series of standards chromatographed simultaneously or spectrophotometrically after elution [127]. The separated pigments can be eluted from impregnated layers using dimethylformamide and freed from interfering oil by washing with petrol ether [36].

3. Vitamin A Group

The socalled "vitamin A" or vitamin A_1 — it has an all-*trans* configuration and is physiologically the most active — and vitamin A_2 as well as their derivatives and *cis-trans* isomers, are very unstable. These compounds yield anhydro-vitamin A and more polar decomposition products rapidly after application on active layers; the more polar compounds are more strongly retained during development (cf. Fig. 112). Swift work is thus essential.

a) Separation

Up to now the A-vitamins have been chromatographed largely in adsorption systems. Separations, even of larger amounts, are good and the run durations short. KATSUI [77] has developed with benzene and chloroform on activated silica gel G and alumina G layers and found that separations of vitamin A-alcohol, several of its esters, vitamin A-aldehyde and -acid were similar to those of JOHN et al. [72], VARMA et al. [138] and others who employed two-component solvents (cf. Table 43). The mobility of the spots can be modified at will by change of solvent composition. We have recommended also silica gel G-sec. calcium phosphate (50 + 50) as layer material [8]. Decomposition of vitamin A compounds is appreciably reduced on alkalised silica gel G layers (prepared with 3% aqueous potassium hydroxide) and also by using pyridine-containing solvents. METZGER [in 108] has accomplished the most successful isolation so far of vitamin A- and A_2-alcohols, each in 0.4 µg amount, using petrol ether (BP 40—45° C)-methyl heptenone (85 + 15) on activated silica gel G. Isomers possessing a sterically hindered *cis* double bond (11-*cis* and 11,13-di*cis*) migrate distinctly faster than the unhindered *cis*-forms (9-*cis*, 13-*cis* and 9,13-di*cis*); the A_2-compounds migrate somewhat more slowly than the corresponding A_1-isomers. Separation of the all-*trans* from the 9-*cis*-forms and of the 11-*cis*-from the 13-*cis*-forms is, however, incomplete. The 13-*cis*-vitamin A is a permanent attendant of vitamin A; it is recognisable as a cap to the principal spot of vitamin A in Fig. 112 and Fig. a of Plate II. Adsorptive separation possibilities are compiled in Tables 38 and 43. TLC systems of this sort have been useful in biological investigations on activity, formation, distribution and transformation of ordinary and of labelled vitamin A compounds [e. g. 52, 72, 81]. For example, with their help, vitamin A-aldehyde was first detected in plant and animal tissues [148, 149] and new, active vitamin A metabolites were discovered [33, 151 etc.]. Work without TLC is now inconceivable in the control of purity, form, source and storage stability of the A-vitamins in concentrates like fish liver oils, feeding stuffs, foods and pharmaceuticals [8, 72, 130 etc.].

BRAEKKAN [11] has separated the vitamin A compounds in plasma by partition on kieselguhr G layers which had been partly impregnated with squalane; ethanol-water (85 + 15) was the mobile phase. KATSUI [77] has used polyamide and chloroform or methanol for fractionating

Table 43. hR_f-values of vitamin A compounds using various separation systems (12—18 cm run in 30—120 min)

Layer (ca. 0.25 mm):	S_1	S_1	S_2	S_2	S_3	S_4	S_5
Solvent:	F_1	F_2	F_3	F_4	F_5	F_6	F_7
References:	[72]	[72]	[138]	[138]	[8]	[108]	[77]
Anhydro-vitamin A	93	82	97	—	75	—	—
All-*trans*-vitamin A-palmitate	91	76	—	—	66	—	41
13-*Cis*-vitamin A-palmitate	—	—	—	—	65	—	—
Retro-vitamin A-acetate	—	—	90	—	45	—	—
13-*Cis*-vitamin A-acetate	—	—	—	—	50	—	—
All-*trans*-vitamin A-acetate	70	47	88	—	45	—	—
All-*trans*-vitamin A-aldehyde	53	26	66	—	—	—	—
5,6-Epoxy-vitamin A-aldehyde	44	—	—	—	—	—	—
11,13-Di*cis*-vitamin A-alcohol	—	—	—	—	—	67	—
13-*cis*-vitamin A-alcohol	—	—	—	—	—	53	—
Retro-vitamin A-alcohol	—	—	12	—	—	—	—
All-*trans*-vitamin A-alcohol	20	8	6	45	10	40	69
5,6-Epoxy-vitamin A-alcohol	7	—	3	32	—	—	—
All-*trans*-vitamin A-acid	0	0	—	0	—	—	56
Anhydro-vitamin A_2	87	69	—	—	—	—	—
All-*trans*-vitamin A_2-acetate	69	45	—	—	—	—	—
All-*trans*-vitamin A_2-aldehyde	50	33	59	—	—	—	—
11,13-Di*cis*-vitamin A_2-alcohol	—	—	—	—	—	61	—
13-*Cis*-vitamin A_2-alcohol	—	—	—	28	—	49	—
All-*trans*-vitamin A_2-alcohol	17	8	0	—	—	35	—
All-*trans*-vitamin A_2-acid	0	0	—	—	—	—	—

Layer:
S_1 = Silica gel (Merck)-gypsum (85 + 15), activated, NS
S_2 = Silica gel G, dried 30 min at 100° C
S_3 = Silica gel G (Merck)-sec. calcium phosphate (50 + 50), activated, CS
S_4 = Silica gel G (Merck), activated, CS
S_5 = Polyamide

Solvent:
F_1 = Petrol ether (40—60° C)-acetone (94 + 6)
F_2 = Petrol ether (40—60° C)-ether (85 + 15)
F_3 = Cyclohexane-methanol (99.75 + 0.25)
F_4 = Cyclohexane-ethanol (97 + 3)
F_5 = Cyclohexane-ether (85 + 15)
F_6 = Petrol ether (40—45° C)-methyl heptenone (85 + 15)
F_7 = Methanol

some vitamin A compounds. Radioactive derivatives from animal extracts have been investigated on silica gel G, buffered to pH 8 [98] and on alumina G containing silver nitrate [68].

b) Detection and Determination

Chromatograms can be best interpreted by inspection in UV light. Many vitamin A compounds fluoresce yellow-green in light of wave length 365 nm (limit of detection 0.05 µg). Anhydro-vitamin A, vitamin A-aldehyde and vitamin A_2 derivatives appear as brownish and vitamin A-acid as dark spots. The fluorescence colours depend of course also on the adsorbent, substance concentration and filter for the light source, so that they have often been variously judged [72]. Pure test substances ought thus always to be used for comparison in detection work. Amounts as small as 0.1 µg of A-vitamins can be detected as dark spots on fluorescent layers, viewed in light of 254 nm. Vitamin A yields a blue colour with antimony(III)- and -(V)-chlorides (Rgt. Nos. 15 and 18), deep blue initially with concentrated sulphuric acid and grey-blue with molybdophosphoric acid (Rgt. No. 168) (limits of detection 0.1—0.3 µg). Only the last mentioned coloured complex is stable. Vitamin A-acid reacts to give a violet colour when sprayed with antimony(III)-chloride on silica gel; under these conditions, vitamin A-aldehyde turns dirty brown and it is more specifically detected with rhodanine or indandione reagents (Rgt. Nos. 222 and 138) (see under 2b). Other vitamins than vitamin A and contaminants like oils and antioxidants can be detected alongside vitamin A on the layer by using universal spraying reagents.

A semi-quantitative determination can be made by comparison with a series of standards on the same layer, before or after carrying out a colour reaction. A flawless quantitative determination after elution has not so far been achieved as a result of the extreme instability of the vitamin A compounds, especially on active adsorbents. Success may perhaps be gained by conversion to the more stable anhydro-form and following then with TLC. The radioactive measurement of isotopes has been carried out frequently [68, 98].

4. Vitamin D Group

Vitamin D_2 (ergocalciferol) and D_3 (cholecalciferol) and their esters and provitamins (ergosterol and 7-dehydrocholesterol) easily suffer unwanted changes under various sets of experimental conditions. These conditions must therefore be as mild as possible [10, 15]. Vitamin D solutions[6] are relatively unstable on certain active adsorbents in presence

---

[6] The term "vitamin D" refers to vitamins D_2 and D_3, the physical and chemical properties of which are closely similar.

of alkalies or water-containing media and at higher temperature, even in the absence of air and in the dark [10, 15, 71]. The D-previtamins are formed by thermal isomerisation (cf. Fig. 117) and possess only 40% of the biological activity of the corresponding D-vitamins [10]. The position of the equilibrium vitamin D ⇌ previtamin D and its speed of establishment are dependent on temperature (e. g., 93:7% at 20°C after 1 month and 78:22% at 80° after 2 h). Using TLC-it has been possible to ascertain easily the conversion and decomposition products and to pass judgment on and to improve procedures of purification, separation and evaluation; this was not possible with methods hitherto available [10]. It must be remembered when processing naturally occurring material, that vitamin D often occurs in esterified form or linked to proteins, rather than free.

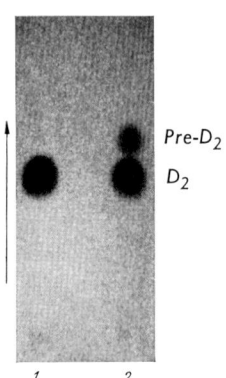

Fig. 117. Evidence of thermal isomerisation in TLC. Separation of 400 µg samples of pure vitamin D_2 before (1) and after warming in chloroform solution (2.5 h at 60°C in the dark) (2); on silica gel GF_{254} with cyclohexane-ether (50 + 50). Photograph in UV light (254 nm) [8]

a) Separation

Many procedures have been described for analytical and preparative purposes; some are given in Tables 38 and 44.

D-vitamins and related compounds can be chromatographed through adsorption on activated layers. Decomposition on the dry layer can be avoided by developing immediately after application. JANECKE and MAASS-GOEBELS [71] were the first to work on the D-group, using hexane, hexane-ethyl acetate (90 + 10) and chloroform on silica gel layers in a stimulating and instructive investigation. In this way, the antirachitic vitamins could be separated from other sterols, degradation products and carotenoids. We have found silica gel G or GF_{254} layers with the solvent mixtures cyclohexane-ether (50 + 50) and cyclohexane-ethyl acetate (75 + 25) to be very satisfactory [8, 10]. A method for determination of vitamin D was then developed, suitable for preparations containing widely varying amounts of vitamin A, even large amounts; it is discussed in more detail below under c). STROHECKER and HENNING [130] have modified the procedure slightly and obtained good results on silica gel HF_{254}, using the same solvents. Since silica gel layers have a large adsorption capacity and can be used for sharp and rapid resolutions of large amounts of substances without overloading the layer (cf. Figs. 119 and 121), the range of their application is broad: e. g., for labelled and unlabelled products of irradiation or synthesis [93, 101, 116, 118]; for D-vitamins and their metabolites in biological material [23, 66, 95, 117]; and for vitamin D-esters of fatty acids [102], sulphuric acid [66] and phophoric acid [23]. Calcium phosphate has been used for the last named also. Alumina has also been valuable in investigations of vitamin D in oily concentrates of vitamins A and D, drugs and hydrogenation products of vitamin D [17, 50, 70].

Table 44. hR_f-values of vitamin D-compounds using various separation systems (12—18 cm run in 40—120 min)

Vitamin D and related compounds	S_1 F_1 [8]	S_2 F_2 [116]	S_3 F_3 [93]	S_4 F_4 [19]	Vitamin D-esters [102]	S_5 F_5 D_2	D_3	S_6 F_5 D_2	D_3
Previtamin D_3	50	70	—	—	-acetate	26	29	—	—
Previtamin D_2	50	—	—	—	-laurate	67	65	—	—
Lumisterol$_3$	45	61	—	—	-myristate	71	69	—	—
Tachysterol$_3$	—	52	64 [a]	—	-palmitate	73	73	—	—
Vitamin D_3	40	50	44	55	-stearate	75	73	79	78
Vitamin D_2	40	—	44	49	-oleate	73	70	57	58
Cholesterol	33	—	41	39	-linoleate	71	69	41	41
Ergosterol	30	—	35	46	-linolenate	61	64	30	30
7-Dehydrocholesterol	30	41	35	45	-arachidonate	69	66	18	20

Layer (ca. 0.25 mm):
S_1 = Silica gel GF$_{254}$ (Merck), activated, CS
S_2 = Silica gel G (Merck), activated (2 mm thick)
S_3 = Cab-O-Sil silica gel (Res. Specialities), activated
S_4 = Kieselguhr G (Merck), dried 15 min at 100° C, impregnated with undecane-petrol ether (40—60° C) (10 + 90), air dried, wedged tip technique, NS
S_5 = Silica gel H (Merck) + 0.02% rhodamine G, activated
S_6 = Silica gel G (Merck)-silver nitrate (96.5 + 3.5) + 0.02% rhodamine G, activated

Solvent:
F_1 = Cyclohexane-ether (50 + 50)
F_2 = Benzene-acetone (90 + 10)
F_3 = Chloroform, 100%
F_4 = Acetic acid-acetonitrile (25 + 75)
F_5 = Hexane-benzene (34 + 66)

[a] Tachysterol$_2$.

Tables 38 and 44 contain examples of partition chromatography of these sterols on silica gel impregnated with paraffin oil [8, 47] and on kieselguhr saturated with undecane [19]. The migration of vitamins D_2 and D_3 differs somewhat on such layers whereas their chromatographic behaviour is the same on the previously mentioned adsorption materials.

The D-group can be separated on silica gel-silver nitrate layers [7, 19, 101]. In particular the socalled "critical pairs" of vitamin D-esters, palmitate-oleate, myristate-linoleate and laurate-linolenate can be separated in this way on the basis of the number of double bonds in the side chain [47, 102].

MURRAY et al. [91] have reported an interesting differentiation and determination of vitamins D_2 and D_3. The principle is to treat a chloroform solution of vitamin D with antimony(III)-chloride (Rgt. No. 15) and then to stop the colour reaction after 1 minute by shaking with 40% aqueous tartaric acid solution. The complexes (presumably isovitamins D_2 and D_3) are extracted with low-boiling petrol ether; they can be separated by gas chromatography, during which they are stable, and amounts down to 0.2 µg can be detected. In the study of animal tissue, saponification must be carried out first, cholesterol precipitated with digitonin and the vitamin A separated thin-layer chromatographically as its anhydro-form. The D-isovitamins must be freed from interfering contaminants on a silica gel G layer after extraction.

b) Detection and Determination

Compounds of the vitamin D family on layers containing a fluorescent additive or which have been sprayed with a fluorescing solution like rhodamine 6G (Rgt. No. 221) [117], are visible in short wave UV light as dark shadows from fluorescence quenching (detection limit 0.5 µg). Many spray reagents have been suggested for identifying vitamin D and distinguishing it from other sterols; the D-previtamins behave then like their D-isomers. With tungstophosphoric acid reagent (No. 256), followed by heating, vitamin D turns grey brown, ergosterol red-violet, 7-dehydrocholesterol dark brown and cholesterol cherry-pink [71]. Antimony(III)- and (V)-chlorides (Rgt. Nos. 15 and 18) and concentrated sulphuric acid similarly yield coloured, unstable complexes with differentiating colours. The last named reagent permits distinction of vitamins D_2 and D_3 as brown or greenish spots respectively, when at least 30 µg are present. The most sensitive test is with antimony(III)-chloride after heating at 120° C (0.025 µg; cf. Table 45 and Fig. 118).

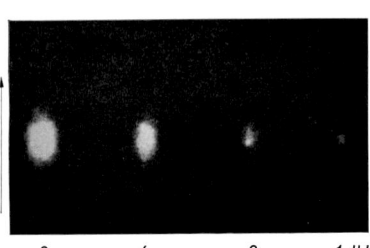

Applied amount of vitamin D_2 in international units (1 IU = 0.025 µg)

Fig. 118. Limit of detection of vitamin D on a 0.25 mm thick silica gel layer after spraying with SbCl$_3$-reagent and warming; inspected in UV-light (365 nm) [8]

CHEN [14] has tested spray solutions of 12 fluorescent dyes like Brilliant yellow 6G, fluorescein, acridine etc.; the chromatogram was then treated with 10% sulphuric acid in methanol, heated and examined in UV light. Iodine [63], dibromofluorescein [19] and permanganate/sodium carbonate solution [95] can be used fairly generally. If the presence of particular substances is suspected, it is advisable always to chromatograph pure specimens of these substances along with the sample studied. These procedures have proved valuable in qualitative testing of naturally occurring and synthetic products.

Table 45. *Detection of vitamin D on silica gel GF_{254} with spray reagents* [8]
(Layer spread 0.25 mm thick and activated)

Visualisation with			Colour of the vitamin D_2 and D_3 spots in		
Reagent and No.	Amount of spray	Treatment and judgement after	UV (365 nm) (Fluorescence)	daylight	Limit of detection in µg
Antimony(III)-chloride (15)	much	5 min/120° C	orange	grey-blue	0.025; 0.3
Antimony(V)-chloride (18)	little	1 min	—	orange-red	0.3
		5 min/120° C	orange	—	0.3
Conc. sulphuric acid	little	3 min	—	D_2 brown D_3 green	30
	medium	5 min/120° C	orange	grey-blue	0.3
Tungstophosphoric acid (256)	medium	20 min/70° C	—	grey-brown	0.2
Molybdophosphoric acid (168 B)	medium	1 min	—	grey-blue	0.3
Trichloroacetic acid (252 B)	medium	5 min/120° C	brown	—	ca. 0.1—0.2
Trifluoroacetic acid (253)	medium	5 min/120° C	brown	—	ca. 0.1—0.2

The approximate amount of active substance can be determined directly on the layer when a clear separation has been obtained. HEAYSMAN and SAWYER [63] have made a visual estimate of vitamin D in pharmaceutical preparations with the help of a series of standards chromatographed on the same plate; they purified with the wedged-tip technique (p. 89) on a second silica gel G chromatogram and visualised with concentrated sulphuric acid. We have been able to determine semi-quantitatively the content of previtamin D in a concentrate from irradiation; this was also done by comparison but without decomposing the substance, by employing layers containing a fluorescent additive and examining in UV light [10]. A further possibility suggested is to determine the amount of vitamin D which is just visible after having visualised a series of various concentrations [71]. Determination of the separated substances after elution from the adsorbent is more accurate and is a universally applicable and the most used method. The solutions are chromatographed on normal or thick layers after spot or band application. The vitamin zone is localised, best with the help of a reference strip which is inspected in UV light or visualised. The equivalent zone on the principal strip is then scraped off, eluted with a suitable solvent like chloroform and the content in the eluent estimated colorimetrically, spectrophotometrically or autoradiographically. In this way, vitamin D-compounds have been determined in irradiation products [10, 49, 93, 101], in concentrates, pharmaceuticals, feeding stuffs etc. [10, 13, 17, 63, 130] and their metabolism and distribution studied in animal tissue [47, 94, 95, 117]. On the preparative scale, vitamin D is isolated similarly from mixtures.

c) Tested Assay of Vitamin D [8, 10]

The method which we have developed, depends on individual processing of the sample, one-dimensional TLC, elution and colorimetry. The vitamin D content can be determined with it in the following preparations which contain from only 5 up to 40 million IU per gram: oily, water-soluble, powder- and resinuous concentrates, crystallised materials, pharmaceutical preparations, feeding stuff mixtures, sweets, chocolate, dietetic slimming preparations and halibut liver oil concentrates. Determinations of this sort have been carried out for 6 years in several laboratories, so that experience about reliability, errors of analysis, efficiency and limits of the method, has been gained. It is specific, sensitive and functions without loss, precisely and accurately. The results agree with the data found biologically. The relative standard deviation amounts to $1.8-2.7\%$, depending on sample source; a duplicate determination lasts $3-7$ hours, depending on the pretreatment.

General instructions are given here. All precautions (see under I) must be taken and the original literature consulted for certain details and explanations.

α) **Preparation of the sample solution for TLC.** The sample for analysis is processed individually in as simple and gentle a way as possible, without saponification[7]; the treatment depends on the composition, amount and form[7] of the vitamin D. This involves direct dilution or extraction, liberation, freezing out, concentration and dissolving in cyclohexane or chloroform[8] (= sample solution with 800—10000 or with 400[9] IU D/ml)[10].

---

[7] Vitamin D can occur free or enclosed for stabilisation. Samples in which vitamin D is esterified or linked to proteins or lipids, have to be saponified, however. CHEN et al [15] have given instructions about this.

[8] If the residue is sparingly soluble in cyclohexane, it can be taken up in chloroform.

[9] For the modified colour reaction to determine smaller amounts.

[10] We consider the suggestion of F. J. MULDER and K. H. HANEWALD (private communication, N. V. PHILIPS-DUPHAR, Weesp) to be valuable for safe protection of vitamin D during TLC: 3,5-ditert.-butyl-4-hydroxytoluene (BHT) is added to the sample solution and to the solvent to yield 0.01% solutions; sample solutions of fat-free extracts are stabilised in addition with 1% squalane.

Oily and, to some extent, water-soluble concentrates can, for example, be directly diluted with cyclohexane or acetone, respectively. Preparations of the former type which contain little vitamin D must be largely freed from the large amount of fat-soluble ballast by dissolving in warm absolute alcohol and freezing out at 0° C. Products in powdered or pulverised form are best submitted to ammoniacal or enzymatic digestion to liberate the vitamin D from the dry powder; this is also done when added as an adsorbate or oily solution. Extraction with petrol ether is then carried out and the extract evaporated and dissolved in cyclohexane or chloroform. The most suitable procedure must be chosen and tested with every preparation.

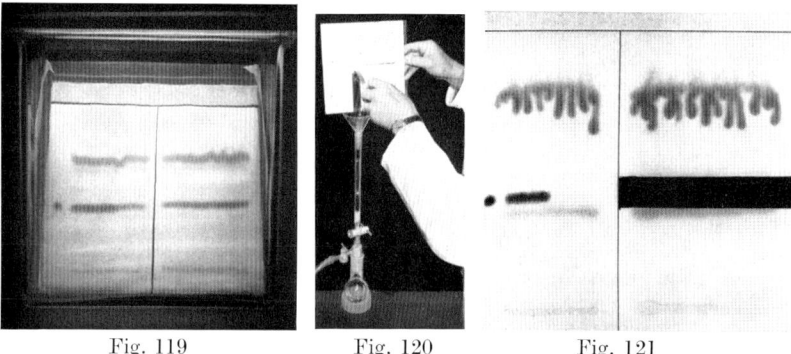

Fig. 119   Fig. 120   Fig. 121

Quantitative Determination of vitamin D by means of TLC (cf. text) [8, 10]

Fig. 119. Thin-layer chromatogram of an oily vitamin A + D concentrate, viewed in UV light (254 nm). Plate illuminated through a quartz cover in the chamber. Reference chromatogram on the left side

Fig. 120. Quantitatively scraping-off the located vitamin D-strip, still moist with solvent, into the elution column

Fig. 121. Thin-layer chromatogram of a halibut liver oil, viewed in UV light (254 nm) after scraping off the vitamin D zone located with the help of the reference chromatogram

β) **TLC and elution.** Vitamin D standard (40000 IU/ml chloroform) and sample solution are applied to the start line using a pipette, avoiding damage to the layer. Activated silica gel GF_{254} is used for the 20×20 cm layer, of thickness 0.5 to 2 mm, depending on the amounts of contaminant to be separated. About 0.05 ml of standard solution is applied as a spot to the left side of the chromatogram (reference part) and then about 0.1 ml as a band alongside it. In the middle, about 0.4—1.0 ml of sample solution is applied with the pipette to this band. Finally, an accurately measured amount of the sample solution containing 800—4000 or 400[9] IU D for the quantitative determination, is applied as a band to the right side of the layer. The plate is placed *immediately* in the separation chamber (CS) and chromatographed[10] for 45—60 min with cyclohexane-ether (50 + 50) or cyclohexane-ethyl acetate (75 + 25)[11]. The still moist layer is then *briefly* inspected in UV light, the right hand vitamin D band located through its absorption or the

---

[11] This solvent mixture is used if vitamin A-alcohol is present.

reference part of the chromatogram and marked round with a spatula at a distance of about 0.5 cm. This marked zone is *immediately* scraped off into a column containing chloroform and Frankonite[12] and fitted with a P_1-sintered glass filter and teflon tap. After evacuating slightly, the vitamin D is eluted with ca. 25 ml chloroform into a 100 ml flask fitted with a ground glass neck, nitrogen swept through, the solution evaporated completely to dryness at ca. 40° C in a vacuum rotary evaporator and the residue dissolved in a measured amount of chloroform at room temperature (= extract solution, containing ca. 200 IU D/ml).

γ) **Colour reaction**[13] **and measurement.** The extract solution with 800 IU or more vitamin D suffices for several colorimetric measurements in the most favourable range of the handy colorimeter "Spectronic 20"[14], thereby increasing the precision. Mixtures of 1.0 ml extract + either 1.0 ml chloroform or 1.0 ml inhibitor[15] are placed in two $^3/_4$ inch cuvettes; to each is added 3.0 ml antimony(III)-chloride-acetyl chloride[16] reagent, then mixed. The extinction (maximum value) of the orange complex is measured at the absorption maximum of 500 nm, 10—20 sec after addition of reagent. 2.0 ml vitamin D_2- or D_3-standard solution (200 IU dissolved in chloroform) are treated in the identical way. Chloroform, with and without inhibitor, + 3.0 ml reagent solution are used as blanks for setting the colorimeter to 100% transmission. The mean extinction value for the mixture with inhibitor is subtracted from that without it and the content of vitamin D calculated using the dilution factor.

In the *modified colour reaction* for small amounts of vitamin D (e. g., extracts with about 400 IU per 2.5 ml), the solutions for a single determination are pipetted into test tubes as described above but are treated with only 2.0 ml reagent. After briefly mixing, the contents are introduced into 1 cm cuvettes and after precisely 20 sec the extinction measured with a spectrophotometer at 500 nm against the blank solution.

This procedure has a few noteworthy advantages over others. The reference chromatogram on each plate shows possible artifacts and gives immediate proof of identity. In our system, vitamin D is separated completely from the other fat-soluble vitamins present, even in large amounts, and from most extraneous contaminants. Traces of lipids which may be scraped off along with the vitamin D, mostly react with the colour reagent only after 30 sec; the colour reaction of almost all other contaminants is prevented by the inhibitor. A degradation product of vitamin A belongs here; it migrates just below vitamin D (cf. Fig. 121) and is partly retained during the washing on the Frankonite. The vitamin D standard needs to be used quantitatively only in the colorimetric stage because no losses occur during the whole separation procedure.

---

[12] 0.5 g granulated Frankonite is scattered into the column when the sample contains vitamin A. For its preparation, see [10].

[13] Spectrophotometric determination is also possible in cleanly separated extracts which contain a large amount of vitamin D.

[14] Firm 17.

[15] Inhibitor (colour inhibitor): chloroform-acetic anhydride (50 + 50).

[16] Antimony(III)-chloride reagent (No. 15) is used for the colour reaction. It contains 2% acetyl chloride and, for stabilisation, 2% acetic anhydride.

Possible interference by scraped-off silica gel has been tested on a large number of equally-sized parallel strips by elution and treatment with colour reagent; it was found always to be zero. The chloroform used in various operations is stabilised with 1% ethanol but need not be purified. The colour intensity from vitamin D_2 is about 3–4% higher than that from an equal amount of vitamin D_3. The calibration curves are linear for amounts up to 500 IU of both active substances. Our standard for each reaction has thus the suitable value of 200 IU. Pure vitamin D is totally and selectively inactivated by the inhibitor so that it need not be added to the standard. The otherwise usual correction for absorption of contaminants, carried out at 550 nm, has been shown to be superfluous; the contaminants which generally interfere have been taken into consideration or eliminated.

STROHECKER [131] has given some useful hints. He has used silica gel HF_{254} as a rule, but for thicker layers on 20 × 40 cm plates, the better adhering silica gel PF_{254}. Up to 5 ml sample solution, containing little vitamin D and larger amounts of fats and oils, are applied as a band alongside the reference chromatogram. He maintains that these interfering substances can be separated through a first development with petrol ether (30–45° C)-chloroform (70 + 30) without freezing out, because they travel with the solvent front. The second chromatographic stage which follows immediately, is then carried out as previously, with cyclohexane-ether (50 + 50). The scraped-off silica gel is freed from traces of moisture by extracting with chloroform and adding anhydrous sodium sulphate; these traces do not however interfere any more in our scheme (evaporation to dryness and solution in chloroform).

5. Vitamin E Group

We possess nowadays a vastly improved idea of the occurrence and chemical changes of the E vitamins, thanks to intensive application and expansion of thin-layer chromatographic methods. PENNOCK et al. [104, 105] have investigated *the naturally occurring tocopherols and tocotrienols* (saturated or unsaturated side-chains, respectively) and proposed the following nomenclature and abbreviations, on the basis of their results:

Ring Substitution	Tocopherols	Tocotrienols
5,7,8-trimethyl	α-T	α-T-3 (ζ_1-, or ζ_2-tocopherol)
5,8-dimethyl	β-T	β-T-3 (ε-tocopherol)
7,8-dimethyl	γ-T	γ-T-3 (η-tocopherol)
8-methyl	δ-T	δ-T-3
		( ) = former term

All these substances of widely varied biological activity contain a methyl group in the 8-position. Representatives which are methylated only in the 5- or 5,7-positions have however never been discovered, so that the former structural formulae for η- and ζ_2-tocopherols have had to be revised. On the other hand, δ-T-3 has been discovered. Whereas the 4 naturally occurring tocopherols and tocotrienols are sterically uniform, synthetic products consist of more or less complex mixtures of diastereoisomers, depending on the method of preparation.

These active substances, which are sensitive to oxidation, occur often only in traces in the sample studied and are accompanied by triglycerides, sterols, lipoquinones, antioxidants etc. Vitamin E is frequently used in esterified and also coated form for reasons of stability. Direct extraction or digestion and, if necessary, preliminary purification or saponification must be carried out, as with vitamin D, as far as possible without losses and changes; the "tocopherols" can then be analysed after a single or multiple chromatographic treatment. LAMBERTSEN [82] has made an excellent survey of this field, with special reference to TLC; the publications quoted in [90] must be mentioned here also.

a) Separation (cf. Tables 38 and 46)

Good and rapid separations of vitamin E can be obtained on 0.2 to 2 mm layers of the adsorbents silica gel G or H [8, 104, 118, 130 etc.], alumina [74, 75, 118 etc.], sec. magnesium phosphate [8], florisil [88] etc. Chloroform, benzene, toluene, benzene-methanol, cyclohexane-ether and many other solvents can be used. Fractionation occurs approximately into groups according to the extent of ring methylation. The double bonds in the side chain have no influence; α-T and α-T-3 have the largest Rf-values. The clear separation of β- and γ-tocopherols and of saturated from the corresponding unsaturated compounds, has been achieved with diisopropyl ether-petrol ether (40–60° C) (20 + 80), a mixture introduced by STOWE [128] and employed in two-dimensional TLC by PENNOCK et al. [104] (cf. Fig. 122). This effect has been explained through a more rapid evaporation of the unpolar component (petrol ether) so that the elution ability of the solvent increases with the length of run; it is claimed that dioxan and ethyl acetate behave similarly [82]. A few examples are cited here, each of which stresses some profitable aspect of using adsorption TLC: testing purity and isolation of synthetic E-preparations [87]; analysis of concentrates of naturally occurring vitamin E [79, 118] (cf. Fig. d in Plate II); determination of tocopherols in pharmaceuticals, in foodstuff concentrates [8, 13, 130] (cf. Fig. e in Plate II), in vegetable oils [111] and in plant tissue in the presence of other quinones [29] etc.; separation of E-oxidation products [119, 120]; studies of origin, occurrence, modifications and functions of E-vitamins and their metabolites in plants [30, 35, 104], animals [107, 145] and humans [5, 88].

Table 46. hR_f-values of vitamin E and related compounds and coloration with antimony(V)-chloride reagent
(12—18 cm run in 40-60 min)

Layer (ca. 0.25 mm):	S₁	S₂	S₃	S₂	S₃	S₂	S₄	Colours[c] with
Solvent:	F₁	F₂	F₃	F₄	F₅	F₆	F₇	SbCl₅
References:	[118]	[8, 145]	[128]	[104,105]	[105,128]	[105]	[105]	[8, 105]
α-T	56	32	93	67	51	57	24	orange-red
β-T	34	30	85	50	47	42	43	light brown
γ-T	31	26	78	53	40	41	42	green
δ-T	21	21	71	37	31	30	57	red-brown
α-T-3	—	—	—	68	43	55	56	
β-T-3	—	—	—	53	40	41	74	brown
γ-T-3	—	—	—	51	35	39	73	orange-brown
δ-T-3	—	—	—	37	27	29	82	brown
Tocol	—	—	—	31	29	27	—	lilac-grey
α-T-acetate	71	40	—	79	76	72	—	—
α-T-succinate	—	30 to 40	—	—	—	—	—	—
α-T-phosphate	—	0	—	—	—	—	—	—
γ-T-3-palmitate	—	—	—	85	97	83	—	—
α-TQ[a]	—	14	—	—	—	—	—	—
α-T-dimer[b]	—	81	—	—	—	—	—	—

Layer:

S₁ = Alumina (Fluka), activated
S₂ = Silica gel G (Merck), activated, CS
S₃ = Silica gel G (Merck), activated, NS
S₄ = Kieselguhr G, activated, impregnated with paraffin oil-petrol ether (40—60°) (5 + 95), air dried, CS

Solvent:

F₁ = Benzene
F₂ = Cyclohexane-ether (80 + 20)
F₃ = Petrol ether (60—80° C)-diisopropyl ether-acetone-ether-acetic acid (85 + 12 + 4 + 1 + 1)
F₄ = Chloroform
F₅ = Diisopropyl ether-petrol ether (40—60° C) (20 + 80)
F₆ = Methanol-benzene (1 + 99)
F₇ = Acetone-water (80 + 20), saturated with paraffin oil

[a] α-TQ = α-Tocopherylquinone.
[b] Oxidation product of α-T with ferricyanide.
[c] Detection on silica gel G: inspected about 3 min after moderate spraying of the dry layer (limit of detection about 2 µg, optimum colour differentiation with 20—50 µg).

Tocopherols can be chromatographed also on silica gel (cf. Table 38) and kieselguhr layers, impregnated with paraffin oil [105]. LAMBERTSEN [82] has obtained splendid results on such layers, using squalane or undecane (BP. 180—200° C) as unpolar stationary phase and ethanol-water (75 + 25 to 95 + 5). During the partition process, the vitamins migrate further, the lower their molecular weight; and unsaturated side-chains increase mobility. The influence of three double bonds corresponds to that of one or two methyl groups. It should be possible theoretically to separate all the 8 naturally occurring "tocopherols" completely from one another by combining adsorption and partition in the two-dimensional system. Success is almost entirely dependent on the amounts of vitamin E and on the influence of contaminants on the shape of the spot.

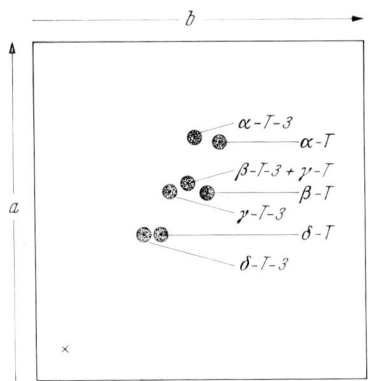

Fig. 122. Two-dimensional separation of the naturally occurring tocopherols (T) and tocotrienols (T-3) on activated, 0.25 mm silica gel G layers. Solvents: *a* chloroform (CS); *b* diisopropyl ether-petrol ether (BP 40—60° C) (20 + 80) (NS). 10 µg of each applied to the start point (×) [104, 105, 146]

EGGER and KLEINIG [41] have succeeded in isolating α-tocopherylquinone from mixtures with related quinones, using polyamide layers. This technique may be suitable for other E-compounds also.

Attempts to separate diastereoisomers by TLC have so far failed.

Some oxidation and metabolic products of tocopherol are mentioned again in the paragraph on the vitamin K group, because these substances often occur together in nature.

b) Detection and Determination

On layers containing an inorganic fluorescent additive, unchanged E-compounds appear as dark spots in UV light (about 20 µg can be detected). They appear violet and detection is appreciably more sensitive (0.02 µg) on layers which contain 0.02% Na-fluorescein; these are moreover visible in daylight as reddish spots (2 µg). The same effect is produced by spraying with fluorescein or dichloro-fluorescein reagent (Nos. 116 and 63), drying and then treating with water vapour [34]; related lipids then show up also.

Tocopherols and tocotrienols yield red with the dipyridyl-iron reagent (No. 91) (0.5 µg α-T can be detected; cf. Fig. d in Plate II), whereas their esters show no reaction. Molybdophosphoric acid (Rgt. No. 168B) shows about the same sensitivity and non-specificity. The free E-vitamins yield grey-blue spots already in the cold (1 µg) and on heating the colours intensify (0.5 µg). Esters and other reducing materials then also react. The background is rendered pale by subsequently treating the layer with ammonia vapour. SEHER [118] has chromatographed a series of standards (10, 20, 40 and 60 µg α-T) simultaneously in order to determine approximately tocopherols in concentrates; the spot sizes were compared by planimetry after having visualised as above. Iodine vapour or ferricyanide-ferric chloride (yielding Turnbull's blue) give non-differentiating colours [120].

Antimony(V)-chloride reagent (No. 18) is especially suitable for qualitative purposes since it yields coloured complexes, dependent however on the layer (cf. Table 46 and Fig. d in Plate II) [8]; β- and γ-tocopherols are in particular well differentiated. Simultaneous chromatography of pure test substances is strongly recommended here. Certain colour nuances of the reaction products are obtained by spraying with concentrated sulphuric, nitric or perchloric acids [77, 120]; with 2,6-dichloroquinonechloroimide reagent (No. 66) (α-T yellow-brown and α-T-acetate pink); and with cerium (IV)-sulphate (Rgt. No. 37) (β-T brown, γ-T blue). Using this last named spray reagent and heating 10 min at 120° C, α-tocopheryl-acetate becomes pink and fluoresces brightly in UV light. T- and T-3-compounds without a methyl group in the 5-position react with diazotised o-dianisidine and those which are unsubstituted in the 5- and 7-positions, with sodium nitrite [104]. Sulphuric acid + heating, antimony(V)-chloride reagent (No. 18) and the Turnbull blue reaction have been utilised to characterise the oxidation products [120].

The tocopherols can be determined planimetrically after having been clearly separated on the plate from contaminants and visualised with molybdophosphoric acid [118]. Direct spectrophotometric measurement of remission of the unsprayed spots seems also possible now.

Quantitative evaluation of spot and band chromatograms by localisation, scraping off, elution with ethanol and photometric determination with the help of the dipyridyl-iron reaction, has been the method most used (cf. Fig. e in Plate II). The easily reduced substances are about 97–100% recovered in the separation operation [8, 74, 111, 131] provided all precautions are respected. Vitamin E-compounds have been thus determined in various materials, both naturally occurring and synthetic. The active substance can also be determined colorimetrically in the eluate by using bathophenanthroline (2.5 times the extinction obtained with dipyridyl-iron) and with other physicochemical methods like spectrofluorometry (at least 0.01 µg/ml) or with the liquid scintillation counter in the case of radioactive compounds [145].

Some authors have obtained low yields of vitamin E, largely due to losses through oxidation during the whole procedure. A certain part of the losses can however be explained or simulated by having used impure test substances. Too high results are due to from reducing contaminants which are not separated in the TLC.

6. Vitamin K Group and Related Quinones

Quinones are widely distributed in nature. They are important in biological systems because they are easily reduced to hydroquinones and can be recovered aga n by oxidation. Members of this group are: Vitamins K_1 and K_2, derived from naphthoquinones; the tocopheryl quinones; and the ubiquinones and plastoquinones which possess the p-benzoquinone parent structure and which do not belong to the vitamins. The synthetically prepared menadione and its derivatives are modified in the intestine by gaining a polyisoprenoid side-chain and then show the complete vitamin K activity. These compounds, which are sensitive and often occur together, must be worked up and handled with care. Chromatography plays an important part in their analysis. One or more separation procedures are usually inevitable since specific methods of determination are lacking, for distinguishing both groups of quinones and quinones within a group. Paper, gas and, above all, thin-layer chromatography are being increasingly used to follow up the column chromatography extensively employed.

We have chosen the first IUPAC suggestion (in [90]) for the consistent nomenclature and abbreviations of some of these quinones, the list of which is continuously growing:

Trivial Name	Suggestion I	Abbreviation	
Vitamin K_1	Phylloquinone	K (K-4)	(n = number of
Vitamin K_2	Menaquinone-n	MK-n	isoprene units (sat. or
Vitamin K_3	Menadione		unsat.) in the side
Plastoquinones	Plastoquinone-n	PQ-n	chain
Ubiquinones, Coenzyme Q	Ubiquinone-n	Q-n	
α-Tocopherylquinone	α-Tocopherylquinone	α-TQ	

a) Separation

A few of the large number of TLC systems hitherto used are compiled in Table 47. They serve mostly for solving particular problems.

The length and degree of saturation of the isoprene side-chain has less influence on the separation of the naphthoquinone derivatives in adsorption TLC than in other procedures. The Rf-values increase with increasing molecular size in accordance with the general rule. Thus all the ubiquinones migrate together [141]. The principal adsorbents used have been silica gel and alumina. Thus menadione in the preparations in which

Vitamins, Including Carotenoids, Chlorophylls and Biologically Active Quinones 289

Table 47. hR_f-values of K-vitamins and related quinones in adsorption and partition chromatography and also on polyamide and silver nitrate-containing layers (15—18 cm runs in 30—60 min; for S_5/F_7, ca. 2 h)

Layer (ca. 0.25 mm):	S_1	S_1	S_1	S_2	S_3	S_4	S_5	S_6
Solvent:	F_1	F_2	F_3	F_4	F_5	F_6	F_7	F_8
References:	[65]	[65]	[64]	[115]	[8, 125]	[41]	[41]	[123]
K (= trans-K-4)	54	69	—	52	—	38	56	72
K-2	50	65	—	—	—	—	—	—
MK-1	41	61	—	—	—	—	—	—
MK-2	46	62	—	—	—	—	—	—
MK-3	—	—	—	—	—	—	—	58
MK-4	53	65	—	—	—	—	—	49
MK-5	—	—	—	—	—	—	—	41
MK-6	58	69	—	52	—	39	51	29
MK-7	—	—	—	43	—	31	46	21
MK-9	64	75	—	33	—	18	35	—
Menadione	29	55	—	—	—	90	76	45
PQ-9 (= PQ-A)	61	74	74	33	—	18	41	—
PQ-10	—	—	—	—	—	12	35	—
PQ-B	—	—	78	—	—	—	—	—
PQ-C	—	—	49	—	—	—	—	—
PQ-D	—	—	40	—	—	—	—	—
Q-1	—	—	—	—	90	—	—	—
Q-2	—	—	—	—	85	—	—	—
Q-3	—	—	—	—	78	—	—	—
Q-4	—	—	—	—	68	—	—	—
Q-5	—	—	—	—	60	—	—	—
Q-6	—	—	—	74	44	58	68	—
Q-7	—	—	—	69	35	—	—	—
Q-8	—	—	—	63	25	44	56	—
Q-9	—	—	—	52	14	—	—	—
Q-10	—	—	—	43	8	30	45	—
α-TQ	—	—	37	—	—	74	71	—
β-TQ	—	—	33	—	—	—	—	—
γ-TQ	—	—	25	—	—	—	—	—

Layer:

S_1 = Silica gel G, activated
S_2 = Silica gel G, activated, impregnated with paraffin oil-petrol ether (40—60°) (5 + 95), air dried, CS
S_3 = Silica gel G-kieselguhr G (50 + 50), act., paraffin oil-impregnated, CS
S_4 = Cellulose MN 300 F_{254}-silica gel HF_{254} (66 + 34), act., par. oil impr., CS
S_5 = Polyamide (Merck)-cellulose MN 300 F_{254} (85 + 15), slurried in methanol, air dried 30 min
S_6 = Silica gel G-silver nitrate (90 + 10); prepared with aq. $AgNO_3$ according to the standard method, act., CS (plates stored in the dark and dry)

Solvent:

F_1 = Benzene
F_2 = Chloroform
F_3 = Chloroform-ether (99 + 1)
F_4 = Acetone-water (95 + 5)
F_5 = Methanol-isopropanol (90 + 10), saturated with paraffin oil
F_6 = Acetone-methanol-water (65 + 33 + 2)
F_7 = Methanol-methyl ethyl ketone-water (63 + 31 + 6)
F_8 = Diisopropyl ether (purified over neutral alumina, activity I)

it is used (h$Rf \simeq$ 38), has been separated from K and MK-7 (h$Rf \simeq$ 61) on silica gel, using cyclohexane-ether (80 + 20) [8]. WALDI [143] has separated menadione (h$Rf \simeq$ 80) from menadione-sodium bisulphite (h$Rf \simeq$ 30) on silica gel HF_{254}, using cyclohexane-chloroform-methanol-acetic acid (10 + 75 + 10 + 5). Silica gel has been likewise employed in biochemical studies listed here: HENNINGER et al. [65], in investigations of algae, have separated 3 new quinones alongside several K, MK-n and Q-9 using chloroform or benzene; the same research team [64] has identified 4 plastoquinones and 3 tocopherylquinones in spinach-chloroplasts, using chloroform-ether (99 + 1) (cf. Table 47). BILLETER et al. [6] have examined the transformation of MK-6 and MK-2 into MK-4 in the organs of birds and mammals using two-stage development with heptane-benzene (50 + 50) and with chloroform. MASITI et al. [86] have chromatographed a fraction from spinach leaves, using 5 different solvents, and discovered the very unstable PQ-3, along with known quinones. Further, adsorption TLC has been employed to isolate plant and synthetic quinones and hydroquinones together with tocopherols [30]; for determination of ubiquinones in plant and animal lipid extracts [140, 141]; for fractionating lipoquinones in microorganisms [112]; for characterising these lipoquinones in the presence of other vitamins in blood [5]; and regularly for testing purity in syntheses of these compounds.

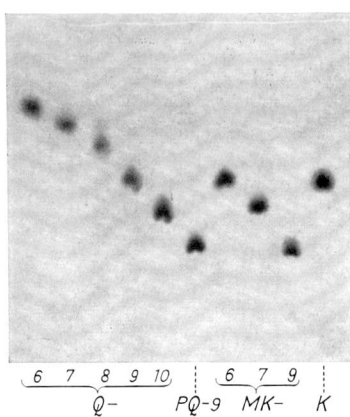

Fig. 123. Separation of ubiquinones (Q-n), plastoquinones (PQ-9), menaquinones (MK-n) and phylloquineon (K) on a paraffin oil-impregnated silica gel G layer, using acetone-water (95 + 5), saturated with paraffin oil (CS). 20 µg of each substance applied. Layer sprayed with conc. sulphuric acid after having heated to 130°C [115]

Homologous series of the quinone groups can be separated especially well by partition chromatography; the mobility falls with increasing chain length, as expected, and the double bonds exert a distinct influence (e. g., K and MK-6). In the main, silica gel, silica gel-kieselguhr or cellulose-silica gel layers, impregnated with paraffin oil, have been used with suitably polar solvents (see Table 47). GLOOR [in 115] has chromatographed various quinones in this way (see Fig. 123) and detected them in tuberculosis bacilli and liver homogenates [in 8]. Using phase reversal, lipoquinones have been determined in fungi [153], bacteria [112], yeast and animal tissues [141].

EGGER and KLEINIG [41] have recently tried out polyamide and SOMMER [123], silver nitrate-silica gel as carriers. As is seen in Table 47, new separation effects have been thus obtained and the scope of the procedures enlarged.

b) Detection and Determination

As a result of their strong light absorption between 240 and 280 nm, all lipoquinones in amounts of 0.5 µg and more are visible as dark spots on layers containing inorganic fluorescent material when illuminated with UV light. By adding Na-fluorescein, rhodamine B or 6G to the adsorbent or by spraying the chromatographed layer with fluorescein or dichlorofluorescein reagents (Nos. 116 and 64 respectively), the compounds can be detected in daylight and, with high sensitivity, in UV light. This non-destructive detection of the active substances is important for the quantitative determination.

All K-vitamins suffer photochemical change and begin to fluoresce yellow when the layer is irradiated for some minutes with a quartz lamp a short distance away (at least 0.3 µg K).

The quinones turn brownish on treatment with iodine vapour, violet with concentrated sulphuric acid followed by heating (ca. 3 µg) and grey-blue with molybdophosphoric acid (Rgt. No. 168B) and subsequent heating (ca. 0.5 µg). Ubiquinones, plastoquinones and tocopherylquinones turn leuco-methylene blue (Rgt. No. 154) immediately blue [53] whereas the K-vitamins scarcely react [30, 112]. Some differentiation of the groups is possible by spraying with antimony(III)-chloride reagent (No. 15) alone or after having pre-treated with dipyridyl-iron (Rgt. No. 91). Little use as a spray test has so far been made of the colour reactions, well known in colorimetry, with alkaline cyanoacetic ester (according to CRAVEN), Na-diethyl dithiocarbamate (according to IRREVERE-SULLIVAN) and Na methylate (DAM-KARRER).

Hydroquinones can be rendered visible by oxidation with dipyridyl-iron or permanganate reagents (Nos. 91 and 199 respectively).

EGGER [40] has directly evaluated in UV light the thin-layer chromatogram containing fluorescent additive, in order to determine the content of K and of PQ-9 in various sorts of leaves. He achieved an accuracy of $\pm 10\%$ by visual comparison of sample and a series of standard amounts. Quinones and hydroquinones are, however, generally determined after elution from the layer. The spots and zones are localised through their fluorescence or quenching as well as through visualisation of reference chromatograms. The isolated substances are then eluted from the adsorbent with suitable solvents, filtered if necessary and characterised and determined by subsequent colorimetric or spectrophotometric procedures. Labelled compounds are evaluated in scintillation counters.

WAGNER et al. [140, 141] (Q-n), REBEL and MANDEL [112] (K, MK-n, Q) and HENNINGER and CRANE [64] (PQ-n, TQ) have described examples of such analyses.

III. TLC of Water-Soluble Vitamins

The solubility of the uncombined vitamins varies considerably. In the sample they may occur also linked as coenzymes, esterified in various ways or as derivatives which develop the same biological activity. Knowledge of this is vital for correct processing, isolation and determination. The water-soluble vitamins are in general more stable than those in the fat-soluble groups; this facilitates chromatography. Nevertheless the precautions recommended under I should be observed.

1. Mixtures of Water-Soluble Vitamins
a) Separation (cf. Table 48)

Adsorption and ion exchange systems are suitable for this. The chromatographic data for the various compound classes do not always agree with theory. Some vitamins are weak acids or bases; the form in which they occur then depends on the dissociation constants and on the pH of the medium.

GÄNSHIRT and MALZACHER [51] have worked out a separation scheme for vitamin mixtures (1—10 µg B-group and 5—30 µg vitamin C) in pharmaceutical preparations. It is generally applicable, provided 50—100 µg of the mixture is spotted (2 spots for each sample) on to a silica gel G or GF_{254} layer containing fluorescent material and acetic acid-acetone-methanol-benzene (5 + 5 + 20 + 70) is used for development. KATSUI and ISHIKAWA [70, 77] have separated 20 hydrophilic vitamins on silica gel G and alumina G 20, using the same solvent mixture. The occasionally marked deviations in Rf-values observed by them, have been confirmed by us; they are probably due to variations in layer activity. We have succeeded in isolating these active substances on the same adsorbent by using water [8] (see Fig. 124). This solvent migrates at about the same speed as the Gänshirt mixture and enables chromatography to be carried out straight after application of aqueous extracts. The chromatogram must, however, usually be briefly dried by warming before visualisation and larger amounts of B_2 and B_6 tend to give tailing. LUDWIG and FREIMUTH [84] have observed a similar fractionation of these mixtures when using water-acetic acid (95 + 5) on layers of super gel-gypsum-fluorescent additive which had been pre-treated with ammonia vapour. HÜTTENRAUCH et al. [67] have resolved the B-complex with ethanol-water (10 + 90) on the exchanger resin Wofatit CP 300,

Table 48. hR_f-values and direct detection of water-soluble vitamins
(12—19 cm run in 40—60 min)

Layer (ca. 0.25 mm): Solvent: References	S_1 F_1 [8]	S_1 F_2 [51]	S_1 F_2 [70]	S_2 F_2 [70]	S_3 F_3 [67]	Visualisation[a] in [8]			Limit of detection (µg)		
						UV (254)	UV (365)	Day-light			
B_1(Thiamine-HCl, -HNO$_3$)	5	0	0	54	0	vio	—	—	2	—	—
B_2 (Riboflavin)	40	35	29	24	42	yel[b]	yel[b]	yel	0.1	0.01	0.3
B_2-5′-phosphate, Na salt	28	0	0	0	100						
Pantothenic acid (Na, Ca salt)	89	57	40	0	—	—	—	—	—	—	—
Nicotinic acid	78	75	—	—	100	dark	—	—	3	—	—
Nicotinamide	49	65	44	62	70						
B_6(Pyridoxin-HCl)	52	15	12	26	35	dk bl	dk bl	—	3	10	—
B_{12}(Cyanocobalamin)	22	0	0	23	100	dark	vio	red	1	0.5	0.3
Folic acid	0	0	7	0	—	dark	dk bl	yel	2	10	10
C(l-ascorbic acid)	96	30	25	0	—	dk bl	—	—	3	—	—
Biotin	70	80	50	54	—	—	—	—	—	—	—
Rutin	[c]	12	10	0	—	dark	dark	yel br	1	5	5

Layer:

S_1 = Silica gel G or GF$_{254}$ (Merck), activated, CS
S_2 = Alumina G (Merck), activated
S_3 = Ion exchanger Wofatit CP 300; for preparation, see [67]

Solvent:

F_1 = Distilled water
F_2 = Acetic acid-acetone-methanol-benzene (5 + 5 + 20 + 70)
F_3 = Ethanol-water (10 + 90) (time of run 2—3 h)
Colours (abbreviations): bl = blue; dk = dark; br = brown; vio = violet; yel = yellow.

[a] On silica GF$_{254}$.
[b] Fluorescence.
[c] Tail formation (from start to front, depending on the amount).

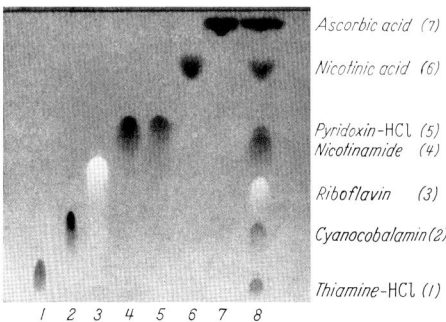

Fig. 124. Separation of water-soluble vitamins on silica gel GF$_{254}$. Solvent: water (CS). 30 µg of *1, 4, 5, 6* and *7* applied; 10 µg of *2*; 5 µg of *3*. *8*: mixture of *1—7* in these amounts. TL-chromatogram photographed in UV (254 nm) [8]

buffered to pH 5.3. Neutral and acid compounds and also cocarboxylase and panthenol migrated with the solvent front on the weakly acid ion exchanger; aquocobalamin remained at the start. The time of migration was 2—3 h.

b) Detection

The characterisation and limits of detection of certain vitamins through inspection of the chromatograms in light of various wave lengths, are quoted in Table 48.

GÄNSHIRT and MALZACHER [51] have detected the components of vitamin preparations which had been spotted in duplicate in two separate rows, by inspection in short- and long-wave UV light; as a result of their fluorescence or light absorbance, respectively, B_2, B_1, B_6, C and nicotinamide (likewise any B_{12} and nicotinic acid present) were visible. To identify biotin, one of the chromatograms was covered and the other sprayed with iodoplatinate reagent (No. 146); biotin showed as a white spot on a pink background. At the same time, B_1 turned grey, nicotinamide pale yellow and C yellow. For identification of B_6 and pantothenic acid, the lower part of the chromatogram which had been covered in the previous test, was sprayed with 2,6-dichloroquinone-chloroimide reagent (No. 66) and exposed to ammonia vapour; B_6 thereby became blue. The chromatogram was then heated 30 min at 160° C and the upper part treated with ninhydrin (Rgt. No. 178A). Pantothenic acid appeared as a violet spot after further brief heating at 160° C. LUDWIG and FREIMUTH [84] and HÜTTENRAUCH et al. [67] characterise the active substances on their layers partly in a similar way and partly also with other reagents. ISHIKAWA and KATSUI [70] have tested multivitamin preparations by chromatographing the extract in 3 different systems (see Table 48 and under "folic acid") and identified the compounds, which were well separated and remote from the start point, like the other authors or with more specific reagents.

Chlorine-tolidine (Rgt. No. 42) has proved to be a universal and sensitive reagent [8]. B_1, B_2, B_6, B_{12}, nicotinic acid, nicotinamide, pantothenic acid, folic acid, biotin and rutin appear as grey blue spots on a white background; after a short while, B_{12} turns violet and B_2 greenish. Vitamin C yields no colour. The reaction depends however on the amount of substance, intensity of spraying, thickness and moisture content of the layer and the time at which the layer is inspected.

As is evident from other examples, certain vitamins can be estimated visually with the help of these methods, when a series of standards is chromatographed at the same time.

Further chromatographic possibilities and selective determination of individual water-soluble vitamins are discussed in the following paragraphs.

2. Vitamin B_1 Group

Almost all thiamine-active compounds from naturally occurring or synthetic products can be extracted with water or aqueous alcohol. The extracts should be worked up under acid conditions, as far as possible at pH 4—6, in order to prevent decomposition of the sensitive substances. B_1-derivatives, linked in biological material to protein, require initial liberation by treatment with enzymes.

a) Separation

Whereas thiamine is strongly adsorbed in some systems (Table 48), it is much more mobile on other layers and can be wholly separated from its esters, derivatives and degradation products. Information about such possibilities is available in Table 49. WALDI [143] has isolated thiamine and its phosphate esters on powdered paper. DAVID and HIRSHFELD [24] have tried out 6 different types of cellulose for the same

Table 49. hR_f-values of thiamine and related compounds
(10—14 cm run in 2—3 h; 10 cm in 1 h with system S_4/F_6)

Layer (ca. 0.25 mm):	S_1	S_2	S_3	S_1	S_4	S_4
Solvent:	F_1	F_2	F_3	F_4	F_5	F_6
References:	[143]	[24]	[24]	[69]	[73]	[8, 130]
Thiamine-HCl and -HNO_3	95	80	10	27	3	56
Thiamine monophosphate	45	47	30	—	—	47[a]
Thiamine diphosphate	25	30	73	—	—	33[a]
Thiamine triphosphate	20	1	80	—	—	—
Thiamine disulphide	—	—	—	51	—	—
O,S-Dicarbethoxythiamine-HCl	—	—	—	90	—	—
Thiamine-thiazolone	—	—	—	67	—	—
α-Hydroxyethylthiamine	—	—	—	—	22	—
Pyrimidine sulphonate	—	—	—	—	30	—
Thiazole-HCl	—	—	—	—	70	—
Thiochrome	—	—	—	53	98	—

Layer:

S_1 = Cellulose MN 300, air dried or 5 min activation, CS
S_2 = Cellulose MN 300 G, air dried
S_3 = Phosphorylated cellulose MN 300 P, air dried
S_4 = Silica gel G, activated, CS

Solvent:

F_1 = Isopropanol-water-trichloroacetic acid-25% ammonium hydroxide (71 + 9 + 20[b] + 0.3)
F_2 = n-Propanol-phosphate buffer, pH 4.9-water (60 + 20 + 20)
F_3 = 0.03 N hydrochloric acid
F_4 = n-Butanol-acetic acid-water (40 + 10 + 50)
F_5 = Pyridine-isobutanol-water (66 + 17 + 17)
F_6 = Pyridine-acetic acid-water (19 + 2 + 79)

[a] Tailing.
[b] Parts by weight.

compounds; the phosphorylated type, i. e., through ion exchange, proved suitable also. INAZU [69] has employed TLC for following the course of the anaerobic decomposition of O,S-dicarbethoxythiamine hydrochloride in solution and published a list of Rf-values for several compounds. TLC on silica gel has proved helpful in studies by JOHNSON and GOODWIN [73] on the formation of thiamine in germinating maize.

b) Detection and Determination

Thiamine and all its esters and also decomposition products can be detected on layers containing fluorescent additives, by inspection in short-wave UV light. B_1 appears violet and amounts down to 2 μg can be detected. A more sensitive and more specific test is to spray with freshly prepared ferricyanide reagent (No. 110); the oxidised thiamine (thiochrome) is recognisable in 365 nm radiation (quartz lamp) through its light blue fluorescence (down to 0.03 μg). The phosphate esters can be detected similarly. The B_1-compounds can be detected also with iodoplatinate reagent (No. 146) (grey; 0.2 μg), chlorine-tolidine (No. 42) or the Dragendorff reagent (No. 96).

An estimate can be made of the B_1-content of the oxidised thiamine by direct comparison on the TLC layer with a series of thiamine standards (0.03 to 2.0 μg). STROHECKER [130] and WALDI [143] have marked the thiamine spots for subsequent quantitative determination by spraying the chromatogram with the ferricyanide reagent (No. 110) and examining it in UV light. The oxidised thiamine was then eluted with methanol from the scraped off adsorbent and the thiochrome phosphate esters extracted with 0.01 N sodium hydroxide solution. Fluorescence intensities were measured after filtration and amounts determined with the help of reference solutions of known concentration which had been similarly chromatographed and treated.

3. Vitamin B_2 Group

Riboflavin readily decomposes through the action of light and of alkalies. It is sparingly soluble in water but can be extracted by adding pyridine or acetic acid or by using dilute mineral acids. It is stable in acid solution. Vitamin B_2 occurs in nature as phosphate ester and, further, as adenine dinucleotide or linked to protein; it must be liberated by enzymatic treatment before determination.

a) Separation

Some ways of separation are compiled in Tables 48 and 50. Suitable spots are obtained on 0.25 mm layers with substance amounts up to 3 μg. TLC is eminently suitable for testing riboflavin and its phosphate ester in pharmaceutical preparations, food- and feeding stuff concentrates [8, 70, 143], for determination in bacteria [80], for purity control in synthesis of B_2 compounds and for preparative isolation. It has been possible to study the products of photochemical degradation of riboflavin on the TLC layer [80, 121].

Table 50. hRf-values of riboflavin and related compounds
(10—12 cm run in 120 min (F_1-F_3) or 60 min (F_4 and F_5))

Layer (ca. 0.25 mm):	S_1	S_1	S_1	S_1	S_2
Solvent:	F_1	F_2	F_3	F_4	F_5
References:	[121]	[121]	[80]	[8]	[143]
Riboflavin	72	30	57	82	75
Riboflavin-5'-phosphate, Na salt	—	—	—	45	38
Lumiflavin	29	30	70	—	—
Lumichrome	42	59	83	—	—

Layer:
S_1 = Silica gel G, activated, CS
S_2 = Silica gel H, activated or air dried, CS

Solvent:
F_1 = Isoamyl alcohol, saturated with water
F_2 = Butanol-ethanol-water (70 + 20 + 10)
F_3 = F_2 (50 + 15 + 35)
F_4 = Pyridine-acetic acid-water (19 + 2 + 79)
F_5 = Acetate buffer, pH 4.62-methanol (50 + 50)

b) Detection and Determination

The yellow flavins and also derivatives and degradation products can be detected specifically and in small amounts in quartz lamp radiation (cf. Table 48). Riboflavin, riboflavin-5'-phosphate and lumiflavin show brilliant yellow fluorescence in long-wave UV light; lumichrome fluoresces blue; and other products of photochemical decomposition likewise yellow and also blue, green and violet [80].

Semi-quantitative determinations can be made using the visual comparison method [8]. WALDI [143] has worked out a quantitative procedure for B_2 and B_2-5'-phosphate as sodium salt; the spots or zones, separated on silica gel H, were detected in the quartz lamp radiation, scraped off, eluted with 0.1 N hydrochloric acid and determined colorimetrically or fluorometrically against a reference solution [130].

4. Pantothenic Acid Group

Pantothenic acid has an optically active centre and only those compounds derived from the D-form possess vitamin activity. Since the free acid is unstable, only its salts, the corresponding alcohol, panthenol, and panthenol ethyl ether are used. These are easily soluble in water or alcohol. In nature, pantothenic acid occurs usually in combined form, as coenzyme A, pantetheine, pantethine etc.

Table 51. hRf-values of pantothenic acid and related compounds
(10—18 cm run in 40—120 min)

Layer (ca. 0.25 mm):	S_1	S_1	S_1	S_2	S_2
Solvent:	F_1	F_2	F_3	F_4	F_5
References:	[8]	[51]	—	[58]	[58]
Pantothenic acid (Na, Ca salt)	89	57	2	—	—
Panthenol	75	57	61	—	—
Panthenol ethyl ether	65	80	66	—	—
Pantetheine	—	—	—	63	—
Pantethine	—	—	—	72	—
4′-Phosphopantetheine	—	—	—	67	—
4′-Phosphopantethine	—	—	—	50	—
Coenzyme A	—	—	—	—	36
Dephosphocoenzyme A	—	—	—	—	48

Layer:
S_1 = Silica gel GF$_{254}$, activated, CS
S_2 = Cellulose MN 300, air dried (0.35 mm thick)

Solvent:
F_1 = Distilled water
F_2 = Acetic acid-acetone-methanol-benzene (5 + 5 + 20 + 70)
F_3 = Absolute ethanol
F_4 = n-Butanol-acetic acid-water (62 + 25 + 13)
F_5 = Ethanol-0.5 N ammonium acetate (66 + 34) (pH 4)

a) Separation

As can be seen from Table 51, a good separation of the members of this group of active substances is possible using TLC. Silica gel layers have proved satisfactory for investigations of extracts of pharmaceuticals, cosmetics or food- and foodstuff concentrates (cf. also Fig. f in Plate II). The products of breakdown of pantothenic acid and panthenol, i. e., β-alanine and aminopropanol, can be isolated at the same time [8]. GÜNTHER and MAUTNER [58], in synthetic studies, have chromatographed coenzyme A, other sulphur-containing pantothenic acid derivatives and the corresponding selenium compounds, using cellulose layers and various solvents.

b) Detection and Determination

Pantothenic acid, panthenol and panthenol ethyl ether are decomposed by heating the plate 30 min at 160° C. After spraying with ninhydrin reagent (No. 178A) and again heating briefly at this temperature, the β-alanine formed yields violet and the aminopropanol wine red. Decomposition is also quantitative by treatment with 10% trichloracetic acid and drying for 10 min at 110° C (limit of detection of β-alanine is about 0.1 µg). Visualisation of the vitamins with the chlorine-tolidine reagent (No. 42) is unspecific and less sensitive.

The sulphur-containing compounds can be detected in UV light after having sprayed with sodium nitroprusside solutions [58].

The vitamins and the amino acids formed by their decomposition, can be semi-quantitatively determined by comparing the ninhydrin-coloured spots with those from a series of standards of the pure substances [8, 103]. Suggestions about evaluation after elution are given in [103]. FREI and FRODYMA [48] have measured the remission spectra of the stable ninhydrin-amino acid complexes on the layer, in order to determine them with high accuracy and precision.

5. Nicotinic Acid and Nicotinamide

Both these so-called vitamin PP-active substances are soluble in water and alcohol.

a) Separation

The vitamins can be isolated on various adsorbents (cf. Table 48). NÜRNBERG [96] has separated pharmaceutical preparations on air-dried silica gel G layers without chamber saturation; the solvent was a freshly prepared mixture of n-propanol and 10% ammonium hydroxide (95 + 5). The solvent front migrates about 8 cm in an hour in this system and the approximate hRf-values are 35 for the acid and 65 for the amide; both can be unambiguously identified and evaluated. STROHECKER and HENNING [130] have used the same solvent mixture or water, silica gel HF_{254} as adsorbent and applied, as bands, larger volumes of the sample solution (0.1–0.4 ml, containing at least) 25 and 50 µg respectively of active material. The quantitative determination was performed after elution. WALSH [144] has investigated tryptophan-nicotinic acid metabolites in urine, with the help of TLC on cellulose.

b) Detection and Determination

Down to 3 µg of both these pyridine derivatives can be detected as dark shadows on the fluorescent layer when irradiated with short-wave UV light. According to NÜRNBERG [96], a specific and sensitive test is sprayng the air-dried layer with p-aminobenzoic acid solution (in total absence of pyridine vapour) and treating subsequently with chlorine cyanide (or bromine cyanide [130]) reagent (No. 23, König reagent). Maximum colour is reached after ca. 6 min. Nicotinamide appears orange-red and nicotine acid red. As little as 0.1 µg of the latter can be detected. Other spray reagents which may be mentioned are iodoplatinate [51] and chlorine-tolidine [8] (Rgt. Nos. 146 and 42 respectively).

A semi-quantitative estimation is possible by visual comparison of the spots with those from a series of standards (nicotinic acid, 0.1 to 2 μg; nicotinamide, 0.5—10 μg) [96]. STROHECKER and HENNING's proposal [130] permits very precise determination. The zones of active substance are localised in UV light, scraped off and determined photometrically or polarographically after elution. It is advisable to treat the pure standards in an exactly parallel way and to take these results into account. According to recent data of STROHECKER [131], nicotinic acid (50—100 μg/10 ml), eluted with 1 N hydrochloric acid, can be conveniently determined by spectrophotometry at 262 nm; a blank must be measured on the same amount of adsorbent.

6. Vitamin B_6 Group

Vitamin B_6-active factors are derived from pyridoxine and can be easily extracted with water. The compounds are destroyed in alkaline or neutral solution under the influence of light.

a) Separation

A selection of TLC systems is given in Table 52. It is evident that the B_6 derivatives can be cleanly separated except pyridoxine and its aldehyde which migrate equally fast on almost all layers. Pyridoxine and pyridoxamine, after having been dissolved in hot methanol, can be

Table 52. hR_f-values of vitamin B_6 compounds
(12—18 cm run in 40—120 min)

Layer (ca. 0.25 mm):	S_1	S_2	S_2	S_3	S_3
Solvent:	F_1	F_2	F_3	F_4	F_5
References:	[97]	[8]	[8]	[152]	[152]
Pyridoxine-HCl	30	53	52	83	73
Pyridoxamine-2 HCl	60	16	19	19	22
Pyridoxal (possibly isomers)	60/66	52	51	83	85
Pyridoxal-5′-phosphate	—	75	64	51	26
Pyridoxal-methylacetal	45	—	—	—	—
Pyridoxal-ethylacetal-HCl	—	44	36	—	—

Layer:
S_1 = Silica gel G (Merck), air dried, CS
S_2 = Silica gel G or GF_{254} (Merck), activated, CS
S_3 = Cellulose powder (over 300 mesh, Toyo Roshi Co.), activated

Solvent:
F_1 = 1 Stage: acetone
    2 Stage: acetone-dioxan-25% ammonium hydroxide (45 + 45 + 10)
F_2 = 0.2% Ammonium hydroxide
F_3 = Distilled water
F_4 = Dioxan-water (70 + 30)
F_5 = n-Butanol-1 N acetic acid (83 + 17)

directly pipetted on to the layer; pyridoxal, also soluble but present as a "cyclic hemiacetal", is unstable. It tends to acetal formation and shows several spots on the chromatogram. The unchanged aldehyde can be detected only after an aqueous solution has been applied. NÜRNBERG [97] has discovered that pyridoxal is quantitatively converted to the methyl acetal by heating in methanol for an hour in the dark; it can then be separated from the other components by stepwise development (S_1/F_1). Silica gel HF_{254} [130] can be used instead of silica gel G or GF_{254}, with the same solvents. YAMADA and SAITO [125] have tested also alumina G, cellulose, kieselguhr G and other adsorbents, using about 20 solvent mixtures; in one case (S_3/F_5) they were able to separate the "critical pair". The method has special practical significance in the analysis of products in which vitamins have been concentrated, in certain biological studies and in control of syntheses.

b) Detection and Determination

The compounds fluoresce dark blue (pyridoxal-5'-phosphate, yellow [152]) on dry silica gel layers containing fluorescent additive, when illuminated with UV light; 3 µg can be detected in 254 nm, 10 µg in 365 nm. These colours depend somewhat on the layer and can be modified by treatment with ammonia and alkalies [152]. All B_6 compounds yield a bluish product after spraying with 2,6-dichloro- or 2,6-dibromoquinone-chloroimide (Rgt. Nos. 66 and 62 respectively) and subsequent treatment with ammonia vapour (limit of detection 0.1 µg; pyridoxal is the weakest, 0.5–1 µg). YAMADA and SAITO [152] have found that the colour shades vary somewhat and depend on the carrier.

These compounds can be extracted from pharmaceutical preparations with water, dilute acids or methanol; pyridoxal present is then converted to the acetal. After spot application and chromatography, the B_6-factors are visualised and estimated by comparison with standard amounts of the pure substances which are chromatographed at the same time [8, 97].

For quantitative evaluation, sample- and standard solution are chromatographed on the same layer, the zones of active substance localised in UV light or after spraying a reference strip, scraped off and extracted with 0.2 N sulphuric acid [130] or 0.05 N hydrochloric acid. The determination is then carried out by photometry, polarography, spectrophotometry [130, 143] or spectrofluorometry.

7. Vitamin B_{12} Group

The most important compound in these socalled corrinoids, is vitamin B_{12} or cyanocobalamin. Aquo- or hydroxocobalamin, in which the cyanide ligand is replaced, exist in equilibrium in solution and possess the same activity. A series of substances, similar to B_{12} but with modified substituents, occur in nature. Some are biologically active (e. g., factor III) whereas others like, cobinamide (factor B), no longer have this property. B_{12} vitamins are soluble in water and 95% alcohol, stable in neutral and weakly acid solution, unstable in strongly acid or alkaline solution and sensitive to both light and oxygen. In the presence of cyanide ions, dissolved cyanocobalamin is converted into the dicyano complex which shows a different absorption spectrum.

a) Separation

Several TLC techniques have been applied here. CIMA and MANTOVAN [16] have investigated the separation of cyanocobalamin and hydroxocobalamin using 20 different solvents on silica gel G which was partly buffered. They isolated photochemical decomposition products at the same time. COVELLO and SCHETTINO [21] have compared paper chromatographic procedures with some fast TLC systems which were especially

Table 53. hRf-values of vitamin B_{12} compounds in various separation systems

Layer (ca. 0.25 mm):	S_1	S_1	S_2	S_2	S_2	S_3	S_4	S_5	S_6
Solvent:	F_1	F_2	F_2	F_3	F_4	F_4	F_5	F_6	F_7
Length of run (cm):	10	10	10	15	11	7.5	12	13	24
Time of run (min):	240	30	30	60	120	60	110	90	240
References:	[16]	[21]	[21]	[62]	[62]	[62]	[62]	[62]	[109][a]
Cyanocobalamin	47	42	32	21	14	37	39	18	62
Aquo-(hydroxo)cobalamin	26	5	45	23	14	29	9	42	—
Factor A	—	—	—	—	2	—	—	—	37
Factor B (cobinamide)	—	—	—	44	28	21	17	46	74
Factor III	—	—	—	—	2	—	—	—	—
Pseudovitamin B_{12} (ψ-B_{12})	—	—	—	3	—	11	21	8	25
Dicyanocobalamin	—	62	19	—	—	—	—	—	—

Layer:
S_1 = Silica gel G (Merck), activated
S_2 = Alumina G (Merck), activated
S_3 = Kieselguhr G (Merck), activated
S_4 = P-Cellulose (MN 300 G/P), dried 1 h at 40° C
S_5 = DEAE-Cellulose (MN 300 G/DEAE), dried 1 h at 40° C
S_6 = Basic alumina, activity II, loose layer of 1 mm thickness

Solvent:
F_1 = Butanol-acetic acid-0.066 M KH_2PO_4-methanol (36 + 18 + 36 + 10)
F_2 = Methanol-water (95 + 5)
F_3 = Chloroform-methanol-water (65 + 25 + 4)
F_4 = Isoamyl alcohol-acetic acid-water (90 + 5 + 5)
F_5 = sec. Butanol, saturated with water-acetic acid (99 + 1)
F_6 = sec.-Butanol-water (83 + 17)
F_7 = Isobutanol-isopropanol-water (33 + 23 + 33)

[a] In the system S_6/F_7, the substances were applied in solution in 5% aqueous NaCN; the dicyano-complexes allegedly change into the monocyano form during chromatography.

suitable for studying therapeutic preparations. HAYASHI and KAMIKUBO [62] have chromatographed several corrinoids on inorganic adsorbents as well as on phosphorylated- and DEAE-cellulose; they established which were the best solvents with the help of the circular technique and achieved new effects by addition of 0.01% potassium cyanide. In this way they identified B_{12} factors in bacteria. Similar schemes have been applied to purity control of the active substances.

b) Detection

Relatively large amounts of B_{12}, as occurring in concentrates or certain medical preparations, can be detected through their red colour (down to 0.3 µg) or in UV light (0.5–1 µg). Even 0.2 µg may be detected through the violet colour with chlorine-tolidine reagent (No. 42).

Quantitative determination can be carried out by direct chromatography of extracts, if necessary after preliminary purification or concentration. Spots of 1 or more µg cyanocobalamin or hydroxocobalamin can be evaluated photodensitometrically on the layer [22]. Elution, followed by spectrophotometry, has been more often used, however [16, 21 etc.]. Bioautography may be used for detection and determination of low concentrations of the B_{12} compounds [62, 99]. The extremely sensitive and specific spectrofluorometry [32] has not so far been used; with the help of a reference chromatogram, with and without addition of B_{12}, the active zone could be scraped off and determined in the eluate in concentrations of only 0.003–0.1 µg/ml.

8. Folic Acid Group

Folic acid (pteroylglutamic acid) and its conjugates occur in very low concentrations, in either free or combined form, in practically every living cell. Folic acid is very sparingly soluble in water and organic solvents. It is amphoteric and thus dissolves in basic or acid solvents. It rapidly decomposes into fluorescent products in light; oxidation, e. g., with permanganate, splits it into the pteridine nucleus and p-aminobenzoylglutamic acid.

a) Separation

Folic acid scarcely migrates with neutral solvents on inorganic adsorbents (cf. Table 48). Decomposition products and contaminants can be easily and quickly separated by TLC with 10% ammonium hydroxide on silica gel G (hR_f = 95) [8] or, according to ISHIKAWA and KATSUI [70], with acetic acid-acetone-methanol-benzene (5 + 5 + 20 + 70) (hR_f = 23) or acetic acid-n-butanol-water (10 + 40 + 50) on alumina G. These procedures are suitable especially for identification in drugs and for purity control of the substance. BAKER et al. [3], in an investigation of transformation of folates in the body, have studied concentrated serum and urine extracts on air dried cellulose MN 300 G, using 5% aqueous citric acid solution which had been adjusted to pH 9.0 with ammonium hydroxide. The approximate hR_f-values found were: folic acid 20; diopterin and teropterin 50; folinic acid and polyglutamates 66. NICOLAUS [92] has checked the purity of several pteridine derivatives on silica gel and alumina layers, using acid, neutral and alkaline solvents.

b) Detection and Determination

Amounts of 2 and 10 μg of folic acid and its derivatives can be detected as dark absorption spots in UV light of 254 and 365 nm, respectively (Table 48). Many pteridines display characteristic fluorescence colours in long-wave UV light, so that substances which are close together on the TLC layer may be distinguished [92].

A highly specific detection of folic acid in very low concentrations and in the presence of all other vitamins is accomplished by spraying with permanganate (Rgt. No. 199) [70, 77]. Even 0.02 μg can be detected through its blue fluorescence in UV radiation of 365 nm. In this way, quantitative determination appears possible directly on the chromatogram or after elution of the spots; we hope to report on this soon. Quantitative determination should likewise be possible with the help of the sensitive spectrofluorometric method [32].

Naturally occurring folic acid compounds which have been separated through TLC are usually evaluated bioautographically on the layer [3].

9. Vitamin C

Vitamin C or free l-ascorbic acid, the active agent against scurvy, occurs in biological material also as dehydroascorbic acid and in combined form as ascorbigen; it is liberated from the latter by hydrolysis. Its salts and esters also are used for pharmaceutical purposes, as food additives and antioxidants. Ascorbic acid is unstable in solution as a result of its powerful reducing character; even in air it can change reversibly into the dehydro form which then suffers degradation. Traces of heavy metals catalyse these changes. It is thus advisable to extract it in an inert gas atmosphere and in the presence of EDTA.

a) Separation

Vitamin C compounds can be separated on various layers (cf. Tables 48 and 54) and separated from further degradation products with, e. g., the systems S_1/F_1 and S_1/F_2 (Table 54). Dehydroascorbic acid yields a compact spot with an hR_f-value of 70—80 on silica gel G, using ethanol or methanol; the other water-soluble C-forms migrate less and tend to tail-formation. With water, dehydroascorbic acid almost reaches the solvent front and the palmitate remains at the start. Several systems are useful especially in drug analysis [8, 51, 70], in studies of isomerisation or degradation [100] and in analysis of potato tubers [61].

Table 54. *hRf-values of ascorbic acid and derivatives*
(12—19 cm run in 40—60 min)

Layer (ca. 0.25 mm):	S_1	S_1	S_2
Solvent:	F_1	F_2	F_3
References:	[8, 51]	—	[130]
l-Ascorbic acid and salts (Na, Ca)	30	50	—
Dehydroascorbic acid	23	73	—
Isoascorbic acid	35	54	—
Ascorbyl palmitate	85[b]	64	—
DNP of dehydroascorbic acid[a]	—	—	24

Layer:
S_1 = Silica gel G or GF_{254}, activated, CS
S_2 = Silica gel H, air dried, CS

Solvent:
F_1 = Acetic acid-acetone-methanol-benzene (5 + 5 + 20 + 70)
F_2 = Ethanol-10% acetic acid (90 + 10)
F_3 = Chloroform-ethyl acetate (50 + 50)

[a] 2,4-dinitrophenylhydrazone of dehydroascorbic acid.
[b] Migrates with the second front.

The separation by TLC of the 2,4-dinitrophenylhydrazone (DNP) of dehydroascorbic acid is extremely specific and has been employed for detection and determination of vitamin C in food and feeding stuffs, fats, fruit juices, wines and bacteria etc. In the method of STROHECKER et al. [129, 130] the ascorbic acid in the extracts is oxidised with 2,6-dichlorophenol-indophenol (VUILLEUMIER and NOBILE [139] use bromine for this); the dehydroascorbic acid formed is then reacted for 3 hours at 70° C with 2,4-dinitrophenylhydrazine in the presence of a little trichloroacetic acid and thiourea [131]. After cooling 10 min in ice, the red or red-brown precipitate is collected on a sintered glass filter, washed with water and dissolved in ethyl acetate or acetone; the solution is evaporated and the residue taken up in acetone. 0.1—1.0 ml (20 to 50 μg vitamin C) of this solution is applied as a band to air dried silica gel H layers[17] and chromatographed with chloroform-ethyl acetate (50 + 50) or, better, with chloroform-ethyl acetate-acetic acid (60 + 35 + 5). The red DNP of the dehydroascorbic acid is thus clearly separated

[17] In cases of complex extracts, larger volumes can be applied to 20 × 40 cm plates, thereby circumventing preliminary purification through column chromatography [131].

from various yellow sugar complexes which would interfere in a direct colorimetric determination.

In experiments on synthesis ascorbigen has also been determined by using TLC [106].

b) Detection and Determination

As a result of its light absorption, amounts of vitamin C down to 3 μg can be detected as dark areas on the layer exposed to short-wave UV light; brief heating to 120° C renders it fluorescent in radiation of 365 nm.

The customary identification of free ascorbic acid depends on its strong reducing properties and any of the reactions known from paper chromatography may be utilised. The limit of detection with indophenol reagent (No. 65) (blue) is around 0.1 μg [77]; dipyridyl-iron (Rgt. No. 91) (red) and molybdophosphoric acid (Rgt. No. 168B) (blue) are almost as sensitive; after brief heating, derivatives and decomposition products yield the colours also. Amounts of 3—5 μg can be visualised with iodoplatinate reagent (No. 146) (yellow) and with alkaline silver nitrate reagent (No. 225).

Phenylhydrazine (Rgt. No. 205) is a specific reagent for dehydroascorbic acid; it yields orange red spots with diketogulonic acid also. The reaction with o-phenylenediamine (Rgt. No. 203) appears to be even more characteristic in yielding an intense blue fluorescence. Ascorbigen can be identified on the layer as free ascorbic acid after acid hydrolysis by spraying with 30% hydrochloric acid and drying at ca. 95° C.

Ascorbic acid can be semi-quantitatively estimated with the comparison method after colour reaction with molybdophosphoric acid [61]. The suggestion of STROHECKER et al. [129, 130] for quantitative determination has met with approval. The red zone of dehydroascorbic acid-DNP is scraped off, eluted with 85% sulphuric acid, filtered or, better, centrifuged, and the light absorption of the solution measured at 520 to 525 nm against water as blank. The analysis result is worked out with the help of a standard solution which is treated identically.

10. Biotin

The naturally occurring d-biotin (vitamin H) is the only biologically active isomer amongst the 8 theoretically possible. It is sparingly soluble in organic solvents and in water but freely soluble in dilute alkali.

a) Separation

As seen from Table 48, biotin can be separated from other water-soluble vitamins on silica gel or alumina, using acetic acid-acetone-methanol-benzene (5 + 5 + 20 + 70) or water.

b) Detection and Determination

Amounts down to 5 µg biotin can be visualised as white spots on a pink background by spraying with iodoplatinate reagent (No. 146). As little as 0.2 µg is rendered visible as a grey-blue spot with the chlorine-tolidine reagent (No. 42) [8].

These procedures are suitable particularly for control of synthesis and purity of the active substance; several impurities and contaminants like dethiobiotin and "γ-biotin" can then be separated and characterised.

Small amounts of biotin, as occurring in natural products and pharmaceuticals, may be detected and determined microbiologically on the layer.

11. Other Vitamins

There are active substances, other than those mentioned, which can similarly be considered as vitamins. They are treated to some extent in the other authors' chapters and are touched on only briefly here. ISHIKAWA and KATSUI [70, 77] have chromatographed and detected some of them.

In particular the vitamin P-active bioflavanoids like rutin [31, 130 etc.] and inositol have often been separated on layers, identified and determined.

Table 55. hRf-values and detection of less well-known water-soluble vitamins, according to ISHIKAWA and KATSUI [70, 77]
(12—19 cm run in 40—60 min)

Layer (ca. 0.25 mm):	S_1	S_2	Detection	Colour	Limit in µg
Solvent:	F_1	F_1	directly or after spraying		
Rutin (vitamin P)	10	0	UV (365 nm)	dk br [a]	0.2
Carnitine (vitamin B_t)	3	20	Dragendorff reagent	red	5
Choline (vitamin Bp)	2	45	Dragendorff reagent	red	1
Inositol	2	0	Silver nitrate-NH_4OH reagent	brown	0.1
p-Aminobenzoic acid	60	54	p-Dimethylaminobenz-aldehyde reagent	yel [b]	0.1
Orotic acid	0	0	$Co(NO_3)_2$ solution + NH_3 vapour	yel [b]	1
Lipoic acid	65	0	Dichromate-conc. H_2SO_4 solution	blue	1

Layer:
S_1 = Silica gel G (Merck), activated
S_2 = Alumina G (Merck), activated

Solvent:
F_1 = Acetic acid-acetone-methanol-benzene (5 + 5 + 20 + 70)

[a] Dark brown.   [b] Yellow.

We thank all colleagues for privately communicated information and for sending reprints of their publications. Further, we should like to thank also Dr. I. ANTENER, Mrs. E. VERES, Mr. W. BÜRKI, Dr. H. GUTMANN, Mr. M. PRÉTÔT, Dr. B. SCHMIDLI, Dr. H. THOMMEN and Dr. H. WEISER who helped us in this work.

Bibliography for Chapter K. Vitamins

1. ALBANESE, A. A.: Newer Methods of Nutritional Biochemistry. New York, London: Academic Press 1965.
2. BACON, M. F.: J. Chromatog. 17, 322 (1965).
3. BAKER, H., O. FRANK, S. FEINGOLD, H. ZIFFER, R. A. GELLENE, C. M. LEEVY, and H. SOBODKA: Am. J. Clin. Nutr. 17, 88 (1965).
4. BENK, E.: Deut. Lebensm.-Rundschau 59, 39 (1963).
5. BIERI, J. G., and E. L. PRIVAL: Proc. Soc. Exp. Biol. Med. 120, 554 (1965).
6. BILLETER, M., u. C. MARTIUS: Biochem. Z. 334, 304 (1961).
7. BOGOSLOVSKY, N. A., L. O. SHNAIDMAN, and E. N. KUZNETOVA: Med. Prom. S.S.S.R. 1965, 41.
8. BOLLIGER, H. R.: In: E. STAHL, Dünnschicht-Chromatographie, p. 217. Berlin-Göttingen-Heidelberg: Springer 1962. English translation "Thin-Layer Chromatography", p. 210. Berlin-Heidelberg-New York: Springer, and New York, London: Academic Press 1965.
9. — A. KÖNIG u. U. SCHWIETER: Chimia (Aarau) 18, 136 (1964).
10. — — Z. Anal. Chem. 214, 1 (1965).
11. BRAEKKAN, O. R.: Int. Z. Vitamin-Forsch. 33, 293 (1963).
12. BUNT, J. S.: Nature 203, 1261 (1964).
13. CASTRÉN, E.: Farm. Aikakauslehti 71, 351 (1962).
14. CHEN jr., P. S.: Anal. Chem. 37, 301 (1965).
15. — A. R. TEREPKA, K. LANE, and A. MARSH: Anal. Biochem. 10, 421 (1965).
16. CIMA, L., e R. MANTOVAN: Farmaco (Pavia), Ed. Prat. 17, 473 (1962).
17. — L. LEVORATO e R. MANTOVAN: Farmaco (Pavia), Ed. Prat. 19, 428 (1964).
18. COLMAN, B., and W. VISHNIAC: Biochim. Biophys. Acta 82, 617 (1964).
19. COPIUS-PEEREBOOM, J. W., and H. W. BEEKES: J. Chromatog. 17, 99 (1965).
20. — — J. Chromatog. 20, 43 (1965).
21. COVELLO, M., e O. SCHETTINO: Farmaco (Pavia), Ed. Prat. 19, 38 (1964).
22. — — Farmaco (Pavia), Ed. Prat. 20, 581 (1965).
23. DAHLQVIST, A., D. L. THOMSON, K. EKBOHM, and B. BORGSTRÖM: Acta Chem. Scand. 18, 1607 (1964).
24. DAVID, S., et H. HIRSHFELD: Bull. Soc. Chim. France 1963, 1011.
25. DAVÍDEK, J., and J. BLATTNÁ: J. Chromatog. 7, 204 (1962).
26. DAVIES, B. H.: Phytochemistry 1, 25 (1961).
27. — D. JONES, and T. W. GOODWIN: Biochem. J. 87, 326 (1963).
28. DEMOLE, E.: J. Chromatog. 1, 24 (1958).
29. DILLEY, R. A., and F. L. CRANE: Anal. Biochem. 5, 531 (1963).
30. — Anal. Biochem. 7, 240 (1964).
31. DRAWERT, F., W. HEIMANN u. A. ZIEGLER: Z. Anal. Chem. 217, 22 (1965).
32. DUGGAN, D. E., R. R. BOWMAN, B. BRODIE, and S. UDENFRIEND: Arch. Biochem. Biophys. 68, 1 (1957).
33. DUNAGIN, P. E., E. H. MEADOWS, and J. A. OLSON: Science 148, 86 (1965).
34. DUNPHY, P. J., K. J. WHITTLE, and J. F. PENNOCK: Chem. and Ind. 1965, 1217.
35. — — and R. A. MORTON: Nature 207, 521 (1965).
36. EGGER, K.: Planta 58, 664 (1962).
37. — Chromatography-Symposium II, Brussels 1962, 75.

38. EGGER, K.: Ber. Deut. Botan. Ges. 77, (145) (1964).
39. — Phytochemistry 4, 609 (1965).
40. — Planta 64, 41 (1965).
41. — u. H. KLEINIG: Z. Anal. Chem. 211, 87 (1956).
42. — u. H. VOIGT: Z. Pflanzenphysiol. 53, 64 (1965).
43. EICHENBERGER, W., u. E. C. GROB: Helv. Chim. Acta 45, 974 (1962).
44. — — Helv. Chim. Acta 48, 1194 (1965).
45. FLEISCHER, S., and G. ROUSER: J. Am. Oil Chemists' Soc. 42, 558 (1965).
46. FRAGNER, J.: Vitamine, Chemie u. Biochemie. Bd. II 1965. Bd. I. Jena: VEB G. Fischer Verlag 1964.
47. FRASER, D. R., and E. KODICEK: Biochem. J. 96, 59 p (1965).
48. FREI, R. W., and M. M. FRODYMA: Anal. Biochem. 9, 310 (1964).
49. FÜRST, W.: Pharm. Zentralhalle 103, 475 (1965).
50. — Pharm. Zentralhalle 104, 381 (1965).
51. GÄNSHIRT, H., u. A. MALZACHER: Naturwissenschaften 47, 279 (1960).
52. GOODMAN, D. S., H. S. HUANG, and T. SHIRATORI: J. Lipid Res. 6, 390 (1965).
53. GOODWIN, T. W.: Lab. Pract. 1964, 295.
54. — Chemistry and Biochemistry of Plant Pigments. London, New York: Academic Press 1965.
55. — and R. J. H. WILLIAMS: Biochem. J. 97, 28 c (1965).
56. GROB, E. C., W. EICHENBERGER u. R. P. PFLUGSHAUPT: Chimia (Aarau) 15, 565 (1961).
57. GSTIRNER, F.: Chemisch-physikalische Vitaminbestimmungsmethoden, 5. Aufl., Stuttgart: F. Enke 1965.
58. GÜNTHER, H. H., and H. G. MAUTNER: J. Am. Chem. Soc. 87, 2708 (1965).
59. HAGER, A., u. T. BERTENRATH: Planta 58, 564 (1962).
60. HALPAAP, H.: Chemiker-Zg. 89, 835 (1965).
61. HASSELQUIST, H., u. M. JAARMA: Acta Chem. Scand. 17, 529 (1963).
62. HAYASHI, M., and T. KAMIKUBO: J. Vitaminol. (Kyoto) 11, 286 (1963).
63. HEAYSMAN, L. T., and E. R. SAWYER: Analyst 89, 1061 (1964).
64. HENNINGER, M. D., and F. L. CRANE: Plant Physiol. 39, 598 (1964).
65. — H. N. BHAGAVAN, and F. L. CRANE: Arch. Biochem. Biophys. 110, 69 (1965).
66. HIGAKI, M., M. TAKAHASHI, T. SUZUKI, and Y. SAHASHI: J. Vitaminol. (Kyoto) 11, 261 (1965).
67. HÜTTENRAUCH, R., L. KLOTZ u. W. MÜLLER: Z. Chem. 3, 193 (1963).
68. HUANG, H. S., and D. S. GOODMAN: J. Biol. Chem. 240, 2839 (1965).
69. INAZU, K.: Ann. Rep. Shionogi Res. Lab. 14, 156 (1964).
70. ISHIKAWA, S., and G. KATSUI: Vitamins (Kyoto) 29, 203 (1964).
71. JANECKE, H., u. L. MAASS-GOEBELS: Z. Anal. Chem. 178, 161 (1960).
72. JOHN, K. V., M. R. LAKSHMANAN, F. B. JUNGALWALA, and H. R. CAMA: J. Chromatog. 18, 53 (1965).
73. JOHNSON, D. B., and T. W. GOODWIN: Biochem. J. 88, 62 p (1963).
74. KATSUI, G., Y. ICHIMURA, and Y. NISHIMOTO: Arch. Pract. Pharm. 23, 299 (1963).
75. — S. ISHIKAWA, M. SHIMIZU, and Y. NISHIMOTO: Vitamins (Kyoto) 28, 41 (1963).
76. — Vitamins (Kyoto) 29, 211 (1964).
77. — Kagaku No Ryoiki Zokan 64, 157 (1964).
78. KNOBLOCH, E.: Physikalisch-chemische Vitaminbestimmungsmethoden. Berlin: Akademie-Verlag 1963.
79. KOFLER, M., P. F. SOMMER, H. R. BOLLIGER, B. SCHMIDLI u. M. VECCHI: Vitamins Hormones 20, 407 (1962).

80. KUWADA, S., and M. HORI: Chem. Pharm. Bull. (Tokyo) **12**, 298 (1964).
81. LAKSHMANAN, M. R., F. B. JUNGALWALA, and H. R. CAMA: Biochem. J. **95**, 27 (1965).
82. LAMBERTSEN, G.: Wiss. Veröffentl. Deut. Ges. Ernährung **16**, 160 (1967).
83. LUDWIG, E., u. U. FREIMUTH: Nahrung **8**, 563 (1964).
84. — — Nahrung **9**, 41 (1965).
85. MERCER, E. I., B. H. DAVIES, and T. W. GOODWIN: Biochem. J. **87**, 317 (1963).
86. MASITI, D., H. W. MOORE, and K. FOLKERS: J. Am. Chem. Soc. **87**, 1402 (1965).
87. MAYER, H., P. SCHUDEL, R. RÜEGG u. O. ISLER: Helv. Chim. Acta **46**, 650 (1963).
88. MIETTINEN, T. A., E. H. AHRENS jr., and S. M. GRUNDY: J. Lipid Res. **6**, 411 (1965).
89. MONTAG, A.: Z. Lebensm.-Untersuch.-Forsch. **116**, 413 (1962).
90. MORTON, R. A.: Biochemistry of Quinones. London, New York: Academic Press 1965.
91. MURRAY, T. K., K. C. DAY, and E. KODICEK: Biochem. J. **90**, 29 p (1965).
92. NICOLAUS, B. J. R.: J. Chromatogr. **4**, 384 (1960).
93. NORMAN, A. W., and H. F. DELUCA: Anal. Chem. **35**, 1247 (1963).
94. — — Biochemistry **2**, 1160 (1963).
95. — J. LUND, and H. F. DELUCA: Arch. Biochem. Biophys. **108**, 12 (1964).
96. NÜRNBERG, E.: Deut. Apotheker-Z. **101**, 142 (1961).
97. — Deut. Apotheker-Z. **101**, 268 (1961).
98. OLSON, J. A., and O. HAYAISHI: Proc. Nat. Acad. Sci. U.S. **54**, 1364 (1965).
99. ONO, T.: Vitamins (Kyoto) **30**, 280 (1964).
100. OTANI, S.: Yakugaku Zasshi **85**, 521 (1965).
101. PAREKH, C. K., and R. H. WASSERMAN: J. Chromatog. **17**, 261 (1965).
102. PASALIS, J., and N. H. BELL: J. Chromatog. **20**, 407 (1965).
103. PATAKI, G.: Dünnschicht-Chromatographie in der Aminosäure- und Peptid-Chemie. Berlin: De Gruyter Verlag 1966.
104. PENNOCK, J. F., F. W. HEMMING, and J. D. KERR: Biochem. Biophys. Res. Commun. **17**, 542 (1964).
105. — and P. J. DUNPHY: Private communication.
106. PIIRONEN, E., and A. I. VIRTANEN: Acta Chem. Scand. **16**, 1286 (1962).
107. PLACK, P. A., and J. G. BIERI: Biochem. Biophys. Acta **84**, 729 (1964).
108. v. PLANTA, C., U. SCHWIETER, L. CHOPARD-DIT-JEAN, R. RÜEGG u. O. ISLER: Helv. Chim. Acta **45**, 548 (1962).
109. POPOVA, Y., K. POPOV, and M. ILIEA: J. Chromatog. **21**, 164 (1966).
110. RANDERATH, K.: Dünnschicht-Chromatographie, 2. Aufl. p. 184—197. Weinheim/Bergstraße: Verlag-Chemie 1965.
111. RAO, M. K. G., S. V. RAO, and K. T. ACHAYA: J. Sci. Food Agric. **16**, 121 (1956).
112. REBEL, G., et P. MANDEL: Biochem. Biophys. Acta **98**, 380 (1965).
113. RILEY, J. P., and T. R. S. WILSON: J. Mar. Biol. Assoc. U.K. **45**, 583 (1965).
114. RISPOLI, G., e A. DI GIACOMO: Boll. Lab. Chim. Provinc. (Bologna) **13**, 587 (1962).
115. RÜEGG, R., u. O. ISLER: Planta med. **9**, 386 (1962).
116. SANDERS, G. M., and E. HAVINGA: Rec. Trav. Chim. **83**, 665 (1964).
117. SCHACHTER, D., J. D. FINKELSTEIN, and S. KOWARSKI: J. Clin. Invest. **43**, 787 (1964).
118. SEHER, A.: Mikrochim. Acta **1961**, 308.
119. SKINNER, W. A., and R. M. PARKHURST: J. Chromatog. **13**, 69 (1964).
120. — — and P. ALOUPOVIC: J. Chromatog. **13**, 240 (1964).

121. SMITH, E. C., and D. METZLER: J. Am. Chem. Soc. **85**, 3285 (1963).
122. SMITH, L. W., R. W. BREIDENBACH, and D. RUBENSTEIN: Science **148**, 508 (1965).
123. SOMMER, P. F.: Private communication, F. Hoffmann-La Roche & Co. AG., Basel.
124. STAHL, E.: Arch. Pharm. **294**/64 ,411 (1959).
125. — H. R. BOLLIGER u. L. LEHNERT: Wiss. Veröffentl. Deut. Ges. Ernährung **9**, 129 (1963).
126. — Chem. Ing.-Tech. **36**, 941 (1964).
127. — Private communication.
128. STOWE, H. D.: Arch. Biochim. Biophys. **103**, 42 (1963).
129. STROHECKER jr., R., u. H. PIES: Z. Lebensm. Untersuch.-Forsch. **118**, 349 (1966).
130. — u. H.. HENNING: Vitamin-Bestimmungen. Weinheim/Bergstraße: Verlag Chemie 1963. English translation "Vitamin Assay — Tested Methods", same publishers 1965.
131. — Private communication.
132. THOMMEN, H.: Naturwissenschaften **49**, 517 (1962).
133. — u. O. WISS: Z. Ernährungswiss. **1963**, Suppl. 3, 18.
134. — u. H. WACKERNAGEL: Biochem. Biophys. Acta **69**, 387 (1963).
135. — U. GLOOR u. O. WISS: Helv. Physiol. Pharmacol. Acta **21**, 345 (1963).
136. — u. H. WACKERNAGEL: Naturwissenschaften **51**, 87 (1964).
137. — u. U. GLOOR: Naturwissenschaften **52**, 161 (1965).
138. VARMA, T. N. R., T. PANALAKS, and T. K. MURRAY: Anal. Chem. **36**, 1824 (1964).
139. VUILLEUMIER, J. P., and S. NOBILE: 12th World's Poultry Congress, Sydney **1962**, 238.
140. WAGNER, H., L. HÖRHAMMER u. B. DENGLER: J. Chromatog. **7**, 211 (1962).
141. — u. B. DENGLER: Biochem. Z. **336**, 380 (1962).
142. WAGNER, F. A., and K. FOLKERS: Vitamines and Coenzymes. New York, London, Sydney: Interscience Publishers 1964.
143. WALDI, D.: Private communication, E. Merck AG., Darmstadt.
144. WALSH, M. P.: Clin. Chim. Acta **11**, 263 (1965).
145. WEBER, F., u. O. WISS: Helv. Physiol. Pharmacol. Acta **21**, 131 (1963).
146. WHITTLE, K. J., P. J. DUNPHY, and J. F. PENNOCK: Biochem. J. **96**, 17 c (1965).
147. WILLIAMS, B. L., and T. W. GOODWIN: Phytochemistry **4**, 81 (1965).
148. WINTERSTEIN. A,, A. STUDER u. R. RÜEGG: Chem. Ber. **93**, 2951 (1960).
149. — u. B. HEGEDÜS: Hoppe-Seyler's Z. physiol. Chem. **321**, 97 (1960).
150. — — Chimia (Aarau) **14**, 18 (1960).
151. YAGISHITA, K., P. R. SUNDARESAN, and G. WOLF: Nature **203**, 410 (1964).
152. YAMADA, M., and A. SAITO: J. Vitaminol. (Kyoto) **11**, 192 (1965).
153. YUSEF, H. M., D. R. THRELFALL, and T. W. GOODWIN: Phytochemistry **4**, 551 (1965).

L. TLC of Steroids and Related Compounds

R. NEHER

Earlier articles or monographs on TLC in general and on steroids or bile acids have been compiled by: BOBBITT, 1963 [23]; GALLETTI [204], HEFTMANN, 1965 [78]; HOFMANN, 1964 [86]; LISBOA [108, 109,

214—216], NEHER 1964 [129, 130]; RANDERATH, 1962 [139]; SJÖVALL [227]; TSCHESCHE, WULFF and RICHERT, 1964 [171] and WALDI, 1962 [181]. Frontier regions have been treated in this book also, in the chapters on lipids by MANGOLD and on clinical diagnosis by ZÖLLNER. Experimental details, e. g., on preparation or quantitative TLC, are given in the different chapters. The literature has been covered up to the end of 1965.

I. Nomenclature

This compound class, widely distributed in nature and intensively studied on the synthetical side, possesses a cyclopentanophenanthrene ring system on which shorter or longer side-chains can be attached. The carbon atoms and rings are numbered and lettered according to the international scheme shown:

The decalin-like ring fusion gives rise to a large number of possible stereoisomers. In the naturally occurring steroids, the fusion of rings B and C is always *trans* and of rings C and D usually *trans* (*cis* in the cardenolides and bufadienolides). Rings A and B are fused in *cis* and *trans* configurations with about equal frequency. The isomer relationships for some substituents on the ring carbon atoms are given in the following two spatial formulae:

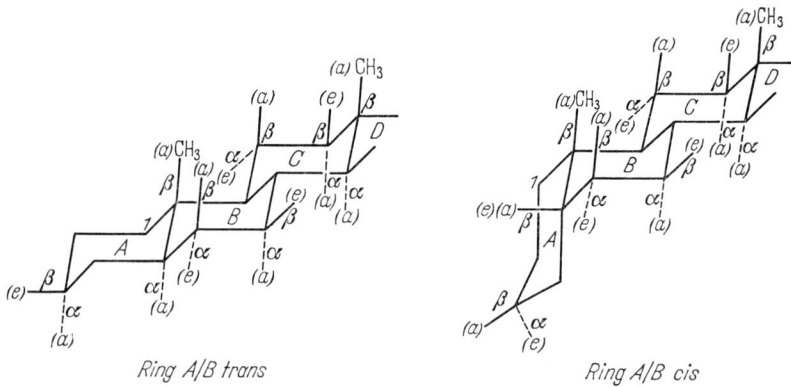

Ring A/B trans   Ring A/B cis

The configuration of these substituents is referred to that of the 19-methyl group on C_{10}, which is conventionally taken as being above the plane of the ring (β-position); substituents of opposite configuration are termed α-substituents (dotted lines). A knowledge of the conformation of the substituents is important especially for the physico-chemical properties; if the substituents lie in the plane of the ring, they are termed equatorial (e) and if at right angles to the ring, axial (a). Thus, e. g., a 3β-hydroxyl group is equatorial in an A/B *trans*-steroid and axial in an A/B *cis*-steroid. Equatorial functional groups are generally appreciably more reactive (and polar!) than axial ones.

Trivial names are used for the ring skeleton:

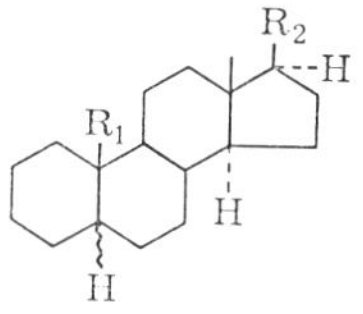

5α/5β

depending on the number and length of the side chains (see Table 56).

Table 56

R_1	R_2		
H	H	5α- or 5β-Oestrane	(Oestrogens)
CH_3	H	-Androstane	(Androgens)
CH_3	COOH	-Etianic acid	
CH_3	C_2H_5	-Pregnane	(Corticosteroids)
CH_3	$CH(CH_3)CH_2CH_2CH_3$	-Cholane	(Bile acids)
CH_3	$CH(CH_3)CH_2CH_2CH_2CH(CH_3)_2$	-Cholestane	(Sterols)

Important, naturally occurring derivatives are quoted in brackets. Sapogenins and steroid alcohols contain hetero atoms linked with the side chain and ring D; the cardenolides and bufadienolides possess an unsaturated 5 or 6-ring lactone instead of the chain at C_{17}, as well as *cis* C/D fusion. In synthetic steroid chemistry, one frequently encounters derivatives which have been formed by shortening, lengthening, opening, transformation or introduction of hetero atoms; they are named after the fundamental skeleton with terms like nor-, homo-, seco-, abeo-, azasteroids etc. Reference must be made to the specialist literature on steroid chemistry for detailed information (e. g., [55, 108]).

II. Range of Application of TLC of Steroids in Comparison with Other Chromatographic Methods

It is not always easy to make the best choice among the multitude of available methods in column, paper, thin-layer and gas chromatography which are based on adsorption and/or partition processes. This choice must depend on the particular problem being studied, on the material at hand and on the special advantages or disadvantages of the individual techniques; maximum efficiency is attained often only by combining techniques and by working on derivatives of the compounds to be investigated. There is no doubt that TLC demands the least time and material for analytical problems in the 0.01—1000 µg region and is thus the first choice. It is often necessary, however, to carry out separations on the preparative scale, quantitative determinations or difficult separations and in the following paragraphs, the essential features of the various chromatographic methods are discussed.

In the choice between *adsorption* or *partition chromatography*, one has to remember that the former is in general preferable for more weakly polar, i. e., more hydrophobic steroids (and derivatives!) and the latter for more strongly polar, i. e., more hydrophilic steroids. As long as PC is carried out according to the partition and TLC according to the adsorption principle, a first basis for decision is given. TLC can, however, very often be carried out just as well as a partition process, so that, a few exceptions apart, TLC yields good results with practically all classes of steroids. It is thus merely a matter of deciding whether to take a piece of cheap filter paper or to prepare a plate with a suitable adsorbent layer (at least as long as no expendable plates are available (see for example, [223]). Experience has shown that partition chromatography is more reproducible [33] and thus preferable where structure analyses are based on the R_M-values (see BUSH [28], NEHER [129], HEFTMANN [78]).

The *capacity* of columns amounts to several kilograms; of analytical PC, up to 1000 µg; of analytical TLC, up to 100 µg; and for GC, up to milligrams. The classical column with continuous development is thus chosen for *preparative* purposes, although the resolution does not attain that of the other methods for various reasons. When silica gel is used as adsorbent, the chromatographic behaviour on the column can be worked out from that on the TLC layer. DUNCAN [47] has applied the relationship $r = a/(b+0.1a)$, where a is the Rf-value of the faster migrating compound, b that of the slower. If $r > 1$, separation is possible in the column also, provided that the ratio of silica gel to substance is 500—1000 and the rate of flow sufficiently small (20—40 ml/h). On the other hand, the properties of the alumina used for column chromatography and that for TLC, vary so much that similar conclusions about conditions in the column cannot be drawn from results on TLC layers.

A compromise solution is to separate up to several grams substance on numerous sheets of paper [129] or on plates with thicker adsorbent layers, using the PC- or TLC-principle respectively (see Chapter D, p. 97; HONEGGER [87], HALPAAP [72]). The column-shaped modifications of DAHN and FUCHS [45] and BALOGH [9] or, in particular, the "dry column" of Loev and Snader [114] have also been recommended for these amounts of material.

In addition to the preparative application, the separation or determination of *very small amounts* in biological material is frequently necessary. The lower limits are set by detection and recovery. GC with amounts of about 0.1 nanogram (ng) offers splendid possibilities and is fully capable of replacing the double isotope technique. TLC with amounts down to 10 ng also fulfils most wishes in contrast to PC with a limit of about 250 ng.

PC and TLC are the methods selected when *many simultaneous analyses* have to be carried out, including serial dilutions of standards; hundreds of such analyses can be performed in a short time and with small demands of space.

The *resolution power* with modern adsorbents is excellent, especially in PC, TLC and GC of the steroids. Partition chromatography is the more profitable type; this applies in particular when, as mentioned above, many polar isomers have to be separated (corticosteroids, glycosides). Impurities interfere less here than in adsorption chromatography.

Side reactions occur rarely in partition chromatography and then only at higher temperature (in GC especially if the system is not entirely of glass) but comparatively often on adsorbents like alumina (see compilation in Neher [129], also [32, 155, 171]). Silica gel is also, however, not without danger for unstable compounds, when they are exposed to the action of air and light on a large surface (e. g., aldosterone or ethyl ketals [153]). This is notable particularly in TLC (see, e. g., [123]).

The *sensitivity of detection* is, with the exception of GC, highest with TLC, provided that corrosive reagents can be used on the adsorbents (alumina, silica gel, kieselguhr; shortcoming of cellulose and polyamide). The highly useful so-called alkali fluorescence [28, 129], specific for the Δ^4-3-ketosteroids, cannot however be carried out on alumina or silica gel but is possible on cellulose or kieselguhr layers. The use of adsorbents with fluorescent additives for non-destructive detection of steroids which adsorb in the UV, is the TLC-equivalent of UV-photocopying in PC. The limit of detection with the most sensitive aids is about 0.20 µg/spot (2.5 cm^2) in PC, 0.01 µg/spot (0.2 cm^2) in TLC and 0.0001 µg (flame ionisation detector) in GC.

GC is the procedure to choose for *quantitative determination*. Unless special equipment is used, direct evaluation of the spots in PC and TLC is not very accurate; in both cases, it is better to elute the substance as quantitatively as possible and to determine in vitro; TLC yields a purer eluate (cf. chapter H).

The *time requirement* amounts to about $10-30-120$ min in GC, $10-30-60$ min in TLC (up to 24 h in difficult separations) and $2-6-48$ h in PC; many analyses can be carried out at the same time in PC and TLC; chromatography on a column of average size requires about 2—24 h (adsorption) or 10—100 h (partition). This last named procedure necessitates the use of a great deal of solvent; rather less for adsorption column chromatography; and little for PC and TLC. On the other hand, GC demands expensive apparatus, PC and column chromatography the least expensive.

The usual conditions for TLC of steroids are discussed in the following sections; information about the application of other useful methods for steroids may be found in the most recent literature: BUSH [28], column and PC; HORNING et al. [89] and in [94], GC; NEHER [129], column, PC, TLC, GC.

III. General Conditions

Adsorbents, solvents, procedures, detection reactions and formation of derivatives are considered here.

1. Adsorbents

The TLC of steroids has been tried on a great variety of adsorbents; loose or adhering adsorbents; silicic acid (silica gel), alumina, magnesium silicate (Florisil [149]), magnesium trisilicate [132], kieselguhr (celite) or cellulose, all with or without binder; even calcium sulphate [119], hydroxylapatite [83], polyamide [57, 193a] or mixtures of silica gel or alumina with kieselguhr [15, 174; 46] and ion exchanger cellulose [135] have been employed for special purposes or have been tested.

Silica gel-gypsum (silica gel G) has been used as adsorbent in the vast majority of *adsorption TLC* procedures; it is expedient to start with it also in separations of steroids. The adsorptive power decreases approximately in the order: silica gel, alumina, magnesium silicate, hydroxylapatite, calcium sulphate, kieselguhr; alumina, however, shows adsorptive power equal to or even higher than silica gel with some substances, e. g., 18-hydroxy-20-ketosteroids, depending also on the solvent. The adsorptive power of silica gel or alumina is reduced by mixing with kieselguhr [15, 174], i. e., the development time with a given solvent is lowered. Activity gradients are formed by judicious mixing of various adsorbents or by partially deactivating them with water [209] (cf. the Section "Gradient-TLC" p. 91).

An addition of 3—10% of silver nitrate to silica gel G or kieselguhr helps considerably in the separation of sterols and steroids which differ only through an unconjugated double bond and which are otherwise difficult to separate (AVIGAN [6], MORRIS [127], HAAHTI [70] and others [37, 50, 41]); this method was introduced in 1962 by B. DE VRIES for separating fatty acids in columns and is based on the stronger retention of unsaturated compounds through complex-formations with silver ions. The technique of deactivation of alumina with acetic acid, previously suggested [150], is probably no longer to be recommended for steroids. Magnesium trisilicate (synthetic) is claimed to show properties differing from those of the typical adsorbents [132]. Starch used as binder interferes when aggressive reagents are used for detection [153]; these reagents attack cellulose and polyamide layers too. On the other hand, cellulose and kieselguhr are better where strong alkalies are needed for the detection reaction (alkali fluorescence, alkaline tetrazolium reagents).

Cellulose and kieselguhr are the carriers which first come into consideration for *partition TLC:* impregnation with the stationary phase is carried out by spraying, or, better, immersion [15, 30, 129, 173, 186], using the Zaffaroni or Bush-type system, known in PC [129]. Smaller spots are obtained with cellulose powder than with kieselguhr or filter paper, due to its very fine structure. Silica gel is, however, also excellently suited as a support for partition chromatography with various PC solvent systems [24, 65, 118, 131]. The carriers mentioned can be used equally well for reversed phase TLC, e. g., after impregnating with paraffin [96, 125] or undecane [40—42, 183]. Information about the properties and capacity of such systems may be found in the PC review of NEHER [129]. Standard size plates are satisfactory for routine work; in special cases, microplates [84, 137] or sheets [103, 223] can be serviceable.

2. Solvents

The choice of solvent for TLC of steroids is governed by their polarity and the nature of the adsorbent. The following exposition relates to silica gel G; R_f-values increase as the activity of the adsorbent falls (alumina, kieselguhr), when the same solvent is used. The solvents most frequently employed for neutral steroids are, in eluotropic or mixotropic order: petrol ether, cyclohexane, carbon tetrachloride, benzene, chloroform, diethyl ether, diisopropyl ether, ethyl acetate, acetone, ethanol, methanol, either alone, in binary or in complex mixtures (see also VAN DAM [175]). The further apart the solvents are from one another in this list, the less miscible they are with one another and the greater the number of differently polar compounds which can be developed with the mixture. Fig. 125 gives hints for the selection of solvent over a wide range of polarity [129, 130]. Between solvents 1 and 6 there are naturally hundreds of possible mixtures, useful for one or other type of neutral, phenolic or weakly basic steroid. A little water, formamide or glycol is sometimes added to prevent tailing or to increase solubility. An acid (acetic acid) or base (pyridine, ammonium hydroxide) is best added for more strongly acid compounds like bile acids or conjugates, or basic steroids, respectively; this suppresses ionisation and tailing. Fig. 126 illustrates a subtle eluotropic gradation of a few useful solvents and binary mixtures of them [129, 130]. This equi-eluotropic scheme is based on the average Rf-values of 20 different steroids on silica gel G layers. At the left of the horizontal mixture-lines

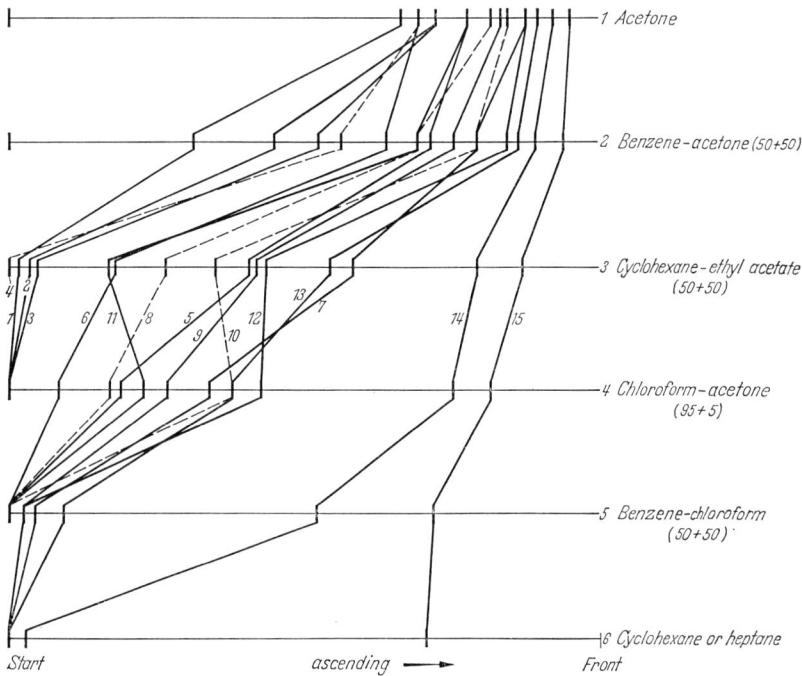

Fig. 125. Lengths of run of sterols and steroids of various polarities in 6 solvent systems of different eluotropic properties (Nos. 1—6 below). Substances: *1* tetrahydrocortisol; *2* cortisol; *3* cortisone; *4* corticosterone; *5* oestradiol-17β; *6* 5β-pregnane-3α,20α-diol; *7* oestrone; *8* testosterone; *9* pregn-5-en-3β-ol-20-one; *10* androst-4-ene-3,17-dione; *11* deoxycorticosterone; *12* progesterone; *13* cholesterol *14* cholesteryl acetate; *15* cholestane

is the more weakly polar pure solvent; at the right, the more strongly polar; the compositions of the mixtures are plotted logarithmically. Methanol is outside the scheme, off the right of the page, corresponding to its high polarity. The pure solvents and binary mixtures of them have been empirically arranged in the figure so that solvent systems of equal average elution power fall on a vertical line. The elution power of the systems thus increases from left to right and remains sensibly constant along vertical lines like X for example. The rings numbered 1 to 6 (from left to right) correspond to the solvent compositions of systems

1–6 in Fig. 125. The choice of a solvent for obtaining a useful R_f-value between 0.2 and 0.7 is not difficult. With more difficult separations, however, it is advantageous to try several equieluotropic solvent systems of widely varying compositions; the resolving power of systems with the same eluotropic properties can vary markedly, thanks to the differing interactions of dissolved material, adsorbent and solvent. The

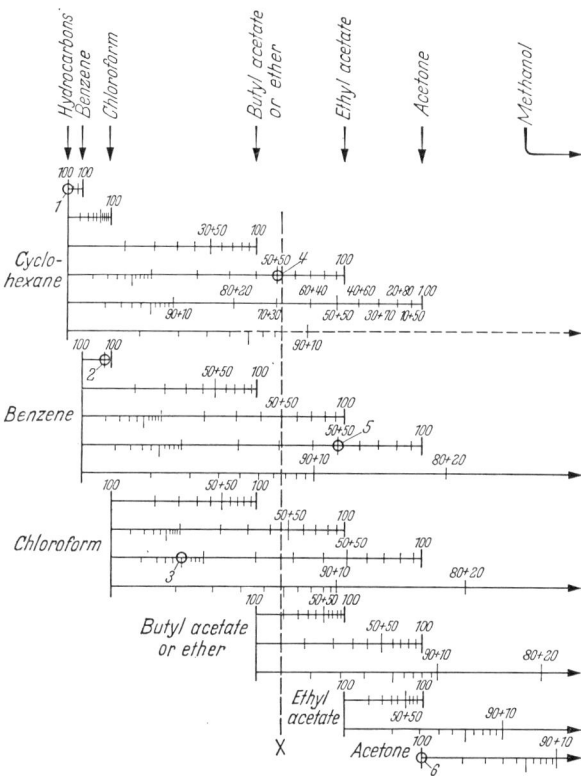

Fig. 126. Arrangement of solvents and their binary mixtures in an equieluotropic series based on the average R_f-values of 20 steroids. Vertical lines join solvent systems of approximately equal eluotropic properties. The first division of the horizontal mixture lines, e.g., cyclohexane-ethyl acetate, from the left corresponds to 99% cyclohexane and 1% ethyl acetate; the first from the right, 10% cyclohexane and 90% ethyl acetate etc. See also text [129, 130]

addition of an ester, ether or ketone can often effect useful modifications. Too large an addition of an alcohol extends the range of polarity, but diminishes the resolving power for substances of similar polarities and causes an interfering second front to be soon formed (see also the next section). See HAGERMAN and SPENCER [71] for the systematic use of

ΔR_{MG}-values in order to discover the optimum solvent systems. The two lipophilic azo dyes, 4-amino-4'-nitro-azobenzene (Oracet Orange 2R) and 4-nitro-2'-methyl-4'-diethanol-aminoazobenzene (Oracet Red 2G) for example, are satisfactory test and reference substances for adsorption TLC using the solvent systems mentioned here. As the proportion of polar solvent is increased, real partition systems are formed which are most useful for steroid TLC, analogous to PC [129]. Some examples of this have been mentioned already in the preceding section. Cellulose powder in particular can be used here as support in addition to silica gel (cf., e. g., [34, 140]); it is best impregnated by immersion in a 10—40% solution of the stationary phase in acetone. Variants can be obtained by using the Bush- and especially the Zaffaroni-type systems, known in PC [129]; these variants can be extended further through the so-called reversed phase systems which are particularly suitable for the very weakly polar steroids. Specific examples are given below; cf. also YAWATA and GOLD [186] for partition systems on kieselguhr-gypsum. A warning may be given here about the excessive use of the toxic benzene (MAC = maximum allowable concentration, 80 mg/m^3 air). It ought to be possible to replace it largely by toluene (MAC 750 mg/m^3).

3. Procedures

Simple development on more or less activated layers will usually be enough for the TLC of steroids, which are best applied to the layer as solutions in alcohol or an alcohol-containing solvent. In the first instance, it is immaterial whether one develops in the normal-, in the S-chamber or in one of the cheaper home made sandwich modifications, provided as good a chamber saturation as possible is aimed at. For difficult separations, the procedures can be improved, or at least rendered possible, through multiple development in the same or different solvents, through continuous development using weakly polar solvent systems (e. g., in the BN-chamber) or through using activity gradients [209]. Further, two-dimensional technique is available as well as the one-dimensional procedure, with or without chemical modification before or between the stages (see p. 325). In any case, it is advantageous to standardise the TLC technique according to the directives in the opening chapters of this book or, e. g. according to COHN and PANCAKE [38] or QUESENBERRY and UNGAR [138] and to employ suitable reference substances consistently.

Additional precautions are needed however to ensure good reproducibility, as shown by the following practical example [131]. Table 57 shows the change in composition of the benzene-methanol (98 + 2) mixture, used for steroid separations, which occurred before, during and after TLC on several silica gel plates (20 × 20 cm) with layers of normal thickness (duplicate determination by gas chromatography with accuracy of ±0.1%). The composition of such a solvent system thus varies appreciably, depending on the operation and brought about by the unequal volatility and, even more, the unequal enrichment on the adsorbent layer; reproducible results cannot be expected in this way. The steroids in 4. migrated in fact furthest on the first layer and least far on the fourth, whereas all migrated equally far in 5., to an extent intermidiate between the second and third layers in 4.

Table 57. *Methanol concentration of the benzene-methanol solvent mixture in the chamber; initial composition (98 + 2)*

	% Methanol in benzene (average)
1. Before use	2.1
2. After introduction into the normal chamber	
a) without paper lining, swilled round gently and kept closed for 10 min	1.8
b) with paper lining, otherwise as a)	1.3
3. As 2a), then left open 5 min	1.5
10 min	0.6
4. As 2a), after running the first plate (15 cm, 60 min)	1.3
after running the second plate (15 cm, 60 min)	0.6
after running the third plate (15 cm, 60 min)	0.3
after running the fourth plate (15 cm, 60 min)	0.1
5. As 2a), after running four plates together	0.5

To ensure high reproducibility in such systems, the same number of plates and always fresh solvent must be used for each run. Systems containing non-hydroxylic components of similar volatility maintain a more stable composition. A tip concerning the *choice of chamber* can be given here in conclusion. The normal chamber is always that first used for ordinary development since it gives good results as a rule. One of the sandwich-chamber types occasionally yields rather smaller spots, however, but this is not the case with every solvent system. We have had quite poor spot formation and wavy fronts using sandwich chambers as soon as the amount of methanol (or homologue) exceeded 3%. Experience with continuous development in these types of chamber (sandwich open at the end, BN) is similar whereas such limitations are found neither with the continuous technique in the normal chamber (with slit cover or blotting paper for absorbing) nor with multiple development in the N-chamber. In fact, the best results have been obtained in continuous development in the BN chamber when using benzene-acetone for example.

The ways of visualising steroids are discussed in the following section. Elution of the unchanged substance in good yield is often a condition for quantitative determination (see chapter H). Since the compounds are more or less strongly adsorbed, it is obvious that they can be desorbed only with hydroxylic solvents, perhaps in mixtures with others; methylene dichloride or ethyl acetate, for example, will not suffice alone [122, 155].

4. Reactions for Detection

Aggressive reagents can be used with inorganic adsorbents, so that all classes of steroid can easily be visualised through colour or fluorescence reactions. The reagents which have been most used and which are most generally applicable are: sulphuric acid-alcohol (No. 214A), chlorosulphonic acid-acetic acid (No. 45), molybdophosphoric acid (No. 168), phosphoric acid (No. 208), aromatic aldehyde-acid (Nos. 11, 48, 133, 218, 260, 262) and antimony(III)-chloride (No. 15). All are in Table 58, with other reagents and some data on application and sensitivity. The colours and especially the fluorescence colours, depend often on concentration. In the following Table, 59, are reagents which react with only a particular grouping and are therefore more specific but usually less sensitive in consequence. Only in exceptional cases are very specific colour or fluorescence reactions available for steroids (cf. the compilation for PC [129]). The intensity of the fluorescence caused by treatment with strong acids, gives no information about the relative amounts of unknown components of a mixture; charring or more specific colour reactions yield more evidence here.

Non-destructive detection, so important for preparative TLC, can be achieved by using known fluorescent adsorbents; the fluorescence is quenched by UV-absorbing substances. The regular use of such adsorbents is recommended. Convenient fluorescent properties can be imparted to the layer *after* TLC, e. g., by spraying with morin [32] (No. 169), 2′,7′-dibromo- or 2′,7′-dichlorofluorescein [43, 127] (No. 63), various dyes [193] or an optical bleach like UVITEX [131] (Nos. 258, 259). Visualisation with iodine is very useful [1, 12, 122, 125, 218] (No. 141), yielding unstable coloured addition products. The zones of lipophilic substances in preparative TLC can be rendered visible also by means of a fine spray of water [10, 69].

Any reagent used in steroid PC can be employed on cellulose layers; some are given in Tables 58 and 59. NEHER [129] gives detailed documentation about application and specificity.

See also the systematic investigations of LISBOA [215, 216] on the application of some of these colour reactions to different groups of steroids.

For completeness, the following reagents may be quoted, although they are not specific and are less important: permanganate-sulphuric acid (No. 200), dichromate-sulphuric acid [70], ferric chloride for bile

Table 58. *Reagents for non-specific detection of steroids*

No.	Reagent	Application, sensitivity limit in μg/spot
241	Sulphuric acid	
	A)	Best with 50% solution in methanol or ethanol; 5—30 min/100—120° C; various colours and fluorescences (365 nm); low specificity but high sensitivity, 0.005; charring with stronger heat
	C) 15% in butanol	Bile acids [5]
	D) 5% in acetic anhydride	Bile acids [5]
	E) 50% in acetic acid	Bile acids [5]; cholesterol 2—4, red
242	Sulphuric acid-hypochlorite	Digitalis glycosides, 0.01, fluorescence [51,52]
45	Chlorosulphonic acid-acetic acid	Cardenolides green, blue violet fluorescence [168]; cholesterol 0.025 [40]
251	p-Toluenesulphonic acid	Similar to sulphuric acid, oestrogens 0.5 [53, 113, 233]
168	Molybdophosphoric acid	Reducing and unsaturated steroids, 0.025, blue on yellow background [13, 68, 111, 153]; cholesterol esters < 0.5 [11, 40, 96]; C_{21}-steroids, pregnanetriols 1—3 [53, 57]; bile acids [62, 172]
208	Phosphoric acid	In water or ethanol, similar to sulphuric acid [129] pregnanetriols [158]; blue spots after spraying with Rgt. No. 168A; cardenolides [181]
209	Phosphoric acid-bromine	Digitalis glycosides 0.001 [52]
11	Anisaldehyde-sulphuric acid	Various colours and fluorescences for many steroids; non-specific but high sensitivity [113, 124]; pregnanes [109]; bile acids]104]
	Anisaldehyde-perchloric acid	Digitalis glycosides 0.1—0.02 [152]
218	Resorcinaldehyde-sulphuric acid	Various steroids [68]
133	p-Hydroxybenzaldehyde-sulphuric acid (Komarowsky reagent)	Ketosteroids; sapogenins 0.1 [164]
48	Cinnamaldehyde-acetic anhydride-sulphuric acid	Steroids; saponins
260	Vanillin-phosphoric acid	Pregnanetriols [142]; sapogenins [148]
262	Vanillin-sulphuric acid	This combination claimed to be better than with weaker acids [121]
	Vanillin-perchloric acid	Pregnanetriols
195	Perchloric acid	Steroids [124]; bile acids [75]
173	Naphthoquinonesulphonic acid -perchloric acid	Sterols 0.03, pink-blue [142]
252	Trichloroacetic acid	Steroids and glycosides
39	Chloramine-trichloro-acetic acid	Cardiac glycosides 0.01 [140, 154]
256	Tungstophosphoric acid	Sterols and esters, red [68, 96]
15	Antimony(III)-chloride	Δ^5-3β-Hydroxysteroids, 0.1, orange-violet; 7-oxygenated Δ^5-steroids, blue-green, etc. cf. [11, 13, 79, 129]
16, 17	Antimony(III)-chloride-acetic acid	Bile acids [5]; dibromosterols, light blue [43, 91]
18	Antimony(V)-chloride	Various steroids [8]
265	Zinc chloride	Similar to No. 15; 0.1 [164]
22	Bismuth chloride	Dibromosterols [42, 43]
221	Rhodamine B	Cholesterol esters; lipids [6, 8, 96]

Table 59. *Reagents for detection of steroids with particular groups*

No.	Reagent	Application, sensitivity limit in µg/spot
	Reducing steroids	
255	Triphenyltetrazolium (chloride) (TTC)	Red 1 [111, 124]
247	Blue tetrazolium (BT)	Blue 0.5 [2, 3, 53, 57, 133]
	High alkali concentration is necessary on silica gel and alumina layers to obtain optimum sensitivity. Alternatively, blue tetrazolium-containing layers can be used [176]	
225	Tollens reagent	Weaker on silica gel or alumina than on cellulose or kieselguhr
122	Folin-Ciocalteu reagent	α-Ketols, α,β-diketones, phenolic steroids, blue [109, 113]
	Ketosteroids	
79	m-Dinitrobenzene	17-Ketosteroids, 0.5, violet [53, 111]; see [129] for behaviour with other ketosteroids
82	2,4-Dinitrophenylhydrazine	Steroids with reactive keto gropus, 2, orange to yellow [111, 148]
150	Isonicotinic acid hydrazide	Δ^4-3-Ketosteroids, 1, yellow [109, 111]
202	Phenylenediamine-phthalic acid	α,β-unsaturated ketones, 2—3, orange-brown [109, 111]
	Phenolic steroids (enols)	
111	Ferricyanide-ferric chloride	Blue, down to 0.2; only 10 or less for other steroids [13, 113]
122	Folin-Ciocalteu	See under "reducing steroids"
180, 182, 238, 240	Diazonium salts	Yellow to red, 2—5 [113]
	Δ^5-3β-hydroxysteroids	
211	Picric acid	Yellow-red, 2—5 µg, can be used also in vitro [198]
	Cardenolides	
80	Dinitrobenzoic acid (Kedde reagent)	Digitaloid 5-ring lactones, blue-violet, 0.005 [108a]
244	Tetranitrodiphenyl	Blue; possibly more specific than No. 80 [20]
254	Trinitrobenzoic acid	Orange-red [219]
	Nitrogen-containing steroids	
146, 147	Iodoplatinate	About 1
96, 97	Iodobismuthate (Dragendorff reagent)	About 1—5
36	Cerium(IV) sulphate	Only on silica gel layers [148]
124	Formaldehyde-acid	Indole derivatives [148]
	Steroid glucuronides	
215	Pyridylazonaphthol	1—2 [44]
	Steroid sulphates	
162	Methylene blue	1—2 [44]

Cf. also the systematic studies of LISBOA [215, 216] in this connection.

acids [5], uranyl nitrate [68], bromothymol blue [14] (No. 31) and iodine-iodide [111] (Rgt. No. 144) for various steroids. See chapter I, p. 157 for measurements of radioactivity.

Other more or less specific but less important reagents are: for unsaturated compounds; molybdoarsenate [111], tetranitromethane [111, 113], osmium tetroxide [111, 113), p-aminodimethylaniline-SO_2 [111, 113]; for phenols, the nitroso reaction (Boute reagent) and Feigl reagent [113]; for o-diphenols, phloroglucinol and vanadic acid [113]; for reactions of the side chain, the Porter-Silber reaction [111], methyl ketones and formaldehydogens [111] and further modified reactions usual in PC [129].

5. Formation of Derivatives (Microreactions)

Derivatives like acetates, propionates or benzoates can occasionally be separated better than the free substances themselves (cf. section IV and special examples). Further, the interplay of chromatographic behaviour, chemical transformations and colour reactions yields appreciably more valuable information about the structure of a compound than do the Rf-values alone. Many examples of this sort are known in PC (BUSH [28], NEHER [129]). MILLER and KIRCHNER [126] have already described TLC-microreactions, in situ or in vitro. It is advisable to control the course and yield of these reactions by using standard steroids under the same conditions. Many reactions can be carried out directly on the layer; reactions in vitro are, however, more reliable (see p. 205).

Acetylation; 10—100 µg substance are esterified by treatment with 0.1 ml acetic anhydride and 0.1 ml pyridine for 8—16 h/room temperature or 1 h/70° C. After reaction, excess of the reagents can be easily removed with a current of nitrogen at 60° C. All primary and secondary hydroxyl groups are acetylated under these conditions, except those sterically hindered (11 β-hydroxyl groups).

Benzoylation is carried out in a parallel way, but using 0.1 ml benzoyl chloride. After reaction, a little ice-water is added, the mixture extracted with methylene dichloride and the organic layer washed with a little 2N hydrochloric acid, with N sodium carbonate solution and finally with water.

Propionylation is performed by dissolving the steroid in 0.3 ml warm propionyl chloride and allowing to stand 10 min at 20° C. The pure ester is obtained by extraction with hexane and washing with water and sodium carbonate solution [37].

Trifluoroacetates are rapidly formed (1 min) by mixing well with a small excess of trifluoroacetic anhydride in hexane or methylene dichloride. The usual working-up procedure follows [15].

Etherification is carried out with diazomethane or dimethyl sulphate.

The separation of so-called "critical pairs" (saturated/unsaturated with one double bond) is difficult but is made possible by *bromination* of the one component. The bromine addition may be done in situ (according to CARGILL [29], by spotting 0.1% bromine solution in chloroform on to the start spot) or, as performed by KAUFMANN, MAKUS and KHOE [40, 97], simply by adding bromine to the developing solvent to give a 0.5% solution. A similar differentiation is accomplished by, e. g., *epoxidation* of the double bond with m-chloroperbenzoic acid in chloroform under mild conditions [8]. The classical hydroxylation with osmium tetroxide leads to far more strongly polar products. *Catalytic hydrogenation* of double bonds can also be carried out on the layer [97]. The *reduction* of keto groups is better performed in vitro by 1 h treatment with a few mg potassium borohydride in methanol [112] or, for example, in situ by spraying with 5% sodium borohydride in 80% methanol [172]. Secondary alcohol groups can be *oxidised* to keto groups in the usual way with either 0.5% chromic acid in 90% acetic acid or the pyridine-chromic acid complex at room temperature (2 h). Ketol- or glycol side chains may be oxidised smoothly with 2% periodic acid in dioxan-water (2–12 h). Other possibilities on the micro scale include: Girard-hydrazone formation from ketones; formation of nitroso complexes from oestrogens [112]; and many other specific reactions already used in PC [28, 54, 129].

IV. Structure and Chromatographic Behaviour
1. Adsorption TLC

As a first approximation, the tendency of a substance to be adsorbed increases with the polarity of its molecule; this is roughly determined by the C/O ratio in the case of the steroids. Thus for example when "equivalent" oxygen-containing groups are present, the polarity increases in the sequence $C_{21}O_2-C_{19}O_2-C_{18}O_2$; propionates therefore migrate faster than acetates. Polarity further depends markedly on the nature of the functional groups; for substituents in the 3β-position on silica gel and alumina, the polarity increases in the order: $-H$, $-Cl$, $-OCOC_6H_5$, $-OCH_3$, $-OCOCH_3$, $=O$, $-OH$, $-N(CH_3)_2$. A phenolic hydroxyl group is more feebly polar than a secondary alcoholic group [167]. The polarity of substituents in the 17β-position has been found to increase in the order: $-COOCH_3$, $-OCOC_6H_5$, $-CN$, $-COCH_3$, $-OCOCH_3$, $=O$, $-OH$ [79]. Isolated double bonds exert only a small influence but conjugated bonds bring about an appreciable increase in polarity, especially in the α,β-unsaturated ketones; further conjugation has then little extra influence. The position of the functional group in the ring structure is of considerable influence, especially for hydroxyl groups, less so for keto groups. The conformation (axial or equatorial) of the former and any possible interactions with neighbouring groups is decisive.

The sequences discussed below are thus of limited validity, the more so because inversions can occur in certain solvent systems. As already known from column chromatography, certain adsorbents can also bring about inversions (cf. p. 57 in [129]). Even though in genuine partition chromatography, an equatorial (e) hydroxyl group is always more polar than an axial (a) in the same position, this does not necessarily always apply in adsorption.

Thus the following sequence of diminishing polarity has been found for the isomeric 3-hydroxycholestanes and 3β-hydroxycholest-5-ene on silica gel and alumina (29, 32):

$$3\beta,5\alpha \geq 3\beta, \Delta^5 > 3\beta,5\beta \geq 3\alpha,5\beta > 3\alpha,5\alpha$$
$$\text{(e)} \qquad\qquad \text{(a)} \qquad\quad \text{(e)} \qquad\quad \text{(a)}$$

It is evident from this, that 5α-H and Δ^5-compounds of the 3β-series and the $3\alpha, 5\beta$ and $3\beta,5\beta$-isomers are difficult to separate; on the other hand, the separation of the equatorial from the axial pairs causes no difficulty in partition chromatography. A combination of both chromatographic types would more easily effect separation here. A further comparison of adsorption (A) and partition (P) leads to the following list [29]:

for A and P, $3\beta,5\alpha > 3\alpha,5\alpha$ holds but
for A $\qquad 3\beta,5\beta \geq 3\alpha,5\beta; 3\beta,5\alpha \geq 3\beta,\Delta^5$
for P $\qquad 3\alpha,5\beta \geq 3\beta,5\beta; 3\beta,\Delta^5 \geq 3\beta,5\alpha.$

The situation is complicated, however, by the partial inversion of sequences which, as already mentioned, can be brought about by certain adsorption systems (ether, ethyl acetate).

The following sequences of decreasing polarity have been established for the hydroxyl groups in other positions [171]:

for 5α-androstanes $\quad 3\beta > 2\alpha > 16\beta > 17\beta > 17\alpha > 15\beta;$
$\qquad\qquad\qquad\qquad 3\beta > 6\alpha > 12\beta > 15\beta;$
for 5α-pregnanes $\quad 3\beta > 11\alpha > 11\beta \geq 21 > 17\alpha;$
for 5β-androstanes and $\quad 11\alpha > 12\beta > 16\beta > 15\alpha > 14\beta;$
$\quad$-pregnanes $\qquad\qquad 3\beta > 7\beta > 12\beta > 7\alpha \geq 12\alpha.$

Here also the equatorial hydroxyl groups are usually more polar than the axial; this could be shown for the bile acid esters too (5β-H) [49, 86]:

$$3\alpha > 3\beta > 7\beta > 12\beta > 7\alpha \geq 12\alpha > 3 = 0 > 7 = 0 \geq 12 = 0.$$

As a result of increased interactions in the more highly substituted derivatives, the situation becomes unclear. The axial 12α-hydroxy isomer of rockigenin shows more polar behaviour than the equatorial isomer, whereas the diacetates show the opposite order [171].

The sequence for esters in 3α- and 6α-position is, e. g., $-OSO_2C_7H_7$ (toluenesulphonate) $> -OCOCH_3$ but in 12α it is the other way round [2]. $2\alpha,3\beta$-Diols shows themselves to be more polar than $3\beta,17\beta$-diols; $1\beta,3\beta$-diols correspond, however, only to a monohydroxy-compound. An 11-keto group, for instance, is less polar than an 11β-hydroxyl group in solvents which consist mainly of benzene or chloroform, but is *more* polar in ethyl acetate; similar inversions are known in PC also. The polarity of oestrogens falls in the order [157]: 16α-hydroxy- $> 6\alpha$-hydroxy- $> 16\beta$-hydroxy- > 16-keto-; a decisive role is played here however by the position or nature of the group in the adjacent 17-position.

ČERNÝ and LÁBLER [105, 106] have carried out a thorough investigation of aminosteroids and steroid alkaloids and found for the dimethylamino group in position 3 almost the same sequence of polarity as for the hydroxyl group (on silica gel):

$$3\beta,5\alpha > 3\beta,\Delta^5 > 3\alpha,5\beta \geqq 3\beta,5\beta > 3\alpha,\Delta^5 > 3\alpha,5\alpha.$$

The dimethylamino group, like the hydroxyl group, is more polar in the 20α- than in the 20β-position. The polarity increases with increasing N-methylation especially for the imino group of the pyrrolidine ring in *Holarrhena* alkaloids. Anti-oximes are said to behave in a more polar fashion than syn-oximes [66].

Some authors have attempted to apply the R_M-function and ΔR_M-values, as conceived by BUSH [28], to TLC [33, 54, 110, 214—216] and to compare in part with the values obtained in PC. This is not discussed further here because of the still scanty material, the sometimes poor reproducibility and the widely fluctuating influence of the various solvents.

2. Partition TLC

The relation between structure and chromatographic behaviour is very much the same here as in paper chromatography; it would be outside the scope of this article to discuss here the extensive material, especially since it has recently been compiled in a monograph [129].

3. Improvement of the Separation of Similar, Polyfunctional Steroids

If only one or two polar groups are present, separation is usually possible as a result of their contribution to the polarity, mentioned above; a better separation is generally possible than of their esters. Two compounds cannot as a rule be separated if they possess the same strongly polar groups and differ only in an additional, more weakly polar substituent; separation becomes possible, however, if the strongly polar groups can be selectively removed or converted into feebly polar groups (see section III 5 on formation of derivatives, p. 325). Strongly polar glycosides, for example, with hydroxyl groups differing in position or even in number in the aglycone part, are separable only after esterification or removal of the sugar moiety [171]. 3β-Hydroxyandrost-5-enes which have a keto-, $-COCH_3$ or $-COOCH_3$ group in the 17-position, cannot be separated in TLC as long as the 3β-hydroxyl group is free; acetylation or benzoylation renders separation easy [79]. On the other hand, the chance of separating polar steroids is frequently greater when based on partition chromatography; PC is then still a valuable tool along with TLC.

V. Sterols

Most sterols in the plant and animal world possess the basic skeleton of cholesterol:

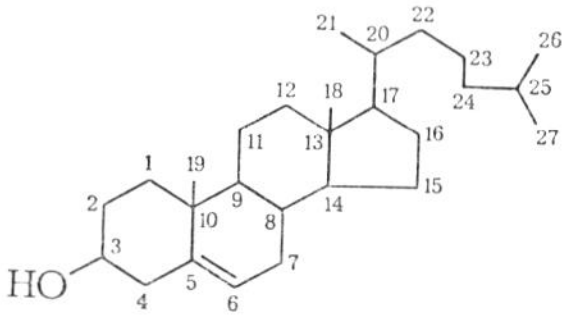

They differ from it partly in the number and/or position of the double bonds, partly through additional methyl groups, for example on C_4, C_{14} and particularly on C_{24}; an ethyl group is occasionally found on the last named. The sterols often occur in complex mixtures, in both free and esterified form. The abbreviated formula, xFC_yzF is a handy representation of the sterols; x gives the number of double bonds in the ring part; y, the total number of carbon atoms; and z, the number of double bonds in the side chain. Thus cholesterol is FC_{27}; 24-dehydrocholesterol (desmosterol) $FC_{27}F$. It is known from PC that the introduction of a double bond into the ring structure or side chain of a sterol, increases its polarity to about the same extent as the introduction of a methyl or ethyl group decreases it. The so-called carbon number can thus be used as a measure of the polarity; it is the difference between the number of carbon atoms and the number of double bonds [43]. Stigmasterol ($FC_{29}F$), campesterol (FC_{28}) and 5α-cholestan-3β-ol (C_{27}), for example, all have the same value. There are numerous such "critical pairs", for the difficult chromatographic separation of which a trick is needed.

Since VAN DAM's [175] investigations on the usual form of sterol adsorption TLC (cf. also Fig. 25 for customary solvents), good progress has fortunately been made because many separation problems, important for practical application, have turned out to be rather thorny:

1. complete separation from one another of the four epimeric 5α- and 5β-cholestan-3-ols;
2. separation of sterols, differing from one another only in one isolated double bond;
3. separation of sterols, differing from one another only in alkyl groups;
4. separation of the many cholesterol esters.

Concerning 1. Fractionation only into groups has been accomplished with adsorption TLC, as investigated in 1961 by ČERNÝ [32] on alumina and in 1962 by CARGILL [29] on silica gel G, using the customary solvent systems like benzene-diethyl ether $(70 + 30)$ or benzene-ethyl acetate $(60 + 30)$ or $(95 + 5)$ or chloroform-acetone $(85 + 15)$; this separation follows the sequence of decreasing Rf-values given above in section IV, namely,

$$3\alpha,5\alpha > 3\alpha,5\beta + 3\beta,5\beta > 3\beta,5\alpha + 3\beta,\Delta^5 \text{ (cholesterol)}.$$

The Rf-values fall between 0.2 and 0.7; values given in this article are only approximate guide values, since good reproducibility between laboratories cannot be guaranteed. Standard substances of appropriate polarity (e. g., cholesterol) should therefore always be chromatographed at the same time, since it is really a question only of the relative sequence (R_{St}-values, St = standard substance). The stanols cannot be completely separated from one another even through partition chromatography, as, for example, using phenyl cellosolve/heptane or undecane/methanol [29]. The sequence is, however, different here (separation into axial and equatorial epimers) so that a combination of both methods must lead to success.

Concerning 2. a) Normal *adsorption TLC* is insufficient here, except with Δ^7-compounds: Separation of 7-dehydrocholestan-3β-ol, 7-dehydrocholesterol and cholesterol, in that order of increasing Rf-values, is possible using cyclohexane-ethyl acetate-water $(60 + 40 + 0.1)$ [16]; oddly enough, the compound with two conjugated double bonds is less polar than that with the single 7(8) double bond (not the case in partition TLC). The pairs cholesterol (FC_{27})-desmosterol ($FC_{27}F$) and β-sitosterol-(F_{29})-stigmasterol($FC_{29}F$) are not separable in the free form; but BENNETT and HEFTMANN [16] have been able to show that their trifluoroacetate esters, only weakly polar, can be separated using the very feebly polar solvent mixture cyclohexane-heptane $(50 + 50)$ on the weakly active adsorbent silica gel G + kieselguhr $(1 + 1)$. AVIGAN [6] has tried to separate cholesterol and desmosterol as acetates on long silica gel G layers, using hexane-benzene $(50 + 10)$.

b) *Partition TLC* has proved better than adsorption TLC. COPIUS-PEEREBOOM [40—43] has investigated it thoroughly in the reversed phase technique (cf. [129] for analogy to PC). The R_{St}-values quoted for various sterol acetates in Table 60 show that fractionation into several zones is possible or a satisfactory separation of pairs like cholesterol/desmosterol, cholesterol/5α-cholestan-3β-ol, stigmasterol/β-sitosterol etc. The stationary phase was undecane on kieselguhr and the mobile phase, a mixture of acetic acid-acetonitrile (1 + 3), saturated with undecane. The TLC, carried out as wedged-tip technique, lasted 2 hours. WOLFMAN and SACHS [183] have separated free cholesterol and desmosterol with the same system on silica gel G. Undecane has the advantage over paraffin, which can also be used with methyl ethyl ketone-acetonitrile (70 + 30) in the reversed phase technique [96], that it can easily be removed after the run and does not interfere in the detection of trace impurities using the usual reagents. Adherence to the directions in the original work [43] is vital for good reproducibility with the undecane system.

Sterols with differing numbers of double bonds can however be more easily separated with the help of various devices.

c) Thus the procedure developed by KAUFMANN [97] for bromination of unsaturated fatty acids by adding bromine to the solvent to give a 0.5 vol% solution, has been applied with success. The bromination is carried out either in situ before chromatography, or during chromatography. The Δ^5-3β-hydroxysterols are thus converted to the more rapidly migrating dibromides while the saturated compounds remain unchanged. Sterols with a double bond in addition to, or otherwise positioned than Δ^5, are decomposed and remain near the start [43] or, in the reversed phase procedure, migrate to the front. CARGILL [29] has been able in this way to detect small amounts of 5α-cholestan-3β-ol in the presence of large excess of cholesterol, using the system benzene-ethyl acetate (100 + 5) on silica gel G. COPIUS-PEEREBOOM [42, 43] has added bromine to the undecane/acetic acid-acetonitrile (1 + 3) system and obtained the R_{St}-values quoted also in Table 60. The sequences change as indicated in relation to the system without bromine and permit further subdivision of the various zones. The dibromides may be characterised additionally with reagents like bismuth chloride or antimony(III)-chloride; the latter gives a light blue colour reaction. AZARNOFF and TUCKER [8] have succeeded in separating saturated from ring-unsaturated sterols by converting the last-named to epoxides with m-chloroperbenzoic acid and following with TLC on silica gel G using benzene-butyl acetate-methyl ethyl ketone (75 + 25 + 10).

d) Probably the currently most elegant method is based on complex-formation of unsaturated compounds with silver. Silica gel containing

5—50% silver nitrate is the adsorbent. AVIGAN [6] has separated mixtures of compounds of different degrees of unsaturation, as acetates on silica gel-silver nitrate, using benzene-ethyl acetate (100 + 20). According to CLAUDE [37], this works even better with propionates and a more weakly polar solvent like hexane-benzene (100 + 20); if necessary, continuous or multiple development is used. A successful separation was achieved in the sequence: 7-dehydrocholesterol ($2FC_{27}$) at the start; desmosterol ($FC_{27}F$); cholesterol(FC_{27}); 5α-cholestan-3β-ol (C_{27}). This represents a good differentiation on the basis of the number of double bonds. The results of COPIUS-PEEREBOOM [43] in Table 61 show this also; the solvents employed here were chloroform-ether-acetic acid (97 + 2.3 + 0.5) (A) for the free sterols and petrol ether-chloroform-acetic acid (75 + 25 + 0.5) (B) for the acetates (1—2 h run). The distinction between one and two double bonds is clearly better with the esters. The separation effect differs from that in the "bromine"-system. TRUSWELL et al. [231], IKAN et al. [211] and DITULLIO et al. [196] quote new examples of good separations of sterols and stanols on silica gel-silver nitrate layers.

Concerning 3. BENNETT and HEFTMANN [16] have been able to separate the acetates of the sterols, β-sitosterol (FC_{29}), cholesterol (FC_{27}) and stigmasterol ($FC_{29}F$) only in a continuous procedure on Anasil B (better here than silica gel G), using hexane-diethyl ether (97 + 3); the separation lasted 2 hours. Here, too, partition chromatography in the undecane system [43] is more successful, as evident from Table 60. The important separation of cholesterol/β-sitosterol, otherwise difficult, must be emphasised.

Separation of the critical pairs, mentioned at the beginning, is, however, comparatively easily accomplished by combining several procedures which differentiate on the one hand according to the number of double bonds and on the other, according to the C-content. COPIUS-PEEREBOOM et al. have presented examples of very complex mixtures of plant sterols and compared also the results with those from PC [40—43].

Concerning 4. Thanks to clinical interest, much work has been done on the separation and determination of cholesterol and its esters in serum or skin fat, so that there are good solutions to this problem.

WALDI [181] has separated the esters on silica gel G, using carbon tetrachloride-chloroform (95 + 5). KAUFMANN et al. [96] have used either tetralin-hexane (25 + 75) or (50 + 50) for adsorption TLC or paraffin/methyl ethyl ketone-acetonitrile (70 + 30) as reversed phase partition system. All the esters from acetate to stearate yield thus a series of spots like a string of beads. Acetic acid is also a suitable mobile phase, as in PC [125]. Unsaturated fatty acid esters can be differentiated from the saturated esters only with difficulty in this way; success is more likely when an adsorption chromatographic separation is carried out

Table 60. R_{St}-values (St = Cholesteryl acetate, Rf ~ 0.28) of sterol acetates in the reversed-phase system: undecane/acetic acid-acetonitrile (20 + 60) [43]

Acetate of	Abbreviation	Zone No.	R_{St}-value	+ 0.5% bromine
5α-Androstan-3β-ol	C_{19}		2.25	
Δ9(11)-Dehydroergosterol	$3FC_{28}F$	6	1.45	
Vitamin D_3	$3FC_{27}$	6	1.41	
Dihydrovitamin D_2	$3FC_{28}$	5/6	1.34	
Zymosterol	$FC_{27}F$	5	1.28	
Vitamin D_2	$3FC_{28}F$	5	1.26	
Desmosterol	$FC_{27}F$	5	1.24	
3-Dehydrocholesterol	$2FC_{27}$	5	1.22	
Ergosterol	$2FC_{28}F$	5	1.19	Front
epi-Cholesterol	FC_{27}	5	1.19	
7-Dehydrocholesterol	$2FC_{27}$	5	1.16	Front
Δ7,9(11),22-Ergostatrienol-3β	$2FC_{28}F$	5	1.11	
Brassicasterol	$FC_{28}F$	4	1.07	1.13
5-Dihydroergosterol	$FC_{28}F$	4	1.07	
22-Dihydroergosterol	$2FC_{28}$	4	1.07	
Cholesterol	FC_{27}	4	1.00 (St)	1.00 (St)
Lanosterol	$FC_{30}F$	4	1.00	Front
Δ3-Cholesten-3β-ol	FC_{27}	4	1.00	
Δ7-Cholesten-3β-ol	FC_{27}	4	0.99	
Δ7-Ergosten-3β-ol	FC_{28}	3	0.92	Front
Campesterol	FC_{28}	3	0.92	0.95
Stigmasterol	$FC_{29}F$	3	0.91	1.06
α-Spinasterol(Δ7,22-Stigmastadienol)	$FC_{29}F$	3	0.91	
5α-Cholestan-3β-ol	C_{27}	3	0.89	0.85
Agnosterol	$2FC_{30}F$	2/3	0.86	
β-Sitosterol	FC_{29}	2	0.83	0.84
Stigmastan-3β-ol	C_{29}	1	0.73	

in the first dimension and a partition chromatographic separation in the second [96]. Application of activity gradients has also been recommended [209]. Excellent separation according to the number of double bonds in the fatty acid moiety has been achieved with silica gel G-silver nitrate, using diethyl ether or diethyl ether-hexane (20 + 80) as by MORRIS [127]; or using benzene or hexane-benzene (10 + 90) as by HAAHTI [70] or, recently, by GOODMAN [207] and by PASCAUD [221]. Fractionation into groups of esters of fatty acids with from 0 to 6 double bonds is brought about. A two-dimensional combination appears here reasonably promising for further subdivision. The work of ZÖLLNER, WOLFRAM and AMIN [189] in the clinical field may be mentioned; they have separated the esters in a single or, better, multiple run on silica gel G, using petrol ether-diisopropyl ether (99 + 1) and photodensitometrically evaluated the coloured spots obtained with antimony(III)-chloride. Reference may be made to the relevant chapter (S) by ZÖLLNER in this book for such clinical applications of TLC.

A few applications of sterol TLC in practice may be added: Determination of dihydrocholesterol in serum [192]; of sterols and vitamin D in cod liver oil [193]; of arachidonates [petrol ether-benzene (60 + 40)] [116] and other synthetic sterols (diisopropyl ether, benzene) [170]; and of sitosteryl esters [cyclohexane-benzene mixtures, chloroform-acetone (95 + 5), heptane-ethyl acetate (95 + 5), carbon tetrachloride-chloroform (95 + 5)] [18]; separations of unsaturated plant sterols from the corresponding hydrogenated compounds, using the bromine-system (separation of $3\beta,5\alpha$-stanols from the $\Delta^5,3\beta$-sterols) [91]; and of skin sterols [88] and milk sterols and their esters [117]. PFEIFER [137] has studied sterol separations on micro plates and their densitometric evaluation.

In particular, sulphuric acid, molybdophosphoric acid, antimony(III)-chloride and fluorescence indicators and others quoted in Table 58 can be used for detection. RICHTER [142] has obtained blue colours with Δ^5- and $\Delta^{5,7}$-3β-hydroxysterols in amounts down to 0.03 μg, using the naphthoquinonesulphonic acid-perchloric acid reagent. Reference may be made to BOLLIGER's chapter (K) for the vitamin D group; to section XI on bile acids for bile alcohols (neutral C_{27}-sterols); and to MANGOLD's chapter (M) for separation of lipids, including cholesterol and its esters.

Table 61. *Separation of sterols and their acetates on silica gel G-silver nitrate layers* [43]
Solvents A and B, see text; detection with 0.2% dibromofluorescein

Sterol	Abbreviation	R_{St}-value of sterol in A	R_{St}-value of sterol acetate in B
Agnosterol	$2\,FC_{30}F$	1.68	0.40
24-Dihydroagnosterol	$2\,FC_{30}$	1.61	
Lanosterol	$FC_{30}F$	1.70	0.78
Dihydro-β-sitosterol	C_{29}	1.14	1.30
5α-Cholestan-3β-ol	C_{27}	1.14	1.25
Cholesterol	FC_{27}	1.00 (St)	1.00 (St)
β-Sitosterol	FC_{29}	1.00	1.00
Δ7-Cholesten-3β-ol	FC_{27}	1.17	1.14
Δ7-Ergosten-3β-ol	FC_{28}	1.22	1.21
5-Dihydroergosterol	$FC_{28}F$	1.13	0.88
Stigmasterol	$FC_{29}F$	0.98	0.87
Vitamin D_2	$3\,FC_{28}F$	0.64	
Dihydrovitamin D_2	$3\,FC_{28}$	0.47	
Δ7,9(11),22-Ergostatrien-3β-ol	$2\,FC_{28}F$	0.83	
Δ9(11)-Dehydroergosterol	$3\,FC_{28}F$	0.69	
Brassicasterol	$FC_{28}F$	0.98 (?)	0.68 (?)

VI. Neutral C_{18}-C_{22}-Steroids
(Androgens, Gestagens, Corticosteroids)

The neutral derivatives of the oestrane, androstane and pregnane basic skeleton are the principal members of this large and important class (see section I). Considerable problems of separation arise here with the plasma- and urine corticosteroids and 17-ketosteroids, progesterone metabolites, steroid extracts from the suprarenal glands, gonads, placenta and tumours and the numerous synthetic steroids with androgenic, gestagenic or adrenal activity.

They extend from the feebly polar monoketones to the highly polar pentahydroxyl derivatives for which partly adsorption-, partly partition-chromatography is very well suited.

A solvent of polarity suitable for *adsorption TLC* is very easy to find on the basis of the data in Figs. 125 and 126 and the composition can be varied at will. Toluene, ethanol and occasionally dioxan, methyl ethyl ketone or methylene dichloride and similar solvents are often used besides those quoted in the two figures. Benefit may accrue from the addition of 0.2–1% water to the systems if tailing occurs due to overloading of the layer or poor solubility (polar corticosteroids).

More or less systematic investigations have been made on silica gel since 1958. Amongst others, ACHREM [4] and BARBIER [11] have worked with mixtures of cyclohexane-ethyl acetate; and BENNETT and HEFTMANN [14] with mixtures containing a small amount of water, such as chloroform-methanol-water (90 + 10 + 1; 94 + 6 + 0.5; 97 + 3 + 0.2) or chloroform-ethyl acetate-water (10 + 90 + 1). Generally silica gel G or silica gel without binder have been used and still are being used nowadays. SMITH and FOELL [153] have, however, found rice starch also to be a good binder but it has the disadvantage of not tolerating aggressive reagents. LISBOA [109, 110, 214–216] has studied systematically the behaviour on silica gel G of many androstane and pregnane derivatives in solvents containing alcohol or ethyl acetate; FEHÉR [53] has used these for practical purposes like urine steroid analyses. QUESENBERRY and UNGAR [138] and GERALI [63] have introduced additional modifications. TAKEUCHI [157] has carried out separations with the usual solvents on Wakogel B_5, a silica gel-gypsum of Japanese origin and of approximately the same quality as silica gel G.

The Czechoslovak teams (32, 79, 149, 150) have found loose alumina to be suitable with the usual solvents; plates with loose adsorbents are rather a disadvantage for routine analyses. Alumina-gypsum could be used here if silica gel was undesirable; the latter is, however, the first choice adsorbent. Separations are indeed possible on calcium sulphate [119] or even polyamide [57] but are scarcely to be recommended. It is stated that abnormal properties are shown on magnesium trisilicate with t.-butanol-hexane solvent [132]. A satisfactory way of modifying silica gel G is to mix it, if necessary, with kieselguhr (1 + 1); this accelerates the run [174].

It is not always simple to ascertain the solvent system with the optimum separation properties from amongst the wealth of possible systems. If the compounds under investigation are known, the decision can be reached with the help of the systematic procedure of BUSH [28], based on R_M and R_{Mg}-values [71]. Since this procedure is time-consuming and unsuitable for mixtures with unknown components, it is simpler to use a few solvent mixtures of similar polarity but widely varying composition or, possibly, continuous or multiple development with correspondingly more weakly polar solvents as explained in section III. After hydrocarbon-alcohol mixtures, it is best to choose those containing a ketone, ester or ether. The last two in particular often effect a sequence change which is useful for separation. For example, in chloroform-methanol, aldosterone, corticosterone and 11-dehydrocorticosterone migrate ahead of cortisol, cortisol and cortisone, respectively; in ethyl acetate-chloroform, the orders are reversed. This difference is striking in general with 11β-hydroxy- and 11-keto derivatives, where the last named are "more polar" in ethyl acetate. Such behaviour is already known in PC (cf. [129]). Incidentally, successful separation of very complex mixtures is likely neither with PC nor TLC, using a single system; two must be used in succession. Even when artificial mixtures can be just separated in one or two dimensions, it is of little help since the components in biological extracts are present in very variable concentrations. Subdivision into C_{19}- and C_{21}-steroids is thus frequently illusory; the polarities of the two groups overlap considerably. Advantage may be gained from using reagents for selective detection, e. g., blue tetrazolium for reducing corticosteroids or m-dinitrobenzene for the 17-ketosteroids. Tabels 58 and 59 contain information about these possibilities, likewise about general detection reagents which are more sensitive but less specific. Combinations can also be employed, e. g., phosphoric acid first and then molybdophosphoric acid [181]; or converting 17β-hydroxy-androstane derivatives, which cannot be selectively detected, by chromic acid oxidation into the 17-ketosteroids which can be [110].

Several practical examples of quantitative evaluation of chromatograms, best done by elution *before* the colour reaction and then subsequent spectrophotometry, are discussed later. MATTHEWS and coworkers [122] have carried out more precise studies of this procedure in comparison with PC and GC and obtained satisfactory results.

The statements in section IV about the relation between structure and chromatographic behaviour apply to the more weakly polar 17-ketosteroids, the polarity of which increases in the order:

3-keto,5α < 3-keto, 5β < 3β,5β < 3α,5α < Δ^5,3β < 3β,5α < 3α,5β < 11-keto, 3α,5α < 11-keto, 3α,5β < 11β-hydroxy, 3α, 5α < 11β-hydroxy, 3α,5β.

Admittedly only the axial and epimeric pairs can be reasonably easily separated from each other; further subdivision is difficult and more is said about this in connection with the urinary 17-ketosteroids. Even the $3\beta,5\alpha$-isomer (epiandrosterone) cannot be separated from the important $\varDelta^5,3\beta$-compound (dehydroepiandrosterone) using normal TLC. As mentioned in the preceding section, complex formation on silver nitrate-containing layers helps in this instance. Many synthetic "C_{19}"-steroids contain a 17α-alkyl as well as the 17β-hydroxyl group; the former reduces the polarity in the order: $-CH_3$, $-C_2H_5$, $-CH=CH_2$ equals $-C\equiv CH$, $-C_3H_7$.

Progesterone and 16-dehydroprogesterone cannot be directly separated. LISBOA [109] has achieved success after prior treatment with isonicotinic acid hydrazide which yields a dihydrazone with the former and a monohydrazone with the latter. Silver nitrate-silica gel would probably be used nowadays for a separation of this sort.

In more polar steroids, especially the corticosteroids containing several keto and hydroxyl groups, interactions between these groups become so significant that the sequences depend more strongly on the solvent and overlap each other. The sequences given in section IV apply only to steroids with isolated hydroxyl groups in various positions.

Partition chromatography of steroids on thin layers has been relatively little used up to now, although various authors [34, 37a, 65, 131, 173, 186, 229] have shown that it yields useful results just as in the long established PC procedures (cf. BUSH [28], NEHER [129]). Adsorption and partition happily complement each other in displaying different sequences to some extent; as mentioned earlier, partition is preferable for the more highly polar compounds or for complex mixtures. That full profit has not been drawn from these advantages of TLC techniques, can be regarded as an unvoiced recognition of the dominance of PC in this domain, especially where it has proved satisfactory. VAEDKE [173] and YAWATA [186] in particular have carried out partition TLC of corticosteroids on kieselguhr-gypsum which had been impregnated with formamide or glycol in acetone solution; the usual PC solvents like hexane, benzene and chloroform and their mixtures (Zaffaroni-type systems) served as mobile phase. YAWATA and GOLD [186] have been able also to apply the Bush-type system to TLC by spraying the stationary phase or spreading it in methanolic solution, a variation known in PC [129]. These systems can be used on cellulose layers just as well as on filter paper [131]. The spots are slightly smaller than those on kieselguhr. On the other hand, the spots in TLC with Zaffaroni-type systems are smaller than with aqueous systems. This procedure still saves time and material. GÖLDEL et al. [65] and SONANINI et al. [229] have successfully employed partition TLC for corticosteroids on silica gel-gypsum based on all variants of the Zaffaroni-type systems. CHANG [34] has used ethylene glycol as stationary phase for C_{19}-steroids and benzene-methylcyclohexane or mixtures with halides as mobile phase.

Practical Applications of Adsorption TLC

Several different teams [35, 36, 48, 53, 74, 141, 160, 232] have investigated the separation and determination of *17-ketosteroids in urine:* for the most part, fractionation only into groups was achieved. STÁRKA and co-workers [160] have separated the fat-free extract on alumina, using solvents like benzene-ethanol (94 + 6), methylene dichloride-ethyl acetate (93 + 7) in one-dimensional, or methylene dichloride in multiple development; they then sprayed only lightly with m-dinitrobenzene reagent for visualisation, eluted with ethanol and finally carried out the determination in vitro with m-dinitrobenzene (recovery about 80%).

HAMMAN and MARTIN [74] give precise directions. They chromatographed the extract of a urine aliquot which had been submitted to enzymatic hydrolysis or solvolysis, on alumina G; in the first direction they developed in a 16 cm run with benzene-methanol (65 + 35) and in the second direction with diethyl ether-ethyl acetate (50 + 50), chromatographing a standard mixture in each dimension in a parallel run on the side of the plate. This part was visualised by spraying with m-dinitrobenzene reagent and the steroids localised on the main part according to the positions of the standards on the reference part. They were then scraped off, eluted and determined colorimetrically using m-dinitrobenzene and with dehydroepiandrosterone (DHEA) as standard for comparison. The separation is represented schematically in Fig. 127; the spot DHEA contains still 3β-hydroxy-5α-androstan-17-one. The sensitivity limit was about 2—3 μg per steroid and the average recovery amounted to 95%. The separation was facilitated by separate determination of those 17-ketosteroids combined as glucuronides and those as sulphate (dehydroepiandrosterone and other 3β-hydroxysteroids occur preferentially as sulphates; 3α-hydroxysteroids, like 3α-hydroxy-5α and -5β-androstan-17-one, as glucuronides). Comparison with PC was satisfactory although optimum TLC separation certainly had not been achieved. Since diethyl ether and ethyl acetate bring about different sequences [131], the outcome of using a mixture of them is uncertain. Information concerning a TLC method for determining dehydroepiandrosterone is given in [232]. GC has recently been competing successfully with TLC in these determinations; it yields still better separations. TLC has then usually been used for preliminary purification.

The situation is similar in the determination of 17-ketosteroids in plasma [53]; the mixtures are then less complex but much less material is available. MATTHEWS et al. [122] have undertaken a more detailed study of the quantitative aspect of determining 17-ketosteroids. TLC has served mainly for preliminary purification in the determination of *testosterone in urine* by GC or by the isotope dilution method [90, 177, 178]. IBAYASHI et al. [90] have used the following sequence of operations: Extraction → 1. TLC (benzene-ethyl acetate, 50 + 50) → acetylation 2. TLC (benzene-ethyl acetate, 60 + 20) → GC. VERMEULEN and VERPLANCKE [177] have worked as follows: Extraction → alumina column

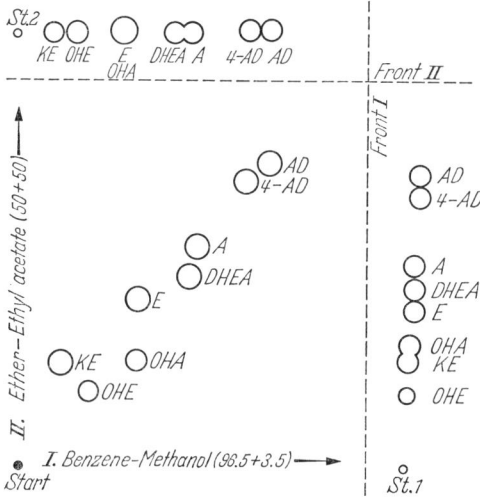

Fig. 127. Separation of a mixture of 17-ketosteroids in two-dimensional TLC on alumina [74]. *A* androsterone $(3\alpha, 5\alpha)$; *AD* 5α-androstane-3,17-dione; *4-AD* Androst-4-ene-3,17-dione; *DHEA* dehydroepiandrosterone $(3\beta, \Delta^5)$; *E* etiocholanolone $(3\alpha, 5\beta)$; *KE* 11-ketoetiocholanolone $(3\alpha, 5\beta)$; *OHA* 11β-hydroxyandrosterone $(3\alpha, 5\alpha)$; *OHE* 11β-hydroxyetiocholanolone $(3\alpha, 5\beta)$

→ 1. TLC (chloroform-ethyl acetate, 80 + 20) → CrO_3-oxidation → 2. TLC (benzene-chloroform-ethyl acetate, 60 + 20 + 20) → micro colour reactions and radiometry. BURGER et al. [27] have determined *testosterone in plasma* with the double isotope technique after having carried out prior purification with TLC. TLC has been used to some extent also for determining C_{19}-16-dehydrosteroids in urine, a determination which has recently become important [25, 68]. GOWER [68] has been able to separate the various epimers by multiple TLC on silica gel G, using benzene-methyl ethyl ketone (90 + 10); the expected sequence $3\beta,5\alpha - \Delta^5,3\beta - 3\alpha,5\beta - 3\alpha,5\alpha$ - 3-hydroxy-aromatic, was found. An easier separation can be accomplished by GC of the trimethylsilyl ethers [25]. Competition and coupling of TLC and GC is likewise encountered in the determination of *progesterone in plasma* or tissue extracts [39, 60, 156, 182], due in part to the better sensitivity of GC. *Pregnanediols* and *pregnanetriols*, present in larger concentrations in urine, have been on the other hand the objects of very satisfactory TLC analyses. Following WALDI's work [180, 182] on rapid determination of pregnanediol as an early test for pregnancy, STÁRKA [159], BANG [10] and LAU [107] have modified or refined (and complicated!) the method; the last named used GC as the final stage. Many TLC variants have been described recently [199, 200, 213, 220, 222, 225, 230]. Further details are given in ZÖLLNER's chapter (S). CHANG-SHEN [35], FEHÉR [53] and STÁRKA [158] have studied the pregnanetriols, a group important for estimating the function of the suprarenal cortex, and their determination in urine with the help of TLC. They did not achieve finally developed methods.

Corticosteroids from urine or tissue extracts, whether in unchanged or metabolised form, have proved to be a profitable class for the application of TLC [2, 3, 19, 31, 123, 133, 138, 190, 191, 194, 205, 206, 217] which competes here with PC. The instability of the corticosteroids must be remembered when using adsorption TLC [123, 138]. ADAMEC et al. [2, 3] have carried out separations using 1, distilled water (preliminary purification), 2. methylene dichloride-ethanol $(95 + 5)$ and 3. chloroform-ethanol $(93 + 7)$ in the same direction on the same silica gel G layer. They then localised the zones with a very thin coating of blue tetrazolium and analysed the eluates from these zones quantitatively according to the PORTER-SILBER method. The zones were not homogeneous, containing particular groups of steroids. NISHIKAZE et al. [133] have applied the two-dimensional technique but succeeded only in fractionating into groups; more could hardly be expected with these complex mixtures.

CAVINA and VICARI [31] have worked on 37 cm long silica gel G layers; they obtained a good separation of metabolised corticosteroids within 2.5 hours, especially with chloroform-methanol $(90 + 10)$. Addition of a little water would presumably eliminate the slight tailing. Solvents containing ethyl acetate led to changes in the sequence. Quantitative determination was carried out UV-spectrophotometrically or through formazan formation with blue tetrazolium. As expected, in the presence of tetrahydrometabolites fractionation was possible only into groups. These can be separated completely by elution and chromatographing again in other solvents, just as in PC. The latter, in its modern variations, can certainly achieve as much here in the same time (cf. [129]), Tetrahydrocorticosterone could not be separated from tetrahydro-S in the first mentioned solvent but this was successfully accomplished in the second: against this, tetrahydrocorticosterone and cortisone migrated together in this second solvent. The angle of inclination seems to be critically important in TLC on long plates [31]. MCCARTHY et al. [123] and QUESENBERRY and UNGAR [138] have recently published their TLC data on the very unstable 18-hydroxycorticosteroids and noted in part the same phenomena which occur in PC [129]. There were certain discrepancies with 18-hydroxy-11-deoxycorticosterone [123]. See [190] and [217] for TLC analyses of corticosteroids in the adrenal cortex. GERDES and STAIB [205] have determined urinary cortisol fluorometrically after separation by TLC.

AUDRIN et al. [7], BENRAAD and KLOPPENBORG [17, 191], BRUINVELS [26], NISHIKAZE and STAUDINGER [134] and GERDES and STAIB [206] have studied theoretically and then practically the application of TLC to the determination of *urinary aldosterone*. The method which was best investigated appears to be that of AUDRIN et al. and GERDES and STAIB; it is, however, neither quicker nor simpler nor appreciably more sensitive than the two stage PC methods [129a], so that for the moment any advantage is scarcely noticeable.

Some additional examples of application in the chemistry of natural products and chemical synthesis may be cited: Removal of pigments from extracts of organs [147]; separation of faecal steroids [144] and of glucocorticosteroids in bile and duodenal juice [146]. TLC has been most useful as a rapid analytical method in microbiological transformations of steroids [124]; it can also be applied conveniently to recovery and separation of the steroids from scintillators after radiometric work [143]. In chemical synthesis, TLC is of particular service for controlling the reactions, homogeneity and stability [194] and for separations on a semi-preparative scale: TSCHESCHE et al. [171] have quoted an instructive example of use in reaction kinetics. 17α-Ethynylsteroids can easily be separated from the saturated analogues on silver nitrate-silica gel layers [50]; differentiation of 19-norsteroids [64] and testing their stability [61], separation and purity control of synthetic corticosteroids [37a, 71a]. Distinction of syn- and antioximes [66]; reactions of $\varDelta^5,3\beta$-hydroxysteroids [63], identification in preparations of sesame oil [101], quantitative analysis of pharmacological active pregnene derivatives [21]. See MATTHEWS et al. [122] concerning the quantitative analysis of steroids by means of TLC.

VII. Cardiac Glycosides and Aglycones
(Cardenolides and Bufadienolides)

Certain plant groups and toad secretions contain poisonous mixtures of glycosides which act on the muscles of the heart; their structures are based on the two aglycone skeletons below:

        Cardenolides   Bufadienolides

The hydroxyl group on the C_3 of the cardenolides is combined with one or more sugars via an ether link; this strongly influences the polarity of these compounds. Table 62 shows this dependence of polarity on structural features for digitalis glycosides, as compiled by WALDI [181]. Apart from *Digitalis purpurea* and *D. lanata*, from which about 90 glycosides are already known, others which contain substances with activity of this sort are: *Strophanthus kombé, St. gratus, Urginea maritima* L. (= syn. *Scilla maritima* L.), *Convallaria majalis* L., *Adonis vernalis* L., *Nerium oleander*, etc. All possible chromatographic methods were used very early on and with good success for separating the very complex mixtures in which these active substances occur (cf. in [129]). Various research teams applied their experience from column and paper chromatography straight away to TLC (cf., e. g., DUNCAN [47]). After TSCHESCHE and co-workers (e. g., [168]) had been able to accomplish at least partial separation of a series of aglycones and their acetates with the help of even simple solvents like ethyl and butyl acetates, the successful application of TLC alongside PC has been reported in many publications [e. g., 95a, 108a]. LEWBART et al. [108a] have used the same solvent in continuous development for separating sarmentogenin- and periplogenin glycosides.

Partition chromatography has generally proved rather more suitable for this large number of fairly polar substances; thus better separations are often found here with paper chromatography. Partition TLC can be carried out on, e. g., silica gel G with benzene or chloroform and an addition of 5–50% of methanol, propanol, isopropanol, butanol, acetone, butanone or ethyl acetate [47]. MANZETTI and REICHSTEIN [118] have used butanone or butanone-benzene (90 + 10), saturated with water on silica gel layers impregnated with acetone-water (70 + 30) for chromatographing more water-soluble materials; cyclohexane-butanone (50 + 50) and (40 + 60) sufficed for the less polar digitoxigenin- and oleandrigenin-glycosides [95a]. DUNCAN [47] has observed that compounds containing hydroxyl groups migrate faster in alcohol-containing systems than in those of the same dielectric constant but without alcohol. The R_f-value

Table 62
Arrangement of digitalis glycosides according to their polarities
D = Digitoxose Ac = Acetyl G = β-Glucose OH = Hydroxyl group

Increasing polarity →

Series:	A	E	B	C	D
Linked to C_3:					
HO- Aglycones	1 Digitoxigenin	2 Gitaloxigenin	5 Gitoxigenin	6 Digoxigenin	9 Diginatigenin
D-D-D-O- Secondary glycosides	3 Digitoxin	4 Gitaloxin	7 Gitoxin	8 Digoxin	12 Diginatin
Ac↓ G-D-D-D-O Primary glycosides of *Digitalis lanata*	10 Lanatoside A	11 Lanatoside E	13 Lanatoside B	15 Lanatoside C	18 Lanatoside D
G-D-D-D-O- Primary glycosides of *Digitalis purpurea*	14 Deacetyl-lanatoside A (=Purpureaglycoside A)	16 [Deacetyl-lanatoside E]	17 Deacetyl-lanatoside B (=Purpureaglycoside B)	19 [Deacetyl-lanatoside C]	20 [Deacetyl-lanatoside D]

of a secondary alcohol decreased in the following sequence of solvents of equal dielectric constant: chloroform-isopropanol, chloroform-methanol, chloroform-acetone.

GÖRLICH [67] has employed ethyl acetate-methanol (95 + 5), methyl ethyl ketone-toluene-methanol-acetic acid-water (80 + 10 + 5 + 2 + 6) and ethyl acetate-chloroform (90 + 10) on silica gel G for *Scilla* glycosides; and STEIDLE [161, 162] has used butanone, saturated with water, on silica gel G, separating into zones, re-chromatographing with similar

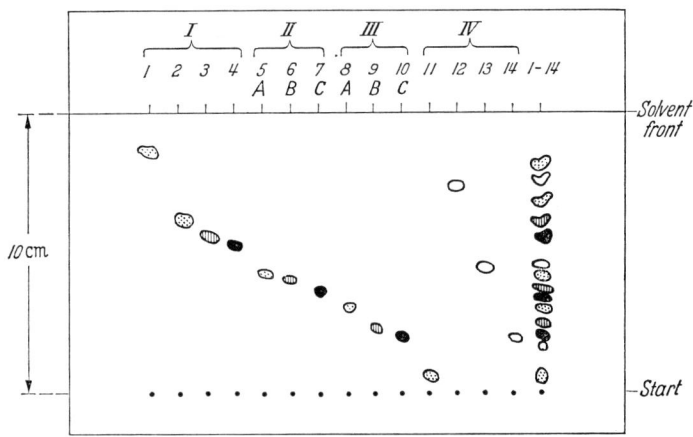

Fig. 128. Separation of cardiac glycosides according to STAHL et al [154]. *I* sec. glycosides; *II* digilanides; *III* deacetyl-diginalides; *IV* digitaloids. *1* acetyldigitoxin; *2* digitoxin; *3* gitoxin; *4* digoxin; *5* digilanide A; *6* digilanide B; *7* digilanide C; *8* deacetyl-diginalide A; *9* deacetyl-digilanide B; *10* deacetyl-digilanide C; *11* k-strophanthoside; *12* cymarin; *13* proscillaridin A; *14* scillaren A; *1—14* mixture. Methylene dichloride-methanol-formamide (80 + 19 + 1) on silica gel G. Detected with reagent No. 39. Colours in UV light (365 nm): ▫ = light yellow; ▤ = brown-yellow; ▦ = light blue; ▪ = violet-blue

mixtures and concluding with spectrophotometry. STAHL and KALTENBACH [154] succeeded already in 1961 in separating complex mixtures of digitalis glycosides by using methylene dichloride-methanol-formamide (80 + 19 + 1) on silica gel (see Fig. 128 and SONANINI [228]). FAUCONNET and WALDESBÜHL [52] turned this experience to account and modified the system to methylene dichloride-methanol-water (87 + 12 + 1; 80 + 19 + 1 and others). They fractionated into groups of acetylated secondary glycosides, aglycones, secondary glycosides, primary glycosides and deacetyl-primary glycosides, with Rf-values decreasing in that order; separation within each group was incomplete. It is evident that when solvent mixtures of this composition are used, marked solvent gradients are formed on silica gel. These systems are quite sensitive to layer overloading in some cases.

Systems like the upper layer of ethyl acetate-pyridine-water (50 + 10 + 40) are also suitable for separation of cardiac glycosides on silica gel, as STEINEGGER and VAN DER WALT [163] have been able to show with the constituents of the white and red sea onion *(Urginea maritima var.)*. REICHELT and PITRA [140] have used loose layers of silica gel which were deactivated with 25% water or 43% acetic acid-water (1 + 1); benzene-ethanol (60 + 20), saturated with water, served as solvent for separation of various cardenolides (pure substances).

SJÖHOLM [152] has made a detailed report on the TLC of 28 digitalis glycosides and aglycones, using two-dimensional technique on silica gel G. Ethyl acetate-methanol-water (80 + 5 + 5) was used in the first direction and chloroform-pyridine (60 + 10) in the second. Naturally, only separation into groups could be achieved in this way. R_{St}-values in both solvents were quoted, likewise the colour reactions with the reagent anisaldehyde-perchloric acid. Butanone-chloroform-formamide (50 + 20 + 10) was used for one-dimensional TLC of the glucose-containing digitalis glycosides. According to WALDI [181], multiple development with less polar solvents is suitable in the TLC of rather larger amounts; he has proposed the following procedure: In a first run with cyclohexane-acetone-acetic acid (49 + 49 + 2), the aglycones of the digitalis glycosides migrate to near the middle (digitoxigenin hRf 68, gitoxigenin 51, digoxigenin 48, digitoxin 42); the secondary glycosides are separated behind these and migrate in the second run to hRf-values between 40 and 70, whereas the primary glycosides have values still below 20; these last named are then separated completely in the third and fourth runs. Other examples of application to digitalis glycosides are to be found in [208, 210, 234] and to *Strophanthus* glycosides in [195].

Information about detection is given in Tables 58 and 59; chloramine-trichloroacetic acid (Rgt. No. 39) and anisaldehyde-perchloric acid [152] have proved especially suitable. FAUCONNET and WALDESBÜHL preferred phosphoric acid-bromine (Rgt. No. 209) and sulphuric acid-hypochlorite (Rgt. No. 242) which also enable the individual groups to be differentiated. Phosphoric acid alone (Rgt. No. 208) or sulphuric acid-acetic anhydride (Rgt. No. 241 D) are also satisfactory. REICHSTEIN'S team has used chiefly a modified Kedde reagent (No. 80) for selective and sensitive detection of cardenolides (blue-violet). Tetranitrobiphenyl yields blue spots; trinitrobenzoic acid, orange-red [219]. The differentiation in Table 63 has been observed in TLC (variations can arise, depending on the adsorbent).

Little work has been done on the TLC of bufadienolides. ZELNIK and ZITI [187, 188] have separated on silica gel-gypsum, using customary solvents like ethyl acetate, ethyl acetate-cyclohexane (40 + 80) or (80 + 20), ethyl acetate-acetone (90 + 10) ethyl acetate-methanol (90

+ 10) or ethyl acetate, saturated with water; visualisation was with antimony(III)-chloride.

See [129] for good ways of using PC with the cardenolides and bufadienolides. The one or other method yields satisfactory results, depending on the separation problem; PC is however usually preferable here.

Table 63

Digitalis-glycoside	Colour of fluo-rescence with chloramine-tri-chloroacetic acid; in UV	Colour of fluo-rescence with phosphoric acid, in daylight (in UV) [123]	Colour with anisaldehyde-perchloric acid, in daylight	Colour of fluo-rescence with phosphoric acid-bromine; in UV	Colour of fluo-rescence with sulphuric acid-hypochlorite; in UV
A-series	brown-yellow	blue-black	blue to blue-black	orange	pink-brown
B-series	light blue	brown-violet (blue)	red-brown	blue	pink-brown
C-series	violet-blue	grey-violet	blue-violet	green-blue	blue-green
D-series		(blue)	red-blue	light blue	grey-blue
E-series		(blue)		grey-green to grey-blue	pink-brown

VIII. Saponins and Sapogenins

The spirostane structure is the basic skeleton of this class of plant constituents which occurs partly together with the cardiac glycosides. Additional isomers are possible here, depending on the α- or β-position of the methyl group at 25 and the tetrahydropyran ring at C_{22}.

In the *saponins*, a sugar residue is combined via an ether linkage with the 3β-hydroxyl group; this renders the molecule strongly polar. As has been previously mentioned, such compounds are best differentiated through partition chromatography (PC or TLC) in which cellulose, kieselguhr or silica gel can function as carrier. CARRERAS-MATAS [30] has chosen silica gel, impregnated with formamide whereas KAWASAKI and MIYAHARA [98] have investigated numerous saponins on silica gel G, using aqueous solvents like chloroform-methanol-water (65 + 35 + 10) (lower phase) or butanol, saturated with water. The Rf-values depend on the number of sugar residues and the number of hydroxyl groups in the sugar and aglycone moieties; generally it is possible to fractionate only into groups. Fully esterified (acetylated) or etherified (methylated) derivatives can be submitted to the usual adsorption TLC since they are only feebly polar. Chloroform-ethanol (100 + 2) or benzene-acetone (80 + 20), for example, can be used as solvents.

Sapogenins, especially monohydroxysapogenins, are much more adaptable to separation, whether by adsorption or partition processes. One of the most important sapogenins is diosgenin, the principal starting material in synthetic steroid chemistry. Many of the solvents customary for steroids of medium polarity (cf. Figs. 125 and 126) have been used for TLC of a large number of sapogenins and their acetates on silica gel [22, 120, 145, 153, 166, 169, 170a] or alumina [32, 79]; a reasonable separation is accomplished in some cases only by combining several systems. BENNETT and HEFTMANN [15] and SCHREIBER and co-workers [148] have made a more precise investigation of the possibility of separating important critical pairs or groups, especially the C_{25}-isomers. They recommend chromatographing an unknown mixture first in its unchanged form, using relatively polar solvents like methylene dichloride-methanol-formamide (93 + 6 + 1), cyclohexane-ethyl acetate (50 + 50) or chloroform-methanol (94 + 6), thus fractionating into groups; the mixture is then acetylated and chromatographed again in a weakly polar system like chloroform-toluene (90 + 10) on silica gel G. Separation, even though only moderately good, of the free C_{25} isomers has been possible with this last system, used on silica gel-kieselguhr (1 + 1). The acetates of the $5\alpha, 5\beta$ and Δ^5-sapogenins could be separated in the same system (on silica gel G) (the sequence found was the reverse of that usually encountered with the free epimers). Nowadays it is possible to separate even the free 5α- and Δ^5-sapogenins (e. g., tigogenin from diosgenin) easily on silver nitrate-containing silica gel; this had been hitherto possible only in the partition system hexane-toluene-ethanol-water (60 + 30 + 3 + 27) and then with difficulty. In the last named system, which is capable of dealing with only small amounts of substances (<0.2 µg), the following sequence of Rf-values has been found: smilagenin $(5\beta, 25\alpha)$ > tigogenin $(5\alpha, 25\alpha)$ > diosgenin $(\Delta^5, 25\alpha)$ and sarsasapogenin $(5\beta, 25\beta)$ > neotigogenin $(5\alpha, 25\beta)$. In contrast, the series found for the acetates in chloroform-toluene (90 + 10) on silica gel G was: tigogenin $(5\alpha, 25\alpha)$ > diosgenin $(\Delta^5, 25\alpha)$ > smilagenin $(5\beta, 25\alpha)$ > neotigogenin $(5\alpha, 25\beta)$ > sarsasapogenin $(5\beta, 25\beta)$. The 25β-sapogenins, whether free or acetylated, thus behave in more polar fashion than the 25α-isomers; on the other hand, the 5β-sapogenins are less polar than the 5α-epimers in the free form and more polar in the acetylated form.

Tables 58 and 59 contain information about detection. The most popular reagents have been aromatic aldehydes-acids (Nos. 11, 48, 133, 219, 260, 262), chlorosulphonic acid-acetic acid (Rgt. No. 45), sulphuric acid (No. 241) and antimony(III)-chloride (No. 15).

IX. Aminosteroids, Steroid Alkaloids and Glycosides

Compounds like those with the structures given below are considered here:

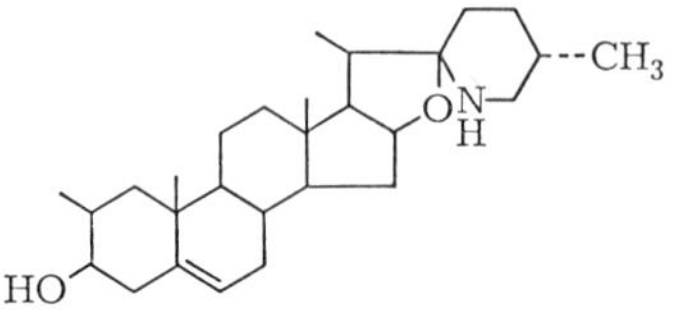

$C_{27}N$, Δ^5-Tomatiden-3β-ol
(5α-H: Tomatidine)

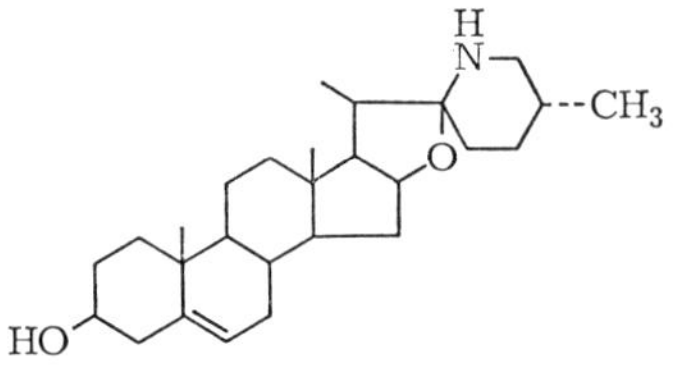

$C_{27}N$, Solasodine
(5α-H: Soladulcidine)

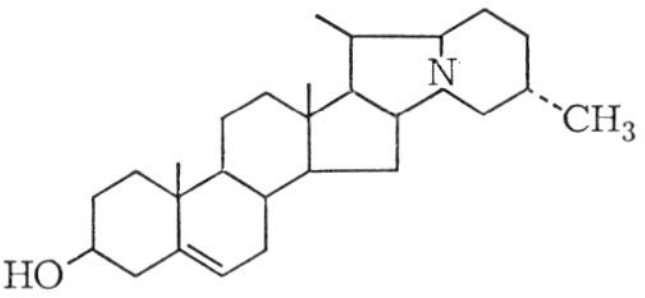

$C_{27}N$, Solanidine

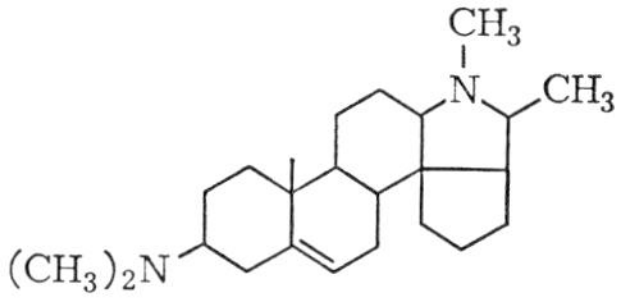

Conessine

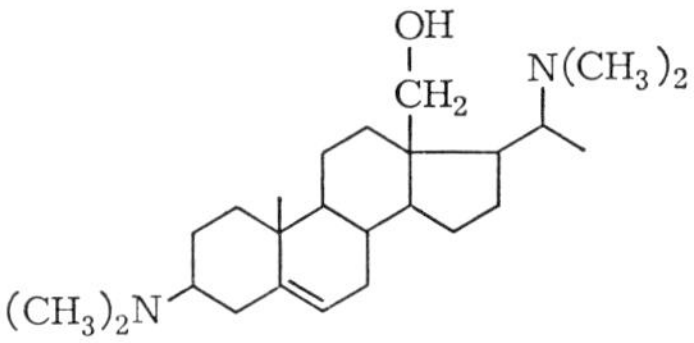

Tetramethylholarrhimine

The *Solanum* glycosides, for example, have the following compositions:

α-Solanine: Solanidine + galactose + glucose + rhamnose
α-Chaconine: Solanidine + glucose + rhamnose + rhamnose
Solasonine: Solasodine + galactose + glucose + rhamnose
Solamargine: Solasodine + glucose + rhamnose + rhamnose
α-Solamarine: Δ^5-Tomatiden-3β-ol + galactose + glucose + rhamnose
β-Solamarine: Δ^5-Tomatiden-3β-ol + glucose + rhamnose + rhamnose

The C_{25}-isomers of the *Solanum* alkaloids are easier to separate than the corresponding isomers of the sapogenins.

As far as the glycosides themselves are concerned, partition systems come into consideration, as with the saponins; bases or acids may sometimes be added. Successful fractionation into groups has been achieved on silica gel, using, for example, acetic acid-95% ethanol

(20 + 60) [136]; butanol, saturated with water [98]; and particularly well, according to the number of sugar residues, using the upper phase of ethyl acetate-pyridine-water (30 + 10 + 30) or lower phase of chloroform-ethanol-1% ammonium hydroxide (40 + 40 + 20) (BOLL [24]). Chloroform-acetone (60 + 20) or chloroform-methanol (95 + 5) have been suggested for the aglycones. SCHREIBER et al. [148] in particular have investigated the separation of aglycones using the long familiar cyclohexane-ethyl acetate mixture (50 + 50) on silica gel G; separation was only into groups and the $3\beta, \varDelta^5$- and $3\beta, 5\alpha$-pairs once again migrated together. They could be distinguished either by selective oxidation of the more sensitive $3\beta, 5\alpha$-isomers to 3-keto-5α-steroids, using chromic acid in pyridine; or by chromatography on silver nitrate-containing silica gel layers. The latter procedure, however, failed with $\varDelta^5$-tomatiden-3β-ol/ tomatidine and solasodine/soladulcidine.

The isomers with the 6-ring nitrogen in α-position on C_{22} are more polar than those with the β-configuration.

LÁBLER and ČERNÝ [105, 106] have worked in detail on the *Kurchi*- or *Holarrhena* alkaloids, including conessine which contains a pyrrolidine ring, and on their derivatives and various aminopregnanes and -cholestanes. Benzene or diethyl ether, saturated with concentrated aqueous ammonium hydroxide, were satisfactory solvents used on silica gel containing gypsum. The systematic study yielded informative data about the relation of structure and chromatographic behaviour; this has been already discussed in section IV.

In other publications, the TLC of iminocholestanes [1] and steroid oximes [66, 102] has been treated; the customary solvents, if necessary with added acetic acid, were used.

Detection of the nitrogen-containing steroids presents no problem. The presence of basic groups enables reagents for visualisation of alkaloids to be used, like the Dragendorff reagent (No. 96), iodoplatinate reagent (No. 146) or cerium(IV)-sulphate (No. 46), in addition to the mainly acid reagents in Tables 58 and 59.

X. Phenolic Steroids (Oestrogens)

LISBOA and DICZFALUSY [112, 113] have carried out the most thorough systematic investigation of the separation and characterisation of phenolic steroids, (24 in all) by means of TLC on silica gel. Various derivatives and microreactions involved in their preparation were also included in the study, as well as a couple of dozen colour reactions of 32 of these compounds.

DIAMANTSTEIN and LÖRCHER [45a], TAKEUCHI [167] and STÁRKA [157] have reported other systematic studies. All this work has, however, been on chromatograms and separations of artificial mixtures, which so far has not been able to provide much information about the suitability of the methods for extracts of biological material. Customary solvent mixtures, as seen in Figs. 126 and 127, have been used for the runs, which were mostly based on adsorption TLC.

Benzene-ethanol (90 + 10), cyclohexane-ethyl acetate-ethanol (45 + 45 + 10) and methylene dichloride-ethanol (82.5 + 7.5), for example, are suitable for the separation on silica gel of mixtures of oestrogens, oestrone, oestradiol-17β and oestriol (16α,17β); these compounds have widely differing polarities and the mixture has been often studied; mixtures of carbon tetrachloride with alcohols, ethers or esters are suitable on alumina [157]. LUISI et al. [115] claim rather better separation in the horizontal BN-chamber when using cyclohexane-ethyl acetate (35 + 65). The TLC of the four epimeric oestriols (16ξ,17ξ) is more difficult. The more weakly polar 16,17-*cis*-glycols (16-epi-oestriol and 17-epi-oestriol) can be separated from the more strongly polar *trans*-compounds (oestriol and 16,17-epi-oestriol) but HERTELENDY and COMMON [80] state that separation of all four from one another necessitates methylation of the phenolic hydroxyl groups and the use of cyclohexane-ethyl acetate-ethanol (45 + 50 + 5) as solvent. Possibly less polar solvent systems containing ethyl acetate and the use of multiple development would accomplish even better separation. Little is known so far about the use of polyamide layers [193a]. PC possibilities and interactions are described in [129]. As already mentioned, LISBOA and DICZFALUSY [112] have investigated a large number of other oestrogen derivatives, like 2-hydroxy-, 2-methoxy-, 6-,7- and 16-keto- and 6α- and 6β-hydroxy-metabolites, many of which may occur in small amounts in biological material. They worked out a separation scheme based on initial TLC with cyclohexane-ethyl acetate-ethanol (45 + 45 + 10) under standardised conditions. A first group of 5 polar oestrogens (Rf up to about 0.37) was eluted and resolved into its components by two-dimensional TLC in hexane-ethyl acetate-ethanol (15 + 80 + 5); the 19 less polar oestrogens from the first TLC stage (Rf between 0.37 and 0.80) were separated into 8 subgroups through two-dimensional TLC in cyclohexane-ethyl acetate (50 + 50); these subgroups could be differentiated in various other systems and also after chemical treatment (reduction by metal hydrides or formation of hydrazones) and using several colour-detection reagents [113]. These and other reagents for detection are given in Tables 58 and 59.

STRUCK [165] and JACOBSOHN [92, 93] have worked on the quantitative determination of oestrogens in TLC, using spectrophotometric and photogram-densitometric methods respectively.

Numerous investigations have been carried out on the application of TLC to determination of *urinary oestrogens*, for years the object of study by the most varied chromatographic techniques, even GC. SIEGEL and DORFMAN [151] replaced the column chromatography in the established method of BROWN by TLC, using an added internal tritium standard. WOTIZ and CHATTORAJ [184] have subjected the urine extract to a preliminary separation on acid-washed silica gel, using benzene-ethyl acetate (50 + 50); four zones were obtained: 1) oestrone + 2-methoxyoestrone; 2) oestradiol + ring D-α-ketols + 17-ketosteroids; 3) oestriols with 16, 17-*cis*-glycols; 4) oestriols with 16, 17-*trans*-glycols. After having chromatographed group 2) again with petrol ether-methylene dichloride-ethanol (50 + 45 + 5) in order to separate out the 17-ketosteroids, all groups were acetylated, again purified on silica gel using petrol ether-methanol (90 + 10) and finally determined gas chromatographically.

JUNG and co-workers [95] have quoted a shortened procedure with only a single TLC stage on silica gel, using benzene-ether (40 + 60) and Kober-reaction on the eluates.

Along with the manifold possibilities for control of reactions and purity in chemical synthesis, the separation of oestrogens from equol [81] and detection of oestrogens in sesame oil preparations [101] may be quoted as further examples of application of TLC.

XI. Bile Acids and Conjugates, Steroid Carboxylic Acids and Steroid Conjugates

It is evident that chromatographic methods of all types were playing a large part already early in the study of the vast number of naturally occurring and synthetic bile acid derivatives; these possess largely the cholane skeleton (C_{24}, *cis* A/B, cf. section I):

	R_1	R_2
Lithocholic acid	H	H
Deoxycholic acid	H	OH
Chenodeoxycholic acid	OH	H
Cholic acid	OH	OH

Along with PC, TLC has now become an indispensable tool. It would appear, however, that the importance of GC will still increase [227].

The bile acids, which occur in the free state or as the more polar taurine- and glycine-conjugates, possess oxygen-containing functional groups in 3-, 6-, 7-, 12-, 16-, or 23-positions (α- or β-isomers) in addition to the C_{24}-carboxyl group. Many isomeric mono-, di- and trihydroxy-cholanic acids and their dehydrogenation products can consequently be present and they can be chromatographed in the free, acetylated or, in particular, esterified form. All conceivable solvent systems of suitable polarity can be used; these must contain at least 1—10% acid (acetic acid), possibly even an additional few % water or methanol, for chromatography of the free bile acids. The acetic acid may be added to the silica gel when the layer is being prepared [5]. TLC of the *methyl esters* is often advantageous. The results of many systematic investigations have been published since 1960, when GÄNSHIRT, KOSS and MORIANZ [62] carried out TLC on silica gel with the system toluene-acetic acid-water (50 + 50 + 10) for the free bile acids and with butanol-acetic acid-water (100 + 10 + 10) for the conjugates; many most useful solvent systems have emerged from these publications (HOFMANN [83, 84, 85], HARA and TAKEUCHI [75], HAMILTON [73], USUI [172], ENEROTH [48]). HOFMANN [85] has recently brought out an excellent review of the TLC of bile acids with tables containing numerous Rf- and R_{St}-values.

Silica gel-gypsum (silica gel G) is the adsorbent which comes first into consideration. HOFMANN [86], however, claims distinctly better performances for another silica adsorbent, Anasil (Firm 7) in the separation of weakly polar acids using ether-heptane mixtures and occasionally in separations of esters of 5β- and 5α-cholanic acids, especially using acetone-di-n-butyl ether (30 + 70; 15 + 85). Alumina, magnesium silicate, hydroxylapatite or kieselguhr can be used but generally furnish no advantages [86, 103, 149]. Kieselguhr has the property of retaining acid components relatively strongly and consequently contributing to the separation of neutral constituents. Modifications on micro plates [84] or aluminium sheets [103] have been described.

Table 64 provides information for selection of suitable solvent mixtures. The following rough rules may be given, without knowledge of the problem: Methyl esters of the bile acids of various polarities can be chromatographed satisfactorily in systems 1—9 and 15 (see reference substances); according to HOFMANN [86], benzene-acetone mixtures, 4, 5 and 7 on silica gel G or 6 and 8 on Anasil B, are especially suitable for such separations. Systems 9—14 are the best for monohydroxy-, monohydroxyketo-, diketohydroxy- and dihydroxy-acids. No. 11 has the virtue of effecting especially good separation of deoxycholic and chenodeoxycholic acids [73], which is important in biological samples. The selectivity for other bile acids is less good, however, than that of the other solvent mixtures in the table. HOFMANN [83] has found the

Table 64. *Solvents for bile acids and derivatives in the order of increasing polarity* [49, 62, 73, 75, 86, 172] (see Text)

The R_f-values of the reference substances quoted are between 0.2 and 0.7

No.	Solvent mixture	Reference substance	Adsorbent
1.	Heptane-diethyl ether (92 + 8)	Methyl oleate	Anasil B
2.	Heptane-diethyl ether (70 + 30)	Methyl oleate	Silica gel G
3.	Benzene-diethyl ether (80 + 20)	Methyl lithocholate	Silica gel G
4.	Benzene-acetone (92 + 8)	Methyl lithocholate	Silica gel G
5.	Benzene-acetone (85 + 15)	Methyl lithocholate	Silica gel G
6.	Di-n-butyl ether-acetone (85 + 15)	Methyl lithocholate	Anasil B
7.	Benzene-acetone (70 + 30)	Methyl deoxycholate	Silica gel G
8.	Di-n-butyl ether-acetone (70 + 30)	Methyl deoxycholate	Anasil B
9.	Benzene-ethyl acetate (20 + 80)	Methyl deoxycholate	Silica gel G
10.	Isooctane-ethyl acetate-acetic acid (50 + 50 + 0.5 to 2.0)	Lithocholic acid	Silica gel G
11.	Isooctane-diisopropyl ether-acetic acid (50 + 25 + 25)	Deoxycholic acid	Silica gel G
12.	Benzene-dioxan-acetic acid (75 + 20 + 2)	Deoxycholic acid	Silica gel G
13.	Benzene-CCl_4-diisopropyl ether-isoamyl acetate-n-propanol-acetic acid (10 + 20 + 30 + 40 + 10 + 5)	Deoxycholic acid	Silica gel G
14.	Benzene-acetic acid (80 + 20)	Deoxycholic acid	Silica gel G
15.	Chloroform-acetone-methanol (70 + 25 + 5)	Methyl cholate	Silica gel G
16.	Cyclohexane-ethyl acetate-acetic acid (7 + 23 + 3)	Cholic acid	Silica gel G
17.	Isoamyl acetate-n-propanol-propionic acid-water (20 + 10 + 15 + 5)	Cholic acid	Silica gel G
18.	Ethyl acetate-methanol-acetic acid (70 + 20 + 10)	Glycocholic acid	Silica gel G
19.	Butanol-acetic acid-water (100 + 10 + 10, upper phase)	Glycocholic acid	Silica gel G

6-component mixture, No. 13, to be usefully versatile. Solvents 13 to 17 are good for trihydroxycholanic acids. The last named separates cholic and hyocholic acids as well as α- and β-muricholic acids. Mixtures like Nos. 17, 18 and 19 serve for fractionating conjugates into groups [58, 62, 172].

Acetates of the bile acids can be separated in systems similar to those employed for the methyl esters, e. g., in Nos. 1—5, or also in mixtures of ethyl acetate with cyclohexane or heptane (for R_f- and R_{St}-values, see HOFMANN [85, 86]). If the necessity arises, additional advantage may be gained from multiple and two-dimensional runs with appropriate solvents.

The chromatographic sequence naturally changes somewhat from system to system but the following approximate rules hold:

The 5β-cholanate esters without a hydroxyl group in the 3-position behave in a less polar fashion than the esters of the 5α-isomers and are therefore more easily separated on anasil B than on silica gel G. If a 3-hydroxyl group is present, the sequence accords with the usual rule that equatorial substituents are more polar than axial ($3\alpha, 3\beta > 3\alpha, 5\alpha$ or $3\beta, 5\alpha > 3\beta, 5\beta$). The polarity sequence of the hydroxyl groups in positions 3, 7 and 12 of the cholanic acids, corresponds: $3\alpha(e) > 3\beta(a) > 7\beta(e) > 12\beta(e) > 7\alpha(a) > 12\alpha(a) >$ 3-keto- > 7-keto- > 12-keto- (cf. also section IV).

With the conjugates, the sequence of Rf-values is: taurocholic acid-taurodeoxycholic acid + taurochenodeoxycholic acid-taurolithocholic acid, then the glycine conjugates in the same order and finally the free bile acids which migrate well ahead [58]. Mild alkaline hydrolysis (0.5 M) is recommended for removing cephalins and lecithins, which migrate similarly; the conjugates are not thereby attacked.

There is still no perfect method for quantitative determination of the bile acids. GÄNSHIRT, KOSS and MORIANZ [62] have used two systems together, one for separation of the free acids and one for the conjugates; the zones were localised by spraying with water, eluted with 65% sulphuric acid and determined spectrophotometrically. FORTH et al. [201] have extended this method. HARA et al. [77] and SEMENUK and BEHER [226] have tried densitometric evaluation; SPRITZ [153a], titrimetric; and IWATA and YAMASAKI [212], enzymatic. FROSCH and WAGENER [59] have made use of the differences in absorption of partially separated mixtures to estimate the components.

Bile acids can be detected without particular difficulty (see Tables 58 and 59). Sulphuric and perchloric acids and antimony(III)-chloride have been the most popular reagents, best used on preheated layers (150° C).

KRITCHEWSKI et al. [104] recommend an anisaldehyde reagent (No. 11) which yields blue to pink spots (1 μg).

Molybdophosphoric acid (Rgt. No. 168) is highly suitable for visualising bile acids containing hydroxyl groups, especially when the plates are preheated to about 150° C; for this test, ketoacids have to be submitted to prior reduction with a spray of 5% sodium borohydride in 80% methanol [172]; they can be visualised directly with dinitrophenylhydrazine (Rgt. No. 82) though with inferior sensitivity.

Iodine is suitable for non-destructive detection.

ANTHONY and BEHER [5] have suggested various modifications, quoting, however, no details of sensitivity.

Some further examples of the application of TLC may be cited: a procedure for two dimensional separation [179]; analysis of bile from various species [76]; fractionation of bile lipids into groups [128]; metabolism [73] and analysis of human faecal bile acids [153a]; quantitative determination of free [203] and bile acid conjugates [202] in serum.

1. Bile Alcohols

These are neutral compounds, based on $3\alpha,7\alpha,12\alpha$-trihydroxy- or 3,7,12-triketocholane, the terminal C_{24}-methyl group of which has been substituted by hydroxyl or hydroxyalkyl groups. They occur chiefly as C_{22}-compounds along with the bile acids and have been classified here for that reason. According to the systematic work of KAZUNO and HOSHITA [100], they can be subjected to TLC on silica gel, using normal solvents as befits their structure; these include benzene-ethyl acetate (60 + 40; 40 + 60), ethyl acetate-acetone (80 + 20); (70 + 30), chloroform-ethanol (90 + 10); (80 + 20) or chloroform-acetone-ethanol (70 + 15 + 15).

The analogous compounds with side chains one methylene group shorter and also those with the same number of carbon atoms but variously positioned hydroxyl group in the side chain, can be satisfactorily separated; this had been impossible by the column chromatography hitherto available. As far as other structural elements are concerned, the bile alcohols behave just like other steroids (see section IV). Strongly acid reagents like sulphuric acid are employed for visualisation.

2. Steroid Carboxylic Acids

Solvent mixtures of usual compositions can be used, depending on polarity or whether these so-called etianic acids (C_{20}) are present in free or esterified form; esters are suitable solvents or solvent components.

BARBIER et al. [11] have chromatographed numerous acids in cyclohexane-ethyl acetate mixtures on silica gel and SCHWARZ and SYKORA [150] on loose alumina. Detection is governed by the nature of the functional groups present in the ring skeleton (see sections III, VI) and can be carried out most simply with concentrated sulphuric acid. DUVIVIER [197] has quoted several systems for the TLC of the free acids.

3. Steroid Conjugates

These compounds, occuring mostly in urine or obtained synthetically, are derivatives of androgens, corticosteroid metabolites and oestrogens; they are principally sulphate or phosphate esters or glucuronides (combined with glucuronic acid via a β-glycosidic link). Their separation in column or paper chromatography has been described earlier (cf. [129]). Synthetic acid esters of steroids with succinic or tetrahydrophthalic acids may be classified here also.

The ionic properties of these substances demand the use of solvents containing added acids or bases or at least of water (partition). Thus, e. g., ethyl acetate-pyridine-acetic acid-water (62 + 21 + 6 + 11), a solvent employed for PC or TLC of amino acids and peptides, has been used for separating steroid phosphates and sulphates on silica gel. Cholesterol sulphate, pregnenolone sulphate, 17-hydroxypregnenolone sulphate separate in that order of decreasing Rf-value [131]. WUSTEMAN et al. [185] have recently chromatographed further steroid sulphates on silica gel G, using benzene-butanone-ethanol-water (30 + 30 + 30 + 10) or isopropanol-chloroform-methanol-10 N ammonium hydroxide (40 + 40 + 20 + 8). KAY and WARREN [99] have used chloroform-methanol-ammonium hydroxide (95 + 5 + 1) for 17-ketosteroid conjugates. Neutral diesters, formed by esterifying the remaining acid groups, e. g., with diazomethane, can be chromatographed in the usual solvents [224]. FISHMAN, HARRIS and GREEN [56] have used stepwise development to separate model mixtures of urine glucuronides of oestradiol and oestriol: first a 15 cm run with system 1 in dimension I, in which the glucuronides themselves did not yet migrate; then two 10 cm runs with system 2 in dimension I and finally a 10 cm run with system 4 in dimension II. The recovery amounted to 85%.

Table 65. *TLC of steroid conjugates on DEAE-cellulose* [135]

Conjugate	hRf-values with solvent: (see footnote Table 66)						
	1	2	3	4	5	6	7
Oestrone sulphate	8	3	7	8	8	2	3
Dehydroepiandrosterone sulphate	19	9	14	20	22	7	6
Androsterone sulphate	23	15	22	24	33	8	7
Pregnenolone sulphate			27				
17-Hydroxypregnenolone sulphate			9				
Dehydroepiandrosterone glucuronide	54	38	41	56	72	29	36
Androsterone glucuronide	59	45	48	62	77	35	46
Etiocholanolone glucuronide	56	40	45	58	75	31	39

Table 66. *TLC of steroid conjugates on ECTEOLA-cellulose* [135]

Conjugate	hRf-values with solvent[a]:						
	1	2	3	4	5	6	7
Oestrone sulphate	8	5	8	7	54	5	4
Dehydroepiandrosterone sulphate	16	14	23	15	79	11	8
Androsterone sulphate	23	18	29	21	85	13	9
Pregnenolone sulphate			34				
17-Hydroxypregnenolone sulphate			12				
Dehydroepiandrosterone glucuronide	65	45	61	64	90	64	48
Androsterone glucuronide	72	49	67	70	91	77	59
Etiocholanolone glucuronide	67	47	64	65	91	71	55

[a] 1 0.5 M acetate buffer, pH 4.25; 2 0.5 M acetate buffer, pH 4.75; 3 1.0 M acetate buffer, pH 4.75; 4 1.5 M acetate buffer, pH 5.00; 5 Isopropanol-water-formic acid (65 + 33 + 2); 6 Ethanol-water-acetic acid (80 + 15 + 3); 7 Methanol-water-acetic acid (75 + 15 + 10).

Mention may be made in conclusion of the work of OERTEL et al. [135]. They have separated sulphates and glucuronides in the S-chamber on DEAE- and ECTEOLA-cellulose (MN 300 G), using on the one hand, acetate buffers of various concentrations and on the other, acid-containing alcohols. The quite interesting results are quoted in Tables 65 and 66. It is evident from them that the Rf-values are larger at lower pH or at higher buffer concentration. Yet increase in the concentration of acid in the organic phase accelerated migration of the glucuronides while having little effect on the Rf-values of the sulphates.

The steroid moiety can be detected once again with sulphuric acid or, for 17-ketosteroid conjugates, with m-dinitrobenzene for example. The acid component can be characterised by either the Folin-Ciocalteu reagent (No. 122) or pyridylazonaphthol reagent (No. 215) for glucuronides and methylene blue (Rgt. No. 162) for sulphates [44].

Bibliography for Chapter L. Steroids

The newest literature sources, worked on while printing was in progress, follow reference [189] in alphabetical order.

1. ADAM, G., u. K. SCHREIBER: Z. Chem. **3**, 100 (1963).
2. ADAMEC, O., J. MATIS, u. M. GALVANEK: Lancet I, **1962**, 7220.
3. — Steroids **1**, 495 (1963).
4. ACHREM, A. A., and A. I. KUZNETSOVA: Proc. Acad. Sci. U.S.S.R. Chem. Section **138**, 507 (1961).
5. ANTHONY, W. L., and W. T. BEHER: J. Chromatogr. **13**, 567 (1964).
6. AVIGAN, J., D. S. GOODMAN, and D. STEINBERG: J. Lipid. Res **4**, 100 (1963).
7. AUDRIN, P., F. C. FOUSSARD, C. BOURGOIN, L. JUNG et P. MORAND: Rev. franc. Etud. clin. biol. **8**, 507 (1963).
8. AZARNOFF, D. L., u. D. R. TUCKER: Biochim. biophys. Acta (Amst.) **70**, 589 (1963).
9. BALOGH, B.: Anal. Chem. **36**, 2498 (1964).
10. BANG, H. O.: J. Chromatogr. **14**, 520 (1964).
11. BARBIER, M., H. JÄGER, H. TOBIAS u. E. WYSS: Helv. chim. Acta **42**, 2440 (1959).
12. BARRETT, G. C.: Nature (Lond.) **194**, 1171 (1962).
13. BEIJEVELD, W. M.: Pharm. Weekblad **97**, 190 (1962).
14. BENNETT, R. D., and E. HEFTMANN: J. Chromatogr. **9**, 348 (1962).
15. — J. Chromatogr. **9**, 353 (1962).
16. — — J. Chromatogr. **9**, 359 (1962); **12**, 245 (1963).
17. BENRAAD, T. J., and P. W. C. KLOPPENBORG: Steroids **3**, 671 (1964).
18. BERGMANN, E. D., R. IKAN, and S. HAREL: J. Chromatogr. **15**, 204 (1964).
19. BERNAUER, W.: Klin. Wschr. **41**, 883 (1963).
20. BINKERT, J., E. ANGLIKER u. A. VON WARTBURG: Helv. chim. Acta **45**, 2122 (1962).
21. BIRD, JR., H. L., H. F. BRICKLEY, J. P. COMER, P. E. HARTSAW, and M. L. JOHNSON: Analyt. Chem. **35**, 346 (1963).
22. BLUNDEN, G., and R. HARDMAN: J. Chromatogr. **15**, 273 (1964).
23. BOBBITT, J. M.: Thin layer Chromatography. New York: Reinhold Co. 1963.
24. BOLL, P. M.: Acta chem. scand. **16**, 1819 (1962).

25. BROOKSBANK, B. W. L., and D. B. GOWER: Steroids **4**, 787 (1964).
26. BRUINVELS, J.: Experientia (Basel) **10**, 551 (1963).
27. BURGER, H. G., J. R. KENT and A. E. KELLIE: J. clin. Endocr. **24**, 432 (1964).
28. BUSH, I. E.: The Chromatography of Steroids. Oxford: Pergamon 1961.
29. CARGILL, D. I.: Analyst **87**, 865 (1962).
30. CARRERAS MATAS, L.: An. Acad. farm. (Madr.) 26, 371 (1960).
31. CAVINA, G., and C. VICARI: In: Marini-Bettolo, ed.: Thin Layer Chromatography. Amsterdam, London, New York: Elsevier 1964; Farmaco Ed. Prat. **19**, 338 (1964).
32. ČERNÝ, V., J. JOSKA, and L. LÁBLER: Coll. Czech. Chem. Comm. **26**, 1658 (1961).
33. CHAMBERLAIN, J., and G. H. THOMAS: J. Chromatogr. **11**, 408 (1963).
34. CHANG, E.: Steroids **4**, 237 (1964).
35. CHANG SHEN, N.-H., F. E. FRANCIS, and R. A. KINSELLA: J. Lab. clin. Med. **60**, 1017 (1962).
36. CHIANG, S. P., and J. S. SCHWEPPE: Fed. Proc. **23**, 270 (1963).
37. CLAUDE, J. R.: J. Chromatogr. **17**, 596 (1965).
37a. CLIFFORD, C. J., J. V. WILKINSON and J. S. WRAGG: J. Pharm. Pharmacol. **16**, Suppl. 11 T (1964).
38. COHN, G. L., and E. PANCAKE: Nature (Lond.) **201**, 75 (1964).
39. COLLINS, W. P., and J. F. SOMMERVILLE: Nature (Lond.) **203**, 836 (1964).
40. COPIUS-PEEREBOOM, J. W., and H. W. BEEKES: J. Chromatogr. **9**, 316 (1962).
41. — Chromatographic Sterol Analysis. Wageningen: Pudoc 1963.
42. — In: MARINI-BÉTTOLO, ed. Thin Layer Chromatography. Amsterdam, London, New York: Elsevier 1964.
43. COPIUS-PEEREBOOM, J. W., and H. W. BEEKES: J. Chromatogr. **17**, 99 (1965).
44. CRÉPY, O., O. JUDAS et B. LACHESE: J. Chromatogr. **16**, 340 (1964).
45. DAHN, H., u. H. FUCHS: Helv. chim. Acta **45**, 261 (1962).
45a. DIAMANTSTEIN, T., u. K. LÖRCHER: Z. analyt. Chem. **191**, 429 (1962).
46. DUMAZERT, C., C. GHIGLIONE et T. PUGNET: Ann. pharm. franç. **21**, 227 (1963).
47. DUNCAN, G. R.: J. Chromatogr. **8**, 37 (1962).
48. DYER, W. G., J. P. GOLD, N. A. MAISTRELLIS, C. T. PENG, and P. OFNER: Steroids **1**, 271 (1963).
49. ENEROTH, P.: J. Lipid Res. **4**, 11 (1963).
50. ERCOLI, A., R. VITALI, and R. GARDI: Steroids **3**, 479 (1964).
51. FAUCONNET, L., et R. FAZAN: Bull. Soc. Vaud. Sci. nat. **66**, 307 (1956).
52. — u. M. WALDESBÜHL: Pharm. Acta Helv. **38**, 423 (1963).
53. FEHÉR, T.: Mikrochim. Acta **1965**, 105.
54. — J. Chromatogr. **19**, 551 (1965).
55. FIESER, L. F., and M. FIESER: Steroids. New York: Reinhold Publ. Corp. 1959.
56. FISHMAN, W. H., F. HARRIS, and S. GREEN: Steroids **5**, 375 (1965).
57. FREIMUTH, U., B. ZAWTA u. M. BÜCHNER: Acta biol. med. germ. **13**, 624 (1964).
58. FROSCH, B., u. H. WAGENER: Z. klin. Chem. **1**, 187 (1963).
59. — Klin. Wschr. **42**, 192 (1964).
60. FUTTERWEIT, W., N. L. MCNIVEN, and R. I. DORFMAN: Biochim. biophys. Acta (Amst.) **71**, 474 (1963).
61. GÄNSHIRT, H., and J. POLDERMAN: J. Chromatogr. **16**, 510 (1965).
62. — F. W. KOSS u. K. MORIANZ: Arzneimittel-Forsch. **10**, 943 (1960).
63. GERALI, G., G. LUGARO e L. FERRARI: Farmaco Ed. Sci. **20**, 148 (1965).
64. GOLAB, T., u. D. S. LAYNE: J. Chromatogr. **9**, 321 (1962).

65. GÖLDEL, L., W. ZIMMERMANN u. D. LOMMER: Z. physiol. Chem. **333**, 35 (1963).
66. GÖNDÖS, GY., B. MATKOVICS u. Ö. KOVACS: Microchem. J. **8**, 415 (1964).
67. GÖRLICH, B.: Arzneimittel-Forsch. **10**, 770 (1960).
68. GOWER, D. B.: J. Chromatogr. **14**, 424 (1964).
69. GRITTER, R. J., and R. J. ALBERS: J. Chromatogr. **9**, 392 (1962).
70. HAAHTI, E., T. NIKKARI, u. K. JUVA: Acta chem. scand. **17**, 538 (1963).
71. HAGERMAN, D. D., and J. M. SPENCER: Steroids **4**, 547 (1964).
71a. HALL, A.: J. Pharm. Pharmacol. **16**, Suppl. 9T (1964).
72. HALPAAP, H.: Chem. Ing. Technik **35**, 488 (1963).
73. HAMILTON, J. G.: Arch. Biochem. **101**, 7 (1963).
74. HAMMAN, B. L., and M. M. MARTIN: J. clin. Endocr. **24**, 1195 (1964).
75. HARA, S., and M. TAKEUCHI: J. Chromatogr. **11**, 565 (1963).
76. — — M. TACHIBANA, and G. CHIHARA: Chem. Pharm. Bull. **12**, 483 (1964).
77. — H. TANAKA, and M. TAKEUCHI: Chem. Pharm. Bull. **12**, 626 (1964).
78. HEFTMANN, E.: Chromatogr. Rev. **7**, 179 (1965).
79. HEŘMÁNEK, S., V. SCHWARZ, and Z. ČEKAN: Coll. Czech. Chem. Commun. **26**, 1669 (1961).
80. HERTELENDY, F., and R. H. COMMON: Steroids **2**, 135 (1963).
81. — J. Chromatogr. **13**, 570 (1964).
82. HODOSAN, F., u. A. POP-GOCAN: Rev. Roumanie de Chim. **9**, 523 (1964).
83. HOFMANN, A. F.: J. Lipid Res. **3**, 127 (1962).
84. — Analyt. Biochem. **3**, 145 (1962).
85. — Acta chem. scand. **17**, 173 (1963).
86. — In: A. T. JAMES and L. J. MORRIS, ed. New Biochemical Separations. London: Van Nostrand 1964.
87. HONEGGER, C. G.: Helv. chim. Acta **45**, 1409 (1962).
88. HORLICK, L., and J. AVIGAN: J. Lipid Res. **4**, 160 (1963).
89. HORNING, E. C., T. LUUKKAINEN, E. HAAHTI, B. G. CREECH, and W. J. A. VAN DEN HEUVEL: Recent Progr. Hormone Res. **19**, 57 (1963).
90. IBAYASHI, H., M. NAKAMURA, S. MURAKAWA, T. UCHIKAWA, T. TANIOKA, and K. NAKAO: Steroids **3**, 559 (1965).
91. IKAN, R., S. HAREL, J. KASHMAN, and E. D. BERGMANN: J. Chromatogr. **14**, 504 (1964).
92. JACOBSOHN, G. M.: Anal. Chem. **36**, 275 (1964).
93. — Anal. Chem. **36**, 2030 (1964).
94. JAMES, A. T., and L. J. MORRIS, ed.: New Biochemical Separations. London: Van Nostrand 1964.
95. JUNG, L., CH. BOURGOIN, J. C. FOUSSARD, P. AUDRIN et P. MORAND: Rev. franç. Étud. clin. biol. **8**, 406 (1963).
95a. KAUFMANN, H., W. WEHRLI u. T. REICHSTEIN: Helv. chim. Acta **48**, 65 (1965).
96. KAUFMANN, H. P., Z. MAKUS u. F. DEICKE: Fette, Seifen, Anstrichmittel **63**, 235 (1961).
97. — — u. T. H. KHOE: Fette, Seifen, Anstrichmittel **64**, 1 (1962).
98. KAWASAKI, T., and K. MIYAHARA: Chem. Pharm. Bull. **11**, 1546 (1963).
99. KAY, H. L., and F. L. WARREN: J. Chromatogr. **18**, 189 (1965).
100. KAZUNO, T., and T. HOHITA: Steroids **3**, 55 (1964).
101. KORZUN, B. P., and S. BRODY: J. Pharm. Sci. **52**, 206 (1963).
102. — L. DORFMAN, and S. BRODY: Analyt. Chem. **35**, 950 (1963).
103. KOSS, F. W., u. D. JERCHEL: Naturwissenschaften **51**, 382 (1964).
104. KRITCHEWSKY, D., D. S. MARTAK u. G. H. ROTHBLATT: Anal. Biochem. **5**, 388 (1963).
105. LÁBLER, L., and V. ČERNY: Coll. Czech. Chem. Commun. **28**, 2932 (1963).

106. — In: MARINI BETTOLO, ed. Thin Layer Chromatography. Amsterdam, London, New York: Elsevier 1964.
107. LAU, H. L., and G. S. JONES: Amer. J. Obstet. Gynec. **90**, 132 (1964).
108. LETTRÉ, INHOFFEN u. TSCHESCHE: Über Sterine, Gallensäuren und verwandte Naturstoffe, Bd. I u. II. Stuttgart: F. Enke (1959).
108a. LEWBART, M. L., W. WEHRLI u. T. REICHSTEIN: Helv. chim. Acta **46**, 505 (1963).
109. LISBOA, B. P.: Acta Endocrin. **43**, 47 (1963).
110. — J. Chromatogr. **13**, 391 (1964).
111. — J. Chromatogr. **16**, 136 (1964).
112. LISBOA, B. P., and E. DICZFALUSY: Acta Endocrin. **40**, 60 (1962).
113. — — Acta Endocrin. **43**, 545 (1963).
114. LOEV, B., and K. M. SNADER: Chem. and Ind. **1965**, 15.
115. LUISI, M., C. SAVI, and V. MARESCOTTI: J. Chromatogr. **15**, 428 (1964).
116. MAHADEVAN, V., and W. O. LUNDBERG: J. Lipid Res. **3**, 106 (1962).
117. MAN, J. M. DE: Z. Ernährungswiss. **5**, 1 (1964).
118. MANZETTI, A. R., u. T. REICHSTEIN: Helv. chim. Acta **47**, 2303 (1964).
119. MATIS, J. O. ADAMEC, and M. GALVÁNEK: Nature (Lond.) **194**, 477 (1962).
120. MATSUMOTO, N.: Chem. Pharm. Bull. **11**, 1189 (1963).
121. MATTHEWS, J. S..: Biochim. biophys. Acta **69**, 163 (1963).
122. — A. L. PEREDA, and A. AGUILERA: J. Chromatogr. **9**, 331 (1962).
123. MCCARTHY, J. L., A. L. BRODSKY, J. A. MITCHELL, and R. F. HERRSCHER: Anal. Biochem. **8**, 164 (1964).
124. METZ, H.: Naturwissenschaften **48**, 569 (1961).
125. MICHALEČ, Č., M. SULC, and J. MĚŠŤAN: Nature (Lond.) **193**, 63 (1962).
126. MILLER, J. M., and J. G. KIRCHNER: Anal. Chem. **25**, 1107 (1963).
127. MORRIS, L. J.: J. Lipid Res. **4**, 357 (1963).
128. NAKAYAMA, F., M. OISHI, N. SAKAGUCHI, and H. MIYAKE: Clin. chim. Acta **10**, 544 (1964).
129. NEHER, R.: Steroid Chromatography. Amsterdam, London, New York: Elsevier 1964.
129a. — 9. Symposium der Deutschen Gesellschaft für Endokrinologie. Berlin-Göttingen-Heidelberg: Springer 1963. S. 21.
130. NEHER, R.: In: MARINI-BÉTTOLO ed. Thin Layer Chromatography. Amsterdam, London, New York: Elsevier 1964.
131. — u. E. VON ARX: Unpublished.
132. NIENSTEDT, W.: Biochem. J. **92**, P 8 (1964).
133. NISHIKAZE, O., R. ABRAHAM, and H.-J. STAUDINGER: J. Biochem. (Tokyo) **54**, 427 (1963).
134. — u. Hj. STAUDINGER: Klin. Wschr. **40**, 1014 (1962).
135. OERTEL, G. W., M. C. TORNERO, and K. GROOT: J. Chromatogr. **14**, 509 (1964).
136. PAQUIN, R., and M. LEPAGE: J. Chromatogr. **12**, 57 (1963).
137. PFEIFER, J. J.: Mikrochim. Acta **51**, 529 (1962).
138. QUESENBERRY, R. O., and F. UNGAR: Anal. Biochem. **8**, 192 (1964).
139. RANDERATH, K.: Dünnschicht-Chromatographie. Weinheim: Verlag Chemie 1962.
140. REICHELT, J., and J. PITRA: Coll. Czech. Chem. Commun. **27**, 1709 (1962).
141. REISERT, P. M., u. D. SCHUMACHER: Experientia (Basel) **19**, 84 (1963).
142. RICHTER, E.: J. Chromatogr. **18**, 164 (1965).
143. RIVLIN, R. S., and H. WILSON: Anal. Biochem. **5**, 267 (1963).
144. SAMUEL, P., M. URIVETZKY, and G. KALEY: J. Chromatogr. **14**, 508 (1964).
145. SANDER, H.: Naturwissenschaften **48**, 303 (1961).

146. SCHEIFFARTH, F., u. L. ZICHA: Acta Endocrin. **43**, 227 (1963).
147. SCHINK, W., u. H. STRUCK: Med. Welt **1964**, 1525.
148. SCHREIBER, K., O. AURICH, and G. OSSKE: J. Chromatogr. **12**, 63 (1963).
149. SCHWARZ, V.: Pharmazie **18**, 122 (1963).
150. — and K. SYKORA: Coll. Czech. Chem. Commun. **28**, 101 (1963).
151. SIEGEL, E. T., u. R. I. DORFMAN: Steroids **1**, 409 (1963).
152. SJÖHOLM, I.: Svensk. Farm. Tidskr. **66**, 321 (1962).
153. SMITH, L. L., and TH. FOELL: J. Chromatogr. **9**, 339 (1962).
153a. SPRITZ, N.: quoted in [86] as private communication.
154. STAHL, E., and U. KALTENBACH: J. Chromatogr. **5**, 458 (1961).
155. STANSFIELD, D. A.: Biochem. biophys. Res. Commun. **16**, 398 (1964).
156. — u. D. I. CARGILL: Biochem. biophys. Res. Commun. **13**, 231 (1963).
157. STÁRKA, L.: J. Chromatogr. **17**, 599 (1965).
158. — and J. MALÍKOVÁ: J. Endocrin. **22**, 215 (1961).
159. — u. J. RIEDLOVA: Endokrinologie **43**, 201 (1962).
160. — J. ŠULCOVÁ, J. RIEDLOVÁ u. O. ADAMEC: Clin. chim. Acta **9**, 168 (1964).
161. STEIDLE, W.: Planta Med. **9**, 435 (1961).
162. — Ann. Chem. **662**. 126 (1963).
163. STEINEGGER, E., u. J. H. VAN DER WALT: Pharm. Acta Helv. **36**, 599 (1961).
164. STEVENS, P. J.: J. Chromatogr. **14**, 269 (1964).
165. STRUCK, H.: Mikrochim. Acta **1961**, 634.
166. TAKEDA, K., S. HARA, A. WADA, and N. MATSUMOTO: J. Chromatogr. **11**, 562 (1963).
167. TAKEUCHI, M.: Chem. Pharm. Bull. **11**, 1183 (1963).
168. TSCHESCHE, R., W. FREYTAG, u. G. SNATZKE: Chem. Ber. **92**, 3053 (1959).
169. — H. SCHWARZ u. G. SNATZKE: Chem. Ber. **94**, 1699 (1961).
170. — u. G. SNATZKE: Ann. Chem. **636**, 105 (1960).
170a. — G. WULFF u. G. BALLE: Tetrahedron **18**, 959 (1962).
171. — — u. K. H. RICHERT: In: A. T. JAMES, L. J. MORRIS, ed. New Biochemical Separations. London: Van Nostrand 1964.
172. USUI, T.: J. Biochem. (Tokyo) **54**, 283 (1963).
173. VAEDTKE, J., and A. GAJEWSKA: J. Chromatogr. **9**, 345 (1962).
174. — — and A. CZARNOCKA: J. Chromatogr. **12**, 208 (1963).
175. VAN DAM, M. J. D., G. J. DE KLEUVER, and J. G. DE HEUS: J. Chromatogr. **4**, 26 (1960).
176. VECSEI, P., V. KEMÉNY, and A. GÖRGÉNYI: J. Chromatogr. **14**, 506 (1964).
177. VERMEULEN, A., and J. C. M. VERPLANCKE: Steroids **2**, 413 (1963).
178. VOIGT, K. D., U. VOLKWEIN u. J. TAMM: Klin. Wschr. **42**, 642 (1964).
179. WAGENER, H., u. B. FROSCH: Klin. Wschr. **41**, 1094 (1963).
180. WALDI, D.: Klin. Wschr. **40**, 827 (1962).
181. WALDI, D.: In: E. STAHL, Thin-Layer Chromatography, 1. Edition, p. 275. Berlin-Göttingen-Heidelberg: Springer 1962.
182. — Ärztl. Lab. **9**, 221 (1963).
183. WOLFMAN, L., and B. A. SACHS: J. Lipid Res. **5**, 127 (1964).
184. WOTIZ, H. H., and S. C. CHATTORAJ: Anal. Chem. **36**, 1466 (1964).
185. WUSTEMAN, F. S., K. S. DODGSON, A. G. LLOYD, F. A. ROSE, and N. TUDBALL: J. Chromatogr. **16**, 334 (1964).
186. YAWATA, M., and E. M. GOLD: Steroids **3**, 435 (1964).
187. ZELNIK, R., and L. M. ZITI: J. Chromatogr. **9**, 371 (1962).
188. — — and C. V. GUIMARÃES: J. Chromatogr. **15**, 9 (1964).
189. ZÖLLNER, N., G. WOLFRAM u. G. AMIN: Klin. Wschr. **40**, 273 (1962).
190. ANGELICO, R., G. CAVINA, A. D'ANTONA, and G. GIOCOLI: J. Chromatogr. **18**, 57 (1965).

191. BENRAAD, TH. J., and P. W. C. KLOPPENBORG: Clin. chim. Acta **12**, 565 (1965).
192. CHATTOPADHYAY, D. P., and E. H. MOSBACH: Analyt. Biochem. **10**, 435 (1965).
193. CHEN, Jr., P. S.: Analyt. Chem. **37**, 301 (1965).
193a. CHIH TUNG, Y., and K. TSUNG WANG: Nature (Lond.) **208**, 582 (1965).
194. COMER, J. P., and P. E. HARTSAW: J. Pharm. Sci. **54**, 524 (1965).
195. CORONA, G. L., and M. RAITERI: J. Chromatogr. **19**, 435 (1965).
196. DITULLIO, N. W., C. S. JACOBS Jr., and W. L. HOLMES: J. Chromatogr. **20**, 354 (1965).
197. DUVIVIER, J.: J. Chromatogr. **19**, 352 (1965).
198. EBERLEIN, W. R.: J. clin. Endocrin. **25**, 288 (1965).
199. EHRLICH, E. N.: J. Lab. clin. Med. **65**, 869 (1965).
200. FISCHER, P.: Schweiz. Apoth. Ztg. **103**, 137, 182 (1965).
201. FORTH, W., P. DOENECKE, u. H. GLASNER: Klin. Wschr. **43**, 20 (1965).
202. FROSCH, B.: Arzneimittel Forsch. **15**, 178 (1965).
203. — Klin. Wschr. **43**, 262 (1965).
204. GALLETTI, F.: Res. Steroids 2, 189 (1966), Rom: Pensiero Scientifico.
205. GERDES, H., u. W. STAIB: Klin. Wschr. **53**, 744 (1965).
206. — — Klin. Wschr. **43**, 789 (1965).
207. GOODMAN, D. S., and T. SHIRATORI: J. Lipid Res. **5**, 578 (1964).
208. HEUSSER, D.: Planta Med. **12**, 237 (1964).
209. HONEGGER, C. G.: Helv. chim. Acta **47**, 2384 (1964).
210. JANSSEN, E. G.: Dtsch. Med. Forsch. **1**, 195 (1963).
211. IKAN, R., and M. CUDZINOWSKI: J. Chromatogr. **18**, 422 (1965).
212. IWATA, T., and K. YAMASAKI: J. Biochem. (Tokyo) **56**, 424 (1964).
213. KULENDA, Z., u. E. HORAKOWA: Z. med. Lab. **4**, 173 (1963).
214. LISBOA, B. P.: Steroids **6**, 605 (1965).
215. — J. Chromatogr. **19**, 81 (1965).
216. — J. Chromatogr. **19**, 333 (1965).
217. LUISI, M., C. SAVI, F. COLI, F. PANICUCCI, V. MARESCOTTI e G. GAMBASSI: Boll. Soc. ital. Biol. sper. **39**, 1264, 1267 (1963).
218. MILBORROW, B. V.: J. Chromatogr. **19**, 194 (1965).
219. MOMOSE, T., T. MATSUKUMA, and Y. OHKURA: J. Pharm. Soc. Japan **84**, 783 (1964).
220. OERTEL, G. W., u. K. GROOT, Clin. Chim. Acta **11**, 512 (1965).
221. PASCAUD, M.: Advances in Tracer Methodology (ed. S. ROTHSCHILD) (1965) in press.
222. PEKKARINEN, A.: Res. Steroids **2**, 223 (1966), Rom: Pensiero Scientifico.
223. QUESENBERRY, R. O., E. M. DONALDSON, and F. UNGAR: Steroids **6**, 167 (1965).
224. RIESS, J.: J. Chromatogr. **19**, 527 (1965).
225. SCHNEIDER, H. P. G., u. Z. SZEREDAY: Klin. Wschr. **43**, 747 (1965).
226. SEMENUK, G., and W. T. BEHER: J. Chromatogr. **21**, 27 (1966)
227. SJÖVALL, J.: Methods Biochem. Analysis **12**, 97 (1964).
228. SONANINI, D.: Pharm. Acta Helv. **39**, 673 (1964).
229. — R. HOFSTETTER, L. ANKER u. H. MÜHLEMANN: Pharm. Acta Helv. **40**, 302 (1965).
230. SULIMOVICI, S., B. LUNENFELD, and M. C. SHELESNYAK: Acta Endocr. **49**, 97 (1965).
231. TRUSWELL, A. S., and W. D. MITCHELL: J. Lipid. Res. **6**, 438 (1965).
232. TAI CHAN, P., R. VENNA, P. OFNER, and P. L. MUNSON: Steroids **6**, 571 (1965).
233. VLASINICH, V., and J. B. JONES: Steroids **3**, 707 (1964).
234. ŽURKOWSKA, J., u. A. OZAROWSKI: Planta Med. **12**, 222 (1964).

M. Aliphatic Lipids

Helmut K. Mangold

I. Introduction

1. Neutral Lipids and Their Hydrolysis Products

Long chain hydrocarbons, alcohols, aldehydes and acids are widely distributed in the plant and animal kingdoms. These aliphatic lipids contain carbon chains which may be straight and saturated or branched or unsaturated, in one or more positions. Compounds with more than one function, such as epoxy- and hydroxy-acids, are also known.

Alcohols, aldehydes and acids occur in nature chiefly in combined form. The esters of long chain alcohols with acids are known as (ester-) waxes.

Aliphatic alcohols and aldehydes occur above all in combination with glycerol; they form alkyl ethers and alk-1-enyl ethers ("vinyl ethers"). Acids are found as esters with glycerol, the mono-, di- and triglycerides.

```
H₂C—O—CH₂—CH₂—R        H₂C—O—CH=CH—R         H₂C—O—C—R
|                      |                      |      ||
HC—OH                  HC—OH                  HC—OH  O
|                      |                      |
H₂C—OH                 H₂C—OH                 H₂C—OH
α-Alkyl glycerol ether  α-Alk-1-enyl glycerol ether  α-Monoglyceride
```

The alkyl glycerol ethers, alkenyl glycerol ethers and monoglycerides can occur as α- and β-isomers; diglycerides may be symmetrical (α,α'-) or asymmetrical (α,β-).

Alkyl and alkenyl glycerol ethers are found naturally as esters of acids, the glycerol ether diesters (alkyl diglycerides) or "neutral plasmalogens" ("aldehydogenic triglycerides"). The triglycerides of long chain (fatty) acids form the principal constituent of fats and oils and are the "fats" in the narrower sense.

It has been discovered only in recent years that esters of various short-chain diols occur in nature as well as the esters of glycerol. The discovery, isolation and characterisation of these "diol lipids" was rendered possible by the application of modern chromatographic methods.

2. Phospholipids, Sulpholipids and Glycolipids

In phospholipids, one of the glycerol hydroxyl groups is esterified with phosphorylcholine, phosphorylethanolamine or phosphorylserine.

Ethanolamine-phosphatides exist in the following three forms:

```
H₂C—O—CH₂—CH₂—R              H₂C—O—CH=CH—R
   |      O                      |      O
   |      ‖                      |      ‖
  HC—O—C—R'                     HC—O—C—R'
   |                             |
  H₂C—PE                        H₂C—PE
Ether-ester phosphatide       Alkenylether-ester
                              phosphatide (plasmalogen)

H₂C—O—C—R
   |  ‖
   |  O                               O
  HC—O—C—R'                           ‖
   |  ‖                    —O—P—O—CH₂—CH₂—NH₂
   |  O                           |
  H₂C—PE                          O—
Diester phosphatide           PE = Phosphorylethanolamine
(phosphoglyceride)
(ethanolamine-cephalin)
```

Presumably choline-phosphatides and serine-phosphatides occur also as ether-esters, alkenylether-esters and diesters.

Glycerol phosphorylcholine, glycerol phosphorylethanolamine and glycerol phosphorylserine, which are esterified with only a single fatty acid molecule and thus contain a free hydroxyl group, are termed "lyso phosphatides". One group of phospholipids which contain no bases, are the phosphatidic acids; these occur naturally, possibly as hydrolysis products of choline-, ethanolamine- and serine-phosphatides:

```
H₂C—O—C—R                    H₂C—O—C—R
   |  ‖                         |  ‖
   |  O                         |  O
  HC—O—C—R'                    HC—OH
   |  ‖                         |  O
   |  O    O⁽⁻⁾                 |  ‖
  H₂C—O—P=O        H₂C—O—P—O—CH₂—CH₂—N⁽⁺⁾(CH₃)₃
         \                      |
          O⁽⁻⁾                  O⁽⁻⁾
α-Phosphatidic acid           Lysolecithin
```

Phosphatidyl glycerol, cardiolipin and phosphatidyl inositols also contain no bases:

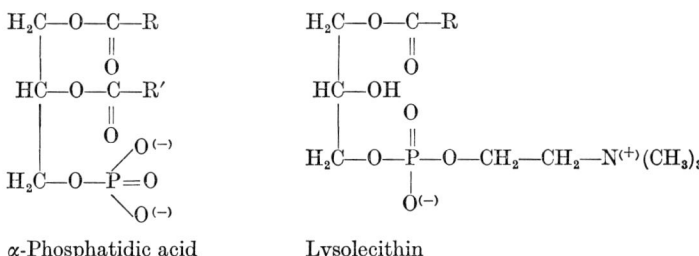

Phosphatidyl glycerol

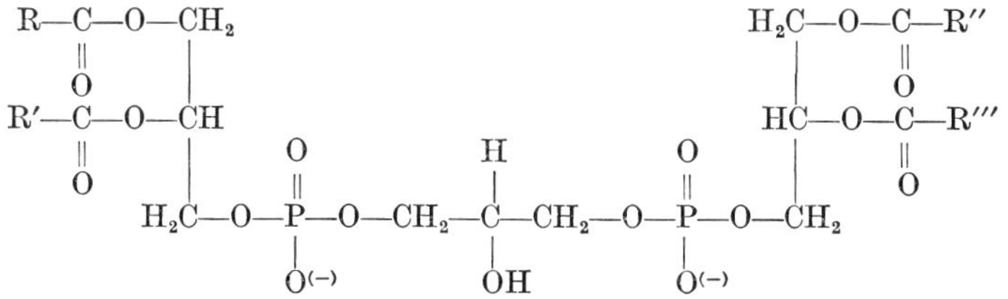

Cardiolipin

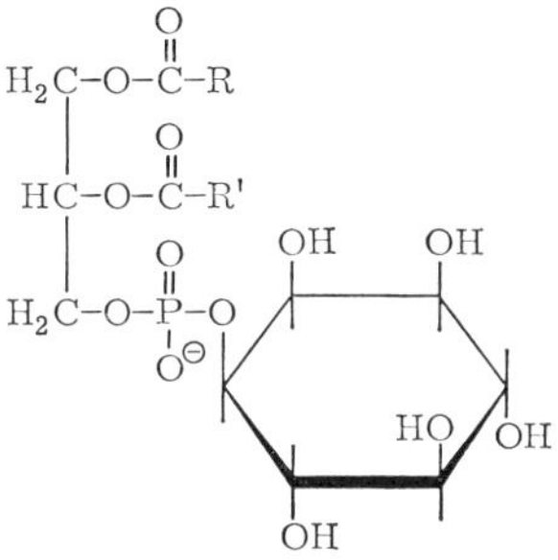

(Structure shown: two diacylglycerol units linked through a glycerophosphate bridge)

Phosphatidyl inositols
(phosphoinositide)   (diphosphoinositide)

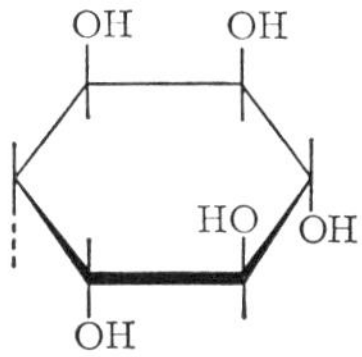

A further group of phospholipids, the sphingomyelins, are built up from a polyhydric amino alcohol like sphingosine, phosphorylcholine and fatty acid. In recent years, a number of long chain amino alcohols of structures similar to that of sphingosine have been detected with the help of chromatographic procedures.

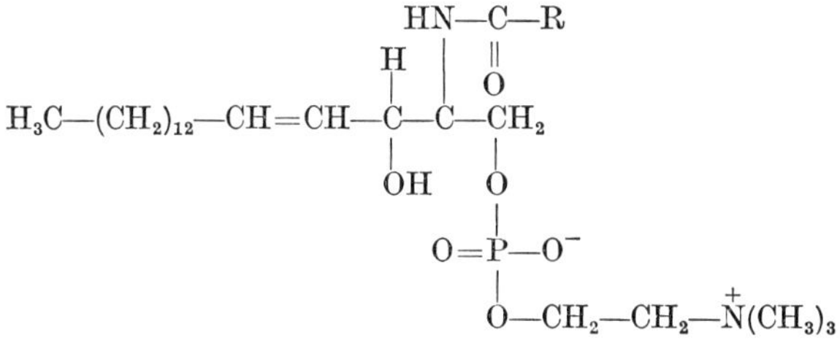

$$H_3C-(CH_2)_{12}-CH=CH-\underset{OH}{\overset{H}{C}}-\underset{}{\overset{HN-C-R}{\underset{\parallel}{C}}}-CH_2-O-\underset{O^-}{\overset{O}{\underset{\parallel}{P}}}-O-CH_2-CH_2-\overset{+}{N}(CH_3)_3$$

Sphingomyelin

Glycolipids, such as cerebrosides and gangliosides, contain sphingosine and one or more sugar molecules:

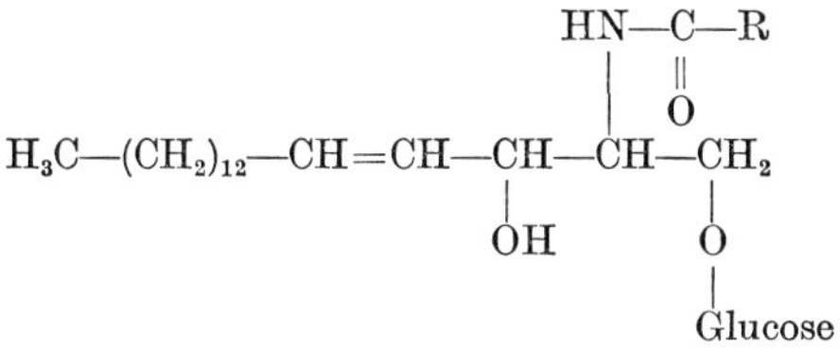

$$H_3C-(CH_2)_{12}-CH=CH-\underset{OH}{CH}-\underset{}{\overset{HN-C-R}{\underset{\parallel}{CH}}}-CH_2-O-Glucose$$

Cerebroside

$$\text{H}_3\text{C}-(\text{CH}_2)_{12}-\text{CH}=\text{CH}-\underset{\underset{\text{OH}}{|}}{\text{CH}}-\underset{\underset{\underset{\text{O}}{\|}}{\underset{\text{HN}-\text{C}-\text{R}}{|}}}{\text{CH}}-\text{CH}_2$$
$$|$$
$$\text{Galactose}$$
$$|$$
$$\text{Glucose-Neuraminic acid}$$
$$|$$
$$\text{Hexosamine}$$

Ganglioside

The structures of the various sulpholipids have not yet been fully elucidated. The following formulae have been proposed:

$$\text{H}_3\text{C}-(\text{CH}_2)_{12}-\text{CH}=\text{CH}-\text{CH}-\text{CH}-\text{CH}_2$$

with OH, HN–C(=O)–R, and O–Glucose–S(=O)(=O)–O⁻ substituents

Cerebroside sulphate

$$\text{H}_2\text{C}-\text{O}-\text{Galactose}-\overset{\overset{\text{O}}{\|}}{\underset{\underset{\text{O}}{\|}}{\text{S}}}-\text{O}^-$$
$$\text{HC}-\text{OH} \qquad \text{O}\text{---}(-\text{O}-\overset{\overset{}{}}{\underset{\underset{\text{O}}{\|}}{\text{C}}}-\text{R}')$$
$$\text{H}_2\text{C}-\text{O}-\overset{\overset{}{}}{\underset{\underset{\text{O}}{\|}}{\text{C}}}-\text{R}$$

Plant sulpholipid

3. Older Methods of Lipid Analysis

Naturally occurring lipid mixtures have a composition of such extreme complexity that the analysis by chemical methods of a group of these substances or the determination of a single compound in such mixtures, appears hopeless. Up to about ten years ago, lipid mixtures were characterised by totals like acid-, saponification-, iodine-, thiocyanogen- and diene-"numbers". The determination of the amount of "non-saponifiable matter" after alkaline hydrolysis was a standard method in the analysis of fats. Phospholipids and sulpholipids were quantitatively determined as inorganic phosphate and sulphate after combustion. These methods were supplemented by detection of subsidiary fat constituents like lipochromes, sterols and resin acids, with the help of colour reactions.

Physical methods for characterising lipids comprised determination of melting and freezing point, density, hardness, viscosity, surface and interfacial tension, solubility, flash point and combustion point. These "classical" methods have been described in detail in the works of T. P. HILDITCH [59] and H. P. KAUFMANN [77].

Several procedures for fractionation of lipid mixtures have been worked out during these last few years. Especially useful have been fractional crystallisation at low temperatures and fractionation through formation of inclusion compounds with urea, e. g., for separating saturated and unsaturated fatty acids or their esters. Vacuum distillation has been employed for isolating the methyl esters of fatty acids having the same chain length. Mono-, di- and triglycerides have been fractionated by molecular distillation. Counter-current extraction between two liquid phases has been applied to separate fatty acids according to chain length or degree of unsaturation and also to separate mono-, di- and triglycerides as well as phospholipids. Neutral and acidic lipids have been fractionated by means of dialysis through rubber membranes.

Lipid samples of 1 to 100 g were required for these separation procedures. They usually achieved no more than the enrichment of a component or semi-quantitative fractionation. The efficiency of the fractionations was generally monitored spectroscopically.

4. Newer Procedures for Separation of Lipids[1]

Chromatographic techniques have largely ousted other methods for lipid fractionation on the analytical and micro preparative scales: adsorption and partition chromatography on silica columns, on (cellulose) paper, impregnated with silica and on glass fibre paper are used in order to separate complex lipid mixtures into compound classes. Reversed phase partition chromatography is used for separating members of a vinylogous series from one another on columns or paper to which hydrophobic properties have been imparted. Gas chromatography is used as a partition chromatographic procedure especially for separating the methyl esters of fatty acids. Lipid classes can be separated according to the degree of unsaturation of their components, by chromatography on silica gel impregnated with silver nitrate. Similar separations can be realised by chromatography of the mercury(II)-acetate adducts of unsaturated lipids. Acids are isolated and strongly polar lipids fractionated by chromatography on ion-exchange columns and -papers. Straight-chain- can be separated from branched-chain fatty acids by chromatography on urea columns. This separation depends on the formation of urea inclusion complexes with the straight-chain fatty acids.

---

[1] See also [63, 158].

Every one of these principles of separation can be applied also in thin-layer chromatography, which usually gives better separations in shorter time; the method has thus rapidly become indispensable in lipid chemistry. The efficiency and scope of TLC may be extended still further by combination with other separation techniques. Adsorption TLC and partition chromatographic procedures, i.e., reversed phase partition TLC, paper or gas chromatography, should be used consecutively. Mixtures of compounds which cannot be fractionated by adsorption chromatography can usually be separated by reversed phase partition chromatography and vice versa.

The following types of lipid fractionation have been accomplished by TLC:

Separation according to compound class, fractionation of vinylogous series, separation according to degree of unsaturation, separation of *cis-trans* isomers and of positional isomers.

5. Processing of the Material for Analysis

The best method of separation is worthless if the material to be analysed has been spoilt by inexpert treatment. Consequently, some methods for isolation of lipids from plant and animal material and some principles for handling the extracts are given here in detail.

a) Homogenisation and Extraction

Extensive investigations of lipid extraction are being currently undertaken. The standard procedures given below, which were introduced many years ago, have been tested thoroughly by TLC. It is known that these methods can lead to artifacts and that they possess other shortcomings [34, 202]; improved instructions have, however, not been published so far.

α) **Isolation of Plant Lipids.** Vegetable oils are obtained from seeds and fruits by chopping into fragments, followed by pressing, grinding and extraction with solvents.

Isolation of Non-Polar Plant Lipids [77]

Procedure: The main part of the oil is first squeezed out of seeds and other plant material of high lipid content; the residue is then ground further and squeezed out once more. The residues from this process are finally extracted for 4—6 h in a Soxhlet apparatus, using petrol ether, B. P. 60—70° C or benzene; chloroform, carbon tetrachloride, trichloroethylene or diethyl ether are also suitable.

Whereas seeds often contain up to 50% fat, leaves and stems of green plants contain only about 1—5% lipids, principally phospholipids, sulpholipids and glycolipids. These compounds have to be extracted with relatively polar solvents. Since aliphatic ethers, ketones and esters activate the enzyme phosphatidase C, which is present in plant tissue, alcohols like n-propanol or isopropanol are used for extraction of polar lipids from green plants. Extraction is carried out in a nitrogen atmosphere to protect unsaturated compounds from oxidation.

Extraction of Polar Plant Lipids [151]

Procedure: Leaves or stems (1 g) are treated for 3—4 min at room temperature in a kitchen mixer with 100 ml isopropanol. The pulpy mass is centrifuged and the residue extracted once with 100 ml isopropanol and once with 100 ml chloroform-isopropanol (2 + 1). The combined extracts are concentrated in vacuo at ca. 35° C and purified by partition between chloroform-methanol (2 + 1) and 0.7% aqueous sodium chloride solution (see procedure of FOLCH, LEES and SLOANE STANLEY [42], p. 370).

The total lipids thus obtained usually still contain free amino acids and peptides, sugars and other hydrophilic, naturally occurring material which are carried into the extract through the action of solubilisers like lecithin. Column chromatography on cellulose or Sephadex is suitable for removal of these contaminants.

β) **Isolation of Animal Lipids.** Detailed descriptions and comparisons of the methods for extracting human and animal lipids may be found in two books which have recently been published [66, 225].

Ordinary kitchen mixers are suitable for homogenising animal tissue in amounts of several hundred grams; special apparatus are commercially available. POTTER and ELVEHJEM's homogeniser is that mostly used for smaller amounts; this consists of a pestle rotating in a thick-walled test tube so that the space between pestle and wall is less than 1 mm. The tissue to be homogenised is mixed with chloroform-methanol (2 + 1) and crushed between pestle and glass wall.

According to ENTENMAN [40], the following basic principles must be observed when processing animal organs:

1. Carry out all operations (homogenisation, extraction, etc.) in a nitrogen atmosphere, so as to prevent autoxidation of unsaturated lipids.
2. Use purified and freshly distilled solvents only. Methanol and ethanol are distilled over potassium hydroxide to remove aldehydes. It is particularly important that only freshly distilled chloroform be used. Peroxides can be removed from diethyl ether by distilling over hydroxylamine or ferrous sulphate; it is kept over iron or sodium wire. Both chloroform and ether are best stored in an explosion-proof refrigerator. Petrol ether should be distilled over concentrated sulphuric acid.
3. Dissect the experimental animal as rapidly as possible after killing and
4. Homogenise immediately the organ to be analysed.
5. Adhere to the established ratio of tissue to extraction solvent.
6. Warm solutions containing lipids only if absolutely necessary.
7. Remove all non-lipids from the lipid extract without loss of lipids themselves.
8. Store the extracts under conditions such that the lipids suffer minimum change, i. e., as far as possible in petroleum ether solution in an atmosphere of nitrogen in a refrigerator.

Extraction of Non-Polar Animal Lipids according to Bloor [15]

Procedure: One part of tissue homogenate is poured into 20 to 30 parts of ethanol-diethyl ether (3 + 1) and left overnight at room temperature in an atmosphere of nitrogen. The extract is filtered from precipitated protein, concentrated in vacuo under nitrogen at a temperature below 50° C (not completely to dryness however) and the lipids reextracted with three portions of petrol ether. Material insoluble in the petrol ether is discarded. The petrol ether solution of lipids may be dried with anhydrous sodium sulphate.

The homogenate must be slightly warmed with ethanol-diethyl ether in order to extract completely the lipids in more fatty tissue. Brain lipids are not wholly extracted with the Bloor-mixture.

Extraction of Polar Animal Lipids according to Folch, Lees and Sloane Stanley [42]

This method gives especially good yields of complex lipids like proteolipids and gangliosides, which cannot be entirely extracted with ethanol-diethyl ether.

Procedure: The mashed tissue is extracted at room temperature with 20 parts of chloroform-methanol (2 + 1). The crude extract is filtered through a coarse, fat-free filter paper into a flask and shaken with one-fifth of its volume of air-free distilled water (boiled out and cooled in a current of nitrogen). After standing some time, the mixture separates into two layers. The upper, aqueous phase, about 40% of the total, is carefully drawn off as far as is possible. The remainder of the upper layer is then rinsed away by repeated washing with small amounts of chloroform-methanol-water (3 + 48 + 47) without stirring up the lower phase. The residual wash liquid and the lower phase are finally rendered homogeneous by adding a little methanol.

Modifications and extensions of the procedure described here have been worked out for isolating proteolipids, phosphatidolipopeptides, gangliosides, sulphatides and triphosphatidyl inositols [42, 110].

b) Separation of Lipids from Non-Lipids

Small amounts of non-lipids can be separated by chromatography on cellulose [199] or Sephadex [217] columns. Counter-current extraction is suitable especially for fractionating amounts of several grams.

c) Degradation of Lipids and Preparation of Derivatives

TLC has found its principal application in the fractionation of complex lipids into compound classes (see p. 374). Further subdivision of these lipid classes is not as a rule carried out, although good methods are available (see p. 394); on the contrary, an aliquot is hydrolysed and the composition of the fatty acids and other components determined gas chromatographically. Even the identification of a pure compound, e. g., a triglyceride, requires qualitative and quantitative analysis of both the lipophilic and hydrophilic hydrolysis products (fatty acids and glycerol, respectively). The most widely used methods for degradation of lipids are described below. Most of the procedures are particularly appropriate for the analysis of long-chain products of hydrolysis; reference is made to methods for identifying the water-soluble moieties (see chapters on amino acids (V) and sugars (X)).

α) **Isolation of Fatty Acids and Non-Saponifiable Matter.** The combined fatty acids in lipids are generally isolated as their methyl esters, following methanolysis (see pp. 372–373). Alkaline hydrolysis of the lipids is almost the only method used for obtaining the non-saponifiable compounds.

Alkaline Hydrolysis of Lipids

Alkaline hydrolysis ("saponification") of fatty materials yields the water-soluble alkali salts of the fatty acids in these lipids, plus non-saponifiable neutral compounds such as hydrocarbons and alcohols. The "non-saponifiable matter" is extracted from the alkaline aqueous-alcoholic phase with non-polar solvents. The fatty acids are then liberated by acidification and likewise extracted with organic solvents.

Procedure: 1 part potassium hydroxide is dissolved in 1 part by weight of distilled water and, after cooling, this concentrated solution is diluted with 3 to 4 parts of methanol. The lipid sample is mixed with about 10 times its amount of this potassium hydroxide solution and left standing overnight at room temperature.

For monitoring the hydrolysis, 1—5 μl of the reaction mixture is applied to silica gel G layers on narrow plates and developed with petrol ether (60—70° C)-diethyl ether-acetic acid (70 + 30 + 2). The lipid fractions are visualised by spraying the chromatogram with the chromic acid-sulphuric acid reagent (No. 46) and heating. At the beginning of saponification, spots of substances of low polarity are visible ahead of the fatty acid spot. After complete hydrolysis, only the spots of the acids, of the non-polar compounds (hydrocarbons) ahead of them and of the strongly polar compounds (mono- and polyalcohols) behind them are visible.

When saponification is complete, the bulk of the methanol is evaporated off in vacuo at below 50° C and the aqueous residue diluted with twice its volume of water. The solution becomes cloudy if larger amounts of non-saponifiable matter are present. This non-saponifiable matter is extracted several times with amounts of diethyl ether of approximately half the volume of the solution. Emulsions can be broken by adding alcohol or by repeated cooling to 5° C and warming up. (The completeness of extraction should be tested by adsorption TLC.) The combined ethereal extracts are washed neutral with distilled water which has been boiled and cooled in a stream of nitrogen. After drying the solution over anhydrous sodium sulphate and evaporating off the ether, the non-saponifiable lipid fraction is sealed into a vessel under nitrogen or dissolved in a suitable solvent and stored in a refrigerator.

The aqueous solution of the potassium salts of the higher fatty acids which has been freed from neutral compounds, is acidified with 5N sulphuric acid under a layer of diethyl ether. The solution of fatty acids thus liberated is washed with distilled, air-free water and then dried over anhydrous sodium sulphate. The ether is evaporated off and the fatty acids sealed into a vessel in an atmosphere of nitrogen or dissolved in petroleum ether and stored in a refrigerator.

Directions for esterifying free fatty acids are given on p. 175.

This "cold saponification" is recommended especially when vitamin-containing lipids or derivatives of fatty acids containing conjugated double bonds, are present. Generally about 1–12 h heating under reflux in a nitrogen atmosphere is required. The duration depends on the lipids to be hydrolysed. 10–20% toluene or xylene can be added to raise the boiling point of the hydrolysis mixture if difficultly hydrolysable compounds are present.

β) **Isolation of Methyl Esters, Dimethylacetals and Alkyl Ethers.** The various lipid fractions can be scraped off the plate after separation and the (ester-)lipids converted by methanolysis to methyl esters without elution from the adsorbent. Alkenyl ethers yield dimethylacetals of the corresponding aldehydes when methanolysis is carried out in the presence of hydrochloric or sulphuric acids; esterified alkyl ethers of polyhydric alcohols are de-acylated. The methyl esters, dimethylacetals and ethers can be separated from the reaction mixture by TLC. Each of these lipid classes may then be analysed separately (see Table 71, p. 399).

Methanolysis with Methanol-Hydrogen Chloride [48, 195]

Procedure: 1—10 mg lipid (or a sample of silica gel on which this amount is adsorbed) and 1—2 ml of a 5% solution of dry hydrogen chloride in absolute methanol, are sealed under nitrogen in an ampoule. This is heated on the water bath at 80° C for 2 h (for glycerides) or 4—6 h (with steryl esters and sphingolipids). After cooling, the ampoule is cautiously opened and 5 ml water added to the reaction mixture. The lipids are then extracted 3 times with 10 ml portions of hexane (the reaction products of alkoxylipids are extracted with diethyl ether). The combined extracts are washed twice with 10 ml N potassium carbonate solution and water and dried over anhydrous sodium sulphate.

Methanolysis with Methanol-Sulphuric Acid [24, 25]

Procedure: 1—2 mg lipid on about 0.2 g silica gel G are heated with 10 ml of a mixture of absolute methanol, benzene and concentrated sulphuric acid (86 + 10 + 4) for 2 h at 80—90° C in a current of nitrogen. After cooling, 15 ml water are added to the reaction mixture and the whole extracted with five 0.5 ml portions of hexane. The combined extracts are dried over sodium bicarbonate-anhydrous sodium sulphate (1 + 3).

Methyl esters are not formed quantitatively by methanolysis. This has, however, no measurable influence on the results of gas chromatographic analysis.

In gas chromatographic analysis of methyl esters of higher fatty acids, substances are often recorded which are "identified" as esters of branched-chain acids, on the basis of their retention times. In reality they are dimethylacetals of aldehydes. The presence of such compounds in methyl esters can be quickly demonstrated through adsorption TLC.

Methyl esters and dimethylacetals are best separated by TLC before gas chromatographic analysis.

Isolation of Methyl Esters, Dimethylacetals and Alkyl Ethers [181, 202]

Procedure: 10—50 mg of the methanolysis products are applied as a band to a 0.5 mm silica gel G layer (20 × 20 cm). The layer is developed two or three times with hexane-diethyl ether (95 + 5) and the fractions visualised in UV with 2',7'-dichlorofluorescein solution (Rgt. No. 63). The methyl esters migrate further than the dimethylacetals and the glycerol alkyl ethers remain at the start. Diethyl ether is used for elution of these lipid classes.

γ) **Isolation of Aldehydes.** Aldehydes can be liberated from neutral and phosphatide plasmalogens by acid catalysed hydrolysis. If this reaction is carried out on the adsorbent layer, the hydrolysis products can be subsequently separated from one another and from other compounds and the aldehydes isolated. Aldehydes may be obtained similarly after reducing plasmalogens with lithium aluminium hydride since this reagent does not attack the alk-1-enyl ether link [184].

Acid-Catalysed Hydrolysis of Plasmalogens [181, 183]

Procedure: 10—100 mg of the lipid sample in hexane solution are applied as a band to a 0.5 mm thick silica gel G layer (20 × 20 cm). The plate is then held about 15 cm above a basin containing warm (50 to 60° C) concentrated hydrochloric acid, so that the layer faces the acid reagent. After 5 min, the plate is developed twice with hexane-diethyl ether (95 + 5). The fractions are visualised by spraying with 2',7'-dichlorofluorescein solution (Rgt. No. 63) and inspecting in UV light and the aldehydes (near the solvent front) eluted with diethyl ether.

Aldehydes may be determined directly (see Table 71, p. 399) or as alkyl acetates after reduction (see below) and acetylation (see p. 175) (see Table 71, p. 399).

δ) **Isolation of the Alkyl Ethers of Polyhydric Alcohols.** The ethers of long-chain alcohols with glycerol are widely distributed in nature in the form of esters with fatty acids; these are the neutral lipids and phospholipids; corresponding derivatives of diols probably occur also. Alkaline hydrolysis of the ester groups in alkoxylipids cannot be recommended since it is difficult to extract glycerol alkyl ethers completely from aqueous solutions of fatty acids. The alkyl ethers of polyhydric alcohols are best liberated from their acyl derivatives by methanolysis (see above) or by reduction with lithium aluminium hydride.

Reduction of Lipids with Lithium Aluminium Hydride [181]

Procedure: 8—10 mg lipid in 1—2 ml absolute diethyl ether are added dropwise to a solution of 0.2 g lithium aluminium hydride in 20 ml absolute diethyl ether; the mixture is stirred magnetically and refluxed 2 h. Excess hydride is decomposed with moist diethyl ether. The reaction mixture is acidified with 2N sulphuric acid and stirred until two clear layers are formed. The aqueous phase is extracted with three 10 ml portions of diethyl ether and the extracts are combined with the ethereal layer. This solution is washed with 10 ml N potassium carbonate solution, three times with 10 ml amounts of water and dried over anhydrous sodium sulphate.

The procedure for isolating alkyl ethers, described on p. 373, may be used to separate the reaction products.

The glycerol 1-alkyl ethers, which are especially extensively distributed in nature, may be analysed gas chromatographically as their isopropylidene derivatives [55, 184] (see Table 71, p. 399) or as alkoxyacetaldehydes after fission with sodium metaperiodate [10] (see Table 71, p. 399). Trifluoroacetates (TFA-derivatives) [222] (see also [62]) (see Table 71, p. 399) and, in particular, trimethylsilyl ethers (TMS-derivatives) [222] (see Table 71, p. 399) are also suitable for gas chromatography of the isomeric glycerol 1- and 2-alkyl ethers or of other alkyl ethers which contain at least one free hydroxyl group.

d) Other Reactions

A few other special procedures are quoted elsewhere in this book: Methods for preparing ozonides (pp. 406–409) and for formation of mercury(II)-acetate adducts of unsaturated lipids (p. 402); procedures for esterifying fatty acids with diazomethane (p. 175) and for acetylation of alcohols with acetic anhydride (p. 175). The procedures described for preparing trifluoroacetates or trimethylsilyl ethers can be applied to aliphatic alcohols.

Detailed procedures can also be found in a recently published book [225].

6. Manufacturers and Suppliers of Pure Lipids

Neutral lipids and fatty acids (Firms 11, 31, 60, 105, 133, 144), phospholipids and other complex compounds (Firms 11, 31, 105, 144) and standard mixtures for TLC and gas chromatography (Firms 11, 144) are commercially available.

Suppliers of radioactively labelled lipids are mentioned in Chapter I, p. 155.

Addresses of firms are listed at the end of this book (p. 918).

II. Thin-Layer Chromatography of Lipids

1. Separation of Lipids According to Compound Class

Over 25 years ago, TRAPPE showed that fractionation of lipids into compound classes was possible by elution from adsorbent columns using a succession of solvents of varying polarity. He classified solvents of differing polarities in the order of their elution effect and named this list "the eluotropic series of solvents". TRAPPE's eluotropic series and further data on solvents are quoted on p. 202.

Several authors have found that mixtures of petroleum ether and diethyl ether in various proportions could be used to advantage instead of a series of individual solvents. Procedures for chromatographic separation of complex lipid mixtures on silica gel columns have been tried out in many places and used as standard methods for routine analysis of human and animal lipid extracts.

After STAHL had introduced TLC [193], several authors [69, 78, 117, 214] applied this new method to lipids and found that it was splendidly suitable for separating compounds in this group of natural products. It is evident from a recently published review [152] that thin-layer and gas chromatography have almost completely displaced other chromatographic procedures for lipid separation within a few years.

Publications on the TLC of lipids which had appeared up to spring 1964 have been listed in two compilations [120, 122].

Fats, oils, waxes and other neutral lipids have been fractionated into compound classes mostly by adsorption TLC on silica gel G. Petrol ether-diethyl ether mixtures have been the solvents generally employed. Alcoholic and aqueous solvents are needed for separating strongly polar lipids on silica gel G layers; the fractionation of such materials into compound classes is consequently often effected by partition rather than by adsorption. Reversed phase partition chromatography on hydrophobic layers has been used in some cases for fractionating lipids into compound classes.

a) Neutral Lipids and Their Hydrolysis Products

Long-chain hydrocarbons, alcohols, aldehydes, acids, mono-, di- and triglycerides and similar lipids can be separated through adsorption TLC into compound classes of differing polarities, according to the

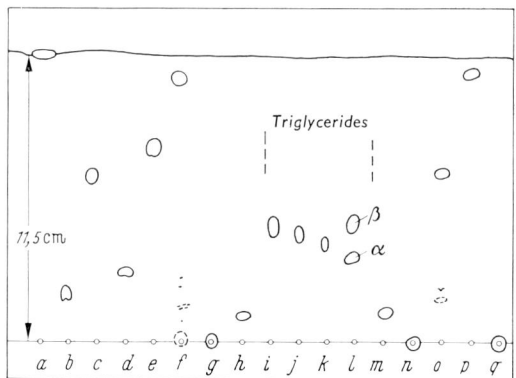

Fig. 129. Separation of neutral lipids and their hydrolysis products by adsorption-TLC on silica gel G [114]. Solvent: petrol ether (BP 60—70° C)-diethyl ether-acetic acid (90 + 10 + 1); time of run: 40 min; spray reagent: 2′,7′dichlorofluorescein in ethanol; 20 μg of each applied. *a* 9-octadecene; *b* oleyl alcohol; *c* oleyl aldehyde; *d* oleic acid; *e* methyl oleate; *f* cholesteryl oleate; *g* monoolein; *h* diolein; *i* triolein; *j* trilinolein; *k* trilinolenin; *l* tricaproin(α) and tristearin(β); *m* cholesterol; *n* selachyl alcohol; *o* selachyl diolein; *p* oleyl oleate; *q* dioleoyl-lecithin

nature and number of their functional groups. Large differences in chain length and degree of unsaturation of the members of a compound class occasionally lead to subfractionation within this class. Such subfractionations are, however, never so marked that they could interfere with the fractionation according to compound class.

The separation of some typical lipids into compound classes is schematically portrayed in Fig. 129. As expected, the separation of different compound types on layers follows the rules drawn up by BROCKMANN and VOLPERS: hydrocarbons are not adsorbed, esters but feebly; aldehydes precede alcohols and acids; short-chain compounds are more strongly adsorbed than long-chain, unsaturated then saturated (see p. 201). The example in Fig. 129 shows the extent of subfractionation within the triglyceride class.

Neutral lipids are applied to the layer as 0.1 or 1% solutions in hexane or hexane-diethyl ether (1 + 1).

Experimental Conditions

Adsorbents. Silica gel has been used more often than any other adsorbent for fractionation of lipids. 10–20 mg of a complex lipid mixture can be separated on a silica gel G layer on a 20 × 20 cm glass plate. Up to 100 mg can be used on a standard layer if only few substances which are of widely varying polarity are to be fractionated. On account of its outstanding properties, silica gel G has recently been recommended also as an adsorbent for column chromatography of lipids [29]. Alumina has been seldom used because lipids are hydrolysed and isomerised on it. Florisil, a synthetic magnesium silicate which is often used in column chromatography of lipids, should be useful in TLC. sec.-Magnesium phosphate is also a highly suitable adsorbent for lipids (see p. 298). Hydroxylapatite is especially good for separating α- and β-monoglycerides [60]. Sugar has been used in columns for fractionating lipids into compound classes. This adsorbent has good separating properties but only low capacity. Its good solubility in water facilitates the recovery of adsorbed lipophilic substances.

Lipid mixtures which can be separated on homogeneous layers only through stepwise development (see p. 87) using two or three solvents, can be fractionated on gradient layers of silica gel G and kieselguhr G [194] using a single solvent.

Solvents. The choice of solvent is governed by the polarity of the components of the lipid mixture and the separation desired. Table 67 contains a list of solvents for fractionation of neutral lipids through TLC. Petroleum ether containing 1–5% benzene or diethyl ether is particularly good for separating hydrocarbons, alkyl esters, steryl esters and polyenol esters into compound classes [52, 118]. Compounds with one or several free hydroxyl or carboxyl groups remain at the starting point when such solvents are employed.

Petroleum ether containing 10–50% diethyl ether has been used for fractionating alcohols and aldehydes, mono-, di- and triglycerides [114, 135, 165, 183]. Weakly polar lipids migrate near the front in this solvent mixture and are scarcely separated from one another. The above-mentioned solvents, plus 1–2% acetic acid, are best for TLC of lipid mixtures containing free fatty acids; this addition prevents streak and tail formation [114, 118].

Table 67. *Solvents for separation of neutral lipids and their hydrolysis products by adsorption TLC on silica gel G*

Solvent	Proportion, v/v	Reference
Petrol ether[a]-diethyl ether	90 + 10	[116]
	80 + 20	[119]
	70 + 30	[120]
Petrol ether[a]-diethyl ether-acetic acid	90 + 10 + 1	[118]
	80 + 20 + 1	[202]
	70 + 30 + 1	[118]
Benzene-diethyl ether-ethanol-acetic acid[b],	50 + 40 + 2 + 0.2	[43]
then hexane-diethyl ether[c]	94 + 6	
Diisopropyl ether-acetic acid[d] then	96 + 4	[190]
petrol ether-diethyl ether-acetic acid[e]	90 + 10 + 1	

[a] Usually petrol ether, BP 60—70° C, i. e., principally hexane.
[b] Migration distance 25 cm.
[c] Migration distance 32 cm.
[d] Migration distance 13—14 cm.
[e] Migration distance 19—19.5 cm.

Since most naturally occurring lipid mixtures contain substances of widely varying polarity, it is rarely possible to accomplish complete separation into compound classes by using a single solvent. A better picture of the composition of a complex lipid mixture is obtained when chromatography is carried out with two or three solvents of differing polarities on separate layers. The same solvents may also be used in stepwise development on one and the same layer [43, 185]. Two-dimensional TLC likewise yields better separations [78]. TLC with a solvent of gradually increasing polarity can be used for achieving better fractionation of lipid classes of considerably differing polarities [179].

Petrol ether-diethyl ether (95 + 5) is especially suitable for chromatography of vegetable and animal waxes on silica gel G. The solvent petrol ether-diethyl ether-acetic acid (90 + 10 + 1) has been that mostly used in the TLC of lipids. It is particularly suitable for fractionating animal fats (Figs. 131 and 135). Vegetable oils which contain epoxy-and hydroxy-compounds are chromatographed with mixtures of petrol ether-diethyl ether-acetic acid (80 + 20 + 1) or (70 + 30 + 2) (Fig. 130). The last mentioned solvent is likewise good for the TLC of alcohols and diols in the "non-saponifiable" fraction of vegetable and animal fats and for separation of fatty acids, epoxy,- hydroxy- and dihydroxy-fatty acids and their methyl esters.

Methods of Detection. Very nearly all indicators commonly used for visualisation of lipids in paper chromatography can be employed for that purpose on coated plates. Moreover, corrosive spray reagents may be used for charring organic substances.

Neutral lipids have been visualised on the layer mostly with: iodine vapour [117, 118] (Rgt. No. 141); 2',7'-dichlorofluorescein [38, 118] (Rgt. No. 63); rhodamine B or 6G [78, 126] (Rgt. Nos. 220, 221); and chromic acid-concentrated sulphuric acid [120, 121] (Rgt. No. 46).

Iodine renders all unsaturated lipids and some saturated, nitrogen-containing lipids dark brown on a pale yellow or white background; less than 1 µg of a simple olefine can be thus detected; most saturated lipids colour only feebly. The brown spots generally disappear by exposing a few minutes to the atmosphere but can be repeatedly formed by re-treatment with iodine vapour. Unsaturated lipids can be visualised also by spraying with a 0.3% solution of Congo red, methyl orange or phenol red and "bleaching" with bromine vapour [176].

Almost all lipids are visualised as light green fluorescent spots on a dark violet background by spraying the chromatogram with 0.2% ethanolic 2',7'-dichlorofluorescein and inspection in UV light (270 nm). 1—5 µg of compound can be detected with this reagent [38, 118]. A similar procedure of spraying with a 0.5% solution of rhodamine B or 6G in 96% ethanol and inspection in UV light, can be used to detect amounts of lipids down to generally about 1 µg; they appear as yellow or blue-violet spots on a pink-red background. As little as 0.1 µg lipid can be detected in UV light as blue spots if a chromatogram which has been sprayed with rhodamine 6G is exposed to iodine vapour [209].

All lipids can be visualised on inorganic adsorbent layers by spraying with 50% sulphuric acid and heating. Isoprenoid compounds display characteristic colours during the carbonisation: cholesterol and its esters turn first red, then violet, brown and finally black; vitamin A and its esters become blue first, turning grey and black at higher temperatures. Chromic acid-sulphuric acid is particularly suitable for charring all involatile organic compounds on heating (180—220° C). Less than 1 µg lipid can be thus detected as a grey or black spot on a white background.

Iodine, 2',7'-dichlorofluorescein and chromic acid-sulphuric acid [121] and also the trio, rhodamine 6G, iodine and sulphuric acid [209] can be used in succession on the same chromatogram. This concecutive use of three indicators increases the probability of discovering every substance.

Less customary spray reagents (Chapter Z) are solutions of bromothymol blue, molybdophosphoric acid, antimony(III)-chloride, antimony(V)-chloride, α-cyclodextrin-iodine, fluorescein-bromine and hydroxylamine-ferric chloride. Lipids containing a conjugated double bond system absorb UV light and can be detected on silica gel layers which contain a fluorescent substance, by illumination with UV light. The radioactively labelled acetyl derivatives of lipids containing hydroxyl or amino groups and labelled esters of fatty acids may be localised by autoradiography or by means of special counters (see p. 160).

Most reactions for detecting lipids are 10–100 times more sensitive on thin-layer than on paper chromatograms. 2′,7′-Dichlorofluorescein and the rhodamines are non-destructive reagents for lipids, permitting their isolation through TLC. Iodine vapour is less suitable. If unsaturated lipids are to be isolated from thin-layer chromatograms, the layers ought not to be exposed to iodine vapour longer than necessary for localising the fractions. The iodine is only physically retained during these first few seconds and the majority of compounds are not drastically changed [39]. Alternatively, the major part of the layer can be covered with a cellophane sheet and only a narrow strip exposed to iodine vapour [148].

Applications and Results

Neutral Lipids in Microorganisms. The amount of a complex lipid mixture needed for fractionation by TLC into compound classes is only a fraction of that required in column chromatography. TLC is thus excellently suited to analysis of the lipophilic constituents of viruses, bacteria, protozoa, yeasts and algae. Attempts have been made to characterise various microorganisms, especially morbific agents which are distinguishable neither morphologically nor immunochemically, through specific "chromatographic patterns". However the isolation of pure viruses and bacteria from a higher organism or from a culture, presents a more difficult problem today than the analysis of their lipids. Thus by means of TLC, substances (siloxanes) have been found in the "endotoxin" from *Eberthella typhi*, which had been added to the growth medium to suppress foaming [28]. The "endotoxin" which had been extracted with the solution of a detergent contained di- and triglycerides, but that obtained by extraction with phenol-water did not [28].

Thin-layer and gas chromatography have been used to characterise the lipids in the protoplast membranes of some streptococci [44] and for analysis of the hydrocarbons from the micrococcus *Sarcina lutea* [2]. Diesters and alkenyl ether-esters of various diols have been found in the "triglyceride" fraction of a yeast, in the same way [12]. Using TLC, it has been possible to detect elementary sulphur in lipid extracts of a bacterium [147].

Vegetable Oils. Triglycerides form the principal component of most edible oils obtained from seeds and fruits; small amounts of free fatty acids and sterols occur also. As shown in Fig. 130, *1*, olive oil, a typical example, can be easily separated by adsorption TLC into these three fractions. Oils containing triglycerides of epoxy- and hydroxy-acids as well as those of the ordinary fatty acids show very characteristic

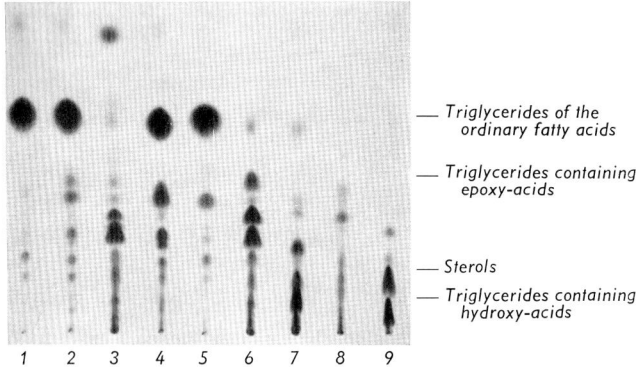

Fig. 130. Thin-layer chromatogram of various seed oils [144]. Adsorbent: silica gel G; solvent: petrol ether (BP 60—70° C)-diethyl ether-acetic acid (70 + 30 + 2); time of run: 1 h; visualisation: carbonisation by heating with chromic acid/sulphuric acid; 200 µg of each. *1 Olea europaea; 2 Malope trifida; 3 Vernonia anthelmintica; 4 Artemisia absinthium; 5 Ceiba pentandra; 6 Cephalocroton cordofanus; 7 Dimorphotheca aurantiaca; 8 Onguekoa Gore; 9 Ricinus communis*

"chromatographic patterns". Ordinary triglycerides, mono-, di- and triepoxy-triglycerides and the three corresponding classes of hydroxy-triglycerides, can be well separated from one another through adsorption TLC.

Some of the frequently used colour reactions of classical fat analysis, like the Halphen-test, can be carried out on the adsorbent layer after thin-layer fractionation; this enables us to discover which substances are responsible for these colour reactions [144]. The sterol fraction is especially easily identified with various spray reagents (see Chapter L) and functions thus as a characterising feature ("marker").

With the aid of adsorption TLC, it has been discovered that the seed oil of the Chinese tallow tree *(Sapium sebiferum)* contains up to about 25% of a fraction only a little more polar than normal triglycerides. These compounds have been characterised as mono-estolide triglycerides, i. e. (mono)hydroxy-triglycerides, esterified with ordinary fatty acids [192]. The following types of estolide-triglyceride have been detected alongside normal triglycerides in ergot oil *(Claviceps purpurea)* [141]:

N: Ordinary fatty acid (esterified with glycerol).
R: Ricinoleic acid (esterified with glycerol).
RN: Ricinoleic acid, esterified with an ordinary fatty acid and with glycerol.

The mono-, di- and tri-estolide triglycerides were isolated by TLC; the fatty acid components of each fraction were determined gas chromatographically [141].

Larger amounts of wax esters occur in some vegetable oils, e. g., jojoba wax (seed oil from *Simmondsia californica*). It is possible to separate these easily from triglycerides, using TLC [86, 118].

Diol lipids have recently been detected in maize oil (seed oil from *Zea mays*) with the help of chromatographic methods [12]. It was evident that adsorption TLC was not capable of separating the esters and alkenyl ether esters of diols from the corresponding glycerol derivatives [12] (see also [22] and [9]). Using TLC, it was nevertheless possible to fractionate the products of alkaline methanolysis of the "triglycerides" of maize oil into three fractions: methyl esters, alkenyl ethers and polyhydric alcohols. The alcohols could be separated and identified as their acetates by gas chromatography [12].

Adsorption TLC is appropriate for checking quality of pharmaceutically important fats and oils [4]. Deteriorated oils can be recognised through detecting their oxidation products [41, 155]. These appear mainly at the start point and behind the sterol fraction; they yield blue with iodide-starch (Rgt. No. 139) and red with thiocyanate-ferrous sulphate solution (Rgt. No. 115).

Adulterations of valuable vegetable oils with inferior animal fats are recognisable by detection of the zoosterols (see Chapter L) in the total oil or, with greater certainty, in the "non-saponifiable material". Added mineral oil can be especially easily detected thin-layer chromatographically through the appearance of hydrocarbons in the solvent front.

Animal Fats. Whereas the lipids of animal and human depot fats consist mainly of triglycerides, blood and various organs contain complex mixtures of hydrocarbons, wax esters, esters of sterols and polyenols, neutral plasmalogens, alkyl diglycerides, triglycerides, free fatty acids and sterols as well as numerous phospholipids and glycolipids. As seen from Fig. 131, a good separation of the non-polar lipids and free fatty acids is achieved through adsorption TLC and each lipid class can be

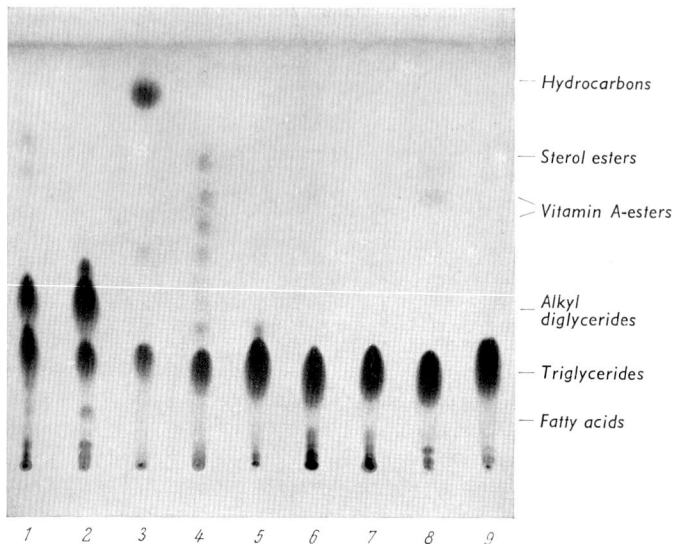

Fig. 131. Thin-layer chromatogram of marine oils [118]. Adsorbent: silica gel G; solvent: petrol ether (BP 60—70° C)-diethyl ether-acetic acid (90 + 10 + 1); time of run 1 h; visualisation: carbonisation by heating with chromic acid/sulphuric acid; 200—300 µg amounts. *1 Squalus acanthias* liver oil; *2 Hydrolagus colliei* liver oil; *3 Cetorhinus maximus* liver oil; *4 Galeorhinus galeus* liver oil; *5 Physeter macrocephalus; 6 Engraulis mordax; 7 Thunnus thynnus; 8 Onocyhynchus gorbuscha* roe; *9 Gadus morrhua* liver oil

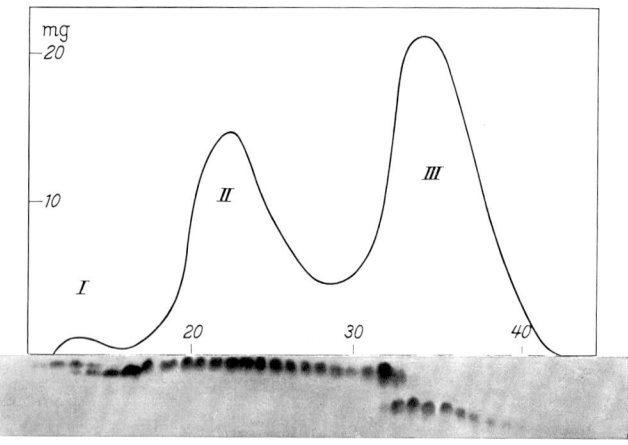

Fig. 132. Analysis of the eluate fractions of a column-chromatographic separation of *Hydrolagus colliei* liver oil. *I* neutral plasmalogens; *II* alkyl diglycerides; *III* triglycerides; (cf with Fig. 131, No. 2)

identified in the presence of the other. Sharp resolutions of this sort cannot be obtained with column chromatography or any other method, even with larger amounts of sample.

TLC has often been employed for monitoring column-chromatographic separations [26, 34, 57, 202]. Fig. 132 gives an example: the liver oil of a shark was chromatographed on a silica gel column and fractions of the eluate analysed by TLC.

Comparison of Fig. 131, 2 with Fig. 132 shows clearly that compound classes with such similar structure and physical properties as possess neutral plasmalogens (I), alkyl diglycerides (II) and triglycerides (III) can be completely separated through adsorption TLC alone (see also Fig. 134, p. 384). This permits thin-layer chromatographic isolation of pure compounds classes. For example, the neutral plasmalogens, alkyl diglycerides and triglycerides have been obtained pure from shark liver oil and the aldehydes [184], glycerol alkyl ethers [115, 184] and fatty acids [115, 185] concerned, determined.

Neutral plasmalogens migrate on adsorbent layers between alkyl diglycerides and dialkyl monoglycerides (Fig. 133). Long-chain aldehydes, which occur, possibly as hydrolysis products of neutral plasmalogens

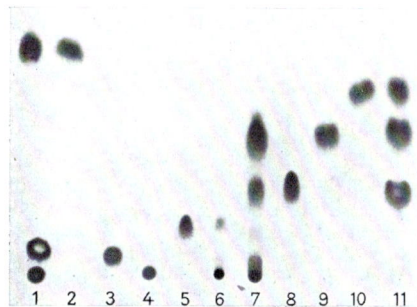

Fig. 133. Thin-layer chromatogram of synthetic and natural alkoxylipids [8]. Adsorbent: silica gel G; solvent: hexane-diethyl ether-acetic acid (90 + 10 + 1); time of run: 40 min; visualisation: charring by heating with chromic acid/sulphuric acid; about 50 µg of each of the synthetic products applied. *1* 1-octadecyl glycerol ether, 1,2-dioctadecyl glycerol ether and 1,2,3-trioctadecyl glycerol ether; *2* 1,2,3-trioctadecyl glycerol ether; *3* 1,2-dioctadecyl glycerol ether; *4* 1-octadecyl glycerol ether; *5* octadecanoic (stearic) acid; *6* non-hydrolysable lipids from the liver oil of a shark *(Hydrolagus colliei)*; *7* liver oil of a shark *(Hydrolagus colliei)*; *8* tristearin; *9* 1-octadecyldistearin; *10* 1,2-dioctadecylstearin; *11* tristearin, 1-octadecyldistearin and 1,2-dioctadecylstearin

and phosphatide plasmalogens in lipid extracts [202], their dimethylacetals and the methyl esters of fatty acids, behave similarly [185].

Aldehydes can be separated from the other compound classes by multiple development with petrol ether-diethyl ether (95 + 5) [183].

This has enabled neutral plasmalogens to be detected in the presence of free aldehydes [183]; combined and free aldehydes to be separated [183]; and aldehydes, dimethylacetals and methyl esters to be fractionated and isolated [185] (see procedures, p. 373). Neutral plasmalogens can be detected by means of the "SRS-technique" (p. 88): after chromatographing in the first direction, the alkenyl ethers are hydrolysed on the layer with hydrogen chloride vapour; the aldehydes and the diglycerides formed are then separated by chromatography in the second direction [183]. Fig. 134 illustrates the detection of neutral plasmalogens in a shark liver oil by means of TLC.

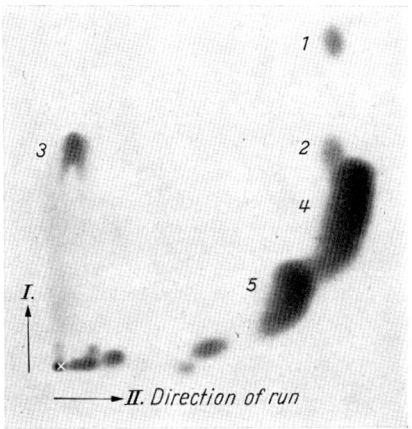

Fig. 134. Detection of neutral plasmalogens in the liver oil of a shark ("Ratfish", *Hydrolagus colliei*) [183]. Adsorbent: silica gel G; solvent: 1 direction: petrol ether (40—60° C)-diethyl ether (95 + 5). After development in the first direction the plate was exposed 5 min to HCl vapour; solvent 2 direction: petrol ether (40—60° C)-diethyl ether (80 + 20); times of run: 40 min in both cases; visualisation: carbonisation by heating with chromic acid/sulphuric acid; amounts: 300 µg liver oil plus 15 µg palmitaldehyde. *1* palmitaldehyde; *2* aldehydes from neutral plasmalogens; *3* diglycerides from neutral plasmalogens; *4* alkyl diglycerides; *5* triglycerides

Two groups have reported that lipid extracts of mouse tumours contain compounds which migrate ahead of the triglyceride fractions on adsorbent layers [113, 191]; these substances were considered at first to be methyl esters of fatty acids but more detailed studies of the material isolated through TLC, showed that they were alkyl diglycerides [191].

The non-polar components of waxes can be isolated also as compound classes by adsorption TLC and analysed in a subsequent gas chromatographic procedure; these are long-chain hydrocarbons and esters of fatty acids with aliphatic or alicyclic alcohols [2, 149].

Esters of long-chain fatty acids with ethanediol and other diols occur in beef lung [22], beef suet and pig lard [22] and in rat liver [12]. Diesters, ether esters and dialkyl ethers of ethanediol cannot be separated from the corresponding glycerol derivatives through adsorption TLC but this separation succeeds with reversed phase partition TLC [9].

Several articles describing the application of adsorption TLC to studies of metabolism are quoted in Chapter S, p. 592—599.

Neutral Lipids of Human Tissue. A thin-layer chromatogram of lipids in human serum and organs is reproduced in Fig. 135.

Several compound classes which could not be found using other methods, have been detected in lipid extracts of human tissue by means of adsorption TLC. These include neutral plasmalogens and alkyl diglycerides, which seem to occur in almost all organs. Small amounts of these substances were isolated from the fat of the perinephrium; the aldehydes and fatty acids from the neutral plasmalogens, the glycerol alkyl ethers and fatty acids of the alkyl diglycerides and the fatty acid components of the triglycerides were determined gas chromatographically [181] (see Fig. 136).

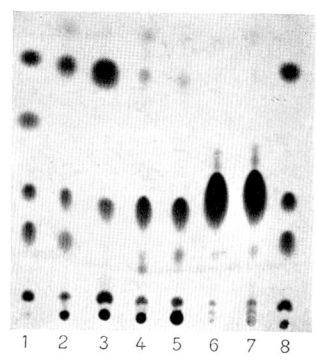

Fig. 135. Thin-layer chromatogram of the lipids from human tissue [202]. Adsorbent: silica gel G; solvent: petrol ether (BP 60—70° C)-diethyl ether-acetic acid (90 + 10 + 1); time of run: 1 h; visualisation: carbonisation by heating with chromic acid/sulphuric acid; amounts: ca 200 µg of each; *1* artificial mixture (cholesterol, oleic acid, triolein, methyl oleate and cholesteryl oleate); *2* serum; *3* aorta "calcification" *4* liver; *5* kidneys; *6* depot fat; *7* bone marrow; *8* artificial mixture (lecithin, cholesterol, oleic acid, triolein and cholesteryl oleate)

Using TLC, several uncommon sterols have been detected in the aorta, especially that of new-born babies [202]. Methyl esters and, in some cases, aldehyde trimers (1,3,5-trioxans) have also been discovered in lipid extracts of human organs. It is assumed, however, that these compounds were formed from free fatty acids or aldehydes, respectively, during extraction of the tissue and storage of the extracts [202].

Adsorption TLC has been used to analyse the products of enzymatic hydrolysis of lipids in both artificial mixtures [61] and the living organism [60]. With the help of chromatography on hydroxylapatite (see Fig. 137), it has been possible to confirm that the lipids in the intestinal contents consist largely of symmetrical monoglycerides [60].

Only a few of the numerous articles on the TLC of blood lipids can be quoted here [e. g. 34, 47, 180, 203]. Pathological changes in metabolism

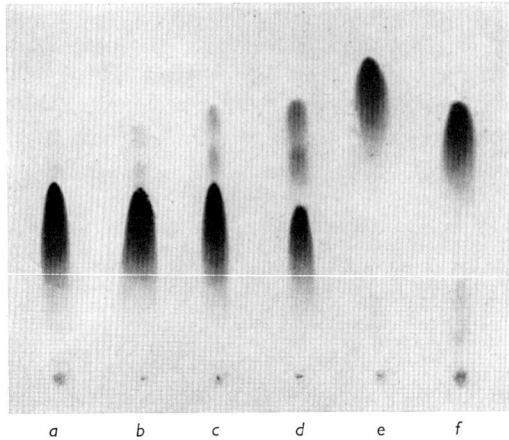

Fig. 136. Isolation of neutral plasmalogens and alkyl diglycerides from the human perinephrium [181]. Adsorbent: silica gel G; solvent: hexane-diethyl ether (99 + 5), double development; times of run: 40 min each; visualisation: carbonisation by heating with chromic acid/sulphuric acid. *a* lipid extract; *b* first concentration stage; *c* second concentration stage; *d* third concentration stage; *e* neutral plasmalogens; *f* alkyl diglycerides

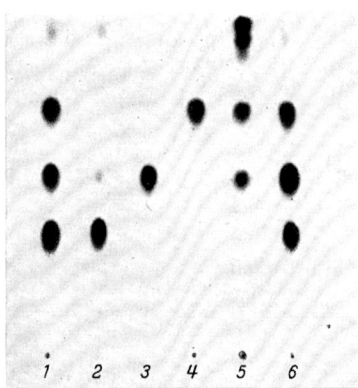

Fig. 137. Separation of isomeric monoglycerides and detection of β-monoglycerides in the human intestine [60]. Adsorbent: hydroxylapatite; solvent: methyl isobutyl ketone; temperature: +10° C; time of run: 1 h; spray reagent: molybdophosphoric acid in ethanol solution; amounts: 10 μg of each. *1* artificial mixture of the compounds *2*, *3* and *4*; *2* α-monoolein; *3* β-monoolein; *4* oleic acid; *5* lipids from the intestinal contents; *6* artificial mixture of the compounds *2*, *3* and *4*

have frequently been detected through chromatographic analysis of the lipids in blood and other body fluids [87, 180, 202, 224]. Examples of practical application of thin-layer and gas chromatography in clinical diagnosis are treated in detail in Chapter S (see also [225]).

Fatty Acids. Ordinary fatty acids, epoxy-, hydroxy- and keto-acids can be separated into compound classes by adsorption chromatography. This applies to the corresponding methyl esters also [78, 82, 134, 135, 205]. Even certain esters which differ only in the position of the functional group in the chain, can be separated [140]. An example of this, a series of hydroxystearate positional isomers, is shown in Fig. 138.

Systematic analyses of the fatty acids in oils from unusual seeds has led to the discovery of several epoxy-acids [134, 135] and hydroxy-acids [94, 134]. With the help of adsorption TLC it has been possible to show that the esters of vicinally unsaturated hydroxy-acids suffer dehydration during gas chromatography [133].

Adsorption TLC has been used with success also for following the metabolism of epoxy-acids [23] and hydroxy-acids [156]. Several publications, based on corresponding investigations on unsubstituted acids, are mentioned in Chapter I and on p. 653.

See p. 399 for applications of TLC to the analysis and isolation of prostaglandins, a class of biologically active fatty acids.

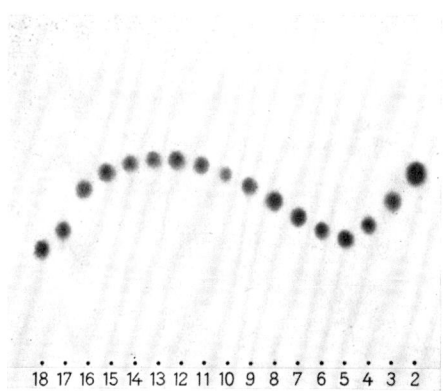

Fig. 138. Thin-layer chromatogram of the methyl esters of isomeric hydroxystearic acids [140]. Adsorbent: silica gel G; solvent: petrol ether (BP 60—70° C)-diethyl ether (50 + 50); time of run: 1 h; visualisation: carbonisation by heating with 50% sulphuric acid. The numbers *18* . . . . . *2* correspond to the position of the hydroxyl group in the esters

Technically important Fatty Acid Derivatives. Nitrogen-containing lipids such as primary, secondary and tertiary amines, quaternary ammonium bases, amides and nitriles, all of which find many varied applications in the plastic and textile industries, can be separated into classes on silica gel G by using ammonia-containing solvents (see Table 68) [19, 121, 162]. The first four solvents in Table 68 have been used in succession in the stepwise development technique (see p. 87) for analysing complex mixtures [121].

Fractionation according to molecular weight, of polyethylene oxide adducts of fatty alcohols and of fatty amines, has been carried out on silica gel G, using butanone-water (50 + 50) and butanone-2.5% aqueous ammonium hydroxide (50 + 50) respectively [18]. Free polyethylene glycols can be detected in adducts by means of TLC [154]. Procedures for separating esters of long-chain fatty acids with sugars, have been developed [46, 92, 93].

25*

A method for the isolation of saturated mono- and dihydric alcohols by chromatography of their nitrates has been described [216].

Strongly polar and ionic surface active substances such as alkyl sulphates and alkyl sulphonates, phosphates and phosphonates, can also be fractionated thin-layer chromatographically [33, 121]. Mixtures of these detergents have been separated on silica gel G containing 10% ammonium sulphate, using the mixtures of chloroform-methanol and sulphuric acid given in Table 68 [121].

Table 68. *Solvents for separation of technically important fatty acid derivatives on silica gel G*

Solvent	Composition, v/v	Reference
Petrol ether[a]-benzene	95 + 5	[121]
Benzene-N ammonium hydroxide[b]	90 + 10	[121]
Ammoniacal chloroform-methanol (chloroform is mixed with N ammonium hydroxide in the proportion 90 + 10)	97 + 3	[121]
Acetone-14 N ammonium hydroxide	90 + 10	[121]
Acid chloroform-methanol[c] (the methanol contains 5% 0.1 N sulphuric acid)	97 + 3	[121]
Acid chloroform-methanol (the chloroform contains 5% 0.1 N sulphuric acid)	80 + 20	[121]

[a] Usually petrol ether, BP 60—70° C, i. e., principally hexane.
[b] The aqueous layer is discarded.
[c] On silica gel G which contains 10% ammonium sulphate.

b) Phospholipids, Sulpholipids and Glycolipids

Phospholipids and similar substances are fractionated into compound classes more according to partition than adsorption effects, in contrast to the separation of non-polar lipids. Alkenyl ether-esters ("plasmalogens") are usually not entirely separated from the corresponding ether-ester and diester-phosphatides. Phosphatide plasmalogens concentrate in thin-layer chromatograms on the forward edge of the phosphatide spots.

Intensive application of TLC has contributed to structural elucidation of complex compounds, such as the gangliosides (see p. 393—394).

Polar lipids are applied to the layers as 0.1 or 1% solutions in chloroform or chloroform-methanol (50 + 50).

Experimental Conditions

Adsorbents. Deactivated adsorbents are the most suitable for fractionating ionic and neutral polar lipids. Examples are silica gel G containing 10% ammonium sulphate [67, 121] or 10% sodium acetate [67]. Especially good separations are achieved on air dried silica gel H or silica gel H containing 10% magnesium silicate [177] or about 0.1% sodium carbonate [187, 188].

Cellulose layers (p. 32) also are suitable for separation of polar lipids and their water-soluble hydrolysis products [126, 221]. The use of ion exchange layers (see p. 44) has a promising future.

A maximum of 10 mg of a mixture of polar lipids can be fractionated on a 20 × 20 cm layer.

Solvents. Mixtures of chloroform and methanol containing a little water have been the principal solvents used for separations of phospholipids [128, 150, 163, 210]. They have been combined in two-dimensional procedures with basic [1, 69, 111, 186] or acidic [1, 111, 127, 177] solvents. The solvents customarily employed for TLC of phospholipids can be used also for separating sulpholipids [69, 210]. Glycolipids have been chromatographed with chloroform-methanol-water [73, 150, 172, 198] but especially with aqueous propanol [106, 107, 163, 199].

Table 69. *Solvents for TLC of phospholipids, sulpholipids and glycolipids on silica gel*

Solvent	Composition, v/v	Reference
Chloroform-methanol-water	65 + 25 + 4	[210]
	60 + 35 + 8	[210]
Chloroform-methanol-10% ammonium hydroxide	60 + 35 + 8	[146]
n-Propanol-water	70 + 30	[106]
n-Propanol-12 N ammonium hydroxide	80 + 20	[69]
n-Butanol-pyridine-water	60 + 40 + 20	[95]
Chloroform-methanol-7 N ammonium hydroxide, then chloroform-methanol-7 N ammonium hydroxide in the second direction	60 + 35 + 5 35 + 60 + 5	[186]
Chloroform-methanol-water, then n-butanol-acetic acid-water in the second direction	65 + 25 + 4 60 + 20 + 20	[177][a]
Chloroform-methanol water-28% ammonium hydroxide, then chloroform-acetone-methanol-acetic acid-water in the second direction	130 + 70 + 8 + 0,5 50 + 20 + 10 + 10 + 5	[159a]

[a] On silica gel H containing 10% magnesium silicate.

Table 69 presents some solvents of good repute for the fractionation of phospholipids sulpholipids and glycolipids on silica gel layers and on paper impregnated with silica gel [126]. Some of these solvents are suitable also for fractionating acyl carnitines [220].

Methods of Detection. Most of the spray reagents usual in paper chromatography may be used for visualisation of phospholipids: ninhydrin solution (Rgt. No. 178) for aminophosphatides [210]; the Dragendorff reagent (No. 97) for choline-containing phosphatides [210]; and molybdate-perchloric acid (Rgt. No. 166) for all phospholipids [210]. Reagents employed for detecting sugars (see Chapter X) can be used also for glycolipids. Universal indicators for lipids are alcoholic rhodamine B solution (Rgt. No. 220), weakly alkaline bromothymol blue solution (Rgt. No. 31) or iodine vapour. Carbonisation with 50% sulphuric acid or with dichromate-sulphuric acid (Rgt. No. 46) has proved particularly satisfactory. These and other spray reagents are described in detail in Chapter Z.

It cannot be too strongly recommended that several reagents be sprayed in succession on to a chromatogram, so that no substance escapes identification. This is evident from Table 70.

Table 70. *Colour reactions for distinguishing various phospholipids and glycolipids* [210] (see also [173])

Substance	hRf-value	Ninhydrin[a] No. 178	Dragendorff Rgt. No. 97	Ammonium molybdate-perchloric acid[a]
Amino acids	0 — 10	+	—	+
Lysolecithin	21 ± 3.7	—	+	+
Lecithin	39 ± 5.5	—	+	+
Sphingomyelin	29 ± 5.5	—	+	+
Ethanolamine-cephalin	57 ± 7.5	+	—	+
Cerebrosides (non-esterified)	78 ± 7.5	—	—	+
Cardiolipin	92 ± 1.5	—	—	+

[a] + positive reaction; — no reaction.

Applications and Results

Polar Lipids in Microorganisms. Bacteria are generally rich sources of uncommon polar lipids. Thus, staphylococci contain large amounts of various O-amino acid esters of phosphatidyl glycerol [16]. The organisms of the PLT-group, closely related to bacteria, contain up to 60% by weight of lipids [74]. These microorganisms, which provoke *p*sittacosis, *l*ymphogranulomatosis and *t*rachoma, are morphologically indistinguishable; they can, however, be characterised thin-layer chromatographically by virtue of the qualitative and quantitative differences in their polar lipids [74].

A two-dimensional separation procedure has been employed for analysing the phospholipids from yeasts; the lipids were first fractionated into classes, the fractions then subjected to methanolysis on the layer and the methyl esters chromatographed in the second dimension [91].

Thin-layer and gas chromatography should find extensive application in microbiology as taxonomic and diagnostic aids.

Polar Plant Lipids. The best known phosphatides can be fractionated into classes with the help of TLC; this is shown in Fig. 139. Even extracts of green plant parts which contain many polar lipids, can be satisfactorily separated if the established experimental conditions are respected [150, 151].

Up to 20 different lipid classes, some of which have not yet been identified, have been detected in lipid extracts of potatoes [111], wheat endosperms [128], lettuce- and cabbage leaves [150] and spinach leaves [57] by means of two-dimensional TLC on silica gel G. TCL proved to be a useful tool for isolating the phosphatidyl glycerols from spinach leaves [57].

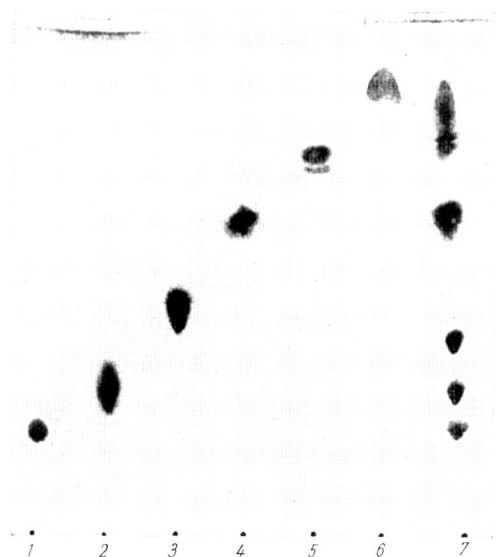

Fig. 139. Thin-layer chromatogram of polar lipids [210]. Layer: silica gel G; solvent: chloroform-methanol-water $(65 + 25 + 4)$; time of run: 2 h; spray reagents: rhodamine B in ethanol (UV light) and Dragendorff-reagent; amounts: 50—100 µg of each. *1* lysolecithin; *2* sphingomyelin; *3* lecithin; *4* ethanolamine-cephalin; *5* cerebrosides; *6* cardiolipin; *7* artificial mixture of compounds 1—6

A detailed account of the TLC of phospholipids, sulpholipids and glycolipids has recently been published [151].

Polar Animal Lipids. The experimental conditions quoted in the first articles [70, 210] on thin-layer chromatographic separation of phospholipids in animal and human organs, have proved satisfactory. They have needed modification, however, in a few special problems [67, 163, 188, 200]. The range of application of TLC has been appreciably extended recently by introduction of the two-dimensional technique [159a, 177, 186] (see Fig. 151).

Animal and human organs contain ether-ester-, alkenyl ether-ester- and diester phosphatides, derived from ethanolamine or choline; diether- and dialkenyl ether phosphatides may occur in smaller amounts.

The available methods are suitable only for fractionation according to the *basic* moieties, whereas phosphatides differing in the lipid moieties are not separated from each other. Compounds which contain differing functional groups in the *lipid moiety* can be indirectly separated from each other using the following procedure [174]: the enzyme phospholipase C from bacteria (*Clostridium perfringens* or *Bacillus cereus*) splits off phosphoryl ethanolamine and phosphoryl choline from the phosphatides. The lipophilic hydrolysis products can then be separated as acetyl derivatives (I, II, III) on adsorbent layers.

$$\begin{array}{ccc}
H_2C-O-R & H_2C-O-CH=CH-R & H_2C-O-\underset{\underset{O}{\|}}{C}-R \\
HC-O-\underset{\underset{O}{\|}}{C}-R' & HC-O-\underset{\underset{O}{\|}}{C}-R' & HC-O-\underset{\underset{O}{\|}}{C}-R' \\
H_2C-OAc & H_2C-OAc & H_2C-OAc \\
\text{I from} & \text{II from alkenyl ether-} & \text{III from diester} \\
\text{ether-ester} & \text{ester phosphatides} & \text{phosphatides} \\
\text{phosphatides} & &
\end{array}$$

The long-chain moieties of the acetyl derivatives I, II and III can be analysed with the usual methods (see procedures, p. 372—374, also Table 71). These procedures have been used for analysing, for example, the phosphatides in hens' eggs and in ox brain [174].

Plasmalogens in mixtures with other phospholipids can be detected and subsequently quantitatively determined using the SRS-technique also (p. 88) [67a, 157].

In recent years, several new glycolipids have been isolated in crystalline form from bovine brain. One of these fractions has been identified as an ether-ester-glycerol galactoside [153]. Alkyl ethers of cerebrosides and the corresponding alkenyl ethers, "sphingoplasmalogens", have been recently isolated for the first time from mixtures of sphingomyelins and sugar-containing sphingolipids in bovine brain, using TLC [101]. Cerebroside fractions of markedly varying fatty acid composition have been found in bovine brain [100] and spinal cord [64]. Fig. 140 reproduces a thin-layer chromatogram of various sphingolipids from beef brain.

Several authors have reported the separation of gangliosides from animal organs, using TLC [32, 107, 146, 218]. TLC has however been devoted in particular to isolation of *human* gangliosides and to the analysis of their degradation products. This is discussed in the following section.

Polar Lipids in Human Tissue. Chromatographic analyses of lipids in blood and cerebrospinal fluid are evaluated for purposes of clinical diagnosis (see Chapter S, p. 592—599).

Investigations of brain lipids count probably amongst the most impressive applications of the Stahl method. For example, differences between the brain lipids of juveniles and of adults have been detected with the help of TLC [177, 199]. Various "lipidoses", diseases associated with marked qualitative and quantitative lipid changes in the white and grey brain matter, have been likewise clearly demonstrated [32, 69,

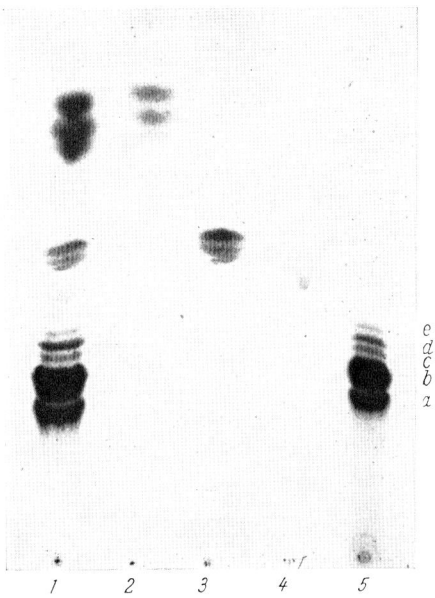

Fig. 140. Thin-layer chromatogram of sphingolipids [210]. Adsorbent: silica gel G; solvent: chloroform-methanol-water (60 + 35 + 8); time of run: 2 h; spray reagents: diphenylamine solution and Dragendorff reagent for detection of sphingomyelin; amounts: 50—100 μg of each: *1* sphingolipid mixture from beef brain; *2* non-esterified cerebrosides; *3* cerebroside sulphates; *4* sphingomyelin; *5* gangliosides (a b c d e)

104, 146]. It has been shown that the lipid composition of the brain of mentally diseased adults is often similar to that of normal infants: much more sulphatide is found in cases of metachromatic leucodystrophy [69, 72, 177]; large amounts of one ganglioside appear in Tay-Sachs disease [96, 177]; and a greatly increased sphingomyelin content of the brain is found with the Niemann-Pick disease [164, 177]. The cerebroside sulphate esters of the brain of normal persons and of those suffering from metachromatic leucodystrophy have been isolated by TLC and shown to be identical [69].

Several gangliosides have been isolated from blood [198] and brain [95, 96, 98, 104, 107]. A thin-layer chromatogram of a ganglioside fraction and the compounds obtained from it, is shown in Fig. 141.

It has been shown with the help of TLC that when brain is stored in formalin, the gangliosides rapidly change chemically [197]. TLC has been valuable in the analysis of hydrolysis and decomposition products in structure elucidation of the gangliosides [95, 96, 106, 107] and similar glycolipids [45, 49, 97, 153, 215] (see p. 598).

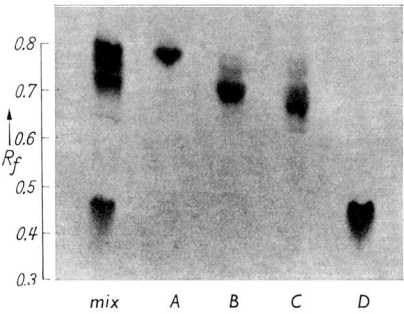

Fig. 141. Thin-layer chromatogram of a mixture of gangliosides (Mix.) and individual gangliosides (A, B, C and D) from human brain [96]. Layer: silica gel G; solvent: n-butanol-pyridine-water $(3 + 2 + 1)$; time of run: 2.5—3 h; amounts: ca 50 µg of each; spray reagent: Bial's reagent

2. Fractionation of Pure Compound Classes

There is no procedure which permits complete separation in a single step, of naturally occurring mixtures of long-chain compounds which differ in the nature and number of their functional groups and in chain length and degree of unsaturation. Several methods, each based on a different principle, are therefore used in succession for fractionating complex mixtures. Lipid extracts are usually first fractionated into compound classes by adsorption chromatography. Each of these classes in then separated subsequently into simpler groups of substances and finally into pure individual compounds, applying two or three of the principles given below. Any of these principles of chromatography may be used in TLC.

1. *Argentation chromatography*. This method depends on the reversible formation of π-complexes of unsaturated lipids with silver ions. Adsorbents containing 5% silver nitrate are used in TLC.

Mixtures of saturated and unsaturated members of a lipid class are fractionated on silver nitrate-impregnated layers *according to number, configuration and position of the double bonds*. Homologous saturated or unsaturated compounds cannot be separated from one another.

Both the saturated and unsaturated lipids can be eluted unchanged from the adsorbent.

2. *Chromatography of mercuric acetate adducts.* Unsaturated lipids form acetoxymercuri-methoxy compounds with methanolic mercuric acetate. These mercuric acetate adducts of unsaturated compounds must be prepared *before* chromatography in contrast to the silver complexes which are formed on the silver nitrate-containing adsorbent *during* chromatography.

Saturated lipids and the adducts of unsaturated substances, are fractionated on adsorbents according to the *number of the acetoxymercurimethoxy groups*. Saturated homologues are not separated from one another. The adducts of homologous, *cis-trans-* and positionally isomeric lipids also cannot be separated.

Saturated compounds can be recovered unchanged directly; unsaturated compounds after decomposing their adducts with hydrochloric acid.

3. *Chromatography of ozonides.* Unsaturated lipids react irreversibly with ozone. The ozonides are prepared *before* chromatography.

Mixtures of saturated lipids and the ozonides of unsaturated lipids are fractionated according to the *number of ozonide groups*. Saturated homologues are not separated from each other, nor, as a rule, are the ozonides of homologues, *cis-trans* and positional isomers.

Saturated compounds can be isolated unchanged after chromatography. The ozonides can be eluted from the adsorbent but the original unsaturated compounds cannot be recovered.

4. *Reversed-phase partition chromatography.* This method utilises differences in the solubilities of the components of a mixture; the substances undergo no chemical change. Adsorbents impregnated with silicones or paraffins are used as stationary phase.

Mixtures of saturated and unsaturated lipids of a class are fractionated *in accordance with the chain length and number of double bonds: cis-trans* isomers and unsaturated positional isomers are generally inseparable (see [160], however). Saturated and unsaturated members of a lipid class which differ from one another by two methylene groups and a double bond, are usually not or only incompletely separated (see p. 410, however).

Adsorption TLC can be combined with argentation-TLC or reversed phase partition TLC on a single layer, using two-dimensional technique. The two last named procedures can be combined also.

These methods are supplemented by gas chromatography, a partition chromatographic procedure. Conversely, adsorption TLC is an ideal method for isolating a compound class for gas chromatographic analysis (see [65] and Chapter F).

a) Argentation Chromatography

The first publications on the separation of lipids on silvernitrate-impregnated adsorbent columns [207] and layers [6, 136, 208] appeared in 1962/63. Argentation-TLC quickly gained favour and has become one of the indispensable aids in the chemistry of lipophilic natural products.

The Brockmann rules (see p. 201) for pure adsorption chromatography are valid with certain limitations for argentation (adsorption) chromatography too. Hydrocarbons are eluted before esters and these before aldehydes, alcohols and acids (cf. Fig. 129, p. 375); the subfractionation of

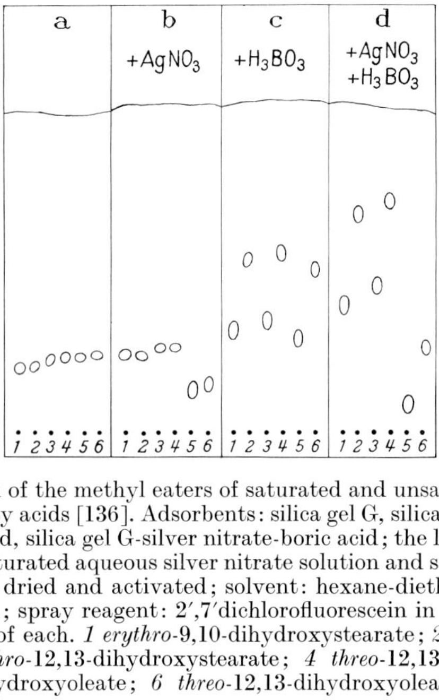

Fig. 142. Separation of the methyl eaters of saturated and unsaturated *erythro*- and *threo*-dihydroxy fatty acids [136]. Adsorbents: silica gel G, silica gel G-silver nitrate, silica gel G-boric acid, silica gel G-silver nitrate-boric acid; the layers were prepared by spraying with saturated aqueous silver nitrate solution and saturated methanolic boric acid solution, dried and activated; solvent: hexane-diethyl ether (40 + 60); time of run: 40 min; spray reagent: 2′,7′dichlorofluorescein in ethanol (UV light); amounts: ca 30 μg of each. *1 erythro*-9,10-dihydroxystearate; *2 threo*-9,10-dihydroxystearate; *3 erythro*-12,13-dihydroxystearate; *4 threo*-12,13-dihydroxystearate; *5 erythro*-12,13-dihydroxyoleate; *6 threo*-12,13-dihydroxyoleate (see also [138])

unsaturated substances within these compound classes is however much more clear-cut on adsorbents containing silver nitrate than on non-impregnated adsorbents. Compounds which differ by only a single double bond and also unsaturated *cis-trans* isomers and positional isomers, can be separated from each other by adsorption TLC. An example is in Fig. 142b-separation of the methyl esters of saturated and unsaturated dihydroxy-fatty acids on silica gel G impregnated with silver nitrate. Isomers of dihydroxyesters are not fractionated on such a layer but they can be separated into the *erythro*- and *threo*-forms on silica gel G impregnated with boric acid (Fig. 142c) (see also Chapter X on sugars).

Saturated and unsaturated compounds are not separated on boric acid-containing adsorbents. From Fig. 142d it is seen that saturated and unsaturated dihydroxy-esters are fractionated according to both the number of double bonds and the configuration of the hydroxyl groups, when chromatographed on a silica gel G layer, impregnated with both silver nitrate *and* boric acid.

On adsorbent layers containing silver nitrate, cyclic olefines and their derivatives migrate less far than the corresponding straight chain compounds. Thus, for example, methyl chaulmoograte and methyl oleate can be clearly separated [124].

The influence of the π-complex on the chromatographic behaviour is so strong that lipids containing several double bonds migrate less far than many saturated compounds containing relatively polar functional groups (cf. Figs. 143 A and B on p. 401).

Lipids are applied to the layers as 0.1 or 1% solutions in hexane or hexane-diethyl ether (50 + 50).

Experimental Conditions

Adsorbents. Silica gel G layers containing 5% silver nitrate are those almost exclusively used [142, 143]. Usually 23.75 g silica gel G are mixed with a solution of 1.25 g silver nitrate in 50 ml water and this slurry spread on to glass plates as in the procedure given on p. 56 [143]. Occasionally, impregnation of silica gel G with 10% instead of 5% silver nitrate is advisable, e. g., for separating at low temperature (see p. 94) the methyl esters of positional isomers of unsaturated fatty acids [142]. Aqueous silver nitrate solutions attack metal parts of most equipment; it is thus best to use a silver plated spreader or an apparatus made of stainless steel (see p. 56).

The following spreading procedure is best for preparing partly impregnated layers: after removing the jacket, the spreader block is divided with a plastic partition into a 5 cm and a 15 cm long compartment; the open ends are closed with rubber stoppers. The slurry for the smaller compartment is then prepared from 8 g silica gel G + 16 ml water and for the larger compartment, from 22 g silica gel G + a solution of 1.5 g silver nitrate in 44 ml water; both compartments are *filled at the same time* and spreading carried out [182].

An already coated plate can be partly or completely impregnated by spraying with 10 to 20% aqueous or methanolic silver nitrate solution [136]. A similarly simple procedure is to allow a 5% aqueous silver nitrate solution to ascend through the still moist layer [30].

Silica gel G layers which have been impregnated with silver nitrate are air dried like ordinary layers and then activated by heating 30 min at 110° C (see p. 60). They are rapidly deactivated by atmospheric moisture and should be stored always over saturated calcium chloride solution [143].

Solvents. Table 71 contains solvent mixtures suitable for fractionating the most important lipid classes. Besides these solvents, the following have also been often used: the methyl esters of unsaturated, positionally isomeric fatty acids have been separated from one another by triple development at $-15°$ C using toluene [142]; triglycerides have been fractionated with chloroform-carbon tetrachloride-methanol-acetic acid $(50 + 50 + 1.5 + 0.5)$ [6, 7]. The separation of lecithins on silica gel H, impregnated with silver nitrate [5], is very satisfactory using chloroform-methanol-water $(65 + 25 + 4)$ [5] (see also [57, 88]).

Methods of Detection. Lipids can be visualised in UV light by spraying with alcoholic 2',7'-dichlorofluorescein solution (Rgt. No. 63) [136]. Aqueous solutions of sulphuric acid (Rgt. No. 241) [136] and ortho-phosphoric acid (Rgt. No. 208) [6] are used for charring organic compounds.

Applications and Results

Argentation-TLC has been often used to isolate groups of compounds from pure compound classes such as methyl esters [136], wax esters [52], cholesteryl esters [137] and triglycerides [75] (see also [108, 109]), which are then fractionated further by gas chromatography (see Figs. 145 and 146, p. 405). Conversely, complex mixtures of methyl esters can be first separated through gas chromatography and their components then characterised by argentation-TLC [178].

In the analysis of triglycerides, generally only the composition of the fatty acids in the individual fractions is determined gas chromatographically after methanolysis [24, 34, 36, 48] (see p. 372). Gas chromatography of intact triglycerides is possible [108, 109].

Coupling of thin-layer and gas chromatography is discussed in detail in Chapter F.

It has been possible to isolate several hitherto unknown fatty acids (mostly as their methyl esters) with the help of argentation-TLC. The relevant work has been summed up in a recent publication [143].

Reference may be made here to a procedure for separating mixtures of mono-olefinic fatty acids and for subsequently determining the structure of the components [11]. This method utilises two-dimensional TLC on silica gel, impregnated with paraffin and silver nitrate; a similar procedure is described in [225]. A scheme for determination of the position and configuration of the double bonds in polyolefinic acids [170] depends on partial hydrogenation with hydrazine, separation of the reduction products through argentation-TLC and determination of their structure by ozonolysis and gas chromatographic analysis of the fission products [170].

Several new epoxy- and hydroxy-acids have been discovered in seed oils. The isolation of a unique compound, 8-(5-hexylfuryl-2)-octanoic acid, from the seed oil of *Exocarpus cupressiformis*, is of particular interest [145].

Procedures for analysing and isolating the various prostaglandins, biologically active cyclic hydroxy-acids, have been published [20, 50].

Argentation-TLC has been especially useful for fractionating the triglycerides of fats and oils. It was recognised early that the chromatographic behaviour of these compounds was not determined alone by the number and configuration of their double bonds [7, 51]. The two double bonds in the linoleic acid group (2) form a more stable complex with silver nitrate than do the two double bonds in two oleic acid moieties (1); and the three double bonds in the linolenic acid group (3) yield stronger complexes than the four double bonds in two linoleic acid groups. Thus, for example, stearodilinolenin (033) with 6 double bonds is more strongly adsorbed than linolenodilinolein (322) with 7. A systematic investigation of synthetic triglycerides yielded the following sequence of elution [51]: 000, 100, 110, 200, 111, 210, 211, 220, 300, 221, 310, 222, 311, 320, 321, 322, 330, 331, 332, 333.

The position of an acyl group in the triglyceride molecule also influences its chromatographic behaviour; thus 1-oleodistearin can be separated from 2-oleodistearin [7].

Table 71. *Solvents for fractionation of pure compound classes using reversed phase partition TLC and argentation-TLC* [123]

Compound class	Adsorption TLC[a] hR_f	Solvents for argentation-TLC (5% AgNO$_3$ in silica gel G)	Solvents for partition TLC in reversed phase (paraffin oil on kieselguhr G)
Hydrocarbons	98	Hexane-ether (95 + 5)	
Alcohols	30	Hexane-ether (50 + 50)	Acetic acid-water (75 + 25)[c]
as acetates	78	Hexane-ether (85 + 15)	Acetic acid-water (85 + 15)[c]
Aldehydes	73	Hexane-ether (80 + 20)	Acetic acid-water (85 + 15)[c]
as dimethyl-acetals	73	Hexane-ether (85 + 15)	Acetone-water (90 + 10)[c]
Methyl ketones	63	Hexane-ether (75 + 25)	Acetic acid-water (90 + 10)[c]
Acids	41	Hexane-ether-acetic acid (50 + 50 + 1)	Acetic acid-water (80 + 20)[c]
as methyl esters	77	Hexane-ether (85 + 15)	Acetic acid-water (85 + 15)[c]
Wax esters	88	Hexane-ether (90 + 10)	
Glycerol alkyl ethers	3	Ether-methanol (90 + 10)	Tetrahydrofuran-water (60 + 40)[b]
as dimethyl-ketals	62	Hexane-ether (75 + 25)	
as acetates	34	Hexane-ether (60 + 40)	Acetic acid-water (60 + 40)[c]

Table 71 (Continued)

Compound class	Adsorption TLC[a] hRf	Solvents for argentation-TLC (5% AgNO$_3$ in silica gel G)	Solvents for partition TLC in reversed phase (paraffin oil on kieselguhr G)
as alkoxyacet-aldehydes	25	Hexane-ether (15 + 85)	Acetic acid-water (85 + 15)[c]
Glycerol dialkyl ethers	30	Hexane-ether (50 + 50)	Chloroform-methanol-water (25 + 75 + 5)[b]
as acetates	63	Hexane-ether (75 + 25)	Acetic acid-water (90 + 10)
Glycerol trialkyl ethers	85	Hexane-ether (80 + 20)	Acetone-ether (50 + 50)[b]
Monoglycerides	2	Ether-methanol (90 + 10)	Tetrahydrofuran-water (50 + 50)[b]
as dimethyl-ketals	60	Hexane-ether (70 + 30)	
as acetates	30	Hexane-ether (60 + 40)	Acetic acid-water (60 + 40)[c]
as acylacet-aldehydes	23	Hexane-ether (15 + 85)	Acetic acid-water (85 + 15)[c]
Diglycerides	21	Hexane-ether (15 + 85)	Chloroform-methanol-water (25 + 75 + 15)[b]
as acetates	45	Hexane-ether (70 + 30)	Acetic acid-water (90 + 10)[c]
Triglycerides	62	Hexane-ether (70 + 30)	Acetic acid-water (99.5 + 0.5) Acetone-acetonitrile (70 + 30)
Alk-1-enyl diglycerides ("Neutral plasmalogens")	82	Hexane-ether (75 + 25)	Acetone-ether (50 + 50)[b]
O-Alkyl diglycerides	78	Hexane-ether (75 + 25)	Acetone-ether (50 + 50)[b]
O,O-Dialkyl glycerides	88	Hexane-ether (80 + 20)	Acetone-ether (50 + 50)[b]
α-Hydroxy-acids	11	Hexane-ether-acetic acid (20 + 80 + 1)	Acetic acid-water (75 + 25)
as methyl esters		Hexane-ether (40 + 60)	Acetic acid-water (80 + 20)[c]
as acetates		Ether-methanol (90 + 10)	Acetic acid-water (80 + 20)[c]
as acetylated methyl esters		Hexane-ether (60 + 40)	Acetic acid-water (80 + 20)[c]
Hydroxyethyl-amides of acids	0		Tetrahydrofuran-water (40 + 60)[b]
as acetates		Ether-methanol (90 + 10)	Tetrahydrofuran-water (50 + 50)[b]
Cholesterol	19	Hexane-ether (10 + 90)	—
Cholesteryl esters	94	Hexane-ether (90 + 10)	Butanone-acetonitrile (70 + 30)
Vitamin A		Hexane-ether (10 + 90)	—
Vitamin A-esters		Hexane-ether (90 + 10)	
Amines	0		Tetrahydrofuran-water (60 + 40)[b]
as acetates			
Amides	5	Ether-methanol (90 + 10)	Acetic acid-water (70 + 30)[c]
Nitriles	62	Hexane-ether (70 + 30)	Acetic acid-water (75 + 25)[c]

[a] Layer: Silica gel G; solvent: hexane-diethyl ether-acetic acid (80 + 20 + 1).
[b] Solvent contains 5% paraffin oil.
[c] Solvent saturated with paraffin oil.

Using argentation TLC, oleodistearin has been isolated from lard and stearodiolein from palm oil, malabar tallow and cocoa butter and it has been shown that these asymmetrical triglycerides are optically active [139].

Triglycerides containing cyclopentenyl acids are appreciably more strongly adsorbed on silver nitrate-containing silica gel G than are those of aliphatic acids and can therefore be easily isolated [124].

Classes of triglycerides which cannot be separated by argentation-TLC are fractionated by reversed phase partition TLC [81, 89, 90, 124] (see Fig. 149, p. 143) or by counter current distribution [124].

The methods worked out for analysis of triglycerides can be used also for investigating acetylated diglycerides from lecithins [174] (see p. 392) Only during recent months have several authors reported the first fractionation of intact lecithins, using argentation-TLC [5, 57, 88, 143]. This method is probably applicable to other ionic lipids.

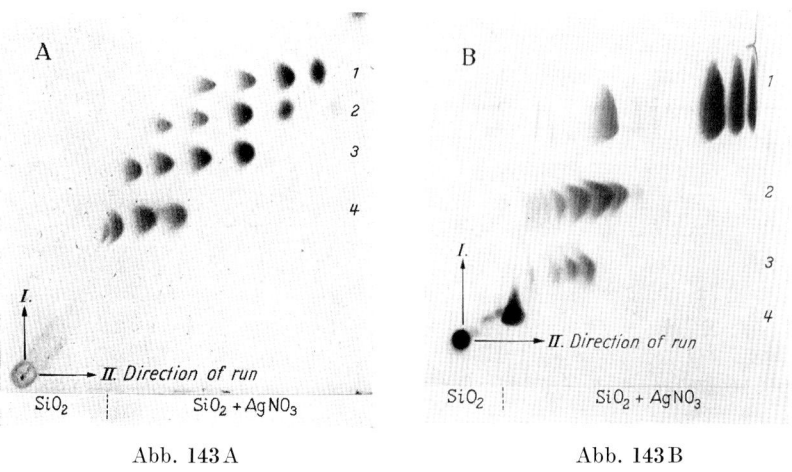

Abb. 143 A   Abb. 143 B

Fig. 143 A. Two-dimensional fractionation of a synthetic mixture of very similar compounds on a layer partially impregnated with silver nitrate [182]. Adsorbents: silica gel G and silica gel G-silver nitrate (5%); solvents: 1 direction: petrol ether (BP 40—60° C)-diethyl ether (90 + 10), 2 direction: petrol ether (BP 40—60° C)-diethyl ether (90 + 10), twice developed. *1* trialkyl glycerol ethers; *2* dialkyl-glycerides; *3* alkyl glycerides; *4* triglycerides; each with 0—3 double bonds

Fig. 143 B. Two-dimensional fractionation of human serum lipids on a layer partly impregnated with silver nitrate [182]. Adsorbents: silica gel G and silica gel G-silver nitrate (5%); solvents: 1 direction: petrol ether (BP 40—60° C)-diethyl ether-acetic acid (85 + 15 + 1), 2 direction: petrol ether (BP 40—60° C)-diethyl ether-acetic acid (75 + 25 + 1); visualisation: charring by heating with chromic acid/sulphuric acid; amount: 300 µg

Adsorbent layers which have been only *partly* impregnated with silver nitrate, may be used in one- or two-dimensional technique [30, 182]; aliquots of a sample solution can be fractionated in *one direction and parallel to one another*, using adsorption and argentation-TLC; in this way, information is obtained about the composition with regard to compound classes, and also about the number and configuration of the double bonds of their components. This procedure is useful particularly for the analysis of intermediate and end products of organic syntheses which aim at changes in the double bond systems or where such changes occur as unwanted side reactions. Complex mixtures of natural products can be submitted to *two-dimensional procedures in succession* on partly impregnated layers, thus separating according to class and number of double bonds. Characteristic "chromatographic patterns" are obtained in this way; classification of the individual fractions is simple. As seen in Fig. 143A, this two-dimensional combination technique permits even mixtures of saturated and unsaturated compounds to be separated on the basis of both class and number of double bonds; with adsorption TLC alone, separation even into classes is not possible. Fig. 143B shows an example of separation of cholesterol esters, triglycerides, fatty acids and sterols in human blood serum.

b) Chromatography of Mercuric Acetate Adducts

There has been no lack of attempts to convert lipids into derivatives which could be chromatographically fractionated and from which the original substances could be subsequently recovered. Acetoxymercurimethoxy adducts are particularly easily accessible derivatives, being formed by reaction of unsaturated compounds with a solution of mercuric acetate in methanol containing a little acetic acid. This addition reaction takes place quantitatively at room temperature [68]. *cis*-Compounds react 10–20 times faster than *trans*-isomers [68]. The unsaturated compounds can be recovered by treating their adducts with hydrochloric acid, without *cis-trans* isomerisation or rearrangement of the double bond occurring [68].

$$\begin{array}{c} \text{H} \quad \text{H} \\ | \quad | \\ -\text{C}=\text{C}- \\ \end{array} \xrightleftharpoons[\text{HCl}]{\text{Hg(OAc)}_2/\text{CH}_3\text{OH}} \begin{array}{c} \text{H} \quad \text{H} \\ | \quad | \\ -\text{C}-\text{C}- \\ | \quad | \\ \text{CH}_3\text{O} \quad \text{HgOAc} \end{array}$$

Saturated lipids and the mercuric acetate adducts of unsaturated lipids of the same compound class can be separated by TLC. In the same way the adducts can be fractionated according to the number of acetoxymercuri-methoxy groups. This is illustrated in Fig. 144 for the example of the methyl esters of long chain fatty acids.

The adducts of methyl chaulmoograte and of the esters of other cyclic acids show chromatographic behaviour which is markedly different from that of the corresponding purely aliphatic esters [124].

Mixtures of *different* classes of lipids can also be fractionated in accordance with the degree of unsaturation of their components, using TLC of the mercuric acetate adducts, provided that these are of not too varied polarity. The chromatographic behaviour of the adducts of, e. g.,

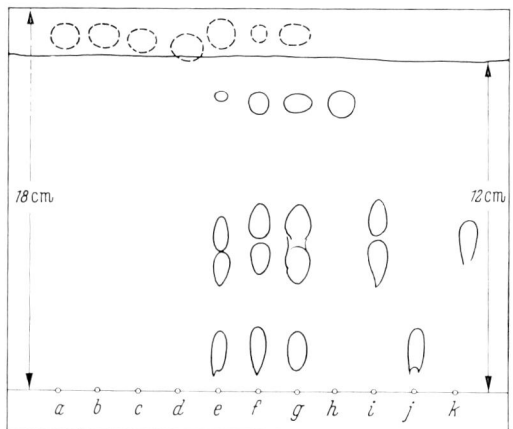

Fig. 144. Fractionation of a lipid class by TLC of the mercuric acetate adducts of its unsaturated components [119]. Adsorbent: silica gel G; solvents: *1* petrol ether (BP 60—70° C)-diethyl ether (80 + 20); *2* n-propanol-acetic acid (100 + 1); times of run: *1* 1.5—2 h; *2* 3—4 h; visualisation: s-diphenylcarbazone in ethanol, then exposure to iodine vapour; amounts: about 20 µg of each ester, 50—100 µg of the adducts. *a* methyl stearate; *b* methyl oleate; *c* methyl linoleate; *d* methyl linolenate; *e—j* mercuric acetate adducts of: *e* methyl esters of C_{16}-acids from *Chlorella pyrenoidosa*; *f* methyl esters of C_{18}-acids from *Chlorella*; *g* methyl esters of all acids from *Chlorella*; *h* methyl oleate; *i* methyl linoleate; *j* methyl linolenate; *k* mercuric acetate

hydrocarbons and wax esters on adsorbent layers, is thus governed mainly by the number of acetoxymercuri-methoxy groups per molecule; i. e. the adducts of octadecene and cetyl oleate or of octadecadiene and cetyl linoleate, migrate together.

Both partition and adsorption effects play a part in the chromatographic separation of mercuric acetate adducts.

Preparation of Mercuric Acetate Adducts [68]

Procedure: The reagent is a solution of 14 g mercuric acetate in 250 ml absolute methanol, containing 2.5 ml water and 1 ml glacial acetic acid. 1 g lipid sample (of low iodine number) is mixed with 40 ml reagent. For lipids of iodine number exceeding 100, a volume of reagent given by 0.4 × iodine number, is taken per g sample. The reaction mixture is kept in the dark in a nitrogen atmosphere.

Reaction is complete at room temperature within 12 h for *cis*-compounds, 2—3 days for *trans*-compounds. After completion of reaction, the major part of the methanol is taken off in vacuo at $< 30°$ C. The residue is dissolved in 10—20 ml chloroform, washed several times with water to remove excess mercuric acetate and the chloroform solution of the adducts finally dried over anhydrous sodium sulphate.

Directions for recovering the unsaturated lipids from the adducts are given on p. 405.

The mercuric acetate adducts of unsaturated lipids are clear, colourless oils. They are applied to thin layers in chloroform solution.

Experimental Conditions

Adsorbents. Silica gel G is suitable for fractionating mercuric acetate adducts of lipids [119]. The adducts of highly unsaturated lipids are chromatographed on layers of silica gel G + kieselguhr G (30 + 70) [211, 212].

The adducts are generally fractionated on a preparative scale; up to 10 mg of these compounds can be separated on a 20 × 20 cm layer of 0.25 mm thickness (see Fig. 145).

Solvents. The chromatograms are developed stepwise as a rule: Petroleum ether (60—70° C)-diethyl ether (80 + 20) is suited to separation of saturated neutral lipids from the mercuric acetate adducts of the corresponding unsaturated compounds [119]. This solvent migrates 15 to 18 cm in 1.5—2 h on silica gel G layers. A second solvent, n-propanol-acetic acid (100 + 1), serves for fractionating the adducts of lipids with one, two, three or more double bonds. It migrates 12—14 cm in 3—4 h. The saturated compounds are found between the two solvent fronts (see Fig. 144). Hexane-dioxan (60 + 40) is better for separating adducts of neutral lipids which contain three, four and more double bonds, on silica gel G; it migrates about 15 cm in an hour [219].

Mercuric acetate adducts of lipids containing more than three double bonds are fractionated on silica gel G-kieselguhr G (30 + 70), using isobutanol-formic acid-water (100 + 0.5 + 15.7) [211, 212]. The solvent should be placed in the chamber 5 hours before chromatography in order to saturate the atmosphere; it migrates about 16—17 cm in ca. 5 h.

Saturated lecithins can be separated from the mercuric acetate adducts of unsaturated lecithins, on silica gel G, using chloroform-methanol-water (70 + 30 + 4) [14].

Methods of Detection. Mercuric acetate adducts can be visualised as violet to purple red spots on a light pink background, by spraying with a solution of s-diphenylcarbazone in 96% ethanol (or Rgt. No. 87) [119, 211]. Saturated lipids are rendered visible in UV light by spraying with 2′,7′-dichlorofluorescein (Rgt. No. 63).

Recovery from the Adducts [119]

Procedure: Adduct fractions are scraped off the plate in strips; each is shaken with 10 ml methanol and 0.5 ml concentrated hydrochloric acid in an atmosphere of nitrogen in a test tube. The adsorbent settles after brief standing and the virtually clear solution is decanted and filtered. The residue in the test tube is treated a second time with methanol-hydrochloric acid. The combined extracts are diluted with 25 ml water and extracted with ether, once with 25 ml and then four times with 10 ml. The ethereal extract is dried over anhydrous sodium sulphate. After decantation and evaporation of the ether, the residue of unsaturated lipids is obtained. It is immediately taken up in petrol ether.

Saturated lipids are eluted from the adsorbent with petrol ether-diethyl ether (50 + 50) or with diethyl ether.

Applications and Results

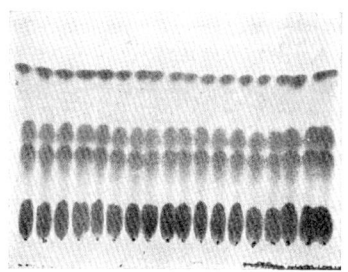

Fig. 145. Thin-layer chromatogram of the mercuric acetate adducts of the fatty acid methyl esters from the alga *Chlorella pyrenoidosa* (cf with Figs. 144 and 146)

Thin-layer chromatographic separation of a lipid class into groups of compounds containing the same number of double bonds, is the ideal

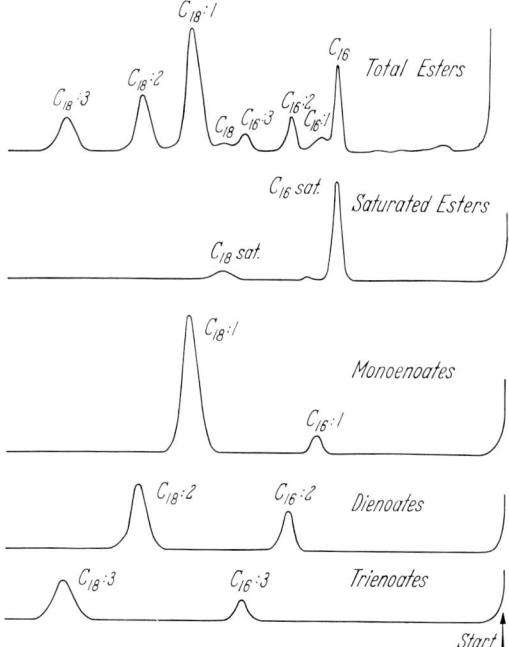

Fig. 146. Gas chromatographic separation of fatty acid methyl esters after prior separation of their mercuric acetate adducts by TLC (see Fig. 145) [119]

complement of gas chromatography: Thus, complex mixtures of the methyl esters of higher fatty acids, which cannot be separated completely through gas chromatography, can be subjected to prior fractionation by TLC of the mercuric acetate adducts. The groups of esters of saturated, mono-, di- and triolefinic acids, are obtained in more than 98% purity. Each group is then gas-chromatographically fractionated on the basis of chain length (see Fig. 146). This facilitates identification of the components of a complex mixture and can reveal traces of esters which cannot be discovered in the chromatogram from the whole sample. The preparative separation of a mixture of methyl esters is shown in Fig. 145. The advantages of coupling thin-layer and gas chromatography are illustrated in Fig. 146.

The sensitivity of detection of traces as well as the accuracy of gas chromatographic analysis is augmented by coupling with TLC of the mercuric acetate adducts; a more reliable identification of the fractions becomes possible.

Examples demonstrating the advantages of consecutive application of TLC and gas chromatography include analyses of: Hydrocarbons from the micrococcus *Sarcina lutea* [2]; fatty acids from the fresh water alga *Chlorella pyrenoidosa* [119] (see Figs. 145 and 146), from several marine algae [212] and from two unusual seed oils [124]. This coupling of methods has been valuable also in the analysis of the long-chain alcohols in fish oils [56].

Thin-layer chromatographic separation of saturated lecithins from the adducts of the corresponding unsaturated compounds has facilitated enzymatic structure analysis of egg-, soya- and milk lecithins and also the gas chromatographic determination of their fatty acid composition [14].

c) Chromatography of Ozonides

Fission of double bonds occurs when unsaturated compounds react with ozone:

$$\begin{array}{c} \text{H} \ \ \text{H} \\ | \ \ \ | \\ -\text{C}=\text{C}- \end{array} \quad \xrightarrow{O_3} \quad \begin{array}{c} \text{H} \ \ \ \ \text{O} \ \ \ \ \text{H} \\ | \ \diagup \ \ \ \diagdown \ | \\ -\text{C} \ \ \ \ \ \ \ \ \ \text{C}- \\ \diagdown \ \ \ \ \diagup \\ \text{O}-\text{O} \end{array}$$

The most favourable experimental conditions for this reaction have been worked out only in recent years; TLC played a decisive role in these investigations [167, 170].

Ozonides can be fractionated from saturated lipids and from one another according to the number of ozonide groups; the ozonides of unsaturated *cis-trans* isomers cannot, however, be separated.

This corresponds to the behaviour of the mercuric acetate adducts (cf. Fig. 144); in contrast to these, however, the ozonides are only slightly more polar than the starting materials.

Ozonisation of Unsaturated Lipids [167]

Procedure: The apparatus necessary for preparing ozone [17] can be constructed by a glass blower. It yields oxygen containing 1—3% ozone. Pentane is used as solvent and is purified as follows:

The gas mixture from the ozoniser is passed through the pentane at —60 to —70° C until it has turned blue. The solution is allowed to warm up slowly to room temperature while a current of pure nitrogen is passed through it to expel oxygen and ozone. Four parts of the pentane are then shaken in a separating funnel with one part concentrated sulphuric acid, the pentane layer washed neutral with water, dried over sodium sulphate and finally distilled over phosphorus pentoxide. For *ozonisation* of unsaturated lipids, 10 ml purified pentane are saturated at —60 to —70° C with the ozone-oxygen mixture from the ozoniser; this requires a flow rate of 100 ml/min, maintained for about 5 min. A solution of 1—50 mg sample in 2—3 ml purified pentane is then cooled as far as possible without crystals separating out, and added to the ozone solution. The ozonides are formed instantly and almost quantitatively. Oxygen and excess ozone are removed as rapidly as possible in a current of nitrogen or by using a rotary evaporator.

The amount of ozone in 10 ml of solution, ca. 0.3 millimole, is sufficient for ozonising about 50 mg methyl oleate or other monoolefine of about the same molecular weight. After the unsaturated compound has been added, the solution should be pale blue or grey. Complete decolorisation indicates that it contained too little ozone.

Note: Ozone attacks the respiratory system and reacts explosively with organic solvents at room temperature. All work with *ozone* and *ozonides* should therefore be conducted in a *fume cupboard*, *protective goggles* should be worn and only *small amounts* used.

Ozonides are colourless crystalline substances which are best kept in solution in pentane. They can be converted to shorter-chain aldehydes by reductive cleavage using Lindlar[2] catalyst; these can then be identified gas chromatographically or thin-layer chromatographically, then usually as derivatives (see also p. 205). This procedure is suitable for determining the structure of unsaturated lipids (cf. p. 408).

Experimental Conditions

Adsorbents. Ozonides are fractionated principally by adsorption chromatography on silica gel G layers.

Solvents. Hexane-diethyl ether mixtures are used as a rule in TLC. Hexane-diethyl ether (90 + 10) is suitable for the ozonides of the methyl esters of higher fatty acids; and hexane-diethyl ether (80 + 20) for those of triglycerides [168].

Methods of Detection. Ozonides may often be detected as white zones on silica gel G layers without using any indicator. 2′,7′-Dichlorofluorescein solution (Rgt. No. 64) is a convenient spray reagent; chromic acid-sulphuric acid (Rgt. No. 46) can be used for charring ozonides [170].

---

[2] See Helv. Chim. Acta **35**, 446 (1952).

Applications and Results

It has been possible to show with the help of adsorption TLC, that isomers are formed when the methyl esters of unsaturated fatty acids are ozonised; each of the isomers from methyl oleate has been isolated in a pure state [175]. Compounds which are not easily fractionated by argentation TLC, like *cis*-methyl octadecenoate (methyl oleate) and *trans-trans*-methyl octadecadienoate (*trans-trans* methyl "linoleate") or the corresponding triglycerides, have been separated as their ozonides [170]. This method is suitable, however, only for lipids with up to four double bonds. Ozone reacts with more highly unsaturated compounds to form short chain ozonides and undefined side products [170].

Small amounts of unsaturated lipids can be isolated using argentation TLC and the structure of these compounds then determined through analysis of their ozonolysis products. This method has been worked out above all for analysing mixtures of the methyl esters of unsaturated fatty acids [169].

The four types of saturated and unsaturated α- and β-monoglycerides can be separated by means of the "ozonisation-reduction" TLC method [165]: first, the α-monoglycerides are broken down with periodate into acetoxyacetaldehydes; the unsaturated acetoxyacetaldehydes and β-monoglycerides are then converted into aldehydic fission products using reductive ozonolysis:

$$\begin{array}{c} H_2C-OH \\ | \\ HC-O-\underset{\underset{O}{\|}}{C}-(CH_2)_x-CH=CH-R \\ | \\ H_2C-OH \end{array} \xrightarrow[\text{Ozonolysis}]{\text{Reductive}} \begin{array}{c} H_2C-OH \\ | \\ HC-O-\underset{\underset{O}{\|}}{C}-(CH_2)_x-C{\overset{H}{\underset{O}{\diagdown}}} \\ | \\ H_2C-OH \end{array}$$

$+$ short chain aldehydes

The "aldehydic cores" are more polar than the acetoxyacetaldehydes and monoglycerides of saturated acids which are unaffected by the ozone; these substances can thus easily be separated into classes through adsorption TLC. As has recently been described, saturated and unsaturated α- and β-monoglycerides can be rapidly and simply separated through TLC on silica gel G which has been impregnated with silver nitrate *and* boric acid; separation is according to the number of double bonds *and* the position of the acyl group [201]. The ozonisation-reduction TLC method may still today be of use, however, for analysis of very complex mixtures, especially those containing unsaturated *cis*- and *trans*-monoglycerides.

Unsaturated di- and triglycerides also can be converted into "aldehydic cores" by reductive ozonolysis; six of the seven diglyceride types and four of the six triglyceride types can then be determined by combination with TLC on silica gel G [165, 168]. This procedure has, in fact, been largely outmoded by argentation-TLC but is still useful for analysing unsaturated *cis-trans* isomeric triglycerides in partly hydrogenated oils.

Saturated lecithins and the three types of "aldehydic cores" obtained by reductive ozonolysis of unsaturated lecithins can be separated through reversed phase partition chromatography [168]. Fig. 147 shows photo-densitometric curves from thin-layer chromatograms of fission products derived from the lecithins of egg, bovine spinal cord, soya bean and wheat germ.

This was the first and only procedure for several years for analysis of the four lecithin types. Only recently have appreciably more subtle methods been developed [14, 174]; they are described on p. 392 and 404.

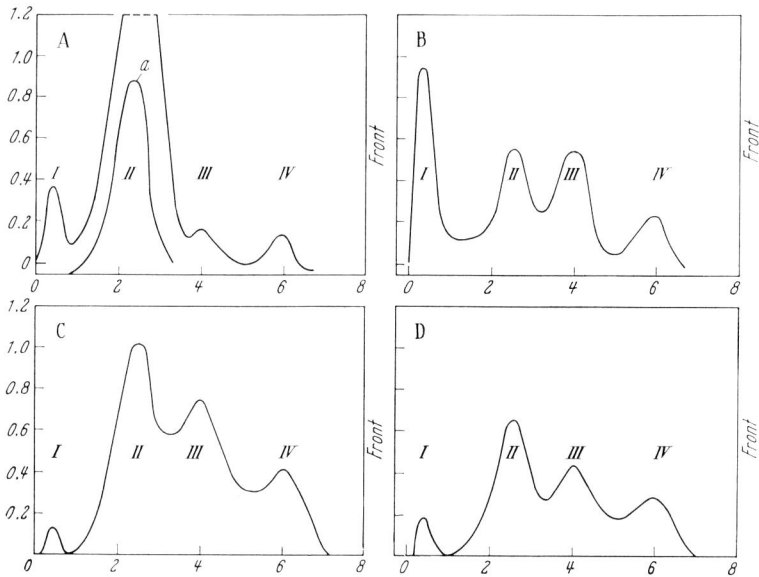

Fig. 147. Densitometric curves of the "aldehydic cores" from naturally occurring lecithin mixtures [*168*]. Stationary phase: silicone on silica gel G; solvent: acetic acid-water (80 + 20); time of run: 1.5 h; visualisation: carbonisation by heating with 50% sulphuric acid. *A* egg-lecithins; *B* lecithins from bovine spinal cord; *C* soya bean lecithins; *D* wheat germ lecithins; *I* saturated lecithins; *II* "aldehydic cores" from saturated(α)-unsaturated(β)- lecithins; *III* "aldehydic cores" from saturated (β)-unsaturated(α)-lecithins; *IV* "aldehydic cores" from lecithins containing two unsaturated fatty acids

d) Reversed Phase Partition Chromatography

This method can be employed for fractionating lipids according to compound *classes* (see p. 408 and 414); it has found greater application, however, for separating mixtures of vinylogous and homologous substances of the same type, based on the number of double bonds *and* chain length. For example, as seen in Fig. 148, the methyl esters of unsaturated C_{18}-acids are separated on the basis of degree of unsaturation, by using reversed phase partition chromatography. Polyolefinic methyl esters migrate ahead of the corresponding monoolefinic and saturated compounds having the same chain length. Even *homologues* are fractionated with the help of this procedure: saturated methyl esters are separated according to chain length; the Rf-values decrease with increasing chain length.

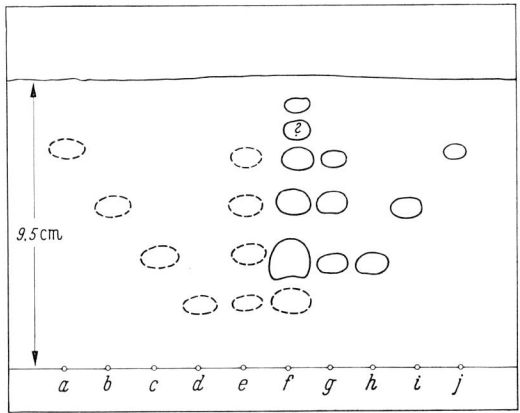

Fig. 148. Fractionation of a lipid class by reversed phase partition TLC [114]. Stationary phase: silicone on silica gel G; solvent: acetonitrile-acetic acid-water (70 + 10 + 25); time of run: 40 min; visualisation: iodine vapour (——) and then α-cyclodextrin-iodine (- - - -); amounts: 20 µg of each; *a* methyl laurate; *b* methyl myristate; *c* methyl palmitate; *d* methyl stearate; *e* methyl esters of the saturated acids; *f* methyl esters of the C_{18}- acids from menhaden oil; *g* methyl esters of the unsaturated C_{18}-acids; *h* methyl oleate; *i* methyl linoleate; *j* methyl linolenate

Certain saturated and unsaturated compounds migrate together in TLC on layers which have been rendered hydrophobic; this is as found in other partition procedures such as liquid-liquid counter-current distribution and partition chromatography in columns or on impregnated paper. Thus, as seen from Fig. 148, methyl palmitate (C_{16}, saturated) and methyl oleate (C_{18}, monoolefine) appear together in one spot as do methyl myristate (C_{14}, saturated) and methyl linoleate (C_{18}, diolefine). Such "critical pairs" are likewise formed by the corresponding free acids or aldehydes and also by tripalmitin/triolein and trimyristin/trilinolein [78].

Most critical pairs may be separated through reversed phase partition chromatography by working at low temperature (see p. 94). Another way is to "develop" the unsaturated lipids through quantitative oxidation with a peroxidic solvent, which has no influence on the separation of the saturated compounds (see p. 412). Argentation-TLC is however the best method to choose (see p. 396–402, also [54]).

Experimental Conditions

Stationary Phase. "Dow Corning 200 Fluid, 10 cs" (Firm 49). A silicone oil which has been of value in the paper chromatography of lipids, can be used to render silica gel G or kieselguhr G layers hydrophobic [114, 168]. Other hydrophobic impregnation agents are hydrocarbons such as undecane [78] (Firm 68), tetradecane [82] (Firms 58, 68) and mixtures of high molecular weight paraffins [4, 81] (Firms 68, 88).

Silicones and paraffins are superior to the more volatile impregnation agents like undecane, since layers treated with them retain their properties even during weeks of storage. Reproducibility suffers when short-chain hydrocarbons are used, due to evaporation of the stationary phase during storage.

Directions for impregnating adsorbent layers are quoted on p. 48. The simplest method is to stand the plates, coated with kieselguhr G, in a 5% solution of *paraffinum subliquidum* (Firm 88) or "*Nujol*" (Firm 11) in petroleum ether (BP 60–70° C) or in benzene and to allow this impregnation solution to ascend through the layer [4, 171]. The impregnated plates are ready for use after the solvent has evaporated. The material for analysis is chromatographed in the same direction as that in which the impregnation solution migrated [4, 171].

Solvents. The behaviour of a lipid class on adsorbent layers yields information about the polarity of these compounds and thus facilitates choice of a solvent which is suitable for their fractionation in reversed phase partition chromatography.

Most of the solvent mixtures used in paper chromatography of lipids are suitable for TLC on hydrophobic layers. Triglycerides and other feebly polar lipid classes can be fractionated on the basis of chain length and number of double bonds by using acetic acid or acetonitrile containing 1–10% water [4, 121, 124, 168]; other good solvents are chloroform-methanol-water (25 + 75 + 5) [78, 121] and mixtures of acetonitrile with acetone or butanone (30 + 70) [78, 80, 81, 85]. Mixtures of acetic acid or acetonitrile with 10–50% water are employed for separating more polar lipids like alcohols [78, 114, 125, 168].

Palmitic and oleic acids, two compounds which migrate together at ambient temperatures, are separable at 4–6° C using acetic acid-formic acid-water (40 + 40 + 20) [114]. Oxidising solvents are also suitable for fractionating these and other critical pairs. Part of the water of aqueous solvent mixtures is replaced by hydrogen peroxide or part of the acetic acid in acetic acid-containing mixtures by peracetic acid (Firm 18): acetic acid-water-hydrogen peroxide (85 + 5 + 10) or acetic acid-peracetic acid-water (75 + 10 + 15) [114, 117]. These solvents oxidise all unsaturated lipids during chromatography; the strongly polar reaction products migrate with the solvent front whereas the unchanged, saturated lipids are separated from one another. Unsaturated lipids can be brominated during development and separated from saturated lipids, by using the solvent propionic acid-acetonitrile-bromine (60 + 40 + 0.5) [83, 84, 102].

Solvents for fractionating the most important lipids are given in Table 71. Three to four hours are usually required for TLC on hydrophobic layers; only solvent systems containing acetonitrile migrate faster.

Methods of Detection. Unsaturated lipids on hydrophobic layers can be visualised with iodine vapour [114]. Most organic compounds can be detected on siliconised layers by carbonisation with dichromate-sulphuric acid (Rgt. No. 46) [121]; this reagent cannot be used on layers impregnated with paraffin. Alcoholic 2′,7′-dichlorofluorescein solution (Rgt. No. 63) can be employed for visualisation on kieselguhr G which has been impregnated with paraffin; the test is especially sensitive when the chromatogram is exposed for a short time to water vapour [38]. This reagent cannot be used on impregnated silica gel G layers [38, 114].

Application and Results

Reversed phase partition TLC may be applied to fractionate alcohols [78, 86], aldehydes and ketones [10, 125], fatty acids [4, 80, 84, 124] and their methyl esters [114, 213], keto-acids, hydroxy-acids and lactones [82]. More complete separations of these compounds or of suitable derivatives, are, however, obtained with gas chromatography and it is appreciably more sensitive than TLC; it is moreover more suitable for quantitative determinations. Reversed phase partition TLC offers substantial advantages only in the analysis of lipids of relatively high molecular weight, like wax esters [86], steryl esters [79] (see also Chapter L), carotenoid esters (see Chapter K) and triglycerides [4, 81, 89, 124]; some derivatives of naturally occurring lipids are also best fractionated on hydrophobic layers [21, 130, 131, 165].

Mixtures of triglycerides, such as triolein, palmitodiolein, oleodipalmitin and tripalmitin, which overlap in chromatography on hydrophobic paper, can be separated by multiple development [85]. Reversed phase partition TLC is suitable also for separation of naturally occurring mixtures, in contrast to paper chromatography: Both olive oil and pig lard yield four triglyceride fractions [83]; soya oil yields thirteen and linseed oil, fifteen spots [85]. The method can be used to test fats and fatty oils from the pharmacopoeia: each individual drug gives a characteristic chromatographic pattern, meeting the requirements for identification; adulterations and substitutes are usually easy to detect

[4]. The partition chromatographic procedure is good for separation also of fish oils, which, because of their high content of triglycerides with five and six double bonds in the fatty acid moieties, are difficult to fractionate using argentation-TLC.

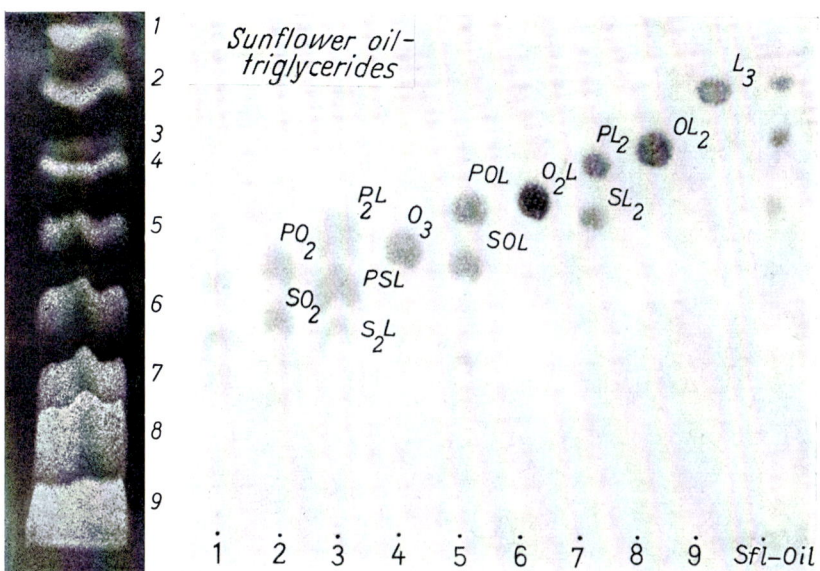

Fig. 149. Separation of the triglycerides of sunflower oil *(Helianthus annuus)* by combination of argentation-TLC and reversed phase partition-TLC [89]. *A* pre-fractionation of the triglycerides through argentation-TLC; adsorbent: silica gel G-silver nitrate (25%); solvent: benzene-diethyl ether (80 + 20); time of run: 3.5 h; spray reagent: 2′,7′-dichlorofluorescein in ethanol (UV light); amount: 20 mg; *B* separation of fractions 1—9 (A) through reversed phase partition-TLC; stationary phase: paraffin on kieselguhr G; solvent: acetone-acetonitrile (80 + 20) (80% saturated with paraffin), developed twice; time of run: 40 min each time; visualisation: aqueous-alcoholic α-cyclodextrin solution-iodine vapour; amounts: ca 12 µg of each of the fractions 1—9 and 20 µg sunflower oil

By adding bromine to the solvent, unsaturated triglycerides can be converted to derivatives which are separable from the saturated triglycerides. Cocoa butter gives five large and two small spots under these conditions, instead of three large and one small; and olive oil, six large and five small instead of four [83].

A very nearly complete fractionation of the triglycerides in naturally occurring mixtures is made possible by coupling argentation- and reversed phase partition TLC. This is shown in Fig. 149. The triglycerides of sunflower oil were first resolved into simple mixtures through argentation-TLC and each of the nine fractions thus obtained was separated further on a layer impregnated with paraffin [89]. Combinations of TLC and GC are useful for the analysis of mixtures of triglycerides [108, 109].

Triglycerides and esters of diols (diol lipids), compounds which cannot be separated by adsorption TLC, can be fractionated into *classes* through reversed phase partition TLC; this applies equally to the corresponding alkoxylipids [9]. Chromatography on hydrophobic layers is also suitable for separating the "aldehydic cores" which are obtained from unsaturated lecithins by reductive ozonolysis [168] (see Fig. 147, p. 409) and for analysing sphingosine bases [129].

3. Quantitative Evaluation of Thin-Layer Chromatograms[3]

a) Neutral Lipids and their Hydrolysis Products

Fractions of neutral lipids can be easily and accurately determined gravimetrically after elution from the adsorbent [37, 103, 112]. A relatively large sample (>50 mg) is admittedly needed. Triglycerides can be determined by IR-spectrophotometry [105]. Smaller amounts are better determined colorimetrically after oxidation with dichromate-sulphuric acid [3, 43, 213]; a separate calibration curve is however necessary for each lipid class (see Fig. 151). The hydroxamic acid method is satisfactory for colorimetric determination of lipid esters [47, 204, 213]. Sugar esters can be treated with resorcinol-hydrochloric acid and determined colorimetrically [46, 93]. Mono-, di- and triglycerides may be quite accurately estimated by alkaline hydrolysis and titrimetric determination of the glycerol [76].

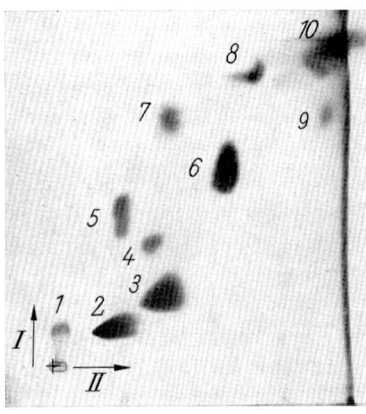

Fig. 150. Thin-layer chromatogram of phospholipids and other polar lipids of bovine milk [159a]. *1* carbohydrate (lactose) and protein, *2* sphingomyelin, *3* phosphatidyl choline, *4* phosphatidyl serine, *5* phosphatidyl inositol, *6* phosphatidyl ethanolamine, *7* cerebroside dihexoside (?), *8* cerebroside monohexoside (?), *9* fatty acids, *10* neutral lipids. Adsorbent: Silica gel HR. Solvents: I, chloroform-methanol-water-28% aqu. ammonia (130 + 70 + 8 + 0.5); II, chloroform-acetone-methanol-acetic acid-water (50 + 20 + 10 + 10 + 5). Time: 40 min in each direction. Indicator: charring with chromic sulphuric acid solution

Photodensitometric evaluation of thin-layer chromatograms of neutral lipids is carried out after having sprayed the layer with 50% orthophosphoric acid [7, 206], 50% sulphuric acid [165, 166] or dichromate-sulphuric acid [13, 121] and

---

[3] cf. Chapter H.

charred the lipids by heating (see also [223]). This method is said to be especially suited to determination of traces.

Thin-layer chromatograms of neutral lipids can be evaluated also by planimetry of the spot areas [19, 163] (see also [171], cf. p. 135)

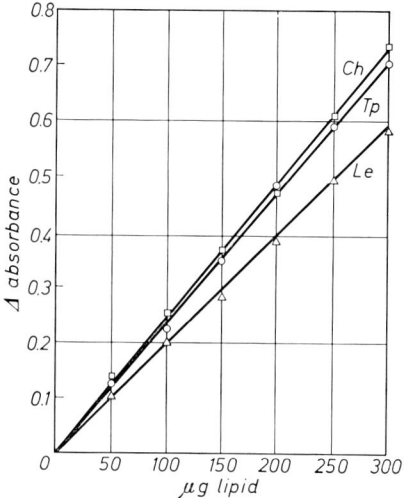

Fig. 151. Calibration curves for the colorimetric estimation of lipids [3]. Ch cholesterol and cholesteryl esters, Tp Tripalmitin and palmitic acid, Le lecithin. Changes in absorbance of the dichromate reagent are plotted against amounts of lipid

b) Phospholipids, Sulpholipids and Glycolipids

Phospholipids are eluted, ignited and colorimetrically determined as phosphate [1, 47, 53, 159, 189]; photodensitometric analysis on the layer is also possible [91, 132, 161]. The experimental conditions given in Fig. 151 are recommended for the two-dimensional separation of phospholipids, sulpholipids and glycolipids, prior to their quantitative analysis. Methods are available for determining gangliosides [73, 196] and other sphingolipids [71, 72, 164].

Procedures for quantitatively evaluating thin-layer chromatograms are described in detail in Chapter H (pp. 133–155).

Note: The thin-layer chromatographic separation of alicyclic constituents of fats is treated in several parts of this book. Reference may be made in particular to the articles on sterols (p. 329), carotenoids and fat-soluble vitamins (pp. 259–292) and terpenes and resins (p. 256 to 258).

Bibliography for Chapter M. Aliphatic Lipids

1. ABRAMSON, D., and M. BLECHER: J. Lipid Res. **5**, 628 (1964).
2. ALBRO, PH. W., and C. K. HUSTON: J. Bacteriol. **88**, 981 (1964).

3. AMENTA, J. S.: J. Lipid Res. **5**, 270 (1964).
4. ANKER, L., u. D. SONANINI: Pharm. Acta Helv. **37**, 360 (1962).
5. ARVIDSON, G. A. E.: J. Lipid Res. **6**, 574 (1965).
6. BARRETT, C. B., M. S. J. DALLAS, and F. B. PADLEY: Chem. & Ind. (London) **1962**, 1050.
7. — — — J. Am. Oil Chemists' Soc. **40**, 580 (1963).
8. BAUMANN, W. J., and H. K. MANGOLD: Biochim. et Biophys. Acta **116**, 570 (1966).
9. — H. H. O. SCHMID, H. W. ULSHÖFER, and H. K. MANGOLD: Biochim. et Biophys. Acta, **144**, 355 (1967).
10. — — and H. K. MANGOLD: J. Lipid Res. (in the press).
11. BERGELSON, L. D., E. V. DYATLOVITSKAYA, and V. V. VORONKOVA J. Chromatog. **15**, 191 (1964).
12. — V. A. VAVER, N. V. PROKAZOVA, A. W. USHAKOV, and G. A. POPKOVA: Biochim. et Biophys. Acta **116**, 511 (1966).
13. BLANK, M. L., J. A. SCHMIT, and O. S. PRIVETT: J. Am. Oil Chemists' Soc. **41**, 371 (1964).
14. — L. J. NUTTER, and O. S. PRIVETT: Lipids **1**, 132 (1966).
15. BLOOR, W. R.: J. Biol. Chem. **77**, 53 (1928).
16. BONSEN, P. P. M., G. H. DE HAAS, and L. L. M. VAN DEENEN: Biochim. et Biophys. Acta **106**, 93 (1965).
17. BONNER, W. A.: J. Chem. Educ. **30**, 492 (1953).
18. BÜRGER, K.: Z. anal. Chem. **196**, 259 (1963).
19. BUSWELL, K. M., and W. E. LINK: J. Am. Oil Chemists' Soc. **41**, 717 (1964).
20. BYGDEMAN, M., and B. SAMUELSSON: Clin. Chim. Acta **13**, 465 (1966).
21. CARTER, H. E., and H. S. HENDRICKSEN: Biochemistry **2**, 389 (1963).
22. — P. JOHNSON, D. W. TEETS, and R. K. YU: Biochem. Biophys. Res. Communs. **13**, 156 (1963).
23. CHALVARDJIAN, A., L. J. MORRIS, and R. T. HOLMAN: J. Nutrition **76**, 52 (1962).
24. — Biochem. J. **90**, 518 (1964).
25. — Can. J. Biochem. **44**, 713 (1966).
26. CHANG, T.-CH. L., and C. C. SWEELEY: Biochemistry **2**, 592 (1963).
27. CHINO, H., and L. I. GILBERT: Biochim. et Biophys. Acta **98**, 94 (1965).
28. CREACH, O., B. ENTRESSANGLES et L. COLOBERT: Biochim. et Biophys. Acta **116**, 80 (1966).
29. CRIDER, Q., P. ALAUPOVIC, J. HILLSBERRY, C. YEN, and R. H. BRADFORD: J. Lipid Res. **5**, 479 (1964).
30. CUBERO, J. M., and H. K. MANGOLD: Microchem. J. **9**, 227 (1965).
31. — Z. BANDI, and H. K. MANGOLD: Separation Sci., (in the press).
32. DAIN, J. A., H. WEICKER, G. SCHMIDT, and S. J. THANNHAUSER: In: Cerebral Sphingolipidoses, A Symposium on Tay Sachs' Disease and Allied Disorders, p. 289, St. M. ARONSON, and B. W. VOLK, Editors. New York: Academic Press 1962.
33. DESMOND, C. T., and W. T. BORDEN: J. Am. Oil Chemists' Soc. **41**, 552 (1964).
34. DHOPESHWARKAR, G. A., and J. F. MEAD: Proc. Soc. Exptl. Biol. Med. **109**, 425 (1962).
35. DOBIÁSOVÁ, M.: J. Lipid Res. **4**, 481 (1963).
36. DOSS, M., u. K. OETTE: Z. klin. Chem. **3**, 125 (1965).
37. DUNN, F., and P. ROBSON: J. Chromatog. **17**, 501 (1965).
38. DUNPHY, P. J., K. J. WHITTLE, and J. F. PENNOCK: Chem. & Ind. (London) **1965**, 1217.

39. ENG, L. F., Y. L. LEE, R. B. HAYMAN, and B. GERSTL: J. Lipid Res. **5**, 128 (1964).
40. ENTENMAN, C.: J. Am. Oil Chemists' Soc. **38**, 534 (1961).
41. FIRESTONE, D.: J. Am. Oil Chemists' Soc. **40**, 247 (1963).
42. FOLCH, J., M. LEES, and G. H. SLOANE STANLEY: J. Biol. Chem. **226**, 497 (1957).
43. FREEMAN, C. P., and D. WEST: J. Lipid Res. **7**, 324 (1966).
44. FREIMER, E. H.: J. Exptl. Med. **117**, 377 (1963).
45. GAVER, R. C., and C. C. SWEELEY: J. Am. Oil Chemists' Soc. **42**, 294 (1965).
46. GEE, M.: J. Chromatogr. **9**, 278 (1962).
47. GLOSTER, J., and R. F. FLETCHER: Clin. Chim. Acta **13**, 235 (1966).
48. GORDIS, E.: Proc. Soc. Exptl. Biol. Med. **110**, 657 (1962).
49. GRANZER, E.: Hoppe-Seyler's Z. physiol. Chem. **328**, 277 (1962).
50. GREEN, K., and B. SAMUELSSON: J. Lipid Res. **5**, 117 (1964).
51. GUNSTONE, F. D., and F. B. PADLEY: J. Am. Oil Chemists' Soc. **42**, 957 (1965).
52. HAAHTI, E., T. NIKKARI, and K. JUVA: Acta Chem. Scand. **17**, 538 (1963).
53. HABERMANN, E., G. BANDTLOW u. B. KRUSCHE: Klin. Wschr. **39**, 816 (1961).
54. HAMMONDS, T. W., and G. SHONE: J. Chromatog. **15**, 200 (1964).
55. HANAHAN, D. J., J. EKHOLM, and C. M. JACKSON: Biochemistry **2**, 630 (1963).
56. HASHIMOTO, A., K. SHIRO, and K. MUKAI: Yukagaku **13**, 586 (1964).
57. HAVERKATE, F., and L. L. M. VAN DEENEN: Biochim. et Biophys. Acta **106**, 78 (1965).
58. HAWTHORNE, B. E., N. TUNA, H. K. MANGOLD, and W. O. LUNDBERG: Federation Proc. **23**, 399 (1964); **27**, 673 (1968).
59. HILDITCH, T. P., and P. N. WILLIAMS: The Chemical Constitution of Natural Fats. 4th Edition, New York: John Wiley & Sons, Inc. 1964.
60. HOFMANN, A. F.: J. Lipid Res. **3**, 391 (1962).
61. —, and B. BORGSTRÖM: Biochim. et Biophys. Acta **70**, 317 (1963).
62. HOLLA, K. S., and D. C. CORNWELL: J. Lipid Res. **6**, 322 (1965).
63. HOLMAN, R. T., W. O. LUNDBERG, and T. MALKIN, Editors: Progress in the Chemistry of Fats and Other Lipids, Vols. I—X, London: Pergamon Press 1952—1968.
64. HOOGHWINKEL, G. J. M., P. BORRI, and J. C. RIEMERSMA: Rec. trav. chim. **83**, 576 (1964).
65. HORNING, E. C., A. KARMEN, and C. C. SWEELEY: In: Progress in the Chemistry of Fats and Other Lipids. R. T. HOLMAN, Editor, Vol. 7, p. 167. London: Pergamon Press 1964.
66. HORNING, M. G.: In: Lipid Pharmacology. R. PAOLETTI, Editor, p. 1, New York and London: Academic Press 1964.
67. HORROCKS, L. A.: J. Am. Oil Chemists' Soc. **40**, 235 (1963).
67a. — J. Lipid Res. **9**, 469 (1968).
68. JANTZEN, E., u. H. ANDREAS: Chem. Ber. **92**, 1427 (1959).
69. JATZKEWITZ, H.: Hoppe-Seyler's Z. physiol. Chem. **320**, 134 (1960).
70. —, u. E. MEHL: Hoppe-Seyler's Z. physiol. Chem. **320**, 231 (1960).
71. — Hoppe-Seyler's Z. physiol. Chem. **326**, 61 (1961).
72. — Hoppe-Seyler's Z. physiol. Chem. **336**, 25 (1964).
73. — H. PILZ u. K. SANDHOFF: J. Neurochem. **12**, 135 (1965).
74. JENKIN, H. M.: Am. J. Ophthalmol. **63**, 1087 (1967).
75. JURRIENS, G., and A. C. J. KROESEN: J. Am Oil Chemists' Soc. **42**, 9 (1965).
76. JURRIENS, G., B. DE VRIES, and L. SCHOUTEN: J. Lipid Res. **5**, 267 (1964).
77. KAUFMANN, H. P. (Editor). Analyse der Fette und Fettprodukte. Berlin, Göttingen, Heidelberg: Springer 1958.

78. KAUFMANN, H. P., u. Z. MAKUS: Fette, Seifen, Anstrichmittel **62**, 1014 (1960).
79. — Z. MAKUS u. F. DEICKE: Fette, Seifen, Anstrichmittel **63**, 235 (1961).
80. — — u. T. H. KHOE: Fette, Seifen, Anstrichmittel **63**, 689 (1961).
81. — — u. B. DAS: Fette, Seifen, Anstrichmittel **63**, 807 (1961).
82. — u. Y. S. KO: Fette, Seifen, Anstrichmittel **63**, 828 (1961).
83. — Z. MAKUS u. T. H. KHOE: Fette, Seifen, Anstrichmittel **64**, 1 (1962).
84. — u. T. H. KHOE: Fette, Seifen, Anstrichmittel **64**, 81 (1962).
85. — u. B. DAS: Fette, Seifen, Anstrichmittel **64**, 214 (1962).
86. — — Fette, Seifen, Anstrichmittel **65**, 398 (1963).
87. —, u. C. V. VISWANATHAN: Fette, Seifen, Anstrichmittel **65**, 538 (1963).
88. — H. WESSELS u. C. BONDOPADHYAYA: Fette, Seifen, Anstrichmittel **65**, 543 (1963).
89. — — Fette, Seifen, Anstrichmittel **66**, 81 (1964).
90. — — Fette, Seifen, Anstrichmittel **68**, 249 (1966).
91. — S. S. RADWAN u. A. K. S. AHMAD: Fette, Seifen, Anstrichmittel **68**, 261 (1966).
92. KINOSHITA, S.: J. Chem. Soc. Japan, Ind. Chem. Sect. **66**, 450 (1963).
93. —, and M. OYAMA: J. Chem. Soc. Japan; Ind. Chem. Sect. **66**, 455 (1963).
94. KISHIMOTO, I., and N. S. RADIN: J. Lipid Res. **5**, 94 (1964).
95. KLENK, E., u. W. GIELEN: Hoppe-Seyler's Z. physiol. Chem. **323**, 126 (1961).
96. — — Hoppe-Seyler's Z. physiol. Chem. **326**, 144 (1961).
97. —, u. M. DOSS: Hoppe-Seyler's Z. physiol. Chem. **342**, 187 (1965).
98. —, W. KUNAU, L. HOF u. L. GEORGIAS: Hoppe-Seyler's Z. physiol. Chem. **346**, 236 (1966).
99. KOCHETKOV, N. K., I. G. ZHUKOVA i I. S. GLUKHODED: Doklady Akad. Nauk. S.S.S.R. **147**, 376 (1962).
100. — — — Biochim. et Biophys. Acta **60**, 431 (1962).
101. — — — Biokhimiya **29**, 570 (1964); Engl. transl. **29**, 487 (1964).
102. KOEHLER, W. R., J. L. SOLAN, and H. T. HAMMOND: Anal. Biochem. **8**, 353 (1964).
103. KOMAREK, R. J., R. G. JENSEN, and B. W. PICKETT: J. Lipid Res. **5**, 268 (1964).
104. KOREY, S. R., C. J. GIDEZ, A. STEIN, J. GONATAS, and K. SUZUKI: J. Neuropathol. Exptl. Neurol. **22**, 2 (1963).
105. KRELL, K., and S. A. HASHIM: J. Lipid Res. **4**, 407 (1963).
106. KUHN, R., H. WIEGANDT u. H. EGGE: Angew. Chem. **73**, 580 (1961).
107. — — Chem. Ber. **96**, 866 (1963).
108. KUKSIS, A., and J. LUDWIG: Lipids **1**, 202 (1966).
109. —, and W. C. BRECKENRIDGE: J. Lipid Res. **7**, 576 (1966).
110. LEES, M., J. FOLCH, G. H. SLOANE STANLEY, and S. J. CARR: J. Neurochem. **4**, 9 (1959).
111. LEPAGE, M.: J. Chromatog. **13**, 99 (1964).
112. LEVIN, E., and C. HEAD: Anal. Biochem. **10**, 23 (1965).
113. LINDLAR, F., u. H. WAGENER: Schweiz. med. Wochschr. **94**, 243 (1964).
114. MALINS, D. C., and H. K. MANGOLD: J. Am. Oil Chemists' Soc. **37**, 576 (1960).
115. — J. C. WEKELL, and C. R. HOULE: Anal. Chem. **36**, 658 (1964).
116. — — — J. Lipid Res. **6**, 100 (1965).
117. MANGOLD, H. K.: Fette, Seifen, Anstrichmittel **61**, 877 (1959).
118. —, and D. C. MALINS: J. Am. Oil Chemists' Soc. **37**, 383 (1960).
119. —, and R. KAMMERECK: Chem. & Ind. (London) **1961**, 1032.
120. — J. Am. Oil Chemists' Soc. **38**, 708 (1961).
121. —, and R. KAMMERECK: J. Am. Oil Chemists' Soc. **39**, 201 (1962).

122. MANGOLD, H. K.: J. Am. Oil Chemists' Soc. **41**, 762 (1964).
123. —, and H. W. ULSHÖFER: Chem. Phys. Lipids (in the press).
124. —, and Z. BANDI: unpublished.
125. MARCUSE, R., U. MOBECH-HANSSEN u. P. O.-GÖTHE: Fette, Seifen, Anstrichmittel **66**, 192 (1964).
126. MARINETTI, G. V.: J. Lipid Res. **3**, 1 (1962).
127. MCKILLICAN, M. E., and R. P. A. SIMS: J. Am. Oil Chemists' Soc. **40**, 108 (1963).
128. — — J. Am. Oil Chemists' Soc. **41**, 340 (1964).
129. MICHALEČ, C.: Biochim. et Biophys. Acta **106**, 197 (1965).
130. — J. Chromatog. **20**, 594 (1965).
131. — Biochim. et Biophys. Acta **116**, 400 (1966).
132. MORIN, R. J.: Clin. Chim. Acta **13**, 395 (1966).
133. MORRIS, L. J., R. T. HOLMAN, and K. FONTELL: J. Lipid Res. **1**, 412 (1960).
134. — — — J. Am. Oil Chemists' Soc. **37**, 323 (1960).
135. — — — J. Lipid Res. **2**, 68 (1961).
136. — Chem. & Ind. (London) **1962**, 1238.
137. — J. Lipid Res. **4**, 357 (1963).
138. — J. Chromatog. **12**, 321 (1963).
139. — Biochem. Biophys. Res. Communs. **20**, 340 (1965).
140. —, and D. M. WHARRY: J. Chromatog. **20**, 27 (1965).
141. —, and S. W. HALL: Lipids **1**, 188 (1966).
142. —, and D. M. WHARRY: private communication, 1966.
143. — J. Lipid Res. **7**, 717 (1966).
144. — R. MAIER, and H. K. MANGOLD: unpublished.
145. — M. O. MARSHALL, and W. KELLY: Tetrahedron Letters **1966**, 4249.
146. MÜLDNER, H. G., J. R. WHERRETT, and J. N. CUMINGS: J. Neurochem. **9**, 607 (1962).
147. MURPHY, M. T. J. (Sister), B. NAGY, G. ROUSER, and G. KRITCHEVSKY: J. Am. Oil Chemists' Soc. **42**, 475 (1965).
148. NEGISHI, T., M. E. MCKILLICAN, and M. LEPAGE: J. Lipid Res. **5**, 486 (1964).
149. NEVENZEL, J. C., W. RODEGKER, and J. F. MEAD: Biochemistry **4**, 1589 (1965).
150. NICHOLS, B. W.: Biochim. et Biophys. Acta **70**, 417 (1963).
151. — In: New Biochemical Separations, A. T. JAMES, and L. J. MORRIS, Editors, p. 321, London: van Nostrand 1964.
152. — L. J. MORRIS, and A. T. JAMES: Brit. Med. Bull. **22**, 137 (1966).
153. NORTON, W. T., and M. BROTZ: Biochem. Biophys. Res. Communs. **12**, 198 (1963).
154. OBRUBA, K.: Coll. Czech. Chem. Communs. **27**, 2968 (1962).
155. OETTE, K.: J. Lipid Res. **6**, 449 (1965).
156. OKUI, S., M. UCHIYAMA, and M. MIZUGAKI: J. Biochem. (Tokyo) **53**, 265 (1963).
157. OWENS, K.: Biochem. J. **100**, 354 (1966).
158. PAOLETTI, R., and D. KRITCHEVSKY (Editors): Advances in Lipid Research, Vols. 1—6, New York, London: Academic Press 1963—1968.
159. PARKER, F., and N. F. PETERSON: J. Lipid Res. **6**, 455 (1965).
159a. PARSONS, J. G., and S. PATTON: J. Lipid Res. **8**, 696 (1967).
160. PAULOSE, M. M.: J. Chromatog. **21**, 141 (1966).
161. PAYNE, S. N.: J. Chromatog. **15**, 173 (1964).
162. PELKA, J. R., and L. D. METCALFE: Anal. Chem. **37**, 603 (1965).
163. PENICK, R. J., M. H. MEISLER, and R. H. MCCLUER: Biochim. et Biophys. Acta **116**, 279 (1966).

164. Pilz, H., and H. Jatzkewitz: J. Neurochem. **11**, 603 (1964).
165. Privett, O. S., and M. L. Blank: J. Lipid Res. **2**, 37 (1961).
166. — — and W. O. Lundberg: J. Am. Oil Chemists' Soc. **38**, 312 (1961).
167. — and E. C. Nickell: J. Am. Oil Chemists' Soc. **39**, 414 (1962).
168. — — J. Am. Oil Chemists' Soc. **40**, 70 (1963).
169. — — and O. Romanus: J. Lipid Res. **4**, 260 (1963).
170. —, and E. C. Nickell: Lipids **1**, 98 (1966).
171. Purdy, S. J., and E. V. Truter: Analyst **87**, 802 (1962).
172. Rapport, M. M., L. Graf, and H. Schneider: Arch. Biochem. Biophys. **105**, 431 (1964).
173. Redman, C. M., and R. W. Keenan: J. Chromatog. **15**, 180 (1964).
174. Renkonen, O.: Biochim. et Biophys. Acta **125**, 288 (1966).
175. Riezebos, G., J. C. Grimmelikhuysen, and D. A. van Dorp: Rec. trav. chim. **82**, 1234 (1963).
176. Rollins, C. B., and R. D. Wood: J. Chromatog. **16**, 555 (1964).
177. Rouser, G., C. Galli, and G. Kritchevsky: J. Am. Oil Chemists' Soc. **42**, 404 (1965).
178. Ruseva-Atanasová, N., and J. Janák: J. Chromatog. **21**, 207 (1966).
179. Rybicka, W. S.: Chem. & Ind. (London) **1962**, 308.
180. Sachs, B. A., and L. Wolfman: Proc. Soc. Exptl. Biol. Med. **115**, 1138 (1964).
181. Schmid, H. H. O., u. H. K. Mangold: Biochem. Z. **346**, 13 (1966).
182. — W. J. Baumann, J. M. Cubero, and H. K. Mangold: Biochim. et Biophys. Acta **125**, 189 (1966).
183. —, and H. K. Mangold: Biochim. et Biophys. Acta **125**, 182 (1966).
184. — W. J. Baumann, and H. K. Mangold: Biochim. et Biophys. Acta **144**, 344 (1967).
185. — L. L. Jones, and H. K. Mangold: J. Lipid Res. **8**, 692 (1967).
186. Skidmore, W. D., and C. Entenman: J. Lipid Res. **3**, 471 (1962).
187. Skipski, V. P., R. F. Peterson, and M. Barclay: J. Lipid Res. **3**, 467 (1962).
188. — — J. Sanders, and M. Barclay: J. Lipid Res. **4**, 227 (1963).
189. — — and M. Barclay: Biochem. J. **90**, 374 (1964).
190. — A. F. Smolowe, R. C. Sullivan, and M. Barclay: Biochim. et Biophys. Acta **106**, 486 (1965).
190a. — —, and M. Barclay: J. Lipid Res. **8**, 295 (1967).
191. Snyder, F., E. A. Cress, and N. Stephens: Lipids **1**, 381 (1966).
192. Sprecher, H. W., R. Maier, M. Barber, and R. T. Holman: Biochemistry **4**, 1856 (1965).
193. Stahl, E.: Pharm. Rundschau **1**, Nr. 2 (1959).
194. — Chemie-Ing.-Techn. **36**, 941 (1964); Angew. Chem., Internat. Edit. **3**, 784 (1964).
195. Stoffel, W., F. Chu, and E. H. Ahrens, Jr.: Anal. Chem. **31**, 307 (1959).
196. Suzuki, K.: Life Sci. **3**, 1227 (1964).
197. — J. Neurochem. **12**, 629 (1965).
198. Svennerholm, E., and L. Svennerholm: Biochim. et Biophys. Acta **70**, 432 (1963).
199. Svennerholm, L.: J. Neurochem. **10**, 613 (1963).
200. — J. Neurochem. **11**, 839 (1964).
201. Thomas, A. E., III, J. E. Scharoun, and H. Ralston: J. Am. Oil Chemists' Soc. **42**, 789 (1965).
202. Tuna, N., and H. K. Mangold: In: Evolution of the arteriosclerotic plaque, R. J. Jones, Editor, p. 83, Chicago and London: The University of Chicago Press (1963).
203. Vacíková, A., V. Felt, and J. Maliková: J. Chromatog. **9**, 301 (1962).

204. VIOQUE, E., and R. T. HOLMAN: J. Am. Oil Chemists' Soc. **39**, 63 (1963).
205. — — Arch. Biochem. Biophys. **99**, 522 (1962).
206. — y A. VIOQUE: Grasas y Aceites **15**, 125 (1964).
207. VRIES, B. DE: Chem. & Ind. (London) **1962**, 1049.
208. —, and G. JURRIENS: Fette, Seifen, Anstrichmittel **65**, 725 (1963).
209. VROMAN, H. E., and G. L. BAKER: J. Chromatog. **18**, 190 (1965).
210. WAGNER, H., L. HÖRHAMMER u. P. WOLFF: Biochem. Z. **334**, 175 (1961).
211. —, u. P. POHL: Biochem. Z. **340**, 337 (1964).
212. — — Biochem. Z. **341**, 476 (1965).
213. WALSH, D. E., O. J. BANASIK, and K. A. GILLES: J. Chromatog. **17**, 278 (1965).
214. WEICKER, H.: Klin. Wschr. **37**, 763 (1959).
215. WEISS, B., and R. L. STILLER: J. Lipid Res. **6**, 159 (1965).
216. WEKELL, J. C., C. R. HOULE, and D. C. MALINS: J. Chromatog. **14**, 529 (1964).
217. WELLS, M. A., and J. C. DITTMER: Biochemistry **2**, 1259 (1963).
218. WHERRETT, J. R., and J. N. CUMINGS: Biochem. J. **86**, 378 (1963).
219. WHITE, H. B., Jr.: J. Chromatog. **21**, 213 (1966).
220. WITTELS, B., and R. BRESSLER: J. Lipid Res. **6**, 313 (1965).
221. WOBER, W., and O. W. THIELE: Biochim. et Biophys. Acta **116**, 163 (1966).
222. WOOD, R., and F. SNYDER: Lipids **1**, 62 (1966).
223. ZÖLLNER, N., G. WOLFRAM u. G. ANIM: Klin. Wschr. **40**, 273 (1962).
224. — — Klin. Wschr. **40**, 1101 (1962).
225. —, u. D. EBERHAGEN: Untersuchung und Bestimmung der Lipoide im Blut. Berlin, Heidelberg, New York: Springer 1965.

N. Alkaloids

F. ŠANTAVÝ

About 5500 alkaloids are known at the present time and new ones are being discovered almost daily. The convenient classification of H.-G. BOIT is adopted for the grouping in the following sections. His book, "Ergebnisse der Alkaloidchemie bis 1960" is an exemplary summary. Reference may be made also to the work of MANSKE in several volumes and to numerous alkaloid chapters in "Fortschritte der Chemie organischer Naturstoffe". A large number of summarising articles on the biology and chemistry of the alkaloids is to be found in the journals "Planta Medica" and "Lloydia". The books and publications mentioned and further relevant ones are quoted in the general bibliography of this chapter on p. 461.

I. Adsorbent Layers and Solvents for TLC

The thin-layer chromatographic separation of alkaloids can be performed on silica gel, alumina, cellulose powder or kieselguhr. Silica gel is the most active stationary phase; good separations are achieved even when several milligrams substance are applied. Alumina is less active and cellulose powder and kieselguhr follow, the last named having the feeblest capacity of only about 25—30 µg alkaloid.

GIACOPELLE [70b] has used Avicel, a microcrystalline cellulose (see p. 34), for TLC of alkaloids. The hRf-values of 48 alkaloids have been measured on these layers, using 5 solvents. The layers are prepared as a rule on flat [258], or, more rarely, grooved glass plates (Fig. 15c) of varying sizes.

PARIS [161b] has recommended coated sheets (see p. 59) for fractionating some alkaloids. Information about the possibilities of preparative TLC is on pp. 97—102.

The layer of silica gel is weakly acid and that of normal alumina, weakly basic. Salts are thus formed during TLC of strong bases on silica gel and they remain at the start when neutral solvents are employed. The acid properties of the silica gel must therefore often be mitigated. Some authors use "basic" or buffered silica gel layers which have been prepared with 0.1—0.5 N NaOH or LiOH, or with a buffer solution instead of with water. Correspondingly, alumina layers are prepared with 0.2 N HCl [245]. Neutral solvents can then be used with such layers. It is advisable to use solvents containing ammonia or organic bases like pyridine, piperidine or frequently also diethylamine, in the TLC of strong bases on silica gel. Cellulose layers are generally impregnated with formamide but chromatography on non-impregnated cellulose layers has been reported. Adsorbent layers containing an added indicator which fluoresces in short wave UV light, are very satisfactory.

During the brief period of its application the TLC of alkaloids has become such a general method that many authors consider it no longer necessary to quote in their publication the adsorbent, whether loose or adhering, the solvent, chamber saturation etc.

There is a growing preference for silica gel G (Firm 88) in the field of alkaloid separation, not only for analytical but also for preparative purposes. TSCHESCHE [244] has worked out an analytical and preparative method for TLC of alkaloids in which the adsorbent contained an added substance fluorescing intensively in long wave UV light. This method is, however, not applicable to the TLC of alkaloids which themselves fluoresce. VON SCHANTZ [202] has described the circular technique for fractionating some opium alkaloids on silica gel layers. STAHL [229], who had earlier studied alkaloid separations, has demonstrated the advantages of gradient layers in this field also (Fig. 152).

WALDI et al. [257] have described a so-called systematic analysis of alkaloids for rapid identification purposes, using TLC on silica gel. It has served as the basis for a great deal of later work. Apart from these authors, TEICHERT et al. [241, 242], KAMP et al. [99], PARIS [160, 161], PFEIFER et al. [267] and SCHWARZ et al. [87, 201b, 201c, 212] have all investigated systematically the TLC of alkaloids. The chromatographic behaviour of the best known alkaloids could thus be established for various TLC techniques and solvents. Most work has been on the TLC of individual alkaloid groups or their derivatives, in contrast to that done formerly. TLC has been applied also in the elucidation of the constitution of alkaloids or in establishing the alkaloids present in the plants under examination [51, 52, 182]. It can be seen from all this that TLC has found far-reaching application in alkaloid chemistry as a whole and in the isolation and use of the alkaloids [103, 185, 188, 189, 238].

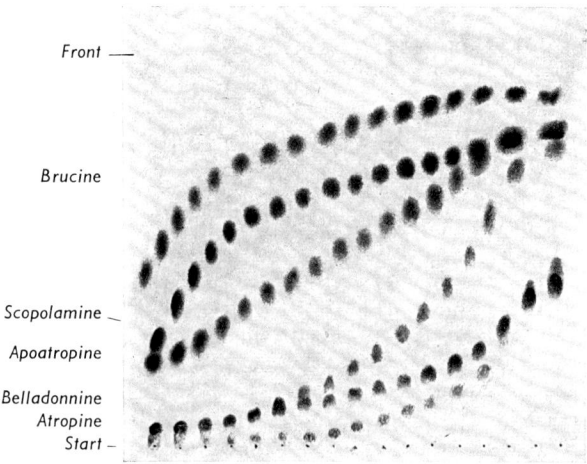

Fig. 152. Separation of an alkaloid mixture on a gradient layer. Left of diagonal spacer: acid alumina for TLC (Firm 153) + 10% gypsum; right: basic alumina for TLC (Firm 153) + 10% gypsum. Solvent: chloroform-methanol (95 + 5); CS; length of run: 15 cm; a contaminant alkaloid (?) has separated below the atropine
(STAHL)

Note: Since differences in adsorbent and solvent affect the reproducibility of Rf-values, it is a good idea to run at the same time one or more alkaloids or even dyes (rhodamine B or butter yellow). Their Rf-values then serve as references for those of the unknown alkaloids. The saturation of the chamber also has a powerful influence on Rf-values. Development only at chamber saturation is thus recommended; this improves the reproducibility of the alkaloid Rf-values, if sometimes at the expense of resolution sharpness. It must consequently be stressed here also that all Rf-values quoted in this chapter are to be regarded only as approximate, "guide" values. Solvent mixtures must be prepared with care and prepared freshly before each use, especially if they contain volatile components. They should preferably contain 1—3 solvent components; too many components militate against good reproducibility of the Rf-values.

II. Visualisation of the Alkaloids

Many alkaloids can be seen on the chromatogram even in daylight. A large number yield typical fluorescence colours in UV light (365 nm).

The agent most commonly used for detecting alkaloids is the Dragendorff reagent in its various modifications[1] (Rgt. Nos. 94, 95, 96, 97, 98). SCHWARZ et al. [201b] recommend an acidified iodine-iodide solution (Rgt. No. 143) for visualisation. If the solvent contains a difficultly volatile base, this must be removed before spraying; for this, the plate is heated to 60–120° C. Colour differentiation is not usually observed with the Dragendorff reagent. Many neutral substances also react with this reagent. More and more reagents have been employed recently for detection of alkaloids, in particular: iodoplatinate (No. 147), antimony-(III)chloride (Nos. 15–17) and cerium (IV) sulphate in sulphuric acid (Nos. 36 and 37) or in phosphoric acid (No. 34). Different alkaloids yield different colours and with some alkaloids, the colour may depend on the time and temperature. This enables substances with the same Rf-value to be distinguished. Alkaloids of the papaverrubine type give a red spot with hydrogen chloride vapour. A mixture of ferric chloride and perchloric acid (Rgt. No. 104) is preferable for *Rauwolfia* alkaloids. Iodine vapour (Rgt. No. 141) is also used. After treatment with formic acid, cinchona bases can be detected as spots which give intense blue fluorescence in UV light. The phenylalkylamines are visualised with ninhydrin (Rgt. No. 178). Additional detection agents are; cinnamaldehyde-hydrochloric acid (Rgt. No. 49) for indole alkaloids; sulphuric acid (Rgt. No. 241) for betaines; and the van Urk reagent (No. 73) for ergot alkaloids. Purine bases can be rendered visible with sulphuric acid, Dragendorff reagent or through sublimation. NEU [145] recommends sodium tetraphenylborate (Rgt. No. 245) for visualising alkaloids of various structures. Other specific reagents are quoted in the sections on the individual groups of alkaloids.

III. Separation Scheme for Alkaloids

WALDI [257] divides the alkaloids into two groups according to Rf-values. Rhodamine B or reserpine are the reference substances. The solvent system cyclohexane-chloroform-diethylamine $(50 + 40 + 10)$, practicable for either alkaloid group, is used in a preliminary run in the TLC of unknown alkaloids. Based on the hRf-values found, the most suitable solvent system for the one or the other group is then chosen for the real run (see 1st Edition, Table 46). For the further identification of the alkaloids, WALDI recommends ascertaining the fluorescence in long wave UV light and the colour after spraying with iodoplatinate reagent (No. 147).

---

[1] See in [189b, 251b] concerning stability and sensitivity of the Dragendorff-reagent. The detection limit for most alkaloids is 0.01 to 0.05 µg.

IV. Quantitative Determination of Alkaloids with TLC[2]

Only rarely does a sample of naturally occurring material contain a single substance. A mixture of related substances or of groups of related substances is encountered as a rule, difficult to fractionate with the methods which were generally available formerly. Many of the methods so far published in the pharmacopoeias are thus still being confronted with an unsolved problem. It is true that alkaloids can be separated from other substances by making use of their basic properties; but the further determination of, e. g., morphine in opium, demands the application of column chromatography which is associated with more or less considerable losses and which moreover requires a larger amount of starting material. Paper chromatography offered the first chance of a microquantitative determination of alkaloids in mixtures. A number of methods has been worked out for paper chromatographic determination of individual alkaloids or other substances; many of these can be adapted to TLC, for which the amounts necessary are even smaller than for PC. TLC-methods for separating alkaloids have also been worked out or modified; these can be characterised as follows:

1. The spot area is evaluated by planimetry [151—153].

2. The spots are photographed (black and white) in daylight or UV light, either directly or after detection with some reagent. The size and intensity of the photographed spots are then evaluated photometrically. POETHKE and KINZE [177] have attempted photometric evaluation directly on the layer but the limits of error are rather broad in this procedure.

3. The zone and adsorbent are scraped off the plate and transferred quantitatively to an extractor. Methanol is generally used for the extraction, the volume being then made up to a definite amount. The substance itself is determined polarographically, colorimetrically (after conversion into a coloured derivative) or, easily the best, spectrophotometrically at the wave length in the UV where the substance has a characteristic adsorption maximum. A blank on the adsorbent is necessary in the spectrophotometric method. The amount of substance can then be calculated from the extinction coefficient. This combination of TLC and spectrophotometric determination of the eluted substance, is a method of wide analytical potentiality in biology and chemistry.

4. Determination of the eluted substance by titration [93].

5. Radioactively labelled substances may be radiometrically evaluated directly or also after scraping off the plate (see Chapter I, p. 155).

V. Special Section
1. Colchicine Alkaloids

The formation of tropolone alkaloids is characteristic of the subfamily Wurmbaeoideae (Liliaceae family). The best known alkaloid

---

[2] Compare Chapter H, "Quantitative Evaluation", p. 133.

of this series is colchicine, found earlier in the genus *Colchicum*. It is a typical mitotic poison, the classical means of creating polyploidal animal and plant organisms. Its less toxic derivatives, e. g., demecolcine, are of interest in the treatment of myeloid leucaemia or skin cancer. The tendency of the tropolone alkaloids to lumitransformation is noteworthy. The plants of this subfamily also form alkaloids without the tropolone ring system.

Silica G layers, prepared using the standard method, have been those preferred for TLC of the tropolone alkaloids. KUHN and co-workers [111, 149] have used ethanol-ethyl acetate (20 + 80) in addition to the solvents given by WALDI and co-workers [257] for colchicine. The one team preferred visualisation with the Dragendorff and iodoplatinate reagents; the other, with antimony(III)-chloride. MÁTHÉ and TYIHÁK [132] have used TLC in an investigation of *Colchicum hungaricum* JANKA.

ŠANTAVÝ et al. [178] have studied in detail the TLC of the alkaloids of this series so far isolated. They employed silica gel layers.

The silica gel was prepared according to [172], grain size 35—75 μm, layer thickness 0.2—0.4 mm, plates of 100 × 200 mm, spots of 2—4 mm diameter and chamber of dimensions 200 × 250 × 35 mm. The individual alkaloids were identified by using four to five plates, each of which was sprayed with a different reagent.

The *solvent systems* used were those proposed by WALDI [257] and slightly modified by POTĚŠILOVÁ [178] (Table 72). The proportions of the components of the solvent mixture can be varied somewhat, depending on the nature of the silica gel, in order to achieve optimum separation. The *best amount to apply* is 2–5 μg of pure substance or 20 μg of crude extract in ethanolic solution. Colchicoside or other glycosidic alkaloids were applied after dissolving in a water-ethanol mixture. The sensitivity of the Oberlin-Zeisel reaction was about 1 μg tropolone substance per spot.

Visualisation; 1. Daylight; 2. long wave UV light; 3. Dragendorff reagent (No. 98); 4. characteristic and sensitive Oberlin-Zeisel reaction, ferric chloride reagent (No. 102); 5. iodoplatinate reagent (No. 147); 6. antimony(III)-chloride reagent (No. 15); and 7. hydrogen chloride vapour.

In addition to the above-mentioned colour reactions, all tropolone alkaloids possess a characteristic UV spectrum (maximum at 352 nm, $\log \varepsilon$ 4.25) which differs from the absorption spectrum of the lumiderivatives with their principal maximum at 266 nm ($\log \varepsilon$ 4.36). All the tropolone alkaloids so far known can be determined in the crude extracts from plants of the subfamily Wurmbaeoideae (after thin-layer chromatographic separation) by subsequent spectrophotometric evaluation at various wave lengths, taking the hR_f-values (see Table 73) and colour reactions into consideration. The phenolic alkaloids can be distinguished from the non-phenolic also (Rgt. Nos. 102 and 238). The lumiderivatives

Table 72. Separation and detection of the colchicum alkaloids on silica gel G layers [178]

No.	Substance	Solvent and hR_f-value [a]					UV-light (365 nm)	Iodoplatinate Reagent (No.147)	Antimony(III)-chloride Reagent (No. 15)	
		I	II	III	IV	V			immediate	after heating to 100°C
1	O-Benzoylcolchiceine	96		96	96		blue-violet	dark red brown	yellow	orange-yellow
2	Cornigerine	58		63			yellow	red brown-beige	yellow	lemon-yellow
3	Colchicine	56		61			yellow-brown			
4	2-Ethyl-2-demethylcolchicine	55		58			light brown			
5	3-Propyl-3-demethylcolchicine	54		67			pink			
6	2-Acetyl-2-demethylcolchicine	53		56			violet			
7	3-Acetyl-3-demethylcolchicine	51		47			pink			
8	3-Ethyl-3-demethylcolchicine	49		68			pink			
9	N-Formyldeacetylcolchicine	44		54			beige	red brown-beige	yellow	lemon-yellow
10	Isocolchicine	38		42			pink			
11	O-Acetylcolchiceine	37		54	95		pink			
12	3-Demethylcolchicine	32		36			dark violet	yellow	yellow	lemon-yellow
13	Colchiceine	27		32	44		brown-yellow			
14	2-Demethylcolchicine	22		27			dark-violet	yellow	yellow	lemon-yellow
15	N-Formyldeacetylcolchiceine	18		26	95		light blue			
16	Colchicoside	7					light blue			
17	Deacetylcolchicine	0		0	52		blue-green			
18	N-Benzoyl-N-deacetylcolchiceine	0		0	48		blue-green			
19	Deacetylcolchicine	35								
20	Deacetylisocolchicine	22								
21	N-Methyldemecolcine		82	87		78	light violet	yellow	yellow	lemon-yellow
22	Speciosine		77			67	grey-violet	yellow-beige	yellow	lemon-yellow
23	Demecolcine		68	70		56	bronze	brown-beige	orange-brown	yellow

Table 72 (Continued)

No.	Substance	Solvent and hRf-value [a]				UV-light (365 nm)	Iodoplatinate Reagent (No.147)	Antimony(III)-chloride Reagent (No. 15)	
		I	II	III	IV			immediate	after heating to 100°C
24	N-Propionyldemecolcine		68	73		violet			
25	N-Acetyldemecolcine		62	65		yellow-brown			
26	3,N-Diacetyl-3-demethyl-demecolcine		55	67		blue			
27	N-Formyldemecolcine		53	55		violet			
28	3-Demethyldemecolcine		51	56	31	dark violet	yellow	orange-brown	yellow
29	3-Ethyl-3-demethyldemecolcine		50	62		light blue			
30	N-Acetylisodemecolcine		48	54		violet			
31	N-Benzoyldemecolcine		38	44		violet			
32	2-Demethyldemecolcine		30	49	22	dark violet	lemon-yellow	yellow	lemon-yellow
33	Demecolceine		9	18		dark violet			
34	β-Lumicolchicine	82				grey	pink	beige-yellow	red-brown-violet
35	γ-Lumicolchicine	70				grey	black	beige-yellow	red-brown-violet
36	N-Acetyllumidemecolcine-β		83			light blue	brown-beige	light beige	red-brown
37	β-Lumidemecolcine		79		74	light grey	brown-beige	light beige	red-brown
38	γ-Lumidemecolcine		65			light pink		light beige	red-brown

Solvent: I = Benzene-ethyl acetate-diethylamine (50 + 40 + 10) + 8% methanol, (NS); II = I but without the methanol addition; III = Chloroform-acetone-diethylamine (70 + 20 + 10), (NS); IV = III but with addition of 8% methanol; V = Benzene-ethyl acetate-diethylamine (70 + 20 + 10), (NS).

Detection: Daylight: Alkaloids 12—15, 17, 18, 28, 32 and 33 showed yellow; Dragendorff reagent (No. 98): all gave orange red; hydrogen chloride vapour: Nos. 1—33 turned yellow; ferric chloride reagent (No. 102): Nos. 1—33 turned brown.

[a] To be regarded as guide values.

Table 73. hRf-values and detection reactions of the subsidiary alkaloids (without tropolone ring) of the sub-family Wurmbaeoideae [178]

Alkaloid	hRf	Fluorescence UV 365 nm	Iodoplatinate (Rgt. No. 147)	SbCl$_3$ (Rgt. No. 15)
Alkaloid AM-4	90	0	Deep yellow-red	—
Alkaloid CC-1	84	Light violet	Orange-yellow; pink edge	—
Alkaloid AM-3	82	0	Orange-yellow	—
Alkaloid O	80	Blue	—	0
Alkaloid CC-15	76	Light grey	Carmine red-grey-pink	—
Isocorydine	73	Light blue	White-blue-green; grey edge	0
Bulbocodine	68	0	White-pink; violet edge	0
Alkaloid CC-2	68	0	Red-brown-beige red	—
Collumellarine	58	Light pink	Yellow; pink edge	—
Androcymbine[a]	53	Dark violet	Dark violet-violet-grey	Momentarily light yellow, then disappears
Melanthioidine	51	Light grey	Light yellow-orange	0
Alkaloid CC-3a	46	0	Violet-brown-red	0
Alkaloid CC-3b	33	0	Brown-brown-red	0
Bechuanine	46	0	White-pale yellow-yellow	0
Alkaloid OGG-3	30	0	Yellow-beige	—
Alkaloid To	18	0	Yellow-brown-red-light brown	Yellow
Alkaloid M	0	Light blue	Brown-red	0
Floramultine[b]	46	0	White-light yellow deep yellow; violet edge	—
Kreysiginine[b]	40	0	Acute green-pink-violet-grey	—

Layer: Silica gel G
Solvent: Benzene-ethyl acetate-diethylamine (70 + 20 + 10) (NS)

[a] Oberlin-Zeisel reaction (Rgt. No. 102) yields brown-blue.
[b] Isolated from *Kreysigia multiflora*.
0 = no reaction.

present, the alkaloids possessing the cyclohexadienone structure[3], the aporphine and homoaporphine alkaloids and alkaloids of unknown constitution without tropolone ring, can also be determined with the help of TLC and UV-spectroscopy. It is difficult to separate colchicine from cornigerine in the crude extracts. However, the antimony(III)-chloride reagent yields orange red with cornigerine and only yellow with colchicine.

Identification of the Alkaloids in the Crude Extracts: 0.5—5 g dried and finely milled plant material is extracted with methanol in a suitably sized chromato-

---

[3] See Chem. Commun. **1965**, 228 and 415 concerning the constitution of androcymbine and melanthioidine.

graphy tube, serving here as a percolator. The methanol extract is evaporated in vacuo at 30° C and the residue dissolved in 1—2 times its own weight of water (based on the weight of the dry sample). After adjusting the pH of the solution to 3 with citric acid, it is extracted in a separating funnel with petroleum ether, ether and then chloroform (= neutral-phenolic chloroform extract). The acidic residue is rendered alkaline with ammonium hydroxide and again extracted with chloroform (= basic chloroform extract) and finally with a chloroform-ethanol (2 + 1) mixture (= glycosidic extract). The glycosidic extract can however be obtained also from a larger amount of starting material (50—100 g) and must then be separated into a neutral and basic portion.

The following must be noted in the separation procedure just described: The extraction with chloroform is carried out with four portions of an amount of solvent equal to 5—10 times the amount of drug. The pH value must be controlled during extraction of the acid and alkaline phases, especially the latter. The chloroform extracts are shaken with small amounts of water until they are neutral. The extracts are concentrated at ca. 35° C in vacuo. The thin-layer chromatographic fractionation of these extracts from partial preliminary separation of the alkaloids then follows (Tables 72 and 73).

Quantitative Determination: The zones of tropolone alkaloids and of their lumi-derivatives can be recognised through their fluorescence in long wave UV light; some appear yellowish in daylight (Table 72). Each separate zone is scraped off, transferred to a 5 ml standard flask, extracted with methanol, the solution centrifuged from silica gel and the extinction measured at the wave lengths stated. Other ways are described in Chapter H, p. 133. It is important to work in subdued light during the quantitative determination, so as to prevent lumitransformation of the tropolone alkaloids.

2. Pyrrolidine, Pyridine and Piperidine Alkaloids

PAILER [154] recommends TLC on silica gel G layers, using chloroform-methanol (33 + 66), for separating the pyrrolidine alkaloid betonicine from glycolbetaine. This had not been possible with PC. The spots were visualised with iodoplatinate (Rgt. No. 147) or by spraying with concentrated sulphuric acid and subsequent heating.

PAPP [157] has succeeded in separating the alkaloids of *Nicotiana tabacum* and *Sedum acre*, using toluene-methanol-chloroform (90 + 30 + 10) on basic silica gel G layers (0.5 N KOH). SPEAKE [220] chromatographs the alkaloids of *N. tabacum* on a silica gel G layer, using pure methanol. TLC has been most valuable also in the study of the demethylation of nicotine [40c].

HODGSON et al. [88d] have made use of two-dimensional TLC in their study of nicotine metabolism. The hR_f-values and experimental conditions are seen in Table 74. DECKER and SAMMECK [42b] have carried out chromatography on silica gel G, using n-butanol-ethanol-0.5 N ammonium hydroxide (66 + 17 + 17) in order to solve similar problems. In both these cases [88d, 42b] the layers were sprayed with a mixture of equal volumes of 2% p-aminobenzoic acid in ethanol and 0.1 M phosphate buffer, pH 7; after drying for 15 min, the plate was exposed to bromine cyanide vapour for visualisation (König reaction, see Rgt. No. 23).

Table 74. hRf-values of the tobacco alkaloids and some of their derivatives on silica gel G layers [88d]

Alkaloid	Solvent I	II
Nornicotine	34	05
Nicotine	77	08
Nicotyrine	87	92
Anabasine	50	06
Nicotine-N-oxide	08	05
Cotinine	75	76
Norcotinine	50	51

Solvent I = Chloroform-methanol-ammonium hydroxide (60 + 10 + 1); II = chloroform-methanol-acetic acid (60 + 10 + 1)

Four alkaloids have been detected in *Gentiana lutea* [38]. One was identified as 3-vinyl-4(2-hydroxy-)ethylpyridine and called gentialutine. The isolation procedure was controlled through TLC on alumina or silica gel G, using the solvents benzene-chloroform-methanol (70 + 30 + 20) or acetone-methanol (50 + 50).

The hemlock alkaloids also can be separated on silica gel G layers [137], using chloroform-absolute ethanol-25% ammonium hydroxide (90 + 10 + 10). The ammonia is expelled by heating 5 min at 110° C before visualisation with iodine vapour in a desiccator; the amines give brown spots. After spraying with 1-chloro-2,4-dinitrobenzene solution (Rgt. No. 44) or with bromothymol blue solution (Rgt. No. 31), the bases appear as blue spots on a yellow background. γ-Coniceine turns red when sprayed with sodium nitroprusside solution (Rgt. No. 186).

PAILER [155] has succeeded in showing that at least twelve alkaloids occur in the seeds of *Evonymus europaea* L. He used chloroform-ethanol (100 + 2) on silica gel G layers, detecting with Dragendorff reagent. This technique was used also for monitoring the separation of the alkaloids by counter-current distribution.

SCHWARTING et al. [150b] have employed TLC in qualitative and quantitative studies of the biosynthesis of hygrine, cuscohygrine, anahygrine, anapherine and isopelletierine.

ROTHER et al. [119, 190] have used TLC (silica gel G/chloroform-methanol-diethylamine (65 + 25 + 5) or alumina G/benzene-methanol-diethylamine (99 + 1 + 5)) for control in structure determination of the alkaloids anapherine and anahygrine and in order to follow their synthesis.

Formerly only the *Lobelia* alkaloids in *Lobelia inflata* L. were considered worth notice; many of these are respiratory stimulants. The other plants of this genus have now been investigated and TLC has been used here, not only as an analytical method but also for preparative isolation as, e. g., by TSCHESCHE et al. [245] with *Lobelia syphilitica*. The adsorbent layer was prepared from alumina G and 0.2 N HCl and chloroform-ethanol (95 + 5) used for development. Chromatography on silica gel G, using chloroform-methanol (75 + 25) was also used. In this way the authors discovered a number of new alkaloids. BORIO and MOREIRA [30b] recommend a silica gel G layer and solvent mixture of cyclohexane-chloroform-diethylamine (50 + 40 + 5) for the thin-layer chromatographic separation of lobeline, lobelanine and lobelanidine (hRf 42, 73, 31). SCHWARZ et al. [212] have separated the *Lobelia* alkaloids on a loose alumina layer. TEICHERT et al. [241] have employed a layer of cellulose powder, impregnated with formamide. PARRÁK et al. [162] have studied the stability of lobeline preparations with the help of TLC on silica gel G or alkalised alumina using as solvent, benzene-chloroform (50 + 50) saturated with ammonia. An old lobeline solution for injection yielded six spots altogether, of which those of lobeline, acetophenone, lobelanidine and lobelanine could be identified (with Rgt. No. 82).

3. Tropane Alkaloids

Although many tropane alkaloids are important in practice as anaesthetics, spasmolytics and parasympathomimetics, it seems only recently that interest has arisen in chromatographic studies of them, either in galenical preparations or for following analytically their biogenesis in plants. In connection with other analgesics or alkaloids, some of these substances have been investigated using TLC.

POETHKE et al. [125, 173] have worked with belladonna herbs and tinctures. They used silica gel G and chloroform-acetone-diethylamine (50 + 40 + 10) as solvent. OSWALD and FLÜCK [151–153] have chromatographed hyoscyamine and other alkaloids from this material, notably belladonnine, apoatropine, aposcopolamine, scopine and scopoline. Six solvents were used, of which the best was found to be butanone-methanol-7.5% ammonium hydroxide (60 + 30 + 10); the Dragendorff reagent was employed for detection. The method was worked out also for quantitative determination (planimetric evaluation) (limit of error $\pm 5.8\%$). A linear relation was found between spot area and amount of alkaloid, provided that the amount of substance did not fluctuate by more than $\pm 20\%$. In this way it was possible to determine the mixture of hyoscyamine and atropine in the presence of scopolamine (hyoscine) in some of the plant organs of *Atropa belladonna*, *Datura stramonium*, *Hyoscyamus niger* and *H. muticus* [153].

IKRAM et al. [93] also have worked out a quantitative TLC method on loose alumina layers. The zone is scraped off, extracted with chloroform, the extract evaporated to dryness, the residue dissolved in ethanol and finally titrated with 0.01 N sulphuric acid (methyl red indicator). SCHNECKENBURGER et al. [204] have used TLC for establishing which pharmacopoeia method is preferable for quantitative determination of the active principles in belladonna leaves; they employed silica gel G or GF_{254} (Firm 81) layers and diethylamine-dimethylformamide-ethanol-ethyl acetate (5 + 5 + 30 + 60) as solvent and detected the spots in UV light or by visualisation with the Dragendorff reagent or the iodoplatinate reagent (No. 147).

Table 75. hR_f-values of tropane-alkaloids

Layer	S	S	S	A	S (0,1 N NaOH)	S	S	S (0,5 N KOH)	C (Formamide)
Solvent	I	II	III	IV	V	VI	VII	VIII	IX
Tropine	—	—	—	—	—	—	16	3	13
Atropine	38	40	16	10	17	17	37	36	15
Homatropine	37	45	15	24	15	—	—	—	—
Apoatropine	54	67	40	40	16	—	44	44	74
Belladonnine	—	—	—	—	—	—	26	17	69
Cocaine	73	90	65	77	62	61	—	—	—
Scopolamine	56	60	19	0	52	—	73	83	53
Scopoline	60	90	44	50	37	—	—	—	—
Tropacocaine	65	90	56	78	35	—	—	—	—
New psicaine	66	90	60	82	59	—	—	—	—

Layer: S = silica gel G; A = alumina G; C = cellulose powder.
Solvent: I = Chloroform-acetone-diethylamine (50 + 40 + 10) [257]; II = Chloroform-diethylamine (90 + 10) [257]; III = Cyclohexane-chloroform-diethylamine (50 + 40 + 10) [257]; IV = Cyclohexane-chloroform (30 + 70) + 0.05% diethylamine [257]; V = Methanol [257]; VI = Methanol-acetone-triethanolamine (50 + 50 + 1.5) [15]; VII = Dimethylformamide-diethylamine-ethanol-ethyl acetate (5 + 5 + 30 + 60) [241]; VIII = 70% Ethanol-25% ammonium hydroxide (99 + 1) [241]; IX = 1 run: 15 cm, heptane-diethylamine (100 + 0.2), 2 run: 10 cm, benzene-heptane-chloroform-diethylamine (60 + 50 + 10 + 0.2) [241].

VÉGH et al. [252] have separated atropine from papaverine in galenical preparations, using TLC on silica gel layers; they developed with methanol-benzene (66 + 33) and visualised the atropine through the Vitali reaction. Proof has been obtained by FROHNE [65] that the drug Radix Scrophulariae is frequently adulterated by the roots of atropine-containing Solanaceae: he used the adsorbent silica gel G and the solvents chloroform-acetone-diethylamine (50 + 40 + 10), chloroform-diethylamine (90 + 10) or cyclohexane-chloroform-diethylamine (50 + 40 + 10). Other work on tropane alkaloids in galenical preparations is described in [34d, 39, 62, 98, 123, 148, and 251].

Stahl and Schorn [229b] have compared the experimental conditions given in the literature for separating tropane alkaloids; their aim was a reliable identification method for Solanaceae drugs in pharmacopoeias. They worked out a rapid extraction method and prefer experimental conditions III (in Fig. 153) as a generally applicable procedure. The Dragendorff reagent of Thies and Reuther, as modified by Vágújfalvi (Rgt. No. 98), is recommended for detection; the limit of detection is about 0.5 µg base.

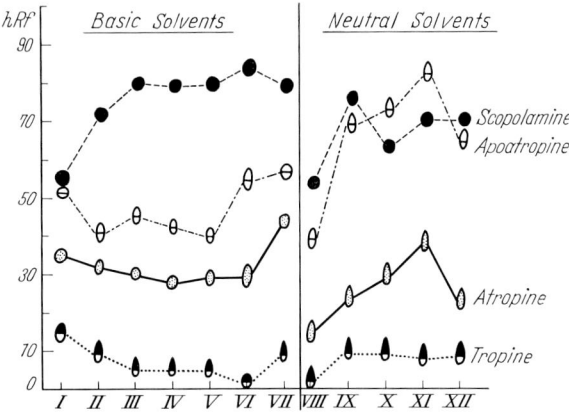

Fig. 153. Comparison of various silica gels and aluminas and solvents with a view to obtaining optimum separation of tropane alkaloids

    I. Silica gel G_{254} (Firm 81); chloroform-acetone-diethylamine (50 + 40 + 10).
   II. Silica gel GF_{254} (Firm 81); butanone-methanol-water-25% ammonium hydroxide (60 + 30 + 7 + 3) [151, 152].
  III. Silica gel G_{254} (Firm 81); acetone-7.5% ammonium hydroxide (90 + 10) [229a].
  IV. Silica gel MN-GHR (Firm 76); solvent III.
   V. Silica gel DGF (Firm 110); solvent III.
  VI. Silica gel (Firm 141); solvent III.
 VII. Silica gel GF_{254} prepared with 0.5N KOH (Firm 81); sclvent III.
VIII. Silica gel GF_{254}, prepared with 0.5N KOH (Firm 81); chloroform-methanol (85 + 15) [229a].
  IX. Alumina GF_{254} (Firm 81); chloroform-methanol (97.5 + 2.5) [229a].
   X. Alumina D (Firm 110); solvent IX.
  XI. Basic alumina (Firm 141); solvent IX.
 XII. Alumina for TLC (Firm 29); solvent IX.

The alkaloids tropine, pseudotropine, 3-α-tigloyloxytropane (3-α-tropanyl tigloate), choline, cuscohygrine (hRf-value 63), isopelletierine and two hitherto unknown alkaloids, anapherine and anahygrine (hRf-value 50) have been isolated from *Withania somnifera*, using control by TLC [101, 102, 120, 211]; benzene-methanol-diethylamine (99 + 1 + 5) was used on alumina G for this purpose. 3-α-Tigloyloxytropane is best separated on silica gel G, using the solvent mixture ethanol-aqueous ammonium hydroxide (80 + 20). *W. ashwagandha* KAUL [101] and *W. americana L.* [156] have also been analysed with TLC.

TLC has been used also by EVANS [50] for isolation of (−)-6β-tigloyloxytropan-3β-ol from *Datura cornigera* HOOK.

With the help of TLC, SULLIVAN and GIBSON [232] have controlled the incorporation of proline-^{14}C into the alkaloids of *Datura strammonium var. tatula* and *D. innoxia*. The biosynthesis of the tropane alkaloids from phenylalanine is also treated in [85]. ROBLES [188] has employed TLC in the preparative separation of tropane alkaloids.

4. Pyrrolizidine Alkaloids

The pyrrolizidine alkaloids form a toxicologically important group today; they occur in some genera of the Fabaceae, Boraginaceae and Asteraceae families. TLC can also be used for identification of these substances. Some hRf-values of the alkaloids occurring most often in these plants are given in Table 76 [200].

Table 76. hRf-values of necine alkaloids

Alkaloid	hRf	Colour[a] in the iodoplatinate reaction (Rgt. No. 147)
Senecionine	57	Red-brown-light beige-violet; white edge
Seneciphylline	55	Red-brown-beige-violet; white edge
Platyphylline	52	Dark brown-beige-white
Rivularine	51	Red-brown grey-grey-violet; white edge
Monocrotaline	38	Red-brown-grey-light beige-white
Ridelline	35	Red brown-beige-white
Retrorsine	31	Dark red-brown-light beige
Jacobine	31	Red-brown-pink-grey pink; white edge
Othosenine	18	Red-brown-beige-white
Retronecine	7	Red-brown-light beige-white

Layer: Silica gel G.
Solvent: Benzene-ethyl acetate-diethylamine (70 + 20 + 10), NS.

[a] All alkaloids also yield an orange-red colour with Dragendorff reagent.

SHARMA et al. [214b] and especially CULVENOR et al. [36b] have recently described the TLC of the pyrrolizidine alkaloids on alkalised silica gel G layers, using methanol as solvent; the last named authors quoted the hRf-values of 58 alkaloids.

5. Quinolizidine Alkaloids

The alkaloids of this group are formed by plants of the Papilionaceae (Fabaceae), Chenopodiaceae, Berberidaceae and Papaveraceae families. Usually alkaloid mixtures occur, as with other plants. They can occur in company with other alkaloid types, e. g., those containing a pyridine nucleus (ammodendrine) or with pyrrolizidine alkaloids (laburnine). The quinolizidine alkaloids can be classified according to the nature of their nucleus, into lupinine, cytisine, sparteine and matrine groups. Most of the substances are toxic and have so far found no medical application.

Table 77. hR_f-values of quinolizidine alkaloids [74]

Alkaloid	I	II	Alkaloid	I	II
Sparteine	93	10	Hydroxysparteine	54	19
L-α-Isosparteine	92	3	Anagyrine	35	60
Retamine	80	30	N-Methylcytisine	30	63
Lupinine	63	12	Hydroxylupanine	19	24
Lupanine	61	38	Cytisine	7	32
Epilupinine	58	35	Calycotomine [a]	5	46

Layer: Silica gel G.   [a] Tetrahydroisoquinoline alkaloid.
Solvent: I. Cyclohexane-diethylamine (70 + 30); II. Chloroform-methanol (80 + 20).
Solvent I is especially suitable for preliminary tests (length of run: 11 cm, time of run: 50 min). Cyclohexane-diethylamine (90 + 10) or (50 + 50) or solvent II is used for further testing for alkaloids.

GILL [71, 72] and STEINEGGER [73–75, 214, 231] have used PC and TLC in qualitative and quantitative (planimetry of the spots after spraying with Dragendorff reagent) investigations of numerous species of the genera *Cytisus*, *Genista* and *Retama*. Adequate material for chemotaxonomic conclusions could be thus obtained [19b]. FAUGERAS [55, 56] has worked on the isolation of the alkaloid anagyrine from the flowers and seeds of *Genista raetum* FORSK; he chromatographed anagyrine and cytisine on silica gel G layers, using benzene-chloroform-diethylamine (20 + 75 + 15).

Sparteine has been isolated also from *Chelidonium majus* (Papaveraceae); with the help of TLC; SCHÜTTE [210] has studied the biogenesis of sparteine from radioactive cadaverine in this plant and concluded that the biosynthesis proceeds as in *Lupinus luteus*. Other work on the TLC of quinolizidine alkaloids has been published [1b, 26, 27].

6. Alkaloids of the Papaveraceae

The following groups of alkaloids occur in the plant family of the Papaveraceae: benzylisoquinoline, cularine, aporphine, proaporphine, morphine, protoberberine, protopine, narceine, benzophenanthridine, phthalide-isoquinoline, rhoeadine, papaverrubine, pavine, isopavine and ochotensimine. Many of these alkaloid groups are formed by one genus only, whereas others occur in many genera and even in plants of other families. Alkaloids of all these groups must thus be taken into consideration when chromatographing crude extracts or testing the purity of the isolated alkaloids.

BORKE et al. [31] ("chromatostrip technique" [104]) and MARIANI and MARIANI-MARELLI [129] in their studies of opium alkaloids, were the first to draw attention to the advantages of TLC over other analytical methods, including PC. The method really developed only after STAHL [227] had introduced standardised conditions and apparatus for separation in 1956.

Various adsorbents, solvents and techniques have been used for the thin-layer chromatographic separation of alkaloid mixtures. The usual technique has been ascending development on silica gel layers. Multiple development with solvents of differing polarity can be advantageous. BÉLA [19] has tested out 28 different solvents and recommends double development in TLC on silica gel G layers, using 1. benzene-methanol (90 + 10) and 2. chloroform-ethanol-ethyl acetate-acetone (60 + 20 + 10 + 10) at 20° C and with a length of run of 15 cm in each case. The following hR_f-values were found: narcotine 80, papaverine 75, narcotoline 67, protopine 62, cryptopine 58, laudanosine 50, thebaine 44, laudanine 39, codeine 20, neopine 18, morphine 10, narceine 0.

The hR_f-values of a number of opium alkaloids and their derivatives have been determined on various adsorbents and using various solvents also by WALDI et al. [257], TEICHERT et al. [241, 242], SCHWARZ et al. [17, 212] and BÄUMLER et al. [15]. According to RAMAUT [182b], a good separation of all these alkaloids is possible on silica gel G layers using hexane-cyclohexane-cyclohexanol (33 + 33 + 33), +5% diethylamine. MACHATA [127] has applied TLC to the detection of opium alkaloids and other medicaments in the stomach contents and bile of a human being who had been poisoned with opium tincture. TEICHERT et al. [241, 242] have found benzene-heptane-chloroform-diethylamine (60 + 50 + 30 + 0.03) preferable with silica gel which had been alkalised with 0.5 N KOH; they thus achieved a good fractionation of the opium alkaloids (morphine remained at the start). TLC on another layer, using chloroform-ethanol (75 + 25), was recommended for separating morphine from contaminants. The Dragendorff reagent (No. 98) or a ferric chloride-perchloric acid mixture (Rgt. No. 104) were used for visualisation.

NEUBAUER and MOTHES [147] have applied TLC more intensively to control of opium, especially in the selection of different types of poppy [25b]. They were able to separate 10 alkaloids from one another, using benzene-methanol (80 + 20) on silica gel layers. These authors employed TLC also for semi-quantitative determination; in this, the spots were visualised with Dragendorff reagent, photographed and the black spots of the positive evaluated photometrically. NEUBAUER [146] has made use of this method in the quantitative determination of alkaloids during the growth of *Papaver somniferum*.

IKRAM et al. [94] describe the separation of morphine, codeine, papaverine and thebaine on a loose alumina layer (Merck), using three solvents. Papaverine and narcotine cannot always be separated satisfactorily and this has led to the use of two-dimensional TLC [99, 176]. With xylene-butanone-methanol-diethylamine $(40 + 60 + 6 + 2)$ in a one-dimensional procedure on activated silica gel G, BAYER [16] has recently succeeded in separating the following alkaloids: narcotine hRf 74, papaverine 59, thebaine 45, codeine 26, morphine 12. The problem of the papaverine-narcotine separation has thus been solved.

PINXTEREN and VAN VERLOOP [170] have separated pseudo-morphine (hRf 4), morphine (hRf 35) and narcotoline (hRf 85) on silica gel, using the solvent carbon tetrachloride-butanol-methanol-6N ammonium hydroxide $(40 + 30 + 20 + 2)$. In later work [171], they determined morphine in poppy heads by converting it on the layer into the intensively fluorescent pseudo-morphine, using ferricyanide reagent (No. 112). KUPFERBERG et al. [116], independently of the last named authors, also have developed this sensitive TLC method for detecting traces of morphium and other analgesics. Only those phenanthrene alkaloids which contain a free phenol group react with reagent No. 112. BEYER [21] has employed TLC on silica gel with 0.1 N ammonium hydroxide as solvent, for determination of acedicone (acetyldihydrocodeinone) and dicodid which is formed from acedicone. A method for detecting this compound in pharmaceutical preparations was also worked out. Visualisation is with the Dragendorff reagent (No. 98) or the Vidic reagent (No. 82).

STAHL and JORK [229c] have critically compared the numerous solvents already quoted in the literature, with a view to introducing TLC into pharmacopoeia procedures and to identifying crude opium and opium preparations.

The standard method was used on silica gel GF_{254} layers, with uniform external conditions. The fluorescence-quenching zones of the six best known opium alkaloids could be visualised with concentrated sulphuric acid, which yielded differentiating colours. Thebaine alone was visible in the cold, as a yellow spot. 10 µg amounts on heating 10 min at 150° C gave the colours: morphine, violet; narceine, brown violet; codeine, reddish blue; thebaine, light violet; papaverine, grey; and narcotine, pale red.

As clearly seen from Fig. 154, solvent selection can be adapted to the particular problem. Thus papaverine and narcotine can be distinguished with neutral solvents which contain benzene as chief component (Nos. I—VI). Morphine and narceine can be separated only in basic solvent mixtures (Nos. X—XIV). If one has no desire to carry out multiple development, solvents X and XIV offer the best separation chance for the 6 alkaloids mentioned, whether in a mixture together or in opium preparations.

POETHKE et al. [174—177] have published a series of articles on the TLC of the opium alkaloids. The alkaloids were determined quali-

Table 78. hRf-values of opium alkaloids on various layers, using solvents I—XVI

Layer	S	S	S	A₂	S 0.1 N NaOH	S	S	S 0.5 N KOH	C	C Form-amide	S	A₁	S	A₁	A₁	A₁
Solvent	I	II	III	IV	V	VI	VII	VIII	IX	X	XI	XII	XIII	XIV	XV	XVI
Narceine	3	0	0	0	0	23	—	—	—	—	—	86	—	—	—	—
Dihydromorphinone (dilaudid)	24	23	8	8	16	28	5	13	27	6	14	—	—	—	—	—
Dihydrocodeinone (dicodid)	51	65	21	43	18	29	10	28	34	63	—	—	—	—	—	—
Morphine	10	8	0	0	34	40	2	2	27	0	24	14	12	3	3	6
Morphine ethyl ether (dionine)	—	—	—	—	—	37	14	37	44	57	—	72	—	—	—	—
Dihydrocodeine (paracodine)	38	54	18	30	25	—	6	22	34	37	26	73	26	77	38	33
Codeine	38	53	16	27	35	43	12	33	41	90	—	—	—	—	—	—
Acetyldihydrocodeinone (acedicone)	—	—	—	—	—	—	24	59	79	75	—	—	—	—	—	—
Dihydro-hydroxycodeinone (eucodal)	—	—	—	—	—	—	47	70	—	—	—	—	—	—	—	—
Thebaine	65	90	51	76	40	—	—	78	86	89	—	91	45	—	—	—
Papaverine	67	90	42	84	70	82	74	78	86	89	74	86	59	89	88	73
Cotarmine	60	90	43	25	0	—	—	—	—	—	69	90	—	—	—	—
Narcotine	72	90	51	79	72	82	78	81	92	94	69	90	74	92	77	77

Layer: S = silica gel G; A₁ = alumina; A₂ = alumina G; C = cellulose powder.
Solvent: I = Chloroform-acetone-diethylamine (50 + 40 + 10) [257]; II = Chloroform-diethylamine (90 + 10) [257]; III = Cyclohexane-chloroform-diethylamine (50 + 40 + 10) [257]; IV = Cyclohexane-chloroform (30 + 70) + 0.05% diethylamine [257]; V = Methanol [257]; VI = Methanol-acetone-triethanolamine (50 + 50 + 1.5) [15]; VII = Chloroform-ethanol (90 + 10) [241]; VIII = Chloroform-ethanol (80 + 20) [241]; IX = Dimethylformamide-diethylamine-ethanol-ethyl acetate (5 + 2 + 20 + 75) [241]; X = Benzene-heptane-chloroform-diethylamine (60 + 50 + 10 + 0.2) [241]; XI = Methanol [127]; XII = Benzene-ethanol (90 + 10) [87]; XIII = Xylene-butanone-methanol-diethylamine (20 + 20 + 3 + 1) [16]; XIV = Anhydrous acetone [94]; XV = Chloroform [94]; XVI = Benzene-chloroform-acetone (70 + 15 + 15) [94].

tatively and quantitatively, using two-dimensional TLC on alumina layers without binder [176]. VIGNOLI et al. [253 b] have used two-dimensional TLC on silica gel G, with the solvents methanol-chloroform-ammonium hydroxide (85 + 15 + 0.7) and diethyl ether, saturated with water-acetone-diethylamine (85 + 8 + 7); they then identified the separate opium alkaloids by means of the different colours yielded with the iodoplatinate reagent (No. 147).

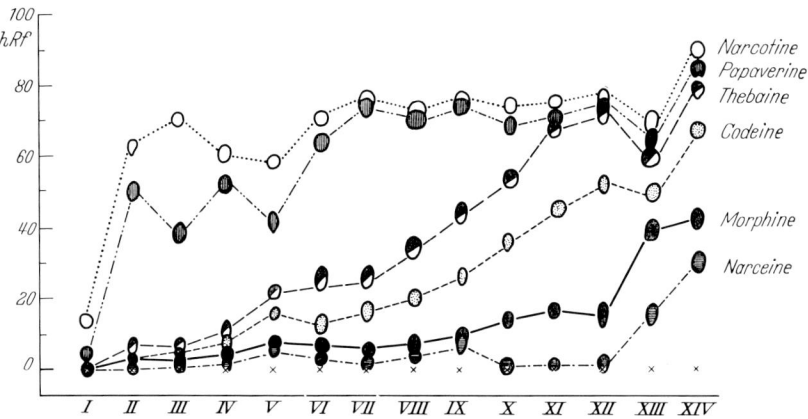

Fig. 154. Comparison of customary solvents for separating 6 important opium alkaloids. *Layer:* Silica gel GF_{254}; chamber with CS; 20° C; 15 cm run.

No.   Solvent, composition, time of run and references

   I. Benzene-tetrahydrofuran (95 + 5); 35 min [265].
  II. Benzene-n-butanol (75 + 25); 30 min [131].
 III. Benzene-tetrahydrofurane (80 + 20); 60 min [131].
  IV. Benzene-n-propanol (80 + 20); 35 min [131].
   V. Benzene-methanol (90 + 10); 30 min [19, 131].
  VI. Benzene-ethanol (80 + 20); 45 min [131].
 VII. Chloroform-isopropanol (80 + 20); 45 min [131].
VIII. Chloroform-n-hexane-methanol (65 + 25 + 10); 30 min [180].
  IX. Chloroform-methanol (90 + 10); 30 min analogous to [265].
   X. Xylene-butanone-methanol-diethylamine (40 + 40 + 6 + 2); 35 min [16, cf 180].
  XI. Benzene-dioxan-ethanol-25% ammonium hydroxide (50 + 40 + 5 + 5); 45 min [39].
 XII. Chloroform-acetone-diethylamine (50 + 40 + 10); 30 min [257].
XIII. Chloroform-acetone-methanol-triethylamine (30 + 40 + 10 + 20); 30 min [88b].
 XIV. Carbon tetrachloride-n-butanol-methanol-6N ammonium hydroxide (40 + 30 + 30 + 2); 90 min [171].

MARY and BROCHMANN-HANSSEN [131] quote eleven solvents with which the most important opium alkaloids can be separated. Methanol-chloroform (10 + 90) is the best for morphine, codeine and thebaine; and ethanol-benzene (20 + 80) good for pavaverine and narcotine (cf.

Fig. 154). In the quantitative determination, the spots were scraped off the plate [silica gel G (Merck)], eluted with methanol, the eluate made up to 5 ml and the concentration of alkaloid determined from the light absorption (wave lengths used were: narcotine 312 nm, morphine 286 nm, thebaine 285 nm, papaverine 279 nm and codeine 215 nm). STEELE [230 b] has recently achieved good separations of opiates in narcotic seizures, using silica gel G layers. The Rf-values of 26 compounds were evaluated in eight solvents; the best solvents were found to be ethyl acetate-benzene-acetonitrile-ammonium hydroxide $(50 + 30 + 15 + 5)$ and acetonitrile-benzene-ethyl acetate-ammonium hydroxide $(40 + 30 + 25 + 5)$.

Other articles in which the TLC of opium alkaloids or their pharmaceutically important derivatives is described, are: [34b, 37, 39, 65b, 110, 150, 163, 180, 188, 191, 202, 208b, 226, 253, 267]. HEUSSER et al. [88b] describe a colorimetric and polarographic determination of morphine in opium, following TLC.

WALDI et al. [256] describe the TLC of quaternary ammonium bases, e. g. of berberine. GERTIG [69, 70] has used TLC in an investigation of the berberine, protopine and chelidonine alkaloids occurring in the plant *Eschscholtzia californica*. DÖPKE [48], SLAVÍKOVÁ [219] and VRUBLOVSKÝ et al. [255] have all recently systematically determined the TLC Rf-values and also studied the colour reactions of a large number of Papaveraceae alkaloids. TSCHESCHE [246] has made use of TLC in his study of the aporphine alkaloids from the bark of *Symplocos celastrinae* MART. (alumina layer; eleven solvents; detection through inspection in UV light or colour reactions with diazotised sulphanilic acid, Rgt. No. 238 or Dragendorff reagent, No. 98). AWE and WINKLER [9, 265] have employed TLC for identification of hydrastinine in a study of the constitution of the rhoeadine and isorhoeadine found in *Papaver rhoeas* L. It was observed that plants of the genus *Papaver* contain also alkaloids which yield a deep red colour with acids (papaverrubines). AWE and WINKLER [8] were the first to use TLC in the investigation of these substances, obtained from the petals of *P. rhoeas* (alumina G, Merck; solvent of benzene-chloroform; detection with hydrogen chloride vapour or spraying with 2N hydrochloric acid). A detailed study of these substances, obtained from plants of species and subspecies of the *Papaver* genus, has been undertaken by PFEIFER [165, 167], using TLC. He found six similar substances which he termed papaverrubines A to F; papaverrubine D is identical with porphyroxine, originally discovered in opium by MERCK. The alkaloids rhoeadine and rhoeagenine, isorhoeadine and isorhoeagenine, glaudine and glaugenine (glaucamine) are chemically very closely related to these papaverrubines [166]. The following hRf-values were found, using butanol-acetone-methanol $(70 + 20 + 10)$ on

Table 79. hR_f-values of alkaloids from the Papaveraceae family

Alkaloids	Solvents				Fluorescence UV 365 nm	Iodoplatinate (Rgt. No. 147)	Cer(IV)-sulphate (Rgt. No. 37)
	III	IV	V	VI			
I. Tetrahydroisoquinoline compounds							
Hydrastinine			46	77	lt bl	r v	lt y
Cotarnine			47	65	y gr	br v	wh
II. Benzylisoquinoline alkaloids							
Armepavine	7	26	54	8	lt bl	v-bl v	y br
Papaverine	9	48	71	18	o	y br	go y
Laudanosine			59	37	—	v-r v	y pk
Latericine			9	0	bl	bl-v	pk
III. Phthalide-isoquinoline alkaloids							
Bicuculine			82	29	lt bl	v	lt y
Adlumidine			90	23	wk lt bl	v	lt y
Capnoidine			90	22	wk lt bl	v	lt y
Corlumine			73	16	lt bl	v	lt y
Corlumidine			73	2	o	r v-y	lt o
D-Adlumine			86	23	lt bl	r v-br	lt y
Hydrastine-β			75	28	bt bl	br v	lt y
α-Narcotine	23	55	84	39	wk bl	r v-o br	lt y
Narcotoline			80	5	—	r v-o	y o
β-Narcotine			85	40	wk bl	r v-o br	lt y
IV. Narceine alkaloids							
Nornarceine			0	0	—	y wh	pk
Narceine			4	0	—	wh	pk-br pk
V. Aporphine and proaporphine alkaloids							
Isothebaine			62	30	o	br r-gr	v-pk
Glaucine	27	62				y br	
Corydine	19	53				bl gr	
Isocorydine	25	56	51	33		v gr	br r
Corytuberine	0	0				cs-gy	
Roemerine	55	71				y with br edge	
Amurine	66[a]						
Domesticine	14	43				cs-r br	
Bulbocapnine	15	48				cs-gr	
Mecambrine	23	50	59	34	r	v-br-dk v	bl v-r br
Mecambroline	8	24				cs	
VI. Protoberberine alkaloids							
Coptisine			53	75	go y	pk br	wk y
Berberine			22	61	y	wk br	br y
Stylopine	56					bg	
Tetrahydro-berberine	51					y br-y	
Tetrahydro-palmatine	36					y br-y	
Tetrahydro-corysamine	68					y br-y	
Thalictricavine	68					y	
Corydaline	50					y	
Hunnefolline	4	22				cs-o	
Mecambridine	12	35				o br	

[a] Solvent II

Table 79 (Continued)

Alkaloids	Solvents					Fluorescence UV 365 nm	Iodoplatinate (Rgt. No. 147)	Cer(IV)-sulphate (Rgt. No. 37)
	I	III	IV	V	VI			
VII. Protopine alkaloids								
Protopine	18	43	64	71	54	o	v	y br
Hunnemannine		0	9				y	
Cryptopine				58	38	—	v	r v
Allocryptopine	7	27	56	53	35	wk o	v-r br	y br
Muramine				52	34	wk o	v-r br	y br
VIII. Benzophenanthridine alkaloids								
Chelidonine	II	30	63	87	40	o	v-y br	pk-o
Norchelidonine		11	42	86	16	gy	v-y br	br pk
Homochelidonine		28	64	84	36	gy-y	v-y br	br r-y
Chelerythrine	59			93	74	go y	bz	r br
Sanguinarine	70			95	78	gy pk	br	wh
Chelilutine	64							
Chelirubine	77							
IX. Pavine and isopavine alkaloids								
Argemonine	55	16	26					
Norargemonine	47	4	16				cs	
Bisnorargemonine		0	2				cs	
Amurensine	26							
X. Alkaloids of so far unknown constitution								
Nudaurine	37							
Amuroline	43							
Amuronine	52							
Roemeridine		2	23				cs	
Platycerine		11	32				cs	

Layer: Silica gel G; (NS).
Solvent: I = Chloroform-benzene (40 + 50), saturated with formamide and mixed with 10 parts of methanol [69, 70]; II = Chloroform-ethyl acetate-methanol (40 + 40 + 20) [48]; III = Cyclohexane-diethylamine (90 + 10) [219]; IV = Cyclohexane-chloroform-diethylamine (70 + 20 + 10) [219]; V = Xylene-butanone-methanol-diethylamine (20 + 20 + 3 + 1) [255]; VI = Cyclohexane-diethylamine (90 + 10) [255].

Abbreviations: bg = beige; bl = blue; br = brown; bt = bright; bz = bronze; cs = colourless; dk = dark; go = golden; gr = green; gy = gray; lt = light; o = orange; pk = pink; r = red; v = violet; wh = white; wk = weak; y = yellow.

silica gel G: isorhoeadine 80, glaudine 73, rhoeadine 70, glaucamine 64, rhoeagenine 60, isorhoeagenine 60; papaverrubine A = 65, B = 59, C = 54, D = 48, E = 34, F = 26. The values, using methanol-chloroform (50 + 50), were: isorhoeadine 88, glaudine 79, alpinine 75, rhoeadine 71, isorhoeagenine 68, rhoeagenine 66, alpinigenine 55, glaugenine 44, papaverrubine A 82, papaverrubine E 59 and papaverrubine D 54.

TLC has shown its usefulness also in the determination of the configuration of the phthalide-isoquinoline alkaloids. BLÁHA et al. [22] have measured R_f-values on silica gel G using normal saturation and benzene-methanol (80 + 20) or xylene-butanone-methanol-diethylamine (40 + 40 + 6 + 2) and compared them with the pK_{MCS}-values[3a] of these alkaloids and their diols. They concluded that the Rf-values of the *erythro*- and *threo*-forms are inversely proportional to the pK_{MCS}-values (excepting α- and β-narcotine), the *erythro* form of the alkaloids possessing a lower Rf-value than the *threo* form. The reverse is the case with the diols. TLC is most useful also for controlling of the isolation procedure [62b and 132b]; in the elucidation of the constitution of the alkaloids found in plants of the genus *Papaver* [179b]; in control of the synthesis of isothebaine [13]; and in the isolation and identification of (+)-reticuline in opium [13b, 34c] and of magnoflorine (silica gel G; solvent of isopropanol-25% ammonium hydroxide-methanol (50 + 25 + 25) or chloroform-methanol (50 + 50)). KNABE [107, 108] has applied TLC in the study of the chemical changes of some tetrahydroisoquinoline alkaloids.

7. Bis(benzylisoquinoline) Alkaloids

DÖPKE [48] has employed TLC for separating the strongly basic alkaloids phaeanthine hRf 55, isotetrandrine 49, pycnamine 47, berbamine 42 and oxyacanthine 35, using chloroform-ethyl acetate-methanol (40 + 40 + 20) on silica gel G (Merck), prepared with 0.1 N NaOH. BOISSIER and co-workers [28] have used this method to detect phaeanthine in the African Menispermaceae *Triclisia patens* OLIVER. FRANCK and BLASCHKE [63] have used TLC likewise for controlling the synthesis of alkaloids of this group. BHATNAGAR et al. [21b] have also worked on the TLC of bis(benzylisoquinoline) alkaloids.

8. Ipecacuanha Alkaloids

The emetine alkaloids and their derivatives from *Uragoga* species are the object of renewed medical interest. STAHL [228] recommends double development with chloroform-methanol (85 + 15) on silica gel G, using chamber saturation, for separating these alkaloids. He found the most sensitive detection to be spraying with about 10 ml of a 0.5% solution of iodine in chloroform, followed by heating 15 min at 60° C (!); the alkaloids then fluoresce intensely yellow or blue in UV light (365 nm).

MACHOVIČOVÁ [128] has compared PC and TLC for testing the stability of emetine hydrochloride solutions for injection. She showed amongst other things that TLC is far better than PC for determining the purity of the emetine.

STAHL [228] describes a rapid test for distinguishing between Rio- and Cartagena-ipecacuanha:

[3a] See p. 455.

100 mg of the finely powdered drug in a small test tube are treated with one drop of concentrated ammonium hydroxide solution; 5.0 ml chloroform are then added and the mixture left 3—4 h, mixing vigorously from time to time with a glass rod. The mixture is then filtered and 5 µl amounts of filtrate used for TLC. 5 µl of a

Table 80. *TLC of the ipecacuanha alkaloids on silica gel G layer according to* STAHL

Alkaloid	h Rf-value	Colour of the iodine-reaction		Limit of detection (µg)
		In daylight	In UV-light	
Cephaeline	13	light brown	light blue	0.006
Psychotrine	16	light brown	yellow with blue edge	0.001
2-Dehydroisoemetine	18	(light beige)	turquoise	0.002
2-Dehydroemetine	21	(light beige)	turquoise	0.002
Emetine	28	yellow	yellow-bluish	0.002
0-Methylpsychotrine	47	yellow	yellow	0.0008
Protoemetine	63	pale beige	blue	0.01
Emetamine	67	light beige	yellow	0.003

0.02% emetine-cephaeline (1 + 1) solution are applied as a reference at the side of the plate. With a Cartagena drug extract, the zone sizes are like those of the reference solution; with a Rio-extract, the cephaeline zone is markedly smaller (Fig. 155).

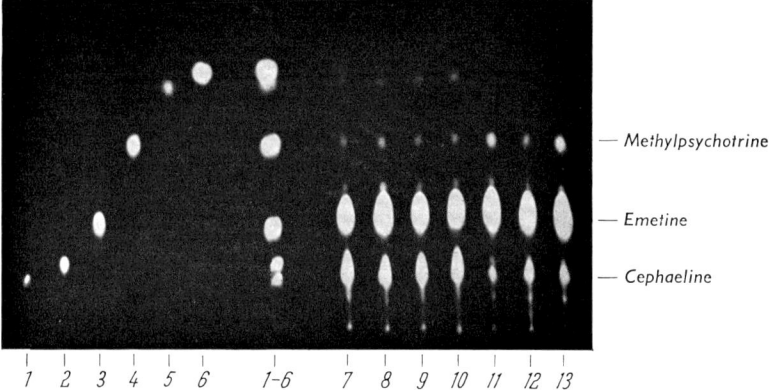

Fig. 155. TLC of ipecacuanha alkaloids (ca 0.1 µg), photographed in UV-light after visualisation with iodine [228]. Further details in text. *1* cephaeline; *2* psychotrine; *3* emetine; *4* 0-methylpsychotrine; *5* protoemetine; *6* emetamine; *7—10* drug extracts from the roots of *Uragoga acuminata* KARSTEN (Cartagena-drug); *11—13* extracts from *Uragoga ipecacuanha* BALL. (Rio-drug)

9. Amaryllidaceae Alkaloids

The belladine, lycorine, lycorenine, crinidine, galanthamine and tazettine alkaloid groups belong to this class.

The Dragendorff reagent (Nos. 96 and 98) is used for detection. LAIHO and FALES [117b] have used silica gel and chloroform-ethyl acetate-methanol (20 + 20 + 60) for preparative separation of the components in a study of the quasi-racemic alkaloid narcissamine.

Table 81. hRf-values of Amaryllidaceae alkaloids according to DÖPKE [48]

Alkaloid	hRf	Alkaloid	hRf	Alkaloid	hRf
Powellamine	81	Krepowine	59	Flexinine	45
Criwelline	74	Parkamine	57	Masonine	44
Clivonine	73	Flexine	56	Caranine	38
Tazettine	72	Nartazine	53	Lycorine	35
Nerispine	71	Hippeastrine	52	Brunsdonnine	32
Nerinine	68	Crinine	51	Narcissamine	29
Crinalbine	66	Crinamidine	50	Lycorenine	28
Neflexine	65	Buphanamine	49	Krelagine	27
Flexamine	63	Galanthine	48	Narcissidine	18
Undulatine	62	Pluviine	47	Annapawine	13

The procedure permits also a ready separation of alkaloids which show only slight structural differences amongst themselves, as, e. g., the following epimers, isomers and dihydro-compounds:

Penarcine	46	Homolycorine	40	Powelline	34
Hippawine	43	Haemanthamine	39	Crinidine	33
Haemanthidine	42	Crinamine	37	Narwedine	26
Oxopowelline	41	Elwesine	36	Galanthamine	24

Layer: Silica gel G (Merck).
Solvent: Chloroform-ethyl acetate-methanol (40 + 40 + 20).

SANDBERG [199] employs TLC in phytochemical investigations of the alkaloid composition in the individual organs of *Pancratium maritimum* L. He uses a two-dimensional procedure on silica gel G, with the solvents ethanol-methanol-diethylamine (65 + 10 + 5) (first dimension) and chloroform-methanol-diethylamine (92 + 3 + 5) (second dimension); the hRf-values of lycorine were 34 and 22, respectively. Altogether 52 bases could be detected in this way and a few new alkaloids were separated and crystallised.

10. Indole Alkaloids

The class of indole alkaloids includes most of the compounds in this chapter. There are some medically important indole alkaloids. They can be subdivided into a number of groups, on the basis of the ring skeleton.

a) Indolyl Alkylamines[4]

The TLC of alcoholic extracts of *Amanita citrina* yields zones corresponding to about 20 different substances; extracts of *A. porphyria* yield

---

[4] Simple indole derivatives are treated in Chapter O, p. 471.

only 12. The presence of bufotenine-N-oxide, serotonine, N-methylserotonine, bufotenine, 5-methoxy-N,N-dimethyltryptamine and N,N-dimethyltryptamine in both fungi was thus established [248]. The solvent chloroform-methanol (95 + 5), saturated with 25% ammonium hydroxide, was used on silica gel G (Merck). The tryptamine derivatives were visualised by spraying with the van Urk reagent (No. 73). Compounds containing hydroxyl groups were detected with the Pauly reagent (No. 240).

Using TLC on alumina or silica gel G, MORIMOTO and OSHIO [139] have succeeded in isolating from *Lespedeza bicolor var. japonica*, a little N,N-dimethyltryptamine and also the new alkaloid 1-methoxy-N,N-dimethyltryptamine, termed "lespedamine".

b) Mavacurine, Fluorocurine, Ellipticine, Eburnamine, Aspidospermine and Strychnine Alkaloids

TLC has proved of great advantage in both preparative and analytical work on this alkaloid group. The layers used have been mostly adhering silica gel and, occasionally, alumina without binder [263, 264].

The Swiss school of KARRER and SCHMID [20, 42, 44, 76, 77, 86, 88, 112—115b, 144, 168, 169, 209, 259—261] shows preference for the solvents chloroform, chloroform-methanol (50 + 10) or benzene-ethyl acetate-diethylamine (70 + 20 + 10). The alkaloids are visualised by spraying with cerium(IV)-sulphate (Rgt. No. 37) or iodoplatinate reagent (No. 147) and then heating. RENNER [184] has been able to detect vobasine also by spraying with 2,6-dibromoquinonechloroimide (Rgt. No. 62). SCHMID [88] cites R_{St}-values for 20 curare alkaloids; he used chloroform-methanol (50 + 10) or benzene-ethyl acetate-diethylamine (70 + 20 + 10) as solvent.

ACHENBACH and BIEMANN [1c] quote the hR_f-values of 10 compounds which they obtained during isolation of the isotuboflavine and norisotuboflavine from *Pleiocarpa mutica*, using chloroform-methanol (90 + 10) or chloroform-ethyl acetate-methanol (75 + 20 + 5) on silica gel G.

DJERASSI's team in America [6, 45—47, 57] has used ethanol-benzene-ethyl acetate (10 + 45 + 45) or the more polar mixture, ethyl acetate-ethanol (50 + 50). WALSER and DJERASSI [257b] combined TLC and column chromatography and succeeded in isolating and in part identifying a total of 28 indole and dihydroindole alkaloids from *Vallesia dichotoma* RUITZ et PAV. Silica gel G was used for the TLC separation, with the solvents methylene dichloride-methanol (95 + 5) and ethyl acetate-methanol (95 + 5). Cerium(IV)-sulphate reagent (No. 37) served for detection, with evaluation at room temperature and after having heated. SCHMUTZ [122] separates the alkaloids of *Aspidosperma ulei* with methyl cellosolve and distinguishes them on the basis of their fluorescence in UV light. PUISIEUX et al. [181] have applied TLC in the analysis and isolation of individual alkaloids in *Voacanga bracteata* STAPF; their adsorbent was silica gel G or alumina G and the solvents recommended were methylene dichloride-methanol, cyclohexane-chloroform and benzene-acetone (without any indication of the proportions).

The TLC of brucine, strychnine and some other alkaloids of this group has been studied by WALDI and co-workers [257]. SCHWARZ et al. [87, 212], RUSIECKI and HENNEBERG [193] and QUIRIN et al. [181b]. ŠARŠÚNOVÁ [201] gives an account of the radiometric determination of strychnine and brucine in galenical preparations. For other publications, see [14, 35 and 183b].

c) Rauwolfia Alkaloids

The experience already gained in PC [97] can be turned to account in the TLC of this alkaloid group.

Table 82. hR_f-values of *Rauwolfia* alkaloids

Layer	S	S	S	A_1	S (0.1 N NaOH)	C (Form-amide)	A_2	A_2	A_2
Solvent	I	II	III	IV	V	VI	VII	VIII	IX
Sarpagine	12	4	0	0	0	3	—	—	—
Serpentine	24	15	0	0	0	6	2	75	34
Ajmaline	47	42	12	13	56	28	2	87	51
Serpentinine	53	56	8	3	12	—	—	86	73
Yohimbine	63	62	18	15	60	33	—	—	—
Rauwolscine	55	63	18	15	68	—	—	—	—
Rescinnamine	—	—	—	—	—	51	—	—	—
Reserpine	72	80	25	35	69	59	60	—	89
Reserpinine	—	—	—	—	—	89	—	—	—
Ajmalicine	—	—	—	—	—	—	77	—	—

Layer: S = Silica gel G; A_1 = alumina G; A_2 = alumina; C = cellulose powder.
Solvent: I = Chloroform-acetone-diethylamine (50 + 40 + 10) [257]; II = Chloroform-diethylamine (90 + 10) [257]; III = Cyclohexane-chloroform-diethylamine (50 + 40 + 10) [257]; IV = Cyclohexane-chloroform (30 + 70) + 0.05% diethylamine [257]; V = Methanol [257]; VI = Heptane-methyl ethyl ketone (50 + 50) in atmosphere of ammonia [241]; VII = Chloroform-acetone (85 + 15) [94]; VIII = Absolute ethanol [94]; IX = Chloroform-ethanol-acetone (90 + 5 + 5) [94].

IKRAM and BAKHSH [94] have separated five of the best known alkaloids from *Rauwolfia serpentina* on a loose alumina layer. Amongst older work, that of WALDI and co-workers [257], TEICHERT et al. [241] and SCHLEMMER and LINK [203] are mentioned here. The last named authors have also worked out a method for quantitative determination of reserpine and other *Rauwolfia* alkaloids; in this, the spot, detectable in UV light (365 nm) on the silica gel layer, was quantitatively scraped off, eluted with a mixture of dioxan and 96% ethanol (1 + 1) and the alkaloid amount estimated from the extinction of the solution at 268 nm. BÄUMLER and RIPPSTEIN [15] separate the *Rauwolfia* alkaloids, also on a silica gel layer, using methanol-acetone-triethylamine (50 + 50 + 1.5).

ULLMANN and KASSALITZKY [249] have established that solubilising agents such as, polyoxyethylene sorbitol monolaurate (*Tween* 20) have no deleterious effect on TLC on silica gel G layers, impregnated with formamide when developing with n-heptane-methyl ethyl ketone (70 + 35) in the dark. Following this preliminary work, they determined the reserpine and rescinamine contents in pharmaceutical preparations.

TLC of *Rauwolfia* alkaloids, carried out under various conditions and with various techniques, is treated in the following articles also: [12, 124, 138, 195—197].

d) Vinca Alkaloids (Catharanthus Alkaloids)

The alkaloids contained in the genus *Vinca* have been studied intensively during the last few years. It has been discovered that some of them are highly promising chemotherapeutic agents against malignant tumours and in treating maladies due to high blood pressure.

The articles of FARNSWORTH et al. [51, 53, 54] and of SVOBODA [235] may be named amongst the older work using TLC. MOKRÝ et al. [133 to 136] and TROJÁNEK and co-workers [141, 142, 243] have also engaged in such work. NEUSS et al. [40] have carried out a notable investigation on the TLC of these alkaloids; they employed silica gel G and alumina layers, testing 11 solvents, of which five proved satisfactory. The colour reaction with cerium(IV)-sulphate (Rgt. No. 34) was quoted for all the alkaloids. In two-dimensional TLC on an alumina layer, pure chloroform (first dimension) and chloroform-ethyl acetate (50 + 50) (second dimension) were used for development.

The most detailed and noteworthy work on the separation of the *Catharanthus* alkaloids on silica gel G layers, has been published by FARNSWORTH et al. [51 b, 53 b]. R_f-values of 63 alkaloids were obtained, using three solvents, and the colours from the test with the cerium(IV)-sulphate reagent (No. 34) cited.

JAKOVLJEVIC et al. [95] have separated eight alkaloids from *Vinca rosea* on an alumina layer which they had prepared with 0.5 N lithium hydroxide; solvents were acetonitrile-ethanol (95 + 5) or benzene-acetonitrile (70 + 30) and visualisation was carried out with cerium(IV)-sulphate (Rgt. No. 34). According to [95] it is better to dilute this reagent solution with an equal volume of water before use. MOZA and TROJÁNEK [142] also comment on the advantages of TLC over PC; on an alumina layer, using petrol ether-anhydrous ether (50 + 50),

Table 83. *hRf-values of some Vinca alkaloids* [40]

No.	Adsorbent layer	A	S	A	S	S	Colour reaction with Cerium(IV)-sulphate, Rgt. No. 34
	Solvent	I	II	II	III	IV	
1	Carosine		71		65	24	Purple-grey
2	Ajmalicine	57	67	72	68	54	Yellow
3	Tetrahydroalstonine	76	60	73	76	69	Yellow-green
4	Catharanthine	77	59	74	58	38	Green (brief period)
5	Carosidine		58		59	10	Yellow
6	Catharine	18	58	76	56	10	Yellow
7	Catharosine		56		58	8	Crimson
8	Virosine	9	54	63	48	9	No colour
9	Pleurosine		51	42	7	3	Yellow
10	Vincarodine		50		50	10	Blue (brief period)
11	Vindolicine	24	46	73			Blue
12	Vindolinine	55	37	70	44	13	Orange-red
13	Vindoline	44		68	57	20	Red
14	Leurosine	27	35				Grey
15	Lochnerine	4	35	70	42	6	Pale grey
15	Sitzirikine		31		45	9	Yellow-green
17	Perivine	5	30	48	39	11	Light brown
18	Neoleurocristine		27		43	3	Blue
19	Lochnericine	15	25	77	46	3	Blue
20	Vinblastine	25	24	66	33	4	Crimson
21	Isoleurosine	35	22				Grey
22	Vindolidine		15		29	0	Blue
23	Vincamicine	1	9	42	20	0	Blue-orange
24	Neoleurosidine		6		17	0	Yellow-brown
25	Serpentine	3	0	11	0	0	—
26	Lochneridine	0	0	21	4	0	Blue-green

Layer: A = Alumina, Fluka, for TLC; S = Silica gel, Merck, for TLC.

Solvent: I = Chloroform-ethyl acetate (50 + 50); II = Ethyl acetate-absolute ethanol (75 + 25); III = Ethyl acetate-absolute ethanol (50 + 50); IV = Pure ethyl acetate.

they accomplished separation of a group of alkaloids which had appeared together as a single spot in PC.

The quantitative separation of the alkaloids contained in *Catharanthus roseus*, using ethyl acetate (or chloroform)-acetone (90 + 10) on an alkalised silica gel G layer, has been described by GRÖGER and STOLLE [82b]. The articles [1, 32, 40, 82b, 179, 234, 237] also deal with the TLC of alkaloids from the genus *Vinca*.

e) Ergot Alkaloids

The significance of the ergot alkaloids in medicine and pharmacy is reflected in the large number of analytical publications concerning them. As long ago as 1959, STAHL [225] separated the principal alkaloids, ergocristine, ergotamine and ergometrine on silica gel G layers, employing chloroform-ethanol (95 + 5). TEICHERT et al. [241] have separated the ergot alkaloids on a layer of cellulose powder, impregnated with formamide. ROCHELMEYER et al. [91, 189] then reported later that a good separation of the hydrogenated ergot alkaloids is also possible on such layers. In most cases, however, the alkaloids have been separated on adhering silica gel G and alumina layers, as, e. g., in the work of KLAVEHN and ROCHELMEYER [105], ZINSER and BAUMGÄRTEL [268], GRÖGER [81], LAUGHLIN [118], SAHLI [198], STAUFFACHER [230] and WOLF [266]. Fluorescence in long-wave UV light has served as a rule for visualisation. Detection with the van Urk reagent (Nos. 73 and 74) may be recommended also.

Numerous methods also for quantitative determination of the individual alkaloids have been worked out. ROCHELMEYER et al. [92, 105 106] have studied this exhaustively. KARÁCSONY and SZARVADY [100] also have developed a procedure which combines the amphi-indicator method with TLC. ZINSER and BAUMGÄRTEL [268] advise photometric evaluation of the zones after coloration with a modified van Urk reagent (No. 74). Band application was used to effect sharper resolution.

PROCHÁZKA et al. [180b] have developed a combined TLC-spectrophotometry procedure for quantitative determination of ergometrine, ergometrinine, ergotamine and ergotaminine from ergot. The extracts had been previously purified with the help of an ion exchanger.

In order to separate the isomers of chanoclavine from *Claviceps purpurea*, STAUFFACHER and TSCHERTER [230] have employed a silica gel G layer and chloroform-butanol (66 + 33) in a chamber filled with ammonia. Further data about TLC of the ergot alkaloids can be found in the publications of WALDI et al. [257] and of SCHWARZ and co-workers [87, 212].

The experimental conditions worked out for the thin-layer chromatographic separation of *Secale* alkaloids have been used also in the investigation of the ergot fungi collected from maize cobs in Mexico [4]. The alkaloids chanoclavine, dihydroelymoclavine, elymoclavine, festuclavine, pyroclavine and agroclavine could be found in this way. Similarly it has been possible to establish the composition of the alkaloids from *Pennisetum*-ergot [4], *Ipomoea* species [78, 79, 128b], *Claviceps paspali* [83] (see GRÖGER [80, 81, 84]), "morning glory" (67b, 239], *Rivea corymbosa* L. [240] and *Penicillium chermenisium* [3]. HOFMANN [89, 91] has succeeded in detecting the alkaloids d-lysergic acid amide, d-isolysergic acid amide, chanoclavine, elymoclavine and lysergol in the seeds of the Mexican "miracle drug", "ololiuqui", (*Rivea corymbosa* (L.) HALL. f. and *Ipomoea violacea* L.), after extraction under mild conditions; TLC on silica gel G using chloroform-methanol (70 + 30) and on alumina using chloroform-methanol (95 + 5) was particularly helpful here.

Table 84. hRf-values of ergot alkaloids

Layer[a]	S	S	C	S	S	C	A	S	S	A—G	Detection[c]	
Solvent[b]	I	II	IV	V	VI	VII	VIII	IX	X	XI	A	B
Lysergic acid	—	—	—	0	—	—	0	—	—	—	←	→
Isolysergic acid	—	—	—	—	—	—	10	—	—	0		
Ergometrine	3	17	—	11	13	—	27	17	12	1		
Ergotamine	13	51	6	43	16	11	58	31	31	1		
Ergosine	13	51	11	43	21	17	68	35	31	1		
Ergometrinine	—	—	—	—	—	—	—	—	38	9	light	
Ergocristine	28	69	67	67	36	41	—	44	56	9		blue
Ergocornine	28	69	73	67	41	50	—	54	59	15		blue
Ergocryptine	28	69	83	67	41	61	—	58	55	6		
Ergotaminine	34	75	50	73	48	—	80	60	64	11		
Ergosinine	34	75	65	—	—	—	73	68	68	38		
Ergocristinine	45	81	90	81	62	—	—	75	74	31		
Ergocorninine	45	81	93	81	—	—	—	80	73	46		
Ergocryptinine	45	81	97	81	—	—	—	83	75	—		
Dihydrolysergic acid	—	—	—	0	—	—	—	85	—	—	←	→
Dihydrolysergic acid amide	0	11	—	6	—	—	—	—	—	—	—[d]	
Dihydroergotamine	—	—	—	35	—	9	—	—	—	—	—[d]	
aci-Dihydrocristine	0	14	—	37	—	—	—	—	—	—	—[d]	
Dihydroergocristine	—	—	s/III	—	—	30	—	—	—	—	—[d]	
Dihydroergocristinine	—	—	—	53	—	—	—	—	—	—	—[d]	
Dihydroergocornine	—	—	—	53	—	38	—	—	—	—	—[d]	
Dihydroergocryptine	—	—	—	53	—	50	—	—	—	—	—[d]	
Chanoclavine	0	2	4	—	—	—	—	—	—	—	—[d]	blue
Elymoclavine	2	13	15	—	—	—	—	—	—	—	—[d]	blue

Peniclavine	4	17	20	—	—	—	—	—	—	—	light blue	light green
Isopeniclavine	8	30	35	—	—	—	—	—	—	—	light blue	light green
Agroclavine	15	38	45	—	—	—	—	—	—	—	—[d]	blue
Setoclavine	18	42	50	—	—	—	—	—	—	—	light blue	light green
Isosetoclavine	20	46	60	—	—	—	—	—	—	—	light blue	light green

[a] *Layer*: S = Silica gel G (Firm 88); C = cellulose powder (Firm 83), impregnated with formamide; A = alumina (firm referred to above); A—G = alumina G (Firm 88).

[b] *Solvent*: I = Chloroform-ethanol (95 + 5) [106]; II = Chloroform-ethanol (90 + 10) [106]; III = Chloroform-ethanol (80 + 20) [106]; IV = Multiple development: 1. Run with heptane-benzene-chloroform (35 + 42.5 + 21.5), 2. Run with heptane-benzene (45.5 + 54.5) [242]; V = Benzene-chloroform-ethanol (28.5 + 57 + 14.5) [268]; VI = n-Heptane-carbon tetrachloride-pyridine (16.5 + 50 + 13.5) [268]; VII = Ethyl acetate-n-heptane-diethylamine (45.5 + 54 + 0.5) [92]; VIII = Chloroform-ethanol (97 + 3) [266]; IX = Ethyl acetate-N,N-dimethylformamide-ethanol (86.5 + 12.5 + 1.0) [118]; X = Benzene-N,N-dimethylformamide (86.5 + 13.5) [118]; XI = Chloroform-diethyl ether-water (70 + 10 + 20) [118].

[c] *Detection*: A = Fluorescence colour in UV-light (365 nm); B = colour reaction with the van Urk-reagent (Nos. 73, 74).

[d] Yellow fluorescence only after intensive irradiation with UV light (without filter).

TLC has been applied to studies of: chemical reactions of ergot alkaloids [90, 223]; the configuration of lysergic acid [224]; biogenesis of the ergot alkaloids [79, 80, 140]; pharmaceutical preparations [68].

KORNHAUSER and PERPAR [108b] have attempted unsuccessfully the thin-layer chromatographic separation of pigments present in ergot drugs and alkaloid extracts.

f) Oxindole Alkaloids

Increased interest is being shown in the oxindole alkaloids from plants of the genus *Mitragyna* (family Rubiaceae). About 10 plants are involved, mostly from tropical Asia. Those invesitgated in most detail up to now have been: *Mitragyna rotundifolia* [217]; *M. parviflora* [218]; *M. stipulosa* [17]; *M. ciliata* [18]; and *M. speciosa* [18b, c]. The alkaloids isolated have been: mitragynine, speciofoline, rhynchophylline, isorhynchophylline, rotundifoline and isorotundifoline. SHELLARD [215—218], using TLC on alumina G with chloroform and on silica gel G with ether-diethylamine (95 + 5) has been able to prove that the rotundifoline and isorotundifoline allegedly isolated from *M. rotundifolia*, really came from *M. parviflora*.

FINCH and TAYLOR [59—61] have also used TLC in establishing the constitution of these alkaloids. They state that the related alkaloids, mitraphylline and isomitraphylline ran as a single spot in ethyl acetate-chloroform (90 + 10) on silica gel layers. KORZUN et al. [109] likewise recommend TLC using ethyl acetate-chloroform (95 + 5) on silica gel G or alumina G for preparative separation of the oxindole alkaloids.

11. Cinchona Alkaloids

The alkaloids from cinchona bark are those longest known and have been often used since 1638 against malarial infections. They are isolated from the bark of *Cinchona* and *Remijia* trees. The Portuguese GOMEZ was the first (in 1810) to obtain a crystalline mixture of alkaloids.

TEICHERT et al. [241], WALDI and co-workers [257] and BÄUMLER and RIPPSTEIN [15] have separated some naturally occurring cinchona alkaloids on silica gel G layers. SCHWARZ et al. [87] have shown that alumina layers without binder can also be used. PREININGER et al. [250] have employed TLC both in the study of *cis-* and *trans-* and of *erythro-* and *threo-*configurations of the quinine and epiquinine bases, and for detecting new alkaloids in the crude extract (Fig. 156). Loose alumina layers proved the best in these investigations. The *erythro-*compounds with the *cis-*configuration (quinidine and cinchonine) yield a higher hRf-value (and higher pK_{MCS}-values) than the analogous *trans-*compounds (quinine and cinchonidine). The epibases (*threo-*

Table 85. hRf- and R_{St}-values in TLC of cinchona alkaloids

Alkaloid	R_{St}-values[a]		hRf-values		Fluorescence on Al_2O_3 [250]
	AI	SII	SIII	SIV	UV (365 nm)
Quininone	112	79	—	—	yellow green
Vuzine	100	100	—	—	no
Cinchonine	80	61	31	40	blue
Dihydrocinchonine	—	—	—	30	—
Cinchotoxine	62	13	—	—	no
Quinotoxine	62	13	—	—	no
Quinidine	54	43, 35	42	47	blue
Dihydroquinidine	—	—	—	39	—
Cinchonidine	54	60	23	40	no
Dihydrocinchonidine	—	—	—	31	—
Quinine	43	43, 35	36	46	blue
Dihydroquinine	—	—	—	40	—
9-Epiquinine	23	31	—	—	blue
9-Epiquinidine	23	31	—	—	blue
Alkaloid X	0	—	—	—	blue
Alkaloid Y	50	—	—	—	blue
Alkaloid Z	106	—	—	—	blue

Layer: A = Alumina; S = Silica gel G.
Solvent: I = n-Hexane-carbon tetrachloride-diethylamine (50 + 40 + 10) [250]; II = Benzene-methanol (80 + 20) [250]; III = Chloroform-ethanol(90+ 10) [241]; IV = Chloroform-methanol-diethylamine (80 + 20 + 10) [233].

[a] Values referred to vuzine as 100.

compounds) show a lower hRf-value but a higher pK_{MCS}-value. Differentiation of cis- from trans-configuration of the epibases has so far failed.

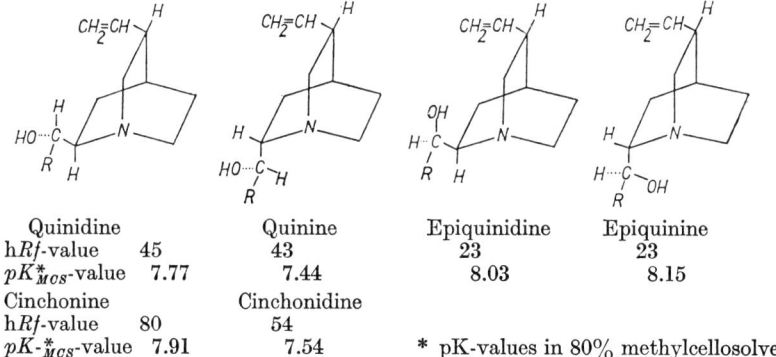

Quinidine	Quinine	Epiquinidine	Epiquinine
hRf-value 45	43	23	23
pK^*_{MCS}-value 7.77	7.44	8.03	8.15
Cinchonine	Cinchonidine		
hRf-value 80	54		
pK^{-*}_{MCS}-value 7.91	7.54	* pK-values in 80% methylcellosolve	

Fig. 156. Constitution, hRf- and pK-values of some cinchona alkaloids

MÜLLER and HONERLAGEN [143] also have worked on the thin-layer chromatographic separation and fluorometric determination of quinine and quinidine. They used silica gel G (Merck) and the solvent petrol ether-diethylamine-acetone (46 + 18 + 18); after treatment with formic acid vapour, the zones were visible in UV-light. A good separation of quinine, quinidine, cinchonine and cinchonidine has been possible only with two-dimensional TLC on silica gel [213]. The presence of epi-bases as well as the alkaloids already known, has been demonstrated in an investigation [250] of crude extracts from various organs of the cinchona tree. Alkaloids of quinotoxine and quininone groups [7] and at least two other alkaloids, previously unknown, have been discovered.

Suszko-Purzycka and Trzebny [233, 233b] have used silica gel G layers in order to separate the four principal alkaloids of the cinchona tree from their dihydro-derivatives. They found that the dihydrobases migrate more slowly than the original vinyl-bases.

Quinine and quinidine show intense blue fluorescence in long-wave UV light, enabling detection on the layer of amounts down to 0.02 μg. As with the other alkaloids, quantitative determination of the individual alkaloids is possible by: a) photography of the spots (coloured with the Dragendorff reagent [151, 152]) and photometry of the negative; or b) scraping off the spot, extraction and spectrophotometry at a suitable wavelength. The relevant UV-spectra can be found in the article [250]. The latter method is particularly suitable since the cinchona alkaloids possess clearly defined maxima in the UV region [33, 143].

Oswald and Flück [151, 152] also have carried out qualitative and quantitative work on the most important alkaloids occurring in cinchona bark. They chromatographed amounts of less than 1 μg on silica gel G, using isopropanol-benzene-diethylamine (20 + 40 + 10). Separation was inadequate with larger amounts; they consequently developed twice with 15 cm runs, using benzene-ether-diethylamine (40 + 24 + 10). The quantitative planimetric determination was accurate to ±8%.

Leete [121] has employed TLC in an investigation of the incorporation of tryptophan into the cinchona alkaloids.

Van Severen et al. [33, 34] have worked out a procedure enabling the quinine-hydroquinine ratio and the amount of quinidine to be determined. Fluorometry was used in the measurement stage. Luckner et al. [126] have described a TLC method for the East German pharmacopoeia [DAB 7 (East)] for identifying the alkaloids contained in cinchona bark; benzene-methanol (80 + 20) was used on a silica gel G layer. The article [236] also deals with this theme.

12. Furanoquinoline Alkaloids

The bases isolated from plants of the genus *Lunasia* (Rutaceae) also belong to the furanoquinoline alkaloids. Rüegger and Stauffacher

[192] have succeeded in identifying the five alkaloids, hydroxylunidine, hydroxylunacridine, lunacrine, lunidine and lunidonine, from *Lunasia amara* BLANCO var. *repanda*. They used TLC on silica gel, with the solvents chloroform-methanol (98 + 2) or cyclohexane-chloroform-diethylamine (60 + 30 + 10). TLC has also been used in the preparative separation of these alkaloids. SCHNEIDER [205] has tested whether the skimmianine, isolated from *Ruta graveolens* L. was homogeneous; he used a silica gel G layer and benzene-ethanol (90 + 10) in a chamber atmosphere of ammonia. WERNY and SCHEUER [262] have employed alumina G layers and benzene-chloroform (50 + 50) or pure chloroform for the same purpose.

13. Purine Alkaloids

Included here are the five plant bases, i. e., heteroxanthine, theophylline, theobromine, caffeine and tetramethyltrihydroxypurine, and the purines from the animal kingdom such as, e. g., adenine, hypoxanthine, guanine, vernine (guanosine), isoguanine, xanthine and uric acid.

Silica gel G layers have been preferred for thin-layer chromatographic separation of the purines [11, 36, 40b, 66, 67, 82, 241, 267]. Loose alumina layers have found only occasional use [201b,c, 212]. Visualisation presents some difficulty. The following spray reagents may be used: 1. acidified iodide solution (Rgt. No. 143), followed with a mixture of 96% ethanol and 25% hydrochloric acid (1 + 1) [242]; 2. permanganate solution (Rgt. No. 200) [36]; 3. mercurous nitrate solution (Rgt. No. 157) which yields colours with aminophylline, theophylline and theobromine with a limit of detection of about 5 µg [201c]; 4. chloramine T (Rgt. No. 38) for caffeine; 5. a solution of tartaric acid, iodine and ferric chloride in acetone (Rgt. No. 103), giving a detection limit of about 1 µg [267]. The purines may also be sublimed from the TLC layer on to a cooled plate [11] (see p. 81). ZARNACK and PFEIFER [267] recommend acetone-chloroform-n-butanol-25% ammonium hydroxide (30 + 30 + 40 + 10) as solvent with silica gel G layers. They quote the following hR_f-values and colours with Rgt. No. 103: theophylline 26 (violet blue); theobromine 47 (intensely grey-blue); hydroxyethyltheophylline 62 (grey-blue); hydroxypropyltheobromine 69 (intensely grey-blue); caffeine 78 (red-brown).

GÄNSHIRT [66] uses this method also for the quantitative spectrophotometric determination of caffeine (accuracy ±12%). Cyclohexane-acetone (40 + 50) was used with silica gel G.

14. Sterol Alkaloids[5]

This group can be subdivided into the real sterol-alkaloids and those containing a benzofluorene skeleton. The alkaloids belonging

---
[5] Further information is in Section IX of Chapter L on Steroids.

to the former subgroup occur in the Solanaceae family (genera *Solanum* and *Lycopersicum*); they are alkaloids with a heterocyclic ring on C_{20}. The genera *Holarrhena* and *Funtumia* of the Apocynaceae family contain alkaloids which are derived from pregnane-(5) or allopregnane and possess a primary, secondary or tertiary amine group on C_3.

Alkaloids with a benzofluorene skeleton are found in the *Veratrum*, *Schoenocaulon*, *Zygadenus* and *Fritillaria* genera of the Liliaceae family. Many of the alkaloids occur as glycosides in the plants. Some are important starting materials for syntheses of steroid hormones.

a) Alkaloids of the Genera Solanum and Lycopersicum

BOLL and ANDERSEN [30] have used TLC in an investigation of the alkaloid composition of *Solanum dulcamara* from different sources. They recommend normal silica gel G layers for alkaloid glycosides and alkalised layers for the aglycones. The Dragendorff reagent (No. 96) or antimony(III) chloride (No. 15) serve for visualisation. Any pyridine from the solvent must be expelled by heating the chromatogram 15 h

Table 86. hRf-values and detection of non-glycosidic steroid-alkaloids [a] and sapogenins [30]

Compounds	hRf	Colour reaction with $SbCl_3$- reagent (No. 15)	
		at 20° C	after 10 min at 105° C
Tomata-3,5-diene	75	red	reddish-violet
Diosgenin	67	red	reddish-violet
Tigogenin	67	colourless	yellow
Yamogenin	67	red	reddish-violet
Tomatidine	47	colourless	grey-red
Δ^5-Tomatiden-3 β-ol	47	red	reddish-violet
Solasodine	23	red	reddish-violet
5 α-Solasodan-3 β-ol	23	colourless	grey-red

Layer: Silica gel G.
Solvent: Chloroform-methanol (95 + 5).

[a] Silica gel G layers and ethyl acetate-pyridine-water (45 + 15 + 45) were used for the TLC of glycosidic steroid alkaloids: γ-solamarine hRf 52; solamargine 44; β-solamarine 38; tomatine 20; α-solamarine 18; solasonine 17.

at 105° C before the Dragendorff reagent can be used. The antimony(III) chloride reagent is suitable for distinguishing saturated and unsaturated alkaloid sapogenins and for differentiating these two groups in the neutral sapogenins (Table 86). TLC on a silica gel G layer using chloroform-ethanol-1% ammonium hydroxide solution (40 + 40 + 20) has been applied in elucidation of the constitution of β- and γ-solamarine [29].

SCHREIBER et al. have also used TLC often in elucidating the constitution of *Solanum* alkaloids. They have reported good separations in more than 17 articles and their experience in TLC is summarised in a special publication [206]. The layers recommended are: 1. silica gel G (Merck); 2. silica gel G, sprayed with 10% silver nitrate solution; 3. silica gel (anhydrous, pure, Firm 88) + 10% gypsum; 4. alumina according to BROCKMANN, +10% gypsum. Solvents preferred are:

1. cyclohexane-ethyl acetate (50 + 50);
2. chloroform-methanol (60 + 40);
3. n-hexane-triethylamine (90 + 6);
4. ethyl acetate-cyclohexane-96% ethanol (50 + 40 + 5);
5. benzine (BP 80–90° C)-benzene-ethyl acetate (85 + 5 + 10).

Detection is carried out with cerium (IV) sulphate (Rgt. No. 37). It cannot be used with layers impregnated with silver nitrate, nor with alumina layers. Iodine vapour or solution (Rgt. No. 143) or paraformaldehyde-phosphoric acid (Rgt. No. 194) can be used also and 2,4-dinitrophenylhydrazine (Rgt. No. 82) with alkaloids containing a keto group. The cerium (IV) sulphate reagent is recommended as a universal detection agent; Δ^5-unsaturated compounds yield a colour reaction even in the cold. Iodine vapour permits non-destructive detection of the sterol alkaloids, so that they may be subsequently isolated unchanged.

Solanidine and demissidine can be separated on a silver nitrate-impregnated silica gel layer. SCHREIBER et al. [207b] have used TLC also for preparative separation of *Solanum* alkaloids. Using 8 solvents on silica gel G layers, PAQUIN and LEPAGE [158] have separated the three principal alkaloids from *Solanum tuberosum*, i. e., solanine, chaconine and solanidine. Antimony(III) chloride reagent was used for visualisation. PÉREZ-MEDINA et al. [164] have investigated many *Solanum* species in the search for a raw material source of solasodine. They used TLC on silica gel G with n-butanol-methanol-diethylamine (85 + 5 + 10) as solvent and antimony(III) chloride (Rgt. No. 16) for visualisation. The TLC of this alkaloid class is described in other publications [2, 25, 183, 208].

b) Alkaloids of the Genera Holarrhena and Funtumia

LÁBLER and ČERNÝ [117] have undertaken a detailed investigation of the TLC of these alkaloids and their derivatives. They separated on silica gel G layers, using benzene or ether, saturated with ammonia. The Dragendorff reagent served for visualisation. Conclusions about constitution and stereochemistry of the compounds could be drawn from the Rf-values.

Table 87. *TLC-conditions for various alkaloids of hitherto unknown structure*

Alkaloid	Alkaloid group or plant	Layer	Solvent	Reference	Remarks
Aconitine and pseudo-aconitine	Diterpene alkaloids	Silica gel G	Isopropanol-methanol-23% ammonium hydroxide (36 + 24 + 1)	[61 b] [35 b, 43, 267 b]	Quantitative determination
Decodine and verticilatine	Steroid alkaloids	Silica gel G	Methanol	[58]	
Deoxyaconitine and hypaconitine	Diterpene alkaloids	Silica gel G	1) Cyclohexane-chloroform-diethylamine (70 + 20 + 10)	[247]	
			2) Benzene-ethyl acetate (80 + 20)	[254]	
Galegine and 4-hydroxygalegine	Aliphatic amines	Silica gel G	Butanol-acetone-water (60 + 15 + 15)	[180 c, 207]	Detection with Rgt. No. 135
Isopilocereine and pilocereine	Tetrahydroisoquinoline alkaloids	Silica gel G	1) Methanol-acetone-2N HCl-acetic acid (70 + 15 + 30 + 15)	[64]	
			2) Methanol + 1% NH$_4$OH	[24]	
Peganine	Quinazoline alkaloids	Al$_2$O$_3$	Chloroform-ethanol (50 + 50)	[81b, 180 c]	Detection with Dragendorff reagent (No. 98)
		Silica gel G	Butanol-acetone-water (60 + 15 + 15)	[207]	
Phyllochrysine and securinine				[159]	
Pseudoephedrine	Phenylalkylamine	Silica gel	Isopropanol + 5% NH$_4$OH (90 + 9)	[187]	Detection with Rgt. No. 178
Rotundine	Rotundine alkaloids	Al$_2$O$_3$-starch	Butanol-water	[41]	
Tylocrebine	Tylocrebine alkaloids	Silica gel-alkali	Methanol	[194]	
Uleine	Elypticine alkaloids	Silica gel G	Ethyl acetate-benzene-methanol-water (40 + 40 + 20 + 1)	[96]	
Cissampelos pereira		Silica gel G	CHCl$_3$-methanol (90 + 10)	[221, 222]	16 substances
Heimia salicifolia		Alumina	4 solvents	[23, 49]	4 substances
Heimia myrtifolia		Silica gel G	3 solvents		
Rhynchosia pyramidalis (Pega palo)		Silica gel	Isopropanol-50% acetic acid (90 + 9)	[186]	1 alkaloid
Frog: *Phyllobates bicolor*		Silica gel G	CHCl$_3$-methanol (60 + 10)	[130]	Preparative TLC

c) Alkaloids of the Benzofluorene Type

DENÖEL and VAN COTTHEM [43] have described the separation of aconitine and pseudoaconitine in galenical preparations. FISCHER and WEIXLBAUMER [61b] have worked on the quantitative determination.

ZEITLER [267c] has separated 9 alkaloids of *Veratrum* species by TLC on silica gel HF_{254}, using cyclohexane-diethylamine (90 + 10) or (70 + 30). TOMKO and VASSOVÁ [242b] have used alumina G and benzene-ethanol for separating the alkaloids in *Veratrum album ssp. lobelianum*.

Note to Table 87: As a supplement, other work may be mentioned in which TLC has been a useful, indeed indispensable, help: on the animal poisons, taricha- and tetrodotoxine [139b]; the lycopodium alkaloid L 9 [10, 49b]; anuloline [209b]; and the biosynthesis of the calycanthus alkaloid, calicanthine [210b].

General Bibliography on the Alkaloids

BENTLEY, K. W.: The Chemistry of the Morphine Alkaloids. Oxford: Clarendon Press 1954.
BOIT, H.-G.: Ergebnisse der Alkaloid-Chemie bis 1960. Berlin: Akademie-Verlag 1961.
GOUTAREL, R.: Les alcaloides stéroidiques des Apocynacées. Paris: Hermann 1964.
HESSE, M.: Indolalkaloide in Tabellen. Berlin, Göttingen, Heidelberg: Springer 1964. Ergänzungswerk 1968.
HOFMANN, A.: Die Mutterkornalkaloide. Stuttgart: Enke 1964.
HOLUBEK, J., O. ŠTROUF et al.: Spectral Data and physical Constants of Alkaloids. Bd. I, II and III. Prague: Publ. House Czech. Acad. Sci. 1965, 1966 and 1968.
KÜHN, L., u. S. PFEIFER: Die Gattung Papaver und ihre Alkaloide. Pharmazie 18, 819—843 (1963).
MANSKE, R. H. F., and H. L. HOLMES: The Alkaloids. Chemistry and Physiology. Bd. I—XI. New York: Academic Press 1950—1968.
MOTHES, K., D. GROSS, M.-W. LIEBISCH u. H.-R. SCHÜTTE: 3. Internationale Arbeitstagung Biochemie und Physiologie der Alkaloide 1965. Berlin: Akademie-Verlag 1966.
— u. H.-B. SCHRÖTER: 1. Internationale Arbeitstagung Biochemie und Physiologie der Alkaloide. 1956. Berlin: Akademie-Verlag 1957.
— — 2. Internationale Arbeitstagung Biochemie und Physiologie der Alkaloide 1960. Berlin: Akademie-Verlag 1963.
PFEIFER, S., I. MANN, and L. KÜHN: Alkaloide der Rhoeadin-Papaverrubin-Klasse. Basen mit einer neuen Grundstruktur. Pharm. Zentralhalle 107, 1—27 (1968).
PINDER, A. R.: Lactonic Alkaloids. Chem. Rev. 62, 551—572 (1964).
SCHULTZ, O. E., u. F. ZYMALKOWSKI: Die quantitative Bestimmung der Alkaloide in Drogen und Drogenzubereitungen. Stuttgart: Enke-Verlag 1960.
SHAMMA, M., and W. A. SLUSARCHYK: The Aporphine Alkaloids. Chem. Rev. 62, 59—79 (1964).
STUART, K. L., and M. P. CAVA: The proaporphine Alkaloids. Chem. Rev. 68, 321—339 (1968).
WALDI, D.: Chromatography of Alkaloids. In A. T. JAMES and L. J. MORRIS: New Biochemical Separations, p. 157—196. London: Van Nostrand, 1964.
Chemie und Biochemie der Solanum-Alkaloide. Tagungsbericht Nr. 27 d. Deutschen Akademie der Landwirtschaftswissenschaften zu Berlin, 1961.
Physical Data of Indole and Dihydroindole Alkaloids. I. and II. Lilly Research Laboratories, Indiana USA, 1960 and 1963.

Special Bibliography for Chapter N. Alkaloids

1. ABDURAKHIMOVA, N., P. KH. YULDASHEV, i S. YU YUNUSOV: Doklady Akad. Nauk U.S.S.R. **21** (2), 29 (1964); C. A. **61**, 9715a (1964).
1b. ABOU-CHAAR, CH. I.: Lebanese Pharm. J. **8**, 82 (1963); C. A. **61**, 14951f (1964).
1c. ACHENBACH, H., and K. BIEMANN: J. Am. Chem. Soc. **87**, 4177 (1965).
2. ADAM, G., u. K. SCHREIBER: Z. Chemie **3**, 100 (1963); C. A. **60**, 4443a (1964).
3. AGURELL, S.: Experientia **20**, 25 (1964).
4. — and E. RAMSTAD: Lloydia **25**, 67 (1962).
5. — u. A. J. ULLSTRUP: Planta Med. **11**, 392 (1963).
6. ANTONACCIO, L. D., N. A. PEREIRA, B. GILBERT, H. VORBRUEGGEN, H. BUDZIKIEWICZ, J. M. WILSON, L. J. DURHAM, and C. DJERASSI: J. Am. Chem. Soc. **84**, 2161 (1962).
7. AUTERHOFF, H., u. K. KALPATHY: Pharm. Acta Helv. **38**, 491 (1963).
8. AWE, W., u. W. WINKLER: Arzneimittel-Forsch. **9**, 773 (1959).
9. — — Arch. Pharm. **294**, 301 (1961).
10. AYER, W. A., A. N. HOGG, and A. C. SOPER: Can. J. Chem. **42**, 949 (1964).
11. BAEHLER, B.: Helv. Chim. Acta **45**, 309 (1962).
12. BARTLETT, M. F., B. F. LAMBERT, H. M. WERBLOOD, and W. I. TAYLOR: J. Am. Chem. Soc. **85**, 475 (1963).
13. BATTERSBY, A. R., and T. H. BROWN: Proc. Chem. Soc. **1964**, 85.
13b. — G. W. EVANS, R. O. MARTIN, M. E. WARREN Jr., and H. RAPOPORT: Tetrahedron Letters **1965**, 1275.
14. BATTERSBY, A. R., and M. GREGORY: J. Chem. Soc. **1963**, 22.
15. BÄUMLER, J., u. S. RIPPSTEIN: Pharm. Acta Helv. **36**, 382 (1961).
16. BAYER, I.: J. Chromatog. **16**, 237 (1964).
17. BECKETT, A. M., E. J. SHELLARD, and A. N. TACKIE: J. Pharm. Pharmacol. **15**, 158T (1963).
18. — — — J. Pharm. Pharmacol. **15**, 166T (1963).
18b. — — — Planta Med. **13**, 241 (1965).
18c. — — J. D. PHILLIPSON, and CALVIN M. LEE: Planta Med. **14**, 266 u. 277 (1966).
19. BÉLA, D.: Acta Pharm. Hung. **34**, 221 (1964).
19b. BERNASCONI, R., ST. GILL u. E. STEINEGGER: Pharm. Acta Helv. **40**, 246 (1965).
20. BERNAUER, K.: Helv. Chim. Acta **46**, 211 (1963).
21. BEYER, K. H.: Dtsch. Apotheker-Ztg. **104**, 697 (1964).
21b. BHATNAGAR, A. K., and S. BHATTACHARJI: Indian J. Chem. **3**, 43 (1965); C. A. **63**, 639h (1965).
22. BLÁHA, K., J. HRBEK Jun., J. KOVÁŘ, L. PIJEWSKA a F. ŠANTAVÝ: Collection Czech. Chem. Commun. **29**, 2328 (1964); 2. communication in preparation.
23. BLOMSTER, R. N., A. E. SCHWARTING, and J. M. BOBBITT: Lloydia **27**, 15 (1964).
24. BOBBITT, J. M., R. EBERMANN, and M. SCHUBERT: Tetrahedron Letters **1963**, 575.
25. BOGNÁR, R., u. S. MAKLEIT: Pharmazie **20**, 40 (1965).
25b. BÖHM, H.: Planta Med. **13**, 234 (1965).
26. BOHLMANN, F., E. WINTERFELDT u. U. FRIESE: Chem. Ber. **96**, 2251 (1963).
27. — G. WINTERFELDT, B. JANIAK, D. SCHUMANN u. H. LAURENT: Chem. Ber. **96**, 2254 (1963).
28. BOISSIER, J. R., A. BOUQUET, G. COMBES, C. DUMONT et M. DEBRAY: Ann. pharm. franç. **21**, 767 (1963).

29. BOLL, P. M.: Acta Chem. Scand. **17**, 1852 (1963).
30. BOLL, P. M., and B. ANDERSEN: Planta Med. **10**, 421 (1962).
30b. BORIO, B. L., and A. MOREIRA: Trib. Farmaceutica Nr. 2—4, 64 (1964).
31. BORKE, M. L., and E. R. KIRSCH: J. Am. Pharm. Assoc. Sci. Ed. **42**, 627 (1953).
32. BORKOWSKI, B., E. BATKIEWICZ and K. DROST: Dissertationes pharm. **16**, 171 (1964).
33. BRAECKMANN, P., R. VAN SEVEREN et L. DE JAEGER-VAN MOESEKE: Pharm. Tijdschr. Belgie **40**, 113 (1963).
34. — — — Dtsch. Apotheker-Ztg. **104**, 1211 (1964).
34b. BROCHMANN-HANSSEN, E., and T. FURUYA: J. pharm. Sci. **53**, 1549 (1964).
34c. — and B. NIELSEN: Tetrahedron Letters **1965**, 2171.
34d. BÜCHI, J., u. A. ZIMMERMANN: Pharm. Acta Helv. **40**, 292 (1965).
35. CAGGIANO, E., e G. B. MARINI-BETTÒLO: Rend. ist. super. sanità **25**, 375 (1962); C. A. **58**, 12847h (1963).
35b. CASTAGNOU, M. R., et S. LARCEBAU: Bull. Soc. Pharm. Bordeaux **103**, 201 (1964); C. A. **62**, 10798f (1965).
36. CERRI, J. O., e G. MAFFI: Boll. chim. farm. **100**, 951 (1961); C. A. **57**, 11467 (1962).
36b. CHALMERS, A. H., C. C. J. CULVENOR, and L. W. SMITH: J. Chromatogr. **20**, 270 (1966).
37. ČIČIRO, V. E.: Apteč. dělo **12**, No. 6, 36 (1963).
38. CIEŚLAK, J., J. KUDUK i. F. RULKO: Acta Polon. Pharm. **21**, 265 (1964).
39. COCHIN, J., et J. W. DALY: Experientia **18**, 294 (1962).
40. CONE, N. J., R. MILLER, and N. NEUSS: J. Pharm. Sci. **52**, 688 (1963).
40b. O'CONNOR, R.: J. Chem. Educ. **42**, 492 (1965).
40c. CRAIG, J. C., N. Y. MARY, N. L. GOLDMAN, and L. WOLF: J. Am. Chem. Soc. **86**, 3866 (1964).
41. DANG HAHN KHOI: Pharm. Zentralhalle **103**, 99 (1964).
42. DASTOOR, N., et H. SCHMID: Experientia **19**, 297 (1963).
42b. DECKER, K., u. R. SAMMECK: Biochem. Z. **340**, 326 (1964).
43. DENÖEL, A., et B. VAN COTTHEM: J. pharm. Belg. **18**, 346 (1963).
44. DEYRUP, J. A., H. SCHMID u. P. KARRER: Helv. Chim. Acta **45**, 2266 (1962).
45. DJERASSI, C., H. W. BREWER, H. BUDZIKIEWICZ, O. O. ORAZI, and R. A. CORRAL: J. Am. Chem. Soc. **84**, 3480 (1962).
46. — Y. NAKAGAWA, J. M. WILSON, H. BUDZIKIEWICZ, B. GILBERT et L. D. ANTONACCIO: Experientia **19**, 467 (1963).
47. — R. J. OWELLEN, J. M. FERREIRA et L. D. ANTONACCIO: Experientia **18**, 397 (1962).
48. DÖPKE, W.: Arch. Pharm. **295**, 605 (1962).
49. DOUGLAS, B., J. L. KIRKPATRICK, R. F. RAFFAUF, O. RIBEIRO, and A. J. WEISBACH: Lloydia **27**, 25 (1964).
49b. DUGAS, H., R. A. ELLISON, Z. VALENTA, K. WIESNER, and C. M. WONG: Tetrahedron Letters **1965**, 1279.
50. EVANS, W. C., and W. J. GRIFFIN: J. Chem. Soc. **1963**, 4348.
51. FARNSWORTH, N. R.: Lloydia **24**, 105 (1961).
51b. — R. N. BLOMSTER, D. DAMRATOSKI, W. A. MEER, and L. V. CAMMARATO: Lloydia **27**, 302 (1964).
52. — and K. L. EULER: Lloydia **25**, 186 (1962).
53. — H. H. S. FONG, R. N. BLOMSTER, and F. J. DRAUS: J. Pharm. Sci. **51**, 217 (1962).
53b. — and I. M. HILINSKI: J. Chromatog. **18**, 184 (1965).
54. — W. D. LOUB, and R. N. BLOMSTER: J. Pharm. Sci. **52**, 1114 (1963).

55. FAUGERAS, G., R. PARIS et M. H. MEYRUEY: Ann. pharm. franç. **21**, 675 (1963).
56. — — — Ann. pharm. franç. **20**, 768 (1962).
57. FERREIRA, J. M., B. GILBERT, R. J. OWELLEN et C. DJERASSI: Experientia **19**, 585 (1963).
58. FERRIS, J. P.: J. Org. Chem. **28**, 817 (1963).
59. FINCH, N., and W. I. TAYLOR: J. Am. Chem. Soc. **84**, 1318 (1962).
60. FINCH, N., and W. I. TAYLOR: J. Am. Chem. Soc. **84**, 3871 (1962).
61. — — Tetrahedron Letters **1963**, 167.
61b. FISCHER, R., u. H. WEIXLBAUMER: Pharm. Zentralhalle **104**, 298 (1965).
62. FLÜCK, H., u. N. BLASCHKE: Lecture 23. Internat. Kongr. Pharmaz. Wiss. Münster Sept. 1963.
62b. FONG, H. H. S., J. BEAL, and M. P. CAVA: Lloydia **29**, 94 (1966).
63. FRANCK, B., u. G. BLASCHKE: Liebigs Ann. Chem. **668**, 145 (1963).
64. — — Tetrahedron Letters **1963**, 569.
64b. FRENCEL, I. M.: Planta Med. **14**, 204 (1966).
65. FROHNE, D.: Dtsch. Apotheker-Ztg. **104**, 1404 (1964).
65b. FUMAGALLI, U., V. AMBROGI e G. BALESTRA: Boll. Chim. Farm. **103**, 911 (1964); C. A. **62**, 8936g (1965).
66. GÄNSHIRT, H.: Arch. Pharm. **296**, 129 (1963).
67. — u. A. MALZACHER: Arch. Pharm. **293**, 925 (1960).
67b. GENEST, K.: J. Chromatog. **19**, 531 (1965).
68. GENEST, K., and C. G. FARMILO: J. Pharm. Pharmacol. **16**, 250 (1964).
69. GERTIG, H.: Acta Polon. Pharm. **21**, 59 (1964).
70. — Acta Polon. Pharm. **21**, 127 (1964).
70b. GIACOPELLE, D.: J. Chromatog. **19**, 172 (1965).
71. GILL, S.: Dissertationes pharm. **16**, 261 (1964).
72. — Acta Polon. Pharm. **21**, 379 (1964).
73. — u. E. STEINEGGER: Sci. Pharm. **31**, 135 (1963).
74. — — Pharm. Acta Helv. **39**, 508 (1964).
75. — — Pharm. Acta Helv. **39**, 565 (1964).
76. GOVINDACHARI, T. R., K. NAGARAJAN u. H. SCHMID: Helv. Chim. Acta **46**, 433 (1963).
77. — B. R. PAI, S. RAJAPPA, N. VISWANATHAN, W. G. KUMP, K. NAGARAJAN u. H. SCHMID: Helv. Chim. Acta **46**, 572 (1963).
78. GRÖGER, D.: Flora (Jena) **153**, 373 (1963); C. A. **60**, 2043e (1964).
79. — Z. Naturforsch. **18**b, 1123 (1963).
80. — Planta Med. **11**, 444 (1963).
81. — u. D. ERDE: Pharmazie **18**, 346 (1963).
81b. — u. S. JOHNE: Pharmazie **20**, 456 (1965).
82. — K. MOTHES, H. SIMON, H. G. FLOSS u. R. WEYGAND: Z. Naturforsch. **16**b, 432 (1961).
82b. — u. K. STOLLE: Arch. Pharm. **298**, 246 (1965).
83. — and V. E. TYLER: Lloydia **26**, 174 (1963).
84. — — and J. E. DUSENBERRY: Lloydia **24**, 97 (1961).
85. GROSS, D., u. H. R. SCHÜTTE: Arch. Pharm. **296**, 1 (1963).
86. GUGGISBERG, A., T. R. GOVINDACHARI, K. NAGARAJAN u. M. SCHMID: Helv. Chim. Acta **46**, 679 (1963).
87. HEŘMÁNEK, S., V. SCHWARZ u. Z. ČEKAN: Pharmazie **16**, 566 (1961).
88. HESSE, M., W. v. PHILIPSBORN, D. SCHUMANN, G. SPITELLER, M. SPITELLER-FRIEDMANN, W. I. TAYLOR, H. SCHMID u. P. KARRER: Helv. Chim. Acta **47**, 878 (1964).

88b. HEUSSER, D., u. E. JACKWERTH: Dtsch. Apotheker-Ztg. **104**, 107 (1964).
88c. — — Dtsch. Apotheker-Ztg. **104**, 107 (1964).
88d. HODGSON, E., E. SMITH, and F. E. GUTHRIE: J. Chromatog. **20**, 176 (1965).
89. HOFMANN, A.: Planta Med. **9**, 354 (1961).
90. — H. OTT, R. GRIOT, P. A. STADLER u. A. J. FREY: Helv. Chim. Acta **46**, 2306 (1963).
91. — et H. TSCHERTER: Experientia **16**, 414 (1960).
92. HOHMANN, T., u. H. ROCHELMEYER: Arch. Pharm. **297**, 186 (1964).
93. IKRAM, M., and M. K. BAKHSH: Anal. Chem. **36**, 111 (1964).
94. — G. A. MIANA, and M. ISLAM: J. Chromatog. **11**, 260 (1963).
95. JAKOVLJEVIC, I. M., L. D. SEAY, and R. W. SHAFFER: J. Pharm. Sci. **53**, 553 (1964).
96. JOULE, J. A., and C. DJERASSI: J. Chem. Soc. **1964**, 2777.
97. KAISER, F., u. A. POPELAK: Chem. Ber. **92**, 278 (1959).
98. KAISER, H., F. BIEDEBACH u. C. MANNS: Pharm. Ztg. **108**, 1380 (1963).
99. KAMP, W., W. J. M. ONDERBERG en W. A. VAN SETERS: Pharm. Weekblad **98**, 993 (1963); C. A. **61**, 1707 (1963).
100. KARÁCSONY, E. M., u. B. SZARVADY: Planta Med. **11**, 169 (1963).
101. KHAFAGY, S., A. M. EL-MOGHAZY och F. SANDBERG: Svensk. Farm. Tidskr. **66**, 481 (1962); C. A. **58**, 10514b (1963).
102. KHANNA, K. L., A. E. SCHWARTING, A. ROTHER, and J. M. BOBBITT: Lloydia **24**, 179 (1961).
103. KINZE, W.: Pharm. Zentralhalle **103**, 715 (1964).
104. KIRCHNER, J. G., J. M. MILLER, and G. J. KELLER: Anal. Chem. **23**, 420 (1951).
105. KLAVEHN, M., u. H. ROCHELMEYER: Dtsch. Apotheker-Ztg. **101**, 477 (1961).
106. — — u. J. SEYFRIED: Dtsch. Apotheker-Ztg. **101**, 75 (1961).
107. KNABE, J., u. J. KUBITZ: Arch. Pharm. **296**, 591 (1963).
108. — u. N. RUPPENTHAL: Arch. Pharm. **297**, 141 (1964).
108b. KORNHAUSER, A., u. M. PERPAR: Arch. Pharm. **298**, 312 (1965).
109. KORZUN, B. P., L. DORFMAN, and S. M. BRODY: Analyt. Chem. **35**, 950 (1963).
110. KOZUKA, H.: Kagaku Keisatsu Kenkyusho Hôkoku **16**, 39 (1963); C. A. **59**, 15121h (1963).
111. KUHN, H. J.: Dissertation, Göttingen 1964.
112. KUMP, W. G., u. H. SCHMID: Helv. Chim. Acta **44**, 1503 (1961).
113. — — Helv. Chim. Acta **45**, 1090 (1962).
114. KUMP, C., J. SEIBL u. H. SCHMID: Helv. Chim. Acta **46**, 498 (1963).
115. — — — Helv. Chim. Acta **47**, 358 (1964).
115b. KUMP, C., J. SEIBL u. H. SCHMID: Helv. Chim. Acta **48**, 1002 (1965).
116. KUPFERBERG, H. J., A. BURKHALTER, and E. L. WAY: J. Chromatog. **16**, 558 (1964).
117. LÁBLER, L., a V. ČERNÝ: Collection Czech. Chem. Commun. **28**, 2932 (1963).
117b. LAIHO, S. M., and H. M. FALES: J. Am. Chem. Soc. **86**, 4434 (1964).
118. MCLAUGHLIN, J. L., J. E. GOYAN, and A. G. PAUL: J. Pharm. Sci. **53**, 306 (1964).
119. LEARY, J. D., J. M. BOBBITT, A. ROTHER, and A. E. SCHWARTING: Chem. & Ind. (London) **1964**, 283.
120. — K. L. KHANNA, A. E. SCHWARTING, and J. M. BOBBITT: Lloydia **26**, 44 (1963).
121. LEETE, G.: J. Am. Chem. Soc. **84**, 4919 (1962).
122. LEHNER, H., u. J. SCHMUTZ: Helv. Chim. Acta **44**, 444 (1961).
123. LIST, P. H., S. HANAFI u. E. STEIN: Dtsch. Apotheker-Ztg. **103**, 1314 (1963).

124. LIUKONNEN, A.: Farm. Aikakauslehti **71**, 329 (1962).
125. LUCKNER, M., K. WINKLER, O. BESSLER, J. HOFFMANN u. W. POETHKE: Pharm. Zentralhalle **103**, 484 (1964).
126. — — — P. SCHRÖDER, J. HOFFMANN u. W. POETHKE: Pharm. Zentralhalle **103**, 660 (1964).
127. MACHATA, G.: Microchim. Acta **1960**, 79.
128. MACHOVIČOVÁ, F., a V. PARRÁK: Českoslov. farm. **13**, 200 (1964).
128b. MARDEROSIAN, A. D., and H. W. YOUNGKEN Jr.: Lloydia **29**, 35 (1966).
129. MARIANI, A., e O. MARIANI-MARELLI: Rend. ist. super. sanità **22**, 759 (1959).
130. MÄRKI, F., et B. WITKOP: Experientia **19**, 329 (1963).
131. MARY, N. Y., and E. BROCHMANN-HANSSEN: Lloydia **26**, 223 (1963); C. A. **60**, 7871 (1963).
132. MÁTÉ, I., és E. TYIHÁK: Herba Hung. **2**, 35 (1963).
132b. MATUROVÁ, M., D. PAVLÁSKOVÁ u. F. ŠANTAVÝ: Planta Med. **14**, 22 (1966).
133. MOKRÝ, J., L. DÚBRAVKOVÁ et P. ŠEFČOVIČ: Experientia **18**, 564 (1962).
134. — a I. KOMPIŠ: Chem. zvesti **17**, 852 (1963).
135. — u. I. KOMPIŠ: Naturwissenschaften **50**, 93 (1963).
136. — — P. ŠEFČOVIČ a Š. BAUER: Collection Czech. Chem. Commun. **28**, 1309 (1963).
137. MOLL, F.: Arch. Pharm. **296**, 205 (1963).
138. MOREIRA, E. A.: Tribuna farm. No. 3—4, 57 (1964).
139. MORIMOTO, H., u. H. OSHIO: Liebigs Ann. Chem. **682**, 212 (1965).
139b. MOSHER, H. S., F. A. FUHRMAN, H. D. BUCHWALD, and H. G. FISCHER: Science **144**, 3622 (1964).
140. MOTHES, K., K. WINKLER, D. GRÖGER, H. G. FLOSS, U. MOTHES, and B. WEYGAND: Tetrahedron Letters **1962**, 933; C. A. **58**, 8251d (1963).
141. MOZA, B. K., and J. TROJÁNEK: Chem. & Ind. (London) **1962**, 1425.
142. — — Collection Czech. Chem. Commun. **28**, 1419, 1427 (1963).
143. MÜLLER, K. H., u. H. HONERLAGEN: Mitt. dtsch. pharm. Ges. **30**, 202 (1960).
144. NAGARAJAN, K., CH. WEISSMANN, H. SCHMID u. P. KARRER: Helv. Chim. Acta **46**, 1212 (1963).
145. NEU, R.: J. Chromatog. **11**, 364 (1963).
146. NEUBAUER, D.: Planta Med. **12**, 43 (1964).
147. — u. K. MOTHES: Planta Med. **9**, 466 (1961).
148. NEUMANN, D., and H.-B. SCHRÖTER: J. Chromatog. **16**, 414 (1964).
149. NEUMÜLLER, O. A., H. J. KUHN, G. O. SCHENCK u. F. ŠANTAVÝ: Liebigs Ann. Chem. **674**, 122 (1964).
150. NÜRNBERG, E.: Arch. Pharm. **292**, 610 (1959).
150b. EL-OLEMY, M. M., A. E. SCHWARTING, and W. J. KELLEHER: Lloydia **29**, 58 (1966).
151. OSWALD, N.: Diss. E. T. H., Zürich 1963.
152. — u. H. FLÜCK: Pharm. Acta Helv. **39**, 293 (1964).
153. — — Sci. Pharm. **32**, 136 (1964).
154. PAILER, M., u. W. H. KUMP: Arch. Pharm. **293**, 645 (1960).
155. — u. R. LIBISELLER: Monatsh. Chem. **93**, 403, 511 (1962).
156. PAIS, M., J. MAINIL et R. GOUTAREL: Ann. pharm. franç. **21**, 139 (1963).
157. PAPP, E., és Z. SZABO: Herba Hung. **2**, 383 (1963).
158. PAQUIN, R., and M. LEPAGE: J. Chromatog. **12**, 57 (1963).
159. PARELLO, J., A. MELERA et R. GOUTAREL: Bull. soc. chim. France **1963**, 898.
160. PARIS, R.: J. pharm. Belg. **18**, 401 (1963).
161. PARIS, R. R., et M. PARIS: Bull. soc. chim. France **1963**, 1597.
161b. PARIS, R., R. ROUSSELET, M. PARIS et J. FRIES: Ann. pharm. franç. **23**, 473 (1965).

162. PARRÁK, V., E. RADĚJOVÁ a F. MACHOVIČOVÁ: Chem. zvesti **18**, 369 (1964).
163. PENNA-HERREROS, A.: J. Chromatog. **14**, 536 (1964).
164. PÉREZ-MEDINA, L. A., E. TRAVECEDO u. J. E. DEVIA: Planta Med. **12**, 478 (1964).
165. PFEIFER, S.: Pharmazie **19**, 678 (1964).
166. — Pharmazie **19**, 724 (1964).
167. — u. S. K. BANERJEE: Pharmazie **19**, 286 (1964).
168. PINAR, M., W. VON PHILIPSBORN, W. VETTER u. H. SCHMID: Helv. Chim. Acta **45**, 2260 (1962).
169. — u. H. SCHMID: Helv. Chim. Acta **45**, 1283 (1962).
170. PINXTEREN, J. A. C., en M. E. VAN VERLOOP: Pharm. Weekblad **97**, 1 (1962).
171. — — Pharm. Acta Helv. **38**, 437 (1963).
172. PITRA, J., a J. ŠTĚRBA: Chem. listy **57**, 389 (1963).
173. POETHKE, W., u. J. HOFFMANN: Pharm. Zentralhalle **103**, 731 (1964).
174. — u. W. KINZE: Pharm. Zentralhalle **101**, 685 (1962).
175. — — Pharm. Zentralhalle **102**, 692 (1963).
176. — — Arch. Pharm. **297**, 593 (1964).
177. — — Pharm. Zentralhalle **103**, 577 (1964).
178. POTĚŠILOVÁ, H., J. HRBEK Jr. a F. ŠANTAVÝ: Collection Czech. Chem. Commun. **32**, 141 (1967).
179. POTIER, P., R. BEUGELMANS, J. LE MEN et M. M. JANOT: Ann. pharm. franç. **23**, 61 (1965).
179b. PREININGER, V., A. D. CROSS a F. ŠANTAVÝ: Collection Czech. Chem. Commun. **31**, 3345 (1966).
180. PREININGER, VL., u. P. VRUBLOVSKÝ: Pharmazie **20**, 439 (1965).
180b. PROCHÁZKA, V., F. KAFKA, M. PRŮCHA a J. PITRA: Česk. Farm. **14**, 154 [1965].
180c. PUFAHL, K., u. K. SCHREIBER: Züchter **33**, 287 (1963).
181. PUISIEUX, F., M. B. PATEL, J. M. ROWSON et J. POISSON: Ann. pharm. franç. **23**, 33 (1965).
181b. QUIRIN, M., J. LÉVY et J. LE MEN: Ann. pharm. franç. **23**, 93 (1965).
182. RAFFAUF, R. F.: Lloydia **25**, 255 (1962).
182b. RAMAUT, J. L.: Bull. soc. chim. Belg. **72**, 406 (1963).
183. RENAULT, J. L.: Bull. soc. chim. Belges **72**, 406 (1963); C. A. **59**, 10449f (1963).
183b. RENNER, U.: Lloydia **27**, 406 (1964).
184. RENNER, U., D. A. PRINS, A. L. BURLINGAME u. K. BIEMANN: Helv. Chim. Acta **46**, 2186 (1963).
185. REUTER, G.: Pharm. Zentralhalle **102**, 573 (1963).
186. RISTIĆ, S., u. A. THOMAS: Arch. Pharm. **295**, 510 (1962).
187. — — Arch. Pharm. **295**, 524 (1962).
188. ROBLES, M. A., en R. WIENTJES: Pharm. Weekblad **96**, 379 (1961).
189. ROCHELMEYER, H.: Pharm. Ztg. **103**, 1269 (1958).
189b. ROPER, E. C., R. N. BLOMSTER, N. R. FARNSWORTH u. F. J. DRAUS: Planta Med. **13**, 98 (1965).
190. ROTHER, A., J. M. BOBBITT, and A. E. SCHWARTING: Chem. & Ind. (London) **1962**, 654.
191. ROVIGATI DA SILVA JARDIM, I.: Arbeit zu Erlangung des Titels: Professor an der Univ. Rio de Janeiro 1961, S. 52.
192. RÜEGGER, A., u. D. STAUFFACHER: Helv. Chim. Acta **46**, 2329 (1963).
193. RUSIECKI, W., et M. HENNEBERG: Ann. pharm. franç. **21**, 843 (1963).
194. RUSSEL, J. H.: Naturwissenschaften **50**, 443 (1963).

195. RUTKOWSKA, U., i. K. WOJSA: Biul. Inst. Przemystu Ziolkarckiogo w Poznaniu **9**, 192 (1964).
196. SAHLI, M.: Arzneimittel-Forsch. **12**, 55 (1962).
197. — Arzneimittel-Forsch. **12**, 155 (1962).
198. — u. M. OESCH: Pharm. Acta Helv. **40**, 25 (1965).
199. SANDBERG, F., and K. H. MICHEL: Lloydia **26**, 78 (1963).
200. ŠANTAVÝ, F.: unpublished
201. ŠARŠÚNOVÁ, M., J. TÖLGYESSY u. M. HRADIL: Pharmazie **19**, 336 (1964).
201b. — u. V. SCHWARZ: Pharmazie **18**, 34 (1963).
201c. — — Pharmazie **18**, 207 (1963).
202. SCHANTZ, M. v.: In: MARINI-BETTÒLO: TLC. Amsterdam: Elsevier 1964.
203. SCHLEMMER, F., u. E. LINK: Pharm. Ztg. **104**, 1349 (1959).
204. SCHNECKENBURGER, J., u. I. HARTIKAINEN: Dtsch. Apotheker-Ztg. **104**, 1402 (1964).
205. SCHNEIDER, G.: Arzneimittel-Forsch. **14**, 435 (1964).
206. SCHREIBER, K., O. AURICH, and G. OSSKE: J. Chromatog. **12**, 63 (1963).
207. — — u. K. PUFAHL: Arch. Pharm. **295**, 271 (1962).
207b. SCHREIBER, K., C. HORSTMANN u. G. ADAM: Chem. Ber. **98**, 1961 (1965).
208. — u. H. RIPPERGER: Chem. Ber. **96**, 3094 (1963).
208b. SCHULTZ, O. E., u. J. SCHNEKENBURGER: Arch. Pharm. **298**, 548 (1965).
209. SCHUMANN, D., u. H. SCHMID: Helv. Chim. Acta **46**, 1996 (1963).
209b. SCHUNACK, W., u. H. ROCHELMEYER: Arch. Pharm. **298**, 572 (1965).
210. SCHÜTTE, H. R., u. H. HINDORF: Naturwissenschaften **51**, 463 (1964).
210b. SCHÜTTE, H. R., u. B. MAIER: Arch. Pharm. **298**, 459 (1965).
211. SCHWARTING, A. E., J. M. BOBBITT, A. ROTHER, C. K. ATAL, K. L. KHANNA, J. D. LEARY, and W. G. WALTER: Lloydia **26**, 258 (1963).
212. SCHWARZ, V., u. M. ŠARŠÚNOVÁ: Pharmazie **19**, 267 (1964).
213. SEVEREN, R. VAN: J. pharm. Belg. **17**, 40 (1962).
214. SHALABY, A. F., u. E. STEINEGGER: Pharm. Acta Helv. **39**, 752 (1964).
214b. SHARMA, R. K., G. S. KHAJURIA, and C. K. ATAL: J. Chromatog. **19**, 433 (1965).
215. SHELLARD, E. J., u. J. D. PHILLIPSON: 23. Intern. Kong. Pharmaz. Wiss. Münster 1963. S. 205.
216. — — 23. Intern. Kongr. Pharmaz. Wiss. Münster 1963, S. 209.
217. — — Planta Med. **12**, 27 (1964).
218. — — Planta Med. **12**, 160 (1964).
219. SLAVÍKOVÁ, L., u. J. SLAVÍK: unpublished.
220. SPEAKE, T., P. MCCLOSKEY, W. K. SMITH, T. A. SCOTT, and H. HUSSEY: Nature **201**, 614 (1964).
221. SRIVASTAVA, R. M., u. M. P. KHARE: Chem. Ber. **97**, 2732 (1964).
222. — — Current Sci. (India) **32**, 114 (1963).
223. STADLER, P. A.: Helv. Chim. Acta **47**, 756 (1964).
224. — u. A. HOFMANN: Helv. Chim. Acta **45**, 2005 (1962).
225. STAHL, E.: Arch. Pharm. **292**, 411 (1959).
226. — Angew. Chem. **73**, 646 (1961).
227. — Pharmazie **11**, 633 (1956).
228. — In: K. PAECH u. M. V. TRACEY: Moderne Methoden der Pflanzenanalyse, Bd. V. Berlin-Göttingen-Heidelberg: Springer-Verlag 1962.
229. — Chem.-Ing.-Tech. **36**, 941 (1964).
229b. — u. P. J. SCHORN: unpublished.
229c. — u. H. JORK: unpublished.
230. STAUFFACHER, D., u. H. TSCHERTER: Helv. Chim. Acta **47**, 2187 (1964).
230b. STEELE, J. A.: J. Chromatog. **19**, 300 (1965).

231. STEINEGGER, G., R. BERNASCONI u. G. OTTOVIANO: Pharm. Acta Helv. **38**, 371 (1963).
232. SULLIVAN, G., and M. R. GIBSON: J. Pharm. Sci. **53**, 1058 (1964).
233. SUSZKO-PURZYCKA, A., and W. TRZEBNY: J. Chromatog. **16**, 239 (1964).
233b. SUSZKO-PURZYCKA, A., and W. TRZEBNY: J. Chromatog. **17**, 114 (1965).
234. SVOBODA, G. H.: Lloydia **24**, 173 (1961).
235. — and A. J. BARNES JR: J. Pharm. Sci. **53**, 1227 (1964).
236. SYPER, L.: Dissertationes pharmac. **15**, 411 (1963).
237. SZABO, Z.: Herba Hung. **2**, 101 (1963).
238. SZÁSZ, G., L. KHIN és Z. BUDVÁRI: Acta Pharm. Hung. **33**, 245 (1963).
239. TABER, W. A.: Phytochemistry **2**, 65 (1963).
240. — Phytochemistry **2**, 99 (1963).
241. TEICHERT, K., E. MUTSCHLER u. H. ROCHELMEYER: Z. anal. Chem. **181**, 325 (1961).
242. — — — Dtsch. Apotheker-Ztg. **100**, 282, 477 (1960).
242b. TOMKO, J., u. A. VASSOVÁ: Pharmazie **20**, 385 (1965).
243. TROJÁNEK, J., O. ŠTROUF, J. HOLUBEK a Z. ČEKAN: Collection Czech. Chem. Commun. **29**, 433 (1964).
244. TSCHESCHE, R., G. BIERNOTH, and G. WULFF: J. Chromatog. **12**, 342 (1963).
245. — K. KOMETANI, F. KOWITZ u. G. SNATZKE: Chem. Ber. **94**, 3327 (1961).
246. — P. WELZEL, and G. LEGLER: Tetrahedron **20**, 1435 (1964).
247. TSUDA, Y., O. ACHMATOWITCZ JR. u. L. MARION: Liebigs Ann. Chem. **680**, 88 (1964).
248. TYLER, V. E., u. D. GRÖGER: Planta Med. **12**, 397 (1964).
249. ULLMANN, E., u. H. KASSALITZKY: Arch. Pharm. **295**, 37 (1962).
250. VÁCHA, P., P. ČUBA, VL. PREININGER, L. HRUBAN u. F. ŠANTAVÝ: Planta Med. **12**, 406 (1964).
251. VÁGUJFALVI, D.: Herba Hung. **3**, 267 (1964).
251b. — Planta Med. **13**, 79 (1965).
252. VÉGH, A., R. BUDVÁRI, G. SZÁSZ, A. BRANTNER és P. GRACZA: Acta Pharm. Hung. **33**, 67 (1963).
253. VIDIC, E., u. J. SCHÜTTE: Arch. Pharm. **295**, 342 (1962).
253b. VIGNOLI, L., J. GUILLOT, F. GOUEZO et J. CATALIN: Ann. pharm. franç. **24**, 461 (1966).
254. VORBRUEGGEN, H., and C. DJERASSI: J. Am. Chem. Soc. **84**, 2990 (1962).
255. VRUBLOVSKÝ, P., H. POTĚŠILOVÁ a F. ŠANTAVÝ: unpublished.
256. WALDI, D.: Naturwissenschaften **50**, 614 (1963).
257. — K. SCHNACKERZ, and F. MUNTER: J. Chromatog. **6**, 61 (1961).
257b. WALSER, A., u. C. DJERASSI: Helv. Chim. Acta **48**, 391 (1965).
258. WASICKY, A.: Anal. Chem. **34**, 1346 (1962).
259. WEISSMANN, CH., H. SCHMID u. P. KARRER: Helv. Chim. Acta **43**, 2201 (1960).
260. — — — Helv. Chim. Acta **44**, 1877 (1961).
261. — — — Helv. Chim. Acta **45**, 62 (1962).
262. WERNY, F., and P. J. SCHEUER: Tetrahedron **19**, 1293 (1963).
263. WINKLER, W.: Naturwissenschaften **48**, 694 (1961).
264. — Arch. Pharm. **295**, 895 (1962).
265. — u. W. AWE: Arch. Pharm. **294**, 301 (1961).
266. WOLF, L., B. SZARVADY és E. M. KARÁCSONY: Acta Pharm. Hung. **34**, 131 (1964).
267. ZARNACK, J., u. S. PFEIFER: Pharmazie **19**, 216 (1964).
267b. ZENDA, H.: Kagaku No Ryoiki, Zokan, No. 64, 133 (1964); C. A. **62**, 10800b (1965).

267c. ZEITLER, H. J.: J. Chromatog. **18**, 180 (1965).
268. ZINSER, M., u. CH. BAUMGÄRTEL: Arch. Pharm. **297**, 158 (1964).

Addendum to Bibliography of the English edition:

ADAMSKI, R., J. LUTOMSKI u. J. WISIEWSKI: Schnellmethode zur Bestimmung der Tropa-Alkaloide in Injektionslösungen. Deut. Apotheker-Ztg. **107**, 185 (1967).

BRAEKMAN, J. C., M. KAISIN, J. PECHER, and R. MARTIN: Indole alkaloids. XI. Absolute configuration of voacerpine. Bull. Soc. Chim. Belges **75**, 465 (1966); C. A. **65**, 2677 (1966).

CIONGA, E., E. NICHIFORESCO, V. MASCOV, N. URICARU et S. ARIZAN: Considération sur l'hydrolyse des glucoalcaloïdes du *Solanum laciniatum* AIT. Ann. Pharm. Franç. **25**, 139 (1967).

DEBRAY, M., M. PLAT et J. LEMEN: Alkaloides des Ménispermacées africaines. II. *Stephania dinklagei* (Engl.): Isolement de la (+)-corydine, de la (+)-isocorydine et de la (—)-roemerine. Ann. Pharm. Franç. **25**, 327 (1967).

DOMAGLINA, E., i J. OCHYNSKA: Thin-layer and column chromatography of benzophenanthridine alkaloids of *Chelidonium majus*. Products of oxidation of chelidonine and homochelidonine. Chem. Anal. (Warsaw) **12**, 267 (1967).

FEJÉR-KOSSEY, O.: The separation of ten tobacco alkaloids by thin-layer chromatography. J. Chromatog. **31**, 592 (1967).

GENEST, K.: Changes in ergoline alkaloids in seeds during ontogeny of *Ipomea violacea*. J. Pharm. Sci. **55**, 1284 (1966).

HAEFELFINGER, P.: Quantitative Bestimmung eines quaternären Alkaloids nach der dünnschichtchromatographischen Trennung. J. Chromatog. **33**, 370 (1968).

HUANG, J.-T., H.-CH. HSIU, and K.-T. WANG: Polyamide layer chromatography of opium alkaloids. J. Chromatog. **29**, 391 (1967).

LEHMAN, G., and P. MARTINOD: Untersuchung über Coffein aus Synthese und Biosynthese. Arzneimittel-Forsch. **17**, 35 (1967).

LUNDSTRÖM, J., and S. AGURELL: Thin-layer chromatography of peyote alkaloids. J. Chromatog. **30**, 271 (1967).

MATTOCKS, R. R.: Detection of pyrrolizidine alkaloids on thin-layer chromatograms. J. Chromatog. **27**, 505 (1967).

NOIRFALISE, A., et G. MEES: Chromatographie sur couche mince de quelques alkaloides et basis aminées. J. Chromatog. **31**, 594 (1967).

PARIS, R., u. M. ŠARŠUNOVÁ: Dünnschichtchromatographische Verteilung von Opiumalkaloiden und einigen partialsynthetischen Abwandlungsprodukten. Pharmazie **22**, 483 (1967).

PFEIFER, S.: Papier- und dünnschichtchromatographische Charakterisierung von Alkaloiden der Gattung *Papaver*. J. Chromatog. **24**, 364 (1966).

PHILLIPSON, J. D., and E. J. SHELLARD: The correlation between the stereochemistry of some indole and oxindole alkaloids and their behaviour on thin layer chromatograms. J. Chromatog. **24**, 85 (1966).

— — The thin layer chromatographic behaviour of some E seco oxindole alkaloids, and their relationship with indolizidine and some simple oxindoles. J. Chromatog. **32**, 692 (1968).

PIJEWSKA, L., J. L. KAUL, R. K. JOSHI, and F. ŠANTAVÝ: The presence of alkaloids in some tribes of the subfamily Wurmbaeoideae. Collection Czech. Chem. Commun. **32**, 158 (1967).

PÖTTER, H., u. R. VOIGT: Versuche zur quantitativen Bestimmung von Reserpin und 3-Isoreserpin durch Dünnschichtchromatographie. Pharmazie **22**, 198 (1967).

RÖDER, K., E. MUTSCHLER u. H. ROCHELMAYER: Über die quantitative Bestimmung der Mutterkornalkaloide nach dünnschichtchromatographischer Trennung. Pharm. Acta Helv. **42**, 407 (1967).
RÖNSCH, H., u. K. SCHREIBER: Solanum-Alkaloide. LXXXIII. Analytische und präparative dünnschichtchromatographische Trennung von 5α-gesättigten bzw. Δ^5-ungesättigten Steroidalkaloiden und Sapogeninen an silbernitrathaltigen Adsorptionsschichten. J. Chromatog. **30**, 149 (1967).
SENANAYAKE, U. M., and R. O. B. WIJESEKERA: A rapid micro-method for the separation, identification and estimation of purine bases: Caffeine, Theobromine and Theophylline. J. Chromatog. **32**, 75 (1968).
SHELLARD, E. J., and M. Z. ALAM: The quantitative determination of some mitragyna oxindole alkaloids after separation by thin layer chromatography. Part IV. Comparison of ultra-violet spectrometry, colorimetry and densitometry as methods for the quantitative determination of oxindole alkaloids in plant material. J. Chromatog. **35**, 72 (1968).
— — The quantitative determination of some mitragyna oxindole alkaloids after separation by thin layer chromatography. Part I. Ultraviolet spectrophotometry. Part II. Colorimetry, using the Vitali-morin reaction. J. Chromatog. **32**, 472 and 489 (1968).
— J. D. PHILLIPSON, and D. GUPTA: The effect of methoxy substitution and configuration on thin layer chromatographic behaviour of some closed E ring oxindole. alkaloids. J. Chromatog. **32**, 704 (1968).
TSCHESCHE, R., R. WELTERS u. H.-W. FEHLHABER: Alkaloide aus Rhamnaceen. I. Scutianin, ein cyclisches Polypeptidalkaloid aus *Scutia buxifolia* REISS. Chem. Ber. **100**, 323 (1967).
WINKLER, B. C., W. J. DUNLAP, L. M. ROHRBAUGH, and S. H. WENDER: A thin-layer chromatograpghy-fluorometry method for quantitative analysis of scopoline and scopoletine in tobacco. J. Chromatog. **35**, 570 (1968).

O. "Simple" Indole Derivatives and Plant Growth Regulators
Urine Metabolites, Auxins, Gibberellins and Cytokinins[1]

HARALD KALDEWEY

I. Introduction

Indole derivatives possessing no ring system in addition to the indole ring (see Table 88A) may be entitled "simple", in contrast to the more complex-structured indole alkaloids (p. 446) and indole dyes.

Phenol and quinaldine derivatives (Table 88,B,C) as well as the indole derivatives are discussed here under urine metabolites. The expression "plant growth regulators" has become adopted in recent years for auxins, gibberellins and cytokinins, which are really quite distinguishable from each other both chemically and in their physiological

---
[1] Synonymous with the term "phytokinin", proposed by MOTHES [82].

activity. The endogenous plant auxins hitherto identified are indole derivatives. Gibberellins however possess the gibbane ring system (see [106]), which contains no nitrogen, as basic skeleton. The cytokinins so far known belong to the purines.

As a "simple" indole derivative, the aromatic amino acid tryptophan (2-amino-3-(3-indolyl)-propionic acid) plays an important part in animal and plant metabolism. Its degradation via kynurenine leads, e. g., to formation of nicotinic acid and its amide or of ommochromes which have been found as pigments especially in crustaceans and insects. Observations made on mutant strains of insects and of the mould *Neurospora crassa*, in which the biosynthetic routes mentioned are blocked, have given a profound insight into the physiology of gene activity [6].

5-Hydroxytryptamine (serotonine), which occurs in both the animal and plant kingdom, is formed by hydroxylation and decarboxylation of tryptophan. Serotonine has attracted attention above all in human physiology. It is regarded along with adrenaline and acetylcholine as the third active substance in the transfer of the nerve stimulus in synapses or from nerve ends to the effector.

One of its metabolites, 5-hydroxyindole-3-acetic acid, is gaining more and more significance in clinical diagnosis. Increased amounts of it are eliminated in the urine of cancer cases. Nothing is yet known about the function of serotonine in plant metabolism.

Extensive work has been carried out on the activity of indole-3-acetic acid (IAA)[2] in the plant world. It is certainly formed in vivo from tryptophan in plant micro-organisms, probably also in higher plant forms. KÖGL et al. [56] gave IAA the trivial name of "Heteroauxin" when it was isolated from urine during the search for the supposed plant auxin. In the meantime, results of many investigations suggest that IAA is probably *the* genuine active auxin. Many other simple indole derivatives (and also numerous non-indoles) likewise show auxin activity. A few have been isolated from plant extracts but some may be artefacts.

Interest has recently been shown in conjugates of IAA and tryptophan which are perhaps to be regarded as irreversibly inactivated products of metabolism of plant growth regulators, no longer needed [134, 136]. In contrast, the so-called bound auxin (auxin precursor) can be activated as required. Glucobrassicin, an indole thioglucoside, has been discussed as a possible auxin precursor in Cruciferae [21, 22, 23, 61, 63, 107, 108]. Tryptophan and auxins can also intervene in the biogenesis of the indole alkaloids [99, 128, 129].

The auxins were first regarded as hormones for cell extension [101, 119]. The investigations of the past decade have however shown that a more central significance in plant growth must be ascribed to them [25, 49, 50, 123]. They appear together with gibberellins[3] principally to influence cell extension [25, 54, 90a, 91, 126, 127]; and with cytokinins, cell division [17, 82]. They thereby control decisively processes of

---

[2] These and later used abbreviations are based on the terms in English in the hope of contributing to uniformity; cf. also nomenclature suggestions of the IUPAC-IUB in J. Biol. Chem. **241**, 2491 (1966). Unnumbered substituents of indole derivatives are in the $_3$C position (cf. Table 88).

[3] In connection with the chemistry of the gibberellins, their derivatives and degradation products, reference is made to the comprehensive, tabulated review of SCHNEIDER, SEMBDNER and SCHREIBER [106] and to the bibliography already published and planned [114].

Table 88. "Simple" indole derivatives and urine metabolites: hR_f-values, colours, fluorescence and lower limit of detection in ng ($= 10^{-9}$ g) on silica gel layers of 250 μm thickness

Ring systems:
- A: indole (positions 1–7, N at 1)
- B: benzene (positions 1–6)
- C: quinoline (positions 1–8, N at 1)

Compound	Ring system and substituents (substituents not given are —H)	Abbreviation (see also Table 89 for TLC-separation of compounds denoted with ●)	hR_f I [46, 68, 131]	II [46, 78]	III [1, 46, 102]	IV [19]	V [19, 46]	VI [14, 46]	VII [46, 60]	VIII [46, 134, 136]	IX [46, 90]	X [28, 46]	XI [28, 46]	XII [2, 46]	Quenching in UV_{254} [46]	Fluorescence colours in UV_{366} and detection limit in ng [2, 9, 12, 46, 104]	van Urk reagent (No. 73)[4], colours[1] and detection limit in ng [1, 8, 9, 12, 19, 35, 46, 60, 78, 102, 120]	4-Dimethylaminocinnamaldehyde reagent (No. 76), colours[1] and detection limit in ng [30, 32, 33, 35, 46]	Procházka Reagent (No. 123) — Colours and detection limit in ng [8, 19, 35, 46, 60, 102, 120]	Procházka — Fluorescence colours in $UV_{366, 254}$ and detection limit in ng	Salkowski reagent (Nos. 104, 108)[4] — $FeCl_3/HClO_4$ [24, 46]	Salkowski — $FeCl_3/H_2SO_4$ [8, 46, 96]	Fluorescence colours in $UV_{366, 254}$ [46]	Ninhydrin reagent (No. 178), colours[1] and detection limit in ng [46, 104]
Acetoxy-indole [14] (cf. also indoxylacetate)	A4, 5, 6 or 7: —O·CO·CH₃	e. G. I-4-OAc																						
N-α-Acetyl-3-hydroxy-kynurenine [2]	B1: —CO·CH₃·CH(NH·CO·CH₃)·COOH + B2: —NH₂ + B3: —OH	Kyn-Nα,Ac-3-OH												68		yellow-green	—	—	—	—	—	—		
3-Acetyl-indole [60]	A3: —CO·CH₃	IAc							52								no	yellow	red-brown	dark				
1-Acetyl-indoline [60]	A1: —CO·CH₃ + A2 + A3: —H₂ (₂C—₃C)	Indolin-1-Ac							70								yellow							
N-α-Acetyl-kynurenine [2]	B1: —CO·CH₂·CH(NH·CO·CH₃)·COOH + B2: —NH₂	Kyn-Nα,Ac												78		azure								
N-α-Acetyl-5-methoxy-tryptamine [78]	A3: —CH₂·CH₂·NH·CO·CH₃ + A5: —O·CH₃	TryAm-N, Ac-5-OMe ●	50														blue							
N-α-Acetyl-tryptophan [8, 19, 134, 135]	A3: —CH₂·CH(NH·CO·CH₃)·COOH	Try-N,Ac ●				37⁺	22⁺		60								violet		grey-yellow	pink-orange		violet		
Adrenaline [104]	B1: —CH(OH)·CH₂·NH·CH₃ + B3 + B4: —OH	Adr ●																					light brown >1000	
2-Amino-acetophenone [33]	B1: —CO·CH₃ + B2: —NH₂	AcPhe-2-NH₂																						
3-(2-Aminoethyl)-indole (see tryptamine)																								
2-Amino-hippuric acid [2, 9]	B1: —CO·NH·CH₂·COOH + B2: —NH₂	HipA-2-NH₂ ●												73		violet	yellow							
4-Amino-hippuric acid [9]	B1: —CO·NH·CH₂·COOH + B4: —NH₂	HipA-4-NH₂ ●																yellow						
2-Amino-3-hydroxy-acetophenone [12]	B1: —CO·CH₃ + B2: —NH₂ + B3: —OH	AcPhe-3-OH-2-NH₂ ●														blue	yellow-brown							
4-Amino-salicylic acid [9]	B1: —COOH + B2: —OH + B4: —NH₂	SalA-4-NH₂ ●															yellow							
Anthranilic acid [9, 12, 46, 120]	B1: —COOH + B2: —NH₂	AntA ●	0	04	47↓		19↓	05	0	53	80	95	95	19	50	light blue 50	yellow 100	pink-red[n] 50	100; blue	no	no	dark; dark	pink[b] 100
Ascorbigen [22, 23, 63, 95]		Ascor																						
Bufotenine [8, 131]	A3: —CH₂·CH₂·N(CH₃)₂ + A5: —OH	Bufo ●	70														grey-blue		grey-yellow	dark		grey-blue		
Bufotenine-N-oxide [131]	A3: —CH₂·CH₂·N(O)(CH₃)₂ + A5: —OH	Bufo-N→O	20																					
3-Butyl-indole [60]	A3: —CH₂·CH₂·CH₂·CH₃	IBut								88							blue							
N,N-Diethyl-tryptamine [87]	A3: —CH₂·CH₂·N(C₂H₅)₂	TryAm-N, DiEt																						
3,3′-Diindolyl-methane [22, 23, 95]	A3: —CH₂·(3′) indolyl	—								—							red							
1,1′-Dimethyl-3,3′-diindolyl-methane [23]	A1: —CH₃ + A3: —CH₂·(3′)-1′-methyl·indolyl	—																						
1,2-Dimethyl-5-hydroxy-indole-3-carboxylic acid, ethyl ester [6]	A1 + A2: —CH₃ + A3: —CO·O·C₂H₅ + A5: —OH	ICA-OEt,1,2-DiMe-5-OH							60								no		no	violet				
1,2-Dimethyl-indole [87]	A1 + A2: —CH₃	I-1,2-DiMe							—															
1,3-Dimethyl-indole [60]	A1 + A3: —CH₃	I-1,3-DiMe								92							blue							
2,3-Dimethyl-indole [87]	A2 + A3: —CH₃	I-2,3-DiMe																						
N,N-Dimethyl-tryptamine [8, 87, 131]	A3: —CH₂·CH₂·N(CH₃)₂	TryAm-N,DiMe ●	94														grey-blue		yellow-orange	grey-yellow		grey-violet		
Dopamine [104]	B1: —CH₂·CH₂·NH₂ + B3 + B4: —OH	DopAm ●																					light brown >1000	
N-Formyl-2-amino-acetophenone [33]	B1: —CO·CH₃ + B2: —NH·CHO	AcPhe-2-NH-For																						
Glucobrassicin [22, 63]	A3: —CH₂·C(=N·OSO₃⁻)(—S·glucose)	Glubra															—[y] 20	green-violet[n] 500						
Gramine [8, 46, 60, 120]	A3: —CH₂·N(CH₃)₂	Gram ●	13	0	03		0	0	45	05	31	39	55	90	50	blue 10	pink-yell.[n] 1000	green-violet[n] 500	grey-pink 500	grey-green 50; yellow	grey-violet 50	brown-violet 50	no; no	violet[n] 500
Histamine [104]		Hist ●															violet-blue						pink-blue >1000	
3-Hydroxyanthranilic acid [2, 9, 12]	B1: —COOH + B2: —NH₂ + B3: —OH	AntA-3-OH ●												50			yellow-orange							
2-Hydroxy-indole [14, 19]	A2: —OH	I-2-OH					60	49	12								→yellow 5000							
5-Hydroxy-indole [8, 14, 19]	A5: —OH	I-5-OH ●					47	57	23								violet		grey-pink 500	brown 500		grey-brown		
Hydroxy-indoles (other) [14]																								
5-Hydroxy-indole-3-acetamide [8]	A3: —CH₂·CO·NH₂ + A5: —OH	IAAm-5-OH ●		—													blue		grey-brown	dark blue		grey-blue		
5-Hydroxy-indole-3-acetic acid [8, 9, 19, 28, 46, 78, 80, 104, 120]	A3: —CH₂·COOH + A5: —OH	IAA-5-OH ●	0	0	22↓		36⁺	01	0	47	76	80	80	31	50	pink 1000	blue[n] 50	grey-violet[n] 50	yellow-pink 500	dark violet 100; violet	grey-violet 50	grey-blue 50	no; no	violet[n] 500
5-Hydroxy-indole-3-ethanol [78]	A3: —CH₂·CH₂·OH + A5: —OH	IEtOH-5-OH ●		30													grey-blue							
3-Hydroxy-kynurenine [2, 9, 12]	B1: —CO·CH₂·CH(NH₂)·COOH + B2: —NH₂ + B3: —OH	Kyn-3-OH ●												86		yellow-green	orange							
3-Hydroxymethyl-indole (see indole-3-methanol)																								
5-Hydroxy-skatole [32, 33, 34, 35]	A3: —CH₃ + A5: —OH	Ska-5-OH															blue-violet	violet	red-brown					
Hydroxy-skatoles (other) [32, 33, 34, 35]																								
5-Hydroxy-tryptamine (see serotonine)																								
5-Hydroxy-tryptophan [8, 9, 14, 28, 46, 87, 90, 104, 120]	A3: —CH₂·CH(NH₂)·COOH + A5: —OH	Try-5-OH ●	0	0	06		0	0	0	05	33	24	24	87	50	reddish 1000	blue-grey[n] 50	violet[n] 50	grey-yellow 500	grey-brown 10; dark	grey-pink 100	grey-green 100	no; no	brown-red 50
Hydroxy-tryptophans (other) [14]																								
Indole [8, 9, 12, 14, 19, 22, 46, 60, 87, 120]	A	Indol ●	64	63	73	81	70	53	83	69*	81	97	100	08	100	grey 50	red[n] 50	green-blue[n] 10	grey-pink 500	brown-pink 100; dark	grey-violet 50	brown-pink 50	yellow; gold	brown[n] 500
Indole-3-acetaldehyde [8, 19, 46, 120]	A3: —CH₂·CHO	IAAld ●	67	40	67	67	58	29	58	64*	83	96		05*	no	grey 100	b'n[n] violet 1000	pink-violet[y] 50	grey-pink 10	grey-yellow 10; yell.	brown 500	brown 500	brown 500	brown-violet[n] 50
Indole-3-acetamide [1, 8, 9, 19, 46, 63, 87, 102, 120, 134]	A3: —CH₂·CO·NH₂	IAAm ●	08	12	51	23	08	02	21	43	70	83	86	47	100	grey 1000	blue[n] 50	violet[n] 50	grey-yellow 500	grey-yellow 50; grey	red-violet 50	red-violet 50	yellow; green	no[n]
Indole-3-acethydrazine [60]	A3: —CH₂·CO·NH·NH₂	IAHydrazin							40							yellow-green								
Indole-3-acetic acid [1, 8, 9, 19, 28, 46, 63, 86, 87, 90, 92, 102, 120, 134]	A3: —CH₂·COOH	IAA ●	0	01	34↓	68⁺	06	02	0	56*	78	93*	93	13	100	grey 1000	pink-violet[n] 30	brown-violet[n] 50	brown-yellow 500	light yel.-green 5; yel.	brown-violet 50	red-violet 10	yellow; gold	pink[y] 100
Indole-3-acetic acid, ethyl ester [1, 46, 87, 134]	A3: —CH₂·COOC₂H₅	IAA-OEt ●	63	62	70		68	39	73	67	80	98	100	10	500	grey 500	red-violet[n] 100	red-violet[n] 50	yellow-brown 500	pink 50; gold	yellow-brown 50	blue-violet 50	brwn.; brwn.	pink[n] 1000
Indole-3-acetic acid, methyl ester [19, 46]	A3: —CH₂·COOCH₃	IAA-OMe ●	59	61	68		68	65	35	72	65	79	88	100	15*	500	grey 500	red-violet[n] 100	red-violet[n] 50	yellow-brown 500	pink 50; gold	yellow-brown 50	blue-violet 50	brwn.; brwn.	pink[n] 1000
Indole-3-acetonitrile [1, 8, 19, 22, 23, 46, 63, 87, 102, 120]	A3: —CH₂·CN	IAN ●	52	53	72		63	61	34	67	63	82			12	500	pink 100	grey-violet[n] 100	red-violet[b]	yellow-pink 500	yellow-pink 50; gold	grey-violet 10	grey-blue 50	pink; gold	pink[n] 1000
N-(Indole-3-acetyl)-aspartic acid [19, 63, 134]	A3: —CH₂·CO·NH·CH(COOH)(CH₂COOH)	IAAspA				22⁺	17⁺		35															
N-(Indole-3-acetyl)-D-glucose [134]	A3: —CH₂·CO·O·Glucose	IAGlc							10															
N-(Indole-3-acetyl)-glucuronide [28]	A3: —CH₂·CO·O·Glucuronic acid	IAGlcUA										45	45											
N-(Indole-3-acetyl)-glutamine [28]	A3: —CH₂·CO·NH·CH(COOH)([CH₂]₂·CO·NH₂)	IAGlu(NH₂)										68	68											
N-(Indole-3-acetyl)-glycine [28]	A3: —CH₂·CO·NH·CH₂·COOH	IAGly										82	82											
Indole-3-acrylic acid [8, 19, 28, 46, 120]	A3: —CH:CH·COOH	IAcrA ●	0	03	50↓	67⁺	56⁺	03	0	53	82	98	98	02	100	grey 2000	brown-pink[y] 100	brown-red[v] 50	brown-yellow 500	grey-green 50; brown	grey-orange 50	yell.-orange 100	no; no	brown-pink[n] 500
N-(Indole-3-acryloyl)-glycine [28, 36]	A3: —CH:CH·CO·NH·CH₂·COOH	IAcrGly										83	83											
Indole-3-aldehyde [1, 19, 22, 33, 46, 60, 63, 102, 120]	A3: —CHO	IAld ●	23	27	61	58	36	12	42	49	75	90	95	13	50	grey 1000	→pink[b] 1000	pink-yell.[r] 1000	pink 1000	grey 500; dark	brown 1000	brown-yellow 500	grey; green	no[n] —
Indole-3-aldoxime [46]	A2: —CH:NOH	IAldoxim	10	18	70				14↓	47	82	93	93	09	50	on —	no[n]	pink[n] 500	no	yellow 1000; dark	yellow-brown 500	yell.-brown 500	no; no	no[n] —
Indole-3-butyric acid [1, 8, 19, 46, 87, 120]	A3: —CH₂·CH₂·CH₂·COOH	IBA ●	0	04	48	72⁺	40⁺	04↓	0	60	83	98	98	05	500	grey 500	blue[n] 100	grey-violet[n] 50	grey-yellow 500	yellow-pink 10; gold	yellow-yellow 50	yellow-yellow 50	yellow; gold	pink[n] 1000
Indole-carbinol (see indole-3-methanol)																								
Indole-3-carboxylic acid [1, 19, 33, 46, 63, 102]	A3: —COOH	ICA ●	0	04	60	71⁺	58⁺	03	0	57	81	98	98	06	500	red[n] 100	blue-grey[v] 100	grey-pink 500	brwn.-violet; brwn.	green-orange 50	yellow; gold	pink[n] 100
Indole-1,3-dipropionitrile [60]	A1 + A3: —CH₂·CH₂·CN	I-1,3-DiproN								68							no	blue-violet 100	no	no				
Indole-3-ethanol [1, 8, 19, 46, 87]	A3: —CH₂·CH₂·OH	IEtOH ●	21	28	66	52	41	11	42	52*	79	93	96	21		grey 100	grey-yell. 500	red-violet 5	yellow-org. 500	pink-yell. 10; orange	yellow-violet 100	grey-violet 100	orange; gold	violet[n] 1000

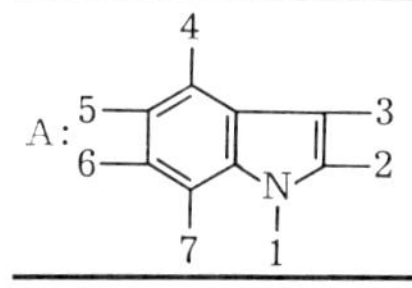

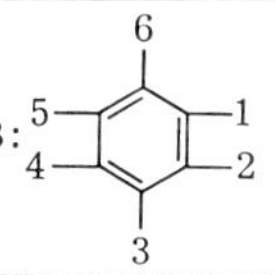

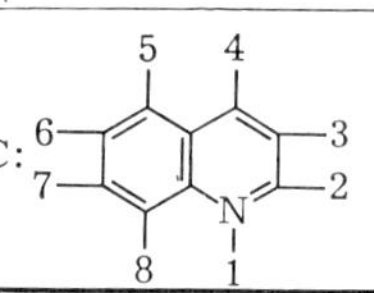

Thin-Layer Chromatography, 2nd Edition

Table 88 (Continued)

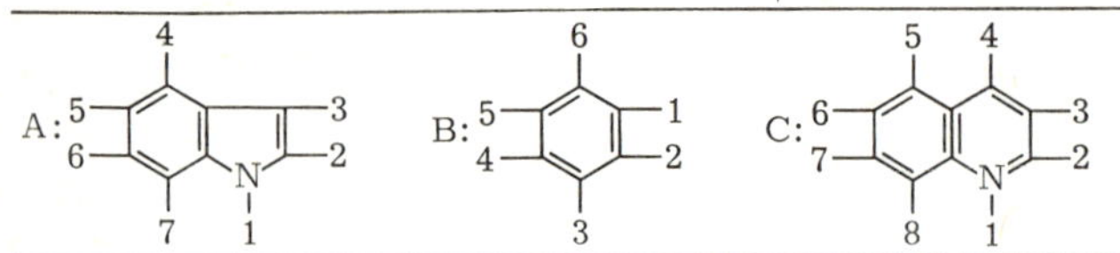

Compound	Ring system and substituents (substituents not given are —H)	Abbreviation (see also Table 89 for TLC-separation of compounds denoted with ●)	hRf-value for a 10 cm run in the solvents below												Quenching in UV$_{254}$ [46]^2	Fluorescence colours in UV$_{366}$ and detection limit in ng	van Urk reagent (No. 73^4), colours^1 and detection limit in ng	4-Dimethylamino-cinnamaldehyde reagent (No. 76), colours^1 and detection limit in ng	Procházka Reagent (No. 123)^5		Salkowski reagent (Nos. 104, 108)^6		Fluorescence colours in UV$_{366/254}$	Ninhydrin reagent^6 (No. 178), colours^1 and detection limit in ng
			I [46, 68, 131]	II [46, 78]	III [1, 46, 102]	IV [19]	V [19, 46]	VI [14, 46]	VII [46, 60]	VIII [46, 134, 136]	IX [46, 90]	X [28, 46]	XI [28, 46]	XII [2, 46]	[2, 9, 12, 46, 104]	[1, 8, 9, 12, 19, 35, 46, 60, 78, 102, 120]	[30, 32, 33, 35, 46]	Colours and detection limit in ng [8, 19, 35, 46, 60, 102, 120]	Fluorescence colours in UV$_{366/254}$ and detection limit in hg	Colours and detection limit in ng FeCl$_3$/HClO$_4$ [24, 46]	FeCl$_3$/H$_2$SO$_4$ [8, 46, 96]	[46]	[46, 104]
Indole-3-glycollic acid [19, 46, 87]	A3: —CH(OH)·COOH	IGlyeA ●	0 0 04			43+ 40+ 0			0 35* 59		63 63 35			100	grey 100	violet-red^n 50	brwn.-violet^{n}50	greyish-pink 500	green-yellow 50; olive	grey-violet 100	brown-violet 50	no; no	yell.-orange 100
Indole-3-glyoxylamide [19, 46]	A3: —CO·CO·NH$_2$	IGlyoxAm ●	09 16 63			37 27* 05			21 48 76		85 85 12			50	no —	light yell.^{n}5000	pink 1000	pink 5000	grey 500; no	grey-green 500	grey-green 500	no; no	no^n —
Indole-3-glyoxylic acis [8, 19, 46]	A3: —CO·COOH	IGlyoxA ●	0 0 07			59+ 45+ 0			0 16 48		56 56 11			100	yellowish 500	yell.-pink^n 100	pink 500	green-violet500; grey	grey-yell.. 500	green-yell.. 500	brwn.; brwn.	pink^n 500
Indole-3-lactic acid [8, 19, 28, 46, 86, 87, 92]	A3: —CH$_2$·CH(OH)·COOH	ILA ●	0 0 06			49+ 44+ 0			0 40 66		75 75 22			100	grey 100	blue^n 50	pink-violet 5	yell.-orng. 100	yell.-pink 10; orange	brwn.-violet50	brwn.-violet12	brwn.; gold	pink^n 500
Indole-3-methanol [19, 22, 23, 33, 46, 87, 95, 120]	A3: —CH$_2$OH	IMeOH ●	62 57 70			72 61 32			65 70* 81*		97*100 16			100	grey 100	grey-pink^{n}100	brwn.-violet50	yell.-pink 500	brown-yell. 5; brown	brwn.-violet50	red-violet 50	no; no	pink-brwn.^{n}100
N-(Indole-3-methyl)aniline [60]	A3: —CH$_2$·NH·C$_6$H$_5$	IMeAnil							68								pink-yellow							
Indole-3-nitrile [60]	A3: —CN	ICN							48								violet		pink					
Indole-3-propionic acid [8, 9, 19, 28, 46, 87, 90, 120]	A3: —CH$_2$·CH$_2$·COOH	IPA ●	0 03 43↓			72+ 67+ 03			0 58 82		96 96 07			100	grey 50	grey-blue^n 50	brwn.-violet50	grey-pink 100	yell.-brown10; orange	yell., brown 50	brown-orange 50	yellow; gold	pink^n 100
Indole-3-pyruvic acid [19, 46, 87, 92]	A3: —CH$_2$·CO·COOH	IPyA ●	* 0 04			56+ 59+ 0			0 49 73		27 * 04			150	grey 50	grey-violet^n 50	brown-pink^{r}50	grey-pink 100	yellow-brown 5; olive	brwn.-violet50	grey-violet 50	brwn.; green	pink^n 100
Indoline [60]	A2: —H$_2$ + A3: —H$_2$; ($_2$C—$_2$C)	Indolin							77								yellow							
Indoxyl-acetate [8] (cf. also acetoxy-indole)	A3: —O·CO·CH$_3$	IOAc ●	— — —								33 73 —				brown-grey	blue		brown-blue		blue-violet —	— —	brwn.-violet —	— —
Indoxyl-sulphate [9, 12, 28]	A3: —O·SO$_3$H	IOSulf ●															brown-red							
Isatin [46, 120]	A2 + A3: =O; ($_2$C—$_2$C)	Isat	30 36 65			38 15			0 73		— 40			500	no	orange^n	yellow^n 500	dark 500; violet	yellow 500	yellow 500	yellow 500	no; no	yellow^n 500
Kynurenic acid [2, 9, 12]	C2: —COOH + C4: —OH	KynA ●												36		blue-green	no —							
Kynurenine [2,12]	B1: —CO·CH$_2$·CH(NH$_2$)·COOH + B2: —NH$_2$	Kyn											— 90			azure	yellow-brown							
Kynurenine-sulphate [9, 28]	B1: —CO·CH$_2$·CH(NH$_2$)·CO·O·SO$_3$H + B2: —NH$_2$	KynSulf											25 30			blue-green	yellow-orange							
N-Malonyl-tryptophan [134, 136]	A3: —CH$_2$·CH(NH·CH$_2$·COOH)COOH	Try-N,Mal ●								40														
Metanephrine [104]	B1: —CH(OH)·CH$_2$·NH·CH$_3$ + B3: —O·CH$_3$ + B4: —OH	MetNeph ●																						violet >1000
5-Methoxy-N,N-dimethyl-tryptamine [68, 131]	A3: —CH$_2$·CH$_2$·N(CH$_3$)$_2$ + A5: —O·CH$_3$	TryAm-N,DiMe-5-OMe	94														red-violet							
5-Methoxy-indole [60]	A5: —O·CH$_3$	I-5-OMe							70								blue							
5-Methoxy-indole-3-acetic acid [78]	A3: —CH$_2$·COOH + A5: —O·CH$_3$	IAA-5-OMe		10													violet		red-brown					
5-Methoxy-indole-3-aldehyde [60]	A3: —CHO + A5: —O·CH$_3$	IAld-5-OMe							54								no —		light-brown	blue-grey				
5-Methoxy-ingole-2-carboxylic acid, ethyl ester [60]	A2: —CO·O·C$_2$H$_5$ + A5: —O·CH$_3$	I-2-CA-OEt-5-OMe							70								green-blue							
5-Methoxy-indole-3-ethanol [78]	A3: —CH$_2$·CH$_2$OH + A5: —O·CH$_3$	IEtOH-5-OMe ●		60																				
5-Methoxy-N-methyl-tryptamine [68]	A3: —CH$_2$·CH$_2$·NH·CH$_3$ + A5: —O·CH$_3$	TryAm-N,Me-5-OMe																						
5-Methoxy-tryptamine [87]	A3: —CH$_2$·CH$_2$·NH$_2$ + A5: —O·CH$_3$	TryAm-5-OMe														blue								
8-Methoxy-xanthurenic acid [2]	C2: —COOH + C4: —OH + C8: —O·CH$_3$	XantA-8-OMe											30											
2-Methyl-3-chloro-indole [60]	A2: —CH$_3$ + A3: —Cl	I-Cl-2-Me							77								no —		green-yellow					
2-Methyl-5-hydroxy-3-acetyl-indole [60]	A2: —CH$_3$ + A3: —CO·CH$_3$ + A5: —OH	IAc-2-Me-5-OH							50								no —		blue-grey					
2-Methyl-5-hydroxy-indole-3-carboxylic acid, ethyl ester [60]	A2: —CH$_3$ + A3: —CO·O·C$_2$H$_5$ + A5: —OH	ICA-OEt-2-Me-5-OH							38								no —		no	violet				
1-Methyl-indole [23]	A1: —CH$_3$	I-1-Me							—															
2-Methyl-indole [60]	A2: —CH$_3$	I-2-Me							83								red-violet							
1-Methyl-indole-3-methanol [23]	A1: —CH$_3$ + A3: —CH$_2$OH	IMeOH-1-Me							—															
2-Methyl-5-methoxy-indole-3-carboxylic acid, ethyl ester [60]	A2: —CH$_3$ + A3: —CO·O·C$_2$H$_5$ + A5: —O·CH$_3$	ICA-OEt-2-Me-5-OMe							68								no —							
3-Methyl-oxindole [33]	A2: =O + A3: —CH$_3$; ($_2$C—$_2$C)	Ska-2-OH	—																					
N-Methyl-serotonine [131]	A3: —CH$_2$·CH$_2$·NH·CH$_3$ + A5: —OH	Sero-N,Me	34																					
N-Methyl-tryptamine [68, 74]	A3: —CH$_2$·CH$_2$·NH·CH$_3$	TryAm-N,Me																						
5-Methyl-tryptamine [87]	A3: —CH$_2$·CH$_2$·NH$_2$ + A5: —CH$_3$	TryAm-5-Me																						
5-Methyl-tryptophan [8, 120]	A3: —CH$_2$·CH(NH$_2$)·COOH + A5: —CH$_3$	Try-5-Me															blue-grey 50		turquoise	yellow-green 5			violet	
Neoglucobrassicine [23]	A1: —O·CH$_3$ + A3: —CH$_2$·C(=N·OSO$_3^-$)(—S·Glucose)	NeoGlubra							—								—		—					
1-Nitroso-gramine [60]	A1: —NO + A3: —CH$_2$·N(CH$_3$)$_2$	Gram-1-N=O							45								no —		purple					
1-Nitroso-2-methyl-indoline [60]	A1: —NO + A2: —HCH$_3$ + A3: —H$_2$; ($_2$C—$_2$C)	Indolin-1-N—O-2-Me							85								grey-yellow							
Noradrenaline [104]	B1: —CH(OH)·CH$_2$·NH$_2$ + B3 + B4: —OH	NoAdr ●																					brown >1000	
Normetanephrine [104]	B1: —CH(OH)·CH$_2$·NH$_2$ + B3: —O·CH$_3$ + B4: —OH	MoMetNeph ●																					grey-brn.>1000	
Oxindole (see 2-Hydroxy-indole)																								
3-Phenylacetyl-indole [60]	A3: —CO·CH$_2$·C$_6$H$_5$	IAcPhe							64								pink							
Serotonine [8, 9, 28, 46, 87, 90, 104, 120, 131]	A3: —CH$_2$·CH$_2$·NH$_2$ + A5: —OH	Sero ●	30 0 03			0 0			03 42		38 46 86			no	no —	grey^n 100	violet^n 500	brown-yell 1000	grey-yewoll 50; grey	grey 500	grey 500	no; no	brwn.-violet^{n}100
Skatole [8, 9, 19, 22, 23, 28, 32, 33, 46, 60, 87, 120]	A3: —CH$_3$	Ska ●	73 74 77			86 74 56			76 70 83		100 100 04			500	grey 500	grey-blue^{n}100	grey-violet^n 50	grey-yellow 500	yellow 5; orange	yellow-brown500	brown 500	no; no	violet^n 500
Tryptamine [8, 9, 19, 28, 46, 63, 87, 120]	A3: —CH$_2$·CH$_2$·NH$_2$	TryAm ●	0 07 02			05 01 0			03 07 46		51 64 85			100	white 50	green-blue^{n}100	brown-red^y 50	yellow 100	yell.-pink5; orange	yellow-brown 50	brwn.-violet50	yell.; gold	brown-violet^n 50
Tryptophan [8, 9, 12, 19, 46, 63, 90, 92, 120, 134, 136]	A3: —CH$_2$·CH(NH$_2$)·COOH	Try ●	0 0 07			0 0			0 08 42		38 38 77			500	grey 500	green-blue^n 50	grey-violet^n 50	yellow 500	brown-orange 50	brwn.-violet50	orge.; gold	brown-red^n 5
Tryptophol (see indole-3-ethanol)																								
Urinary indican (see indoxyl-sulphate)																								
Vanillin-mandelic acid [104]	B1: —CHO + B3: —O·CH$_3$ + B4: mandelic acid	Van-Mand ●																					no —	
Xanthurenic acid [2, 9, 12]	C2: —COOH + C4 + C8: —OH	XantA ●											17			blue-green	no —							

Signs: * compound decomposes; ↑↓ compound shows tailing; + methyl ester of the quoted compound; → colour develops slowly

Colour abbreviations: gold = brilliant gold-brown; no = no colour

No.	Solvent Composition (v/v)	No.	Fields of Application and Comments
I	Chloroform-methanol (93 + 7), saturated with 25% NH$_4$OH	I	Neutral indole derivatives, indole alkaloids
II	Chloroform-methanol (93 + 7)	II	Neutral indole derivatives, urine metabolites
III	Ethyl acetate-isopropanol-water (65 + 24 + 11)	III	Acid auxins
IV	Diisopropyl ether-dimethylformamide (80 + 20)	IV	Two-dimensional technique; acids previously esterified with diazomethane
V	Benzene-dioxan (66 + 35)	V	
VI	Benzene-acetone (90 + 10)	VI	Neutral indole derivatives
VII	Benzene-ethanol (90 + 10) on neutral alumina	VII	Neutral indole derivatives
VIII	Chloroform-ethyl acetate-formic acid (35 + 55 + 10)	VIII	
IX	n-Butanol-acetic acid-water (65 + 13 + 22)	IX	Basic indole derivatives, urine metabolites
X	Acetone-chloroform-acetic acid (96%)-water (40 + 40 + 20 + 5)	X	Urine metabolites; after thorough drying, chromatographed again in solvent XI
XI	Acetone-25% ammonium hydroxide (100 + 1)	XI	in order to separate the amines from the indole amino acids
XII	Methanol-water-formic acid (37.5 + 60 + 2.5) on polyamide layer	XII	Tryptophan metabolites; very well-defined spots; sensitive layer

[1] The indices of the colours refer to the fluorescence of the reaction product with the test reagent; b = blue; y = yellow to yellow-green; h = grey; r = red; v = violet fluorescence colour; n = no.

[2] Dark violet quenching zones when using F$_{254}$-layers (with fluorescent additive).

[3] Fluorescence yielded or intensified only after irradiation with UV$_{254}$.

[4] Layers warmed for 10 min at 60° C after spraying; colours are intensified and sometimes altered by exposure to a slight current of aqua regia vapour.

[5] Layers warmed for 20–30 min at 100° C after spraying; colours and fluorescence are intensified by exposure to a slight current of aqua regia vapour (fluorescence is stable for weeks).

[6] Layers warmed for 1 h at 110° C after spraying; all the indole derivatives tested (except isatin) showed fluorescence after standing several days in the laboratory (visible with amounts down to 10 ng).

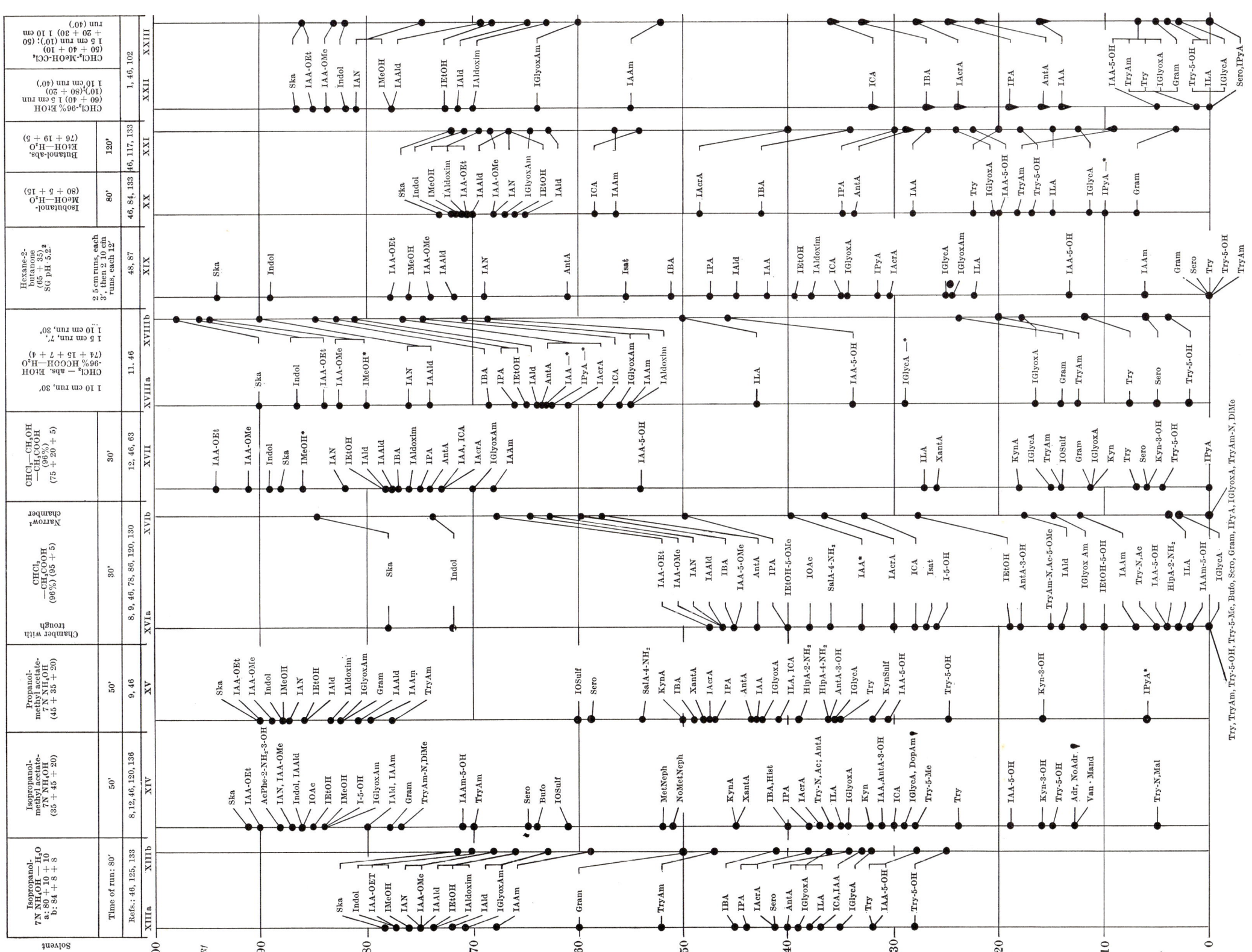

Table 89. *Thin-layer chromatographic separation of indole derivatives and urine metabolites; hRf-values and times for 10 cm run on silica gel G, H or P*

plant development. Observations of further interactions between the various plant growth regulators are, however, accumulating.

Reference may be made in particular to the participation of growth regulators in plant movements [27], apical dominance [25, 26], leaf and fruit abscission [25, 26], bud dormancy [26, 132], vernalisation and photo-periodic phenomena [26, 65, 109], flower and fruit development [10, 25, 26, 65, 109], seed germination [26, 59], tumour and gall formation [17, 26] and plant senescence [26, 81].

The compounds mentioned are present in the organisms mostly in very low concentration and are often chemically unstable, all of which renders difficult investigations of occurrence, biogenesis and metabolism.

PC and electrophoresis [73, 125] have already enriched the methods. A further step forward has come with TLC. Since the first TLC-analyses of growth regulators [22, 44, 120], the method has found wide application, recently in combination with GC [124]. As well as advantageously short times of run (15—50 min) with the consequently reduced decomposition of less stable substances, TLC usually yields better separations than do methods hitherto employed. Moreover, nanogram amounts (1 ng = 10^{-6} mg) can be detected with suitable spray reagents. The sensitivity is thus of the same order of magnitude as that of biological tests.

II. Preparation of the Material for Analysis
1. General Information

The processing technique is largely to be adapted to the particular problem and the material under investigation. Scarcely more than 50 µl of a solution of the substances, as far as possible in a volatile solvent, are applied to layers of normal thickness. Extracts must thus usually be appreciably concentrated and pre-purified.

The commonest procedures are separations into hydrophilic and lipophilic compounds, followed by extraction in stages of the neutral and acid materials from suitably buffered aqueous solution using ether, chloroform or ethyl acetate [3, 55, 66, 67, 79].

It is sometimes especially difficult to separate chlorophyll and other plant colouring matter. A procedure which has proved helpful for prior purification of methanolic extracts, very rich in chlorophyll, is given here: the extract is concentrated to dryness under reduced pressure in a rotary evaporator, the residue taken up in a little dilute HCl (pH ca. 5) and this solution filtered through a short (1—2 cm) silica gel column [133]. Pre-purification by chromatography with water in an atmosphere of acetic acid ([3], p. 508), if necessary on thicker (preparative) layers, is a promising but so far scarcely used procedure.

Reference may be made to the possibility of separation into groups in neutral solvents ([1], cf. p. 478), the use of ion exchange columns [5] or of activated charcoal [9] for preliminary purification. Rotary evaporators with cylinders or ground glass centrifuge tubes of 10—20 ml capacity, are suitable for final concentration of the pre-purified extracts.

Concentration is usually carried out to dryness, the residue taken up several (2–3) times with 1–3 drops of a volatile solvent and each solution applied with a 10 µl pipette. Aqueous solutions should be applied in a nitrogen atmosphere in order to keep the start zone small and to suppress decomposition through oxidation. Start zones which are too large may be "closed up" by brief preliminary chromatography in a strongly polar solvent (like water-methanol (50 + 50)).

2. Free Auxins

Free auxins should generally be obtained by short-time extraction of the plant material; the activity of the endogenous enzymes should be eliminated. By observing these precautions it has been shown that numerous components of *Brassica* species, considered to be natural indole derivatives, were extraction artefacts [22, 23, 63, 107]. The procedure used by GMELIN and VIRTANEN [22] appears suitable, namely, extraction of the fresh, whole plant organs with boiling methanol (10 min) followed by grinding or squeezing out or by further extraction for 1–2 h at room temperature [63]. Small samples of not too woody material are best covered with liquid nitrogen in a mortar and ground up; the powder is then brushed through a plastic funnel into a centrifuge glass and boiling methanol added; the glass is maintained a few minutes in a sand or water bath at 70–80° C, shaken up several times and centrifuged. The whole procedure lasts but a few minutes. Before use, all the apparatus is cooled with liquid nitrogen in a covered beaker.

Large yields of auxins have been obtained by prolonged extraction with ether at room temperature. Probably "bound auxins" are liberated during extraction.

3. Diffusible Auxins

"Diffusible" auxins may be collected at cut surfaces of the plant organs in a moist atmosphere, using 1–3% aqueous agar blocks [3, 66, 67, 119]. The polar and mainly basipetal transport of growth regulators in plants is utilised here. For chromatography, the compounds diffused into the agar are extracted by shaking (at least 30 min) 2–3 times with a volume of 50–70% methanol or ethanol about 5–10 times that of the agar. Thin silica gel layers (on microscope slides) can also be used as a diffusion medium. The material which has penetrated into the layer is easily obtained in clear solution by briefly (3 min) shaking the scraped-off silica gel with methanol in a centrifuge tube and then centrifuging [45].

4. Indole Derivatives in Urine

It has been possible to detect indole derivatives in urine through TLC of 50–100 µl of the untreated samples [12]. Ether- or ethyl acetate-extraction of urine, saturated with NaCl and acidified [104]; and elution with phenol-distilled water (8 + 92) of the substance, adsorbed on inactive charcoal in a previous stage [9], have also been recommended.

5. Gibberellins

Gibberellins are extracted from fresh or dried plant material using acetone, methanol or ethanol (long-time extraction at low or at room temperature). The substances which have been adsorbed if necessary on charcoal and eluted with acetone [55], can be extracted from the crude aqueous extract at pH 3 by means of ethyl acetate. It appears usually necessary to follow this with purification by extracting with a buffer solution at pH 6, renewed extraction with ethyl acetate at pH 3 and prior separation on columns of celite/charcoal, celite/silica gel, celite or silica gel [51, 55, 75, 76].

SEMBDNER and co-workers have extracted the crude extract with ethyl acetate and then with n-butanol, thereby obtaining the more polar gibberellins from the flowers of *Phaseolus* [113] and shoots of *Nicotiana* [115]. JONES and PHILLIPS [40] have detected "diffusible gibberellins" in agar after 18 hours diffusion (see "diffusible auxins" above). The compounds were extracted with methanol at room-temperature the agar had been frozen at $-15°$ C.

6. Cytokinins

Cytokinins have been so far detected in aqueous ethanolic extracts of various plant materials.

The cytokinins hitherto found can be separated from gibberellins and acid auxins by extraction of the aqueous extracts at pH 3, using ethyl acetate. The organic phase has been shown not to be active in cytokinin tests. The summary of MILLER [79] and the contributions of NITSCH [85] may be consulted for information about experience gained up to now (cf. [58, 88] for new bio-assays).

Only LETHAM and MILLER [68a, 69] appear to have used TLC so far. They compared the zeatin obtained from immature fruit of *Zea mays* with the "maize factor", isolated likewise from maize grains. The preparations, characterised through their kinetin activity, are evidently identical. They accordingly show the same Rf-values in TLC on alumina G with butanone, saturated with water (hRf 58); with ethyl acetate saturated with water (hRf 14); and with chloroform-95% ethanol (10 + 90, hRf 86, 90 + 10, hRf 61). The compound is 6-(4-hydroxy-3-methylbut-*trans*-2-enyl) aminopurine.

III. Adsorbent and Solvents
1. General Information

The layers are best prepared by the standard procedure, using a spreader (p. 56, 85). Pouring procedures are used successfully in many laboratories; a definite amount of a suspension of the adsorbent is then spread over the carrier plate [111]. Spread layers should be preferable on account of their more uniformly distributed adsorbent. Layers of about 250 μm thickness, drying to 150 μm, are usually suitable for analytical work.

The substance mixtures may be applied as spots (<5 mm diameter) or bands (<5 mm height) (cf. p. 63). Separation of bands is generally superior. Band application is especially advantageous when the separated substances are to be tested biologically (cf. p. 487). Layers with fluorescent indicators should be chosen here [45].

Separations are generally carried out in chambers with CS. Better resolution is occasionally achieved in S-chambers which are more economical in solvent. In general, the chamber atmosphere should be enriched with the less volatile components of the solvent mixture. This can be done by attaching a strip of filter paper, saturated with these components, on to the cover plate [45]; or by using a coated cover plate [37] which has been sprayed with the particular component (about 10 ml) and which dips into the solvent during chromatography [46] (cf. Table 89, XVI b). The hRf-values in the S-chamber are normally larger; they are strongly influenced by the chamber atmosphere.

Runs of 10 cm length usually suffice. Runs of 6—8 cm in auxin analysis have yielded equally good separations in shorter time and with lowered limits of detection [44, 45]. Since many materials are sensitive to light, the chambers should be covered with a black cloth or a box during chromatography.

2. Auxins and Urine Metabolites

Adsorbents. The first separations were carried out on silica gel G [21, 120]. Since added gypsum has an unfavourable influence on many biological tests (cf. p. 488), adhering silica gel, free of gypsum, is preferable; the times of run are then shorter. Silica gel PF_{254} (Firm 88), really developed for preparative TLC, has been recently used with success in our laboratory. Layers prepared with it are harder, an advantage which becomes evident particularly in analyses of plant extracts requiring concentration through multiple application on a start point or a starting line. OBREITER and STOWE [87] have obtained harder layers which could be written on with a pencil, by mixing purified and sieved (44 μm diameter) silica with carboxymethyl cellulose (28.5 g silica + 1.5 g + 60 ml warm water). Silica gel-starch mixtures, which must be stirred at the boiling point, yield layers which can be wiped and possess especially good separating properties; times of run are, however, up to twice as long. It has been possible to separate at least 17 indole derivatives on layers of extremely pure silica gel (Firm 83), prepared with M/30 phosphate buffer of pH 5.3 or with 0.075 M H_3PO_4, and using hexane-butanone-2 with a length of run 11 cm (cf. p. 478), Table 89, XIX).

Some authors have used cellulose [103] or alumina layers [60] (Table 88, VII), partly in mixtures with silica gel [19], in order to separate basic and neutral substances. Indole derivatives on alumina- and especially on cellulose layers, generally tend to tail-formation. BENASSI et al. [2] have accomplished good separations of urine metabolites on 100 μm thick polyamide layers. The completely smooth layers prepared with polyamide powder DF (Firm 118), fluorescing in UV_{254}, yield sharply defined zones; they are, however, very sensitive to rubbing so that there are limits to multiple application of an extract [46] (Table 88, XII).

Solvents. The acid and ammoniacal solvents (Table 89, XIV and XVI), suggested by STAHL and KALDEWEY [120], have been in the meantime (with some modifications – Table 89, XV and XVII) applied successfully to separation of auxins and urine metabolites.

More strongly polar, acid solvents (Tables 88 and 89, VIII, IX, X, XVII; [28]) have proved satisfactory for the mostly basic urine metabolites of serotonine and tryptophan; these have been used partly in combination with basic systems and two-dimensional separation (Tables 88 and 89, XIV and XVI, XV and XVI, XIII and IX) or multiple development (Table 88, X and XI; cf. here p. 481).

The formic acid-containing solvents (Table 88/89, VIII and XVIII), so far not used for this purpose, should be suitable for separating urine metabolites especially in multiple development.

When using the mixtures mentioned, the compounds show essentially the same sequence after separation; a more or less inverted separation pattern is shown in solvent XII, Table 88, on polyamide layers.

HEACOCK and MAHON [32, 33, 34, 35] have carried out TLC analyses of the hydroxyskatoles occurring during urine metabolism; they used special solvents and reagents for detection. EICH and ROCHELMEYER [14] report TLC-separations of hydroxyindoles (Table 88, VI) and hydroxy-tryptophans (cf. Table 90).

Table 90. hR_f-values of the metabolites formed during hydroxylation of skatole and indole (Silica gel G, 10 cm run)

Solvent	A	B		A	B		C	D
Time of run (h)	1.5	2		1.5	2		1/3	
Ska	68	72	IMeOH			Indole	59	80
Ska-2-OH	18	51	AcPhe-2-NH_2	57	68	I-2-OH	12	45
Ska-4-OH	71	60	AcPhe-2-NH—	28	68	I-4-OH	29	43
Ska-5-OH	56	49	For			I-5-OH	23	37
Ska-6-OH	47	29	IAld	15	32	I-6-OH	17	22
Ska-7-OH	55	14	ICA	06	00	I-7-OH	23	18

A Diisopropyl ether; B 1,2-Dichloroethane-diisopropylamine (90 + 15) [34]; C Benzene-acetone (90 + 10); C Methylene dichloride-ethyl acetate (65 + 35) [14]. A → B and C → D are suitable for two-dimensional separation. See Table 88 for abbreviations.

Caution is demanded when ammoniacal or acid solvents are used in investigations on plant auxin metabolism. Indole-3-pyruvic acid and glucobrassicin decompose during TLC in ammonia-containing media. Good separation of a number of indole derivatives is possible in solvents containing acetic and formic acids but they are attacked, especially by the latter, turning reddish as soon as they come into contact with air after chromatography (the asterisked (*) compounds in Table 88 and 89). Indole-3-acetic acid thereby loses its biological activity [8, 120].

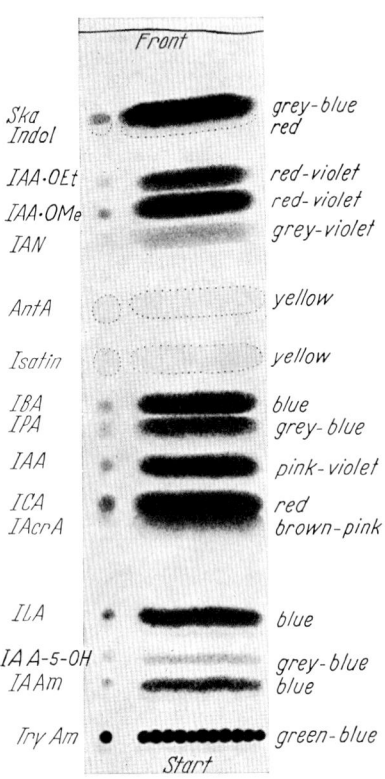

Fig. 157. TLC-separation of 16 indole derivatives. Solvent: hexane-butanone-2 (65 + 35); layer: silica gel HR$_{254}$ (Firm 83), made up with M/30 phosphate buffer, pH 5.2; length of run: 10 cm; multiple development: 2 runs of 6 cm (each of 5 min) and 2 of 10 cm (each of 12 min). Ca 0.1 μg substance per point. Visualisation with 4-dimethylaminobenzaldehyde (van Urk reagent (No. 73)

There is almost no decomposition in neutral solvents; it is true, however, that the resolutions are less sharp. The solvents proposed by BALLIN [1] (Table 89, XXII and XXIII) are particularly suitable for separating into groups of basic, acid and neutral compounds; and solvent III, Table 88, for separating acid compounds. The acids tend to tailing; this can be reduced by one or two preliminary chromatography stages over a run of about 5 cm. A quite good separation of neutral derivatives can be achieved in the solvents II, VI, VII, XXII and XXIII of Tables 88 and 89, using reduced alcohol content [1]; a sharp resolution is possible with hexane-butanone-2 (82 + 18) on silica gel containing carboxymethylcellulose [87]. Acid and basic derivatives hardly migrate at all in these solvents. The acids, too, can be separated with the hexane-butanone mixture on acidified layers. The best separations, with sharply defined zones, have been thus obtained with multiple development using the solvent combination XIX, Table 89 (Fig. 157) [48].

Sharp resolutions of the acids and to some extent of the basic indole derivatives also, have been accomplished using solvents XX and XXI of Table 89, although migration speeds are low; pyruvic acid decomposes and tryptamine is attacked in XXI.

The TLC of glucobrassicin and ascorbigen has been carried out in various acid and neutral solvents (Table 91) [21, 22, 23, 61, 63].

Table 91. hRf-values of some products from the metabolism of glucobrassicin in Cruciferae (Silica gel G)

Solvent	A	B	C
Time for a 10 cm run	20′	30′	15′
Ascorbigen	—	0	80
3,3′-Diindolylmethane	55	—	—
Glucobrassicin	—	0	50
Indole	66	85	100
Indole-3-acetonitrile	42	48	100
Indole-3-aldehyde	14.5	10	100
Indole-3-carboxylic acid	0	24	100
Indole-3-acetic acid	0	31	100
Indole-3-methanol	9.5	53	100[a]
Skatole	70		100
Tryptophan	0	0	20

A Chloroform-ethanol (99 + 1) [22, 23, 46]; B Carbon tetrachloride-chloroform-acetic acid (47.5 + 47.5 + 5) [46, 63]; C Propanol-methanol-ethyl acetate-acetic acid (25 + 25 + 50 + 5) [46, 63].

[a] Substance decomposes.

3. Gibberellins

Table 92 contains a compilation of gibberellin separations on various layers and with various solvents. The Rf-values fluctuate considerably from run to run. Many authors thus prefer to quote R_{St}-values, based on those of gibberellin A_3, instead of Rf-values [15, 39, 111, 115]. Silica gel G, first suggested by SEMBDNER et al. [111], has been the principal adsorbent used. The gibberellins $A_{1-3, 8}$ migrate slowly on activated adsorbents; the authors thus used inactive silica gel G layers for them (Table 92, 20—22, 24, 27, 29). MACMILLAN and SUTER [77] and, later, KAGAWA and co-workers [35a, 43] have chosen kieselguhr G as adsorbent for this purpose (Table 92, 4—6, 25). Continuous development on alumina layers [64] seems to offer no advantages over the techniques mentioned.

Solvents used have been chiefly mixtures of chloroform, carbon tetrachloride, benzene or diisopropyl ether with acetic acid, occasionally with added ethyl acetate, butanol or water to increase polarity. The ammoniacal and especially the neutral solvents cited (Table 92, 11—15, 27—29), are less suitable for analyses of the gibberellins; they have, however, yielded successful separations of hydrolysis products (e. g., 1 → 3-lactones from GA_1 and GA_7 [15]) and of the methyl esters of gibberellins-A_{1-9} (Table 92, solvent 1).

Table 92. *Thin-layer chromatographic separation of gibberellins*

No.	Solvent	References	Time of Run [h]	Run [cm]	Layer	Separation Pattern
1	CHCl$_3$-ethyl acetate-CH$_3$COOH (70 + 30 + 5)	[111, 113]	15		d'	a	A8 ... A3 A1 A4—7,9
2	CHCl$_3$-ethyl acetate-CH$_3$COOH (60 + 40 + 5)	[111, 113]	2		i	A8 A3 A1 ... A7 A4,5 A9
3	Benzene-butanol-CH$_3$COOH (70 + 25 + 5)	[43]	1	15	a	A8 A2 A1 A3 ... A5,6 A7 A4,9
4	CCl$_4$-CH$_3$COOH-H$_2$O (50 + 19 + 31) lower phase + 20% ethyl acetate*	[43]	1	15	g	A8 ... A2 A3 A1 ... A6 A5 A7 A4,9
5	CCl$_4$-CH$_3$COOH-H$_2$O (50 + 19 + 31) lower phase + 10% ethyl acetate*	[43]	1	15	g	A8 A3 A2 A1 ... A5,7 A6 A4 A9
6	Benzene-acetic acid-H$_2$O (50 + 19 + 31) upper phase**	[39, 77]	1	15	g	A8 A3 A1 A2 ... A6 A5 A4,7,9
7	CHCl$_3$-ethyl acetate-CH$_3$COOH (50 + 40 + 10)	[112, 115]	3		a	DA A8 A1,3,GE A6 A5 A4,7 A9 AG Gl
8	Benzene-butanol-CH$_3$COOH (80 + 15 + 5)	[43]	1	15	a	A8 A2 A3 A1 ... A5 A6 A7 A4 A9
9	Diisopropyl ether-CH$_3$COOH (95 + 5)	[39, 51, 53, 77, 83]	0,6	15	a	A2,8 A1,3 A6 A5 A4,7 AG' A9,Gl
10	Benzene-CH$_3$COOH-H$_2$O (50 + 19 + 31) upper phase**	[15, 53, 77, 115]	1	15	a	A1,2,3,8 A6 A5 A7 A4 A9
11	Propanol-5N NH$_4$OH (85 + 17)	[113, 115]	10		a, basic	GE DA A8 A1,3 A6,AG A4 A7 A5 Gl A9
12	Butanol-4.5N NH$_4$OH (75 + 25)	[15]		10	a, basic	GE A8 A6 A1,3 A2 A5 A4,7 A9
13	Isopropanol-water (75 + 25)	[15]		10	a, basic	GE A8 A2 A1,3,5,6 A4,7 A9
14	Water, after previous run in ethyl acetate	[15, 39]		14	a, neutral	A9 A5 A2,4,7 A6 A1,3 A8, GE
15	Phosphat buffer 0,1 M, pH 6,3; layer previously chromatographed in 7% capryl alcohol in petroleum ether	[15]		10	i, neutral	A4,7,9 ... A5 A2 A6 A1,3 GE A8
16	Diisopropyl ether-CH$_3$COOH (98 + 2)	[77, 115]	0,6	15	a, ester-separation	E2,8 E3 E1 E6 E5 E7 E4 E9
17	Ethyl ether-Benzene (80 + 20)	[43]	1	15	a, ester-separation	E8 E2 E1 E3 E5 E6 E7 E4 E9
18	Benzene-CH$_3$COOH-water (50 + 19 + 31) upper phase**	[15, 77, 115]	1	15	a, ester-separation	E8 E3 E1 E2 E4,5,6,7,9

Solvent Variations for Special Purposes***

No.	Solvent	References	Time of Run [h]	Run [cm]	Layer	hRf
19	Benzene-CH$_3$COOH (70 + 30)	[115]	3		a	
20	CHCl$_3$-ethyl acetate-CH$_3$COOH (90 + 10 + 5)	[111, 115]	3		i	(8)
21	CHCl$_3$-ethyl acetate-CH$_3$COOH (70 + 30 + 5)	[111, 113]	2		i	(8)
22	CHCl$_3$-ethyl acetate-CH$_3$COOH (80 + 20 + 5)	[111]	2		i, acid solvents	(8)
23	Benzene-CH$_3$COOH (77 + 23)	[111]	3		a	(8)
24	CHCl$_3$-ethyl acetate-CH$_3$COOH (80 + 20 + 1)	[111]	18		d' i	(10)
25	CCl$_4$-CH$_3$COOH-H$_2$O (50 + 19 + 31) lower phase without ethyl acetate*	[43]	1	15	g	(10)
26	CCl$_4$-CH$_3$COOH-H$_2$O (50 + 19 + 31) lower phase + 10% ethyl acetate*	[43]	1	15	a	(10)
27	Propanol-3N NH$_4$OH (85 + 17)	[111]	7		i, basic	(11)
28	Isopropanol-4.5N NH$_4$OH (75 + 25)	[15]		10	a, basic	(12)
29	Butanol-3N NH$_4$OH (85 + 17)	[111]	7		i	(12)
30	Ethyl ether-petrol ether (80 + 20), double run	[43]	1	15	a, ester	(17)

Abbreviations:
A1—9 = Gibberellin A1—9
E1—9 = Gibberellin-A1—9, methyl esters
GI = Gibberic acid
GE = Gibberellenic acid
AG = Allogibberic acid
DA = Dicarboxylic acid A
* plates equilibrated before the run by about 16 h in the upper phase
** plates equilibrated before the run by about 16 h in the lower phase
*** the separation pattern is similar to that in the solvent denoted by the number in brackets in the last column (number refers to the above series 1—18)
d = continuous development
a = activated silica gel
i = inactive silica gel
g = kieselguhr
▼ = position of indole-3-acetic acid

It has not been possible to separate all the known gibberellins in a single operation, using one of the solvents suggested. Combination of solvents 10 and 3 (Table 92) should give good results in two-dimensional and perhaps also in stepwise development technique (cf. p. 87).

SCHNEIDER et al. [105] have used thin-layer electrophoresis on silica gel G with a field of 3.8—5 V/cm for further characterisation of various gibberellins. The separation patterns after thin-layer electrophoresis and after TLC in water (Table 92, 14) are similar. Determination of the "free electrophoretic mobility" of a substance permits conclusions to be drawn about the number of carboxyl groups it contains; this constitutes the special value of the method.

IV. Multiple Development, Stepwise Development and Two-Dimensional Separation
(cf. also p. 86)

Highly volatile solvents are particularly suitable for these procedures since the layer has to be dried carefully after each run (test by smell!). If the substances to be separated are easily oxidised or if higher-boiling solvents are used, the plates should be dried in an evacuated desiccator, if necessary in a nitrogen atmosphere.

Compared with two-dimensional development, multiple and stepwise development have the advantage of allowing analysis of several fractions on the same layer, even after band application.

Indole derivatives. Better separations than in a single run have been obtained with: hydroxyindoles using solvents C and D (Table 90) [14]; various indole derivatives and their methyl esters in the solvents V and VI (Table 88) [19, 20], using double development; and hydroxyskatoles through four runs in chloroform-cyclohexane-diethylamine (50 + 40 + 10) [33].

Generally, tailing is less noticeable in multiple development. Further, better resolution is accomplished of substances which remain near the start after a single run. In contrast, substances near the front move together. This latter disadvantage can be largely eliminated by first developing once or several times using a short length of run and then finally over the full distance (Fig. 158, cf. data for solvent XIX, Table 89). The method may be varied by changing the solvent composition for the different runs. Using the solvent combinations XXII and XXIII (Table 89), the first run separates chiefly the acids; the second, non-acidic compounds. This variation is a stage of transition to stepwise development technique (cf. p. 87) in which quite different solvents are used for the successive runs (X and XI, Table 88).

In contrast to the techniques so far described, two-dimensional separation (cf. p. 88) permits use of the complete plate square for the components of a fraction, assuming appropriate solvent combinations are chosen. In the analysis of a naturally occurring mixture of compounds it supplements usefully one-dimensional chromatography in different solvents. Comparison of the spot pattern found here with a two-dimensional chromatogram of as many as possible "suspected" authentic

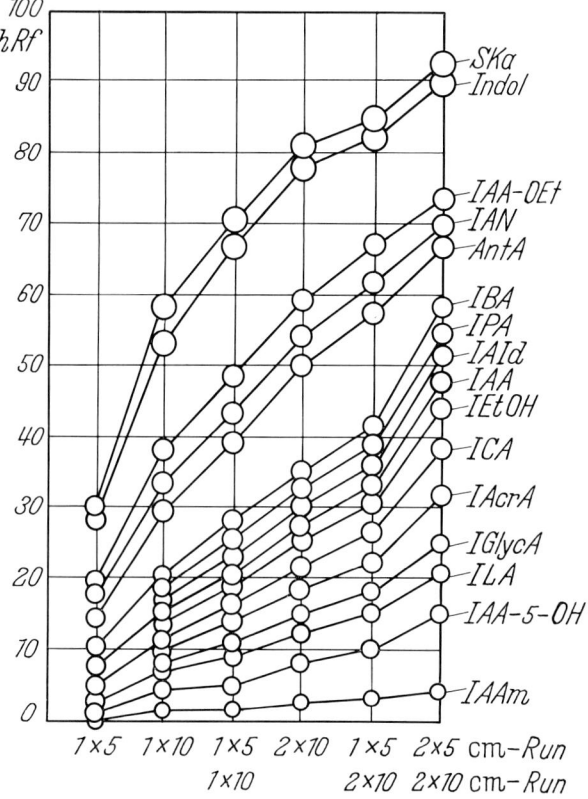

Fig. 158. Migration of various indole derivatives (0.2—0.3 µg of each) during multiple development in hexane-butanone-2 (65 + 35) on silica gel HR_{254} (Firm 83), prepared with M/30 phosphate buffer, pH 5.2; 3—5 min time of run for 5 cm, 12—15 cm for 10 cm. Visualisation with van Urk reagent (No. 73). Spot sizes copied from the original

substances, often renders possible a classification of the separated components. One must guard against "chromatographically related" derivatives (e. g., non-polar compounds, acids, acids with an amino- or hydroxyl-group, amines, amides) falling together as groups in the solvent mixture chosen. Sometimes data concerning at least the compound group to which a substance belongs, can be thus obtained. Care must be taken that the substances to be separated are not decomposed by the solvent, particularly in the first run. This can be tested by two-dimensional separation in the solvent in question, provided enough material is available. If no decomposition takes place, all the components lie on a diagonal; any decomposition products very probably fall outside the diagonal (= SRS-technique, p. 88; cf. [8]).

The following solvent combinations, in order of use, have been so far proposed (Roman numerals refer to Tables 88/89, capital letters to Table 90): XIV → XVIa [120] and V → IV [19] for separating simple indole derivatives, particularly from auxin metabolism (the second combination is suitable for the methyl esters of the acids also); XVII → XIV [12], XV → XVIa [9]; isopropanol-25% ammonium hydroxide-water (80 + 4 + 8) → IX [90] for separation of urine metabolites; A → B for hydroxyskatoles; and C → D for hydroxyindoles.

Gibberellins. No experience using stepwise technique or two-dimensional development and only that from occasionally used multiple development (solvent 30, Table 92) is available (cf. however, the hint on p. 481).

V. Visualisation and Identification
1. Chemical and Physical Methods of Detection

General Information. Clues for characterising the separated compounds are given by Rf-values, fluorescence quenching, fluorescence in short- and long-wave UV light, colour with various reagents and the behaviour of reaction products in UV light. These criteria alone do not, however, usually suffice for certain identification, since they are subject to marked fluctuations through uncontrollable factors.

The procedure described below has proved useful for the analysis of naturally occurring mixtures on UV_{254}-fluorescent layers: The sample is applied as a band, 2—4 cm long; about 0.1—0.5 µg of genuine reference substances (the compounds probably expected) in methanolic solution (about 0.1%) which can be stored in a refrigerator for weeks if necessary, are applied as spots away from the centre of the band but not right at the end. After chromatography, the areas of quenched fluorescence (in UV_{254}) of the references substances are marked round with a sharp needle and any zones from the sample showing quenching are marked with square brackets outside the separation path. Dotted lines are used to mark zones of fluorescence in UV_{366} light; this sometimes occurs or is intensified only after activation with UV_{254} light (Fig. 159). Since many compounds are sensitive to UV light, the layers should be exposed to it for only a minute or two [47]. After this marking has been carried out, the zones observed in the half of the migration path containing no chromatographed reference substances, are scraped off and submitted to further chemical or biological tests (see below). The other half remaining on the plate is then sprayed with a suitable reagent and the colour and any fluorescence of the reaction products in short and long wave UV light is observed (cf. Table 88). One must bear in mind the fact that colour and fluorescence often do not immediately appear and that the colours can change with time (sprayed layers should be covered with glass plates after drying).

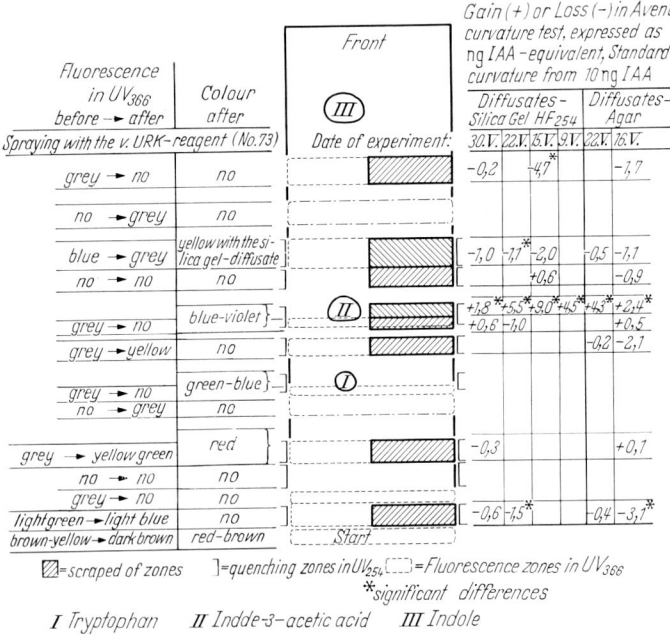

Fig. 159. TLC-separation of the diffusates from sections of fruit axes from *Fritillaria meleagris* L. and treatment of the chromatogram in chemical (left) and biological (right) characterisation of the separated substances. Layer: silica gel HF_{251}; solvent: isopropanol-25% ammonium hydroxide-water (85 + 5 + 15); 60 mm run in S-chamber, saturated with isopropanol (cf p. 476) (according to [45])

When a coloured spot of the authentic reference substances lies in an identically coloured band of the naturally occurring mixture, the identity of the substances is suspected. If this is repeated in various solvents of widely differing separation properties, the identity is probable. In any case, reference substances and mixture of naturally occurring substances should be chromatographed together (co-chromatography) as shown, since colour and Rf-value can be altered considerably by contaminants.

The procedure described is satisfactory for one-dimensional separations, also with multiple development or when stepwise technique is used (see p. 87). In two-dimensional separation, suitable bands may be applied outside the separation square itself. Further, reference substances which come into consideration after the first run has been completed, can be applied directly alongside the spots observed by quenching in UV light of 254 nm or by fluorescence in 366 nm light; they are then chromatographed with the sample in the second run.

If there is enough substance available, the eluates of the scraped-off zones can be qualitatively and quantitatively examined by the usual spectrophotometric, colorimetric and fluorometric methods.

SEILER et al. [110] describe a direct quantitative determination of anthranilic acid and some indole derivatives, without elution. Layers on 5×20 cm plates were chromatographed using solvent XIV, Table 89, sprayed with the Procházka reagent (No. 123, Table 88), illuminated with UV light of wave-length 365 nm and the fluorescence measured with a slightly modified Zeiss-Spektralphotometer PMQ II (Firm 155). The surface areas under the lines registered with a recorder were found to be directly proportional to the amount of material between 10 and 2000 ng. A TLC-attachment for the Zeiss Spektralphotometer has been developed with which direct routine qualitative and quantitative measurements on normal TLC-plates have become possible [41, 42].

WAKHLOO [133] has obtained information about tryptophan, separated by TLC from *Solanum nigrum;* he chromatographed the plant extract under investigation and 3—4 known amounts of tryptophan on the same layer and made densitometric measurements on a contact diapositive of the TLC-layer.

Auxins and Urine Metabolites. Table 88 contains the most generally used reagents for detecting auxins and urine metabolites, their quenching effect on UV_{254}-fluorescent layers and their fluorescence in UV_{366}, together with data on the lower limits of detection. For the best colour formation, the experimental conditions given for use of the reagents, should be adhered to. Even then, there are considerable fluctuations of colour, depending on the concentration of the compounds, on the adsorbent and layer thickness, on whether observed in transmitted or reflected light and, above all, on the influence of contaminants. Delicate colour nuances have consequently not been quoted in Table 88.

The van Urk-reagent (No. 73) yields relatively well differentiated colours with simple indole derivatives and sensitivity is quite high (cf. [13, 97] concerning the chemistry of the reactions).

4-Dimethylaminocinnamaldehyde (Rgt. No. 76) reacts in particular with a series of indole auxin metabolites more than twice as sensitively and to some extent with characteristically varied colours. Ammonia must be entirely removed from the layer before this reagent is used, otherwise the whole layer turns red-brown. Neither of these two reagents gives fluorescent derivatives except in a few cases (indole-3-acrylic acid, glucobrassicin). In contrast, the Procházka reagent (No. 123) is characterised by the powerful fluorescence of the reaction products in UV_{366}, to some extent in UV_{254} also; sensitivity is often ten or more times higher. The visible colours are feeble and little differentiated.

Both the Salkowski reagents (Nos. 104 and 108) yield reaction products with differentiating colours; many of the products fluoresce in long and short wave UV light. The colours obtained with the two reagents differ amongst themselves in some cases but the fluorescence colours are the same. It may be pointed out that some compounds which display the same fluorescence colour in UV_{366}, show differing colours in UV_{254} and vice versa. The lower limit of detection in the colour reaction corresponds approximately to that with the van Urk reagent; it is much lower in fluorescence detection.

The ninhydrin reagent (No. 178), in customary use for detection of amino acids, is usually suitable for visualising urine metabolites, even of non-indole nature. The colours are not very specific. Most indole derivatives have a high limit of detection. After keeping for some time (one to several days), amounts of reaction product down to 10 ng fluoresce in UV_{366} without characteristic colour.

In addition to the reagents for detection which are quoted in Table 88, many others have been mentioned in the literature: [89, 90, 104] for urine metabolites; [78] for serotonine metabolites; [14] for hydroxyindoles and -tryptophans; [35] with comprehensive data for hydroxyskatoles; [19] for the methyl esters of indolecarboxylic acids; [22, 23, 63] for glucobrassicin. 2,4-Dinitrophenylhydrazine (Rgt. No. 82) is suitable for the often feebly reactive indole-aldehydes.

None of these reagents is, however, so specific that the appearance of a colour testifies the certain presence of an indole derivative.

Gibberellins. A most sensitive test for gibberellins and their methyl esters is spraying with 70—95% or ethanolic sulphuric acid (Rgt. No. 241),

Table 93. *Fluorescence colours of various gibberellins in UV_{366}, on silica gel G or kieselguhr*

Substance	70—75% H_2SO_4 (Rgt. No. 241); plates heated 10 min at 120° after spraying [77, 113, 115]	Conc. H_2SO_4-ethanol (5 + 95) (Rgt. No. 241); plates heated to 120° after spraying [39, 77]	
	Fluorescence colour	Duration of heating (min)	Fluorescence colour
Gibberellin A_1	green-blue	30—40	blue
A_2	purple-blue	4—8	purple
A_3	turquoise blue; yellow green without heating	1—3	green-blue, turning blue
A_4	purple-blue	4—8	purple
A_5	green-grey	10—20	blue
A_6	green-blue	10—20	blue
A_7	blue-grey (yellow-brown edge); yellow-green without heating	1—2	bright yellow, turning pale yellow
A_8	green-blue	10—20	blue
A_9	purple-blue	4—8	purple
Gibberic acid	yellow brown	1—3	green-blue, turning blue
Gibberellenic acid		1—3	
Acetylgibberellinic and diacetyl-gibberellinic acids			green-blue

followed by 10 min heating at 110–120° C (cf. [119a] for the chemistry of the reaction); the product fluoresces in UV_{366} and the lower limit of detection has been quoted as 0.3 ng for GA_3 to 10 ng for GA_6 [39]. The methyl esters fluoresce more weakly. Semiquantitative determination is possible when known amounts of reference substances are subjected to parallel chromatography [16, 83]. The fluorescence spectra are in part characteristic and can be utilised for identification purposes [15, 39]. Further distinction is possible through graded periods of heating [39] (cf. Table 93).

A mixture of saturated aqueous ceric sulphate solution and concentrated sulphuric acid $(1 + 1)$ is a sensitive reagent for detecting and partly differentiating (fluorescence) the gibberellins $A_{3, 4, 7 \text{ and } 9}$ [111]. Gibberellins $A_{1 \text{ to } 9}$, gibberic and allogibberic acids yield yellow-brown reaction products with 0.5% potassium permanganate solution; the lower limit of detection is between 2 and 5 µg [15, 64, 77, 111]. With 20% antimony(III) chloride solution in chloroform, followed by heating 10 min at 120° C, gibberellins A_3 and A_7 give a purple-black colour; A_4 and A_8, orange-brown; A_1, A_5, A_6 and A_9, pale orange; and A_2 no colour. 1 µg of A_3 can be detected [43].

2. Biological Methods of Detection

Auxins. Growth reactions of various plant organs, principally from *Avena* or *Triticum* coleoptiles (oat or wheat coleoptiles) are used for bioassay of the auxin activity of a substance.

In the so-called *Avena* curvature test, the angle of curvature caused when an auxin-containing agar block is attached to one side of the decapitated coleoptile, is taken as a measure of the auxin concentration in the agar. In the straight growth or section test, the elongation of the organ cylinders (sections) placed in the aqueous solution to be investigated, is measured (see [66, 67, 72, 119] etc. for further details). Straight growth tests are suitable also for detecting inhibitors (see [70, 80] etc.). After growth substances have been added to the agar blocks, antagonists can be recognised also from the curvature test (cf. Fig. 159). The tests are highly sensitive; a single oat coleoptile responds for example to 0.5 ng of indole-3-acetic acid in a 5 mm^3 agar block with a clearly measurable curvature.

The following procedure has proved satisfactory in the bioassay of substances separated by TLC:

After chromatography, the zones in the part of the migration path free from reference substances (cf. p. 483 and Fig. 159), are located in UV light and marked round with a needle, using the marking template. The silica gel of the zone to be tested is then scraped off into a small heap, using a metal or plastic spatula, 3–4 mm broad. The plate is then held vertically over a sheet of celluloid (35 mm film strip) and the powdered silica gel is brushed on to the sheet (Fig. 160A). Better still, the silica gel can be scraped off with a piece of razor blade, fixed in a wooden holder; the powder collects on the blade and transfer to the celluloid is not necessary.

For the curvature test, the powder is spread as uniformly as possible over a set of 6—12 agar blocks (B). This is best done by adding 1—2 drops of water to form a suspension and then spreading with a narrow spatula in a back and forth movement over the agar surface (C). Care must be taken that the suspension is not carried over the edge of the blocks. After leaving 30 min in a humid atmosphere in the dark, the separate agar blocks with the adhering silica gel can be used for the test [44, 45]. Indole-3-acetic acid can be recovered practically without loss by using this procedure [47].

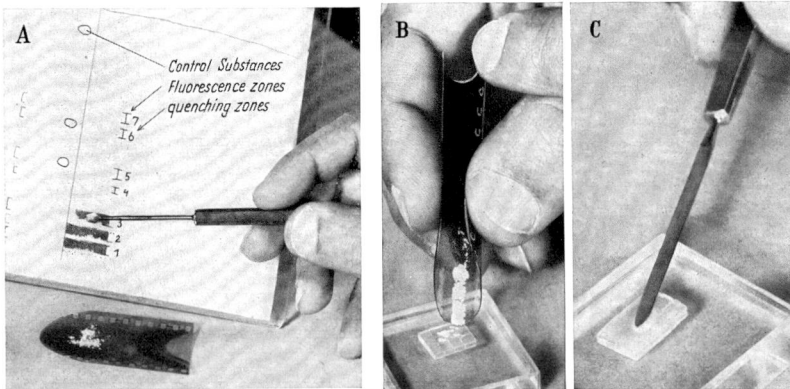

Fig. 160. Treatment of a thin-layer chromatogram in biological detection using the *Avena*-curvature test. *A*. Transfer of the silica gel from the layer to a celluloid sheet. *B*. Bringing the silica gel powder on to the 12 agar blocks. *C*. Spreading the powder evenly after having added a drop of water [45]

For the straight growth test, the scraped-off silica gel is added directly to the test solution. Alcoholic eluates of the silica gel are occasionally used, to prevent growth inhibition [134]; no such interference by the presence of silica gel has been however noted in our laboratories (cf. also [7, 8]). The procedure mostly used is that with test sections of a definite length (3—5 mm); these are floated in the solution which may contain sugar and/or buffer and their elongation is measured after 16—20 h. Inactive auxin precursors which can be activated by the test sections, perhaps under the influence of epiphytic bacteria [71], are indistinguishable here from active auxins and are also assayed. Richer information is yielded by the straight growth test proposed here: in this, the elongation of a coleoptile cylinder, standing inversely (upside down) in the test solution, is recorded photographically at brief time intervals [44, 45].

Gibberellins. Straight growth tests with hypocotyls and dwarf mutants have been adopted along with various germination tests for biological detection of gibberellin activity [4, 55, 57, 91, 94]. The *Hordeum* (barley) endosperm test, in which the influence of the gibberellins on the amylase activity is measured, is rapid and sensitive [24a, 82a, 91a]. A new chlorophyll assay appears promising [15a]. Since the response to the individual gibberellins varies in the different test objects, the activity spectrum gained from several biotests permits more precise characterisation of the compound to be identified [3a, 4, 24a, 51, 91a, 112, 116, 133a].

The substances separated by TLC are transferred to the substrate for the biotest after elution from the scraped-off silica gel using methanol, acetone or ethyl acetate (saturated with water), evaporation of this eluate to dryness, and taking the residue up in water, if necessary after addition of 0.05% Tween 20 (Firm 119) [39, 51, 112, 115].

Reference substances should be co-chromatographed so as to enable the zones to be located as described above for the biological auxin test. The gibberellins, in the amounts usually found, give no quenching zones in UV_{254} on layers containing fluorescent indicators. If one wishes neither to scrape off "blindly" (i. e., successive zones of definite length, usually one-tenth of the length of run) nor to make use of fluorescing or quenching contaminants in the extract as clues, the part of the chromatogram containing the reference substances must be sprayed with reagent before the zones are scraped off. The other part of the layer containing the substances for the biotest, must be protected from the reagent; this is done by pressing a glass plate firmly on to it and removing a narrow strip of adsorbent from between the two parts of the layer. If required, the sprayed part of the layer alone is heated by placing the plate only halfway on to a heater.

The separation layer can be divided into two halves when purchased "single use plates" or coated sheets are used. The suggestion has been made also of chromatographing dyes with the same Rf-values as particular gibberellins, so as to facilitate location [100].

With the help of TLC it has been possible to detect, in addition to gibberellins already known, numerous substances with gibberellin activity and even gibberellin precursors [93] as genuinely present in various plant genera (in *Bryophyllum* [118a]; *Citrus* [52]; *Echinocystis* [15]; *Fusarium* [16, 29, 39a, 83]; various Gramineae [38, 121]; *Hordeum* [39, 93, 98a]; *Malus* [11a]; *Nicotiana* [115, 118]; *Phaseolus* [38, 113]; *Pisum* [51, 109a]; *Rudbeckia* [97a]; *Solanum tuberosum* [31, 31a]; *Trifolium* [122]; *Zea mays* [38]).

I wish to thank cordially Dr. G. SEMBDNER of the Institut für Kulturpflanzenforschung, Gatersleben, for looking through my manuscript and for valuable hints.

Bibliography for Chapter 0. "Simple" Indole Derivatives

1. BALLIN, G.: J. Chromatog. **16**, 152 (1964).
2. BENASSI, C. A., F. M. VERONESE, and E. GINI: J. Chromatog. **14**, 517 (1964).
3. BENTLEY, J. A.: In: Handbuch Pflanzenphysiol. Herausg.: W. RUHLAND, Bd. **14**, S. 501. Berlin-Göttingen-Heidelberg: Springer 1961.
3a. BENTLEY-MOWATT, J. A.: Ann. Botany (London) N. S. **30**, 165 (1966).
4. BRIAN, P. W., H. G. HEMMING, and D. LOWE: Ann. Botany (London) NS. **28**, 369 (1964).

5. Burnett, D., L. J. Audus, and H. D. Zinsmeister: Phytochemistry **4**, 891 (1965.)
6. Butenandt, A.: Über die Analyse der Erbfaktorenwirkung und ihre Bedeutung für biochemische Fragestellungen. Schriftenreihe Arbeitsgem. f. Forsch. d. Landes Nordrhein-Westfalen, H. **62**. Köln: Westdeutscher Verlag 1960.
7. Collet, G.: Compt. Rend. **259**, 871 (1964).
8. — J. Dubouchet et P. E. Pilet: Pysiol. Vég. **2**, 157 (1964).
9. Cotte, I., M. Chetaille, F. Poulet et I. Christiansen: J. Chromatog. **19**, 312 (1965).
10. Crane, J. C.: Ann. Rev. Plant Physiol. **15**, 303 (1964).
11. Das, V. S. R., I. V. S. Rao, and K. U. K. Murthy: Current Sci. (India) **34**, 94 (1965).
11a. Dennis, F. G., and J. P. Nitsch: Nature (London) **211**, 781 (1966).
12. Diamantstein, T., u. H. Ehrhart: Hoppe-Seylers Z. Physiol. Chem. **326**, 131 (1961)
13. Dibbern, H. W., u. H. Rochelmeyer: Arzneimittel-Forsch. **13**, 7 (1963).
14. Eich, E., u. H. Rochelmeyer: Pharm. Acta Helv. **41**, 109 (1966).
15. Elson, G. W., D. F. Jones, J. MacMillan, and P. J. Suter: Phytochemistry **3**, 93 (1964).
15a. Fletcher, R. A., and D. J. Osborne: Nature (London) **211**, 743 (1966).
16. Focke, J., G. Sembdner u. K. Schreiber: Biol. Zentr. **84**, 309 (1965).
17. Gautheret, R. J.: Rev. Cytol. Biol. Végetales **27**, 99 (1964).
18. Glombitzka, K.-W.: J. Chromatog. **19**, 320 (1965).
19. — J. Chromatog. (in the press).
20. — u. T. Hartmann: Planta **69,** 135 (1966).
21. Gmelin, R.: 5th Int. conf. on plant growth subst. Paris 1963. Coll. Intern. CNRS Nr. **123**, 159 (1964).
22. — u. A. I. Virtanen: Ann. Acad. Sci. Fennicae Ser. A. **107**, 25 S. (1961).
23. — — Acta Chem. Scand. **16**, 1378 (1962).
24. Gordon, S. A., and R. P. Weber: Plant Physiol. **26**, 192 (1951).
24a. Griffith, C. M., J. C. MacWilliam, and T. Reynolds: Nature (London) **202**, 1026 (1964).
25. Handbuch der Pflanzenphysiologie (Editor W. Ruhland). Bd. XIV. Wachstum und Wuchsstoffe Redig. v. H. Burgström. Berlin-Göttingen-Heidelberg: Springer 1961.
26. Handbuch der Pflanzenphysiologie (Editor W. Ruhland). Bd. XV. Differenzierung und Entwicklung. Redig. v. A. Lang. Berlin-Göttingen-Heidelberg: Springer 1965.
27. Handbuch der Pflanzenphysiologie (Editor W. Ruhland). Bd. XVIII. Physiologie der Bewegungen. Redig. v. E. Bünning, Berlin-Göttingen-Heidelberg: Springer 1959 u. 1962.
28. Hansen, I. L., and M. A. Crawford: J. Chromatog. **22**, 330 (1966).
29. Harada, H., and A. Lang: Plant Physiol. **40**, 176 (1965).
30. Harley-Mason. J., and A. A. P. G. Archer: Biochem. J. **69**, 60p. (1958).
31. Hayashi, F., and L. Rappaport: Nature **205**, 414 (1965).
32. Heacock, R. A., and M. E. Mahon: Can. J. Biochem. Physiol. **41,** 2381 (1963).
33. — — Can. J. Biochem. Physiol. **41**, 487 (1963).
34. — — Can. J. Biochem. Physiol. **42**, 813 (1964).
35. — — J. Chromatog. **17**, 338 (1965).

35a. IKEKAWA, N., T. KAGAWA, and Y. SUMIKI: Proc. Japan. Acad. **39**, 507 (1963).
36. INHOFFEN, H. H., K.-H. NORDSIEK u. H. SCHÄFER: Liebigs Ann. Chem. **668**, 104 (1963).
37. JÄNCHEN, D.: J. Chromatog. **14**, 261 (1964).
38. JONES, D. F.: Nature **202**, 1309 (1964).
39. — J. MACMILLAN, and M. RADLEY: Phytochemistry **2**, 307 (1963).
39a. JONES, K. C.: Diss. Univ. California, Los Angeles 1965.
40. JONES, R. L., and J. D. J. PHILLIPS: Nature **204**, 497 (1964).
41. JORK, H.: 3. Symp. Chromatogr. Soc. Belge Sci. Pharm., S. 295. Brussels 1964.
42. — Z. Anal. Chem. **221**, 17 (1966); **236**, 310 (1968).
43. KAGAWA. T,, T. FUKINBARA, and Y. SUMIKI: Agr. Biol. Chem. **27**, 598 (1963).
44. KALDEWEY, H.: Habilitation, Saarbrücken 1961.
45. — 5th Internat. Conf. on plant growth subst., Paris 1963. Coll. Internat. CNRS, **Nr. 123,** 421 (1964).
46. — unpublished experiments.
47. — u. E. STAHL: Planta **62**, 22 (1964).
48. — — u. J. FUCHS: in preparation.
49. KEFFORD, N. P.: Science **142**, 1495 (1963).
50. — and P. L. GOLDACRE: Am. J. Botany **48**, 643 (1961).
51. KENDE, H., and A. LANG: Plant Physiol. **39**, 435 (1964).
52. KHALIFAH, R. A., L. N. LEWIS, and C. W. COGGINS JR: Plant Physiol. **40**, 441 (1965).
53. — — and P. C. RADLICK: J. Exp. Botany **16**, 511 (1965).
54. KNAPP, R. (edit.): Eigenschaften und Wirkungen der Gibberelline. Berlin-Göttingen-Heidelberg: Springer 1962.
55. — In: Moderne Methoden der Pflanzenanalyse. Edit. H. F. LINSKENS u. M. V. TRACEY, Bd. 6, S. 203. Berlin-Göttingen-Heidelberg: Springer 1963.
56. KÖGL, F., A. J. HAGEN-SMIT u. H. ERXLEBEN: Hoppe Seylers Z. Physiol. Chem. **228**, 90 (1934).
57. KÖHLER, D., and A. LANG: Plant Physiol. **38**, 555 (1963).
58. KÖHLER, K. H., u. K. CONRAD: Biol. Rdsch. **4**, 36 (1966).
59. KOLLER, D., A. M. MAYER, A. POLJAKOFF-MAYBER, and S. KLEIN: Ann. Rev. Plant Physiol. **13**, 437 (1962).
60. KOST, A. N., T. V. KORONELLY i R. S. SAGITULLIN: Akad. Nauk. USSR **19**, 125 (1964). (Russian with English summary)
61. KUTÁČEK, M.: Biol. Plant. Acad. Sci. Bohemoslov. **6**, 88 (1964).
62. — private communication.
63. — et Ž. PROCHÁZKA: 5th Internat. congr. on plant growth subst., Paris 1963. Coll. Internat. CNRS Nr. **123**, 445 (1964).
64. — J. ROSMUS, and Z. DEYL: Biol. Plant. Acad. Sci. Bohemoslov. **4**, 226 (1962).
65. LANG, A.: In: Physiology of reproduction Proc. 22, Biol. Coll. Oregon State Univ. 1961, S. 53.
66. LARSEN, P.: In: Moderne Methoden der Pflanzenanalyse. Ed. K. PAECH u. M. V. TRACEY. Bd. **3**, S. 565. Berlin-Göttingen-Heidelberg: Springer 1955.
67. — In: Handbuch der Pflanzenphysiologie (Editor W. RUHLAND), Bd. **14**, S. 521. Berlin-Göttingen-Heidelberg: Springer 1961.
68. LEGLER, G., u. R. TSCHESCHE: Naturwissenschaften **50**, 94 (1963).
68a. LETHAM, D. S.: Phytochemistry **5**, 269 (1966).

69. LETHAM, D. S., and C. O. MILLER: Plant Cell Physiol. **6**, 355 (1965).
70. LEWIS, L. N., R. A. KHALIFAH, and C. W. COGGINS JR.: Plant Physiol. **40**, 500 (1965).
71. LIBBERT, E., S. WICHNER, U. SCHIEWER, H. RISCH, and W. KAISER: Planta **68**, 327 (1966).
72. LINSER, H., u. O. KIERMAYER: Methoden zur Bestimmung pflanzlicher Wuchsstoffe. Vienna: Springer 1957.
73. LINSKENS, H. F.: Papierchromatographie in der Botanik. 2. Aufl. Berlin-Göttingen-Heidelberg: Springer 1959.
74. LOU, V., W. Y. KOO, and E. RAMSTAD: Lloydia **28**, 207 (1965).
75. MACMILLAN, J., J. C. SEATON, and P. J. SUTER: Tetrahedron **11**, 60 (1960).
76. — — Tetrahedron **18**, 349 (1962).
77. — and P. J. SUTER: Nature **197**, 790 (1963).
78. MCISAAC, W. M., G. FARELL, R. G. TABORSKY, and A. N. TAYLOR: Science **148**, 102 (1965).
79. MILLER, C. O.: In: Moderne Methoden der Pflanzenanalyse. Ed. H. F. LINSKENS u. M. V. TRACEY. Bd. **6**, S. 194. Berlin-Göttingen-Heidelberg: Springer 1963.
80. MITSUHASHI, M., and H. SHIBAOKA: Plant Cell Physiol. **6**, 87 (1965).
81. MOTHES, K.: Naturwissenschaften **47**, 337 (1960).
82. — 5th Internat. congr. on plant growth subst. Coll. Internat. CNRS Nr. **123**, S. 131 (1964).
82a. NICHOLLS, P. B., and L. G. PALEG: Nature (London) **199**, 823 (1963).
83. NINNEMANN, H., J. A. D. ZEEVAART, H. KENDE, and A. LANG: Planta **61**, 229 (1964).
84. NITSCH, J. P.: In: The chemistry and mode of action of plant growth substances. (Ed. R. L. WAIN, and F. WIGHTMAN) p. 3. London: Butterworth 1956.
85. — (Edit.): Regulateurs naturels de la croissance végétale. 5th Internat. conf. on plant growth subst. Paris 1963, Coll. Internat. CNRS Nr. **123**, Paris 1964.
86. NOVAT, N.: Compt. Rend. Soc. Biol. **158**, 1458 (1964).
87. OBREITER, J. B., and B. B. STOWE: J. Chromatogr. **16**, 226 (1964).
88. ONCKELEN, H. A. VAN, R. VERBEEK, and L. MASSART: Naturwissenschaften **52**, 561 (1965).
89. OPIÉNSKA-BLAUTH, J., M. CHAREZINSKI, and H. BERBÉĆ: Anal. Biochem. **6**, 69 (1963).
90. — H. KRACZKOWSKI, H. BRZUSZKIEWICZ, and Z. ZAGÓRSKI: J. Chromatog. **17**, 288 (1965).
90a. OVERBEEK, J. VAN: Science **152**, 721 (1966).
91. PALEG, L. G.: Ann. Rev. Plant Physiol. **16**, 291 (1965).
91a. — D. ASPINALL, and P. B. NICHOLLS: Plant Physiol. **39**, 286 (1964).
92. PERLEY, J. E., and B. B. STOWE: Plant Physiol. **41**, 234 (1966).
93. PETRIDIS, C., R. VERBEEK, and L. MASSART: Naturwissenschaften **53**, 331 (1966).
94. PHINNEY, B. O., and C. A. WEST: In: Handbuch der Pflanzenphysiologie (Editor: W. RUHLAND), Bd. **14**, S. 1185. Berlin-Göttingen-Heidelberg: Springer 1961.
95. PIIRONEN, E., and A. I. VIRTANEN: Acta Chem. Scand. **16**, 1286 (1962).
96. PILET, P. E.: Rev. Gén. Bot. **64, 1** (1957).
97. PÖHM, M.: Arch. Pharm. **286**, 509 (1953).
97a. PONTLEZIA, R. F.: Bull. Soc. Roy. Sci. Liège **34**, 49 (1965).

98. Procházka, Ž.: In: J. M. Hais u. K. Macek: Handbuch der Papierchromatographie. Jena: VEB Fischer 1958.
98a. Radley, M.: Nature (London) **210**, 969 (1966).
99. Ramstad, E., and S. Agurell: Ann. Rev. Plant Physiol. **15**, 143 (1964).
100. Reinhard, E., W. Konopka u. R. Sacher: J. Chromatogr. **16**, 99 (1964).
101. Sachs, R. M.: Ann. Rev. Plant Physiol. **16**, 73 (1965).
102. Schiewer, U., u. E. Libbert: Planta **66**, 377 (1965).
103. Schlossberger, H. G., H. Kuch u. I. Buhrow: Hoppe Seylers Z. Physiol. Chem. **333**, 152 (1963).
104. Schmid, E., L. Zicha, J. Krautheim u. J. Blumberg: Med. Exptl. **7**, 8 (1962).
105. Schneider, G., G. Sembdner u. K. Schreiber: J. Chromatogr. **19**, 358 (1965).
106. — — — Gibberelline — ihre Derivate und Abbauprodukte. Berlin: Akademie-Verlag 1966; see also Kulturpflanze **13**, 267 (1965).
107. Schraudolf, H.: Experientia **21**, 520 (1965).
108. — u. F. Bergmann: Planta **67**, 75 (1965).
109. Searle, N. E.: Ann. Rev. Plant Physiol. **16**, 97 (1965).
109a. Šebánek, J.: Flora, Abt. A. (Jena) **156**, 303 (1965).
110. Seiler, N., G. Werner u. M. Wiechmann: Naturwissenschaften **50**, 643 (1963).
111. Sembdner, G., R. Gross u. K. Schreiber: Experientia **18**, 584 (1962).
112. — G. Schneider u. K. Schreiber: Planta **66**, 65 (1965).
113. — — J. Weiland u. K. Schreiber: Experientia **20**, 89 (1964).
114. — — — — Kulturpflanze **13**, 137 (1965).
115. — u. K. Schreiber: Phytochemistry **4**, 49 (1965).
116. — — Flora, Abt. A (Jena) **156**, 359 (1965).
117. Sen, S. P., and A. C. Leopold: Physiol. Plantarum **7**, 98 (1954).
118. Sirois, J. C., and E. V. Parups: Physiol. Plantarum **18**, 70 (1965).
118a. Skene, K. G. M., and A. Lang: Plant Physiol. **39** (Suppl.) XXXVII (1964).
119. Söding, H.: Die Wuchsstofflehre. Stuttgart: G. Thieme 1952.
119a. Speake, R. N.: J. chem. Soc. (London) **1963**, 7 (1963).
120. Stahl, E., u. H. Kaldewey: Hoppe Seylers Z. Physiol. Chem. **323**, 182 (1961).
121. Stoddart, J. L.: Ann. Botany (London) N. S. **29**, 741 (1965).
122. — J. Exptl. Bot. **17**, 96 (1966).
123. Stowe, B. B.: In: Fortschritte Chemie organischer Naturstoffe. Bd. **17**, S. 249. Vienna: Springer 1959.
124. — and J. F. Schilke: 5th Internat. conf. on plant growth subst. Paris 1963, Coll. CNRS Nr. **123**, S. 409 (1964).
125. — and K. V. Thimann: Arch. Biochem. Biophys. **51**, 499 (1954).
126. — and T. Yamaki: Ann. Rev. Plant Physiol. **8**, 181 (1957).
127. Stowe, B. B., and T. Yamaki: Science **129**, 807 (1959).
128. Teuscher, E.: Flora (Jena) **155**, 80 (1964).
129. — Phytochemistry **4**, 341 (1965).
130. Teuscher, G., u. E. Teuscher: Phytochemistry **4**, 511 (1965).
131. Tyler, V. E., u. D. Gröger: Planta Med. **12**, 397 (1964).
132. Vegis, A.: Ann. Rev. Plant Physiol. **15**, 185 (1964).
133. Wakhloo, J. L.: Planta **65**, 301 (1965).
133a. Wittwer, S. H., and M. I. Bukovac: Amer. J. Botany **49**, 524 (1962).
134. Zenk, M. H.: 5th Internat. conf. on plant growth subst. Paris 1963. Coll. Internat. CNRS Nr. **123**, 241 (1964).
135. — u. H. Scherf: Biochim. Biophys. Acta **71**, 737 (1963).
136. — — Planta **62**, 350 (1964).

P. Amines and Tar Bases

Egon Stahl and P. J. Schorn

I. Amines

The chromatographic behaviour of amines is closely related to that of the alkaloids. Amongst others, Stein von Kamienski [43] has summarised the possibilities of paper chromatographic separation. In the same article he discussed also the isolation of amines from plant material and the preparation of characteristic derivatives for further identification.

Primary, secondary and tertiary amines are distinguished in the aliphatic series and are classified separately from the aromatic amines. The free amine bases, their salts or suitable derivatives may be chromatographed; the last-named procedure is particularly to be recommended.

1. Aliphatic Amines

Teichert et al. [45] were the first to carry out TLC of amines. The hydrochlorides were dissolved in 70% alcohol and applied in 1 to 10 μg amounts. The hR_f-values and experimental conditions for the separation are summarised in Table 94. As can be seen from the table, only a partial improvement in separation is achieved by buffering silica gel G layers with a mixture of 0.2M primary potassium phosphate and 0.2M secondary sodium phosphate (1 + 1) or with 0.15M sodium acetate solution.

Table 94. hR_f-values of amines on various layers, using different solvents [45]

Amine	Silica gel G-layer			Buffered silica gel G layer		Cellulose powder layer
	I[a]	II	III	IV	V	VI
Methylamine	—	11	10	10	7	12
Ethylamine	—	15	13	20	14	19
n-Propylamine	—	30	19	30	23	31
i-Amylamine	—	49	39	45	44	58
Cadaverine	3	6	2	1	1	—
Putrescine	3	4	2	1	1	—
Ethanolamine	30	18	11	10	10	7
Histamine	41	26	2	3	3	2
Tyramine	56	44	38	55	40	28
Phenylethylamine	66	56	37	55	42	—
Benzylamine	70	49	36	50	42	—
Tryptamine (s. p. 471)	60	54	43	90	45	—

[a] Solvents: I = 96% Ethanol-25% ammonium hydroxide (80 + 20); II = Phenol-water (80 + 30); III = Upper phase of butanol-acetic acid-water (40 + 10 + 50); IV = 70% Ethanol on phosphate-buffered layer; V = solvent III on acetate-buffered layer; VI = Upper phase of amyl alcohol-acetic acid-water (40 + 10 + 50).

Grasshof [8] quotes similar thin-layer chromatographic behaviour on magnesium silicate (Firm 153), using chloroform-methanol (50 + 50) or n-propanol-water-chloroform (66 + 22 + 11) for amines or the analogous aminoalcohols, respectively. On the other hand, the TLC-separation of aminoalcohols on neutral alumina (Firm 153) and of amines on silica gel (Firm 153) and on neutral alumina, using the same solvents, has proved unsatisfactory.

Primary and secondary amines are identified with certainty by preparing the 3,5-dinitrobenzamides (DNBs) and separating them on normal silica gel G layers with chloroform-ethanol (99 + 1). The hR_f-values found for the DNBs of the following amines are [14, 45]: methylamine 14, dimethylamine 47, ethylamine 68, n-propylamine 38, i-butylamine 42, i-amylamine 50.

Preparation of the 3,5-dinitrobenzamides (DNBs)

50 mg of the amine hydrochloride or 25 mg of free amine, dissolved in 5 ml water in a separating funnel, are treated with 15 ml ether, 0.25 ml pyridine and 250 mg 3,5-dinitrobenzoyl chloride, dissolved in 1 ml benzene. Potassium carbonate (5.5 g) is added with cooling and continuous stirring. After 20 min, the aqueous layer is discarded and the ether phase shaken with two 5 ml portions of 1% sulphuric acid and finally with water. The ethereal solution is then dried over sodium sulphate, filtered and evaporated to dryness. If necessary, the DNB product is crystallised from 50% ethanol. For TLC, 5—25 µg amounts are applied as a 1% solution in ether or chloroform.

Successful separations of the red 4'-nitroazobenz-(4)-amides[1] on silica gel G layers using chloroform, have been accomplished by Neurath and Doerk [20]. Multiple development and/or two-dimensional TLC with the same solvent, improved separation further. As with the benzamides mentioned above, the hR_f-values increased with increasing number of C atoms in the amine. Seiler and Wiechmann [36] separate primary and secondary amines after reaction with 1-dimethylamino-naphthalene-5-sulphonyl chloride (= DANS, Firm 60). They used the solvents: I. ethyl acetate-cyclohexane (60 + 40); II. benzene-methanol-cyclohexane (85 + 5 + 10) or III: benzene-triethylamine (83 + 17) in one- and/or two-dimensional procedures (Fig. 161).

The DANS amides show yellow-green fluorescence in long wave UV light (365 nm), whereas that of the amines, the phenol group of which reacted, is intensely yellow or yellow-orange; about 10^{-10} mole of amide can be thus detected.

A partition chromatographic procedure for TLC of homologous 2,4-dinitrophenylamines is also of interest. Schwartz et al. [39] prepare the layer as follows: 14 g Seasorb and 7 g Celite 545 are shaken with 8 ml polyethylene glycol 400 and 48 ml 95% ethanol to yield a homogeneous

---

[1] 4'-nitroazobenzoyl(4)chloride (Firm 88).

slurry which is then spread in the usual way. The development solvent is n-heptane, saturated with the polyethylene glycol serving as stationary phase. The dinitrophenyl-derivatives of the primary and secondary amines are separated on this layer according to the number of C atoms in the amine, e. g., methylamine hRf 5 up to decylamine hRf 95.

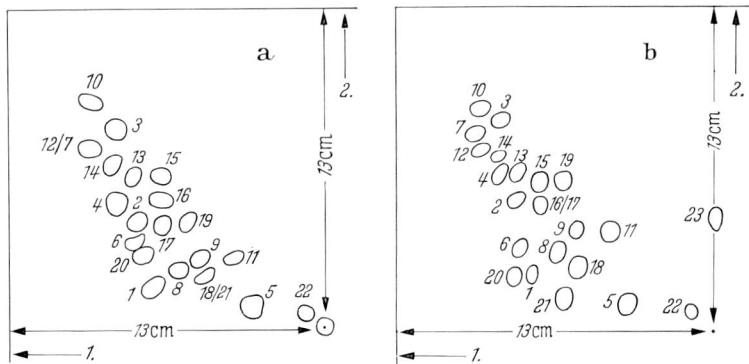

Fig. 161a and b. Two-dimensional thin-layer chromatogram of DANS-amides [36] a) Ethyl acetate-cyclohexane (60 + 40) (Run 1); benzene-methanol-cyclohexane (85 + 5 + 10) (Run 2); b) ethyl acetate-cyclohexane (60 + 40) (Run 1); benzene-triethylamine (83 + 17) (Run 2).

1 ammonia
2 methylamine
3 dimethylamine
4 ethylamine
5 ethanolamine
6 cysteamine
7 isoamylamine
8 spermidine
9 spermine
10 piperidine
11 histamine
12 β-phenylethylamine
13 4-hydroxy-β-phenylethylamine
14 4-methoxy-β-phenylethylamine
15 3,4-dihydroxy-β-phenylethylamine
16 3-methoxy-4-hydroxy-β-phenylethylamine
17 3,4-dimethoxy-β-phenylethylamine
18 noradrenaline
19 adrenaline
20 tryptamine
21 serotonine
22 pyridoxamine
23 hordenine

TLC on silica gel G has been used to test the stability of Tris (tris-(hydroxymethyl)aminomethane) buffers [49].

LANE [17] has separated long chain tertiary amines on alumina G layers, using isobutyl acetate or isobutyl acetate-acetic acid (98 + 2). The tertiary amines and also any primary or secondary amines present as impurities, are detected with cobalt(II) thiocyanate (Rgt. No. 53). The amines appear as blue spots on a pink background. The colours fade after about 2 hours but can be restored by spraying with water or placing the plate in a humid atmosphere. Long chain quaternary ammonium salts and amine oxides react similarly.

Detection. *Primary aliphatic amines* are generally detected with ninhydrin (Rgt. No. 178). Like secondary amines, they react giving a violet-red colour. Primary amines yield a characteristic blue colour with the Folin reagent (No. 172). Other tests which are used are with vanillin (Rgt. No. 261) and diazotisation (of aromatic amines), followed by coupling with suitable phenols [9].

Secondary amines react only feebly with ninhydrin. The nitroprusside-acetaldehyde reagent (No. 185) is preferable, giving a blue colour.

Tertiary amines yield an orange-red colour with the Dragendorff reagent (No. 98); the red-brown with iodine (Rgt. No. 144) and blue with molybdophosphoric acid (Rgt. No. 168A) are even less specific. They can be distinguished from primary and secondary amines on the TLC layer by treatment with hydrogen peroxide. The last two amine classes do not react and migrate near the solvent front; they thus separate from the reaction products of the tertiary amines, the strongly polar amine oxides $R_3N \to O$, which, on silica gel G layers, using chloroform saturated with ammonia-methanol (97 + 3), remain in the lower to middle Rf-range [25].

3,5-Dinitrobenzamides can be detected either on fluorescent layers (e. g., silica gel GF_{254}) or through the brown colour with iodine. They may be visualised also with α-naphthylamine (Rgt. No. 177). Another test is to expose for 15 min to the radiation of a mercury lamp without a filter, which yields violet spots on a white background.

The possibility is not excluded that diphenylpicrylhydrazyl [21], which has been used as a colorimetric reagent, may find useful application in the detection of amines in TLC.

2. Nitrosamines

Aliphatic and aromatic nitrosamines can be separated on silica gel G layers. PREUSSMANN et al. [19, 32, 33] use appropriately modified solvents for TLC of the various nitrosamine classes:

	Hexane-ether-CH_2Cl_2
Symmetrical dialkyl-nitrosamines	40 + 30 + 20
Methyl alkyl nitrosamines	40 + 30 + 20
Cyclic nitrosamines	25 + 35 + 40
Aryl-alkyl and diaryl nitrosamines	66 + 20 + 15

Detection. Violet spots are obtained with the diphenylamine-palladium chloride reagent (No. 84). The limit of detection is about 1.2 µg for volatile nitrosamines and 0.5 µg for solids. Detection with the sulphanilic acid-α-naphthylamine reagent (No. 239) is still more sensitive; aliphatic nitrosamines yield a red-violet colour and aromatic nitrosamines, green to blue (0.2—0.5 µg detected). YASUDA [50] has separated 20 different N-nitroso- and nitro-diphenylamines, using two-dimensional TLC; he utilised the results in the analysis of explosives (see p. 672, Fig. 188).

3. Aminoalcohols and Quaternary Ammonium Salts

The thin-layer separation of these physiologically important compounds is carried out with strongly polar acid solvents. BAYZER [1a] has separated the hydrochlorides of choline, chlorocholine and acetylcholine and tetramethylammonium chloride on cellulose layers with the help of the wedged tip technique; solvents were the "Partridge mixture" (Table 95, C) or chloroform-methanol-water (75 + 22 + 3). Choline hydrochloride was used as reference substance ($R_{St} = 1.00$). This group of compounds can be separated also in combination with thin-layer electrophoresis [1b]; a prior separation with TLC is first carried out on silica gel G layers, using methanol-acetone-hydrochloric acid (90 + 10 + 4)); this is followed by electrophoresis in the second dimension, using a pyridine acetate buffer, pH 3.6 (30—40 V/cm for 2 h). On cellulose layers an advantageous combination is electrophoresis using a pyridine acetate buffer, pH 6.5 (30—40 V/cm for 40 min) and TLC using the Partridge mixture (Table 95, C). TAYLOR [44] fractionates the same group of compounds on alumina layers.

Table 95. hRf[a]-values and colour reactions of choline derivatives on alumina G layers [44]

Hydrochloride of	Solvent			Iodoplatinate (Rgt. No. 147)		Dragendorff reagent (No. 98)	
	A	B	C				
Choline	41	50	49	Blue	1 µg	Red	1 µg
Acetylcholine	55	65	66	Blue	1 µg	Orange	1 µg
Carbamylcholine	31	57	46	Blue	1 µg	Orange	1 µg
Succinylcholine	32	39	35	Pink	1 µg	Orange	2 µg
Succinyldicholine	5	10	25	Blue-black	0.5 µg	Orange	0.5 µg
β-Methylcholine	52	59	56	Blue	1 µg	Red	1 µg
Bethanechol	46	61	63	Pink	1 µg	Orange	1 µg
Methacholine	59	70	67	Pink	1 µg	Orange	1 µg

Solvents: A: n-Butanol-water-formic acid (60 + 35 + 15) (upper layer); B: n-Butanol-water-acetic acid (66 + 17 + 17); C: n-Butanol-water-acetic acid (40 + 50 + 10) (upper phase).

[a] The hRf-values must be regarded only as guide values.

SULLIVAN and BRADY [37] have likewise used alumina G layers for separating betaine, choline and muscarine with the solvent methanol-carbon tetrachloride-acetic acid (70 + 30 + 2.5); the hRf-values were 57, 77 and 87 respectively. TYIHÁK [46], however, works on silica gel G layers, using methanol-water (50 + 50); hRf-values: 6 for choline and 45 for betaine.

Detection. The alkaloid reagents (Nos. 98, 147, 245, for example) have been those principally used for detecting aminoalcohols and quaternary ammonium salts. The dipicrylamine reagent (No. 90) is still more sensitive, detecting 0.5–1 µg of choline or its derivatives (as red spots on a yellow background).

4. Catechol Amines (Phenylalkylamines)

DESIMIO [5] has reported that biogenic amines can be separated on silica gel G layers, using the solvent mixture, commonly employed in PC, of n-butanol-acetic acid-water (66 + 17 + 17). 6-Hydroxycatechol amines and metabolites can be separated also with the same solvent or with n-butanol-25% ammonium hydroxide-ethyl acetate (60 + 20 + 20) [4a].

SEILER and WIECHMANN [35] separate mescaline and other phenylethylamines on silica gel G layers with isopropanol-chloroform-concentrated ammonium hydroxide (80 + 5 + 15). Through condensation with formaldehyde in the presence of concentrated ammonium hydroxide and subsequent heating with hydrochloric acid or with the Procházka reagent (No. 123), derivatives are obtained which show intense green fluorescence, adaptable to fluorometric quantitative determination or to further separation with ethyl acetate-methanol-formic acid (60 + 35 + 5 or 80 + 10 + 10) (limit of detection for mescaline = 0.01 µg).

SEGURA-CARDONA and SOEHRING [34] have made a detailed study of the TLC of the catechol amines and their derivatives. They separate these physiologically active compounds on polyamide layers (Firm 153), using the solvent mixture isobutanol-acetic acid-cyclohexane (80 + 7 + 10). These compounds may be detected through reactions of the phenolic $-OH$ or the $-NH_2$ group (see also Clinical diagnosis, Chapter S, pp. 578–612).

A good separation of catechol amines and their metabolites is possible also on cellulose MN 300 layers, using various solvents [31]; n-butanol, saturated with 3N hydrochloric acid, is suitable. The following hR_f-values are quoted for a 15 cm run:

Noradrenaline	31	Metanephrine	58
Adrenaline	38	3,4-Dihydroxymandelic acid	80
Normetanephrine	48	3-Methoxy-4-hydroxymandelic acid	89

BECKETT and CHOULIS [2], using cellulose, silica gel G and alumina G layers and the solvents n-butanol-acetic acid-water (40 + 10 + 50) or water-saturated n-butanol, separate histamine, noradrenaline, adrenaline, isoprenaline, ephedrine and β-phenylethylamine.

Note: In TLC on cellulose, the free amines or the corresponding salts in the presence of acids yield two substance spots. A single spot appears only when basic solvents are used. It has been found that this is due to the hydroxyl groups of the cellulose, since the effect vanishes when it is methylated with diazomethane. On silica gel G or alumina G layers, whether using acid or neutral solvents, only one spot is always obtained [2].

5. Aromatic and Heterocyclic Amines

The chromatographic behaviour of the aromatic amines is closely related to that of the aliphatic amines. A good, sharply defined separation is especially valuable here also for following rapidly chemical changes in organic syntheses.

Silica gel G layers and benzene-methanol (95 + 5) solvent have proved the best for TLC of the amines mentioned below. The following hR_f-values have been obtained: aniline 38, p-xylidine 44, β-naphthylamine 44, acridine 50, carbazole 58, diphenylamine 69 [38].

HANSSON and ALM [10] have separated diphenylamine, triphenylamine and the corresponding nitroso- and nitro derivatives on silica gel G layers, using toluene, benzene or chloroform.

MISTRYUKOV [18] has chromatographed *trans*-decahydroquinoline, *cis*-decahydroquinoline, (1)-*cis*-perhydropyridine, their derivatives and some aromatic amines; he used loose alumina layers and various neutral or basic solvents such as acetone-methanol-water (72 + 18 + 9) and chloroform (saturated with ammonia)-benzene (50 + 50).

The thin-layer chromatographic separation of positionally isomeric primary amines is of interest. WAKSMUNDZKI and co-workers [48] have separated o-, m- and p-nitroaniline, for example, with chloroform; the hR_f-values were 84, 73, and 58, respectively. Amine isomers have been separated also by GILLIO-TOS et al. [7], using standard conditions.

All the compounds quoted in Table 96, except the aminobenzoic acids, show the sequence of hR_f-values of ortho > meta > para. PASTUSKA and PETROWITZ [24] have found the same chromatographic separation behaviour in TLC on silica gel G layers, using the solvents benzene-methanol (80 + 20) or benzene-dioxan-acetic acid (79 + 19 + 1). The same sequence has been observed also in TLC of the isomeric nitrophenols on silica gel G layers, using benzene-ether (70 + 30) [42].

FELTKAMP [6] has separated stereoisomeric cyclohexylamines also on silica gel G; solvents were acetone-concentrated ammonium hydroxide-petrol ether (BP 50—70° C) (66.5 + 1.5 + 32) or (49 + 1 + 50). In TLC of the individual pairs of isomers, the compound with axial —NH_2 groups has usually the higher hR_f-value.

BOOTH and BOYLAND [3] have separated 18 different arylamine metabolites on silica gel G layers, using petrol ether (40—60° C)-acetone (70 + 30) or chloroform-ethyl acetate-acetic acid (60 + 30 + 10). The Folin-Ciocalteu reagent (No. 122) is an especially suitable spray reagent for detection; after further spraying with 10% sodium carbonate solution, the separated arylamines yield green-blue, grey or violet spots. PARIHAR and co-workers [21a] separate amines after conversion into p-toluenesulphonamides. This conversion can be carried out directly on the starting line; for this purpose, 1—2 µg amine, dissolved in pyridine-

Table 96. hRf-values[a] of primary aromatic amines on silica gel G layers [7]

Amine	Solvent			Amine	Solvent		
	A	B	C		A	B	C
o-Toluidine	62	17	84	o-Nitroaniline	69	52	93
m-Toluidine	54	10	83	m-Nitroaniline	64	36	92
p-Toluidine	40	5	80	p-Nitroaniline	58	29	91
o-Aminophenol	34	—	80	o-Phenylenediamine	—	—	63
m-Aminophenol	29	—	75	m-Phenylenediamine	—	—	53
p-Aminophenol	6	—	62	p-Phenylenediamine	—	—	40
o-Aminobenzoic acid	62	44	98	o-Bromoaniline	81	69	95
m-Aminobenzoic acid	50	12	95	m-Bromoaniline	70	44	93
p-Aminobenzoic acid	59	29	97	p-Bromoaniline	61	27	89
o-Anisidine	60	15	81	o-Chloroaniline	78	66	96
m-Anisidine	51	9	80	m-Chloroaniline	68	40	94
p-Anisidine	11	2	58	p-Chloroaniline	60	22	89

Solvents: A: Dibutyl ether-ethyl acetate-acetic acid (50 + 50 + 5); B: Dibutyl ether-n-hexane-acetic acid (80 + 16 + 4); C: n-Butanol-acetic acid-water (40 + 10 + 50) (upper phase).

[a] The hRf-values must be regarded only as guide values.

petrol ether (1 + 1), and 1.1 μmole p-toluenesulphonyl chloride (Firm 88) (in mixtures, 1.5 μmole) in pyridine solution, are applied to the TLC layer. The plate is maintained 4 h at 60° C and chromatography then carried out with chloroform or chloroform-xylene (95 + 5) or (80 + 20) on silica gel G layers for example. Neutral, acid or basic alumina can serve also as adsorbent, using the same solvents. The spots can be detected in long wave UV light. A good separation of isomeric naphthylamine monosulphonic acids is possible on cellulose MN adsorbent (300 G, Firm 83), using the solvent n-butanol-propanol-water-concentrated ammonium hydroxide (50 + 25 + 20 + 5) [1]. The spots are detected in long wave UV light.

Hetarines (halogen substituted piperidinoquinolines) can be separated on alumina G layers with benzene-chloroform (90 + 10) [15]. Coloured products are formed by irradiating the still moist chromatogram with short wave UV light (254 nm).

Heterocyclic nitrogen compounds can be separated on silica gel G layers according to PETROWITZ [29], using solvents of the type benzene-methanol (90 + 5), chloroform-methanol (90 + 10) or benzene-acetone (50 + 50); the following approximate hRf-values are quoted for the last named solvent mixture: imidazole 6; 1,2,4-triazole 12; benzimidazole 17; pyrazole 34; 1,2,3-benzotriazole 47, indazole 50; indole 64. Visualisation was accomplished by spraying with ammoniacal silver nitrate solution (Rgt. No. 225), followed by 5 min exposure to short wave UV radiation (254 nm); this yields white to light grey spots on a dark background.

VERNIN and METZGER [47] have separated 4-aryl-thiazoles through adsorption TLC, using solvents of the same type on silica gel G or alumina G layers. They used also partition chromatography on silica gel G layers which had been impregnated with dimethylformamide; cyclohexane saturated with this impregnation agent served as solvent.

Polyamine products of reaction between ethylene dichloride and ammonia can be separated on silica gel G layers, using concentrated ammonium hydroxide-ethanol (66 + 33); the following hR_f-values have been obtained [22]: $C_2H_8N_2$ (ethylenediamine) 44; $C_4H_{13}N_3$ 41; $C_6H_{18}N_4$ 32; $C_8H_{23}N_5$ 27. Iodine vapour was used for detection.

Δ^3-Pyrroline-2-carboxylic acid can be separated from Δ^1-pyrroline-5-carboxylic acid on silica gel G layers, using chloroform-methanol-concentrated ammonium hydroxide (50 + 25 + 25) [51].

The Pyrrole-C-carboxylic acids have been separated likewise on silica gel G layers with the help of two-dimensional TLC [4]; solvents were n-butanol-ethanol-concentrated ammonium hydroxide-water (40 + 40 + 4 + 8) and ethyl acetate-ethanol-acetic acid (60 + 12 + 20).

Detection. Aromatic amines can be visualised with the iodine-chloroform reagent (No. 144), with iodine vapour or with the reagents employed for detecting aliphatic amines and quoted on p. 496-7. Other suitable reagents are cerium (IV) sulphate-sulphuric acid (No. 47) or antimony(V)chloride (No. 18): acridine yields yellow; carbazole, green; aniline, pink; p-xylidine, weakly violet; diphenylamine, blue; and β-naphthylamine, grey. Special reference may be made to the advantages of the "fluorescent layers".

6. Thin-Layer Electrophoresis of Amines

Equipment for thin-layer electrophoresis is described on p. 109. PASTUSKA and TRINKS [23] have separated a number of amines on silica gel G layers which had been impregnated with borax. The amines listed below migrated towards the cathode and the M_G-values refer to methylamine as 1.00: ethylamine 0.69; n-butylamine 0.82; ethanolamine 1.09; ethylenediamine 0.32 and triethanolamine 0.51. These M_G-values were obtained under the following experimental conditions:

The silica gel G was shaken up with a 3% borax solution and poured on to glass plates. After 20—25 min, 5—10 μg of each of the amine hydrochlorides were applied to the still moist layer at about 7 cm from the strip on the anode side. The electrolyte consisted of a mixture of 80 ml ethanol, 30 ml distilled water and 2 g anhydrous sodium acetate which had been brought to pH 12 with 40% sodium hydroxide. A duration of 2 h is quoted at a field strength of 10 V/cm. It is important to renew the electrolyte solution for each experiment.

Reference may be made to the combination of thin-layer electrophoresis with TLC in a two-dimensional procedure, described by HONEG-

GER [11]; and to the possibility of coupling TLC with GC [12, 13] (see also p. 114).

II. Tar Bases

Pyridine and its homologues, especially lutidines and collidine, occur in the base fractions of tar from hard coal and lignite. The individual bases can be isolated in some cases by fractional distillation.

TLC and GC can be successfully applied to test the purity of these bases and also to monitor their chemical reactions (cf. p. 124). Separation of the aniline group from the quinoline and pyridine groups and of the isomers from one another is thus possible.

Early reference was made to the thin-layer chromatographic fractionation of tars from hard coal and wood [40, 41]. Using silica gel G layers (ca. 750 μ thick), PETROWITZ [26, 27] has chromatographed some of these bases, including the methylquinoline isomers, with a view to

Table 97. hR_f-values of 18 tar bases

Layer	Silica gel G	Silica gel G + 0,1 m Na-acetate		Group
Solvent	I	II	III	
Aniline	38	—	—	
p-Xylidine	44	—	—	
β-Naphthylamine	44	72	86	I
Acridine	50 (45)[a]	38	89	
Carbazole	58 (97)	85	92	
Diphenylamine	69	87	91	
2-Methylquinoline (Quinaldine)	40 (27)	18	74	
4-Methylquinoline (Lepidine)	33 (22)	25	74	II
Isoquinoline	30 (22)	35	74	
Quinoline	35 (30)	39	79	
2,4,6-Trimethylpyridine (2,4,6-Collidine)	24	7	23	
2,6-Dimethylpyridine (2,6-Lutidine)	28	6	36	
2,4-Dimethylpyridine (2,4-Lutidine)	22	8	38	III
2,5-Dimethylpyridine (2,5-Lutidine)	26	10	48	
4-Methylpyridine (γ-Picoline)	21	10	49	
Pyridine	22	13	55	
3-Methylpyridine (β-Picoline)	24	15	59	
2-Ethylpyridine	30	12	62	

Run: 10 cm

Solvent: I, Benzene-methanol (95 + 5); II, Ethyl acetate-methanol-formic acid (80 + 10 + 10); III, Ethyl acetate-methanol-acetic acid (75 + 20 + 5).

[a] Values given by PETROWITZ [26] are type-set in ( ).

analysing tar oils. The possibilities of separating and detecting other tar bases (Firm 65) have been investigated by SCHORN [38], using standard conditions on silica gel layers. The hR_f-values are quoted in Table 97.

Normal silica gel G layers and solvent I have proved the most suitable for TLC of the bases listed in Group I. (KUCHARCZYK and co-workers [16] have separated the same group of substances with similar success on silica gel- and alumina G layers, using neutral, weakly polar solvents.) The best separation of a number of quinoline and pyridine derivatives has been obtained on silica gel G layers which had been prepared with 0.1 M sodium acetate solution instead of with water. The most favourable differences in hR_f-values of the quinoline derivatives quoted, were achieved with the acid solvent II; for the pyridine derivatives, with solvent III.

It is important to apply only very little substance. 0.5—3 mm³ of 0.1% solutions of the bases are used. Strongly activated layers should be avoided and good results have been obtained only with layers which were air dried.

Table 98. *Structure and hR_f-values of various pyridine derivatives* [28, 30]

Solvent	α	hR_f	β	hR_f	γ	hR_f
E	—COOH	2	—COOH	6	—OH	0
A	—COOH	4	—COOH	6	—OH	2
E	—OH	6	—CH$_2$OH	13	—COOH	5
A	—OH	20	—CH$_2$OH	39	—COOH	5
E	—CH$_2$OH	18	—NH$_2$	18	—NH$_2$	5
A	—CH$_2$OH	45	—NH$_2$	45	—NH$_2$	14
E	—NH$_2$	27	—OH	23	—CH$_2$OH	4
A	—NH$_2$	50	—OH	53	—CH$_2$OH	39
E	—CH$_3$	30	—CHO	33	—CH$_3$	27
A	—CH$_3$	54	—CH$_3$	55	—CH$_3$	48
E	—CHO	51	—CH$_3$	35	—CHO	36
A	—CHO	67	—CHO	58	—CHO	56
E	—Hal	61	—Hal	57		
A	—Hal	70	—Hal	70		

Layer: Silica gel G.
Solvents: E = Ethyl acetate; A = acetone.

PETROWITZ [28, 30] has carried out a thorough investigation of the adsorption affinity of heterocyclic compounds; he established that the differences in R_f-values between the α- and β-substituted derivatives are usually small in comparison with those between these and γ-substituted derivatives. The hR_f-values of the individual pyridine compounds are quoted in Table 98 in the order of diminishing adsorption.

The Rf-values on silica gel G layers, using chloroform, of quinoline and isoquinoline compounds show only small differences, with the sole exception of those between monomethyl- and dimethyl-quinolines, which are rather larger [28, 30]. A methyl group in the 8-position, however, reduces the adsorption affinity strongly (e. g., 2-methylquinoline hRf 16; 8-methylquinoline 31; 2,4-dimethylquinoline 8; 2,8-dimethylquinoline 47).

Detection; The less volatile tar bases can be detected on fluorescent layers (e. g., silica gel GF_{254}). Other possibilities are using the spray reagents listed for the amines, e. g., iodine solution (Rgt. No. 144), permanganate solution (Rgt. No. 200), antimony(V) chloride (Rgt. No. 18) or the modified Dragendorff reagent for alkaloids (Rgt. No. 98).

Bibliography for Chapter P. Amines and Tar Bases

1. ASMUS, E., u. G. SCHULZE: Z. anal. Chem. **217**, 180 (1966).
1a. BAYZER, H.: Experientia **20**, 233 (1964).
1b. — J. Chromatog. **24**, 372 (1966).
2. BECKETT, A. H., u. N. H. CHOULIS: Lecture: 23. Internat. Kongr. Pharmaz. Wissensch., Münster, Sept. 1963, S. 627.
3. BOOTH, J., and E. BOYLAND: Biochem. J. **91**, 364 (1964).
4. CHIERICI, L., e M. PERANI: Ricerca sci., Rc., A., **6**, 168 (1964).
4a. DALY, J. W., J. BENIGNI, R. MINNIS, Y. KANAOKA and B. WITKOP: Biochemistry **4**, 2518 (1965).
5. DESIMIO, M.: Boll. soc. ital. farm. ospi. **8**, 155 (1962); ref.: Pharm. Acta Helv. **38**, 383 (1963).
6. FELTKAMP, H.: Lecture: 23. Internat. Kongr. Pharmaz. Wissensch., Münster, Sept. 1963, S. 741.
7. GILLIO-TOS, M., S. A. PREVITERA, u. A. VIMERCATI: J. Chromatog. **13**, 571 (1964).
8. GRASSHOF, H.: J. Chromatog. **20**, 165 (1965).
9. HAIS, I. M., u. K. MACEK: Handbuch der Papierchromatographie, Band I, S. 752, Reag. D 108c, Jena: VEB Gustav Fischer Verlag, 1958.
10. HANSSON, J., and A. ALM: J. Chromatog. **9**, 385 (1962).
11. HONEGGER, C. G.: Helv. Chim. Acta **44**, 173 (1961).
12. KAISER, R.: Chromatographie in der Gasphase, Band IV, BI-Hochschultaschenbücher (92/92a), Mannheim: Bibliographisches Institut AG., 1965.
13. — Z. anal. Chem. **205**, 284 (1964); Lectures during the conference on modern methods of analysis of organic compounds, 20.—23. V. 1964 in Eindhoven/Holland.
14. KALTENBACH, U.: Dissertation, Saarbrücken 1964.
15. KAUFFMANN, TH., J. HANSEN u. R. WIRTHWEIN: Ann. Chem. Liebigs **680**, 31 (1965).
16. KUCHARCZYK, N., J. FOHL, u. J. VYMÉTAL: J. Chromatog. **11**, 55 (1963).
17. LANE, E. S.: J. Chromatog. **18**, 426 (1965).
18. MISTRYUKOV, E. A.: J. Chromatog. **9**, 314 (1962).
19. MOHR, U., A. AUTHALER u. J. ALTHOFF: Naturwissenschaften **52**, 188 (1965).
20. NEURATH, G., u. E. DOERK: Chem. Ber. **97**, 172 (1964).
21. PAPARIELLO, G., and M. A. M. JANISH: Anal. Chem. **37**, 899 (1965).
21a. PARIHAR, D. B., S. P. SHARMA u. K. C. TEWARI: J. Chromatog. **24**, 443 (1966).

22. PARRISH, J. R.: J. Chromatog. **18**, 535 (1965).
23. PASTUSKA, G., u. H. TRINKS: Chem. Ztg. **86**, 135 (1962).
24. —, u. H. J. PETROWITZ: Chem.-Ztg. **88**, 311 (1964).
25. PELKA, J. R., u. L. D. METCALFE: Anal. Chem. **37**, 603 (1965).
26. PETROWITZ, H. J.: Materialprüfung **2**, 309 (1960).
27. — Chemiker-Ztg. **85**, 143 (1961), see also Erdöl und Kohle **14**, 923 (1961).
28. — Chimia (Schweiz) **18**, 137 (1964).
29. — Chimia (Schweiz) **19**, 426 (1965).
30. — G. PASTUSKA u. S. WAGNER: Chemiker-Ztg. **89**, 7 (1965).
31. POTTER, W. P. DE, R. F. VOCHTEN u. A. F. DE SCHAEPDRYVER: Experientia **21**, 482 (1965).
32. PREUSSMANN, R., D. DAIBER, and H. HENGY: Nature **201**, 502 (1964).
33. — G. NEURATH, G. WULF-LORENTZEN, D. DAIBER u. H. HENGY: Z. anal. Chem. **202**, 187 (1964).
34. SEGURA-CARDONA, R., u. K. SOEHRING: Med. exp. **10**, 251 (1964).
35. SEILER, N., u. M. WIECHMANN: Z. physiol. Chem. Hoppe-Seyler's **337**, 229 (1964).
36. — — Experientia **21**, 203 (1965).
37. SULLIVAN, G., and L. R. BRADY: Lloydia **28**, 68 (1965).
38. SCHORN, P. J.: unpublished.
39. SCHWARTZ, D. P., R. BREWINGTON, and O. W. PARKS: Microchem. J. **8**, 402 (1964).
40. STAHL, E.: Parf. u. Kosm. **39**, 564 (1958).
41. — Arch. Pharm. **292**. 411 (1959).
42. — unpublished.
43. STEIN VON KAMIENSKI, E.: In: H. F. LINSKENS: Papierchromatographie in der Botanik. Berlin-Göttingen-Heidelberg: Springer-Verlag 1959.
44. TAYLOR, E. H.: Lloydia **27**, 96 (1964).
45. TEICHERT, K. H., E. MUTSCHLER u. H. ROCHELMEYER: Deut. Apotheker-Ztg. **100**, 283 (1960).
46. TYIHÁK, E.: Naturwissenschaften **51**, 315 (1964).
47. VERNIN, G., et J. METZGER: Chim. anal. **46**, 487 (1964).
48. WAKSMUNDZKI, A., J. RÓŻYŁO i J. OŚCIK: Chem. Anal. (Warsaw) **8**, 965 (1963).
49. WINKLER, H.: Pharm. Ztg. **109**, 217 (1964).
50. YASUDA. ST. K.: J. Chromatog. **14**, 65 (1964).
51. ZBIRAL, E.: Monatsh. Chem. **94**, 639 (1963).

Q. Synthetic Pharmaceutical Products

HERBERT GÄNSHIRT

Synthetic pharmaceutical products are classified in this chapter according to their pharmacological activity; within each such class, they are subdivided on the basis of their chemical structure. The chromatography of the naturally occurring substances which are used in a parallel way may be referred to in the relevant chapters of this book.

Some overlapping is unavoidable in order to preserve continuity, e. g. in the case of the hypnotically functionating analgesics.

The methods given can generally be employed for both toxicological testing and analytical control. Since toxicological investigation demands in most cases prior isolation and purification in order to be able to apply the substances for separation in adequate amounts and purity, each section contains hints on extraction from the organs. Toxicological orientation analysis is possible[1] when use is made of several of the solvents and spray reagents quoted in the pertinent section. The application of TLC to clinical diagnosis and the study of metabolites is described in Chapter S, p. 578.

Unequivocal separation of the known active substances using the full length of the plate is required in working out methods for routine control analysis in pharmaceutical laboratories. Information about the necessary conditions for accomplishing this are to be found in particular in the section on the TLC of pharmaceutical commercial preparations (p. 554 of this chapter). TLC is excellently suited for the "accelerated" test of stability of pharmaceutical active materials and preparations, since it is possible with its help, rapidly to separate the reaction partners from the medium and to evaluate them; a compilation of such investigations is therefore made at the end of this chapter.

TLC may be applied with success also to drug recognition and to detection of adulteration (cf. Chapter U, II).

As far as possible, the international non-proprietary names proposed by the WHO (World Health Organisation) have been chosen, as in the compilation of NEGWER [107a].

Where the commercial form of the product is a salt, the salt form has deliberately not been quoted; this is because several salts are often commercially available and chromatography has usually been carried out with the active compounds isolated in a preliminary stage — just as they occur in the course of studies of toxicology, physiological metabolism and pharmaceutical analysis.

1. Antihistamines, Anti-Allergics and Structurally Related Compounds with Psychic Activity

The pharmacological properties of the chemical compound classes in this section overlap to some extent those quoted in section 2. Pheno-

---

[1] MACEK et al. [97] have worked out a procedure for systematic toxicological analysis of 160 known pharmaceutical products. The substances are divided into 3 main groups by extraction at low and high pH or by using ion exchangers and are subsequently identified by PC and TLC.

thiazines and substances with a related ring structure are quoted here together with diazepines and the "antihistamine" medicaments, the antihistamine activity of which predominates and which are, chemically, often derivatives of ethylenediamine and monoethanolamine.

The non-proprietary names suggested by the WHO have been used as far as possible, as in the compilations of NEGWER [107a] and BLAŽEK et al. [20].

a) Phenothiazines and Diazepines

For the chromatography of phenothiazine derivatives, silica gel [6, 10, 47, 101, 124, 145], various aluminas [28, 47] and cellulose powder [47, 112] have been used as adsorbents. The phenothiazine derivatives studied by EIDEN and STACHEL [47] could be separated better on alumina layers and silica gel paper than on silica gel adsorption layers. Nevertheless, most investigations of phenothiazines and substances with related ring structure have been carried out on silica gel layers, using neutral and basic solvents; adsorption chromatographic effects then predominate (cf. Tables 99 and 100).

Thiazinamine, prothipendyl, prochlorperazine, triflupromazine, chlorprothixene, fluphenazine, thiopropazate and phenothiazine can for example, be separated using the chromatographic conditions E from Table 100. Amongst others, perazine, prochlorperazine, prothipendyl, promazine, isothipendyl, pecazine, chlorpromazine, levomepromazine and diethazine can be separated under conditions G. 2- and 4-Chlorophenothiazines have been separated on silica gel layers, using petrol ether-ether (60 + 40) and determined spectrophotometrically after elution [157a].

The phenothiazine bases can be set free from acid solution with 5% sodium hydroxide solution and extracted with ether. Buffer solutions of pH 10 are better since, e. g., chlorpromazine decomposes already in normal alkali solutions [51]. It is imperative to use peroxide-free ether in order to avoid oxidation. The phenothiazine hydrochlorides may be extracted directly from hydrochloric acid solution with chloroform, in which they are easily soluble [46]. Since phenothiazines easily undergo change in daylight or UV light, especially in acid solution [101, 150], they must be protected against this influence.

Neither the R_f-value in a solvent nor the selectivity of a colour reaction suffices for unambiguous identification of a compound of the phenothiazine series or of a series with a related ring system. AWE and SCHULZE [6, 145] have therefore employed a systematic analytical procedure using several solvents with which definite dependences of migration distance on structure could be discerned (cf. Table 99). Other authors have oxidised the separated phenothiazines with 3% hydrogen peroxide and then extracted and transferred the sulphoxides thus formed to a second layer for identification [88a]. Some solvents and spray reagents have also been proposed for distinguishing the phenothiazines from other central depressives and stimulants in toxicological tests [111].

MAROZZI and FALZI [100] have compared the R_f-values of pure phenothiazines and those isolated from biological material, finding statistically detectable differences.

Chromatography on layers which have been impregnated with fluorescent material and then irradiation in short wave UV light can be used also for non-specific detection of phenothiazines and related ring systems. Detection has been carried out often with the Dragendorff reagent also and occasionally with the iodoplatinate reagent [35, 111] and iodine-potassium iodide [46] (cf. Table 100).

More selective localisation of numerous phenothiazines is possible through their own fluorescence in short wave UV light [101]. Phenothiazines, substituted with chlorine or unsubstituted in the 2-position, show no fluorescence. Oxidising agents often yield characteristic colours with phenothiazines; phenothiazones and other reaction products may be formed, depending on the substituent on the nitrogen atom [107]. Colorations of this type are given after spraying with 65% nitric acid [46], 10% aqueous molybdophosphoric acid [42, 46], 10% hydrogen peroxide and aqueous or ethanolic sulphuric acid [101,124] (cf. Table 100). Compounds in which carbon replaces the heterocyclic nitrogen (chlorprothixene) or sulphur (imipramine) or both (amitriptyline) react more feebly and slowly with sulphuric acid than do the phenothiazines themselves, or no reaction takes place at all, as with the last named. Phenothiazines and substances of related ring structure can be differentiated through the different colours obtained by spraying with formalin-sulphuric acid in the cold and after heating to 120–150° C [6, 145]. Phenothiazines can be distinguished from their sulphoxides with 2% ferric chloride solution; the former yield a red to violet colour, the latter remain colourless [35]. Palladium chloride solution serves to distinguish the sulphur-containing ring system from that without sulphur; only the former phenothiazines give violet or yellow complexes [10, 12]. Reference may be made to other literature on the TLC of phenothiazines [99, 138].

The dibenzazepine, imipramine, can be detected as a blue spot after spraying with dichromate-sulphuric acid (0.5% dichromate in 20% sulphuric acid) [2]. Imipramine and the dibenzocycloheptadiene, amitryptilene, fluoresce in UV light after having been sprayed with sulphuric or phosphoric acid [144]. Chlordiazepoxide can be identified by acid hydrolysis on the layer to 2-amino-5-chlorobenzophenone, followed by diazotisation and coupling [12, 133]. Non-specific detection is possible with iodoplatinate (Rgt. No. 147) and Dragendorff reagent (Nos. 96–98) [133]. Amounts of over 8 µg of diazepam can be likewise detected with Dragendorff reagent or through the yellow colour of the ketone yielded by hydrolysis with hydrochloric acid [116].

Table 99. *Phenothiazines, phenothiazine-like compounds and diazepines*[a] *functioning as antihistamines, anti-allergics and antipsychotics*

WHO-Term[d]	Substituents[b] R₁	R₂	X	hRf-values A	B	C	D	E	F	G	H	I	K	L
Phenothiazin		H	CH				81	94	98	98			90	
Diethazine	—CH₂CH₂N(C₂H₅)₂	H	CH		45		32	40	44	91		81	54	
Promethazin	—CH₂CHN(CH₃)₂ | CH₃	H	CH		41	22 (30)[c]	32	43	44	96		52	56	59 (57)
Thiazinamine	—CH₂CHN⁺(CH₃)₃ | CH₃	H	CH		7		00	00	00	00				
Profenamine	—CH₂CHN(C₂H₅)₂ | CH₃	H	CH		43		32	40	35	65		94	58	
Promazine	—CH₂CH₂CH₂N(CH₃)₂	H	CH	11	23	37 (15)	17	18	16	43		37	38	38 (46)
Methopromazine Methoxyproma-zine	—CH₂CH₂CH₂N(CH₃)₂	CH₃O	CH	09		24								58
Levomepromazine	—CH₂CHCH₂N(CH₃)₂ | CH₃	CH₃O	CH		45		32					72	63	
Chlorpromazine	—CH₂CH₂CH₂N(CH₃)₂	Cl	CH	23	33	30 (19)	25	31	28	73	44	5	50	70 (47)
Triflupromazine	—CH₂CH₂CH₂N(CH₃)₂	CF₃	CH	22		40 (24)	31	37	32	88	50	62		79 (48)

Synthetic Pharmaceutical Products

Compound	Side chain	R'	R''										
Acepromazine	—CH$_2$CH$_2$CH$_2$N(CH$_3$)$_2$	CH$_3$CO	CH				14					34	
Propionylpromazine	—CH$_2$CH$_2$CH$_2$N(CH$_3$)$_2$	C$_2$H$_5$CO	CH		30								
Aminopromazine	—CH$_2$CHCH$_2$N(CH$_3$)$_2$ / N(CH$_3$)$_2$	H	CH		17								
Perazine	—CH$_2$CH$_2$CH$_2$–N⟨⟩N–CH$_3$	H	CH	08			14	26	32	30	31	17	
Prochlorperazine	—CH$_2$CH$_2$CH$_2$–N⟨⟩N–CH$_3$	Cl	CH	12	32 (13)		29	23	24	21	31	26	27 (28)
Trifluoperazine	—CH$_2$CH$_2$CH$_2$–N⟨⟩N–CH$_3$	CF$_3$	CH	10	40		34	33	32	41	33		33
Butyrylperazine	—CH$_2$CH$_2$CH$_2$–N⟨⟩N–CH$_3$	C$_3$H$_7$CO	CH	13			29				31		
Thiethylperazine	—CH$_2$CH$_2$CH$_2$–N⟨⟩N–CH$_3$	C$_2$H$_5$S	CH								50		
Perphenazine	—CH$_2$CH$_2$CH$_2$–N⟨⟩N–CH$_2$CH$_2$OH	Cl	CH	24	48		43	32	26	35	57	9	44
Fluphenazine	—CH$_2$CH$_2$CH$_2$–N⟨⟩N–CH$_2$CH$_2$OH	CF$_3$	CH	27	68		50	50	32	48	56	12	58
Acetophenazine	—CH$_2$CH$_2$CH$_2$–N⟨⟩N–CH$_2$CH$_2$OH	CH$_3$CO	CH	18							36		
Proketazine	—CH$_2$CH$_2$CH$_2$–N⟨⟩N–CH$_2$CH$_2$OH	C$_3$H$_7$	CH	25							45		

Table 99 (Continued)

| WHO-Term[c] | Substituents[b] R_1 | R_2 | X | \multicolumn{12}{c|}{hRf-values} |
				A	B	C	D	E	F	G	H	I	K	L
Thiopropazate	$-CH_2CH_2CH_2-N\diagup\diagdown N-CH_2CH_2COOCH_3$	Cl	CH	53		70	71				79			67
Pipamazine	$-CH_2CH_2CH_2-N\diagup\diagdown-CONH_2$	Cl	CH	37		56					71			72
Pecazine, Mepazine (Pacatal)	$-CH_2-\text{(piperidine-}CH_3)$	H	CH	13		29 (24)[c]	25				46	45		57 (59)
Thioridazine	$-CH_2CH_2-\text{(piperidine-}CH_3)$	CH_3S	CH	15		20	13				39	45	48	65
Selvigon	$-COOCH_2CH_2OCH_2CH_2-N\diagup\diagdown$	H	N		11									
Isothipendyl	$-CH_2CHN(CH_3)_2$ $\quad\;\;CH_3$	H	N	11	26		22	22	25	62	39			
Prothipendyl	$-CH_2CH_2CH_2N(CH_3)_2$	H	N	08	18		15	16	15	35	21			
Trimepazine	$-CH_2CHCH_2N(CH_3)_2$ $\quad\;\;\;CH_3$	H	CH			30								64
Methdilazine	$-CH_2-\text{(pyrrolidine-}N-CH_3)$	H	CH			16								61

Synthetic Pharmaceutical Products

Phenothiazine-like compounds and diazepines

Name	Formula									
Chlorprothixene	CH–CH$_2$–CH$_2$–N(CH$_3$)$_2$ (thioxanthene)	36	46	25			78	69		73
Imipramine	CH$_2$–CH$_2$–N(phenyl)–CH$_2$CH$_2$CH$_2$N(CH$_3$)$_2$	16	29				55	47	45	77
Amitryptyline	CH$_2$–CH$_2$–C=CHCH$_2$CH$_2$N(CH$_3$)$_2$		39							65
Chlordiazepoxide	NHCH$_3$, N=C–CH$_2$·HCl, C=N→O, C$_6$H$_5$, Cl		73						85	67

Numbering according to [165]

Basic formula: phenothiazine with positions 1,2,3,4 on one ring, 5=S, 6,7,8,9 on other ring, 10=N, R$_1$ on N, R$_2$ on position 2.

a See Table 100 for explanations.
b Basic formula:
c Values in brackets ( ) are the Rf-values for the corresponding sulphoxides.
d Where no WHO-term was known the trade name of the substance used in the investigation is given.

X = group in position 9 in formula.

33 Thin-Layer Chromatography, 2nd Edition

Table 100. *Explanations for the hRf-values in Table 99 for phenothiazines, compounds of similar structure to phenothiazines, diazepines, carbamate esters and compounds of antihistamine activity*

Compound class	Term in hRf-data Table (99)	Chromatographic conditions	Reagents for detection	Reference
Phenothiazines	A	Silica gel G, Merck, 500 μ; n-propanol-water (85 + 15); salt form applied	1. Fluorescence in UV 263 nm	[101]
	H	n-Propanol-N ammonium hydroxide (88 + 12); salt form applied	2. 40% sulphuric acid	
Phenothiazines	B	Silica gel G, Merck, air dry; acetone-water (85 + 15); salt form applied	1. Dragendorff reagent (Nos. 96—98)	[47]
Phenothiazines, tranquillisers and antihistamines in biological material		Silica gel G, Merck, prepared accdg. to BRENNER and NIEDERWIESER [24]. Bases applied after isolation from urine	1. Iodoplatinate (Rgt. No. 147)	[35]
	C	Methanol-butanol (60 + 40)	2. 50% sulphuric acid	
	L	Ethanol-acetic acid-water (50 + 30 + 20)		
	N	Benzene-dioxan-aq. ammonium hydroxide (60 + 35 + 5)		
Phenothiazines, substances with ring structure similar to that of phenothiazine		Silica gel, 250 μ; start spot applied 1 cm from lower plate edge; 2 μl (ca. 4 μg) applied in the base form; 15 cm run	1. Dragendorff reagent accdg. to Vágujfalvi (No. 98)	[6]
	D	Methanol, normal chamber saturation	2. Formaldehyde-sulphuric acid (Rgt. No. 125)	[145]
	E	Methanol-methylal (50 + 50), NS		
	F	Methanol-methylal-chloroform (15 + 30 + 45), chamber supersaturation		
	G	Methanol-methylal-NH$_4$OH (d = 0.963) (50 + 50 + 1)		
	M	Methanol-methylal-NH$_4$OH (d = 0.963) (50 + 50 + 0.6) (G and M can be used only once)		
Phenothiazines and similarly acting compounds	I	Silica gel G, Merck; 11.5 cm run; some applied as bases, some as salts; Benzene-acetone-25% NH$_4$OH (75 + 15 + 7.5)	1. Dragendorff reagent (Nos. 96—98) 2. 10% H$_2$SO$_4$ in ethanol 3. Various other reagents	[124]
Phenothiazines and similarly acting compounds	K	Silica gel G, Merck, 250 μ; coarser particles sieved out; methanol-acetone-triethanolamine (50 + 50 + 1.5)	1. UV 2. Dragendorff reagent (Nos. 96—98) 3. Palladium chloride (No. 193)	[10]
Antihistamines in biological material	O	Silica gel G; cyclohexane-benzene-diethylamine (75 + 15 + 10)	1. Dragendorff reagent (Nos. 96—98)	[52]
	P	Silica gel G, prepared with 0.1 N NaOH; methanol solvent; plates stored at 52% relative humidity and 22° C for 16 h in both cases	2. 1% Ammonium vanadate in sulphuric acid	

Table 101. *hRf-values of antihistamines and substances with anti-allergic activity, in various solvents*[a]

Basic formula	R₁	R₂	WHO-term[b]	D	E	F	M	N	C	P	O
$CH_3\!\!>\!\!N-CH_2-CH_2-N\!\!<\!\!^{R_1}_{R_2}$	phenyl	2-thenyl	Methaphenilene	34	47	49	86			46	55
	pyridyl(2)	2-thenyl	Methapyrilene	20	25	17	65	83	30	36	47
	pyridyl(2)	3-thenyl	Thenfanil (Thenyldiamine)	16	25	17	55			32	47
	pyridyl(2)	4-methoxybenzyl	Mepyramine (Pyrilamine)	16	20	16	53	82	27	33	42
	pyridyl(2)	benzyl	Tripelenamine	17	23	15	57	92	23	35	50
	pyridyl(2)	4-chlorobenzyl	Chloropyramine	17	24	17	60				
	pyridyl(2)	4-bromobenzyl	Bromopyramine	17	24	15	61				
	pyrimidyl(2)	4-methoxybenzyl	Thonzylamine					65	32	38	41
	pyridyl(2)	5-chloro-2-thenyl	Chloropyrilene (Chlorothen)					70	29	37	43
$\square\!\!>\!\!N-CH_2-CH_2-N\!\!<\!\!^{R_1}_{R_2}$	benzyl	benzyl	Myostimin	15	24	19	66				
	benzyl	phenyl	Histapyrrodine	31	30	57	86				
	phenyl	phenyl	Aspasan	34	49	48	94				
	phenyl	tolyl-(4)	Pragman	25	31	29	57				
$CH_3\!\!>\!\!N-CH_2-CH_2-CH\!\!<\!\!^{R_1}_{R_2}$	pyridyl(2)	chlorophenyl-(4)	Chlorphenamine (Chlorpheniramine)					40	20	19	38
	pyridyl(2)	bromophenyl-(4)	Bromphenir-amine	9	12	10	14	42	16		
	pyridyl(2)	phenyl	Pheniramine	12	13	8	11	68	19	18	40

[a] See Table 100 for explanatory notes.
[b] Where no WHO-term was found, the trade name of the substance used in the investigation is given.

Table 101 (Continued)

Basic formula	Substituents		WHO-term[b]	D	E	F	M	N	C	P	O
$\text{N-CH}_2-\text{CH}=\text{C}\langle^{R_1}_{R_2}$	4-chlorobenzyl	phenyl	Pyrrobutamine	25	32	37	84	84	32	39	62
	pyridyl(2)	tolyl-(4)	Tripolidine					40	26	40	41
$\text{CH}_3\rangle\text{N}-\text{CH}_2-\text{CH}_2-\text{O}-\text{CH}\langle^{R_1}_{R_2}$	phenyl	phenyl	Diphenhydra-amine	22	29	24	57	91	25	37	52
	phenyl	tolyl-(2)	Orphenadrin	27	31	29	75				
	phenyl	bromophenyl-(4)	Bromazine (Bromdiphen-hydramine)	23	31	29	74	86	33	37	50
	pyridyl(2)	chlorophenyl-(4)	Carbinoxamine					33	17	21	29
	pyridyl(2)	phenyl	Doxylamine					52	17		
$\text{CH}_3\rangle\text{N}-\text{CH}_2-\text{CH}_2-\text{O}-\text{C}\langle^{R_1}_{R_2}-\text{CH}_3$	phenyl	chlorophenyl-(4)	Chlorphenox-amine	25	33	29	67				
$\text{C}_2\text{H}_5\rangle\text{N}-\text{CH}_2-\text{CH}_2-\text{O}-\text{C}\langle^{R_1}_{R_2}-\text{CH}_3$	phenyl	chlorophenyl-(4)	Keithon	25	35	36	90				
$R_1-\text{N}\underset{}{\bigcirc}\text{N-CH}\langle^{R_2}_{\text{(naphthyl)}}$	methyl	H	Cyclicine							46	55
	methyl	Cl	Chlorcyclicine	40	48	54	86	90	40	44	49
	hydroxyethoxy-ethyl	Cl	Hydroxyzine	58	61	63	92	27	62	59	8
	methylbenzyl-(3)	Cl	Meclozine	81	88	95	98			71	69
	tert.-butylben-zyl-(4)	Cl	Buclizine	81	88	95	98			72	73
$\text{CH}_3-\text{N}\underset{}{\bigcirc}\text{N}-\text{CH}_2-R$	benzyl		Soventol	23	30	33	43				
	2-thenyl		Thenalidine	22	29	32	51				

	8	42	55		33	58	46
	11	25	53		67	54	51
	40		48				
	31		90				
	9	24	90	83	94	90	
	5	17	60	51	89	70	
	6	23	51	43	80	52	
	5	17	41	34	76	44	
	Antazoline	Diphenylpyra-line	Phenindamine	Mebhydroline	Clemizole	Captodiamine (Covatin)	Phenyltoxol-amine

b) Antihistamines

Like the phenothiazine derivatives, substances with antihistamine activity have been separated largely by adsorption chromatography on silica gel layers [6, 35, 52, 145] (cf. Tables 100, 101). Silica gel layers that had been prepared with 0.1 N sodium hydroxide, have been used in one instance [52]. AWE and SCHULZE [6, 145] have used a systematic analysis procedure with several solvents for antihistamines also.

Antazoline, myostimin,e keithon, mebhydroline, captodiamine, orphenadrine, clemizole and buclizine, for example, can be separated by using conditions D in Table 100. Under conditions F, antazoline, brompheniramine, mepyramine, diphenhydramine, chlorphenoxamine, pyrrobutamine, mebhydroline, histapyrrodine, phenindamine, hydroxyzine, captoindamine, clemizole and buclizine, etc., can be separated.

Many antihistamines possess no groups conferring the property of absorption in the UV region and cannot therefore be located on fluorescing adsorbent layers. The Dragendorff reagent (Nos. 96—98) and iodoplatinate reagent (No. 147) can be used again here as general, non-specific reagents for detection. The possibility of detection with the chlorine-tolidine reagent (No.42) or chlorine-benzidine, may be pointed out also [70]. Many antihistamines can be identified also through the characteristic colours or colour changes with ammonium vanadate-sulphuric acid [52], before and after heating to 85° C. Various antihistamine types may be similarly distinguished through the different colours yielded by treatment with formalin-sulphuric acid and by heating 10 min at 120° C. MORRISON and CHATTEN [104] have determined some antihistamines quantitatively by comparing the sizes of chromatogram spots obtained by spraying with cerium(IV) sulphate (Rgt. similar to No. 36) and heating 2 h at 120° C (see Chapter H, p. 135). Quaternary ammonium compounds, impurities in some antihistamines of the aminoalkoxybenzhydryl type, could be detected through PC and TLC [39].

2. Analeptics, Psychotherapeutic Agents with Antidepressive Activity, Appetite Depressants and Some Carbamate Esters of Varied Activity

Analeptics

The possibility of separation by TLC and the detection of some classical analeptics may be seen in Table 103 and Fig. 162 [63].

Psychotherapeutic Agents and Appetite Depressants

Most of the monoaminoxidase inhibitors with antidepressive function, listed in Table 104, are hydrazine derivatives and can be separated by TLC on silica gel layers, using the basic solvents quoted, and detected with the Folin-Ciocalteu reagent or with molybdophosphoric acid [144]. The substances can be extracted from aqueous solution with n-butanol.

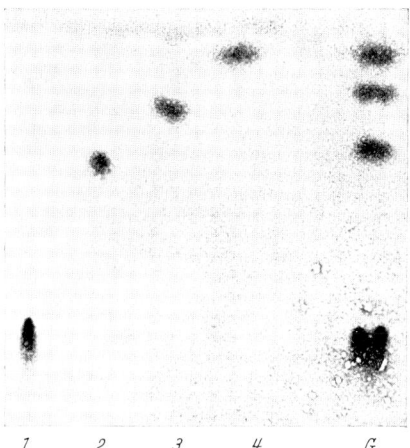

Fig. 162. Separation of some analeptics. Nikethamide (*1*), pentetrazole (*2*), benegride (*3*), camphor (*4*), mixture (*G*). Layer: silica gel H; solvent: methyl ethyl ketone; CS; 10 cm run; detection: exposed to iodine vapour, excess iodine removed by leaving in the air; sprayed with Na-fluorescein solution and inspected in UV-light. Amounts applied: (*1*) 10 μg; (*2*) 30 μg; (*3*) 50 μg; (*4*) 50 μg

Data on the TLC of some hydrazine-psychotherapeutic agents may be found elsewhere also [111]. Various such agents, appetite depressants and sympathomimetics have been separated on silica gel layers with a neutral solvent, as seen in Table 102 [44]. Detection was carried out

Table 102. hRf-values of some antidepressive psychotherapeutics, appetite depressants and sympathomimetics

Free (non-registered) Name[a]	Chemical term	hRf[b]
Methamphetamine	1-Phenyl-2-methylaminopropane	9
Amphetamine	1-Phenyl-2-aminopropane	16
Phenmetrazine	2-Phenyl-3-methyltetrahydro-1,4-oxazine	26
Captagon	7-[2'-(1''-methyl-2''-phenylethylamino)-ethyl]-theophylline	54
Tradon	5-Phenyl-2-imino-4-oxazolidinone	60
Methylphenidate	Methyl α-phenyl-piperidyl-(2)acetate	70
Prolintane	1-Phenyl-2-(1-pyrrolidinyl)pentane	78
Diethylpropione	1-Phenyl-2-diethylamino-1-propanone	90

[a] Where unknown, trade name used.
[b] *Layer:* silica gel GF$_{254}$; *Solvent:* dimethylformamide-ethyl acetate (10 + 90), plus 30 drops n-octanol.

with Fast Blue salt (Rgt. No. 100) which requires subsequent spraying with N sodium hydroxide solution and then again with the Fast Blue salt. Further information about the TLC of amphetamine, dexamphetamine and methylphenidate may be found in [111].

The substances in Table 102 were isolated from urine by adjusting its pH to 9—10 and extracting with methyl acetate-ether (1 + 1). After purifying the extract over alumina, it was concentrated in vacuo, the residue taken up in a little ether or methanol and this solution applied to the TLC layer.

Table 103. hR_f-values and detection of some analeptics [63]

Analeptic	hR_f-value in solvent:		Detection			
	Butanone	Cyclohexane-acetic acid-chloroform (40 + 10 + 50)	UV 254 nm	Iodine vapour	Active hydrogen Rgt. (No. 42)	Dragendorff reagent (No 96—98)
Micoren, SpotA	42⎫ 51⎭ double spot	30⎫ 37⎭ double spot	+	(+)	—	—
SpotB	05 elongated	00	—	+	—	+
Lobeline hydrochloride	10 elongated	00	+	+	—	+
Nikethamide (coramine)	22	13	+	+	+ (blue)	+
Pentetrazole (cardiazole)	50	20	—	+	—	—
Bemegride (eukraton)	67	31	—	+	+ (dark blue)	—
Camphor	76	—	—	+	—	(+)

Layer: Silica gel HF$_{254}$; 10 cm run.
Amounts applied: 20—50 µg in the form of extracts and diluted solutions of the relevant medicaments.

Carbamate Esters

The experimental conditions for chromatographing some carbamate esters may be seen in Table 105.

The carbamates can be detected specifically with a mixture of formalin and hydrochloric or sulphuric acid (Rgt. Nos. 123, 124) [78]. Other literature sources [53, 63, 94] contain information about the TLC of meprobamate alone and in the presence of various sedatives.

3. Sympathomimetics of the Adrenaline Type

POTTER and co-workers [132] have separated adrenaline, noradrenaline and some of their metabolites on cellulose layers [7.5 g cellulose MN-300 (cf. p. 34) + 45 ml methanol; layers dried 10 min at 105° C] at chamber saturation, using the solvent n-butanol, saturated with 3N HCl. Noradrenaline, adrenaline, isoprenaline, ephedrine and amphetamine have been chromatographed similarly on cellulose layers [20 g cellulose MN-300 G (cf. p. 499) + 100 ml water; layers dried 2 h at

Table 104. *Psychotherapeutic anti depressives*

WHO-Term	Formula	hRf-values I	hRf-values II
Isoniazide	N-pyridyl-CO-NH-NH$_2$	45	0
Iproniazide	N-pyridyl-CO-NH-NH-CH(CH$_3$)$_2$	51	6
Nialamide	N-pyridyl-CO-NH-NH-CH$_2$-CH$_2$-CO-NH-CH$_2$-phenyl	40	0
Tervasid	phenyl-CH$_2$-NH-NH-CO-CH(CH$_3$)$_2$	86	51
Isocarboxazide	phenyl-CH$_2$-NH-NH-CO-(5-methylisoxazole)	84	42
Phenelzine	phenyl-CH$_2$-CH$_2$-NH-NH$_2$	85	57
Pheniprazine	phenyl-CH$_2$-CH(CH$_3$)-NH-NH$_2$	88	62
Aetryptamine	indolyl-CH$_2$-CH(C$_2$H$_5$)-NH$_2$		13
Tranyl-cypromine	phenyl-CH(-CH$_2$-)CH-NH$_2$ (cyclopropyl)	88	

Explanatory notes:
I. Adsorbent: Silica gel G. Solvent: Chloroform-acetone-diethylamine (50 + 40 + 10), CS.
II. Adsorbent: Silica gel G. Solvent: Chloroform-cyclohexane-diethylamine (50 + 40 + 10), CS.
Spray reagents: 1. Folin-Ciocalteu (No. 122), diluted 1 + 1 with water; with and without heating 10—20 min at 110° C. 2. 5% molybdophosphoric acid in ethanol; subsequent treatment with ammonia.

Table 105. *Carbamate esters*

WHO-term	Chemical term	hRf-values I	hRf-values II
Meprobamate (tranquilliser)	2-Methyl-2-n-propyl-1,3-propandiol dicarbamate	18	36
Carisoprodol (muscle relaxant)	N-Isopropyl-2-methyl-2-propyl-1,3-propandiol carbamate	33	—
Hexapropymate (hypnotic)	1-Propinylcyclohexyl carbamate	—	72

I: *Layer:* Silica gel G. Solvent: Cyclohexane-ethanol (80 + 20) [161].
II: *Layer:* Silica gel G. Solvent: Acetone-chloroform (50 + 50) [78].
Detection: Furfuraldehyde-HCl or -H$_2$SO$_4$ (Rgt. No. 126).

100° C] with the solvent n-butanol-acetic acid-water (40 + 10 + 50); the substances migrated in the order quoted, regarded from start to solvent front. So-called double spot formation, as known in paper chromatography, occurs at the start when salts of the bases are applied and the above acetic acid-containing solvent or neutral solvents are employed; and also in the presence of acids stronger than that used in the solvent mixture. BECKETT and CHOULIS have again discussed the cause of this phenomenon [15] (cf. p. 499).

This double spot formation is not observed on silica gel or alumina layers. The sensitivity to oxidation of the adrenaline derivatives is, however, increased by the metal or heavy metal impurities present in such layers. Adrenaline and noradrenaline have therefore been chromatographed on buffered silica gel layers (Sörensen buffer, pH 6.8), prepared with addition of sodium bisulphite and using 10% ethanol as solvent [163]. The sympathomimetics and the secondary products formed by oxidation tend moreover to form complex salts; this manifests itself in elongated spots. Silica gel layers which had been prepared with 0.1 M EDTA have been used with the solvent acetone-formic acid-water (70 + 10 + 20) for the separation of adrenaline, noradrenaline and various substances of similar structure [110]; silica gel HR (Firm 88) ought to be especially advisable for this purpose. On the other hand, HALMEKOSKI [74] has made use of this very ability of the adrenaline-type of sympathomimetic to form complexes, in order to accomplish fractionation on buffered layers containing molybdate, tungstate or borax components

Table 106a. *Structure of the adrenaline-type sympathomimetics quoted in Table 106b*

Name of the adrenaline derivative	Substituents				
	R_1	R_2	R_3	R_4	R_5
1. Adrenaline	OH	OH	OH	H	CH_3
2. Synephrine (Oxedrine)	OH	H	OH	H	CH_3
3. Noradrenaline	OH	OH	OH	H	H
4. Adrenalone (Adrenone)	OH	OH	O=	H	CH_3
5. Corbadrine	OH	OH	OH	CH_3	H
6. Hydroxyamphetamine	OH	H	H	CH_3	H
7. Isoprenaline	OH	OH	OH	H	$CH(CH_3)_2$
8. Phenylephrine (Metaoxedrine) . .	H	OH	OH	H	CH_3
9. 3-Hydroxytyramine	OH	OH	H	H	H

(cf. Tables 106a and b). Sympathomimetics containing two hydroxyl groups in o-position to each other in the benzene ring, are more firmly held to the layer as complexes than are monophenols; this is clearly seen from comparison of adrenaline and synephrine (cf. Table 106b).

Table 106b. hR_f-values of some adrenaline-type sympathomimetics [74]

Solvent	Complexing agent	hR_f of sympathomimetic No:.								
		1	2	3	4	5	6	7	8	9
I	None	27	35	41	38	57	69	45	44	53
	Na_2MoO_4	06	40	12	06	14	71	11	47	17
	Na_2WO_4	05	48	06	05	12	72	14	53	12
	Borax	15	38	14	07	18	51	20	40	28
II	None	33	42	52	41	56	64	48	48	54
	Na_2MoO_4	10	40	13	10	21	66	21	50	25
	Na_2WO_4	18	45	17	08	27	71	34	53	29
	Borax	27	56	30	17	48	73	40	61	46
III	None	27	30	34	27	38	48	31	34	37
	Na_2MoO_4	05	35	14	06	19	56	17	39	19
	Na_2WO_4	09	36	17	08	22	60	20	40	21
	Borax	13	34	20	16	22	60	20	36	23
IV	None	42	45	52	45	60	67	53	51	56
	Na_2MoO_4	22	47	25	16	31	64	24	47	24
	Na_2WO_4	24	50	30	19	36	64	31	52	31
	Borax	28	55	36	24	42	73	46	61	42

Explanatory Notes to Table 106b, Sympathomimetics

Adsorption layer, method of application, length of run and other chromatographic conditions	Solvents	Detection
30 g Silica gel G + 45 ml distilled water + 15 ml buffer, pH 4 [92], with or without 0.01 mole of one of the following complexing agents: sodium molybdate, sodium tungstate or borax. The adrenaline derivatives were applied as 1% solutions in a buffer of pH 4 [92]. Temperature during separation was 19—21° C. 1 hour equilibration with the organic phase before chromatography. 10 cm length of run.	Organic phases of the following mixtures: I. n-Butanol, saturated with aqueous SO_2-solution (H_2SO_3) II. n-Butanol-acetic acid-H_2SO_3 (40 + 10 + 50) III. n-Amyl alcohol-acetic acid-ethanol-H_2SO_3 (40 + 10 + 10 + 50) IV. n-Butanol-n-propanol-acetic acid-H_2SO_3 (40 + 10 + 10 + 50)	Folin-Denis reagent [155] and then kept 10 min in a chamber containing 25% ammonium hydroxide

Oxidation during chromatography has been prevented in these studies by using essentially solvents containing sulphur dioxide. Catechol amines and their metabolites have been chromatographed also on polyamide layers [7 g polyamide (Firm 88) + 45 ml methanol; layer of 250 µ thickness, air dried 2 h], using the solvent isobutanol-acetic acid-cyclohexane (80 + 7 + 10) [148]. The TLC of adrenaline, noradrenaline and dopamine has been described elsewhere also [31]. Adrenaline has been quantitatively determined by preparing the triacetyl derivative, extracting it with methylene dichloride, chromatographing on silica gel G layers with chloroform-methanol (90 + 10), visualising with molybdophosphoric acid and evaluating the spot area [169]. Others have prepared the relatively stable tetraphenylborate derivatives in order to isolate adrenaline and noradrenaline from biological material; these were then separated by TLC, visualised with ferricyanide-ferric chloride reagent (No. 111) and quantitatively determined by recording the light absorption [76a].

Sympathomimetics have been detected with: iodine vapour; Folin-Denis reagent [74, 155]; ferricyanide solution [16, 110, 132]; diazonium salt solutions [132, 148]; and dichloroquinone-4-chloroimide solution [148]. Ephedrine has been visualised by spraying with 0.2% ninhydrin in n-butanol.

4. Analgesics, Antipyretics and Antirheumatic Agents
a) p-Aminophenol Derivatives, Pyrazolones etc.

Paracetamol, phenazone, ethenzamide and phenylbutazone can be separated on silica gel with the pyridine-containing solvent used in experimental conditions III in Table 107 [63]. Clear separation of paracetamol, phenazone and aminopyrine has been achieved with another basic solvent, butanone-diethylamine (85 + 5) [65]. The separation of phenylbutazone, acetylsalicylic acid, phenacetin, phenazone, novalgin and aminopyrine is possible on silica gel, using the acid solvent V from Table 107 [177]. As is seen in Fig. 163, some of the substances cited in Table 107 may be separated with a neutral solvent mixture also [63]. TLC of a mixture of salicylamide, acetylsalicylic acid and free salicylic acid is more difficult; success has however been achieved on silica gel G, using methanol-acetic acid-ether-benzene (1 + 18 + 60 + 120) and a 10 cm length of run [65]. TLC on a polyamide layer with one of the usual simple solvents is certainly better in this example; the separation is then accomplished as a result of the differing strengths of the hydrogen bonds between the adsorbent and the various substances to be separated. Various pyrazolones have been separated on silica gel with the solvent ethanol-acetone-chloroform (3 + 30 + 70) [5]. TLC has been employed also in investigating the oxidative degradation of 4-aminophenazone

Table 107. hRf-values of some substances used as analgesics, antipyretics and antirheumatic agents

Free (non-registered) Name	hRf-values				
	I	II	III	IV	V
Novalgin	00	00	00	10	3
Cincophene	elongated	15	elongated	—	—
Acetylsalicylic acid	elongated	50	elongated	83	19
Phenazone (antipyrine)	9	5	19	21	60
Aminopyrine (pyramidone)	16	3	24	4	75
Phenacetin	27	15	18	57	81
Ethenzamide(o-ethoxy-benzamide)	39	42	35	—	—
Paracetamol (N-acetyl-p-aminophenol)	44	00	6	46	67
Phenylbutazone	63	4 spots	63	90	40
Salicylamide	59	26	12	51	59

Explanatory notes:

I, II and III. Layer: silica gel HF_{254}, Merck, prepared using the standard method; length of run: 10 cm; 10 μg applied (30 μg of ethenzamide).

Solvent I: butanone; II: cyclohexane-chloroform-acetic acid (40 + 50 + 10); III: cyclohexane-chloroform-pyridine (20 + 60 + 5) [63].

IV and V Layer. silica gel G, Merck, prepared by pouring by hand. Chromatograms developed on slanted plates in shallow dishes; length of run: 13 cm; 10 μg of each applied, dissolved in chloroform or chloroform-methanol (1 + 1).

Solvent IV: butyl acetate-chloroform-85% formic acid (60 + 40 + 20).

Solvent V: butyl acetate-acetone-n-butanol-10% ammonium hydroxide (50 + 40 + 30 + 10) [177].

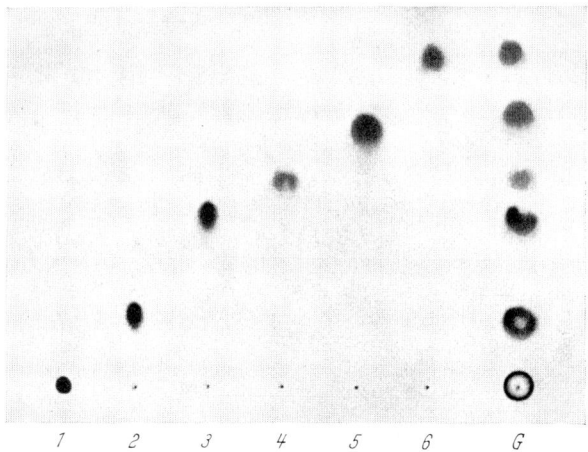

Fig. 163. Separation of novalgin (1), phenazone (2), aminopyrine (3), ethenzamide (4), salicylamide (5), phenylbutazone (6), mixture (G). Layer: silica gel H; solvent: cyclohexane-acetone (40 + 50); CS; 10 cm run; detection: exposure to iodine atmosphere; amounts applied: 10 μg of 1, 2, 3, 5 and 6; about 30 μg of 4

[126]; in testing processes of hydrolysis of metamizole-type pyrazolones [127]; and in studying the metabolites of aminopyrine [128, 146] and nicotinoylaminophenyldimethylpyrazolone [79] after passage through the body. The detection of phenacetin and N-acetyl-γ-aminophenol and of their metabolites in urine has likewise been reported [139].

It has been possible with the help of TLC, to detect diplosal (salicylsalicylic acid) in free salicylic acid [9]. The impurities, chloroacetanilide and acetanilide, could similarly be detected in phenacetin [57a, 143, 164a].

FRESEN [57a] has hydrolysed with acid, treated the mixture with ammonium hydroxide, extracted the hydrolysis products (phenetidine, p-chloroaniline and possibly aniline) with ether, applied aliquots of the ether extract to a silica gel layer and developed with methylene dichloride. A 3% solution of permanganate in concentrated sulphuric acid was used for detection. THOMA's method [164a] is even more sensitive; he carries out acid hydrolysis, renders basic with ammonium hydroxide, dilutes with methanol and applies this solution to the thin layer. The p-chloroaniline and any other amines present are detected by diazotisation and coupling. Down to 10 ppm of p-chloroacetanilide can be detected in this way.

The fractions obtained from the analysis procedure of MÜHLEMANN and BÜRGIN in toxicological and analytical investigations of pharmaceutical interest, can also be further examined with TLC [57]. Data on the chromatography of pharmaceutical commercial preparations containing the substances quoted in Table 107, can be found in the relevant section I of this chapter.

The substances in Table 107 can be detected on fluorescent layers by irradiation with UV light or by exposure to iodine vapour (cf. Table 108). Treatment with the chlorine-o-tolidine reagent (No. 42) is worth consideration as a further, though non-specific test [63]. The test depends sensitively on the duration of chlorination, which should last only 10 sec. The other reagents from Table 108 can also be used for some of the substances [65]. After thin-layer chromatographic separation, pyrazolones can be detected very sensitively with ferricyanide-ferric chloride reagent (No. 111) [65, 117]. As seen from Table 108, paracetamol, phenacetin and salicylamide also respond well to this reagent. Acetylsalicylic acid itself does not react; but it can often be identified through the colour given by the salicylic acid, which is formed from it by thermal decomposition during evaporation of the solvent at higher temperature [65]. As well as ferric chloride solution [127, 128, 146], a mixture of 10% aqueous ferric chloride solution and 1% p-dimethyl-aminobenzaldehyde solution in N hydrochloric acid (1 + 2) has been used to detect pyrazolone derivatives [5]; various colours are yielded, depending on the structure of the compound. Pyrazolones have been identified also by spraying first with the Dragendorff reagent of THIES and REUTHER (No. 98), and then with 2% sulphuric acid and also by spraying with Fast Blue salt solution (Rgt. No. 100) [128]. Nicotinoylaminophenyldimethylpyrazolone has been detected through the colour with 1.5% ethanolic picryl chloride solution; the colour can be intensified by exposure to ammonia vapour [79].

Table 108. *Methods of detection of some of the substances quoted in Table 107* [65]

Substances	Permanganate-reagent (No. 199)	Ferricyanide-FeCl$_3$ (No. 111)	Silver nitrate-NH$_4$OH (No. 225)	Iodine vapour (No. 141)	UV$_{254}$ nm
Acetylsalicylic acid	(+)	—	—	+	+
Phenazone	—	+ (brown)	—	+	+
Aminopyrine	(+)	+ (blue)	+	+	+
Phenacetin	—	+ (blue)	—	+	+
Paracetamol	+	+ (blue)	+	+	+
Salicylamide	+	+ (violet)	—	+	+

b) Analgesics with Narcotic Activity

This group of substances comprises opiates and synthetic drugs with similar activity. Knowledge of the chromatographic behaviour of these substances amongst themselves is important especially in toxicological investigations; the narcotics which occur naturally and those synthetically obtained are therefore described together below. The TLC of the opiates in connection with other alkaloids is mentioned in the appropriate chapter, N, pp. 436—444).

A procedure for isolating narcotic analgesics from biological material has been worked out by MULÉ [106].

A 6 ml sample of urine, plasma or tissue homogenate (10% in 0.1 N HCl) in a 40 ml glass stoppered centrifuge tube is adjusted to pH 10.0 with 2.5 N NaOH and buffered with 3 ml potassium phosphate buffer of pH 10.4. The solution is saturated with 2 g NaCl and mixed with 15 ml ethylene dichloride, containing 25% isobutanol by volume, for 30 min on a shaking machine. After subsequent centrifuging, a 4 ml aliquot of the upper organic layer is evaporated to dryness in a 15 ml centrifuge tube on a water bath in an atmosphere of nitrogen. The residue is dissolved in 25 to 100 μl methanol and the solution applied to the layer. In order to isolate the iminoethanophenanthrenes, 15 ml pure ethylene dichloride is used instead of the solvent mixture mentioned.

Other investigators have extracted narcotics from urine in a similar way [33]. Since most morphine derivatives and synthetic narcotics are eliminated as glucuronides by the human body, hydrolysis has been carried out by heating the urine with one-tenth of its volume of concentrated hydrochloric acid for 1 h at 100° C [33]. Acid hydrolysis is also described elsewhere [50].

EBERHARDT and NORDEN [45] have extracted narcotics from urine by making alkaline with ammonium hydroxide and shaking with ethyl acetate. Interfering phenothiazines and their metabolites could be previously removed by acidifying the urine strongly with hydrochloric acid and extracting with chloroform.

TLC of analgesics with narcotic action has usually been carried out on silica gel G layers (cf. Table 109a). Alumina G [33] and cellulose

Table 109a. *Experimental condition for TLC of analgesics with narcotic activity*

Cond. No.	Layer, preparation, etc.	Solvent	Refs.
I	Silica gel G. Coarser particles sieved out with fine sieve of Swiss Pharmacopoeia. Layer prepared by standard method, p. 85	Methanol-acetone-triethanolamine (50 + 50 + 1.5)	[10]
II	Silica gel G	0.1 N ammonia in CH_3OH	[168]
III a	Silica gel G. 10 cm run. 5—10 µg hydrochloride salt applied	Ethanol-pyridine-dioxan-water (50 + 20 + 25 + 5)	[33]
IV a	as III a	Ethanol-dioxan-benzene-NH_4OH (5 + 40 + 50 + 5)	[33]
V a	as III a	Methanol-n-butanol-benzene-water (60 + 15 + 10 + 5)	[33]
III b	Silica gel G. Standard method. 10 cm run. Reference substances applied as free bases: 1—15 µg as 0.1% solutions in methanol or ethyl acetate. 1—25 µg in 25—100 µl methanol of the extracted substances applied	as III a	[106]
IV b	as III b	as IV a	[106]
V b	as III b	as V a	[106]
VI	Silica gel G. Pure substances applied as free bases	Dimethylformamide-ethyl acetate (25 + 75)	[45]
VII	Silica gel; standard method, CS. 1 hour equilibration. 30 ml beaker containing 10 ml 28% NH_4OH placed in chamber 20 min before TLC. Chamber newly equilibrated for each run. 25 µg salt of the analgesics applied in 5 µl absolute ethanol. 10 cm run	Acetone	[49]
VIII	Silica gel G; 30 g slurried with 60 ml 0.5 N LiOH and spread. Heated 1 h at 110° after drying. No ammonia in the chamber, otherwise as in VII	Acetone	[49]
IV c	Silica gel G layers. Standard method. Activated 2 h at 100°. Freshly prepared solvent for development of each new chromatogram. 10 cm run	as IV a	[157]

layers (MN-cellulose, Firm 83), adjusted to pH 8 with phosphate buffer [106] have occasionally been used also.

The active substances are generally isolated from biological material or from pharmaceutical preparations in the form of bases, whereas the reference substances are usually salts. If basic solvents are selected, the bases are formed from these salts without any extra preliminary treatment and hR_f-values can be compared. Solvents of this sort have thus been used in many investigations, as seen in Table 109a. A further general technique, used also in TLC of narcotic analgesics, is to use neutral solvents but basic layers [49, 91]. EMMERSON and ANDERSON [49] have compared this procedure, using layers alkalised with lithium hydroxide, with another method; in this latter method, TLC was carried out in chambers saturated with a neutral solvent and in which a small beaker containing ammonium hydroxide was placed shortly before introducing the plate with silica gel layer (cf. Table 109a). The relative migration distances in the two methods agreed well. The ammonia-method has the advantage that, after evaporation of the ammonia, chromatography with a neutral solvent in a second dimension can be carried out.

Some authors have used the solvent mixture IV, Table 109a, which contains ammonia [33, 50, 106, 157]. Agreement between the hR_f-values is adequate, as seen from Table 109b. In two of the publications mentioned [33, 106], the use of a pyridine-containing mixture, III (Table 109a) and the acid solvent, n-butanol-n-dibutyl ether-acetic acid (40 + 50 + 10), was quoted. Reference may be made also to neutral solvents (Table 109a and [25]) which have been employed in the chromatography of narcotic bases. Other more or less modified solvents can be found in the articles quoted and in [41].

It has been possible to separate morphine, pethidine, codeine, methadone, nalorphine and propoxyphene in two-dimensional TLC on silica gel G with solvent V of Tables 109a and b in the first dimension and solvent IV in the second [33]. The frequently abused narcotics morphine, codeine, oxycodone and pethidine have likewise been chromatographed in two dimensions [106]; solvent IV (Table 109a) was used in the first direction, mixture III in the second. Morphine, codeine, thebaine, papaverine and noscapine (2 µg amounts or 1 µl of opium tincture) could be separated in a one-dimensional procedure on silica gel G, using the solvent xylene-butanone-methanol-diethylamine (40 + 40 + 6 + 2) and a 13.5 cm length of run [13]. KUPFERBERG et al. [91] have used solvent V (Table 109a) in order to separate normorphine, paramorphan, morphine, 6-monoacetylmorphine and N-allylnormorphine on silica gel G layers which had been brought to pH 8.5 with 5N sodium hydroxide. A thin-layer chromatographic or electro-

Table 109b. *hRf-values of various analgesics with narcotic activity*

Free (non-registered) Name	Chemical term	I	II	III a	III b	IV a	IV b	IV c	V a	V b	VI	VII	VIII
	Iminoethanophenanthrofurans												
Normorphine	3,6-Dihydroxy-4,5-epoxy-morphin-7-ene			14	8	5	4		16	7		5	0
Morphine	3,6-Dihydroxy-4,5-epoxy-N-methyl-morphin-7-ene	40	45	40	29	17	11	18	34	21	24	12	4
Nalorphine[a]	3,6-Dihydroxy-4,5-epoxy-N-allylmorphin-7-ene			82	71	34	35		75	67			
Codeine	3-Methoxy-6-hydroxy-4,5-epoxy-N-methylmorphin-7-ene	43		40	30	46	39	28	32	25	37	30	9
Codethyline	3-Ethoxy-6-hydroxy-4,5-epoxy-N-methylmorphin-7-ene	37			33	46	40		27				
Diamorphine	3,6-Diacetoxy-4,5-epoxy-N-methylmorphin-7-ene				37	76			35				
Thebacon	3-Methoxy-6-acetoxy-4,5-epoxy-N-methylmorphin-6-ene	31							55				
Thebaine	3,6-Dimethoxy-4,5-epoxy-N-methyl-morphina-6,8-diene	41					65						
Paramorphan	3,6-Dihydroxy-4,5-epoxy-N-methylmorphinan			15		10	9		10				
Hydromorphone	3-Hydroxy-6-oxo-4,5-epoxy-N-methyl-morphinan	28		19	11	22	17	18	18	13	16		
Hydrocodone	3-Methoxy-6-oxo-4,5-epoxy-N-methyl-morphinan	29	26				36		22				
Oxycodone	14-Hydroxy-3-methoxy-6-oxo-4,5-epoxy-N-methyl-morphinan			60		46	87		29	80			

Table 109b. (Continued)

Free (non-registered) Name	Chemical term	I	II	III		IV			V		VI	VII	VIII
				a	b	a	b	c	a	b			
Isoquinolines													
Noscapine	2-Methyl-8-methoxy-6,7-methylene-dioxy-1(6′,7′-dimethoxy-phthalidyl)-(3′)-)-1,2,3,4-tetrahydroiso-quinoline	82						80					
Papaverine	1-(3′-4′-Di-methoxy-benzyl)-6,7-dimethoxy-isoquinoline	82						73					
Iminoethanophenanthrenes													
Levor-phanol	L-3-Hydroxy-N-methyl-morphinan	28		34	11	62	80	57	29	10	20		
Levo-methorphan	L-3-Methoxy-N-methyl-morphinan				13		91			8			
Dextro-metho-morphan	D-3-Methoxy-N-methyl-morphinan	23	26									47	7
Levallor-phan[a]	L-3-Hydroxy-N-allyl-morphinan				65		98			41			
Benzomorphans													
Phenazo-cine	D,L-2′-Hydroxy-5,9-dimethyl-2-phenethyl-6,7-benzo-morphan			90	88	93	97	96	59	82			
Arylpiperidines													
Pethidine	Ethyl 1-methyl-4-phenyl-piperidino-4-carboxylate	56	55	51	42	90	97	63	44	36	62	61	19
Anileridine	Ethyl 1-(4′-aminophene-thyl)-4-phenyl-piperidino-4-carboxylate							95				77	62
Alpha-prodine	Ethyl D,L-1,3-dimethyl-4-phenylpiper-idino-4-carboxylate				39		93	83		34			

34*

Table 109b. (Continued)

Free (non-registered) Name	Chemical term	I	II	III a	III b	IV a	IV b	IV c	V a	V b	VI	VII	VIII
Keto-bemidone	1-Methyl-4-(3'-hydroxy-phenyl)-piperidyl-(4)-ethyl ketone	56	48		31		47			24	41		
	Diarylalkoneamines												
Methadone	D,L-6-Dimethyl-amino-4,4-di-phenylheptan-3-one	48	42	53	34	96	99	79	20	17	86	79	48
Nor-methadone	6-Dimethyl-amino-4,4-di-phenylhexan-3-one		59								71		
Acetyl-methadol	6-Dimethyl-amino-4,4-diphenyl-3-acetoxyheptane				64		99			40			
Propoxy-phene	D-4-Dimethyl-amino-3-methyl-2-propionyloxy-1,2-diphenyl-butane			80	73	97	97		53	54		80	64
Dextromor-amide	D-2,2-Diphenyl-3-methyl-4-morpholino-butyryl-pyrrolidine	87	85								96		

[a] Morphine antagonists.

phoretic method for detection of ketobemidone in a toxicological examination may also be mentioned [76].

The narcotic analgesics have been detected principally with the non-specific alkaloid reagents, iodoplatinate (No. 147) and Dragendorff reagent (Nos. 96—98). Narcotics with a phenol group and phenanthrene structure may be specifically detected on alkalised silica gel G layers by spraying with a solution of 57 mg potassium ferricyanide and 7.8 mg potassium ferrocyanide in 100 ml water and inspection in UV light [91]. Fluorescent pseudomorphine analogues are thus formed by oxidation. The concentration of the reagents and the pH value of the silica gel layer have a profound influence in determining the best fluorescence development.

5. Anticoagulants of the 4-Hydroxycoumarin Group

The conditions used in TLC of these substances are seen in Table 110.

Table 110. *TLC of some anticoagulants* [136]

Free (non-registered) name	Chemical term	hRf	Detection a	b	c	d	e
Ethyl biscoumacetate	Ethyl bis-(4-hydroxy-3-coumarinyl)-acetate	67—69	+	+	—	+[a]	+
Phenprocoumon	1-(4'-hydroxy-3'-coumarinyl)-1-phenylpropane	50—53	+	+	+[a]	+[a]	+
Acenocoumarol	1-(4'-hydroxy-3'-coumarinyl)-1-(4'-nitrophenyl)-butan-3-one	19—20	+	—	—	+[a]	+
Dicoumarol	Bis-(4-hydroxy-3-coumarinyl)-methane	10	—	—	+[a]	+[a]	+
Warfarin	1-(4'-hydroxy-3'-coumarinyl)-1-phenyl-butan-3-one	25—27	+	+[a]	—	+[a]	+[a]

Layer: Silica gel G.
Solvent: 80 ml benzene, saturated with 99% formic acid at room temperature and then mixed with 4 ml butanone.
Detection: a) Tincture of iodine (German Pharmacopoeia, DAB VI)-ethanol (1 + 5)
b) 5% aqueous ferric chloride solution
c) Ehrlich's reagent (No. 72)
d) Reagent b + reagent c (1 + 1), freshly prepared for each use
e) 5% aqueous iron acetate solution
+ Positive test.
— Negative test.

[a] Heated 20 min at 100° C after spraying.

6. Hypnotics

a) Barbiturates

In order to detect barbituric acids in organs or body fluids, they must first be concentrated and purified in procedures preceding TLC separation. Processing can be as follows:

The material to be examined is extracted with ethanol on the water bath; after evaporating off the ethanol, the residue is taken up in water, the solution acidified with tartaric acid and the barbituric acids extracted with ether. Any strongly acid contaminants which are extracted at the same time may be separated from the barbituric acids by shaking the ether solution with a buffer of pH 4.4.

The pre-purification of barbiturates to be isolated from urine can likewise be carried out by acid ether extraction [43, 56, 166]. Methylene dichloride has been employed also [34]. Acidification with hydrochloric acid is preferable since, if this is performed with weaker, organic acids, amphoteric medicinal components like sulphonamides, are extracted at the same time [166]. After drying the organic phase with sodium sulphate, further purification on a charcoal or alumina column can be carried out [43, 166]. Urine colouring materials can be removed by shaking the ethereal solution with 5% lead acetate solution [56]. After purification, the organic phases are evaporated to dryness in vacuo, the residue dissolved in ethyl acetate or ethanol and applied to the thin layer. Before application, the barbiturates can, if necessary, be further purified by sublimation.

Blood and tissue are brought to pH 5 with hydrochloric acid and the barbituric acids and their metabolites extracted with methylene dichloride [34]; prior homogenisation of tissue is carried out with isotonic potassium chloride solution. Barbituric acids can be isolated from serum by adding 0.1 ml concentrated hydrochloric acid and 2 g anhydrous sodium sulphate to 3 ml serum and extracting the mixture with 15 ml chloroform. 10 ml of this extract are evaporated down, the residue dissolved in 0.2 ml 70% ethanol and an aliquot applied to the thin layer [93].

In contrast to these purely acid processing procedures, SUNSHINE [160] has extracted the barbiturates from stomach fluid and urine under alkaline conditions; the solutions were filtered and acidified and the barbituric acids obtained by shaking with chloroform.

Silica gel has been the adsorbent mainly used in TLC of barbituric acids [10, 34, 43, 56, 77, 93, 122, 140, 152, 160, 166, 174]. Chromatography has been carried out also on so-called mixed layers (silica gel G

Table 111. *hRf-values of barbituric acids under various conditions*

Free (non-registered) name	Chemical term	I-hRf				I-R_{st}[a]	II-hRf		
		a	b	c	d	e	a	b	c
1. Barbital	5,5-Diethyl-B	49	24	28	50	1.00	45	47	49
2. Phenobarbital	5-Ethyl-5-phenyl-B	54	25	30	50	1.00	36	37	37
3. Hexethal	5-Ethyl-5-hexyl-B	—	—	—	—	1.56	—	—	—
4. Butobarbital	5-Ethyl-5-n-butyl-B	—	—	39	62	1.41	—	64	60
5. Butabarbital	5-Ethyl-5-sec.-butyl-B	—	32	34	—	1.28	—	—	64
6. Amobarbital	5-Ethyl-5-isopentyl-B	56	38	41	60	1.44	—	67	66
7. Pentobarbital	5-Ethyl-5-(1'-methyl-butyl)-B	63	36	41	57	1.42	65	73	66
8. Melidorm	5-Ethyl-5-(2-butenyl)-B	—	—	41	—	—	—	—	61
9. Vinbarbital	5-Ethyl-5-(1'-methyl-Δ^1-butenyl)-B	—	—	—	—	1.19	—	—	—
10. Heptabarbital	5-Ethyl-5-(Δ^1-cyclo-heptenyl)-B	55	—	37	—	1.32	55	—	55
11. Cyclobarbital	5-Ethyl-5-(Δ^1-cyclo-hexenyl)-B	55	32	35	68	F	50	54	52
12. Allobarbital	5,5-Diallyl-B	63	32	36	55	1.30	—	50	50
13. Aprobarbital	5-Allyl-5-isopropyl-B	57	31	38	—	1.30	—	67	57
14. Butalbital	5-Allyl-5-(sec.-butyl)-B	60	38	41	67	—	—	—	62
15. Secobarbital	5-Allyl-5-(1'-methyl-butyl)-B	70	41	44	64	1.67	—	73	67
16. Talbutal	5-Allyl-5-(3'-isobutyl)-B	—	—	40	—	1.41	—	—	64
17. Cyclopal	5-Allyl-5-cyclopenten-(2')-yl-B	—	36	35	—	1.38	—	—	57
18. Vinylbital	5-Vinyl-5-(methyl-butyl)-B	—	—	38	—	—	—	—	56
19. Dormovit	5-Isopropyl-5-furfuryl-B	—	—	35	—	—	—	—	57
20. Axeen	5-Allyl-5-(2-hydroxy-propyl)-B	—	—	3	—	—	—	—	27
21. Propallylonal	5-Isopropyl-5-(2-bromoallyl)-B	—	—	34	—	—	—	—	53

Table 111 (Continued)

Free (non-registered) name	Chemical term	I-hRf a	b	c	d	I-R_{St}[a] e	II-hRf a	b	c
22. Butallylonal	5-Sec-butyl-5-(2-bromoallyl)-B	—	—	42	—	1.50	75	—	60
23. Sigmodal	5-Sec-amyl-5-(2-bromoallyl)-B	—	—	44	—	1.77	—	—	66
24. Methylphenobarbital	5-Ethyl-5-phenyl-1-methyl-B	—	53	53	98	—	62	76	62
25. Metharbital	5,5-Diethyl-1-methyl-B	—	—	—	85	2.35	—	—	—
26. Methohexital	5-Allyl-5-(1'-methyl-2'-pentinyl)-1-methyl-B	—	—	—	77	—	—	—	—
27. Hexobarbital	5-Methyl-5-(1-cyclohexen-1-yl)-1-methyl-B	—	46	50	92	2.06	70	76	76
28. Narcobarbital	5-Isopropyl-5-(2-bromoallyl)-1-methyl-B	—	—	62	—	—	79	—	82
29. Inactin	5-Ethyl-5-(1'-methylpropyl)-2-thio-B	—	—	0	—	—	59	—	5
30. Thiopental	5-Ethyl-5-(1'-methylbutyl)-2-thio-B	—	68	58	94	F	67	—	70
31. Buthalital	5-Allyl-5-(2'-methylpropyl)-2-thio-B	—	—	63	—	—	—	—	65
32. Thiamylal	5-Allyl-5-(1'-methylbutyl)-2-thio-B	—	—	—	95	F	—	—	—
33. Methitural	5-(2-Methylthioethyl)-5-(1'-methylbutyl)-2-thio-B	—	—	66	95	—	—	—	75

[a] Referred to phenobarbital = 100.
F = in the solvent front.
-B = -barbituric acid.

Solvent I: Chloroform-acetone (90 + 10).
Layer a: Silica gel G; coarser particles removed with the fine sieve of the Swiss Pharmacopoeia; standard method (cf. p. 85); 10 cm run; method for toxicological studies [10].
Layer b: Silica gel G; standard method: 20 μg of each barbituric acid applied in acetone solution; method for testing pharmaceutical preparations [140].
Layer c: Silica gel G; standard method; chamber saturation; study carried out with pure substances [106].
Layer d: Silica gel G, prepared according to [24]; activated by heating 30 min at 80° C, then stored in a desiccator; 10 cm run; 1—20 μg barbituric acid applied in solution in 5—10 μl chloroform or ethanol; method aimed at analysis of organ extracts, metabolites also mentioned [34].
Layer e: Silica gel G; standard method; 10—12 cm run; method for identifying barbiturates after extraction from organ constituents [160].
Solvent II: Isopropanol-chloroform-25% ammonium hydroxide (45 + 45 + 10).
Layer a: Silica gel; standard method; chamber saturation; 10 cm run; 20 to 50 μg barbituric acids or their Na-salts, dissolved in ethyl acetate or in water, applied as standard; method for identifying hypnotics and their metabolites in aqueous solutions and urine [56].
Layer b: Silica gel G; layers air dried 30 min, then heated 1 h at 150° C; 2 μg of each barbituric acid applied in ether solution at 1 cm from the lower edge of the plate; 10 cm run; method for pharmaceutical testing of the barbiturates commonly used in Great Britain [152].
Layer c: see Ic [166].

+ alumina G, 1 + 1) [170] and on alumina[2] layers without binder [142]. The principal solvents have been neutral and basic mixtures (cf. Table 111). According to FRAHM and co-workers [56], barbituric acids can be separated with acid solvents also.

As is seen in Table 111, there are large differences in the hRf-values quoted by different authors for the same solvents.

Despite conflicting findings [56] it seems evident that the hRf-values are not, as is usual, determined alone by the experimental, chromatographic conditions like thickness and activation of the layer, chamber saturation etc., but depend on whether a single barbituric acid or a mixture is being chromatographed [152]. The migration distances can thus be compared only when referred to a standard substance which is chromatographed simultaneously. This yields the corresponding R_{St}-values (cf. Table 111 and p. 127). Those in Table 111 are referred to phenobarbital as 1.00 [160]. Other authors have chosen barbital [152] or a non-barbiturate [166] as reference standard.

Phenobarbital, butabarbital, secobarbital or barbital and aprobarbital have been separated for example with solvent I in Table 111 [140]. Phenobarbital, cyclopal, pentobarbital or barbital, phenobarbital, butalbital or cyclobarbital and hexobarbital have been separated on silica gel, using benzine (Swiss Pharmacopoeia)-dioxan (75 + 30) [140]. Barbital, butobarbital and methylphenobarbital could be separated with chloroform-ether (75 + 25) on silica gel [32]. Among the eleven barbituric acids in column IIb of Table 111, phenobarbital, barbital, cyclobarbital, butobarbital and pentobarbital could be separated [152]. It has been possible to separate up to eight of the barbituric acids in column IIb, Table 111, using two-dimensional chromatography with the solvents II (first dimension) and diisopropyl ether-chloroform-benzene (13 + 8 + 4) (second) [152]. Some barbituric acids could be separated also on microscope slides which had been coated with silica gel [83]. Barbiturates, N_1-methylated barbiturates and hypnotics with other structures, have been fractionated into groups by chromatography on silica gel, using ether-butanol-25% ammonium hydroxide (90 + 10 + 10) [174].

Detection of barbituric acids is possible non-specifically by chromatography on so-called fluorescent layers, e. g., silica gel GF_{254}. If not present in too small amounts, they appear in short wave UV light as spots of quenched fluorescence. A popular method of detection has been with mercury compounds, especially mercurous nitrate (Rgt. No. 157) which yields insoluble barbiturates [10, 34, 43, 56, 77, 140, 166]. The sensitivity of response varies with the barbituric acid. A combined spray reagent of a mercuric salt and diphenylcarbazone solutions (No. 156) has been often used: the spots then stand out better against the background [32, 93, 160]. Cobalt salts have been used with various bases for detection (Zwikker-reaction and modifications, Rgt. No. 51) [56, 152, 170, 174]. N-methylated barbituric acids do not react since the second nitrogen atom cannot take part in enolisation [98]. A further, non-specific method

---

[2] Alumina for analysis, "Lachema" (CSN 68 51 31; grain size 0.075).

of detection is with silver nitrate-dichromate (Rgt. No. 229) [174]. Barbituric acids with double bonds in the C_5 side chain react with 1% permanganate solution. Allobarbital reacts immediately whereas cyclobarbital and hexobarbital are sterically hindered and react more slowly [32]. 2-Mercapto-barbituric acids also react with permanganate.

A more specific method of detecting barbituric acids is through the murexide reaction [86a, 173]; and of thiobarbituric acids, through the iodine-azide reaction (Rgt. No. 142) [173]. Bromine-containing barbituric acids may be characterised by oxidation yielding bromine which then converts fluorescein into eosin (Rgt. No. 119). Larger amounts of barbituric acids and ureides, separated by TLC, can be removed from the layer by microsublimation and identified through their crystal form [8].

Quantitative evaluation of barbituric acid chromatograms is similarly possible when the instability of these compounds in alkaline medium is taken into consideration [156]. The author [64] has determined methylphenobarbital, after separation from phenobarbital, by elution and measuring the UV-absorption of the solution (see Table 14, p. 152). This method can be used only when the material under investigation is extracted with alcohol or another neutral solvent and chromatography is carried out with a neutral or acid solvent. The barbituric acid spots must then be eluted also with a neutral solvent and measurement made directly after adding a buffer (40.5 ml 0.1 N NaOH + 59.5 ml 0.05 M borax). Since MORRISON and CHATTEN [104] extracted barbituric acids from the drug under alkaline conditions and chromatographed with isopropanol-chloroform-25% ammonium hydroxide on silica gel layers which had been prepared with 0.1 N NaOH, they were unable to use the UV spectrophotometric method with success. After chromatographic separation, they therefore sprayed with mercuric nitrate, sucked the spots, after drying, on to small sintered glass filters, eluted with water and added mercuric chloride solution (300 mg + 1 ml N HCl, made up with water to 100 ml) and phosphate buffer of pH 8.0. The resulting cloudy solution was extracted with chloroform and dithizone solution added, yielding an orange mercury derivative ($Hg(HDz)_2$), soluble in chloroform, which was colorimetrically determined at 475 nm.

b) Hydantoins

The hydantoin derivatives in Table 112 have been chromatographed on silica gel. Detection is possible on fluorescent layers, using short wave UV light [98] or with the mercury reagents already described for barbituric acid detection [140]; the limit of detection is about 10 μg. The combined mercuric salt-diphenylcarbazone reagent, mentioned in connection with barbituric acids, has also been employed [129].

Table 112. *hRf-values of hydantoin derivatives*

Free (non-registered) name and chemical term	hRf-values		
	I	II	III
Phenytoin: 5,5-Diphenylhydantoin	19	16	33
Mephenytoin: 3-Methyl-5-ethyl-5-phenylhydantoin	43	29	35
Phenyl-dibromoethyl-methylhydantoin	16	15	21

Solvents: I Chloroform-acetone (90 + 10); II Benzine (Swiss Pharmacopoeia)-dioxan (75 + 30); III Benzene-ether (50 + 50).
See Table 111, I b [140] for other chromatographic conditions.

c) Bromoureides

During the customary processing of organ extracts, the bromoureides, as neutral substances, pass along with the barbituric acids from the acid stock solution into the ether or chloroform phase. The barbituric acids can be removed from the organic phase with weakly basic aqueous solutions and thus separated from the bromoureides.

Bromisoval and carbromal and the corresponding metabolites have been separated on silica gel, using chloroform-acetone (90 + 10) [11, 95]. They can be differentiated in other solvents also (cf. Table 113). The experimental conditions under IV, Table 113, are suitable for detecting these two substances along with acetylcarbromal in pharmaceutical preparations [63] (cf. Fig. 71, p. 129). Sensitive but non-specific *detection* can be achieved with the chlorine-tolidine reagent (No. 42), reacting with N-active hydrogen; carbromal reacts the most feebly [63]. Silver nitrate-ammonium hydroxide (Rgt. No. 225) can also be used for non-specific detection [65]. The method described for detecting bromine-containing barbituric acids, depending on conversion of fluorescein into eosin (Rgt. No. 119), ought to be a more specific test for bromoureides. An identification of carbromal has been described which depends on reaction of the bromine content to form "Wurster's red" (Rgt. No. 77) [11].

d) Other Hypnotics

Table 113 also contains experimental conditions for TLC of other frequently used hypnotics. Some of these compounds may be detected (non-specifically) with the mercurous nitrate reagent (No. 157), quoted for the barbituric acids; sensitivity is generally low, however. In this way, pyrithyldione [56], methyprylone [34, 56], glutethimide [32, 34, 56, 122, 174], ethinamate [34, 56, 174] and the barbiturate antagonist benegride [56] have been detected on silica gel layers. The test is so insensitive for pyrithyldione and methylprylone, however, that it has been quoted as negative by some authors [122, 174]; thalidomide (contergan) does not react. The non-specific detection with the chlorine-tolidine reagent (No. 42) has been cited as quite sensitive for thalidomide [123], glutethimide [94, 123], meprobamate [94], hexapropymate [94] and ethinamate [94]. PAULUS and KEYMER [123] have used chlorine which had been dried over calcium chloride, in order to prevent the whole plate from darkening. Intensive chlorination is needed for the identification of thalidomide.

Thalidomide cannot be extracted from acid solution. A solution in N sodium hydroxide can be applied to the plate. TLC can be carried

Table 113. hRf-values of various non-barbiturates used as hypnotics

Free (non-registered) name[a] and Chemical term	hRf-values						
	Ia	Ib	Ic	II	III	IV	V
1. Apronal 2-Isopropyl-4-pentenoylurea	—	—	—	—	—	—	22
2. Bromisoval 2-Bromo-3-methyl-butyrylurea	—	—	—	—	60	27	23
3. Carbromal 2-Bromo-2-ethyl-butyrylurea	—	—	—	—	75	68	54
4. Acetylcarbromal 1-Acetyl-3-(2-bromo-2-ethylbutyryl)-urea	—	—	—	—	—	48	39
5. Ethinamate 1-Ethinylcyclohexyl carbamate	44	—	81	82	45	—	50
6. Hexapropymate 1-(2-propinyl-)cyclohexyl carbamate	—	—	—	—	—	—	61
7. Meprobamate 2-Methyl-2-n-propyl-1,3-propanediol dicarbamate	—	—	—	—	—	—	5
8. Centalun 2-Methyl-1-phenyl-3-butyne-1,2-diol	37	—	—	99	—	—	—
9. Methyprylone 2,4-Dioxo-3,3-diethyl-5-methylpiperidine	23	35	50	84	50	—	—
10. Dihydroprylone 2,4-Dioxo-3,3-diethyl-tetrahydropiperidine	40	58	—	82	40	—	—
11. Glutethimide 2-Ethyl-2-phenyl-glutarimide	57	78	80	88	90	—	66
12. Thalidomide N-(2,6-Dioxo-3-piperidyl)-phthalimide	—	45	—	—	—	—	—
13. Methaqualone 2-Methyl-3-o-tolyl-4(3H)-quinazolinone	59	—	—	90	80	—	—

Key to Table 113

I. Solvent: Chloroform-acetone (90 + 10); Conditions: a — see Table 111, Ic [166]; b — see Table 111, Ia [10]; c — see Table 111, Id [34]; d — see Table 111, Ib [140].

II. Solvent: Isopropanol-chloroform-25% ammonium hydroxide (45 + 45 + 10). See Table 111, Ic for conditions [166].

III. Solvent: Petrol ether (BP 50—75°)-pyridine (75 + 15); silica gel G layer; standard method (cf. p. 85). Testing soporifics and their metabolites in urine [43].

IV. Solvent: Chloroform-cyclohexane-pyridine (60 + 20 + 5); silica gel G layer; standard method; chamber saturation; 10 cm run. Solvent removed after TLC by 15 min/150° C. Method for testing pharmaceutical preparations [63].

V. Solvent: Chloroform-diethyl ether (85 + 15); silica gel G layer; standard method. Aim of the work was forensic investigation [94].

[a] If not known, commercial name is given.

out with solvent I in Table 113 or with dimethylformamide-methanol-water (25 + 70 + 5). Benzine-pyridine (80 + 20) has also been used for the separation of thalidomide and glutethimide [123]. Detection is possible with the non-specific methods mentioned and also with Zwikker's reagent (No. 51) or after conversion into the corresponding hydroxamic acid (Rgt. No. 137) [56, 153]. Glutethimide may in addition be detected with the Dragendorff reagent (No. 96—98), formaldehyde-sulphuric acid (Rgt. No. 125) [167] or through its strong UV-absorption [69]. Methaqualone can be extracted from acid solution with ether. It can be localised with modified Dragendorff reagent or with p-dimethylaminobenzaldehyde in hydrochloric acid [69]; with the latter, it yields an intensive blue-red spot after 48 h. Methyprylone may be visualised specifically by spraying alternately with 20% KOH and 1% methanolic m-dinitrobenzene [40]. BURGER [29] has given information about the TLC of centalun(2-methyl-1-phenylbutyne-1,2-diol).

7. Bactericidal and Bacteriostatic Substances
a) Phenols of Pharmaceutical Interest

Some compounds used as preservatives (see Chapter TN, p. 636) and many which occur in dermatological preparations belong to this group.

Since the lower phenols are volatile, they are best separated after conversion into involatile, coloured derivatives (azo dyes) (see p. 664) [37, 87, 154]. The TLC of the azo compounds derived from other simple alkyl phenols also has been described in these articles [37, 87]. Kieselguhr layers, impregnated with formamide, have been used as well as strongly alkaline silica gel layers. After separation, the coloured derivatives can be quantitatively eluted from the layer and determined [154]. Mixtures of alcohols and phenols can be satisfactorily chromatographed as their dinitrobenzoates [38]. Numerous phenolcarboxylic acids have been separated on mixed layers of silica gel G and cellulose, using the solvents toluene-ethyl formate-formic acid (50 + 40 + 10) and chloroform-acetic acid-water (80 + 20 + 20). Phenolcarboxylic acids differing little in polarity, can be better separated if the layers are saturated with moisture before use [158].

The chromatographic behaviour of homologous series of mono- and dihydric phenols with a single nucleus, on silica gel G and polyamide layers, has also been described [73]; solvents were benzene-methanol (95 + 5) and n-butyl ether (saturated with water)-acetic acid (90 + 9). Mono- and polynuclear phenols, phenol aldehydes and phenolcarboxylic acids have been chromatographed on silica gel G layers, using various solvents; thin-layer electrophoretic investigations have also been conducted [119, 120, 121]. Using silica gel G layers and, e. g., benzene-dioxan-acetic acid (90 + 25 + 4), it has been possible to separate mononuclear, polyhydric phenols into the groups catechol, resorcinol, hydroquinone; and pyrogallol and phloroglucinol [121]. Silica gel and the solvent benzene-methanol (95 + 5) have been quoted also for separating the three first mentioned dihydric phenols [147]. Polyhydric phenols may be chromatographed also on polyamide layers using acetone- or methanol-water mixtures [48, 71].

Resorcinol and its mono- and diacetates can be separated from one another and from hexachlorophene by TLC after extraction from dermatological preparations; silica gel layers were used, prepared with 0.01 M aqueous sodium tungstate as complexing agent, benzene-dioxan-acetic acid (90 + 10 + 2) as solvent and ferricyanide-ferric chloride (Rgt. No. 111) for detection [64]. Hexylresorcinol and hexachlorophene can be separated on silica gel G layers, using methyl isobutyl ketone [64]. A separation of dichlorophene and hexachlorophene has been possible with n-heptane, saturated with acetic acid, and layers of a hand prepared mixture of silica gel and starch binder [26]. Some of the components of tar oils and their TLC are also of pharmaceutical interest [130]. TLC-data for dithranol have also been described [15]. Iodochlorhydroxyquin(5-chloro-8-hydroxy-7-iodoquinoline) has been separated from possible contaminants originating from its synthesis, by TLC in methanol on polyamide-calcium sulphate layers (5 g polyamide + 3.5 g calcium sulphate + 10 ml water) [90].

Detection; The phenols can be detected on layers containing a fluorescent indicator by inspection in short wave UV light. Visualisation is usually possible with diazonium salts (Rgt. No. 100). Ferricyanide-ferric chloride (Rgt. No. 111) and chlorine-tolidine (Rgt. No. 42) may also be employed. The action of the latter reagent depends on the easy formation of polychlorocyclohexadienones from chlorination of the phenols; the active chlorine in these compounds oxidises the o-tolidine to a dye of the diphenoquinone-diimine type (KI is not necessary) [171].

b) Sulphonamides

Many sulphonamides can be extracted with acetone from pharmaceutical preparations.

For tablets, an amount of powdered sample containing 10 mg sulphonamide is extracted with 50 ml acetone and an amount of solution containing 1—3 μg of the individual sulphonamides applied to the start. This amount should not be exceeded if critical sulphonamide pairs are to be separated. A method has been described also for concentrating smaller amounts of sulphonamides present in feeding stuffs [3].

Silica gel, occasionally also alumina has been the adsorbent principally used in TLC of sulphonamides (cf. Tables 114 and 115). Since many sulphonamides are amphoteric, both acid and basic solvents have found considerable use in their TLC, just as in PC (cf. Tables 114 and 115) [3, 4, 9, 15, 16]. The relatively small chamber volume in TLC, with consequent rapid saturation of the gas phase, leads to more reproducible chromatograms than in PC with its large chambers; TLC is moreover faster.

Several investigators have used a solvent containing diethylamine [63, 68, 102, 134]. In this way, the sulphonamides in the appendix of the German Pharmacopoeia (DAB VI) could be separated [134] (cf. Table 114/4). Numerous sulphonamide mixtures can be separated, however, by using adsorption TLC also with neutral solvents (cf. Table 114/1, 2, 6, 7, 8, 10, 12, 13, 14 and Fig. 164). Good separations are often accomplished with solvents which contain two or more components, as a result of the composition gradients which are then set up. Especially uniform solvent fronts and reproducible hRf-values have been obtained in sulphonamide separations, using a neutral solvent mixture when chromatography was carried out on trapezoidally shaped silica gel layers [82] (cf. Table 114/6 and 115).

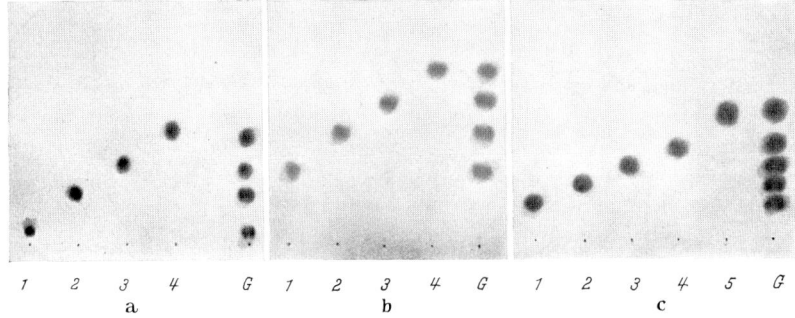

Fig. 164. Separation of sulphonamides; 10 cm run; detection with 4-dimethylaminobenzaldehyde reagent (No. 72). a) sulphaguanidine (1), sulphanilamide (2), sulphapyridine (3), sulphaphenazole (4), mixture (G); layer: silica gel H; solvent: chloroform-ethanol (80 + 10): CS. b) sulphadimidine (1), sulphadiazine (2), sulphamethoxypyridazine (3), sulphadimethoxine (4), mixture (G); layer: silica gel, Woelm; solvent: chloroform-pentane-ethanol (35 + 25 + 30); CS. c) sulphadiazine (1), sulphamethoxypyridazine (2), sulphapyridine (3), sulphaguanidine (4), sulphanilamide (5), mixture (G); layer: silica gel H; solvent: 5% ammonium hydroxide-n-butanol (50 + 50); CS

For separating sulphonamides in trisulphapyridine syrup (US Pharmacopoeia XVI), a neutral solvent has been used which contained a definite amount of water, checked by titration with Karl Fischer reagent [86] (cf. Table 114/5 and 115). The water content appears to be less critical with another neutral solvent mixture [172] (cf. Table 114 and 115/11). The sulphonamides commonly used in Canada have been separated in this way. Sulphonamide separations have been carried out also on inclined plates in appropriately shaped chambers [177]. Sulphanilamide, sulphacetamide, sulphacarbamide, sulphaguanidine, sulphaphenazole and sulphamethoxypyridazine could be thus separated in a 13 cm run, using the solvent butyl acetate-n-butanol-acetone-10% ammonium hydroxide (30 + 30 + 40 + 10); and sulphathiazole, sulphaethidole, sulphisomidine and sulphamethoxypyridazine, using chloroform-n-butanol-acetone-85% formic acid (80 + 20 + 20 + 20). It has been possible to separate N_4-phthalic acid and succinic acid derivatives from sulphanilamide, sulphacetamide and sulphathiazole in a two stage procedure on silica gel (cf. Table 114/9). The acid derivatives could be separated from the corresponding original sulphonamides by this method. Mixtures of sulphathiourea, sulphathiazole, sulphisomidine, sulphacarbamide, maphenide, sulphaphenazole, sulphanilamide, sulphaguanidine and phthalyl-sulphathiazole can be separated on silica gel using the two-stage technique [102]; the first development is with butanol-acetone-methanol-diethylamine (90 + 10 + 10 + 10), the second with chloroform-methanol (80 + 15), after having dried the layer for 30 min at 50° C. Others have carried out two-dimensional separations on alumina [131]. A dissertation [170a] contains another detailed description of the TLC of the sulphonamides named in Table 115 under the experimental conditions 17—19, and also of sulphacarbamide, pallidin and carbutamide, a sulphonamide which is an active anti-diabetic agent (cf. section 10, p. 550); neutral and buffered silica gel layers and neutral, acid and basic alumina layers were used. TLC, PC and thin-layer electrophoresis of the sulphonamides mentioned were compared in this work also. Mention may be made of some other literature on the TLC of sulphonamides [58, 60, 103, 117, 118].

Detection of sulphonamides can be performed non-specifically on fluorescent layers, making use of the compounds' absorption in the UV. More sensitive detection of those containing a free p-amino group, is, however, with a 1% solution of p-dimethylaminobenzaldehyde-hydrochloric acid (Ehrlich's reagent, No. 72) or by diazotisation and coupling with N-(1-naphthyl)-ethylenediammonium dichloride (Bratton-Marshall reagent, No. 61); excess nitrous acid may be expelled from the layer by heating to 100° C before spraying with the coupling component; if a definite excess remains on the layer, however, colour changes characteristic of the particular sulphonamide are observed [172]. A 2% solution of vanillin in acetic acid has also been employed for detecting sulphonamides [137a]. Another spray reagent for sulphonamide chromatograms is a mixture of equal volumes of a 0.2M solution of 1-phenyl-3-methyl-2-pyrazoline-5-one in pyridine and of aqueous potassium cyanide, applied after having exposed the plate to chlorine [20a].

Maphenide responds to the Ehrlich reagent only in amounts exceeding 25 μg, after longer heating at 90° C. It is better detected by spraying with 0.3% ninhydrin solution in butanol and then heating 30 min at 90° C [102, 134]. N_4-succinic and phthalic acid derivatives of sulphonamides have been detected through their ability to quench the fluorescence of suitable layers and also with a 0.05% ethanolic solution of bromocresol purple [84].

Quantitative evaluation of separated sulphonamides also has been described in detail [168a]. The best procedure was found to be elution of the sulphonamides with N hydrochloric acid, diazotisation and coupling with N,N-diethyl-N'-(1-naphthyl)-ethylenediammonium oxalate.

Table 114. *Experimental conditions for TLC of sulphonamides*

No.	Details	Nos. of the sulphonamides separated (Table 115)	Ref.
1	Layer: Silica gel G; standard method. 1 μg amounts of sulphonamides applied in 10 μl acetone. Solvent: chloroform-heptane-ethanol (30 + 30 + 30)	21, 17, 13 and sulphanilic acid	[175]
2	Layer: Silica gel G; standard method. 1—5 μg amounts of sulphonamide in acetone applied. Solvent: chloroform-methanol (80 + 15)	9, 6, 10, 4 and 9, 17, 5 and 6, 14, 1	[63]
3	Layer: Silica gel G; standard method. 1—5 μg amounts of each sulphonamide in acetone applied. Solvent: acetone-methanol-diethylamine (90 + 10 + 10)	3, 17, 9, 1	[63]
4	Layer: Silica gel G; CS; 10 cm run; 5—15 μg of each applied. Solvent: n-butanol-methanol-acetone-diethylamine (90 + 10 + 10 + 10)	1, 10, 5, 14, 4 German Pharm. (DAB VI)-sulphonamides cf. also Table 115	[134]
5	Layer: Silica gel G; standard method. 1 μg of each sulphonamide applied. 15 cm run. Solvent: chloroform-abs. ethanol-heptane (30 + 30 + 30) + a little water (1.5 volume % in the solvent). CS. 3 hours run	3, 7, 8, 9. Cf. also Table 115	[86]
6	Layer: Silica gel G; standard method. Wedge-shaped portions of layer scraped off, leaving a section shaped like a trapezium, yielding uniformly ascending solvent front. 2—4 μg of the sulphonamides in acetone applied. Solvent: chloroform-n-butanol-petrol ether (BP 60—80° C) (30 + 30 + 30)	Cf. Table 115 for hRf-values. Rf-data for other sulphonamides in original work	[82]
7	Layer: Silica gel G; standard method. 4 μg of each sulphonamide in ethanol applied. Solvent: ether for narcotic purposes	Cf. Table 115 for hRf-values	[18]
8	Like 7 but with solvent chloroform-methanol (100 + 10) (cf. also No. 2 in this table)	Cf. Table 115 for hRf-values	[18]
9	Layer: Silica gel G and GF_{254}; standard method. 1—3 μg sulphonamide applied. Separated by stepwise development (p. 87). 1st solvent: anhydrous ethanol-methanol (50 + 50) without chamber saturation. 2nd solvent: n-propanol-0.05N HCl (80 + 20) with CS. 5 cm run in 1st solvent, then 5 min evaporation at 100°, cooling and 10 cm run in 2nd solvent	Identification of the N_4-subd. sulphonamides of USP XVI and NF XI. Possible separations: a) phthalylsulphacetamide, phthalylsulphathiazole, succinylsulphathiazole b) phthalylsulphanilamide, suc-	[84]

Table 114 (Continued)

No.	Details	Nos. of the sulphonamides separated (Table 115)	Ref.
		cinylsulphanilamide, succinylsulphacetamide	
10	Layer: Alumina G (30 g + 60 ml water rapidly mixed in a 250 ml erlenmeyer flask and spread). Air-dried 15 min, then 30 min/100° C. Solvent: methanol-chloroform (30 + 70). Chamber saturation > 30 min. 1—2 µg of each sulphonamide applied	3, 7, 20, 8, 11, 9, 1	[3]
11	Layer: Silica gel G; standard method. Activated again by 30 min/110° before use. Solvent: chloroform-methanol-dist. water (80 + 20 + 1.25)	15, 3, 1, 14, 7, 6, 8, 9 (cf. Table 115)	[172]
12	Layer: Silica gel (Firm 153) (30 g silica gel + 40 ml water slurried); standard method (p. 85) but 1 h activation at 150° C. Multiple development. Solvent: chloroform-acetonitrile (50 + 50). Two 10 cm runs. CS. 1 µg of each sulphonamide applied	Especially for separating sulphonamide-diazine derivs. E. g., 13, 12, 8, 11, 10 cf. Table 115)	[64]
13	Layer prepared as 12. Solvent: pentane-ethanol-chloroform (25 + 30 + 35). CS. 10 cm run. 1—2 µg of each sulphonamide applied	E. g., 13, 11, 7, 10; cf. Fig. 164	[64]
14	Layer: Silica gel HF_{254}; standard method. Solvent: chloroform-96 vol.-%-ethanol (80 + 10). CS. 10 cm run. 1—2 µg of each sulphonamide applied	E. g., 19, 6, 1, 4; cf. Fig. 164	[64]
15	Layer: Silica gel HF_{254}; standard method. Solvent: 5% (vol.) ammonium hydroxide-n-butanol (50 + 50). CS. 10 cm run. 1—2 µg of each sulphonamide applied	E. g., 1, 4, 6, 11, 7; cf. Fig. 164	[64]
16	Layer: Silica gel G; standard method. Solvent: cyclohexane-acetone-acetic acid (40 + 50 + 10). CS. 10 cm run. 1—2 µg sulphonamide applied	Cf. Table 115 for hRf-values	[63]
17	Adsorbent: Alumina (VEB Chemiewerk Greiz-Dölan, GDR), containing about 15% gypsum. 15 g mixed with 45 ml water to prepare the layers. Up to 12 cm run. CS. Solvent: chloroform-methanol (60 + 30)	Separation of long- and medium duration acting sulphonamides, e. g., 11, 13, 18, 19 and pallidin; cf. Table 115	[170 a]
18	Conditions as 17 but with solvent methanol-water (96 + 8)	Cf. Table 115 for hRf-values	[170 a]
19	Conditions as 17 but with solvent n-butanol-water (90 + 9)	Cf. Table 115 for hRf-values	[170 a]

Table 115. *hRf-values of frequently used sulphonamides*

No.	Free (non-registered) name[a]	Chemical formula[b]	4	5	6	7	8	11	12	16	17	18	19
1	Sulphanilamide	R–H	68	53	43	61	36	52	42	43		86	82
2	Maphenide	$H_2N-CH_2-\!\!\!\bigcirc\!\!\!-SO_2NH_2$	56										
3	Sulphacetamide	$R-CO-CH_3$	38	42	31			47		49		48	28
4	Sulphaguanidine	$R-\underset{\underset{NH}{\|\|}}{C}-NH_2$	27		15	4	15		11	28		77	66
5	Sulphathiourea	$R-CS-NH_2$	48		34					56			
6	Sulphapyridine	R–(pyridyl)			37			69	31	54			
7	Sulphadiazine	R–(pyrimidyl)	39	47	39	53	50	65	50			30	35
8	Sulphamerazine	R–(4-methylpyrimidyl)		57	44	60	59	72	48			44	56
9	Sulphadimidine (Sulphamethazine)	R–(4,6-dimethylpyrimidyl)	52	64	52	72	63	76	52	56		70	72
10	Sulphisomidine	R–(2,6-dimethylpyrimidyl)			19				6	33		62	53
11	Sulphamethoxypyridazine	R–(methoxypyridazinyl)			61	38	68		41		51	55	67
12	Durenat	R–(methoxypyrimidyl)			56				55				
13	Sulphadimethoxine	R–(dimethoxypyrimidyl)			67				68		56	55	63
14	Sulphathiazole	R–(thiazolyl)	39	50	41	14	38	60		38		37	62
15	Sulphamethizole	R–(methylthiadiazolyl)						41					
16	Sulphaethidole	R–(ethylthiadiazolyl)	41									31	12

Table 115 (Continued)

No.	Free (non-registered) name[a]	Chemical formula[b]	\multicolumn{11}{c}{hRf-values under the conditions given in Table 114}
			4	5	6	7	8	11	12	16	17	18	19
17	Sulphafurazole	R—[isoxazole with CH₃ CH₃]			51	81	51			55		64	22
18	Sulphuno (Sulphadimethyl-oxazole)	R—[oxazole with CH₃, CH₃]							35		64	64	65
19	Sulphaphenazole	R—[pyrazole-phenyl]			77					52		69	74	76
20	Sulphaquinoxalin	R—[quinoxaline]											
21	Acetylgantrisin	N¹-Acetylsulphafurazole				75	81						

[a] If not known, commercial name quoted.

[b] R = $NH_2-\langle\rangle-SO_2-NH-$

c) Chemotherapeutic Agents of the Nitrofuran Series, etc.

The nitrofuran derivatives nitrofurantoin, nitrofurfuraldehyde and furazolidine and the nitroimidazole derivative metronidazole, can be separated on silica gel G using chloroform-diethylamine (90 + 10). The spots are visible in daylight or UV light without using any reagent [23]. In the presence of acridine derivatives (vioform and acriflavine), nitrofurans have been chromatographed on silica gel GF, using the solvents acetone-chloroform-ether (50 + 20 + 30) or acetone-chloroform-ether-n-butanol (45 + 20 + 30 + 5) [107b].

WALLHÄUSER discusses the TLC of the antibiotics in the next chapter (R) (p. 566).

8. Diuretics

The chromatographic behaviour of various chlorothiazide and hydrochlorothiazide derivatives in neutral solvents on silica gel G and alumina G layers has been described [1] and is summarised in Table 116. Hydrochlorothiazide, hydroflumethiazide, chlorazanil, acetazolamide, quinethazone, fursemide and chlorthalidone have been separated on silica gel GF_{254} (Merck) in a 13 cm run, using the basic solvent toluene-xylene-dioxan-isopropanol-25% ammonium hydroxide (10 + 10 + 30 + 30 + 20); 50 µg amounts of the substances were applied in solution in acetone [108].

All these diuretics can be detected on fluorescent layers as dark spots from quenching when exposed to short wave UV light. Quinethazone and hydroflumethiazide fluoresce themselves. Most of the compounds can be visualised with a mixture of 5 ml 20% sodium hydroxide, 15 ml 1% disodium pentacyanoammineferrate(II) and a drop of hydrogen peroxide (Fearon's reagent). The reagent can be kept for 24 h.

Table 116. *hRf-values of chlorothiazide- and hydrochlorothiazide derivatives, used as diuretics*

Free (non-registered) name	Basic formula	Substituents			hRf-values		
		R_1	R_2	R_3	a	b	c
Chlorothiazide ...	I	—Cl	—H	—H	2	22	17
Hydrochlorothiazide	II	—Cl	—H	—H	24	35	29
Flumethiazide ...	I	—CF_3	—H	—H	2	40	33
Polythiazide	II	—Cl	—CH_2—S—CH_2—CF_3	—CH_3	10	51	47
Hydroflumethiazide .	II	—CF_3	—H	—H	36	46	41
Methyclothiazide ..	II	—Cl	—CH_2—Cl	—CH_3	46	50	47
Benzthiazide	I	—Cl	—CH_2—S—CH_2—C_6H_5	—H	55	58	55
Thiabutazide	II	—Cl	—CH_2—CH(CH_3)$_2$	—H	55	59	57
Trichloromethiadiazide	II	—Cl	—$CHCl_2$	—H	29	59	57
Cyclopenthiazide ..	II	—Cl	—CH_2—⬠	—H	55	60	58
Epithiazide	II	—Cl	—CH_2—S—CH_2—CF_3	—H	54	61	59
Bendroflumethiazide	II	—CF_3	—CH_2—C_6H_5	—H	64	65	66

^a Alumina G, ethyl acetate, 16 cm run.
^b Silica gel G, ethyl acetate, 15 cm run.
^c Silica gel G, ethyl acetate-benzene (80 + 20), 15 cm run.
Layer 150 µ thick; 5—10 µg applied.

Chlorothiazide-Derivatives

Hydrochlorothiazide-Derivatives

9. Purine Derivatives of Various Activities

The most suitable conditions for TLC-separation of the naturally occurring xanthine derivatives, caffeine, theobromine and theophylline, are basic silica gel layers with neutral solvents or neutral silica gel with basic solvents (Table 117/I and II) [163, 177]. Separation can be achieved also on neutral layers with acid solvents (Table 117/III) [7]. Using the experimental conditions quoted in Table 117/II, it has been possible to separate five xanthine derivatives from one another, namely, the three mentioned above, hydroxyethyltheophylline (cf. Table 118, No. 4) and hydroxypropyltheobromine (cf. Table 118, No. 6) [177]. All these and other synthetic xanthine derivatives have been separated by two-dimensional chromatography on silica gel GF_{254} layers, using benzene-acetone (30 + 70) in an ammonia atmosphere in the one direction and chloroform-ethanol-formic acid (88 + 10 + 2) in the other (Table 118) [145a]. The purines, uric acid, xanthine, hypoxanthine and 6-mercaptopurine, have also been separated by TLC [30]. Chromatography of hypoxanthine, xanthine, uric acid, guanine, adenine, theophylline, theobromine and caffeine has been carried out on silica gel HF_{254} layers which had been prepared with piperazine solution; the solvent used was isopropanol-chloroform-10% aqueous piperazine (60 + 20 + 20) [137b]. Xanthine derivatives, mostly derived from theophylline, have been submitted also to TLC on silica gel HF_{254}, using the neutral solvent chloroform-acetone-methanol (30 + 30 + 30) [161a].

Table 117. *hRf-values of theobromine, theophylline and caffeine*

	I	II	III
Theobromine	22	47	36
Theophylline	37	26	50
Caffeine	57	78	41

I = Silica gel G layer, buffered to pH 6.8 (25 g silica gel G slurried with 50 ml of a mixture of equal amounts of 0.2M primary potassium phosphate and 0.2M secondary sodium phosphate solutions). Chloroform-96% ethanol (90 + 10) as solvent.
II = Silica gel G layer. Acetone-chloroform-n-butanol-25% ammonium hydroxide (30 + 30 + 40 + 10) solvent. Chromatography in shallow dishes as chambers, at normal saturation.
III = Silica gel G layer. Ethyl acetate-methanol-acetic acid (80 + 10 + 10) as solvent.

The purines can be detected on fluorescent layers in UV light, through their quenching of the fluorescence. Detection is also possible by spraying with iodine solution (Rgt. Nos. 141 and 144) [145a] or by exposure to iodine vapour. Sensitive visualisation through spraying with a solution of iodine and ferric chloride in acetone containing tartaric acid (Rgt. No. 103) has also been described [145a, 177]. Caffeine and probably numerous other purines can be detected by chlorination and then spraying with a solution containing pyridine, potassium cyanide and 1-phenyl-3-methyl-2-pyrazolin-5-one (Rgt. No. 41) [20a]. Larger amounts of chromatographically separated purine derivatives can be sublimed on to a cooled glass plate from the chromatogram [7].

Table 118. *TLC of xanthine derivatives* [145a]

No.	Substance name	Chemical term	hR_f-values A	B
1	Theophylline	1,3-Dimethylxanthine	6	45
2	Theobromine	3,7-Dimethylxanthine	31	34
3	Caffeine	1,3,7-Trimethylxanthine	70	57
4	Hydroxyethyl-theophylline	1,3-Dimethyl-7-(2'-hydroxyethyl)-xanthine	65	52
5	Proxyphylline	1,3-Dimethyl-7-(2'-hydroxypropyl)-xanthine	55	37
6	Hydroxypropyl-theobromine	1-(2' Hydroxypropyl)-3,7-dimethyl-xanthine	50	39
7	Dihydroxypropyl-theophylline	1,3-Dimethyl-7-(2',3'-dihydroxypropyl)-xanthine	17	16
8	Dihydroxypropyl-theobromine	1-(2',3'-Dihydroxypropyl)-3,7-dimethyl-xanthine	17	14
9	Hexyltheobromine	1-Hexyl-3,7-dimethylxanthine	85	72
10	Diethylaminoethyl-theophylline	1,3-Dimethyl-7-(2'-diethylaminoethyl)-xanthine	81	3
11		1,3-Dimethyl-7-[2'-(1''-methyl-2''-phenyl-ethylamino)-ethyl]-xanthine	76	9
12		1,3-Dimethyl-7-[2'-(1''-methyl-2''-hydroxy-2''-phenylethylamino)-ethyl]-xanthine	61	7
13		1,3-Dimethyl-7-[2'-hydroxy-2''-(3''', 4'''-dihydroxyphenyl)-ethylamino)-ethyl]-xanthine	0	0

Layer: Silica gel GF$_{254}$ (Firm 88). 3 µg of each substance applied, in methanol or methylene dichloride solution.

Solvent A: Benzene-acetone (30 + 70): chamber saturation achieved by placing a small dish of 25% ammonium hydroxide in the chamber.

Solvent B: Chloroform-ethanol-formic acid (88 + 10 + 2).

10. Oral Antidiabetic Agents

Oral antidiabetic agents can be chromatographed with either of the basic solvents given in Table 119 [109, 137]; substances No. 3, 5, 6 and 7 are separated with solvent II. The antidiabetics containing an aromatic amino group can be detected with Ehrlich's reagent (No. 72); the tolbutamides also respond to this reagent if present in higher concentration. All the substances in Column I, Table 119, can be visualised with Ehrlich's reagent in which the volatile hydrochloric acid has been replaced by phosphoric acid; the plate is then heated 10 min at 150° C after spraying and the layer sprayed with ninhydrin solution while still hot [137]. The oral antidiabetics, excepting silubin, can of course be detected through their property of quenching the fluorescence in UV light of layers containing a fluorescent additive.

Table 119. hR_f-values of some oral antidiabetic agents

Free (non-registered) name[a]	Chemical term	hR_f-values[b]	
		I	II
1	1-Butyl-3-(m-aminobenzenesulphonyl)-urea	14	—
2 Glybuthiazole	N'-(5-tert. Butyl-1,3,4-thiadiazol-2-yl)-sulphanilamide	15	—
3 Carbutamide	1-Butyl-3-(p-aminobenzenesulphonyl)-urea	27	45
4 Chlorpropamide	1-[(p-Chlorophenyl)sulphonyl]-3-propylurea	47	—
5 Tolbutamide	1-Butyl-3-(p-tolylsulphonyl)-urea	57	54
6 Glycodiazine	Na N-[5-(2-Methoxy-ethoxy)-2-pyrimidinyl]-benzenesulphonamide	—	50
7 Buformin	1-n-Butylbiguanide	—	36

[a] If not known, trade name.
[b] I = Silica gel G "Merck". n-Butanol-chloroform-diethylamine (45 + 45 + 5). CS [137]; II = Silica gel GF "Merck". Butanol-chloroform-methanol-25% ammonium hydroxide (40 + 15 + 15 + 15) [109].

11. Laxatives

Phenisatin (triacetyl-diphenylisatin) and bisacodyl (4,4'-(2-pyridyl-methylene)-diphenol diacetate) can be separated on silica gel HF_{254} layers, using the solvent chloroform-cyclohexane-butanone (30 + 30 + 30) [64]. Detection is possible in short wave UV light.

12. Local Anaesthetics

Chromatography of local anaesthetics has been carried out up to now on silica gel layers, prepared with sodium hydroxide solution [22, 63, 159] and on loose alumina and silica gel layers (cf. Table 120) [27, 141]. Various mixtures of local anaesthetics can be separated using the experimental conditions described under I—IV, Table 120. Procaine and benzocaine have been separated with the system V. Further solvents for chromatography on loose alumina layers [141] and two other publications on the chromatography of local anaesthetics [59, 176] may be referred to here. With the help of TLC, the relationships between properties of the substances which are significant in the distribution within organisms (such as pK, water- and lipid solubility, partition coefficient) can be determined. Reference may be made here to the example of partition chromatographic study of numerous local anaesthetics with oleyl alcohol on cellulose powder as stationary phase and aqueous buffer solutions as solvents [27a]. In this investigation, chromatography was carried out on benzocaine homologues with alkyl chains of various lengths; benzocaine analogues with methyl-substituted ethyl group; procaine analogues with branched side chain; procaine analogues in which the p-amino group had been replaced by various

Table 120. *hRf-values of local anaesthetics*

Name	Chemical formula	hRf-values				
		I	II	III	IV	V
Anaesthesine (Benzocaine)	$R_1-CH_2-CH_3$	48	67	6	74	74
Orthoform	$H_2N-\bigcirc-COO-CH_3$, OH	0	—	—	—	—
Butoform (Butamben)	$R_1-CH_2-CH_2-CH_2-CH_3$	—	69	7	78	—
Procaine	$R_1-CH_2-CH_2-R_2$	31	36	5	18	60
Butacaine	$R_1-CH_2-CH_2-CH_2-N(C_4H_9)_2$	—	61	8	49	—
Butethamine	$R_1-CH_2-CH_2-NH-CH(CH_3)_2$	—	48	5	28	—
Dimethocaine	$R_1-CH_2-CH(CH_3)-CH_2-R_2$, with CH$_3$	47	—	—	—	—
Tutocaine	$R_1-CH(CH_3)-CH(CH_3)-CH_2-R_3$	43	—	—	—	—
Tetracaine	$C_4H_9-NH-\bigcirc-COO-CH_2-CH_2-R_3$	40	27	18	11	65
Chloroprocaine	$H_2N-\bigcirc(Cl)-COO-CH_2-CH_2-R_2$	—	38	3	21	—
Metabutoxycaine	$\bigcirc(NH_2)(OC_4H_9)-COO-CH_2-CH_2-R_2$	—	49	30	41	—
Proxymetacaine (Proparacaine)	$C_3H_7O-\bigcirc(NH_2)-COO-CH_2-CH_2-R_2$	—	38	31	18	—
Meprylcaine	$\bigcirc-COO-CH_2-C(CH_3)_2-NH-C_3H_7$	—	54	58	31	—
Piperocaine	$\bigcirc-COO-CH_2-CH_2-CH_2-N\langle\text{piperidine-CH}_3\rangle$	—	31	63	17	—
Cyclomethycaine	$\text{(4-cyclohexyloxyphenyl)}-COO-CH_2-CH_2-CH_2-N\langle\text{piperidine-CH}_3\rangle$	—	32	66	14	
Pramocaine (Pramoxine)	$C_4H_9O-\bigcirc-O-CH_2-CH_2-CH_2-N\langle\text{morpholine}\rangle$	—	43	52	29	

Synthetic Pharmaceutical Products 553

Table 120 (Continued)

Name	Chemical formula	hRf-values				
		I	II	III	IV	V
Dyclonine	C_4H_9O—⟨ ⟩—$CO-CH_2-CH_2-N$⟨ ⟩	—	34	54	16	
Lidocaine	2,6-(CH_3)$_2$-C_6H_3—$NH-CO-CH_2-R_2$	50	66	39	62	
Hostacaine	2,6-(CH_3)$_2$-C_6H_3—$NH-CO-CH_2-NH-C_4H_9$	26	—	—	—	
Mepivacaine	2,6-(CH_3)$_2$-C_6H_3—$NH-CO$—(N-CH_3 piperidyl)	—	52	37	29	
Trimecainum	2,4,6-(CH_3)$_3$-C_6H_2—$NH-CO-CH_2-R_2$	35	—	—	—	
β-Eucaine	(benzoyloxy-trimethyl-piperidine)	28	—	—	—	
Phenacaine	$CH_3-C(=N-C_6H_4-OC_2H_5)(NH-C_6H_4-OC_2H_5)$	52	63	12	64	
Diocaine	$CH_3-C(=N-C_6H_4-O-CH_2-CH=CH_2)(NH-C_6H_4-O-CH_2-CH=CH_2)$	55	—	—	—	
Cinchocaine	quinolinyl-OC_4H_9, $CO-NH-CH_2-CH_2-R_2$	46	—	—	—	
Amolanone	(benzofuranone with $CH_2-CH_2-R_2$ and phenyl)	—	61	64	60	
Cocaine		65	43	50	41	
Dextrocaine	d—ψ—Cocaine	46	—	—	—	

other substituents; procaine-N-dialkyl homologues; procaine-alkylene homologues; and parethoxycaine- and cinchocaine analogues with alkoxy chains of various lengths.

The local anaesthetics may be detected on adsorbent layers prepared with fluorescent additives, by illumination with short wave UV light [63]. Reagents which can be used for visualisation are: 4-dimethylaminobenzaldehyde solution (Rgt. No. 72) [27, 63]; Dragendorff reagent, prepared in various ways [63, 141, 159]; iodine-iodide solution (Rgt. No. 144) [141]; and a 0.5% solution Fast Red GG in water [159]. Meprylcaine, which does not react with the Dragendorff reagent, can be located by spraying with a 5% aqueous solution of sodium nitrite [159]. SUNSHINE and FIKE [159] have described a procedure for isolating local anaesthetics from biological material.

13. Miscellaneous Other Active Substances

The TLC of amine mixtures, e. g., piperazine and various piperazine derivatives, on silica gel G layers, using the solvent chloroform-methanol-17% ammonium hydroxide (40 + 40 + 20) and the quantitative evaluation of such chromatograms through measurement of the spot sizes (cf. p. 135) has also been described [70a].

I. Analysis of Various Drug Forms and of Commercial Preparations

Knowledge of a thin-layer chromatographic method of separating the active materials present in a pharmaceutical preparation is not alone enough for establishing a procedure for control of the preparation. With solid drugs, an extraction procedure has first to be worked out, aiming at preventing interfering contaminants from reaching the thin

---

Explanatory notes to Table 120

$R_1 = H_2N-\langle\bigcirc\rangle-COO-$ $\qquad$ $R_2 = -N\langle{}^{C_2H_5}_{C_2H_5}$ $\qquad$ $R_3 = -N\langle{}^{CH_3}_{CH_3}$

I. *Layer:* Alumina without binder, analytical grade, manufactured by Lachema CSN. 68513.1 activity grade III, pH 8.6; grain size 0.075 mm; 0.6 mm thick layer. Solvent: benzene-95% ethanol (95 + 5) [141].

II, III and IV. *Layer:* Silica gel G, "Merck" with 0.1 N NaOH (30 g + 60 ml). Layers air-dried and kept at least 6 h at 22° in 50% relative humidity. 10—12 cm run. 20—40 µg of each applied. Solvent for II, ethanol; III, cyclohexane-benzene-diethylamine (75 + 15 + 10); IV, methyl acetate [159].

V. *Layer:* Silica gel G + alkali + fluorescent additive (25 g silica gel + 0.5 g ZS-Super, Firm 118 + 50 ml 0.5N NaOH). Layers dried 2 h at 120° C before use. 10 cm run. Chamber saturation. Solvent: chloroform-methanol (80 + 10) [63].

layer. NUSSBAUMER [115] has discussed this aspect in detail for the examination of penicillin preparations. The extraction procedure of a commercial suppository preparation in Table 121/3 may be regarded as another example of such a pre-purification; in it, the suppository mass is frozen out after extraction of the active material.

With liquid drug preparations, separation of the additives and base before chromatography is in most cases impossible or only partly possible. These substances are thus carried over into the chromatographic separation. Examples of this are to be found in Table 121/5 and 6. As shown in Fig. 165, the experimental conditions for the steroid emulsion, Table 121/5, are chosen in such a way that, apart from the emulsion base remaining at the start, two other emulsion constituents are separated from the two active steroid components during development. In the example of sesame oil preparations, Table 121/6, a suitable solvent displaces the lipophilic oil, serving as excipient, to the solvent front and

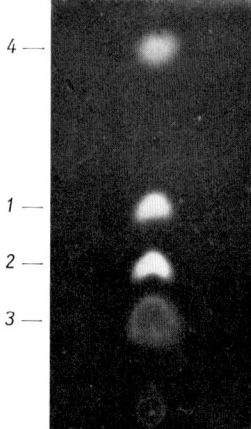

Fig. 165. Chromatogram of a steroid emulsion. Progesterone (*1*), oestradiol benzoate (*2*), auxiliary materials (*3* and *4*). See Table 121/5 for experimental conditions

Fig. 166. Chromatogram of a tablet extract for detecting small amounts of ethinyl oestradiol. See Table 121/2 for explanatory data

the active substances remain on the rest of the migration path. Since the oil is present in considerable excess of the active substances, the choice of solvent is critical and must be adapted also to small changes in the structure of these substances. Thicker layers are an advantage in such cases, since this diminishes the saturation density of the layer with oil and increases the separation effect. Mention may be made also of the possibility of separating steroids and interfering glycerides, especially diglycerides, by drawing off the solvent from the solvent front [29a]. Chromatography has been carried out in this way on preparations of testosterone propionate, progesterone, 19-nortestosterone propionate and oestradiol cyclopentylpropionate in olive oil containing 1% benzyl alcohol; the solvents were petrol ether (BP 65°)-ether-acetic acid (70 + 30 + 1) and (50 + 50 + 1) and the steroids were quantitatively determined after elution.

The amount applied is decisive in perfecting the separation of active substance mixtures from pharmaceutical preparations. Since the composition of the preparation is fixed, enough must be applied to permit certain identification of the most difficultly detectable active component. If the composition is as unfavourable as in the case of the steroid tablets in Table 121/2, chromatography cannot be carried out in a single operation. The two steroids were present in the ratio 1000:1 in these tablets. To be able to detect the small amount of ethinyl oestradiol, an extraction agent was chosen in which the second steroid, present in excess, was

Table 121. *Analysis of various drug preparations*

Drug form and components	Extraction and experimental chromatographic conditions	Detection
1. *Tablets* Active materials: 50 mg caffeine; 200 mg phenacetin; 50 mg salicylamide; 200 mg acetylsalicylic acid; small amounts of salicylic acid, formed by hydrolysis [63]	Two tablets are finely powdered, extracted with 5 ml methanol and 1 μl of this solution applied. Silica gel HF layer. Methanol-acetic acid-ether-benzene (1 + 18 + 60 + 120) solvent. CS. hR_f-values: caffeine 16, phenacetin 45, salicylamide 57, acetyl-salicylic acid 76, salicylic acid 85	All identifiable as quenching spots in UV 254 nm. Salicylamide fluoresces itself in UV light. Phenacetin, salicylic acid and salicylamide yield colour with ferricyanide-FeCl$_3$ reagent (No. 111); acetyl-salicylic acid gives colour with this reagent only after heating and thermal decomposition
2. *Tablets* Active materials: 10 mg ethisterone and 10 μg ethinyl oestradiol [64]	Detection of ethisterone: 1 tablet extracted with 2 ml CHCl$_3$ and centrifuged; 20 μl (ca. 100 μg) applied. Detection of ethinyl oestradiol: 10 tablets extracted with 1.5 ml benzene and centrifuged; 30 μl of this solution applied three times to the same start spot (ca. 6 μg). Silica gel H. Chloroform-acetone (90 + 10) for TLC of ethisterone; benzene-ethyl acetate (80 + 20) for TLC of ethinyl oestradiol. 15 cm run (cf. Fig. 166)	Antimony-(III)-chloride-acetic acid (1 + 1); heated 15 min/110° C. Inspected in UV light, 366 nm

Table 121 (Continued)

Drug form and components	Extraction and experimental chromatographic conditions	Detection
3. *Suppositories* Active materials: 150 mg theophylline, 100 mg papaverine and 50 mg phenobarbital [62]	1 suppository shaken with hot water-methanol (20 + 80) until fully disintegrated, mass frozen out in refrigerator, centrifuged and 5 µl clear liquid applied. Silica gel HF$_{254}$. Benzene-ethanol-acetic acid (80 + 12 + 5). hRf-values: theophylline 17, papaverine 50, phenobarbital 78	All substances visible in UV light of 254 nm, as quenching spots
4. *Ointment or liniment* Active materials: 3-butoxyethyl nicotinate and nonylvanillamide [62]	Method of extraction depends on base. Extraction with methanol is often possible, then centrifuging and TLC with this extract, using an amount corresponding to 100 µg active mixture. Silica gel HF$_{254}$. Butanol-acetic acid (90 + 10). hRf-values: 3-butoxyethyl nicotinate 64, nonylvanillamide 93	Both materials can be detected in UV 254 nm. 3-Butoxyethyl nicotinate also with Königs reagent (No. 23) and nonylvanillamide with Fast Blue Salt B (Rgt. No. 100)
5. *Emulsion* Active materials: 2 mg 17-β-oestradiol benzoate and 10 mg progesterone per ml [64]	Emulsion diluted with ethanol, 1 to 10. 10 µl then applied. Silica gel H. Carbon tetrachloride-methanol (95 + 5), shaken with 1 ml 25% ammonium hydroxide and organic phase taken as solvent after phase separation and filtration; freshly prepared each time. hRf-values: progesterone 30, oestradiol benzoate 20 (cf. Fig. 165). The spots with the highest and lowest hRf-value are components of the emulsion base	Antimony(III)-chloride-acetic acid (1 + 1); heated 15 min/110° after spraying
6. *Steroid-sesame oil* preparations [89]	An amount of oil equivalent to 100 µg steroid applied as a 2 cm band. Silica gel G. Chloroform for testosterone propionate (hRf 40) and deoxycorticosterone acetate (hRf 15); benzene-methylene dichloride (40 + 16) for oestradiol dipropionate (hRf 45); chloroform-ethyl acetate (70 + 30) for aldosterone acetate (hRf 30); chloroform-methanol (98 + 2) for methandrostenolone (hRf 40). The oil migrates in the neighbourhood of the solvent front. CS. Two 15 cm runs	0.8 ml 37% formaldehyde mixed with 10 ml H$_2$SO$_4$ and diluted with 3 ml water. Heated 5—10 min/80°. Inspected in daylight and UV-light

sparingly soluble. In this way and through careful choice of a suitable solvent for chromatography, ready semi-quantitative detection was achieved (cf. Fig. 166). The quantitative determination of ethinyl oestradiol and its 3-methyl ether, after preliminary isolation through TLC, has been described also in the meantime [77a].

Several authors have worked on TLC of the more frequently occurring mixed pharmaceutical preparations such as, the analgesic-antipyretic mixture of caffeine, phenacetin, acetylsalicylic acid and salicylamide (Table 121/1) or the anti-rheumatic preparation in Table 121/4. In this connection, one may mention here also the separations and quantitative evaluations of the analgesic mixtures, caffeine-benzyl mandelate-aminopyrine-phenacetin and caffeine-propylphenazone-phenacetin-pyrithyldione [66, 72, 96] (cf. p. 147). Other TLC separation procedures which have been described are of the components of: an analgesic powder on ion exchange layers [135]; an anti-influenza elixir, with semi-quantitative evaluation [114]; the water-soluble vitamins in a commercial vitamin preparation [67]; and the fat-soluble vitamins in such preparations [21]. Further extensive information about the analysis of commercial preparations is to be found elsewhere [19, 63, 65 and 75].

II. Stability Tests on Pharmaceutical Materials

The development of a new medicament requires as thorough a knowledge as possible of the physico-chemical properties of the chosen active substances and auxiliary materials. The composition of the pharmaceutical product must be selected so that its properties will remain as long as possible unchanged during storage under the climatic conditions expected. In order to be able to make a quick forecast of the stability, suitably packed samples are subjected to rigorous conditions, e. g., higher temperatures, various humidities, influence of light etc. and tested at regular time intervals for possible changes of physical and chemical properties.

When TLC is applied to testing stability, it must be previously established whether the compounds to be separated really suffer no change under the chosen experimental conditions for the chromatography. Changes due to the influence of light and air can take place quickly in the dried-out start spot on the large adsorbent surface. Rapid work, if necessary in subdued light, is consequently often imperative. On the other hand, the large surface of the adsorbent can be utilised for fast orientation experiments, investigating the influence on the pharmaceutically active substances, of air, of the adsorbent itself and of the additives in the adsorbent. An example is shown in Fig. 167. Oestrenols (cf. p. 152) with different substituents were applied to silica gel layers which had been prepared with solutions of various metal salts; these were then stored for a definite time in the dark under inert gas in order to test the influence of a diluent powder containing a heavy metal salt of particular moisture content, on these steroids [64]; chromatography was then carried out with heptane-acetone (80 + 40) and visualisation by spraying with sulphuric acid and inspection in UV light. The advantage of the method is that reaction products formed under the test conditions can be separated by the development and subsequently detected without further working-up. Of course it is possible in such experiments to test only those diluent powders which can be used as adsorbents.

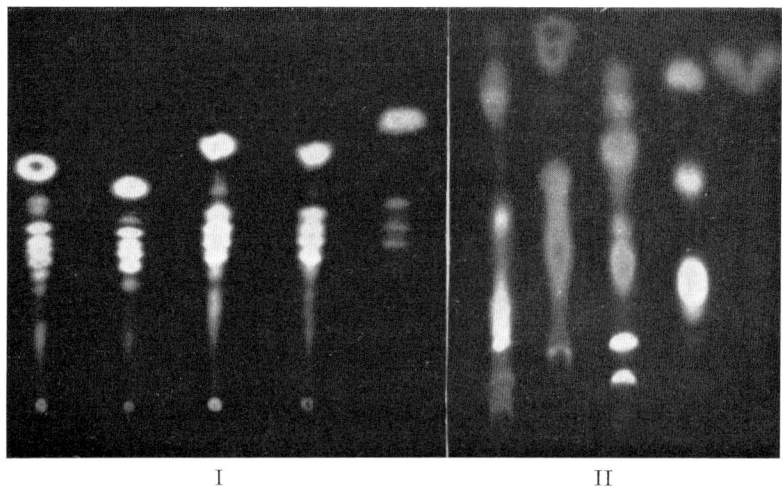

Fig. 167. Rapid testing of the influence of heavy metals on steroids on adsorbent layers. *I* silica gel G layer, prepared with a 0.5% solution of $Pb(NO_3)_2$; 5 variously substituted oestrenols applied; nitrogen atmosphere in the chromatography chamber; plate kept 12 weeks at room temperature in this chamber, under exclusion of light; then chromatographed with heptane-acetone (80 + 40); dried, sprayed with sulphuric acid (Rgt. No. 241 B) and inspected in 366 nm UV-light. *II* as *I* but using layers prepared with 10% $HgCl_2$ solution

The amount to be applied to the layer in stability testing is governed in particular by the sensitivity of the method of detection and the related question, beyond what degree of decomposition the activity of the preparation becomes really impaired or the decomposition products show an undesireable side effect. An example may be seen in Fig. 168. The resolved steroid mixture can be detected with either dinitrophenylhydrazine (Rgt. No. 82) or sulphuric acid (Rgt. No. 241 B), followed by inspection in UV light. In this particular case, the sulphuric acid reaction is too sensitive, since it detects impurities and degradation products which are without therapeutic interest [64].

TLC has been often employed for qualitative stability testing. Thus, preparations containing nicotinate esters have been tested for free nicotinic acid [62]; a laxative phenol ester, bisacodyl(4,4'-(2-pyridylmethylene)-diphenol diacetate) for deacetylated degradation products [62]; 4-aminosalicylic acid solutions for 3-aminophenol possibly formed from it [164]; and solutions containing the phenothiazine, perphenazine, for the corresponding sulphoxide [113]. Investigations of the stability of solutions of tetracaine [22], pyrazolone [5, 126, 127] and sulphadimethyloxazole [151] and of tablets containing hydroxyoestrene [55]

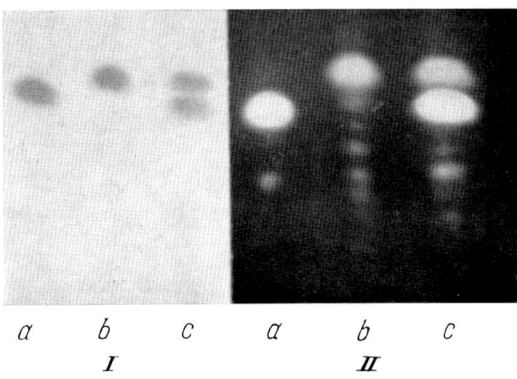

Fig. 168. Separation of progesterones. a progesterone; b alkylsubstituted progesterone c mixture. I detection with 2,4- dinitrophenylhydrazine (Rgt. No. 82), observing in daylight; II detection with sulphuric acid (Rgt. No. 241 B), inspecting in UV-light. Layer: silica gel H, 500 μ thick; solvent: cyclohexane-chloroform-acetic acid (70 + 20 + 10); CS; multiple development: 3 runs of 15 cm; 100 μg of each substance applied

may be mentioned also. TLC has been applied to check qualitatively the reaction processes, worked out from other methods, in investigating the kinetics of reactions of substituted dihydrotriazines [162].

Unknown products or substances which interfere with available quantitative methods of determination occur frequently during the degradation of active compounds in drugs. In such cases, it is advantageous to isolate the undecomposed active compound with the help of TLC, following then with quantitative evaluation (cf. Chapter H).

Eye drops containing fluprednisolone acetate have been tested [80] in a way similar to that previously described for hydrocortisone acetate preparations [63]. After submitting to appropriate test conditions, the steroid extracts were chromatographed on fluorescent silica gel layers, the separated steroid ester and alcohol spots localised in UV light, eluted and determined quantitatively by UV spectrophotometry or colorimetrically with tetrazolium compounds. The stability of tablets containing oestr-4-en-17 β-ols and the influence of the quality of the auxiliary material on the decomposition of these active compounds, has likewise been described [61] (cf. Fig. 169). The tablet extracts were developed with the solvent, the spots visualised with iodine vapour, the iodine evaporated off, the still unchanged fraction of active substance eluted and the colour obtained with sulphuric acid or vanillin/sulphuric acid, used for quantitative determination.

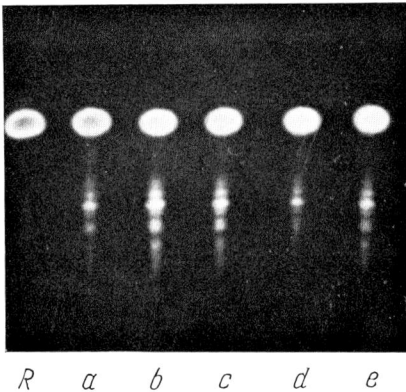

Fig. 169. Influence of variation in quality of auxiliary material on the storage properties of oestrenol tablets. a—e tablet extracts with an auxiliary material from various sources; R pure substance ; Layer: silica gel H; solvent: heptane-acetone (80 + 40); CS; detection: sulphuric acid reagent (No. 241 B), inspecting in UV-light; quantitative evaluation through sulphuric acid reaction also possible after localisation with iodine vapour (cf p. 147)

A ^{14}C-labelled steroid has been used in testing the storage properties of a cream formulation containing 6α-fluoro-16α-hydroxyhydrocortisone-16,17-acetonide [36]. The values which were found from spectrophotometric determination of the steroid content of a chloroform extract of cream samples, stored at room temperature and at 37°, agreed well with those obtained from radioactivity measurements on decomposition products, separated by TLC. The radioactive method was superior to the spectrophotometric method in this test when the active compound decomposed via another route on storing the cream at 50°. The complex and relatively fast photochemical degradation of prochlorperazine has been similarly investigated with the help of TLC [150]. By using prochlorperazine-^{35}S for preparing the test solutions, the individual products of degradation could be evaluated quantitatively. The spots, localised by autoradiography, were scraped off and determined after suspension in 15 ml of a gel designed for evaluation with a liquid scintillator. Ten decomposition products were thus determined quantitatively during a period of 35 hours.

The possibility of carrying out quantitative stability tests on aqueous preparations of ergot alkaloids, using TLC, has also been reported [85, 178]; reference may be made too to the semi-quantitative detection of free nicotinic acid in polyvitamin preparations containing nicotinamide [113].

The amount of m-hydroxyphenyltrimethylammonium salt, present in neostigmine methylsulphate ampoule solutions as impurity or degradation product, has been determined by TLC [4a] also.

Bibliography for Chapter Q. Synthetic Pharmaceutical Products

1. ADAM, E., et C. L. LAPIÈRE: J. Pharm. Belg. **19**, 79 (1964).
2. ADANK, K., u. W. HAMMERSCHMIDT: Chimia **18**, 361 (1964).
3. ALEXANDER, L. R., and E. R. STANLEY: J. Ass. Off. Agr. Chem. **48**, 278 (1965).
4. ALHA, A. R., u. R. LINDFORS: Ann. Med. exp. Fenn. **37**, 149 (1959).
4a. ANAND, D. R.: Pharmaz. Zentralhalle **104**, 370 (1965).
5. AWE, W., u. H.-G. TRACHT: Pharm. Ztg. (Frankfurt) **108**, 1365 (1963).
6. — u. W. SCHULZE: Pharm. Ztg. (Frankfurt) **107**, 1333 (1962).
7. BAEHLER, BR.: Helv. chim. Acta **45**, 309 (1962).
8. — Pharm. Acta Helv. **29**, 457 (1964).
9. BAILEY, R. W.: Analyt. Chem. **36**, 2021 (1964).
10. BÄUMLER, J., u. S. RIPPSTEIN: Pharm. Acta Helv. **36**, 382 (1961).
11. — — Arch. Pharm. **296**, 301 (1963).
12. — Chimia **17**, 257 (1963).
13. BAYER, J.: J. Chromatog. **16**, 237 (1964).
14. BEYER, K.-H.: Angew. Chem. **75**, 1136 (1963), Lecture.
15. BECKETT, A. H., and N. H. CHOULIS: J. Pharm. Pharmacol. **15**, 236 T (1963).
16. — M. A. BEAVEN, and A. E. ROBINSON: J. Pharm. Pharmacol. **12**, 204 (1960).
17. BEYRICH, TH.: Pharmazie **17**, 280 (1962).
18. BICÁN-FISTER, T., and C. KAJGANOVIĆ: J. Chromatog. **11**, 492 (1963).
19. — Acta pharm. jugosl. **12**, 73 (1962); Ref. C. A. **58**, 13721 (1963).
20. BLAŽEK, J., V. ŠPINKOVÁ, u. Z. STEJSKAL: Pharmazie **17**, 497 (1962).
20a. BOHNSTEDT, G., and M. R. F. ASHWORTH: Talanta **16**, 1631 (1966).
21. BOLLIGER, H.-R., u. A. KÖNIG: Z. analyt. Chem. **214**, 1 (1965).
22. BREINLICH, J.: Pharm. Ztg. (Frankfurt) **110**, 579 (1965).
23. — Dtsch. Apoth.-Ztg. **104**, 535 (1964).
24. BRENNER, M., u. A. NIEDERWIESER: Experientia (Basel) **16**, 378 (1960).
25. BROCHMANN-HANSSEN, E., and T. FURUYA: J. Pharm. Sci. **53**, 1549 (1964).
26. BRAVO, R. O., and F. A. HERÁNDEZ: J. Chromatog. **7**, 60 (1962).
27. BRUD, i W. DANIEWSKI: Chem. Analit. (Warszawa) **9**, 267 (1964).
27a. BÜCHI, J., u. J. A. FRESEN: Pharm. Acta Helv. **41**, 551 (1966).
28. BULENKOW, T. J.: Med. Prom. S.S.S.R. **17**, 26 (1963); ref. C. A. **60**, 5280e (1964).
29. BURGER, E.: Arch. Toxikol. **21**, 121 (1965).
29a. CAVINA, G., and G. MORETTI: J. Chromatog. **22**, 41 (1966).
30. CERRI, O., e G. MAFFI: Boll. chim. farm. **100**, 940 (1961).
31. CHOULIS, N. M.: Chimica Chronika **30**A, 37 (1965).
32. CHRISTENSEN, E. K. J., TH. VOS en T. HUIZINGA: Pharm. Weekbl. **100**, 517 (1965).

33. COCHIN, J., and J. W. DALY: Experientia (Basel) **18**, 294 (1962).
34. — — J. Pharm. exp. Ther. **139**, 154 (1963).
35. — — J. Pharm. exp. Ther. **139**, 160 (1963).
36. COMER, J. P., and P. E. HARTSAW: J. Pharm. Sci. **54**, 524 (1965).
37. CRUMP, G. B.: Analyt. Chem. **36**, 2447 (1964).
38. DHONT, J. H., and C. DE ROOY: Analyst **86**, 527 (1961).
39. DOORENBOS, H. J., R. F. REKKER, J. GOOTJES, J. R. A. SIMOONS en W. TH. NAUTA: Pharm. Weekbl. **98**, 1037 (1963).
40. DRESSLER, A.: Arch. Toxikol. **17**, 293 (1959).
41. DRYON, L.: J. Pharm. Belg. **19**, 19 (1964).
42. EBERHARDT, H., O. W. LERBS u. K. J. FREUNDT: Naunyn-Schmiedebergs Arch. exp. Path. Pharmak. **245**, 136 (1963).
43. — K. J. FREUNDT u. J. W. LANGBEIN: Arzneimittel-Forsch. **12**, 1087 (1962).
44. — u. M. DEBACKERE: Arzneimittel-Forsch. **15**, 929 (1965).
45. — u. D. NORDEN: Arzneimittel-Forsch. **14**, 1354 (1964).
46. — O. W. LERBS u. K. J. FREUNDT: Arzneimittel-Forsch. **13**, 804 (1963).
47. EIDEN, F., u. H.-D. STACHEL: Dtsch. Apoth.-Ztg. **103**, 121 (1963).
48. ENDRES, H., u. H. HÖRMANN: Angew. Chem. **75**, 288 (1963).
49. EMMERSON, J. L., and R. C. ANDERSON: J. Chromatog. **17**, 495 (1965).
50. ELLIOT, W. H., N. NOMOF, K. PARKER, L. DEWEY, and E. L. WAY: Clin. Pharm. Ther. **5**, 405 (1964).
51. FELS, G. I., M. KAUFMANN, and A. G. KARCZMAR: Nature (Lond.) **181**, 1266 (1958).
52. FIKE, W. W., and I. SUNSHINE: Analyt. Chem. **37**, 127 (1965).
53. FIORI, A., and M. MARIGO: Nature (Lond.) **182**, 943 (1958).
54. FISCHER, R., u. W. KLINGELHÖLLER: Arch. Toxikol. **19**, 119 (1961).
55. FOKKENS, J., u. J. POLDERMAN: Pharm. Weekbl. **96**, 657 (1961).
56. FRAHM, M., A. GOTTESLEBEN u. K. SOEHRING: Pharm. Acta Helv. **38**, 785 (1963).
57. FRESEN, J. A.: Pharm. Weekbl. **100**, 532 (1965).
57a. — Pharm. Weekbl. **99**, 829 (1964).
58. FUWA, T., T. KIDO, and H. TANAKA: Yakuzaigaku (Arch. Pract. Pharm. Japan) **23**, 102 (1963), ref. C. A. **60**, 3951 g (1964).
59. — — — **24**, 123 (1964), ref. C. A. **61**, 15934 f (1964).
60. GAJDOŠ, M.: Česk. Farm. **14**, 70 (1965).
61. GÄNSHIRT, H. G., and J. POLDERMAN: J. Chromatog. **16**, 510 (1964).
62. — u. F. MALZACHER: Arch. Pharm. **293/65**, 925 (1960).
63. — In: E. STAHL: Thin-Layer Chromatography. Berlin-Göttingen-Heidelberg: Springer 1965.
64. — unpublished.
65. — Arch. Pharm. **296**, 73 (1963).
66. — Arch. Pharm. **296**, 129 (1963).
67. — u. F. MALZACHER: Naturwissenschaften **47**, 279 (1960).
68. GRÄFE, G.: Dtsch. Apoth.-Ztg. **104**, 1763 (1964).
69. GELDMACHER-MALLINCKRODT, M., u. L. LAUTENBACH: Arch. Toxikol. **20**, 31 (1963).
70. GENDI, S. E., W. KISSER u. G. MACHATA: Mikrochim. Acta **1965**/1, 120.
70a. GNEHM, R., H. REICH u. P. GUYER: Chimia **19**, 585 (1965).
71. GRAU, W., u. H. ENDRES: J. Chromatog. **17**, 585 (1965).
72. HAEFELFINGER, P. B., SCHMIDLI u. H. RITTER: Arch. Pharm. **297**, 641 (1964).
73. HALMEKOSKI, J., and H. HANNIKAINEN: Suom. Kemist B **36**, 24 (1963).
74. — Suom. Kemist B **36**, 58 (1963).

75. HARA, S., and H. TANAKA: Kagaku No Ryoiki, Zokan **64**, 141 (1964), Ref. C. A. **62**/9, 10799 g (1965).
76. HARDMEIER, E., u. J. SCHMIDLEIN-MÈSZÁROS: Arch. Toxikol. **20**, 102 (1963).
76a. HAUPTMANN, S., and J. WINTER: J. Chromatog. **21**, 341 (1966).
77. HENRICHS, J.: Dissertation, Bonn, 1962.
77a. HEUSER, D.: Dtsch. Apoth.-Ztg. **106**, 411 (1966).
78. HEYNDRICKX, A. M., SCHAUVLIEGE et A. BLOMME: J. Pharm. Belg. **1965**/3—4, 117.
79. HOFFMANN, H., u. K. ROLLER: Arzneimittel-Forsch. **14**, 1001 (1964).
80. JENSEN, E. H., and D. J. LAMB: J. Pharm. Sci. **53**, 402 (1964).
81. JOMMI, G., P. MANITTO, and M. A. SILANOS: Arch. Biochem. **108**, 334 (1964).
82. KARPITSCHKA, N.: Mikrochim. Ichnoanalyt. Acta **1963**/1, 157.
83. KELLEHER, J., and J. G. ROLLASON: Clin. chim. Acta **10**, 92 (1964).
84. KHO, B. T., and S. KLEIN: J. Pharm. Sci. **52**, 404 (1963).
85. KLAVEHN, M., H. ROCHELMEYER u. J. SEYFRIED: Dtsch. Apoth.-Ztg. **101**. 75 (1961).
86. KLEIN, S., and B. T. KHO: J. Pharm. Sci. **51**, 967 (1962).
86a. KLÖCKING, H.-P.: Pharmazie **20**, 737 (1965).
87. KNAPPE, E., u. I. ROHDEWALD: Z. analyt. Chem. **200**, 9 (1964).
88. KÖNIG, J., I. HYNIE u. K. KAĆL: Pharmazie **20**, 242 (1965).
88a. KORCZAK-FABIERKIEWICZ, C., J. KOFOED, and G. H. W. LUCAS: J. Forensic Sci. **10**, 308 (1965).
89. KORZUN, B. P., and S. BRODY: J. Pharm. Sci. **52**, 206 (1963).
90. — — and F. TISHLER: J. Pharm. Sci. **53**, 976 (1964).
91. KUPFERBERG, J. H., A. BURKHALTER, and E. L. WAY: J. Chromatog. **16**, 559 (1964).
92. LANGE, N. A.: Handbook of Chemistry, 10th Ed. S. 951. New York: McGraw-Hill Book Comp. 1961.
93. LEHMANN, J., and V. KARAMUSTAFAVĞLU: Scand. J. Clin. Lab. Invest. **14**, 554 (1962).
94. LINDFORS, R.: Ann. Med. exp. Fenn. **41**, 355 (1963).
95. — u. A. RUOHONEN: Arch. Toxicol. **19**, 402 (1962).
96. LÜDY-TENGER, F.: Pharm. Acta Helv. **37**, 770 (1962).
97. MACEK, K., J. VEČERKOVÁ u. J. STANISLAVOVÁ: Pharmazie **20**, 605 (1965).
98. MACHATA, G., u. W. KISSER: Arch. Toxikol. **19**, 327 (1962).
99. MARGASINSKI, Z., et al.: Acta Polon. Pharm. **21**, 5 (1964) und **21**, 253 (1964); ref. C. A. **62**, 16784b (1965).
100. MAROZZI, E., et G. FALZI: Farmaco, Ed. Pr. **20**, 302 (1965).
101. MELLINGER, T. J., and C. E. KEELER: J. Pharm. Sci. **51**, 1169 (1962).
102. MEYER-DULHEUER, K.-H., u. R. RITTER: Pharm. Ztg. (Frankfurt) **110**, 260 (1965).
103. MOREIRA, E. A.: Tribuna Farm. **31**, 49 (1964).
104. MORRISON, J. C., and L. G. CHATTEN: J. Pharm. Pharmacol. **17**, 655 (1965).
105. — — J. Pharm. Sci. **53**, 1205 (1964).
106. MULÉ, S. J.: Analyt. Chem. **36**, 1907 (1964).
107. NANO, G. M., P. SANCIN u. G. TAPPI: Pharm. Acta Helv. **38**, 623 (1963).
107a. NEGWER, M.: Organisch-chemische Arzneimittel und ihre Synonyma. Stuttgart: Kunst und Wissen 1961.
107b. NEIDLEIN, R., E. HOHNDORF u. J. D. ROSENBLATH: Pharm. Ztg. (Frankfurt) **111**, 874 (1966).
108. — H. KRÜLL u. M. MEYL: Dtsch. Apoth.-Ztg. **105**, 481 (1965).
109. — G. KLÜGEL u. U. LEBERT: Pharm. Ztg. (Frankfurt) **110**, 651 (1965).

110. Nishimoto, Y., and S. Toyoshima: J. Pharm. Soc. Japan **85**, 327 (1965).
111. Noirfalise, A.: J. Chromatog. **20**, 61 (1965).
112. — and M. H. Grosjean: J. Chromatog. **16**, 236 (1964).
113. Nürnberg, E.: Dtsch. Apoth.-Ztg. **101**, 142 (1961).
114. — Arch. Pharm. **292/64**, 610 (1959).
115. Nussbaumer, P. A.: Pharm. Acta Helv. **37**, 161 (1962).
116. Oehlschläger, H., J. Volke u. E. Kurek: Arch. Pharm. **297**, 431 (1964).
117. Pastor, J., et R. Raimondi: Bull. Soc. chim. Fr. **1965/9**, 2426.
118. — — Trav. Soc. Pharm. Montpellier **23**, 220 (1963), Ref. C.A. **62**, 11634b (1965).
119. Pastuska, G., u. H. Trinks: Chemiker-Ztg. **85**, 535 (1961).
120. — Z. anal. Chem. **179**, 427 (1961).
121. — u. H.-J. Petrowitz: Chemiker-Ztg. **86**, 311 (1962).
122. Paulus, W.: Arch. Toxikol. **20**, 191 (1963).
123. — u. R. Keymer: Arch. Toxikol. **20**, 38 (1963).
124. — W. Hoch u. R. Keymer: Arzneimittel-Forsch. **13**, 609 (1963).
125. — et al.: Arch. Kriminol. **135**, 84 (1965).
126. Pechtold, F.: Arzneimittel-Forsch. **14**, 258 (1964).
127. — Arzneimittel-Forsch. **14**, 1056 (1964).
128. — Arzneimittel-Forsch. **14**, 972 (1964).
129. Pennington, G. W., and D. Smyth: Arch. int. Pharmacodyn. **152**, 285 (1964).
130. Petrowitz, H.-J.: Materialprüfung **2**, 309 (1960).
131. Poethke, W., u. W. Kinze: Pharmaz. Zentralhalle **103**, 95 (1964).
132. Potter. W. P. de, R. F. Vochten, J. F. de Schaepdryver, and C. Heymanns: Experientia (Basel) **21**, 482 (1965).
133. Pribilla, O.: Arzneimittel-Forsch. **14**, 723 (1964).
134. Reisch, J., H. Bornfleth u. J. Rheinbay: Pharm. Ztg. (Frankfurt) **107**, 920 (1962).
135. — — — Pharm. Ztg. (Frankfurt) **108**, 1182 (1963).
136. — — — Pharm. Ztg. (Frankfurt) **108**, 1183 (1963).
137. — — u. G. L. Tittel: Pharm. Ztg. (Frankfurt) **109**, 74 (1964).
137a. Ritschel-Beurlin: Arzneimittel-Forsch. **15**, 1247 (1965).
137b. Rink, M., and A. Gehl: J. Chromatog. **21**, 143 (1966).
138. Rusiecki, W., u. M. Henneberg: Acta Polon Pharm. **21**, 23 (1964).
139. Rüdiger, W., u. H. Büch: Arch. exp. Path. Pharm. **251**, 107 (1965).
140. Sali, M., and M. Oesch: J. Chromatog. **14**, 526 (1964).
141. Šaršúnová, M.: Pharmazie **18**, 748 (1963).
142. — u. V. Schwarz: Pharmazie **18**, 207 (1963).
143. Savidge, R. A., and J. S. Wragg: J. Pharm. Pharmakol. **17**, Suppl. 60 S (1965).
144. Schmid, E., E. Hoppe, Chr. Meythaler jr. u. L. Zicha: Arzneimittel-Forsch. **13**, 969 (1963).
145. Schulze, W.: Dissertation, Braunschweig 1962.
145a. Schunack, W., E. Mutschler u. H. Rochelmeyer: Dtsch. Apoth.-Ztg. **105**, 1551 (1965).
146. Schüppel, R., u. Kl. Soehring: Pharm. Acta Helv. **40**, 105 (1965).
147. Seeboth, H., u. H. Görsch: Chem. Techn. **15**, 294 (1963).
148. Segura-Cardona, R., u. K. Soehring: Med. exp. **10**, 251 (1964).
149. Seiler, N., u. M. Wiechmann: Hoppe-Seylers Z. physiol. Chem. **337**, 229 (1964).
150. Seno, S., W. V. Kessler, and J. E. Christian: J. Pharm. Sci. **53**, 1101 (1964).

151. SEYDEL, J. H., BUETTNER u. F. PORTWICH: Klin. Wschr. **43**, 1060 (1965).
152. SHELLARD, E. J., and I. V. OSISIOGU: Lab. Pract. **13**, 516 (1964).
153. SHEPPARD, H., B. S. D'ASARO, and A. J. PLUMMER: J. Pharm. Sci. **45**, 681 (1956).
154. SMITH, G. A. L., and P. J. SULLIVAN: Analyst. **89**, 312 (1964).
155. SNELL, F. D., and C. T. SNELL: Colorimetric. Methods of Analysis. Vol. II, p. 623. Princeton, New Jersey: Third Ed., D. van Nostrand Company, Inc. 1955.
156. STAINIER, C., J. BOSLY, FR. DUTRIEUX et R. STAINIER: Pharm. Acta Helv. **38**, 587 (1963).
157. STEELE, J. A.: J. Chromatog. **19**, 300 (1965).
157a. SUGITA, J., and Y. TSUJINO: J. Chromatog. **21**, 341 (1966).
158. SUMERE, C. F. VAN, G. WOLF, H. TEUCHY, and J. KINT: J. Chromatog. **20**, 48 (1965).
159. SUNSHINE, I., and W. W. FIKE: New Engl. J. Med. **271**, 487 (1964).
160. — E. ROSE, and J. LE BEAU: Clinical-Chemistry **9**, 312 (1963).
161. — Amer. J. Clin. Path. **40**, 576 (1963).
161a. SZENDEY, G. L.: Arch. Pharm. **299**, 527 (1966).
162. SZULCZEWSKI, D. H., C. M. SHEARER, and A. J. AGUIAR: J. Pharm. Sci. **53**, 1156 (1964).
163. TEICHERT, K., E. MUTSCHLER u. H. ROCHELMEYER: Dtsch. Apoth.-Ztg. **100**, 283 (1960).
164. THOMA, F.: Tuberk.-Arzt **16**, 362 (1962).
164a. — Arzneimittel-Forsch. **16**, 771 (1966).
165. *The ring Index, A.C.S. Monograph Series* 84,252 No 1860 (1940).
166. UHLMANN, H.-J.: Pharm. Ztg. (Frankfurt) **109**, 1998 (1964).
167. VERCRUYSSE, A.: J. Pharm. Belg. N.S. 18, 569 (1963).
168. VIDIC, E.: Arch. Toxikol. **19**, 254 (1961).
168a. WAGNER, G., u. J. WANDEL: Pharmazie **21**, 105 (1966).
169. WALDI, D.: Arch. Pharm. **295**, 125 (1962).
170. WALDI, D.: In: STAHL: Thin-Layer Chromatography. Berlin-Göttingen-Heidelberg: Springer 1965.
170a. WANDEL, J.: Dissertation, Leipzig 1965, „Beiträge zur DC von Sulfonamiden".
171. WEBER, S. H., u. A. LANGEMANN: Helv. chim. Acta **48**, 1 (1965).
172. WEHRLI, A.: Canad. Pharm. J. Sci. Sect. **97**, 208 (1964).
173. WEICHSEL, H.: Mikrochim. Ichnoanal. Acta **1965**/2, 325.
174. WEIDMANN, H.: Dissertation, Berlin 1961.
175. WOLLISH, E. G., M. SCHMALL, and M. HAWRYLYSHYN: Analyt. Chem. **33**, 1138 (1961).
176. YATABE, M., and H. OKI: Kagaku Keisatsu Kankynsho Hokoku **17**/2, 167 (1964) ref. C.A. **61**, 13, 15933f (1964).
177. ZARNACK, J., u. S. PFEIFER: Pharmazie **19**, 216 (1964).
178. ZINSER, M., u. CH. BAUMGÄRTEL: Arch. Pharm. **297**, 158 (1964).

R. Antibiotics

K. H. WALLHÄUSSER

According to WAKSMAN's definition, antibiotics, which strictly speaking belong to the chemotherapeutical agents, are substances which are formed by microorganisms and which kill off other microorganisms or inhibit their growth. This narrow interpretation is nowadays often overstepped and anti-microbial compounds occurring in higher plants and some animals are included. A total of nearly 800 antibiotically active compounds is known, of which only about 45 are however in use. Even though the expansion of research on antibiotics is no longer so hectic as 10 years ago, this in no way means that the antibiotic era has come to an end. Alongside the search for new antibiotics, modifications of known substances, e. g., the "semi-synthetic preparation" of new penicillins starting from 6-aminopenicillanic acid, is becoming more and more important.

TLC is providing valuable help in the domains of both research and production, since it enables rapid and dependable results to be obtained. Its chief spheres of application are:

a) in research on antibiotics
1. for identifying known antibiotics in the screening programme and in searching for new compounds
2. for separation of substance mixtures
3. for isolating individual components (preparative methods)

b) in production of antibiotics
1. for quality control and testing for impurities
2. for evidence of identity in the control laboratory
3. for testing stability of preparations

It is not possible to consider all the known antibiotics in the following tables. The inclusion of all substances in use has, however, been thought worth while. There is anyhow no information about the behaviour in TLC of the compounds which were described early but which are no longer used. Wherever the Rf-values of new compounds are known, they are incorporated into the tables. Some products of decomposition of the known antibiotics are included in so far as they play a part in quality control.

Classification of Antibiotics according to their Biosynthetic Structural Units

Structural Unit				
Acetic acid (Propionic acid)	Sugar	Nucleic acid derivatives	Amino acids	Miscellaneous
I. Polyenes and polyacetylenes II. Macrolides III. Tetracyclines IV. Substances with similar structural unit	V. Basic, *non*-extractable substances	VI	VII. Acyclic compounds VIII. Heterocyclic compounds IX. Macrocyclic peptides X Other peptides	XI. Antibiotics not belonging to the previous groups are treated here

Detection of Antibiotics on the Thin-Layer Chromatograms

In the evaluation of thin-layer chromatograms of antibiotic mixtures, it is not only a question of visualising the spots but rather of detecting the antibiotically active and inactive components. Two methods are available here:

a) *microbiological detection* (bioautograph),
b) *chemical detection* through specific colour reactions.

These methods must be complementary. The bioautograph shows only the active spot and does not detect the inactive impurities. The chemical test on the other hand provides no information about the biological activity of a component; this is possible only through comparison with the bioautograph. Chemical detection is generally non-specific and has to be tried out first with every substance to be tested. Only when the most suitable reagent has been found and side reactions can be genuinely excluded, can the chemical test replace the bioautograph. It is an advantage that the chemical test can be carried out appreciably faster than the microbiological procedure; in the latter, results are obtained only after 6–16 h, depending on the growth of the test organism. Yet the bioautograph is a universal test which is available for all antibiotics, even unknown substances, when various test species are used (gram positive and negative bacteria, yeasts, fungi etc.).

1. Execution of the Microbiological Test

Both culture media must be sterilised by heating 20 min at 120° C in an autoclave. The Sabouraud medium is suitable especially for culture of fungi. In the preparation of the agar culture medium, 10 g meat extract and 1000 ml distilled water can be used instead of broth.

The organisms recommended for testing a particular antibiotic are seen from the following tables.

Composition of the Culture Medium (g/litre)

Component	Agar culture	Sabouraud medium
Dextrose	—	40
Meat peptone	10	10
Broth (ml)	1000	—
NaCl	3	—
Na_2HPO_4	2	—
Agar	20	15
Distilled water (ml)	—	1000
pH	7.2	5.7

Sources of Ready-Prepared Dry Culture Media

Suppliers	More detailed description in ordering:	
	Agar culture	Sabouraud medium
BBL Biotest-Serum-Institut, Frankfurt, W. Germany	Nutrient-Agar 03-124T	Sabouraud-Maltose-Agar[a] 03-142T
Difco: O. Nordwald, Hamburg-Altona, W. Germany	Bacto Nutrient-Agar[a] B 1	Bacto Sabouraud-Dextrose Agar B 109
E. Merck, Darmstadt, W. Germany	Standard-II-Nähragar[a]	Sabouraud-Agar modifiziert Merck[a]
Oxoid: Nährboden und Chemie, Wesel/Rhein, W. Germany	Nutrient-Agar[a] CM 3	Sabouraud-Dextrose-Agar CM 41

[a] These cultures have a somewhat different composition but can still be used for bioautographs.

Frequently Used Test Organisms

Species	Species No. of the ATCC	Sources[b]
Staphylococcus aureus	ATCC 6538P	American Type Culture
Micrococcus flavus	ATCC 10240	Collection (ATCC), 12,301
Sarcina lutea	ATCC 9341	Park Lawn Drive
Staphylococcus epidermidis	ATCC 12228	Rockville
Bacillus subtilis	ATCC 6633	Maryland
Bac. cereus var. mycoides	ATCC 11778	20,852 USA
Bac. megaterium	ATCC[a] 9885	
Saccharomyces cerivisiae	ATCC 2601	
Candida albicans		
Candida pseudotropicalis		
Microsporium gypseum	ATCC 14683	

[a] Instead of this organism, *Bac. subtilis* or *Bac. cereus var. mycoides* can be used.
[b] Most of these test organisms can be obtained in Germany also from pharmaceutical branches of the chemical industry which supply antibiotics.

The growth inhibition zones yielded by the active substances are difficult to recognise on the thin-layer chromatogram because it is opaque. Visualisation is then accomplished with TTC(2,3,5-triphenyl-2H-tetrazolium chloride) [43, 27] which is converted to a red formazan by the dehydrogenase of the multiplying organisms. TTC may be added directly to the agar medium or, better, the incubated plate is sprayed over with a 0.1% TTC solution and allowed to incubate for a further 30 min. A 0.5% solution of 2,6-dichlorophenol-indophenol can be used instead of TTC; the inhibition zones then appear blue against a colourless (decolourised) background [5].

a) Direct Method

The TLC plate is coated over with agar culture or Sabaroud agar which has been previously liquefied in a heated vessel; after cooling to 40° C, it is inoculated with the desired test organism.

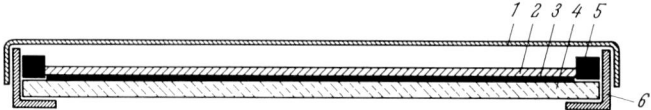

Fig. 170. Set-up for obtaining a TLC-bioautograph. *1* aluminium cover; *2* agar layer; *3* TL; *4* 20×20 cm glass plate; *5* V$_2$A-steel frames laid on after TLC; *6* aluminium frames

b) Contact Method (Reprint or Filter Paper Print Process)

A strip of filter paper (Whatman No. 1 (Firm 115)), 2–3 cm × 17 cm, is moistened with a suitable buffer solution or solvent and pressed firmly on to the thin-layer chromatogram by means of another glass plate. After 10–15 min, the strip is removed, laid immediately on to an inoculated agar plate (after drying if it had been moistened with an organic solvent) and left for 10–15 min. It is then removed and the agar plate allowed to incubate for 6–16 h, after which the inhibition zones are located and their positions marked in the corresponding places on the thin-layer chromatogram. According to [6], it is better to place filter paper No. 403 (Firm 83) on the agar test plate and then to press the TLC plate on to the paper for about 20 min with a heavy weight. TLC plate and filter paper are then removed and the test plate incubated.

2. Execution of the Chemical Test

About 20% of the antibiotics are coloured and their spots on thin-layer chromatograms can thus be detected without any special aid. Some others fluoresce in UV light. The overwhelming majority, however, are visualised only by spraying with a (specific) reagent. This reagent must be previously determined, in accordance with the properties of the substance to be detected. Specific or group reagents used for the individual antibiotics are seen in the following tables.

3. General Information about Layers and Solvents Used

Generally one of two principal adsorbents has been used for separating antibiotics with the help of TLC; these are silica gel G (Firm 88), and, less often, alumina. Impregnated layers, e. g., with buffers, have occasionally been used. The solvents frequently used are compiled in Table 122.

Table 122. *Solvents for TLC of mixtures of antibiotics*

No.	Composition	V/V Proportion
1	Methanol	
2	Methylene dichloride-methanol-benzene-formamide	65 + 16.3 + 16.3 + 2.4
3	Chloroform-methanol	98 + 2
4	Chloroform-methanol	97 + 3
5	Chloroform-methanol	95 + 5
6	Chloroform-methanol	90 + 10
7	Chloroform-methanol	50 + 50
8	Chloroform-methanol-17% ammonium hydroxide	40 + 40 + 20
9	Chloroform-methanol-17% ammonium hydroxide	50 + 25 + 25
10	Chloroform-acetone	50 + 50
11	Ethanol-water	80 + 20
12	Ethanol-conc. ammonium hydroxide-water	80 + 10 + 10
13	Ethyl acetate	
14	Ethyl acetate-methanol	66.7 + 33.3
15	Ethyl acetate-methanol	87 + 13
16	Ethyl acetate-methanol	97 + 3
17	Ethyl acetate-methanol	95 + 5
18	Ethyl acetate-sym-tetrachloroethane-water (lower layer)	42.8 + 14.4 + 42.8
19	Ethyl acetate-di-n-butyl ether-water (upper layer)	42.8 + 14.4 + 42.8
20	Ethyl acetate-di-n-butyl ether-water (upper layer)	40 + 20 + 40
21	n-Propanol-pyridine-acetic acid-water	39.5 + 26.3 + 7.9 + 26.3
22	n-Butanol, saturated with water	
23	Butanol-buffer of pH 4.6	
24	Butanol-methanol-10% citric acid	57.2 + 14.2 + 28.6
25	Butanol-acetic acid-water	49.4 + 1.2 + 49.4
26	Butanol-acetic acid-water	60 + 20 + 20
27	n-Butanol-oxalic acid-water	50 ml + 2.5 g + 50 ml
28	n-Butanol-tartaric acid-water	50 ml + 3 g + 50 ml
29	Butyl acetate-n-butanol-acetic acid-phosphate buffer of pH 5.8	50.3 + 9.4 + 25.2 + 15.1
30	10% Tartaric acid	
31	10% Citric acid	
32	10% Citric acid, saturated with butanol	

No. 26 (butanol-acetic acid-water) has been the most widely employed; it effects separation in 5 of the 11 antibiotic groups quoted. Other "standard solvents" are chloroform-methanol, ethyl acetate-methanol and n-propanol-pyridine-acetic acid-water, which have been used in mixtures of various proportions.

The results so far obtained in TLC of antibiotic mixtures are tabulated below; the hR_f-values and experimental conditions are given.

I. Polyenes and Polyacetylenes

These unsaturated compounds are, with few exceptions, active mainly against fungi and especially against yeasts.

Antibiotic	hRf-value with solvent		Ref.	Test organism
	12[a]	26		
Amphotericin A	33	33		*Candida albicans*
Nystatin (Fungicidin)	18	18		*Candida albicans*
Pimaricin	34	34		*Sachcaromyces*
Trichomycin A—C	45	17	[17]	*Candida albicans*
Variotin				*Microsporium* ATCC 14683
Pentamycin	67	67		
Unamycin A	—	36		

Layer: Silica gel G (Firm 88).

[a] Silica gel previously treated with phosphate buffer, pH 8.

II. Macrolides

These antibiotics, which contain one or several sugars, can be extracted and are distinguished from the representatives of group V through their large lactone ring. They are active principally against gram positive bacteria.

Antibiotic	hRf-values with solvent				Ref.	Detection[a]	Microbiological detection
	1	5	7	26			
Acumycin	66	35	82		[6]	weakly pink	*Staph. aureus*
Amaromycin				38	[17]	yellow	
Angolamycin	65	18	82		[6]		
Brefeldin A					[40]		
Carbomycin	75	40	88	55	[6, 17]	blue	*B. megaterium*
Niddamycin							
Erythromycin	16	03	29	39	[3, 6, 17]	brownish green	*B. subtilis*
Narbomycin	22	12	41		[6]		
Leucomycin				58	[17]	brownish red	
Lancamycin	74	37	87		[6, 21]		
Oleandomycin				29	[17]	purple	*B. megaterium*
Picromycin	22	7	36	41	[6, 17]	pink	*B. megaterium*
Spiramycin I—III				8	[17]	red-brown	or *Staph.*
Tertiomycin A				86	[17]	pale red	*epidermidis*
Tertiomycin B				63	[17]	pale red	ATCC 12228
Tylosin	68	7	81	59	[6, 17]	olive green	
Foromacidin A	31	2	59		[6]		
Foromacidin B	32	5	61		[6]		
Foromacidin C (identical with spiramycin)	37	6	64		[6]		

Layer: Silica gel G (Firm 88).

[a] The spots are visualised by spraying with 10% sulphuric acid and heating 5—10 min at 80° C; most substances of this group show a characteristic colour. Spraying with 50% aqueous sulphuric acid and subsequent charring by heating

was carried out by [3]. Treatment with 10% molybdophosphoric acid in alcohol, followed by heating, has proved better. Erythromycin appears as a blue spot on a yellow background; in 2 h, however the colour disappears. According to [3], solvent 2 gives a good separation of erythromycin A and B, anhydro-2'-acetylerythromycin and mixtures of these components; separation is complete in a run of 10 cm, lasting 30—40 min. Brefeldin A (identical with decubin) has been detected in [40] with a chromate-reagent (26.7 g CrO_3 + 21.3 ml conc. sulphuric acid, diluted to 100 ml with water).

III. Tetracyclines

These have a wide range of activity against gram positive and gram negative bacteria and against large viruses.

Antibiotic	hR_f-value with solvent					References	Chem. test [a]	Microbiol. test [b]
	22	24	27	28	31			
Chlortetracycline	30	43	49	35	30	[20, 27, 42]	dark grey	Bac. cereus var. mycoides ATCC 11778
Demethylchlor-Tc.								
Oxytetracycline	27	41	46	31	58	[20, 27, 42]	dark grey	
Tetracycline (Tc.)	23	38	38	26	36	[20, 27, 42]	dark grey	
Anhydrochlor-Tc.		46				0-20 [27, 42]		
Anhydrotetra-cycline		45				0-20 [27, 42]		

Layer: Silica gel G (Firm 88); a mixture of 30 g silica gel G + 9 g di-Na EDTA + 60 ml water used with solvents 22, 27 and 28 [20].

[a] see a) below.
[b] see b) below.

Radial (circular) technique has been employed as well as the usual procedure [20, 42]. Two or three chromatographic runs on the same layer have been recommended for better separation of tetracycline and 6-demethyltetracycline [42]; the layer is dried between each run.

Application of the sample; about 20 μg in methanol solution.

Visualisation

a) *chemical detection;* spraying with a 5% solution of $FeCl_3$ in methanol.

Most of the substances fluoresce yellow-green in UV light; ammonia vapour intensifies the fluorescence.

Spraying with a 50% solution of $SbCl_3$ in acetic acid and then heating 5—10 min at 110° C: Tc, Cl-Tc and 6-demethyl-Tc then appear as reddish spots in daylight and brick red in UV light; oxy-Tc is yellow in daylight and blue in UV light.

Spraying with conc. H_2SO_4 and heating to 110° C: the spots are yellow brown in daylight; only oxy-Tc shows fluorescence (yellow) in UV light.

Spraying with a solution of Fast Blue B in water: Tc, chlor-Tc and oxy-Tc give a reddish colour, epi-Tc a violet tinge and anhydro-Tc blue.

b) *microbiological detection;* test organisms are *Bac. subtilis* ATCC 6633 or *Sarcina lutea* ATCC 9341. Limit of detection is 0.01–0.1 µg.

IV. Substances with Similar Structural Units

Antibiotic	hR_f-values with solvent			Ref.	With iodine vapour	Colour in UV light
	3	3^a	4			
Griseofulvin				[8, 23]		
Epi-griseofn.				[23]		
Dehydrogriseofn.				[23]		
Roridin A	18	70	21		yellow-brown	dark
Roridin B	26	55	49		green-black	dark
Roridin C			41		yellow-brown	dark
Verrucarin A	28	70	59		yellow-brown	light
Verrucarin B	47	83	69		yellow-brown	light
Verrucarin C	28	74	52	[15]	yellow-brown	light
Verrucarin D	28	70	55		yellow-brown	dark
Verrucarin E	0	0	9		violet	dark
Verrucarin F	54	—	—		yellow-brown	dark
Verrucarin G	49	—	—		brown	not visible

Layer: Silica gel G (Firm 88).

^{a} Alumina (Firm 88); *Microbiological detection: Microsporum gypseum* used for griseofulvin; *Candida albicans* or *Saccharomyces cerevisiae* ATCC 2601 for verrucarin A and B.

V. Basic, Water-Soluble, Non-extractable Antibiotics

Antibiotic	hR_f-values with solvent		References
	9	21	
Aminocidin		40	[17]
Amminosidin	68		[17]
Catenulin	60	54	[17]
Dihydrostreptomycin			[34]
Kanamycin A	65	55	[17]
Neomycin B	51	46	[7, 17]
Paromomycin	68	40	[17]
Zygomycin A	62	56	[17]
Streptomycin			[34]
Streptothricin I and II (identical with neomycin B and C)	26	52	[17]

Layer: Silica gel G (Firm 88).

VI. Nucleic Acid Derivatives

Antibiotic	hRf-values with solvent		Reference
	9	14	
Angustmycin C		+	
Blasticidin S	70		[17]
Toyacamycin		+	
Tubercidin		+	

+ Rf-value not quoted.

Layer: Silica gel G (Firm 88).

VII. Acyclic Compounds

Antibiotic	hRf-value with solvent 26	Reference
Enteromycin	73	[17]

Layer: Silica gel G (Firm 88).

VIII. Heterocyclic Compounds

Antibiotic	hRf-values with solvent					References
	15	23	25	26	29	
Actithiazic acid				74		[17]
Ampicillin						
6-Aminopenicillanic acid		9	26			[27]
Aureothricin	63			58	57	[17]
Methicillin						
Oxacillin						
Penicillin G					65	[30]
Penicillin V		23	56			[27, 30, 31, 32]
Phenethicillin K						[29]
Thiolutin	70			65	64	[17]

Layer: Silica gel G (Firm 88).

Microbiological detection with *Staph. aureus* ATCC 6538P or *Sarcina lutea* ATCC 9341.

IX. Macrocyclic Peptides

Antibiotic	hRf-values with solvent							Refs.	Microbiological detection
	11	12	16	18	19	20	26		
Actinomycin C 1				44	40	28			
C 2				51	46	30			
C 3				58	53	33		[9]	Bac. subtilis
F 1				21	23	10			ATTC 6633
F 3				35	29	13			
Actinomycin C							68		
J							73		
Althiomycin	78								
Amphomycin							53		
Bacitracin	13	58						[17]	Micrococcus
Etamycin	80						66		flavus
Mikamycin A			+						
Pyridomycin	18						38		
Telomycin							44		

Layer: Silica gel G (Firm 88); alumina (Firm 88) with solvents 18, 19 and 20[1]
+: No hRf-value quoted.

X. Other Peptides

Antibiotic	hRf-values with solvent 9	Reference
Ferrimycin		
Phleomycin		
Viomycin	11	[17]

Layer: Silica gel G (Firm 88).

XI. Miscellaneous Antibiotics

(not classifiable in the preceding groups)

Antibiotic	hRf-values with solvent							References
	4	8	12	15	17	26	29	
Antimycin A			72				81	[17]
Blastmycin		78	82					[17]
Chloramphenicol								
Chromomycin A3				65				[17]
Cyanein	48[a]							[5]
Gliotoxin	88[a]							[5]
Homomycin			42					[17]

[a] Alumina (Firm 88).

(Continued)

Antibiotica	hRf-values with solvent							References
	4	8	12	15	17	26	29	
Nonactin								[44]
(Werramycin)								
Monactin					48			[4, 44]
Dinactin					32			[4, 44]
Trinactin					15			[4, 44]
Novobiocin			82					[17]
Moenomycin A		92				20		
B								
C								
Porfiromycin					48	50	50	[17]
Rifomycin B								[27]
O								[27]
S								[27]
SV								[27]

Layer: Silica gel G (Firm 88).
Detection: Nonactin by spraying with conc. H_2SO_4 and heating to 150° C, or with the Dragendorff reagent. Microbiological detection of nonactin with *Staph. aureas.* ATTCC 6538 P; cyanein with *Candida pseudotropicalis;* rifomycins with *Sarcina lutea* ATCC 9341; novobiocin with *Bac. subtilis* ATCC 6633; chloramphenicol with *Sarcina lutea* ATCC 9341.

Bibliography for Chapter R. Antibiotics

1. ACHENBACH, H., u. H. GRIESEBACH: Z. Naturforsch. **19**b, 561 (1964).
2. AKITA, E.: J. Antibiotics (Japan) Ser. A **17**, 200 (1964).
3. ANDERSON, T. T.: J. Chromatog. **14**, 127 (1964).
4. BECK, J.: Helv. Chim. Acta **45**, 620 (1962).
5. BETINA, V., and Z. BARATH: J. Antibiotics (Japan) Ser. A **17**, 127 (1964).
6. BICKEL, H.: Helv. Chim. Acta **45**, 1396 (1962).
7. BRODASKY, T. E.: Anal. Chem. **35**, 343 (1963).
8. BROSSI, A.: Helv. Chim. Acta **45**, 1292 (1962).
9. CASANI, G.: J. Chromatog. **13**, 238 (1964).
10. CEDER, O.: Acta Chem. Scand. **18**, 83 (1964).
11. CIFERRI, O.: Biochem. J. **90**, 82 (1964).
12. COMIN, J.: Helv. Chim. Acta **46**, 409 (1963).
13. FISCHBACH, H., and J. LEVIN: Antibiotics & Chemotherapy **5**, 640 (1955).
14. FISCHER, R., u. H. LAUTNER: Arch. Pharm. **294**, 1 (1961).
15. HÄRRI, E.: Helv. Chim. Acta **45**, 839 (1962).
16. HÜTTENRAUCH, R., u. J. SCHULZE: Pharmazie **19**, 334 (1964).
17. IKEKAWA, T.: J. Antibiotics (Japan) Ser. A **16**, 56 (1963).
18. — J. Antibiotics (Japan) Ser. A **17**, 194 (1964).
19. ISHII, S., and B. WITKOP: J. Am. Chem. Soc. **86**, 1848 (1964).
20. KAPADIA, G. J., and G. S. RAO: J. Pharm. Sci. **53**, 223 (1964).
21. KELLER-SCHIERLEIN, W., u. G. RONCARI: Helv. Chim. Acta **47**, 78 (1964).
22. KONDO, S.: J. Antibiotics (Japan) Ser. B **14**, 1 (1964).
23. KYBURZ, E.: Helv. Chim. Acta **45**, 813 (1962).
24. LIBOSVAR, J.: Českoslov. farm. **11**, 73 (1962).
25. MASSE, J.: Ann. pharm. franç. **22**, 349 (1964).
26. MISTRYUKOV, E. A.: J. Chromatog. **9**, 311 (1962).

27. NICOLAUS, B. J. R.: Pharmaco 8, 349 (1961).
28. — Experientia 17, 473 (1961).
29. NUSSBAUMER, P. A.: Pharm. Acta Helv. 37, 65 (1962).
30. — Pharm. Acta Helv. 37, 161 (1962).
31. — Pharm. Acta Helv. 38, 245 (1963).
32. — Pharm. Acta Helv. 38, 758 (1963).
33. — Pharm. Acta Helv. 39, 647 (1964).
34. — et M. SCHORDERET: Pharm. Acta Helv. 40, 205 (1965).
35. — Pharm. Acta Helv. 40, 210 (1965).
36. OKUDA, T.: J. Antibiotics (Japan) Ser. A 17, 218 (1964).
37. SENSI, P.: J. Chromatog. 5, 519 (1961).
38. — J. med. Chem. 7, 586 (1964).
39. SIGG, H. P.: Helv. Chim. Acta 46, 1061 (1963).
40. — Helv. Chim. Acta 47, 1401 (1964).
41. SMITH, R. H., and W. MCKERNAN: Nature 195, 1303 (1962).
42. SONANINI, D., u. L. ANKER: Pharm. Acta Helv. 39, 518 (1964).
43. WALLHÄUSSER, K. H., u. A. RIPPEL-BALDES: Naturwissenschaften 37, 450 (1950).
    — Naturwissenschaften 38, 190 (1951).
44. — Arzneimittelforsch. 14, 356 (1964).
45. WAISVISZ, J. M.: Acta Chem. Scand. 18, 83 (1964).

S. TLC in Clinical Diagnosis

NEPOMUK ZÖLLNER and GÜNTHER WOLFRAM

Introduction

It is difficult to write a chapter on application in a book which is devoted to methods without repeating what has already been said (and presumably better said) about thin-layer techniques and methods of separation. Without exaggeration it can be said that most of the TLC methods are applicable to medicine; for the solution of any particular medical problem nowadays, the multitude of possibilities offered by TLC should be considered. No attempt is made here to give detailed methods, as in other parts of this book. That would have involved simply repetition. We have, however, quoted original sources for procedures which are given in other chapters.

It could not have been our task to comment on the applicability to medicine of all the methods described in this book. Such a list would be very soon out of date, apart from the fact that the specialist investigator on each separate compound class could have given fuller information; nor could special features of the analysis of biological material be the subject of this chapter, since other contributors are foreseen for this. We have deemed it right to discuss below only work in which the clinical applicability of TLC has been put to the test; we have done this less with a view to making a review compilation than to demonstrate the potential versatility of TLC in clinical medicine and to show that it might be used in many places. Only occasionally have we ventured our own opinion and offered hints for work in hitherto unexplored regions where the application of TLC is promising.

TLC represents one of the most important steps forward, not only in the field of analytical chemistry but also particularly in clinical chemistry. It is regrettable that the method is being so slowly introduced into routine analysis and clinical research. In our opinion, it is certain that, in many places, analysis would be faster and more accurate with the help of TLC. Its systematic use as an analytical tool would enable new data to be gained in all spheres of clinical activity.

I. Investigation of Endogenous Substances

1. Sugars, Their Derivatives and Metabolites

Numerous diseases are accompanied by increased elimination of various sugars in the urine or faeces. One may mention here the different forms of glucosuria, pentosuria, fructosuria, galactosaemia, the group of disaccharidase deficiencies or hereditary fructose intolerance.

TLC is a simple technique, available for clinical diagnosis, investigations of metabolism or control of appropriate therapy. Another example is the determination of fructose in human ejaculate [120]; this can give important information in the diagnosis of postpuberal Leydig cell insufficiency even when the spermiogram is normal.

Many solvents are at our disposal for separating sugars in the free state, their phenylosazones or 2,4-dinitrophenylhydrazones (see p. 827). The method of KÄSER and MASERA [85] may be quoted as an example of the determination of sugars in biological material. The authors point out the clinical applicability of sugar detection in urine and faeces in cases of progressive muscular dystrophy and ribosuria, in essential pentosuria and other maladies with increased sugar elimination.

A sugar extract is obtained by suspending samples of faeces in distilled water (1/3, w/v), mixing vigorously for 5 min, centrifuging and filtering the supernatant. The filtrate is used directly for the chromatography. The 24 h-urine is collected in the presence of 5 ml of both toluene and chloroform or of a few crystals thymol and stored in a refrigerator until worked up further. The urine must be demineralised before submitting to the chromatographic separation. An ion exchange procedure with the synthetic resin Biodeminrolit (The Permutit Company, London) is used. This is shaken 15—20 min in distilled water, saturated with carbon dioxide; the water is then discarded and equal amounts of urine sample and the moist ion exchanger, now saturated with carbon dioxide, are stirred together in a centrifuge tube for 20 min. The sample is then centrifuged, yielding a clear supernatant sensibly free of salts yet containing all the sugars. The adsorbent for the TLC-layers is prepared from 1 part by weight of silica gel G ("Merck") and 2 parts by volume of

0.1 N boric acid solution. The plates are ready for use after activating the layers by heating 30 min at 110° C. 10 µl amounts of the demineralised urine or the filtrate from the faeces sample, are applied as spots. If the concentrations in the sample are unknown, it is advisable to apply progressively varying amounts along the start line, always keeping a free route for the standard solution containing several known sugars (3 µg of each).

The chromatography is complete within 4 h at chamber saturation, using isopropanol-n-butanol-distilled water (50 + 30 + 20). The plates are then air-dried and sprayed with ethanolic naphthoresorcinol solution and 1 ml 95% sulphuric acid. After heating 15 min at 110° C, the sugars are visible as differently coloured spots. Palatinose, fructose, ribose, lactose, galactose and sucrose are adequately separated from one another; only glucose and maltose have the same Rf-values. This pair can be resolved with the same solvent on a layer of silica gel G/alumina G (1 + 1). Semi-quantitative determination is possible by visual comparison of the size and colour intensity of the individual spots with those from standard solutions of known concentration (see p. 134). It is best to express the data as the amount of sugar eliminated in urine or faeces during a 24 h period.

BICKEL and SCHWINDT [11] have separated the sugars from urine on silica gel G layers, buffered with sodium acetate. They developed twice in ascending chromatography, using ethyl acetate-isopropanol-water (65 + 23 + 12). 10 µl amounts of demineralised urine and increasing amounts of standard sugars were applied and the developed chromatograms were semi-quantitatively evaluated after spraying with naphtholindole-phosphoric acid (10 parts of a 0.2% solution of naphtholindole in ethanol + 1 part 85% orthophosphoric acid) and drying 15 min at 100° C. In work of importance for application to medicine, SCHERZ and co-workers [139] have determined aldo- and keto-hexoses, separated by TLC, on the basis of their colour reaction with diphenylamine. The oligosaccharides built up from these sugars can be determined according to the same principle since they are hydrolysed in the acid (hydrochloric acid) reaction mixture.

A reagent prepared from 10 parts conc. hydrochloric acid, 8 parts acetic acid and 2 parts of a 10% solution of diphenylamine in ethanol, is added to the sugar-containing fractions in test tubes. The mixture is heated in a bath of boiling water, yielding a blue complex. Maximum colour intensity is attained after 5—10 min with ketoses and 30 min with aldoses. The mixture is cooled, made up to a fixed volume with acetic acid, centrifuged from silica gel and the light absorption measured at 640 nm. Standards and blanks are treated in the same way.

RINK and HERRMANN [132] separate the sugars in 10 ml urine as their phenylosazones, using borate-buffered kieselguhr G layers and chloroform-dioxan-tetrahydrofuran-0.1 M sodium tetraborate (40 + 20 + 20 + 1,5).

a) Glycoproteins

With the help of TLC, WEICKER and co-workers [183] have studied the composition of the sugar components of a low molecular weight, perchloric acid-soluble glycoprotein in human urine. They established that there was no difference in concentration between that from patients with tumours, from patients with acute inflammation and from healthy individuals (Fig. 171). Xylose was found in all the glycoproteins investigated.

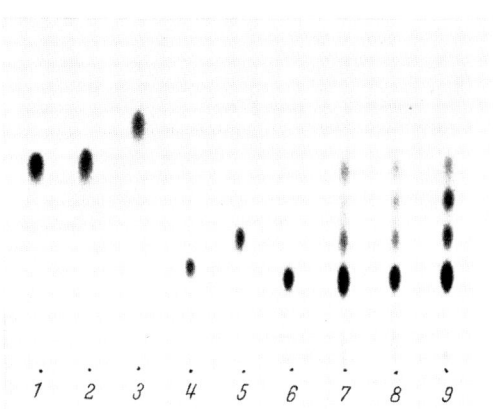

Fig. 171. TLC of sugars on cellulose-MN layer using the solvent water-butanone-formic acid-*tert*. butanol (15 + 30 + 15 + 40); time of run: 2 h; visualisation with aniline phthalate; *1* ribose; *2* fucose; *3* rhamnose; *4* glucose; *5* mannose; *6* galactose; *7* urine glycoprotein from inflammations, hydrolysed for 4 h with Amberlite pH 1; *8* urine glycoprotein from tumours, hydrolysed as in *7*; *9* sugar standard of galactose, mannose, xylose, fucose

b) Ketone Bodies

A sequel to the disturbance of metabolism in diabetes is acidosis of the blood and elimination of ketone bodies in the urine, provided the kidneys are in order. RINK and HERRMANN [131] have converted acetone and acetoacetic acid contained in 5 ml filtered urine into the corresponding hydrazones by dropwise addition of 2,4-dinitrophenylhydrazine (0.4% in 2N HCl). After standing 30 min in the cold, these derivatives were extracted with 1—2 ml ethyl acetate and the organic phase applied to a thin layer of cellulose MN 300. After developing with methanol-water-25% ammonium hydroxide (90 + 10 + 3), both substances could be recognised as yellow spots in daylight and dark spots in UV light.

2. Amino Acids, Their Derivatives and Metabolites

In order to be able to detect and determine quantitatively amino acids in biological material (serum, urine, sperm, lymph, milk, tissue) other components of the body fluids and organ extracts, such as proteins, peptides, lipids, carbohydrates, salts and urea, must be removed (see p. 736). If the amino acids are to be separated as their dinitrophenyl derivatives, these may be efficiently separated from the seriously interfering salts by virtue of their solubility in organic solvents (see p. 758). An excellent survey of the TLC of amino acids and related compounds in biological material has been made by PATAKI [118, 123].

a) Amino Acids in Urine

About 1.1 g of free amino acids is eliminated in urine daily. This represents about 1.2% of the total nitrogen in urine. The relative proportions of amino acids in urine differ from those in serum; accordingly, the clearance values vary from amino acid to amino acid. The amounts of amino acids eliminated vary greatly from one individual to another, even when on the same diet. A hyperaminoaciduria can be brought about by a disturbance in the intermediary metabolism which leads to increased amounts of one or several amino acids in the blood and hence to an amount exceeding the renal threshold; or by interferences with the transport mechanisms in the kidneys themselves. Examples of prerenal disturbances are phenylketonuria, accompanied by increased elimination of phenylalanine; and aminoacidurias occurring with serious liver affections or during rapid parenteral amino acid infusion. Examples of renal disturbances are the Fanconi syndrome, the aminoaciduria of Wilson's disease and in galactosaemia. Determination of the relative amounts of the individual amino acids as well as the total amount eliminated is evidently necessary for diagnostic purposes.

In some diseases, e. g., cystinuria, an unequivocal diagnosis is possible only on the basis of the pathological pattern of amino acid distribution in the urine. The reference amount is important for evaluating the results of thin-layer chromatography of urine amino acids. Use of standard volumes in TLC is unsatisfactory since the daily urine volumes passed can vary widely, e. g., from 600—2000 ml for adults. In order to circumvent this difficulty, some investigators refer the results to the amount of creatinine contained in the same sample; the elimination of creatinine is largely independent of the volume of urine. Others use volumes per time unit; thus JEPSON [81] has proposed 2-second portions for TLC in cases of severe aminoaciduria. For a period of collection of, say, 2 hours and a urine volume of 98 ml for example, the amount used would be $\frac{98 \times 2}{2 \times 60 \times 60}$ = 27.2 µl, or 30.0 µl if the urine sample is diluted to 108 ml.

OPIENSKA-BLAUTH and co-workers [114, 115] and ROKKONES [133, 134] have investigated the thin-layer chromatographic separation of free amino acids in urine. Using 10 µl of urine which had not been previously treated, the former [114] were able to detect 3 to 6 ninhydrin-positive spots in two-dimensional TLC on silica gel G with the solvents butanol-acetic acid-water (60 + 15 + 15) and phenol-water (75 + 25); 100 µl of the same but demineralised urine yielded 14—16 visible spots. ROKKONES [134] has obtained 9—12 ninhydrin-positive spots from an amount of urine containing 20 μg creatinine from 2—3 year old children; he separated on silica gel G layers with chloroform-methanol-17% ammonium hydroxide (40 + 20 + 20) and phenol-water (75 + 25). He carried out also semi-quantitative determination of the urinary amino acids and found differences in the urinary amino acid elimination, depending on age group. Glycine, cystine, histidine, alanine, lysine, serine, glutamine, phenylalanine and tryptophan are eliminated in large amounts during the first months of life, attaining the values for adults after 2—4 years. Proline and hydroxyproline are eliminated in scarcely measurable amounts during the first days after birth, reach maximum amounts before the end of the first month of life, thereafter falling to unmeasurably small concentrations in about 4—5 months. The elimination of 3-methylhistidine and taurine is as high a few days after birth as with adults. Increased cystine elimination is proportional as a rule to increase in lysine just as higher phenylalanine, tyrosine and tryptophan eliminations are related with one another. In a comparison of amino acid eliminations in the urine of adults of various ages, it has been found that merely the histidine concentration diminished somewhat with older people. CRAWHALL and co-workers [22] have studied urinary amino acids in cases of cystinuria.

DITTMANN [30, 31] has given procedures for TLC of amino acids without prior demineralisation. He used cellulose layers and triple development. The solvents must be buffered if compounds are present like phosphoethanolamine which are unstable in acid solution [31].

WALZ and co-workers [180] have carried out TLC-separation of the DNP-derivatives of urine amino acids. PATAKI and KELLER [119] have been able to detect a total of 45 DNP-compounds in the ether-soluble fraction of the DNP-amino acids from human urine; 24 of these had Rf-values agreeing with those of authentic amino acids; the others were designated by the symbols X_1 to X_{19}. BÜRGI and co-workers [19] have supplemented this "spot map" by detecting the compounds $X_{20}-X_{34}$. A number of unidentified compounds (Y_1-Y_{32}) has been found also among the acid-soluble DNP-compounds; $Y_{21}-Y_{32}$ occur regularly only with infants. Homoarginine and homocitrulline could be identified in the urine of two children during tolerance tests with lysine. Another degradation route for lysine thus appears to exist, in addition to that hitherto known via pipecolic acid and α-aminoadipic acid. The elimination of amino acids by adults increased only modestly in comparison with that by children after protein hydrolysates had been administered orally. In one case, increased amounts of sarcosine were eliminated and two unidentified substances, X_{31} and X_{32} appeared. KELLER and PATAKI [92] have identified 18 water-soluble DNP-compounds in human urine of which 5 agreed with authentic DNP- derivatives of amino acids.

TANCREDI and CURTIUS [165] have investigated the ether-soluble DNP-amino acids in the urine of healthy and sick children. In a case of hyperglycinaemia, one of hypoaminoaciduria and one of renal insufficiency, they were able to discover several amino acids which could not be found in the urine of a healthy girl, using the same technique.

α) **Tryptophan and Metabolites.** The separation of tryptophan and 10 of its metabolites from biological material has been described by DIAMANTSTEIN and EHRHARD [28]. They identified a number of tryptophan metabolites in the urine of patients suffering from chronic myelosis who had been fed with tryptophan-containing diet. Samples of 0.05—0.08 ml urine were used, without prior concentration. Salts and other foreign matter did not interfere in the separation. SCHLOSSBERGER and co-workers [141] have separated indole derivatives from urine using Sephadex G 25, and 0.5% aqueous ascorbic acid as eluent; they then chromatographed the individual substances on a cellulose MN-300 layer. OPIENSKA-BLAUTH and co-workers [116] have separated indole derivatives from urine on silica gel G and on cellulose layers, employing a modified Adamkiewicz-Hopkins reagent for quantitative determination of tryptophan. Pyridoxine has an important function as co-ferment in the degradation of tryptophan. DAHLER [26] has investigated the elimination of xanthurenic acid and kynurenine in the urine of babies and infants which had received a tryptophan-containing diet. He interpreted the pathologically increased values as due to a latent deficiency of vitamin B_6; doses of this vitamin normalised the augmented xanthurenic acid elimination. A sharp increase in kynurenine is found further in cases of vitamin B_2 deficiency. This implies that vitamin B_2 participates in the oxidation of kynurenine to 3-hydroxykynurenine. Kynurenine is determined quantitatively by separation on silica gel layers using n-butanol-acetic acid-water (60 + 20 + 20); the zones, recognised by their green fluorescence in UV light, are scraped off, 4-dimethylaminobenzaldehyde reagent is added, the solution filtered and evaluated photometrically. BENASSI and co-workers [7] separate tryptophan and its metabolites on a polyamide layer.

β) **β-Aminoisobutyric Acid (β-AIB).** About 90% of the adult, white population eliminate not more than 52 mg β-aminoisobutyric acid per 24 h; values of up to 300 mg/24 h however, have been measured on the remaining 10% [16]. This widely differing elimination by normal persons is probably genetically determined. Pathologically increased β-AIB elimination is found in cases of leucaemia, diseases of protein deficiency, following operations and radiation treatment and also in particular forms of mental deficiency. GOEDDE and BRUNSCHEDE [59] have developed a simple method in which the β-AIB is quantitatively converted to its dinitrophenyl-derivative, this compound isolated by TLC, eluted from the adsorbent and determined photometrically. The limit of error of the method is stated by the authors to be below $\pm 5\%$ for 0.8—80 µg amounts of β-AIB per sample. In a study of a number of patients with liver diseases, the authors were unable to confirm the hypothesis that increased β-AIB elimination is due to the damaged liver being incapable of degrading β-AIB further [16]. Three out of six patients with haemochromatosis had increased β-AIB elimination whereas the members of two families suffering from maple-syrup disease showed normal β-AIB values in their 24-hour urine.

b) Amino Acids in Blood and Organs

About 5.5—8 mg-% amino acid nitrogen is found in the plasma of a fasting human. The concentrations lie between 0.03 mg-% for aspartic acid and 8.3 mg-% for glutamine. Oral administration of individual amino acids leads to increases of these amino acids in peripheral blood.

The free amino acids in plasma have been separated by OPIENSKA-BLAUTH [114] on cellulose MN-300 layers, using the solvents butanol-acetic acid-water (60 + 15 + 15) and phenol-water (75 + 25). 14 ninhydrin-positive spots were isolated from 20 µl plasma. PATAKI and KELLER [117] have been able to detect 21 ether- and water-soluble dinitrophenyl-amino acids in 60 µl serum; no histidine was, however, present.

With the help of one- and two-dimensional TLC on silica gel G, DIMILLIER and TROUT [29] have observed increases in concentration of histidine, proline, hydroxyproline, serine, threonine, aspartic acid, glutamic acid and tryptophan in blood during extracorporal circulation; this increase was proportional to the duration of the extracorporal circulation. Through TLC on silica gel layers with the solvent n-butanol-acetic acid-water (60 + 15 + 15), V. EULER and co-workers (40) have been able to isolate from the tumour serum of the rat, after precipitation of proteins with methanol, a substance of hRf-value of 23 which was present at much lower concentration in the normal serum. By chromatographing again in two different solvents it could be identified as glycine.

WERNZE and FUJII [187] have employed TLC on cellulose layers for quantitative determination of the valine liberated by enzymatic hydrolysis as a measure of the angiotensinase activity in plasma and renal tissue. The data were stated to be reproducible.

The use of labelled substances is an indispensable technique in research on metabolic routes. DRAWERT and co-workers [33] have added ^{14}C-labelled glutamic acid to fermentation mixtures and found that the dinitrophenyl derivatives of glutamic acid, alanine, aspartic acid, proline and glycine, separated by TLC, were radioactive.

SQUIBB [154, 155] has separated the free amino acids in bird liver extracts, using silica gel G layers on plastic sheets. The coloured reaction products with ninhydrin were evaluated photometrically at 525 nm, directly on the layer. Lysine, histidine, arginine, asparagine, alanine, valine and leucine were determined in this way. Recoveries lay between 90 and 106%.

γ-Aminobutyric acid is, along with glutamic acid and glutamine, highly important in the metabolism of the free amino acids of the brain. VOIGT and co-workers [171] have separated the free amino acids in extracts of rat brain, using isopropanol-water (70 + 30) on silica gel G layers; they determined the content of γ-aminobutyric acid by spraying with a cadmium acetate-ninhydrin mixture, eluting from the silica gel and measuring the light absorption at 500 nm.

c) Amino Acids in Other Body Fluids

Amino acids in milk have been separated by BUJARD and MAURON [17] on a cellulose MN-300 layer. According to KELLER and PATAKI [91], 24 amino acids can be detected in sperm fluid. Since human ejaculate contains several proteolytic enzymes, the authors have drawn conclusions about the individual proteolytic activities from this result. MOUHG-RABI [109a] reports the separation of amino acids from the perilymph of the inner ear, using ion exchange TLC with gradient elution.

d) Iodoamino Acids (cf. p. 779)

Triiodothyronine and thyroxine are the physiologically active hormones of the thyroid gland. Tyrosine is the starting material for their synthesis. GRIES and co-workers [62] have chromatographed thyroid-active iodoamino acids on cellulose G layers and obtained good separations of DL-thyroxine, DL-triiodothyroxine, DL-diiodothyronine, DL-monoiodothyronine, diiodotyrosine and monoiodotyrosine.

SCHNEIDER and SCHNEIDER [145] have quoted solvent systems for separation of iodoamino acids on silica gel G layers and employed this method for detecting iodo compounds in urine of patients with hyperthyroidism who had been given ^{131}I. 48 hours after this dose, the iodoamino acids, concentrated over Dowex 50, were separated by TLC and, using autoradiography, seven ^{131}I-labelled fractions could be detected and an activity spectrum demonstrated. HERBERHOLD and NEUMÜLLER [68] report a thin-layer chromatographic investigation of the contents of the lymph tract of human thyroid glands. Amounts averaging 2 mm³ of thyroid lymph were obtained from autopsy material by means of glass capillaries; these amounts were extracted and separated on silica gel HF_{254} layers. Mono- and diiodotyrosine and tri- and tetra-iodothyronine could be detected along with iodine-containing proteins and inorganic iodine; TLC of native thyroid gland lymph shows that the inorganic iodide, diiodotyrosine and tetraiodothyronine are present in the free state. The same results were obtained in studies of the lymph obtained during operations.

WEST et al. [188] have used TLC to determine the thyroxine level in human serum. For 43 persons with normal thyroid function they found an average of 5.4 µg thyroxine per 100 ml serum with a normal range from 3.2 to 7.9 µg per 100 ml. Altogether 339 patients were examined. SCHORN and WINKLER [148] have reported the radiochromatographic determination of labelled tri- and tetraiodothyronine in the serum of a patient who had been dosed with carrier-free $Na^{131}I$ for therapy of hyperthyroidism (Fig. 172). This separation is important because triiodothyronine is about four times as active as tetraiodothyronine. Determination of total hormone iodine in serum with the usual methods does not reveal any pathologically modified ratio of tri- to tetraiodothyronine.

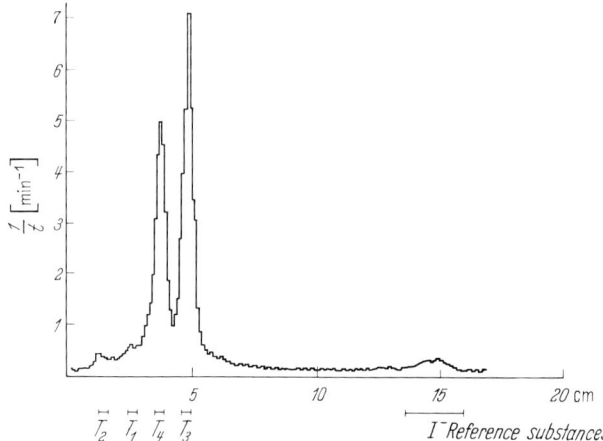

Fig. 172. Activity distribution on a chromatogram. Iodoamino acids labelled with ^{131}I were separated from the serum of a patient [148]. Layer: silica gel G; solvent: butanone-ethanol-N ammonium hydroxide (80 + 10 + 10), double development

e) ε-Aminocaproic Acid

As a result of its inhibiting influence on the activation of plasminogen, a precursor of the fibrinolytically active plasmin, ε-aminocapronic acid (EACA) has recently acquired growing importance in treatment of increased fibrinolysis with haemorrhage or danger of it. KÄSER and GUGLER [86] have developed a simple TLC-method for determining EACA in serum and urine. According to the authors, the mean error of the analyses was below 3%. The smallness of the sample required renders possible continuous control of the EACA concentration in serum or urine in therapeutic applications or metabolic studies, even with infants.

f) Creatinine

PATAKI [121] has detected creatinine in urine on cellulose-D-layers. This procedure enables the creatinine to be separated from other Jaffé-chromogens which interfere in the colour reactions on which many methods of determination are based. Quantitative evaluation of the creatinine spots, obtained using picric acid/sodium hydroxide, is possible since there is a linear relation between the square root of the spot surface area and the logarithm of the amount of substance applied (see p. 136) [122]. The reliability of the creatinine determination has been tested by parallel determinations of various amounts of creatinine in standard solutions and in urine. Tests on recovery gave a value of 95.7 ± 3.4% of the creatinine added. Up to 20 µg creatinine or ca. 20 µl urine can be used in one determination; larger amounts can lead to overloading of the layer.

The slurry for TLC is prepared from 10 g cellulose-D (Firm 33) and 65 ml distilled water and is spread in a layer 0.5 mm thick. 10 µl urine (A), 10 µl urine/distilled water (1 + 1) (B) and 10 µl creatinine standard (= 10 µg creatinine) (S) are applied to the air-dried layer. Chromatography is carried out with n-butanol-acetic acid-water (60 + 15 + 15) at chamber saturation. The layers are then sprayed with 5% alcoholic picric acid solution and 10% sodium hydroxide (lower limit of detection is 0.4 µg) and the outlines of the orange spots immediately marked round with a needle. The areas are then measured planimetrically (five measurements per spot) or by counting squares on millimeter paper.

From the equation:

$$\log G_A = \log G_S + \frac{\sqrt{F_A} - \sqrt{F_S}}{\sqrt{F_B} - \sqrt{F_A}} \cdot \log v$$

where
F_A = spot surface area from A
F_B = spot surface area from B
F_S = spot surface area from S
G_A = creatinine (µg) in A
G_S = creatinine (µg) in S
v = dilution factor

the creatinine concentration (G_A) of the urine sample can be determined.

The determination of creatinine in urine and blood enables the creatinine clearance to be calculated, a value of significance in renal diagnostics.

In PATAKI's method [123] for this purpose, 10 µl urine or 300 µl serum (deproteinised with an equal volume of acetone) and a standard solution are applied to the layer, chromatographed and the creatinine zones localised with the help of visualised reference runs at the side of the plate. The creatinine is then eluted from the scraped-off cellulose by shaking 3 h with 2 ml 0.1 N HCl and the solution neutralised with 2 ml 0.1 NaOH. 2 ml of colour reagent (100 ml 1% picric acid solution mixed with 40 ml 10% NaOH and made up to 1000 ml with distilled water) are then added and the absorbance measured within an hour at 490 nm with a cuvette of thickness 2 cm, against a blank on the cellulose alone.

g) Amines and Metabolites

Adrenaline. WALDI [177] describes the semi-quantitative determination of acetylated adrenaline and noradrenaline, using TLC. SEGURA-CARDONA and SOEHRING [149] have separated very small amounts of catechol amines and derivatives on polyamide layers. The lower limit of detection for adrenaline is about 0.003 µg.

Vanillin mandelic acid (4-hydroxy-3-methoxymandelic acid) (VMA), *vanillic acid* (VA) and *homovanillic acid* (HVA). Vanillin mandelic acid is quantitatively the most important degradation and elimination product of catechol amine metabolism. Elimination of over 15 mg VMA in the 24 hour-urine is specific proof of excessive adrenaline or noradrenaline production in cases of phaoechromocytoma. The endogenous VA, amounting to about 10% of the VMA elimination, is to be regarded as the degradation product of the VMA formed from the catechol amines. Since the total vanillic acid eliminated by humans is largely determined by the diet, exact separation of VA and VMA is very important. SANKOFF and SOURKES [137] report the TLC-determination of HVA in the urine of healthy and sick persons. SCHMID and HENNING [142] describe the thin-layer chromatographic detection of VMA from urine. A procedure for quantitative determination of VMA, VA and HVA in urine is due to TAUTZ et al. [166] who belong to the same team.

Reagents: 36.4% HCl; NaCl[1]; diethyl ether[1], free of peroxides; absolute ethanol[1]; silica gel G and kieselguhr G (Firm 88); fluorescent pigment ZS super (Firm 118); isopropanol for chromatography; ethyl acetate[1], refluxed for 12 h over NaOH (25 g/l) and distilled; 25% ammonium hydroxide; crystallisable benzene[1]; 96% acetic acid for chromatography; alkalised methanol solution (2% aqueous Na_2CO_3 + methanol[1], 1:3); diazotised Rose reagent (0.25% p-aminophenyl-β-diethylaminoethyl sulphone in 1% HCl + 0.5% aqueous $NaNO_2$ solution, mixed at 0° C in the proportion 3:1 immediately before use and kept in the dark); the test substances (1 mg/ml HVA, VMA, VA and 5-HIA (5-hydroxyindole acetic acid), chromatographically pure) are obtainable from the California Corporation Biochem. Res., Lucerne, Switzerland; 0.1% dichloroquinonechloroimide in methanol; the TLC layers are treated with ammonia vapour before spraying; diazotised p-nitroaniline (5 ml 0.1% nitroaniline solution + 5 ml 0.2% $NaNO_2$ solution in distilled water + 1 ml N NaOH, to be used immediately after mixing).

10—20 ml of the 24 hour-urine are adjusted to pH 0 with conc. HCl, saturated with NaCl and extracted by rhorough shaking with three 50 ml portions of diethyl ether. The ether extract is evaporated to dryness, the residue dissolved in 1 ml ethanol and 0.1 ml of this solution is applied to the layer (adsorbent = 25 g silica gel G + 25 g kieselguhr G + 2.5 g fluorescent pigment ZS super + 90 ml water; layer thickness 250 µm; dried 1 h at 105° C). This is developed at chamber saturation with isopropanol-ethyl acetate-ammonium hydroxide-water (45 + 30 + 17

---

[1] Analytical grade.

+ 8) to a height of 8 cm, then dried and twice developed with benzene-acetic acid (90 + 10) in 15 cm runs. The substances are clearly recognisable in short wave UV light (254 nm) through their fluorescence quenching and can be easily identified with the help of parallel runs of standard mixtures. The corresponding zones are scraped off and eluted with 2.5 ml alkalised methanol solution. After centrifuging at $+5°$, 2 ml of the supernatant liquid are mixed with 0.2 ml freshly prepared Rose reagent and left 30—60 min. The light absorption is measured at 490 nm for VMA and VA and at 370 nm for HVA, using a 1 cm cuvette thickness. The relevant details for a Zeiss Spektralphotometer PMQ III are: VA and VMA—slit width 0.02 mm, 490 nm, amplification 1/1/0; HVA—370 nm, amplification 1/10/0.

HVA absorbs appreciably less than VMA and VA. Calibration values, reagent and absorbent blanks are subjected to the same procedure. Under the conditions described, the lower limit of detection is about 0.05 µg/ml. Recoveries of the pure substances from thin-layer chromatograms were 97% for HVA, 90% for VA and 78% for VMA. The relatively low value for VMA is believed due to its greater polarity. An average error of $\pm 2.6\%$ was found for duplicate determinations with separate working up but chromatography on the same plate; and $\pm 3.5\%$ when using different layers. In newer work [143a], the authors have eluted with 7 ml of Titrisol buffer, pH 1 and added diazotised p-nitroaniline to 5 ml of the clear supernatant from centrifuging. The absorbance was then measured in a 2 cm cuvette at 510 nm. The HVA was eluted with distilled water and the colour reaction carried out with Pauly's reagent because Rose's reagent was not commercially available at the time.

Using TLC, SCHMID and co-workers [143, 143a] have discovered increased elimination of HVA, VMA, dihydroxyphenylacetic acid and VA in cases of malignant glomus tomour and neuroblastoma. DITTMANN [31a] has suggested applying to cellulose layers an amount of the 24 hour-urine (2.5 µl per 1 of the urine eliminated daily) which, with normal urine, contains an amount of VMA just below the limit of detection with diazotised p-nitroaniline. A positive reaction would then be obtained on the thin layer in cases of pathologically increased elimination of the VMA. A negative result can be further substantiated by comparison with a standard. A simple, direct photometric evaluation of urinary VMA on chromatograms has been given by KÖHLER and BAUFELD [96].

5-Hydroxyindoleacetic acid. This acid (5-HIA) is the degradation product of serotonine (5-hydroxytryptamine). Urinary elimination of over 15 mg per day renders probable the diagnosis of a metastasising carcinoid, provided the influence of diet (bananas, walnuts etc.) is excluded. Standard amounts of 5-HIA and increasing amounts of the urine extract are applied to the adsorbent in the semi-quantitative determination of SCHMID and KUSCHKE [143]. After the TLC-separation, the plate is sprayed with 0.1% 2,6-dichloroquinonechloroimide solution in methanol and the spot sizes compared in the approximate evaluation. Especially distinct, brownish-red spots are obtained by spraying with diazotised p-nitroaniline (Fig. 173). This reaction also can be adapted to quantitative determination [143a].

Methylhistamine. FRAM and GREEN [47] have detected methylhistamine as histamine metabolite in normal urine. The dinitrophenylmethylhistamine was separated by TLC on a silica gel G layer and photometrically determined at 358 nm after elution.

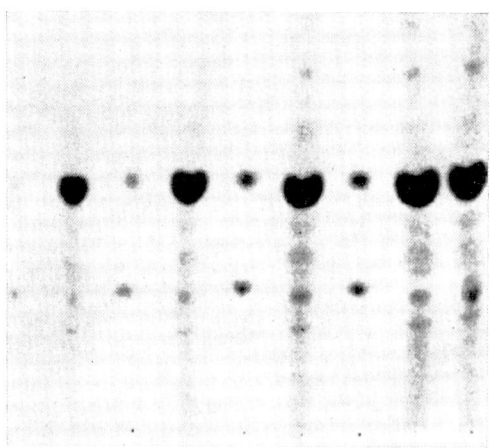

Fig. 173. TLC of an extract of the urine of a patient with carcinoid syndrome. 2.5—10 µl extract applied, with 0.25—1.0 µg VMA or 5-HIA as standards. The 5-HIA concentration is greatly increased in the urine, that of VMA is normal [143a]. Layer: silica gel G; solvent: isopropanol-ethyl acetate-ammonium hydroxide-water (45 + 30 + 17 + 8); visualisation with diazotised p-nitroaniline

h) Serum Proteins

JOHANSSON and RYMO [82, 83] describe protein separation through TLC-gel filtration. Sephadex G 50 and G 75 (Firm 102) proved to be especially suitable for proteins of relatively low molecular weight and Sephadex G 100 and G 200 for separating higher molecular weight proteins and therefore serum proteins. Using this method, with Sephadex G 200 "superfine", the authors investigated a normal serum, a myeloma serum with γ_{SS}-globulin, two myeloma sera with γ_{1A}-globulin and a serum with a pathological macroglobulin. FASELLA and co-workers [43] have likewise reported the TLC-separation of proteins on Sephadex. ANDREWS [2] has based statements concerning the molecular weights of individual proteins on their rates of migration in TLC-gel filtration on Sephadex G 100 "superfine". Similar investigations have been undertaken by MORRIS [108].

WIELAND and DETERMANN [191] have separated lactate-dehydrogenase isoenzyme and AMP, ADP and ATP from one another on DEAE-Sephadex layers, using gradient elution chromatography (p. 90).

WAGNER-ROMERO [176] has been able to accomplish separation of a serum into 15 fractions, using TL-starch gel electrophoresis. REISSELL and co-workers [130] have separated serum lipoproteins electrophoretically on a thin layer of starch and used TLC to investigate the lipid content of the single fractions.

V. EULER and co-workers [41] have separated the protein fractions in serum by thin-layer electrophoresis and compared those of healthy rats and of rats with tumours. Albumin and the α-globulins were present in distinctly smaller amounts in the serum of rats with larger tumours.

3. Lipids and Related Compounds

In many ways, clinical lipid chemistry has only become possible at all since the advent of TLC. An early application was the analytical separation of serum lipids by WEICKER [181]. The pathological lipid in Refsum's disease was first observed as a result of the investigation of serum lipids through TLC [201]. Further, it seems that the analysis of small tissue samples is becoming important, especially in the lipid domain.

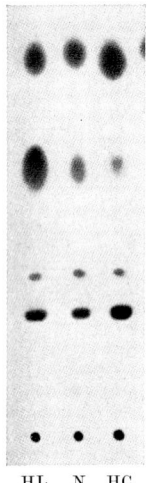

Fig. 174. TLC of the plasma lipids of a healthy individual(N), a patient with hyperlipaemia (HL) and with hypercholesterolaemia (HC) according to [140]. Layer: silica gel G; solvent: petrol ether (BP 30—60° C)-diethyl ether-acetic acid (82 + 18 + 1); CS

The application of TLC in clinical lipid chemistry is indicated in at least three types of problem. First, a thin-layer chromatogram using a suitable solvent, permits a rapid survey of which lipids have increased in amount as a result of hyperlipaemia and consequently the diagnostically especially important distinction between hypercholesterolaemia with and without hyperlipaemia (HUHNSTOCK and WEICKER [73], SACHS and WOLFMAN [135], SCHLIERF and WOOD [140]) (Fig. 174). Secondly, the amounts of lipid eluted from thin-layer chromatograms of many fractions suffice for GC of the fatty acids contained in these fractions. BOWYER and co-workers [13] were among the first to develop a practicable method with which fatty acids could be trans-esterified in the presence of silica gel. Loss of unsaturated fatty acids by oxidation during the TLC could not be demonstrated. Thirdly and most important, however, is the possibility of quantitative analysis of lipid fractions, separated by TLC, with or without elution.

Detailed information about the extraction, investigation and determination of lipids in blood has been given by ZÖLLNER and EBERHAGEN

[202]. Several TLC-systems are suitable for a clear-cut separation of the most important lipid fractions of serum (see p. 389). We have found the solvent petrol ether (BP 50—70° C)-butanone-acetic acid (95 + 4 + 1) on silica gel G layers to be especially satisfactory for enabling the lipid composition to be quickly estimated [201]. Separation of the polar lipids can be still further improved by increasing the ketone content of the solvent.

a) Neutral Fat in Plasma

SCHLIERF and WOOD [140] have carried out semi-quantitative determination of the triglycerides and free fatty acids in serum after thin-layer chromatographic separation on silica gel G; it was based on the linear relation between the square root of the area of the substance spot and the logarithm of the amount of substance. The reproducibility of replicate determinations on the same serum was good, with a standard deviation of 8.1—8.5%. The method is complicated in relation to the limited information yielded.

An accurate determination after isolation through TLC has been accomplished by KRELL and HASHIM [98] with the help of IR-spectrophotometry. The recovery is about 96% and serum concentrations as low as 20 mg-% can still be determined reasonably well. A mean triglyceride level of 93 mg-% with a standard deviation of ± 43 mg-% was established for a group of 51 healthy men aged between 40 and 59. A method quoted by PRIVETT and BLANK [128] yields still more analytical information, especially about the composition of the glycerides; distinction is made between neutral fat with and without unsaturated fatty acids. The enzymatic determination of glyceride glycerol [38] following TLC-separation into mono-, di- and triglycerides, due to ZÖLLNER and co-workers [203], requires much simpler apparatus. The results agree very well with those from chemical analyses. The ratio of triglyceride glycerol to the sum of mono- and diglyceride glycerol is fairly constant at 90:10 for healthy individuals; the ratio remains constant after a meal but the mono- diglyceride glycerol ratio is changed. Using TLC, WOOD and co-workers [194] have separated the lipids in the plasma chylomicrons after consumption of a fat meal. The composition of the fatty acids, which was studied by GC, was very similar in the triglycerides, diglycerides and the butter of the meal.

b) Cholesteryl Esters in Plasma

TLC of serum cholesteryl esters on silica gel G using triple development with carbon tetrachloride at chamber saturation yields up to 5 zones; the esters are fractionated merely according to the number of double bonds in the fatty acid moiety (cf. also Chapter J). Direct photometric evaluation of a chromatogram of the cholesteryl esters which has been sprayed with antimony(III) chloride (Rgt. No. 15) is possible according to ZÖLLNER and co-workers [200], provided that the largest single fraction does not contain more than 3 μg cholesterol. Prior chemical determination of total cholesteryl esters is necessary in order to be able to observe this condition and to calculate the absolute values of the individual fractions.

Reagents: Carbon tetrachloride; 25% antimony(III)chloride in reagent grade chloroform (freshly prepared each day).

If the starting material is a Sperry extract [202], 1 ml of extract is concentrated to 0.14 ml and the amount of this concentrated extract to be applied is calculated from the equation: ml lipid concentrate = 1.67/mg-% ester cholesterol in the serum. Immediately after the TLC-separation, the 200 × 38 mm plate is sprayed evenly with antimony(III)chloride reagent and then heated ca. 5 min at 110°C. The characteristic coloured reaction product changes from red to blue at a rate which varies from fraction to fraction; correction factors must be thus determined for the particular technique used, with the help of standard substances. Table 123 contains such factors for the Beckmann spectrophotometer DU G 4700 at 575 nm. Some electrophoresis scanners with adequate light intensity are also suitable for photometric evaluation of the chromatograms (cf. also p. 139).

With the help of this method some important observations have been made concerning the cholesteryl esters of serum (Table 123). The linoleic acid content of umbilical cord blood is distinctly lower than that of the adult [199]; the baby attains the adult value by the end of the second year of life at the latest, depending on the linoleic acid amount in its diet [204]. Cases of liver disease are accompanied by a clear reduction in the content of linoleate ester [198]; this has not so far been explained. In familial hypercholesterolaemia the increase in total cholesterol is made up of uniform increases in the amounts of each individual cholesterol ester [200]; the relative proportions of the various cholesteryl esters in 20 patients with peripheral arterial sclerosis (detected angiographically), have been likewise found not to be significantly altered [195]. The distribution spectrum of serum cholesteryl esters in a case of chylous ascites, showed only a slight deviation [195]. The pathological cholesteryl phytanate was discovered through simple thin-layer chromatographic separation of the serum cholesteryl esters [201] (Fig. 175). KLENK and KAHLKE [95] subsequently identified it. If there is clinical suspicion of a Refsum syndrome, a four-fold amount of cholesteryl esters must be used in the TLC so as not to overlook the phytanate ester which is present only in small amounts in the serum cholesteryl esters.

According to MORRIS [107] a better separation on the basis of the number of double bonds is possible on silica gel G layers containing added silver nitrate. KAUFMANN and co-workers [87] have described a partition chromatographic procedure. Both procedures may be utilised in two-dimensional chromatography. MOSER and co-workers [109], using

TLC-analyses, report the occurrence of cholesteryl sulphate in the serum and various organs of humans.

Table 123. *The cholesteryl esters in serum, expressed as relative %. The standard deviation of the mean (δx) is less than 1.1 for all values*
See Fig. 175 concerning the evaluation

Serum from	Fatty acid moieties						References
	n	16:0 18:0	16:1 18:1	18:2	20:3 20:4	20:5 22:5 22:6	
Adults	14	16.1	23.6	46.7	11.8	2.0	ZÖLLNER and co-workers [200]
New-born	27	23.5	30.6	23.6	20.6	1.2	ZÖLLNER and co-workers [204]
Typical case of liver disease [a]	1	21.9	28.2	37.4	10.3	2.3	ZÖLLNER and WOLFRAM [198]
Arteriosclerosis	20	16.1	25.2	44.6	12.8	1.6	WOLFRAM [195]
Correction factors for the quantitative evaluation		1.1	1.1	1.4	1.3	0.8	ZÖLLNER and co-workers [200]

[a] The average from a wider range of patients is not representative.

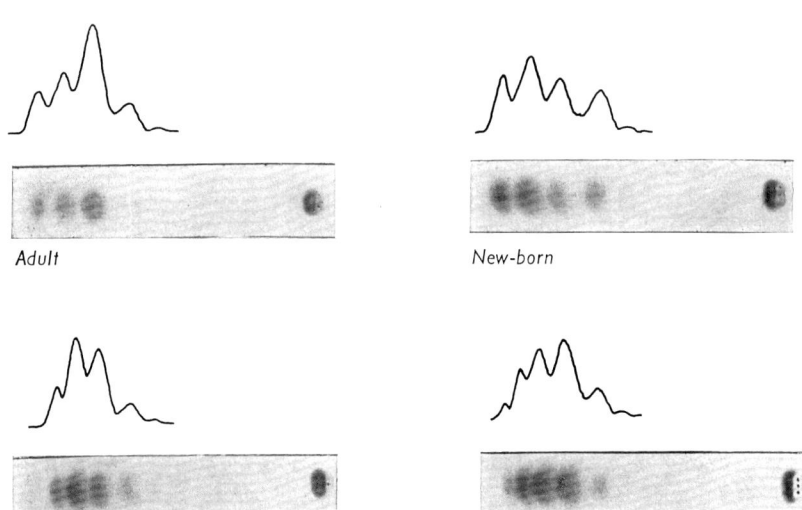

Adult

New-born

Liver cirrhosis

Refsum syndrom

Fig. 175. Thin-layer chromatograms of the cholesteryl esters of various sera and the corresponding absorption curves from photometric evaluation. Reading from left to right (decreasing Rf-value): esters of acids with no; one; two; three and four; and more double bonds; cholesterol on the right near the start

c) Phospholipids in Plasma

The solvent chloroform-methanol-water (65 + 25 + 4), described by WAGNER et al. [175] has found extensive application in the thin-layer chromatographic separation of serum phospholipids on silica gel G layers. HABERMANN and co-workers [67] have published in this connection a method for quantitative determination of the single phospholipid fractions. The methods so far known for separating phospholipids fractionate them only into groups of compounds; this contrasts with the TLC of cholesteryl esters, where separation into individual substances is possible. The results hitherto obtained are summarised in Table 124.

Table 124. *Phospholipids and phospholipid fractions in serum*

Serum from	n	mg-% Total Phospholipid	% Cephalin	% Lecithin	% Sphingomyelin	% Lysolecithin	References
Adults	30	223	3.1	71.9	16.4	8.9	ZÖLLNER and co-workers [204]
Adults	20	199	3.0	66.0	21.5	9.4	WAGENER and co-workers [172]
Pregnancy	5	328	5	73	21	1.3	VIKROT [169]
Children (3—7 years)	10	193	8.5[b]	68.8	17.5	5.2	CHRISTIAN and co-workers [23]
New-born	19	114	8.8	55.3	19.8	16.2	ZÖLLNER and co-workers [204]
Arteriosclerosis	20	255	3.0	66.4	22.6	8.0	WAGENER and co-workers [172]
Liver diseases[a]	14	269	2.2	80.5	12.3	5.3	ZÖLLNER and WOLFRAM (unpublished)
Hypercholesterolaemia	1	694	3.5	72.5	20.7	3.3	SCHOLDERER [147]
Hypercholesterolaemia	1	315	6.9[b]	67.4	22.5	3.2	CHRISTIAN and co-workers [23]
Carbohydrate-induced hyperlipaemia	1	390	7.4[b]	75.0	9.7	7.9	CHRISTIAN and co-workers [23]

[a] The behaviour of the phospholipids in various liver diseases appears to be so uniform that an average value can be worked out.
[b] Total of cephalin, phosphatidyl inositol and phosphatidic acid.

Sphingomyelin in Plasma. MICHALEC and KOLMAN [105] have been able to detect C_{16}- and C_{17}-sphingosine as well as C_{18}-sphingosine and small amounts of C_{18}-dihydrosphingosine in the sphingomyelin fractions of human serum, using two-dimensional TLC. WOOD and HOLTON [193], employing GC after preparative TLC, have found predominantly C_{16}-fatty acids in the more polar fractions and C_{22}- and C_{24}-fatty acids in the less polar fractions.

d) Methods for Separating Other Medically Important Lipids

GRÉEN and SAMUELSSON [61] quote solvent systems for TLC separation of all known prostaglandins, both as fatty acids and as methyl esters. BYGDEMAN and SAMUELSSON [20] have published work on the quantitative determination of prostaglandins in pooled human semen.

GOODMAN [60] has investigated the level of squalene in plasma and its metabolism in humans; he used ^{14}C-mevalonic acid and found the major part of the labelled squalene in the lipoproteins of lower density and a concentration of $30-35$ μg-% in plasma. Small amounts of lanosterol probably also occur in plasma.

SVENNERHOLM and SVENNERHOLM [161, 162] have extracted the neutral glycolipids from human serum, spleen and liver and were able to separate 4 different glycolipids, ascertaining their structure and distribution. FAILLARD and CABEZAS [42] give a method for isolation of N-acetyl- and N-glycolylneuraminic acid from calf and chicken serum.

e) Lipids in Secreted and Excreted Material

The lipids in faeces are well known, especially through column chromatographic separations (e. g., AYLWARD and WOOD [5]); application to TLC should present no difficulties. Appropriate investigations are important for the differential diagnosis of digestive insufficiencies.

WEICKER and collaborators [182] have studied the lipids in normal and in strongly chylous duodenal secretions obtained from a patient with exsudative enteropathy by means of a double balloon tube; principally neutral fats along with phospholipids and cholesterol could be detected with the help of TLC. Oleic acid, labelled with ^{131}I was introduced into the jejunum and 15 min later it was already possible with the help of TLC to detect active phospholipids in the duodenal juice. With two healthy subjects, this increase in activity in the duodenal secretion took place only after 2 hours and was much less than with the patients suffering from enteropathy. WILLIAMS and co-workers [192] have investigated the lipid composition of faeces and faecoliths from the appendix and found chiefly free fatty acids. The faecal sterols are mentioned below in section 4.

KAUFMANN and VISWANATHAN [88] have found over 2.5 mg-% cholesteryl ester and 1.0 mg-% free cholesterol in the urine of patients with lipoid nephrosis and a preponderance of triglycerides (348 mg per 100 ml urine) in a case of lipiduria. Using TLC we have similarly been able to find predominantly neutral fats in another chyluria case.

The phospholipids in pooled cerebrospinal fluid have been investigated by PHILLIPS and ROBINSON [127]. CZEGLÉDI-JANKÓ [25] has separated the lipoids of human milk, using preparative TLC. The methyl esters of fatty acids were subsequently studied by GC.

f) Lipids in Tissue

The analysis of tissue lipids has always been important for elucidation of lipid storage diseases. TLC has greatly simplified the necessary investigations and permits the range of application to be extended also to punch biopsy material and the smallest organs of experimental animals.

α) **Brain Lipids.** A great deal of work has been done on the qualitative and quantitative analysis of brain lipid fractions after separation by TLC [e. g., 4, 70, 71, 72, 76, 84, 90, 94, 110, 125, 185, 190]. JATZKEWITZ [78] has developed a micro-method for quantitatively determining brain sphingolipids. It depends essentially on TLC in connection with the "ultramicro analytical system" of M. C. SANZ. It could be demonstrated, using this procedure, that sulpholipids of the kerasine and cerebron types accumulate in cases of metachromatic leucodystrophy. Cerebrosides of the kerasine type are more rapidly degraded than those of the cerebron type in cases of demyelinating disorders, whether through metachromatic leucodystrophy or multiple sclerosis. SVENNERHOLM [163] has determined the gangliosides in the brain of foetuses, new born babies and adults. He found no differences between those of the foetus and the adult. Whereas only 3–6% of the ganglioside acyl-sphingosine-N-triose-N-acetylneuraminic acid is found in a normal brain, 90% was found in a case of infantile amaurotic idiocy. PENICK and co-workers [126] report the thin-layer chromatographic separation of 9 gangliosides from normal human brain. WAGNER [173] has separated the gangliosides from the brain of a patient with Tay-Sachs disease, using TLC on a silica gel G layer. The ganglioside with a 1:1 molar galactose-glucose ratio, typical of this disease, could be isolated. A non-typical case of infantile amaurotic idiocy is reported by JATZKEWITZ et al. [77, 79]. The normal ganglioside A (KLENK) had been accumulated whilst only traces of the ganglioside typical for the Tay-Sachs disease could be detected. BOOTH and co-workers [12] have investigated the composition of the gangliosides in the white and grey matter of fresh, unfixed brain from patients with Tay-Sachs-, Niemann-Pick- and Pfaundler-Hurler's diseases; they used silica gel G layers and were able to determine accurately the molar distribution of the sugars.

SEIFERT and UHLENBRUCK [150], with the help of a TLC-technique, have been able to isolate the principal ganglioside of the meningiomata and to elucidate its constitution. It contains one molecule each of fatty acid, sphingosine, glucose, galactose and N-acetylneuraminic acid and occurs only in small amounts in normal brain. By using polyzonal TLC, SEIFERT [151] has been able to detect a second ganglioside that is not always found in meningiomate but is always present in gliomata, ependymomata and medulloblastomata. Its molecule contains an additional neuraminic acid unit.

β) **Lipids in Other Tissue.** Liver lipids have been investigated by, e. g., SCHÖN and co-workers [146], DOBIÁŠOVÁ [32], SKIPSKI and co-workers [153] and ZÖLLNER [197]. Studies of fatty and regenerating liver are of especial interest. Depot fat [32] and the adrenal glands [3] are other tissues which have been more often investigated. In the first edition of this book it was possible already to report analysis of lipids in vascular walls. GLUCK and co-workers [58] have carried out direct photometric determination of the phospholipids in lung tissue after thin-layer chromatographic separation and spot carbonisation.

MASORO et al. [104] have studied the lipids of muscular tissue with the help of TLC. WAGNER and WEICKER [174] have been able to obtain from normal human spleen tissue a ganglioside-like substance (0.06% of the dry material). The spleen ganglioside could also be separated by TLC into 6 fractions, differing in the sugar moiety of the molecule. The fraction which was present in largest amounts migrated fast and contained no hexosamine; it is built up of a molecule of each of sphingosine, fatty acid, galactose, glucose and neuraminic acid. SUOMI and AGRANOFF [160] have studied the lipids in the spleen of 8 patients with Gaucher's disease. They could find no increase in the phospholipid fraction accompanying the known changes in the cerebrosides but a moderate increase in neutral lipids was established. EBERHAGEN [36] has investigated the lipids of human placenta.

Experimental conditions for TLC-studies of lipids of skin and hair have been given by KAUFMANN and VISWANATHAN [89]. HAAHTI and co-workers [65, 66] also have used TLC in investigations of the cholesteryl esters and waxes of sebum. STÜTTGEN and VOGELBERG [159] have compared the qualitative and quantitative composition of the xanthoma-lipids of 7 patients with tuberous xanthomata, with the values for normal skin and fat of the skin surface. The deposited cholesterol was largely in ester form. No increased deposition of neutral fats occurs in the xanthomata in cases of hyperlipaemia. An extra-cellular accumulation of free cholesterol could be established in a case of cytomycotic, histiocytic granulomatosis.

Using TLC, HOLCZABEK [69] has investigated lipid extracts from depot fat, normal lung tissue and from lungs with fat embolism. DOBIÁŠOVÁ [32] and also EBERHAGEN [37] in particular, have worked on special questions of analysis of very small amounts of tissue. ARNOLD [3a] has reported the TLC-separation of lipids in cholesteatomata of the middle ear.

4. Steroids
(see also Chapter L in this connection)

Compounds with steroid structures, such as adrenocortical hormones, sex hormones, cholesterol and bile acids, occupy a central position in the human organism. The level of the active hormones in serum, their elimination products in urine and the reserve capacity of the endocrine organs are of principal interest in clinical diagnosis of hormone disturbances. Increase in the cortisol level in blood or increased elimination of it and its metabolites in urine, is characteristic of the Cushing syndrome. Increased elimination of the total 17-ketosteroids does not always occur. Chromatographic separation of the 17-ketosteroids can provide information, since the 11-hydroxy-17-ketosteroids are metabolites of cortisol and cortisone and are thus present in increased amounts. Hyperaldosteronism leads to augmented elimination of aldosterone in urine while the elimination of 17-ketosteroids and 17-hydroxysteroids remains normal. The adrenogenital syndrome is characterised by increased production of androgenic hormones. Elimination of the 17-ketosteroids, pregnanetriol, androsterone and dehydroepiandrosterone yield important diagnostic information. Reduced elimination of 17-ketosteroids and 17-hydroxysteroids takes place in cases of hypoadrenocorticism.

Steroid concentrations in cerebrospinal fluid are lower than in serum. OERTEL and BRÜHL [113] have investigated pooled cerebrospinal fluid of patients and, after having extracted the free steroids, they used ion exchange chromatography and solvent partition to separate the steroid conjugates into steroid sulphates, sulpholipids and glucuronosides. Following fission of the conjugates, the steroid groups (oestrogens, 17-ketosteroids, 17-hydroxycorticosteroids) or single steroids were isolated by means of TLC and PC.

a) C_{21}-Steroids

In WALDI's simple, semi-quantitative determination of progesterone in serum [179], it is separated from cholesterol by TLC after extraction from 3.5 ml of serum. Comparison with standards is the basis of the evaluation. LUISI et al. [101] have extracted from the serum, purified by liquid/liquid partition and TLC and determined the progesterone quantitatively by gas chromatography. The progesterone content of the serum of 10 normally menstruating women fluctuated during the menstrual cycle between 0.25 and 3.5 µg/100 ml.

TLC may be used for easy and rapid determination of modifications in the hormone regulation, brought about by pregnancy or pathological disturbances of the cycle which are accompanied by a change in the urinary elimination of pregnanediol. WALDI's pregnancy test [178] depends on the determination of progesterone metabolites in urine. Maximum elimination of pregnanediol in urine during the secretion phase is as a rule 3—5 mg/l, exceptionally as high as 7 mg/l. Urinary pregnanediol exceeds the normal maximum value by about 2 mg/l already some 10 days after conception. Pregnanediol determination may be applied also for control of the cycle and of the hormone excretion when oral contraceptives are administered. If an adrenogenital syndrome is suspected, an aliquot of the 24 hour urine can be tested for pregnanetriol. BANG [6] has applied WALDI's extraction procedure and solvents to determine pregnanediol quantitatively in urine. SCHNEIDER and SZEREDAY [144] have spectrophotometrically determined pregnanediol from urine. Other authors also quote quantitative determinations of pregnanediol after TLC-separation [39, 99, 156]. ADAMEC and co-workers [1] have employed TLC for purification of a crude extract of 17-hydroxycorticosteroids from urine; quantitative colorimetric determination was with the help of the Porter-Silber reaction. BERNAUER [9] has evaluated corticosteroids quantitatively on thin-layer chromatograms by photometric measurements on their blue-violet formazan complexes. BRUINVELS [15] gives a simple procedure for TLC-separation and quantitative determination of aldosterone, hydrocortisone and corticosterone. A fluorometric determination of free cortisol after isolation from urine by means of TLC is described by GERDES and STAIB [55].

Increased amounts of 6-β-hydroxycortisol are found in human urine during pregnancy, in cases of hyperadrenocorticism and normally with new born babies, possibly also in cases of disturbed microsomal liver function. BERTHOLD and STAUDINGER [10] have isolated 6-β-hydroxycortisol using two-dimensional TLC-technique on silica gel HF_{254} with cyclohexane-isopropanol (50 + 50) and chloroform-acetic acid-ethanol (65 + 30 + 5). After elution from the silica gel, quantitative determination was carried out with the help of triphenyltetrazolium chloride reagent. The average daily elimination of 10 healthy middle-aged males was 525 µg $\pm$ 123(s); of 10 females, 534 µg $\pm$ 156(s).

NISHIKAZE and STAUDINGER [111] have employed a two-dimensional TLC-procedure to isolate aldosterone after extraction from urine. The steroid was eluted and determined quantitatively with blue tetrazolium. The normal aldosterone elimination in the 24 hour-urine was established as 10 µg. The aldosterone, isolated by the same TLC-method, has been determined fluorometrically in newer work of NOWOTNY and STAUDINGER [112]. BENRAAD and KLOPPENBORG [8] have separated aldosterone from other urinary steroids by TLC and then determined it quantitatively. The recovery of aldosterone added to the urine was 74.2%. GERDES and STAIB [56] have had similar results in their extraction procedure for fluorometric determination of aldosterone in human urine.

SCHEIFFARTH and co-workers [138] have used TLC to investigate the elimination of prednisolone and its metabolites in duodenal juice. Prednisolone acetate is eliminated by healthy livers but not by patients with liver damage. The formation of glucocorticoid acetates can again be demonstrated after normalisation of the pathological liver function tests.

b) C_{19}-Steroids

TLC-separations of individual 17-ketosteroids have been given by REISERT and SCHUMACHER [129], SHEN and co-workers [152], DYER et al. [34], STÁRKA and co-workers [157] and WEINAND et al. [184]. KIRCHNER and LIPSETT [93] have combined TLC and GC for determination of several 17-ketosteroids, pregnanediol and pregnanetriol in the 24 hour urine of men and women. PENG and co-workers [125] give a simple TLC-method for determining dehydroepiandrosterone in urine.

SZEREDAY and SACHS [164] have isolated testosterone from urine as its coloured dinitrophenylhydrazone and then quantitatively determined it. The recovery of ca. 70% is relatively good. VERMEULEN and VERPLANCKE [168] and also BROOKS [14] consider that prior column chromatographic purification over alumina is necessary in the determination of urinary testosterone; the testosterone is then isolated by TLC and determined quantitatively with the help of the Zimmermann reaction. About 40 and less than 12 μg are eliminated in the 24 hour urine by men and women, respectively; 34 μg was found in the urine of a woman suffering from hirsutism. FUTTERWEIT and co-workers [54] have established that increased testosterone elimination occurs in cases of idiopathic hirsutism, Stein-Leventhal syndrome with hirsutism, congenital adrenocortical hyperplasia and Cushing syndrome with virilisation. IBAYASHI and co-workers [75] have had similar results. KORENMAN and co-workers [97] have used isotope dilution and determined fluorometrically both the testosterone in urine and its rate of metabolism, after thin-layer chromatographic separation. VOIGT and co-workers [170] likewise have made use of TLC in determining the testosterone eliminated in urine. GUERRA-GARCIA et al. [64] have found a testosterone concentration of 0.3—1.27 μg/100 ml plasma for young men; the recoveries from this coupled TLC-GC method were, however, only 39%. A similar method has been given by BURGER and co-workers [18].

c) C_{18}-Steroids

The quotient oestriol / (oestradiol + oestrone) in urine normally has the value of unity for men and women. This value changes in some diseases and during pregnancy. DETTER and co-workers [27] have investigated these changes, using a semi-quantitative TLC-method. STRUCK [158] gives a quantitative determination of the oestrogens after separation on silica gel. LADANY and FINKELSTEIN [100] have carried out separations by TLC and PC in studies of the urinary oestrogens of healthy women between the 5th and 9th and the 20th and 24th days of the menstrual cycle and also of a female patient suffering from secondary amenorrhoea with normal elimination of follicle-stimulating hormone (FSH) in urine. The oestrogen elimination was almost nil in the urine of a patient with secondary amenorrhoea and increased FSH-elimination and of a patient after bilateral ovariectomy. The reliability of this fluorometric determination was checked with labelled oestradiol and oestrone and recovery was found to be about 50%. FISHMAN and co-workers [44] have used a two-dimensional technique on silica gel H to separate oestradiol and oestriol from urine as their glucuronide conjugates. After elution from the silica gel they were hydrolysed with the help of β-glucuronidase. A densitometric determination of an oestrogen mixture after thin-layer chromatographic separation is given by JACOBSON [74]. WOTIZ [196] has analysed the oestrogens in urine by GC after a TLC-separation.

d) Bile Acids

CURTIUS [24] proposes a semi-quantitative determination of bile acids in serum, duodenal juice and faeces, based on the method of the limiting concentration just visible on the TLC-layer; he demonstrated the effectiveness of this procedure by analyses on the serum of 20 healthy men and women. FROSCH and WAGENER in several publications [49–52] have reported the quantitative determination of free and combined bile acids in serum and duodenal juice, following separation by TLC. In more recent work [53], FROSCH has investigated the behaviour of bile acid conjugates of serum in liver diseases. He was able to show that the concentration of trihydroxy- and dihydroxycholanic acid conjugates remain high for some time in the serum of patients with hepatitis, even after the acute inflammation has subsided, as recognised by the normalisation of transaminase values. A more costly but more sensitive method is the spectrofluorometric determination of the bile acids, separated by TLC, due to FORTH and co-workers [46]. This enables cholic acid to be reliably determined in concentrations as low as 0.02 µg/ml cuvette content. The authors determined bile acids in bile [45], in preparations isolated from rat intestine [57] and in the bile and serum of humans. Neither free cholic nor free deoxycholic acid could be detected in normal sera; the ratio of the conjugate of cholic acid with glycine to that with taurine was about 0.4. The concentration of both conjugates was elevated in patients with obstructive jaundice; similar features were found up to the 20th day of illness with epidemic hepatitis. Free cholic acid could be detected in 3 of 17 sera from patients with obstructive jaundice or hepatitis but free chenodeoxycholic acid was not detectable in the serum of patients.

e) Sterols in Faeces

SAMUEL and URIVETZKY [136] have separated cholesterol-7α^3H and coprosterol-4-^{14}C from human faeces by double development with toluene-ethyl acetate (90 + 10) on silica gel layers. MIETTINEN and co-workers [106] report the determination of neutral steroids in faeces; after prior separation by TLC they were analysed gas chromatographically as their trimethylsilyl ethers. The method is especially suitable for balance studies. The same authors also report analysis of bile acids in faeces [63].

5. Porphyrins and Metabolites

Interferences in the biosynthesis of protoporphyrin, a percursor of the red blood colouring matter haem, lead to qualitative or quantitative changes in the elimination of porphyrin metabolites in the urine. According to JENSEN [80], coproporphyrins I and III can be separated on silica gel using 2,6-lutidine-water (70 + 21) and a chamber atmosphere saturated with ammonia. Detection of 0.01 µg coproporphyrin in UV light is possible. The chromatograms can be quantitatively evaluated by comparison with increasing amounts of a standard or by elution from the silica gel with 2N HCl and fluorometry.

The experimental conditions for TLC of protoporphyrins have been investigated in detail by STAHL and co-workers in connection with the "magnolipin" test. The products to be tested for purity were applied in solution in very pure pyridine (2—20 µl of 0.1—0.5% solutions). The best separation conditions were on alkalised silica gel G layers (slurried with 0.5 N KOH instead of water), developing five times with chloroform-methanol-tetrahydrofuran (30 + 30 + 30) on the 10 cm run at chamber saturation. A protoporphyrin, presumed to be pure, was resolved into 6 zones. They are most easily recognised through their red fluorescence in long wave UV light and a high pressure lamp (p. 78, Firm 44) is suitable for this.

TENHUNEN [167] quotes a simple TLC-procedure for separation of bile pigments from human bile. The diazotised pigments were separated on silica gel G layers with the solvent butanone-propionic acid-water (60 + 15 + 15); bilirubin sulphate, bilirubin glucuronide, bilirubin monoglucuronide, free bilirubin and biliverdin were found in this order of increasing Rf-value and were eluted and determined quantitatively by a photometric method. The glucuronic acid content can be determined also after acid hydrolysis. In serum, proteins must be precipitated beforehand.

II. Investigations of Exogenous Substances

The detection of exogenous substances and their metabolites in human organs and excreta is not only theoretically interesting but is important in many ways in practical medicine. Function tests, diagnoses of poisoning and control of drug therapy are clinically important. Pharmacologists are interested in metabolic routes via which drugs are broken down in healthy and in diseased organisms. The possibilities offered by TLC in this sphere are discussed in Chapter Q.

1. Function Tests

The rate of breakdown or elimination of foreign material by a particular organ is a measure of the efficiency of that organ. Several function tests in clinical diagnosis depend on this principle.

The bromosulphalein (BSP) test, i. e., the elimination of disodium phenoltetrabromophthalein-sulphonate from plasma by the liver, is in general use as a liver function test. According to WHELAN and PLAA [189], the metabolites of BSP, produced during its passage through the liver parenchyma and detectable in the bile, plasma, lymph and urine, can be separated and quantitatively determined by TLC; cellulose thin layers and the solvent butanol-acetic acid-water (40 + 10 + 50) (upper phase) are used. WERNZE [186] has determined 6 ninhydrin-positive BSP-metabolites in rat bile using the same solvent; the largest of these fractions is referred to as BSP-glutathione (GSH) complex. The unchanged BSP has the highest Rf-value under the experimental conditions chosen (Fig. 176); the quantitative evaluation was by elution with 3 ml 0.1 N NaOH and determination of the absorbance at 570 nm versus 0.1 N NaOH. According to the author's data, liver damage due to carbon tetrachloride, thioacetamide, thyroxine or protein deficiency is accompanied by qualitative changes in the BSP-metabolites of the bile. Direct qualitative and quantitative resolution of the BSP-metabolites promises more precise information about the liver cell function than does the mere overall determination of the rate of elimination of BSP from serum. In certain cases of liver damage, a back-flow of BSP-conjugates into the blood is brought about with increased elimination of these compounds in the urine [103].

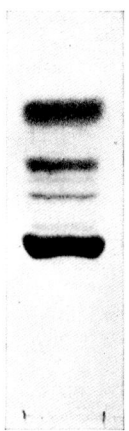

Fig. 176. Thin-layer chromatogram of rat bile (1 hour collection) after intravenous injection of 20 mg/kg bromosulphalein. Layer: cellulose; solvent: butanol-acetic acid-water (40 + 10 + 50) [186]

2. Poisoning

The doctor is repeatedly confronted in his clinical practice with cases of acute poisoning where an unambiguous diagnosis, based on case history or clinical symptoms, is not possible. In such cases, rapid identification of the poison or its metabolites in the gastric or intestinal juices or in urine is important for the therapy and for medico-legal reasons. TLC is markedly superior here to PC as a result of the much shorter separation times and the greater sensitivity. The TLC-methods of the forensic scientist have been reported in detail by MACHATA [102] for investigations of tissue, blood, stomach and intestine contents; they may be applied virtually unchanged for clinical purposes. Reference may be made to the special chapters in this book for individual methods for extraction and separation of alkaloids and their synthetic substitutes, sedatives, tranquillisers, antidepressants, analgesics, insecticides etc.

3. Control of Therapy

The doctor can usually judge the success of a therapeutic regimen by clinical criteria. If the desired success is not forthcoming, the cause is sought in the remedy or in the patient. It is not uncommon for patients to take their medicine irregularly, either deliberately or through negligence or superstition; they force the doctor to adopt the job of a detective. A check on whether sufficient doses have been prescribed or taken is possible through determination of the blood drug level. With many substances this level is too low for a chemical test. In such cases, the doctor can control whether the patient has been taking the medicament at all or in satisfactory amounts, by detection of the eliminated drug either unchanged or as a degradation product. Through correlation with the clinical picture, true and false therapeutic failures can be distinguished.

Clinicians have, surprisingly, not availed themselves so far of this objective method, even though the chemical analysis of urine, e. g., for sulphonamides or salicylates, is quite usual. We are therefore limited here to pointing out applications which appear interesting and possible, bearing in mind the present state of TLC-technique. The main problem would seem to be not the TLC-detection but the separation from contaminants, particularly if present in excess.

During prolonged treatment with digitoxin, 30—45% of the dose taken is eliminated unchanged through the kidneys. Isolation by TLC and quantitative determination of the digitoxin (p. 341) in urine provides information about the degree of digitalisation; according to FRIEDMAN [48], elimination of 32—44 µg in 24 h corresponds to a complete digitalisation. Similar control determinations are conceivable for treatments with other drugs, e. g., *Rauwolfia* alkaloids, thyreostatic drugs, anti-rheumatics and antibiotics.

In the therapeutic treatment of gout and urate nephrolithiasis with allopurinol, xanthine and hypoxanthine are eliminated instead of uric acid as end-products of purine metabolism. A good separation of these compounds from each other and from uric acid is possible using TLC [21]; their increase in urine is proof that the therapy is being observed.

The augmented elimination of β-aminoisobutyric acid (β-AIB) in urine in cases of highly increased degradation of nuclear material has been mentioned already (see p. 584). It seems logical to take the rate of elimination of β-AIB as a measure of the effectiveness of antineoplasmatics. Pilot experiments [16] have shown however that endoxan, myleran and nitrogen-mustard, used with various tumours, had differing influences on the elimination of β-AIB.

In cases of phenylketonuria, an enzyme defect leads to large increases of phenylalanine in blood and tissue, with consequent damage to the brain. The only possible therapy is a diet of low phenylalanine content. "Phenistix", used for diagnosis of phenylketonuria, are too inaccurate for adequate control; moreover it is a matter of normalising not only the phenylalanine but the whole amino acid pattern in the serum and elimination in urine. Only column chromatography has so far been capable of fulfilling these requirements. The introduction of TLC could effect a real simplification.

Continued post-operative control of the relevant hormone metabolites being eliminated in the urine, is necessary after removal of hormone-active tumours in cases of phaeochromocytoma, Cushing's disease and metastasising carcinoid. Only in this way can the complete success of the operation be shown and relapses recognised sufficiently early.

Bibliography for Chapter S. TLC in Clinical Diagnosis

1. ADAMEC, O., J. MATIS, and M. GALVÁNEK: Steroids **1**, 495 (1963).
2. ANDREWS, P.: Biochem. J. **91**, 222 (1964).
3. ANGELICO, R., G. CAVINA, A. D'ANTONA, and G. GIOCOLI: J. Chromatog. **18**, 57 (1965).
3a. ARNOLD, V.: Dissertation, Düsseldorf 1966.
4. AUSTIN, J. H.: J. Neurochem. **10**, 921 (1963).
5. AYLWARD, F., and P. D. S. WOOD: Brit. J. Nutr. **16**, 345 (1962).
6. BANG, H. O.: J. Chromatog. **14**, 520 (1964).
7. BENASSI, C. A., F. M. VERONESE, and E. GINI: J. Chromatog. **14**, 517 (1964).
8. BENRAAD, TH. J., and P. W. C. KLOPPENBORG: Steroids **3**, 671 (1964).
9. BERNAUER, W.: Klin. Wschr. **41**, 883 (1963).
10. BERTHOLD, K., u. HJ. STAUDINGER: Z. klin. Chem. **4**, 130 (1966).
11. BICKEL, H.: Universitäts-Kinderklinik Marburg, private communication (1966).
12. BOOTH, D. A., H. GOODWIN, and J. N. CUMINGS: J. Lipid Res. **7**, 337 (1966).
13. BOWYER, D. E., W. M. F. LEAT, A. N. HOWARD, and G. A. GRESHAM: Biochim. Biophys. Acta **70**, 423 (1963).
14. BROOKS, R. V.: Steroids **4**, 117 (1964).
15. BRUINVELS, J.: Experientia (Basel) **19**, 551 (1963).
16. BRUNSCHEDE, H., R. HOFFBAUER u. H. W. GOEDDE: Klin. Wschr. **43**, 93 (1965).
17. BUJARD, EL., and J. MAURON: J. Chromatog. **21**, 19 (1966).
18. BURGER, H. G., J. R. KENT, and A. E. KELLIE: J. Clin. Endocrinol. Metabolism **24**, 432 (1964).

19. Bürgi, W., J. P. Colombo u. R. Richterich: Klin. Wschr. **43**, 1202 (1965).
20. Bygdeman, M., and B. Samuelsson: Clin. Chim. Acta **10**, 566 (1964).
21. Cerri, O., e G. Maffi: Boll. chim. farm. **100**, 940 (1961).
22. Crawhall, J. C., E. Saunders u. C. J. Thompson: quoted from [*123*].
23. Christian, J. C., S. Jakovcic, and D. Yi-Young Hsia: J. Lab. Clin. Med. **64**, 756 (1964).
24. Curtius, H. Ch.: Z. klin. Chem. **4**, 27 (1966).
25. Czeglédi-Jankó, G.: Z. klin. Chem. **3**, 14 (1965).
26. Dahler, R. P.: Ann. paediat. **200**, 5 (1963).
27. Detter, F., J. Dietrich u. V. Klingmüller: Klin. Wschr. **44**, 100 (1966).
28. Diamantstein, T., u. H. Ehrhard: Z. physiol. Chem. **326**, 131 (1961).
29. Dimillier, I., u. R. G. Trout: quoted from [*123*].
30. Dittmann, J.: Z. klin. Chem. **4**, 8 (1966).
31. — Z. klin. Chem. **4**, 10 (1966).
31a. — Z. klin. Chem. **4**, 265 (1966).
32. Dobiášová, M.: J. Lipid Res. **4**, 481 (1963).
33. Drawert, F., O. Bachmann, and K. H. Reuther: J. Chromatog. **9**, 376 (1962).
34. Dyer, W. G., J. P. Gould, N. A. Maistrellis, T. C. Peng, and P. Ofner: Steroids **1**, 271 (1963).
35. Eberhagen, D., u. N. Zöllner: Z. klin. Chem. **5**, 149 (1963).
36. — Z. physiol. Chem. **333**, 179 (1963).
37. — Z. analyt. Chem. **212**, 230 (1965).
38. Eggstein, M., u. F. H. Kreutz: Klin. Wschr. **44**, 262 (1966).
39. Ehrlich, E. N.: J. Lab. Clin. Med. **65**, 869 (1965).
40. Euler, H. von, H. Hasselquist och I. Limnell: Arkiv Kemi **21**, 259 (1963).
41. — — — Z. Krebsforsch. **65**, 404 (1963).
42. Faillard, H., u. J. Cabezas: Z. physiol. Chem. **333**, 266 (1963).
43. Fasella, P., A. Giartosio, and C. Turano: In: Thin layer Chromatography. Ed. by G. B. Marini-Bettólo. Amsterdam-London-New York: Elsevier 1964, S. 205.
44. Fishman, W. H., F. Harris, and S. Green: Steroids **5**, 375 (1965).
45. Forth, W., W. Rummel u. H. Glasner: Naunyn-Schmiedebergs Arch. exp. Pathol. Pharmakol. **247**, 382 (1964).
46. — P. Doenecke u. H. Glasner: Klin. Wschr. **43**, 1102 (1965).
47. Fram, D. H., and J. P. Green: J. biol. Chem. **240**, 2036 (1965).
48. Friedman, M., S. St. George, and R. Bine jr.: Medicine **33**, 15 (1954).
49. Frosch, B., u. H. Wagener: Z. klin. Chem. **1**, 7 (1964).
50. — — Klin. Wschr. **42**, 192 (1964).
51. — — Klin. Wschr. **42**, 901 (1964).
52. — Klin. Wschr. **43**, 262 (1965).
53. — Verhandl. deut. Ges. inn. Med. **72**, 697 (1966).
54. Futterweit, W., N. L. Mc Niven, R. Guerra-Garcia, N. Gibree, M. Drosdowsky, G. L. Siegel, L. J. Soffer, J. M. Rosenthal, and R. I. Dorfman: Steroids **4**, 137 (1964).
55. Gerdes, H., u. W. Staib: Klin. Wschr. **43**, 744 (1965).
56. — — Klin. Wschr. **43**, 789 (1965).
57. Glasner, H., u. W. Forth: Naunyn-Schmiedebergs Arch. exp. Pathol. Pharmakol. **250**, 325 (1965).
58. Gluck, L., M. V. Kulovich, and S. J. Brody: J. Lipid Res. **7**, 570 (1966).
59. Goedde, H. W., and H. Brunschede: Clin. Chim. Acta **11**, 485 (1965).
60. Goodman, D. S.: J. Clin. Invest. **43**, 1480 (1964).
61. Gréen, K., and B. Samuelsson: J. Lipid Res. **5**, 117 (1964).
62. Gries, G., K. H. Pfeffer u. E. J. Zappi: Klin. Wschr. **43**, 515 (1965).

63. GRUNDY, S. M., E. H. AHRENS JR., and T. A. MIETTINEN: J. Lipid Res. **6**, 397 (1965).
64. GUERRA-GARCIA, R., S. C. CHATTORAJ, L. J. GABRILOVE, and H. H. WOTIZ: Steroids **2**, 605 (1963).
65. HAAHTI, E., u. T. NIKKARI: Acta Chem. Scand. **17**, 536 (1963).
66. — — u. K. JUVA: Acta Chem. Scand. **17**, 538 (1963).
67. HABERMANN, E., G. BANDTLOW u. B. KRUSCHE: Klin. Wschr. **39**, 816 (1961).
68. HERBERHOLD, C., u. O. A. NEUMÜLLER: Klin. Wschr. **43**, 717 (1965).
69. HOLCZABEK, W.: Klin. Med. **19**, 483 (1964).
70. HONEGGER, C. G.: Helv. Chim. Acta **45**, 2020 (1962).
71. — u. T. A. FREYVOGEL: Helv. Chim. Acta **46**, 2265 (1963).
72. HORROCKS, L. A.: J. Am. Oil Chemists Soc. **40**, 235 (1963).
73. HUHNSTOCK, K., u. H. WEICKER: Klin. Wschr. **38**, 1249 (1960).
74. JACOBSON, G. M.: Anal Chem. **36**, 275 (1964).
75. IBAYASHI, H., M. NAKAMURA, S. MURAKAWA, T. UCHIKAWA, T. TAMIOKA, and K. NAKAO: Steroids **3**, 559 (1964).
76. JATZKEWITZ, H., u. E. MEHL: Z. physiol. Chem. **320**, 251 (1960).
77. — u. K. SANDHOFF: Biochim. Biophys. Acta **70**, 354 (1963).
78. — Z. physiol. Chem. **336**, 25 (1964).
79. — H. PILZ, and K. SANDHOFF: J. Neurochem. **12**, 135 (1965).
80. JENSEN, J.: J. Chromatog. **10**, 236 (1963).
81. JEPSON, J. B.: In: The Metabolic Basis of Inherited Disease, p. 1283. New York-Toronto-Sydney-London: McGraw-Hill Book Company 1966.
82. JOHANSSON, B. G., and L. RYMO: Acta Chem. Scand. **16**, 2067 (1962).
83. — — Acta Chem. Scand. **18**, 217 (1964).
84. JOHNSON, G. A., and R. H. McCLUER: Biochim. Biophys. Acta **70**, 487 (1963).
85. KÄSER, H., u. G. MASERA: Schweiz. med. Wschr. **94**, 158 (1964).
86. — u. E. GUGLER: Z. klin. Chemie **3**, 33 (1965).
87. KAUFMANN, H. P., Z. MAKUS u. F. DEICHE: Fette, Seifen, Anstrichmittel **63**, 235 (1961).
88. — u. C. V. VISWANATHAN: Fette, Seifen, Anstrichmittel **65**, 538 (1963).
89. — — Fette, Seifen, Anstrichmittel **65**, 607 (1963).
90. KEAN, E. L.: J. Lipid Res. **7**, 449 (1966).
91. KELLER, M., u. G. PATAKI: Helv. Chim. Acta **46**, 1687 (1963).
92. — — Klin. Wschr. **44**, 99 (1966).
93. KIRCHNER, M. A., and M. B. LIPSETT: Steroids **3**, 277 (1964).
94. KLENK, E., u. W. GIELEN: Z. physiol. Chem. **326**, 144 (1961).
95. —, u. W. KAHLKE: Z. physiol. Chem. **333**, 133 (1963).
96. KÖHLER, P., u. H. BAUFELD: Ärztl. Lab. **10**, 224 (1964).
97. KORENMAN, S. G., H. WILSON, and M. B. LIPSETT: J. Clin. Invest. **42**, 1753 (1963).
98. KRELL, K., and S. A. HASHIM: J. Lipid Res. **4**, 407 (1963).
99. KULENDA, Z., u. E. HORÁKOVÁ: Z. med. Lab. Techn. **4**, 173 (1963).
100. LADANY, S., and M. FINKELSTEIN: Steroids **2**, 297 (1963).
101. LUISI, M., G. GAMBASSI, V. MARESCOTTI, C. SAVI, and F. POLVANI: J. Chromatog. **18**, 278 (1965).
102. MACHATA, G.: Methods of Forensic Science, Vol. IV, p. 229. Ed. by A. S. CURRY. London-New York-Sydney: Interscience Publ. 1965.
103. MANDEMA, E., W. H. DE FRAITURE, H. O. NIEWEG, and A. ARENDS: Am. J. Med. **28**, 42 (1960).
104. MASORO, E. J., L. B. ROWELL, and R. M. McDONALD: Biochim. Biophys. Acta **84**, 493 (1964).
105. MICHALEČ, C., u. Z. KOLMAN: Naturwissenschaften **53**, 254 (1966).

106. MIETTINEN, T. A., E. H. AHRENS JR., and S. M. GRUNDY: J. Lipid Res. **6**, 411 (1965).
107. MORRIS, L. J.: J. Lipid Res. **4**, 357 (1963).
108. MORRIS, C. J. O. R.: J. Chromatog. **16**, 167 (1964).
109. MOSER, H. W., A. B. MOSER, and J. C. ORR: Biochim. Biophys. Acta **116**, 146 (1966).
109a. MOUHGRABI, A.: Dissertation, Düsseldorf 1966.
110. MULDNER, H. G., J. R. WHERETT, and J. N. CUMINGS: J. Neurochem. **9**, 607 (1962).
111. NISHIKAZE, O., u. HJ. STAUDINGER: Klin. Wschr. **40**, 1014 (1962).
112. NOWOTNY, E., u. HJ. STAUDINGER: Z. klin. Chem. **4**, 203 (1966).
113. OERTEL, G. W., u. P. BRÜHL: Z. klin. Chem. **4**, 66 (1966).
114. OPIENSKA-BLAUTH, J.: quoted from [*123*].
115. — H. KRACZKOWSKI, and H. BRZUSZKIEWICZ: In: Thin layer Chromatography, p. 165. Ed. by G. B. MARINI-BETTÓLO. Amsterdam-London-New York: Elsevier 1964.
116. — — u. Z. ZAGÓRSKI: J. Chromatog. **17**, 288 (1965).
117. PATAKI, G., u. M. KELLER: Z. klin. Chem. **1**, 157 (1963).
118. — Z. klin. Chem. **2**, 129 (1964).
119. — u. M. KELLER: Helv. Chim. Acta **47**, 787 (1964).
120. — — Gynaecologia **158**, 129 (1964).
121. — Schweiz. med. Wschr. **94**, 1789 (1964).
122. — u. M. KELLER: Klin. Wschr. **43**, 227 (1965).
123. — Dünnschichtchromatographie in der Aminosäure- und Peptidchemie. Berlin: Walter de Gruyter 1966.
124 PAYNE, S. N.: J. Chromatog. **15**, 173 (1964).
125 PENG, T. C., R. VENA, P. OFNER, and P. L. MUNSON: Steroids **6**, 571 (1965).
126. PENICK, R. J., M. H. MEISLER, and R. H. MCCLUER: Biochim. Biophys. Acta **116**, 279 (1966).
127. PHILLIPS, B. M., u. N. ROBINSON: Clin. Chim. Acta **8**, 832 (1963).
128. PRIVETT, O. S., and M. L. BLANK: J. Lipid Res. **2**, 37 (1961).
129. REISERT, P., u. D. SCHUMACHER: Experientia (Basel) **19**, 84 (1963).
130. REISSELL, P. K., L. M. HAGOPIAN, and F. T. HATCH: J. Lipid Res. **7**, 551 (1966).
131. RINK, M., and S. HERRMANN: J. Chromatog. **12**, 249 (1963).
132. — — J. Chromatog. **12**, 415 (1963).
133. ROKKONES, T.: Scand. J. Clin. Lab. Invest. **16**, 149 (1964).
134. — quoted from [*123*].
135. SACHS, B. A., and L. WOLFMAN: Proc. Soc. Exp. Biol. Med. **115**, 1138 (1964).
136. SAMUEL, P., M. URIVETZKY, and G. KALEY: J. Chromatog. **14**, 508 (1964).
137. SANKOFF, I., and T. L. SOURKES: Canad. J. Biochem. **41**, 1381 (1963).
138. SCHEIFFARTH, F., L. ZICHA, F. W. FUNCK, and M. ENGELHARDT: Acta Endocrinol. **43**, 227 (1963).
139. SCHERZ, H., W. RUCKER u. E. BAUCHER: Mikrochim. Acta (Wien) **1965**, 876.
140. SCHLIERF, G., and P. WOOD: J. Lipid Res. **6**, 317 (1965).
141. SCHLOSSBERGER, H. G., H. KUCH u. J. BUHROW: Z. physiol. Chem. **333**, 152 (1963).
142. SCHMID, E., u. N. HENNING: Klin. Wschr. **41**, 567 (1963).
143. — u. H. J. KUSCHKE: Ergeb. Labor. Med. **2**, 175 (1965).
143a. — B. LAUDI, J. KRAUTHEIM u. N. A. TAUTZ: Z. klin. Chem. **4**, 250 (1966).
144. SCHNEIDER, H. P. G., u. Z. SZEREDAY: Klin. Wschr. **43**, 747 (1965).
145. SCHNEIDER, G., u. C. SCHNEIDER: Z. physiol. Chem. **332**, 316 (1963).
146. SCHÖN, H., u. N. KRAUSE: Klin. Wschr. **41**, 743 (1963).
147. SCHOLDERER, D.: Dissertation, München 1965.
148. SCHORN, H., u. C. WINKLER: J. Chromatog. **18**, 69 (1965).

149. SEGURA-CARDONA, R., u. K. SOEHRING: Med. exp. (Basel) **10**, 251 (1964).
150. SEIFERT, H., u. G. UHLENBRUCK: Naturwissenschaften **52**, 190 (1965).
151. — Klin. Wschr. **44**, 469 (1966).
152. SHEN, N.-H. C., F. E. FRANCIS, and R. A. KINSELLA: J. Lab. Clin. Med. **60**, 1017 (1962).
153. SKIPSKI, V. P., R. F. PETERSON, and M. BARCLAY: Biochem. J. **90**, 374 (1964).
154. SQUIBB, R. L.: Nature **198**, 317 (1963).
155. — Nature **199**, 1216 (1963).
156. STÁRKA, L., u. J. RIEDLOVÁ: Endokrinologie **43**, 201 (1962).
157. — J. SULCOVÁ, J. RIEDLOVÁ u. O. ADAMEC: Clin. Chim. Acta **9**. 168 (1964),
158. STRUCK, H.: Mikrochim Acta **1961**, 634.
159. STÜTTGEN, G., u. K. H. VOGELBERG: Arch. klin. u. exp. Dermatol. **222**, 43 (1965).
160. SUOMI, W. D., and B. W. AGRANOFF: J. Lipid Res. **6**, 211 (1965).
161. SVENNERHOLM, E., and L. SVENNERHOLM: Biochim. Biophys. Acta **70**, 432 (1963).
162. — — Nature **198**, 688 (1963).
163. — J. Neurochem. **10**, 613 (1963).
164. SZEREDAY, Z., u. L. SACHS: Experientia (Basel) **21**, 166 (1965).
165. TANCREDI, F., u. H. C. CURTIUS: quoted from [*123*].
166. TAUTZ, N. A., G. VOLTMER u. E. SCHMID: Klin. Wschr. **43**, 233 (1965).
167. TENHUNEN, R.: Acta Chem. Scand. **17**, 2127 (1963).
168. VERMEULEN, A., and J. C. M. VERPLANCKE: Steroids **2**, 453 (1963).
169. VIKROT, O.: Lancet **1963** II, 891.
170. VOIGT, K. D., N. VOLKWEIN u. J. TAMM: Klin. Wschr. **42**, 642 (1964).
171. VOIGT, S., M. SOLLE, and K. KONITZER: J. Chromatog. **17**, 180 (1965).
172. WAGENER, H., D. LANG u. B. FROSCH: Z. ges. exp. Med. **138**, 425 (1964).
173. WAGNER, A.: Klin. Wschr. **44**, 398 (1966).
174. — u. H. WEICKER: Z. klin. Chem. **4**, 73 (1966).
175. — Fette, Seifen, Anstrichmittel **62**, 1115 (1960).
176. WAGNER ROMERO, F.: Experientia (Basel) **20**, 588 (1964).
177. WALDI, D.: Arch. Pharm. **295**/32, 125 (1962).
178. — Klin. Wschr. **40**, 827 (1962).
179. — Ärztl. Lab. **9**, 221 (1963).
180. WALZ, D., A. R. FAHMY, G. PATAKI, A. NIEDERWIESER u. M. BRENNER: Experientia (Basel) **19**, 213 (1963).
181. WEICKER, H.: Klin. Wschr. **37**, 763 (1959).
182. — A. WAGNER u. H. SCHÖNTHAL: Tagung der Deutschen Gesellschaft f. Fettwissenschaft e.V. Karlsruhe Okt. 1963. Lochham b. München: Pallas-Verlag Dr. Edmund Gans.
183. — D. KUHN u. H. STEGMANN: Klin. Wschr. **43**, 1215 (1965).
184. WEINAND, K., W. RINDT u. G. W. OERTEL: Acta Endocrinol. **51**, 210 (1966).
185. WELLS, M. A., and J. C. DITTMER: J. Chromatog. **18**, 503 (1965).
186. WERNZE, H.: Z. ges. exp. Med. **138**, 485 (1964).
187. — u. J. FUJII: Z. ges. exp. Med. **140**, 128 (1966).
188. WEST, C. D., V. J. CHAVRÉ, and M. WOLFE: J. clin. Invest **44**, 1109 (1965).
189. WHELAN, F. J., and G. L. PLAA: Toxicol. Appl. Pharmacol. **5**, 457 (1963).
190. WHERETT, J. R., and J. N. CUMINGS: Biochem. J. **86**, 378 (1963).
191. WIELAND, T., u. H. DETERMANN: Experientia (Basel) **18**, 431 (1962).
192. WILLIAMS, J. A., A. SHARMA, L. J. MORRIS, and R. T. HOLMAN: Proc. Soc. exp. Biol. Med. **105**, 192 (1960).
193. WOOD, P. D. S., and S. HOLTON: Proc. Soc. exp. Biol. Med. **115**, 990 (1964).
194. WOOD, P., K. IMAICHI, J. KNOWLES, G. MICHAELS, and L. KINSELL: J. Lipid Res. **5**, 225 (1964).
195. WOLFRAM, G.: Dissertation, München 1965.

196. Wotiz, H. H.: Biochim. Biophys. Acta **74**, 122 (1963).
197. Zöllner, N.: Rev. Intern. Hepatol. **15**, 283 (1965).
198. — u. G. Wolfram: Klin. Wschr. **39**, 817 (1961).
199. — — Klin. Wschr. **40**, 267 (1962).
200. — — u. G. Amin: Klin. Wschr. **40**, 273 (1962).
201. — — Klin. Wschr. **40**, 1101 (1962).
202. — u. D. Eberhagen: Untersuchung und Bestimmung der Lipoide im Blut. Berlin-Heidelberg-New York: Springer 1965.
203. — H. Hündorf u. G. Wolfram: unpublished.
204. — G. Wolfram, W. Londong u. K. Kirsch: Klin. Wschr. **44**, 380 (1966).

TF. Synthetic Colouring Materials[1]

H. Schweppe

Some years ago, mixtures of dyes were separated on alumina which had been strewn in a thin layer on to glass plates. Mottier and coworkers [47, 48] were those who used this procedure most. They employed alcohol-water mixtures as solvents. Lagoni and Wortmann [36] have described a circular technique, likewise on loose layers, for detecting food colorants.

It has become apparent that, in addition to the known disadvantages of loosely spread layers, the solvents quoted [71, 82] cannot necessarily be used without difficulty. The possibilities of separating numerous dyes, using TLC, have been subsequently investigated. Along with silica gel G and alumina G and H (Firm 88), 1:1 mixtures of both, MN-cellulose powder, acetylated MN-cellulose powder (Ac) (Firm 83) and polyamide powder (Firm 153) have proved useful in thin-layer chromatographic separation.

I. Solvent Dyes

Solvent dyes are employed for colouring naturally occurring and synthetic oils, fats and waxes, benzine and mineral oils and also plastics. Dyes of this type are mono- and bisazo dyes, azines, indophenols and anthraquinone derivatives containing no sulphonate groups. They may be separated on alumina layers, using a variety of organic solvents. Separations through PC are possible only by reversed phase chromatography on pre-treated paper or acetyl-paper [22, 73, 81].

The first separations of solvent dyes were carried out by Stahl [75] on standardised silica gel G layers. He separated butter yellow (C. I. 11020)[1a], Sudan R (C. I. 12150) and indophenol (C. I. 49700), using benzene as solvent. Montag [46] has reported the detection

---

[1] The possibilities of thin-layer chromatographic separation of naturally occurring mixtures of colouring matter are described in the pertinent chapters; carotenoids (p. 259) and anthocyanins (p. 705).

[1a] Colour Index (1956).

of some synthetic and naturally occurring solvent dyes in foodstuffs, with the help of TLC. He employed benzene-carbon tetrachloride (50 + 50) on silica gel G. Experimental details for identifying carotene and bixin in margarine and cheese and of paprika and turmeric pigments in sausage were included. Similar work on the use of TLC in identification of annatto and synthetic solvent dyes has been carried out by RAMA-MURTHY and BHALERAO [59]. HEŘMÁNEK and co-workers [27] have determined the activity of aluminas with the help of the Rf-values found in separations of various fat-soluble azo dyes, using carbon tetrachloride on layers of the aluminas. WALKER and BEROZA [83] have used the solvents chloroform or benzene or their mixtures with diethyl ether, acetone, ethyl acetate, methanol or acetic acid in order to separate butter yellow, Sudan Orange R (C. I. 12055), Sudan III (Fat Red HRR) (C. I. 26100) and Sudan Red BB (C. I. 26105) on silica gel.

PEEREBOOM [52], using various solvents, has separated numerous synthetic and naturally occurring fat-soluble colouring materials on silica gel G, alumina G and kieselguhr G. The results are given in Table 125.

NEHER [50] uses a mixture of Oracet Orange 2R (C. I. Disperse Orange 3, C. I. 11005), Oracet Red 2G (C. I. Solvent Red 12) and Oracet Scarlet 2B (C. I. Solvent Red 50) as reference solution in TLC-separations of steroids on silica gel G. The best separations of the dye mixture were obtained with the solvents benzene-acetone (50 + 50), chloroform-methanol (90 + 10), dioxan and acetone. RUIZ and LAROCHE [62], using alumina G which had been impregnated with a 2% sodium carbonate solution, have obtained excellent separations of Yellow AB and Yellow OB with benzene-carbon tetrachloride (50 + 50) as solvent, and of Sudan III and Sudan Red BB with carbon tetrachloride. FUJII and KAMIKURA [16] have described the separation of 15 solvent dyes on silica gel, using 12 different solvents. Apparently all the yellow and orange fat-soluble compounds of the mono- and bisazo dye types can be separated in this way. RETTIE and HAYNES [61] have separated the solvent dyes of the anthraquinone series, Waxoline Blue A (C. I. Solvent Blue 36), Waxoline Purple A (C. I. 60725) and Waxoline Green G (C. I. 61565) on silica gel G, using toluene-cyclohexane (50 + 50). DAVÍDEK and co-workers [10] use alumina (activity grade III according to BROCKMANN) and petrol ether-carbon tetrachloride (50 + 50) for separating some solvent dyes. TOPHAM and WESTROP [80] have separated dimethylaminoazobenzene and some of its metabolites on silica gel G, using the solvent chloroform-methanol (95 + 5); after development, the chromatogram was exposed to hydrogen chloride vapour, the yellow spots thereby turning orange to red. DAVÍDEK and JANÍČEK [11] have studied the TLC of fat-soluble dyes on layers of starch, prepared by spreading on glass plates a suspension of the starch in

Table 125. hRf-values of solvent dyes [52]

Dye	Colour index No. (1956)	Layer: Silica gel G				S_2	S_3
		F_1	F_2	F_3	F_4	F_5	F_6
Sudan Orange R	12055	68	60	77	70	56	63
Sudan Orange RR	12140	72	58	78	67	62	44
Sudan III	26100	56	52	68	61	41	15
Sudan Red BB	26105	56	53	68	61	38	15
Martius yellow	10315	0	0	28	0	0	0
Sudan yellow GG (butter yellow)	11020	68	62	68	57	59	85
Sudan Red G	12150	18	46	30	36	19	16
Sudan Orange G	11920	14	24	36	37	0	0
Sudan Yellow 3G	12700	54	61	74	75	56	54
Sudan Yellow G	12740	60	64	81	80	68	40
Yellow OB	11390	27	82	50	49	27	87
Yellow AB	11380	25	80	46	41	22	88

Solvents: F_1 = hexane-ethyl acetate (90 + 10)
F_2 = chloroform
F_3 = petrol ether-ether-acetic acid (70 + 30 + 1)
F_4 = petrol ether-ether-ammonium hydroxide (70 + 30 + 1)
F_5 = hexane-ethyl acetate (98 + 2)
F_6 = cyclohexane

Layer: S_2 = alumina G
S_3 = kieselguhr G

a 10% solution of paraffin oil in petrol ether; suitable solvents for this method are methanol-water-acetic acid (80 + 15 + 5) and (80 + 10 + 10).

The coloured plate III, between pages 616 and 617 shows examples of separation of solvent dyes in mixtures [37]. Many of them can be clearly distinguished already after a run of 5—6 cm. It is, however, expedient to adhere to the standard conditions (p. 85) and to chromatograph simultaneously the Desaga test mixture, likewise made up of solvent dyes. The hRf-values and colour index (C. I.) numbers of numerous dyes of this type are given in the following Table 126 [82].

Table 126. hRf-values of solvent dyes on silica gel G, using benzene as solvent [82]

Dye	Colour index No. (1956)	hRf, main spot	Subsidiary spot (s)	Colour of the main spot
Sudan Orange G	11920	4	—	yellow-orange
Sudan III	26100	12	30, 41 [a]	carmine red
Sudan Red G	12150	13	32	carmine red
Sudan Blue G	61525	20	12, 35	blue
Sudan Deep Black BB	26150	28	0, 4, 16 [a], 63 [a]	steel blue
Sudan Orange RR	12140	29	14	light red
Sudan Yellow GG (butter yellow)	11020	40	—	orange-yellow
Sudan Violet BR	61705	53	—	violet-blue

[a] Faintly visible.

Fig. 21 on p. 65 demonstrates how the sharpness of resolution depends on the technique of application.

SCHORN and STAHL [69] have established the hRf-values of a further series of dyes, under the same conditions. 0.2 µg of each dye was applied to a silica gel G layer and development performed with benzene under standard conditions (chamber saturation). Spraying with concentrated hydrochloric acid afforded further characterisation; the colours thus yielded are also quoted in Table 127.

Table 127. *hRf-values of azobenzene derivatives on silica gel G, using benzene as solvent* [69]

Dye	hRf-value	Colour after spraying with conc. HCl
p-Hydroxyazobenzene, *trans*	12	orange-yellow
p-Aminoazobenzene, *trans*	19	orange
Benzeneazonaphthol, *trans*	44	orange
p-Dimethylaminoazobenzene, *trans*	55	red
p-Dimethylaminoazobenzene, *cis*	39	red
p-Methoxyazobenzene, *trans*	64	orange-yellow
p-Methoxyazobenzene, *cis*	12	orange-yellow
Azobenzene, *trans*	72	yellow
Azobenzene, *cis*	30	yellow
Guaiazulene ⎫	76	
Sudan Red G ⎬ test mixture	22	
Indophenol ⎭	8	

Complex mixtures of solvent dyes can also be separated better with the help of gradient TLC than on the normal uniform layers; this is seen in Plate I (Figs. b and c) (between p. 96 and p. 97).

Special Applications

a) Dyes in Motor Fuels

Manufacturers of different motor fuels have to add dyes to their petrol as a means of characterisation. According to HÄUSSER [26] and MACHATA [39], these dyes are best detected and compared in the residues from distillation of the petrol. HÄUSSER [26] carries out intermediate purification. The 5—10 ml sample is concentrated to about one-third and then chromatographed with petrol ether on active alumina in an Allihn tube. The column is extruded, the dye zones extracted with acetone, the extracts filtered and evaporated to dryness. The dyes thus obtained are dissolved in a few drops of acetone and applied to a silica gel G thin layer. Benzene is employed for development. Apart from BP and NORDÖL, all the petrol sorts on sale in Germany can be thus distinguished (ARAL, ARALIN, BP Super, DEA, DEA-Super, ESSO, ESSO Extra, FANAL, FANAL-Super, NORD-ÖL Spezial, SHELL, SHELL-Super). The Rf-values are quoted in the work. This procedure is valuable in clearing up cases of petrol theft.

The difficultly volatile petrol fractions may, in addition, be compared in filtered UV light.

b) Colouring Matter in Naturally Occurring Fats and Oils

In order to isolate the colouring material of fats, 25 g of the fat is saponified with 200 ml 50% alcoholic potassium hydroxide solution by refluxing 30 min on the water bath (DAVÍDEK and JANÍČEK [11]). 160 ml water is then added and the solution extracted with three 50 ml portions of pentane. The combined pentane extract is shaken with water, dried and the solvent distilled off in vacuo. The solution of the residue in 2 ml ethanol is used for TLC.

c) Dyes in Polystyrene

The following method may be used for isolating the colouring material in a transparent, coloured, polystyrene sample [71]: About 5 g sample is dissolved in 25 ml benzene or chloroform at room temperature by shaking several hours on a machine or by allowing to stand overnight. The solution is poured in a thin stream into double its volume of methanol, with stirring. Polystyrene precipitates whereas the solvent dyes remain in solution. The polystyrene is filtered off and the filtrate brought to dryness in a porcelain dish on the water bath. The residue is taken up in a few ml acetone and this solution used for TLC.

II. Disperse Dyes

Disperse colouring materials are used principally for colouring cellulose acetate, polyamide and polyester fibres. Their chemical structure is similar to that of the fat-soluble dyes. They belong to the compound classes of nitrodiphenylamine derivatives and azo- and anthraquinone dyes without a sulphonate group. They are more or less easily soluble in organic solvents but insoluble or difficultly soluble in water. Mixtures of disperse dyes can be separated on alumina columns, using solvents such as ether, methylene dichloride, ethyl acetate and tetrahydrofuran [71]. PC-separations on cellulose paper are incomplete [14, 30, 31, 73, 85]. Separations are better on acetyl-paper [29, 30, 43] or pre-treated paper [19, 20, 25, 44, 73].

There are only a few publications in the field of TLC of disperse dyes, although this method provides better resolution and is more time-economical than PC. WOLLENWEBER [84] has separated Celliton Fast Red Violet RN (C. I. 61100), Celliton Fast Pink B (C. I. 60710) and quinizarin (C. I. 58050) through TLC on acetyl-cellulose powder of 10% acetyl content; he used the solvent ethyl acetate-tetrahydrofuran-water (6 + 35 + 47) which is suitable also for PC-separations on acetyl-paper (43). Many azo- and anthraquinone disperse dyes have been separated by RETTIE and HAYNES [61] with the help of TLC on the same adsorbent, using tetrahydrofuran-water-4N acetic acid (80 + 54 + 0.05) and mixtures of similar composition. Better separations of disperse dyes of the anthraquinone class can be obtained by TLC on silica gel G, using chloroform-acetone (90 + 10) [61]; 1-Amino-, 2-amino-, 1,2-diamino- and 1,4-diamino-anthraquinones could be thus separated.

Our own experience [34, 71] shows that the solvent chloroform-methanol (95 + 5) is suitable for separating many disperse dyes on

Plate III. TLC of Synthetic Dyes (see p. 617 for explanatory notes)

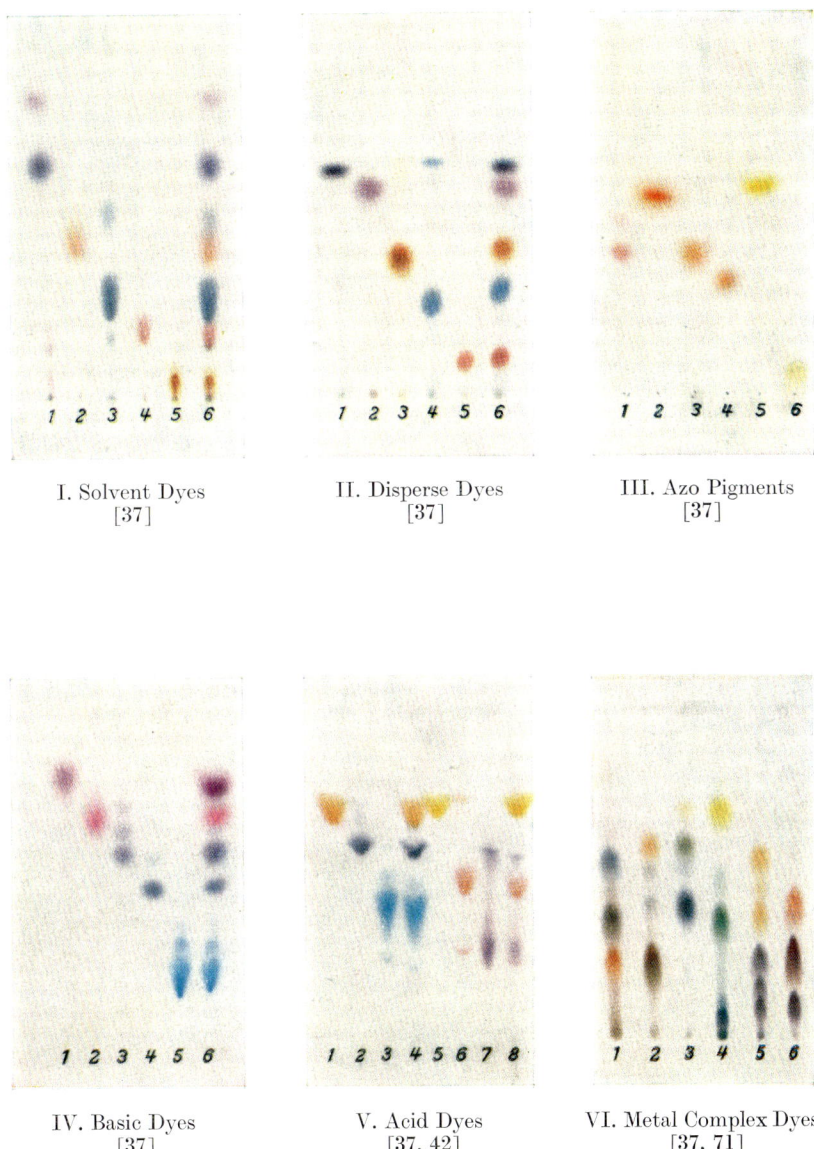

I. Solvent Dyes
[37]

II. Disperse Dyes
[37]

III. Azo Pigments
[37]

IV. Basic Dyes
[37]

V. Acid Dyes
[37, 42]

VI. Metal Complex Dyes
[37, 71]

Thin-Layer Chromatography, 2nd Edition

silica gel G. A chromatogram is given on Plate III (between p. 616 and p. 617), showing the separation of some in this way.

III. Organic Pigments

Organic pigments belong mainly to the azo- and anthraquinone dyes, with and without sulphonate groups, the phthalocyanines and the vat dyes. Many organic pigments are so poorly soluble in water and organic solvents that chromatographic separations are difficult. Some have been separated by PC [45, 70].

FUJII and KAMIKURA [17] have separated various organic pigments through TLC. Our own experiments [37, 71] have shown that some azopigments can be separated with TLC on silica gel G when hexane-benzene-pyridine (65 + 20 + 15) is used as solvent (see also Plate III).

Explanatory notes for Plate III

I. Solvent Dyes [37]
1. Sudan violet BR (C. I. 61,705); 2. Sudan Orange RR (C. I. 12,140); 3. Sudan Blue G (C. I. 61,525); 4. Sudan Red R (C. I. 12,155); Sudan Orange G (C. I. 11,920); 6. Mixture of 1—5.
Layer: Silica gel G (standard method); solvent: benzene, 5—6 cm run, lasting 20 min.

II. Disperse Dyes [37]
1. Celliton Fast Blue B (C. I. 61,500); 2. Celliton Fast Pink B (C. I. 60,710); 3. Celliton Fast Brown 3R (C. I. 11,100); 4. Celliton Fast Blue FFG (C. I. 62,050); 5. Celliton Fast Red GG (C. I. 11,210); 6. Mixture of 1—5.
Layer: Silica gel G (standard method); solvent: chloroform-methanol (95 + 5), 5—6 cm run, lasting 15 min.

III. Azo Pigments [37]
1. Permanent Red FR extra (C. I. 12,300); 2. Amaplast Orange ORC 6673 (C. I. 12,100); 3. Autol Red BL (C. I. 12,070); 4. Permanent Orange Toner (C. I. 12,060); 5. Helio Fast Yellow R (C. I. Pigment Yellow 26); 6. Permanent Yellow NCR (C. I. 12,780).
Layer: Silica gel G (standard method); solvent: hexane-benzene-pyridine (65 + 20 + 15), 5—6 cm run of duration 30 min.

IV. Basic Dyes [37]
1. Fuchsine (C. I. 42,510); 2. Rhodamine B (C. I. 45,170); 3. Crystal Violet (C. I. 42,555); 4. Malachite Green (C. I. 42,000); 5. Methylene Blue (C. I. 52,015); 6. Mixture of 1—5.
Layer: Silica gel G (standard method); solvent: n-butanol-acetic acid-water (50 + 10 + 20), 5—6 cm run lasting 90 min.

V. Acid Dyes [37, 42]
1. Orange II (C. I. 15,510); 2. Amido Black 10b (C. I. 20,470); 3. Patent Blue AE (C. I. 42,090); 4. Mixture of 1—3; 5. Metanil Yellow (C. I. 13,065); 6. Ponceau R (C. I. 16,150); 7. Acid Violet 4BL (C. I. 42,580); 8. Mixture of 5—7.
Layer: Silica gel G + 2.5% sodium carbonate; solvent: butyl acetate-pyridine-water (30 + 45 + 25), 5—6 run, lasting 60 min.

VI. 1:2-Metal Complex Dyes [37, 71]
1. Isolan Brown BLS; 2. Irgalan Brown FL; 3. Irgalan Olive BGL (C. I. Acid Black 64); 4. Lanasyn Green 5GL; 5. Cibalan Brown VRL (C. I. Acid Brown 225); 5. Ortolan Brown 3R (C. I. Acid Brown 33).
Layer: Polyamide powder; solvent: methanol-water-conc. ammonium hydroxide (80 + 16 + 4), 5—6 cm run of duration 30 min.

IV. Basic Dyes

Basic dyes are water-soluble in the form of their salts and are used for colouring paper, leather, cellulose- and polyacrylonitrile fibres. The free bases dissolve in many organic solvents and find application similar to that of the solvent dyes. Basic dyes can function as pigments in lacquer form. Mixtures of basic dyes are usually responsible for brown, green and black tones.

Basic dyes are triarylmethane, xanthene, azine, oxazine, thiazine and acridine derivatives. Azo and methine dyes are encountered particularly as colouring agents for polyacrylonitrile fibres.

Mixtures of basic dyes may be separated by column chromatography on polyamide powder, using water or ethanol-water (80 + 20) or (60 + 30) as solvents [71]. Separations are possible by PC on impregnated paper [6, 18] or acetyl-paper [70]. TLC separations are advantageously shorter than these PC procedures. Many authors have worked on the TLC of basic dyes. Experimental conditions for separations are summarised in the following Table 128.

Table 128. *Solvents for TLC of basic dyes on silica gel G layers*

Solvent	References
1. Chloroform-methanol (90 + 10)	[83]
2. Benzene-methanol (90 + 10)	[83]
3. Chloroform-methanol (80 + 20)	[61]
4. n-Butanol-ethanol-water (90 + 10 + 10)	[62]
5. n-Butanol-acetic acid-water (40 + 10 + 50)	[28]
6. n-Butanol-acetic acid-water (20 + 10 + 50)	[38]
7. Butanone-acetic acid-isopropanol (40 + 40 + 20)	[49]
8. n-Propanol-formic acid (80 + 20)	[78]
9. Chloroform-acetone-isopropanol-sulphurous acid (5—6% SO_2) (30 + 40 + 20 + 10)	[78, 82]
10. 0.75% Sodium acetate-1% HCl-methanol (40 + 10 + 40)	[78]

Solvents 1—3 in Table 128 are suitable for separating bismarck-brown (C. I. 21000) and other basic azo dyes [61, 83]. The triphenylmethane dyes, malachite green (C. I. 42000) and methyl violet (C. I. 42535) have been separated with solvent 4 [62]. Fuchsine (C. I. 42510), rhodamine B (C. I. 45170) and rhodamine 6G (C. I. 45160) can be separated with solvent 5. Rhodamine B, malachite green, crystal violet (C. I. 42555), methylene blue (C. I. 52015) and Victoria blue B (C. I. 44045) have been separated with solvent 7, using silica gel layers on microscope slides [49]. Many basic dyes of the xanthene class which are used for histological staining, e. g., Acridine Red 3B (C. I. 45000), Pyronine G (C. I. 45005), rhodamine S (C. I. 45050), rhodamine G (C. I. 45150) and rhodamine B, can be separated with solvents 8—10 [78]. Numerous subsidiary conta-

minants have been detected in basic triphenylmethane dyes by TLC using solvent 6 [38].

Since the dyes used for microscopic purposes are largely basic, they may be treated here. Table 129 contains the hR_f-values of dyes frequently employed in histology, bacteriology and biology, obtained by WALDI [82] using standard conditions. The solvent chloroform-acetone-isopropanol-sulphurous acid (5–6% SO_2) (30 + 40 + 20 + 10) was used on silica gel G layers. 2 mm^3 amounts of 0.25% solutions of the dyes in methanol were applied.

Table 129. hR_f-values of dyes used in microscopy [82]

Dye	Schultz No.	Colour index No. (1956)	hR_f	Colour
Acridine orange	902	46005	41	yellow
Alkali blue	811	42765	16 and 34	blue (blue)[a]
Brilliant green	760	42040	59 (0)[a]	green
Brilliant cresol blue	992	51010	21 and 52	green (green)[a]
Eriochromazurol S				
Chromazurol S	841	43825	39	dark violet
Gentian violet				
Methyl violet 2B	783	42535	43 and 48	red (violet)[a]
Crystal violet	785	42555	43	violet
Light green, yellowish	765	42095	11 (0)[a]	green
Malachite green	754	42000	35 (0)[a]	green
Metanil yellow	169	13065	39	yellow
Methylene blue B	1038	52015	9	blue-green
Methylene green	1040	52020	18	green
Victoria blue B	822	44045	51	blue

[a] The hR_f-values of the "subsidiary spots" and their colours are quoted in brackets.

According to SCHORN and STAHL [69], the dyes quoted in Table 130 are best separated with n-propanol-formic acid (80 + 20) on silica gel layers; development lasts 70–90 min for a 10 cm run at chamber saturation. The dyes denoted with[a] fluoresce intensively in UV light (365 nm) and very small amounts can be detected in this way.

Table 130. hR_f-values of some fluorescent dyes [69]

Dye	Colour index No. (1956)	hR_f	Dye	Colour index No. (1956)	hR_f
Methyl green	42590	0	Nile blue	51180	49
Diazine green[a]	30295	12	Fuchsine	42510	55
Pyronine G	45005	20	Rhodamine B[a]	45170	62
Acridine orange G[a]	46005	40	Fluorescein	45350	74

[a] Fluorescent dyes.

Plate III (between p. 616 and 617) displays separations of basic dyes [37, 71].

V. Acid Dyes

Acid dyes are employed above all for colouring wool, polyamide fibres, paper, leather and inks. Only certain acid dyes are permitted for colouring foodstuffs and such food colorants are therefore treated in a special section below (p. 623). The difficultly soluble salts of some acid dyes are used as pigments. Most acid dyes are azo-, triarylmethane-, xanthene-, or anthraquinone derivatives. Various investigators have reported PC-separations of acid dyes in detail [12, 31, 33, 40]. Such separations can be carried out more quickly with TLC and often with superior resolution. Table 131 summarises experimental conditions of the abundant work in this field.

Acid triarylmethane dyes, such as those used for example in inks, can be separated according to JAMIESON [28] with solvent 3 in Table 131. Solvents 1 and 4 are suitable also for separating various ink dyes [13, 61]. PERKAVEC and PERPAR [53] have described in detail a TLC method for separating dyes in writing inks, using solvent 2.

Table 131. *Solvents for TLC of acid dyes on silica gel G layers*

Solvent	References
1. Ethanol (95%)	[13]
2. n-Butanol-acetic acid-water (20 + 10 + 50)	[38, 53]
3. n-Butanol-acetic acid-water (40 + 10 + 50)	[28]
4. n-Butanol-ethanol-water-acetic acid (60 + 10 + 20 + 0.5)	[61]
5. Benzene-dioxan-acetic acid (90 + 25 + 4)	[28]
6. Chloroform-acetic acid (90 + 10)	[83]
7. Benzene-acetic acid (90 + 10)	[83]
8. Toluene-acetic acid (65 + 35)	[49]
9. Benzene-propionic acid (80 + 20)	[61]
10. Benzene-chloroform-propionic acid (40 + 40 + 20)	[61]
11. Benzene-isopropanol-acetic acid (60 + 40 + 1)	[61]
12. Ethyl acetate-pyridine-water (60 + 30 + 10)	[61]
13. Amyl alcohol-ethanol-conc. NH_4OH (50 + 45 + 5)	[61]
14. n-Butanol-ethanol-conc. NH_4OH-pyridine (40 + 10 + 30 + 20)	[4]
15. n-Butyl acetate-pyridine-water (40 + 40 + 20)[a]	[61]
16. n-Butyl acetate-pyridine-water (30 + 45 + 25)[b]	[42]

[a] Or cellulose.
[b] Silica gel or alumina G, + 2.5% Na_2CO_3.

Fluorescein (C. I. 45350) and its halogen substitution products such as eosin (C. I. 45380) may be separated with the help of solvents 5—10 [28, 49, 61, 83]. According to RETTIE and HAYNES [61], various acid anthraquinone dyes can be separated on silica gel G or cellulose layers, using solvent 15.

MECKEL and co-workers [42] have succeeded in separating the substantive wool dyes Benzyl Fast Red GRG (C. I. 22245), Palatine Fast

Green GN and Palatine Fast Red on alkalised silica gel G and alumina G layers, as had been first suggested by STAHL [76]. The slurry was prepared by mixing the adsorbent with 2.5% sodium carbonate solution instead of with water. Solvent 15 (Table 131) is employed for development; it must be freshly prepared for each use. Resolution is superior on the alkalised silica gel G layers. The limit of detection is a tenth of that in the PC-procedures so far used. Most of the dyes (acid and substantive) consist of several components of different colours.

Various sulphonphthalein dyes which are used as acid-base indicators, e. g., thymol blue, bromophenol blue and phenol red, may be separated using solvents 11—13 (Table 131). WALDI [82] has determined the hR_f-values of indicator dyes with a view to identification and purity testing; he used standard conditions on alumina G-silica gel G [1:1] layers, with ethyl acetate-methanol-5N ammonium hydroxide solution (60 + 30 + 10). The time of run was 30 min. Results are in Table 132.

Table 132. *hR_f-values of indicator dyes on alumina G-silica gel G (1:1) layers* [82]

Dye	hR_f	Colour of the spots
Chlorophenol red	27 (78)[a]	violet (yellow)[a]
Bromocresol purple	40 (0)	violet (yellow-brown)
Cresol red	42 (60; 77)	orange-yellow (light red and yellow)
m-Cresol purple	43 (71)	orange-yellow (yellow)
Bromophenol blue	48 (39)	blue-violet (red-violet)
Bromochlorophenol blue	48	blue-violet
Bromothymol blue	65	green-brown
Benzyl orange	67	yellow
Methyl orange	67	yellow-orange
Thymol blue	74	orange-yellow
Phenolphthalein	83	colourless; red with alkali
p-Ethoxychrysoidine	84	orange

[a] The hR_f-values and colours of the contaminant dyes are given in brackets; 2 mm³ of 0.25% solutions of the dyes in methanol were applied.

Stahl's gradient-TLC (p. 91) offers a new and interesting way of separating and identifying basic and acid dyes. An example is in Plate I (between p. 96 and 97).

PASTUSKA and TRINKS [51] report the thin-layer electrophoresis of indicator dyes.

Plate III (between p. 616 and 617) reproduces some separations of mixtures of acid dyes.

VI. Direct Dyes

Direct dyes are used for colouring cellulose fibres, paper and leather. They are principally polyazo dyes of comparatively high molecular weight. Various authors have reported the separation of direct dyes through PC [3, 31, 32, 37, 74]. Meckel [41] has tried to circumvent tail formation and the low migration velocity of direct dyes in PC on cellulose paper by chromatographing on acetyl- or glass fibre paper.

These difficulties of tailing and too small R_f-values do not generally arise in the TLC of direct dyes. Moreover, the resolution of direct dyes in TLC usually surpasses that in PC. Table 133 contains experimental conditions for TLC of direct dyes, compiled from numerous publications.

Gasparič and Cee [21] have used solvents 2 and 3 (Table 133) for the separation into zones of various direct azo dyes such as, Dianil Blue G (C. I. 24340), Chicago Blue B (C. I. 24380), Sella Fast Brown DGR (C. I. Acid Brown 235), Solophenyl Grey 4 G and Diphenyl Fast Green BE. Meckel and co-workers [42] have used solvent 6 on alkalised layers of silica gel G or alumina G[2] for separating Benzamin Green 3GS (C. I. 28280), Sirius Supra Scarlet BN (C. I. Direct Red 95), Sirius Supra Brown 6 RL (C. I. 29166) and other direct dyes, into several components. Congo red (C. I. 22120), Trypan Blue (C. I. 23850), Direct Blue 2B (C. I. 22610), Direct Green B (C. I. 30295) and Direct Red F (C. I. 22310) have been separated into several zones by Raban [58], using solvent 1 on alumina.

Table 133. *TLC of direct dyes*

Layer	Solvent	Ref.
1. Alumina	Ethanol-water (various proportions)	[58]
2. Silica gel G	n-Propanol-ammonium hydroxide (60 + 30)	[21]
3. Silica gel G	Pyridine-n-amyl alcohol-NH$_4$OH (30 + 30 + 30)	[21]
4. Silica gel G	n-Butanol-acetone-water-NH$_4$OH (d = 0.88) (50 + 50 + 10 + 20)	[4]
5. Silica gel G	n-Butanol-ethanol-NH$_4$OH-pyridine-water (40 + 15 + 20 + 20 + 15)	[4]
6. Silica gel G[a] or alumina G[a]	n-Butyl acetate-pyridine-water (30 + 45 + 25)	[42]
7. Alumina G[a]	Diethylene glycol monoethyl ether-2% NH$_4$OH (80 + 20)	[62]
8. Alumina G[a]	n-Butanol-ethanol-water (50 + 25 + 25)	[62]

[a] With addiiton of 2.5% Na$_2$CO$_3$.

Meckel [42] applied 2—3 mm^3 amounts of 0.15% dye solutions; larger amounts led to undesirable tailing. Gasparič and Cee [21] used about 5 mm^3 of 5% aqueous dye solutions. Evidently larger amounts of dyes can be cleanly separated using alkaline solvents.

VII. Reactive Dyes

Reactive dyes are attached to cellulose or wool with a covalent link formed from a reactive group. They are azo dyes, anthraquinone

---
[2] See section "Acid Dyes" for preparation.

and phthalocyanine derivatives containing sulphonate and reactive groups.

Various reactive dyes may be separated by PC [60, 72]. TLC-separations have some advantages: Resolution is better and the dyes do not remain at the starting point because the substantivity to cellulose is lacking. PERKAVEC and PERPAR [55] apply 0.1% aqueous solutions of reactive dyes to silica gel G layers and develop with the solvents:

1. Isobutanol-n-propanol-ethyl acetate-water (20 + 40 + 10 + 30) [suitable for separating Procion dyes (ICI)].

2. Dioxan-Acetone (50 + 50) [suitable for separation of Cibacron dyes (CIBA) and Procion-H dyes (ICI)].

3. n-Propanol-ethyl actetate-water (60 + 10 + 30) [suitable for separating Drimaren dyes (Sandoz), Reacton dyes (Geigy), Remazol dyes (Hoechst), Levafix dyes (Bayer) and Primazin dyes (BASF)].

VIII. Metal Complex Dyes

Acid metal complex dyes are used for dyeing wool and polyamide fibres. They are suitable also for preparing coloured, transparent cellulose nitrate varnishes, partly then as their dye acids.

Two chemical types of acid metal complex may be distinguished:

1. 1:1 chromium complex dyes always have sulphonate groups in the molecule. The stoichiometric ratio of metal: dye is 1:1.

2. Technical 1:2-metal complex dyes contain, with few exceptions, no sulphonate group. The stoichiometric ratio chromium (or cobalt): dye is 1:2.

The 1:1 chromium complexes can be separated in PC and TLC under the conditions suitable for acid dyes [42]. PC-separation of 1:2 metal complexes has succeeded in some cases on acetyl-paper [70] or on DEAE-cellulose paper [3]. Various isomeric 1:2-chromium and cobalt complexes of the azo-, azomethine- and formazan series have been separated by SCHETTY and co-workers [2, 64, 65, 66, 67), using methanol on alumina layers. POLLARD and co-workers [57] have reported the TLC-separation of some dyes, complexable with chromium, using silica gel G layers containing starch. HÄFELINGER and BAYER [24] have separated some structural isomers of cobalt chelates of 2-hydroxy-azobenzenes with the help of TLC on silica gel G, benzene being the solvent. Many 1:2-metal complex dyes can be separated on columns of polyamide powder [70]. Separations of 1:2-metal complexes can be carried out under the same conditions, through TLC on polyamide layers with the solvent methanol-water-ammonium hydroxide (80 + 16 + 4) [71]. The usefulness of this method has been demonstrated by separating various 1:2-mixed metal complexes, as often found in brown and green dye shades, with the aid of TLC on polyamide powder. A chromatogram of this type is seen in Plate III (between p. 616 and 617).

IX. Synthetic Food Colorants

Paper chromatographic methods are among those employed for characterising the synthetic food colorants proposed by the Dyestuff Commission of the German Research Council (Deutsche Forschungsgemeinschaft) (DFG). A number of solvents for developing these chromatograms is quoted in the 8th communication of this commission. Samples of the colorants and coloured reproductions of the chromatograms as well as their spectra, are quoted also[3].

Synthetic food colorants have the chemical structure of the acid dyes. Their separation through TLC is thus largely possible under the conditions given in the section on "acid dyes". Work specifically on the TLC of food colorants is summarised in Table 134.

Table 134. *TLC of food colorants*

Layer	Solvent	Ref.
1. Alumina G	n-Butanol-ethanol-water (e. g. 50 + 25 + 25)	[62]
2. Alumina G	Isopropyl acetate-pyridine-water (30 + 40 + 20)	[62]
3. Calcium carbonate	n-Butanol-ethanol-10% NH$_4$OH (50 + 25 + 25)	[79]
4. Silica gel G	Ethyl acetoacetate-pyridine-conc. NH$_4$OH (50 + 20 + 10)	[1]
5. Silica gel G	Ethyl acetate-pyridine-water (70 + 30 + 10)	[1]
6. Silica gel G	Benzene-dioxan-acetic acid (90 + 25 + 4)	[28]
7. Silica gel G	n-Butanol-ethanol-water-conc. NH$_4$OH (50 + 25 + 25 + 10)	[5]
8. Silica gel G	Benzene-methanol-conc. NH$_4$OH (65 + 30 + 4)	[7]
9. Silica gel G	Benzene-n-propanol-conc. NH$_4$OH (60 + 30 + 10)	[7]
10. Silica gel G	Benzene-n-amyl alcohol-conc. HCl (65 + 30 + 5)	[7]

Using solvents 1 and 2 (Table 134), RUIZ and LAROCHE [62] separate the food colorants permitted in Switzerland. BARRET and RYAN [1] have used solvents 4 and 5 and others for separating the food colorants allowed in New South Wales, Australia. JAMIESON [28] separates dyes used in lipsticks with the help of solvent 6. COTSIS and GAREY [7] have shown that all dyes used for colouring lipstick could be separated by TLC on silica gel G, using solvents 8—10. CANUTI and LUBOZ MAGRASSI [5] use solvent 7 for separating the food colorants authorised in Italy.

WOLLENWEBER [84] and WALDI [82] have tried out TLC for chromatographic characterisation of the food colorants allowed in the Federal German Republic. It was found that the dyes listed in Table 135, could

---

[3] Comm. No. 8 of the Dyestuff Commission of the DFG, 2nd Edition, published by Franz Steiner, Wiesbaden, W. Germany 1957.

Table 135. hRf-values of food dyes on cellulose MN layers [84]

DFG-8th Comm.	Commercial name(s)	Schultz No.	Colour index No. (1956)	DFG-6th Comm.	hRf in solvent 1	hRf in solvent 2
Gelb 1 (Yellow 1)	Acid or Fast Yellow Säuregelb Echtgelb (also extra) Jaune solide	172	13015	23	53 (50)[a]	58 (—)[a]
Gelb 2 (Yellow 2)	Tartrazine Hydrazine Yellow O FD & C yellow No. 5 Jaune tartarique	737	19140	64	25 (—)	72 (—)
Gelb 3 (Yellow 3)	Quinoline Yellow Chinolingelb (also water-soluble or extra)	918	47005	97	39 (27)	12 (20)
Gelb 4 (Yellow 4)	Chrysoine S Resorcinol Yellow Tropaeolin O	186	14270	26	88 (—)	4 (—)
Gelb 5 (Yellow 5)	Gelb 27175 N	—	—	30	31 (—)	29 (—)
Orange 1 (Orange 1)	Orange GGN or GGL	—	15980	32	56 (69)	45 (—)
Orange 2 (Orange 2)	Sunset Yellow FCF FD & C Yellow No. 6 Gelborange S	—	15985	29	55 (69)	42 (24)
Rot 1 (Red 1)	Azorubine	208	14720	38	64 (75)	12 (—)
Rot 2 (Red 2)	Echtrot E (= Fast Red E)	210	16045	39	57 (71)	20 (—)
Rot 3 (Red 3)	Amaranth S Naphtholrot S (= Naphthol Red S)	212	16185	40	27 (—)	31 (—)
Rot 4 (Red 4)	Brilliant Ponceau 4 RC Cochillenrot A (= Cochineal Red A)	213	16255	41	33 (—)	55 (—)
Rot 5 (Red 5)	Scarlet 6 R Ponceau 6 R	215	16290	42	16[b] (—)	76 (—)
Rot 6 (Red 6)	Scharlach GN	—	—	34	64 (73)	90 (94)
Blau 1 (Blue 1)	Indanthrene Blue RS Indanthrenblau RS	1228	69800	104	0	0
Blau 2 (Blue 2)	Indigo Carmine FD & C Blue No. 2 Indigotin I or Ia	1309	73015	105	26 (—)	19 (7)
Schwarz 1 (Black 1)	Brillantschwarz BN (= Brilliant Black BN)	—	28440	58	20[b](—)	10 (—)

[a] hRf-Values of subsidiary spots are quoted in brackets; [b] elongated spots (tailing).

be clearly separated on layers of cellulose MN 300G powder (Firm 83), using a single solvent. The solvents, which were tested by these investigators independently of each other, were:

1. n-Propanol-ethyl acetate-water (60 + 10 + 30) [82].
2. 2.5% aqueous sodium acetate solution-25% ammonium hydroxide solution (80 + 20) [84].

Their separation patterns differ so that they are useful for two-dimensional TLC. The migration times were 90 min for solvent 1 and only 30 min for solvent 2.

SALO and SALMINEN [63] likewise use cellulose MN 300 powder to separate the food colorants authorised in Finland. Their solvents were:

1. 2% Trisodium citrate in 5% ammonium hydroxide.
2. *tert.*-Butanol-propionic acid-water (50 + 12 + 38) with 0.4% potassium chloride.

SCHNEIDER and HOFSTETTER [68] have separated various food colorants in pill coatings used in pharmaceutical preparations, on cellulose powder layers.

CRIDDLE et al. [8, 9] report the thin-layer electrophoresis (see pp. 105—114) of food colorants.

X. Dye Intermediates

FRANC and HAJKOVÁ [15] report the TLC-separation of 1- and 2-amino-, 1,2-, 1,4-, 1,5-, 1,6-, 1,7-, 1,8- and 2,6-diaminoanthraquinones; they used alumina layers and cyclohexane-ether (50 + 50).

GILLIO-TOS and co-workers [23] have chromatographed isomeric toluidines, aminophenols, aminobenzoic acids, anisidines, nitroanilines, phenylenediamines, bromo- and chloroanilines on silica gel G using, for example, the solvents dibutyl ether-ethyl acetate-acetic acid (50 + 50 + 5), (75 + 25 + 5) or (25 + 75 + 5). Detection was carried out by first spraying with 5% sodium nitrite in 0.2N hydrochloric acid, dryng the plates at 50° C and spraying then with 5% α-naphthol in methanol.

Two dye components are needed for dyeing with azo dyes which are synthesised on the fibre: a diazonium salt and the coupling compound, abbreviated as the "naphthol". Diazonium components are aromatic amines, the fast bases, which must be previously diazotised; or stabilised diazonium salts, the fast salts. Coupling components are arylamides of aromatic o-hydroxyacids and of acylacetic acids. PERKAVEC and PERPAR [54] report the TLC-separation of these products.

The *fast bases* can be separated on silica gel G using the solvents: 1. n-butanol-acetic acid-water (64 + 16 + 20) or 2. n-butanol-pyridine-water (25 + 50 + 25). 4-Dimethylaminobenzaldehyde (Rgt. No. 72) is used for visualisation. hR_f-values obtained are in Table 136.

The *fast salts* are applied as solutions in 30% acetic acid to silica gel G and developed with n-butanol-acetic acid-water (10 + 10 + 50). A solution of naphthol AS-LR is sprayed on for detection (0.15 g dissolved in 0.3 ml 28% sodium hydroxide, 2 ml 96% ethanol added and the mixture diluted to 50 ml with water). Table 137 shows the hR_f-values found.

PC methods are more suitable for separation of the naphthol components.

According to a private communication of KUHN [35], various naphthalenesulphonic acids can be separated by TLC on silica gel G if they have been previously converted into the corresponding naphthols by fusion with alkali.

Table 136. hR_f-values of the fast bases [54]; details in text

Fast bases	C. I. No. 1956	hR_f 1	2
Fast Yellow GC Base	37000	96	89
Fast Orange GC Base	37005	94	91
Fast Scarlet LG Base	37145	90	86
Fast Scarlet TR Base	37080	93	86
Fast Red TR Base	37085	89	82
Fast Red RC Base	37120	94	84
Fast Bordeaux GP Base	37135	88	87

Table 137. hR_f-values of the fast salts [54]; details in text

Fast salts	hR_f	Colour of the spot
Fast Orange GGD Salt	46	red-yellow
Fast Orange GR Salt	23	orange
Fast Red ITR Salt	18	red
Fast Red RC Salt	14	red
Fast Scarlet GG Salt	17	orange
Fast Brown VA Salt	97; 38	yellow-brown; grey
Fast Blue B Salt	4	violet-blue
Variamine Blue Salt FGC	33	reddish blue
Fast Black K Salt	97; 92	violet; yellow

Special Application: TLC of 2,7-naphthalenedisulphonic acid. 1 g 2,7-naphthalenedisulphonic acid and 10 g potassium hydroxide (flakes) are weighed into a 100 ml erlenmeyer flask fitted with a ground glass opening. The mixture is melted together under a reflux condenser and boiled 15 min. After cooling, the mixture is dissolved in 80 ml distilled water, washed into a 200 ml erlenmeyer flask, also with a ground glass opening, and the solution treated with 1:1 sulphuric acid until acid to Congo red. It is allowed to cool completely, 50 ml ether are added, the flask stoppered and the contents vigorously shaken. 10 mm³ of the ether layer, which contains the 2,7-dihydroxynaphthalene, is applied to a silica gel G layer and developed with benzene-methylene dichloride-ether (70 + 10 + 20). Detection is carried out by spraying with 1% aqueous Fast Blue BB salt (diazotised 4-benzoyl-amino-2,5-diethoxyaniline). The contaminants in the 2,7-naphthalenedisulphonic acid can be detected by TLC in this way. Table 138 contains the hR_f-values of the hydroxynaphthalenes, determined under these conditions.

Table 138. *hR_f-values of the hydroxynaphthalenes* [35]

Substance	hR_f	Spot colour with Fast Blue BB salt
2,7-Dihydroxynaphthalene (2,7-naphthalenedisulphonic acid)	24	brown-violet
2,6-Dihydroxynaphthalene (2,6-naphthalenedisulphonic acid)	31	violet
1,6-Dihydroxynaphthalene (1,6-naphthalenedisulphonic acid)	39	dark brown
1,5-Dihydroxynaphthalene (1,5-naphthalenedisulphonic acid)	50	dark brown-violet
2-Hydroxynaphthalene (2-naphthalenesulphonic acid)	65	red-violet
1-Hydroxynaphthalene (1-naphthalenesulphonic acid)	75	blue-violet

Bibliography for Chapter TF. Synthetic Colouring Materials

1. BARRETT, J. F., and A. J. RYAN: Nature (Lond.) **199**, 372 (1963).
2. BEFFA, F., P. LIENHARD, E. STEINER u. G. SCHETTY: Helv. chim. Acta **46**, 1369 (1963).
3. BROWN, J. C.: J. Soc. Dyers and Colourists **76**, 536 (1960).
4. — J. Soc. Dyers and Colourists **80**, 185 (1964).
5. CANUTI, A., e B. LUBOZ MAGRASSI: Chim. Ind. (Milano) **46**, 284 (1964).
6. CIGLAR, J., J. KOLŠEK u. M. PERPAR: Chem. Z. **86**, 41 (1962).
7. COTSIS, T. P., u. J. C. GAREY: Proc. Sci. Sect. Toilet Goods Ass. **41**, 3 (1964), abs. C. A. **61**, 6855a (1964).
8. CRIDDLE, W. J., G. J. MOODY, and J. D. R. THOMAS: J. Chromatog. **16**, 350 (1964).
9. — — — Nature (Lond.) **202**, 1327 (1964).
10. DAVÍDEK, J., J. POKORNÝ u. G. JANÍČEK: Z. Lebensmitt.-Untersuch. **116**, 13 (1961).
11. — u. G. JANÍČEK: J. Chromatog. **15**, 542 (1964).
12. DOBAS, J.: Coll. Čs. Chem. Commun. **23**, 146 (1958).
13. DRUDING, L. F.: J. Chem. Educ. **40**, 536 (1963).
14. ELLIOT, K., and L. A. TELESZ: J. Soc. Dyers and Colourists **73**, 8 (1957).
15. FRANK, J., and M. HAJKOVA: J. Chromatog. **16**, 345 (1964).
16. FUJII, S., u. M. KAMIKURA: Shokuhin Eiseigaku Zasshi **4**, 96 (1963).
17. — — Shokuhin Eiseigaku Zasshi **4**, 125 (1963).
18. GASPARIČ, J., and M. MATRKA: Coll. Čs. Chem. Commun. **24**, 1943 (1959).
19. — and I. TÁBORSKÁ: J. Soc. Dyers and Colourists **77**, 160 (1961).
20. — and I. GEMZOVÁ-TÁBORSKÁ: Coll. Čs. Chem. Commun. **27**, 2996 (1962), abs. J. Chromatog. **12**, 10 D (1963).
21. — u. A. CEE: J. Chromatog. **14**, 484 (1964).
22. — u. M. MATRKA: Coll. Čs. Chem. Commun. **25**, 1969 (1960).
23. GILLIO-TOS, M., S. A. PREVITERA, and A. VIMERCATI: J. Chromatog. **13**, 571 (1964).
24. HÄFELINGER, G., u. E. BAYER: Naturwissenschaften **51**, 136 (1964).

25. HARRIS, P., and F. W. LINDLEY: Chem. and Ind. **1956**, 922.
26. HÄUSSER, H.: Arch. Kriminol. **125**, 72 (1960).
27. HEŘMÁNEK, S., V. SCHWARZ u. Z. ČEKAN: Coll. Čs. Chem. Commun. **26**, 3170 (1961).
28. JAMIESON, G. R.: Lecture on "Thin-layer chromatography and its application to dyes and plasticisers" given at symposium organised by Scottish Section of the Society for Analytical Chemistry and the Institute of Chemistry of Ireland on "Modern Aspects of Chromatography" (Dublin, Sept. 1963).
29. JANOUSEK, J.: J. Soc. Dyers and Colourists **73**, 328 (1957).
30. JOHNSON, C. D., u. L. A. TELESZ: J. Soc. Dyers and Colourists **78**, 496 (1962).
31. JUNGBECK, J.: SVF Fachorgan **15**, 417 (1960).
32. KIEL, E. G., u. G. H. A. KUYPERS: Tex. **22**, 779 (1963).
33. KOLŠEK, J.: Chem. Z. **82**, 35 (1958); **82**, 457 (1958).
34. — Chem. Z. **83**, 478 (1959).
35. KUHN, A., Basel (Switzerland): private communication.
36. LAGONI, H., u. A. WORTMANN: Milchwissenschaft **10**, 360 (1955); **11**, 206 (1956).
37. LICHTENBERGER, W.: private communication and see Z. anal. Chem. **185**, 111 (1962).
38. LOGAR, S., J. PERKAVEC, and M. PERPAR: Mikrochim. Acta **1964**, 712.
39. MACHATA, G.: Arch. Toxikol. **127**, 1 (1961).
40. MCNEIL, C.: J. Soc. Dyers and Colourist **76**, 272 (1960).
41. MECKEL, L.: Textil-Rundschau **15**, 353 (1960).
42. — H. MILSTER u. U. KRAUSE: Textil-Praxis **16**, 1032 (1961).
43. MICHEEL, F., and H. SCHWEPPE: Mikrochim. Acta **1954**, 53.
44. MITCHELL, L. C.: J. Ass. Off. Agric. Chem. **36**, 943 (1953).
45. MOLOSTER, Z.: Ann. Chim. **3**, 771 (1958).
46. MONTAG, A.: Z. Lebensmitt.-Untersuch. **116**, 413 (1962).
47. MOTTIER, M.: Mitt. Lebensmitt.-Hyg. **47**, 372 (1956).
48. — u. M. POTTERAT: Analyt. Chim. Acta **13**, 46 (1955).
49. NAFF, M. B., and A. S. NAFF: J. chem. Educ. **40**, 534 (1963).
50. NEHER, R.: J. Chromatog. **1**, 205 (1958).
51. PASTUSKA, G., u. H. TRINKS: Chem.-Ztg. **86**, 135 (1962).
52. PEEREBOOM, J. W. C.: Chem. Weekblad **57**, 625 (1961), abs. J. Chromatog. **11**, D 2 (1963).
53. PERKAVEC, J., u. M. PERPAR: Kem. Ind. (Zagreb) **12**, 829 (1963).
54. — — Mikrochim. Acta **1964**, 1029).
55. — — Z. anal. Chem. **206**, 356 (1964).
57. POLLARD, F. H., G. NICKLESS, T. J. SAMUELSON, and R. G. ANDERSON: J. Chromatogr. **16**, 231 (1964).
58. RABAN, P.: Nature (Lond.) **199**, 596 (1963).
59. RAMAMURTHY, M. K., u. V. R. BHALERAO: Analyst **89**, 740 (1964).
60. REIF, J.: Dtsch. Textil-Techn. **13**, 86 (1963).
61. RETTIE, G. H., and C. G. HAYNES: J. Soc. Dyers and Colourists **80**, 629 (1964).
62. RUIZ, S. L., et C. LAROCHE: Bull. Soc. chim. Fr. **1963**, 1594.
63. SALO, T., u. K. SALMINEN: Suomen Kemistilehti **35**, Nr. 9, 146, (1962); abs. Z. anal. Chem. **200**, 160 (1964).
64. SCHETTY, G., u. W. KUSTER: Helv. chim. Acta **44**, 2193 (1961).
65. — Helv. chim. Acta **45**, 809 (1962).
66. — Helv. chim. Acta **45**, 1095 (1962).
67. — Helv. chim. Acta **46**, 1132 (1963).
68. SCHNEIDER, H., u. J. HOFSTETTER: Dtsch. Apoth.-Ztg. **103**, 1423 (1963).
69. SCHORN, P. J., u. E. STAHL: unpublished work.
70. SCHWEPPE, H.: Paint Technol. Vol. **27**, Nr. 8, 12 (1963).
71. — unpublished work.

72. Šrámek, J.: Textil (Praha) **13**, 387 (1958).
73. — J. Soc. Dyers and Colourists **78**, 326 (1962).
74. — J. Chromatog. **15**, 57 (1964).
75. Stahl, E.: Chem.-Ztg. **82**, 323 (1958).
76. — Arch. Pharm. **292**, 411 (1959).
77. — Arch. Pharm. **293**, 531 (1960).
77a. — Chem. Ing. Tech. **36**, 941 (1964) or Angew. Chem. Internat. Edit. **3**, 784 (1964) and Z. analyt. Chem. **221**, 3 (1966).
78. Stier, A., u. W. Specht: Naturwissenschaften **50**, 549 (1963).
79. Synodinos, E.: Chim. cronica (Athens) **28**, 77 (1963), abs. C. A. **60**, 1089 e (1964).
80. Topham, J. C., and J. W. Westrop: J. Chromatog. **16**, 233 (1964).
81. Verma, M. R., u. R. Dass: Naturwissenschaften **44**, 351 (1957).
82. Waldi, D.: unpublished work.
83. Walker, K. C., u. M. Beroza: J. Ass. Off. Agric. Chem. **46**, 250 (1963).
84. Wollenweber, P.: J. Chromatog. **7**, 557 (1962).
85. Zahn, H.: Textil-Praxis **6**, 127 (1951).

Some recent noteworthy articles which appeared during preparation of the 2nd German edition for publication:

86. Cotsis, T. P., and J. C. Garey: Drug and Cosmetic Ind. **95**, 172, 291 and 294 (1964) (determination of dyes in lipstick by means of TLC.)
87. Fujii, S., and M. Kamikura: Kagaku no Ryôiki, Zokan No. **64**, 173 (1964); abstract C. A. **62** 10800b (1965) (TLC of pigments).
88. Gasparič, J.: Z. anal. Chem. **218**, 113 (1966) (PC and TLC of the fast bases).
89. Gemzová, L., and J. Gasparič: Coll. Čs. Chem. Commun. **31**, 2527 (1966) (TLC of primary aromatic amines).
90. Peereboom, J. W. C., and H. W. Beekes: J. Chromatog. **20**, 43 (1965) (TLC of dyes on polyamide and "silver nitrate" layers).
91. Pietsch, H. P., and R. Meyer: Nahrung **9**, 154 (1965) (TLC-separation of food colorants on silica gel D).
92. Purzycki, J., A. Szwark i M. Owoc: Chem. Anal. (Warsaw) **10**, 485 (1965) (TLC of inks).
93. Synodinos, E., G. Kotakis, and E. Kokoti-Kotakis: Riv. ital. sost. grasse **40**, 674 (1963): abstract C. A. **61** 16189 h (1964) (separation of synthetic dyes by means of TLC).

TN. Foodstuffs and Their Additives

J. W. Copius-Peereboom

I. General Applications

TLC is applied to the analysis of foodstuffs in the following ways:

a) Chromatographic separation and determination of the chief components of foodstuffs and separation and identification of trace components which may be of special interest for the characterisation or effectiveness.

b) Separation and analysis of the additives, e. g., antioxidants, preservatives and protective agents against insects.

The chromatography of the principal food components is treated in various chapters: triglycerides, fatty acids etc. in Chapter M; sugars and derivatives in Chapter X; amino acids and peptides in Chapter V; and naturally occurring colouring materials and vitamins in Chapter K.

TLC is especially advantageous for detecting foodstuff components which are present only as traces, e. g., substances in the non-saponifiable residue of fats and oils. It is often difficult to prepare suitably purified concentrates of these trace constituents. This is, however, far less necessary for TLC than for PC. In TLC, interfering components, e. g., triglycerides, can be displaced to the region of the front in a first development using hexane; the real analysis is then carried out in a second development. LIBBY and DAY [56] give an example of the identification of trace components of this sort. They detected 3–30 ppb of the strongly smelling compound, methyl mercaptan, in Cheddar cheese, using TLC of the corresponding derivatives with 2,4-dinitrofluorobenzene.

The TLC of such trace substances is, in any case, treated in detail in other chapters, e. g., of the sterols in Chapter L, and carbonyl compounds in Chapter TS.

A special province of TLC application is the "finger-printing" of some foodstuffs through characteristic trace components, termed *"guide substances"*. COPIUS-PEEREBOOM [17, 18] has characterised mixtures of animal and vegetable fats in this way through the TLC of their sterol fractions. With the help of TLC, MEYER [61] detected a characteristic phenol contaminant in pure cocoa press butter, disclosing its adulteration with cocoa extraction fat. Blending of dried egg powder with vegetable lecithin has been revealed by ACKER et al. [3] through thin-layer chromatographic analysis of the phytosterol glycosides.

TLC has proved its worth in separation and detection of various additives in the foodstuff industry. Vitamin additives may be mentioned here and other examples are the discovery of pesticide residues in foods, the recognition of authorised and unauthorised synthetic dyes and the detection of polyphosphate additives in foodstuffs. The analysis of the last-named class is treated in Chapter Y, "TLC of Inorganic Ions".

II. Antioxidants

These must be first separated from the substrate. The following procedure is satisfactory for fat-containing foodstuffs [99]:

A solution of 40 g fat or substance of high fat content in 250 ml petrol ether is extracted with 40 ml and then 20 ml absolute methanol. 20 ml water is then added to the combined methanol extract. The petrol ether layer is discarded and the methanol phase concentrated in vacuo (15 mm mercury) at 40°, cooled to 0° C, filtered and the filtrate extracted with 20 ml ethyl acetate. The ethyl acetate solution is dried with sodium sulphate and concentrated to 0.5 ml. This solution can be applied to the TLC layer.

CASSIDY and FISHER [14] quote a similar procedure. The samples for analysis must be stored in containers with ground glass stoppers to prevent contact with rubber or plastics; the latter often contain certain plasticisers and/or antioxidants. It must be noted also that some antioxidants, e. g., BHT (= butylhydroxytoluene = 2,6-di-*tert.*-butyl-p-cresol) cannot be obtained quantitatively with this method; BHT can, however, be separated quantitatively through distillation in steam (apparatus on p. 208) or column chromatography [60, 73].

Simultaneous chromatography of reference substances or the use of a "transparent template" according to SEHER [84] (Firm 44) are necessary for reliable identification of antioxidants. Components of essential oils and flavours can otherwise easily simulate added antioxidants. SEHER [84] has developed the procedure below for the special problem of detecting synthetic antioxidants in food fats and energy feeds. It can be seen from the two-dimensional chromatogram reproduced in Fig. 177.

Method: The silica gel G layers, prepared by the standard method, are dried at 120° C. Chloroform alone is first allowed to ascend to a height of 12 cm in order to establish definite conditions. The layers are re-dried and three start spots of the material under investigation and two spots of the reference mixture of three dyes (p. 84) applied according to the scheme seen in Fig. 177. Development is then carried out with chloroform in direction 1 and benzene in direction 2, both runs being 10 cm long.

Detection is with molybdophosphoric acid (Rgt. No. 168). The strongly reducing antioxidants appear within 1–2 min as blue spots against a yellow background. The plate is then exposed to ammonia vapour which renders the background white again and the substances clearly visible as blue or violet spots. Less powerfully reducing compounds may be visualised similarly by heating subsequently for 10 min at 120° C.

The substances remaining at the start (cf. Fig. 177), in particular the gallates, nordihydroguaiaretic acid (NDGA) and components of guaiacum resin, can be separated through further two-dimensional chromatography, though with a certain amount of difficulty. SEHER [84] has investigated various commercially available mixed antioxidant preparations with this procedure. The commercial butylhydroxyanisoles (BHA) were found to contain 2,5-di-*tert.*-butyl-4-hydroxy-anisole, hydroquinone monomethyl ether and 4-*tert.*-butoxy-anisole in addition to the butylhydroxyanisole isomers which cannot be separated by this procedure (i. e., 2-*tert.*-butyl-4-hydroxyanisole and 3-*tert.*-butyl-4-hydroxyanisole). SEHER [85] has also quantitatively

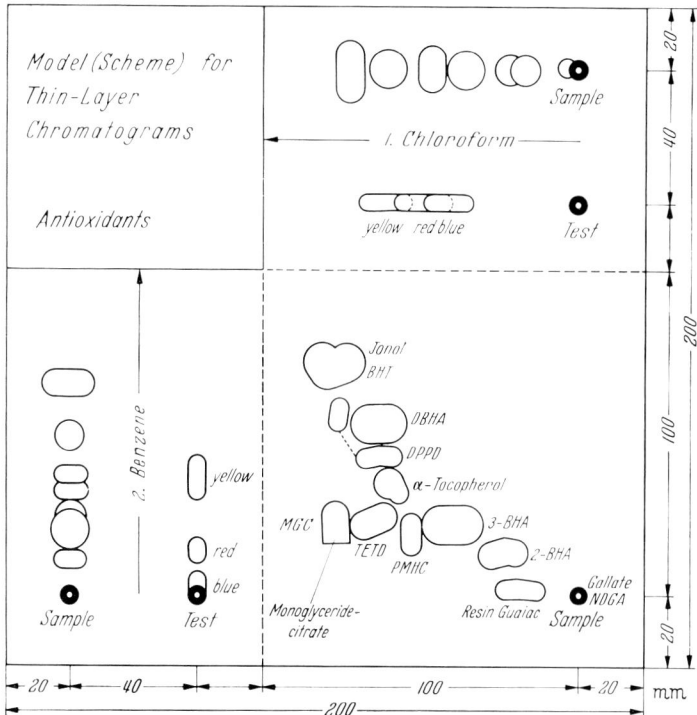

Fig. 177. Scheme for evaluating thin-layer chromatograms of antioxidants according to SEHER [84]. Distances in mm. Silica gel G layer. BHT = butylhydroxytoluene; $DBHA$ = dibutylhydroxyanisole; $DPPD$ = diphenyl-p-phenylenediamine; $TETD$ = tetraethylthiuram disulphide; $PMHC$ = pentamethylhydroxychroman; BHA = butylhydroxyanisole; MGC = monoglyceride citrate; $NDGA$ = nordihydroguaiaretic acid

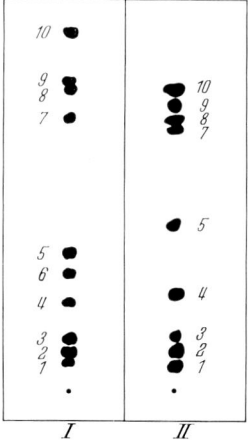

Fig. 178. Thin-layer chromatograms of antioxidants according to [60]. *I.* Adsorbent: silica gel-kieselguhr (25 + 5); solvent: hexane-acetic acid (80 + 20). *II.* Adsorbent: silica gel-kieselguhr (20 + 10); solvent: hexane-acetic acid (85. 7 + 14.3). Detection with 5% molybdophosphoric acid. *1* NDGA; *2* propyl gallate; *3* butyl gallate; *4* octyl gallate; *5* dodecyl gallate; *6* vanillin[1]; *7* butylhydroxyanisole; *8* eugenol[1]; *9* thymol[1]; *10* butylhydroxytoluene

[1] *6*, *8* and *9* were chromatographed along with the antioxidants since these can easily simulate the presence of added antioxidants

determined the purity of these BHA samples or the amounts of their impurities; contact photocopies of the chromatogram were evaluated planimetrically (cf. p. 135). A better separation of the gallates is accomplished on acid layers [85]; these are prepared from silica gel containing 4.5% oxalic acid.

MEYER [60] has separated the various gallates and also other antioxidants, using hexane-acetic acid mixtures on mixed silica gel-kieselguhr layers (Fig. 178). Double development was shown to be advantageous. The method has been checked on examples from practice [89]. Fat-soluble dyes and other alcohol-soluble substances sometimes obstruct a one-dimensional separation and identification of the antioxidants, especially the gallates. Success has then been attained by two-dimensional TLC with the solvents chloroform and hexane-acetic acid.

DAVÍDEK and POKORNÝ [24] have studied TLC on loose polyamide layers. NDGA, BHA, propyl gallate and ascorbyl palmitate were separated by developing with the solvents methanol-acetone-water $(60 + 20 + 20)$ or $(60 + 10 + 30)$. Carbon tetrachloride-ethanol $(70 + 30)$ has been used to separate the gallates [25]. Detection is difficult on the layers which are easily damaged.

COPIUS-PEEREBOOM [19] has employed adhering polyamide layers for the TLC of antioxidants.

Polyamide layers may be prepared in various ways. In the first procedure, 5 ml of 10% starch solution is added to a suspension of 10 g polyamide (Firm 83) in 50 ml methanol and this mixture subsequently spread. A solution of polyvinyl acetate also makes a good binder. It can be prepared from Mowolith CT 5A (Firm 56), for example. In the second procedure, 45 ml of a 10% solution of this Mowolith in methanol is mixed with 7 g polyamide (Firm 153) and the plates are coated with this mixture.

The hRf-values of the alkyl gallates decrease regularly as the chain length increases, using the solvent methanol-acetone-water $(60 + 20 + 20)$[1] (Table 139). This system should be regarded as a reversed-phase system, since fatty acids and esters show similar TLC behaviour. JONAS [41] has found a similar sequence in the separation of antioxidants on paraffin-impregnated layers. A reverse order of hRf-values is found with the solvent benzene-petrol ether-acetic acid-dimethylformamide $(38 + 38 + 22 + 2)$; the hRf-values of the longer chain alkyl homologues are higher than those having shorter chains. A similarly excellent separation of antioxidants is accomplished by this solvent on silica gel G layers also (Table 139). It was therefore assumed that in the system polyamide/petrol ether-benzene-acetic acid-DMF, the stationary phase is formed through a polar polyamide-acetic acid complex [19].

---

[1] This system is very suitable for separation of NDGA and propyl gallate, compounds which are difficult to separate under usual conditions [83].

Table 139. *hRf-values of various antioxidants on polyamide and silica gel G layers (chamber saturation) according to* [19]

Layer	Polyamide[a]	Polyamide[b]	Silica gel G
Solvent	Methanol-acetone-water (60 + 20 + 20)	Petrol ether-benzene-acetic acid-dimethylformamide (38 + 38 + 22 + 2)	Petrol ether-benzene-acetic acid (40 + 40 + 20)
Chromatography	Reversed phase	Polar, stationary phase	
Methyl gallate	68	6	6
Ethyl gallate	55	8	9
Propyl gallate	52	10	11
Butyl gallate	48	14	16
Octyl gallate	18	27	23
Dodecyl gallate	6	47	32
BHA	30	67	63
BHT	24	96	94
Sesamol	—	40	50
NDGA	26	8	9
Ascorbyl palmitate	4	58	8

[a] Polyamide (Firm 83) with starch binder.
[b] Polyamide (Firm 153) with Mowolith CT 5A binder (Firm 56).

Substances on polyamide layers may be detected by spraying with a ferricyanide-ferric sulphate reagent (similar to No. 111). The limit of detection of BHT is about 1 µg (on silica gel).

According to SALO et al. [83], the antioxidants NDGA, dodecyl gallate, BHA, BHT and tocopherol can be separated on silica gel G layers[2] with the solvent Shell Sol A-n-propanol-acetic acid-formic acid (75 + 10 + 5 + 10). Since NDGA and propyl gallate cannot be separated in this way, SALO estimated them UV-spectrophotometrically.

SAHASRABUDHE [77] has separated the mixture of antioxidants in a two-dimensional procedure, then scraped the substances off and determined them colorimetrically. In his procedure, 10 g of the fat sample, e. g., lard, are dissolved in 100 ml hexane and the solution extracted four times with 25 ml amounts of 80% ethanol and then eight times with 25 ml portions of acetonitrile. The acetonitrile phase is brought to dryness, the residue dissolved in a few drops of ethanol and applied to the silica gel G layer. Development is carried out first with benzene and then in the second dimension with acetonitrile. The reducing antioxidants are visualised by spraying with 2,6-dichloroquinonechloroimide (Rgt. No. 66). Only 3-BHA, 2-BHA and BHT give small, distinct zones; propyl gallate and NDGA yield larger and more diffuse spots. With the help of a second reference chromatogram, the substance zones are then located, scraped off and determined quantitatively: 3-BHA with 2,6-dichloroquinonechloroimide; 2-BHA and BHT

[2] Prepared by suspending silica gel G in methanol and spreading.

with α,α'-dipyridyl-ferric chloride; and propyl gallate and NDGA with a ferrous sulphate reagent.

STROHECKER [97] has worked out a procedure for quantitative determination of ascorbyl palmitate in oils and fats; he oxidises with 2,6-dichlorophenol-indophenol and treats the reaction product with 2,4-dinitrophenylhydrazine. The 2,4-dinitroosazone of ascorbic acid is formed under these conditions and may be easily separated from other compounds containing phenolic groups such as tocopherols. He uses a silica gel H layer and chloroform-ethyl acetate (50 + 50) for TLC. After development, the brick red zone of the osazone is scraped off and determined colorimetrically in solution in sulphuric acid. Down to 0.001% of ascorbyl palmitate in antioxidant mixtures and in oils and fats can be determined with this procedure.

III. Preservatives

The articles of JARCZYNSKI [39] and JOUX [43] furnish information about the possibilities of isolating preservatives from foodstuffs.

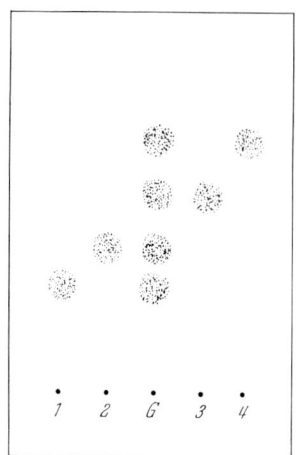

Fig. 179. Separation of p-hydroxybenzoate esters [32]. Silica gel G layer containing 2% Leuchtstoff ZS-Super (highly active). Solvent: pentane-acetic acid (88 + 12). 13 cm run. *1* methyl ester; *2* ethyl ester; *3* n-propyl ester; *4* n-butyl ester; *G* mixture of *1*, *2*, *3* and *4*

The p-hydroxybenzoate esters can be separated with pentane-acetic acid (88 + 12) on silica gel layers containing a fluorescence indicator [32]. These compounds are then detected in short wave UV radiations as dark spots of fluorescence quenching (Fig. 179). The layers are prepared from silica gel G by adding 2% of the fluorescent material "Leuchtstoff ZS-Super" (Firm 118); or by using a silica gel, such as silica gel GF_{254} (Firm 88), which already contains a fluorescent additive.

The four esters in Fig. 179 can be separated only when the following conditions are strictly respected: the plates must be dried for 2 h at 160° C and then stored over potassium hydroxide in an evacuated desiccator. The methyl and propyl esters can be relatively easily separated. Mixtures of these two can be determined quantitatively after separating, by scraping off the zones, eluting and measuring the absorbance in the UV region; the reproducibility is ±3—4% (see p. 152—153).

Separation of methyl, ethyl and propyl p-hydroxybenzoates succeeds on rigorously dried, highly active silica gel GF_{254} layers (3 h at 160° C) or by triple development with the solvent petrol ether-diethyl ether-acetic acid (81 + 5 + 14) [21].

On the other hand, Salo [80] has separated numerous p-hydroxybenzoates (C_1–C_{12}) on mixed layers of polyamide powder (Firm 153) and acetylated cellulose powder (Firm 83) (90 + 10), using Shell Sol A-acetic acid (83 + 17) as solvent.

As well as the p-hydroxybenzoate esters, sorbic and benzoic acids are also often used in food technology as preservatives. According to Copius-Peereboom [20], these preservatives may be satisfactorily separated on cellulose layers (Firm 83) using n-butanol-35% ammonium hydroxide-water (70 + 20 + 10) at chamber saturation. Visualisation is performed with a bromophenol blue-methyl red solution (Rgt. No. 29), followed by spraying with permanganate solution.

Tests on various inorganic adsorbents have shown that benzoic and sorbic acids are best separated on a mixture of silica gel G and kieselguhr G (1 + 1) (Fig. 180); hexane-acetic acid (96 + 4) is the solvent [20].

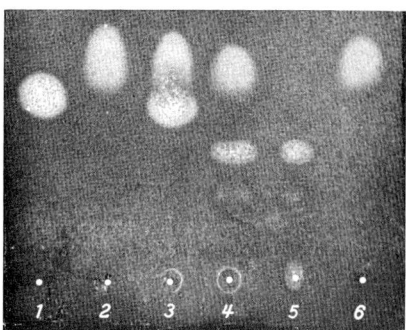

Fig. 180. Separation of preservatives [20, 21]. Layer: silica gel G-kieselguhr G (1 + 1); solvent: hexane-acetic acid (96 + 4). Chamber saturation. 17 cm run, with 1 h development time. Visualisation: alkaline permanganate (Rgt. No. 197). Ca. 100 µg applied. *1* sorbic acid; *2* benzoic acid; *3* mixture of *1* and *2*; *4* mixture of brominated sorbic and benzoic acids (*5* + *6*); *5* brominated sorbic acid; *6* benzoic acid (is not brominated)

Identification is improved if the mixture of sorbic and benzoic acids is brominated at the start point by adding a few drops of 5% bromine solution [21, 58]; the brominated sorbic acid then yields two spots of lower Rf-value. The benzoic acid is not brominated (Fig. 180). Lück and Courtial [58], however, claim that the two bromine derivatives of sorbic acid can easily interfere in the detection; they thus propose bromination with potassium bromate in aqueous solution, whereby sorbic acid yields only a single spot in TLC.

These acids, their esters and also o-phenylphenol, can be detected by introducing 0.02% of an optical bleach such as Ultraphor WT (Firm 16) into the adsorbent layer [22]. The substances may further be visualised by spraying with a rhodamine B solution (Rgt. No. 220) or a bromocresol green solution (Rgt. No. 25). Special spray reagents have been described for specific detection of the different types of preservative: e.g., thiobarbituric acid (Rgt. No. 248) or thymol-sulphuric acid for sorbic acid; hydrogen peroxide-ferric chloride for dehydroacetic acid; and ferric chloride for salicylic acid. o-Phenylphenol, a fungicide, can be visualised with 2,6-dichloroquinonechloroimide (Rgt. No. 66), diazotised 4-nitroaniline solution (Rgt. No. 182) or cerium-(IV) sulphate-trichloroacetic acid (Rgt. No. 37).

The fungicides, biphenyl, o-phenylphenol and 2,4-dichlorophenoxyacetic acid, used for treating citrus fruit peel, have been separated on silica gel G layers by Salo and Mäkinen [81]; the following hR_f-values were obtained, using the solvent Shell Sol A-acetic acid (96 + 4): biphenyl 81; o-phenylphenol 34; 2,4-dichlorophenoxyacetic acid 10. The sensitivity of detection may be appreciably increased by nitrating the biphenyl.

IV. Pesticides

Analytical detection of the mostly fairly small amounts of toxic plant protection agents is of general interest. Not only plant protection authorities and the Plant Protection Section of the World Health Organisation are concerned here but also institutes dealing with food control. It is further of great interest to know how long the material sprayed on to fruit and vegetables needs for conversion into non-toxic derivatives. It must be mentioned here that the incorrect use of too large amounts of such agents has already led to numerous cases of poisoning; their detection is thus also a concern of toxicology.

A good separation and identification of, in particular, insecticides with predominantly lipophilic properties, can be realised with TLC. PC (see work of Mitchell [64]) and GC with highly sensitive detectors [15, 40] are used in practice as alternative analytical procedures. In PC, the lower limit of detection of insecticides is about 15—50 µg; in TLC, as little as 0.05 µg. Special reference may be made to IR-spectroscopy for identifying insecticides (see under Morris [66]).

Special "clean-up" procedures are necessary in the isolation of pesticide residues from foodstuffs. Thornburg [100], Eder [28] and Mckinley [59] have reviewed the various possibilities. After having first extracted the foodstuff with, e.g., methylene dichloride, the extract can be further purified by treatment with concentrated sulphuric acid or through column chromatography on celite-sulphuric acid ("Davidow-column")

[26], magnesium oxide-celite, "Florisil" [65] or alumina. Partition in a two-phase system like petrol ether-acetonitrile [42], petrol ether-dimethylformamide [29] or petrol ether-dimethyl sulphoxide [36] can be undertaken with the same goal.

1. Phosphate Esters

The most important thiophosphate esters can be separated through TLC on various types of layer, using various solvents.

BÄUMLER and RIPPSTEIN [6] have chromatographed on silica gel G, using solvent I (Table 140). They detected with a palladium chloride solution (Rgt. No. 193). SALO [79] has developed with toluene.

FISCHER and KLINGELHÖLLER [31] describe the toxicological detection and the necessary isolation from organs and body fluids. They chromatographed also on silica gel G layers, using the standard method with chamber saturation but at higher temperature (30–31° C). The length of run was 12 cm and solvents were methylene dichloride-methanol-10% ammonium hydroxide solution (60 + 35 + 5) and (80 + 20 + 3), i. e., solvent II of Table 140.

The analysis of breakdown products and metabolites of insecticides is also important. With the help of TLC, KATZ [44] has identified the products of degradation (e. g., benzamide) of phosphate esters. KOVÁČ [50] has utilised preparative TLC to detect some contaminants of

Table 140. hR_f-values of 11 phosphate esters on silica gel layers

Solvent[a]	I [6]	II [31]	III	IV	V	VI	VII	VIII	IX
			←		accdg. to [78]				→
Diazinon	76—82	44	75	66	86	37	43	66	25
E 605 Parathion	65—68	42	70	94	87	77	63	82	75
S 1572 Mercaptophos	—	—	66	90	91	66	76	85	90
Malathion	52—54	32; 44, 58; 66	55	83	89	30	33	64	35
E 1513 (Azinphos ethyl; ethyl guthion)	—	—	42	85	85	18	25	27	36
E 1582 (Azinphos methyl; guthion)	—	—	37	75	80	15	16	25	30
FAC	20—26	—	35	25	61	5	5	10	8
Methyl demeton (metasystox)	62—64	30, 40 and 48	32	24	45	6	10	14	10
DDVP	—	—	30	37	60	13	15	17	34
Dimethoate (Rogor)	4—7	—	15	5	12	0	2	3	2
Chlorthion	43—45	31	—	—	—	—	—	—	—

[a] I = Hexane-acetone (80 + 20); II = Methylene dichloride-methanol-10% ammonium hydroxide (80 + 20 + 3); III = Petrol ether-acetone (75 + 25); IV = Petrol ether-chloroform (10 + 90); V = Petrol ether-ethyl acetate (50 + 50); VI = Petrol ether-ethanol (97.5 + 2.5); VII = Petrol ether-methanol (98 + 2); VIII = Petrol ether-methanol (95 + 5); IX = Benzene-chloroform (50 + 50).

Sumithion (BAYER 41831) (see also PETSCHIK [72]). BLINN [9] has undertaken the identification of oxidation products of phorate (in the "Thimet" mixture) in plant residues; he chromatographed on silica gel G layers, using chloroform-methanol (98.2 + 1.8) and toluene-acetonitrile-nitromethane (45 + 40 + 15).

SALAMÉ [78] has studied the separation of ten thiophosphate esters on silica gel G layers. A selection of the solvents which he used is given in Table 140 (III–IX).

WOGGON [104] has used silica gel-water glass layers for analysis of "Tinox" mixtures. He chromatographed in the S-chamber with toluene-isopropanol-methanol-acetonitrile-water (40 + 16 + 16 + 20 + 9).

EDER et al. [28] first fractionate the phosphate esters by partition between petrol ether and aqueous methanol. They then chromatograph the substances in the petrol ether phase, e. g., parathion, in a two-dimensional procedure on silica gel layers, using as first solvent, hexane-acetone (80 + 20) and as second, hexane-methylene dichloride (50 + 50), at chamber saturation; the phosphate esters in the aqueous phase such as phosdrin, phosphamidon, Sulfotepp, Rogor and Metasystox, can be separated in a one-dimensional procedure with methylene dichloride-ethyl acetate (50 + 50).

WALKER and BEROZA [102] have carried out chromatography of, altogether, 49 phosphate esters on silica gel G layers, at 30° C and chamber saturation, testing thereby the separating efficency of 19 solvents. Four of these (I–IV) have been selected for citation in Table 141. 62 insecticides (46 phosphate esters and 16 organic halides) could be visualised with a bromine-fluorescein-silver nitrate reagent (No. 24). Only some (33 of them) are visible immediately after spraying; the others have to be irradiated with UV light for a short time. The limit of detection of phosphate esters is in the 1–5 µg region.

STANLEY [95] uses the so-called micro-plate method, i. e., he spreads silica gel G on microscope slides. The length of run is then only 5 cm. Solvents used are cyclohexane, benzene, acetone, ethyl acetate and isopropanol (Nos. VI–X in Table 141).

BUNYAN [13] has used alumina layers (Firm 153) as well as silica gel H layers (Firm 88). Benzene-acetone (90 + 10) served as solvent for the TLC of 17 phosphate esters.

KOVACS [52] also has separated the thiophosphate esters on alumina G layers (Firm 117), previously impregnated, however, with dimethylformamide (20% solution in diethyl ether). Development was carried out with methylcyclohexane at chamber saturation (Table 141). Using TLC, extracts of fruits and vegetables also can be tested for insecticide residues. The procedure of, e. g., STORHERR [96] can be employed for preliminary purification (clean-up).

Table 141. hRf-values of insecticides, especially phosphate esters

Layer	Silica gel G [102]				Al$_2$O$_3$ [52]	Micro-plates with silica gel G [95]				
Solvent[a]	I	II	III	IV	V	VI	VII	VIII	IX	X
Aramite	81	56	15	28	—	—	—	—	—	—
Bayer 25141	65	18	0	2	—	—	—	—	—	—
Carbophenothion (Trithion)	76	78	43	38	59	76—86	100	100	92	100
Chlorthion	2	23;61	0;13	2;22	—	—	—	—	—	—
Ciodrin	—	—	—	—	—	0;5	96	66;75	77	94
CORAL	64	46	6	14	15	—	—	—	—	—
Delnav	10;82	0;44;53	0;7	0;26	24	0;33	0;100	0;100	0;88	96
Demeton (Systox)	74;80	35;60	0;34	14;34	—	0;54	0;95;100	0;77;98	85;90	87
Demeton, thiono isomer	0;80	0;33;60	0;33	32	67	—	—	—	—	—
Demeton, thiol isomer	72	34	0	13	32	—	—	—	—	—
DEF	—	—	—	—	—	0;4;100	0;100	0;100	93	100
DDVP	—	—	—	—	—	0	0;93	0;73	0;75	0;82
Disan (Betasan)	—	—	—	—	—	3	100	100	95	100
Diazinon	13	0;28;35	0	0;10	78	8	100	100	83	92
Dicapton (Isochlorthion)	79	65	18	45	—	64	100	100	89	95
Dimethoate (Rogor)	59	12	0	0	1	0	88	39	77	92
Di-syston (Dithiosystox)	83	72	41	41	72	68	100	100	92	100
EPN	81	70	24	29	33	64—74	100	100	87	86-100
Eradex	74	64	26—40	35	—	—	—	—	—	—
Ethion	83	74	34;43	37;44	63	—	—	—	—	—
Guthion (Azinphos methyl)	78	45	0	7	6	—	—	—	—	—
Malathion	73	44	4	22	22	0;7	0;100	0;94	0;88	86;94
Menazon	28	8	0	0	—	—	—	—	—	—
Methyl parathion	73	61	13	22	11	55	100	100	89	96
Methyl demeton (meta-systox)	—	—	—	—	—	0	0;93	0;66	4;74—86	91
Merphos	—	—	—	—	—	0;9;100	100	100	92	100
Methyl trithion	—	—	—	—	36	68—78	100	100	84—94	100
Naled (dibrom)	71	39	30	0;11	—	0;8	0;96	0;85	0;80	0;90
Paraoxon	—	—	—	—	—	0	94	77	79	96
Parathion	74	67	23	30	27	60	98	98	87	89
Phorate (thimet)	75	72	41	42	71	0;68	0;100	0;100	89	100
Phosdrin	59	11;18	0	2	—	0	88	45;59;82	68	85
Phostex	75	76	39	42	—	—	—	—	—	—
Phosphamidon	55	7	0	0	—	0	0;82	0;26	2;60	9;85
Ronnel (trolene)	—	—	—	—	62	86	100	98	83	96
Ruelene	—	—	—	—	—	0	86	43	79	93
Schradan (OMPA)	—	—	0	—	—	0	4	1	19—36	76

Table 141. (Continued)

Layer	Silica gel G [102]				Al$_2$O$_3$ [52]	Micro-plates with silica gel G [95]				
Solvent[a]	I	II	III	IV	V	VI	VII	VIII	IX	X
Sevin	58	35	0	6	—	—	—	—	—	—
Sulfotepp	62	57	31	32	55	—	—	—	—	—
Sulphenone	66	51	8	18	—	—	—	—	—	—
TEPP (tetron-100)	—	—	—	—	—	0	0;81	0;35	7;84	92
Tetradifon	77	75	34	31	—	—	—	—	—	—
Trichlorfon (dipterex)	—	—	—	—	—	0	88;98	100	82;97	88;98
VC-13	61	65	46	38	—	86	100	100	90	100
Zectran	65	2	1	0	—	—	—	—	—	—
Zinophos	—	—	—	—	—	0—14	96	90	71—83	80—96

[a] *Solvents:* I = Chloroform-methanol (90 + 10); II = Benzene-acetic acid (90 + 10); III = Hexane-diethyl ether (90 + 10); IV = Hexane-acetic acid (90 + 10); V = Methylcyclohexane, used on an alumina layer which had been impregnated with a 20% solution of dimethylformamide in diethyl ether; VI = Benzene; VII = Acetone; VIII = Ethyl acetate; IX = Isopropanol; X = Methanol.

Detection of Phosphate Esters

The following reagents may be used: ferric chloride-sulphosalicylic acid solution (Rgt. No. 106) [78, 95]; palladium chloride (Rgt. No. 193) [6, 78]; and dibromoquinonechloroimide (Rgt. No. 62) [10]. The bromine-fluorescein-silver nitrate reagent (No. 24) has also proved satisfactory [102]; only some of the insecticides are visible directly after spraying; others are visualised by subsequent exposure to UV light.

KOVACS [52] has suggested a further improvement of the fluorescein-silver nitrate reagent (No. 24), namely, replacing the fluorescein with the ethyl ester of tetrabromophenolphthalein. Amounts as low as 0.05 µg insecticide and therefore traces of 0.02—0.05 ppm in extracts can then be detected. The TLC-procedure is consequently 20 times more sensitive than similar methods using PC. STANLEY [95] uses iodine vapour, platinum chloride, silver nitrate solution (Rgt. No. 232), fluorescein (Rgt. No. 116), methylumbelliferone (Rgt. No. 163) and ammonium molybdate-perchloric acid (Rgt. No. 166) for detection.

BUNYAN [13] has introduced into TLC COOK'S [16] biological detection of insecticides, i. e., the inhibition of the cholinesterase activity.

2. Chlorinated Hydrocarbons

BÄUMLER and RIPPSTEIN [6] have separated chlorinated hydrocarbons on alumina layers (solvent I in Table 142). PETROWITZ [69] has used silica gel G layers and developed with hexane, cyclohexane or petroleum ether, for example. YAMAMURA and NIWAGUCHI [106] have separated aldrin, dieldrin, endrin and thiodan on silica gel-starch layers,

Table 142. hR_f-values of 18 chlorine-containing insecticides

Layer[a]	Alumina		M	Silica gel G									
Solvent	I[b]	II[c]	III	IV	V	VI	VII	VIII	IX[d]	X	XI	XII	XIII
Reference	[6]	[51]	[1]	[101]	[102]	[102]	[102]	[102]	[51]	[1]	[1]	[1]	[1]
Aldrin	80	167	98	61	82	80	64	56	200	70	58	64	67
Chlordan	—	—	—	—	81	78	62	41—53	—	—	—	—	—
p,p'-DDE	—	162	98	—	66	68	66	54	190	65	74	57	65
o,p'-DDT	—	—	90	—	66	68	56	48	—	50	50	46	59
p,p'-DDT	60	141	91	54	65	68	55	46	145	42	52	39	57
p,p'-Dichlorobenzophenone	—	—	—	—	—	—	—	—	—	14	—	27	59
Dieldrin	18	53	58	15	79	71	34	37	21	12	30	48	65
Endosulfan(Thiodan) I	—	—	—	—	77	75	51	40	—	17	—	35	58
Endosulfan II	—	—	—	—	73	64	8	27	—	2	—	—	12
Endrin	—	54	—	—	82	73	37	39	22	13	—	26	49
Heptachlor	—	159	98	—	80	80	63	57	174	58	48	53	65
Heptachlor epoxide	—	74	—	—	79	75	44	41	36	17	—	—	39
Kelthane	—	6	—	—	73	70	31	33	21	—	—	—	—
HCH,γ-BHC(Lindane)	40	103	58	27	73	73	35	36	48	—	—	18	46
Methoxychlor	11	34	—	9	75	68	23	33	5	—	28	10	—
Perthane	49	117	—	—	—	—	—	—	79	—	—	—	—
TDE (DDD)	—	≡100	77	—	78	83	45	41	≡100	25	67	26	52
Toxaphene	—	—	—	—	73	81	60	45	—	—	—	—	—

[a] Alumina = alumina layer for TLC; M = mixed layer of silica gel G-alumina G (1 + 1).
[b] Highly active, heated 2 h at 200° C.
[c] R_{St}-values, referred to o,p'-DDD = 100.
[d] R_{St}-values, like 3 on Adsorbosil 1-layer.

Solvents: I = n-Hexane; II and IX = n-Heptane; III = Cyclohexane-silicone oil (92 + 8); IV = Cyclohexane-chloroform (80 + 20); V = Chloroform-methanol (90 + 10); VI = Benzene-acetic acid (90 + 10); VII = n-Hexane-ether (90 + 10); VIII = n-Hexane-acetic acid (90 + 10); X = n-Hexane; XI = Cyclohexane-benzene-liquid paraffin (45 + 45 + 10); XII = Petrol ether-liquid paraffin (80 + 20) XIII = Petrol ether-liquid paraffin-dioxan (94 + 5 + 1).

using cyclohexane-acetone (90 + 10) or (92 + 8). WALDI [101] has studied the separation on silica gel G layers (standard method and solvent IV of Table 142). Loose layers have been employed by TAYLOR and FISHWICK [98].

WALTER and BEROZA [102] have investigated the separation of 16 chlorinated hydrocarbons on silica gel G layers (chamber saturation, 30° C), using 19 different solvents. The hR_f-values obtained with four of these solvents (V to VIII) are quoted in Table 142. Visualisation was with the bromine-fluorescein-silver nitrate reagent (No. 24), with a limit of detection of about 5 µg.

The possibilities of detecting chlorinated hydrocarbons in extracts of foodstuffs have been studied by KOVACS [51]. He used prewashed layers of alumina G and silica gel G with the solvents n-heptane and n-heptane-acetone (98 + 2) (see Nos. II and IX, Table 142). Silver nitrate solutions (Rgt. Nos. 231, 232) served for visualisation, with a limit of detection of 0.05 to 0.01 µg.

Using TLC, Kovacs [51] could detect residual chlorinated hydrocarbons in amounts down to 0.002 ppm in extracts of foodstuffs prepared according to the procedure of Mills [62]. The method has proved valuable in practice as can be seen from the work of Salo et al. [79] and Onley [67]. The latter was able to detect amounts as low as 0.005 ppm of insecticides in milk (see Fig. 181).

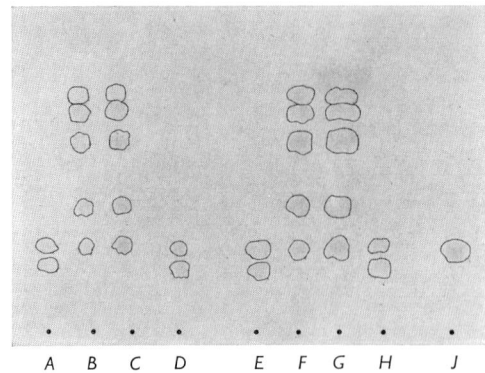

Fig. 181. Thin-layer chromatogram of chlorinated pesticides [67][1]. Detection of pesticide residues in milk. Layer: alumina G; solvent: n-heptane (according to [51]). A and E 0.2 µg endrin and aldrin; B and F aldrin, heptachlor, DDT, lindane and heptachlor epoxide; C 2 g milk with added aldrin, heptachlor, lindane, heptachlor epoxide (all 0.1 ppm) and DDT (0.5 ppm); D 2 g milk with added endrin and dieldrin (0.1 ppm); G 5 g milk with added aldrin, heptachlor, lindane, heptachlor epoxide (all 0.04 ppm) and DDT (0.2 ppm); H 5 g milk with added endrin and dieldrin (0.04 ppm); J 50 g milk with spot of heptachlor epoxide; I milk blank

[1] With kind permission of the author and the AOAC publishers.

The preparation of hexachlorocyclohexane (HCH or BHC) yields a mixture of isomers. The γ-isomer, known as "lindane", is the most effective. The six isomers can be separated on silica gel G layers with petrol ether-carbon tetrachloride (50 + 50) and cyclohexane-chloroform (80 + 20). Waldi [101] found the following hRf-values when using the latter solvent mixture: α-BHC 40—43; β- 25—28; γ- 33—36; and δ- 14—17.

A layer impregnated with fluorescein (Rgt. No. 121,6) has been used for detection. Pink spots are formed by amounts exceeding 5 µg after standing for 1—2 h. Dark blue-violet spots on the green fluorescent layer are seen in long-wave UV radiation. The rhodamine B reagent (No. 120) can also be used, followed by spraying with a 10% solution of sodium carbonate [101] (see also Rgt. No. 77).

ABBOTT et al. [1] have undertaken an exhaustive investigation of the separation of 16 chlorine-containing insecticides. They used kieselguhr G, alumina G and mixed layers of silica gel G and alumina G $(1 + 1)$ as well as the ordinary silica gel G. The hRf-values found in five of the 15 solvents tried out are given in Table 142 (Nos. III and X—XIII). Colour differentiation was obtained with a silver nitrate-bromophenol blue reagent.

A two-dimensional procedure for separating chlorinated hydrocarbons has been worked out by EDER et al. [28]. They worked with silica gel G layers, using hexane in the first dimension and isooctane in the second, at chamber saturation. Details were given also of the clean-up needed for determination of insecticide residues in fruit and vegetables.

KAWASHIRO and HOSOGAI [44a] have investigated many spray reagents as part of a study of the detection of pesticide residues in food. They found the following procedure to be highly satisfactory for visualising halogen-containing organic pesticides:

A 0.5% alcoholic solution of a suitable aromatic amine is sprayed on to the developed chromatogram. After drying, the layer is exposed to short-wave UV radiation (254 nm). Green spots appear after only 1 min. The limit of detection is 0.5—1 µg when o-toluidine or o-dianisidine are used as amines.

3. Pyrethrins and Synergists

Pyrethrins and rotenones, compounds of plant origin, are finding increasing use in addition to the synthetic insecticides. These so-called synergists are often added to suitable preparations in order to augment the activity of the plant insecticides. STAHL [93] has applied the SRS-technique (see p. 88) to study the inactivation of pyrethrins through light. He chromatographed the pyrethrin concentrate on a silica gel layer, first in direction 1 (Fig. 182); he then irradiated the layer with UV light or sunlight (Fig. 182, shaded zone), following with development in direction 2 using the *same* solvent. The more strongly polar photochemical decomposition products were then found below the respective pyrethrins. Conclusions about the rate of decomposition could be drawn from the sizes of the zones. The pyrethrin peroxides, inactive as insecticides, and the so-called lumipyrethrins were found as intermediates.

For separating the pyrethrins and synergists, silica gel layers (standard method, chamber saturation) and various so-called "chloroform-isoeluotropic" solvents have been employed for development: 1. benzene-butanone $(90 + 10)$; 2. benzene-ethyl acetate $(85 + 15)$; 3. carbon tetrachloride-ethyl acetate $(80 + 20)$; 4. hexane-butanone $(80 + 20)$; 5. hexane-ethyl acetate $(75 + 25)$.

It must be pointed out nevertheless, that the pyrethrin I spot contains a mixture of "pyrethrin I-compounds"; the pyrethrin II spot also contains the II-compounds. For this reason, STAHL and PFEIFLE [92] took up the separation of the pyrethrins again later. They found that it was possible to separate the pyrethrin I compounds into 3 zones on silica gel HF_{254} layers in the BN-chamber, using hexane-ethyl acetate (95 + 5); the more polar solvent hexane-heptane-ethyl acetate

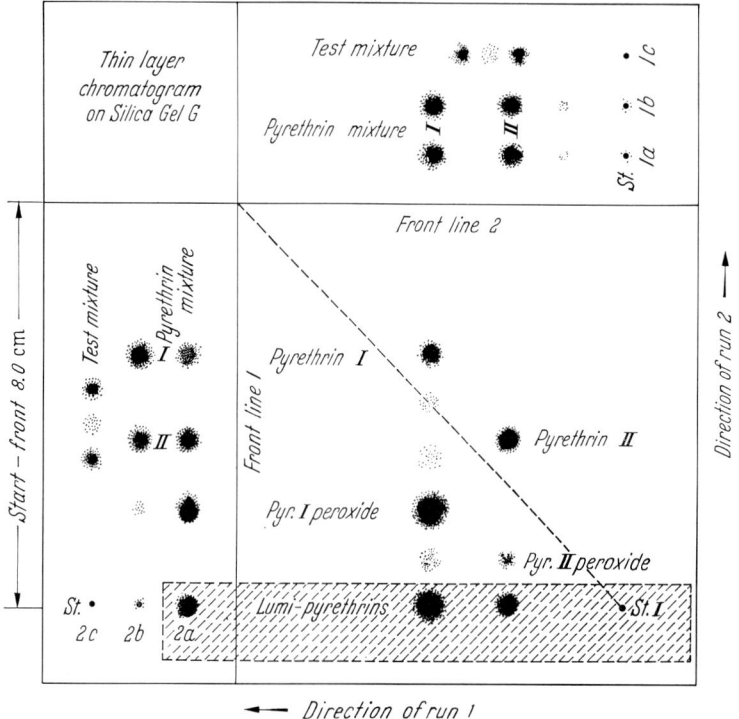

Fig. 182. Thin-layer chromatogram (SRS-technique) of a pyrethin concentrate according to STAHL [93]. The irradiation (shaded area) was carried out after separation in direction 1. The colour reactions with antimony(III)chloride, with 2,4-dinitrophenylhydrazine and with iodide/starch, are superimposed on the single diagram (for details see text and original article)

(48 + 40 + 12) was used to separate the II-compounds. Details are given in Fig. 183.

Detection: Silica gel HF_{254} (Firm 88) is a convenient adsorbent, since the compounds mentioned can be detected without decomposition through their quenching of fluorescence in short-wave UV light. The zones may then be scraped off and submitted to GC. The anisaldehyde-sulphuric acid reagent (No. 11) has proved especially good for visualisation. Pyrethrins I and II yield a dark grey colour after heating; the cinerins and jasmolins turn brown. Only the pyrethrins I and II show a colour reaction (brown-grey) with antimony(III) chloride (Rgt. No. 15). All six compounds give brown-grey spots with antimony(V)chloride (Rgt. No. 18) after heating. Molybdophosphoric acid (Rgt. No. 168) and 2,4-dinitrophenylhydrazine (Rgt. No. 82) can also be used for visualisation.

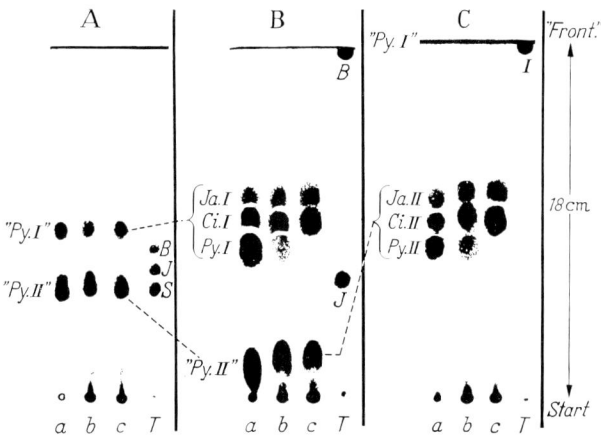

Fig. 183. Thin-layer chromatograms of ordinary commercially available pyrethrum extracts of different ages (a, b, c), photographed in short wave UV light [92]. A on silica gel HF_{254} under normal conditions; B in the BN-chamber using continuous development with solvent I (concluded when the butter yellow attained the evaporation zone); C continuous development with solvent II (concluded when the indophenol reached the evaporation zone); Ja jasmolin; Ci cinerin; Py pyrethrin; T standard Desaga mixture (B butter yellow; I or J indophenol; S sudan red)

Iodide-starch reagent (No. 139) is employed for visualising pyrethrin peroxides. Biological detection of insecticidal activity after chromatography is especially interesting. STAHL [93] has used for this the highly sensitive *Aedes aegyptici*-larvae test (age 5—8 days, length 2—4 mm, detection within 12 h) [12] and also the less sensitive *Drosophila* test.

According to STAHL, the following synergists can be separated on silica gel with hexane-ethyl acetate $(75 + 25)$: synergist S_{421} BASF (= octachlorodipropyl ether) hRf 67; piperonyl butoxide hRf 35; bucarpolate hRf 23. BEROZA [8] has separated synergists like piperonyl butoxide, sesamin, sesamex etc., also on silica gel G layers, using benzene-acetone $(75.5 + 2.5)$.

4. Herbicides

TLC has been used with success also in detection of residues of plant protection substances with herbicidal activity. BACHE [4] has been able to detect amiben in tomatoes. HENKEL and EBING [37] have separated the triazine group, e. g., prometryn and propazine, on silica gel G layers, using ethyl acetate-methanol (80 + 20); and the phenoxyalkanecarboxylate ester group such as hexyl 4-chloro-2-methylphenoxypropionate (MCCP-hexyl) and the corresponding butoxyethyl ester (MCPP-butoxyethyl), using cyclohexane-diisopropyl ether (84 + 16).

ABBOTT et al. [1] have separated six herbicides, e. g., 2,4-, 2,4,5-T and 4-chloro-2-methylphenoxyacetic acid (MCPA), 2,2-dichloropropionic acid (dalapon) in the S-chamber on mixed layers of silica gel G-kieselguhr G (2 + 3); liquid paraffin-benzene-acetic acid-cyclohexane (4 + 11 + 8 + 77) was the solvent. It has been possible also to detect herbicide residues in extracts of soil and water, using TLC.

Pentachlorophenol, likewise a herbicide and a component of oily wood preservatives, can be separated from various other contact insecticides on acid silica gel G layers (addition of 1% oxalic acid for example), using benzene and chloroform as solvents [70].

V. Artificial Sweeteners

Saccharin and dulcin can be separated by TLC on silica gel G layers, using chloroform-acetic acid (90 + 10) [101]. For detection, the layer is sprayed first with rhodamine B (Rgt. No. 220) and then with silver nitrate (Rgt. No. 225). The hRf-values of saccharin and dulcin are about 30 and 50 respectively.

In order to isolate saccharin and/or dulcin from an aqueous solution, it is acidified and extracted with ethyl acetate. The ethyl acetate solution is then concentrated and saccharin in it detected through TLC. The acid aqueous phase is then made alkaline and dulcin similarly extracted with ethyl acetate and subsequently chromatographed. See SCHILDKNECHT and KÖNIG's [87] comment on this.

SALO and co-workers have separated artificial sweeteners on mixed layers of acetylated cellulose and polyamide [82]. A good separation of dulcin (hRf 66), saccharin (47) and cyclamate (28) was possible using the solvent Shell Sol A-n-propanol-acetic acid-formic acid (75 + 10 + 12 + 3). Rhodamine B (Rgt. No. 220) or dichlorofluorescein (Rgt. No. 63) were used for detection. The layers were prepared from 9 g acetylated cellulose powder (MN 300 Ac, Firm 83) and 6 g polyamide powder for TLC (Firm 153). These were mechanically mixed into a homogeneous suspension with 60 ml methanol and then spread; the layers were dried for 10 min at 70° C. A 10 cm run took about 25 min at chamber saturation.

VI. Emulsifiers and Swelling Agents

Surface active substances, such as the commercially available preparations of "sucrose palmitate" and other esters, have been separated into their individual components by MIMA and KITAMORI [63] and by GEE [33], using TLC on silica gel layers. Solvents were benzene-ethanol (75 + 25) or toluene-ethyl acetate-95% ethanol (50 + 25 + 25) (see also LINOW [57]).

Other emulsifiers also can be analysed by TLC. Thus, for example, "Planta-Emulgator" ME 18 has been separated into 6 constituents on silica gel G layers, using isooctane-ethyl acetate (85 + 15) [21]. KRÖLLER [53] has separated the components of this emulsifying agent in a two-dimensional procedure on silica gel G, using methylene dichloride-acetic acid (99 + 1) first and then benzene-methanol (90 + 10). He worked out in addition some procedures for isolating various emulsifiers from foodstuffs like margarine, pastry and fruit syrups [54].

The emulsifier stearyl tartrate has been detected in bread by WILLIAMS [103]. Chromatography was carried out on silica gel layers with diethyl ether-petrol ether (90 + 10) in order to identify stearyl alcohol present in the non-saponifiable part of the extracted bread fat.

Partially esterified glycerol and polyglycerol are employed in the food industry, e. g., the monoglyceride of diacetyltartaric acid, polyglycerides esterified with tartaric acid or almost pure mixtures of polyglycerols. Polyglycerol-based emulsifiers are not permitted as additives in all countries and their detection is therefore of interest. SEHER [86] has developed a procedure for detection of polyglycerols in mixtures. He separates on silica gel G layers, using the solvent ethyl acetate-isopropanol-water (65 + 22.7 + 12.3), suggested by STAHL and KALTENBACH [94] for the TLC of sugars (see Chapter X). He was thus able to separate mixtures of glycerol, diglycerol, triglycerol and tetraglycerol. Sodium periodate solution was used for detection, followed by spraying with benzidine or ammoniacal silver nitrate.

Thickening or swelling agents are used a great deal in the food industry. For purposes of detection, they are hydrolysed according to the procedure of BECKER and EDER [7]. The decomposition products, namely sugars and related compounds like uronic acids, have been separated by GRAU [35] on cellulose layers, using SCHWEIGER's [90] solvent, ethyl acetate-pyridine-water (40 + 20 + 40). Swelling agents, such as carob bean flour, agar, alginate and pectin, yield characteristic series of spots which can be utilised for identification. Cellulose ethers, like methyl cellulose (tylose), hydroxyethylcellulose, carboxymethylcellulose, etc., can also be identified by this "finger-print" procedure [21].

VII. Alcohols and Glycols

The methods used in the TLC of alcohols and glycols are described on p. 660. It may be mentioned here that various diols, such as propane-1,2-diol or butane-2,3-diol, are used in practice as tobacco humectants. According to WRIGHT [105], they can be separated on silica gel G layers with acetone or butanol-acetone-water (44 + 55 + 1) and detected with a 1% solution of lead tetraacetate in benzene (Rgt. No. 152). Worthy of mention in this connection are the experiments of DUMAZERT et al. [27] who have separated ethylene glycol diacetate, propylene glycol diacetate, glycerol triacetate, erythritol tetraacetate etc., on silica gel G layers, using benzene-ethyl acetate-ethanol (89 + 10 + 1). They employed the hydroxylamine-ferric chloride reagent (No. 134) for visualisation. See also the work of PREY et al. [74], HROMATKA and AUE [38] and KNAPPE, PETERI and ROHDEWALD [47].

KUČERA [55] has separated several diols on alumina layers. He used layers impregnated with 3% ammonium borate solution in order to separate the 1,2-diols from other glycols. Silver nitrate solution was used for visualisation. TLC-separation of stereoisomeric diols also has been achieved by FISCHER and KOCH [30] on layers of fibrous alumina.

The terpene- and sesquiterpene alcohols, used as food flavouring agents in the food industry, are treated in Chapter J on p. 225 and the sugar alcohols in Chapter X.

VIII. Organic Acids

Organic acids can be separated by TLC under various sets of conditions. According to BRAUN and GEENEN [11], it is convenient to prepare a 2% solution of the ammonium salts of the acids in ethanol-water (1 + 1) and to apply 2 mm^3 (= 40 µg). A selection is given below of the adsorbent layers and corresponding solvents which have been tried out so far:

A. *Silica gel layers*

    I. 96% Ethanol-water-25% ammonium hydroxide (78 + 9.5 + 12.5); without chamber saturation; 10 cm run in 120 min [11].

    II. Benzene-methanol-acetic acid (79 + 14 + 7); chamber saturation; 110 min time of run [71].

    III. Propanol-28° Bé ammonium hydroxide (70 + 30) [68].

    IV. Ethanol-chloroform-28° Bé ammonium hydroxide-water (53 + 30.3 + 15.2 + 1.5) [68].

    V. Diisopropyl ether-formic acid-water (90 + 7 + 3) [48].

    VI. Methanol-5N ammonium hydroxide (80 + 20) [101].

VII. Benzene-dioxan-acetic acid (75.6 + 21 + 3.4); separates like II [71].
B. *Silica gel G, impregnated with propionic acid* [76]
VIII. Petrol ether-ethyl formate (65 + 35).
C. *Silica gel G-kieselguhr* (1 + 1) *layers*
IX. Benzene-ethanol-conc. ammonium hydroxide (28.6 + 57 + + 14.4); chromatographed on microscope slides [5].
D. *Kieselguhr layers*, impregnated with polyethylene glycol [45, 46]
X. Diisopropyl ether-formic acid-water (90 + 7 + 3), saturated with polyethylene glycol M 1000; use of the M 4000 gives rise to small deviations.
E. *Cellulose layers* [91]
XI. Pentanol-formic acid-water (48.8 + 48.8 + 2.4).
F. *Acetylcellulose layers* [75]
XII. n-Propanol-n-butanol-10% ammonium carbonate-5N ammonium hydroxide (45 + 22 + 22 + 11) or (33 + 33 + 23 + 11) (or 39 + 28 + 22 + 11).
XIII. n-Propanol-10% ammonium carbonate-5N ammonium hydroxide (67 + 22 + 11).
G. *Polyamide layers* [48, 49]
XIV. Diisopropyl ether-petrol ether-carbon tetrachloride-formic acid-water (50 + 20 + 20 + 8 + 1).
XV. Acetonitrile-ethyl acetate-formic acid (81.8 + 9.1 + 9.1).
XVI. Butyl formate-ethyl acetate-formic acid (81.8 + 9.1 + 9.1).

The procedure developed by KNAPPE and PETERI [45, 46] appears to be especially useful for TLC of organic acids. Silica gel G layers are impregnated with polyethylene glycol M 1000 (Table 143). The hR_f-values of the various unsaturated dicarboxylic acids on such layers resemble one another much more than those of the saturated acids. KNAPPE has accordingly worked out a special SRS procedure (see p. 88) for complex mixtures of these dicarboxylic acids. After first separating with solvent X, the acids are visualised with bromocresol purple (Rgt. No. 28) and then hydrogenated using a colloidal palladium solution as catalyst. Development is then carried out in the second direction with the same solvent. In this way, fumaric, maleic, oxalic, pimelic, citraconic, itaconic, mesaconic, suberic and glutaconic acids can be identified [46].

Preparation of the layers according to [46]: 30 g kieselguhr G (Firm 88) + 0.05 g sodium diethyldithiocarbamate are carefully stirred into a mixture of 45 ml water and 15 g polyethylene glycol and the mixture spread on to 5 plates using the standard method; these are dried for 30 min/100° C. After developing in a 12 cm run, they are heated 10 min at 100° C (!) and, when cooled, sprayed with a solution of 0.04 g bromocresol purple in 100 ml of 50% ethanol, adjusted to pH 10.

KNAPPE and PETERI [48, 49] use polyamide layers also for separating organic acids. Many of the acids listed in Table 143 can be identified by alternately using solvents XIV, XV and XVI in combination with the systems (A, V) and (D, X), and hydrogenation of unsaturated acids directly on the layer. TLC on polyamide with solvent XIV has proved highly suitable for separating the biochemically important acids of the citric acid cycle [49]. Treating with iodic acid achieves further differentiation; some acids are then modified or decomposed in a characteristic fashion [49]. The treated acids are best chromatographed with the solvents acetonitrile-propyl formate-propyl acetate-formic acid (41 + 41 + 9 + 9) or amyl alcohol-carbon tetrachloride-formic acid (47 + 35 + 18).

Strongly adhering polyamide layers can be prepared by adding formic acid. A suspension of 5 g polyamide for TLC purposes (Firm 153) in 50 ml of a mixture of ethanol-chloroform-formic acid (32 + 15 + 3) is spread on the plates. These are dried 20 min at room temperature and then 30 min at 105° C.

Table 143 shows that the hR_f-values of the dicarboxylic acids increase with increasing chain length, both in basic and acid solvents and also on kieselguhr layers which have been impregnated with polyethylene glycol. The *cis-trans* pair, maleic-fumaric acid, has been separated with solvents II, V and X amongst others. Citric acid may be separated from tricarballylic and methylsuccinic acids (hR_f-values 15, 35 and 80 respectively) using solvent VI on silica gel G layers [101].

RONKAINEN [76] has chromatographed the 2,4-dinitrophenylhydrazones of some keto acids on acid-impregnated silica gel layers. The layers were prepared by spreading a slurry of 30 g silica gel G in 60 ml water and 5 ml propionic acid. A good separation of the hydrazones of a number of keto acids can then be accomplished using solvent VIII.

DANCIS et al. [23] also have worked on the separation of 2,4-dinitrophenylhydrazones of the keto acids. They used silica gel G layers and isoamyl alcohol-0.25N ammonium hydroxide (95.2 + 4.8). RINK [75] has succeeded in separating the rhodanine derivatives of 16 ketocarboxylic acids, using acetylated cellulose layers and solvents XII and XIII.

Footnotes for table 143:
   [a] The compositions of layers and solvents are to be found on p. 650.
   [b] The acids marked "hy" can be hydrogenated on a Pd catalyst with stirring.
   [c] Standard method, without saturation (NS).
   [d] Standard method with chamber saturation (CS).
   [e] R_M-values, referred to malic acid = 100.
   [f] Polyamide layer prepared with formic acid according to [49]; NS.
   [g] Polyamide layer prepared without formic acid according to [48]; NS.

Table 143. hR_f-values of carboxylic acids on various layers (A-G) using various solvents[a]

Layer[a]	A. Silica gel				C.	D.	E.	G. Polyamide		
Solvent[a]	I[c]	II	IV	V	IX	X	XI	XIV	XV	XVI
Tartaric acid (DL)	8[d]	—	—	4	4	4	66[e]	4[f]	28[g]	15[g]
Cis-aconitic acid	—	—	—	—	—	—	—	4	—	—
Isocitric acid	—	—	—	—	—	—	—	4	—	—
Citric acid	5[d]	—	—	4	4	3	75	4	17	10
Tartronic acid	—	—	—	—	—	—	—	7	—	—
Oxalic acid	5	0	0	4	6	9	—	9	17	11
Malic acid (DL)	—	—	4	8	8	9	100	11	45	29
Tricarballylic acid	—	—	—	—	—	—	—	11	—	—
α-Ketoglutaric acid	—	—	10	—	—	—	127	15	—	—
Malonic acid	14	13	—	16	13	23	129	22	52	39
Succinic acid	30	28	9	44	20	31	145	33	67	56
Maleic acid, hy[b]	—	7	—	22	—	18	—	18	35	27
Fumaric acid, hy[b]	—	23	12	72	28	60	—	47	46	49
Pyruvic acid	—	—	3	—	24	—	—	—	—	—
Glutaric acid	39	35	—	46	27	45	163	50[g]	74	68
Adipic acid	43	42	16	49	32	51	177	57	75	73
Pimelic acid	53	47	—	54	—	61	—	72	76	79
Suberic acid	54	50	—	62	—	82	—	84	79	83
Azelaic acid	56	53	—	67	—	91	—	95	80	86
Sebacic acid	67	55	—	72	—	96	—	98	82	88
Lactic acid	—	—	35	—	46	—	161	—	—	—
Glycollic acid	—	—	22	—	32	—	—	—	—	—
β-Ketobutyric acid	—	—	45	—	—	—	—	—	—	—
Dehydroascorbic acid	—	—	46	—	—	—	—	—	—	—
Ascorbic acid	—	—	14	—	—	—	77	—	—	—
Laevulinic acid	—	—	52	—	60	—	—	—	—	—
Benzoic acid	76[d]	—	—	—	—	—	—	—	—	—
p-Toluic acid	76[d]	—	—	—	—	—	—	—	—	—
Methylsuccinic acid	—	—	—	55	—	52	—	57	69	63
Pyromellitic acid	—	—	—	0	—	2	—	2	0	0
Trimellitic acid	—	—	—	41	—	13	—	14	9	13
Phthalic acid	26[d]	—	—	51	—	30	—	39	41	36
Citraconic acid, hy[b]	—	—	—	34	—	31	—	36	48	39
Glutaconic acid, hy[b]	—	—	—	52	—	32	—	44	64	56
Itaconic acid, hy[b]	—	—	—	52	—	33	—	45	61	53
Endomethylene-tetra-hydrophthalic acid, hy[b]	—	—	—	41	—	48	—	60	68	65
Tetrahydrophthalic acid, hy[b]	—	—	—	60	—	54	—	68	65	63
Isophthalic acid	—	—	—	75	—	64	—	71	46	59
Endomethylene-hexa-hydrophthalic acid	—	—	—	60	—	65	—	71	65	66
Hexahydrophthalic acid	—	—	—	60	—	65	—	79	65	66
Terephthalic acid	73[d]	—	—	0	—	69	—	81	0	0
Mesaconic acid, hy[b]	—	—	—	78	—	85	—	82	56	62
Tetrachlorophthalic acid	—	—	—	69	—	88	—	80	9	23
Hexachloro-endomethylene-tetrahydrophthalic acid	—	—	—	82	—	96	—	98	30	51

GOEBELL and KLINGENBERG [34] have separated several acids of the tricarboxylic acid cycle, using two-dimensional TLC on cellulose layers and the solvents, ethanol-25% ammonium hydroxide-water (72.6 + 18.2 + 9.2) and isobutanol-5M formic acid (40 + 60). They concluded with quantitative evaluation by autoradiography. The combination of TLC with enzyme test methods permitted accurate quantitative determination of these acids in the nanomole region.

PREY and co-workers [74] chromatograph formic, lactic, acetic and pyruvic acids on silica gel G layers with pyridine-petrol ether (33.4 + 66.6) or ethanol-ammonium hydroxide-water (80 + 4 + 16). They recommend detection with dihydroindanthroazine disulphate (9,14-dihydroxy-5,18-anthrazinedione disulphate) [88].

Detection of the acids or their salts is carried out as a rule with a suitable pH-indicator solution; bromocresol green (Rgt. No. 25) has been often used. Chromatograms developed with acid solvents must be previously heated for 60 min at 120° C to remove all traces of (acetic) acid. After spraying, the acids appear as blue spots on a yellow background with detection limit between 0.8 and 8 µg. The colour intensity decreases with increasing chain length. Light red spots on a blue violet background are obtained with the mixed indicator methyl red-bromophenol blue. Exposure to ammonia vapour is sometimes helpful.

Note: The TLC of the fatty acids is treated in Chapter M, p. 363.

Bibliography for Chapter TN. Foodstuffs and Their Additives

1. ABBOTT, D. C., H. EGAN, and J. THOMSON: J. Chromatog. **16**, 481 (1964).
2. — — E. W. HAMMOND, and J. THOMSON: Analyst **89**, 480 (1964).
3. ACKER, L., H. GREWE u. H. O. BEUTLER: Dtsch. Lebensmitt. Rdsch. **59** 231 (1963).
4. BACHE, C. A.: J. Ass. Off. Agr. Chem. **47**, 355 (1964).
5. BANCHER, E., H. SCHERZ u. V. PREY: Mikrochim. Acta **1963**, 712.
6. BÄUMLER, J., u. S. RIPPSTEIN: Helv. chim. Acta **44**, 1162 (1961).
7. BECKER, E., u. M. EDER: Z. Lebensmitt. Unters. **104**, 187 (1956).
8. BEROZA, M.: J. Agric. Food Chem. **11**, 51 (1963).
9. BLINN, R. C.: J. Ass. Off. Agr. Chem. **46**, 952 (1963).
10. BRAITHWAITE, D. P.: Nature (Lond.) **200**, 1011 (1963).
11. BRAUN, D., u. H. GEENEN: J. Chromatog. **7**, 56 (1962).
12. BRUCHFIELD, H. P., and A. HARZELL: J. Econ. Entomol. **48**, 210 (1955).
13. BUNYAN, P. J.: Analyst **89**, 615 (1964).
14. CASSIDY, W., and A. J. FISHER: Analyst **85**, 295 (1960).
15. CASSIL, C. C.: In: F. A. GUNTHER: Residue Reviews, vol. 1, S. 37. Berlin-Göttingen-Heidelberg: Springer 1962.
16. COOK, J. W.: J. Ass. Off. Agr. Chem. **38**, 150 (1955).
17. COPIUS PEEREBOOM, J. W.: Chromatographic Sterol Analysis, as applied to the investigation of milk fat and other oils and fats. Wageningen: Pudoc 1963.
18. — J. Chromatog. **17**, 99 (1965).

19. COPIUS-PEEREBOOM: Nature (Lond.) **204**, 748 (1964).
20. — J. Chromatog. **14**, 417 (1964).
21. — unpublished results.
22. — J. Chromatog. **4**, 323 (1960).
23. DANCIS, J., J. HUTZLER u. M. LEVITZ: Biochim. biophys. Acta (Amst.) **78**, 85 (1963).
24. DAVÍDEK, J., u. J. POKORNÝ: Z. Lebensmitt. Unters. **115**, 113 (1961).
25. — J. Chromatog. **9**, 363 (1962).
26. DAVIDOW, B.: J. Ass. Off. Agr. Chem. **33**, 130 (1950).
27. DUMAZERT, CH., C. GHIGLIONE et T. PUGNET: Bull. Soc. chim. Fr. **1963**, 475.
28. EDER, F., H. SCHOCH u. R. MÜLLER: Mitt. Lebensmitt. Hyg. **55**, 98 (1964).
29. DE FAUBERT MAUNDER, M. J.: Analyst **89**, 168 (1964).
30. FISCHER, F., and H. KOCH: J. Chromatog. **16**, 246 (1964).
31. FISCHER, R., u. W. KLINGELHÖLLER: Arch. Toxikol. **19**, 119 (1961).
32. GÄNSHIRT, H., u. K. MORIANZ: Arch. Pharm. **293**, 1065 (1960).
33. GEE, M.: J. Chromatog. **9**, 278 (1962).
34. GOEBELL, H., u. M. KLINGENBERG: In: Chromatographie, Symposium II, Société Belge des Sciences Pharmaceutiques Bruxelles 1962, publ. 1963, S. 153.
35. GRAU, R., u. A. SCHWEIGER: Z. Lebensmitt. Unters. **119**, 213 (1963).
36. HAENNI, E. O., J. W. HOWARD, and F. L. JOE: J. Ass. Off. Agr. Chem. **45**, 67 (1962).
37. HENKEL, H. G., u. W. EBING: J. Chromatog. **14**, 283 (1962).
38. HROMATKA, O., u. W. A. AUE: Mh. Chem. **93**, 503 (1962).
39. JARCZYNSKI, R., u. F. KIERMEIER: Z. Lebensmitt. Unters. **99**, 91 (1954).
40. JOHNS, T., and C. H. BRAITHWAITE: In: F. A. GUNTHER: Residue Reviews, vol. 5, S. 45. Berlin-Göttingen-Heidelberg: Springer 1964.
41. JONAS, J.: J. Pharm. Belg. **17**, 103 (1962).
42. JONES, L. R., and J. A. RIDDICK: Anal. Chem. **24**, 569 (1952).
43. JOUX, J. L.: Ann. Fals. Fraudes **50**, 205 (1957).
44. KATZ, D., u. I. LEMPERT: J. Chromatog. **14**, 133 (1964).
44a. KAWASHIRO, I., and Y. HOSOGAI: J. Food Hyg. Soc. Japan **5**, 54 (1964).
45. KNAPPE, E., u. D. PETERI: Z. anal. Chem. **188**, 184 u. 352 (1962).
46. — — Z. anal. Chem. **190**, 380 (1962).
47. — — u. I. ROHDEWALD: Z. anal. Chem. **199**, 270 (1964).
48. — — Z. anal. Chem. **210**, 183 (1965).
49. — — Z. anal. Chem. **211**, 49 (1965).
50. KOVÁČ, J.: J. Chromatog. **11**, 412 (1963).
51. KOVACS, M. F.: J. Ass. Off. Agr. Chem. **46**, 884 (1963).
52. — J. Ass. Off. Agr. Chem. **47**, 1097 (1964).
53. KRÖLLER, E.: Fette, Seifen, Anstrichmittel **64**, 85 (1962).
54. — Fette, Seifen, Anstrichmittel **65**, 482 (1963).
55. KUČERA, J.: Coll. Czech. Chem. Commun. **28**, 1341 (1963).
56. LIBBY, L. M., and E. A. DAY: J. Dairy Sci. **46**, 859 (1963).
57. LINOW, F., H. RUTTLOFF u. K. TÄUFEL: Naturwissenschaften **50**, 689 (1963).
58. LÜCK, E., u. W. COURTIAL: Dtsch. Lebensmitt.-Rdsch. **61**, 78 (1965).
59. MCKINLEY, W. P., D. E. COFFIN, and K. A. MCCULLY: J. Ass. Off. Agr. Chem. **47**, 863 (1964).
60. MEYER, H.: Dtsch. Lebensmitt. Rdsch. **57**, 170 (1961).
61. — Süßwaren **6**, 645 (1962), und Revue internationale de la chocolaterie **17**, 290 (1962).

62. MILLS, P. A., J. H. ONLEY, and R. A. GAITHER: J. Ass. Off. Agr. Chem. **46**, 186 (1963).
63. MIMA, H., and N. KITAMORI: J. Am. Oil Chem. Soc. **39**, 546 (1962).
64. MITCHELL, L. C.: J. Ass. Off. Agr. Chem. **40**, 294 (1957); **41**, 781 (1958); **42**, 684 (1959); **43**, 810 (1960); **44**, 643 (1961).
65. MOATS, W. A.: J. Ass. Off. Agr. Chem. **46**, 172 (1963).
66. MORRIS, W. W., and E. O. HAENNI: J. Ass. Off. Agr. Chem. **46**, 964 (1963).
67. ONLEY, J. H.: J. Ass. Off. Agr. Chem. **47**, 317 (1964).
68. PASSERA, C., A. PEDROTTI, and G. FERRARI: J. Chromatog. **14**, 289 1(964),
69. PETROWITZ, H. J.: Chem. Ztg. **85**, 867 (1961).
70. — Chem. Ztg. **86**, 815 (1962).
71. — and G. PASTUSKA: J. Chromatog. **7**, 128 (1962).
72. PETSCHIK, H., and E. STEGER: J. Chromatog. **9**, 307 (1962).
73. PHILIPS, M. A., and R. D. HINKEL: J. Agric. Food Chem. **5**, 379 (1957).
74. PREY, V., H. BERBALK u. M. KAUSZ: Mikrochim. Acta **1962**, 449.
75. RINK, M., and S. HERRMANN: J. Chromatog. **14**, 523 (1964).
76. RONKAINEN, P.: J. Chromatog. **11**, 228 (1963).
77. SAHASRABUDHE, M. R.: J. Ass. Off. Chem. **47**, 888 (1964).
78. SALAMÉ, M.: J. Chromatog. **16**, 476 (1964).
79. SALO, T., K. SALMINEN u. K. FISKARI: Z. Lebensmitt. Unters. **117**, 369 (1962).
80. — — Z. Lebensmitt. Unters. **124**, 448 (1964).
81. — u. R. MÄKINEN: Z. Lebensmitt. Unters. **125**, 170 (1964).
82. — E. AIRO u. K. SALMINEN: Z. Lebensmitt. Unters. **125**, 20 (1964).
83. — R. MÄKINEN u. K. SALMINEN: Z. Lebensmitt. Unters. **125**, 450 (1964).
84. SEHER, A.: Fette, Seifen, Anstrichmittel **61**, 345 (1959).
85. — Nahrung **4**, 466 (1960).
86. — Fette, Seifen, Anstrichmittel **66**, 371 (1964).
87. SCHILDKNECHT, E., u. H. KÖNIG: Z. anal. Chem. **207**, 269 (1965).
88. SCHLÖGL, K.: Naturwissenschaften **46**, 447 (1959).
89. SCHNEIDER, E.: Lebensm. chem. gerichtl. Chem. **17**, 172 (1963).
90. SCHWEIGER, A.: J. Chromatog. **9**, 374 (1962).
91. — Z. Lebensmitt. Unters. **124**, 20 (1963).
92. STAHL, E., u. J. PFEIFLE: Naturwissenschaften **52**, 620 (1965).
93. — Arch. Pharm. **293**, 531 (1960).
94. — and U. KALTENBACH: J. Chromatog. **5**, 351 (1961).
95. STANLEY, C. W.: J. Chromatog. **16**, 467 (1964).
96. STORHERR, R. W.: J. Ass. Off. Agr. Chem. **47**, 1087 (1964).
97. STROHECKER, R.: Fette, Seifen, Anstrichmittel **66**, 787 (1964).
98. TAYLOR, A., and B. FISHWICK: Lab. Practice **13**, 525 (1964).
99. TER HEIDE, R.: Fette, Seifen, Anstrichmittel **60**, 360 (1958).
100. THORNBURG, W.: In: G. ZWEIG: Pesticides, Planth growth regulators and Food additives, vol. I, p. 87. New York: Academic Press 1963.
101. WALDI, D.: unpublished results, in: STAHL, Thin-Layer Chromatography, 1. Engl. Edition, p. 357—365. Berlin-Göttingen-Heidelberg: Springer 1965.
102. WALKER, K. C., and M. BEROZA: J. Ass. Off. Agr. Chem. **46**, 250 (1963).
103. WILLIAMS, E. I.: Analyst **89**, 289 (1964).
104. WOGGON, H., D. SPRANGER u. H. ACKERMANN: Nahrung **7**, 612 (1963).
105. WRIGHT, J.: Chem. and Ind. (London) **1963**, 1125.
106. YAMAMURA, J., and T. NIWAGUCHI: Proc. Japan Academy **38**, 129 (1962).

TS. Synthetic Organic Products
H.-J. Petrowitz

I. Plastics and Plasticisers
1. Polymers and Polymerisable Compounds

The analysis of plastics comprises the identification of the monomers which participate in polymer formation and of products of degradation reactions [49]; and also the detection of plastic additives like plasticisers, stabilisers and catalysts. Any simple alcohols, aldehydes, ketones or fatty acids occurring amongst these may be detected with the TLC procedures described elsewhere in the relevant chapters of this book.

Braun and Vorendohre [7] have carried out TLC of unsaturated polymerisable compounds which are found also in the pyrolysis products of high polymers. They chromatographed the addition compounds obtained by reacting 15—20 mg of the monomers with 120 mg mercuric acetate in methanol. Excess mercuric acetate can be reduced with hydrazine without the adducts' suffering attack (those of α-methylstyrene, vinyl acetate, vinylcarbazole and vinyl chloride are unstable

Table 144. hRf-values of polymerisable compounds

Compound	hRf	Compound	hRf
Styrene	43—56	Dicyclopentadiene	58—68
α-Methylstyrene [a]	47—58	Isoprene	35—43
2,4-Dimethylstyrene	55—65	Ethylene	7—10
p-Chlorostyrene	49—56	Acrylic acid	0—5
Ethylstyrene	54—63	Methacrylic acid	8—12
Divinylstyrene	53—59	Methyl acrylate [b]	—
Vinylcarbazole [a]	66—73	Methyl methacrylate	27—35
Vinyl acetate [a]	17—24	Butyl methacrylate	54—64
Vinyl chloride	18—20	Acrylonitrile	44—51
Cyclohexene	24—32		

Layer: silica gel G; solvent: Butanone-n-propanol-ethanol-conc. ammonium hydroxide (45.5 + 4.5 + 18 + 32) [7].

[a] Unstable to hydrazine sulphate.
[b] Marked tailing.

exceptions). Separation was on silica gel G (Firm 88) layers, using butanone-n-propanol-ethanol-conc. ammonium hydroxide (45.5 + 4.5 + 18 + 32). Visualisation was carried out after prior removal of all traces of solvent from the layer, by exposing it for 5—10 min to hydrogen chloride and then spraying with a 0.1% solution of dithizone in carbon tetrachloride; red or yellow spots were thus formed.

The novolaks, formed by acid condensation of phenols with formaldehyde, are closely related to the plastics. HAUB and KÄMMERER [30] have fractionated condensation products of p-cresol and formaldehyde by TLC on silica gel G. They visualised the compounds through spraying with antimony (V) chloride reagent (No. 18) and subsequently heating to 100°C. The solvents quoted in Table 145 and also chloroform-methanol-water (95 + 4 + 1-emulsion) were used for separating the phenol dialcohols; the hR_f-values found with it for n = 0, 1 and 2 (Table 145) were 13, 22 and 32 respectively. The homologous series of the compound class in the upper half of Table 145 could be separated completely by two-dimensional chromatography.

Table 145. hR_f-values of some p-cresol-formaldehyde condensation products (see [30] for others) on silica gel layers, using solvents: I: Benzene-methanol-acetic acid (95 + 2,5 + 2,5); II: Benzene-methanol (75 + 25); III: Chloroform-methanol (96 + 4).

Structural formula	n	hR_f-values with		
		I	II	III
H—[OH-C₆H₂(CH₃)-CH₂-]ₙ—OH-C₆H₂(CH₃)—H H-n-H	0	35	61	45
	1	30	65	52
	2	32	69	65
	3	41	—	74
	4	50	—	78
	5	61	—	—
	6	72	—	—
Cl—[OH-C₆H₂(CH₃)-CH₂-]ₙ—OH-C₆H₂(CH₃)—Cl Cl-n-Cl	0	61	67	65
	1	59	69	67
	2	59	73	71
	3	60	—	75
	4	65	—	79
	5	71	—	—
	6	76	—	—
	7	82	—	—

1a. Urethanes

Cyclic, homologous oligo- and polyurethanes can be satisfactorily analysed by TLC. KUNTZ [48] has been able to separate the oligomers of molecular weight between 40 and 4220 on silica gel G layers, using a solvent containing cyclohexanone as principal component. KERN and co-workers [36] have employed TLC for purity testing of cyclodiurethanes which had been synthesised by the high dilution principle of RUGGLI and ZIEGLER. They, too, used silica gel G layers and their solvent was benzene-ethyl acetate-cyclohexanone (62.5 + 12.5 + 25). After the solvent had been expelled by heating 2 h at 120°C, the plate was sprayed with a saturated solution of chlorine in carbon tetrachloride. Chlorine adsorbed by the layer was removed by leaving 1 h in the fume cupboard and the layer was then sprayed with iodide-starch solution. The limit of detection of the oligo-urethanes was ca. 0.1 µg.

2. Plasticisers

Plastic products are prepared by addition of plasticisers. Esters of phthalic, of phosphoric and of aliphatic dicarboxylic acids, epoxides and some high molecular weight compounds are especially suitable. Since some of these substances are highly toxic, a sensitive chemical method of detection was desirable. GC analysis is rendered difficult by the high boiling points of the plasticisers but identification with the help of TLC is most successful. Plasticisers authorised in the USA for use in plastic package

Table 146. R_{St}-values of plasticisers in various solvents, referred to dibutyl sebacate $= 100$ [12]

No.	Compound	hRf[a]		
		I	II	III
1	Triacetin (glycerol triacetate)	18	34	17
2	Ethylphthalyl ethyl glycollate	22	66	30
3	Triethyl acetylcitrate	26	51	29
4	Triphenyl phosphate	33	80	50
5	Tricresyl phosphate	42	86	69
6	Butylphthalyl butyl glycollate	43	90	65
7	2-Ethylhexyl diphenyl phosphate	46	77	58
8	Diethyl phthalate	51	79	60
9	Tributyl acetylcitrate	53	85	70
10	Di-n-butyl phthalate	74	103	84
11	Diisobutyl adipate	83	86	85
12	Dibutyl sebacate	100	100	100
13	Dinonyl phthalate	101	118	114
14	Di(2-ethylhexyl) phosphate	114	116	115
15	Butyl stearate	161	123	128
16	Paraflex G 62 (epoxides of naturally occurring glycerides)	5 spots	9 spots	4 spots

Solvents: I Isooctane-ethyl acetate $(90 + 10)$; II Benzene-ethyl acetate $(95 + 5)$; III Dibutyl ether-hexane $(80 + 20)$.

[a] Layer: Silica gel G.

materials for foodstuffs, have been analysed by Copius-Peereboom [12] on silica gel G layers containing 0.005% of the water-soluble fluorescence indicator Ultraphor-WT[1]. 10 µl amounts of approximately 5% solutions in ether of each plasticiser or equivalent extraction product

[1] Optical bleach of Firm 16.

were applied and developed with the solvents given in Table 146. Various colour reactions are suitable for visualising plasticisers; in UV light (365 nm), some fluoresce and the phthalate derivatives may be detected as dark spots of quenched fluorescence. The three critical pairs 5—6, 5—7 and 7—9 (Table 146), which cannot be separated in the solvents quoted, may be identified through colour reactions.

Also using silica gel G layers, BRAUN [4, 5] has separated numerous plasticisers with methylene dichloride, after they had been extracted from the plastic material with benzene or ether (provided the polymer itself was not soluble). Antimony(V)chloride (Rgt. No. 18) is a generally applicable spray reagent; it yields brown spots with most of the plasticisers after the plate is heated to 120° C. Phthalate esters can be detected in addition with resorcinol solution (Rgt. No. 218) and phosphate esters with a diazonium salt reagent (No. 238).

As can be seen from the hR_f-values, plasticisers of low molecular weight are more easily separated than esters of longer-chain alcohols.

Ester plasticisers can be directly analysed or acid and alcohol (or phenol) products of saponification may be detected by TLC.

Table 147. hR_f-values of important plasticisers on silica gel G; solvent: methylene dichloride [5]

Compound	hR_f	Compound	hR_f
Dimethyl phthalate	51	Di-(2-ethylhexyl) adipate	44
Dibutyl phthalate	69	Dinonyl adipate	44
Dihexyl phthalate	80	Adipic acid polyester	2
Dioctyl phthalate	86	Dibutyl sebacate	41
Di-(2-ethylhexyl) phthalate	85	Dioctyl sebacate	61
Didecyl phthalate	85	Di-(2-ethylhexyl) sebacate	61
Diisodecyl phthalate	84	Sebacic acid polyester	2
Trioctyl phosphate	23	Triethyl citrate	12
Diphenyl octyl phosphate	42	Tributyl citrate	14
Triphenyl phosphate	47	Triethyl acetylcitrate	15
Diphenyl cresyl phosphate	51	Tributyl acetylcitrate	24
Tricresyl phosphate	53	Tri(2-ethylhexyl) acetylcitrate	46
Dioctyl adipate	42	Glycerol triacetate	13

3. Alcohols

a) Simple Alcohols

Volatile, simple alcohols may be submitted to TLC only in the form of their 3,5-dinitrobenzoate esters (DNB-esters). These can be satisfactorily separated on silica gel G layers, using cyclohexane-carbon tetrachloride-ethyl acetate (10 + 75 + 15) under standard conditions (Fig. 184). Detection is carried out by spraying the layer with rhodamine

B reagent and then inspecting in UV light. The DNB-esters are conveniently prepared by the procedure worked out by WALDI [89]:

If, as often the case, the alcohols occur in aqueous solution, they are first extracted with alcohol-free ether[2]. The extract is dried over freshly ignited sodium sulphate, 3,5-dinitrobenzoyl chloride added and refluxed for 30 min. Unreacted acid chloride is then removed by adding water and 5—10% sodium hydroxide is added to adjust the pH to 9—10. The ether layer, containing the DNB-ester, is separated in a separating funnel and the aqueous layer extracted three to four times with alcohol-free benzene[2]. The combined ether and benzene extracts are dried over sodium sulphate and evaporated to dryness. The residue is taken up in a small, accurately measured amount of benzene; this solution can be chromatographed directly.

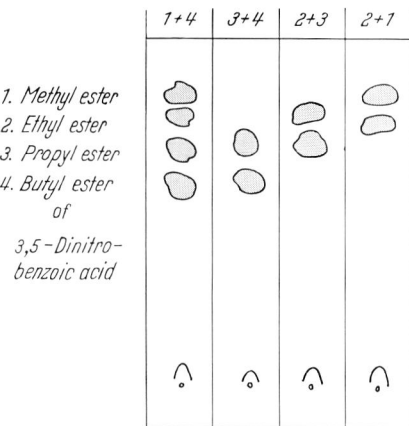

Fig. 184. DNB-esters of lower alcohols, developed on a silica gel G layer with cyclohexane-carbon tetrachloride-ethyl acetate (10 + 75 + 15). Detection: rhodamine B (Rgt. No. 220) [89]

For the analysis of plasticisers, BRAUN [5] has recommended transesterification of the esters with 3,5-dinitrobenzoic acid, so as to obviate saponification and isolation of the alcohol.

Procedure: 2 ml of the plasticiser under investigation are mixed with 2 drops concentrated sulphuric acid and 1.5 g 3,5-dinitrobenzoic acid and heated 30 min on an oil bath at 150° C. After cooling, the mass is dissolved in 25 ml ether and the resultant solution washed with 25 ml 5% sodium carbonate and then with water. The ether is evaporated off, whereby the 3,5-dinitrobenzoate sometimes crystallises out and can be recrystallised if necessary. Otherwise the dark, oily residue after evaporation of the ether can be dissolved in a little ether and chromatographed directly.

Table 148 contains hR_f-values obtained with the solvent benzene-methyl acetate (150 + 1).

---

[2] Commercially available ether or benzene are freed from alcohols by refluxing with excess 3,5-dinitrobenzoyl chloride for about 60 min and then distilling.

Table 148. hRf-values of 3,5-dinitrobenzoate esters on silica gel G, using benzene-methyl acetate solvent (150 + 1) [5]

3,5-DNB Ester	hRf	3,5-DNB Ester	hRf
Methyl-	40	2-Ethylhexyl-	81
Ethyl-	49	n-Decyl-	81
Butyl-	65	Cyclohexyl-	68
n-Hexyl-	73	Benzyl-	63
n-Octyl	80	Glyceryl-	5

b) Polyalcohols

Glycerol, ethylene glycol and propane-1,2-diol can be separated on air-dried silica gel G layers, using chloroform-acetone-5N ammonium hydroxide solution (10 + 80 + 10) in a 10 run at chamber saturation [89]; the hRf-values are 35, 70 and 85, respectively. The benzidine-periodate reagent (Nos. 158 and 159) is used for detection.

PREY and co-workers [70] have separated glycerol (hRf 38) from ethylene glycol (hRf 45) on normal silica gel G layers, using butanol-water (90 + 10); the separation may be improved by impregnation with 0.1N boric acid. They used dichromate-sulphuric acid (Rgt. No. 46) for detection, obtaining white spots on a yellow-brown background.

HROMATKA and AUE [34] have established for diols a linear relation between the log of the Rf-value and the number of carbon atoms when using absolute ethanol or dioxan on silica gel G layers; they studied ethylene glycol, propane-1,3-diol, hexane-1,6-diol, heptane-1,7-diol, nonane-1,9-diol, decane-1,10-diol and tridecane-1,13-diol.

KNAPPE et al. [40] separate technically important polyalcohols on alumina G layers, using chloroform-toluene-formic acid (80 + 17 + 3); on silica gel G layers with n-butanol, saturated with 1.5N ammonium hydroxide; or on kieselguhr-polyamide layers with chloroform. For detection, they recommend a number of spray reagents made up of a strong oxidising agent and an oxidisable base.

Polyalcohols can be separated on cellulose-gypsum layers (100 + 6) also, for which suitable solvents are n-butanol-25% ammonium hydroxide-water (85 + 5 + 10), n-butanol-pyridine-water (46 + 31 + 23) or n-butanol-ethanol-25% ammonium hydroxide-water (40 + 15 + 5 + 40) [20].

ULLMANN and co-workers [95] have worked extensively on the TLC of polyethylene glycols. They showed that these compounds of molecular weight between 200 and 6000 can be separated on silica gel G layers with the solvent chloroform-methanol-water (6 + 50 + 24). In later work [96], the same authors showed that the surface-active polyethylene glycol derivatives (esters and ethers) could also be separated on silica gel G layers, using n-butanol-ethanol-25% ammonium hydroxide (70 + 15 + 25). In order to separate a mixture of different polyethylene glycol stearates, about 400 μg are applied to the start and two-dimensional TLC is carried out; in direction I with n-butanol-ethanol-25% ammonium hydroxide (70 + 15 + 25) and in direction II with chloroform-methanol-water (30 + 50 + 24). Alternatively in direction II, butanone saturated with water may be used with continuous development in the BN-chamber. *Visualisation* can be with the Dragendorff reagent; for larger amounts, the layer is saturated by spraying with 0.005N iodine solution, dried and then sprayed with 0.2% starch solution. This detection is, however, not very sensitive.

4. Phenols

BRAUN and VOHRENDOHRE [8] describe the analysis of phenolic components of plasticisers. It is carried out after prior saponification by boiling for 3 hours with 1N alcoholic potassium hydroxide solution. The salts formed during the hydrolysis are best removed before chromatography. Silica gel G layers, impregnated with formamide[3], and methylene dichloride-cyclohexane (55 + 45) are used. PETROWITZ [64] has used benzene and benzene-methanol (95 + 5) in TLC on normal silica gel G layers.

Table 149. *hRf-values of some phenols*

Compound	hRf-values[a]			Compound	hRf-values[a]		
	I	II	III		I	II	III
Phenol	21	17	43	2,4-Dimethylphenol	76	24	51
o-Cresol	51	26	52	2,5-Dimethylphenol	75	28	58
m-Cresol	39	19	45	2,6-Dimethylphenol	93	39	67
p-Cresol	40	20	49	3,4-Dimethylphenol	56	15	44
2,3-Dimethylphenol	68	27	59	3,5-Dimethylphenol	61	17	44

[a] I on silica gel G, impregnated with formamide, using the solvent methylene dichloride-cyclohexane (55 + 45) [8]; II on silica gel G using benzene [64]; III on silica gel G using benzene-methanol (95 + 5) [64].

As seen from the *Rf*-values, the adsorption affinity of the phenols depends on the position of the methyl group(s) in the nucleus.

KNAPPE and ROHDEWALD [41 a] have shown that an especially good separation of the red-brown coupling products of phenols with Fast Red AL Salt is possible. Phenols and the isomeric cresols and xylenols could be separated from one another on basic or acidic silica

---

[3] The cold plates are impregnated by immersing them vertically for a few seconds in acetone-formamide (2 + 1), laying them at once horizontally, removing a margin 0.5 cm broad on both sides and drying in a current of warm air.

gel G layers (prepared with 0.5N potassium carbonate or 0.5 N oxalic acid, respectively). Solvents were methylene dichloride-ethyl acetate-diethylamine (92 + 5 + 3) and chloroform-ethyl acetate-diethylamine (93 + 5 + 2) at chamber saturation and pure benzene at normal saturation.

Preparation of the azo dyes: 10 mg phenol are dissolved in 1—2 ml 0.1 N sodium hydroxide; 150 mg Fast Red AL Salt (Firm 56) are then added, followed by 20 ml water. After 30 min, the reaction mixture is acidified with a little 2N hydrochloric acid and extracted with 20 ml chloroform. The extract is washed with water and can then be used directly in TLC.

SMITH and SULLIVAN [78a] separate the same groups of compounds after coupling with diazotised p-nitroaniline. They use kieselguhr G layers, impregnated with formamide.

5. Other Auxiliary Materials in the Plastics Industry

Peroxides of different types are important as polymerisation initiators in the manufacture of plastics. KNAPPE and PETERI [39] have separated various technically important organic peroxides on silica gel G layers, using toluene-carbon tetrachloride (67 + 33) and toluene-acetic acid (95 + 5).

Substituted 2-hydroxybenzophenones, used as UV-light absorbers in the plastics and varnish industries, have been separated by KNAPPE and co-workers [41] on layers impregnated with adipic acid triethylene glycol polyester, using the solvent m-xylene-formic acid (98 + 2). The substances show characteristic fluorescence colours and can be visualised in addition as brick-red spots (only salol gives violet) by spraying with Fast Red AL Salt solution.

II. Organo-Metallic Compounds
1. Organo-Tin Compounds

The organo-tin compounds are associated closely with the plastics; they are especially important as stabilisers in the PVC (polyvinyl chloride) industry. They may be satisfactorily separated by TLC on silica gel layers [9, 31, 56, 87]; solvents so far used for development are:

    I. n-Butanol-acetic acid (98.3 + 1.7) [87].
   II. Water-n-butanol-ethanol-acetic acid (48 + 24.5 + 24.5 + 2.5) [87].
  III. Isopropanol-10% ammonium carbonate solution-5N ammonium hydroxide (67 + 22 + 11) [9].
  IV. n-Butanol-2.5% ammonium hydroxide (80 + 20) [9].
   V. Isopropyl ether-acetic acid (98.5 + 1.5) [56].
  VI. Hexane-acetic acid (92 + 8) [31].

The spots are visualised by spraying the layers with a solution of dithizone in chloroform or by exposure to UV-radiation for 30 min and subsequently spraying with catechol violet solution. Organo-tin derivatives of the four alkylation stages can be separated and amounts down to 1 µg can be detected. The toxicologically important distinction

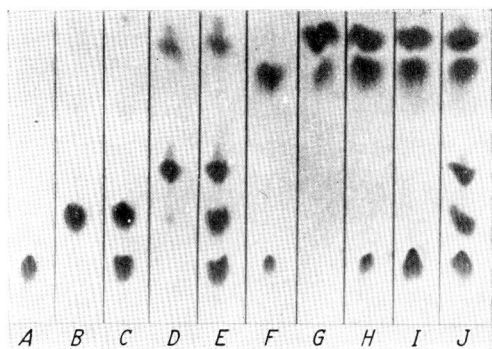

Fig. 185. TLC of various organo-tin compounds [56]. Layer: silica gel G; solvent: isopropyl ether containing 1.5% acetic acid.

A: Di-n-butyl-tin dichloride
B: Di-n-octyl-tin dichloride
C: Separation of $A + B$
D: Di-isooctyl-tin dichloride
E: Separation of $A + B + D$
F: Tri-n-butyl-tin chloride
G: Tetrabutyl-tin
H: Separation of $F + G$
I: Separation of $A + F + G$
J: Separation of all compounds

between dibutyl- and dioctyl-tin and the identification of sulphur-containing compounds are possible; according to NEUBERT [56], the chromatographic behaviour of the last-named is determined by the sulphur-containing acidic group and no longer by the alkyl group linked to tin. BÜRGER [9] has shown that phenyl-tin compounds suffer decomposition in UV light; the phenol thus formed can serve as the basis of a rapid detection method.

Other stabilisers encountered in PVC, such as 2-phenylindole or urea derivatives, may be separated on silica gel layers, using chloroform [32, 43].

2. Ferrocenes

SCHLÖGL and co-workers [80] have employed TLC on both analytical and preparative scale in their extensive work on syntheses and properties of ferrocene derivatives. It has been used in investigations like: following chemical changes, such as oxidations [81]; checking the purity of products of synthesis; or separating isomers [21, 84]. TLC has further enabled compound classes such as the alkyl ferrocenes to be separated from acyl ferrocenes [82] and correlations between molecular structure and

R_f-values to be demonstrated. Thus, e. g., the R_f-values of the polyethylferrocenes decrease as the number of ethyl groups increases [83]; and the R_f-values of heterocyclic ferrocene derivatives increase linearly in the order β-pyridyl-, α-pyridyl-, quinolyl-[84].

TLC has been carried out on silica gel G layers, using the solvents I: benzene; II: benzene-ethanol (90 + 3); III: benzene-ethanol (90 + 6); IV: propane-1,2-diol-methanol (50 + 50); and V: chlorobenzene-propane-1,2-diol-methanol (33 + 33 + 33). In this way, ferrocenyl-methanol, ferrocene ketones, esters, ethers and nitrogen-containing derivatives can be separated, whereas ferrocene and ferrocenyl-hydrocarbons migrate with the front; n-hexane is suitable for separating these last named.

Special procedures for visualising ferrocene derivatives are unnecessary since they are coloured. Alkylferrocenes and derivatives without chromophore groups in conjugation with the ferrocene nucleus are yellow; monoacylferrocenes are orange; diacyl derivatives, red; and decidedly conjugated systems, red-violet. Less intensively coloured ferrocene derivatives may be oxidised to blue or blue-green ferrocinium derivatives. The lower limit of detection of most ferrocence derivatives is 2–3 µg.

3. Other Organo-Metallic Compounds

According to VOBECKY et al. [88], the triphenyl derivatives of antimony, bismuth, phosphorus and arsenic and also ditolyl-tellurium isomers can be separated on alumina layers using petrol ether (hR_f-values of the first two, 69 and 62 respectively). They may be visualised by spraying with permanganate solution.

III. Polyphenyls and Fused Polynuclear Aromatic Hydrocarbons

1. Polyphenyls

Mixtures of linear and branched polyphenyls are important refrigerating agents for reactors. GEISS and co-workers [25, 26] report the TLC-separation of the complex mixtures derived from biphenyl and the three terphenyls through radiolysis and pyrolysis during operation of a reactor. Alumina G proved to be the best adsorbent and n-heptane the best solvent. Separation of the individual compounds depends markedly on the relative humidity prevailing during pre-treatment and development of the chromatogram; the development chamber must thus be air-conditioned. Separation is improved by "hot elution"; chromatography is then carried out on still warm layers which are more active and, further, bring about increased solvent evaporation. The polyphenyls are visualised by spraying with a 0.5% solution of ceric sulphate in concentrated nitric acid (Rgt. No. 35) which yields spots of various colours, strongly fluorescent. Quantitative spectrophotometric determination of the separated spots is possible following elution with n-heptane or methylene dichloride.

2. Polynuclear Aromatic Hydrocarbons

Large numbers of polynuclear aromatic compounds occur in the tar from hard coal distillation, in products of pyrolysis and in a variety of residues from combustion. These compounds have been separated by TLC on silica gel [46, 47], alumina [46, 59] and acetyl-cellulose [1, 78].

Table 150. hR_f- and R_{St}-values of polyphenyl compounds, arranged in descending order

No.	Compound	hR_f	R_{St}[a]
1	Biphenyl	75	1.21
2	o-Terphenyl	62	1.00
3	o',o''-Quaterphenyl	56	0.90
4	m-Terphenyl	54	0.87
5	o',m''-Quaterphenyl	46	0.74
6	p-Terphenyl	45	0.73
7	1,2,3-Triphenylbenzene	40	0.65
8	1,2,4-Triphenylbenzene	38	0.61
9	o',p''-Quaterphenyl	38	0.61
10	1,3,5-Triphenylbenzene	37	0.60
11	Triphenylene	36	0.58
12	m',m''-Quaterphenyl	34	0.55
13	m',p''-Quaterphenyl	27	0.44
14	p',p''-Quaterphenyl	16	0.26

Layer: Alumina G; *Solvent:* n-Heptane; *Temperature:* 23—24° C; *Relative humidity:* 40—50% [75].

[a] Referred to o-terphenyl.

Further improvements in separation have been achieved using two-dimensional TLC on impregnated layers (BERG and LAM [3]); on mixed layers (KÖHLER et al. [42]); and using multiple development (PETROWITZ [67, 68]). Suitable solvents for use with silica gel and alumina layers are hexane, heptane or carbon tetrachloride, also with small additions of a polar solvent, as in the mixtures listed in Table 152.

The substances are detected by inspection of the layer in UV light or by spraying with reagents such as antimony(V)chloride (No. 18), tetracyanoethylene (No. 243) or formaldehyde-sulphuric acid (No. 125).

As both PETROWITZ [68] and MATSUSHITA and co-workers [52] have been able to show, a linear relation exists for polynuclear aromatic

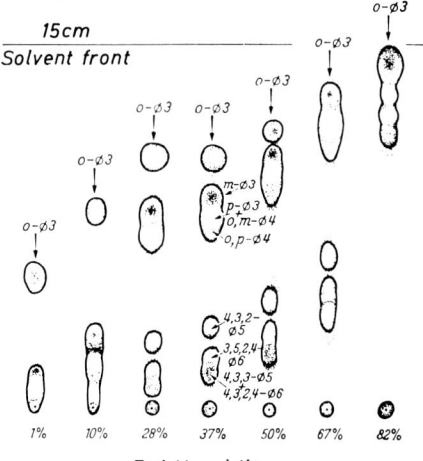

Fig. 186. Influence of relative humidity in TLC of a polyphenyl mixture on alumina G (Merck), using n-heptane. o- ⌀3, m- ⌀3, p- ⌀3 = o-, m-, p-terphenyl; o,m- ⌀4 and o,p- ⌀4 = o,m- or o,p-quaterphenyl respectively; 4,3,2- ⌀5 and 4,3,3- ⌀5 = p,m,o- or p,m,m-quinquaphenyl respectively; 3,5,2,4- ⌀6 = m,m,o,p-hexaphenyl; 4,3,2,4- ⌀6 = p,m,o,p-hexaphenyl

Table 151. hRf-values of polynuclear aromatic hydrocarbons

Compound	Silica gel			Alumina		
	I	II	III	II[a]	III[b]	II[b]
Indene	40	55	84	67	81	67
Naphthalene	36	59	77	63	85	63
1-Methylnaphthalene	32	54	74	55	78	64
2-Methylnaphthalene	—	50	87	52	85	64
2,3-Dimethylnaphthalene	28	48	75	39	81	—
2,6-Dimethylnaphthalene	31	50	72	52	85	—
2,7-Dimethylnaphthalene	30	53	—	42	80	—
Biphenyl	28	45	74	58	79	59
Acenaphthene	32	44	83	44	70	56
Fluorene	24	32	66	35	73	42
Anthracene	27	37	65	35	59	—
Phenanthrene	24	33	65	20	66	33
Pyrene	23	32	64	10	65	26
Fluoranthene	21	29	66	10	55	—
Chrysene	16	0	11	0	0	11

Solvents: I = Heptane (Silica gel G, Merck) [67]: II = Hexane [46, 59]; III: Carbon tetrachloride [46].

[a] Neutral.
[b] Activity I—II.

compounds, between the Rf-values and the size of the molecule, expressed as the log of the molecular weight or as the number of carbon atoms in the molecule. This dependence is found also with the methylnaphthalenes, provided the methyl C-atoms are included in the sum of aromatic carbon atoms. According to the data of KUCHARCZYK and FOHL [47], the Rf-values depend linearly on the π-electron energy for linear and angular fused hydrocarbons.

SAWICKI and co-workers [78] have separated in addition to the compounds in Table 152, a series of aza compounds of structures resembling those of polynuclear aromatic compounds; they used dimethylformamide-water (35 + 65) and ethanol-water (30 + 70) on cellulose layers and pentane-ether (95 + 5) on alumina layers.

Table 152. R_{St}-values of polynuclear aromatic hydrocarbons [77]

Compound	R_{St}-value under conditions[a]:		
	IV	V	VI
Phenanthrene	1.99	3.74	1.13
Anthracene	1.99	3.33	1.14
Fluoranthene	1.89	2.92	1.09
Chrysene	1.75	—	1.10
Pyrene	1.72	3.16	1.25
Triphenylene	1.49	—	1.07
Benz(a)anthracene	1.47	2.70	1.03
11H-Benzo(b)fluorene	1.33	3.54	1.08
Benzo(e)pyrene	1.16	2.94	1.04
Perylene	1.14	2.86	0.91
Benzo(k)fluoranthene	1.03	2.40	0.98
Benzo(a)pyrene	1.00	1.00	1.00
Anthanthrene	0.70	2.17	0.71
Benzo(ghi)perylene	0.69	3.04	0.89
Dibenz(a,h)anthracene	0.66	2.92	0.74
Naphtho(1,2,3,4-def)-chrysene	0.48	1.85	0.78
Benzo(rst)pentaphene	0.45	2.41	0.68
Coronene	0.37	2.87	0.46
Benzo(a)coronene	0.15	2.48	0.10
Dibenzo(h,rst)pentaphene	0.14	2.35	0.12

[a] R_{St} = Benzo(a)pyrene.
IV = Dimethylformamide-water (50 + 50); cellulose layer. V = Ethanol-toluene-water (68 + 16 + 16); acetylated cellulose layer. VI = Pentane-ether (95 + 5); alumina layer.

IV. Explosives

HARTHON [28] was the first to apply TLC to problems of control of the nitration of hexamethylenetetramine to yield hexogen (RDX) (Bachmann synthesis) and of detection of the side products which are formed. 30 µg amounts in methanolic solution of each of the nitroamines in Table 153 were applied to layers of silica gel G (Firm 88); petrol ether (BP 40—60°C)-acetone (62.5 + 37.5) was used for development. The nitroamines were visualised by spraying with the fluorescence indicator solution (Rgt. No. 121), the spots then appearing in UV light as dark areas of absorption and consequent quenching. It would appear better to use layers which already contain a fluorescent indicator, e. g., silica gel GF_{254}. Very small amounts (0.5 μg) can be detected as blue-violet spots on a weakly brown background by spraying the layers with a 1% solution of diphenylamine in ethanol and then exposing them to a 125 watt high-pressure mercury lamp.

Table 153. *hRf-values of nitroamine explosives*

Explosive	hRf-value
1-Acetyloctahydro-3,5,7-trinitro-1,3,5,7-s-tetrazine (SEX)	16
Octahydro-1,3,5,7-tetranitro-s-tetrazine (Octogen, HMX)	48
3,7-Dinitro-1,3,5,7-tetrazabicyclo-3,3,1-nonane (DPT)	62
Hexahydro-1,3,5-trinitro-s-triazine (Hexogen, RDX)	71
2,4,6-Trinitro-2,4,6-triazaheptane-1,7-diol acetate (BSX)	93
2,4,6,8-Tetranitro-2,4,6,8-tetrazanonane-1,9-diol acetate (AcAn)	93

Layer: Silica gel G; *Solvent:* Petrol ether-acetone (20 + 12) [28].

Using silica gel G (Firm 88) which had been activated at 110° C, HANSSON [27] has separated some of the most important explosives belonging to various compound classes. Amounts of 1.5—4 μl of a solution of 0.01 g explosive in 1 ml acetone were applied (explosives containing ammonium salts which do not dissolve in the acetone, are best first extracted with water). Benzene, chloroform or a mixture of petrol ether (BP 30—50° C)-acetone (62.5 + 37.5) are suitable solvents for development. A 10 cm run suffices in most cases. Continuous development is necessary only for final separation of picric acid, dipicrylamine and ammonium nitrate (hRf-values 49, 19 and 4 respectively) with a time of run of 2 h. The substances, mostly colourless or faintly coloured, can be visualised by spraying with a 5% solution of diphenylamine in ethanol, followed by exposure to UV light (Table 154).

With the help of two-dimensional TLC, YASUDA [92] has been able to separate and identify a number of compounds occurring as impurities in 2,4,6-trinitrotoluene; and has also identified N-nitroso and nitro-diphenylamines which are formed by reaction of diphenylamine, used as a stabiliser for cellulose nitrate, with oxides of nitrogen (formed during the slow decomposition of cellulose nitrate). All the separations were carried out on silica gel G layers containing zinc dust.

Table 154. hRf-values of various explosives on silica gel G [27]

Compound	hRf-value with			Colour after spraying with 5% diethylamine in ethanol, before and after exposure to UV	
	C_6H_6	$CHCl_3$	X[a]	before	after
Ammonium nitrate	0	0	0 (4)	no colour	sea green
Dipicrylamine	0	0	6 (19)	red-orange	red-orange
Picric acid	0	0	9 (49)	yellow	yellow
Octogen (HMX)	4	0	23	no colour	slate violet
Hexogen (RDX)	5	10	39	no colour	violet
DINA	16	42	56	no colour	neutral violet
Tetryl	26	46	62	olive yellow	olive yellow
Trinitrobenzene	40	62	71	brown-red	red-brown
Penthrite	41	61	74	no colour	green
Trotyl (TNT)	48	68	73	orange	orange

[a] X = Petrol ether-acetone (50 + 30).
The values in brackets were obtained after continuous development lasting two hours.

30 g silica gel G and 3 g zinc dust are added to 65 ml of distilled water with vigorous stirring. The layers are then spread and activated by heating 1—2 h at 110° C.

The R_{St}-values of the nitro-, dinitro- and trinitrotoluene isomers and of some oxidation and reduction products of α-(2,4,6-)trinitrotoluene (= St) are referred to this last named compound. Ethyl acetate-petrol ether (15 + 85) and 1,2-dichloroethane-petrol ether (25 + 75) were the solvents used in the 1st and 2nd directions respectively. The substances were visualised by spraying with the 4-diethylaminobenzaldehyde reagent, yielding yellow, brown or red spots (Fig. 187).

The N-nitroso and nitrodiphenylamines (Fig. 188 and Chapter P) were separated by chromatographing with acetone-benzene-petrol ether (1 + 49.5 + 49.5) in the first direction and ethyl acetate-petrol ether (20 + 80) in the second. Visualisation was likewise with 4-diethylaminobenzaldehyde reagent (similar to No. 72).

For the *analysis* of explosives, a 0.4 g sample is extracted with 25 ml methylene dichloride for 2 h at room temperature, the extract filtered and the sample again extracted with methylene dichloride for 1.5 h: the combined extracts are evaporated to dryness in vacuo and the residue taken up in 0.2 ml acetone; this solution is applied to the layer.

Mixtures of the nitrate esters of glycol, glycerol, diethylene glycol and diglycerol can be separated on silica gel G with benzene-petrol ether (50 + 50) [71].

DICARLO and co-workers [18] have carried out thin-layer chromatographic analysis of the hydrolysis products of pentaerythritol tetranitrate, important as an explosive and of interest in the metal working

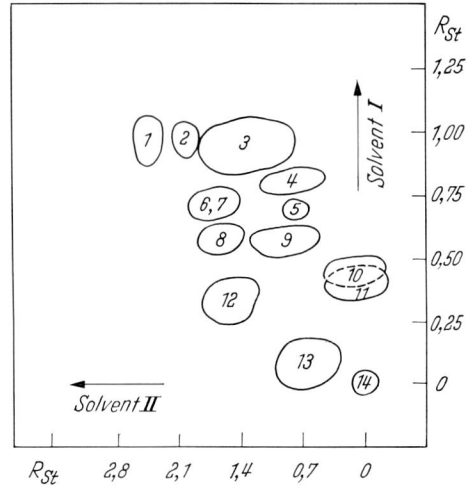

Fig. 187. Two-dimensional TLC of impurities in trinitrotoluene on silica gel G-zinc layers, using the solvents I = ethyl acetate-petrol ether (15 + 85) and II = 1,2-dichloroethane-petrol ether (25 + 75)

1 m-nitrotoluene; *2* 2,5-dinitrotoluene; *3* 2,4,6-trinitrotoluene; *4* 1,3,5-trinitrobenzene; *5* 4,6-dinitroanthranil; *6* 3.5-dinitrotoluene; *7* 2,6-dinitrotoluene; *8* 2.4-dinitrotoluene; *9* 2,4,5-trinitrotoluene; *10* 2,4,6-trinitrobenzaldehyde; *11* 2,4,6-trinitrobenzyl alcohol; *12* 3,4-dinitrotoluene; *13* 2,3,4-trinitrotoluene; *14* 2,x,6-trinitrobenzoic acid; reference substance for R_{S_r} values: 2,4,6-trinitrotoluene

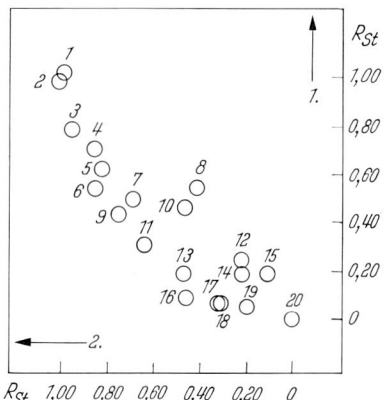

Fig. 188. Two-dimensional TLC of diphenylamine derivatives on silica gel G-zinc layers using the solvents I = acetone-benzene-petrol ether (1 + 49.5 + 49.5) and II = ethyl acetate-petrol ether (20 + 80). *1* 2-nitroDPA; *2* diphenylamine(DPA); *3* N-nitrosoDPA; *4* N-nitroso-4-nitroDPA; *5* 2,4-dinitroDPA; *6* 2,4,6-trinitroDPA; *7* 2,4′-dinitroDPA; *8* 2,2′-dinitroDPA; *9* N-nitroso-4,4′-dinitroDPA; *10* N-nitroso-2-nitroDPA; *11* 4-nitroDPA; *12* N-nitroso-2,4′-dinitroDPA; *13* 2,4,4′-trinitroDPA; *14* 2,2′,4-trinitroDPA; *15* N-nitroso-2,2′-dinitroDPA; *16* 4-nitrosoDPA; *17* 2,2′, 4,4′-tetranitroDPA; *18* 2,2′,4,4′,6-pentanitroDPA; *19* 4,4′-dinitroDPA; *20* 2,2′,4,4′,6,6′-hexanitroDPA

industry and also in medicine on account of its relaxing effect. They used ^{14}C-labelled compounds on silica gel G layers. Two solvents had to be used to separate the reaction mixture completely: I. toluene-ethyl acetate (50 + 50) and II. ethyl acetate, saturated with water. Pentaerythritol remained at the start with each; it has an hR_f-value of 18 when ethanol-chloroform (50 + 50) (III) is used. Table 155 contains the hR_f-values obtained with these three solvents.

Table 155. *hR_f-values of the hydrolysis products of pentaerythritol tetranitrate on silica gel G* (solvents see text) [18]

Compound	hR_f-value with solvent		
	I	II	III
Pentaerythritol (PE)	0	0	18
PE-mononitrate	5	37	61
PE-dinitrate	20	69	74
PE-trinitrate	50	80	79
PE-tetranitrate	70	84	79

V. Industrial Additives

An account of the TLC of miscellaneous, industrially important additives is given in the following section; it illustrates especially the versatility of the procedure.

1. Inhibitors and Antioxidants

Inhibitors are added to retard the ageing processes of insulating oils such as those used in high tension transformers. TLC is especially convenient for detecting such inhibitors because it is sensitive and simple to operate. The insulating oils themselves consist of petroleum fractions and can be fractionated by TLC into paraffins, naphthenes and aromatics. REY [72, 72a] has carried out TLC of a number of commercially available inhibitors, using benzene and hexane on silica gel layers (cf. Table 156). In tests of stability of insulating oils, the technique allows the behaviour of the inhibitors to be examined without prior separation of the insulating oil.

Some antioxidants in polyethylene are identical with or similar to the inhibitors (e. g., Nos. 2, 7, 10 and 11); they have been separated by TLC on silica gel using petrol ether-ethyl acetate (90 + 10) after having been extracted by ether from polyethylene samples [33]. 2,6-Dichloroquinonechloroimide or diazonium salt reagent were used for detection.

Table 156. *hRf-values and colour reaction of some usual commercial inhibitors on silica gel G using benzene* [72, 72a]

No.	Chemical name of the inhibitor	hRf-value	Colour reaction with SbCl$_5$ at 120° C
1	2,6-Di-tert.-butyl-4-methylphenol	81 (36)	violet
2	2,6-Di-tert.-butylphenol	79 (33)	yellow
3	4,4'-Bis(2,6-di-tert.-butylphenol)	79 (11)	canary yellow
4	4,4'-Methylene-bis-(2,6-di-tert.-butylphenol)	78 (10)	red
5	Dodecyl-o-cresol	71 (7)	dark yellow
6	o-tert.-Butylphenol	65	yellow ochre
7	N-phenyl-2-naphthylamine	63	brown
8	Dodecylphenol	56	brown-red
9	Diphenylpicrylhydrazyl	55	light yellow
10	4,4'-Thiobis(6-tert.-butyl-o-cresol)	47	yellow
11	N,N'-Diphenyl-p-phenylenediamine	46	red
12	4,4'-Methylene-bis-(6-tert.-butyl-o-cresol)	46	wine red
13	2,6-Di-tert.-butyl-1-methoxy-p-cresol	41	red
14	2-tert.-Butyl-4-hydroxyanisole	28	yellow-brown
15	2,4,6-Tri-tert.-butylphenol	13	brown
16	2,6-Di-tert.-butyl-1-dimethylamino-p-cresol	0	yellow
17	4,4'-Isopropylidene-diphenol	0	brownish

2. Detergents

The alkyl sulphates, sulphonates, phosphates and phosphonates take pride of place among the multitude of materials used as detergents. MANGOLD and KAMMERECK [51] have separated these compounds through TLC on silica gel containing 10% ammonium sulphate. The suspension must be rapidly spread on the carrier plates. These layers have proved suitable also for analysis of oleic acid esters of hydroxysulphonic acids and N-acylated short-chain amino acids. Mixtures of methanol containing 5% 0.1 N sulphuric acid and chloroform [(97 + 3) and (80 + 20)] were used as solvents. The unsaturated compounds are conveniently visualised with iodine vapour or chromic acid-sulphuric acid (Rgt. No. 46); both saturated and unsaturated lipids may be detected by spraying with 2',7'-dichlorofluorescein (Rgt. No. 63) and then inspecting in UV light.

TLC has been useful in the analysis of polyethylene oxide compounds. Compounds containing a small, medium or large number of ethylene oxide units are formed, depending on the amount introduced in the reaction. BÜRGER [10] has determined the ethylene oxide content and molecular weight distribution of polyethylene oxide "condensates" with fatty acids, fatty alcohols and alkylphenols, using silica gel and the upper layer of a well shaken mixture of methyl ethyl ketone-water (50 + 50) (Fig. 189). Polymeric products ("condensates") from reaction of ethylene oxide with fatty amines can be chromatographed in the upper phase of the mixture methyl ethyl ketone-2.5% ammonium hydroxide

(50 + 50). The separated homologues are visualised with modified Dragendorff reagent (No. 94). The Rf-values fall with increasing content of ethylene oxide units.

OBRUBA [58] has determined free polyethylene glycols in non-ionic surface-active ethylene oxide adducts. The glycols were chromatographically separated on loose silica gel layers, using the upper phase

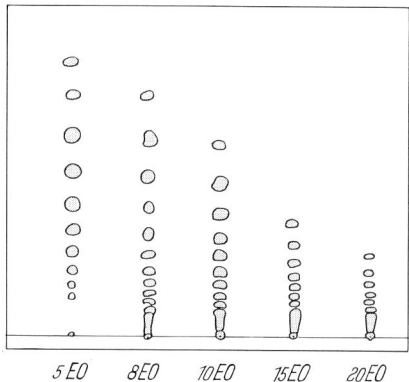

Fig. 189. Thin-layer chromatogram of ethylene oxide "condensates" with phenol with various numbers of ethylene oxide (EO) units. Layer: silica gel G; solvent: see text [10]

of the mixture diethyl ether-methanol-conc. ammonium hydroxide (67 + 22 + 11); the separated glycols were then decomposed with hydriodic acid and the iodine thus liberated was determined volumetrically.

3. Optical Brighteners

The TLC of optical brighteners or bleaches, derived from 4,4'-diaminostilbene-2,2'-disulphonic acid, has been described by THEIDEL [86]. He was able to separate the *cis* and *trans* isomers on polyamide layers, using methanol-ammonium hydroxide-water (66 + 27 + 7); surprisingly the *cis*-compound migrated ahead of the *trans*-. The opposite order was found in TLC experiments of LATINÁK [50] on silica gel G, using the solvents n-propanol-5% sodium bicarbonate solution (66 + 33), n-butanol-pyridine-water (33 + 33 + 33) and amyl alcohol-pyridine-25% ammonium hydroxide (33 + 33 + 33). The substances were detected by observation in UV light.

4. Wood Protective Agents

In addition to the old established tar oil from hard coal distillation, newer wood protective oils, containing insecticidal and fungicidal additives, are now commercially available. Their detection is especially important for testing the wood protection work or for estimating the duration of activity. High sensitivity is demanded of the analytical procedure. TLC has proved valuable here and it is to be included in the DIN (Deutsche Industrie Norm) German official procedures [19]. The most important active components of wood protection oils are α-monochloronaphthalene, γ-hexachlorocyclohexane (HCH, BHC), E 605, chlorinated phenols, aldrin and DDT. They are chromatographed on neutral silica gel layers; if phenolic components are present, acidic layers are better. PETROWITZ [65, 66] favours petrol ether or heptane as solvent for separating BHC isomers from other active agents in a single analytical step. DETERS [17] has used chloroform and benzene, which enabled γ-BHC and monochloronaphthalene to be separated.

Contaminants adversely affect the sensitivity of detection of the active agents, especially in wood extracts; the limits of detection are consequently rather high (5 μg for γ-BHC and pentachlorophenol, 2 μg for monochloronaphthalene). As has been shown by PETROWITZ, the inclusion of new types of fluorescent dye[4] in the layers also does not increase sensitivity. The procedures in Section IV, p. 77 may be used for visualising the individual substances.

5. Chemicals for Photography

Little detailed work has so far been published on the application of TLC in investigations of the numerous chemicals needed in photography, although experience has been gained in the industry. The work of EGGERS [22] contains diagrams illustrating separations of the aminophenol isomers, N-methyl-p-aminophenol and some 1-naphthylamine-mono- and disulphonic acids on silica gel G layers but without experimental details. PASTUSKA and PETROWITZ [61] have similarly used silica gel G for separating o-, m- and p-aminophenols; their solvent was benzene-methanol (80 + 20) and the hR_f-values were 50, 47 and 41 respectively, metol(N-methyl-p-aminophenol) and amidol (2,4-diaminophenol remaining at the start point. The separated compounds were visualised with a solution of 1 g p-dimethylaminobenzaldehyde in a mixture of 30 ml ethanol, 3 ml hydrochloric acid (d = 1.19) and 180 ml n-butanol.

6. Special Mineral Oil Analyses

Mineral oils are analysed mainly by gas chromatography. There are, however, some important examples of the use of TLC in detecting traces of mineral oils.

---

[4] E. g. trisodium salt of 3-hydroxypyrene-5,8,10-trisulphonic acid.

CRUMP [13] has analysed edible oils which had been contaminated or adulterated with mineral oil-containing lubricants and with the help of TLC was able to distinguish between synthetic ester oils, glycerides and mineral oils. He used silica gel G layers and chloroform-benzene (70 + 30) as solvent. The oil spots can be located in UV light or by spraying with concentrated sulphuric acid and subsequently heating to 160° C [14].

KRIEGER [45] has applied TLC to investigations of soil strata containing water sources used for drinking water supply. He succeeded in detecting traces of mineral oil which had soaked into the soil, e. g., from a leaky container. Using the circular technique with silica gel layers and a constant amount of hexane as solvent, amounts down to 50 μg of mineral-, heating- or tar oil could be detected in extracts of soil samples. The mineral oils are visualised by spraying the chromatograms with 0.03% fluorescein solution (Rgt. No. 116) and then inspecting in UV light.

VI. Intermediates in Organic Synthesis

TLC has become a valuable and indispensable aid especially to the chemist working on organic preparations: for example, to check the course of a reaction or to test the composition and purity of reaction products. The changes taking place in a synthesis of one or several stages may be followed very rapidly because the individual reaction stages usually lead to changes of polarity. As is known, chromatographic behaviour is determined by the number and position of multiple bonds, the presence of functional groups and their polarity, heteroatoms in ring structures, hydrogen bonds and steric features, to mention only some of the factors. Opinions concerning the constitution of compounds can therefore be based on results obtained from TLC. Only work of this type which is of general interest can be considered to be within the scope of a laboratory handbook.

WESSELY and co-workers [90] have applied TLC on a preparative scale instead of fractional distillation in order to isolate the hydroquinone acetate formed during treatment of *bromophenols* with lead tetraacetate. They worked with benzene on silica gel G layers and visualised by spraying with diazotised sulphanilic acid or a 1% solution of p-dimethylaminobenzaldehyde (Rgt. No. 72).

TAKACS [85] has separated the o- and p-benzoquinol acetates and o-quinone diacetate resulting from oxidation with lead tetraacetate of *phenols* containing an o-isopropyl, sec-butyl or tert.-butyl group. The o-derivatives were separated and purified on silica gel G. A 1% solution of p-dimethylaminobenzaldehyde in concentrated sulphuric acid was used as a specific reagent for detection, yielding grey-brown to yellow spots; o-quinone diacetate reacts immediately after spraying but the benzoquinol acetates only after warming. The reaction mixture yielded, for example, by oxidation of 2,4-dimethylphenol could be separated using petrol ether (BP 60—80° C) on layers of cellulose-MN 300 G powder, impregnated with formamide; the hR_f-values for o-quinone diacetate, p-benzoquinol acetate and o-benzoquinol acetate were 63, 52 and 45, respectively.

Tropones may be synthesised by ring expansion of benzoquinol acetates and have been detected and purified by ZBIRAL et al. [94] on fluorescent silica gel G layers. Chloroform-methanol (95 + 5) or benzene-ethyl acetate (50 + 50) are suitable solvents. Quenching of fluorescence and also spraying with a 0.5% iodine solution in chloroform can be used for detection.

As STAHL showed already in 1958, TLC is a valuable aid in the investigation of peroxidic intermediates and synthetic *peroxides*. RIECHE and co-workers [73, 74] have chromatographed with benzene-acetone (50 + 50) on silica gel G layers. The separated hydroperoxides and easily hydrolysable peroxides can be visualised by spraying with an aqueous solution of potassium iodide containing acetic acid (Rgt. No. 139). Dialkyl peroxides and difficultly hydrolysable peroxides may be visualised only when the layer is subsequently sprayed with concentrated hydrochloric acid.

NEALEY [55] has been able to show that *polyphenyl ethers* can be separated by multiple development on silica gel G with cyclohexane-benzene (95 + 5). m-Phenoxyphenol remains at the start and the other compounds migrate in the order: m-diphenoxybenzene, bis(m-phenoxyphenyl) ether and m-bis-(m-phenoxyphenoxy-)benzene; the migration distances thus diminish with increasing molecular size.

KLEMENT and WILD [37] have investigated numerous *phosphate esters*, using TLC in order to identify them with certainty in starting and end products of reactions. Good separation of even non-polar phosphorus compounds was also possible. Silica gel G layers were used in the separation of aliphatic and aromatic esters of various phosphoric and phosphoramidic acids and of the ammonium salts of dialkylphosphoric acids. Suitable solvents are listed below:

    I: Hexane-benzene-methanol (50 + 25 + 25),
   II: Benzene-chloroform-methanol (50 + 25 + 25),
  III: Benzene-hexane (50 + 50),
  IV: Hexane-chloroform (50 + 50),
   V: Dimethylformamide-ethanol (66.5 + 33.5),
 VI: Methylene dichloride-methanol (50 + 50),

VII: Benzene-acetic acid-ethanol (40 + 40 + 20),
VIII: n-Butanol-acetic acid (80 + 20).

Detection was by spraying the chromatograms with ammonium molybdate-perchloric acid reagent (No. 166).

Table 157 contains the hR_f-values of the individual compounds obtained with the various solvents.

Table 157. *hR_f-values of phosphate esters on silica gel G layers using miscellaneous solvents* (see p. 678)

Compound	hR_f	Solvent	Compound	hR_f	Solvent
$(C_2H_5O)_3PO$	94	I	$(C_6H_5O)_3P$	61	IV
	94	V		77	III
$(C_4H_9O)_3PO$	93	I		81	II
$(C_6H_{11}O)_3PO$	83	I	$(C_6H_5O)_5P_2ON$	95	I
$(C_2H_5)_2(NH_2)PO$	40	I	$(C_2H_5O)_4P_2S_5$	87	II
$(C_6H_5O)_3PO$	61	I	Triethyl ester of triethyl-	80	II
	95	VI	trimetaphosphinic acid		
	94	V	C_6H_5OH [a]	44	III
	85	VII		73	I
$(C_6H_5O)_2(NH_2)PO$	40	I	$(C_2H_5O)_2(ONH_4)PO$	59	VIII
$(C_6H_5O)(NH_2)_2PO$	19	I		56	VI
				45	V
$(C_2H_5O)_3PS$	50	I	$(C_4H_9O)_2(ONH_4)PO$	64	VIII
$(C_6H_5O)_3PS$	80	I	$(C_6H_5O)_2(ONH_4)PO$	76	VI
	91	III		81	V
$(C_6H_5O)_2(NH_2)PS$	36	I		67	VII
$(C_6H_5O)(NH_2)_2PS$	21	I	$(C_6H_{11}O)_2(ONH_4)PO$	46	I
$(C_6H_5O)_3PSe$	86	II	$(CH_3O)_2(ONH_4)PS$	41	VIII
$(CH_3O)_3P$	15	II	$(NH_2)_3PO$	0	I
$(C_2H_5O)_3P$	67	II	$(NH_2)_3PS$	0	I
	42	I			

[a] Phenol gives a brown spot with molybdate.

Alkyl-, aryl- and steroid *sulphate esters* can likewise be separated on silica gel G. WUSTEMAN and co-workers [91] have used the following solvents for chromatography:

I: Benzene-butanone-ethanol-water (30 + 30 + 30 + 10),
II: Isopropanol-chloroform-methanol-water (37 + 37 + 19 + 7),
III: Isopropanol-chloroform-methanol-10 N ammonium hydroxide (37 + 37 + 19 + 7) and
IV: n-Butanol-acetic acid-water (60 + 20 + 20).

Hydroxyamino acids, their sulphate esters, L-tyrosine and the corresponding p-hydroxyphenyl derivatives including their esters, were chromatographed with solvent IV; solvent II was used for separating the sulphate esters of aliphatic alcohols. Solvent III is suitable like solvent I for analysing steroid esters. The hR_f-values of the sulphate esters of substituted phenols, separated with solvents I and II, are in Table 158.

Table 158. hRf-values of sulphate esters of substituted phenols on silica gel G [91]

Sulphate ester of	hRf		Sulphate ester of	hRf	
	I[a]	II[a]		I[a]	II[a]
Phenol	46	45	4-Methoxyphenol	45	44
2-Chlorophenol	48	47	4-Hydroxy-3-nitrophenol	46	48
3-Chlorophenol	50	49	4-Hydroxy-2-nitrophenol	60	53
4-Chlorophenol	50	48	2-Hydroxy-5-nitrophenol	51	41
2-Methylphenol	47	47	2,3-Dichlorophenol	54	53
3-Methylphenol	47	51	2,4-Dichlorophenol	54	53
4-Methylphenol	47	49	3-Nitrophenol	53	50
2-Methoxyphenol	40	41	4-Nitrophenol	57	52
3-Methoxyphenol	45	46	2-Hydroxy-4-chlorophenol	56	47

[a] See text for solvents I and II.

Detection was carried out with the spray reagents customarily used for the individual compound classes.

DÉNES et al. [16] have identified the products formed by intermolecular rearrangement of *thiocarbonate esters* on heating. They were thus able to elucidate the mechanism of the reaction. They used loose layers of alumina (activity III according to BROCKMANN) and chloroform-carbon tetrachloride-diethyl ether (30 + 30 + 40) as solvent.

NEURATH and co-workers [57] have worked on the TLC of 5-nitro-2-hydroxybenzalhydrazines, showing thereby a means of identifying *nitrosoamines* and the unsymmetrical *dialkylhydrazines* derived from them by reduction with lithium aluminium hydride. Fig. 190 reproduces the positions of the substance spots after TLC on silica gel G with the solvent carbon tetrachloride-ethyl acetate (95 + 5). The colour of the spots deepens to yellow brown on spraying with 3% potassium hydroxide in ethanol. Spraying with a 1% potassium ferricyanide solution in 5% aqueous hydrochloric acid, followed by heating, is a particularly sensitive, although non-specific test; amounts down to 0.05 μg can be thereby detected (blue spots). Chromatographic separation procedures have been of valuable service in the analysis of mixtures of isomers, which were otherwise not at all or only partly separable and then with wearisome procedures. Thus, isomeric dibromodecalins [35], the isomeric trithiofluorobenzaldehydes [11], diastereoisomeric diols [24] or cycloalkane- and *threo-erythro* isomer pairs [53] have been separated with the help of TLC. HRANISAVLJEVIĆ-JAKOVLJEVIĆ and co-workers [29, 62] have been able to separate some oxime isomers on silica gel G layers, using benzene-ethyl acetate (83 + 17). Table 159 contains the hRf-values of the α- and β-forms.

A 0.5% aqueous cupric chloride solution is an especially suitable spray reagent.

The TLC of numerous heterocyclic compounds has been described in detail; silica gel G layers have been those predominantly used.

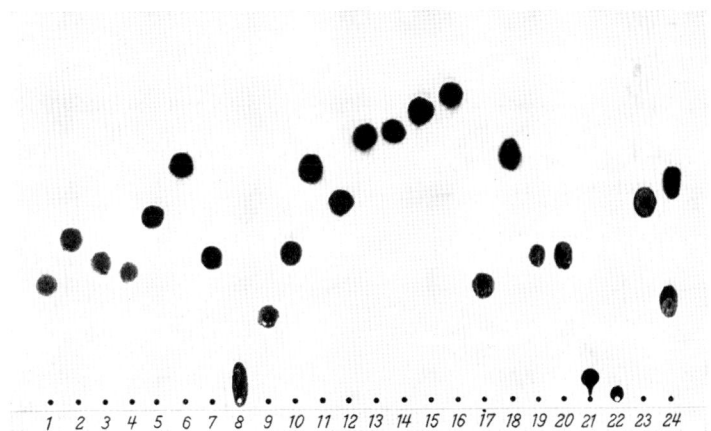

Fig. 190. TLC of benzal derivatives of unsymmetrical hydrazines on silica gel using the solvent carbon tetrachloride-ethyl acetate (95 + 5) [57]

(hydrazine = —h) *1* methyl-ethyl-h; *2* methyl-n-butyl-h; *3* methyl benzyl-h; *4* methyl allyl-h; *5* methyl n-pentyl-h; *6* ethyl n-heptyl-h; *7* ethyl phenyl-h; *8* 5-nitro-2-hydroxybenzaldehyde; *9* dimethyl-h; *10* diethyl-h; *11* di-n-propyl-h; *12* diisopropyl-h; *13* di-n-butyl-h; *14* diisobutyl-h; *15* di-n-pentyl-h; *16* di-n-hexyl-h; *17* methyl phenyl-h; *18* diphenyl-h; *19* 1-amino-pyrrolidine; *20* 1-amino-piperidine; *21* 4-aminomorpholine; *22* 4-amino-1-methylpiperazine; *23* diallyl-h; *24* standard mixture

Table 159. hR_f-values of isomeric oximes on silica gel G using benzene-ethyl acetate (83 + 17) solvent [29, 62]

Oxime	hR_f-value of the	
	α-Form	β-Form
Benzaldoxime	50	32
Benzoin-oxime	15	37
Anisoin-oxime	5	23
p-Tolualdoxime	54	33
p-Anisaldoxime	42	27
p-Cuminaldoxime	54	37
o-Nitrobenzaldoxime	47	40
m-Nitrobenzaldoxime	52	35
p-Nitrobenzaldoxime	53	34

CURTIS and PHILLIPS [15] have used alumina G also for separating *thiophene derivatives;* they obtained the hR_f-values quoted in Table 160, using the solvent petrol ether (BP 40–60° C). The results in Table 161 were obtained on the other hand on silica gel G layers. The substances were located by inspection in UV light and by spraying with a 0.4% solution of isatin in concentrated sulphuric acid (Rgt. No. 148).

Table 160. *hRf-values of thiophene derivatives on alumina G, using petrol ether (BP 40—60° C)*[a] [15]

Compound	hRf	Compound	hRf
2-Methylthiophene	76	Bithienyl	80
3-Methylthiophene	82	α-Terthienyl	57
2-Ethylthiophene	92	α-Quaterthienyl	26
2,5-Dimethylthiophene	95	5,5'-Dichlorobithienyl	89
2,3,5-Trimethylthiophene	87	5,5'-Dimethylbithienyl	76
2,3-Dimethyl-4-ethylthiophene	84	α-Phenyl-α-bithienyl	50
Tetramethylthiophene	89	5,5'-Diphenylbithienyl	16
Methyl 3-thienyl sulphide	92	5,5''-Dimethylterthienyl	48
n-Decyl 3-thienyl sulphide	96		

[a] The volatile compounds were chromatographed at 4° C.

Table 161. *hRf-values of thiophene derivatives on silica gel G, using benzene-chloroform (90 + 10) (I) and methanol (II)* [15]

Compound	hRf (I)	Compound	hRf (II)
Thiophenecarboxaldehyde	34	Thiophenecarboxylic acid	65
2-Nitrothiophene	61	β-(α-Thienyl)-acrylic acid	57
Phenyl 2-thienyl ketone	39	4-(α-thienyl)-butyric acid	60
Methyl 2-thienyl ketone	25	Bithienyl-5-carboxylic acid	63
		2,2'-Bithienyl-methylamine (HCl)	35

With the help of TLC on silica gel G, MAYER et al. [54] have investigated many *heterocyclic sulphur compounds* of the trithione (I) (1,2-dithiol-3-thiones) and dithione (II) (1,2-dithiol-3-ones) types, their anils (III), the 1,2-thiazoline-5-thiones (IV) and xanthane hydride (V). Petrol ether-benzene (50 + 50) or carbon disulphide are good solvents.

```
    R—C—C⟨X        R'—C—C⟨S         N—C⟨S
    ‖    ⟩S         ‖    ⟩S         ‖    ⟩S
    R'—C—S/        R''—C—N/        H₂N—C—S/
                         |
                         R

I:  X=S
II: X=O                 IV              V
III: X=N—C₆H₅
```

Chromatographic separation of more strongly polar compounds is better with benzene-ethyl acetate (75 + 25); and of heterocyclic compounds containing hydroxyl or carboxyl groups, with acetone. The compounds may be easily detected on the layer since they are coloured. In addition, compounds of groups I, II and IV yield colours with tetracyanoethylene. As seen from the Rf-values, polar groups reduce markedly the adsorption capacity.

RUNGE and co-workers [76] have prepared a series of unsymmetrical heterocyclic *disulphides* and found a relation between the constitution and the Rf-values. Chromatography was carried out with chloroform-hexane (30 + 70) on silica gel G (Table 162) and visualisation by spraying with a 0.1% aqueous-alcoholic solution of fluorescein, followed by exposure to bromine vapour (Rgt. No. 118). The thiazole disulphides appear as yellow spots on a reddish background. The Rf-values increase with increasing atomic weight of the halogen substituent. The compounds with a phenyl substituent in the thiazole ring are more weakly adsorbed than those with an alkyl substituent.

Table 162. hRf-values of unsymetrical heterocyclic disulphides on silica gel G, using chloroform-hexane (30 + 70) [76]

Basic formula	Substituents		hRf-values
	R	R′	
(structure 1)	H	—	34
	F	—	38
	Cl	—	42
	Br	—	54
(structure 2)	Phenyl	H	34
	2′-Thienyl	H	31
	Phenyl	Phenyl	32
	Methyl	H	17
	Methyl	Methyl	8
	Methyl	Carbethoxy	7
(structure 3)	o-Cl	o-Cl	95
	p-Cl	o-Cl	93
	o-Cl	p-Cl	70
	p-Cl	p-Cl	60
(structure 4)	H	2-Nitrophenyl	34
	H	2,4-Dinitrophenyl	15
	H	2-Nitro-4-chlorophenyl	40
	H	2-Chlorophenyl	52
	H	4-Chlorophenyl	52
	H	4-Methylphenyl	70
	H	Trichloromethyl	75
	Br	Trichloromethyl	67

KORTE and VOGEL [44] have worked on the TLC of *lactones*, *lactams* and *thiolactones*. They were able to separate a large number of compounds belonging to these groups, using silica gel G and the solvents diisopropyl ether, diisopropyl ether-ethyl acetate (80 + 20) and diisopropyl ether-isooctane (20 + 80) and (60 + 40). The lactams are best visualised with Dragendorff reagent. Thiolactones are first treated with alkali and the −SH groups thus formed yield colour with nitroprusside (2% in 75% ethanol).

Heterocyclic nitrogen compounds of the pyridine and quinoline series, sometimes technically important in organic synthesis, have been chromatographed by PETROWITZ and co-workers [69] on silica gel G layers, using the solvents chloroform, ethyl acetate and acetone. Most of the compounds could be visualised by spraying with Dragendorff reagent (Rgt. No. 98). Hydroxypyridines, pyridyl carbinols and pyridine aldehydes were detected by spraying with permanganate solution. Pyridinecarboxylic acids and 2-halopyridines were located by chromatographing on fluorescent layers (e. g., silica gel HF_{254}) and inspection in UV light.

EISTERT and LANGBEIN [23] have detected the *pyrazolines* yielded by reaction of tetraphenylcyclopentadienone with diazomethane and the tetraphenylbenzene derivatives prepared from decomposition of these pyrazolines via bicyclohexenone; they used silica gel G layers and, as solvent, ethanol-free chloroform containing 0.3–0.4% added methanol. Aqueous permanganate solution (Rgt. No. 199) was suitable for visualisation.

Bibliography for Chapter TS. Synthetic Organic Products

1. BADGER, G. M., J. K. DONNELLY, and T. M. SPOTSWOOD: J. Chromatog. **10**, 397 (1963).
2. BANCHER, E., u. H. SCHERZ: Mikrochim. Acta **1964**, 1159.
3. BERG, A., and J. LAM: J. Chromatog. **16**, 157 (1964).
4. BRAUN, D.: Kunststoffe **52**, 2 (1962).
5. — Chimia **19**, 77 (1965).
6. — and H. GEENEN: J. Chromatog. **7**, 56 (1962).
7. — u. G. VORENDOHRE: Z. analyt. Chem. **199**, 37 (1964).
8. — — Z. analyt. Chem. **207**, 26 (1965).
9. BÜRGER, K.: Z. analyt. Chem. **192**, 280 (1963).
10. — Z. analyt. Chem. **196**, 259 (1963).
11. CAMPAIGNE, E., and M. GEORGIADIS: J. org. Chem. **27**, 135 (1962).
12. COPIUS-PEEREBOOM, J. W.: J. Chromatog. **4**, 323 (1960).
13. CRUMP, G. B.: Analyst **88**, 456 (1963).
14. — Nature (Lond.) **193**, 674 (1962).
15. CURTIS, R. F., and G. T. PHILLIPS: J. Chromatog. **9**, 366 (1962).
16. DÉNES, V. I., G. CIURDARU u. M. FARCASAN: Chem. Ber. **96**, 2691 (1963).
17. DETERS, R.: Holz als Roh- u. Werkst. **21**, 362 (1963).
18. DICARLO, F. J., J. M. HARTIGAN and G. E. PHILLIPS: Anal. Chem. **36**, 2301 (1964).
19. *DIN 52161* (Draft).
20. DYATLOVITSKAYA, E. V., V. V. VORONKOVA u. L. D. BERGELSSON: Ber. Akad. Wiss. UdSSR **145**, 325 (1962).

21. EGGER, H., u. K. SCHLÖGL: Organometal. Chem. **2**, 398 (1964).
22. EGGERS, J.: Phot. u. Wiss. **10**, 40 (1961).
23. EISTERT, B., u. A. LANGBEIN: Justus Liebigs Ann. Chem. **678**, 78 (1964).
24. FISCHER, F., and H. KOCH: J. Chromatog. **16**, 246 (1964).
25. GEISS, F., H. SCHLITT, F. J. RITTER, and M. WEIMAR: J. Chromatog. **12**, 469 (1963).
26. — — Naturwissenschaften **50**, 350 (1963).
27. HANSSON, J.: Explosivstoffe **1963**, 73.
28. HARTHON, J. G. L.: Acta chem. scand. **15**, 1401 (1961).
29. HRANISAVLJEVIĆ-JAKOVLJEVIĆ, M., I. PEJKOVIĆ-TADIĆ, and A. STOJILYKOVIC: J. Chromatog. **12**, 70 (1963).
30. HAUB, H.-G., and H. KÄMMERER: J. Chromatog. **11**, 487 (1963).
31. V. D. HEIDE, R. F.: Z. Lebensmitt.-Untersuch. **124**, 348 (1964).
32. — Z. Lebensmitt.-Unters. **124**, 198 (1964).
33. — u. O. WOUTERS: Z. Lebensmitt.-Unters. **115**, 129 (1962).
34. HROMATKA, O., u. W. A. AUE: Mh. Chem. **93**, 503 (1962).
35. HÜCKEL, W., u. H. WAIBLINGER: Justus Liebigs Ann. Chem. **666**, 17 (1963).
36. KERN, W., K. J. RAUTERKUS u. W. WEBER: Makromol. Chem. **43**, 98 (1961).
37. KLEMENT, R., u. A. WILD: Z. analyt. Chem. **195**, 180 (1963).
38. KNAPPE, E., u. D. PETERI: Z. analyt. Chem. **188**, 184 u. 352 (1962) u. **190**, 380 (1962).
39. — — Z. analyt. Chem. **190**, 386 (1962).
40. — — u. J. ROHDEWALD: Z. analyt. Chem. **199**, 270 (1964).
41. — — — Z. analyt. Chem. **197**, 364 (1963).
41a. — u. I. ROHDEWALD: Z. analyt. Chem. **200**, 9 (1964).
42. KÖHLER, M., H. GOLDER u. R. SCHIESSER: Z. analyt. Chem. **206**, 430 (1964).
43. KORN, O., u. H. WOGGON: Nahrung **8**, 351 (1964).
44. KORTE, F., and J. VOGEL: J. Chromatog. **9**, 381 (1962).
45. KRIEGER, H.: Gas- u. Wasserfach **104**, 695 (1963).
46. KUCHARCZYK, N., J. FOHL, and J. VYMETAL: J. Chromatog. **11**, 55 (1963).
47. — — private communication.
48. KUNTZ, E.: Dissertation, Mainz 1960.
49. KUPFER, W.: Z. analyt. Chem. **192**, 219 (1963).
50. LATINÁK, J.: J. Chromatog. **14**, 482 (1964).
51. MANGOLD, H., and R. KAMMERECK: J. Am. Oil Chem. Soc. **39**, 201 (1962).
52. MATSUSHITA, H. Y. SUZUKI, and H. SAKABE: Bull. Chem. Soc. Japan **36**, 1371 (1963).
53. MAUGRAS, M., M. CH. ROBIN et R. GAY: Bull. Soc. chim. Biol. **44**, 887 (1962).
54. MAYER, R., P. ROSMUS, and J. FABIAN: J. Chromatog. **15**, 153 (1964).
55. NEALEY, R. H.: J. Chromatog. **14**, 120 (1964).
56. NEUBERT, G.: Z. analyt. Chem. **203**, 265 (1964)
57. NEURATH, G., B. PIRMANN u. M. DÜNGER: Chem. Ber. **97**, 1631 (1964).
58. OBRUBA, K.: Collect. Czech. chem. Commun. **27**, 2968 (1962).
59. OGNYANOV, I.: C. R. Acad. Bulg. Sci. **16**, 265 (1963).
60. PASTUSKA, G., u. H.-J. PETROWITZ: J. Chromatog. **10**, 517 (1963).
61. — — Chemiker-Ztg. **88**, 311 (1964).
62. PEJKOVIĆ-TADIĆ, I., M. HRANISAVLJEVIĆ-JAKOVLJEVIĆ, and S. NESIĆ: Thin-Layer-Chromatography, Proc. Sympos. Rome 1963, s. 160. Amsterdam - London - New York: Elsevier Publ. Comp.
63. PETROWITZ, H.-J., u. G. PASTUSKA: J. Chromatog. **7**, 128 (1962).
64. — Erdöl u. Kohle **14**, 923 (1961).
65. — Chemiker-Ztg. **85**, 867 (1961).
66. — Chemiker-Ztg. **86**, 815 (1962).

67. PETROWITZ, H.-J.: Chemiker-Ztg. 88, 235 (1964).
68. — Thin-Layer-Chromatography, Proc. Sympos. Rome 1963, s. 132. Amsterdam-London-New York: Elsevier Publ. Comp.
69. — G. PASTUSKA u. S. WAGNER: Chemiker-Ztg. 89, 7 (1965).
70. PREY, V., H. BERBALK u. M. KAUSZ: Mikrochim. Acta 1962, 449.
71. RAO, K. R. K., A. K. BHALLA, and K. SINHA: Current Sci. (India) 33, 12 (1964).
72. REY, E., u. L. ERHART: Bull. schweiz. elektrotechn. Ver. 52, 401 (1961).
72a. — Elektrotechn. Z., Ausg. B 13, 299 (1961).
73. RIECHE, A., u. M. SCHULZ: Chem. Ber. 97, 190 (1964).
74. — E. HÖFT u. H. SCHULTZE: Chem. Ber. 97, 195 (1964).
75. RITTER, F. J., P. CANONNE u. F. GEISS: Z. analyt. Chem. 205, 313 (1964).
76. RUNGE, F., A. JUMAR u. F. KOEHLER: J. prakt. Chem. 21, 39 (1963).
77. SAWICKI, E., T. W. STANLEY, W. C. ELBERT and J. D. PFAFF: Anal. Chem. 36, 497 (1964).
78. — — J. D. PFAFF and W. C. ELBERT: Anal. Chim. Acta 31, 359 (1964).
78a. SMITH, G. A. L., and P. J. SULLIVAN: Analyst 89, 312 (1964).
79. SCHLÖGL, K., A. MOHAR u. H. PELOUSEK: Naturwissenschaften 46, 447 (1959).
80. — H. PELOUSEK u. A. MOHAR: Mh. Chem. 92, 533 (1961).
81. — u. A. MOHAR: Mh. Chem. 93, 861 (1962).
82. — — u. H. PELOUSEK: Mh. Chem. 92, 921 (1961).
83. — u. M. PETERLIK: Mh. Chem. 93, 1328 (1962).
84. — u. M. FRIED: Mh. Chem. 94, 537 (1963).
85. TAKACS, F.: Mh. Chem. 95, 961 (1964).
86. THEIDEL, H.: Melliand Textilber. 1964, 514.
87. TÜRLER, M., u. O. HÖGL: Mitt. Lebensmitt.-Hyg. 52, 123 (1961).
88. VOBECKY, M., V. D. NEFEDOV u. E. N. SINOTOVA: Z. obšč. Chim. 33, 4023 (1963).
89. WALDI, D.: unpublished work.
90. WESSELY, F., E. ZBIRAL u. J. JÖRG: Mh. Chem. 94, 227 (1963).
91. WUSTEMAN, F. S., K. S. DODGSON, A. G. LLOYD, F. A. ROSE, and N. TUDBALL: J. Chromatog. 16, 334 (1964).
92. YASUDA, ST. K.: J. Chromatog. 13, 78 (1964).
93. — J. Chromatog. 14, 65 (1964).
94. ZBIRAL, E., F. TAKACS u. F. WESSELY: Mh. Chem. 95, 402 (1964).
95. THOMA, K., R. ROMBACH u. E. ULLMANN: Scientia Pharmaceutica 32, 216 (1964).
96. — — — Arch. Pharmaz. 298, 19 (1965).

U. Hydrophilic Plant Constituents and Their Derivatives

The conditions for TLC of the predominantly lipophilic plant constituents have been treated already in preceding chapters. Apart from amino acids, nucleic acids and sugars, likewise discussed in special chapters, plants contain numerous other hydrophilic constituents. Secondary products of metabolism, sometimes important in medicine, are of especial interest here; they can often be utilised to characterise a plant chemically. Some more general considerations concerning separation possibilities for plant phenol derivatives have been worked out in the first part of the present chapter. Other compound classes, in particular those used medicinally, have been treated in a second part. Finally, general possibilities of applying TLC to drug identification in pharmacopoeias, have been compiled and explained.

I. Plant Phenol Derivatives
Kurt Egger
1. Compound Classes and Their Distribution

Surveys of the structures of the individual compounds and their distribution in the plant kingdom are to be found in the work or review articles of Bate-Smith [9], Geissmann [60], Harborne [86], Karrer [118], Bernfeld [12], Thomson [216] and numerous other authors [3, 7, 19, 33, 50, 85, 87, 195, 197].

This group of diverse compounds can be subdivided in various ways. The following classification is useful:

a) Simple phenols and their glycosides, like arbutin, syringin and Salicaceae glycosides [215]. The steam-volatile plant phenols are treated in Chapter J on "Essential Oils" (p. 206). They are obtained partly from degradation of complex natural products, e. g. in the alkali fusion of lignin [55] and of flavonoids [60].

b) Phenol carboxylic acids, predominantly derivatives of benzoic and cinnamic acids, e. g., gentisic, gallic and caffeic acids. A number of tannins is derived from gallic acid [197], e. g., those of the Hamamelidaceae.

c) α-Pyrones: coumarins and isocoumarins [7, 32, 69]. They are biogenetically related to cinnamic acid from which they can be derived. Aesculetin and xanthotoxin may be mentioned as examples in addition to coumarin:

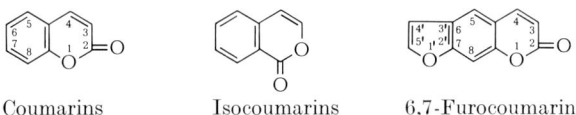

Coumarins   Isocoumarins   6,7-Furocoumarins

d) Depsides and depsidones, belonging to the specific constituents of lichens. They are occasionally associated with the derivatives of vulpinic acid which may be regarded as a cinnamic acid dimer [3, 148].

e) Lignans, likewise dimerisation products; well-known representatives of these are the active principles of *Podophyllum* (see p. 709).

f) Chromones, e. g., eugenin and furochromones like khellin, khellol etc. [37, 39, 195].

g) Flavonoids [60], to which the γ-pyrones, the compound class chiefly discussed here, belong; they are subdivided into flavones and isoflavones. Chalkones and aurones and also the stilbenes [19] are included here. Many tannins (catechin type) are derived from the already mentioned catechins [50, 51, 196].

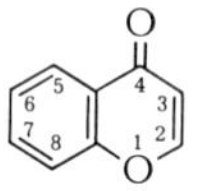

Chromones
(γ-Benzopyrones)

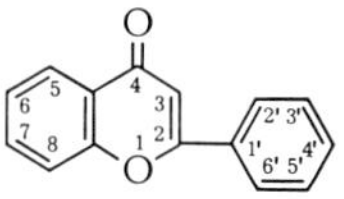

Flavones
(2-phenyl-γ-benzopyrones)
(light to deep yellow)

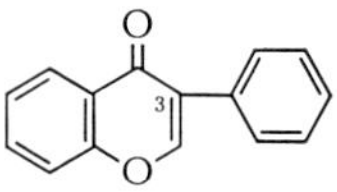

Isoflavones
(3-phenyl-γ-benzopyrones)
(colourless)

The flavones occur in various oxidation stages and the following are distinguished:

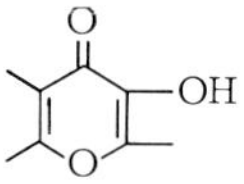

Flavonols
(light to deep yellow)

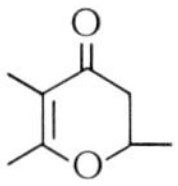

Flavanones
(colourless)

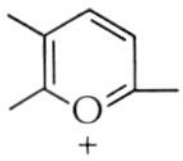

Anthocyanidins
(red, blue, violet)

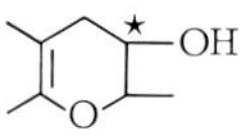

Catechins
(colourless)

Table 163. *Group A, slightly polar phenol derivatives*
Comparison of the hR_f-values obtained under standard conditions with two solvents on both silica gel and polyamide layers

Substance	Silica gel		Polyamide	
	I	II	III	IV
Veratric acid	60	29	90	70
Ferulic acid acetate	26	5	62	20
Dimethoxycinnamic acid	27	2	57	15
3,4-Methylenedioxycinnamic acid	35	8	44	12
Trimethoxycinnamic acid	24	3	64	19
Coumarin	75	32	87	59
Umbelliferone	34	5	33	7
4'-Hydroxy-β-phenylcoumarin acetate	69	33	92	76
4'-Hydroxy-β-phenylcoumarin	54	19	53	15
Khellin	21	4	85	52
4'-Hydroxychalkone acetate	77	45	90	65
4'-Hydroxychalkone	45	14	48	14
Myricetin hexaacetate	28	1	82	53
Quercetin pentaacetate	35	4	84	58
Kaempferol tetraacetate	42	6	85	65
Quercetin-5,7,3',4'-tetramethyl ether	46	16	64	18
Quercetin-7,3',4'-trimethyl ether	51	22	72	31
Alkannin	78	30	79	47
Anthraquinone	90	73	88	70
Menadione	89	68	92	90
Benzoquinone	85	51	90	75
(+)-Pinoresinol dimethyl ether	37	5		
(+)-Catechin pentaacetate	34	4		

Solvents: I: Benzene-acetone (90 + 10); II: Toluene-acetone (95 + 5); III: Petrol ether (high boiling)-benzene-methanol-butanone (50 + 40 + 5 + 5); IV: as III but (60 + 30 + 5 + 5).

h) Quinones, occurring in various oxidation stages, as mono- and polynuclear benzoquinones (e. g., diphenoquinones) or also as naphtho- and anthraquinones [216].

In view of the large number of compounds which come into consideration here it is advisable to subdivide first into three large chromatographic substance groups according to the type and number of the functional groups and to the solubility. This has been done in Tables 163 to 165:

A. Compounds with one phenolic group; those containing no free —OH group; and carboxylic acids containing no free phenolic hydroxyl group (Table 163).

B. Polyphenols, hydroxycarboxylic acids, tannins and other phenol derivatives with fused rings (Table 164).

Table 164. *Group B, substances of medium polarity*
Phenols, phenolcarboxylic acids, hydroxycoumarins, flavone aglycones. Comparison of the hR_f-values obtained under standard conditions on cellulose, silica gel and polyamide thin layers

Substance	Substituents	C	S	Polyamide	
		I	II	III	IV
Catechol	1,2-OH	84	50	66	62
Hydroquinone	1,4-OH	73	39	65	70
Phloroglucinol	1,3,5-OH	28	17	41	70
Pyrogallol	1,2,3-OH	56	28	47	65
Vanillic acid	4-OH, 3-OCH$_3$	91	37	66	58
Protocatechuic acid	3,4-OH	64	19	38	53
Gentisic acid	2,5-OH		27		15
Syringic acid	4-OH,3,5-OCH$_3$	95	36	93	72
Gallic acid	3,4,5-OH	28	7	23	63
o-Coumaric acid	2-OH	92	37	50	41
p-Coumaric acid	4-OH	90	37	55	48
Ferulic acid	4-OH,3-OCH$_3$	95	37	65	53
Caffeic acid	3,4-OH	75	21	51	60
Sinapic acid	4-OH,3,5-OCH$_3$	99	34	75	65
Umbelliferone	7-OH	92	55	67	52
Aesculetin	6,7-OH	73	31	50	52
(+)-Catechin	3,5,7,3',4'-OH			22	70
Galangin	3,5,7-OH	95	62	50	10
Kaempferol	3,5,7,4'-OH	71	39	20	8
Quercetin	3,5,7,3',4'-OH	38	27	8	8
Myricetin	3,5,7,3',4',5'-OH	13	13	4	8
Isorhamnetin	3,5,7,4', 5'-OH,3'-OCH$_3$	83	26	31	8
Apigenin	5,7,4'-OH	84	43	30	9
Luteolin	5,7,3',4'-OH	64	28	19	9
Datiscetin	3,5,7,2'-OH	91	36	36	2
Morin	3,5,7,2',4'-OH	60	6	10	2

Experimental Conditions: I: Cellulose; chloroform-acetic acid-water (50 + 45 + 5). II: Silica gel; toluene-chloroform-acetone (40 + 25 + 35). III: Polyamide; benzene-butanone-methanol (60 + 20 + 20). IV: Polyamide; water-butanone-methanol (40 + 30 + 30).

C. Glycosides and salt-like compounds such as anthocyanidins (Table 165).

Table 165. *Group C, strongly polar compounds*
hRf-values of flavone glycosides on various layers

Substance	Glycosidation	S	Polyamide		C
		I	II	III	IV
Myricitrin	My-3-rh	57	11	54	54
Quercitrin	Q-3-rh	62	9	64	72
Afzelin	K-3-rh	65	7	72	84
Myricetin glucoside	My-3-gluc	46	20	34	37
Quercituron	Q-3-gron	59	5	5	54
Isoquercitrin	Q-3-gluc	51	16	56	56
Kaempferol glucuronide	K-3-gron	63	4	8	70
Astragalin	K-3-gluc	54	14	69	73
Myricetin rutinoside	My-3-rhgluc	21	35	16	25
Rutin	Q-3-rhgluc	30	30	42	43
Nicotiflorin	K-3-rhgluc	36	27	60	57
Kaempferol sophoroside	K-3-glucgluc	21	50	40	53
Paenonoside	K-3-gluc-7-gluc	13	74	28	33
Robinin	K-3-rhgal-7-rh	16	70	40	41
Equisetum glycoside	K-3-rhgluc-7-gluc	7	80	15	20
Helleborus glycoside	K-3-xylgluc-7-gluc	6	84	15	20
Cosmosiin	Ap-7-gluc	53	20	78	65
Rhoifolin	Ap-7-rhgluc	34	38	74	58
Apiin	Ap-7-apiogluc	36	33	72	57
Vitexin	Ap-8-glucosyl	37	25	68	44
Naringin	Nar-7-rhgluc		63	80	54

Experimental conditions: I: Silica gel; CS; solvent VII: ethyl acetate-butanone-formic acid-water $(50 + 30 + 10 + 10)$ [119]. II: Polyamide; solvent: water-ethanol-butanone-acetylacetone $(65 + 15 + 15 + 5)$ [20]. III: Polyamide; solvent; benzene-methanol-butanone $(60 + 20 + 20)$ [25]. IV: Cellulose layer; solvent: Partridge mixture.

rh = rhamnoside; gluc = glucoside; gron = glucuronide; gal = galactoside; xyl = xyloside; apio = apioside.

2. Concentration from Plant Material

The diversity of the compounds, as seen from the preceding section, explains why widely differing procedures are used for extraction. There are, however, four basic types:

a) Acetone-extraction of Fresh Material

The fresh plant parts are minced in a mortar or an electrical mixing machine (Starmix), if necessary after adding acetone. Extraction is carried out by repeated digestion or through percolation. After filtering, the filtrate is treated with about double its volume of petrol ether (BP 40—60° C) and, if two phases are not formed, a little water is added. The lower, aqueous phase contains glycosides and polar aglycones; the petrol ether layer, the fats and lipophilic colouring material. In order to obtain the anthocyanins the extraction must be conducted under weakly acid conditions.

b) Methanol-extraction of Dried Drugs

A methanolic extract is prepared, evaporated to dryness and the residue digested with hot water. The glycosides dissolve in the water and the fats, chlorophylls and other lipids remain in the residue. A milder variant is using petrol ether to remove lipids from the methanolic extract.

c) Extraction in Stages

These extraction stages are in the sequence: petrol ether, chloroform, ether, acetone, methanol; they are carried out in either a percolation tube or a Soxhlet apparatus. After removing fat with petrol ether, the chloroform extract contains chiefly compounds of group A, e. g., daphnoretin [218], lichen components, fully methylated flavones, many furocoumarins etc. Ether then extracts all the aglycones and some glycosides (rhamnosides). Most of the glycosides are then to be found in the acetone and methanol extracts.

d) Extraction with Water

Glycosides may also be extracted with water. The extract is carefully concentrated to a syrupy consistency and then taken up in alcohol. The precipitate formed can often be discarded.

Other suitable concentration procedures are liquid/liquid partition of the extracts, e. g., between water and ether or benzene and formamide. Precipitation with lead and, recently with success, column chromatography on polyamide [60, 171] can be used for purification.

3. Chromatographic Separation (without TLC)

Column Chromatography. The separation of phenolic compounds on cellulose columns [72] has been of some importance but inorganic adsorbents are, however, preferable. They are more easily handled and the solvent can be continuously changed (see Gradient elution, p. 90). Adsorption capacity and rate of flow are greater than in the cellulose column. Silica gel is the principal column filling [11]; magnesol [232] and alumina are sometimes used; polyamide has, however, gained the greatest acclaim [21, 44, 45, 92, 93, 94, 106, 110, 153, 154]. The separation of simple quinones can be hindered by irreversible reaction with the amide groups of the polyamide; acetylated polyamide is suitable for separating such compounds [44, 45].

Ion exchange resins have occasionally been used to adsorb flavones from aqueous solutions; they can then be eluted with alcohol. Weakly polar substances have been separated by partition chromatography on silica gel, impregnated with dimethyl sulphoxide [156]; benzene was used as solvent. An interesting separation is that of betacyanins with an aqueous potassium acetate solution on polyamide [198].

Paper Chromatography. Paper chromatography has been and is still being widely applied to separation of the compound classes treated here; extensive experience has been acquired in this very domain. The fields of application and many solvents correspond to those used in TLC on cellulose layers.

Gas Chromatography. Simple phenols and phenol carboxylic acids can be subjected to GC-analysis, directly or after methylation. Good separations of the trimethyl silyl ethers of higher boiling compounds are often obtained [71].

4. Experimental Conditions for TLC

Three adsorbents have been of especial value in the separation of plant phenolic derivatives: cellulose, silica gel and polyamide. Polyacrylonitrile and ion exchangers like Amberlite CG 50-III have also been used. The three tables 163–165 permit comparison of the hRf-values a larger number of phenol derivatives from plants. These values naturally fluctuate considerably and must be taken only as guide values. In general, experience has shown that silica gel is the best adsorbent for separating substances of group A, i. e., of less polar compounds. A polyamide layer is the best for separation of compounds containing free phenol groups and the corresponding glycosides (flavone compounds). Cellulose layers have been employed in particular for separating mixtures of substances with a high proportion of glycoside units.

a) Cellulose

Resolution on the thin, fine-grained cellulose layers used in TLC is superior to that in PC and is accomplished in shorter time. As has been mentioned already, the solvents used in PC can be applied in TLC and, by and large, the same separation sequences are then obtained. The seven solvents listed below are particularly suitable for development on cellulose layers and they are adequate for most purposes:

1. *Benzene;* with cellulose layers impregnated with formamide (10% solution in ether). Partition chromatographic separation of substances of group A, compounds with no $-OH$ group, oligophenols, methyl ethers, acetates etc., is then possible under these conditions [116]. The same separation can be obtained on correspondingly impregnated silica gel layers [15] for which dibutyl ether appears to be a good solvent.

2. *Benzene-acetic acid-water* (57 + 28 + 15), upper phase [17, 239]; this solvent is convenient for oligophenols and phenolcarboxylic acids which migrate too rapidly in the mixtures below.

3. *Chloroform-acetic acid-water* (50 + 45 + 5); this is suitable for TLC of polyphenols, phenolcarboxylic acids, flavone-aglycones and monoglycosides. The corresponding methyl ethers can be recognised through their much higher hRf-values [36, 48]. It is important to apply

only very little of these substance mixtures in order to obtain good separations. Column C, I, Table 164, contains the relevant hRf-values.

4. *Ethyl acetate-formic acid-water* (66 + 14 + 20); this mixture migrates fast and is appropriate for separating both aglycones and glycoside mixtures [96]. Di- and triglycosides remain in the vicinity of the start, however.

5. *Butanol-acetic acid-water* (40 + 10 + 50); the upper phase of this "Partridge" mixture yields with all glycosides excellently sharp resolution which cannot be surpassed even on polyamide or silica gel. The aglycones have, however, very high hRf-values and the time of run is relatively long. The hRf-values are given in Table 165, column C, IV.

6. *Water*, containing a little acetic acid: this permits differentiation of glycosides, phenolcarboxylic acids, catechins, dihydroflavones and flavan-3,4-diols. It is a particularly good second solvent in two-dimensional chromatography [184].

7. *Propanol-ammonium hydroxide-water;* this is a popular solvent mixture (in various proportions) for quinones. These usually have too high Rf-values with solvents 2—5 in the list. They migrate as phenolates in the basic solvent and are thus separable from one another.

Further information can be found in the specialist PC literature. Along with the general monographs [72, 74] articles on the following compound classes may be mentioned: cinnamic acid glucosides [84]; lignans [116]; stilbenes [88]; phenolcarboxylic acids [17, 239]; flavonoids [36, 80, 81, 83]; anthocyanins [79]; and phenols [53].

The separation effect of cellulose can be combined with that of polyamide in a two-dimensional procedure: A 4-cm wide margin is removed from a ready, dried cellulose layer. The remaining part of the layer is moistened with high-boiling petrol ether and a polyamide layer is then spread on to the free, cleaned plate surface using the pouring procedure. The petrol ether prevents the polyamide slurry from flowing on to the cellulose layer. The two layers are then divided from each other by a 1 mm wide cut. Separation is carried out first on the polyamide layer. The two layers are then united by filling up the cut between them with polyamide and the substances transferred to the cellulose layer by brief development in the second dimension using a 0.5% solution of ammonia in methanol. This is then interrupted and, after drying, the development is continued with the Partridge mixture (see above, solvent 5).

b) Silica Gel

The separation patterns on silica gel layers resemble those in PC; the rules of thumb of BATE-SMITH and WESTALL [8] therefore apply:

hRf (R–CH$_3$) > hRf (R–H) > hRf (R–OCH$_3$) > hRf (R–O-sugar).

If the lipophilic nature of a compound is intensified, e. g., by introducing an alkyl group, the hRf-value rises; polar substituents decrease it. The introduction of a carboxyl group has about the same influence as that of a hydroxyl group; a methoxy and an acetoxy group are roughly equivalent in this respect.

The capacity of the silica gel layer is substantially higher than that in PC. An additional advantage of purely inorganic layers is that the substances may be visualised with aggressive reagents also. It must be borne in mind when using silica gel, however, that substances containing the o-diphenol grouping are easily oxidised, especially on the dried layer; this generally gives rise to a brown coloration. Flavone solutions also turn brown after application. In this case, however, no decomposition has yet taken place; this occurs only after longer standing in the air.

The choice of solvent is governed primarily by the polarity of the substance mixture to be separated; the more polar the mixture, the more polar the solvent must be, so that it is a more powerful eluent. Table 166 contains a selection of the most commonly used and satisfactory solvents. Many others are mentioned in the text and in Table 170.

Table 166. *Customary solvents and their fields of application with silica gel layers.* Additional solvents are in the text and Table 170

No.	Solvent composition	Compound classes	Refs.
I	Benzene-chloroform (50 + 50)	A: Lichen constituents	[203]
II	Benzene-acetone (90 + 10)	A: Coumarins, polyphenol acetates, lignans, methyl ethers	[203, 238]
III	Chloroform-methanol (97 + 3)		[176, 204, 208]
IV	Benzene-ethyl formate-formic acid (75 + 24 + 1)	B: Anthraquinone derivatives[a]	[203]
V	Toluene-ethyl formate-formic acid (50 + 40 + 10)	B: Phenols, flavone aglycones, phenolcarboxylic acids, hydroxycoumarins	[203]
VI	Toluene-chloroform-acetone (40 + 25 + 35)	B: Compound classes as in V	
VII	Ethyl acetate-butanone-formic acid-water (50 + 30 + 10 + 10)	C: Glycosides, anthocyanins	[203]
VIII	n-Butanol-n-propanol-2N ammonium hydroxide (10 + 60 + 30)	Hydroxyquinones (as enolates)	[169]

[a] See Table 170 also.

According to STAHL and SCHORN [203], *constituents of lichens* can be separated on acidic silica gel layers using solvent I in Table 166. Identification is aided when the hydrolysis products of the lichen acids are chromatographed at the same time as the acids themselves.

Since the content of 0.5—5% is comparatively high, about 50 mg of finely powdered dry material is adequate for identification of a lichen. This is extracted with several 0.5 ml portions of boiling acetone. The filtered, combined extract is concentrated to ca. 0.5 ml and 10—100 µl amounts are applied. For the acid hydrolysis, the extract is evaporated to dryness, 1—2 drops concentrated sulphuric acid added, the mixture left 5—10 min, then diluted with 2 ml water and the hydrolysis products extracted with ether [228].

The acidic silica gel layers are prepared by using 0.5 N oxalic acid solution instead of water.

Particularly good colour differentiation is obtained with the anisaldehyde-sulphuric acid reagent (No. 11). The following hR_f-values and colours have been obtained under the conditions mentioned above: vulpinic acid, hR_f 80, yellow; usnic acid, 65, violet; evernic acid, 11, red; and the decomposition product orcinol, 3, red. Further details about lichen constituents are given by NUNO [159] and RAMAUT [174].

Non-polar coumarins like coumarin itself and its precursors [237, 238] can be chromatographed on silica gel layers with benzene or with the solvents II, III or V (Table 166, p. 694). hR_f-values were quoted first by STAHL and SCHORN [203], then by TSCHESCHE and co-workers [218] and HÖRHAMMER et al. [103]. BEYRICH [15] has found that especially good separations of the furocoumarins could be achieved on silica gel layers, impregnated with formamide; he used dibutyl ether as solvent.

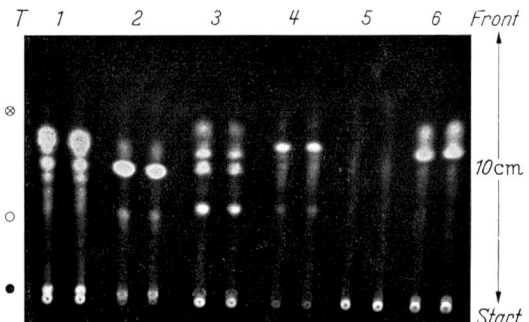

Fig. 191. Distinction of Umbelliferae drugs through TLC. Alcoholic extracts chromatographed with solvent V (Table 166) on a silica gel layer. Photographed in long-wave UV light without using any spray reagent

1 fruit of *Ammi majus* L., *2 Ammi visnaga* L.,; *3* Radix Angelicae; *4* Rhizoma Imperatoriae; *5* Radix Levistici; *6* Radix Pimpinellae [203]

As can be seen in Fig. 191, many Umbelliferae drugs may be distinguished on the basis of their different constituents by using solvent V (Table 166). Other investigators have had comparable results [31, 64, 155]. It may be mentioned that benzotetronic acid and 4-hydroxycoumarin are appreciably more strongly adsorbed than 7-hydroxycoumarin [31].

The mould *Aspergillus flavus* forms carcinogenic "aflatoxins" (furocoumarins) from peanut flour. TLC in combination with column chromatography under similar conditions has proved very useful in the isolation and identification of these compounds [27, 29, 61, 187]. The TLC-separation was carried out with chloroform containing 1—3% methanol, on silica gel G or neutral alumina layers. The aflatoxins display intense blue to blue-green fluorescence, permitting their easy detection; the limit of this detection is 3—6 ng.

Flavonoid acetates and *methyl ethers* have been separated by WEINGES and TORIBIO [236] using solvent II (Table 166). Catechin-resorcinol condensation products in particular can be resolved in this way [236]. Catechin acetate diastereoisomers and the acetates of oligomeric flavonoid tannins such as the acetates of dicatechin and anhydrocatechin can be thus detected with certainty [56]. The acetates of the more highly condensed compounds remain at the start point. Plant extracts containing catechins and flavonoid tannins are best acetylated initially by 15 h treatment with acetic anhydride/absolute pyridine at room temperature. Solvent II or chloroform-ethyl acetate (90 + 10) can be used with silica gel columns for preparative separation of such mixtures.

The acetates, methyl ethers and esters of the remaining flavonoid groups, coumarins and phenol carboxylic acids can be chromatographed on silica gel layers with benzene or benzene-acetone (90 + 10). Silica gel G layers and benzene-ethyl acetate (75 + 25) were used in the discovery and characterisation of digicitrin [145]; the following hR_f-values were found:

5,3'-dihydroxy-3,6,7,8,4',5'-hexamethoxyflavone (digicitrin)	38—40
3,5,6,7,8,3',4',5'-octamethoxyflavone	24—28
5-hydroxy-3,6,7,8,3',4',5'-heptamethoxyflavone	64—66
5,3'-dibenzyloxy-3,6,7,8,4',5'-hexamethoxyflavone	68—71
2-hydroxy-ω,3,4,5,6,-pentamethoxyacetophenone	44—47

Table 167. hR_f-values of some coumarin derivatives on silica gel layers

Substance	Solvent	hR_f	Ref.
Daphnoretin	1	30	[218]
Daphnoretin methyl ether	1	60	[218]
Daphnoretin	2	20	[218]
6,7-Dimethoxycoumarin	2	35	[218]
3-Bromo-6,7-dimethoxycoumarin	2	55	[218]
Umbelliprenin	3	80	[103]
Oxypeucedanin	3	40	[103]
Imperatorin	3	53	[103]
Athamantin	V	68	[203]
Bergapten	V	68	[203]
Xanthotoxin	V	56	[203]
Imperatorin	V	64	[203]

1 = chloroform-acetone (83 + 17); 2 = chloroform-methanol (98 + 2); 3 = chloroform-methanol (99 + 1); V = V in Table 166.

PARIS [163, 164] also reports very good separations of flavonoid mixtures on silica gel-starch layers, using the solvents ethyl acetate-chloroform, ethyl acetate-methanol (both 95 + 5) or hexane-isopentanol-acetic acid, saturated with water.

Methylation of the 5-hydroxyl group which is in peri position to the carbonyl group, lowers the hR_f-value markedly; etherification of the 3-hydroxyl group increases the hR_f-values as expected.

Solvents containing formic acid should be avoided as far as possible in chromatography of acetylated compounds; transesterification and saponification may occur.

Simple phenols can generally be separated by solvent VI, Table 166. Advantage is often gained by prior coupling with a stable diazonium salt [121] (see p. 223). Some hR_f-values of simple phenols are given in Table 164.

Quinones: BARBIER [6] has separated some benzoquinones on silica-starch layers, using hexane-ethyl acetate (85 + 15). PETTERSSON [169] later made a detailed comparison of the chromatographic behaviour of benzoquinones, of which 31 contained hydroxyl group(s) and 21 were without. Neutral solvents, especially mixtures of benzene-chloroform and xylene, were suitable for the last named. The basic solvent VIII (Table 166) was better for the more polar hydroxy-quinones; they migrate as enolates in this system. The hR_f-values on silica gel layers of some recently discovered [170] fungus colouring materials are given below; they were obtained with solvent VIII (values without brackets) and chloroform (values in brackets):

2,3-dimethyl-5,6-dimethoxyquinone	98 (57)
2,3-dimethyl-5-methoxy-6-hydroxyquinone	65 (26)
2,3-dimethyl-5,6-dihydroxyquinone	40 (0)

Polyphenols, flavonoid aglycones: STAHL and SCHORN [203] have shown that solvent V (Table 166) can be used successfully for separating this group of compounds on "basic" silica gel layers. The layers must be prepared with 0.3M sodium acetate solution instead of water. The glycosides remain at the start under these conditions. Good separations can be obtained on non-impregnated layers also with solvent VI. The hR_f-values are given in Table 164, column S, II. HÖRHAMMER and co-workers [97, 98, 102] have tested some solvents customarily used in PC, yet opted for the mixture benzene-pyridine-formic acid (72 + 18 + 10) [111], which permits separation even of critical pairs like genistein/apigenin and rhamnetin/isorhamnetin. The last named has a higher hR_f-value than kaempferol.

Aurones and *chalkones* can likewise be separated by TLC, as has been shown by HÄNSEL and co-workers [75]; using the solvent benzene-ethyl acetate-formic acid (45 + 25 + 20), the aurones are separated according to their polarity, in the hR_f-region between 20 and 90. The glycosides stay near the start. Mixed layers of silica gel and kieselguhr and the solvent cyclohexane-ethyl acetate (70 + 10), saturated with formamide-water (2 + 1), have been recommended for separation of chalkones. Flavokawin A and B [76], differing by only a methoxyl group (in A), have been separated using this solvent on pure silica gel layers; the hR_f-values for A and B were 37 and 51, respectively.

Xanthones from *Gentiana* species have been separated on silica gel with benzene-ethyl acetate-ethanol (50 + 43 + 7); the corresponding acetates have been chromatographed using the same solvent mixture but of composition (72 + 25 + 3) [144].

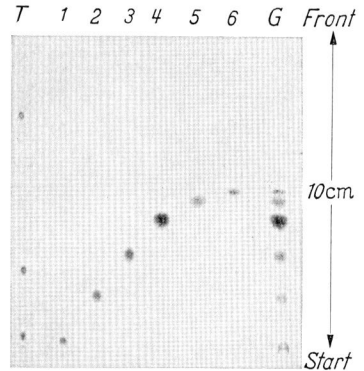

Fig. 192. TLC of some phenolcarboxylic acids on a silica gel G layer using solvent V (Table 166); after visualisation with Rgt. No. 168.

1 chlorogenic acid, hR_f 7; *2* m-digallic acid, hR_f 27; *3* gallic acid, hR_f 39; *4* caffeic acid, hR_f 47; *5* gentisic acid, hR_f 51; *6* ferulic acid, hR_f 56, *G* mixture (1—6) [203]

Phenolcarboxylic acids can be separated on silica gel layers with mixture V (Table 166) as shown by STAHL and SCHORN [203] also (Fig. 192). Other investigators have worked on the separation of this substance class [43, 73, 127, 141, 224]. Isoamyl ether-butanol (75 + 25) has been used with success in the separation of some degradation products of lignin [127]; the following hR_f-values were obtained: syringic acid 24; p-hydroxybenzoic acid 31; vanillin 58; p-hydroxy-benzaldehyde 67. HALMEKOSKI [73] has examined the influence on the hR_f-values of impregnation with chelate-forming

Table 168. hR_f-*values of some phenolcarboxylic acids on silica gel layers.*
F1, F2: values from PASTUSKA [165]; F3, F4: from HALMEKOSKI [73]

Acid	Solvent:	F1	F2	F3				F4			
	Layer:	St	St	St	Mo	Tu	Bo	St	Mo	Tu	Bo
Protocatechuic acid		32	39	38	10	24	10	81	71	15	4
Caffeic acid		24	43	31	14	22	8	65	85	31	2
Vanillic acid		54	61	45	44	39	34	82	93	38	53
Ferulic acid		50	58	35	34	36	28	63	87	37	50
Isoferulic acid		43		30	25	31	25	60	86	33	45

Solvents: F1: Benzene-dioxan-acetic acid (90 + 25 + 4); F2: Benzene-methanol-acetic acid (90 + 16 + 8); F3: n-Butyl ether (water-saturated)-acetic acid (90 + 9); F4: Ethyl acetate-isopropanol-water (65 + 24 + 11).

Layers: St: Standard layer; Mo, Tu, Bo: Layer prepared under standard conditions using 0.01 M solutions of sodium molybdate, tungstate and borate, respectively.

salts; some of these results are quoted in Table 168, together with those of PASTUSKA [165]. It is noteworthy that salicylic acid always has a higher hR_f-value than the m- and p-hydroxybenzoic acids. This is probably connected with inner chelation through hydrogen bond formation. Phenols and phenolcarboxylic acids can be separated also by thin-layer electrophoresis. TLC can be advantageously combined with electrophoresis in a two-dimensional procedure [166] (see p. 113 and Table 12, p. 112).

LYMANN and co-workers [141] have demonstrated in the case of isomeric dihydroxybenzoic acids that the separation sequence can be considerably modified by skilful choice of solvent (Table 169).

Table 169. hR_f-values of isomeric dihydroxybenzoic acids
α-, β- and γ-resorcyclic acids, gentisic (2,5-) and protocatechuic (3,4-) acids

Solvent	V + V	3,5 (α)	2,4 (β)	3,6 (γ)	2,5	3,4
Hexane-diethyl ether	(30 + 70)	21	57	10	35	32
Hexane-ethyl acetate	(25 + 75)	61	85	15	65	55
Hexane-acetone	(75 + 25)	73	19	10	17	10

Glycosides: Flavone-, coumarin- and cinnamic acid glycosides can be clearly separated from one another with solvent VII, tested by STAHL and SCHORN [203]. The hR_f-values are cited in Table 165, column S, I. The sequence of substances on the chromatogram is in accordance with the rules of thumb already given. Anthocyanins also can be chromatographed with this mixture. Their aglycones, the anthocyanidins, can be developed with the solvent ethyl acetate-formic acid-water-conc. hydrochloric acid (55 + 6 + 8 + 1). The hydrochloric acid is added to stabilise the aglycones [34, 160, 164, 212]. Strongly polar solvents like VIII (Table 166) are suitable for separating enolates and have been discussed already under quinones.

c) Polyamide and Other Polymers

The application of polyamide to the chromatography of phenolic plant constituents has brought considerable advantages; it has been studied in particular by the teams of ENDRES [44, 45, 68] and HÖRHAMMER [91, 92, 93, 94]. The properties of this interesting adsorbent are described by ENDRES on p. 41—44.

Our own experiments have shown that the addition of 10—20% cellulose powder to the suspension of polyamide powder in methanol gives a more stable layer, yet does not affect the hR_f-values. Thin layers, between about 0.1 and 0.2 mm thickness, are best. Good separations can be obtained on polyamide layers with both water-alcohol mixtures and lipophilic solvents. The experience accumulated so far is summarised below:

Polar Solvents

The following rough rules have been established for TLC on polyamide layers, using water-alcohol mixtures [45, 93]:

1. The more isolated phenolic hydroxyl groups a substance possesses, the more strongly it is retained on the layer.

2. o-Dihydroxy and vicinal trihydroxy groups in the molecule influence the chromatographic behaviour to about the same extent as a single hydroxyl group.

3. Reaction of a hydroxyl group to form a glycosidic linkage greatly increases the hRf-value of the compound. The value depends, however, on the nature of the sugar. The compound is not retained on the polyamide adsorbent when it no longer contains a free phenolic hydroxyl group.

4. Phenols adhere most strongly when water is the solvent. An "eluotropic series" (p. 43) can thus be established.

In accordance with these rules, many substances fall in the same Rf-region. One group is made up of p-coumaric, ferulic and caffeic acids; others are the flavonols myricetin, quercetin, isorhamnetin and kaempferol and the flavones luteolin and apigenin. This applies equally to the glucosides of each of these groups. Consequently the position on the chromatogram, while yielding no information about the aglycone part, provides a clue to the corresponding sugar moiety [40, 42]. Fig. 193 shows the position of the commonest types of flavonol glycoside. The sequence of increasing hRf-value is: glucuronide < rhamnoside < glucoside < rhamnoglucoside < diglucoside < 3-galacto-7-rhamnoside < 3-glucoside-7-glucoside < 3-rhamnoglucoside-7-glucoside < 3-xyloglucoside-7-glucoside.

It may be mentioned in passing that the individual zones are well formed when some butanone or acetylacetone is added to the solvent. The mixture water-ethanol-butanone-acetylacetone $(65 + 15 + 15 + 5)$ has proved very good (Figs. 193 and 196 and Table 165 [38, 42]). The above-mentioned rules hold when this solvent is used, as has been shown by investigations on Solanaceae [21, 133] and Ranunculaceae [41]. Solvents with more powerful elution capacity, such as water-methanol-butanone $(40 + 30 + 30)$ (Table 164) must be used for separating the free cinnamic acids. The flavonol aglycones require a mixture of methanol-butanone $(60 + 40)$, although this yields only meagre differentiation of the commonly occurring aglycones (see Fig. 194).

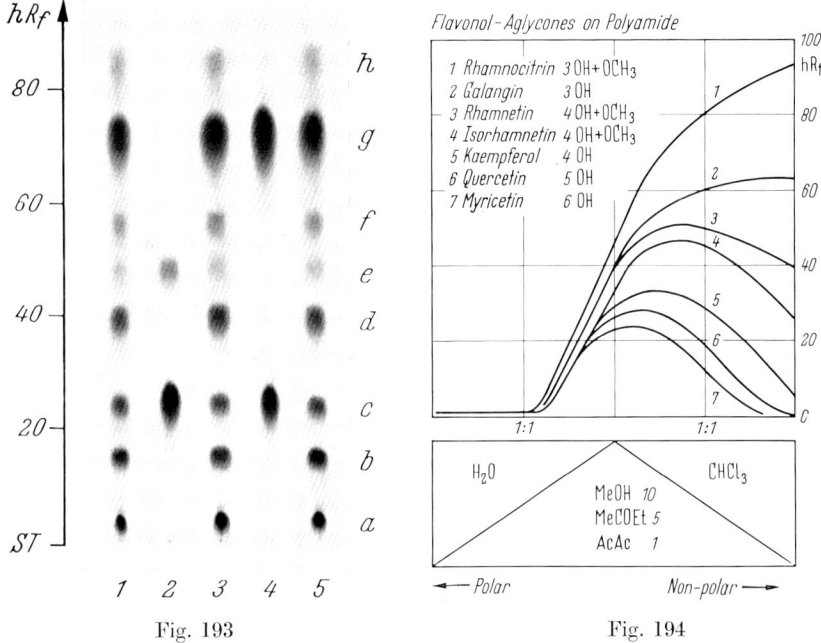

Fig. 193. Fig. 194.

Fig. 193. Thin-layer chromatogram on polyamide showing the sequence of flavonol glycosides. Solvent: water-ethanol-butanone-acetylacetone (65 + 15 + 15 + 5). *a* quercetin-3-glucuronide; *b* kaempferol-3-rhamnoside; *c* K-3-glucoside; *d* rutin-(Q-3-rhamnoglucoside); *e* K-3-glucoside; *f* quercetin-3-galactosido-7-rhamnoside; *g* kaempferol-3-glucoside-7-glucoside; *h* K-3-xyloglucoside-7-glucoside. Start points 1, 3 and 5: a—h; 2: c + e; 4: c + g. Untreated chromatogram photographed in transmitted UV light

Fig. 194. Dependence of the hR_f-values of flavonol aglycones on the composition of the solvent. Water-containing: only slight group separation; chloroform-containing: separation into the individual components. Note that introduction of a methoxyl group raises the hR_f-value

Lipophilic Solvents for Glycosides

As already implied, different separation results are obtained when lipophilic solvents, e. g., ether-96% ethanol (80 + 20) are used. The constituents of arbutin drugs can be separated on polyamide layers, using this solvent [129, 130]: arbutin hR_f 35, hydroquinone hR_f 80. hR_f-Values of 33, 24 and 12 have been found for astragalin, isoquercitrin and rutin, respectively, using a toluene-containing solvent [16]. Systematic investigation of the influence of lipophilic solvents on hR_f-values has shown that partition processes must operate here as well as the adsorption based on hydrogen bond formation [40, 41]. It can be assumed that when a non-polar lipophilic solvent is used, the polyamide acts more or less as a polar stationary phase; partition processes between the phases are then more marked than adsorption phenomena. Rule 3 above

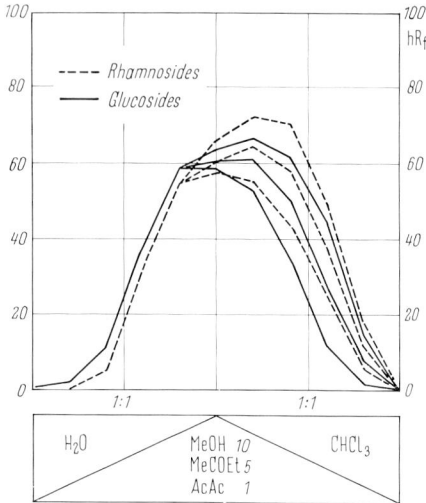

Fig. 195. Dependence of the hR_f-values of flavonol glycosides on the solvent composition. Solvent containing water: separation only into groups, with glucoside > rhamnoside; solvent containing chloroform: separation into components, with glycoside of kaempferol > that of quercetin > that of myricetin

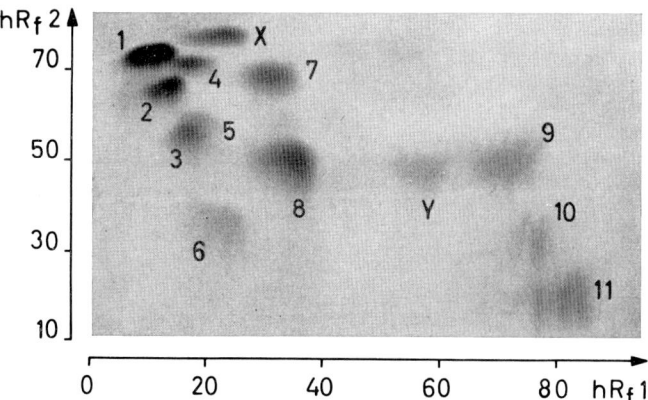

Fig. 196. Two-dimensional chromatogram on polyamide. Solvent 1: water-ethanol-butanone-acetylacetone (65 + 15 + 15 + 5); solvent 2: chloroform-methanol-butanone (60 + 26 + 14) (almost water-saturated).

Substances: *1* kaempferol-3-rhamnoside; *2* quercetin-3-rhamnoside; *3* myricetin-3-rhamnoside; *4* kaempferol-3-glucoside; *5* quercetin-3-glucoside; *6* myricetin-3-glucoside; *7* kaempferol-3-rhamnoglucoside; *8* rutin; *9* robinin; *10* kaempferol-3-glucoside-7-glucoside; *11* kaempferol-3-rhamnoglucoside-7-glucoside; X: unknown compound; Y: quercetin-3-galactoside-7-rhamnoside (formation of groups) then no longer applies and glycosides with similar sugar moieties but different aglycone parts can be separated. Figs. 194 and 195 show the transition from aqueous development (chiefly adsorption) to a more lipophilic development (adsorption + partition).

These two types of separation can also be combined in a two-dimensional chromatogram, as seen in Fig. 196. First, the separation of the glycoside groups is carried out with an aqueous solvent. This may be followed by separation into the derivatives of the various aglycones, using benzene-methanol-butanone (60 + 20 + 20). The shape of the glycoside spots (but not of the aglycone spots) can be still bettered when the solvent used in the second chromatographic direction is saturated with water. It must, however, remain a single phase. The chromatogram must be air-dried for several hours before the second development. Similar results are obtained when the chloroform is replaced by benzene and the butanone by ethyl formate.

Solvents for Aglycones and Compounds containing no Hydroxyl Groups

Both aglycones and substances of group B (see p. 689) can be differentiated on polyamide layers with the solvent chloroform-methanol-butanone (60 + 26 + 14). These separations are possible on a preparative scale also [40, 42]. The introduction of a methoxyl group increases the hR_f-value; thus isorhamnetin, for example, migrates farther than kaempferol.

Compounds containing no or only a single phenolic group are thus subject to only weak adsorption on polyamide and can be separated on these layers by partition chromatography. Care must then be taken that the solvent is sufficiently non-polar. Good spot formation of the substances in group A of Table 163, for example, is thus obtained when petrol ether-benzene-butanone-methanol (50 + 40 + 5 + 5) or (60 + 30 + 5 + 5) is employed. The hR_f-values may be suitably modified by variation of the hydrophilic portion of the solvent.

A satisfactory separation of *quinones* also is possible on polyamide layers. Application and development must be speedily carried out, otherwise they may interact irreversibly with the polyamide [44, 45]. This disadvantage may be countered by acetylating the amide groups [68] and a product of this sort is already available commercially (Firm 83). It may be mentioned that lipoquinones and anthraquinones do not react thus. Aloin and aloeemodin rhamnoside yielded hR_f-values of 71 and 47, respectively, when using methanol [106, 107].

ENDRES [45] has investigated the effect of introducing a carboxyl group into a phenol. He obtained the following hR_f-values, using methanol-water (50 + 50) as solvent:

phenol	63	benzoic acid	56
hydroquinone	57	p-hydroxybenzoic acid	34
resorcinol	58	salicylic acid	60
α- and β-naphthol	28	α-naphthoic acid	27
2,2'-dihydroxybenzophenone	27	4,4'-dihydroxybenzophenone	16

It is evident from these values that the hR_f-value is reduced still more by introducing a carboxyl group than by an additional phenolic hydroxyl group. This is probably because the former can form two hydrogen bonds. The R_f-value is high only where inner chelation is possible, as with salicylic acid, for example. This powerful influence of the carboxyl group is encountered also with the derivatives of glucuronic acid. Flavonol glucuronides are more strongly adsorbed than their corresponding aglucones [42, 183]. These substances are, however, easily desorbed from the layer, like all the other phenols, with a 0.1–1% solution of ammonia in methanol. Any confusion of glucosides with glucuronides can thus be avoided.

d) Polyacrylonitrile

A mixture of polyacrylonitrile and polyamide [70 + 20] enables anthocyanins and anthocyanidins to be separated by TLC; this is not possible on pure polyamide layers. BIRKOFER, KAISER and co-workers [20] describe a special continuous development procedure in this connection. They used n-butanol-n-pentanol-n-propanol-acetic acid-water (20 + 30 + 20 + 20 + 10). The higher alcohols improve spot formation on the chromatogram. The following hR_f-values were quoted for the anthocyanins: delphinidin 21; petunidin 27; cyanidin 31; malvidin 37; paeonidin 39; pelargonidin 41. Cinnamic acid and its glucosides also can be separated on such mixed layers [20]. This is, however, possible also, and more simply, on perlon layers [221]. HÄNSEL and RIMPLER [78] have separated some C_6-substituted 4-methoxy-α-pyrones on pure polyacrylonitrile layers, using cyclohexane-ethyl acetate mixtures. A type of wedged-tip technique (p. 89) was used, employing multiple development. The hR_f-values are influenced by the number and especially the position of the double bonds of the so-called Kawa-lactones (from *Piper methysticum* FORST).

e) Ion Exchangers

Exchange resins of the type Amberlite CG 50-III for TLC (see Table 10, p. 46), containing carboxyl groups, have likewise been used for separation of flavone compounds. Isopropanol-water (40 + 6) was used for development [222, 223]. Good separations are accomplished and the sequence of substances on the chromatogram differs from that on perlon or silica gel layers. The hR_f-values found increased in the order:

kaempferol < quercetin < K-3-rhamnoglucoside < Q-3-rhamnoglucoside < astragalin < isoquercetin. There appears to be no further experience of the use of ion exchangers in these and other domains of separation of natural products.

5. Visualisation of Phenol Derivatives

Only the anthocyanins and some quinone derivatives are sufficiently intensely coloured to be directly detectable on the layer; the other phenolic natural products are colourless and must consequently be rendered visible on the white layer. Many can be detected in short- or long-wave UV light; they display fluorescence or quench fluorescence on layers containing an inorganic fluorescent indicator. Exposure to ammonia vapour or spraying with an alkaline solution usually changes characteristically the fluorescence of many substances. Numerous authors, e. g., GEISSMANN [59, 60], HÄNSEL [74], HAIS/MACEK [72] and SEIKEL (in [86]) have worked extensively on these colour reactions and summarised their results in tables. Reagents containing chelate-forming metal salts are particularly worth considering for visualisation. Thus, e. g., ferric chloride (Rgt. No. 102) yields intense colours in the ochre to violet range with polyphenols; aluminium chloride (Rgt. No. 3), 2% zirconium oxychloride in methanol and also lead acetate (Rgt. No. 151) give, especially with flavone compounds, yellow to orange spots which frequently fluoresce intensely in long-wave UV light. The β-aminoethyl ester of diphenylboric acid (Rgt. No. 86), tested by NEU [152], yields a particularly good differentiation of the flavones. o-Dihydroxy groups can be detected with Benedict's reagent (No. 56); this group is responsible for fluorescence quenching whereas substances which do not contain the group (e. g., coumarins, cinnamic acid, flavonoids [177]) usually fluoresce when sprayed with the reagent; it cannot, however, be used on polyamide layers.

Most phenols yield intensely coloured orange and violet compounds by coupling with suitable diazonium salts (Rgt. Nos. 100, 238, 240). Phenols with reducing properties can be identified also with a silver nitrate reagent (No. 225); dark spots are yielded on a background which is first light, then gradually darkens. The 2,6-dibromoquinonechloroimide reagent (No. 62) also gives intense colours with phenols. Blue spots on a yellow background (see Fig. 192) are obtained by the reaction of phenols and numerous other compounds with the molybdophosphoric acid reagent (No. 168). The Folin-Ciocalteu reagent (No. 122) yields similar colours. Acids (in this case, phenol carboxylic acids) can be detected with the usual pH indicators, e. g., bromocresol green (Rgt. No. 25); acid or basic solvent components must be wholly removed from the layer beforehand. Spray reagents containing strong inorganic acids cannot be employed on a polyamide layer since they dissolve it.

II. TLC in the Characterisation of Animal and Plant Drugs

Egon Stahl and P. J. Schorn

The methods which have been used hitherto for morphological and anatomical drug comparison can yield valuable information concerning the plant or animal of origin but not about the nature and amount of the active substances. Moreover, the drug may be present as an extract; and microscopic analysis fails with very finely powdered drugs. Despite all the value which has been attributed to this classical method, it is really no more than a "finger print" technique. The current goal is to obtain, in a rapid test, the individual substances to which the activity is due; chromatography (TLC, GC) is the appropriate tool in this endeavour.

1. Anthraquinone Drugs

The therapeutic application of some drugs is due to 1,8-dihydroxy-anthraquinones they contain. The single compounds may occur in the quinone, anthranol or anthrone form, or possibly as glucosides or dimers [195, 227]. As a rule, mixtures of many compounds of this class occur in the plants. Since their activities are not identical, the analytical determination of the individual compounds is attempted.

The vast majority of publications in this connection is concerned with the *Aloe* natural drugs. Fundamental work on the TLC of the anthraquinone drugs has been carried out by Hörhammer's team [104, 105, 106, 113] after we had previously pointed out the benefits of a TLC-analysis of such plant extracts (see Fig. 6 in [203]).

Most authors prefer silica gel layers. The first solvent to be considered was the mixture ethyl acetate-formic acid-water (60 + 12 + 18), used in PC. It is, however, best to replace the formic acid by methanol; this mixture is stabler and can be more easily removed from the layer (Table 170, I).

The solvents in Table 170 are suitable for TLC-separation of the constituents of *Aloe*, *Rhamnus* and *Senna* species and also for tinctures and extracts prepared from them [23—26, 108, 114, 142, 143, 178—182, 213]. Solvent I has an especially wide range of application with mixtures of anthraquinone glucosides. Solvent VII is more suitable, however, for the dianthrone glucosides.

Table 170. *Solvents for separating anthraquinone derivatives on silica gel G layers*

For anthraquinone glycosides
    I Ethyl acetate-methanol-water (100 + 16.5 + 13.5) [104]
   II Benzene-ethyl formate-formic acid (75 + 24 + 1) [203]
  III Benzene-acetic acid (66 + 33) [149]
  IV Chloroform-95% ethanol (75 + 25) [63]
   V Chloroform-95% ethanol-water (60 + 30 + 2) [25]
  VI Methylene dichloride-methanol (83.5 + 16.5) [134]
 VII n-Propanol-ethyl acetate-water (40 + 40 + 30) [104]
VIII a) n-Butanol, water-saturated (1. stage)
       b) n-Propanol-ethanol-chloroform-water-acetic acid (40 + 40 + 12 + 16 + 4) (2. stage) [135]

For the aglucones
  IX Benzene [109]
   X Benzene-methanol (90 + 10) [110]
  XI Isopropyl ether [106]
 XII Heptane-benzene-chloroform (33 + 33 + 33) [114]

The spot positions of numerous anthraquinone derivatives after development on silica gel layers, is seen in Fig. 197.

Single compounds in drug extracts have been determined quantitatively by separating with solvents I, IV or V on silica gel layers, scraping off the zones, eluting and comparing light absorption of the eluates with that of standard solutions [25, 63, 104].

The individual *Aloe* anthraquinones are usually obtained in preparative amounts on polyamide columns, the purity being monitored by TLC [105, 106]

The same conditions can be chosen in principle for TLC of the *Frangula* and *Rheum* constituents as for *Aloe* (Fig. 197). KORTE and co-workers [134, 189] have separated *Frangula* anthraquinones on silica gel G layers using solvent VI or on acid layers (silica gel G impregnated with 0.5 N oxalic acid) using methylene dichloride-methanol (100 + 5). The methylated aglucones can be satisfactorily separated on normal silica gel G layers using solvents IX—XII. Quantitative determination has also been described [134].

POETHKE and co-workers [172, 173] have succeeded in separating and identifying the anthraquinone derivatives in *Oreoherzogia fallax* (BOISS.) W. VENT (syn *Rhamnus fallax* BOISS.) on silica gel layers using solvents I, II and XI.

TLC of the *Senna* aglucones (= 10,10'-dianthrones) has been carried out with solvent III on silica gel G layers; sennidin A and B migrated together (hR_f 65) but were separated from rhein (hR_f 72). The glucosides can be separated with solvent VII; the hR_f-values quoted were: sennoside A 30; sennoside B 17; rhein-8-glucoside 58; and rhein 68 [104]. LONGO and co-workers [135] have employed stepwise development on silica

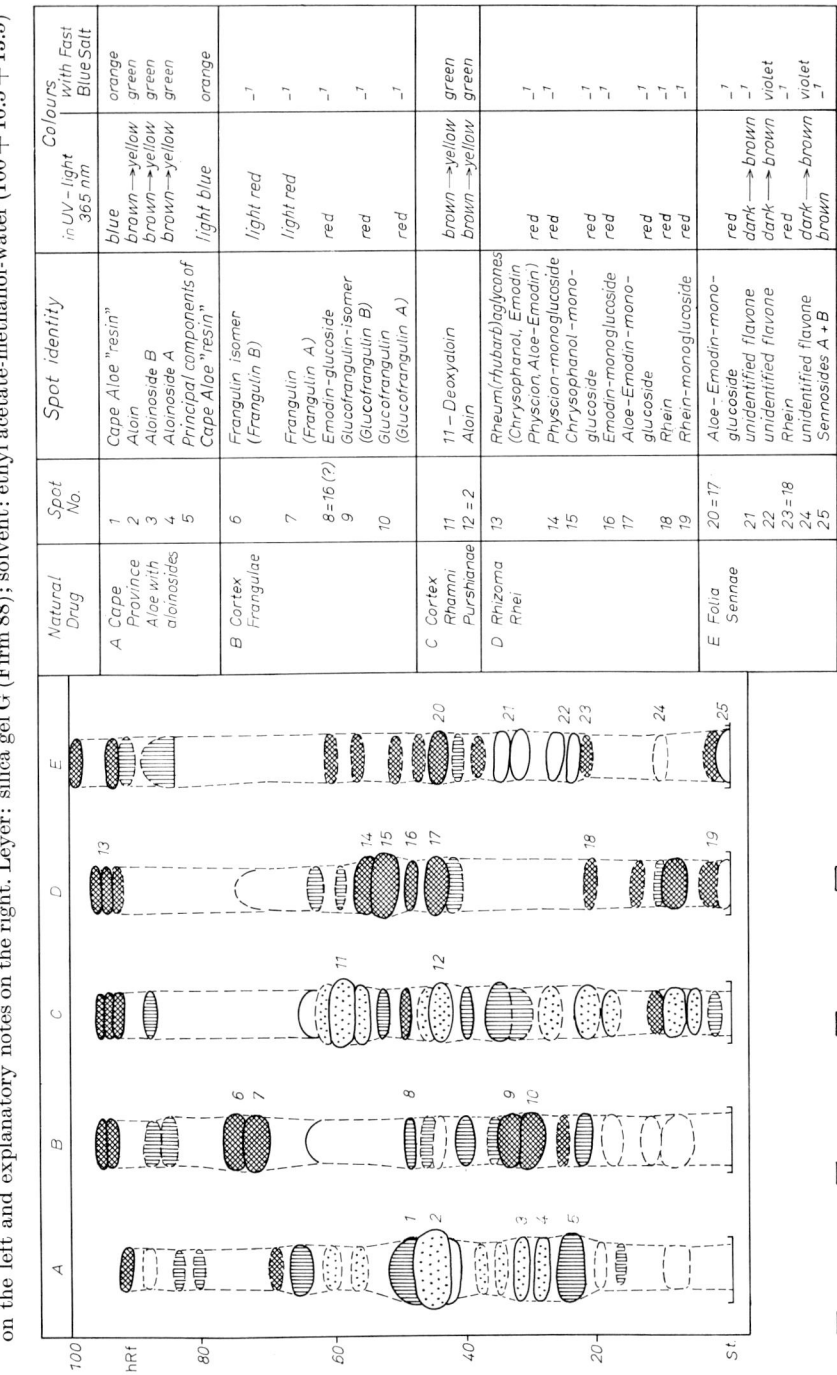

Fig. 197. A = Cape Aloe; B = Cortex Frangulae; C = Cortex Rhamni Purshianae; D = Rhizoma Rhei; E = Folia Sennae. The spot colours quoted were obtained in daylight with Fast Blue Salt (Rgt. No. 100) with the untreated chromatogram by inspection in UV light.

TLC of important anthraquinone drugs according to Hörhammer, Wagner and Bittner [104]; schematically drawn chromatogram on the left and explanatory notes on the right. Layer: silica gel G (Firm 88); solvent: ethyl acetate-methanol-water (100 + 16.5 + 13.5)

Natural Drug	Spot No.	Spot identity	Colours	
			in UV-light 365 nm	with Fast Blue Salt
A Cape Province Aloe with aloinosides	1	Cape Aloe "resin"	blue	orange
	2	Aloin	brown→yellow	green
	3	Aloinoside B	brown→yellow	green
	4	Aloinoside A	brown→yellow	green
	5	Principal components of Cape Aloe "resin"	light blue	orange
B Cortex Frangulae	6	Frangulin isomer (Frangulin B)	light red	-?
	7	Frangulin (Frangulin A)	light red	-?
	8 = 16 (?)	Emodin-glucoside	red	-?
	9	Glucofrangulin-isomer (Glucofrangulin B)	red	-?
	10	Glucofrangulin (Glucofrangulin A)	red	-?
C Cortex Rhamni Purshianae	11	11-Deoxyaloin	brown→yellow	green
	12 = 2	Aloin	brown→yellow	green
D Rhizoma Rhei	13	Rheum (rhubarb aglycones (Chrysophanol, Emodin Physcion Aloe-Emodin)	red	-?
	14	Physcion-monoglucoside	red	-?
	15	Chrysophanol-monoglucoside	red	-?
	16	Emodin-monoglucoside	red	-?
	17	Aloe-Emodin-monoglucoside	red	-?
	18	Rhein	red	-?
	19	Rhein-monoglucoside	red	-?
E Folia Sennae	20 = 17	Aloe-Emodin-monoglucoside	red	-?
	21	unidentified flavone	dark→brown	-?
	22	unidentified flavone	dark→brown	violet
	23 = 18	Rhein	red	-?
	24	unidentified flavone	dark→brown	violet
	25	Sennosides A + B	brown	-?

-? = no characteristic colour

▨ = yellow   ▦ = blue or green   ▩ = red   ☐ = brown or dark gel H layers for separation; they developed with solvent VIIIa in the first step of 7 cm and with solvent VIIIb in the second step.

The plant excretion chrysarobin is a mixture of various anthracene derivatives and can be chromatographed just like the synthetic 1,8-dihydroxyanthranol (cignolin, anthralin, dithranol) on silica gel G layers with solvent XII; it may be detected in this way in the corresponding preparations [14].

Cassiamin, a 2,2'-dianthraquinone derivative, has been isolated from the bark of *Cassia siamea* LAM. and separated from other anthraquinones on silica gel G layers using benzene-acetone (80 + 20) [35]. Similar compounds have been found as metabolites of *Penicillium* species; thus, with the help of TLC, skyrin, a 1,4'-dianthraquinone derivative, has been detected in the mycelium of *Preussia multispora* [150].

TLC has been useful also for following the fermentative oxidation of rhein-anthrone to rhein and rhein-dianthrone [136]; solvent II on silica gel G was used.

It has been possible to separate mixtures of the aminoanthraquinones which are intermediates in dye preparations; neutral alumina layers were used with cyclohexane-ether (50 + 50). The individual components could then be determined directly by remission measurements at 510 nm with the absorption-recording apparatus ERI-10 (Firm 156) [49].

Visualisation: Anthraquinone derivatives may be detected on silica gel GF_{254} layers through their quenching of short-wave UV light. They fluoresce brown-yellow to red in long-wave UV light (see Fig. 197). After spraying with alkali solution, a more intense yellow, orange or red fluorescence is observed in long-wave UV light. If this is followed by further spraying with Fast Blue B salt (Rgt. No. 100), the fluorescent zones of many compounds become visible in daylight (orange-yellow, violet or green — see explanatory notes to Fig. 197). 2,6-Dichloroquinonechloroimide (Rgt. No. 66), followed by spraying with a 10% solution of sodium carbonate in 30% methanol, yields brown with the "resins" and blue-green or violet zones with the anthraquinones.

Detection of the brown-fluorescing sennosides in long-wave UV light, is non-specific. It is thus expedient to spray the chromatogram with 25% nitric acid and then to heat the plate 10 min at 120° C. This causes decomposition and oxidation to the corresponding anthraquinone derivatives which can be rendered visible in UV light after spraying with alkali; or in daylight when this is succeeded by spraying with the Fast Blue B salt (Rgt. No. 100).

2. Lignan Drugs

The compounds termed lignans by HAWORTH [89] are formed by dimerisation of two phenylpropane derivatives at the β-carbon atom of their propane side-chain. The *Podophyllum* lignans, some of which have a laxative and antineoplasmatic affect, aroused special interest in recent years; HARTWELL and SCHRECKER [87] and others have surveyed their chemistry.

As STAHL and KALTENBACH [204] showed in 1961, the *Podophyllum* lignans and their glycosides can be quickly separated through stepwise development on a silica gel G layer; experimental details are given in the title to Fig. 41, p. 87. STEINEGGER and GEBISTORF [208] took up later the quantitative determination and laid particular store by the separation of α- and β-peltatin. They used silica gel G layers for this purpose also and developed with chloroform-methanol (97 + 3) to a height of 11 cm; the lignan- and flavone glycosides remained unseparated at the start.

Visualisation has been carried out by spraying with a 0.1 N silver nitrate solution. α-Peltatin and demethylpodophyllotoxin yield jet-black spots already in the cold and β-peltatin turns brown-black. Only after spraying with large amounts of reagent do the other compounds appear as white, fatty spots on the transparent layer; they turn brownish after several hours. It is simpler to develop on silica gel GF_{254} layers, so that these substances may be detected through their fluorescence quenching in short-wave UV light.

KUHN and VON WARTBURG [131, 230] have chromatographed *podophyllum* lignans and their glycosides on silica gel G layers, using the solvents:

chloroform-methanol-water (70 + 25 + 5)
ethyl acetate-isopropyl acetate-ethanol-water (40 + 40 + 15 + 15)
dimethyl sulphoxide-chloroform-acetic acid (10 + 60 + 10)

The same team [176] worked later on the synthesis of neopodophyllotoxin and on the transesterification products of podophyllic acid. Silica gel layers were again used for the TLC, the solvent being chloroform which contained from 1 to 6% methanol, depending on the separation problem. Visualisation was carried out by spraying with 50% sulphuric acid containing 0.2% ceric sulphate and heating to 120° C.

SCHORN [200] has shown that both lignans and their glycosides can be separated in a single development by using chloroform-methanol (95 + 5) on a gradient layer from kieselguhr G to silica gel GF_{254} (length of run 13 cm). The sequence is that of Fig. 41. TLC on magnesol layers with benzene-chloroform mixtures ought to be successful also, analogous to the column chromatographic separation of podophyllin described by RÜTTIMANN and FLÜCK [186].

GENSLER and GATSONIS [62] have used TLC in the study of the isomerisation of podophyllotoxin to picropodophyllin which is without laxative activity. AUTERHOFF and THEILACKER [5] have chromatographed the semi-synthetic azo- and aminopeltatins on silica gel G layers, using methylene dichloride-ether (80 + 20) or methylene dichloride-ethyl acetate-ether (50 + 40 + 10). The azopeltatins are red and the nitro- and aminopeltatins yield yellow to orange colours with the 4-dimethylaminobenzaldehyde-hydrochloric acid reagent (No. 72).

Pyrethrin synergists of the lignan type, notably sesamin and sesamolin, were separated some years ago by BEROZA [13, 115], using silica gel layers and 14 different solvents; he preferred benzene containing 2.5% acetone. JORK [117] also reports the TLC of these lignans on silica gel G layers, with the solvent chloroform-methylene dichloride (50 + 50) at chamber saturation; he quotes the following colour reactions (Table 171).

Table 171. *Colour reactions with various spray reagents* [a]

Lignan derivatives	hRf-values	I	II	III	IV
Asarinin	58	ochre with blue border	blue	orange	brown-grey, blue border
Sesamolin	56	brick red	blue	olive green, blue border	violet-grey
Sesamin	43	ochre with blue border	blue	orange-red	brown-grey, blue border

[a] I = Anisaldehyde reagent (No. 11); colour after heating 5 min at 110° C; II = Molybdophosphoric acid reagent (No. 168); only sesamolin turns blue before heating to 100° C; III = Conc. sulphuric acid; colour after heating 5 min at 110° C; IV = Antimony(III)chloride reagent (No. 15); only sesamolin turns light blue before heating to 110° C.

BEROZA [13] prefers visualising with a mixture of chromotropic and sulphuric acids (Rgt. No. 47) and was able to detect amounts of these lignans down to 0.1–0.2 µg by the purple colour they give after treating with this reagent and heating 30 min at 105° C.

HÄNSEL and co-workers [77] have studied the lignan glycoside arctiin as a chemotaxonomic characteristic of the Compositae family. Arctiin sould be separated from arctigenin on silica gel G layers with ethyl acetate-methanol (95 + 5), the hRf-values being 23 and 77, respectively. Both yield red-violet colours with antimony (III) chloride (Rgt. No. 15) or with concentrated sulphuric acid.

TLC has been finding increasing application in the sphere of wood research. Using chloroform-ethyl acetate (90 + 10) on silica gel G layers, FREUDENBERG and SIDHU [54] have separated (+)-sesamin, hRf 78; (+)-asarinin, 87; (+)-epiasarinin, 93 from (+)-pinoresinol dimethyl ether, 47 and (+)-epipinoresinol dimethyl ether, 55. After spraying with formalin-sulphuric acid (1 + 9) and heating to 110° C, the first three compounds turn green and the last two, red. KRATZL and MIKSCHE [128] have been able to resolve a pre-purified, synthetic DL-pinoresinol into four fractions by TLC on silica gel G layers with benzene-acetic acid-water (60 + 30 + 15). VON RUDLOFF and SATO [185] have employed two-dimensional TLC on silica gel G layers to fractionate lignin-, flavone- and stilbene-containing wood extracts from *Pinus banksiana* LAMB; development was with chloroform-acetic acid (90 + 10) in the first run and toluene-dioxan-water (33 + 33 + 33) in the second.

WEINGES [234—236] and FREUDENBERG [51, 52, 56] draw attention to a further way of separation. They methylate the lignan mixture in the total extract and separate the ethers thus obtained on a silica gel layer using chloroform-ethyl acetate (90 + 10).

3. Drugs with Phloroglucinol Derivatives
a) Filix-Phloroglucinol Butanones

The constituents of the male fern rhizomes [*Dryopteris filix-mas* (L) SCHOTT] which are active against tapeworms, are phloroglucinol derivatives. These active substances can be separated as their water-soluble barium or magnesium phenolates from ether extracts of the drugs. The mixture known as crude filicin is precipitated by adding acid [2]. A procedure for preparing a solution of the drug, suitable for TLC-identification, is given on p. 722. The filix phloroglucides cited in Table 172 differ greatly in their taenicidal activity.

TLC-separation is carried out mostly on acid or basic impregnated silica gel GF_{254} layers. The reversed phase technique has also been used, for which normal silica gel G layers were impregnated with paraffin-petrol ether (5 + 95) (p. 48) and methanol-formic acid-water (75 + 10 + 15) was used as solvent. The results are in Table 172.

The influence of pH is clearly seen by comparing the hRf-values on the weakly acid layer (I) with those on the weakly basic layer (III). T-gradient TLC (weakly acid → weakly basic) illustrates this (Fig. 198); differences in the acidity of the filix-phloroglucides are thereby demonstrated [200]. On more strongly acid layers (II), all the compounds with one ring are in the lowest third of the chromatogram while flavaspidic acid, the principal substance in the crude filicin, is to be found along with the other two-ring substances.

The relation between the number of phloroglucinol rings and the hRf-value, observed in reversed phase procedures, is of some interest. The hRf-value falls as the molecular weight rises; for one ring it lies between 78 and 100 and for two rings, either ca. 70 (24 carbon atoms) or ca. 10 (with one methyl group more).

Visualisation: Phloroglucinol derivatives may be detected through their quenching on fluorescence layers in short-wave UV light. Coupling with the stable Fast Blue B salt (Rgt. No. 100) yields the colours quoted in Table 172. VON SCHANTZ [191] gives data about the reaction products

Table 172. hRf-values of phloroglucinol derivatives isolated from Dryopteris fern species

Phloroglucinol derivatives	Formula	Buffered silica gel G			"Reversed phase"	Colour with Fast Blue B salt
		I	II	III	IV	(Rgt. No. 100)
Filicinic acid (1)[a]	$C_8H_{10}O_3$	—	3	0	100	red-violet
Desaspidinol (1)	$C_{11}H_{13}O_4$	—	22	75	87	orange-red
Methylphloro-butyrophenone[b] (1)	$C_{11}H_{14}O_4$	—	14	55	83	red-violet
Butyrylfilicinic acid (1)	$C_{12}H_{16}O_4$	—	5	5	81	red-orange
Aspidinol (1)	$C_{12}H_{16}O_4$	41	25	73	78	red-violet
Phloropyrone (2)	$C_{21}H_{26}O_7$	—	60	50	72	orange
Flavaspidic acid (2)	$C_{24}H_{30}O_8$	7	53	9	70	orange-red
Desaspidin (2)	$C_{24}H_{30}O_8$	82	53	12	70	orange
Aspidin (2)	$C_{25}H_{32}O_8$	85	75	33	11	yellow
Albaspidin (2)	$C_{25}H_{32}O_8$	87	79	26	11	orange-red
Filixic acid (3)	$C_{36}H_{44}O_{12}$	90	83	16	3	orange-red

I = Silica gel G, acid-buffered with McIlvaine buffer, pH 6; solvent: petrol ether-chloroform-ethanol (47.5 + 47.5 + 5) [191]; II = Silica gel G, acid-buffered with 0.5 N oxalic acid; solvent: benzene-chloroform (50 + 50), chamber saturation [199]; III = Silica gel GF_{254}, alkali-buffered (0.3M sodium acetate); solvent: ethyl acetate, double development at chamber saturation [205]; IV = Silica gel GF_{254}, impregnated with paraffin; solvent: methanol-formic acid-water (75 + 10 + 15), chamber saturation [205].

[a] Number of rings in the molecule is given in brackets ( ).
[b] DF_x [199].

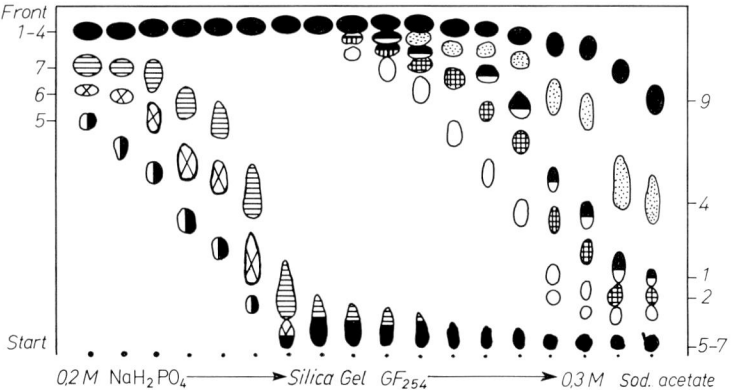

Fig. 198. T-Gradient chromatogram of crude filicin [200]. Solvent: chloroform-methanol (95 + 5); chamber saturation; for substances 1—7, see Fig. 199

with other diazonium salts. He used also the spray reagent 1% ferric chloride solution-1% potassium ferricyanide solution (1 + 1) and added 10 drops concentrated nitric acid per 10 ml of the mixture (see Rgt. No. 111); this reagent yields deep blue spots. The Folin-Ciocalteu reagent (No. 122), which we prefer, also gives a blue reaction product. Quantitative evaluation after TLC has been described in other work; the scraped-off zones were extracted and determined spectrophotometrically [199] or the eluate was treated with Fast Blue B salt solution and the coloured reaction product evaluated photometrically [192, 193].

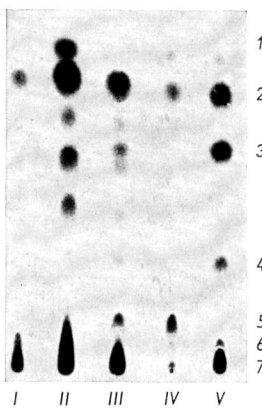

Fig. 199. TLC of crude filicin and of commercial preparations containing phloroglucinol [206]. *I* crude filicin (old); *II* crude filicin (new); *III* filmaron (new); *IV* filmaron (old); *V* taeniver (Belgian preparation); *1* albaspidin; *2* filixic acid; *3* aspidinol; *4* methylphlorobutyrophenone; *5* butyrylfilicinic acid; *6* filicinic acid; *7* flavaspidic acid

Applications: With the help of TLC, Pentillä and Sundman [168] in particular have been able, in a comparatively brief period, to discover and identify new filix phloroglucides [22, 46, 47, 48]. The quantitative evaluations after TLC-separation furnished the first useful data about the composition of the crude filicins obtained in different ways [192, 199].

Reference may be made to the possibility of a TLC-identification of Rhiz. Filicis or crude filicin (see p. 722).

b) Hop Bitter Principles

The bitter substances from hops, which are important in brewing, are also phloroglucinol derivatives. The humulones, lupulones and corresponding iso-compounds can be separated on silica gel G layers, using weakly polar solvents [4, 132, 211, 214]. The method proposed by Grant [67] for detection of hop constituents in beer or in extracts, appears most serviceable:

Beer samples and extracts are extracted with n-hexane. The extract is carefully concentrated and taken up in methanol and the solution cooled to about 1° C so that waxes separate out. After centrifuging, the clear liquid is concentrated and can be applied directly to the silica gel GF_{254} layer. A mixture of 2,2,4-trimethylpentane-isopropanol-formic acid (83.5 + 16.5 + 0.5) is employed as solvent. The substances can be detected through their quenching of the fluorescence and can thus be isolated and submitted to further identification procedures. All absorption maxima are near 274 nm yet the specific bitter values vary greatly. The usual phenol reagents, such as Fast Blue B salt for example (Rgt. Nos. 100 or 122), may be used for chemical detection.

4. Drugs containing Bitter Principles

Mixtures of bitter substances of unknown structure can likewise be separated by TLC. The "bitter value" is estimated according to WASICKY [231] by determining the "bitter substance threshold"; this is the dilution (with water) at which a bitter taste can just still be detected.

Alcoholic extracts (1:10) of two different bitter woods have been compared by TLC on silica gel G layers, using chloroform-methanol (90 + 10) at chamber saturation (Fig. 200). The bitter substances were located on the chromatogram by scraping off both the fluorescing and other zones in successive millimeter lengths and determining the bitter value of each; this was done by suspending each of the silica gel portions in 0.5 ml pure ethanol and preparing a dilution series with water. In this way it was established that, in contrast to former belief, Jamaican quassia wood (*Picrasma excelsa* PLANCH.) contains a number of bitter principles and that they are not identical with those in a Brasilian wood from *Aeschrion crenata* VEL. [233].

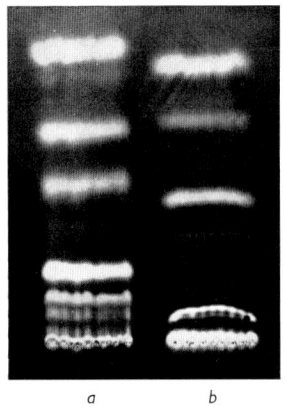

Fig. 200. Thin-layer chromatogram of two different bitterwoods; photographed in long-wave UV light without prior spraying with a reagent; *a* Brasilian bitterwood; *b* Jamaican quassia wood [233]. See text for details

5. Constituents of Hashish

KORTE and co-workers [28, 123—126] in particular have worked on the TLC of the narcotic drug hashish, also known as marihuana. This drug is the resinous constituent of the female inflorescence of *Cannabis sativa*, a plant which occurs in various varieties or chemical races. The solution for investigation is obtained by extraction at high rotation speed with petrol ether (BP 40—60° C). Separation is carried out on silica gel G layers impregnated by dipping in dimethylformamide-carbon tetrachloride (60 + 40) and allowing the mixture to ascend 15 cm in a suitable chamber. After 45 min, when the solvent and some of the impregnation liquid have evaporated, 1—5% solutions in n-hexane of the cannabis and hashish extracts are applied. Ascending development is carried out 2–3 times with cyclohexane to a height of 10 cm, at chamber saturation. Fast Blue B salt (Rgt. No. 100) is the most suitable reagent for visualising the separated substances, permitting detection of amounts down to 0.01 µg. Fig. 201 shows that TLC is just as suitable for distinguishing hashish extracts of different origins and possibly chemical races as it is for ascertaining the pyrolysis products arising when hashish is smoked. MIRAS and co-workers [146] also have worked on this and established that only the cannabidiolic acid is decomposed during smoking of hashish.

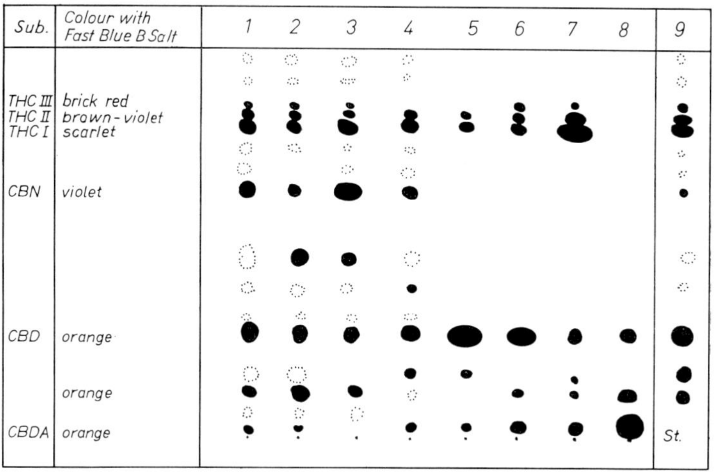

Fig. 201. TLC of hashish extracts and CBD pyrolysis products [124]. THC tetrahydrocannabinols; CBN cannabinol; CBD cannabidiol; CBDA cannabidiolic acid; *1—4* hashish of oriental origin; *5 Cannabis* non indica, cultivated in Karlsruhe, Germany, 1956; *6* as 5, but 1957; *7 Cannabis indica*, Karlsruhe 1957; *8 Cannabis* non indica, Karlsruhe 1962; *9* pyrolysis product of CBD; the zones with dotted outlines are only faintly visible

The chief substances in Fig. 201 can be quantitatively determined by either spraying the chromatogram with alkali and the Fast Blue B salt mentioned or by visualisation with blue tetrazolium (Firm 88) [125, 126]; in each case the zones are than scraped off, eluted with acetic acid-methanol (1 + 1) and the solution evaluated photometrically in the visible spectral region.

6. Other Drugs and Mixtures of Natural Products

The TLC of drug extracts has been applied in the detection of adulteration and sometimes also to differentiate closely related species.

It has thus been possible to recognise an adulteration of Radix Pimpinellae with the roots of *Heracleum sphodylium* L. [95] (see Table

173). Folia Farfarae and Petasites have also been chromatographically distinguished [101]. NOVOTNÝ and co-workers [157, 158] have applied TLC with considerable success in chemotaxonomic studies of European *Petasites* species; extracts of the Petasites-rhizomes were separated by multiple development in the S-chamber (p. 69) on silica gel G layers and visualisation was carried out by spraying with concentrated sulphuric acid and heating. Numerous *Tilia* species have been separated by TLC, profiting from the different flavonoid compositions in the inflorescences [209]. The barks of *Viburnum prunifolium* L. and *V. opulus* L. may be easily distinguished with the help of TLC. This is important since confusions and adulterations of the expensive *V. prunifolium* drug with the other are repeatedly encountered [112]. Differences in the compositions of various *Crataegus oxyacantha* drugs and their preparations can be established through TLC [102]. The drugs from the flowers of *Arnica montana* L. and *Arnica chamissonis* MAG. can also be distinguished [139]. TLC has been employed also for identifying Lichen islandicus [137], Rad. Gentianae [140], Rad. Liquiritiae (p. 723), Rhiz. Filicis (p. 722), hydroquinone-drugs (Table 174), cannabis (Fig. 201, p. 716), anthraquinone drugs (Fig. 197, p. 708), drugs containing lignans and stilbenes (p. 710) and many others. Further information is found in Tables 173 and 174 and in the chapters in which the separation of natural products is discussed.

Table 173. *Experimental conditions for TLC of some drugs or their extracts*

Drug	Layer	Solvent	Detection	Refs.
Arnica montana L. *Arnica chamissonis* MAG. (Flores)	Silica gel G	Ethyl acetate-formic acid-water (80 + 10 + + 10)	Rgt. No. 86	[139]
Crataegus oxyacantha L. (drugs)	Silica gel	Ethyl acetate-methanol-water (100 + 20 + 10)	151, 260	[102]
Petasites-species in *Tussilago farfara* L. (Folia)	Silica gel G	Chloroform	11	[101]
Petasites-species (*Rhizomes*)	Silica gel G	Methylene dichloride (S-chamber; multiple development)	241	[157]
Pimpinella saxifraga L. *Heracleum sphondylium* (Radix)	Silica gel G	Chloroform	18	[95]
Tilia species (Flores)	Silica gel G	Toluene-methyl acetate-formic acid-water (30 + 50 + 14 + 6)	3	[209]
Viburnum prunifolium L. *Viburnum opulus* L. (Cortex)	Silica gel G	Chloroform-acetic acid-acetone (75 + 25 + 10)	100, 213, 262	[112]

Table 174. Conditions for separation and detection of other groups of plant constituents

Group	Plant of origin	Layer	Solvent	Detection (Rgt.-No.)	Comments	Refs.
Aristolochic acid (Nitrophenanthrene-carboxylic acids) and their methyl esters	Aristolochia clematitis L. (root)	Silica gel G Cellulose	Benzene-methanol-acetic acid (85 + 10 + 5) Benzene-heptane-chloroform-acetic acid (51 + 15 + 70 + 3)		quantitative determination	[167] [201]
Aurantiacin (1,4-benzo-quinone derivative)	Hydnellum caeruleum (Basidiomycetae)	Alumina G Silica gel G	Benzene Benzene-ethyl acetate-acetic acid (75 + 24 + 1)		qualitative qualitative	[162] [147]
Bitter principles	Physalis franchetti Mast. (berries)	Silica gel G	Benzene-chloroform-methanol (67 + 16 + 16)	15 + 241	qualitative	[226]
	Cnicus benedictus L. (plant)	Silica gel G	Chloroform-acetone (80 + 20)	15	qualitative	[220]
Capsaicin (vanillyl-amide of 7-methyloct-5-ene-1-carboxylic acid)	Capsicum annuum L. (fruit)	Silica gel G Polyamide	Cyclohexane-chloroform-acetic acid (70 + 20 + 10) Chloroform-methanol-acetic acid (95 + 1 + 5) Water-dioxan [66 + 34]	147 62 100	qualitative quant. detn. separation from vanillylamide of nonanoic acid	[229] [90] [57]
Ergot pigments	Claviceps purpurea Tul. (Secale cornutum)	Silica gel G	Benzene-chloroform-ethanol (40 + 40 + 10)	UV_{366}	qualitative	[122]
Dhurrin and taxiphyllin	Taxus species (leaves)	Silica Gel G	Chloroform-acetic acid (90 + 10) Butanone-ethyl acetate-formic acid-water (50 + 30 + 20 + 10)	102 200	qualitative glucosides of p-hydroxymandelo-nitrile	[1] [217]

Table 174 (Continued)

Group	Plant of origin	Layer	Solvent	Detection (Rgt.-No.)	Comments	Refs.
Exogonic acid	*Exogonium purga* BENTH (Jalapa resin)	Silica gel G	Isopropanol-10% ammonium hydroxide (66 + 33)	262	qualitative	[66]
Hydroquinones and quinones	Arthropods (insect repellents)	Silica gel G	Chloroform-ether (66 + 33)	111	qualitative	[194]
		Alumina G, impregnated	Petrol ether (BP 30—50°C)		qualitative	
Hydroquinone glucosides (arbutin + methyl-arbutin)	*Arctostaphylos uva ursi* L. (leaves)	Silica gel GF$_{254}$	Toluene-ethyl acetate-formic acid (50 + 40 + 10)	165	qualitative	[203]
		Silica gel	Ethyl-acetate-methanol-water (100 + 16.5 + 13.5)	165, 168	qualitative	[100] [138] [190]
Jalapinolic acid (4 acylated glucosides of it)	*Ipomea parasitica* DON. (seed)	Silica gel G	Chloroform-methanol (90 + 10)	241	qualitative	
Paeonol glucoside paeoniflorin	*Paeonia* species (root)	Silica gel GF$_{254}$	Chloroform-methanol (85 + 15)	11, 165	BN-chamber (2 h)	[30]
Rotenone	*Derris* species (root)	Alumina G	Benzene-ethanol-water (60 + 30 + 15)	UV$_{366}$, 111	qualitative	[240]
Sinalbin	*Sinapis alba* L. (seed)	Silica gel G	Water-methanol-ethyl acetate (75 + 5 + 5)	225	qualitative	[120]
Stilbene (pinosylvin)	*Pinus sylvestris* L. (wood)	Silica gel G	Benzene-methanol (90 + 10)	Autoradiography	preparative	[18]
Tormentoside (triterpene glycoside)	Rosaceae (whole plant)	Silica gel G	Benzene-n-butanol-acetic acid-water (50 + 25 + 25 + 5)	241	qualitative	[210]
Ustilaginoidines (Naphthopyrone derivatives)	*Ustilaginoidea virens* TAK.	Silica gel G, prepared with oxalic acid	Benzene-acetone (80 + 20)			[188]

III. TLC as a Legally Binding Method for Characterisation of Drugs

Egon Stahl and P. J. Schorn

Special demands are currently being made of those who have to work out a chromatographic method for legally binding compilations of procedures such as pharmacopoeias. There are more points to be observed than is generally supposed.

A detailed description of the general method in which the standard conditions are stated (see p. 85) is required for the acceptance of a chromatographic procedure for characterising drugs. The introduction of paragraphs on chromatographic characterisation of drugs into specialist monographs is possible only on this basis. Such a paragraph is most conveniently classified into:

a) preparation of the solution for investigation and the standard (reference) solution.

b) chromatography, i. e., the special experimental conditions for TLC

c) evaluation.

Information worthy of note is given below, concluding with two examples from practice.

1. General Information for Practical Directions

Solution for Investigation (drug extract)

a) Sample: 0.1–1.0 g of the powdered drug.

b) Choice of the best extraction procedure: the solvent should be as selective as possible, capable of dissolving out the active substances already in a short time through digestion or maceration. If necessary, a simple intermediate purification can be carried out by liquid/liquid partition in a separating funnel.

c) The extract of active substances, which may be concentrated to dryness, is dissolved in a fixed, definite amount of solvent (usually 0.1–1.0 ml) and particular amounts of this (usually 1, 5 and 10 μl) are applied in parallel.

d) With a view to better separation, it must be tested whether start bands, 1–2 mm broad, are better than the usual start points.

Standard Solution (artificial mixture)

A so-called standard (reference) solution must always be chromatographed at the same time; it should contain from one to four of the principal components present in the solution under investigation. This procedure enables the positions of substance zones to be compared and also, if necessary, their sizes and colour reactions. Further, the positions of zones of unknown substances on the chromatogram of the solution being investigated can be described better and, above all, with greater certainty by using such a reference chromatogram than by Rf-values. Co-chromatography of a standard solution is an important control for the experimentor to judge whether he has properly adhered to the experimental conditions.

The following is required of a standard solution for analysing a drug:

a) It must contain the chief active substance(s) in a pure state.

b) The ratios of the amounts of these reference substances to one another should be approximately the same as those in an ordinary drug.

c) The chosen reference substances should be commercially available. If not, a "test mixture" of one to three chemically distinct but commercially available substances can be made; the chromatographic behaviour of these substances should, however, be the same as or closely similar to those of the principal active components of the drug.

d) The detailed procedure for preparing the standard solution and the amount to be applied must be given and its stability must be tested.

Chromatography

Although the technique in general, here TLC, is usually treated under "general methods", the following data should be quoted in the drug monograph:

Nature of the layer (e. g., silica gel GF_{254}) and thickness in the dry state; volume composition (totalling 100) of the solvent; normal or chamber saturation; length of run in cm; single or multiple development; visualisation (detection), for which simple but as specific procedures as possible should be chosen (inspection of the chromatogram in short- and/or long-wave UV light in order to detect compounds which absorb UV light or fluoresce should not be the sole test; a colour reaction frequently follows).

Note: solvent and reagents selected for pharmacopoeia procedures should be, as far as possible, normally available in chemists' shops or at least readily obtainable.

Evaluation

This involves comparison of the chromatogram zones, if necessary of the spot areas and the behaviour during visualisation. The chromatogram of the solution under investigation is described with reference to that of the standard solution.

Since definite amounts are applied, semi-quantitative estimation is possible through visual comparison of the sizes of spots. Added information is obtained by applying progressively increasing amounts of the solution studied (e. g., 1, 5 and 10 µl). Thus, for example, application of 1 µl may reveal only the principal component whereas with 10 µl, additional substances zones are usually observed.

The chromatograms of the solution under investigation and of the standard solution can be diagrammatically reproduced instead of quoting a longer comparative description (cf. here Fig. 71c, p. 129); this has been done in the East German Pharmacopoeia 7.

Little or nothing is known about the type and nature of the real active principles in a number of drugs. TLC can be of help here also, providing a chromatographic finger print. Metabolic products which play no part in the activity can be utilised for identification. One is thinking especially here of the differing α- and/or γ-pyrone composition of plants. The possibility of carrying out a chemical reaction before the real test of identity has been little or not at all explored; an example is to carry out acid hydrolysis of a drug extract (see Liquiritia below) and to characterise the hydrolysis products chromatographically. Particular reference may be made here to the section "Reactions at the start point" (p. 206).

2. Two Examples of Special Procedures [207]
a) Rhizoma Filicis (Male Fern Rhizome)

The identification described below depends on chromatographic detection of active phloroglucinol derivatives, especially flavaspidic acid, obtained by concentration using the "baryta" method. The position of the active substances on the chromatogram is established by co-chromatography of a standard solution of resorcinol and phloroglucinol; both are commercially available substances which are not present in the drug extract, however.

Preparation of the Solution for Investigation and of the Standard Solution

Solution for investigation: 1.0 g of the finely powdered drug is allowed to stand 30 min with 10 ml saturated barium hydroxide solution, shaking frequently. After filtration, the filtrate is acidified with dilute hydrochloric acid and extracted with two 5 ml portions of peroxide-free ether. The combined ether extract is filtered through a smooth filter (7 cm diameter) to which a pinch of dry sodium sulphate has been added. The filtrate is concentrated on the water bath and taken up in 2.0 ml pure chloroform; 10 µl of this solution are applied on a start band 1 cm long.

Standard solution: 50 mg of both resorcinol and phloroglucinol are dissolved in 10 ml methanol and 1 µl of this solution applied near to the start band of the solution under investigation.

Experimental Conditions

Normal TLC-conditions apply (p. 85).

The layer is prepared by mixing silica gel GF_{245} (Merck) with 0.3 M sodium acetate solution, spreading into layers of about 250 μm thickness. Development is carried out twice to a height of 10 cm in a saturated chamber atmosphere, using chloroform-methanol (85 + 15). The spots are visualised by spraying with 10 ml of a freshly prepared 0.5% aqueous solution of Fast Blue B salt, followed some minutes later with a few ml of aqueous sodium hydroxide to intensify the colours.

Evaluation

The chromatogram of the *standard solution* shows two coloured spots: red brown from resorcinol at Rf 0.45–0.50 and blue-violet from phloroglucinol at Rf 0.20–0.25.

Flavaspidic acid, principal component of the *solution being examined*, has an Rf-value only slightly larger than that of phloroglucinol and yields an orange-red colour like the phloroglucides above it on the chromatogram. Further orange-red zones are found also in the space between the two standard substances. More similarly coloured or yellow spots are above the resorcinol spot. There should, however, be no blue to violet zones in the chromatogram of the investigated solution (decomposition products).

b) Radix Liquiritiae (Liquorice Root)

The TLC-identification of Radix Liquiritiae, described below, depends on detecting glycyrrhetinic acid following acid hydrolysis of the drug extract. It is present in the original drug as glucuronide (glycyrrhizic acid). Hydrolysed and non-hydrolysed drugs are compared, resulting in certain identification of the drug.

Preparation of the Solution for Investigation and of the Standard Solution

Solution for investigation: 1.0 g of the finely powdered drug is refluxed 15 min with 20.0 ml methanol. The filtrate is cooled and its volume made up to 20.0 ml with methanol.

a) 10.0 ml of the filtrate is evaporated to dryness. The residue is dissolved in 1.0 ml of a mixture of chloroform-methanol (1 + 1); any insoluble remainder can be neglected.

b) 10.0 ml filtrate is evaporated to dryness and the residue refluxed for 1 h with 20.0 ml 5% sulphuric acid. After cooling, the solution is extracted with 20.0 ml chloroform. This extract is dried with sodium sulphate, concentrated to dryness and the residue taken up in 1 ml of a mixture of chloroform-methanol (1 + 1). 1,5 and 10 µl amounts of both solutions a) and b) are applied as spots to the start.

Standard solution: A solution of 50 mg purified glycyrrhetinic acid in 5 ml chloroform-methanol (1 + 1) is prepared and 1 and 5 µl amounts applied alongside the solution under examination.

Experimental Conditions

Normal TLC-conditions apply (p. 85). The slurry is prepared from silica gel GF_{254} (Merck) and 0.25% orthophosphoric acid instead of water, spread to give layers about 250 µm thick and dried. The solvent is chloroform-methanol (95 + 5) and development is carried out in the saturated chamber with a 10 cm length of run. For visualisation the layer is sprayed with about 10 ml anisaldehyde-sulphuric acid reagent and heated 10 min at 100–110° C.

Evaluation

The glycyrrhetinic acid in the standard solution appears as a violet-blue spot in the Rf region of 0.3. Since the acid does not occur in the free state in the plant, it should appear only in the chromatogram of extract b) and not a).

Glycyrrhetinic acid may be detected in short-wave UV light through its property of quenching fluorescence, without using any spray reagent. Two yellow zones are visible at Rf 0.20–0.25 on the untreated chromatogram in daylight.

Semi-quantitative estimation of the glycyrrhetinic acid in the drug is possible by comparing the spot size with that from the standard solution.

Bibliography for Chapter U. Hydrophilic Plant Constituents

1. ABERHART, D. J., S. CHEN, P. DE MAYO, and J. B. STOTHERS: Tetrahedron **21**, 1417 (1965).
2. ACKERMANN, M., u. M. MÜHLEMANN: Pharm. Acta Helv. **21**, 157 (1946).
3. ASAHINA, Y.: In: L. ZECHMEISTER: Fortschritte der Chemie org. Naturstoffe, Bd. 6. Vienna: Springer 1961.
4. ASHURST, P. R., and D. R. J. LAWS: J. Chem. Soc. (C) **1966**, 1615.
5. AUTERHOFF, H., u. G. THEILACKER: Arch. Pharm. **297**, 88 (1964).
6. BARBIER, M.: J. Chromatog. **2**, 649 (1959).
7. BARRY, R. D.: Chem. Rev. **64**, 229 (1964).
8. BATE-SMITH, E. C., and R. G. WESTALL: Biochim. et Biophys. Acta **4**, 427 (1950).
9. — J. Linnean Soc. London **58**, 95 (1962).
10. BAUMGARTNER, R., u. K. LEUPIN: Pharm. Acta Helv. **36**, 445 (1961).
11. BELIČ, I., and J. BERGANT-DOLAR: J. Chromatog. **5**, 455 (1961).
12. BERNFELD, P.: Biogenesis of Natural Compounds. Oxford, London, New York, Paris: Pergamon Press 1963.

13. BEROZA, M.: Agric. Food Chemistry **11**, 51 (1963).
14. BEYRICH, T.: Pharmazie **17**, 280 (1962).
15. — Planta Med. **13**, 439 (1965).
16. BHANDARI, P. R.: J. Chromatog. **16**, 130 (1964).
17. BILLEK, G., u. H. KINDL: Monatsh. Chem. **92**, 493 (1961).
18. — u. W. ZIEGLER: Monatsh. Chem. **93**, 1430 (1962).
19. — In: L. ZECHMEISTER: Fortschritte der Chemie org. Naturstoffe. Bd. 22. Vienna: Springer 1964.
20. BIRKOFER, L., u. CH. KAISER: Z. Naturforsch. **17**b, 352 (1962).
21. — — Z. Naturforsch. **17**b, 359 (1962).
22. BLAKMORE, R. C., K. BOWDEN, J. L. BROADBENT, and A. C. DRYSDALE: J. Pharm. and Pharmacol. **16**, 464 (1964).
23. BOGS, U., u. G. ZESSIN: Pharmazie **21**, 547, 550, 553 (1966).
24. BÖHME, H., u. L. KREUTZIG: Deut. Apotheker-Ztg **103**, 505 (1963).
25. — — Arch. Pharm. **297**, 681 (1964).
26. — — Arch. Pharm. **298**, 262 (1965).
27. BROADBENT, J. H., J. A. CORNELIUS, and G. SHONE: Analyst **88**, 214 (1963).
28. CLAUSSEN, U., u. F. KORTE: Naturwissenschaften **53**, 541 (1966).
29. COOMES, T. J., P. C. CROWTHER, B. J. FRANCIS, and G. SHONE: Analyst **89**, 436 (1964).
30. COOPER, S. F.: Dissertation Saarbrücken 1967.
31. COPENHAVER, J. H., and M. J. CARVER: J. Chromatog. **16**, 229 (1964).
32. CROMBIE, L.: In: L. ZECHMEISTER: Fortschritte der Chemie org. Naturstoffe. Bd. 21, Vienna: Springer 1963.
33. DEAN, F. M.: In: L. ZECHMEISTER: Fortschritte der Chemie org. Naturstoffe. Bd. 9. Vienna: Springer 1952.
34. DRAWERT, F.: Vitis **4**, 42 (1963).
35. DUTTA, N. L., A. C. GHOSH, P. M. NAIR, and K. VENKATARAMAN: Tetrahedron Letters **40**, 3023 (1964).
36. EGGER, K.: J. Chromatog. **5**, 74 (1961).
37. — Z. Naturforsch. **16**b, 697 (1961).
38. — Z. anal. Chem. **182**, 161 (1961).
39. — Planta **58**, 326 (1962).
40. — Planta Med. **12**, 265 (1964).
41. — u. M. KEIL: Ber. deut. botan. Ges. 78, 153 (1965).
42. — — Z. anal. Chem. **210**, 201 (1965).
43. EL-BASYOMI, S. Z., and G. H. N. TOWERS: Can. J. Biochem. **42**, 203 (1964).
44. ENDRES, H.: Z. anal. Chem. **181**, 331 (1961).
45. — u. H. HÖRMANN: Angew. Chem. **73**, 288 (1963).
46. FIKENSCHER, L. H., and M. R. GIBSON: Lloydia **25**, 196 (1962).
47. — u. R. HEGNAUER: Planta Med. **11**, 348 (1963).
48. — — Planta Med. **11**, 355 (1963).
49. FRANC, J., and M. HÁJKOVÁ: J. Chromatog. **16**, 345 (1964).
50. FREUDENBERG, K., u. K. WEINGES: In: L. ZECHMEISTER: Fortschritte der Chemie org. Naturstoffe. Bd. 16. Vienna: Springer 1958.
51. — — Tetrahedron 8, 336 (1960).
52. — — Tetrahedron **15**, 115 (1961).
53. — — CH.-L. CHEN u. G. CARDINALE: Chem. Ber. **95**, 2814 (1962).
54. — u. G. S. SIDHU: Chem. Ber. **94**, 851 (1961).
55. — In: L. ZECHMEISTER: Fortschritte der Chemie org. Naturstoffe. Bd. 20. Vienna: Springer 1962.
56. — u. K. WEINGES: Ann. Chem. Liebigs **668**, 92 (1963).
57. FRIEDRICH, H., u. R. RANGOONWALA: Naturwissenschaften **52**, 514 (1965).

58. GAGE, T. B., Q. L. MORRIS, W. E. DETTY, and S. H. WENDER: Science **113**, 522 (1951).
59. GEISSMANN, T. A.: In: K. PAECH u. M. V. TRACEY: Moderne Methoden der Pflanzenanalyse, Bd. III. Berlin, Göttingen, Heidelberg: Springer 1955.
60. — The Chemistry of Flavonoid Compounds. Oxford, London, New York, Paris: Pergamon Press 1962.
61. GENEST, C., and D. M. SMITH: J. Assoc. Off. Agr. Chemists **46**, 817 (1963).
62. GENSLER, W. J., and C. D. GATSONIS: J. org. Chem. **31**, 3224 (1966).
63. GERRITSMA, K. W., u. M. C. B. v. RHEEDE VAN OUDTSHOORN: Pharm. Weekblad **97**, 765 (1962).
64. GHOSAL, C. R.: Chem. & Ind. (Lond.) **1963**, 1430.
65. GOODWIN, T. W.: Chemistry and Biochemistry of Plant Pigments. London, New York: Academic Press 1965.
66. GRAF, E., E. DAHLKE u. H. W. VOIGTLÄNDER: Arch. Pharm. **298**, 81 (1965).
67. GRANT, H. L.: Proc. Americ. Soc. Brew. Chem. **1965**, 208.
68. GRAU, W., and H. ENDRES: J. Chromatog. **17**, 585 (1965).
69. GRISEBACH, H., u. W. D. OLLIS: Experientia **17**, 4 (1961).
70. — Z. Naturforsch. **18**b, 466 (1963).
71. — u. K. O. VOLLMER: Z. Naturforsch. **19**b, 781 (1964).
72. HAIS, I. M., u. K. MACEK: Handbuch der Papierchromatographie. Jena: VEB G. Fischer 1958.
73. HALMEKOSKI, J.: Acta chem. Fennica **35**B, 39 (1962).
74. HÄNSEL, R.: In: H. F. LINSKENS: Papierchromatographie in der Botanik. Berlin, Göttingen, Heidelberg: Springer 1959.
75. — L. LANGHAMMER, J. FRENZEL, and G. RANFT: J. Chromatog. **11**, 369 (1963).
76. — G. RANFT u. P. BÄHR: Z. Naturforsch. **18**b, 370 (1963).
77. — H. SCHULZ u. CH. LEUCKERT: Z. Naturforsch. **19 b**, 727 (1964).
78. — u. H. RIMPLER: Z. anal. Chem. **207**, 270 (1965).
79. HARBORNE, J. B.: J. Chromatog. **1**, 473 (1958).
80. — Biochem. J. **70**, 22 (1958).
81. — Chromatog. Rev. **1**, 209 (1959).
82. — J. Chromatog. **2**, 581 (1959).
83. — Chromatog. Rev. **2**, 105 (1960).
84. — and I. J. CORNER: Biochem. J. **81**, 242 (1961).
85. — In: L. ZECHMEISTER: Fortschritte der Chemie org. Naturstoffe. Bd. 20. Vienna: Springer 1962.
86. — Biochemistry of Phenolic Compounds. London, New York: Acad. Press 1964.
87. HARTWELL, J. L., u. A. W. SCHRECKER: In: L. ZECHMEISTER: Fortschritte der Chemie org. Naturstoffe. Bd. 15. Vienna: Springer 1958.
88. HATHWAY, D. E.: Biochem. J. **83**, 80 (1962).
89. HAWORTH, R. D.: J. Chem. Soc. (London) **1942**, 448.
90. HEUSSER, D.: Plante Med. **12**, 237 (1964).
91. HÖRHAMMER, L., H. WAGNER u. W. LEEB: Naturwissenschaften **44**, 513 (1957).
92. — — J. ISQUIERDO u. H. ENDRES: Arch. Pharm. **291**, 269 (1958).
93. — — Pharm. Ztg. **104**, 783 (1959).
94. — — u. H. GRASMEIER: Naturwissenschaften **45**, 388 (1959).
95. — — u. B. LAY: Pharmazie **15**, 645 (1960).
96. — — In: W. D. OLLIS: Recent Developments in the Chemistry of Natural Phenolic Compounds. Oxford, London, New York, Paris: Pergamon Press 1961.

97. HÖRHAMMER, L., u. H. WAGNER: Arzneimittel-Forsch. **12**, 1002 (1962).
98. — Deut. Apotheker-Ztg. **102**, 759 (1962).
99. — — u. B. SALFNER: **13**, 33 (1963).
100. — — u. H. KÖNIG: Deut. Apotheker-Ztg **103**, 1 (1963).
101. — — u. B. LAY: Deut. Apotheker-Ztg **103**, 429 (1963).
102. — — u. M. SEITZ: Deut. Apotheker-Ztg **103**, 1302 (1963).
103. — — u. W. EYRICH: Z. Naturforsch. **18**b, 639 (1963).
104. — — u. G. BITTNER: Pharm. Ztg. **108**, 259 (1963).
105. — — — Arzneimittel-Forsch. **13**, 537 (1963).
106. — — — Z. Naturforsch. **19**b, 222 (1964).
107. — In: Methods in Polyphenol Chemistry. Oxford: Pergamon Press 1964. p. 89.
108. — G. BITTNER u. H. P. HÖRHAMMER jr.: Naturwissenschaften **51**, 310 (1964).
109. — L. FARKAS, H. WAGNER u. S. IMRE: Acta Chim. Acad. Sci. Hung. **40**, 309 (1964).
110. — — — u. E. MÜLLER: Chem. Ber. **97**, 1662 (1964).
111. — H. WAGNER, and K. HEIN: J. Chromatog. **13**, 235 (1964).
112. — — u. H. REINHARDT: Deut. Apotheker-Ztg **105**, 1371 (1965).
113. — — u. E. GRAF: Deut. Apotheker-Ztg **105**, 827 (1965).
114. JANIAK, B., u. H. BÖHMERT: Arzneimittel-Forsch. **12**, 431 (1962).
115. JONES, W. A., M. BEROZA, and E. D. BECKER: J. org. Chem. **27**, 3232 (1962).
116. JÓRGENSEN, C., and H. KOFOD: Acta Chem. Scand. **8**, 941 (1954).
117. JORK, H.: Dissertation Saarbrücken 1963.
118. KARRER, W.: Konstitution und Vorkommen organischer Pflanzenstoffe. Basel: Birkhäuser 1958.
119. KAWANO, N., H. MIURA, and H. KIKUCHI: J. Pharm. Soc. Japan **84**, 469 (1964).
120. KINDL, H.: Monatsh. Chem. **95**, 439 (1964).
121. KNAPPE, E., u. I. ROHDEWALD: Z. anal. Chem. **200**, 9 (1964).
122. KORNHAUSER, A., S. LOGAR u. M. PERPAR: Pharmazie **20**, 447 (1965).
123. KORTE, F., u. H. SIEPER: Ann. Chem. Liebigs **630**, 71 (1960).
124. — — J. Chromatog. **13**, 90 (1964).
125. — — J. Chromatog. **14**, 178 (1964).
126. — H. SIEPER, and S. TIRA: Bull. Narcotics. UN. Dep. Social Affairs **17**, 35 (1965).
127. KRATZL, K., u. G. PUSCHMANN: Holzforschung **14**, Heft 1, 1 (1960).
128. — u. G. E. MIKSCHE: Monatsh. Chem. **94**, 434 (1963).
129. KRAUS, L., u. D. DUPAKOVÁ: Pharmazie **19**, 41 (1964).
130. — Farm. Obzor. **33**, 309 (1964).
131. KUHN, M., u. A. VON WARTBURG: Helv. Chim. Acta **46**, 2127 (1963).
132. KUROIWA, Y., and H. HASHIMOTO: J. Inst. Brewing **67**, 347, 352 (1961).
133. LIST, P. H., u. S. HANAFI: Arch. Pharm. **298**, 107 (1965).
134. LONGO, R., G. MEINARDI u. F. KORTE: Arch. Pharm. **297**, 248 (1964).
135. — — Bull. Chim. France **104**, 503 (1965).
136. LOTH, H., G. SCHENCK u. K. KOSSMANN: Arch. Pharm. **297**, 331 (1964).
137. LUCKNER, M., O. BESSLER u. P. SCHRÖDER: Pharmazie **20**, 203 (1965).
138. — — — Pharmazie **20**, 300 (1965).
139. — — u. R. LUCKNER: Pharmazie **20**, 681 (1965).
140. — — u. P. SCHRÖDER: Pharmazie **20**, 16 (1965).
141. LYMANN, R. L., A. L. LIVINGSTON, E. M. BICKOFF, and A. N. BOOTH: J. Org. Chem. **23**, 756 (1958).
142. MCCARTHY, T. J., and C. H. PRICE: Pharm. Weekblad **100**, 761 (1965).
143. — u. M. C. B. VAN RHEEDE VAN OUDTSHOORN: Planta Med. **14**, 62 (1966).
144. MARKHAM, K. R.: Tetrahedron **20**, 991 (1964).

145. MEIER, W., u. A. FÜRST: Helv. Chim. Acta **45**, 232 (1962).
146. MIRAS, C., S. SIMONS and J. KIBURIS: Bull. Stupefiants **16**, 13 (1964).
147. MONTFORT, M. L., V. E. TYLER jr., and L. R. BRADY: J. Pharm. Sci. **55**, 1300 (1966).
148. MOSBACH, K.: Biochem. Biophys. Research Communs **17**, 363 (1964).
149, MÜLLER, K. H., B. CHRIST u. G. KÜHN: Arch. Pharm. **295**, 41 (1962).
150. NATORI, S., F. SATO, and S. UDAGAWA: Chem. & Pharm. Bull. (Tokyo) **13**, 385 (1965).
151. NEU, R.: Naturwissenschaften **44**, 181 (1957).
152. — Microchim. Acta **1957**, 196.
153. — Nature **182**, 660 (1958).
154. — Arch. Pharm. **293**, 169 (1960).
155. NIELSEN, B. E., and J. LEMMICH: Acta Chem. Scand. **18**, 932 (1964).
156. NILSSON, M.: Acta Chem. Scand. **15**, 154 (1961).
157. NOVOTNÝ, L., u. F. ŠORM: Beiträge zur Biochemie und Physiologie von Naturstoffen, S. 327. Festschrift MOTHES, K., Jena: VEB Fischer 1965.
158. — J. TOMAN, F. STARÝ, A. D. MARQUEZ, V. HEROUT, and F. ŠORM: Phytochemistry **5**, 1281 (1966).
159. NUNO, M.: J. Japan Bot. **39**, 97 (1964).
160. NYBOM, N.: Fruchtsaft-Ind. **8**, 205 (1963).
161. — Physiol. Plantarum **17**, 157 (1964).
162. PAILER, M., P. BERGTHALLER u. G. SCHADEN: Monatsh. Chem. **96**, 863 (1965).
163. PARIS, R.: Pharm. Acta Helv. **36**, 176 (1961).
164. — et M. PARIS: Bull. soc. chim. France **1963**, 1597.
165. PASTUSKA, G.: Z. anal. Chem. **179**, 355 (1961).
166. — u. H. TRINKS: Chemiker-Ztg. **85**, 535 (1961).
167. PATT, P.: Arzneimittel-Forsch. **15**, 90 (1965).
168. PENTILLÄ, A., u. J. SUNDMAN: Planta Med. **14**, 157 (1966).
169. PETTERSSON, G.: J. Chromatog. **12**, 352 (1963).
170. — Acta Chem. Scand. **18**, 2303 (1964).
171. PRIDHAM, J. B.: Methods in Polyphenol Chemistry. Oxford, London, New York, Paris: Pergamon Press 1964.
172. POETHKE, W., H. BEHRENDT u. E. MATSCHKE: Pharm. Zentralhalle **102**, 492 (1963).
173. — — Pharm. Zentralhalle **104**, 549 (1965).
174. RAMAUT, J. L.: Bull. soc. chim. Belges **72**, 316 (1963).
175. REIO, L.: J. Chromatog. **13**, 475 (1964).
176. RENZ, J., M. KUHN u. A. VON WARTBURG: Ann. Chem. Liebigs **681**, 207 (1965).
177. REZNIK, H., u. K. EGGER: Z. anal. Chem. **183**, 196 (1961).
178. RHEEDE VAN OUDTSHOORN, V. M. C. B.: Planta Med. **11**, 332 (1963).
179. — u. K. W. GERRITSMA: Pharm. Weekblad **99**, 1425 (1964).
180. — Phytochemistry **3**, 383, 390 (1964).
181. — u. K. W. GERRITSMA: Naturwissenschaften **52**, 35 (1965).
182. — Planta Med. **14**, 72 (1966).
183. RIBÉREAU-GAYON, P.: Compt. rend. **258**, 1335 (1964).
184. ROUX, D. G., and S. R. EVELIN: J. Chromatog. **1**, 537 (1958).
185. RUDLOFF, E. VON, and A. SATO: Can. J. Chem. **41**, 2165 (1963).
186. RÜTTIMANN, O., u. H. FLÜCK: Pharm. Acta Helv. **39**, 417 (1964).
187. SARGEANT, K.: Chem. & Ind. (London) **1963**, 53.
188. SHIBATA, S., and Y. OGIHARA: Chem. Pharm. Bull. (Tokyo) **11**, 1576 (1963).
189. SIEPER, H., R. LONGO u. F. KORTE: Arch. Pharm. **296**, 403 (1963).
190. SMITH, C. R. jr., L. H. NIEZE, H. F. ZOBEL, and I. A. WOLFF: Phytochemistry **3**, 289 (1964).

191. Schantz, M. v.: Planta Med. **10**, 22 (1962).
192. — Planta Med. **10**, 98 (1962).
193. — L. Ivars, I. Lindgren, L. Laitinen, E. Kukkonen, H. Walenius u. C. J. Widén: Planta Med. **12**, 112 (1964)
194. Schildknecht, H., u. H. Krämer: Z. Naturforsch. **17**b, 701 (1962).
195. Schmid, H.: In: L. Zechmeister: Fortschritte der Chemie organischer Naturstoffe. Bd. 11. Vienna: Springer 1954.
196. Schmidt, O. Th.: In: K. Paech u. M. Tracey: Moderne Methoden der Pflanzenanalyse. Bd. III. S. 517. Berlin, Göttingen, Heidelberg: Springer 1955.
197. — In: L. Zechmeister: Fortschritte der Chemie org. Naturstoffe. Bd. 13. Vienna: Springer 1956.
198. — u. W. Schönleben: Z. Naturforsch. **12**b, 262 (1957).
199. Schorn, P. J.: Dissertation, Saarbrücken 1963.
200. — Vortrag: Chromatographie-Symposium III, Brüssel 1964.
201. Schunack, W., E. Mutschler u. H. Rochelmeyer: Pharmazie **20**, 685 (1965).
202. Stafford, H. A.: Plant Physiol. **40**, 130 (1965).
203. Stahl, E., u. P. J. Schorn: Z. physiol. Chem., Hoppe-Seyler's **325**, 263 (1961).
204. — u. U. Kaltenbach: J. Chromatog. **5**, 458 (1961).
205. Stahl, E., u. P. J. Schorn: Naturwissenschaften **49**, 14 (1962).
206. — — Scient. Pharm. **31**, 157 (1963).
207. — — unpublished.
208. Steinegger, E., u. J. Gebistorf: Pharm. Acta Helv. **38**, 840 (1963).
209. — — Scient. Pharm. **31**, 298 (1963).
210. — u. K. Peters: Pharm. Acta Helv. **41**, 102 (1966).
211. Stocker, H. R.: Schweiz. Brau.-Rundschau **72**, 243, 267 (1961).
212. Tanner, H., H. Rentschler u. G. Senn: Mitt. Klosterneuburg Ser. A Rebe, Wein **13**, 156 (1963).
213. Teichert, K., E. Mutschler u. H. Rochelmeyer: Z. anal. Chem. **181**, 325 (1961).
214. Teuber, M.: Dissertation München 1962.
215. Thieme, H.: Pharmazie **19**, 471 (1964).
216. Thomson, R. H.: Naturally Occuring Quinones. New York: Academic Press 1957.
217. Towers, G. H. N., A. G. McInnes, and A. C. Neish: Tetrahedron **20**, 71 (1964).
218. Tschesche, R., U. Schacht u. G. Legler: Ann. Chem. Liebigs **662**, 113 (1963).
219. — — — Naturwissenschaften **50**, 521 (1963).
220. Tyihak, E., és I. Palvy: Herb. Hung. **3**, 469 (1964).
221. Urion, E., M. Metche u. J. P. Haluk: Brauwissenschaft **16**, 211 (1963).
222. Vancraenenbroeck, R., A. Rogirst, H. Lemaitre et R. Lontie: Bull. soc. chim. Belges **72**, 619 (1963).
223. — A. Vanclef u. R. Lontie: Chromatographie-Symposium III. Brussels 1964.
224. van Summére, C. F.: Arch. intern. physiol. et biochim. **72**, 709 (1964).
225. Venkataraman, K.: In: L. Zechmeister: Fortschritte der Chemie org. Naturstoffe. Bd. 17. Vienna: Springer 1959.
226. Völksen, W.: Arch. Pharm. **294**, 337 (1961).
227. Vorträge (Lectures): Tagung Dtsch. Ges. Arzneipflanzenforsch. In: Planta Med. **7**, 336—449 (1959).
228. Wachtmeister, C. A.: In: H. F. Linskens: Papierchromatographie in der Botanik. 2. Aufl. Berlin, Göttingen, Heidelberg: Springer 1959.
229. Waldi, D.: In: E. Stahl: Dünnschicht-Chromatographie. Ein Laboratoriumshandbuch. 1. Aufl. Berlin, Göttingen, Heidelberg: Springer 1962.

230. WARTBURG, A. VON, M. KUHN u. H. LICHTI: Helv. Chim. Acta **47**, 1203 (1964).
231. WASICKY, R.: Leitfaden für die pharmakognostischen Untersuchungen im Unterricht und Praxis. Leipzig, Wien: Deuticke 1936.
232. WATKIN, J. E.: Chem. & Ind. (London) **1960**, 378.
233. WEIGERT, E., u. P. J. SCHORN: Tribuna Farmaceutica **30**, 48 (1962).
234. WEINGES, K.: Chem. Ber. **94**, 3032 (1961).
235. WEINGES, K.: Phytochemistry **3**, 263 (1964).
236. — u. F. TORIBIO: Ann. Chem. Liebigs **681**, 161 (1965).
237. WEYGAND, F., u. H. WENDT: Z. Naturforsch. **14**b, 421 (1959).
238. — H. SIMON, H. G. FLOSS u. U. MOTHES: Z. Naturforsch. **15**b, 765 (1960).
239. WONG, E., and A. O. TAYLOR: J. Chromatog. **9**, 449 (1962).
240. YOSHITAKE, D.: Kagaku Keisatsu Kenkyusho Hokoku **16**, 51 (1963).

V. Amino Acids and Derivatives

M. BRENNER, A. NIEDERWIESER and G. PATAKI*

I. Introduction

Free amino acids and peptides are markedly hydrophilic compounds, only slightly soluble in non-aqueous solvents. This must be borne in mind when sampling and preparing materials for TLC as well as when choosing the solvent. Some data on solubilities of a few amino acids in various solvents which illustrate this are:

Table 175. *Solubility of some amino acids in different solvents*
Moles per liter at 25° C[a]

	Glycine	DL-Norleucine	L-Aspartic acid
Water	2.886	0.0866	0.0375
20% Ethanol	1.343	0.0516	0.0149
80% Ethanol	0.0278	0.0130	0.00070
Ethanol	0.00039	0.00104	0.0000116
Methanol	0.00426	0.00854	
n-Butanol	0.0000959	0.000336	
Acetone	0.0000305	0.0000793	

[a] E. J. COHN and J. T. EDSALL [1].

Further, amino acids are amphoteric and tend to complex formation. It therefore makes a difference whether they are free, occur as hydrochlorides or as alkali salts or as complexes with heavy metal ions when applied to the layer. It is also of importance whether they are chromatographed in an acid, neutral or basic medium and at what pH the colour reactions are carried out for visualising the spots. Particular attention

* Revised and supplemented by J. JENTSCH and E. STAHL in the 2nd edition.

must be paid to the danger of hydrolysis of amino acid salts; depending on concentration, an amino acid hydrochloride decomposes in water more or less completely into the zwitter ion amino acid and HCl; this change is complete at high dilution. All chromatographic techniques involve dilution; layer and solvent must thus be appropriately buffered if salts are to be chromatographed without decomposition. Attention may be drawn also to chemical changes which, under certain conditions, can occur more rapidly and more specifically on chromatograms than in vitro. The difference between the ninhydrin reaction carried out on paper and in solution constitutes a well-known example of this.

The problem of amino acid separation was the starting point for the development of PC. Separation was considered first to be based on partition of the substances between water, bound to cellulose by imbition causing swelling, and a "mobile" phase, immiscible with water (e. g., phenol saturated with water). This attitude has changed somewhat since it was found that single-phase solvents could also be used; with these, the less mobile liquid involved in the cellulose swelling can be regarded as a type of phase, separate from the more mobile solvent-liquid. A partial transition to Freundlich adsorption on the cellulose surface may even occur sometimes.

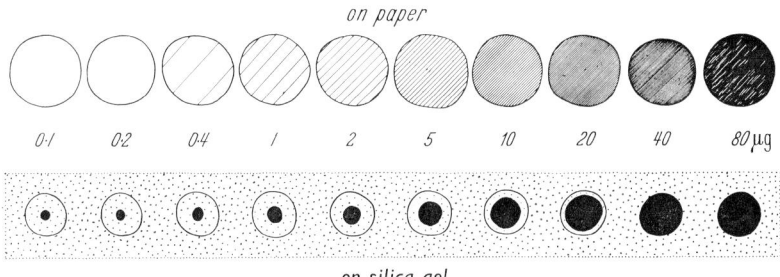

Fig. 202. Spreading of spots on silica gel G (0.25 mm thick layer, air-dried) and Whatman No. 1 paper from application of acetone solutions of DNP-serine. 10 µl amounts with increasing DNP-serine content (0.1 to 80 µg) were applied in equal pipetting times. At smaller concentrations, the material applied to silica gel covers only a fraction of the area wetted by the solvent whereas that applied to paper covers the whole wet area

TLC on silica gel, developed especially for separating lipophilic materials, seemed at first less suitable for hydrophilic substances. Silica gel however, like cellulose, contains imbibed water (depending on how it has been dried) so that its excellent suitability for chromatography of amino acids, now recognised, is scarcely surprising. There is also a basic similarity to PC in practice and the great wealth of experience gained in PC can be drawn on when applying TLC to free amino acids. As long as there is no evidence to the contrary, the specifications in PC[1] concerning preparation of the material, solvents and reactions for detection, can be taken over as a rough rule. This analogy is restricted to the use of silica gel, however.

TLC on silica gel has the two known, decisive advantages over PC:

1. Less spreading of substance zones, namely:
   a) when applying solutions at the start (cf. Fig. 202),
   b) during chromatography (slower diffusion, cf. Fig. 203).
2. Saving of time.

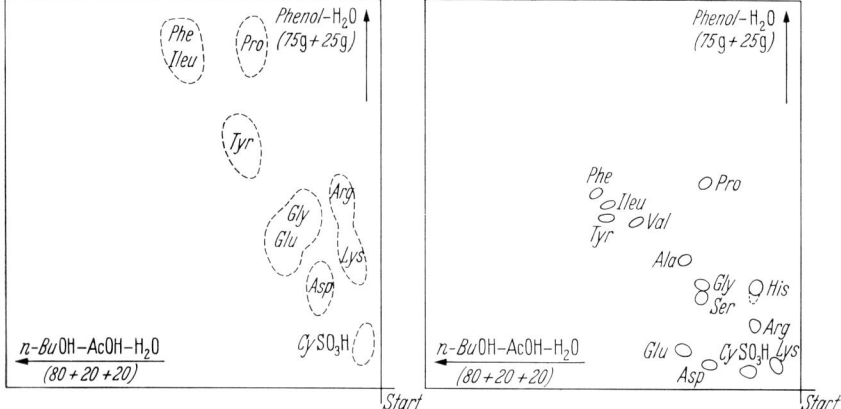

Fig. 203. Comparison of TLC with PC. Two-dimensional chromatogram reduced from original size by the factor 0.57
Left: 10 amino acids on Whatman No. 1 paper; times of run about 2 and 2.5 h.
Right: 14 amino acids on silica gel G; times of run about 1.5 and 2 h.
1 μl aqueous solution containing 1 μg of each amino acid was applied: ascending technique; phenol-water was used in the second dimension; detection with the ninhydrin reagent (No. 179), modified by MOFFAT and LYTLE [3]

II. General Technique
1. Preparation of the Layer

The preparation of layers in general is described on pp. 52–59. Activation of the layer is undesirable, *at least* for the compound classes treated here.

When the layer is heated even to a little over 100° C, the gypsum in the silica gel G ceases to fulfil properly its function as a binder and the layer becomes soft and almost powdery. Layers "activated" at high temperatures (e. g., 2 h at 140° C according to CHERBULIEZ et al. [4] or even 4 h at 140° C according to NICOLAUS [5]) necessarily become more sensitive to atmospheric moisture[2]; it is not surprising that larger fluctuations of Rf-values occur on such layers.

---

[1] I. M. HAIS and K. MACEK in [2].
[2] Cf. here R. E. KIRK and D. F. OTHMER in [6].

In contrast to the general procedure (p. 60) we air-dry moist layers by leaving overnight. The separations described below (amino acids, peptides, DNP-amino acids and PTH-amino acids) have all been obtained on air-dried layers [7—9]. Narrow plates (50 × 200 mm) are suitable only for orientation experiments since layers on such plates are as a rule uneven.

Further pre-treatment of the layers may be needed when certain solvents are used. An example of this is referred to later, in the section on chromatography of DNP-amino acids (p. 768).

We have carried out preliminary experiments with "Alox for TLC" (Firm 60) (a suspension of 20 g Alox in 60 ml water was spread with the usual spreader and the layer thus obtained was dried as for silica gel); they were unsatisfactory because of the slow rate of flow of the mobile phase, a mixture of n-butanol-acetic acid-water (80 + 20 + 20); 4 hours were required for a 10 cm run! The amino acid separation was, however, comparable to that achieved on silica gel.

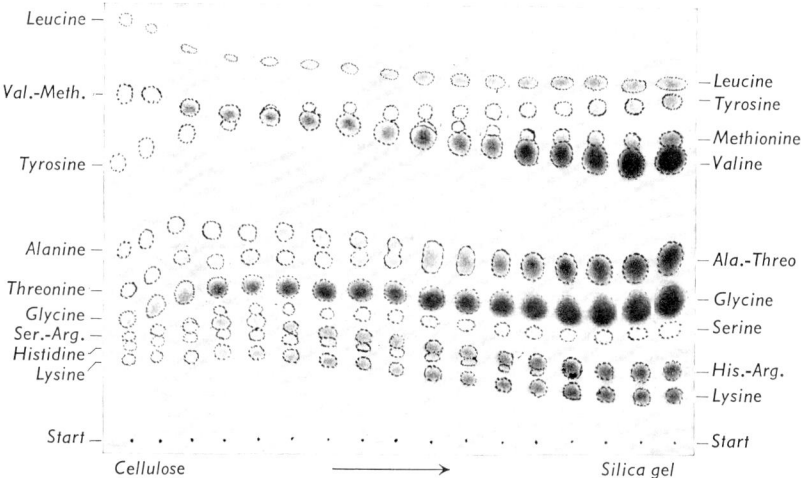

Fig. 204. Behaviour of an amino acid mixture in T-gradient cellulose/silica gel; solvent: upper phase of butanol-acetic acid-water (50 + 10 + 40); CS; (STAHL)

MUTSCHLER and ROCHELMEYER [10] take the particular properties of amino acids into consideration, preparing a layer from 25 g silica gel and 50 ml of a mixture of equal amounts of 0.2 M KH_2PO_4 and of 0.2 M Na_2HPO_4 instead of water; after spreading, they dry for only 30 min at 110° C, i. e., not really "activating". As far as can be gauged from our own experience, such buffering is superfluous and may even be disadvantageous if elution is planned after the chromatographic separation.

MOTTIER [11] has developed a TLC-variant. He activates "Merck" alumina for chromatography (No. 1097) by heating 45 min at 300—500° C and applies it to the plate without a binder. He chromatographs the sodium salts instead of the free amino acids and his approach differs from that of STAHL in other respects also.

In recent years, cellulose has been used increasingly and successfully as well as silica gel for layers in TLC of amino acid mixtures. The corresponding hRf-values and solvents are compiled in Tables 179 and 181. There has been no lack of attempts to use other adsorbents, e. g., basic alumina or kieselguhr (celite). Many investigators favour so-called mixed layers, like silica gel-kieselguhr (4:11) [203] or silica gel-cellulose (1:1). TURNER and REDGWELL [230] have obtained a better separation on this latter mixture than on the simple layers. Using gradient TLC (p. 91), STAHL [227] has, however, been able to show that even in this case there is no optimum mixture composition and that it is therefore advantageous, in particular with unknown mixtures of amino acids, to employ cellulose-silica gel gradient layers (Fig. 204).

The mixture is then applied as 15 or 20 equally sized start points in a row and developed at right angles to the direction of the gradient ([227] and p. 91). It is evident also from Fig. 204 that the colour with ninhydrin becomes weaker as the proportion of cellulose in the layer increases.

2. Chromatographic Technique (7, 8, 12)

a) Sample Application. Samples (optimum amount 0.5—2 µg in 0.5 µl solution) are applied on a line exactly 15 cm from the lower edge of the plate. The outermost spots should be at least 15 mm from the layer sides and the interval between spots at least 8 mm. The plate is left a few minutes before being placed in the chamber so that solvent can evaporate completely from the surface; special care is needed with involatile and acid solvents.

b) Solvent. Only freshly prepared mixtures of components of highest possible purity should be used.

c) Ascending Technique. 100—120 ml solvent are placed in the chamber which should be completely lined with filter paper and shaken thoroughly (chamber saturation, p. 66) before the plates (200 mm × 200 mm) are inserted. The chamber should close tightly (weights placed on the lid if necessary) and must not be opened during chromatography.

d) Horizontal Technique [13]: A description is given in the general section (p. 75). Chamber saturation is unnecessary here.

e) Temperature. The temperature must be maintained constant during chromatography; its value is usually of secondary importance. A higher temperature (50—60° C) may help to speed up solvent flow with viscous solvents like phenol-water.

f) Length of Run. 10 cm is usually adequate. The length of longer or shorter runs must be noted when quoting the Rf-values (except for single-component solvents). If the run is limited by a dividing line across the layer, the plate must be removed from the chamber as soon as the solvent reaches this line.

III. Amino Acids

1. Preparation of the Solution for Investigation

Amino acids should be as free as possible from contaminants. If standards for comparison of Rf-values etc. are required, it is advisable to use a commercially available set of pure amino acids[3] and to prepare

---

[3] Firms 60, 86, 88, 124, 127, 130.

solutions containing 1 mg per 1 ml of each standard substance. The best solvent is water containing about 10% added n-propanol (by volume); such solutions may be kept in the refrigerator or even at room temperature for 2—4 weeks. The more sparingly soluble amino acids tyrosine and cystine must be dissolved in 0.1 N hydrochloric acid. Generally 0.5 or 1 μl is applied, containing 0.5 or 1 μg amino acid. If solutions of standards in hydrochloric acid are used, it is best to dissolve and apply the unknown sample in the same solvent; before chromatography the plate should be aerated for 15—20 min to remove the excess hydrochloric acid. In paper chromatography, hydrochloric acid is occasionally buffered by exposure to ammonia vapour; care should be taken with silica gel layers since ammonia is taken up relatively strongly and chromatography with a neutral solvent may turn out to be under basic conditions. Amino acids in acid hydrolysates (see below) of proteins or peptides are mostly present as hydrochlorides. Their aqueous solution corresponds approximately to a solution of the standard amino acids in 0.1 N hydrochloric acid. Amino acids in animal and plant extracts as well as in body fluids such as urine, serum etc., should be freed from contaminants before application or chromatographed as their DNP-derivatives. The experimental procedures given below were developed for paper chromatography; they are not necessarily reliable for quantitative purposes.

2. Hydrolysis of Proteins and Peptides

a) Acid Hydrolysis. Hydrolysis carelessly carried out may lead to considerable losses, particularly of tyrosine. Special care should be taken if carbohydrates are present (darkening of the hydrolysate through humin formation). Acid hydrolysis destroys especially trytophan (see under "Strongly acid ion exchangers" below for an exception) and, to some extent, serine and threonine also (the original content of hydroxyamino acids may be estimated approximately in quantitative determinations by comparing the hydrolysates after 24 and 72 hours). Reproducible results may be obtained by the following hydrolysis procedure used by us in quantitative amino acid analyses[4]:

6 N Hydrochloric Acid. The substance is mixed with a 200—500 fold excess of 6 N hydrochloric acid (distilled two or three times) in a thick glass ampoule designed for evacuation and sealing (the excess acid is related to the amount of protein present and should be particularly large if the material contains carbohydrates). The contents of the ampoule are then frozen in an acetone/solid CO_2 mixture. After the air in the ampoule has been displaced by pure nitrogen, the ampoule is evacuated to about 1 mm mercury and then sealed. For this, the ampoule is held in a wooden beaker filled with solid CO_2 and possessing a handle. Hydrolysis is carried out by heating 18—72 h in a thermostat at $110° \pm 1$. The hydrochloric acid is best removed after hydrolysis by standing over NaOH in an evacuated desiccator.

Strongly Acid Ion Exchangers. A method has been discovered only recently for acid hydrolysis which allows quantitative recovery of tryptophan and, for

---

[4] Chromatography on ion exchangers according to STEIN and MOORE [14].

example, lysergic acid also. It has the additional advantage that humin formation is practically eliminated. According to M. Pöhm [15], about 0.05—0.2 g of the peptide, 1 g of Amberlite IR-112 (H) per millival amide nitrogen and 3—10 ml 80% ethanol are heated in a nitrogen atmosphere in a sealed tube for 6—10 h at 90—95° C. After cooling, the amino acids are eluted from the exchanger with 10% ammonium hydroxide solution.

Earlier experiments with Dowex-50 in 0.05N hydrochloric acid [16] had shown that aspartic acid, serine and threonine are liberated very rapidly but valine and isoleucine more slowly in comparison to hydrolysis in 6N hydrochloric acid. Peptide bonds with cystine and cysteic acid are said to be very resistant to this type of hydrolysis. About 25% of the glutamic acid present is converted into pyrrolidone carboxylic acid. Basic amino acids are only incompletely removed from the ion exchange resins by means of dilute ammonium hydroxide solution.

Further details about hydrolysis methods with particular reference to the use of formic acid, acetic acid, stannous chloride, sulphuric acid and hydriodic acid are to be found in the books of Hais and Macek[5] and Linskens[6].

b) Basic Hydrolysis. 5—10 mg material, 65 mg $Ba(OH)_2 \cdot 8H_2O$ and 1 ml water are heated for 24 h at 125—130° C in a sealed tube. The cooled reaction mixture is adjusted to pH 6 with $2N\ H_2SO_4$, heated to boiling and centrifuged to separate $BaSO_4$. The latter is washed with a little water, the combined supernatant liquid and washings are evaporated to dryness and the residue dissolved in 0.5—1 ml water or 0.1N HCl. Cystine and β-hydroxyamino acids are partly destroyed during this procedure while tryptophan remains unattacked.

3. Free Amino Acids in Biological Material

Aqueous extracts of animal or plant organs, fruit juices, homogenates and body fluids generally contain peptides[7], proteins, carbohydrates, urea, salts and lipoids in addition to free amino acids in water-soluble or emulsified form.

a) Separation of amino acids from proteins and polysaccharides: *Alcohol* [18] or acetone are best used since they do not interfere in the chromatography and are moreover easily removed. Awapara [19] mixes aqueous solutions with 4—8 parts by volume of alcohol, centrifuges after a few hours and washes the residue with 80% alcohol.

According to Hais and Macek[8], "1 ml plasma is evaporated to dryness in a vacuum desiccator over concentrated H_2SO_4. The residue is allowed to stand 2 h with 8 ml *acetone* containing 1% conc. HCl. The liquid is centrifuged, the precipitate washed and centrifuging repeated two or three times. The liquid is then evaporated at 37° C in a stream of dry air. The residue is dissolved in 0.5 ml water, the solution extracted two or three times with an equal volume of ether and the aqueous solution again brought to dryness in a desiccator containing H_2SO_4. The residue is taken up in 20—100 µl water and applied".

Ion exchangers which adsorb amino acids are also suitable for deproteinisation. Proteins are only weakly adsorbed for steric reasons and consequently largely pass

---

[5] Hais and Macek (pp. 415, 482 in [2]).

[6] Linskens (p. 149 in [17]).

[7] Removal of smaller peptides from the amino acids is usually impossible. See section on peptides, p. 751.

[8] Hais and Macek (p. 780 in [2]).

through along with carbohydrates and inorganic cations and anions (cf. demineralisation).

An eminently suitable method for separating compounds of high molecular weight is *gel-filtration with Sephadex* [20][9, 10, 11] a cross-linked dextran with swelling properties (see p. 40). Large molecules are unable to penetrate into the pores but inorganic salts and smaller molecules diffuse into the gel without hindrance. If an aqueous solution of such substances is allowed to pass through a Sephadex column, the high molecular weight parts appear in the first fractions; smaller molecules are retained for a time but can be eluted quantitatively with more water or dilute salt solution. The main advantage of this elegant "dialysis" procedure is the time it saves; only about 60 min are required.

b) Destruction of Urea. With a trace of urease, urea in urine is converted within 24 h into CO_2 and NH_3. These gases are evolved during subsequent evaporation [25].

c) Demineralisation. Either an electrodialysis apparatus, such as that described by CONSDEN, GORDON and MARTIN[12] [26] or a suitable ion exchange system[13] [27—29] may be used.

Electrolytic demineralisation partially converts arginine into ornithine and 10—30% of the histidine, lysine, methionine, proline and tyrosine are lost [30]. *Losses also ensue occasionally from demineralisation with ion exchangers:* arginine and possibly lysine are not retained by strongly basic exchangers and are difficult to elute from acid exchangers.

The following ion exchange procedure, based on a method of DRÈZE et al. [32, 33], has proved successful in our laboratory [31] for *removing both salts and soluble carbohydrates:*

A 1×10 cm exchanger tube is filled to a height of 2 cm with:

α) Dowex-50,8% DVB, 200—400 mesh (H^+-form) for *demineralisation of basic amino acids or tryptophan*. It is treated with 10 ml 1 N HCl and washed until neutral. 1—4 ml of sample solution (the pH of which is adjusted to <6 with HCl in order to lower the CO_2-content) are brought on to the column and the rate of flow set at 1 ml/3 min. When the sample has penetrated into the resin, it is washed with 20 ml 0.5N HCl. The amino acids are then eluted with 4N HCl, discarding the first 3 ml eluate. The following 5 ml are collected and repeatedly evaporated to dryness in vacuo, adding a little water each time. The residue is dissolved in water and chromatographed.

β) Dowex-2,10% DVB, 200—400 mesh (OH^-)-form) for *demineralisation of neutral and acid amino acids*. The column is pre-treated with 20 ml 2N NaOH (CO_2-free) and washed neutral with boiled, CO_2-free water. 1—4 ml sample solution are then added to the column. After this solution has penetrated into the column, it is washed with 20 ml water and elution then carried out with 10 ml 1N acetic acid which causes the colour of the resin to change from brown to light yellow. When the acid front reaches the bottom of the resin bed, collection of 5 ml eluate is begun. It is evaporated and the subsequent procedure is as described under α).

d) Removal of Lipoids [19]: After proteins have been removed, the clear supernatant liquid from centrifuging is extracted by vigorous shaking with 3 parts of chloroform by volume. The layers are separated in a separating funnel, the major part of the amino acids being in the aqueous (upper) phase.

---

[9] General procedure and demineralisation: [21].
[10] Gel filtration of proteins, peptides and amino acids: [22, 23].
[11] Peptide separation : [24].
[12] Cf. LINSKENS (p. 32 in [17]), etc.
[13] Cf. HAIS and MACEK (p. 415 in [2]).

e) Examples. A new procedure from the literature for detecting amino acids in serum is quoted below [34].

In order to test recovery, analysis of amino acids after treatment with an ion exchanger was carried out on (i) an aliquot of serum; (ii) an aliquot of the same serum after known amounts of amino acids had been added; and (iii) an artificial amino acid mixture of known composition, similar to that occurring in serum.

1.5 ml fresh serum is passed slowly through a 5 × 0.4 cm bed of a strongly acid ion exchanger [Zeocarb 225 (H)]. Inorganic ions, proteins, sugars and other uncharged contaminants are removed by washing with 50 ml water. The amino acids are then eluted with 20 ml 10% ammonium hydroxide and the extract evaporated to dryness in vacuo. The residue consists of free neutral amino acids and acid amino acids as their monoammonium salts or as salts with the basic amino acids; the last-named suffer some loss due to incomplete elution.

Table 176 is taken from the work cited and gives a good idea of the inaccuracies which have to be reckoned with; the authors attribute shortcomings of the method explicitly to the amino acid separation.

Table 177 gives some further applications of the methods mentioned.

Table 176. *Recovery of amino acids after filtration through a column of an acid ion exchanger* (Zeocarb 225) (Cook and Luscombe [34])

Amino acid (abbreviations cf. Table 180)	Recovery of amino acids added to serum %	Recovery from a synthetic amino acid mixture %
Glu (NH$_2$)	113	—
Cit	102	—
Thr	74	93
Pro	76	85
Met	—	41
Phe	84	107
Ser	91	110
Val	90	93
Leu	—	88
Ileu	79	—
Ala	101	122
Gly	—	124
Orn	—	77
Arg	82	53
Average	89	83

Table 177. *Literature on the extraction of amino acids from biological material*

Material	Authors
Potatoes	Dent et al. [35]
Rice	Parihar [36]
Leaf juice (potato plants)	Reindel and Bienenfeld [37]
Biological material	Biserte et al. [38]
Fungi and micro-organisms	Close [39]
Ascites and pleuracentesis	Knauff et al. [40]

4. Solvents and Separation Efficiency

One-phase Systems. In view of the extremely limited solubility of free amino acids in organic solvents (cf. Table 175), chromatographic solvent systems for them must generally contain water. Markedly polar organic solvents such as methanol, ethanol or acetone may be used as components. Fairly satisfactory separations can often be accomplished with such systems but they give relatively diffuse spots and tend to cause tailing. This tendency to tailing may occasionally be checked by adding a few volume percent of acetic acid (cf. solvent G in Table 178); the adsorption capacity for free amino acids remains, however, low (ca. 2 µg per acid). Smaller spots are obtained when higher alcohols are used but this advantage is offset by slower rates of flow, primarily as a result of increased viscosity. The larger the viscosity of the solvent, the smaller the spots, i. e., the resolution is better; phenol is a good example of this effect. If alcohols are replaced by non-polar liquids of lower viscosity, a solubiliser like methanol, pyridine or acetic acid must be added to restore miscibility with water. Separation may be very satisfactory and fast with systems of this latter type.

Two-phase Systems. These are best avoided since they cannot be used straight after mixing and are, moreover, extremely sensitive to temperature changes. In most cases, it suffices to use a solvent mixture in which the organic phase is nearly saturated with water. We have found, for instance, that the Rf-values in the phenol-water system change only slightly if water-saturated phenol, containing about 71% phenol, w/w, is replaced by 80% phenol; we therefore use 75% phenol (Table 178).

Some solvents suitable for separation of amino acids are listed in Table 178. Some are known already from paper chromatography. The corresponding Rf-values are in Table 180.

Selection of the solvent suitable for a particular purpose is facilitated by using the diagrams in Fig. 205. It is interesting that in neutral solvents such as ethanol- or n-propanol-water mixtures, the acid amino acids migrate much faster than lysine and arginine, which show very snall Rf-values (cf. Table 180). This difference might be due to cation exchange. AHRLAND et al. [43] have found that the titration curve of silica gel *resembles that of a weakly acidic ion exchanger* (cf. Chapter Y on Inorganic TLC). It is noticeable also that a hydroxyl group in the molecule does not necessarily reduce the Rf-value; the reverse may even be true, depending on the interaction with the solvent (cf. ser/gly, thr/ala, hypro/pro and tyr/phe in solvents A and E). By far the best separation efficiency is with chloroform-methanol-17% ammonium hydroxide (40 + 40 + 20), n-butanol-acetic acid-water (80 + 20 + 20) and phenol-water (75 + 25, w + w).

Table 178. *Solvents*[a] *for TLC of amino acids on silica gel G*

	Components	Proportions w/w	v/v
A	96% Ethanol-water	63 + 37	70 + 30
B	n-Propanol-water	64 + 36	70 + 30
C	n-Butanol-acetic acid-water	60 + 20 + 20	80 + 20 + 20
D	Phenol-water[b]	75 + 25	—
E	n-Propanol-34% ammonium hydroxide	67 + 33	70 + 30
F	96% Ethanol-34% NH$_4$OH	77 + 23	70 + 30
G	Butanone-pyridine-water-acetic acid		70 + 15 + 15 + 2
H	Chlorofom-methanol-17% NH$_4$OH		40 + 40 + 20

[a] E. MUTSCHLER and H. ROCHELMEYER [10] use on buffered layers (see p. 733) the solvents: 70% ethanol; 96% ethanol-25% ammonium hydroxide (80 + 20); and 96% ethanol-25% ammonium hydroxide-water (70 + 10 + 20).

E. NÜRNBERG [41] uses n-propanol-water (50 + 50) and phenol-water (100 + 40) for two-dimensional chromatograms.

According to JOST, RUDINGER and ŠORM [42], phenol-water (90 + 30) is suitable for separation of tyrosine, N-methyltyrosine and O-methyltyrosine.

[b] 75 g molten phenol are mixed with 25 ml water and about 20 mg NaCN or other antioxidant added. The mixture should be cooled to room temperature before use.

Table 179. *Some solvents*[a] *used in TLC of amino acids on cellulose layers*

	V/V Proportion	Ref.
Acid solvents		
a) Butanol-acetic acid-water	40 + 10 + 50	[232]
b) Butanol-acetic acid-water	63 + 27 + 10	[218]
c) Butanol-formic acid-water	75 + 15 + 10	[232]
d) Butanol-ethanol-propionic acid-water	40 + 40 + 8 + 20	[232]
e) Water-saturated phenol		[206]
f) Phenol-water (vapour phase saturated with 3% NH$_4$OH	75 + 25 w/w	[202]
Basic solvents		
g) Pyridine-butanone-water	15 + 70 + 15	[232]
h) Pyridine-methanol-water	4 + 20 + 80	[232]
i) Pyridine-isoamyl alcohol-water	35 + 30 + 30	[215]
j) Pyridine-water	80 + 20	[226]
k) Propanol-8.8% ammonium hydroxide	80 + 20	[232]
Neutral solvents		
l) Isopropanol-water	80 + 20	[209]
m) tert. Amyl alcohol-butanone-water	60 + 20 + 20	[207]
n) sec. Butanol-tert. butanol-butanone-water	25 + 25 + 25 + 25	[216]

[a] Other solvents which may be used are those given under B, G and H in Table 178.

The combination of solvents C and D (Fig. 203, [7]) and, in particular, the combination of H and D (Fig. 206, [9]) is suitable for two-dimensional chromatography. The latter enables all protein amino acids + β-alanine

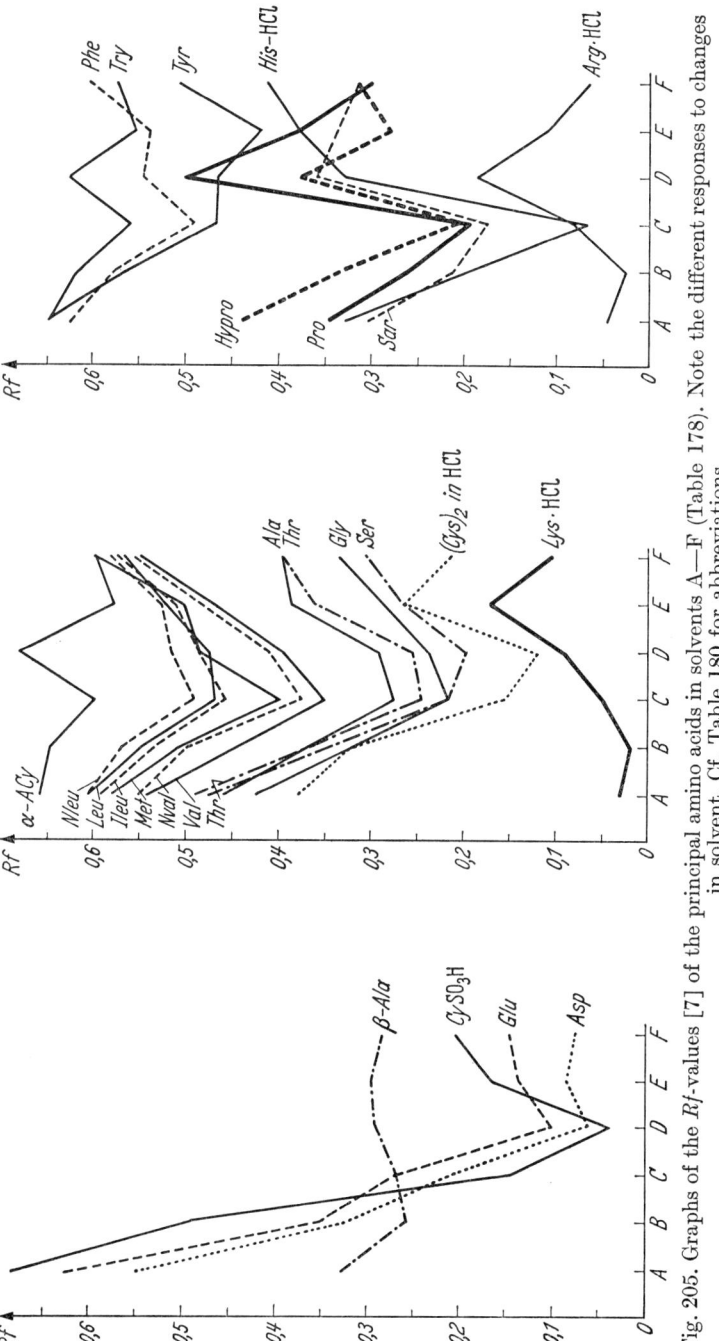

Fig. 205. Graphs of the R_f-values [7] of the principal amino acids in solvents A—F (Table 178). Note the different responses to changes in solvent. Cf. Table 180 for abbreviations

+ γ-amino-n-butyric acid to be separated, excepting leucine and isoleucine. These isomers can be differentiated in a parallel experiment using the continuous technique [13] with solvent G (cf. Fig. 207).

It is striking that the Rf-values generally do not exceed 0.7 in one-dimensional thin-layer chromatography of amino acids. Only about 2/3 of the distance from start to front is thus utilised in separation in all solvents investigated by us. Increasing the water content of the solvent raises small Rf-values more than larger, so that the effective separation path is diminished.

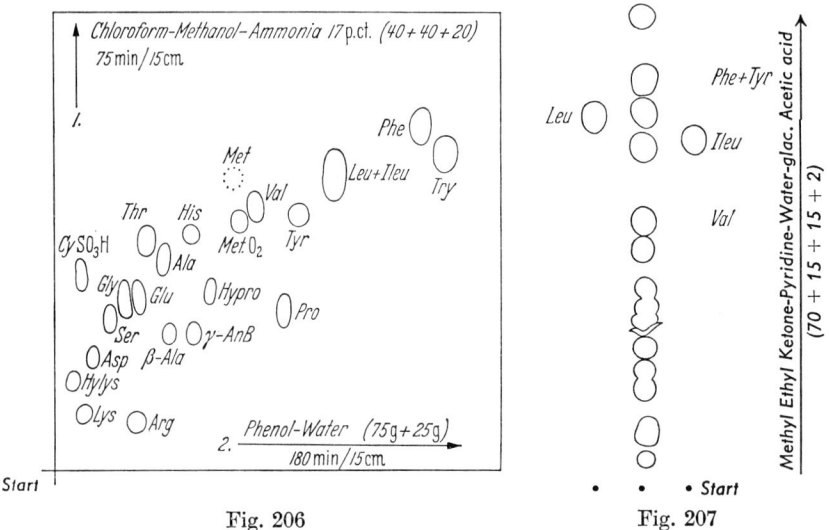

Fig. 206. Two-dimensional chromatographic separation of a mixture of 20 protein amino acids + β-alanine + γ-amino-n-butyric acid (γ-AnB) after oxidation with performic acid [44]; ascending technique.
0.5 μg of each amino acid applied in a total of 0.5 μl 0.1N hydrochloric acid. The layer is aerated 20 min in a fume cupboard after the first run. The methionine spot (dotted) appears only when the performic acid oxidation is not carried out.
Visualisation with ninhydrin (Rgt. No. 178)

Fig. 207. One-dimensional separation in a chamber for continuous development [13] (see p. 76) for detection of leucine and isoleucine in the presence of 18 protein amino acids + β-alanine + γ-amino-n-butyric acid
0.5 μg of each amino acid applied in a total of 0.5 μl 0.1N hydrochloric acid: 4.5 h run; detection with ninhydrin; oxidation with performic acid [44] is necessary if methionine is present. Identification of leucine and isoleucine is unambiguous if a standard sample of each is chromatographed in a parallel run on the same layer

Supplementary Notes. It should be remembered that the Rf-value of a pure substance is not a constant but is influenced by a number of factors [12]. Fig. 208 illustrates, for example, the dependence on the amount of substance applied. Further, the Rf-values can depend slightly on other materials present. Acidic amino acids, chromatographed in n-propanol-water (70 + 30), thus have higher Rf-values in the presence of other amino acids than in their absence. Rf-values of DNP-amino acids in mixtures also depend somewhat on the ratio of the components; dinitrophenol has a particularly marked influence, as has already been noticed in paper chromatography [45]. In the second dimension of two-dimensional chromatograms, Rf-values are affected not so much by the relative proportions of components as by previous treatment of the layer (chromatography in the first dimension, intermediate drying). Such "pre-treatment" can be largely standardised and Rf-values in the second dimensions are therefore generally quite reproducible. Mixtures yield characteristic spot patterns in this way in two-dimensional chromatography, which are scarcely obscured by fluctuations of Rf-values. It is advisable for this reason not to chromatograph an *unknown sample* on its own but together with a standard mixture containing just enough of each substance for spots to be still detectable after two-dimensional chromatography. Compounds present in the unknown sample are thus quickly and satisfactorily recognisable through the intensity of the relevant spot.

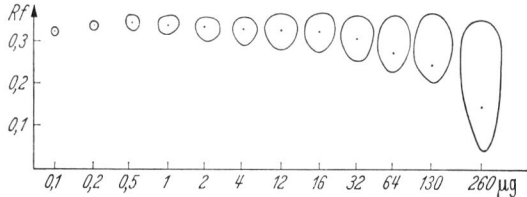

Fig. 208. Dependence of the Rf-value of a pure substance (glycine) on the amount applied

The point of highest substance density in the spot is marked with a dot; the amount quoted (in µg) was applied in solution in 1 µl; solvent was n-propanol-water (70 + 30), using ascending technique

The method of preparation of the adsorbent layer is also not without influence: amino acid pairs which are difficult to separate on layers of silica gel from one particular manufacturer can be separated on another preparation (PATAKI [219]).

The products of hydrolysis of sulphur-containing amino acids are liable to yield interfering subsidiary spots; it is thus advisable to oxidise with performic acid before hydrolysis [211].

VON ARX and NEHER [202] have separated a total of 52 amino acids by two-dimensional chromatography on 3 cellulose layers (see p. 88). Solvent I was the only solvent used in the first dimension and solvents II,

III and IV in the second. The hRf-values quoted in Table 181 define the position on the chromatograms and a "position map" can be established.

Table 180. *Rf-values*[a] *of the main amino acids in solvents A—F of Table 178* (ascending technique, solvent migration 10 cm.), *and* $R_{Leucine}$-*values*[b] *in solvent G of Table 178* (horizontal technique[c], 4 hours duration). Ca 0.5 µg of each spotted

Amino acid	Abbreviation	hRf-value in solvent						$R_{Leucine}$
		A	B	C	D	E	F	G[c]
Alanine	Ala	47	37	22	29	39	40	51
β-Alanine	β-Ala	33	26	22	30	30	29	35
α-Amino-i-butyric acid	α-AiB			27				61
α-Amino-n-butyric acid	α-AnB			27				59
β-Amino-i-butyric acid	β-AiB			25				48
γ-Amino-n-butyric acid	γ-AnB			27				45
ε-Amino-n-caproic acid	ε-ACo			34				68
α-Amino-n-caprylic acid	α-ACy	66	65	59	69	58	60	143
Arginine	Arg	04	02	06	19	10	06	19
Aspartic acid	Asp	55	33	17	06	09	07	18
Cysteic acid	CySO$_3$H	69	50	10	04	17	21	53
Cystine	(Cys)$_2$	39	32	09	12	27	22	18
Dibromotyrosine	DBT			60				168
Glutamic acid	Glu	63	35	24	10	14	15	32
Glycine	Gly	43	32	18	24	29	34	45
Histidine	His	33	20	05	32	38	42	18
Homocystine	(Hcys)$_2$			17				28
Hydroxyproline	Hypro	44	34	16	38	28	31	50
Isoleucine	Ileu	60	53	43	49	52	58	92
Leucine	Leu	61	55	44	48	53	58	100
Lysine	Lys	03	02	03	09	18	11	11
Methionine	Met	59	51	35	49	51	60	92
Methionine sulphone	Met.O$_2$							66
1-Methylhistidine	Mehis							09
Norleucine	Nleu	61	57	45	52	53	59	102
Norvaline	Nval	56	50	36	42	49	57	77
Ornithine	Orn			04				14
Phenylalanine	Phe	63	58	43	55	54	60	109
β-Phenylserine	β-Φ-Ser			41				108
Proline	Pro	35	26	14	50	37	30	40
Sarcosine	Sar	31	22	12	37	34	31	32
Serine	Ser	48	35	18	20	27	31	47
Taurine	Tau			18				79
Threonine	Thr	50	37	20	26	37	40	51
Tryptophan	Try	65	62	47	63	55	58	122
Tyrosine	Tyr	65	57	41	47	42	51	107
Valine	Val	55	45	32	40	48	56	72
Asparagine	Asp(NH$_2$)			14				43
Glutamine	Glu(NH$_2$)			15				47
Glycinamide	Gly(NH$_2$)			16				62

[a] Each figure given is the arithmetic mean of the results of 4 (solvents A, B, D, E, F) or of 9 separate measurements (solvent C). Factors affecting Rf-values are discussed by M. BRENNER, A. NIEDERWIESER, G. PATAKI and A. R. FAHMY [12].

[b] $R_{Leucine}$ = Rf-value referred to leucine. Each figure given is the arithmetic mean of the results of 9 separate measurements.

[c] M. BRENNER and A. NIEDERWIESER [13], cf. p. 76.

Table 181. *hRf-values of amino acids on cellulose layers (2-dimensional chromatogram)*
First direction: I_1; second direction: II_2—IV_2 [202]

Amino acid	hRf-value in solvent				Colour reaction	
	I_1	II_2	III_2	IV_2	Isatin-Rgt.	Ninhydrin-Rgt. (No. 178)
Glycocyamine	4	34	8	71	pink	—
Arginine	5	13	2	90	pink	violet
Creatine	5	34	8	86	yellow	—
α,α-Diaminopimelic acid	6	5	6	36	red	violet
Aspartic acid	11	21	20	20	violet	green
Lanthionine	14	8	22	35	orange	violet
Canavanine sulphate	15	7	11	66	brownish	violet
Dihydroxyphenylalanine	17	16	10	13	lilac	grey
Glutamic acid	18	30	21	25	lilac	violet
Hydroxyglutamic acid	19	30	15	25	pink	violet
Cystine	20	4	26	35	pink	brown
Citrulline	22	16	15	68	pink	violet
Asparagine	22	12	21	47	pink	yellow
Cysteic acid	26	6	41	9	yellow	violet
Methionine sulphoxide	26	19	26	82	pink	violet
Glutamine	27	14	18	59	pink	violet
Hydroxyproline	29	22	19	66	blue	yellow
Glycine	32	22	19	46	pink	brown
Ornithine	32	8	5	82	red	violet
Hydroxylysine	33	8	11	71	lilac	violet
Lysine	36	10	6	82	red	violet
Alanine	37	34	22	61	violet	violet
β-Alanine	38	37	14	68	lilac	green
Creatinine	38	36	25	97	yellow	—
Sarcosine	40	29	17	80	yellow	grey
α,γ-Diaminobutyric acid	40	8	20	73	pink	violet
Histidine	40	11	34	87	lilac	grey
Methionine sulphone	42	20	36	73	pink	violet
Dimethylcysteine	44	10	53	73	pink	violet
Proline	46	35	19	87	blue	yellow
β-Aminobutyric acid	47	43	21	87	yellowish	lilac
α-Amino-i-butyric acid	47	50	25	78	yellow	violet
α-Amino-n-butyric acid	48	42	31	78	lilac	violet
γ-Aminobutyric acid	48	46	13	78	lilac	violet
β-Amino-i-butyric acid	48	47	22	78	lilac	violet
Serine	48	21	38	41	orange	violet
p-Aminohippuric acid	49	65	74	89	yellow	violet
Tyrosine	51	38	56	71	red	brown
Taurine	52	21	44	51	yellow	violet
Valine	54	55	44	87	pink	violet
ε-Aminocaproic acid	58	55	20	87	pink	violet
Norvaline	59	55	45	87	red	violet
Kynurenine	60	39	47	80	pink	brown
Allothreonine	60	31	55	59	pink	violet
Tryptophan	62	41	60	88	lilac	violet
Methionine	64	50	54	88	pink	violet
Alloisoleucine	64	65	62	88	pink	violet
Norleucine	66	65	60	83	lilac	violet
Threonine	66	31	67	56	pink	violet
α-Phenylalanine	67	54	73	88	lilac	violet
Diiodotyrosine	68	50	73	71	lilac	violet

Table 181. (Continued)

Amino acid	hRf-value in solvent				Colour reaction	
	I_1	II_2	III_2	IV_2	Isatin-Rgt.	Ninhydrin-Rgt. (No. 178)
Thyronine	69	62	82	88	brown	brown
α-Phenylglycine	69	76	81	77	yellow	yellow
β-Hydroxyvaline	69	41	81	65	pink	violet
Thyroxine	75	73	84	86	yellow	brown

Layer: Cellulose MN 300 (Firm 83); 16 cm ascending development without CS.

Solvents: I: n-Butanol-acetone-diethylamine-water (30 + 30 + 6 + 15), pH 11.05—11.4; II: Isopropanol-formic acid (99%)-water (80 + 4 + 20), pH 2.7—2.8; III: sec. Butanol-butanone-dicyclohexylamine-water (30 + 30 + 6 + 15), pH 10.9 to 11.0; IV: Phenol-water (75 + 25); equilibration of the vapour phase carried out with 3% ammonium hydroxide; pH 7.1—7.3.

Note on intermediate drying of two-dimensional chromatograms.

Intermediate drying presents no problem where highly volatile solvents are used; the plate is allowed to stand for about 15 min in a current of air (well ventilated fume cupboard) after which it is immediately submitted to the second run. Sensitive substances may suffer decomposition on the chromatogram when heating is needed to remove less volatile solvents, e. g., phenol. It would be useful if small and inexpensive vacuum equipment[14] was designed for rapid and less drastic drying of chromatograms. For the time being, any involatile solvent has to be used only in the second dimension or the plate must be left overnight in a current of air if sensitive substances are present; oxidation may then occur and appreciable amounts of the involatile solvent may remain behind and interfere in the second run (cf. "indirectly" obtained Rf-values of DNP-amino acids, Table 188). Another possibility is to leave the plate 10 min in a current of air, heat it 15 min at 60° C in a drying oven and allow it to cool for 15 min in an air current.

If the thin-layer chromatogram has to be kept for a longer period after the first run and intermediate drying, it must be covered with a glass plate and is best stored in the dark.

TLC-Electrophoresis

Silica gel layers were used already in 1946 by CONSDEN, GORDON and MARTIN [99] for electrophoretic separation of amino acid mixtures. PASTUSKA and TRINKS [195] and HONEGGER [123] were the first to draw attention to the advantages in TLC. The combination of normal TLC in one direction and electrophoretic separation in the second (Fig. 59), as described on pp. 105—114 by HANNIG and PASCHER in the chapter on electrophoresis, is especially interesting. Individual results are thus not discussed here.

5. Detection of Amino Acids on the Chromatogram

At present there is no known specific reagent for amino acids alone. Some non-specific reagents for nitrogen-containing compounds and

---

[14] The usual vacuum-drying ovens with heating are quite expensive.

several which are specific for one or a few amino acids are available. Some details worth knowing about the most important procedures are described below.

a) Ninhydrin. Ninhydrin is still the most popular reagent for amino acids. Although the colour reaction of indane-1,2,3-trione hydrate with amino acids and lower peptides has been known since 1910 (RUHEMANN), there is still doubt about the course of the reaction, CALDIN [46] has summarised the many theories about the reaction mechanism. Treatment with complex-forming cations (Cu, Cd, Ca) changes the colour obtained with ninhydrin bathochromically towards red and considerably increases the colour stability. Whereas the normal method (cf., e. g., [26, 47, 48]) for colour development yields yellow (proline and hydroxyproline) or violet (all other α-amino acids) products, more specific colours may be obtained by addition of bases such as collidine or benzylamine; although this may facilitate the identification of individual amino acids, it is offset by a loss in sensitivity of detection.

Practical details for carrying out the ninhydrin reaction are given in Chapter Z, Reagent No. 178. The acetic acid added to the reagent ensures that the pH of about 5, as required for the reaction, is attained even where basic solvents have been employed to run the chromatogram. Prolonged heating at higher temperatures (110° C) renders the chromatogram background pink. The sensitivity of detection on thin-layer chromatograms for some amino acids is given in Table 182.

α) *Stabilisation of the coloured spots obtained with ninhydrin:* After visualisation with the ninhydrin reagent, which should not contain any complex-forming buffer such as citrate, KAWERAU and WIELAND [49] spray the chromatogram with a cupric nitrate solution (Rgt. No. 178). The Cu-complex thus obtained is, however, stable only in the absence of free acid. The chromatogram should thus be briefly exposed to ammonia vapour immediately after spraying. Moreover, the chromatogram must be protected from moisture since the Cu-ninhydrin complex dissociates reversibly between pH 7 and 9 and irreversibly at pH-values above 9. A collodion film, as proposed by BARROLLIER [50], may be used for this purpose (see also under DNP-amino acids, documentation, p. 771).

β) *Polychromatic ninhydrin reaction:* We have found the reagent described by MOFFAT and LYTLE [3] (Rgt. No. 179) to be satisfactory. Similar effects have been achieved by WOIWOD [51], using collidine; and by HARDY et al. [52] with cyclohexylamine or dicyclohexylamine.

ESSER [210] describes a quantitative determination of the ninhydrin-positive compounds separated on cellulose powder MN 300.

The dried plates are briefly immersed in a 0.5% solution of ninhydrin in acetone and then maintained for 105 min at 70° C. The zones are immediately scraped off, mixed with 0.4 ml of a 0.5% solution of cadmium acetate in methanol and brought into suspension by a current of air. After 2 h, the suspension is again formed, centrifuged and the light adsorption of the clear liquid measured at 494 nm. The average error is quoted as $\pm 0.5\%$.

Table 182. *Limits of detection (µg.) of each of 19 amino acids on thin-layer chromatograms (silica gel G) using the ninhydrin reaction* [9]

Amino acids (cf. Table 180 for abbreviations)	One-dimensional chromatogram in n-propanol-water (70 + 30)	Two-dimensional chromatogram of the amino acid mixture as shown in Fig. 206
Ala	0.009	0.05
β-Ala	0.01	0.06
Arg	0.01	0.06
Asp	0.1	0.4
CySO$_3$H	0.01	0.1
Glu	0.04	0.2
Gly	0.001	0.006
His	0.05	0.5
Hypro	0.05	0.1
Leu	0.01	0.2
Lys	0.005	0.03
Met	0.01	0.4
Phe	0.05	0.2
Pro	0.1	0.5
Ser	0.008	0.1
Thr	0.05	0.1
Try	0.05	0.5
Tyr	0.03	0.1
Val	0.01	0.2

b) **Chlorine-Tolidine** (Rgt. No. 42). Free amino acids respond to this test which was developed principally for detecting substances containing the grouping —NH—CO— and which is described in the section on peptides; relatively high concentrations of amino acids are, however, needed.

c) **Other Reagents.** Tables 183 and 184 summarise experience gained in paper chromatography. Appropriately modified techniques should be applicable in TLC. The excellent separation effects furnished by TLC render specific reagents less important for the detection of individual amino acids but they may still be useful for detecting such amino acids in intact peptides.

The fluorescence test customarily used in PC depends, according to work of OPIENSKA-BLAUTH et al. [53], on a reaction taking place on heating between the amino groups of the amino acids and aldehyde groups formed from carbohydrates in the paper; it is therefore inapplicable to chromatograms on silica gel layers.

Table 183. *Amino acid detection with predominantly non-specific reagents*

Type	Rgt. No.	Reagent	Authors
polychromatic	178	Ninhydrin	Caldin [46] Tsukamoto and Komori [48] Patton and Chism [54] Toennies and Kolb [55] Consden et al. [47] Ruhemann [56—60]
		Ninhydrin + cobaltous chloride	Wiggins and Williams [61]
	179	Ninhydrin + copper nitrate + collidine	Moffat and Lytle [3]
		Ninhydrin + collidine	Woiwod [51]
		Ninhydrin + cyclohexylamine or dicyclohexylamine	Hardy et al. [52]
		Ninhydrin + phenol	Hais [a]
		Isatin	Saifer and Oreskes [62] Noworytko and Sarnecka-Keller [63] Acher et al. [64] Grassmann and v. Arnim [65]
polychromatic		Isatin + zinc acetate + pyridine	Barrollier et al. [66]
		Alloxan	Saifer and Oreskes [62] Rosebeek [67]
		Na-1,2-naphthoquinone-4-sulphonate	Kofrányi [68] Consden [69]
polychromatic	172	....	Müting [70] Giri and Nagabhushanam [71]
		Na-1,2-naphthoquinone-4-sulphonate + zinc acetate + quinoline	Barrollier et al. [66]
		4,5-Diacetyl-cyclohexene-(1)	Riemschneider and Preuss [72]
	42	Chlorine/tolidine	Reindel and Hoppe [73]

[a] Cf. p. 419, 753 in [2].

Table 184. *Amino acid detection with relatively specific reagents*

Amino acid	Rgt. No.	Reagent	Authors
Arginine	136	α-Naphthol (or oxine)/NaOBr (or NaOCl) "Sakaguchi reaction"	Bhattacharya et al. [74] Roche et al. [75] Jepson and Smith [76] Acher and Crocker [77] Sakaguchi [78—80]
	187	Na-nitroprusside/potassium ferricyanide	Roche et al. [75]
		Na-α-naphtholate/diacetyl	Tuppy [81]
Cysteine	184	Sodium nitroprusside	Toennies and Kolb [55] Winegard et al. [82]
	cf. 147	Iodoplatinate	Toennies and Kolb [55] Winegard et al. [82]

Table 184 (Continued)

Amino acid	Rtg. No.	Reagent	Authors
Cysteine	142	Iodine-azide	KIRBY-BERRY et al. [83]
			AWE et al. [84]
			CHARGAFF et al. [85]
	255	Tetrazolium salts	BURTON et al. [86]
		p-Aminodimethylaniline/potassium ferricyanide	TOYADA [87]
Cystine	184 cf.	Sodium nitroprusside/NaCN	WINEGARD et al. [82]
	147	Iodoplatinate	WINEGARD et al. [82]
	142	Iodine-azide	KIRBY-BERRY et al. [83]
			AWE et al. [84]
			CHARGAFF et al. [85]
		p-Aminodimethylaniline/potassium ferricyanide	TOYADA [87]
Glycine		o-Phthaldehyde	PATTON and FOREMAN [88]
Histidine	238	Diazotised sulphanilic acid or sulfanilamide ("Pauly reagent")	FRANK and PETERSEN [89]
			BRAY et al. [90]
			KIRBY-BERRY et al. [83]
			PAULY [91, 92]
	100	Fast Blue B salt (diazo-reagent)	
		Diazotised p-chloroaniline	EDLBACHER [93, 94]
		Diazotised p-bromoaniline	
		Diazotised p-anisidine	SANGER and TUPPY [95]
		Bromine	KIRBY-BERRY et al. [83]
		o-Phthaldehyde	PATTON and FOREMAN [88]
Hydroxy-proline		Isatin/Ehrlich reagent	JEPSON and SMITH [76]
		Ninhydrin modification	CLARKSON [96]
		Periodate/acetylacetone + ammonium acetate	SCHWARTZ [97]
		Ninhydrin	
Lysine	261 cf.	Vanillin	
Methionine	147	Iodoplatinate	WINEGARD et al [82]
	142	Iodine-azide	KIRBY-BERRY et al. [83]
			AWE et al. [84]
			CHARGAFF et al. [85]
		Potassium permanganate	DALGLIESH [98]
Ornithine	261	Vanillin	
Proline	178	Isatin	ACHER et al. [64]
		Ninhydrin	
Serine		Periodate/acetylacetone + ammonium acetate	SCHWARTZ [97]
	160	Periodate/Nessler reagent	CONSDEN [69], CONSDEN et al. [99]
		Periodate/KI + starch	METZENBERG and MITCHELL [100]
		1,2-dinitrobenzene-enediol-reaction	FEARON and BOGGUST [101]
Threonine		Periodate/Na-nitroprusside + piperidine	SCHWARTZ [97]
		Periodate/Na-nitroprusside + piperazine ("Rimini reagent")	EDWARD and WALDRON [102]
	160	Periodate/Nessler reagent	CONSDEN [69], CONSDEN et al. [99]
		Periodate/KI + starch	METZENBERG and MITCHELL [100]
		1,2-Dinitrobenzene-enediol-reaction	FEARON and BOGGUST [101]

Table 184 (Continued)

Amino acid	Rgt. No.	Reagent	Authors
Tryptophan	72	p-Dimethylaminobenzaldehyde + HCl (Ehrlich reagent)	Smith [103] Dalgliesh [104] Pachéco [105]
	49	Cinnamaldehyde + HCl	Wieland and Bauer [106] Jerchel and Müller [107]
		NaNO$_2$ + HCl	Fischer [108]
	123	Formaldehyde reagent modified Salkowski reagent	Procházka [109] Linser et al. [110]
	238	Diazotised sulphanilic acid Diazotised p-nitroaniline Diazotised benzidine Diazotised ethyl-α-naphthylamine	Erspamer [111] Erspamer [111] Clerk-Bory et al. [112] Dalgliesh [104] Ekman [113]
		o-Phthalaldehyde	Patton and Foreman [88]
Tyrosine		α-Nitroso-β-naphthol/HNO$_3$	Acher and Crocker [77] Gerngross et al. [114]
	238	Diazotised sulphanilic acid (Pauly reagent)	Frank and Petersen [89] Bray et al. [90] Kirby-Berry et al. [83]
	100	Fast Blue B salt (diazo reagent)	
	122	Tungstomolybdophosphoric acid (Folin-Ciocalteu reagent)	Kudzin et al. [115] Folin and Ciocalteu [116]
		Folin-Denis reagent	Kudzin et al. [115] Folin and Denis [117]
		Millon reagent	Durant [118] Millon [119]

d) Fluorodinitrobenzene. According to Pataki [220], amino acids separated in one-dimensional chromatography can be visualised by reaction with fluorodinitrobenzene on the layer, followed by removal of excess reagent (limit of detection is 10^{-2} to 10^{-3} μmole). They may be eluted and re-chromatographed or directly chromatographed in the second dimension.

IV. Peptides

Peptides are generally hydrophilic like amino acids. Thin-layer chromatographic techniques developed for amino acids are therefore applicable in principle to peptides. There are, however, limits to this analogy. The number, nature and sequence of the amino acid units have a bearing on solubility and adsorption behaviour of higher peptides. Other chromatographic conditions or even other methods of separation may thus have to be employed. Peptides with masked functional groups, such as intermediates in peptide synthesis, are less hydrophilic than those without protective groups.

Peptides exist along with amino acids in biological material[15], usually, however, in small amounts and often as conjugates (phosphopeptides, peptidyl nucleotides, glucopeptides, lipopeptides, peptide-protein complexes). Smaller, free peptides accompany the amino acids in extraction; their separation from the amino acids may be difficult.

There may be no other way than to run a two-dimensional chromatogram of the amino acid-peptide mixture, elute each of the separated substances and to test their homogeneity by further chromatography and, if necessary, to establish peptide character with certainty through hydrolysis. A combination of PC and paper electrophoresis has often been used successfully to separate such mixtures (cf., e. g., ANFINSEN et al. [122]). According to HONEGGER [123], TLC may be combined with thin-layer electrophoresis (pp. 111—114 and Fig. 59).

Column chromatography on cellulose[16]- or dextran[17]-based ion exchangers offers new possibilities of efficient pre-fractionation.

DEAE-cellulose, for example, contains diethylaminoethyl groups; it does not adsorb neutral and basic amino acids but does adsorb neutral peptides which are eluted with water or water, saturated with carbon dioxide [125].

Gel filtration with Sephadex[18] separates substances on the basis of their molecular weight [21, 24, 126, 127]. Its effect corresponds roughly to that of dialysis but it works appreciably faster. LINDNER et al. [128] have extracted a dry preparation from the posterior lobe of hog pituitary glands (oxytocin and vasopressin activity of 2—3 units/mg) with a pyridine acetate buffer, neutralised, filtered through a column of Sephadex G 25, eluted with the same buffer and obtained two ninhydrin-positive fractions. The first, faster migrating fraction contained oxytocin and vasopressin as peptide-protein complexes; the second, containing only inactive material of low molecular weight, was rejected. The complex was split up by treating the first fraction with 1M formic acid for 10 min at 70° C; vasopressin and oxytocin, each with activities of about 100 units/mg, were then obtained in a slowly migrating fraction by re-filtration through Sephadex G-25 and elution with 1M formic acid.

Separation of amino acids or of salts from peptides on Sephadex is less effective. The situation is complicated by the fact that migration rates depend more on experimental conditions than on molecular size, so that the former play a vital part [22].

Pre-fractionation by Craig counter-current distribution requires more complicated apparatus[19].

Mixtures of peptides result from partial degradation of proteins. Separation of the individual components and establishment of their amino acid sequence form the basis for elucidating the chemical structure of proteins; for information about the methods of degradation which yield larger and smaller peptides, reference may be made to: SANGER [130] (partial hydrolysis); CRAIG et al.[20] (oxidative fission of S—S

---

[15] Cf. [120, 121], for example.

[16] DEAE-Cellulose, ECTEOLA-Cellulose, Carboxymethyl-cellulose, Phosphoryl-cellulose; cf. pp. 39, 43, 45, 47 in [124].

[17] DEAE-Sephadex (Firm 102).

[18] Sephadex consists of cross-linked dextran (p. 40).

[19] CRAIG, L., and D. CRAIG (p. 290 in [129]).

[20] CRAIG, L. C., W. M. KÖNIGSBERG and T. P. KING (p. 70 in [124]).

bridges); H. ZUBER [131] (enzymatic hydrolysis); and SJÖQUIST [132, 133] (peptides as products of the Edman degradation).

Synthetic peptides are gaining importance as newer synthetic methods are being developed; one may mention bradykinin and analogues [134—136], oxytocin analogues [137], polymyxin [138, 139] and corticotropically active polypeptides [139a].

PC was formerly extensively employed in all these domains but TLC has partly taken over its role as an analytical aid today. Some examples of TLC of peptides are given in Tables 185 and 186.

Table 185. hR_f-values [a] of pairs of isomeric dipeptides on silica gel G in two solvents [b] [140]. For abbreviations cf. Table 186

Dipeptide pair	I	II
H · Ala-Gly-OH/H · Gly-Ala · OH	15—16	15—14
H · Gly-Hypro-OH/H · Hypro-Gly · OH	08/13	12—13
H · Gly-Leu · OH/H · Leu-Gly · OH	37—35	27/33
H · Ala-Leu · OH/H · Leu-Ala·· OH	45/37	30—31
H · Gly-Phe · OH/H · Phe-Gly · OH	36—38	32/37
H · Gly-Pro · OH/H · Pro-Gly · OH	08—09	08—07
H · Gly-Ser · OH/H · Ser-Gly · OH	12—14	12—14
H · Phe-Ala · OH/H · Ala-Phe · OH	48—45	42—40
H · Gly-Val · OH/H · Val-Gly · OH	32/27	22—24

I = n-Butanol-acetic acid-water (80 + 20 + 20);
II = n-Propanol-water (70 + 30).

[a] hR_f-values joined by a hyphen can be distinguished in parallel runs but do not differ enough to permit separation in the usual procedure without continuous development.

[b] Other suitable solvents are: ethanol-34% ammonium hydroxide (70 + 30); n-propanol-34% ammonium hydroxide (70 + 30). Our experience with about 100 peptides shows that the chromatographic behaviour is generally unambiguous up to the nonapeptides.

For chromatographing protected, larger peptides (up to 10 amino acid units), GUTTMANN [143] recommends silica gel or Alox layers and dimethylformamide or acetic acid containing 5—10% water as solvents. Detection is with the chlorine-iodine reaction and R_f-values lie between 0.3 and 0.9. These are said to be not very reproducible but the method has been used successfully for purity testing. SCHELLENBERG [144] recommends the solvents: chloroform-acetone (90 + 10) and (80 + 20); cyclohexane-ethyl acetate (50 + 50); and chloroform-methanol (90 + 10).

The possibility of separating similar, complex peptides from one another is limited according to VOGLER [145]. VOGLER et al. [146] prepared synthetically four isomeric cyclodecapeptides in the course of the structural elucidation of polymyxin B_1; W. HAUSMANN [147] and BISERTE and DAUTREVAUX [148] had suggested them as possible structures for the naturally occurring product. These cyclic, isomeric oligopeptides, denoted by the symbols 8γ, 8α, 7γ and 7α, differ from one another partly in the number of ring amino acids and in the way in

Table 186. *TLC of peptides and protected peptides on silica gel G in the solvents A, B and* [a,b,c] *according to* RINIKER [141]. *For abbreviations cf.* KAPPELER *and* SCHWYZER [142]

Peptide and protected peptide	hRf A	hRf B
H · Pro · OH	10	11
Z · Pro · OH	39	81
H · Pro · OtBu	81	34
Z · Pro · OtBu	86	87
Z · Val-Lys (BOC) · OH	57	87
Z · Val-Lys · OH	26	49
Z · Val-Tyr · OCH₃	86	82
Z · Val-Tyr · OH	48	78
Z · Val-Tyr-Pro · OtBu	84	81
Z · Val-Tyr-Pro · OH	45	75
Z · Val-Tyr-Pro · OtBu	73	58
H · Val-Tyr-Pro· OH	30	46
H · Val-Tyr-Val-His-Pro-Phe · OCH₃	76	22
H · Val-Tyr-Val-His-Pro-Phe · OH	47	18
H · Val-tyr-Val-His-Pro-Phe · OH	51	24
H · Asp-Arg-Val-Tyr-Val-His-Pro-Phe · OH (= Hypertensin II)	21	03
H · Asp-Arg-Val-tyr-Val-His-Pro-Phe · OH	26	05
H · Asp(NH₂)-Arg-Val-Tyr-Val-His-Pro-Phe · OH	20	02
H · Asp(NH₂)-Arg-Val-Tyr-Val-His-Pro-Phe · NH₂	19	03
H · Asp(NH₂)-Arg-Val-Tyr-Val-His-Pro-Phe · OCH₃	21	05
e-[Val-Orn-Leu-phe-Pro-Phe-phe-Asp(NH₂)-Glu(NH₂)-Tyr] (= Tyrocidine A)	32	40
(Val-Orn-Leu-phe-Pro)₂ (= Gramicidin)	45	36
(Val-Lys-Leu-phe-Pro)₂ Lysin-(= Gramicidin)	38	31
H · Arg-Pro-Pro-Gly-Phe-Ser-Pro-Phe-Arg · OH (= Bradykinin)	12	01
Z · Glu(OtBu)-His-Phe-Arg(NO₂)-Try-Gly · OH	53	53
Z · Glu-His-Phe-Arg(NO₂)-Try-Gly · OH	24	42
BOC · Ser-Tyr-Ser-Met-Glu(OtBu)-His-Phe-Arg-Try-Gly · OH	38	28
H · Ser-Tyr-Ser-Met-Glu-His-Phe-Arg-Try-Gly · OH	18	
H · Lys(BOC)-Pro-Val-Gly-Lys(BOC)-Arg-Arg-Pro-Val-Lys(BOC)-Val-Tyr-Pro · OtBu	32	22
Z · Glu(OtBu)-His-Phe-Arg(NO₂)-Try-Gly-Lys(BOC)-Pro-Val-Gly-Lys(BOC)-Lys(BOC)-Arg-Arg-Pro-Val-Lys(BOG)-Val-Tyr-Pro · · OtBu	39	43
Z · Val-Tyr-Val-His-Pro-Phe · OCH₃	77[a]	81[b]
H · Val-Tyr-Val-His-Pro-Phe · OCH₃	59[a]	66[b]

A: sec-Butanol-3% ammonium hydroxide (100 + 44);
B: n-Butanol-glacial acetic acid-water (100 + 10 + about 30), upper phase.

[a] Dioxan-water (90 + 10).
[b] Methanol.
[c] Water, methanol, acetone, dioxan and dimethylformamide, either alone or in mixtures with each other, are generally less useful for separation of free peptides and of protected higher peptides because of tailing and formation of diffuse spots.

which the side chains are linked to the ring; they have been the object of precise thin-layer chromatographic studies [145]:

Ascending Technique:

No difference could be established between 7γ, 7α and polymyxin B_1 in the hRf-region 50—90 when using the solvents:

n-butanol-pyridine-acetic acid-water (v/v)
30	20	6	24
15	5	8	12
15	3	10	12
30	3	23	24

n-butanol-pyridine-acetic acid-water-ethyl acetate (v/v)
| 5 | 1 | 2 | 2 | 2 |

n-butanol-acetic acid-water (v/v) (upper phase)
| 40 | 10 | 50 |

isopropanol-pyridine-acetic acid-water (v/v)
| 10 | 5 | 4 | 4 |
| 4 | 8 | 1 | 1 |

ethanol-pyridine-acetic acid-water-ethyl acetate (v/v)
| 5 | 1 | 2 | 2 | 2 |

Horizontal technique, continuous development:

The chromatograms from continuous development [13] in n-butanol-pyridine-acetic acid-water (30 + 20 + 6 + 24) showed no significant Rf differences after the longest possible run (8—10 h) (spots at the far end of the layer). All three substances migrated about 20 mm overnight (14 h) in ethyl acetate-pyridine-acetic acid-water (50 + 10 + 10 + 10) and all yielded sharply-edged, round spots at exactly the same height.

Peptides may be detected best using the methods found suitable in PC. The popular ninhydrin reaction is not always sufficiently sensitive with higher peptides; it fails utterly with cyclic peptides unless these contain free amino groups in side chains. More generally available and more sensitive (limit of detection about 0.1 µg) is the N-halogenation of REINDEL and HOPPE [73], modified as follows [8, 149]:

About equal volumes (20—50 ml) of 1.5% $KMnO_4$ solution and 10% HCl are placed in a suitably sized photographic development tank which contains a grating of glass rods, 2—3 cm above the bottom. The plate with the chromatographed substances to be detected is placed on the grating and the tank is covered with a large glass plate. After 15—20 min reaction time, the plate is aerated for 2—3 min in a well ventilated fume cupboard; the smell of chlorine should have completely disappeared before the subsequent spraying. The spray reagent (No. 42) should be applied with the greatest care since this increases sensitivity; in this way, even poorly separated substances can be seen as separate spots for at least a moment. The washing process with 2% acetic acid, usual in PC, is omitted. If a chlorine cylinder is available, it is simpler to fill a Desaga chamber or similar vessel with chlorine gas and to place the plate for 5—10 min in this chlorine atmosphere.

Iodine also can be used for halogenation and the peptides then detected by subsequent spraying with a starch solution (modification of Rgt. No. 141).

A combination of the ninhydrin- and chlorination techniques is especially good. The layers are sprayed first with ninhydrin and the positions of the spots marked. Treatment with chlorine gas is then carried out as described above; the previous treatment with ninhydrin does not affect the results. Spots appearing only on halogenation are thus easily distinguished from ninhydrin-positive compounds.

SCHELLENBERG [144] uses a variation of the well-known fluorescence test with morin; sensitivity is relatively low (limit of detection 2 µg). SANGER and TUPPY's [95] fluorescence test seems to us to be rather uncertain in the absence of cellulose. Naturally, all procedures which depend on destruction of the organic compounds yielding brown or black products (Rgt. Nos. 46, 241) can be used here, in contrast to PC.

For *elution of the substances* [150] the corresponding area of the layer is carefully scraped off, suspended in an appropriate solvent and filtered. Water should not be used with silica gel G layers since it dissolves the gypsum, which may interfere later. n-Butanol and acetic acid have been found to be satisfactory.

V. N-(2,4-Dinitrophenyl)-amino Acids and 3-Phenyl-2-Thiohydantoins

Dinitrophenylamino acids (DNP-amino acids) and phenylthiohydantoins (PTH-amino acids) are formed when proteins or peptides are treated with dinitrofluorobenzene [151—154] or phenyl isothiocyanate [155], respectively, and the reaction product suitably broken down. Their separation from the reaction mixture and their identification is of considerable practical importance since systematic application enables sequence analysis of peptide structures to be performed. Numerous investigators have worked on this problem[21].

Substitution in the amino or carboxyl group or in both transforms amino acids into acids, bases or neutral compounds, i. e., the zwitter ion character is lost. The derivatives remain more or less polar, depending on the type of substituent. This must be borne in mind when the solvent is chosen. Substitution products which still possess a free carboxyl and free amino group (e. g., monoacyl derivatives of basic amino acids, monoesters of acidic amino acids or ethers of hydroxyamino acids) behave chromatographically like free amino acids.

A. Dinitrophenylamino Acids

The formation of dinitrophenylamino acids (DNP-amino acids), as mentioned above, is portrayed in the reaction scheme below.

---

[21] Literature compilation in [8]; BISERTE et al. [156] also have made an excellent survey.

Free amino acids instead of proteins or peptides can of course be reacted with dinitrofluorobenzene (DNFB). This can arise when an amino acid mixture cannot be chromatographed because of contaminants; the DNP-amino acids can often be more easily converted into a form capable of being chromatographed, than can the free amino acids. The fact that these derivatives are coloured facilitates quantitative analysis since all the difficulties associated with subsequent colour reactions are obviated.

DNP-amino acid derivatives are sensitive to light. They must be kept in the dark and should be exposed for only short periods to light (and only indirect light).

End-group determination according to SANGER

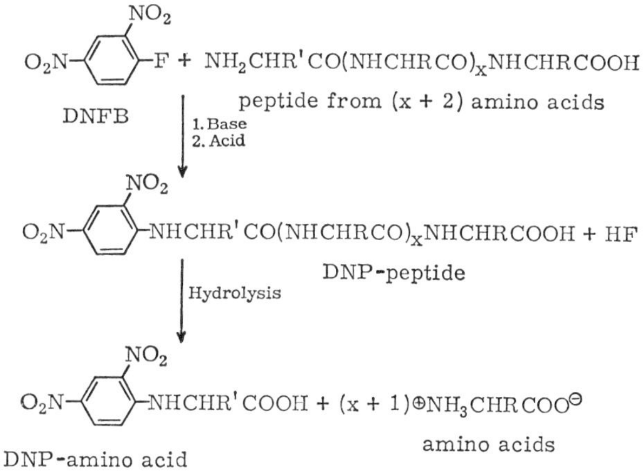

$O_2N-C_6H_3(NO_2)-F$ + $NH_2CHR'CO(NHCHRCO)_xNHCHRCOOH$

DNFB   peptide from (x + 2) amino acids
1. Base
2. Acid

$O_2N-C_6H_3(NO_2)-NHCHR'CO(NHCHRCO)_xNHCHRCOOH$ + HF

DNP-peptide

Hydrolysis

$O_2N-C_6H_3(NO_2)-NHCHR'COOH$ + $(x + 1)^{\oplus}NH_3CHRCOO^{\ominus}$

DNP-amino acid   amino acids

1. Dinitrophenylation

The customary procedures for dinitrophenylation differ amongst themselves in the experimental conditions and outlay of apparatus. BISERTE et al. [156] have composed an excellent review. A brief survey is given here.

a) Amino Acids

α) **Preparation of DNP-amino acids**[22]: According to LEVY and CHUNG [157], 5 mmoles of the amino acid and 1 g of anhydrous Na_2CO_3 in 20 ml water are mixed with 5 mmoles 2,4-dinitrofluorobenzene (DNFB) as a 10% solution in acetone (twice this amount is needed for amino acids with two reactive groups and $2^1/_2$ times for histidine). The suspension is shaken well for 30—90 min at 40° C in the dark, during which the drops of DNFB slowly disappear. Any residual DNFB is then extracted with ether and the aqueous solution is carefully acidified with 1—2 ml conc. hydrochloric acid, whereby the DNP-amino acids separate as oils or crystals; they are isolated by extraction with ether or by filtration, respectively. Derivatives which have separated as oils usually crystallise during evaporation

---

[22] A collection can be obtained from Firm 86, for example.

of the ether solution (DNP-glutamic acid presents particular difficulties). The derivatives can generally be successfully recrystallised by taking up in benzene containing a little ethanol and adding petrol ether to the hot solution. Highly polar DNP-amino acids are best recrystallised from aqueous methanol; those insoluble in ether may be purified by dissolving in dilute hydrochloric acid and precipitating with a base, e. g., pyridine.

Reference may be made to the review of BISERTE [156], mentioned above, for information about preparation of DNP-cysteic acid and the various mono-derivatives of cysteine, cystine, histidine, lysine, ornithine and tyrosine; the water-solubility of these products creates some special problems.

Melting-point data can be found in the publications of PORTER and SANGER [154], LEVY and CHUNG [157], DU VIGNEAUD et al. [158] and others.

β) **Quantitative dinitrophenylation of a mixture of amino acids** (e. g., a protein hydrolysate): In WALLENFELS' [159] procedure, 2—5 mg of the air-dried product from oxidation of proteins by performic acid is hydrolysed and the dried residue dissolved with vigorous stirring (magnetic stirrer) in 2 ml of CO_2-free water at room temperature. An aliquot (1.2 ml) is transferred to a small reaction vessel equipped with a magnetic stirrer and diluted with 1.8 ml of CO_2-free water + 0.1 ml of 3.1 N KCl; the temperature is raised to $40.0 \pm 0.1°$ C (thermostat). The pH is then adjusted to 8.9 by adding 0.2 N NaOH with an autotitrator, stirring energetically. Under exclusion of light about 0.1 ml of 2,4-dinitrofluorobenzene ("Merck" analysis grade, No. 2966; amount representing a slight excess) is added and the pH maintained at 8.9 for 100 min by means of the auto-titrator. A recorder connected to the auto-titrator follows the uptake of alkali. Dinitrophenylation is practically complete after 50 min; the remaining 50 min period is mainly for establishing the rate of hydrolysis of the dinitrofluorobenzene, yielding dinitrophenol. Excess of dinitrofluorobenzene is then removed by extracting twice with 5 ml of peroxide-free ether [160, 161] and the aqueous phase is acidified with 0.5 ml hydrochloric acid (1 part HCl, density 1.19, + 1 part water). The ether-soluble DNP-amino acids are extracted with five 4 ml portions of peroxide-free ether; the combined extracts are made up with ether to exactly 25 ml. For chromatography, 1 ml of the solution is concentrated to a small volume and quantitatively applied with a capillary pipette.

The aqueous phase still contains the acid-soluble DNP-amino acids. Dissolved ether is removed in vacuo and the volume made up to exactly 10 ml with CO_2-free water. A 0.5 ml aliquot of this solution is evaporated in vacuo to dryness and the residue is dissolved in the minimum amount of acidified acetone (2 ml 6N hydrochloric acid made up to 25 ml with acetone); the solution is applied quantitatively to the layer.

The acid-soluble DNP-amino acids are better extracted with a mixture of equal volumes of ethyl acetate and n-butanol (6 extractions, volume ratio 3:1) since the salts then remain largely in the aqueous phase (WALZ et al. [231]).

b) Peptides

α) **Dinitrophenylation according to Lockhard and Abraham**[23]. 50—150 µg of the peptide are dissolved in 0.1 ml of a 1.5% aqueous solution of trimethylammonium carbonate (pH 9.3) and 0.2 ml of a 5% alcoholic solution of dinitrofluorobenzene is added. The mixture is left in the dark for $2^1/_2$ hours. The ethanol is then evaporated off in vacuo, 0.24 ml more of trimethylammonium carbonate

---

[23] Micromethod according to [162].

solution and 1 ml ether are added and the mixture thoroughly shaken. The phases are separated by centrifuging, the ether layer discarded and the aqueous solution evaporated to dryness in vacuo.

β) **Total hydrolysis of a DNP-peptide:** The residue from the aqueous layer [see *α*)] is dissolved in 0.1 ml of 6N hydrochloric acid. The solution is heated for 9 h at 105° C in nitrogen in a sealed tube and the hydrolysate then diluted with two volumes of water. The ether-soluble DNP-amino acids are obtained from their solution by extracting three times with an equal volume of ether or ethyl acetate (di-DNP-histidine).

c) Polypeptides and Proteins

α) **Dinitrophenylation:** LEVY and LI [163] dissolve at least 0.2 μmole of the material in 3 ml of 0.05N aqueous KCl at 40° C and adjust the pH to 8 with 0.05 potassium hydroxide using an auto-titrator. About 0.1 ml of dinitrofluorobenzene is added and the solution energetically stirred in the dark at constant pH and temperature. The end of the reaction is indicated when alkali consumption ceases. The solution is then extracted three times with ether and the aqueous phase acidified in order to precipitate the dinitrophenyl derivative. The precipitate is centrifuged off, washed with water, acetone and ether and dried in a desiccator over P_2O_5.

β) **Partial hydrolysis of a DNP-protein:** The partial hydrolysis of a DNP-protein can be carried out with hydrochloric acid or enzymatically, in the usual way. Hydrolysis is appreciably slower than that of naturally occurring material but it yields important analytical information. During electrophoretic or chromatographic treatment of the degradation products, the N-terminal fragments on which the interest is centred are already recognisable through their colour and can thus be eluted effortlessly and with certainty from two-dimensional chromatograms. However, non-terminal peptide fragments which contain lysine, for example, are also coloured yellow through their *ε*-DNP-group. Whereas separation of DNP-peptides from free peptides and amino acids is not problematic (talcum pretreated with hydrochloric acid adsorbs only DNP-compounds; see under *δ*), extraction from acid aqueous solution by means of organic solvents often does not satisfactorily differentiate between N-terminal-*α*-DNP-peptides and non-terminal DNP-peptides; the last named should remain in the aqueous phase because they possess a free NH_2-group [164].

γ) **Total hydrolysis of a DNP-protein:** For qualitative end-group determination, the protein is hydrolysed in a sealed tube for 16 h at 105° C with 100 times its amount of 5.7N hydrochloric acid (twice distilled). DNP-glycine [165, 166] and DNP-proline [167] are partially destroyed especially if tryptophan is present [168]; di-DNP-tyrosine partly loses its O-DNP group [166]; DNP-cystine should be removed by prior treatment with performic acid [44].

δ) **Separation of DNP-amino acids from a total hydrolysate:** The hydrolysate is diluted until it is about 1N in hydrochloric acid. This solution is extracted five times with peroxide-free ether [160, 161] and, if histidine is present, five times with ethyl acetate; the extracts are washed three times with 0.1N hydrochloric acid. All extracts are then combined (Fraction A: ether-soluble DNP-amino acids and dinitrophenol) and the aqueous phase is combined with the washings (Fraction B: free amino acids and acid-soluble dinitrophenyl derivatives like DNP-arginine, DNP-cysteic acid, mono-DNP derivatives of cysteine, cystine, histidine, lysine, ornithine and tyrosine; if the extraction is performed only with ether, a part of the di-DNP-histidine may also be found in this fraction).

Fraction A: The removal of dinitrophenol is recommended if much is present; this may be done in one of two ways:

Sublimation. Dinitrophenol largely sublimes when heated in high vacuum at 70—80° C in a special apparatus designed by MILLS [169]. Some DNP-methionine and, as observed in our laboratory, also di-DNP-cystine may be lost. If necessary, the sublimation procedure may have to be interrupted, the substance dissolved in a little acetone and redeposited as a thin film on the walls of the vessel by evaporation.

Adsorption (ion exchange): [170]. A few μmoles of DNP-amino acids are dissolved in 0.5 ml methanol and brought on to a column of anionotropic alumina[24] (1 × 10 cm). After rinsing the column with 0.5 ml methanol, the dinitrophenol is quantitatively eluted with 2% acetic acid. The DNP-amino acids are desorbed by first eluting with a small amount of 0.1 N NaOH (to avoid CO_2-evolution and consequent rupture of the column) and then with 1% $NaHCO_3$ solution. The DNP-amino acids can be isolated from the eluate by carefully acidifying with HCl, extracting with ether and evaporating the extract.

The residue from removal of dinitrophenol is dissolved in acetone (ca. 1 ml per 10 μmole protein) and amounts of about 1 μl can be chromatographed directly.

Fraction B: The acid-soluble DNP-amino acids can often be directly identified by chromatography even when free amino acids are present. The solution is repeatedly evaporated to dryness, some water being added each time. The residue is taken up on 0.5N hydrochloric acid or in acetic acid (about 1 ml per 10 μmole protein) and 1 μl amounts applied[25].

In order to remove the free amino acids, the residue from evaporation (see above) is dissolved in 2 ml 1N hydrochloric acid and this solution passed through a column (diameter 2.5 cm) of a mixture of 20 g Hyflo-Super-Cel and 50 g talcum (pretreated with 0.01N and 1N hydrochloric acid [171]). All the DNP-amino acids (apart from DNP-cysteic acid) are adsorbed, in contrast to the free amino acids. The column is rinsed with 100 ml 1N hydrochloric acid and finally the DNP-amino acids are eluted with alcoholic hydrochloric acid (alcohol-1N hydrochloric acid, 40 + 10) or, better, with ammoniacal alcohol (alcohol-0.3% ammonium hydroxide, 40 + 10). The eluate is evaporated to dryness and the residue chromatographed as indicated above.

2. Solvents and Separation Efficiency

Solvents used in PC for separating free amino acids are mostly directly applicable in TLC. This does not apply or applies only to a very limited degree to DNP-amino acids. DNP-amino acids have a bad reputation in PC since they tend greatly to tailing and their Rf-values are highly dependent on the quantity applied and on the presence of other DNP-amino acids. The "toluene" system of BISERTE and OSTEUX [45], n-butanol-0.1% ammonium hydroxide of BRAUNITZER [172] and 1.5M phosphate buffer of LEVY [173] have proved good solvents in this sphere.

In TLC, the "toluene" system shows thin "beards"[26] and a far too meagre elution effect; DNP-leucine has an Rf-value of 0.25. If the layer is inactivated by exposing it to the vapour of the aqueous phase

---

[24] "Merck" alumina is shaken 10 min with excess 1N hydrochloric acid and washed free of acid by decantation with water.

[25] The acid solvent must be completely removed from the layer before chromatography.

[26] Cf. HAIS and MACEK (p. 147 in [2]) for this expression.

of this solvent system at least overnight before application, the R_f-value of DNP-leucine rises to 0.66; excellent separations are then accomplished in two-dimensional chromatograms despite the "beards" mentioned. BRAUNITZER's [172] solvent gives long spots and the phosphate buffer [173] is completely useless for TLC; all separations are rendered impossible through elongated spots and an often considerable diffusion. Buffering the layer does not help and addition of a few percent acetic acid makes little difference in this case. In addition to the modified "toluene" system, the solvents cited under a) and b) below have proved useful. Solvents of defined standard quality must be used in the preparation of all systems.

Quality of the Solvent Components

0.8 N Ammonium hydroxide	25% "Merck" ammonium hydroxide, diluted with distilled water.
34% Ammonium hydroxide	Commercially available quality.
t-Amyl alcohol	Fraction boiling between 100.5 and 102° from distillation, with a short column, of "Fluka"[27] "pract." grade t-amyl alcohol.
Ethylene chlorohydrin	"Fluka"[27] puriss. grade 2-chloroethanol.
Benzene	Shaken 3 times with 10% its volume of conc. sulphuric acid, washed with water, 2N sodium carbonate solution and water, dried over calcium chloride and distilled through a short column.
Benzyl alcohol	Shaken with saturated bisulphite solution, washed with 2N sodium carbonate, dried over sodium sulphate and vacuum distilled through a short column in a nitrogen atmosphere (benzaldehyde contaminant modifies R_f-values considerably).
n-Butanol	n-Butanol for chromatography.
Chloroform	Distilled twice through a short column or passed through an alumina column[28] and used immediately (phosgene is rapidly formed from alcohol-free chloroform on standing).
Acetic acid Methanol n-Propanol	Commercial product distilled through a short column.
Pyridine	Refluxed for 24 h over barium oxide and distilled through a short column.
Toluene	As for benzene.

---

[27] Firm 60.
[28] Cf. G. WOHLLEBEN, Angew. Chem. 68, 752 (1956).

a) Solvents for Chromatography of Acid- and Water-soluble DNP-Amino Acids, not Extractable with Ether

n-Propanol-34% ammonium hydroxide (70 + 30). DNP-arginine, DNP-cysteic acid, mono-DNP-cystine, α-DNP-histidine, di-DNP-histidine[29],

Table 187. *Identification of acid- and water-soluble DNP-amino acids*[a], by TLC on silica gel G in the system n-propanol-34% ammonium hydroxide (70 + 30). Ascending technique, solvent migration 10 cm, 0.5—1 µg spotted

DNP-Amino acids	hR_f[b]	Colour	UV-Absorption (360 nm)	Colour with ninhydrin
Mono-DNP-(Cys)$_2$	29	yellow	+	brown
DNP-CySO$_3$H	29	yellow	+	yellow
α-DNP-Arg	43	yellow	+	yellow
ε-DNP-Lys	44	yellow	+	brown
O-DNP-Tyr	49	colourless	+[c]	violet
α-DNP-His	57	yellow	+	yellow
Di-DNP-His	65	yellow	+	yellow

[a] For abbreviations of amino acids cf. Table 180.
[b] Average of the results of 6 separate measurements.
[c] Cf. "documentation", p. 771.

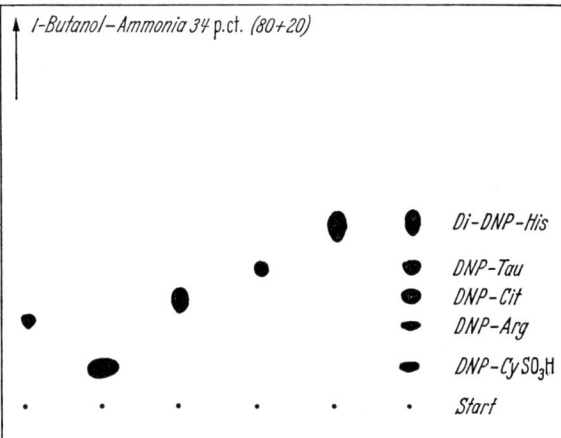

Fig. 209. Separation of an artificial mixture of urinary, water-soluble DNP-amino acids using ascending technique [175]
0.5 µg amounts applied; UV-photocopy (see p. 772); see Table 180 for abbreviations of the amino acids; Tau = taurine, Cit = citrulline

---

[29] im-DNP-histidine also belongs theoretically to the group of acid-soluble DNP derivatives. ZAHN and PFANNMÜLLER [174], however, state that it cannot be detected in hydrolysates of the corresponding DNP-peptides on account of its instability. We have therefore restricted our study to the α-DNP- and di-DNP-histidine derivatives.

ε-DNP-lysine and O-DNP-tyrosine can be identified through ascending chromatography in this system, either when alone or in the presence of one another should this occur as a practical problem. The time of run is about 2 h. Table 187 gives Rf-values and means of differentiating the spots. Although DNP-arginine and ε-DNP-lysine are not wholly separated, each can be detected in the presence of the other through the different colours in the ninhydrin reaction. DNP-cysteic acid and mono-DNP-cystine should never occur together in practice. It is important to remove excess acid after having applied the sample solution to the layer; this is done by warming the plate for about 10 min at ca. 60° C in a current of air and allowing 15 min for subsequent cooling.

n-Butanol-34% ammonium hydroxide (80+20). DNP-arginine, -citrulline, -cysteic acid and -taurine and di-DNP-histidine are water-soluble-DNP-derivatives formed from dinitrophenylation of urinary amino acids with excess dinitrofluorobenzene. The separation of an artificial mixture is illustrated in Fig. 209.

b) Solvents for Chromatography of Acid-insoluble DNP-Amino Acids, Extractable with Ether

Since DNP-amino acids like other carboxylic acids, tend to association in organic solvents, the addition of some acetic acid to the solvent is usually advantageous, suppressing the association and thereby removing the chief cause of tailing. It is noteworthy that this "acetic acid effect" is observed also in solvents which contain pyridine and are thus to be thought of as basic. Further, acetic acid increases the elution power of the solvent to a quite astonishing extent. This favourable influence of acetic acid in TLC of ether-soluble DNP-amino acids on silica gel layers is limited, however, to concentrations between 0.5 and 5%; larger concentrations act like high water content in diminishing the separating ability of a solvent.

α) Solvents for Separation in General

No. 1: Toluene-pyridine-ethylene chlorohydrin-0.8N ammonium hydroxide (100 + 30 + 60 + 60) ("toluene"-system) [45].

The upper layer is the solvent used in the chromatography, the lower for pre-treatment of the layer (see p. 768). This system yields spots with long "beards"[30] and is thus associated with a certain loss of substance. On account of its excellent separation power we employ it in the first dimension of two-dimensional chromatography. We consider this system to be indispensable at the present time and consequently tolerate the necessity of a pre-treatment of the layers. It separates, for example: DNP-phenylalanine from DNP-methionine; DNP-norleucine

---

[30] Cf. I. M. HAIS (p. 147 in [2]) for this expression.

from DNP-valine; DNP-β-alanine from the DNP-leucines; DNP-α-amino-n-butyric acid from DNP-proline; DNP-alanine from DNP-sarcosine; the group of di-DNP-amino acids (except di-DNP-cystine) from the smaller DNP-amino acids; and these last named from the DNP-derivatives of the acidic amino acids which remain at the start.

No. 2: Chloroform-benzyl alcohol-acetic acid (70 + 30 + 3).

This solvent yields symmetrical spots and separates 2,4-dinitrophenol[30a] and 2,4-dinitroaniline[30a] from all the DNP-amino acids.

No. 3: Chloroform-t-amyl alcohol-acetic acid (70 + 30 + 3).

This solvent has separation properties similar to those of No. 2. DNP-valine and 2,4-dinitrophenol[30a] migrate more closely to DNP-leucine but as a compensating advantage, t-amyl alcohol is more stable and more volatile than benzyl alcohol.

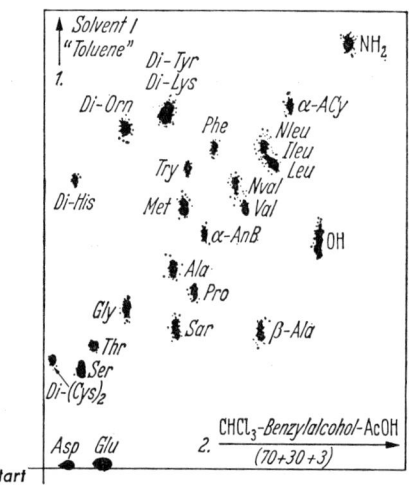

Fig. 210. Two-dimensional chromatogram of a standard mixture of 0.2 µg of each amino acid, using ascending technique with the solvents 1 and 2.
Symbols: Only the symbol for the relevant amino acid is given for mono-DNP derivatives (see Table 180 for these); Di = di-DNP derivatives; OH = 2,4-dinitrophenol; NH_2 = 2,4-dinitroaniline; UV-photocopy, original 12 × 10 cm [8]

No. 4: Benzene-pyridine-acetic acid (80 + 20 + 2).

In continuous development [13] (BN-chamber, cf. p. 75) this system is highly suitable for separating the less polar DNP-amino acids, e.g. the isomeric leucine derivatives. 2,4-Dinitroaniline[30a] migrates fastest ($R_{\text{DNP-leucine}} = 1.28$).

[30a] 2,4-Dinitrophenol and 2,4-dinitroaniline are by-products of the synthesis and acid hydrolysis, respectively, of DNP-peptides.

No. 5: Chloroform-methanol-acetic acid (95 + 5 + 1).

Di-DNP-tyrosine and di-DNP-lysine, which are not separated by any of the above systems 1—4, can be clearly distinguished using continuous development with No. 5 [13] (BN-chamber).

Table 188 contains Rf-values in the solvents mentioned. Migration times and a few facts worthy of special mention are given in Table 189.

Separations in Two-dimensional Procedures. Two-dimensional technique is usually necessary for unambiguous identification of a DNP-

Table 188. hRf-values of ether-soluble DNP-amino acids[a] observed after one-dimensional ascending or horizontal [13] TLC on silica gel G in solvents Nos. 1—5
Each figure represents average result of 6 observations. 0.5—1 μg spotted

DNP-Amino acids	1[b]	2		3		4[c]		5[c]			
		ascending		ascending				horizontal		horizontal	
	as-cend-ing		indi-rect[d]		indi-rect[d]	as-cend-ing	indi-rect[d]	as-cend-ing	indi-rect[d]		
solvent migration →	15 cm	10 cm	10 cm	10 cm	10 cm	15 cm	DNP-Leu: 10 cm		15 cm	DNP-Leu: 10 cm		
DNP-α-AnB	46	72	44	73	42	52	52	55	79	85	75
DNP-α-ACy	79	92	66	83	57	105	108	109	108	101	106
DNP-Ala	34	54	35	60	34	32	33	38	59	66	58
DNP-β-Ala	27	71	57	73	50	89	98	100	99	95	102
DNP-Asp	02	13	08	09	13	06	05	11	07	06	06
DNP-Glu	01	26	17	31	21	12	12	23	12	12	14
DNP-Gly	27	32	22	40	23	17	18	22	31	38	31
DNP-Ileu	64	83	63	81	57	107	107	107	100	100	104
DNP-Leu	66	82	62	80	54	100	100	100	100	100	100
DNP-Nleu	69	82	60	80	52	86	90	88	101	100	98
DNP-Met	55	70	39	69	38	43	43	47	72	81	74
DNP-Met.O$_2$	17	—	—	—	04	03	03	02	10	10	07
DNP-Phe	67	75	46	74	41	44	46	52	81	86	76
DNP-Pro	29	65	41	67	38	58	59	62	78	84	75
DNP-Sar	23	56	35	57	32	34	35	41	59	65	60
DNP-Ser	15	11	10	11	10	09	10	14	07	08	07
DNP-Thr	20	17	13	15	12	12	14	20	09	11	11
DNP-Try	65	69	38	69	31	23	25	33	54	61	49
DNP-Val	53	79	56	77	51	76	81	85	91	98	86
DNP-Nval	56	77	52	76	48	65	70	75	86	95	89
Di-DNP-(Cys)$_2$	—	03	02	01	01	00	00	02	00	02	02
Di-DNP-His	53	11	09	08	04	05	04	08	12	16	14
Di-DNP-Lys	74	56	35	60	30	12	13	19	66	73	65
Di-DNP-Orn	70	34	23	40	20	06	06	10	39	46	39
Di-DNP-Tyr	76	58	35	60	30	17	16	19	57	65	57
2,4-DNP-OH[e]	41	100	76	83	55	22	21	23	148	102	111
2,4-DNP-NH$_2$[f]	90	90	84	72	63	115	128	129	131	101	115

[a] For abbreviations of amino acids, cf. Table 180.
[b] Cf. remarks on the use of the "toluene" system.
[c] Rf-values with reference to DNP-leucine.
[d] After "pre-treatment" of the layer by chromatography in the "toluene" system (No. 1) and intermediate drying (see text).
[e] 2,4-DNP-OH = 2,4-dinitrophenol.
[f] 2,4-DNP-NH$_2$ = 2,4-dinitroaniline.

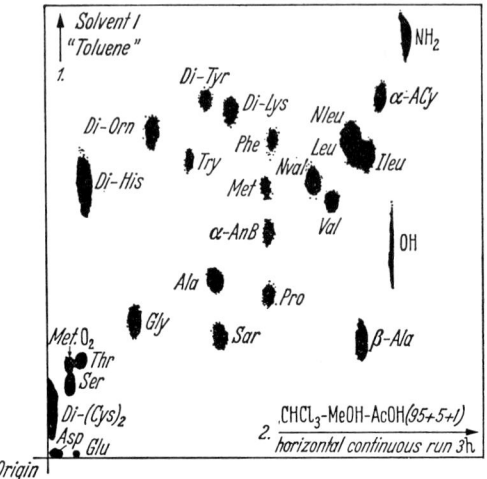

Fig. 211. Two-dimensional chromatogram containing 1 μg of each DNP-amino acid in solvents 1 (ascending technique) and 5 (horizontal technique with continuous development [13]).

See Fig. 210 for symbols; UV-photocopy with longer exposure than Fig. 210; original size 13 × 13 cm [8]

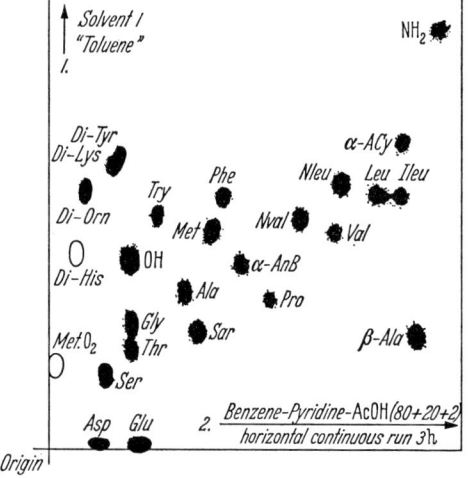

Fig. 212. Two-dimensional chromatogram of a mixture containing 1 μg of each DNP-amino acid in solvents 1 (ascending technique) and 4 (horizontal technique with continuous development [13]).

See Fig. 210 for symbols; Di-His and Met. O_2 were omitted from this mixture; UV-photocopy, original size 15 × 14 cm [8] derivative belonging to the relatively large group of acid-insoluble DNP-amino acids, extractable with ether. We use the "toluene" system of BISERTE and OSTEUX [45] (Solvent No. 1) in the first dimension and one of solvents No. 2—5 in the second. Fig. 210 shows the separation of a standard mixture of 2 μg of each of the DNP-amino acids, using a combination of solvent Nos. 1 and 2. The leucine and valine groups are not separated, nor are di-DNP-lysine and di-DNP-tyrosine. The last named are separated by a combination of Nos. 1 and 5 (Fig. 211) or perhaps in No. 5 alone. The combination of No. 1 with No. 4 (Fig. 212) enables the isomeric leucine derivatives and isomeric valine derivatives to be separated.

Table 189. *Times of run and notable separation effects observed in one-dimensional TLC on silica gel G in solvents Nos. 1—5 (p. 763—765)*

Solvent No.	Separation of the DNP-amino acids from		Separation of DNP-derivatives of		Separation of the di-DNP-derivatives of Tyr, Lys	time of run
	Dinitro-aniline	Dinitro-phenol	Leu, Ileu, Nleu	Val, Nval		
1	+	—a	—	—	—	1 h/15 cm
2	+	+	—	—	—	1½ h/10 cm
3	—e	—b	—	—	—	1 h/15 cm
4 f	+	—c	+	+	—	2—3 h
5 f	—d	—d	—	+	+	2—3 h

^a Spot of 2,4-dinitrophenol is between spots of DNP-Val and DNP-Ala.
^b Spot of 2,4-dinitrophenol is near spot of DNP-Leu.
^c Spot of 2,4-dinitrophenol is between spots of DNP-Ala and DNP-Gly.
^d Spots of 2,4-dinitrophenol and 2,4-dinitraniline are just above spot of DNP-Leu.
^e Spot of 2,4-dinitraniline is between spots of DNP-Phe and DMP-Met.
^f Horizontal chromatography [13]; DNP-Leu migrates about 10 cm in 2½ hrs.

The characteristic spot patterns are scarcely distorted by fluctuations in R_f-values. An unknown sample can therefore be identified by chromatography with a standard mixture which contains enough (0.2 μg) of each DNP-amino acid in question, to be just detectable after two-dimensional separation[31] (Fig. 210). A compound present in the sample is then revealed rapidly and unambiguously in most cases through the intensity of the spot ascribed to it. Table 188 incidentally contains R_f-values obtained in one-dimensional chromatography and also those values found in the second dimension after prior chromatography with the "toluene" system *("indirectly" obtained R_f-values)*. The reproducibility of these values is good, provided the instructions for working with the

---

[31] A solution of 1 mg of each DNP-amino acid in a total of 5 ml acetone keeps in a refrigerator for at least 4 weeks. 1 μl is needed for the experiment.

"toluene" system and for intermediate drying, given below in the next paragraph, are followed.

Note on the Use of the "Toluene" System

Pre-treatment of the thin-layer (equilibration): The lower phase of the "toluene" system is placed in a chamber lined with filter paper. A thick, bent, glass rod on the bottom of the chamber serves as a grating. Two glass plates coated with silica gel G are laid on the middle of the grating with their coated sides outwards, each leaning with its upper edge against a chamber wall. To prevent solvent passing from filter paper to layer, the latter is divided in two by a heavy pencil stroke, parallel to the upper edge of the plate. The system is left overnight. The silica gel takes up a great deal of moisture; its appearance does not show this but substances spotted on it during this treatment diffuse considerably.

Chromatography on a layer pre-treated in this fashion probably depends on partition between two liquid phases. The effect of the pre-treatment is practically lost if the plates are left in the open air for even 45 min, a fact which should not be overlooked when employing the layers.

Application of samples: If a plate is left unprotected in the air for various periods between pre-treatment and chromatography, the logarithm of these time periods in excess of 2 min shows a linear relation to the logarithm of the Rf-values subsequently found. The pre-treated layer should consequently be covered immediately after removal from the chamber; a glass plate is used, leaving a margin of about 1.7 cm at the lower edge. The samples can be applied on this margin at leisure. If this "leisure period" does not exceed 5 min, interference is negligible. As far as possible, no more than 1 µl should be applied per spot since the rate of evaporation of the applied solutions is less on the pre-treated moist layers than on dry layers. After spotting is complete, the cover plate is carefully removed and the chromatography begun without delay, using 120 ml of the upper phase of the "toluene" system (ascending technique).

Intermediate drying: The plate is left for 10 min in an air current (well-ventilated fume cupboard), heated in a drying oven for 10 min at 60° C and allowed to cool for 15 min in an air current; chromatography in the second dimension can then be carried out immediately. Drying for too long a period is not advisable because exposure to the air may lead to some decomposition of the DNP-amino acids (oxidation of DNP-methionine can result in a product behaving in the subsequent chromatography like di-DNP-histidine). If the plate has to kept a longer time after the intermediate drying, the layer must be covered with a glass plate and stored in the dark.

β) Solvents for Special Separations

The diagrams reproduced in Figs. 213—217 may be useful in the quest for solvents with particular properties.

Urinary amino acids and by-products can be separated after dinitrophenylation, by two-dimensional chromatography on silica gel G using the following solvent combinations:

Ether-soluble Amino acids:

I. Toluene-ethylene chlorohydrin-pyridine-25% ammonium hydroxide (50 + 35 + 15 + 7).
II. Chloroform-benzyl alcohol-acetic acid (70 + 30 + 3).

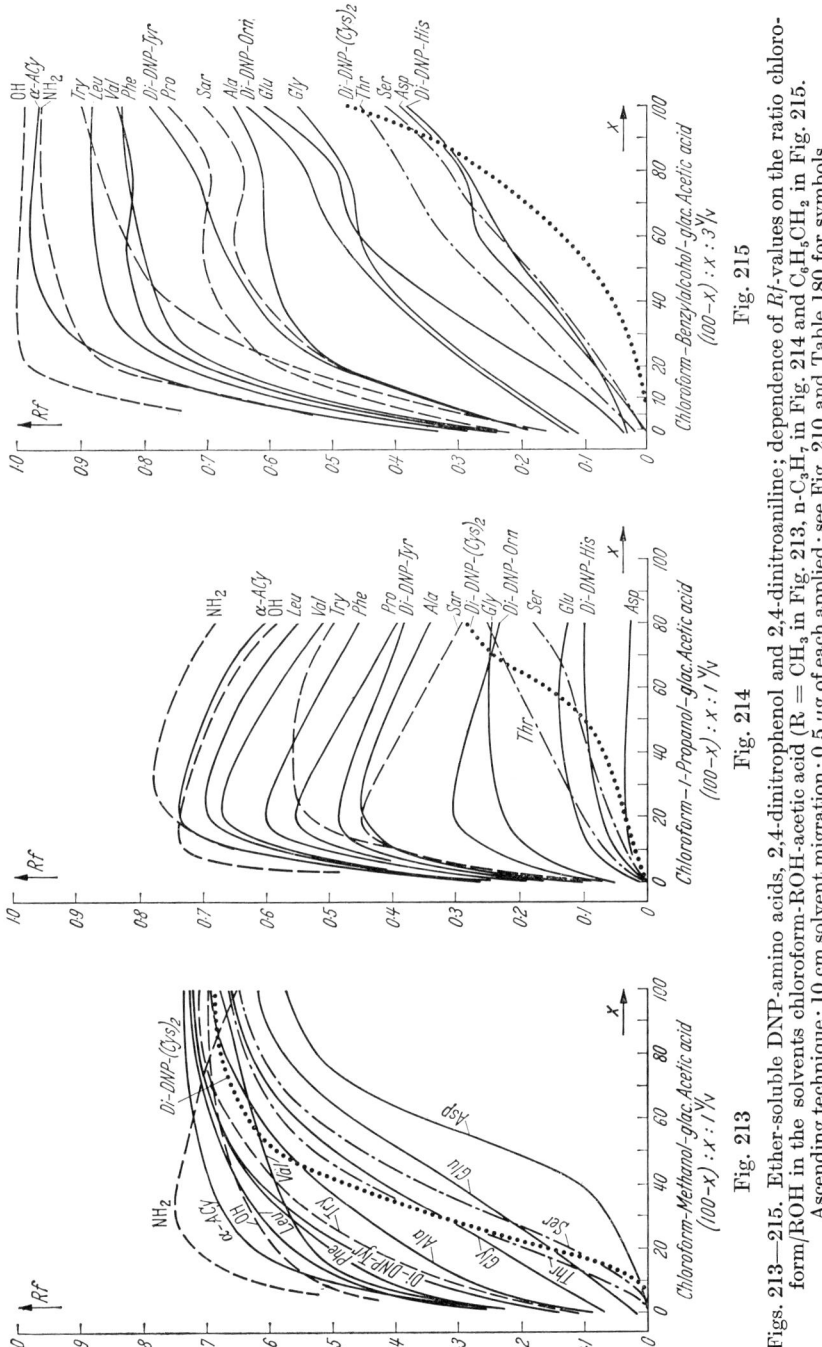

Figs. 213—215. Ether-soluble DNP-amino acids, 2,4-dinitrophenol and 2,4-dinitroaniline; dependence of R_f-values on the ratio chloroform/ROH in the solvents chloroform-ROH-acetic acid (R = CH_3 in Fig. 213, n-C_3H_7 in Fig. 214 and $C_6H_5CH_2$ in Fig. 215. Ascending technique; 10 cm solvent migration; 0.5 µg of each applied; see Fig. 210 and Table 180 for symbols

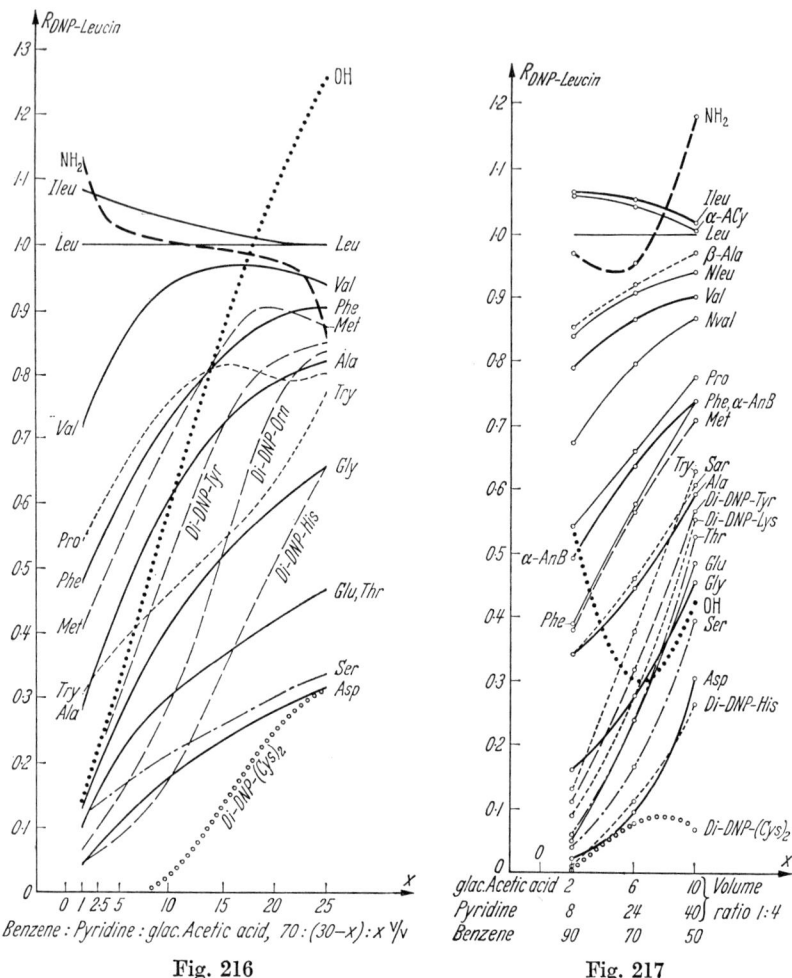

Fig. 216. Ether-soluble DNP-amino acids, 2,4-dinitrophenol, 2,4-dinitroaniline; dependence of the $R_{\text{DNP-leucine}}$-values[32] on the ratio acetic acid/pyridine in solvents with constant benzene content.

Ascending technique; 14 cm solvent migration; 0.5 µg of each applied; cf. Fig. 210 and Table 180 for symbols. Di-DNP-lysine has always the same migration speed as di-DNP-tyrosine; DNP-β-alanine shows $R_{\text{DNP-leucine}}$ of ca. 0.9 everywhere

Fig. 217. Ether-soluble DNP-amino acids, 2,4-dinitrophenol and 2,4-dinitroaniline; dependence of the $R_{\text{DNP-leucine}}$-values[33] on the ratio benzene/pyridine/acetic acid in solvents containing a constant acetic acid/pyridine ratio.

Ascending technique; 14 cm solvent migration; about 0.5 µg of each applied; cf. Fig. 210 and Table 180 for symbols.

[32, 33] See Footnotes on p. 771.

III. Chloroform-methanol-acetic acid (95 + 5 + 1).
IV. Chloroform-methanol-acetic acid (70 + 35 + 5).

Acid-soluble amino Acids:

V. Pyridine.
VI. n-Butanol, saturated with 25% ammonium hydroxide [231].

3. Documentation

All the DNP-derivatives investigated here are yellow, with the exception of O-DNP-tyrosine. Amounts of 0.1 µg (one-dimensional chromatogram) or 0.5 µg (two-dimensional chromatogram) yield spots which are still easily visible if the layer is examined in transmitted daylight. A copy of the chromatogram must be made within a few hours because the spots gradually fade. The layer may be preserved by spraying with a mixture of paraffin and ether [176] or, better, with a plastic dispersion [34] [50, 177] (p. 127). The resulting (plastic) film is easily peeled off the glass plate and may be handled like a paper chromatogram.

We usually stretch a sheet of transparent paper across the layer with two paper clips, examine in transmitted light and trace the spots in ink with a soft pen. Even very feeble, pale yellow spots can be detected on the white silica gel layer with a little practice, It is often easier to inspect the layer in transmitted UV light (light source → layer → glass plate → protective goggles → eye). The DNP-amino acids are then recognisable as dark spots. Weakly absorbing substances in low concentrations are difficultly visible since the silica gel-gypsum layer itself absorbs and therefore appears dark. Layers containing a suitable fluorescent indicator can be profitably used; O-DNP-tyrosine can then be detected in amounts down to 0.06 µg through its fluorescence quenching of short-wave UV-light.

---

[32] hR_f-values referred to DNP-leucine. As x (acetic acid content) increases, the R_f-value of the reference substance varies as follows:

x	0	1.5	3	5	10	15	20	25
hR_f-values of DPN-leucine	3	28	37	46	62	68	76	60

In solvents of x < 5, only continuous development [13] is thus practicable (see p. 76, Fig. 34).

[33] hR_f-values referred to DNP-leucine. The R_f-value of this substance varies with the benzene content of the solvent as follows:

benzene content v/v	90%	70%	50%
hR_f-value of DPN-leucine	17%	51%	70%

Continuous development is thus as a rule necessary in solvents with over 70% benzene (see [13] concerning the method)

[34] A material specially selected for preserving thin-layer chromatograms is available under the name of "Neatan" (Firm 88).

A UV-photostat is especially to be recommended. The sensitive side of Gevaert "Gevacopy" paper is pressed directly against the layer with a glass plate. UV light (365 nm) is then allowed to pass through the layer on to the paper for a few seconds and a positive is developed in the ususal way. Figs. 210—212 were prepared in this way. Maximum sensitivity results from an exposure time which renders the negative background grey but not black. A faithful reproduction is thus obtained. The negative background becomes darker with increasing exposure and the light spots decrease in diameter, finally disappearing altogether on over-exposure. This fact can be utilised to break up partly overlapping large spots into smaller, separated spots.

B. Phenylthiohydantoins

The formation of phenylthiohydantoins (PTH-amino acids), mentioned in the introduction, is based on the reaction series below [178, 179]: EDMAN [155] proposed the scheme and a generally applicable working procedure has been subsequently developed in other laboratories [180, 181].

Scheme of the EDMAN-degradation

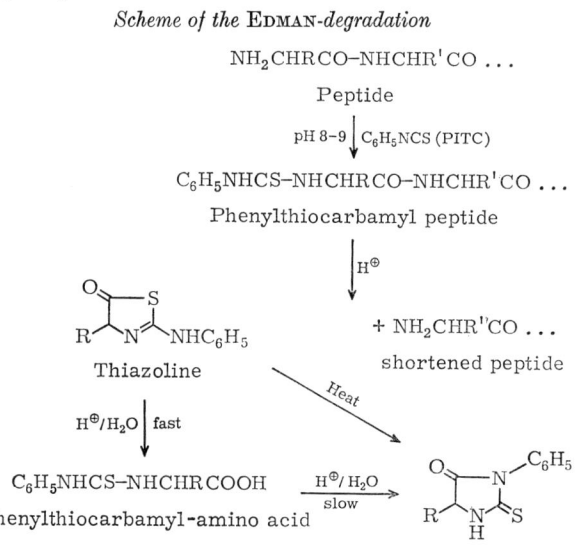

Free amino acids as well as proteins or peptides can also, of course, be reacted with phenyl isothiocyanate (PITC) [182]. This can arise like dinitrophenylation when an amino acid mixture cannot be directly chromatographed as a result of contaminants; the PTH-amino acids are often more easily converted into a form suitable for chromatography than are the free amino acids. Since the PTH-amino acids absorb in the UV region, the quantitative determination of amino acids is facilitated [182] (cf. pp. 142—144).

1. Preparation of the Phenylthiocarbamyl-derivatives and their Conversion into PTH-Amino Acids
Amino Acids

SJÖQUIST describes a micro-method [181]; for preparing larger amounts this method may be suitably modified or EDMAN's [183] procedure used:

α) **Preparation of PTH-amino acids**[35]. 10 mmoles amino acid are dissolved in 25 ml water and 25 ml pyridine, the pH of the solution is brought to ca. 9 by adding 1N sodium hydroxide (indicator paper), and its temperature raised to 40°. 2.4 ml phenyl isothiocyanate are added and the pH maintained at 9 by successive additions, with stirring, of 1N sodium hydroxide. After about 30 min, the reaction is complete and NaOH consumption ceases. Unreacted phenyl isothiocyanate and the bulk of the pyridine are removed by several extractions with benzene. An amount of hydrochloric acid equivalent to the sodium hydroxide is then added, the solution concentrated if necessary and the phenylthiocarbamyl-amino acid separating out is filtered off. *Note:* phenylthiocarbamyl-arginine and -histidine precipitate at pH 7 and 3.5, respectively.

The phenylthiocarbamyl derivatives are converted into PTH-amino acids by refluxing for 2 h with 30 ml N hydrochloric acid, evaporating the solution several times to dryness after repeated additions of water. The crude PTH-amino acid (80—90% yield) is recrystallised from acetic acid-water, alcohol or water. Melting points of 18 PTH-amino acids are to be found in the original article [181]. (*Note:* the tryptophan derivative is prepared by reflux with acetic acid instead of hydrochloric acid.) The PTH-derivatives of the diamino-monocarboxylic acids possess the phenylthiocarbamyl substituent on the amino group which is not in the α-position. β-Hydroxy- and β-mercapto-amino acid derivatives have a marked tendency to eliminate water or H_2S respectively; PTH-threonine is, for example, the phenylthiohydantoin of α-aminocrotonic acid. All PTH-amino acids are sensitive to light. Optically active derivatives racemise easily, yet are sufficiently stable for their configuration to be determined by optical rotatory dispersion measurements [184].

β) **Quantitative conversion of an amino acid mixture (e. g., a peptide hydrolysate) into PTH-amino acids** [182]: 0.5—1 mg peptide or protein is hydrolysed in a sealed quartz tube by heating with 0.3 ml of constant boiling (5.7N) hydrochloric acid (twice distilled in glass) for 22 h at 110° C in an atmosphere of nitrogen. The hydrolysate is repeatedly evaporated to dryness in vacuo over potassium hydroxide with intermediate additions of water. The residue is mixed thoroughly with 250 µl of triethylamine-acetic acid buffer (see "Reagents") and 250 µl phenyl isothiocyanate solution in acetone (corresponding to 6 µl of reagent) and left in a closed vessel for $2^1/_2$ h on a water bath at 25°. The solvent is removed by 15 min evacuation with a water pump and then leaving overnight over phosphorus pentoxide in high vacuum. The residue (phenylthiocarbamyl derivatives of the amino acids) is taken up in 100 µl water (distilled in glass) and 200 µl acetic acid, saturated with hydrogen chloride, and the resulting solution maintained 6 h on a water bath at 25°; solvent and acid are then removed over potassium hydroxide as described above. The PTH-amino acids are in the residue. A parallel experiment without

---

[35] A collection may be obtained from Firm 86, for example.

amino acids is carried out for chromatographic comparison. Cysteine and cystine do not give PTH-derivatives under these conditions; peptide- or protein-bound cysteine or cystine may be detected by oxidation with performic acid before hydrolysis; cysteic acid is then yielded after the hydrolysis and its PTH derivative is formed moderately well. SJÖQUIST uses a slightly modified method of HIRS [44] for oxidation.

Reagents

Buffer:	2 ml 2N acetic acid (analytical reagent grade) and 1.2 ml triethylamine (Eastman-Kodak, refluxed 3 h over 5% (w/w) phthalic anhydride and distilled through a short column immediately before use) are diluted to 25 ml with distilled water and this solution mixed with 25 ml acetone (analytical grade, refluxed over potassium permanganate and distilled through a short column immediately before use); pH 10.1
Phenyl isothiocyanate:	Distilled under reduced pressure.
HCl-acetic acid:	Acetic acid (analytical grade) in a quartz vessel is saturated with HCl gas (washed with conc. sulphuric acid) at room temperature.

The "normal" degradation procedures of SJÖQUIST et al. [132, 133] and SJÖQUIST's special procedure [133] are outstanding among the many procedures in the literature [132, 133, 155, 185—188]; the special procedure ensures maximum yields of PTH-amino acids. Our comments must be restricted to the following:

The customary by-products mono- and diphenylthiourea (MPTU and DPTU respy.) are formed in especially large amounts in the special procedure mentioned. DPTU does not interfere in the chromatographic identification of the PTH-amino acids (cf. Fig. 220). MPTU, however, has a chromatographic behaviour (TLC and PC [189]) similar to that of PTH-glycine (cf. Fig. 219). We consequently use a specific colour reaction for detecting glycine, which is not affected by the presence of MPTU (see under "Detection", p. 777).

In the separation of each PTH-amino acid from the shortened peptide by ethyl acetate extraction, PTH-histidine, PTH-arginine and the products formed from any cysteic acid present, remain in the acid, aqueous solution. PTH-histidine may be practically completely extracted when the bulk of the acid is first removed by cautious evaporation, the residue taken up in a little water and the pH of the solution adjusted to about 7; triethylamine is suitable for neutralisation.

We have found that difficulties arise when cysteic acid occurring in a sequence, enters the degradation cycle[36].

2. Solvents and Separation Efficiency

Solvents employed in paper chromatographic separation of PTH-amino acids [182, 190—192] cannot be used on silica gel G. We have, however, achieved good separations using the solvents given in Table 190. Chloroform-methanol-formic acid (70 + 30 + 2) is suitable for separating PTH-aspartic acid and PTH-glutamic acid (Fig. 218).

---

[36] Cf. D. G. SMYTH, W. H. STEIN and S. MOORE (J. Biol. Chem. **237**, 1845 (1962) concerning the degradation of glutamine peptides.

Table 190. *hRf-values of PTH-amino acids and of mono- and diphenylthiourea*
For abbreviations, cf. Table 180 and [a,b,c]

Amounts spotted: 0.5 μg in 0.5 μl of methanol or acetone. Ascending technique, solvent migration 10 cm. Each figure represents average of 6 determinations.

PTH-derivatives of	Solvent			
	I	II	III	IV
Ala	16	68	39	11
Arg	00	01	00	00
Asp	00	01	13	00
Asp(NH$_2$)	00	23	07	00
Glu	01	04	17	00
Glu(NH$_2$)	01	28	08	00
Gly	10	56	33	05
His	01	29	00	02
Ileu	40	77	57	37
Leu	40	77	60	37
Lys	12	71	34	03
Met	33	75	51	13
Met.O [a]	01	40	12	01
Phe	28	74	50	18
Pro	60	82	65	21
Thr	04	45	15	00
Try	13	62	39	10
Tyr	03	47	21	01
Val	32	74	55	23
MPTU [b]	12	54	31	03
DPTU [c]	43	76	67	22

I: *Chloroform;*
II: *Chloroform-methanol* (90 + 10);
III: *Chloroform-formic acid* (100 + 5);
IV: *"Heptane"-system:* n-heptane-ethylene dichloride-formic acid-propionic acid (90 + 30 + 21 + 18), 100 ml of upper phase used.

Specification of chemicals used:

Chloroform:	stabilised with 1.5% ethanol (Firm 60)
Formic acid:	purest analytical grade, anhydrous
Heptane:	n-heptane ASTM pure grade (Firm 60)
Propionic acid:	pure grade (Firm 60)
Ethylene chloride:	1,2-dichloroethane, purest analytical grade (Firm 60)
Methanol:	commercial grade, distilled once with a short column.

[a] Methionine sulphoxide.
[b] Monophenylthiourea.
[c] Diphenylthiourea.

CHERBULIEZ et al. [4] recommend heptane-pyridine-ethyl acetate (50 + 30 + 20) in the separation of PTH-glycine, -proline and -leucine.

hRf-Values: Table 190 provides information about hRf-values in some of the solvents which we have tested. The values are independent of the amount applied within wide limits; for example, this is valid for PTH-proline for amounts between 0.05 and 72 μg (chloroform-methanol (90 + 10)). If substance amounts within these limits are applied in

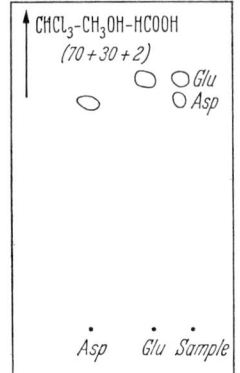

Fig. 218. One-dimensional chromatogram for detection of PTH-aspartic acid and PTH-glutamic acid.

Ascending technique; 11 cm run; 0.5 μg in 0.5 μl methanol applied; visualisation with chlorine-tolidine or inspection in UV light of 270 nm

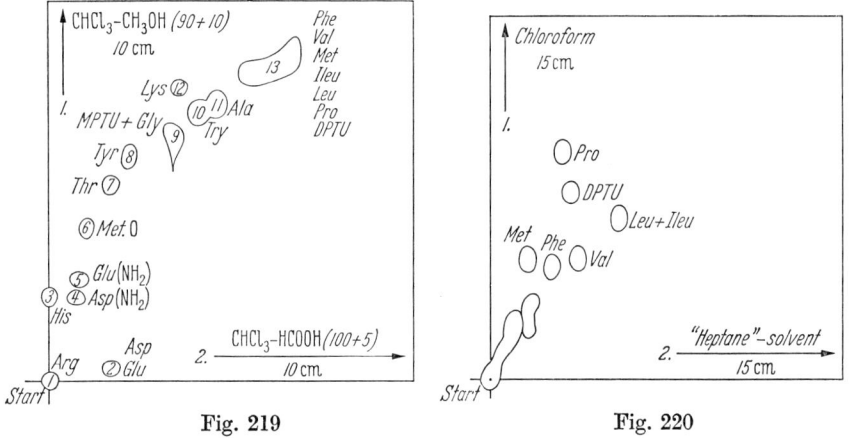

Fig. 219

Fig. 220

Fig. 219. Two-dimensional chromatogram of a mixture containg 0.5 μg of each PTH-amino acid + MPTU + DPTU in 0.5 μl methanol or acetone; ascending technique.

Detection: chlorine-tolidine or UV light (270 nm); note that only 3 out of the total of 13 spots contained more than one component; see text, Fig. 218 and Fig. 220 for the examination of spots 2, 9 and 13

Fig. 220. Two-dimensional chromatogram of a mixture containing 0.5 μg of each PTH-amino acid + MPTU + DPTU in 0.5 μl methanol or acetone; ascending technique.

Detection: UV light (270 nm); note that the components of spot 13 in Fig. 219 have been separated except for PTH-leucine and PTH-isoleucine; see text for the identification of these last two substances a fixed volume of solvent (0.5 µl methanol), the spot area (evaluated on squared paper) is roughly proportional to the logarithm of the amount of substance. The table above has been expanded in a publication by PATAKI [221].

Separation and Identification: The following procedure [193] is preferable to that described previously [8]: Two two-dimensional (Figs. 219 and 220) and one one-dimensional (Fig. 218) chromatogram(s) are run simultaneously; the pairs PTH-glycine/monophenylthiourea (MPTU) and PTH-leucine/PTH-isoleucine are not then separated. We recommend a modification of the colour reaction with ammonia, described by SCHRAMM et al. [189], for detection of PTH-glycine in presence of MPTU (Fig. 219); precise details are given below. PTH-leucine and PTH-isoleucine can be separated in a 20 cm run through multiple development with chloroform-methanol (50 + 50). A mixture of the two PTH-amino acids is applied immediately alongside the sample to be identified. Development is repeated (2—4 times) with intermediate drying (40° C) until the faster migrating PTH-leucine is clearly distinguished from the slower migrating PTH-isoleucine [213].

3. Detection of the Phenylthiohydantoins

The chlorine/tolidine test (Rgt. No. 42), described already for peptides (p. 755), is valuable; down to about 0.05 µg PTH-amino acid may be detected. Layers containing a fluorescence indicator, like silica gel GF_{254} (Firm 88), can be used here with advantage. As little as 0.1 µg PTH-amino acid is then detectable on the chromatogram through fluorescence quenching.

In the *specific detection of PTH-glycine* (mentioned above), the layer is slightly moistened by spraying with water and then exposed to the vapour of concentrated ammonia; a stable, deep red spot is yielded, with limit of detection at about 0.08 µg PTH-glycine. Threonine gives a much feebler red colouration under these conditions.

The iodine-azide reagent (No. 142), used by SJÖQUIST [191] and recently by CHERBULIEZ [4] for visualisation, is for from ideal for PTH-amino acids, in our opinion. It yields white spots on a light blue background with the silica gel G layers used by us: these are difficult to see and disappear rapidly. The Grote-reagent [194] (No. 189) would be preferable but its application [190] is laborious and time-consuming.

C. Other Amino Acid Derivatives
1. Dinitropyridyl-Amino Acids

Amino acids and peptides can be reacted with 2-chloro-3,5-dinitropyridine like with 1-fluoro-2,4-dinitrobenzene. Terminal substituted amino acids of this sort are easily broken down by acid hydrolysis (15—20 min with 6N hydrochloric acid-30% formic acid). The following solvents are recommended for chromatography on silica gel G:

A) Chloroform-methanol-acetic acid (95 + 5 + 1);

B) n-Propanol-33% ammonium hydroxide (70 + 30);

C) Toluene-pyridine-ethylene chlorohydrin-0.8N ammonium hydroxide (40 + 12 + 24 + 24);

D) Benzene-pyridine-acetic acid (80 + 20 + 2);

E) Chloroform-formic acid (100 + 5);

F) Butanone-pyridine-water-acetic acid (70 + 15 + 15 + 2).

The amino acid derivatives are visible in daylight and in UV light (DI BELLO and SIGNOR [204]).

2. 1-Dimethylamino-5-Naphthalenesulphonyl-Amino Acids (DANS-Amino Acids)

The free amino groups of amino acids and peptides react with 1-dimethylamino-5-naphthalenesulphonyl chloride (DANS-Cl) yielding intensively fluorescing poducts.

Procedure: 10^{-3} μmole peptide is dissolved in 15 μl 0.1M $NaHCO_3$ solution and mixed with 15 μl of a solution of 1 mg DANS-Cl in 1 ml acetone. After 3 h at room temperature, the mixture is evaporated to dryness in vacuo and hydrolysed with 20 μl 6N HCl by heating 6—12 h at 105°. The acid is evaporated off under reduced pressure and the residue submitted to chromatography (GRAY and HARTLEY [212]).

Separation on silica gel G (SEILER and WIECHMANN [225]) can be carried out with the following solvents:

A) Methyl acetate-isopropanol-conc. ammonium hydroxide (45 + 35 + 20);

B) Chloroform-methanol-acetic acid (75 + 20 + 5);

C) Chloroform-ethyl acetate-methanol-acetic acid (30 + 50 + 20 + 1).

The chromatograms are inspected in UV light, best while still moist since the fluorescence intensity of the spots decreases markedly on drying; the limit of detection is 10^{-10} mole.

3. Carbobenzoxy Compounds

TLC has proved useful for following the course of reaction during peptide synthesis. The reaction mixture is applied to silica gel G layers and chromatographed with one of the solvents below, for example:

A) Ethanol-water (70 + 30) [222];

B) n-Butanol-acetone-acetic acid-5% ammonium hydroxide-water (45 + 15 + 10 + 10 + 20) [208];

C) n-Butanol-acetic acid-water-pyridine (45 + 9 + 36 + 20) [208].

Chlorine-tolidine or ninhydrin reagents are used for detection.

D. Iodoamino Acids and Similar Compounds

According to SCHNEIDER [229], monoiodotyrosine (hR_f 25), diiodotyrosine, diiodothyronine, triiodothyronine and thyroxine (hR_f 46) can be separated in that order of increasing hR_f-values on silica gel G layers using 95% acetic acid-benzene-xylene (60 + 20 + 20). For the most part however, only mono- and diiodotyrosine have been chromatographed along with tyrosine [217]. These three compounds can be satisfactorily separated with butanol-acetic acid-water (40 + 10 + 50) or (100 + 10 + 10), butanol-methanol-20% ammonium hydroxide (80

Table 191. *hR_f-values of organic iodine-containing compounds in various solvents* [228]

No.	Iodine compound	Commercial name	I	II	III	IV	V	VI
1	3,3'-(Adipoyldiimino)bis[2,4,6-triiodobenzoic acid]	Biligrafin	9	27	33	0	0	33
2	3,5-Diacetamido-2,4,6-triiodobenzoic acid	Gastrografin	12	30	37	40	0	37
3	1-Ethyl-2-(2,4,6-triiodo-3-hydroxyphenyl)-propionic acid	Teridax	19	30	36	0	33	37
4	3-Acetamino-2,4,6-triiodobenzoic acid	Triopac	26	40	47	0	0	44
5	Monoiodomethanesulphonic acid	Abrodil	28	43	48	0	0	44
6	3,5-Diiodo-4-oxo-1(4H)-pyridineacetic acid	Ioduron	29	43	50	33	0	43
7	N-(3-amino-2,4,6-triiodobenzoyl)-N-phenyl-2-aminopropionic acid	Osbil	31 (37)	42 (45)	49 (57)	10	15 (16)	44
8	2-(3-Dimethylamino-methylene-amino-2,4,6-triiodophenyl)-propionic acid	Biloptin	35	43	50	40	28	50
9	2-(2,4,6-Triiodophenoxy)-butyric acid	Baygnostil	37	43	50	27	7	57
10	n-Propyl 3,5-diiodo-4-oxo-1(4H)-pyridineacetate	Propyliodan	87	63	71	0	60	67
11	3-Iodo-L-tyrosine		7(10)	20	23	0	0	8
12	Iodogorgo(n)ic acid(3,5-diiodo-L-tyrosine)		7	20	23	0	0	8
13	3,5-Diiodo-L-thyronine		31	42	38	0	0	13
14	3,3',5-Triiodo-L-thyronine		30	39	33	0	0	17
15	Thyroxine (D, L)		22	35	30	0	0	12
16	Potassium iodide		33	48	47	0	0	40
17	Potassium iodate		0	0	0	0	0	0

Layer: Silica gel HF$_{254}$; 25 g stirred into a homogeneous suspension with 65 ml water in which 0.5 g starch amylum solubile analytical grade, (Firm 88) has been dissolved with warming; spread and dried 30 min at 110° C.

Solvents: I = Ethyl acetate-isopropanol-25% ammonium hydroxide (55 + 35 + 20); II = Acetone-isopropanol-25% ammonium hydroxide (40 + 40 + 20); III = Isopropanol-25% ammonium hydroxide (80 + 20); IV = Acetic acid-chloroform (5 + 95); V = Chloroform-methanol-pyridine (85 + 5 + 10); VI = Ethyl acetate-methanol-diethylamine (50 + 40 + 20).

+ 20 + 20) or phenol-water (75 + 25). The solvents recommended for chromatographing iodine-containing Röntgen contrast agents are unsuitable for this separation but may be used to separate the iodothyronines (see Table 191).

Using, for example, *tert.*-butanol-2N ammonium hydroxide-chloroform (75 + 14 + 12) on cellulose layers, the iodoamino acids can be separated in the sequence (start to front): diodotyrosine, monoiodotyrosine, (tyrosine), thyroxine, (thyronine), tri + diiodothyronine [214].

BERGER and co-workers [205] have shown that the mixture of mono- and diiodotyrosine, labelled with ^{131}I, and thyroxine, can be separated also on Dowex 1 × 2 (OH$^-$) layers; they proposed chromatography on discontinuous layers.

The hRf-values of organic iodine-containing compounds in various solvents are compiled in Table 191; the accent is on X-ray contrast agents.

Detection of Organic Iodine Compounds. The most specific possible detection of organically bound iodine is of particular interest. According to STAHL and PFEIFLE [228] this is possible directly on a silica gel-starch layer by means of photochemical deiodination. The liberated iodine yields the blue-violet starch inclusion compound and amounts down to 0.5 µg may be thus visualised.

The solvent is removed from the chromatogram by heating at 100°. After cooling, the layer is sprayed with a little 50% acetic acid and is exposed for a few minutes to unfiltered UV light of ca. 254 nm wave length. For this purpose, 4 UVIS reflectors (Fig. 35, Firm 44) are equipped with germicidal-fluorescence tubes (Firm 134) and the layers irradiated at a distance of 5 cm (protect the eyes). Iodine-containing compounds then become weakly violet to brown. The colour may be intensified by spraying with 10% acetic acid to incipient transparence and re-exposing to the UV radiation. The typical blue colour then usually appears suddenly.

This detection does not, however, succeed on cellulose layers. The procedure of PATTERSON and CLEMENTS [224] can be used here:

Solution a: 2.7% FeCl$_3$ · 6H$_2$O in 2N HCl.
Solution b: 3.5% aqueous solution of K$_3$[Fe(CN)$_6$].
Solution c: 3.8 g As$_2$O$_3$ are dissolved in 25 ml 2N NaOH with heating and the solution cooled to 5° C; 50 ml 2N H$_2$SO$_4$ (at 5° C) are then added and the solution diluted to 100 ml with water.

5 ml a + 5 ml b + 1 ml c are mixed just before use and the mixture sprayed on to the dry layer. After the colour-producing reaction has taken place, excess reagent can be carefully washed out with water in a developing tank.

Iodoamino acids can be detected also with ninhydrin but it is appreciably feebler than with the corresponding iodine-free compounds. The detection of radioiodinated substances is possible through autoradiography (p. 157) or through direct determination of the activity (p. 160).

Bibliography for Chapter V. Amino Acids and Derivatives

1. COHN, E. J., and J. T. EDSALL: "Proteins, Amino Acids and Peptides as Ions and Dipolar Ions", p. 203 und 201. New York: Reinhold Publishing Corporation 330 West Forty-second St. 1943.
2. HAIS, I. M., u. K. MACEK: Handbuch der Papierchromatographie. Jena: VEB Gustav Fischer Verlag 1958.
3. MOFFAT, E. D., and R. I. LYTLE: Analyt. Chem. 31, 926 (1959).
4. CHERBULIEZ, E., BR. BAEHLER u. J. RABINOWITZ: Helv. Chim. Acta 43, 1871 (1960).
5. NICOLAUS, B. J. R.: J. Chromatog. 4, 384 (1960).
6. KIRK, R. E., and D. F. OTHMER: Encyclopedia of Chemical Technology. Vol. 2, p. 773, New York: Interscience Publishers Inc. 1948.
7. BRENNER, M., u. A. NIEDERWIESER: Experientia 16, 378 (1960).
8. — — u. G. PATAKI: Experientia 17, 145 (1961).
9. FAHMY, A. R., A. NIEDERWIESER, G. PATAKI u. M. BRENNER: Helv. Chim. Acta 44, 2022 (1961).
10. MUTSCHLER, E., u. H. ROCHELMEYER: Arch. Pharm. 292, 449 (1959).
11. MOTTIER, M.: Mitt. Lebensm. u. Hyg. 49, 454 (1958).
12. BRENNER, M., A. NIEDERWIESER, G. PATAKI u. A. R. FAHMY: Experientia 18, 101 (1962).
13. BRENNER, M., u. A. NIEDERWIESER: Experientia 17, 237 (1961).
14. MOORE, S., D. H. SPACKMAN, and W. H. STEIN: Analyt. Chem. 30, 1185 (1958); 30, 1190 (1958).
15. PÖHM, M.: Naturwiss. 48, 555 (1961).
16. PAULSON, C., F. E. DEATHERAGE and E. F. ALMY: J. Am. Chem. Soc. 75, 2039 (1953).
17. LINSKENS, H. F.: Papierchromatographie in der Botanik, 2. Aufl. Berlin-Göttingen-Heidelberg: Springer-Verlag 1959.
18. DE VERDIER, C.-H., u. G. ÅGREN: Acta Chem. Scand. 2, 783 (1948).
19. AWAPARA, J.: Arch. Biochem. 19, 172 (1948).
20. TISELIUS, A.: Experientia 17, 433 (1961).
21. FLODIN, P.: J. Chromatog. 5, 103 (1961).
22. PORATH, J.: Biochim. Biophys. Acta 39, 193 (1960).
23. — Clin. Chim. Acta 4, 776 (1959).
24. STEPANOV, V., D. HANDSCHUH u. F. A. ANDERER: Z. Naturforsch. 16 b, 626 (1961).
25. BOISSONNAS, R. A., u. S. LO BIANCO: Experientia 8, 425 (1952).
26. CONSDEN, R., A. H. GORDON, and A. J. P. MARTIN: Biochim J. 41, 590 (1947).
27. REDFIELD, R. R.: Biochim. Biophys. Acta 10, 344 (1953).
28. CARSTEN, M. E.: J. Am. Chem. Soc. 74, 5954 (1952).
29. PIEZ, K. A., E. B. TOOPER, and L. S. FOSDICK: J. Biol. Chem. 194, 669 (1952).
30. STEIN, W. H., and S. MOORE: J. Biol. Chem. 190, 103 (1951).
31. BUCHNER, H.: Dissertation Basel 1957, S. 84.
32. DRÈZE, A., et A. DE BOECK: Arch. intern. physiol. et biochem. 60, 201 (1952).
33. — S. MOORE, and E. J. BIGWOOD: Analyt. chim. Acta 11, 554 (1954).
34. COOK, E. R., and M. LUSCOMBE: J. Chromatog. 3, 75 (1960).
35. DENT, C. E., W. STEPKA, and F. C. STEWARD: Nature 160, 682 (1947).
36. PARIHAR, D. B.: Naturwiss. 41, 502 (1954).
37. REINDEL, F., u. W. BIENENFELD: Z. physiol. Chem. 305, 123 (1956).
38. BISERTE, G., P. BOULANGER et P. PAYSANT: Bull. soc. chim. biol. 40, 2067 (1958).
39. CLOSE, R.: Nature 185, 609 (1960).

40. KNAUFF, H. G., H. SELMAIR u. H. ZICKGRAF: Z. physiol. Chem. **318**, 73 (1960).
41. NÜRNBERG, E.: Arch. Pharmaz. **292**, 610 (1959).
42. JOST, K., J. RUDINGER, and F. ŠORM: Collection Czech. Chem. Commun. **26**, 2496 (1961).
43. AHRLAND, S., I. GRENTHE u. B. NORÉN: Acta Chem. Scand. **14**, 1059, 1077 (1960).
44. HIRS, C. H. W.: J. biol. Chem. **219**, 611 (1956).
45. BISERTE, G., et R. OSTEUX: Bull. soc. chim. biol. **33**, 50 (1951).
46. CALDIN, D. J. MC.: Chem. Rev. **60**, 39 (1960).
47. CONSDEN, R., A. H. GORDON, and A. J. P. MARTIN: Biochem. J. **38**, 224 (1944).
48. TSUKAMOTO, T., and T. KOMORI: Chem. Pharm. Bull. (Tokyo) **8**, 913 (1960).
49. KAWERAU, E., and TH. WIELAND: Nature **168**, 77 (1951).
50. BARROLLIER, J.: Naturwissenschaften **48**, 404 (1961).
51. WOIWOD, A.: J. Chromatog. **3**, 278 (1960).
52. HARDY, T. L., D. O. HOLLAND, and J. H. C. NAYLER: Analyt. Chem. **27**, 971 (1955).
53. OPIENSKA-BLAUTH, J., M. SANECKA, and M. CHAREZINSKI: J. Chromatog. **3**, 415 (1960).
54. PATTON, A. R., and P. CHISM: Analyt. Chem. **23**, 1683 (1951).
55. TOENNIES, G., and J. J. KOLB: Analyt. Chem. **23**, 823 (1951).
56. RUHEMANN, S.: J. Chem. Soc. **99**, 792 (1911);
57. — J. Chem. Soc. **99**, 1306 (1911);
58. — J. Chem. Soc. **99**, 1486 (1911);
59. — J. Chem. Soc. **97**, 1483 (1910);
60. — J. Chem. Soc. **97**, 2025 (1910).
61. WIGGINS, L. F., and J. H. WILLIAMS: Nature **170**, 279 (1952).
62. SAIFER, A., and I. ORESKES: Analyt. Chem. **28**, 501 (1956).
63. NOWORYTKO, J., i M. SARNECKA-KELLER: Acta Biochim. Polon. **2**, 91 (1955).
64. ACHER, R., C. FROMAGEOT et M. JUTISZ: Biochim. Biophys. Acta **5**, 81 (1950).
65. GRASSMANN, W., u. K. V. ARNIM: Ann. Chem. **519**, 192 (1935).
66. BARROLIER, J., J. HEILMAN u. E. WATZKE: Z. physiol. Chem. **304**, 21 (1956).
67. ROSEBEEK, S.: Chem. Weekblad **46**, 813 (1950).
68. KOFRÁNYI, E.: Z. physiol. Chem. **299**, 129 (1955).
69. CONSDEN, R.: Nature **162**, 359 (1948).
70. MÜTING, D.: Naturwissenschaften **39**, 303 (1952).
71. GIRI, K. V., u. A. NAGABHUSHANAM: Naturwissenschaften **39**, 548 (1952).
72. RIEMSCHNEIDER, R., u. K. PREUSS: Mh. Chem. **90**, 924 (1959).
73. REINDEL, F., u. W. HOPPE: Chem. Ber. **87**, 1103 (1954).
74. BHATTACHARYA, K. R., J. DATTA, and D. K. ROY: Arch. Biochem. **84**, 377 (1959).
75. ROCHE, J. N., VAN THOAI, and J. L. HATT: Biochim. Biophys. Acta **14**, 71 (1954).
76. JEPSON, J. R., and I. SMITH: Nature **172**, 1100 (1953).
77. ACHER, R., and CH. CROCKER: Biochim. Acta **9**, 704 (1952).
78. SAKAGUCHI, S.: J. Biochem. (Tokyo) **5**, 25 (1925).
79. — J. Biochem. (Tokyo) **5**, 133 (1925).
80. — J. Biochem. (Tokyo) **37**, 231 (1950).
81. TUPPY, H.: Sitzungsberichte (Vienna) **162**, 342 (1953).
82. WINEGARD, H. M., G. TOENNIES, and R. J. BLOCK: Science **108**, 506 (1948).
83. KIRBY-BERRY, H., H. E. SUTTON, L. CAIN, and J. S. BERRY: Univ. Texas Pubs. **5109**, 22 (1951).
84. AWE, W., I. REINECKE u. J. THUM: Naturwissenschaften **41**, 528 (1954).

85. CHARGAFF, E., C. LEVINE, and C. GREEN: J. Biol. Chem. **175**, 67 (1948).
86. BURTON, R. B., A. ZAFFARONI, and E. H. KENTMANN: J. Biol. Chem. **188**, 763 (1951).
87. TOYADA, H.: J. Bull. Chem. Soc. (Japan) **9**, 263 (1954).
88. PATTON, A. R., and E. M. FOREMAN: Science **109**, 339 (1949).
89. FRANK, H., u. H. PETERSEN: Z. physiol. Chem. **299**, 1 (1955).
90. BRAY, H. G., W. V. THORPE, and K. WHITE: Biochem. J. **46**, 271 (1950).
91. PAULY, H.: Z. physiol. Chem. **42**, 517 (1904).
92. — Z. physiol. Chem. **94**, 288 (1915).
93. EDLBACHER, S., H. BAUR, H. R. STAEHELIN u. A. ZELLER: Z. physiol. Chem. **270**, 158 (1941).
94. — Z. physiol. Chem. **157**, 106 (1926).
95. SANGER, F., and H. TUPPY: Biochem. J. **49**, 463 (1951).
96. CLARKSON, T. W.: Biochim. Biophys. Acta **18**, 453 (1955).
97. SCHWARTZ, D. P.: Analyt. Chem. **30**, 1855 (1958).
98. DALGLIESH, C. E.: Nature **166**, 1076 (1950).
99. CONSDEN, R., A. H. GORDON, and A. J. P. MARTIN: Biochem. J. **40**, 33 (1946).
100. METZENBERG, R. L., and H. K. MITCHELL: J. Am. Chem. Soc. **76**, 4187 (1954).
101. FEARON, W. R., and W. A. BOGGUST: Analyst **79**, 101 (1954).
102. EDWARD, J. T., and D. M. WALDRON: J. chem. Soc. **1952**, 3631.
103. SMITH, I.: Nature **171**, 43 (1953).
104. DALGLIESH, C. E.: Biochem. J. **52**, 3 (1952).
105. PACHÉCO, H.: Bull. Soc. Chim. Biol. **33**, 1915 (1951).
106. WIELAND, TH., u. L. BAUER: Angew. Chem. **63**, 511 (1951).
107. JERCHEL, D., u. R. MÜLLER: Naturwissenschaften **38**, 561 (1951).
108. FISCHER, A.: Planta **43**, 288 (1954).
109. PROCHÁZKA, Ž.: Chem. Listy **47**, 1643 (1953).
110. LINSER, H., H. MAYR u. F. MASCHEK: Planta **44**, 103 (1954).
111. ERSPAMER, V., and G. BORETTI: Arch. Intern. Pharmacodyn. **88**, 296 (1951).
112. CLERC-BORY, M., H. PACHÉCO et CH. MENTZER: Compt. Rend. **238**, 525 (1954).
113. EKMAN, B.: Acta Chem. Scand. **2**, 383 (1948).
114. GERNGROSS, O., K. VOSS u. H. HERFELD: Ber. dtsch. Chem. Ges. **66**, 435 (1933).
115. KUDZIN, S. F., R. M. DE BAUN, and F. F. NORD: J. Am. Chem. Soc. **73**, 4615 (1951).
116. FOLIN, O., and V. CIOCALTEU: J. Biol. Chem. **73**, 617 (1927).
117. — and W. DENIS: J. Biol. Chem. **12**, 239 (1912).
118. DURANT, J. A.: Nature **169**, 1062 (1952).
119. MILLON, E.: Compt. Rend. **28**, 40 (1949).
120. BRICAS, E., and CL. FROMAGEOT: Advances in Protein Chem. **8**, 4 (1953).
121. SCHACHTER, M.: Polypeptides which affect smooth muscles and blood vessels. Oxford: Pergamon Press 1960.
122. KATZ, A. M., W. J. DREYER, and C. B. ANFINSEN: J. Biol. Chem. **234**, 2897 (1959).
123. HONEGGER, C. G.: Helv. Chim. Acta **44**, 173 (1961).
124. Biochem. Preparations 8 (1961).
125. YANARI, S.: Biochim. Biophys. Acta **45**, 595 (1960).
126. PORATH, J., and P. FLODIN: Nature **183**, 1657 (1959).
127. GELOTTE, B. J.: J. Chromatog. **3**, 330 (1960).
128. LINDNER, E. B., A. ELMQUIST, and J. PORATH: Nature **184**, 1565 (1959).
129. Weissberger Technique of Organic Chem. Vol. III, part I, 2nd Ed. New York: Interscience 1956.
130. SANGER, F.: Advances in Protein Chem. **7**, 1 (1952).

131. ZUBER, H.: Chimia (Aarau) 14, 405 (1960).
132. SJÖQUIST, J., B. BLOMBÄCK, and P. WALLÉN: Arkiv Kemi 16, 425 (1961).
133. — Arkiv Kemi 14, 291 (1959).
134. VOGLER, K., R. O. STUDER u. W. LERGIER: Helv. Chim. Acta 44, 1495 (1961).
135. GUTTMANN, ST., u. R. A. BOISSONNAS: Helv. Chim. Acta 44, 1713 (1961).
136. — J. PLESS u. R. A. BOISSONNAS: Helv. Chim. Acta 45, 170 (1962).
137. HUGUENIN, R. L., u. R. A. BOISSONNAS: Helv. Chim. Acta 44, 213 (1961).
138. VOGLER, K., R. O. STUDER, W. LERGIER u. P. LANZ: Helv. Chim. Acta 43, 1751 (1960).
139. STUDER, R. O., K. VOGLER u. W. LERGIER: Helv. Chim. Acta 44, 131 (1961).
139a. SCHWYZER, R., u. H. KAPPELER: Helv. Chim. Acta 44, 1991 (1961).
139b. — u. H. DIETRICH: Helv. Chim. Acta 44, 2003 (1961).
140. BRENNER, M., u. G. PATAKI: Helv. Chim. Acta 44, 1420 (1961).
141. Private communication from Dr. R. RINIKER, CIBA AG., Basel.
142. KAPPELER, H., u. R. SCHWYZER: Helv. Chim. Acta 44, 1137 (1961).
143. Private communication from Dr. ST. GUTTMANN, SANDOZ AG., Basel.
144. SCHELLENBERG, P.: Angew. Chem. 74, 118 (1962).
145. Private communication from Dr. K. VOGLER, Hoffmann-La Roche AG., Basel.
146. VOGLER, K., R. O. STUDER, P. LANZ, W. LERGIER u. E. BÖHNI: Experientia 17, 223 (1961).
147. HAUSMANN, W.: J. Am. Chem. Soc. 78, 3663 (1956).
148. BISERTE, G., et M. DAUTREVAUX: Bull. Soc. Chim. Biol. 39, 795 (1957).
149. BRENNER, M., J. P. ZIMMERMANN, J. WEHRMÜLLER, P. QUITT, A. HARTMANN, W. SCHNEIDER u. U. BEGLINGER: Helv. Chim. Acta 40, 1497 (1957).
150. NIEDERWIESER, A., and G. PATAKI: Chimia (Aarau) 14, 378 (1960).
151. SANGER, F.: Biochem. J. 39, 507 (1945).
152. — Biochem. J. 40, 261 (1946).
153. — Biochem. J. 45, 562 (1949).
154. PORTER, R., and F. SANGER: Biochem. J. 42, 287 (1948).
155. EDMAN, P.: Acta Chem. Scand. 4, 283 (1950).
156. BISERTE, G., J. W. HOLLEMAN, J. HOLLEMAN-DEHOVE, and P. SAUTIÈRE: J. Chromatogr. 2, 225 (1959); 3, 85 (1960).
157. LEVY, A. L., and D. CHUNG: J. Am. Chem. Soc. 77, 2899 (1955).
158. DAVOLL, H., R. A. TURNER, J. G. PIERCE, and V. DU VIGNEAUD: J. Biol. Chem. 193, 363 (1951).
159. WALLENFELS, K., u. A. ARENS: Biochem. Z. 33, 217 (1960).
160. LI, C. H., and L. ASH: J. Biol. Chem. 203, 419 (1953).
161. TUPPY, H., u. G. BODO: Mh. Chem. 85, 807 (1954).
162. LOCKHART, I. M., and E. P. ABRAHAM: Biochem. J. 58, 633 (1954).
163. LEVY, A. L., and C. H. LI: J. Biol. Chem. 213, 487 (1955).
164. DESNUELLE, P., and C. FABRE: Biochim. Biophys. Acta 18, 49 (1955).
165. PORTER, R.: Methods Med. Res. 3, 256 (1951).
166. MIDDLEBROOK, W. R.: Biochem. J. 59, 146 (1955).
167. SCANES, F. S., and B. T. TOZER: Biochem. J. 63, 282 (1956).
168. THOMPSON, A.: Nature 168, 390 (1951).
169. MILLS, G. L.: Biochem. J. 50, 707 (1952).
170. TURBA, F., u. G. GUNDLACH: Biochem. Z. 326, 322 (1955).
171. BAILEY, K., and F. R. BETTELHEIM: Biochim. Biophys. Acta 18, 495 (1955).
172. BRAUNITZER, G.: Chem. Ber. 88, 2025 (1955).
173. LEVY, A. L.: Nature 174, 126 (1954).
174. ZAHN, H., u. H. PFANNMÜLLER: Biochem. Z. 330, 97 (1958).
175. FAHMY, A. R.: unpublished.
176. HEFENDEHL, F. W.: Planta Med. 8, 65 (1960).

177. Lichtenberger, W.: Z. Anal. Chem. **185**, 111 (1962).
178. Edman, P.: Nature **177**, 667 (1956).
179. — Acta Chem. Scand. **10**, 761 (1956).
180. Fraenkel-Conrat, H., and J. J. Harris: J. Am. Chem. Soc. **76**, 6058 (1954).
181. Sjöquist, J.: Arkiv Kemi **11**, 129 (1957).
182. — Biochem. Biophys. Acta **41**, 20 (1960).
183. Edman, P.: Acta Chem. Scand. **4**, 277 (1950).
184. Djerassi, C., K. Undheim, R. C. Sheppard, W. G. Terry u. B. Sjöberg: Acta Chem. Scand. **15**, 903 (1961).
185. Fraenkel-Conrat, H.: Methods of Biochem. Analysis, Vol. 2, p. 383. New York: Interscience Publishers 1955.
186. Cherbuliez, E., Br. Baehler, M. C. Lebeau u. A. R. Sussmann: Helv. Chim. Acta **43**, 896 (1960).
187. — A. R. Sussmann u. J. Rabinowitz: Pharmac. Acta Helv. **36**, 131 (1961).
188. Harris, J. J., u. P. Roos: Biochem. J. **71**, 434 (1959).
189. Schramm, G., J. W. Schneider u. A. Anderer: Z. Naturforsch. **11**b, 120 (1956).
190. Landmann, W. A., M. P. Drake, and J. Dillaha: J. Am. Chem. Soc. **75**, 3638 (1953).
191. Sjöquist, J.: Acta Chem. Scand. **7**, 447 (1953).
192. Edman, P., u. J. Sjöquist: Acta Chem. Scand. **10**, 1507 (1956).
193. Pataki, G.: Dissertation, Basel (1962).
194. Grothe, J. W.: J. Biol. Chem. **93**, 25 (1931).
195. Pastuska, G., u. H. Trinks: Chemiker-Ztg. **85**, 535 (1961).
196. Stahl, E.: Pharmazie **11**, 633 (1956).
197. — Chemiker-Ztg. **82**, 323 (1958).
198. — Parfüm. Kosmetik **39**, 564 (1958).
199. — Arch. Pharmaz. **292**, 411 (1959).
200. — Pharm. Rdsch. **1**, Nr. 2, 1 (1959).
201. Weber, R.: Helv. Chim. Acta **36**, 424 (1953).
202. Arx, E. von, and R. Neher: J. Chromatog. **12**, 329 (1963).
203. Bancher, E., H. Scherz u. V. Prey: Mikrochim. Acta 712 (1963).
204. Bello, D. di, and A. Signor: J. Chromatog. **17**, 506 (1965).
205. Berger, J. A., G. Meynel, J. Petit et P. Blanquet: Bull. Soc. Chim. France 2662 (1963).
206. Brandner, G., and A. I. Virtanen: Acta Chem. Scand. **17**, 2563 (1963).
207. Bujard, El.: quoted from [*223*].
208. Cramer, F., and E. Ehrhardt: J. Chromatog. **7**, 405 (1962).
209. Dittmann, J.: Z. klin. Chem. **1**, 190 (1963).
210. Esser, K.: J. Chromatog.**18**, 414 (1965).
211. Fahmy, A. R., A. Niederwieser, G. Pataki u. M. Brenner: Helv. Chim. Acta **44**, 2022 (1961).
212. Gray, W. R., and B. S. Hartley: Biochem. J. **89**, 59P (1963).
213. Habermann, E.: unpublished.
214. Hollingsworth, D. R., M. Dillard, and P. K. Bondy: J. Lab. Clin. Med. **62**, 346 (1963).
215. Hörhammer, L., H. Wagner u. F. Kilger: Deut. Apotheker **15**, 164 (1962).
216. Llosa, P. de la, C. Tertrin, and M. Jutisz: J. Chromatog. **14**, 136 (1964).
217. Massaglia, A., and U. Rosa: J. Chromatog. **14**, 516 (1964).
218. Myhill, D., and D. S. Jackson: Analyt. Biochem. **6**, 193 (1963).
219. Pataki, G.: J. Chromatog. **17**, 580 (1965).
220. — J. Chromatog. **16**, 541 (1964).

221. PATAKI, G.: Chimia (Aarau) **18**, 24 (1964).
222. — J. Chromatog. **16**, 553 (1964).
223. — Dünnschichtchromatographie in der Aminosäure- und Peptid-Chemie. Berlin: W. de Gruyter u. Co. 1966.
224. PATTERSON, S. J., and R. CLEMENTS: Analyst **89**, 328 (1964).
225. SEILER, N., u. J. WIECHMANN: Experientia **20**, 559 (1964).
226. SJÖHOLM, I.: Acta Chem. Scand. **18**, 889 (1964).
227. STAHL, E.: Z. anal. Chem. **221**, 3 (1966).
228. — u. J. PFEIFLE: Z. anal. Chem. **200**, 377 (1964).
229. SCHNEIDER, G., u. C. SCHNEIDER: Hoppe-Seylers Z. physiol. Chem. **332**, 316 (1963).
230. TURNER, N. A., and R. J. REDGWELL: J. Chromatog. **21**, 129 (1966).
231. WALZ, D., A. R. FAHMY, G. PATAKI, A. NIEDERWIESER u. M. BRENNER: Experientia **19**, 213 (1963).
232. WOLLENWEBER, P.: J. Chromatog. **9**, 369 (1962).

W. Nucleic Acids and Nucleotides

HELMUT K. MANGOLD

I. Introduction

1. Nucleic Acids and their Hydrolysis Products

The "genuine" high polymer nucleic acids form a part of all living matter.

Two types of polynucleotides may be distinguished through their structure; deoxyribonucleic acids (DNA) with molecular weights as high as 100 million; and ribonucleic acids (RNA) with molecular weights up to 500,000.

Deoxyribonucleic acids are much more alkali-resistant than ribonucleic acids and require other enzyme systems for hydrolitic degradation. Hydrolysis of deoxyribonucleic acids with mineral acids splits off purine bases, yielding "thymic acid"; and "oligonucleotides" can be prepared through the action of pancreatic enzymes.

Both types of nucleic acid are composed of mononucleotides which are built according to the scheme:

base − carbohydrate − o-phosphoric acid

Nucleosides are formed from mononucleotides (sometimes referred to as "simple nucleic acids") by elimination of the phosphoric acid moiety. Those derived from deoxyribonucleic acid contain the purine bases guanine and adenine and the pyrimidine bases cytosine and thymine; the bases are linked to deoxyribose via a N-glycosidic bond. 5-Methylcytosine, 5-hydroxymethylcytosine and 5-hydroxymethyluracil also occur in some deoxyribonucleic acids. The four nucleosides which can be isolated from ribonucleic acids contain the same purines and the pyrimidine bases cytosine and uracil, linked to ribose.

Purines are bound to the sugar phosphates as glycosides at N_9, pyrimidines at N_3; the phosphoric acid moiety may be linked to either C_2 or C_3 of the carbohydrate.

The components of the high molecular weight nucleic acids are given below:

Bases

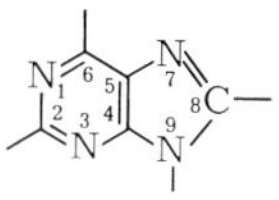

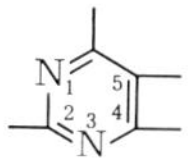

Purines
Adenine (6-Amino-purine)
Guanine (2-Amino-6-hydroxy purine)

Pyrimidines
Cytosine (2-Hydroxy-6-amino-pyrimidine)
Uracil (2,6-Dihydroxy-pyrimidine)
Thymine (2,6-Dihydroxy-5-methyl-pyrimidine)
Also 5-Methylcytosine and 5-Hydroxymethyl cytosine

Carbohydrates

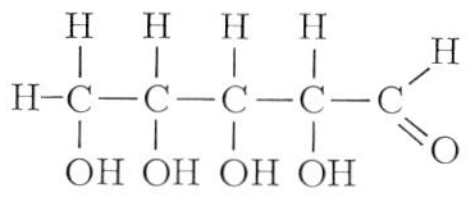

Ribose

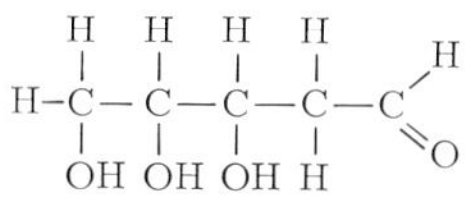

2-Deoxyribose

Examples of *Nucleoside* and a *Nucleotide*:

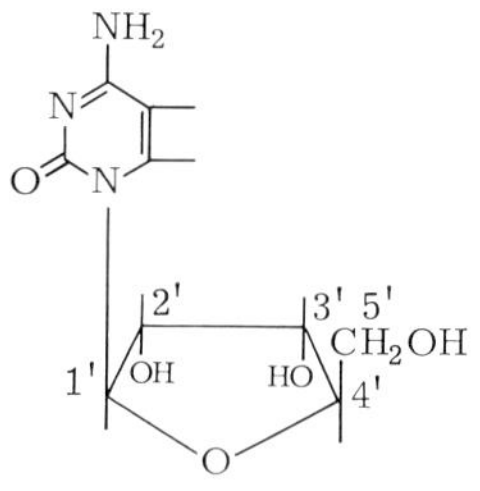

Cytidine

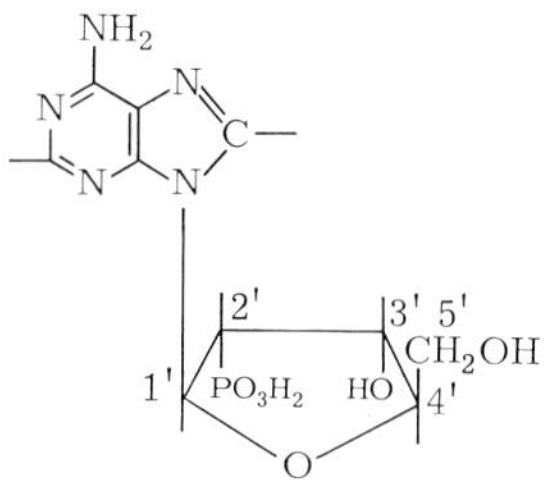

Adenosine-2'-phosphate

The "soluble ribonucleic acids", including the (amino acid) transfer ribonucleic acids, have molecular weights of about 30,000. They contain alkylated purines and pyrimidines in addition to adenine, guanine, cytosine and uracil; pseudouridine, a C-nucleoside of uracil, is another characteristic component. In addition, at least three sulphur-containing bases have been identified in transfer ribonucleic acids.

2. Nucleotide-Coenzymes

The nucleotide coenzymes are structurally related to the mononucleotides. Typical nucleotide coenzymes are adenosine triphosphate (ATP), flavin-adenine-dinucleotide (FAD) and numerous other phosphate esters of complex structure, containing adenosine, guanosine, cytidine or uridine. Five coenzymes are known for example, which are derived from cytidine diphosphate (CDP); CDP-choline, CDP-ethanolamine, CDP-diglyceride, CDP-glycerol and CDP-ribitol.

3. Older Methods of Nucleic Acid Analysis

Determination of phosphorus and nitrogen content was always important in nucleic acid analysis. Purines and pyrimidines in nucleic acid hydrolysates were identified and determined quantitatively by isolation on a small scale and also with the help of various colour reactions, polarography and microbiological methods. These techniques are, almost without exception, only of historical interest today.

4. Newer Methods for Isolation of Nucleic Acids and Separation of their Constituents

The different procedures for isolating nucleic acids yield products which vary considerably in composition and properties. One reason for this is the presence of nucleolytic enzymes in most plant and animal tissues. Work is always carried out at as low a temperature as possible to retard this enzyme activity and sodium citrate is used in an attempt to inhibit the action of deoxyribonucleases; ribonucleases are inactivated with guanidine hydrochloride or dodecylsulphate.

Nucleic acids are degraded by concentrated salt solutions and heat is also harmful. Even dilute acids attack nucleic acids and dilute alkali hydrolyses ribonucleic acids rapidly.

It is generally assumed that deoxyribonucleic acid solutions of high viscosity correspond most closely to the natural product. Good preparations of ribonucleic acids are not necessarily viscous. Pure nucleic acids contain about 9.2% phosphorus and possess an absorption maximum in the UV between 257 and 261 nm at pH 7: the should be free of proteins, polysaccharides, lipids and inorganic salts.

Deoxyribonucleic acids are usually prepared by the procedures of SIGNER and SCHWANDER [92], MARMUR [46a] or MASSIE and ZIMM [47]. VOLKIN and CARTER'S [100] method is suitable for isolation of high molecular weight ribonucleic acid. Soluble ribonucleic acid is frequently isolated by ZUBAY'S [112] procedure.

During the last 20 years, various chromatographic and electrophoretic procedures have been developed which permit high molecular nucleic acids to be fractionated. Complex hydrolysates of nucleic acids can be separated through PC and other partition procedures and through chromatography on ion exchangers. Quantitative analyses of very small amounts of such hydrolysates can be carried out nowadays by determining the concentrations of bases, nucleosides, mono- and oligonucleotides after separation.

Detailed descriptions of these modern separation procedures are to be found in the first volume of the standard work on nucleic acid chemistry, edited by CHARGAFF and DAVIDSON [9] in Vol. XII of the series "Methods in Enzymology" [13]; and in a number of review articles [12, 85, see also 16].

Procedures for isolation and analysis of nucleic acids are given in CANTONI and DAVIS's monograph [8].

5. Colour Reactions for Distinguishing Ribo- and Deoxyribonucleic Acids

The biuret reaction [alkaline Cu(II)] and the arginine test [102] are used to detect peptides in nucleic acid preparations. Two other colour reactions permit either type of nucleic acid to be detected in the presence of the other; one is due to DISCHE and the other is the phloroglucinol test of v. EULER and HAHN.

Dische-Reaction [19]

Procedure: A solution of 50—500 µg nucleic acid in 1 ml water is heated 10 min at 100° C with 2 ml of a solution of 1 g diphenylamine and 2.75 ml conc. sulphuric acid in 100 ml acetic acid. The solution turns blue (abs. max. 595 nm) if deoxyribonucleic acid is present. The difficultly hydrolysable pyrimidine-deoxyribonucleosides and -nucleotides do not react under these conditions.

Reaction of v. Euler and Hahn [21]

Procedure: A sample containing about 2 mg ribonucleic acid per ml is heated on the water bath for 50 min with 8 ml of a 0.1% solution of ferric chloride in conc. hydrochloric acid-acetic acid (1 + 6). The reaction mixture is cooled to room temperature and 1 ml of a 25% solution of phloroglucinol in conc. hydrochloric acid-acetic acid-water (25 + 50 + 25) is added. After standing 20 min, the mixture is heated 4 min on the water bath. The colour (abs-max. 680 nm) attains maximum intensity after about 10 h at room temperature. Deoxyribonucleic acid yields no colour.

6. Hydrolysis of Nucleic Acids

Hydrolysis of nucleic acids may lead to oligonucleotides, mononucleotides, nucleosides, purine and pyrimidine bases, sugar phosphates and free carbohydrates. Secondary products always result also, e. g., from deamination of the aminopurines and aminopyrimidines or their nucleosides and nucleotides.

a) Alkaline Hydrolysis

Ribonucleic acids are broken down into mononucleotides by dilute aqueous solutions of sodium or potassium hydroxide. Deoxyribonucleic acids are attacked only slightly and can thus be separated from ribonucleic acids by treatment with dilute sodium hydroxide [88].

The following procedure has been useful:

Conversion of Ribonucleic Acids into Mononucleotides, according to Schmidt and Thannhauser [88]

Procedure: 10 mg ribonucleic acid are maintained for 20 h at 37° with 1 ml 0.2 N sodium hydroxide. After cooling, the hydrolysate is acidified with 0.5 N hydrochloric acid (pH 4—6) and the acid solution used for analysis of the mononucleotides.

b) Acid Hydrolysis

The N-glycosidic purine-carbohydrate bond in both types of nucleic acid is particularly susceptible to acids whereas the pyrimidine nucleosides and -nucleotides are quite resistant. Acid removes the phosphoric acid moiety much more easily from purine nucleotides than from pyrimidine nucleotides.

Dilute hydrochloric or sulphuric acid splits guanine and adenine from deoxyribonucleic acids without appreciably altering the high molecular structure of the polynucleotide. The purine-free product, termed "thymic acid" has a molecular weight of about 15,000; it can be purified by dialysis with dilute hydrochloric acid as receiving solution and is isolated as "apurinic acid" [95]. Cytosine- and thymine deoxyribodiphosphoric acids are obtained from apurinic acid or directly from deoxyribonucleic acid by heating with methanolic hydrogen chloride solution for 1—5 h [41]. The procedures below have proved satisfactory for liberating the purine and pyrimidine bases from nucleic acids:

Cleavage of Purines and Pyrimidines from Deoxyribonucleic Acids according to Vischer and Chargaff and to Wyatt [99, 105]

Procedure: 0.5 ml 98% formic acid is added to 10—20 mg of deoxyribonucleic acid which has been dried in vacuo at 40—60° C and the mixture heated for 30 min at 175° C in a thick-walled, sealed glass tube, placed in an electric oven. After cooling, the tube is carefully opened and the formic acid removed in a desiccator over potassium hydroxide. The residue is taken up in 3—4 ml N hydrochloric acid and the solution used for analysis of the purine and pyrimidine bases.

Ribonucleic acids are not completely hydrolysed under these conditions.

Cleavage of Purines and Pyrimidines from Ribo- and Deoxyribonucleic Acids according to Marshak and Vogel [46, 105]

Procedure: About 10 mg nucleic acid, dried in vacuo at 40—60° C, are weighed into a tube fitted with a glass stopper and dissolved in 0.2 ml 72% perchloric acid. The tube is closed and heated 1 h at 100° on the water bath. After cooling, 0.2 ml water is added and the diluted hydrolysate centrifuged. The clear solution is decanted from the small amount of black residue and used for analysing the purine and pyrimidine bases.

c) Enzymatic Hydrolysis

Enzymatic degradation has been an especially valuable tool in studying the structure of nucleic acids.

The enzyme ribonuclease I from pancreas brings about hydrolysis of ribonucleic acids to mono-, di-, tri- and tetranucleotides; a large part of the ribonucleic acid remains always unattacked. Ribonuclease I is inactive towards deoxyribonucleic acid but thymic acid suffers partial hydrolysis.

Deoxyribonuclease I from pancreas breaks down high molecular weight deoxyribonucleic acids into oligonucleotides and a small amount of mononucleotides; large parts of the polynucleotide structure are unaffected. Ribonucleic acids are unattacked. Deoxyribonucleases from snake venom are capable of breaking down deoxyribonucleic acids more completely than is the deoxyribonuclease I from pancreas. These enzymes have, however, not yet been adequately investigated.

Reference may be made to specialist literature concerning the properties and use of the various nucleases [8, 9, 13, 28].

7. UV-Spectra of Nucleic Acid Constituents

Table 192. *The UV spectra of nucleic acid derivatives*

	Molecular weight	Normality of solution	Absorption maximum nm	Millimol Extinction-coefficient
Purines and Pyrimidines				
Adenine	135.13	0.1 HCl	262	13.1
Guanine	151.13	0.1 HCl	249	11.1
Cytosine	111.10	0.1 HCl	276	10.0
Uracil	112.09	Water	260	8.2
Thymine	126.11	0.1 HCl	265	7.95
5-Methylcytosine	125.13	0.1 HCl	283	9.8
5-Hydroxymethyl cytosine	141.14	0.1 HCl	279	9.7
Ribonucleosides				
Adenosine	267.25	Water	259	15.4
Guanosine	283.24	Water	252	13.7
Cytidine	243.22	0.1 HCl	262	10.0
Uridine	244.20	Water	262	10.0
Ribonucleoside-2'- and -3'-monophosphates				
Adenosine-2'-monophosphate	347.23	Water	260	15.0
Adenosine-3'-monophosphate				
Guanosine-2'-monophosphate	363.24	0.01 HCl	280	12.9
Guanosine-3'-monophosphate				
Cytidine-2'-monophosphate	323.21	0.01 HCl	278	12.7
Cytidine-3'-monophosphate				
Uridine-2'-monophosphate	324.20	Na salt in water	260	10.0
Uridine-3'-monophosphate				

Knowledge of their UV spectra enables the individual nucleic acid derivatives to be identified and quantitatively determined after separation and elution. The absorption maxima and molar absorption coefficients of a number of substances are compiled in Table 192.

The UV spectra of methylated purines and pyrimidines, of other uncommon bases, and of their nucleosides and nucleotides, are described in a monograph [98].

8. Manufacturers and Suppliers of Pure Preparations

Purines, pyrimidines and their nucleosides and nucleotides are available commercially (Firms 26, 31, 80, 86, 98, 124, 130). Enzymes for degradation of nucleic acids are also obtainable (Firms 26, 154).

Imformation about manufacture and sources of radioactively labelled compounds is to be found on p. 167 and in the list of firms at the end of this book.

II. Thin-Layer Chromatography of Nucleic Acids and their Constituents

Nucleobases, nucleosides, mononucleotides and oligonucleotides can be separated with the help of TLC. The method is suitable also for fractionating and characterising soluble ribonucleic acids as well as high molecular weight nucleic acids. Silica gel, cellulose, dextran gel and ion exchangers are used as adsorbents. An especially good separation of oligonucleotides is accomplished through thin-layer electrophoresis and thin-layer electrophoresis-chromatography (see Chapter E, p. 105-114).

1. Purines, Pyrimidines and Nucleosides

PC was being used already 25 years ago for fractionation and quantitative analysis of these substances in nucleic acid hydrolysates [45, 46, 99, 105]. RANDERATH was the first to report the thin-layer chromatographic separation of nucleic acid constituents on silica gel G [59] and cellulose [60]. He found TLC on cellulose layers to be superior to PC [59, 61]. TLC on finely powdered cellulose layers yields smaller and more sharply defined spots than PC on the fibrous paper under comparable conditions [60]; moreover, thin-layer chromatographic separation of nucleic acid derivatives requires only a fraction of the time needed in PC [60, 61].

Experimental Conditions

a) Adsorbents. Any of the following can be used: silica gel G (F) and silica gel H(F) [59, 80, 86, 87]; pure cellulose powder [14, 26, 31, 36];

cellulose powder containing gypsum [10, 27, 38, 60]; mixtures of cellulose and silica gel (90 + 10) [24]; and celite-starch [91]; pure cellulose is mostly preferred.

Ion exchange-layers like DEAE- and ECTEOLA-cellulose [11, 35, 59, 78] and PP-cellulose [55] also are suitable for separating bases and nucleosides.

Procedures for preparing the various layers are given in Chapter B, p. 47.

b) Solvents. Distilled water is suitable for separation of purine and pyrimidine bases and of their nucleosides on paper [96]. These substances may be fractionated using water as solvent on silica gel G and cellulose layers also [60, 61]; separations require only 45 min. Especially good separations of the bases have been obtained by two-dimensional chromatography on cellulose [26, 74]; solvents were, for example, methanol-conc. hydrochloric acid-water (70 + 20 + 10) in the first dimension and n-butanol-methanol-water-conc. ammonium hydroxide (60 + 20 + 20 + 1) in the second [74]. After development with the first solvent, hydrochloric acid was removed by placing the layer for a few minutes in a current of air, first in the cold and then at 60–70° C.

Water has been used as solvent on DEAE- and ECTEOLA-cellulose layers [11, 59] and also 0.005N hydrochloric acid [11] and isobutyric acid-conc. ammonium hydroxide-water (66 + 2 + 32) on the former [11]. Saturated ammonium sulphate solution-N sodium acetate-isopropanol (80 + 18 + 2), a solvent which was developed for the PC [45] and TLC (cellulose) [60] of nucleotides, is particularly good for separating purine and pyrimidine bases on layers of DEAE-cellulose [11].

c) Methods of Detection. Purine and pyrimidine bases and their nucleosides can be detected on thin-layer chromatograms in the light of a short-wave UV lamp (maximum emission 254 nm) (p. 77-79). This method is less sensitive on inorganic layers than on cellulose. The lower limit of detection may be somewhat reduced by adding a fluorescent material to the silica gel or by spraying the chromatogram with a solution of fluorescein in dilute ammonium hydroxide (Rgt. No. 117) [23, 36]. Amounts down to about 10^{-3} μmole of adenine derivatives can be detected on cellulose in short-wave UV light without using a fluorescence indicator; the limit of detection is rather higher for compounds which contain cytosine or uracil [66].

The "photoprint" procedure for photographic detection in UV light cannot of course be used with coated glass plates.

Adenine may be detected specifically on silica gel and cellulose by using the Dragendorff-reagent (No. 98). All the reactions for detection of pentoses in nucleosides are useful for identifying these compounds on thin-layer chromatograms.

Applications and Results

Table 193 gives a comparison of the separations of nucleobases and nucleosides on paper and on layers of cellulose, silica gel G and ECTEOLA-cellulose. The extensive agreement of the hR_f-values of the purine bases and purine nucleosides on the cellulose and ECTEOLA-cellulose layers and on paper, is striking; appreciably higher hR_f-values were found on silica gel G. The various pyrimidines and pyrimidine-ribonucleosides have very nearly identical hR_f-values on the three layers and on paper. A better separation of the purine and pyrimidine bases from the corresponding ribonucleosides is achived on cellulose and ECTEOLA cellulose layers than on silica gel G. These three adsorbents are about equally useful for fractionating the nucleosides.

Table 193. *hR_f-values of some purines, pyrimidines and their nucleosides from paper- and thin layer-chromatographic separations using water as developing solvent* [59]

Substance	Silica Gel G [a]	Cellulose [b]	ECTEOLA [c]	Paper [d]
Adenine	57	30	29	38
Adenosine	75	43	56	56
Guanine	66	37	33	38
Guanosine	80	58	50	57
Cytosine	—	—	—	—
Cytidine	76	80	82	77
Uracil	78	72	73	75
Uridine	85	81	84	84

[a] *Layer:* Silica gel G (Firm 88).
[b] *Layer:* MN 300 (Firm 83).
[c] *Layer:* the capacity of the exchanger is 0.26 m.eq.N/g (Firm 127).
[d] "Ederol 202" from the Firm of Binzer, Hatzfeld/Eder, W. Germany.

TLC on cellulose and silica gel layers has been applied to separate the fluorinated [30] and iodinated pyrimidines and pyrimidine ribonucleosides [23, 47] which are important in cancer therapy. Other domains in which TLC has been applied with profit are analyses of products of organic synthesis [7, 42, 44] and of enzymatic reactions [38, 110].

Purine and pyrimidine bases and their nucleosides can be determined quantitatively on the (cellulose) layer by a fluorometric method [53]; or spectrophotometrically in the UV after elution [10, 36, 37, 43] (cf. Chapter H, p. 133-155).

2. Nucleotides and Nucleotide-Coenzymes

Nucleotides contain both acid and basic groups; mixtures of these substances may therefore be separated by anion- *or* cation exchange chromatography [12, 85]. Column chromatography on Dowexionexchangers has been used successfully for years in the fractionation, isolation and characterisation of nucleic acid structural components [9, 12, 13, 85]; the method has been rendered automatic for routine investigations [1].

The sequence of elution of the nucleotides should be determined by their net charge (Fig. 221) but in practice these theoretical expectations are not usually fulfilled. An *adsorptive activity* of the ion exchanger is given as the reason [12]. Purine derivatives are generally adsorbed more strongly than the corresponding pyrimidine compounds; thus, for example, uridine-3'-phosphate is eluted before guanosine-3'-phosphate from Dowex columns although the reverse order would be expected from their net charge [12].

RANDERATH was the first to describe fractionations of nucleotides on ECTEOLA- [58, 59] and DEAE-cellulose layers [64, 66]. He found that TLC on ion exchangers renders possible better and faster separations than column chromatography and is also considerably more sensitive than PC. In 1962 RANDERATH described the preparation of PEI-cellulose and the application of this new anion exchanger in thin-layer chromatography of nucleotide mixtures [63, 65].

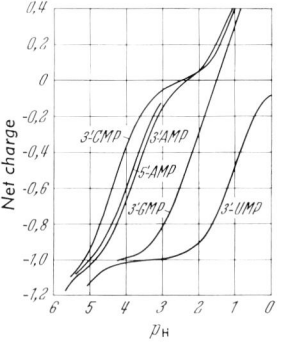

Fig. 221. Relation between net charge and pH for the mononucleotides of ribonucleic acid [12]

The chromatographic behaviour of the various nucleotides can be forecast with fair certainty from physical data [64]. This represents a marked advantage of ion exchange TLC over partition chromatographic procedures. Ion exchange TLC on modified celluloses may be used also for separations on the micro-preparative scale [64].

Experimental Conditions

a) Adsorbents. Separation of nucleotides and nucleotide-coenzymes can be performed on layers of: silica gel G [86, 87]; cellulose (mostly without added gypsum) [10, 26, 36, 61]; DEAE-cellulose [11, 20, 22, 94] and DEAE-sephadex [29, 104]; ECTEOLA-cellulose [52, 58, 59, 66]; and, principally, PEI-cellulose ion exchangers [51, 56, 70, 103].

Procedures for preparing these layers are given on p. 37-39 and 45-47.

Plates coated with ion exchangers are best used as soon as possible after preparation. PEI-cellulose layers become unusable within a few days at ordinary temperature and in daylight [77]; but they keep several weeks if stored at ca. 5° in the dark [77].

b) Solvents. Pyridine-nucleotides can be chromatographed on silica gel G using isobutyric acid-conc. ammonium hydroxide-water (66 + 1 + 33) [86]; n-butanol-acetone-acetic acid-5% ammonium hydroxide-water (35 + 25 + 15 + 15 + 10) is recommended for purine- and pyrimidine-nucleotides [60]. Basic solvents understandably cannot be used with layers containing gypsum.

Saturated ammonium sulphate-M sodium citrate-isopropanol (80 + 18 + 2), a solvent developed for PC of mononucleotides [45], can be employed also for separating isomeric purine-mononucleotide on cellulose layers [60]; it migrates about 10 cm in 90 min. Excellent separations are obtained on cellulose layers with the solvent *tert.* amyl alcohol-formic acid-water (30 + 20 + 10), likewise applied in paper chromatographic separation of mononucleotides [48]; the time of run is about 120 min. It is superior to all other solvent systems described here (Table 194).

Nucleoside-monophosphates, -diphosphates and -triphosphates can be separated on paper or on cellulose layers using the solvent n-butanol-acetone-acetic acid-5% ammonium hydroxide-water (45 + 15 + 10 + 10 + 20) [61].

Hydrochloric acid (0.02–0.04 N) is suitable for fractionating mononucleotides on layers of DEAE- and ECTEOLA-cellulose [26, 66] (see also [104]); the runs are rather longer on the former. The mononucleotides can be separated on PEI-cellulose by stepwise development using N acetic acid and 0.1–0.4N sodium chloride or lithium chloride [57, 70]. Complex mixtures of these compounds may be fractionated by two-dimensional TLC on PEI-cellulose: stepwise development is carried out in the first direction with 0.2M, 1M and 1.6M lithium chloride; the lithium salt is then washed out of the layer by laying the chromatogram for 15 min in a flat dish containing 0.5–1 litre absolute methanol; finally, stepwise development is performed in the second direction using 0.5M, 2M and 4M sodium formate buffer, pH 3.4 [56] (Fig. 222).

Aqueous 0.15M sodium chloride solution is convenient for fractionating mono-, di- and triphosphates on DEAE- and ECTEOLA-cellulose [59]. 0.3–0.5M lithium chloride may be used in TLC on PEI-cellulose of these compounds as well as of pyridine-nucleotides; the separations do not take longer than 15 min [51, 63, 70, 75]. 0.3–0.8M ammonium sulphate is also suitable for fractionation of mono-, di- and triphosphates [77].

A successful separation of ribomononucleotides from the corresponding deoxyribomononucleotides has been achieved by TLC on PEI-cellulose using solvents containing boric acid, such as 2% aqueous boric acid-2M lithium chloride (50 + 25) [69, 70]. Mixtures of ribo- and deoxyribonucleoside triphosphates also can be separated on PEI-cellulose: stepwise development is first carried out with 2N acetic acid-2M lithium chloride (50 + 50) (4 cm) and 4N acetic acid-2.5M lithium chloride (50 + 50) (15 cm); the layer is then freed from acetic acid by a stream of warm air (40°) and from lithium chloride by washing with absolute methanol; stepwise development follows in the second

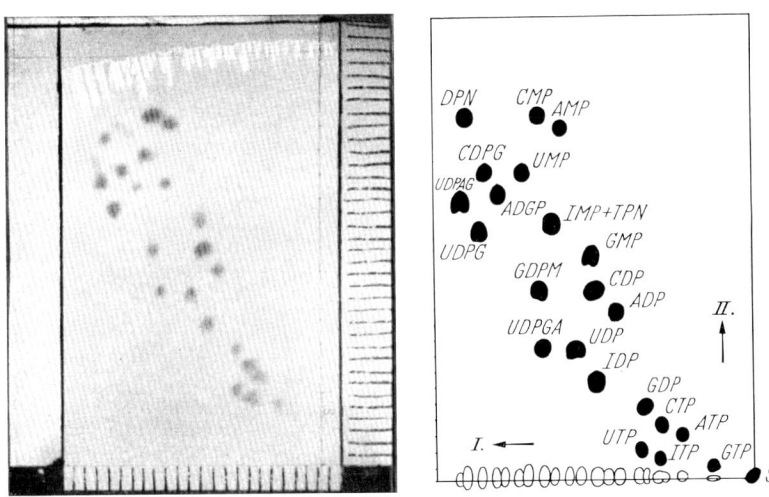

Fig. 222. Separation of a mixture of mononucleotides through two-dimensional ion-exchange TLC [56]

Layer: PEI-cellulose (0.5 mm); solvent: 1. direction: stepwise development with lithium chloride solutions: 0.2M (2 min), 1.0M (6 min) and 1.6M (13 cm); lithium chloride removed with methanol (see p. 796) and then stepwise development in the 2. direction with sodium formate buffers, pH 3.4: 0.5M (0.5 min), 2.0M (2 min) and 4.0M (15 cm). Visualisation in UV light; on the left the chromatogram; on the right, the identified individual fractions (cf. Table 195)

direction using the solvents 2M ammonium acetate containing 3% boric acid (4 cm) and 3.5M ammonium acetate, 4% in boric acid (14 cm) (the two last named solvents are adjusted to pH 7 with ammonium hydroxide [51]). Fig. 98 (p. 179) reproduces the autoradiograph of a chromatogram of ribo- and deoxyribonucleoside triphosphates containing ^{32}P, which were prepared in this way.

c) **Methods of Detection.** Nucleotides and nucleotide-coenzymes are most simply detected on thin-layer chromatograms by exposure to the radiation of a short-wave UV lamp (maximum emission 259 nm) (see p. 77-79). The lower limit of detection depends on the adsorbent and the solvent used for separation. Amounts of adenylic acid down to 5×10^{-4} µmole can be detected on DEAE- and ECTEOLA-cellulose.

Adenine, cytosine and uracil derivatives fluoresce dark blue and guanine derivatives light blue [64].

Applications and Results

The separations which are realisable with TLC can be seen in Table 194 and Fig. 226 A.

Table 194. *hRf-values of 5'-mono-, di- and triphosphates on cellulose*[a] *layers, using various solvents* [60, 61]

Mono-, di- and triphosphates	Solvent 1	Solvent 2	Solvent 3
Adenosine-5'-monophosphate	52	35	38
Adenosine diphosphate	29	17	26
Adenosine triphosphate	16	8	16
Guanosine-5'-monophosphate	—	—	—
Guanosine diphosphate	—	—	—
Guanosine triphosphate	—	—	—
Cytidine-5'-monophosphate	48	30	34
Cytidine diphosphate	27	13	22
Cytidine triphosphate	13	7	13
Uridine-5'-monophosphate	47	30	37
Uridine diphosphate	26	14	25
Uridine triphosphate	13	8	17

Solvent 1 = *tert*. amyl alcohol-formic acid-water (30 + 20 + 10); 120 min run.

Solvent 2 = n-butanol-acetone-acetic acid-5% NH_4OH-water (35 + 25 + 15 + 15 + 10); 90 min time of run.

Solvent 3 = n-butanol-acetone-acetic acid-5% NH_4OH-water (45 + 15 + 10 + 10 + 20); 50—60 min time of run.

[a] Cellulose MN 300 (Firm 83).

Table 195 enables the separation efficiencies of DEAE-, ECTEOLA- and PEI-celluloses to be compared. On these ion exchange layers, mono-, di- and triphosphates are separated primarily into groups on the basis of their differing negative charges; monophosphates migrate fastest and triphosphates the most slowly. The net negative charge within each of these groups is in the order: cytosine- < adenine- < guanine- < uracil derivatives (cf. Fig. 221). A corresponding reduction in migration speed is indeed observed on ion exchange layers: in dilute hydrochloric acid, cytosine- and adenine-nucleotides have consistently higher hRf-values than the corresponding guanine derivatives which in turn exhibit higher values than the uracil derivatives (cf. Fig. 222). Cytosine- and adenine nucleotides, a group which can be fractionated only with difficulty using hydrochloric acid, may be fully separated from each other with aqueous sodium chloride solutions; the spots are slightly more diffuse, however.

Table 195. *hRf-values of 5'-mono, di- and triphosphates on ion exchange layers* [59, 66, 70]

Mono-, di- and triphosphates	ECTEOLA-Cellulose		DEAE-Cellulose		PEI-Cellulose	
	solvent 1	2	solvent 3	4	solvent 5	6
Adenosine-5'-monophosphate	57	26	45	65	>80	>80
Adenosine diphosphate	36	8	24	48	29	70
Adenosine triphosphate	21	—	6	11	4	33
Guanosine-5'-monophosphate	55	14	36	60	50	72
Guanosine diphosphate	37	3	9	27	13	61
Guanosine triphosphate	17	—	5	7	2	24
Cytidine-5'-monophosphate	74	31	46	65	>80	>80
Cytidine diphosphate	51	11	31	53	35	73
Cytidine triphosphate	34	—	9	13	4	37
Uridine-5'-monophosphate	80	13	31	49	64	>80
Uridine diphosphate	63	0	7	15	11	60
Uridine triphosphate	44	—	4	4	2	20

Solvent 1: 0.15 M sodium chloride; time of run: 15 min.
Solvent 2: 0.01 N hydrochloric acid; time of run: 15 min.
Solvent 3: 0.01 N hydrochloric acid; time of run: 40 min.
Solvent 4: 0.02 N hydrochloric acid; time of run: 40 min.
Solvent 5: 2.0 N formic acid-0.5 M lithium chloride; time of run: 45 min.
Solvent 6: 2.0 N formic acid-2.0 M lithium chloride; time of run: 45 min.

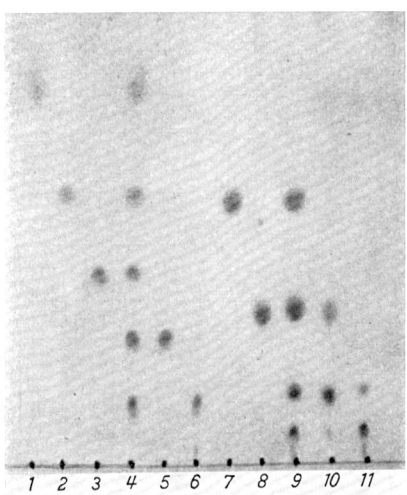

Fig. 223. Separation of nucleotides through ion exchange TLC [77].
Layer: PEI-cellulose (0.5 mm); solvent: stepwise development with lithium chloride solutions: 0.5 M (5 min), 1.0 M (10 min) and 1.5 M (35 min). Visualisation by spraying with 0.002% fluorescein solution in methanol and inspection in UV light. *1* uridine diphosphate-glucose; *2* uridine monophosphate; *3* uridine diphosphate-glucuronic acid; *4* compounds 1, 2, 3, 5 and 6; *5* uridine diphosphate; *6* uridine triphosphate; *7* guanosine diphosphate-mannose; *8* guanosine monophosphate; *9* compounds 7, 8, 10 and 11; *10* guanosine diphosphate; *11* guanosine triphosphate

The separation efficiency is virtually the same on all three ion exchangers; the spots are especially sharply outlined on DEAE- and PEI-celluloses [56, 64] (cf. Fig. 222).

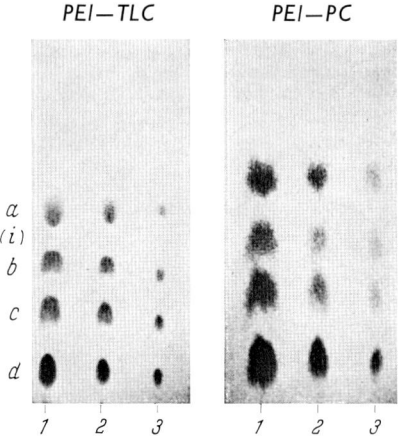

Fig. 224. Comparison of thin-layer and paper chromatographic fractionations of nucleotide coenzymes [73]. PEI-TLC: separation on a PEI-cellulose layer (0.5 mm); PEI-PC: separation on PEI-paper.
Solvents: in both cases, 1.0N acetic acid (2 cm) and 1.0N acetic acid-3.0M lithium chloride (90 + 10) (15 cm); times of run: 120 min (TLC) and 60 min (PC). Visualisation in UV light; amounts: 10—100 μmoles. *a* uridine monophosphate; *b* uridine-diphosphate-N-acetylglucosamine; *c* uridine diphosphate-glucose; *d* uridine diphosphate. An unknown substance (*i*) appears on the thin-layer chromatogram which is not found on the paper chromatogram

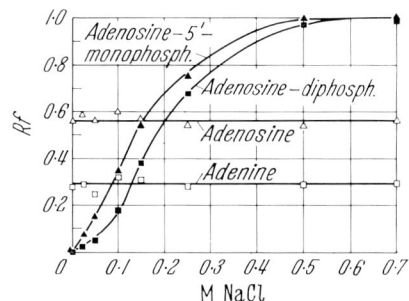

Fig. 225. Separation of nucleic acid derivatives on an ion exchange layer [59]. Rf-values as a function of the concentration of sodium chloride in the solvent. Ion exchanger: ECTEOLA-cellulose, capacity of 0.26 mval/g; solvent: 0.1—0.7M sodium chloride solutions; length and time of run: ca. 10 cm and 15 min respectively

Fig. 224 shows that ion exchange chromatography on layers of PEI-cellulose yields better resolutions than on PEI-paper.

Depending on the type of solvent, ECTEOLA-, DEAE- and PEI-celluloses can function as stationary phases for partition chromatography or as ion exchangers; this is seen from Fig. 225.

Nucleotides and sugar-nucleotides can be determined spectrophotometrically in the UV after thin-layer chromatographic separation and elution [18, 24, 36, 43]; mixtures of mono-, di- and triphosphates may be analysed in the same way [5, 50, 57].

Mononucleotides in alkaline hydrolysates of ribonucleic acids are best separated on PEI-cellulose and quantitatively determined after elution. In a recommended procedure, the individual fractions are eluted with M lithium chloride on to a narrow paper strip, the strip is dried, the nucleotides are eluted from it with water and the concentrations of the aqueous solutions determined by spectrophotometric measurements in the UV [57]; considerably more accurate analyses are possible with this procedure than by direct extraction from the adsorbent [57] (see also Fig. 54, p. 104).

Sugar phosphates may be analysed through determination of their phosphate content after thin-layer chromatographic separation and elution [14] (see also [18, 101]).

TLC has proved suitable for analysis of the products of enzymatic degradation of nucleotide coenzymes [38, 72] and of soluble ribonucleic acids [3]; it has been demonstrated that enzymatic reactions can be carried out on layers of PEI-cellulose [72]. The biosynthesis of nucleosides [110], ribonucleoside- and deoxyribonucleoside triphosphates [52], diribonucleoside di-, tri- and tetraphosphates [76, 108, 111] and polynucleotides [62] has been followed with the help of TLC.

3. Oligonucleotides and Nucleic Acids

Mixtures of oligonucleotides can be separated with the help of electrophoresis (see pp. 105–114), through partition chromatography on cellulose columns [106] and through ion exchange chromatography on columns of DEAE-cellulose [93]. Countercurrent distribution in particular is suitable for fractionating and isolating transfer ribonucleic acids [32, 107]; the same can be accomplished by chromatography on cellulose columns [106]. Ribo- and deoxyribonucleic acids of high molecular weight are usually chromatographed on columns of DEAE- or ECTEOLA-cellulose [39, 81, 82] (see also [12, 85]).

Little work has been done on the TLC of oligonucleotides, transfer ribonucleic acids and ribo- and deoxyribonucleic acids of high molecular weight. Synthetic deoxyribo-oligonucleotides have been separated on layers of PEI-cellulose, using stepwise development with 0.2–1.2 M sodium chloride; DEAE-cellulose proved less suitable [103]. Chromatography of oligonucleotides on silica gel H layers has also been possible [87]. Complex formation between deoxyribonucleotides and polyribonucleotides has been demonstrated with the help of ion exchange TLC on PEI-cellulose [68].

MORTON and ROGERS [49] have been able to distinguish various ^{14}C-labelled "soluble ribonucleic acids" through TLC on PEI-cellulose and other ion exchangers. Their solvents were the upper phase of the system isopropanol-formamide-phosphate buffer, pH 6.2 (20 + 5 + 50) [32] or the lower phase of the mixture n-butanol-water-tri-n-butylamine-acetic acid-di-n-butyl ether (100 + 130 + 10 + 2.5 + 29) [107]; the fractions were detected by autoradiography [49].

BAUER and MARTIN [2] have chromatographed ribo- and deoxyribonucleic acids on ECTEOLA-cellulose, using as solvents aqueous solutions of ammonium hydroxide (M), sodium chloride (2M) and phosphate buffer (0.01 M, pH 11). The authors showed that deoxyribonucleic acids which had been degraded by heating migrated further on ion exchanger layers than did preparations of high molecular weight (cf. Table 196).

Table 196. *Sedimentation coefficients and hRf-values of deoxyribonucleic acid which had been heated in aqueous solution* [2]

Duration of heating (min)	Sedimentation coefficient	hRf-value on ECTEOLA-cellulose
0	24.2	0
5	12.1	46
10	11.5	53
15	10.1	54
30	6.4	61
60	5.3	76

III. Thin-Layer Electrophoresis of Hydrolysis Products of Nucleic Acids

Nucleobases and nucleosides [19] and mixtures of mononucleotides [15] can be separated by paper electrophoresis; and oligonucleotides by two-dimensional paper electrophoresis-chromatography [80] or by two-dimensional electrophoresis on cellulose acetate and DEAE-paper [84]. Chromatographic and electrophoretic techniques are complementary in that compounds which cannot be fractionated suitably by the former can usually be completely separated by the latter.

Thin-layer electrophoresis and thin-layer electrophoresis-chromatography (see Chapter E, p. 105) are also suitable for fractionating complex mixtures. Purine and pyrimidine bases, for example, can be fractionated by electrophoresis on cellulose in 0.05 M formate buffer, pH 3.4 at 0° C

(1500 V, 25 mA), followed by TLC with methanol-conc. hydrochloric acid-water (64 + 17 + 18) [37]. The bases may be eluted and quantitatively determined by UV-spectrophotometry; about three hours is required for determining the composition of the bases in a deoxyribonucleic acid [37].

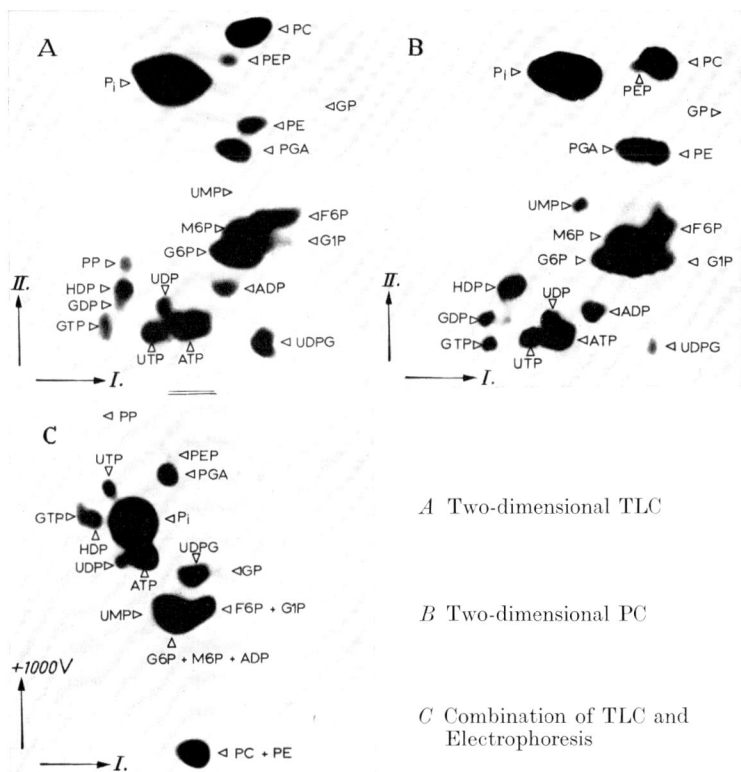

Fig. 226. Comparison of the separation of ^{32}P-labelled nucleotides and other phosphate esters from plant tissue of *Brassica pekinensis* [4]
A Two-dimensional TLC; layer: cellulose-MN-300; solvents: 1. direction: n-propanol-conc. ammonium hydroxide-water (60 + 30 + 10, +2.0 g EDTA per l), twice developed; 2. direction: n-propyl acetate-90% formic acid-water (55 + 25 + 15), twice developed. Times of run: twice 180 min (1 direction) and twice 80 min (2. direction); visualised by autoradiography. *B* Two-dimensional paper chromatography with the solvents given under *A*. *C* TLC-electrophoresis; layer: cellulose MN 300; solvents: n-propanol-conc. ammonium hydroxide-water (60 + 30 + 10, +2.0 g EDTA per l), twice developed; electrophoresis: 1000 V, 35 mA, 0.28 M ammonium acetate buffer (0.1 g EDTA per l), pH 3.6; 16 min

Ribo- and deoxyribonucleosides can be separated [37] on cellulose by electrophoresis and subsequent chromatography with saturated ammonium sulphate-M sodium citrate-isopropanol (80 + 18 + 2) [45].

51*

Complete separation of the mononucleotides in alkali hydrolysates of ribonucleic acids is possible in 75 min through electrophoresis (450 V) on a cellulose layer, sprayed with sodium formate buffer of pH 3.4 [17] (see also [97]); 10–20 h are needed for paper electrophoresis of these compounds [15]. Mixtures of mononucleotides have also been separated by two-dimensional thin-layer electrophoresis-chromatography on cellulose [3].

Thin-layer electrophoresis is suitable for fractionating mono-, di- and triphosphates [90]. Fig. 226 compares the separations achieved by two-dimensional TLC on a cellulose layer (226 A), two-dimensional PC (226 B) and thin-layer electrophoresis-chromatography on a cellulose layer (226 C); experimental details are in the legend to Fig. 226.

It has recently been shown that complex mixtures of oligonucleotides also can be separated by thin-layer electrophoresis. One may anticipate still further development of this method.

Bibliography for Chapter W. Nucleic Acids and Nucleotides

1. ANDERSON, N. G., J. G. GREEN, M. L. BARBER, and (Sister) F. C. LADD: Anal. Biochem. **6**, 153 (1963).
2. BAUER, R. D., and K. D. MARTIN: J. Chromatog. **16**, 519 (1964).
3. BERGQUIST, P. L.: Biochim. Biophys. Acta **103**, 347 (1965).
4. BIELESKI, R. L.: Anal. Biochem. **12**, 230 (1965).
5. BOERNIG, H., u. C. REINICKE: Acta Biol. Med. Ger. **11**, 600 (1963).
6. BOVÉ, J. M.: Bull. soc. chim. Biol. **45**, 421 (1963).
7. BROOM, A. D., L. B. TOWNSEND, J. W. JONES, and R. K. ROBINS: Biochemistry **3**, 494 (1964).
8. CANTONI, G. L., and D. R. DAVIS: Procedures in Nucleic Acid Research. New York-London: Harper and Row, Publishers 1966.
9. CHARGAFF, E., and J. N. DAVIDSON (Editors): The Nucleic Acids, Chemistry, and Biology, Vols. I, II, III. New York-London: Academic Press 1955, 1960.
10. CHMIELEWICZ, Z. F., and M. ACARA: Anal. Biochem. **9**, 94 (1964).
11. COFFEY, R. G., and R. W. NEWBURGH: J. Chromatog. **11**, 376 (1963).
12. COHN, W. E.: In: Chromatography, 2. Edition, p. 627, E. HEFTMANN, Editor. New York: Reinhold Publishers 1967.
13. COLOWICK, S. P., and N. O. KAPLAN (Editors): Methods in Enzymology, Vol. 12. New York-London: Academic Press 1967.
14. DAVIDSON, I. W. F., and W. G. DREW: J. Chromatog. **21**, 319 (1966).
15. DAVIDSON, J. N., and R. M. S. SMELLIE: Biochem. J. **52**, 594 (1952).
16. — and W. E. COHN (Editors): Progress in Nucleic Acid Research and Molecular Biology, Vols. 1—8. New York-London: Academic Press 1962—1968.
17. deFILIPPES, F. M.: Science **144**, 1350 (1964).
18. DIETRICH, C. P., S. M. C. DIETRICH, and H. G. PONTIS: J. Chromatog. **15**, 277 (1964).
19. DISCHE, Z., u. K. SCHWARZ: Microchim. Acta **2**, 13 (1937).
20. DYER, T. A.: J. Chromatog. **11**, 414 (1963).
21. v. EULER, H., och L. HAHN: Svensk Kem. Tidskr. **58**, 251 (1964).
22. FAHN, S., R. W. ALBERS, and G. J. KOVAL: Anal. Biochem. **10**, 468 (1965).

23. GARRETT, E. R., T. SUZUKI, and D. J. WEBER: J. Am. Chem. Soc. **86**, 4460 (1964).
24. GEBICKI, J. M., and S. FREED: Anal. Biochem. **14**, 253 (1966).
25. GERBER, N. N.: J. Med. Chem. **7**, 204 (1964).
26. GRIPPO, P., M. IACCARINO, M. ROSSI, and E. SCARANO: Biochim. Biophys. Acta **95**, 1 (1965).
27. HANSBURY, E., and D. G. OTT: Los Alamos Scientific Laboratory, Biol. and Med. Res. Group (H-4), Annual Report 1961—1962, LAMS-2780, p. 268.
28. HARBERS, E., G. E. DOMAGK u. W. MÜLLER: Die Nucleinsäuren. Eine einführende Darstellung ihrer Chemie, Biochemie und Funktionen. Stuttgart: Georg Thieme 1964.
29. HASHIZUME, T., and Y. SASAKI: J. Agr. Biol. Chem. (Tokyo) **27**, 881 (1963).
30. HAWRYLYSHYN, M., B. Z. SENKOWSKI, and E. G. WOLLISH: Microchem. J. **8**, 15 (1964).
31. HOLDGATE, D. P., and T. W. GOODWIN: Biochim. Biophys. Acta **91**, 328 (1964).
32. HOLLEY, R. W., and S. H. MERRILL: J. Am. Chem. Soc. **81**, 753 (1959).
33. — J. APGAR, P. B. DOCTOR, J. FARROW, M. A. MARINI, and S. H. MERRILL: J. Biol. Chem. **236**, 200 (1961).
34. JACOBSON, K. B.: Science **138**, 515 (1962).
35. — J. Chromatog. **14**, 542 (1964).
36. JOSEFSSON, L.: Biochim. Biophys. Acta **72**, 133 (1963).
37. KECK, K., u. U. HAGEN: Biochim. Biophys. Acta **87**, 685 (1964).
38. KESSELRING, K., u. G. SIEBERT: Hoppe-Seyler's Z. physiol. Chem. **337**, 79 (1964).
39. KLOUWEN, H. M., and H. WEIFFENBACH: J. Chromatog. **7**, 45 (1962).
40. LETHAM, D. S.: J. Chromatog. **20**, 184 (1965).
41. LEVENE, P. A., u. H. MANDEL: Ber. dtsch. chem. Ges. **41**, 1905 (1908).
42. LOHRMANN, R., and H. G. KHORANA: J. Am. Chem. Soc. **86**, 4188 (1964).
43. LOMAKINA, T. S., L. I. GUSKOVA, and N. I. GRINEVA: Khim. Prirodn. Soedin. Akad. Nauk. Uz. SSSR **1965**, 335.
44. MAHAPATRA, G. N., and O. M. FRIEDMAN: J. Chromatog. **11**, 265 (1963).
45. MARKHAM, R., and J. D. SMITH: Biochem. J. **49**, 401 (1951).
45a. MARMUR, J.: J. Mol. Biol. **3**, 208 (1961).
46. MARSHAK, A., and H. J. VOGEL: J. Biol. Chem. **189**, 597 (1951).
47. MASSIE, H. R., and B. H. ZIMM: Proc. Natl. Acad. Sci. U.S. **54**, 1641 (1965).
48. MICHELSON, A. M.: J. Chem. Soc. **1959**, 1371.
49. MORTON, M. J., and W. I. ROGERS: Anal. Biochem. **13**, 108 (1965).
50. NAYAR, M. N. S.: Life Sci. **3**, 1307 (1964).
51. NEUHARD, J., E. RANDERATH, and K. RANDERATH: Anal. Biochem. **13**, 211 (1965).
52. PANTELEEVA, N. S.: Vestn. Leningr. Univ. **19**, Ser. Biol. **2**, 73 (1964).
53. PATAKI, G., and A. KUNZ: J. Chromatog. **23**, 465 (1966).
54. PETERSON, E. A., and H. A. SOBER: J. Am. Chem. Soc. **78**, 751 (1956).
55. RANDERATH, E., and K. RANDERATH: J. Chromatog. **10**, 509 (1963).
56. — — J. Chromatog. **16**, 126 (1964).
57. — — Anal. Biochem. **12**, 83 (1965).
58. RANDERATH, K.: Angew. Chem. **73**, 436 (1961).
59. — Angew. Chem. **73**, 674 (1961).
60. — Biochem. Biophys. Res. Communs. **6**, 452 (1961/62).
61. — u. H. STRUCK: J. Chromatog. **6**, 365 (1961).
62. — and F. CRAMER: Biochim. Biophys. Acta **61**, 346 (1962).
63. — Biochim. Biophys. Acta **61**, 852 (1962).
64. — Angew. Chem. **74**, 484 (1962); Internat. Edit. Engl. **1**, 435 (1962).
65. — Angew. Chem. **74**, 780 (1962); Internat. Edit. Engl. **1**, 553 (1962).

66. RANDERATH, K.: Nature **194**, 768 (1962).
67. — J. Chromatog. **10**, 235 (1963).
68. — and G. WEIMANN: Biochim. Biophys. Acta **76**, 129 (1963).
69. — Biochim. Biophys. Acta **76**, 622 (1963).
70. — and E. RANDERATH: J. Chromatog. **16**, 111 (1964).
71. — Experientia **20**, 406 (1964).
72. — u. E. RANDERATH: Angew. Chem. **76**, 494 (1964).
73. — — Anal. Biochem. **13**, 575 (1965).
74. — Nature **205**, 908 (1965).
75. — and E. RANDERATH: J. Chromatog. **22**, 110 (1966).
76. — C. M. JANEWAY, M. L. STEPHENSON, and P. C. ZAMECNIK: Biochem. Biophys. Res. Communs. **24**, 98 (1966).
77. — private communication, 1966.
78. RATAPONGS, C.: Naturwissenschaften **53**, 252 (1966).
79. REINAUER, H., and F. H. BRUNS: J. Chromatog. **19**, 453 (1965).
80. REMENCHIK, A. P., and I. BERNSOHN: Anal. Biochem. **18**, 1 (1967).
81. RINK, M., u. A. GEHL: J. Chromatog. **21**, 143 (1966).
82. RUSHIZKY, G. W., and C. A. KNIGHT: Virology **11**, 236 (1960).
83. SANDER, E. G., D. B. McCORMICK, and L. D. WRIGHT: J. Chromatog. **21**, 419 (1966).
84. SANGER, F., G. G. BROWNLEE, and B. G. BARRELL: J. Mol. Biol. **13**, 373 (1965).
85. SAUKKONEN, J. J.: Chromatog. Revs. **6**, 53 (1964).
86. SCHEIG, R. L., A. ANNUZIATA, and L. A. PESCH: Anal. Biochem. **5**, 291 (1963).
87. SCHEIT, K. H.: Biochim. Biophys. Acta **134**, 217 (1967).
88. SCHMIDT, G., and S. J. THANNHAUSER: J. Biol. Chem. **161**, 83 (1945).
89. SCHWARTZ, A. N., A. W. G. YEE, and B. A. ZABIN: J. Chromatog. **20**, 154 (1965).
90. SCHWEIGER, A., u. H. GÜNTHER: J. Chromatog. **19**, 201 (1965)
91. SHASHA, B., and R. L. WHISTLER: J. Chromatog. **14**, 532 (1964).
92. SIGNER, R., u. H. SCHWANDER: Helv. Chim. Acta **33**, 1521 (1950).
93. STAEHELIN, M,: Biochim. Biophys. Acta **49**, 11 (1961).
94. STICKLAND, R. G.: Anal. Biochem. **10**, 108 (1965).
95. TAMM, C., M. E. HODES, and E. CHARGAFF: J. Biol. Chem. **195**, 49 (1952),
96. — H. S. SHAPIRO, R. LIPSHITZ, and E. CHARGAFF: J. Biol. Chem. **203**, 673 (1953).
97. TOMETSKO, A. M., and N. DELIHAS: Anal. Biochem. **18**, 72 (1967).
98. VENKSTERN, T. V., and A. A. BAEV: Absorption Spectra of Minor Bases. Their Nucleosides, Nucleotides, and Selected Oligoribonucleotides. New York: Plenum Press 1966.
99. VISCHER, E., and E. CHARGAFF: J. Biol. Chem. **176**, 715 (1948).
100. VOLKIN, E., and C. E. CARTER: J. Am. Chem. Soc. **73**, 1516 (1951).
101. WARING, P. P., and Z. Z. ZIPORIN: J. Chromatog. **15**, 168 (1964).
102. WEBER, C. J.: J. Biol. Chem. **86**, 217 (1930).
103. WEIMANN, G., and K. RANDERATH: Experientia **19**, 49 (1963).
104. WIELAND, T., G. LÜBEN u. H. DETERMANN: Experientia **18**, 430 (1962).
105. WYATT, G. R.: Biochem. J. **48**, 584 (1951).
106. ZACHAU, H. G.: Hoppe-Seyler's Z. physiol. Chem. **342**, 98 (1965).
107. — M. TADA, W. B. LAWSON, and M. SCHWEIGER: Biochim. Biophys. Acta **53**, 221 (1961).
108. ZAMECNIK, P. C., M. L. STEPHENSON, C. M. JANEWAY, and K. RANDERATH: Biochem. Biophys. Res. Communs. **24**, 91 (1966).
109. ZARNACK, J., u. S. PFEIFER: Pharmazie **19**, 216 (1964).
110. ZIMMERMAN, M., and D. HATFIELD: Biochim. Biophys. Acta **91**, 326 (1964).

111. Ziporin, Z. Z., and R. W. Hanson: Anal. Biochem. **14**, 78 (1966).
112. Zubay, G.: J. Mol. Biol. **4**, 347 (1962).

X. Sugars and Derivatives

B. A. Lewis and F. Smith

I. Introduction

Thin-layer chromatography affords a simple, rapid and sensitive method for the qualitative and quantitative analysis of low molecular weight sugars and their derivatives. In the few years since Stahl and Kaltenbach [1] extended TLC to the separation of sugar mixtures, the method has been adopted for a wide spectrum of analytical problems. This chapter considers two classes of compounds: the hydrophilic carbohydrates and their derivatives which are hydrophobic in nature.

Carbohydrates, being strongly hydrophilic, require polar solvent systems. These solvents have relatively slow migration rates requiring from 0.5 to 3 hours for one ascent of the plate. A limited resolution of the simple sugars is attained on silica gel G. The hR_f-values decrease as the number of hydroxyl groups and the molecular weight increase (pentose > hexose > disaccharide). Impregnating the silica gel and kieselguhr layers with salts such as sodium acetate enhances the resolution and simple mixtures of the sugars can be resolved [1, 2]. Cellulose thin layers [3, 4] offer the resolution of paper chromatography combined with the speed and sensitivity of TLC and it is anticipated that this method will assume more importance for the simple sugars.

A significant feature of inorganic layers is that organic compounds can be detected with sulfuric acid. TLC is of particular value, therefore, for derivatized carbohydrates which are not readily detected by conventional chromatography. The highly substituted carbohydrates migrate rapidly on silica gel with non-polar solvents. By this means, reactions in carbohydrate chemistry involving the preparation or transformation of derivatives can be followed with facility and, as a consequence of the sensitivity of the technique, traces of side products or impurities can be detected. A survey of the literature suggests that TLC is now routine procedure for examining the purity of synthetic carbohydrate compounds.

II. Preparation of Plates

Several adsorbents have been found to give satisfactory resolution of the sugars and their derivatives on thin-layer plates. A brief description of the preparation of these plates follows.

1. Silica Gel G and Kieselguhr G Layers

A slurry of silica gel G according to Stahl[1] (30 g) in water (60 ml) is applied to 5 plates (20 × 20 cm) by means of the Desaga applicator to give a layer 250 µ thick. The plates are dried at 110° for 30 min prior to use. Kieselguhr G[1] layers are prepared in the same way.

The carbohydrate mixture to be chromatographed is dissolved in water or other suitable solvent and applied by means of a capillary to the starting point 15 mm from the end of the plate. For kieselguhr G layers the optimum quantity of each sugar for application is 0.5–2 µg per spot, whereas with silica gel G from 5–50 µg can be applied.

2. Kieselguhr G, Impregnated with Sodium Acetate [1]

Kieselguhr G (30 g) is mixed with 0.02 M sodium acetate (60 ml) and the plates are prepared as in [1]. An amount of 0.5–2 µg of each sugar is applied to the origin.

3. Kieselguhr G, Impregnated with pH 5 Phosphate Buffer [5]

Kieselguhr G (20 g) is mixed with 40 ml of pH 5 phosphate buffer (prepared by mixing equal volumes of 0.1 M phosphoric acid and 0.1 M disodium hydrogen phosphate). The plates are spread to a thickness of 250 µ and dried overnight in the air. From 5 to 25 µg of carbohydrate is applied to the starting point.

4. Impregnated Silica Gel G Layers

The plates are prepared as described in [1] using 60 ml of the following solutions instead of water:

a) 0.02 M sodium borate buffer, pH 8.0 (100 ml of 0.02 M boric acid and 3 ml of 0.02 M sodium tetraborate) [6].

From 1–30 µg of each sugar is applied to the origin.

b) 0.02 M sodium acetate [2].
c) 0.02 M boric acid [7] or 0.1 M boric acid [8].
d) 0.1 M sodium bisulfite [9].

5. Cellulose Layers (a) Cellulose MN 300 [3]

Cellulose MN 300[2] powder (15 g) is homogenized in a blender with distilled water (90 ml) for 30 sec [3]. The suspension is spread onto the plate to a thickness of 250 µ with the Desaga apparatus and the moist plate is dried at 100° for 10 min. The addition of calcium sulphate or other

---

[1] Firm 88.
[2] Firm 83.

binders is not necessary for durable layers; however, a slightly faster development time is obtained with calcium sulfate-bound cellulose layers [10].

(b) "Avirin" ("Avicel")[3] [4]

The microcrystalline cellulose ("Avirin" or "Avicel", 100 g) is blended in a mixer for 15–45 sec with distilled water (430 ml). The amount of water required may vary with the particular lot of cellulose. An insufficient amount of water results in cracking of the layer. The cellulose suspension is transferred to a filter flask and placed under vacuum for a few min to remove air bubbles. The plates are spread by means of a Desaga apparatus (Firm 44) (drawn slowly) to a thickness of 1.0 mm and dried overnight at room temperature or at 80° for 30–60 min. Thin layers (250 μ) are not as suitable for either analytical or preparative work. The performance of the plates is also influenced by the time of blending and for good smooth plates the mixture should "peak" somewhat.

Cellulose layers are very durable and the plates may be stacked together for storage or written on without crumbling. They need not be desiccated.

6. ECTEOLA-Cellulose Layers for Sugar Phosphates [11]

A suspension of sieved ECTEOLA-cellulose powder (Firm 127) (2 g) in 0.004 M ethylenediaminetetraacetic acid, pH 7.0 (18 ml) is shaken vigorously for 5 min. The slurry is spread uniformly on a glass plate (20 × 20 cm) and dried overnight at room temperature giving a layer about 200 μ thick. The dried layer is then sprayed with 0.1 M ammonium tetraborate, pH 9.0, and dried at 50° for 30 min. No changes are found on storing the plates for 3 days.

7. Celite Layers (a) Filter-Cel[4] and Hyflo Super-Cel[4] [12]

Calcium sulfate hemihydrate (0.8 g) is ground for 1 min in a mortar with distilled water (5 ml). The Celite adsorbent (Filter Cel or Hyflo Super-Cel) (15 g) and additional water (60 ml) are added and the mixture is ground for 1–2 min. The smooth slurry is spread with an applicator in the usual way to give five 8 × 8 in (20 × 20 cm) plates which are air-dried for 10–15 min and then dried at 100° for 30 min.

(b) Celite 535[4]-Starch [13]

A suspension of 200-mesh Celite 535 (18 g) in 0.25 N sodium hydroxide is mixed with a suspension of powdered potato starch (1.5 g) in water

---

[3] "Avirin" is a microcrystalline cellulose supplied by Firm 5. Avicel is the pharmaceutical product for the same material.

[4] Firm 77.

(20 ml) and homogenized for 1—2 min. The mixture is applied to glass plates (20 × 20 cm) to a thickness of 500 μ. The plates are dried at 25° for 16 hr.

III. Visualization

The use of a non-specific destructive reagent such as sulfuric acid in conjunction with inorganic layers for TLC permits the detection of highly substituted sugar derivatives which cannot be visualized by other methods. Sulfuric acid alone or admixed with nitric acid [14] or permanganate [15] is used. The air-dried plate is sprayed with concentrated sulfuric acid (5—50% acid in alcohol or water is also used) and heated at 100—150° for 5—10 min to char the organic substances. Sugar alcohols, glyconic acids and the inositols are less sensitive to this reagent than the reducing sugars and their derivatives [16]. It is frequently advantageous to detect these polyols with alkaline permanganate (Rgt. No. 198).

Table 197. *Color reactions of sugars on thin-layers*

Sugar	Anisaldehyde-sulfuric acid [a] (No. 11) [1]	Naphthol-resorcinol-sulfuric acid [b] (No. 175) [8]	Aniline-diphenylamine [c] (No. 8) [7]
D-Digitoxose	blue	—	—
L-Rhamnose	green	green	pale green
D-Ribose	blue	—	—
D-Xylose	grey	light blue	bright blue
L-Arabinose	yellow-green	blue-green	bright blue
L-Sorbose	violet	red	—
D-Fructose	violet	red-black	red scarlet
D-Mannose	green	light blue	—
D-Glucose	light blue	blue-violet	grey-green
D-Galactose	green-grey	blue-violet	grey-green
Sucrose	violet	red	lilac
Maltose	violet	—	—
Lactose	greenish	red-violet	blue violet
D-Glucuronic acid	—	blue	—
D-Galacturonic acid	—	blue	—

[a] *Adsorbent:* Kieselguhr G-0.02M sodium acetate impregnated. *Solvent:* Ethyl acetate-65% isopropanol (65 + 35).
[b] *Adsorbent:* Silica gel G-0.1N boric acid impregnated. *Solvent:* Benzene-glacial acetic acid-methanol (20 + 20 + 60).
[c] *Adsorbent:* Silica gel G buffered with boric acid or sodium acetate.

The anisaldehyde-sulfuric acid reagent of STAHL (Rgt. No. 11) is extremely sensitive (0.05 μg of sugar) and gives characteristic colors for the sugars (Table 197). The reagent does not give good results with boric acid-impregnated layers. Characteristic colors are also obtained with naphtholresorcinol-sulfuric acid (Rgt. No. 175) [8] and with aniline-diphenylamine (Rgt. No. 8) [7, 17] (Table 197). Variations in color observed with reagents No. 8 and No. 175 may result from differences in TLC conditions or in the temperature used for developing the color.

α-Naphthol-sulfuric acid reagent (Rgt. No. 170) gives a blue color with most sugars [6], whereas thymol-sulfuric acid (Rgt. No. 250) gives a dark pink color [9] and carbazole-sulfuric acid (Rgt. No. 32) gives violet spots on a blue background [9]. Phenol-sulfuric acid (Rgt. No. 201) and o-aminobiphenyl-phosphoric acid (Rgt. No. 5) give brown colors [9]. Ketoses are detected preferentially with dimedon-phosphoric acid (Rgt. No. 69) [9] and 2-deoxy-sugars by the quinaldine reaction (Rgt. No. 59) [2]. The ketoses appear grey-yellow with Reagent 69 under daylight and show strong dark pink fluorescence under UV. Reagents 5, 69, 201 and 250 detect as little as 0.1 µg of sugar [9] and under UV light 0.05–0.25 µg of 2-deoxy-sugars are detected with Reagent 59 [2].

p-Anisidine phthalate (Rgt. No. 12) gives characteristic colors on cellulose thin-layers: hexoses (green), 6-deoxy-hexoses (yellow-green), pentoses (red-violet) and uronic acids (brown) [3]. The reagent is very sensitive, detecting 0.5 µg of hexoses and 0.1–0,2 µg of pentoses and uronic acids.

Other reagents suitable for paper chromatography are also applicable to TLC (Nos. 10, 21, 134, 159, 178, 234, 255) and some other reagents containing periodic acid. Of these, iodine vapor [18] has the most universal application since it detects a variety of carbohydrate derivatives including the free sugars, partially and fully substituted methyl and benzyl ethers, esters and acetals. Although iodine is less sensitive than sulfuric acid, it is non-destructive in the short exposure time required (5–20 min) and can therefore be used to detect components on preparative plates or for quantitative analysis. The adsorbed iodine disappears when the plate is exposed to the air.

IV. Chromatography of Sugars and Derivatives

Specific conditions for TLC of sugars and their common derivatives are described below. The unsubstituted sugars, sugar alcohols and acids which require polar solvents are considered first. Polar solvent systems are relatively slow and cause some flaking of the layers but this is not a serious problem. Since silica gel layers dehydrate these solvents, repeated use results in a marked lowering of the hR_f-values.

Derivatized carbohydrates possessing hydrophobic substituents are considered in the final sections. These compounds migrate rapidly on silica gel G with relatively non-polar solvents and resolution of α,β-anomers and pyranose and furanose sugars is observed.

1. Sugars

The mobility of the sugars on silica gel depends primarily on the molecular weight and the number of hydroxyl groups and consequently the diastereoisomers are poorly resolved [16, 19]. Resolution is improved by impregnating silica gel G and kieselguhr G with salts of weak acids or by the use of cellulose layers.

A good resolution of the deoxy-sugars rhodinose (hR_f 73), D-digitoxose (hR_f 63) and 2-deoxy-L-fucose (hR_f 45) has been reported [20] on silica gel G layers with chloroform-acetone (50 + 50).

a) Separation of Sugars on Buffered Kieselguhr G Layers

STAHL and KALTENBACH [1] observed a marked improvement in resolution of simple sugars on kieselguhr G, weakly buffered with 0.02 M sodium acetate. A mixture of eight simple sugars was resolved in 25 to 30 min with solvent I as shown in Table 198. With acetone as solvent, glucose and arabinose have hR_f-values of 50 and 71, respectively [21].

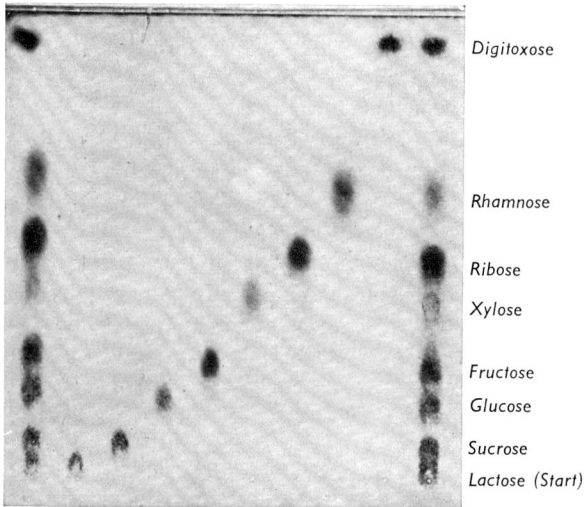

Fig. 227. Separation of sugars (0.5 µg of each) on a buffered silica gel G layer. Visualization with anisaldehyde-sulphuric acid [1]

Other solvent systems are also satisfactory for the sugars [17]. Acetate-buffered kieselguhr G layers have an optimum capacity of 0.5—2.0 µg of each sugar [1].

WALDI [5] and KRINGSTAD [22] have shown that impregnating kieselguhr G layers with 0.1 M sodium dihydrogen phosphate buffer (pH 5) increases the capacity of the layer to about 25 µg of each sugar.

Table 198. hRf-values of sugars on buffered layers [1, 5, 7, 8, 9]

Sugar	Kieselguhr G		Silica Gel G					
	Sodium acetate	Sodium phosphate, pH 5	Boric acid	Sodium acetate			Sodium bisulfite	
	I	II	III	IV	V	VI	VII	VIII
Rhamnose	62	93	47	54	71	61	57	62
Ribose	49	75	—	—	—	—	50	57
Xylose	39	73	40	46	65	55	34	59
Arabinose	28	65	37	41	53	46	32	51
Sorbose	26	68	—	—	—	—	43	47
Fructose	25	60	14	30	47	34	28	48
Mannose	23	63	—	—	—	—	41	53
Galactose	18	36	25	27	45	35	32	39
Glucose	17	55	20	37	55	42	28	48
Sucrose	8	40	—	19	48	24	20	40
Maltose	6	30	—	—	—	—	11	35
Lactose	4	17	6	12	29	13	8	23
Trehalose	—	23	—	—	—	—	5	23
Raffinose	—	5	—	—	—	—	4	13

Solvents: I—VIII (see Table 216 for composition). Ca. 30 min run for I—V; ca. 40 min for VI.

The sugars migrate rapidly in solvent II and a good resolution is obtained in one development [5] (see Table 198). The phosphate buffer in the solvent may be replaced by water with little effect on the hRf-values [5, 22].

A mixture of kieselguhr G, aluminum oxide G and polyacrylonitrile also serves as an effective adsorbent [23].

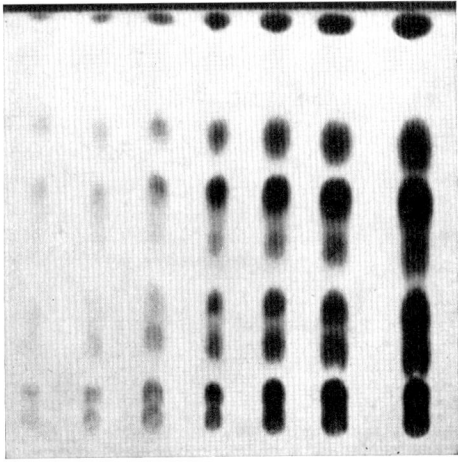

Fig. 228. Ascertaining the optimum amount for application. Increasing amounts of the sugar mixture (Fig. 227) were applied from left to right [1]

b) Separation of Sugars on Buffered Silica Gel G

The resolution of simple sugars on silica gel G is enhanced by impregnating the plates with 0.02 M sodium acetate [7], 0.02 M [7] and 0.1 N [8, 24] boric acid, 0.02 M sodium borate [6] or pH 8 phosphate buffer [25]. The preparation of these plates is described in section II. Table 198 summarizes the hR_f data obtained with a variety of solvents on buffered silica gel G layers.

In addition, an excellent resolution of 2-deoxy-sugars is obtained with solvent I on silica gel G buffered with sodium acetate. hR_f-Values: 2-deoxy-galactose (24), 2-deoxy-glucose (32—35) and 2-deoxy-ribose (43—45) [2]. The 2-deoxy sugars are readily detected as green-yellow fluorescent spots by the quinaldine reaction (Rgt. No. 59) which is more sensitive on thin-layer plates (limit of sensitivity 0.05—0.25 μg) than on paper.

Chromatography on acetate and borate-buffered silica gel G layers [26] has been used for the identification of sugars in blood and urine.

ADACHI [9] examined silica gel G impregnated with 0.1 M sodium bisulfite as an adsorbent for thin-layer chromatography of sugars (Table 198) and has reported hR_f-values for the sugars, including various ketoses, in several solvent systems.

WEICKER and BROSSMER [27] have observed the formation of ninhydrin-positive spots during thin-layer chromatography of pure sugars on silica gel G with solvents containing ammonium hydroxide.

c) Separation of Sugars on Cellulose

Although TLC of simple sugars on cellulose layers has had limited use, it is apparent [3] that the solvent systems and spray reagents most suitable for paper chromatography are directly applicable to cellulose thin-layer chromatography. The separations are considerably faster than those attained by paper chromatography and smaller amounts of sugars can be detected. In general much longer development times are required on cellulose layers than on buffered silica gel G or kieselguhr G. An advantage for preparative work is the greater capacity of cellulose compared with inorganic adsorbents.

Excellent resolutions have been obtained on either cellulose MN 300 [3] or "Avirin" cellulose [4] with solvent systems previously used for paper chromatography. The data for the common sugars are summarized in Table 199. Where the sugars have low hR_f-values multiple development is required (solvents IX and X) or the plate can be modified for continuous flow [4].

Multiple development (2 ascents) in solvent IX [3] or in solvent X [10] gives a good separation of the common sugars in about 4 hrs and 6 hrs, respectively.

Table 199. *TLC of sugars on cellulose layers* [3, 4, 10]

Sugar	Cellulose MN 300			Avirin cellulose	
	hR_{St}[a] IX	X	hRf XI	hRf XII	XIII
L-Rhamnose	152	—	—	60	46
D-Ribose	142	191	57	59	39
D-Lyxose	—	170	46	—	—
D-Xylose	125	160	41	52	33
L-Arabinose	111	151	51	46	31
D-Fructose	—	130	47	—	29
L-Sorbose	—	123	37	—	—
D-Mannose	109	123	40	44	30
D-Glucose	100	100	35	39	25
D-Galactose	90	91	40	36	21
Sucrose	—	65	37	—	—
Maltose	—	38	34	29	15
Cellobiose	—	32	32	25	13
Lactose	—	26	37	—	—
Total time	4 hr (2 ascents)	6 hr (2 ascents)	6 hr	1.75 hr (13.3 cm)	3 hr (14.2 cm)

Solvents: IX—XIII: see Table 216 for composition.

[a] hR_{St} = hRf relative to glucose after 2 ascents.

d) Chromatography of Sugars on Miscellaneous Adsorbents

Various inexpensive adsorbents have been examined briefly for their ability to resolve sugars on thin-layer plates. Some of these show promise for hydrophilic compounds but as yet there are few reports describing their use. TORE [28] obtained a resolution of simple sugars on layers of Silene E. F. (Firm 40) (a hydrated calcium silicate) and Silene E. F.-Celite 535 mixtures. Magnesium silicate layers have been used [29] as well as gypsum [30].

SHASHA and WHISTLER [13] prepared an inexpensive coating of Celite 535 utilizing potato starch as a binder. Chromatoplates prepared with the Celite 535-starch coating gave an excellent resolution of a simple mixture of L-rhamnose (hRf 92), D-ribose (hRf 50), D-fructose (hRf 31), D-glucose (hRf 28) and maltose (hRf 12) with isopropanol-water (90 + 10) (solvent XIV).

A comparison of the chromatographic resolution of three simple sugars on thin-layers of Filter-Cel and Hyflo Super-Cel is shown in Table 200 [12]. With solvent XV a good separation was obtained on the Filter-Cel layers. The development time is decreased by incorporation of Hyflo Super-Cel and the resulting mixture still has good resolving properties.

Table 200. *Comparison of Celite adsorbents for TLC of carbohydrates* [12]

Compound	hRf-values with solvent XV (see Table 216)			
	Filter-Cel	Hyflo-Super-Cel	Hyflo Super-Cel: Filter-Cel (6 + 4)	Kieselguhr G
Xylose	85	94	90	95
Glucose	67	94	76	95
Maltose	48	84	51	83
Development time in min (8—10 cm)	80	30	50	50

2. Oligosaccharides

Solvents II, V, VIII, X and XII described previously for the monosaccharides are also satisfactory for smaller oligosaccharides. The lower molecular weight maltodextrins (DP 2—6)[5] are resolved on kieselguhr G plates with solvent XVI (Table 201) whereas the higher maltodextrins (DP 5—10) have good mobility in n-butanol-ethanol-water (50 + 30 + 20) [31]. Mixtures of n-butanol-pyridine-water are effective for the intermediate dextrins (Table 201). The mobilities of the higher oligosaccharides are increased by decreasing the butanol and increasing the water content of the solvent. The differences in mobility primarily reflect differences in molecular weight.

Table 201. *TLC of maltodextrins on kieselguhr G* [31]

Solvent	hRf							
	Glucose	DP of maltodextrins						
		2	3	4	5	6	7	8
XVI	88	76	55	32	13	5	—	—
XV	94	84	73	54	33	19	11	—
XVII	95	88	74	64	51	41	29	18
Solvents: See Table 216.								

KOLLER and NEUKOM [32] analyzed enzymic hydrolyzates of pectin directly on silica gel G using butanol-formic acid-water (33 + 50 + 17) with no interference from acetate, phthalate and phosphate ions. The following hRf values were obtained for the homologous series of α 1 → 4 linked oligosaccharides: D-galacturonic acid (43), digalacturonic acid (34), trigalacturonic acid (26), tetragalacturonic acid (21) and pentagalacturonic acid (16).

---
[5] DP = Degree of polymerisation (average).

Silica gel G layers have been used also for following enzymic transglycosylation reactions [33]. The α- and β-Schardinger dextrins are differentiated on silica gel G with n-butanol-glacial acetic acid-water-pyridine-N,N-dimethylformamide (37 + 19 + 6 + 13 + 25) in which the β-cyclodextrin has hR_f 50 whereas the α-dextrin remains at the origin [34]. Additional examples of thin-layer resolutions of oligosaccharides are worthy of note [17, 19, 24, 35, 36]. PREY et al. [17] used a layer of silica gel G and kieselguhr G (1 + 4) buffered with 0.02 M sodium acetate and the solvent system ethyl acetate-methanol-water (68 + 23 + 9) to separate the following oligosaccharides: sucrose (hR_f 48), maltose (42), lactose (32), melizitose (28), melibiose (28) and raffinose (21).

3. Amino Sugars

The most effective separations of amino sugars have been obtained with cellulose thin-layers. FAILLARD and CABEZAS [37] report the separation of N-acetylneuraminic acid from the N-glycolyl derivative on cellulose MN 300 with the solvent systems ethanol-water-concentrated ammonium hydroxide (80 + 20 + 1) (hR_f 37 and 30 respectively) and n-butanol-propanol-0.1 N hydrochloric acid (25 + 50 + 25) (hR_f 49 and 39 respectively). Glucosamine and muramic acid are resolved on the same solid phase by double development with butanol-pyridine-glacial acetic acid-water (60 + 45 + 4 + 30) [38].

A complete resolution [39] of glucosamine, galactosamine and their N-acetyl derivatives was obtained by two-dimensional chromatography on cellulose MN 300 using solvents XII, XVIII, XIX and XX (see Table 202).

Table 202. *TLC of amino sugars on cellulose MN 300* [39]

Amino sugar	hR_{St}-values [a]		
	Solvent: XII	XIX	XX [b]
2-Amino-2-deoxy-D-glucose	100	100	100
2-Amino-2-deoxy-D-galactose	83	88	—
2-Acetamido-2-deoxy-D-glucose	162	182	295
2-Acetamido-2-deoxy-D-galactose	153	170	165
Solvents: See Table 216.			

[a] hR_{St} = hR_f relative to glucosamine after two ascents. Each ascent requires 2—3 hrs.
[b] Layer had been sprayed with borate buffer and dried prior to use.

For good resolution two ascents in solvent XII or XVIII followed by two ascents in solvent XIX or XX are required. After the second ascent in solvent XII or XVIII, the plate is sprayed with borate buffer of pH 8.0

(0.2 M boric acid, 0.05 M NaCl and 0.05 M sodium tetraborate) and dried prior to chromatography in solvent XIX or XX. The total time required for the two-dimensional chromatography is 6–8 hrs. Table 202 gives the hR_{St}-values for these hexosamines.

"Avirin" cellulose is also effective for the resolution of these common amino sugars. The mobilities of several amino sugars and their derivatives in solvent XII are given in Table 203 [4].

Table 203. hRf-values for amino sugars and their derivatives on "Avirin" (Firm 5) cellulose [4]

Compound [a]	hRf
1,2-Diamino-1,2-dideoxy-D-glucitol [b]	11
1,2-Diamino-1,2-dideoxy-D-mannitol	12
2-Amino-2-deoxy-D-galactose	18
2-Amino-2-deoxy-D-glucose	22
3-Amino-3-deoxy-D-mannose	24
2-Amino-2-deoxy-D-ribose	25
2-Amino-2-deoxy-D-lyxose	28
2-Amino-2-deoxy-L-xylose	31
1,2-Diacetamido-1,2-dideoxy-D-glucitol	38
2-Acetamido-2-deoxy-D-glucose	54
2-Acetamido-2-deoxy-D-glucose diethyl dithioacetal	85

Solvent XII: See Table 216 for composition (15.5 cm ascent requires 115 min).

[a] Amino sugars were applied as the hydrochloride.
[b] Dihydrobromide.

The 2,4-dinitrophenyl derivatives of glucosamine and galactosamine are resolved on silica gel G layers buffered with 0.1 M potassium tetraborate (pH 10.8) with the solvent n-propanol-ethyl acetate-water (70 + 10 + 20) [40]. One ascent (10–12 cm) on a 250 µ thick layer requires 60–90 min; hRf-values: DNP-glucosamine, 46; DNP-galactosamine, 33. The DNP-derivatives are detected by their yellow color.

The ninhydrin reagent (No. 178) is the most sensitive for detection of the free amino sugars (lower limit of detection, 0.5 µg), whereas a periodic acid reagent, followed by thiobarbituric acid, although less sensitive (3 to 5 µg), gives characteristic colors (red for the N-acetyl compounds). The Elson-Morgan test can be used on cellulose MN 300 A but the characteristic color formation is hindered on cellulose MN 300 [39].

TLC on silica gel G has been useful particularly in the organic synthesis of amino sugars for the characterization of derivatives which are not readily detected on paper. It has found application in the synthesis of 4-acetamido-4-deoxy-D-*threo*furanose [41], 2,6-diamino-2,6-dideoxy-D-mannose dihydrochloride [42], 2,3-epimino-hexopyranosides [43], 2-amino-2,6-dideoxy-D-galactose [44], the 1-O-acyl-2-acylamido-

2-deoxy-D-glucopyranoses [45], and 3-O-(2-acetamido-2-deoxy-β-D-galactopyranosyl)-α-D-galactose [46].

4. Acids (Aldonic, Aldaric, Uronic and Saccharinic)

Thin-layer chromatography of a number of aldonic acids, their lactones and phenylhydrazides on silica gel G with XXI has been reported (Table 204) [16]. The mobilities of the γ-lactones have been reported also in the neutral solvent XXII (Table 204) [47]. While the γ- and δ-lactones and the free acids are well differentiated, the differences in hR_f-values between the free acids are slight [16].

Table 204. hR_f-values of aldonic acids and their derivatives on silica gel G [16, 47]

Compound	hR_f					
	XXI				XXII	XXIII
	Acid	Lactone		Hydrazide	γ-lactone	
		γ	δ			
D-Ribonic	38	61	—	—	55	60
L-Arabonic	35	64	58	62	—	—
Ca Salt	33	—	—	—	—	—
D-Gluconic	40	68	—	60	69	62
D-Galactonic	38	61	47	—	—	—
Ca Salt	38	—	—	—	—	—
D-Gulonic	31	53	43	—	32	52
L-Rhamnonic	—	64	—	—	—	—
D-Glucoheptonic	—	43	—	—	—	—

Solvents: See Table 216.

It should be noted that the aldonic acids need not be deionized prior to TLC since the same mobility is observed when either the acid or the calcium salt is applied. In each sugar acid investigated with solvent XXI, the γ-lactones had a higher hR_f than the corresponding δ-lactone and both moved faster than free acid. This relative mobility of the two lactones may be dependent on the solvent, however, since in neutral solvents the inverse order ($\delta > \gamma$) is observed [47].

WALDI [5] has separated the isomeric D-*arabo*- and D-*xylo*-hexulosonic acids on phosphate-buffered kieselguhr G with solvent II (hR_f 10 and 19, respectively).

The hR_f values of the common uronic acids and their lactones on silica gel G with solvent XXI [16] and on cellulose layers (Camag) with solvents XII and XXIV [47] are shown in Table 205.

An excellent separation of the 1,4- and 6,3-lactones of D-glucaric (glucosaccharic) acid has been observed on silica gel G (Table 205) [16].

Preparative TLC of the isomeric "α"- and "β"-D-glucoisosaccharino-1,4-lactone tribenzoates on silica gel HF_{254} with ethyl acetate-petroleum ether (b. p. 40–60°) (17 + 83) has afforded the "β"-isomer for the first time [48].

Table 205. hRf-values of uronic and aldaric acids and their lactones [16, 47]

Compound	hRf		
	Silica gel G XXI	Cellulose XII	XXIV
D-Glucuronic acid	—	29	37
D-Glucurono-γ-lactone	58	—	—
D-Galacturonic acid	32	25	33
D-Mannuronic acid	36	—	—
D-Mannurono-γ-lactone	53	—	—
D-Glucaric acid	—	23	29
D-Glucaro-1,4-lactone	43	—	—
D-Glucaro-6,3-lactone	85	—	—

Solvents: See Table 216.

5. Sugar Alcohols

WASSERMANN and HANUS [49] resolved a mixture of glucose (hR_{St} 100), glucitol (hR_{St} 51) and mannitol (hR_{St} 70), a difficult separation in paper chromatography, using an activated kieselguhr-silica gel G (60 + 40) layer and isopropanol-ethyl acetate-water (83 + 11 + 6). Glycerol (hRf 90), mannitol (hRf 52), galactitol (hRf 45) and glucitol (hRf 39) are resolved on sodium phosphate-buffered kieselguhr G with solvent II [5].

In general the chromatographic conditions used for the sugars (section IV, 1) are also suitable for the alditols which migrate in compact zones on buffered [5, 17] or non-buffered [16, 49] layers. Incorporation of boric acid into the silica gel G layers retards the migration of the sugar alcohols [50].

6. Methyl Glycosides

The chromatographic conditions used for the sugars (section IV, 1) are suitable for the methyl glycosides which migrate somewhat faster than the free sugars. It is apparent that the α and β anomers are readily differentiated and in the few glycosides examined the β anomer has the higher hRf value [4, 16].

SHASHA and WHISTLER [13] have reported a rapid (12 cm ascent in 10 min) separation of the following methyl α-glycopyranosides on Celite 535-starch with 90% aqueous butanone: rhamnose (hRf 88), xylose (hRf 78), galactose (hRf 54), glucose (hRf 50) and mannose (hRf 46).

Table 206. hR_f-values for methyl glycosides on "Avirin" cellulose [4]

Methyl glycoside	hR_f
Methyl α-D-lyxopyranoside	75
Methyl β-D-xylopyranoside	70
Methyl β-D-arabinopyranoside	65
Methyl β-D-glucopyranoside	61
Methyl α-D-glucopyranoside	57
Methyl β-D-galactofuranoside	72
Methyl β-D-galactopyranoside	56
Methyl α-D-galactopyranoside	54
Methyl β-cellobioside	44

Solvent XII: See Table 216 for composition (15.6 cm ascent requires 120 min).

Table 206 gives the mobilities of some common methyl glycosides on "Avirin" cellulose [4]. A comparable resolution on Celite 535-starch thin-layers requires much less time.

7. Sugar Phosphates

The sugar phosphates are separated on ECTEOLA-cellulose thin-layers impregnated with ammonium tetraborate (see section II, 6) with solvent XXV [11]. The mobility of the sugars is altered by preparing the solvent and plates with pH 10 ammonium tetraborate buffer. Table 207 gives the mobilities of some common sugar phosphates.

Table 207. *TLC of sugar phosphates* [11, 51]

Compound	ECTEOLA-cellulose (Firm 83) (0.1 M ammonium tetraborate)		Cellulose MN 300 (Firm 83)	
	hR_{St}-values[a]		hR_f-values	
	XXV (pH 9.0)	XXV (pH 10.0)	XXVI	XXVII
N-Acetyl-glucosamine 1-phosphate	129	137	—	—
N-Acetyl-galactosamine 1-phosphate	112	116	—	—
Glucose 1-phosphate	120	115	32	27
Mannose 1-phosphate	90	88	—	—
Galactose 1-phosphate	77	80	—	—
Mannose 6-phosphate	70	57	—	—
Fructose 6-phosphate	68	68	41	20
Fructose 1-phosphate	59	54	—	—
Glucose 6-phosphate	56	39	29	17
Fructose 1,6-diphosphate	33	20	34	13

Solvents XXV-XXVII: See Table 216 for composition. Note the differences in pH value.

[a] hR_{St} = hR_f-values of the sugar phosphates referred to inorganic phosphate (7.2 cm in 2 h).

Phosphate esters have been resolved also by two-dimensional chromatography on cellulose MN 300 layers (Avicel and other grades of cellulose gave inferior resolutions) [51]. The plates are developed for 6—8 hrs (16—18 cm ascent) with solvent XXVI in the direction of slurry application, air-dried overnight, and then developed in solvent XXVII at right angles to the first ascent (2—4 hr for 14—18 cm ascent). The mobilities of the sugar phosphates are given in Table 207 [51]. TLC has also been used in the preparation of 2-deoxy-D-glucopyranoside 6-phenyl phosphate [52].

For visualization of the sugar phosphates, the plates are sprayed successively with benzidine trichloroacetate (Rgt. No. 21) to detect the hexose 6-phosphates, and with the molybdate reagent (Rgt. No. 166) which detects all phosphates. The lower limits of detection of the two reagents are 10 and 5 μmoles, respectively. Stannous chloride reagent also reveals the sugar phosphates.

8. Acetates and Benzoates

The acylated derivatives of the sugars are readily characterized by TLC on silicic acid which affords a rapid method (15—30 min) for resolving anomeric mixtures [53]. The anomers are not easily separated by paper chromatography except by a two-phase solvent system in which the paper is impregnated with a polar solvent. Chromatography on columns of inorganic adsorbents has given satisfactory resolutions of the acetates of mono- and oligo-saccharides and gas-liquid chromatography has been used for the low molecular weight sugar (or sugar alcohol) acetates.

Thin layers of silicic acid containing 10% starch binder [53], silica gel G [19, 54, 55], and Magnesol (Firm 150) containing calcium sulfate (13%) [56] give good resolutions. Magnesol must be washed thoroughly with dilute acetic acid before use to avoid deacylation catalyzed by traces of alkali.

Benzene and mixtures of benzene with methanol or ethyl acetate resolve the acylated sugars with the acetates requiring more polar solvents than the benzoates. The hR_f values of some peracetylated sugars are given in Table 208.

DEFERRARI et al. [53] noted that with one exception the monosaccharide acetates with a 1,5-*trans* configuration have the higher R_f value. However the anomeric sequence ($\beta > \alpha$) has been observed for the 2,4-dinitrophenyl derivatives of D-glucosamine tetraacetate [57] and for other mono- and disaccharide acetates [54, 58]. The α-linked oligosaccharide acetates migrate faster than the corresponding β- on both Magneso (Table 209) [56] and silica gel G [54, 55].

Two phase paper chromatography with dimethyl sulfoxide (DMSO) as the stationary phase was used successfully by WICKBERG [59] for

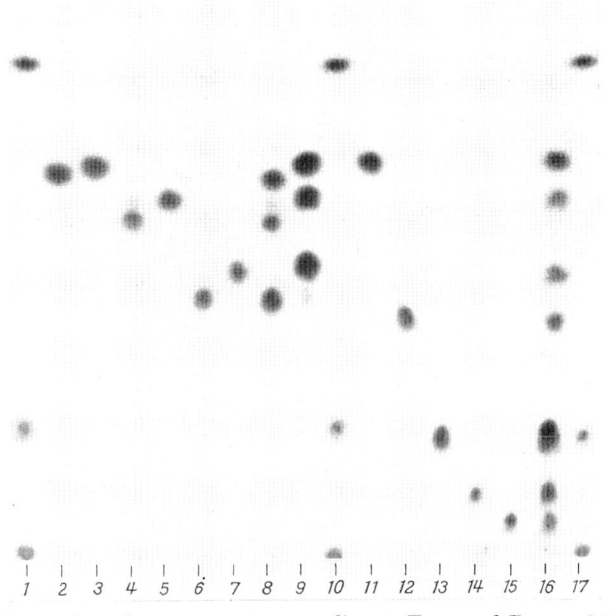

Fig. 229. Separation of sugar acetates according to TATE and BISHOP [54]; double development (17 cm runs) on silica gel G layer using benzene-methanol (96 + 4).

1, 10, 17 Standard dyes (Sudan III, methyl red, methyl orange)
2 α-D-glucopyranose pentaacetate
3, 11 β-D-glucopyranose pentaacetate
4 α-Maltose octaacetate
5 β-Maltose octaacetate
6 α-Cellobiose octaacetate
7 β-Cellobiose octaacetate
8 Mixture of 2, 4 and 6
9 Mixture of 3 (11), 5 and 7
12 β-Laminaribiose octaacetate
13 β-Laminaritriose undecaacetate
14 β-Laminaritetrose tetradecaacetate
15 β-Laminaripentaose heptadecaacetate
16 Mixture of 3 (11), 5, 7, 12—15

Table 208. *TLC of anomeric monosaccharide peracetates and perbenzoates on silica gel-10% starch layers* [53]

Compound	hRf-values Anomer		Solvent
	α	β	
D-Glucopyranose pentabenzoate	56	49	XXVIII
D-Galactopyranose pentabenzoate	52	0	XXIX
D-glycero-L-manno-Heptose hexabenzoate	42	68	XXX
D-glycero-D-gulo-Heptose hexabenzoate	40	69	XXX
D-glycero-D-galacto-Heptose hexabenzoate	26	0	XXIX
D-Glucopyranose pentaacetate	66	65	XXXI
D-Galactopyranose pentaacetate	64	57	XXXI
D-Galactofuranose pentaacetate	52	49	XXXI
D-glycero-L-manno-Heptose hexaacetate	49	52	XXXI

Solvents: See Table 216 for composition.

resolving anomeric acetates. INGLIS [60] has adapted the technique to thin-layers with similar results.

The plate coated with silica gel G is sprayed until damp with a mixture of DMSO and sulfur-free toluene (50 + 50) and dried in an oven until the dampness has just disappeared leaving the silica layer opaque white in appearance (this process requires about 5 min at 130°). The

Table 209. *hRf-values of the β-acetates of oligosaccharides on thin-layers of Magnesol-13% calcium sulfate* [55] *with ethyl acetate-benzene* (50 + 50)[a]

Compound	hRf
β-Maltose octaacetate	75
β-Gentiobose octaacetate	56
β-Isomaltose octaacetate	64
β-Cellobiose octaacetate	62
β-Cellotriose hendecaacetate	44
β-Cellotetraose tetradecaacetate	32

[a] Methanol-benzene (3 + 97) can be used also.

plates are cooled in an air-tight chamber and the sugar acetates applied rapidly avoiding excessive exposure of the plates to moisture. Development of the plate is carried out in the usual manner in a saturated chamber. Excess DMSO is removed from the plate by heating at 150° before applying the detection reagent.

This method gives good resolution of the anomeric acetates in 15–20 min but the hRf values are variable unless the plates are carefully standardized (Table 210). The α-anomers have higher hRf values by this technique.

Table 210. *Chromatography of sugar acetates on silica gel G thin-layers with dimethyl sulfoxide as stationary phase* [60]

Compound	Mobile phase	hRf
α-D-Glucopyranose pentaacetate	Diisopropyl ether-ether (50 + 50)	45—50
β-D-Glucopyranose pentaacetate		35—40
Methyl α-D-glucopyranoside tetraacetate	Ether	80—95
Methyl β-D-glucopyranoside tetraacetate	Ether	65—75
Methyl 2-deoxy-α-D-glucopyranoside triacetate	Diisopropyl ether	45—50
Methyl 2-deoxy-β-D-glucopyranoside triacetate	Diisopropyl ether	30—40
Methyl 2-deoxy-α-D-galactopyranoside triacetate	Diisopropyl ether	45—80
Methyl 2-deoxy-β-D-galactopyranoside triacetate	Diisopropyl ether	30—60

The acetates are readily visualized with sulfuric acid, by the hydroxamic acid reaction (Rgt. No. 134), with silver nitrate (Rgt. No. 228) or with iodine vapor which is somewhat less sensitive (lower limit of detection 100 μg). Strongly hydrophobic compounds such as oligosaccharide peracetates are detected by spraying with water to give opaque white spots on the translucent background (lower limit of detection 100 μg) [54]. Although the latter technique is not as sensitive, it is convenient for preparative plates.

9. Hydrazones and Osazones

The sugars are occasionally identified by conversion to the crystalline hydrazones or osazones which are characterized in turn by their melting points and infrared spectra. Chromatography on columns of calcium carbonate [61] and alumina [62] resolves mixtures of the osazones and some success has attended paper chromatographic methods [62].

Thin-layer chromatography of the osazones can be effected on a variety of adsorbents. TORE [63] reported the chromatography of the phenylosazones on unbound Silene E. F. plates (activated at 110° for 24 hours) with solvent systems containing mixtures of chloroform, acetone, ethanol and water (solvents XXXII and XXXIII). The hRf values increase rapidly with an increase in the water content of the solvent as shown in Table 211.

Table 211. hRf-values of some sugar phenylosazones [63, 65]

Phenylosazone of	hRf		
	Silene E. F. [63]		Kieselguhr G-0.5M sodium tetraborate [65]
	XXXII	XXXIII	XXXIV
Arabinose	65	—	91
Xylose	—	—	72
Rhamnose	78	—	—
Glucose	22	—	39 [a]
Galactose	29	—	52
Sorbose	—	—	21
Lactose	—	50	2
Maltose	—	57	12
Cellobiose	—	57	—

Solvents: See Table 216 for composition.

[a] The same value for fructose.

HAAS and SEELIGER [64] investigated the resolution of phenylosazones on polyamide (Firm 153). The hRf values are primarily dependent on the molecular weight (pentoses > hexoses) and little differentiation between stereoisomers was obtained. With pyridine and water (15 + 85), a reversal in the order of migration occurred and the oligosaccharide osazones migrated while the monosaccharide derivatives remained at the origin.

BANCHER et al. [47], in an extensive study of TLC of the sugars and derivatives on silica gel G, observed that benzene-acetone (45 + 55) afforded an excellent resolution of the phenylosazones of the sugars varying in molecular weight from the hexoses to glycerose and, in addition, glyoxal and methylglyoxal.

RINK and HERRMANN [65] have recommended that the sugars in urine be identified by chromatography of their phenylosazones on kieselguhr G layers, impregnated with 0.05 M sodium tetraborate.

Procedure [65]: 10 ml of urine are heated with 0.4 g phenylhydrazine hydrochloride and 0.6 g sodium acetate for 30 min in a boiling water bath. The sugar phenylosazones, which crystallize on cooling, are filtered, washed with water and dissolved in dioxane-methanol (1 + 1) for application to the plate. For comparison, a 1% sugar solution is treated with 0.2 g phenylhydrazine hydrochloride and 0.3 g sodium acetate in the same manner.

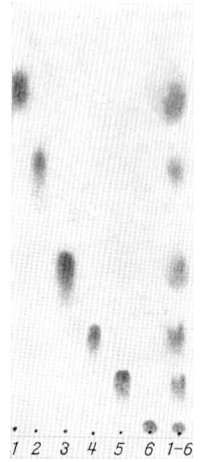

Fig. 230. Separation of phenylosazones on a polyamide layer, using benzene-dimethylformamide (97 + 3). 14 cm run at chamber saturation [64].
1 glycolaldehyde; *2* glyceraldehyde; *3* D-erythrose; *4* D-xylose; *5* D-glucose; *6* D-lactose

The usefulness of this method is limited by the fact that sugars which differ only in the stereochemistry at C_2 give rise to the same osazone. The hRf values of the phenylosazones in solvent XXXIV are given in Table 211.

Chromatography of the sugars and their phenylhydrazones on acidic aluminum oxide has been reported [66], but under the conditions selected only small differences in hRf values between pentoses and hexoses were obtained.

ANET [67], during studies on the degradation of carbohydrates, examined the mono- and bis-2,4-dinitrophenylhydrazones derived from hydroxy-carbonyl and α,β-dicarbonyl compounds by TLC on thin layers of aluminum oxide G and silica gel G using solvent systems of toluene and ethyl acetate in varying proportions (see Table 212). The hRf values, which are dependent on the number of hydroxyl groups, the adsorbent and the solvent, can be increased by deactivating the adsorbent or by increasing the proportion of ethyl acetate. The yellow 2,4-dinitrophenylhydrazones are visible in quantities of 0.1 µg and the intensity of the spot is increased by spraying the air-dried plate with a 2% solution of sodium hydroxide in 90% ethanol which gives blue or purple colors for the 1,2-bis-hydrazones.

Table 212. Adsorbents and solvent systems for TLC of sugar hydrazones and osazones

Parent Compound	Derivative	Adsorbent	Solvent	Ref.
Sugars (triose, tetrose, pentoses, hexoses)	Phenylosazone	Polyamide (Firm 153)	N,N-Dimethylformamide-benzene (7 + 93)	[64]
Hexoses, pentoses	Phenylosazone	Polyamide (Firm 153)	Ethanol-chloroform (3 + 97)	[64]
Oligosaccharides	Phenylosazone	Polyamide (Firm 153)	Pyridine-water (15 + 85)	[64]
Pentoses, hexoses	Phenylosazone	Silene E. F. (Firm 40) (activated at 110°)	Solvent XXXII	[63]
Hexoses, disaccharides	Phenylosazone	Silene E. F. (Firm 40) (activated at 110°)	Solvent XXXIII	[63]
Sugars (triose, tetrose, pentoses, hexoses)	Phenylosazone	Silica gel G (Firm 88) (unactivated)	Benzene-acetone (45 + 55) Chloroform-DMF[a] (87 + 13) n-Propanol-n-hexane (63 + 37) n-Butanol-n-hexane (50 + 50)	[47]
Pentoses, hexoses	Phenylosazone	Kieselguhr G–0.5 M sodium tetraborate	Solvent XXXIV	[65]
Pentoses, hexoses	Phenylhydrazone	Aluminum oxide (acidic)	n-Butanol-acetone-water (40 + 50 + 10)	[66]
Pentoses, hexoses	Phenylhydrazone	Aluminum oxide (acidic)	n-Butanol-acetone-water (70 + 20 + 10)	[66]
Pentoses, hexoses, methyl ethers	p-Bromophenylosazone	Silica gel G (Firm 88)	Benzene-methanol (90 + 10)	[69]
Glyoxal, glycerose, erythrose	2,4-Dinitrophenylosazone	Silica gel	Toluene-glacial acetic acid-water (50 + 37 + 13) Methylcyclohexane-ethyl acetate-acetonitrile-water (57 + 33 + 9 + 1)	[68]
Hydroxy-carbonyl and α,β-dicarbonyl compounds	2,4-Dinitrophenyl-hydrazone and osazone	Aluminum oxide G (deactivated) Aluminum oxide G (activated) Silica gel G (activated)	Toluene-ethyl acetate (50 + 50) Toluene-ethyl acetate (50 + 50) and (75 + 25) Toluene-ethyl acetate (50 + 50) and (75 + 25)	[67]

Solvents: XXXII–XXXIV: See Table 216 for composition.

[a] DMF = N,N-Dimethylformamide.

Excellent resolution of the 2,4-dinitrophenylosazones of glyoxal, glycerose and erythrose is provided by silica gel layers with the solvent systems toluene-glacial acetic acid-water (50 + 37 + 13) or methyl-cyclohexane-ethyl acetate-acetonitrile-water (58 + 32 + 9 + 1) [68].

The p-bromophenylosazones of the hexoses (hR_f 4), pentoses (10), mono-O-methyl-hexoses (18), mono-O-methyl-pentoses (24) and di-O-methyl-pentoses (29) are resolved in 40 min on silica gel (calcium sulfate-bound) with benzene-methanol (90 + 10). Under these conditions the disaccharide osazones remain at the origin [69].

All osazones and hydrazones are detected by their yellow color directly or under UV light but the intensity of the spots is enhanced by spraying with alcoholic sodium hydroxide. Alternatively the plate can be sprayed with diazotized sulfanilic acid in 2 N sodium carbonate which transforms the yellow color into brown or red [64]. Free sugars or other colorless impurities are readily detected by the sulfuric acid sprays.

A summary of the adsorbents and solvent systems applicable to the hydrazones and osazones is given in Table 212.

10. Methyl Ethers

Thin layer chromatography on silica gel G affords a rapid method for either qualitative or quantitative determination of methylated sugars. Not only is a good resolution obtained based on the degree of substitution but the isomeric ethers are resolved also. Benzene-acetone (50 + 50) is particularly effective in the identification of the tri-methyl ethers of glucose and galactose (see Table 213) [70]. Using the same solvent system BISHOP et al. (72) have separated the methyl sugars derived from a methylated galactomannan by preparative TLC, applying 50—75 mg of the hydrolyzate to each plate (20 × 20 cm, 0.6 mm thick). The tri-methyl ethers of mannose have also been separated by preparative TLC with butanone as solvent [73].

TLC of the di-O-methyl tetroses [74], tri-O-methyl apiose [75], mycinose [76], and the methyl ethers of N-acetyl-D-glucosamine [77, 78] on silica gel G has been reported. The mono-O-methyl-fructoses are readily separated on boric acid impregnated silica gel G using n-butanol-acetone-water (40 + 50 + 10) [79, 80]. TLC of several methylated sugars on Magnesol containing calcium sulfate as binder has also been reported [56].

Examination [81] of a variety of fully methylated monosaccharides (ethyl ether-toluene, 67 + 33) (Table 214) and disaccharides (butanone-toluene, 50 + 50) on silica gel G reveals that the α- and β-anomers are resolved in most cases, the order of mobility paralleling that attained by gas-liquid chromatography.

Table 213. *TLC of methyl ethers on silica gel G and cellulose* [4, 70, 71]

Compound	Silica gel G		Avirin
	hR$_{St}$^{a}	hRf-values	
	XXXV	XXXVI	XXXVII
2,3,4,6-Tetra-O-methyl-D-glucose	100	45	86
2,3,4-Tri-O-methyl-D-xylose	111	50	—
2,3,4,6-Tetra-O-methyl-D-galactose	81	33	78
2,3,4-Tri-O-methyl-D-glucose	—	34	—
2,3,6-Tri-O-methyl-D-glucose	—	25	60
2,4,6-Tri-O-methyl-D-glucose	30	23	—
3,4,6-Tri-O-methyl-D-glucose	38	27	—
2,3,6-Tri-O-methyl-D-mannose	—	—	57
2,3,4-Tri-O-methyl-D-galactose	25	16	—
2,3,6-Tri-O-methyl-D-galactose	59	21	—
2,4,6-Tri-O-methyl-D-galactose	45	20	48
3,4,6-Tri-O-methyl-D-galactose	23	—	—
4,6-Di-O-methyl-D-glucose	—	17	—
2,4-Di-O-methyl-D-galactose	—	—	23
2,6-Di-O-methyl-D-galactose	—	—	15

Solvents: See Table 216 for composition. 17.1 cm migration in 2 h with solvent XXXVII.

^{a} hR$_{St}$ = Values referred to 2,3,4,6-tetra-O-methyl-D-glucose (= 100).

Table 214. hRf-values of permethylated sugars on silica gel G
(Ethyl ether-toluene, 67 + 33) [81]

Methyl glycoside of	Anomer:	hRf	
		α	β
2,3,5-Tri-O-methyl-D-arabinose		51	38
1,3,4,6-Tetra-O-methyl-D-fructose		39	39
1,3,4,5-Tetra-O-methyl-D-fructose		23	16
2,3,4,6-Tetra-O-methyl-D-glucose		29	46
2,3,5,6-Tetra-O-methyl-D-glucose		31	39
2,3,5,6-Tetra-O-methyl-D-galactose		29	44
2,3,4,6-Tetra-O-methyl-D-galactose		21	28

Paper chromatography and more recently gas liquid chromatography have been the most convenient methods for analysis of the methylated sugars. Butanone-water azeotrope, one of the most effective solvents for paper chromatography, is equally effective on cellulose thin-layers (see Table 213) [4]. Although chromatography on cellulose-plates is more rapid than on paper it is significantly slower than on inorganic layers.

Quantitative Analysis

TLC on silica gel G permits a rapid determination of the molar ratio of the mono-, di-, tri-, and tetra-O-methyl components in the hydrolyzate of a methylated polysaccharide [16]. The resolved components are quantitatively removed from the chromatoplate and analyzed by an appropriate colorimetric procedure such as the phenol-sulfuric acid method [16]. Silica gel G contains small amounts of impurities which may interfere in such colorimetric analyses unless corrected for by a blank. A simple device has been suggested [82] for the removal of these impurities by which the plate is subjected to ascending chromatography in ethyl ether-methanol (20 + 80) prior to loading with the mixture to be analyzed. The plate is removed from the development chamber, dried at 110° for 25 minutes, loaded with the sugars and chromatographed in the usual way at right angles to the first ascent.

The reducing methylated sugars are detected by the same reagents used for paper chromatograms such as aniline phthalate (No. 10), p-anisidine phthalate (No. 12), 2,3,5-triphenyltetrazolium chloride (No. 255), alkaline silver nitrate (No. 234) and, in addition, the sulfuric acid sprays and iodine vapor. Brief exposure to iodine vapor does not cause a detectable destruction of the methyl ethers of the sugars [83] and is therefore a convenient method for visualizing the components on preparative plates [73] or for quantitative analysis.

11. Miscellaneous Derivatives

At the present time, TLC is an indispensable tool for synthetic carbohydrate chemistry. This method together with gas-liquid chromatography permits the analysis of compounds which cannot be detected on paper chromatograms. GLC, however, is restricted to compounds which can be volatilized without degradation, whereas TLC affords a rapid analysis under mild conditions. Column chromatography employing silica gel or alumina is a standard method for purification and fractionation of hydrophobic derivatives of carbohydrates, but thin-layers afford a sharper resolution of similar compounds.

Suitable conditions for chromatography of these derivatives have already been discussed in the sections dealing with acetates, osazones and methyl ethers (section IV, 8—10) and section V contains further examples. A few applications of TLC to the analysis of still other carbohydrate derivatives have been selected here to illustrate the scope and utility of this technique.

Silica gel G is the most frequently used adsorbent together with solvent systems of benzene, petroleum ether or chloroform containing varying proportions of solvents such as methanol, acetone or ethyl acetate. The hR_f-values increase as the degree of substitution increases.

The derivatized sugars are detected with iodine vapor or with sulfuric acid. TATE and BISHOP [84] found iodine vapor to be the most convenient and sensitive for detection of benzyl ethers on silica gel G and they were able to detect 0.4 µg of benzyl tetra-O-benzyl-D-glucopyranoside in this way. Using silica gel G containing 1% fluorescent zinc silicate[6], 4 µg of the benzyl ether could be detected under short wave UV light. The mobility of the benzylated sugars in the solvent system of petroleum ether (b. p. 65–110°) containing 3–5% methanol is dependent on the methanol content with the more polar compounds requiring the higher methanol levels. Since benzylated carbohydrates are usually sirups, TLC offers an excellent criterion of purity.

Acetonation of 3-O-methyl-D-glucitol affords the isomeric 1,2:5,6- and 2,4:5,6-di-O-isopropylidene derivatives which are differentiated on silica gel layers [hR_f 50 and 25, respectively, in benzene-methanol (90 + 10)] with visualization by iodine vapor [85]. The four stereoisomers of 1,2:5,6-di-O-trichloroethylidene-α-D-glucofuranose, which differ in the configuration of the acetal carbons atoms, are resolved on silica gel with ethyl acetate-petroleum ether (b. p. 40–50°) (50 + 50) [86]. Chloroform-acetone (97 + 3) and benzene-ether (90 + 10) have been used to separate mixtures of acylated di-O-isopropylidene-hexitols [87, 88].

Preparative TLC has afforded the two pure isomeric dichlorides obtained by addition of chlorine to tri-O-acetyl-D-glucal. The capacity of silica gel G plates is favorable for preparative purposes. The 250 µ thick layers (20 × 20 cm plate) permitted separation of 25 mg of the mixture of dichlorides with the solvent toluene-ether (67 + 33) [89]. TATE and BISHOP [84] report the separation of up to 100 mg of the benzylated sugars on one plate.

Methyl β-D-xylothiopyranoside triacetate (hR_f 71) was separated from its oxidation products, the isomeric "α" and "β"-sulfoxides (hR_f 28 and 18) and the sulfone (hR_f 46), on silica gel G in chloroform-acetone (93 + 7) [90].

ROSENTHAL and coworkers [91, 92] have described the TLC of some acylated anhydro-hexitols and heptitols. WOLFROM et al. [93, 94, 95, 96, 97] characterized various derivatives of D-glucosamine and acyclic sugar nucleoside analogs by TLC. The α- and β-anomers of tri-O-benzoyl-3-O-methyl-D-glucopyranosyl fluoride have been separated on silica gel H with benzene-acetone (90 + 10) [98].

V. Monitoring Reactions by TLC

The adaptability of TLC to the analysis of a wide range of carbohydrate derivatives [16] suggested the feasibility of using this method for monitoring reactions in carbohydrate chemistry. The value of this technique has been illustrated by DUTTON et al. [14] who used TLC on

---

[6] ZA2, S 104 Fluorescent green, Firm 84.

Table 215. *Summary of reactions which have been monitored by TLC on silica gel*

Reaction	Reactant	Product	hRf of Product	Solvent	Ref.
Tritylation	1,6-Anhydro-β-maltose	6'-O-trityl	30	XXXVII	[14]
Detritylation	1,6-Anhydro-6'-O-trityl-β-maltose pentaacetate	detritylated Reactant Triphenylcarbinol	4 58 75	XXXVIII	[14]
Tosylation	1,6-Anhydro-β-maltose 2,2',3,3',4'-pentaacetate	6'-O-tosyl	22	XXXVIII	[14]
Displacement	1,6-Anhydro-6'-O-tosyl-β-maltose pentaacetate	6'-thioacetate	30	XXXVIII	[14]
Acetolysis	1,6-Anhydro-β-maltose hexaacetate	β-octaacetate α-octaacetate Reactant	37 34 24	XXXVIII	[14]
Thioacetalation	D-Glucose	diethyl dithioacetal Reactant	50 0	XXXVII	[14]
Benzylidenation	2,3,5-Tri-O-benzyl-D-arabinose Methyl α-D-glucopyranoside	diethyl dithioacetal 4,6-O-benzylidene Reactant	— 59 6	XXXIX XXXVII	[99] [14]
Debenzylidenation	Methyl 4,6-O-benzylidene-2,3-di-O-methyl-α-D-glucoside	Methyl 2,3-di-O-methyl-α-D-glucoside	33	XXXVII	[14]
Isopropylidenation	Methyl α-D-mannoside	2,3-O-isopropylidene	57	XXXVII	[14]
Deisopropylidenation	1,2:5,6-Di-O-isopropylidene-D-glucose	1,2-O-isopropylidene Reactant	50 70	XXXVII	[14]
	Methyl 2-benzamido-2-deoxy-5,6-O-isopropylidene-β-D-glucoside	Methyl 2-benzamido-2-deoxy-β-D-gluco-furanoside	10	XL	[100]
Deacetylation	Allyl 2,3,4,6-tetra-O-acetyl-β-D-galactoside	Allyl β-D-galactoside	—	XLI	[101]
Periodate Oxidation	2,3,4-Tri-O-benzyl-D-galactitol	2,3,4-Tri-O-benzyl-L-lyxose	85	XLII	[101]
Glycoside Formation	2,3,4-Tri-O-benzyl-L-lyxose	Ethyl glycosides	75, 85	XLI	[101]
Isomerization	Allyl 6-O-allyl-2,3,4-tri-O-benzyl-α-D-galacto-pyranoside	Prop-1-enyl derivative	80	XLIII	[101]
Deamination	D-Lyxonamide tetraacetate	D-Lyxonic acid tetraacetate		XLIV	[102]
Pyrolysis	Methyl 4,6-O-benzylidene-2-deoxy-α-D-glucopyranoside 3-xanthate	Methyl 4,6-O-benzylidene-2,3-dideoxy-α-D-*erythro*-hexoside		XLV	[103]
Epimerization	L-Xyloascorbic acid	L-Araboascorbic	38	XLVI	[104]

Solvents: XXXVII–XLVI: See Table 216 for composition. Silica gel G impregnated with metaphosphoric acid used with solvent XLVI silica gel (Camag D_5) to monitor a variety of reactions, in particular the displacement of tosyl groups by nucleophiles and the opening of 1,6-anhydro rings by acetolysis (Table 215). In most cases the reaction mixture can be applied directly to the plate without interference from the catalysts (for example, zinc chloride used to catalyze acetal formation), but in the case of acetolysis mixtures containing acetic anhydride, acetic acid and sulfuric acid it it necessary to neutralize the reaction mixture. A spot of pyridine applied to the origin before and after the reaction mixture eliminates interference [14]. Reference compounds must be applied in the same solvent as that used in the reaction mixture since solvents such as dimethylformamide may greatly alter the hRf values.

A few examples of the applications of TLC to the monitoring of carbohydrate reactions are given in Table 215. Microchromatoplates prepared by coating microscope slides [105] are convenient for this purpose since the analysis is complete in approximately 10 min.

In some instances the chemical reaction has been performed directly on the plate. HANESSIAN and HASKELL [106], using the ninhydrin degradation to distinguish between 2-amino and 3-amino sugars, carried out the reaction in the following manner.

One mg samples of 3,6-diamino-3,6-dideoxy-D-idose, 2,6-diamino-2,6-dideoxy-D-galactose, glucosamine and galactosamine were each dissolved in a mixture of 0.96 ml of water and 0.04 ml of pyridine containing 2% ninhydrin and applied to cellulose MN 300 G thin-layer plates. The plates were heated at 100° for 80 min to effect the reaction, cooled and developed with t-butyl alcohol-acetic acid-water (40 + 40 + 20). The 2-amino-sugars were degraded whereas the 3-amino-sugar was mostly unchanged.

Table 216. *Solvents used in TLC of sugars and their derivatives*

No.	Mixture	Composition	Refs.
I	Ethyl acetate-65% isopropanol	65 + 35	[1]
II	n-Butanol-acetone-phosphate buffer, pH 5	40 + 50 + 10	[5]
III	Methanol-chloroform-acetone-conc. ammonium hydroxide	42 + 16.5 + 25 + 16.5	[7]
IV	Chloroform-methanol	60 + 40	[7]
V	Acetone-water	90 + 10	[7]
VI	Acetone-water-chloroform-methanol	75 + 5 + 10 + 10	[7]
VII	Ethyl acetate-acetic acid-methanol-water	60 + 15 + 15 + 10	[9]
VIII	n-Propanol-water	85 + 15	[9]
IX	Ethyl acetate-pyridine-water	40 + 20 + 40, upper phase	[3]
X	Formic acid-butanone-*tert.*-butanol-water	15 + 30 + 40 + 15	[10]

Table 216 (Continued)

No.	Mixture	Composition	Refs.
XI	Aqueous phenol (ca. 90%)-water + 0.002% oxine	89 + 11	[10]
XII	Pyridine-ethyl acetate-acetic acid-water	36 + 36 + 7 + 21	[4]
XIII	n-Butanol-acetic acid-water	60 + 20 + 20	[4]
XIV	Isopropanol-water	90 + 10	[13]
XV	n-Butanol-pyridine-water	75 + 15 + 10	[12]
XVI	n-Butanol-2,6-lutidine-water	60 + 30 + 10	[31]
XVII	n-Butanol-pyridine-water	70 + 15 + 15	[31]
XVIII	Ethanol-pentanol-ammonium hydroxide-water	62 + 15 + 15 + 8	[39]
XIX	Ethyl acetate-pyridine-tetrahydrofuran-water	50 + 22 + 14 + 14	[39]
XX	Ethyl acetate-isopropanol-pyridine-water	50 + 22 + 14 + 14	[39]
XXI	n-Butanol-acetic acid-water	50 + 25 + 25	[16]
XXII	Ethyl acetate-acetone-water	40 + 50 + 10	[47]
XXIII	n-Butanol-acetic acid-water	48 + 31 + 16 + 5	[47]
XXIV	Ethyl acetate-formic acid-water	60 + 20 + 20	[47]
XXV	95% Ethanol-0.1 M ammonium tetraborate, pH 9.0	60 + 40	[11, 51]
XXVI	t.-Amyl alcohol-water-p-toluene-sulphonic acid	65 + 33 + 2 (v/v/w), upper phase	[11, 51]
XXVII	Isobutyric acid-ammonium hydroxide-water	66 + 1 + 33	[11, 51]
XXVIII	Benzene-chloroform	70 + 30	[53]
XXIX	Benzene	100	[53]
XXX	Benzene-methanol	99.5 + 0.5	[53]
XXXI	Benzene-ethyl acetate	70 + 30	[53]
XXXII	Chloroform-acetone-95% ethanol	38 + 38 + 24	[63, 65]
XXXIII	Chloroform-acetone-ethanol-water	37 + 37 + 23 + 3	[63, 65]
XXXIV	Chloroform-dioxan-tetrahydrofuran-0.1 M sodium tetraborate buffer	50 + 24 + 24 + 2	[63, 65]
XXXV	Benzene-acetone	50 + 50	[70]
XXXVI	Diisopropyl ether-methanol	83 + 17	[71]
XXXVII	Butanone-water azeotrope		[4]
XXXVIII	Diethyl ether-toluene	67 + 33	[14]
XXXIX	Benzene-diethyl ether	80 + 20	[99]
XL	Ethyl acetate		[100]
XLI	Diethyl ether-petrol ether (B. P. 60—80° C)	50 + 50	[101]
XLII	Diethyl ether		[101]
XLIII	Petrol ether (B. P. 60—80°)-diethyl ether	67 + 33	[101]
XLIV	Chloroform-methanol	90 + 10	[102]
XLV	Benzene-methanol	99 + 1	[103]
XLVI	Acetonitrile-butyronitrile-water	65 + 33 + 2	[104]

Bibliography for Chapter X. Sugars and Derivatives

1. STAHL, E., u. U. KALTENBACH: J. Chromatog. **5**, 351 (1961).
2. WEIDEMANN, G., u. W. FISCHER: Z. physiol. Chem. Hoppe-Seyler's **336**, 189 (1964).

3. SCHWEIGER, A.: J. Chromatog. **9**, 374 (1962).
4. WOLFROM, M. L., D. L. PATIN, and R. M. DE LEDERKREMER: J. Chromatog. **17**, 488 (1965).
5. WALDI, D.: J. Chromatog. **18**, 417 (1965).
6. JACIN, H., and A. R. MISHKIN: J. Chromatog. **18**, 170 (1965).
7. PIFFERI, P. G.: Anal. Chem. **37**, 925 (1965).
8. PASTUSKA, G.: Z. anal. Chem. **179**, 427 (1961).
9. ADACHI, S.: J. Chromatog. **17**, 295 (1965).
10. VOMHOF, D. W., and T. C. TUCKER: J. Chromatog. **17**, 300 (1965).
11. DIETRICH, C. P., S. M. C. DIETRICH, and H. G. PONTIS: J. Chromatog. **15**, 277 (1964).
12. GARBUTT, J. L.: J. Chromatog. **15**, 90 (1964).
13. SHASHA, B., and R. L. WHISTLER: J. Chromatog. **14**, 532 (1964).
14. DUTTON, G. G. S., K. B. GIBNEY, P. E. REID, and K. N. SLESSOR: J. Chromatog. **20**, 163 (1965).
15. ERTEL, H., and L. HORNER: J. Chromatog. **7**, 268 (1962).
16. HAY, G. W., B. A. LEWIS, and F. SMITH: J. Chromatog. **11**, 479 (1963).
17. PREY, V., H. SCHERZ u. E. BANCHER: Mikrochim. Acta **1963**, 567.
18. GREENWAY, R. M., P. W. KENT, and M. W. WHITEHOUSE: Research (London) **6**, Suppl. No. 1, 6S (1953).
19. GUILLOUX, E., et S. BEAUGIRAUD: Bull. soc. chim. France **1965**, 261
20. BROCKMANN, H., u. T. WAEHNELDT: Naturwissenschaften **50**, 43 (1963).
21. CLAISSE, J., L. CROMBIE, and R. PEACE: J. Chem. Soc. **1964**, 6032.
22. KRINGSTAD, K.: Acta Chem. Scand. **18**, 2399 (1964).
23. BIRKOFER, L., C. KAISER, H.-A. MEYER-STOLL u. F. SUPPAN: Z. Naturforsch. **17b**, 352 (1962).
24. PREY, V., H. BERBALK u. M. KAUSZ: Mikrochim. Acta **1961**, 968.
25. RAGAZZI, E., e G. VERONESE: Farmaco (Pavia), Ed. pract. **18**, 152 (1963)
26. COTTE, J., M. MATHIEU et C. COLLOMBEL: Pathol. et. biol. Semaine hôp. **12** 747 (1964).
27. WEICKER, H., u. R. BROSSMER: Klin. Wochschr. **39**, 1265 (1961).
28. TORE, J. P.: J. Chromatog. **12**, 413 (1963).
29. GRASSHOF, H.: J. Chromatog. **14**, 513 (1964).
30. ZHDANOV, YU. A., G. N. DOROFEENKO i S. V. ZELENSKAYA: Doklady Akad. Nauk S.S.S.R. **149**, 1332 (1963).
31. WEILL, C. E., and P. HANKE: Anal. Chem. **34**, 1736 (1962).
32. KOLLER, A., and H. NEUKOM: Biochim. et Biophys. Acta **83**, 366 (1964).
33. WHEELER, M., P. HANKE, and C. E. WEILL: Arch. Biochem. Biophys. **102**, 397 (1963).
34. WIEDENHOF, N.: J. Chromatog. **15**, 100 (1964).
35. GEE, M.: J. Chromatog. **9**, 278 (1962).
36. PREY, V., W. BRAUNSTEINER, R. GOLLER u. F. STRESSLER-BUCHWEIN: Z. Zuckerind. **14**, 135 (1964).
37. FAILLARD, H., u. J. CABEZAS: Z. physiol. Chem. **333**, 266 (1963).
38. ESSER, K.: J. Chromatog. **18**, 414 (1965).
39. GÜNTHER, H., and A. SCHWEIGER: J. Chromatog. **17**, 602 (1965).
40. MARCUS, D. M., E. A. KABAT, and G. SCHIFFMAN: Biochemistry **3**, 437 (1964).
41. SZAREK, W. A., and J. K. N. JONES: Can. J. Chem. **43**, 2345 (1965).
42. WOLFROM, M. L., P. CHAKRAVARTY, and D. HORTON: J. Org. Chem. **30**, 2728 (1965).
43. BUSS, D. H., L. HOUGH, and A. C. RICHARDSON: J. Chem. Soc. **1963**, 5295; **1965**, 2736.
44. ZEHAVI, U., and N. SHARON: J. Org. Chem. **29**, 3654 (1964).

45. HARRISON, R., and H. G. FLETCHER JR.: J. Org. Chem. **30**, 2317 (1965)
46. FLOWERS, H. M., and D. SHAPIRO: J. Org. Chem. **30**, 2041 (1965).
47. BANCHER, E., H. SCHERZ, and K. KAINDL: Mikrochim. Acta **1964**, 1043.
48. FEAST, A. A. J., B. LINDBERG, and O. THEANDER: Acta Chem. Scand. **19**, 1127 (1965).
49. WASSERMANN, L., u. H. HANUS: Naturwissenschaften **50**, 351 (1963).
50. PREY, V., H. BERBALK, and M. KAUSZ: Mikrochim. Acta **1962**, 449.
51. WARING, P. P., and Z. Z. ZIPORIN: J. Chromatog. **15**, 168 (1964).
52. WOLFROM, M. L., and N. E. FRANKS: J. Org. Chem. **29**, 3645 (1964).
53. DEFERRARI, J. O., R. M. DE LEDERKREMER, B. MATSUHIRO, and J. F. SPROVIERO: J. Chromatog. **9**, 283 (1962).
54. TATE, M. E., and C. T. BISHOP: Can. J. Chem. **40**, 1043 (1962).
55. DUMAZERT, C., C. GHIGLIONE, et T. PUGNET: Bull. soc. pharm. Marseille **12**, 337 (1963).
56. WOLFROM, M. L., R. M. DE LEDERKREMER, and L. E. ANDERSON: Anal. Chem. **35**, 1357 (1963).
57. HORTON, D.: J. Org. Chem. **29**, 1776 (1964).
58. WOLFROM, M. L., and R. M. DE LEDERKREMER: J. Org. Chem. **30**, 1560 (1965).
59. WICKBERG, B.: Acta Chem. Scand. **12**, 615 (1958).
60. INGLIS, G. R.: J. Chromatog. **20**, 417 (1965).
61. JØRGENSEN, P. F.: Dansk Tidsskr. Farm. **24**, 1 (1950).
62. BARRY, V. C., and P. W. D. MITCHELL: J. Chem. Soc. **1954**, 4020.
63. TORE, J. P.: Anal. Biochem. **7**, 123 (1964).
64. HAAS, H. J., and A. SEELIGER: J. Chromatog. **13**, 573 (1964).
65. RINK, M., and S. HERRMANN: J. Chromatog. **12**, 415 (1963).
66. STROH, H. H., u. W. SCHUELER: Z. Chemie **4**, 188 (1964).
67. ANET, E. F. L. J.: J. Chromatog. **9**, 291 (1962).
68. BLUMENFELD, O. O., M. A. PAZ, P. M. GALLOP, and S. SEIFTER: J. Biol. Chem. **238**, 3835 (1963).
69. APPLEGARTH, D. A., G. G. S. DUTTON, and Y. TANAKA: Can. J. Chem. **40**, 2177 (1962).
70. TSCHESCHE, R., and G. BALLE: Tetrahedron **19**, 2323 (1963).
71. — and G. WULFF: Tetrahedron **19**, 621 (1963).
72. BISHOP, C. T., M. B. PERRY, F. BLANK, and F. P. COOPER: Can. J. Chem. **43**, 30 (1965).
73. BOUVENG, H. O., I. BREMNER, and B. LINDBERG: Acta Chem. Scand. **19**, 967 (1965).
74. DUTTON, G. G. S., and K. N. SLESSOR: Can. J. Chem. **42**, 614 (1964).
75. HULYALKAR, R. K., J. K. N. JONES, and M. B. PERRY: Can. J. Chem. **43**, 2085 (1965).
76. BRIMACOMBE, J. S., M. STACEY, and L. C. N. TUCKER: J. Chem. Soc. **1964**, 5391.
77. WOLFROM, M. L., J. R. VERCELLOTTI, and D. HORTON: J. Org. Chem. **29**, 547 (1964).
78. — — — J. Org. Chem. **29**, 540 (1964).
79. GRUNDSCHOBER, F., u. V. PREY: Monatsh. Chem. **92**, 1290 (1961).
80. PREY, V., H. BERBALK, and M. KAUSZ: Mikrochim. Acta 449 (1962).
81. GEE, M.: Anal. Chem. **35**, 350 (1963).
82. BROWN, T. L., and J. BENJAMIN: Anal. Chem. **36**, 446 (1964).
83. HANDA, N., and F. SMITH: unpublished.
84. TATE, M. E., and C. T. BISHOP: Can. J. Chem. **41**, 1801 (1963).
85. FOSTER, A. B., M. H. RANDALL, and J. M. WEBBER: J. Chem. Soc. **1965**, 3388.

86. FORSÉN, S., B. LINDBERG, and B. SILVANDER: Acta Chem. Scand. **19**, 359 (1965).
87. BAKER, B. R., and D. H. BUSS: J. Org. Chem. **30**, 2304 (1965).
88. BUKHARI, M. A., A. B. FOSTER, and J. M. WEBBER: J. Chem. Soc. **1964**, 2514.
89. LEFAR, M. S., and C. E. WEILL: J. Org. Chem. **30**, 955 (1965).
90. WHISTLER, R. L., T. VAN ES, and R. M. ROWELL: J. Org. Chem. **30**, 2719 (1965).
91. ROSENTHAL, A., and D. ABSON: Can. J. Chem. **42**, 1811 (1964).
92. — and H. J. KOCH: Can. J. Chem. **43**, 1375 (1965).
93. WOLFROM, M. L., D. HORTON, and D. H. HUTSON: J. Org. Chem. **28**, 845 (1963).
94. — W. A. CRAMP, and D. HORTON: J. Org. Chem. **29**, 2302 (1964).
95. — H. G. GARG, and D. HORTON: J. Org. Chem. **29**, 3280 (1964).
96. — W. VON BEBENBURG, R. PAGNUCCO, and P. MCWAIN: J. Org. Chem. **30**, 2732 (1965).
97. — H. G. GARG, and D. HORTON: J. Org. Chem. **30**, 1556 (1965).
98. LUNDT, I., C. PEDERSEN, and B. TRONIER: Acta Chem. Scand. **18**, 1917 (1964).
99. FLETCHER, H. G. JR., and H. W. DIEHL: J. Org. Chem. **30**, 2312 (1965).
100. GIGG, R., and C. D. WARREN: J. Chem. Soc. **1965**, 1351.
101. — — J. Chem. Soc. **1965**, 2205.
102. WOLFROM, M. L., and R. B. BENNETT: J. Org. Chem. **30**, 1285 (1965).
103. FERRIER, R. J.: J. Chem. Soc. **1964**, 5443.
104. BRENNER, G. S., D. F. HINKLEY, L. M. PERKINS, and S. WEBER: J. Org. Chem. **29**, 2389 (1964).
105. PEIFER, J. J.: Mikrochim. Acta **1962**, 529.
106. HANESSIAN, S., and T. H. HASKELL: J. Org. Chem. **30**, 1080 (1965).

Y. Inorganic Ions

H. SEILER

As long ago as 1949, MEINHARD and HALL [16] applied the method termed "surface chromatography" by them, for separation of mixtures of simple ferric and zinc salts.

Fears that the layers would retain their adhesion only when relatively non-polar solvents were used and that consequently the TLC of inorganic ions would be accompanied by difficulties, have proved groundless. Aqueous solvents may be used without loosening of the layer, whether starch, gypsum or agar-agar is used as binder.

Impurities present in ordinary silica gel have, however, created difficulties in separation and detection reactions. Adsorbents destined for use in separating inorganic ions should therefore be submitted to thorough prior washing with acid and distilled water. Purified silica gels of this sort are nowadays commercially available, e. g., under the term MN silica gel HR (Firm 83). Experiments aimed at establishing the influential factors in inorganic TLC [21] have shown that migration and separation are determined by the ion exchange character of the silica gel, by complex formation of the various ions with the solvents and by the solubility of the applied salts in the solvents.

I. Preparation of the Solutions for Analysis

If a total analysis is to be carried out, the substance mixture under investigation is best separated in a preliminary stage into the classical analytical groups. The separation procedure in the scheme below has proved most satisfactory; a more detailed description was given earlier [24] (1st Edition).

Scheme for Separation into the Analytical Groups

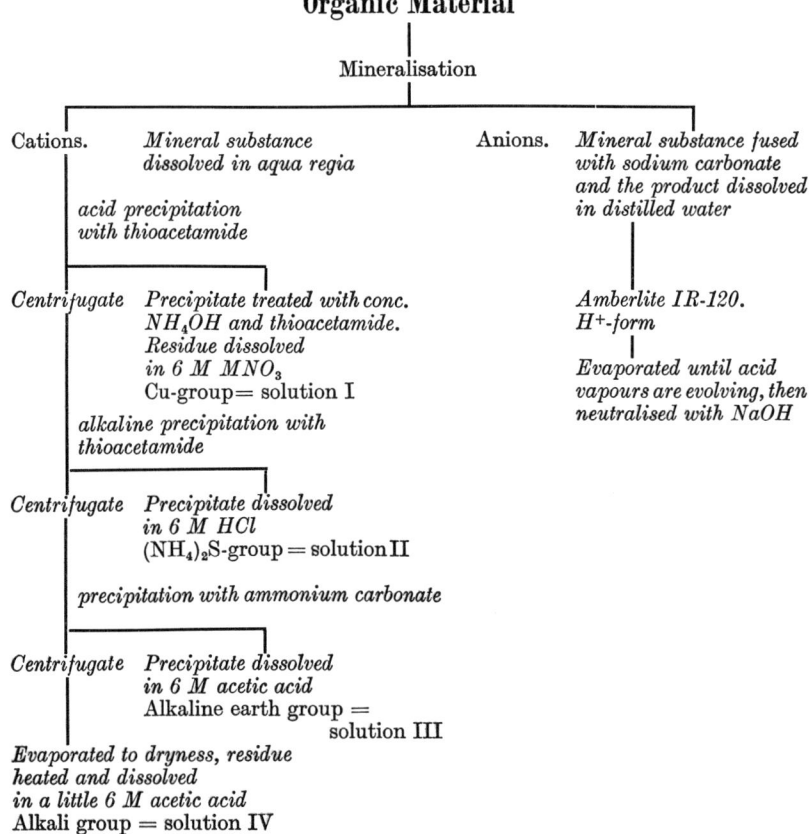

II. TLC of the Cations Separated in this Preliminary Stage

One-dimensional ascending technique has been used, except where otherwise stated. It is evidently best not to quote Rf-values in the TLC of inorganic ions since they depend on too many factors. The *sequence of the migration distances of the ions, which is always the same*, is thus quoted below. The individual ions can be easily identified through specific tests when the chromatograms are evaluated. Preliminary separation into the analytical groups is taken for granted.

The "group solution" for analysis is best applied in 2 or 3 different concentrations alongside one another at the start points. It is possible in this way to detect traces of an element and also to eliminate occlusions through excessive concentrations of another element. It is also always an advantage to carry out parallel chromatography of the ions expected in the chromatogram. The chromatography chambers in all work are lined with filter paper to ensure a constant vapour phase (CS).

1. Separation of the Cu-group (Solution I) [23]

The Cu-group is separated with n-butanol-1.5 N HCl-acetonyl-acetone (100 + 20 + 0.5). The sequence of migration distances in this *solvent* is Hg > Bi > Cd > Pb > Cu (see Fig. 231).

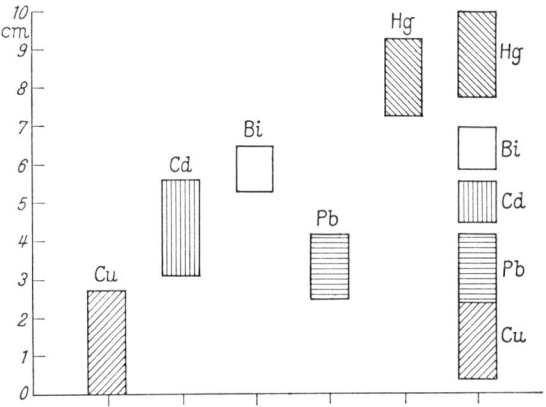

Fig. 231. Separation of the Cu-group

Layer: MN silica gel S-HR.

Standard samples: 2 µl of 0.1 M solutions of $Hg(NO_3)_2 \cdot 2H_2O$; $Cd(OOCCH_3)_2 \cdot 2H_2O$; $BiONO_3$; $Pb(NO_3)_2$; $Cu(OOCCH_3)_2 \cdot H_2O$ as separate spots in a row.

Duration of run: About 2 h for a 15 cm run.

Detection: The layer is spread with a 2% solution of KI, dried, exposed to NH_3 vapour and placed in a chamber filled with H_2S (Table 217).

Table 217. *Colour reactions of ions in the Cu-group*

Ion	KI solution	H_2S
Hg^{II}	red	brown-black
Bi^{III}	brown-yellow	brown-black
Cd^{II}		yellow
Pb^{II}	yellow-brown	brown
Cu^{II}	brown	dark brown

CANIĆ and PETROVIĆ [5] have employed layers of maize starch for separation of Pb^{2+}, Ag^+ and Hg^{2+}. Their solvent was acetone-3 N HNO_3 (50 + 50) and the sequence of migration distances was:

$$Hg^{2+} > Ag^+ > Pb^{2+}.$$

2. Separation of the $(NH_4)_2S$-group (Solution II) [23]

Acetone-conc. HCl-acetonylacetone (100 + 1 + 0.5) is used as *solvent* to separate this group. Separations are improved if a small bowl of water is placed in the chamber. The sequence of migration distances of the ions is:

$$Fe > Zn > Co > Mn > Cr > Ni > Al \text{ (see Fig. 232)}.$$

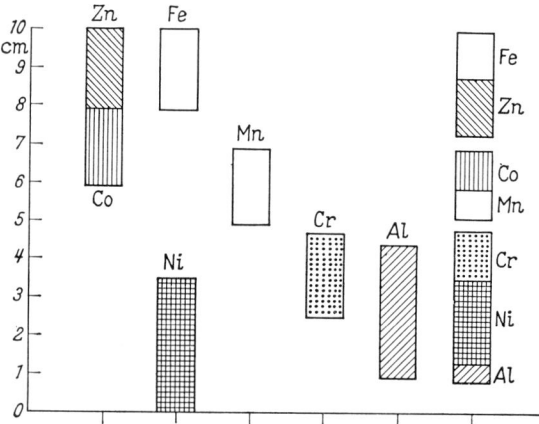

Fig. 232. Separation of the $(NH_4)_2S$-group

Layer: MN silica gel S-HR (Firm 83).
Standard samples: 2 µl of 0.1M solutions of $NiSO_4 \cdot 7H_2O$; $Co(NO_3)_2 \cdot 6H_2O$; $ZnSO_4 \cdot 7H_2O$; $MnSO_4 \cdot H_2O$; $Al(OOCCH_3)_3$; $CrCl_3 \cdot 6H_2O$; and $FeCl_3$ as separate spots in a row.
Detection: The layers are held over NH_3 vapour and then sprayed with a solution of 0.5 g 8-hydroxyquinoline in 100 ml 60% ethanol. The spots are evaluated in UV light of 365 nm (Table 218).

Table 218. *Colour reactions of the ions in the $(NH_4)_2S$-group*

Ion	NH_3	Oxine	UV
Fe^{III}		brown	dark
Zn^{II}		pink	yellow
Co^{II}	blue	yellow	dark
Mn^{II}		orange	dark
Cr^{II}	green		dark
Ni^{II}			dark
Al^{III}			light yellow

3. Separation of the Ammonium Carbonate-group (Solution III)

Starch must be used as binder for this group since insoluble sulphates would be formed from gypsum in the layer and the ions are detected with violuric acid. The anions present must be acetates also; this is easily effected by dissolving the carbonates in 6N acetic acid.

Marked tailing may be observed in the presence of Ca. This can be suppressed by spotting 1 µl acetic acid on to each start spot. The following sequence of migration distances has been found in the *solvent* ethanol-n-propanol-acetic acid-acetylacetone-distilled water (37.5 + + 37.5 + 5 + 1 + 20):

$$Ca > Sr > Ba \text{ (see Fig. 233).}$$

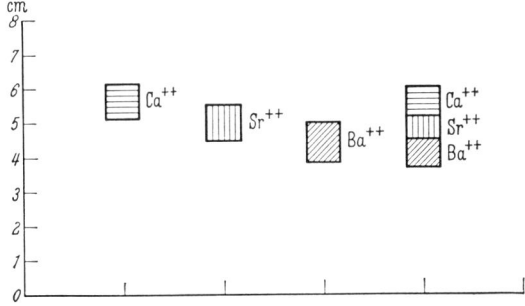

Fig. 233. Separation of the ammonium carbonate-group [24]

Layer: MN-silica gel S-HR (Firm 83).

Standard samples: 1 µl of 1M solutions of Ca, Sr and Ba acetate applied as separate spots in a row, followed in each case with 1 µl acetic acid.

Duration of run: 80—90 min for a 15 cm run.

Detection: The layer is sprayed with a freshly prepared 1.5% solution of violuric acid in distilled water and then heated for 20 min in a drying oven (100°) (Table 219).

Table 219. *Colour reactions of ions in the ammonium carbonate-group*

Ion	Ca^{2+}	Sr^{2+}	Ba^{2+}
Color with violuric acid	yellow-orange	pink	red-violet

DRUDING [7] has employed acetic acid-ethanol (5 + 95) and 2N HCl-*tert.*-butanol (5 + 95) as solvents for the separation of alkali and alkaline earth ions. Violuric acid and lithium tetracyanoquinodimethanide were used for detection.

BERGER et al. [4] have separated the ions of this group on layers of ion exchanger resins (see p. 46, Table 10).

GAGLIARDI and LIKUSSAR [8] have separated Be, Mg, Ca, Sr and Ba on layers of cellulose MN 300 HR; especially good solvents were: I, dioxan-conc. HCl-water (58 + 12 + 30) and II, methanol-conc. HCl-water (73 + 12 + 15). The ions were detected by spraying first with conc. NH_4OH and then with a 2% solution of oxine in ethanol. The order of migration distances was:

$$Be > Mg > Ca > Sr > Ba.$$

According to CANIĆ and PETROVIĆ [5], Mg, Ca, Sr and Ba can be separated on layers of maize starch, using acetone-3N HCl (40 + 60). The sequence of migration distances is:

$$Mg > Ca > Sr > Ba.$$

Detection is carried out by spraying with oxine solution and subsequent inspection in UV light of 250 nm.

4. Separation of the Alkali-group (Solution IV) [25]

Adsorbents with starch as binder must be used here too as a result of interference by gypsum; acetate anions are necessary in this group also. Alkali salts of strong acids can be converted into acetates with the help of an anion exchanger in the acetate form but this is comparatively protracted. Exchange on the layer itself is simpler: Ba ions remain at the start in the solvent used. An amount of barium acetate approximately equivalent to the expected amount of alkali ions is first applied, followed by the alkali sulphates. Double decomposition occurs and the alkali ions studied can now be separated and detected with violuric acid. A red spot from Ba^{2+} is yielded at the start. The solvent is allowed to ascend rather higher (15 cm) than when the acetates are directly applied (10 cm).

Solvent: Absolute ethanol-acetic acid (100 + 2).

Order of migration distances of the ions:

$$Li > Mg > Na > K \text{ (see Fig. 234)}.$$

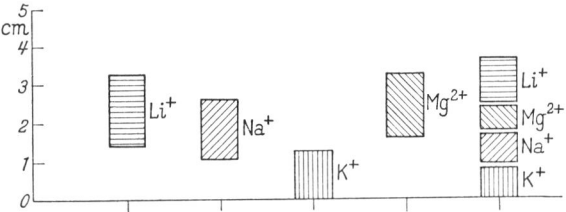

Fig. 234. Separation of the alkali-group

Layer: MN silica gel S-HR (Firm 83).

Standard samples: 1 µl of M solutions of Li, Na, K and Mg acetates, all slightly acidified with acetic acid, are applied as separate spots in a row.

Time of run: acetates, 50 min; sulphates, 70 min.

Detection: The layers are sprayed with 1.5% violuric acid in distilled water and heated 20 min at 100° afterwards in a drying oven (Table 220).

Table 220. *Colour reactions of ions in the alkali-group*

Ion	Li^+	Mg^{2+}	Na^+	K^+
Color with violuric acid	light red	yellow-orange	red-violet	blue-violet

LESIGANG [13] has separated the ions of this group on layers of the salts of heteropoly acids.

III. Separation of Special Cation Mixtures

It is frequently of interest to detect only a particular group of cations or even individual ions, irrespective of the other ions present. Separations of this type are described below. Preliminary separation into the analytical groups would be superfluous and time-consuming in such cases.

1. UO_2^{2+} in a Mixture of Cations [23, 28]

UO_2^{2+} ions migrate towards the upper part of the chromatogram in the solvent quoted (below) while Fe, Co, Cu, Ni, Al and Th remain at or near the start. If a $UO_2(NO_3)_2$ solution is applied, the daughter nuclide ^{234}Th can be separated from it and determined at the start by activity measurements.

Layer: MN silica gel S-HR (Firm 83).
Solution for application: $UO_2(NO_3)_2$ in 4.7N HNO_3.
Solvent: Freshly distilled ethyl acetate-ether saturated with water-tri-n-butylphosphate (50 + 50 + 2).
Detection: 0.25% solution of pyridylazonaphthol in ethanol.
Duration of run: 10—15 min for 15 cm.

MARKL and HECHT [14] have separated U^{6+}, Mo^{6+} and Fe^{3+} from mixtures of other ions using silica gel G layers and triisooctylamine as solvent.

Layers with exchanger properties can be obtained by impregnating the adsorbent with triisooctylamine [15]; these may be used with various acids as solvents for separating mixtures such as U, Co, Cu, Zn; Fe, Co, Ni etc.

2. Ga^{3+} in Presence of Large Excess of Al^{3+} [22]

Ga migrates near the front and Al remains behind at the start in this separation. Down to 0.5% Ga in Al may be detected unequivocally and 1 µg of Ga is still clearly identifiable.

Layer: MN silica gel S-HR (Firm 83).
Solvent: Freshly distilled acetone-conc. HCl (100 + 0.5).
Detection: Spraying with 0.5% 8-hydroxyquinoline in 60% ethanol, followed by exposure to NH_3 vapour and inspection in UV light of 365 nm.
Duration of run: 10—15 min for 15 cm.

3. Sn, Cu, Hg, Pb, Bi, Cd and Zn as Dithizonates [11]

The cations are applied as their dithizone derivatives in chloroform solution. No detection reaction is needed since the dithizonates have characteristic colours.

Layer: Silica gel G (Firm 88).
Solvent: Benzene.
Duration of run: 40 min for 10 cm.

4. Ag, Pd, Au and Pt as Dithizonates [12]

The ions are extracted from aqueous solution at different pH values into a solution of dithizone in benzene: AgHDz, Au(HDz)$_2$ and Au$_2$Dz$_3$

from alkaline solution; $Pd(HDz)_2$ and traces of Au_2Dz_3 from weakly acid solution; and $Pt(HDz)_2$, from strongly acid solution.

The sequence of migration distances of the dithizonates is:

$$Pd(HDz)_2 > Pt(HDz)_2 > Au_2Dz_3 > Au(HDz)_2 > AgHDz = 0.$$

Layer: Silica gel G (Firm 88).
Solutions for application: 0.1% solutions of H_2PtCl_4, $AuCl_3$, $PdCl_2$ and $Ag(OOCCH_3)$; 0.1% solution of dithizone in benzene.
Solvent: Benzene-methylene dichloride (50 + 50).
Duration of run: 40 min for 10 cm.

5. Separation of Cations on Layers of Ion Exchangers
(cf. pp. 44–48)

BERGER [4] and SHERMA [32] describe separations of various cations on layers of ion exchangers. The following separations are possible according to SHERMA: As^{3+}, Fe^{3+} and Bi^{3+} on layers of Amberlite CG-120 in the NH_4^+-form, using a 0.5M solution of NH_4F; As^{3+}, Cd^{2+} and Ba^{2+} on layers of Amberlite CG-120 in the Na^+-form, with 0.5M NaCl as solvent; and Mg^{2+}, Tl^+ and Ce^{4+} on Amberlite CG-400 in the thiosulphate form, using 0.1M $(NH_4)_2S_2O_3$ and 0.1M NH_4OH solvents.

Mn^{2+}, Cd^{2+} and Ag^+ are separated in 0.05M malonic acid on layers of Amberlite CG-400 in the malonate form.

6. Circular TLC of Cations

HASHMI et al. [10] have separated a large number of cations by circular TLC (cf. p. 73), after a preliminary separation into the classical analytical groups. They developed a special apparatus for the work. Both alumina and silica gel were used as adsorbents. Migration times were exceptionally short, averaging 2 min.

Migration sequence	Layer	Solvent
$Hg^{2+} > Ag^+ > Pb^{2+}$	Alumina	n-Butanol-acetone-conc. HNO_3 (23 + 23 + 4)
$Cu^{2+} > Cd^{2+} > Hg^{2+} > Bi^{3+} > Pb^{2+}$	Alumina	n-Butanol-8M HCl-acetylacetone (43 + 4 + 3)
$Sb^{3+} > As^{3+} > Sn^{4+}$	Silica gel	Acetone-8M HCl-acetylacetone (48 + 1 + 1)
$Fe^{3+} > Al^{3+} > Cr^{3+}$	Silica gel	Acetone-4M HCl-acetylacetone (97 + 2 + 1)
$Fe^{3+} > Co^{2+} > Mn^{2+} > Ni^{2+}$	Alumina	Acetone-4M HCl-acetylacetone (46 + 2 + 2)
$Cu^{2+} > Zn^{2+} > Ni^{2+}$	Alumina	Acetone-4M HCl-acetylacetone (97 + 2 + 1)
$Ca^{2+} > Sr^{2+} > Ba^{2+}$	Silica gel	Acetone-4M HCl (97 + 3)

7. Separation and Detection of Toxic Metals

a) Qualitative Separation [17] of Tl, Ni, Cu, Bi and Hg and of Ce, Ni, Cu, Be, Bi and Hg

The same solvent is used in each case; the orders of migration distances are:

$$Hg > Bi > Cu > Ni > Tl,$$
$$Hg > Bi > Be > Cu > Ni > Ce.$$

Layer: Cellulose MN 300 (Firm 83).
Solvent: Acetone-25% HNO_3 (70 + 30).
Detection: 1 group: 0.5% aqueous Na_2S solution; 2 group: saturated solution of alizarin in 96% ethanol. The layers are exposed to NH_3 vapour after spraying.

b) Determination of Hg [3]

Since the spectrophotometric determination of Hg by means of dithizone can be falsified by other heavy metals and oxidation products of the dithizone, it is best to purify the dithizone extract by TLC beforehand. The various metal dithizonates, dithizone and its oxidation products are separated from one another on the chromatogram. In particular the ubiquitous Cu-dithizonate can be separated. The spots of Hg-dithizonate may be scraped off, eluted and evaluated colorimetrically. The order of migration distances is:

oxidation products > Hg-dithizonate > Cu-dithizonate > dithizone.

Layer: Silica gel G (Firm 88); it is purified by allowing methanol-hydrochloric acid to ascend about 18 cm before separation.
Solvent: Benzene.

8. Separation of *cis-trans* Co-Complex Isomers [29]

Good separations of the *cis-trans* isomers of complexes of the type $(Coen_2Cl_2)X$ ($X = Cl^-$, NO_3^-, SCN^-) and of nitro-compounds of Co are possible. The *cis*-forms always migrate further than the corresponding *trans*-forms.

Layer: MN silica gel S-HR (Firm 83).
Solvents: a) for $(Coen_2Cl_2)X$: methanol-0.5% methanolic sodium acetate-1N acetic acid-water (90 + 10 + 0.5 + 1).
b) for nitro-compounds: 96% ethanol-methanol-25% aqueous ammonium acetate-1N methanolic acetic acid (30 + 70 + 5 + 0.3).
Detection: Sprayed with 2N NH_4OH, dried and then sprayed with 1% dithiooxamide in ethanol.

9. Separation of Radionuclides [18]

Mixtures of radioelements like Ba—La, Ba—Cs, Ca—Sc, Sr—Y, Zn—Ga, Nb—Ta, $I^- - IO_3^- - TeO_3^-$, $PO_4^{3-} - SO_4^{2-}$ etc., can be separated using suitable solvent systems with various adsorbents such as silica gel, kieselguhr and lanthanum oxide.

Separations of this sort have been carried out by thin-layer electrophoresis also [19].

IV. Separation of Anions
1. Separation of the Halides [26]

The alkali halide salts are those applied. Detection can be undertaken with a pH-indicator or with a fluorescence indicator. Fluoride is detected with the zirconium-alizarin lake. The anions appear as light yellow spots on a blue background when the pH indicator (below) is used. The migration sequence is:

$$I^- > Br^- > Cl^- > F^- \text{ (see Fig. 235)}.$$

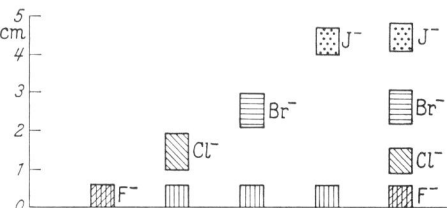

Fig. 235. Separation of the halides

Layer: MN silica gel S-HR (Firm 83).

Standard samples: 1 µl of 1M solutions of NaF, NaCl, KBr and KI.

Solvent: Acetone-n-butanol-conc. NH_4OH-distilled water (65 + 20 + 10 + 5).

Duration of run: 30—40 min for 15 cm.

Detection: a) 0.1% solution of bromocresol purple in ethanol, brought just to the colour change with a few drops of NH_4OH.

b) 1% solution of ammoniacal silver nitrate and 0.1% ethanolic fluorescein.

c) 0.1% solution of zirconium-alizarin lake in concentrated hydrochloric acid.

GAGLIARDI et al. [9] have separated halides and pseudohalides on MN silica gel S-HR (Firm 83). The same order of migration distances of the ions were observed in all the solvents used:

$I^- > SCN^- > Br^- > Cl^- > N_3^- > Fe(CN)_6^{3-}, Fe(CN)_6^{4-} > CN^- > F^-$.

K salts were applied in all cases except for N_3^- and F^- which were applied as their Na salts.

Good separations were obtained with the solvents:
a) n-butanol-n-propanol-di-n-butylamine (45 + 45 + 10);
b) n-butanol-benzylamine (90 + 1);
c) n-propanol-chloroform-benzylamine (60 + 30 + 10).

The run durations for 10 cm were 95, 90 and 55 min for a), b) and c) respectively.

BERGER et al. [4] have separated the halides on layers of ion exchange resins.

2. Separation of Phosphates [30]

Sodium pyro- and orthophosphates, ortho- and hypophosphites have been separated.

Layer: MN silica gel S-HR (Firm 83).

Standard samples: 1 µl each of 0.1 M solutions of $Na_2H_2P_2O_7$, NaH_2PO_4, NaH_2PO_3 and NaH_2PO_2.

Solvent: Methanol-conc. NH_4OH-10% trichloroacetic acid-water (50 + 15 + 5 + 30).

Duration of run: 50–60 min for 15 cm.

The sequence of migration distances found, is:

$$H_2PO_2^- > H_2PO_3^- > H_2PO_4^- > H_2P_2O_7^{2-}.$$

Detection: Blue spots were yielded by spraying the dried chromatograms first with 1% aqueous ammonium molybdate and then with 1% stannous chloride in 10% HCl. The blue may appear rather later with hypophosphite.

CLESCERI and LEE [6] separate ortho- and pyrophosphate on cellulose layers using the solvent dioxan-distilled water-trichloroacetic acid-conc. NH_4OH (67 + 27.5 + 5 g + 0.25); the order of migration distances is:

$$P_2O_7^{4-} > PO_4^{3-}.$$

Mono- and diphosphoric acids have been separated on layers of cellulose MN 300 HR (Firm 83) by BAUDLER et al. [2]; good results were obtained in particular with the solvents:

a) methanol-conc. NH_4OH-water-trichloroacetic acid (55 + 5 + 40 + 3 g);

b) methanol-conc. NH_4OH-water-trichloroacetic acid (55 + 5 + 35 + 3.5 g).

3. Separation of Condensed Phosphates

AURENGE et al. [1] have separated both linear and cyclic condensed phosphates on cellulose layers. The best separation of linear polyphosphates was accomplished with water-ethanol-isobutanol-isopropanol-conc. NH_4OH-trichloroacetic acid (30 + 35 + 15 + 20 + 0.4 + 5 g). The best solvent for the cyclic polyphosphates was conc. NH_4OH-isobutanol-water-methanol-formic acid (9 + 10 + 31 + 50 + 0.3).

Detection was carried out with the help of the molybdophosphate reaction with subsequent reduction to "molybdenum blue" using H_2S.

RÖSSEL [20] suggests using layers containing no gypsum, since, on both organic and inorganic adsorbents containing it, only the metaphosphates and not the linear condensed phosphates can be separated. Maize or soluble starch in 2% amount have proved suitable binders. The starch is mixed in with the water used to prepare the slurry. The best separations of meta- and linear condensed phosphates are obtained on layers of acid-washed cellulose powder (Type S & S 142 dg, Firm 121) (Fig. 236). He uses mainly a solvent (MD 2) prepared from: 75 ml analytical grade methanol; 20 ml of a solution of 700 ml analytical grade isopropanol and 100 ml distilled water; 25 ml of a solution of 125 g analytical grade trichloroacetic acid and 32 ml 25% ammonium hydroxide (analytical grade), made up to 100 ml with distilled water; and 6 ml of a solution of 200 ml 96% acetic acid and 800 ml distilled water.

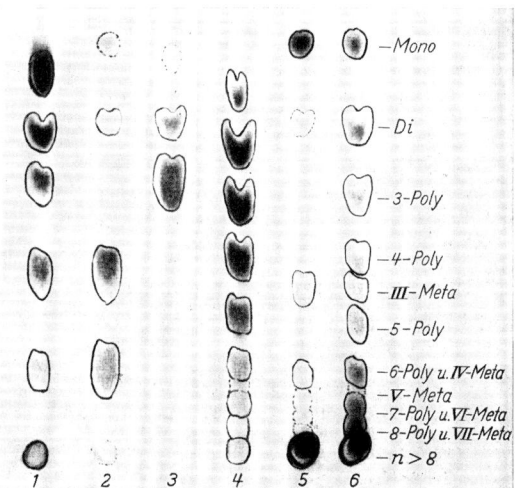

Fig. 236. Separation of condensed phosphates on a cellulose layer. Detailed information is in the text (RÖSSEL [20])

A mixture containing cyclic as well as the linear condensed polyphosphates can be successfully separated in a two-dimensional chromatogram (Fig. 237). In this, the acid solvent (MD2), mentioned above, is employed in the first direction; after drying, the basic solvent AD 2, is used at right angles to the first direction. AD 2 consists of: 67.5 ml analytical grade methanol; 22.6 ml of a mixture of 700 ml analytical grade isopropanol and 100 ml distilled water; 50 ml of a mixture of 75 g analytical grade trichloroacetic acid and 80 ml 25% ammonium hydroxide (analytical grade), made up tp 1000 ml with distilled water; and 6 ml of a mixture of 200 ml 96% analytical grade acetic acid and 800 ml distilled water.

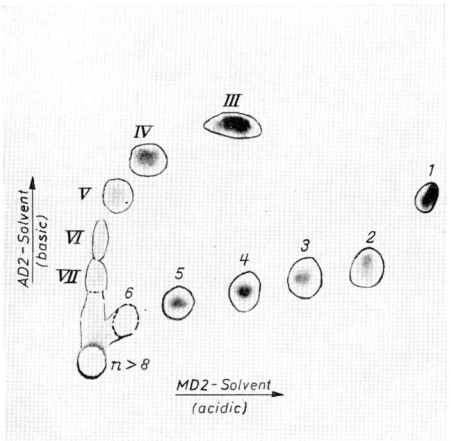

Fig. 237. Two dimensional chromatogram of a 1,1-basic phosphate glass on a cellulose layer. See text for details. *1* to *6* = linear phosphates, mono- and medium sized. III—VII = cyclic condensed phosphates from trimeta- to heptametaphosphate (RÖSSEL [20])

4. Separation of Sulphates and Polythionates [27, 28]

Separation is possible of mixtures of sulphates, sulphites, persulphates and thiosulphates; of thiosulphates, di-, tri-, tetra- and pentathionates; and of the anions present in Wackenroder's solution. The various oxidation products from reactions of thiosulphate with Cl_2, Br_2, I_2 and SO_2 can be identified with the help of these separations. The following sequences of migration distances are obtained using the solvents quoted below:

a) $S_2O_8^{2-} > SO_3^{2-} > S_2O_3^{2-} > SO_4^{2-}$,
b) $S_5O_6^{2-} > S_4O_6^{2-} > S_3O_6^{2-} > S_2O_6^{2-} > S_2O_3^{2-}$.

Layer: MN silica gel S-HR (Firm 83).

Standard samples: 1 µl of 1% solutions of the Na or K salts. The solutions should be freshly prepared each time since they change to some extent on standing.

Solvents: a) methanol-n-propanol-conc. NH_4OH-water(50 + 50 + 5 + 10).

b) methanol-dioxan-conc. NH_4OH-water (30 + 60 + 10 + 10).

Duration of run: 45 min for 15 cm.

Detection: a) 0.1 M $AgNO_3$ to which 2 N NH_4OH has been added so as just to dissolve the first formed precipitate. All anions with reducing properties or containing sulphide yield dark spots.

b) 0.1% aqueous bromocresol green solution, adjusted just to the colour change with dilute NH_4OH. The other anions give pale spots on a blue green background.

V. Quantitative Determination [31]

Two principles of quantitative determination may be distinguished.

A. The separated substances are extracted from the adsorbent and determined by a conventional method.

B. Quantitative determination directly on the chromatogram.

Although accurate micromethods are available for determining the substances according to principle A, appreciable errors can result during the scraping-off and elution of the spots.

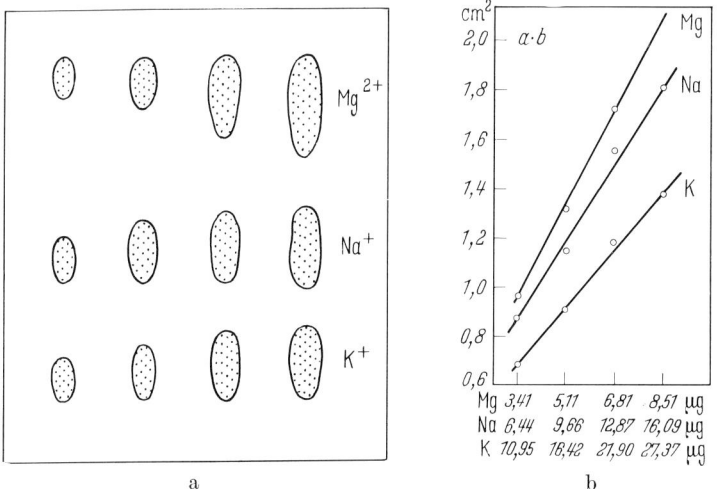

Fig. 238a. Separation of the metal ions, Na, K and Mg

Fig. 238b. Determination of Na, K and Mg, based on spot size

Several methods of direct determination (B) are possible (see also Chapter H, pp. 133—135):

1. Semi-quantitatively by visually comparing the spots (size and colour intensity) with those from known amounts; the error is ca. ± 30%.

2. Planimetric evaluation of the spot area, which is proportional to the concentration; it is difficult to decide exactly where the borders of the spots are; the error is about ±10% (Figs. 238a, b).

3. Photometric evaluation of the spots, which is more accurate; it can be carried out in transmitted and in reflected light; the latter

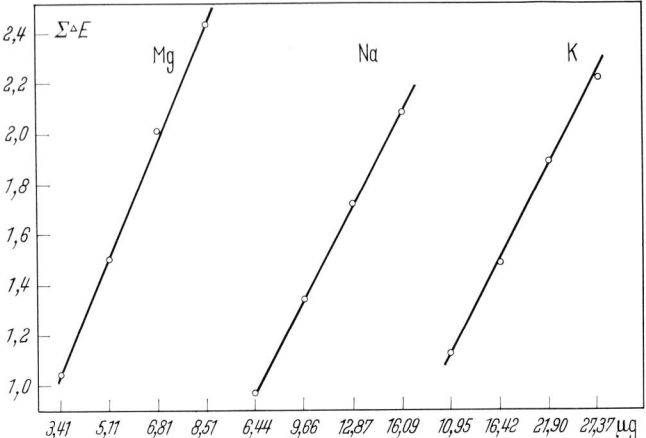

Fig. 239. Photometric determination of Na, K and Mg

yields better results but the former requires less elaborate apparatus and various instruments are commercially available (see pp. 140—144); the error is about ±4% (Fig. 239) (Apparatus of Firm 104).

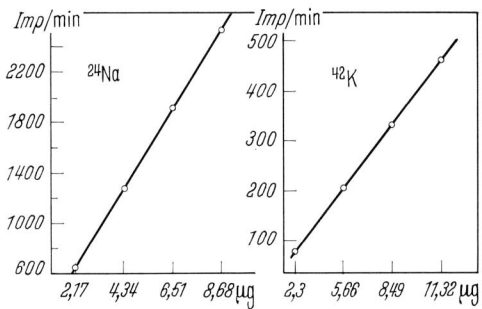

Fig. 240. Radiometric determination of Na and K

4. Using radioactive isotopes, an accurate and sensitive method (cf. p. 155); measurement of the total radioactivity of the whole spot or the "scanning" technique may be carried out; the error is less than ±1%, the principal error resulting from the spotting procedure (Fig. 240).

It has been shown that a small aperture is best in the photometric determination in transmitted light; the whole ray of light can then be passed through the substance spot in the "scanning" technique. Measurement is along the axis of symmetry of the spot, in the direction of migration and extinction values are recorded. The extinction is referred to the blank on the layer treated with solvent and detection reagent. The area under the extinction curves is proportional to the spot concentration; $\Sigma \Delta E = k \cdot c$.

Incident stray light must be excluded. The layer must withstand rubbing so that the scanning head can glide over it. Very stable layers can be prepared with starch as binder (MN silica gel S-HR, Firm 83). The entire measuring part must be shielded from lateral incident light since glass plates are highly transparent.

Bibliography for Chapter Y. Inorganic Ions

1. Aurenge, J., M. Degeorges et J. Normand: Bull. Soc. chim. Fr. **1964**, 508.
2. Baudler, M., u. M. Mengel: Z. analyt. Chem. **206**, 8 (1964); Baudlee, M., u. F. Stuhlmann: Naturwissenschaften **51**, 57 (1964)
3. Bäumler, J., u. S. Rippstein: Mitt. Lebensmitt.-Hyg. **54**, 57 (1963)
4. Berger, J. A., G. Meyniel et J. Petit: C. R. Acad. Sci. (Paris) **259**, 2231 (1964).
5. Canić, V. D., u. S. M. Petrović: Z. analyt. Chem. **211**, 321 (1965).
6. Clesceri, N. L., and G. F. Lee: Analyt. Chemistry **36**, 2207 (1964).
7. Druding, L. F.: Analyt. Chemistry **35**, 1582 (1963).
8. Gagliardi, E., u. W. Likussar: Mikrochim. Acta **1965**, 765.
9. — u. G. Pokorny: Mikrochim. Acta **1965**, 699.
10. Hashmi, M. H., M. A. Shahid u. A. A. Ayaz: Talanta **12**, 713 (1965).
11. Hranisavljević-Jakovljević, M., et al.: Thin-layer Chromatography. p. 221. Ed. G. B. Marini-Bettòlo. Amsterdam - London - New York: Elsevier 1964.
12. — — Mikrochim. et Ichnoanalyt. Acta **1965**, 141.
13. Lesigang, M.: Mikrochim. Ichnoanalyt. Acta 508 (1964).
14. Markl, P., u. F. Hecht: Mikrochim. Acta **1964**, 889.
15. — — Mikrochim. Acta **1963**, 970.
16. Meinhard, J. E., and N. F. Hall: Analyt. Chemistry **21**, 185 (1949).
17. Merkus, F. W. H. M.: Pharm. Weekblaad **98**, 947 (1963).
18. Moghissi, A.: J. Chromatog. **13**, 542 (1964).
19. — Analyt. chim. Acta **30**, 91 (1964).
20. Rössel, T.: Z. analyt. Chem. **197**, 333 (1963).
21. Seiler, H.: Helv. chim. Acta **45**, 381 (1962).
22. — u. M. Seiler: Helv. chim. Acta **44**, 939 (1961).
23. — — Helv. chim. Acta **43**, 1939 (1960).
24. — This book, 1. Edition.
25. — u. W. Rothweiler: Helv. chim. Acta **44**, 941 (1961).
26. — u. T. Kaffenberger: Helv. chim. Acta **44**, 1282 (1961).
27. — u. H. Erlenmeyer: Helv. chim. Acta **47**, 264 (1964).
28. — u. M. Seiler: Helv. chim. Acta **48**, 117 (1965).
29. — Chr. Biebricher u. H. Erlenmeyer: Helv. chim. Acta **46**, 2636 (1963).

30. SEILER, H.: Helv. chim. Acta **44**, 1753 (1961).
31. — Helv. chim. Acta **46**, 2629 (1963).
32. SHERMA, J.: J. Chromatog. **19**, 458 (1965).

The TLC of inorganic ions has been summarised by
F. H. POLLARD, K. W. C. BURTON, and D. LYONS: Lab. Pract. **13**, 505 (1964)

Z. Spray Reagents

K. G. KREBS, D. HEUSSER and H. WIMMER

The list below contains the most important of the known spray reagents. These are given in alphabetical order (of a principal component) in Part I. Part II is classified alphabetically according to compounds or compound classes to be identified and lists the reagents customary for each of these. Spray reagents which are named after authors or have well known abbreviations are compiled in Part III.

Many reagents are quoted for individual compound classes to enable the analyst, through judicious choice, to classify a compound in a definite class with more certainty. A warning must be given, however, against over-estimating the specificity of the reagent concerned. Thus reducing compounds such as ascorbic acid react as well as amino acids with ninhydrin. Unambiguous characterisation is therefore possible only by using several different reagents.

Only pure chemicals and solvents should be used for *preparation* of the spray reagents. "Ethanol" means 96% ethanol.

Relevant information is given about *toxic* or otherwise dangerous reagents or spray solutions. The usual laboratory rules and precautions apply to all work with spray reagents.

An analytical quartz lamp with a wave length maximum of 366 nm is used for inspection of chromatograms in long-wave UV light; the maximum wave length of the short wave UV lamp is ca. 254 nm (cf. p. 78).

Spraying. The reagent must be applied as an aerosol or fine spray in order to distribute it uniformly over the whole layer. Only propellent sprays fulfil this requirement, as mentioned on p. 80. Optimum pressure conditions must be observed with sprays of higher specific gravity, e. g., sulphuric acid reagents and aqueous solutions.

Some of the commonest reagents, e. g., ninhydrin, bromocresol green, aniline phthalate, are commercially available as aerosol sprayers. These atomise the reagent most efficiently with the simplest operation.

Reagent vessel and propellent gas container are separate in the combined sprayers offered by several firms (Fig. 38). The relevant reagent vessel is fixed to the container before use, using a plastic adapter. The spray solutions are thus exchangeable.

D. WALDI's scheme provides even spraying (see Fig. 241). The spray jet is swept several times over the layer in the way portrayed, at a distance of 30—40 cm.

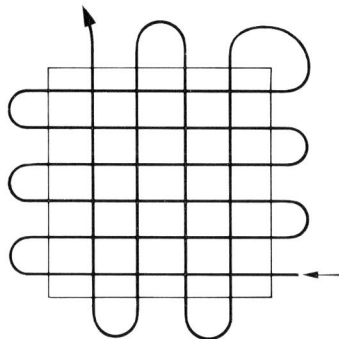

Fig. 241. Scheme for spraying. The jet of solution is swept over the layer surface in the direction indicated by the arrows

Note: Chromatograms must be sprayed in an efficient fume cupboard (cf. p. 81) especially as some reagents are toxic.

I. Preparation and Application of the Spray Reagents

1. **Acetic anhydride-sulphuric acid** (Liebermann-Burchard reagent): for Δ^5-3-sterols (cholesterol and esters) and for a number of steroids and triterpene glycosides.
   Spray reagent: 5 ml acetic anhydride are carefully mixed under cooling with 5 ml conc. sulphuric acid; this mixture is added cautiously to 50 ml abs. ethanol with cooling. Freshly prepared before use.
   Treatment after spraying: Heated 10 min at 100° C; fluorescing spots are observed in long-wave UV light.
   C. MICHALEČ: Biochem. et biophys. Acta **19**, 187 (1956).
   R. TSCHESCHE: J. Chromatog. **5**, 217 (1961).
   K. TAKEDA, S. HARA, A. WADA, and N. MATSUMOTO: J. Chromatog. **11**, 562 (1963).

2. **Alizarin:** for cations.
   Spray reagent: Saturated solution of alizarin in ethanol. The plate, while still moist, is placed in a chamber containing 25% ammonium hydroxide solution.
   G. DE VRIES, G. P. SCHÜTZ, and E. VAN DALEN: J. Chromatog. **13**, 119 (1964).

3. **Aluminium chloride:** for flavonoids.
   Spray reagent: 1% aluminium chloride solution in ethanol. Yields yellow fluorescence in long-wave UV light.
   T. G. Gage, C. D. Douglas, and S. H. Wender: Anal. Chem. **23**, 1582 (1951).

4. **4-Aminoantipyrine-potassium ferricyanide:** for phenols (Emerson reaction).
   Spray reagent I: 2% 4-aminoantipyrine solution in ethanol.
   Spray reagent II: 8% aqueous potassium ferricyanide solution.
   Procedure: After spraying first with I and then with II, the chromatogram is placed in a chamber containing 25% ammonium hydroxide; this yields red-orange to salmon pink spots
   G. Gabel, K. H. Müller u. J. Schoknecht: Dtsch. Apotheker-Ztg. **102**, 293 (1962).

5. **o-Aminobiphenyl-phosphoric acid** (modified according to Lewis-Smith): for sugars.
   Spray reagent: 0.3 g o-aminobiphenyl and 5 ml 85% phosphoric acid are dissolved in 95 ml ethanol.
   Treatment after spraying: Heated 15—20 min at 110° C. Sugars yield brown spots.
   T. E. Timell, C. P. J. Glaudemans, and A. L. Currie: Anal. Chem. **28**, 1916 (1956).

6. **4-Aminohippuric acid:** for reducing sugars.
   Spray reagent: 0.3% 4-aminohippuric acid in ethanol.
   Treatment after spraying: Heated 8 min at 140° C; the spots fluoresce in long-wave UV light.
   L. Sattler, and F. W. Zerban: Anal. Chem. **24**, 1862 (1952).

7. **Ammonium hydroxide:** for tetracyclines.
   Procedure: The plate with thin layer is placed in a vessel containing 25% ammonium hydroxide; the tetracyclines fluoresce yellow in long-wave UV light.
   M. Urx, J. Vondráčková, L. Kovařík, O. Horský, and M. Herold: J. Chromatog. **11**, 62 (1963).

8. **Aniline-diphenylamine-phosphoric acid:** for reducing sugars.
   Spray reagent: 4 g diphenylamine, 4 ml aniline and 20 ml 85% phosphoric acid are dissolved in 200 ml acetone.
   Treatment after spraying: Heated 10 min at 85° C; the reagent yields various colours; 1,4-aldohexose oligosaccharides turn blue.
   R. W. Bailey, and E. J. Bourne: J. Chromatog. **4**, 206 (1960).
   J. L. Buchan, and R. J. Savage: Analyst **77**, 401 (1952).
   S. Schwimmer, and A. Bevenne: Science **123**, 543 (1956).

9. **Aniline-phosphoric acid:** for sugars.
   Spray reagent: 2N aniline solution in n-butanol, saturated with water + 2N orthophosphoric acid in n-butanol (1 + 2, v/v).
   Treatment after spraying: Heated 10 min at 105° C.
   I. L. Bryson, and T. I. Mitchell: Nature **167**, 864 (1951).

10. **Aniline phthalate:** for reducing sugars and anions of halogen oxyacids.
    Spray reagent: 0.93 g aniline and 1.66 g o-phthalic acid are dissolved in 100 ml n-butanol, saturated with water.

Treatment after spraying: Heated 10 min at 105° C.
S. M. PARTRIDGE: Nature **164**, 443 (1949).
W. PESCHKE: J. Chromatog. **20**, 572 (1965).

11. **Anisaldehyde-sulphuric acid:** for sugars, steroids, terpenes, etc.
    Spray reagent: 1 ml conc. sulphuric acid is added to a solution of 0.5 ml anisaldehyde in 50 ml acetic acid. Freshly prepared before use.
    Treatment after spraying: Heated at 100—105° C until the spots attain maximum colour intensity. The pink background can be bleached by exposure to steam (water-bath).
    Lichen constituents, phenols, terpenes, sugars and steroids yield violet, blue, red, grey or green products, depending on the compound.
    Modified spray reagent: A freshly prepared mixture of 0.5 ml anisaldehyde, 9 ml ethanol, 0.5 ml conc. sulphuric acid and 0.1 ml acetic acid can be used for visualising sugars.
    Treatment after spraying: Heated 5—10 min at 90—100° C.
    E. STAHL, u. U. KALTENBACH: J. Chromatog. **5**, 351 (1961).
    B. P. LISBOA: J. Chromatog. **16**, 136 (1964).

12. **p-Anisidine phthalate:** for reducing sugars.
    Spray reagent: 0.1 M solution of p-anisidine and phthalic acid in 96% ethanol.
    Treatment after spraying: Heated 10 min at 100° C.

13. **Anthrone:** for ketoses.
    Spray reagent: 0.3 g anthrone are dissolved in 10 ml acetic acid and 20 ml ethanol; 3 ml 85% phosphoric acid and 1 ml water are added. The solution can be kept several weeks in a refrigerator.
    Treatment after spraying: Heated 5—6 min at 110° C. Ketoses and oligosaccharides containing ketoses appear as yellow spots.
    R. JOHANSON: Nature **172**, 956 (1953).

14. **Antimony(III) chloride:** for flavonoids.
    Spray reagent: 10% solution of antimony(III) chloride in chloroform; yields spots which fluoresce in long-wave UV light.
    L. HÖRHAMMER, H. WAGNER u. K. HEIN: J. Chromatog. **13**, 235 (1964).
    R. NEU u. P, HAGEDORN: Naturwissenschaften **40**, 411 (1953).

15. **Antimony(III) chloride** (Carr-Price reagent): for vitamins A and D, carotenoids, steroid sapogenins, steroid glycosides and terpene derivatives.
    Spray reagent: 25 g antimony(III) chloride are dissolved in 75 g chloroform; generally a saturated solution of the antimony(III) chloride in chloroform or carbon tetrachloride is used.
    Treatment after spraying: Heated 10 min at 100° C. The chromatogram is also inspected in long-wave UV light.
    E. STAHL: Chemiker-Ztg. **82**, 323 (1958).
    K. TAKEDA, S. HARA, A. WADA, and N. MATSUMOTO: J. Chromatog. **11**, 562 (1963).

16. Antimony(III) chloride-acetic acid: for steroids and diterpenes
Spray reagent: 20 g antimony(III) chloride are dissolved in a mixture of 20 ml acetic acid and 60 ml chloroform.
Treatment after spraying: Heated 5 min at 100° C. The diterpenes appear as reddish yellow to blue-violet zones. Inspected in long-wave UV light.
H. P. KAUFMANN u. A. K. SEN GUPTA: Chem. Ber. **97**, 2652 (1964).

17. Antimony(III) chloride-sulphuric acid: for bile acids.
Spray reagent: 20 g antimony(III) chloride are dissolved in 50 ml anhydrous n-butanol and mixed with 10 ml conc. sulphuric acid and 20 ml acetic acid. Freshly prepared before use.
Treatment after spraying: Dried 15 min in the air, then heated at 110° C for 25—30 min (with conjugated bile acids) or 45 to 50 min (free bile acids).
Colour reaction from yellow to green.
W. L. ANTHONY, and W. T. BEHER: J. Chromatog. **13**, 567 (1964).

18. Antimony(V) chloride: for vitamins A, D and E, terpenes, oils, resins, phenoxyalkane carboxylic esters, steroid sapogenins.
Spray reagent: Antimony(V) chloride + chloroform or carbon tetrachloride (1 + 4). Freshly prepared before use.
Treatment after spraying: Heated at 120° C until the spots appear. Evaluated also in long-wave UV light.
J. M. MACMAHON, R. B. DAVIS, and G. KALNITZKY: J. Am. Chem. Soc. **74**, 4483 (1952).
E. STAHL: Chemiker-Ztg. **82**, 323 (1958).
R. IKAN, J. KASHMAN, and E. D. BERGMANN: J. Chromatog. **14**, 275 (1964).
H. G. HENKEL, and W. EBING: J. Chromatog. **14**, 285 (1964).

19. Benzidine: for persulphates.
Spray reagent: 0.05 g benzidine is dissolved in 100 ml N acetic acid. Blue spots are formed immediately on spraying.
Y. SERVIGNE, et C. DUVAL: Compt. Rend. **245**, 1803 (1957).

20. Benzidine, diazotised: for phenols.
Stock benzidine solution: 5 g benzidine and 14 ml 36% hydrochloric acid are diluted to 1000 ml with water.
Nitrite solution: 10% solution of sodium nitrite in water. Prepared freshly before use.
Spray reagent: 20 ml of the benzidine solution are mixed with 20 ml of the nitrite solution at 0° C, stirring continuously.
Note: The reagent can be kept 2—3 h. The colour may appear very rapidly or after several hours, depending on the phenol.
J. SHERMA, and L. V. S. HOOD: J. Chromatog. **17**, 307 (1965).

21. Benzidine-trichloroacetic acid: for sugars.
Spray reagent: 0.5 g benzidine is dissolved in 10 ml acetic acid. 10 ml 40 % aqueous trichloroacetic acid added and the mixture diluted to 100 ml with ethanol.

Treatment after spraying: Exposed 15 min to unfiltered UV light. The sugars appear as grey-brown to deep red-brown spots.
Heating to 110° C also yields dark spots.

J. S. D. BACON, and J. EDELMANN: Biochem. J. **48**, 114 (1951).
G. HARRIS, and I. C. MACWILLIAM: Chem. & Ind. (London) **1954**, 254.

22. Bismuth(III) chloride: for sterols.
Spray reagent: 33% ethanolic solution of bismuth(III) chloride.
Treatment after spraying: Heated at 110° C until optimum fluorescence of the spots in long-wave UV light.

J. W. COPIUS-PEEREBOOM: in: Thin-Layer Chromatography, p. 199, edited by G. B. MARINI-BETTOLO, Amsterdam: Elsevier Publ. Co. 1964.

23. Bromine cyanide-4-aminobenzoic acid (König reagent): for pyridine compounds with at least one free α-position.
Pretreatment: Before spraying, the chromatogram is placed for 1 h in a chamber containing a beaker of bromine cyanide solution (toxic!).
The bromine cyanide solution is prepared by adding 10% aqueous sodium cyanide solution to saturated bromine water, cooled in ice, until it is decolorised.
Spray reagent: 2 g 4-aminobenzoic acid are dissolved in 75 ml 0.75 N hydrochloric acid and the solution made up to 100 ml with ethanol.

E. KODICEK, and K. K. REDDI: Nature **168**, 475 (1951).

Variation:
Spray reagent: 2% ethanolic 4-aminobenzoic acid solution + 0.1 M phosphate buffer, pH 7.0 (1 + 1).
Procedure: The plate is dried for 15 min at room temperature after spraying and then placed in a chamber containing a few crystals of bromine cyanide.

E. HODGSON, E. SMITH, and F. E. GUTHRIE: J. Chromatog. **20**, 176 (1965).

24. Bromine-fluorescein-silver nitrate: for insecticides.
Bromine solution: 5% in carbon tetrachloride.
Spray reagent I: 1 ml of a 0.25% solution of fluorescein in dimethylformamide is diluted to 50 ml with ethanol.
Spray reagent II: 1.7 g silver nitrate is dissolved in 5 ml water, 10 ml phenyl cellosolve(2-phenoxy-ethanol) added and the solution diluted to 200 ml with acetone.
Procedure: The plate is placed for 30 sec in a chamber containing the bromine solution and then sprayed first with solution I and then with II; it is subsequently exposed to long-wave UV light for 7 min.

K. C. WALKER, and M. BEROZA: J. Assoc. Off. Agr. Chemists **46**, 250 (1963).

25. Bromocresol green: indicator.
Spray reagent: 0.04 g bromocresol green is dissolved in 100 ml ethanol and 0.1 N sodium hydroxide added until the blue colour just appears.

F. BRYANT, and B. T. OVERELL: Biochim. et Biophys. Acta **10**, 471 (1953).

26. Bromocresol green-bromophenol blue-potassium permanganate: for organic acids.

Solution a: 0.075 g bromocresol green and 0.025 g bromophenol blue are dissolved in 100 ml absolute ethanol.

Solution b: 0.25 g potassium permanganate and 0.5 g sodium carbonate ($Na_2CO_3 \cdot 10\,H_2O$) are dissolved in water and the volume made up to 100 ml.

Spray reagent: Solution a and b are mixed 9:1 (volume) and used immediately for spraying; the mixture can be kept only 5—10 min.

J. Pásková, and V. J. Munk: J. Chromatog. **4**, 241 (1960).

27. Bromocresol purple: for halide ions when the solvent acetone-n-butanol-25% ammonium hydroxide-water (65 + 20 + 10 + 5) is used. Indicator reagent, since the anions migrate as ammonium salts.

Spray reagent: 0.1% bromocresol purple solution in ethanol is treated with 10% ammonium hydroxide dropwise until the colour change just appears.

H. Seiler u. T. Kaffenberger: Helv. chim. Acta **44**, 1282 (1961).

28. Bromocresol purple: for dicarboxylic acids on layers impregnated with polyethylene glycol.

Spray reagent: 0.04 g bromocresol purple is dissolved in 100 ml 50% ethanol and the pH adjusted to 10.0 (glass electrode) using 0.1 N sodium hydroxide.

Procedure: The TLC layers which have been developed with diisopropyl ether-formic acid-water (90 + 7 + 3) are heated for exactly 10 min at precisely 100° C. They are sprayed after having completely cooled to room temperature, and yield yellow spots on a blue background.

E. Knappe u. D. Peteri: Z. anal. Chem. **188**, 184 (1962).

29. Bromophenol blue-methyl red-Pauly reagent: for phenols.

Spray reagent I: A mixture is made of 100 ml amounts of 0.12% aqueous bromophenol blue solution, 0.06% ethanolic methyl red solution and Sörensen-phosphate buffer, pH 7.2.

Spray reagent II: see Rgt. No. 238 (diazotised sulphanilic acid).

Procedure: The layer is sprayed first with I and then with II.

J. W. Copius-Peereboom, and H. W. Beekes: J. Chromatog. **14**, 417 (1964).

30. Bromosuccinimide-fluorescein: for vulcanisation accelerators.

Spray reagent I: 0.5 g N-bromosuccinimide is dissolved in 100 ml acetic acid.

Spray reagent II: 0.01 g fluorescein is dissolved in 100 ml ethanol.

Procedure: The chromatograms are sprayed first with I and then with II. The sprayed layers are observed in daylight and in long-wave UV light.

A. Popov, and V. Gadeva: J. Chromatog. **16**, 256 (1964).

31. Bromothymol blue: for lipoids.

Spray reagent: 0.04 g bromothymol blue is dissolved in 100 ml 0.01 N sodium hydroxide.

H. Jatzkewitz, u. E. Mehl: Hoppe-Seylers Z. physiol. Chem. **320**, 251 (1960).

32. Carbazole-sulphuric acid: for sugars.
Spray reagent: 0.5 g carbazole is dissolved in 95 ml ethanol and 5 ml conc. sulphuric acid added. Freshly prepared before use.
Treatment after spraying: Heated 10 min at 120° C, yielding violet spots on a blue background.
S. ADACHI: J. Chromatog. **17**, 295 (1965).

33. Ceric(cerium(IV)) nitrate, ammonium-N,N-dimethyl-p-phenylenediammonium dichloride: for polyalcohols.
Solution a: 1% solution of ammonium hexanitratocerate (IV) in 0.2 N nitric acid.
Solution b: 1.5 g N,N-dimethyl-p-phenylenediammonium dichloride is dissolved in a mixture of 128 ml methanol, 25 ml water and 1.5 ml acetic acid.
Spray reagent: Solutions a and b are mixed 1:10 (v/v) before use.
Treatment after spraying: Heated 10 min at 105° C, giving yellowish green spots on a red background.
E. KNAPPE, D. PETERI, and J. ROHDEWALD: Z. anal. Chem. **199**, 270 (1964).

34. Ceric(cerium(IV)) sulphate, ammonium: for *Vinca*-alkaloids.
Spray reagent: 1% solution of ammonium ceric sulphate in 85% phosphoric acid.
I. M. JAKOVLJEVIC, L. D. SEAY, and R. W. SHAFFER: J. Pharm. Sci. **53**, 553 (1964).

35. Ceric (cerium(IV))sulphate-nitric acid: for polyphenyls.
Spray reagent: 0.3 g ceric sulphate is dissolved in 100 ml 65% nitric acid.
Treatment after spraying: Heated 15—20 min at 120° C. Inspection is carried out in long-wave UV light.
F. GEISS u. H. SCHLITT: Euratom-Report EUR-I-19d (Nov. 1961).

36. Ceric sulphate-sulphuric acid: for *Solanum* steroid alkaloids and steroid sapogenins.
Spray reagent: Saturated solution of ceric sulphate in 65% sulphuric acid.
Treatment after spraying: Heated 15 min at 120° C.
Note: Cannot be used on alumina layers.
K. SCHREIBER, O. AURICH u. G. OSSKE: J. Chromatog. **12**, 63 (1963).

37. Ceric sulphate-sulphuric acid (modified according to Sonnenschein): for alkaloids and iodine-containing organic compounds.
Spray reagent: 0.1 g ceric sulphate is suspended in 4 ml water; 1 g trichloroacetic acid is added, the solution boiled and conc. sulphuric acid added dropwise until turbidity disappears.
Treatment after spraying: Heated at 110° C for some minutes until the spots appear.
Note: Apomorphine, brucine, colchicine, papaverine and physostigmine respond to the reagent; organic iodides also can be detected by it.
O.-E. SCHULTZ u. D. STRAUSS: Arzneimittel-Forsch. **5**, 342 (1955).

38. **Chloramine-T:** for caffeine.
    Spray reagent I: 10% aqueous chloramine T solution.
    Spray reagent II: N hydrochloric acid.
    Procedure: The layer is sprayed with I, briefly dried and then sprayed with II. After having heated to 96—98° C until the smell of chlorine has disappeared, the plate is exposed to ammonia (ca. 5 min) in a chamber containing 25% ammonium hydroxide and finally warmed again until the pink-red colour of the spots attains its maximum intensity.
    H. GÄNSHIRT u. A. MALZACHER: Arch. Pharm. **293**, 925 (1960).

39. **Chloramine T-trichloroacetic acid:** for digitalis glycosides.
    Solution a: Freshly prepared 3% aqueous solution of chloramine T.
    Solution b: 25% ethanolic solution of trichloroacetic acid (which can be kept for a few days).
    Spray reagent: 10 ml a and 40 ml b are mixed before use.
    Procedure: Heated 7 min at 110° C; bluish or yellow fluorescence is observed in long-wave UV light.
    D. WALDI: Arch. Pharm. **292**, 206 (1959).

40. **Chlorine cyanide-4-aminobenzoic acid:** for pyridine compounds with at least one free α-position.
    Spray reagent: 5% methanolic solution of 4-aminobenzoic acid.
    Procedure: The sprayed plate is introduced into a chamber containing a freshly prepared mixture of 20 ml 28% chloramine suspension in water, 20 ml N hydrochloric acid and 10 ml 10% aqueous potassium cyanide solution. The spots appear after brief exposure (poisonous vapour and solution).
    E. NÜRNBERG: Dtsch. Apotheker-Ztg. **101**, 142 (1961).

41. **Chlorine-pyrazolinone-cyanide:** for indoles, amides and sulphonamides.
    Chlorination: The plate is placed for 2—3 min in an atmosphere of chlorine, prepared from potassium permanganate and 25% hydrochloric acid; it is then allowed to stand 5 min in the air or, better, heated in a drying oven at 100° C to remove excess chlorine.
    Solution a: 0.2 M 1-phenyl-3-methyl-2-pyrazolin-5-one solution in pyridine.
    Solution b: N potassium cyanide solution in water.
    Spray reagent: Equal volumes of a and b are mixed.
    Procedure: After removal of the excess chlorine, the plate is sprayed until just appearing transparent (caution! poison!). The spots are bright red, turning blue after ca. 2 min.
    M. R. F. ASHWORTH, and G. BOHNSTEDT: Talanta **13**, 1631 (1966).

42. **Chlorine-tolidine:** for nitrogen containing compounds which can be converted into chloramines.
    Chlorination: The plate is placed in an atmosphere of chlorine. 5—10 min exposure suffices if the chlorine is

Spray reagent: obtained from a cylinder; 15—20 min if it is prepared from a mixture of 1.5% potassium permanganate solution and 10% hydrochloric acid (1 + 1). Excess chlorine is then removed by allowing the plates to stand some 5 min in the air.

Spray reagent: 160 mg o-tolidine are dissolved in 30 ml acetic acid, the solution diluted to 500 ml with distilled water and 1 g potassium iodide then added.

Note: A corner of the chromatogram is first sprayed cautiously to establish if chlorine has been completely removed (no blue colour should appear). Only when this condition is fulfilled can full spraying be carried out.

F. REINDEL u. W. HOPPE: Chem. Ber. **87**, 1103 (1954).
G. PATAKI: J. Chromatog. **12**, 541 (1963).

43. Chlorine-tolidine (modified by Greig and Leaback).
Spray reagent I: 2% aqueous potassium hypochlorite solution.
Spray reagent II: Equal volumes of saturated o-tolidine solution in 2% acetic acid and 0.85% aqueous potassium iodide are mixed before use.
Procedure: The plate is lightly sprayed with I, left 1—1.5 h at room temperature and then sprayed evenly with II.

C. G. GREIG, and D. H. LEABACK: Nature **188**, 310 (1960).

44. 1-Chloro-2,4-dinitrobenzene: indicator.
Spray reagent: 0.5% of 1-chloro-2,4-dinitrobenzene in ethanol

45. Chlorosulphonic acid-acetic acid: for triterpenes, sterols and steroids.
Spray reagent: 5 ml chlorosulphonic acid are dissolved in 10 ml acetic acid under cooling.
Treatment after spraying: Heated 5—10 min at 130° C and fluorescence in long-wave UV light noted.

R. TSCHESCHE u. G. WULF: Chem. Ber. **94**, 2019 (1961).
R. TSCHESCHE: J. Chromatog. **5**, 217 (1961).
K. TAKEDA, S. HARA, A. WADA, and N. MATSUMOTO: J. Chromatog. **11**, 562 (1963).

46. Chromic acid-sulphuric acid: reagent for general detection of organic substances.
Spray reagent: 5 g potassium dichromate are dissolved in 100 ml 40% sulphuric acid.
Note: The reagent is especially suitable for charring organic substances (in particular, lipids); the plate is heated to 150° C for this purpose.

J. BERTETTI: Ann. Chim. (Rome) **44**, 495 (1954).

47. Chromotropic acid: for compounds capable of splitting off formaldehyde, e. g., those containing the methylenedioxy group (narcotine, hydrastine, sesamine etc.).
Solution a: 10% aqueous solution of sodium 1.8-dihydroxynaphthalene-3,6-disulphonate.
Solution b: Conc. sulphuric acid and water are mixed in the volume ratio of 5:3 and cooled to room temperature.

Spray reagent: Solutions a and b are mixed (1 + 5, v + v) before use.
Treatment after spraying: Heated 30 min at 105° C.

M. BEROZA: Agricult. and Food Chemistry **11**, 51 (1963).

48. Cinnamaldehyde-acetic anhydride-sulphuric acid: for steroid-sapogenins.
Spray reagent I: 1% ethanolic solution of cinnamaldehyde.
Spray reagent II: A mixture of acetic anhydride and conc. sulphuric acid, 12:1 by volume, freshly prepared before use.
Procedure: The layer is sprayed with I, dried 5 min at 90° C and sprayed with II. After 1—2 min at room temperature, the chromatogram is heated at 90° C in a drying oven until the spots appear.

49. Cinnamaldehyde-hydrochloric acid: for indole derivatives.
Spray reagent: 5 ml cinnamaldehyde are diluted to 100 ml with ethanol and 5 ml 36% hydrochloric acid added. Freshly prepared before use.
Treatment after spraying: The layers are exposed to hydrogen chloride yielding red spots.

D. JERCHEL u. R. MÜLLER: Naturwissenschaften **38**, 561 (1951).

50. Cobalt(II) chloride: for organic phosphate esters.
Spray reagent: 1% solution of anhydrous cobalt(II) chloricde in acetone.
Treatment after spraying: Warmed to 40—50° C. Blue spots are yielded with this rather insensitive reagent.

R. DONNER u. K. LOHS: J. Chromatog. **17**, 349 (1965).

51. Cobalt(II) nitrate-ammonia (Zwikker reagent): for barbiturates.
Spray reagent: 1% solution of cobalt(II)nitrate in absolute ethanol.
Treatment after spraying: The layer is dried at room temperature and the plate introduced into a chamber, saturated with water-vapour and containing 25 % ammonium hydroxide solution.

E. J. SHELLARD, and J. V. OSISIOGU: Lab. Practice **13**, 516 (1964).

52. Cobalt(II) nitrate-lithium hydroxide: for barbiturates.
Spray reagent I: 2% cobalt(II)nitrate solution in anhydrous methanol.
Spray reagent II: 0.5% lithium hydroxide in methanol.
Procedure: The layer is sprayed with I, dried in the air and sprayed with II.

H. WEIDMANN: Dissertation, Berlin 1961.

53. Cobalt(II) thiocyanate: for alkaloids and for primary, secondary and tertiary amines.
Spray reagent: 3 g ammonium thiocyanate and 1 g cobalt (II) chloride are dissolved in 20 ml water.
Note: Alkaloids and amines appear as blue spots on a white to pink background.
The colours fade after about 2 h but can be revived by spraying with water or introducing the plate into a moisture-saturated atmosphere.

E. S. LANE: J. Chromatog. **18**, 426 (1965).

54. Cupric chloride: for oximes.
Spray reagent: 0.5% aqueous solution of cupric chloride,
Note: Complexes of β-oximes appear as green zones immediately after spraying; those of α-oximes only after 10 min heating at 110° C, then as green-brown spots.

M. Hranisavljević-Jakovljević, I. Pejković-Tadić, and A. Stojiljković: J. Chromatog. **12**, 70 (1963).

55. Cupric sulphate-benzidine: for pyridine monocarboxylic acids.
Spray reagent I: 0.3 g cupric sulphate is dissolved in 100 ml of a mixture of water-ethanol (5 + 4).
Spray reagent II: 0.1% solution of benzidine in 50% ethanol.
Procedure: The layer is sprayed with I, dried at 60° C and sprayed then with II. Blue spots are obtained.

56. Cupric sulphate-citrate (sodium) (Benedict's reagent): for flavonoids and coumarins with the o-dihydroxy group.
Spray reagent: 1.73 g cupric sulphate ($CuSO_4 \cdot 5H_2O$), 17.3 g sodium citrate and 10 g anhydrous sodium carbonate are dissolved in water and the volume made up to 100 ml with water.
Note: The fluorescence in long-wave UV light of compounds with the o-dihydroxy group is reduced or totally quenched by the reagent; that of compounds of these classes, but without this particular group, is unchanged or may be intensified (in which case a change of fluorescence colour often occurs).

H. Reznik u. K. Egger: Z. anal. Chem. **183**, 196 (1961).

57. Cupric sulphate-quinine-pyridine: for barbiturates and thiobarbiturates
Spray reagent I: 0.2 g cupric sulphate and 0.02 g quinine hydrochloride are dissolved in 50 ml water, 2 ml pyridine added and the mixture made up to 100 ml with water.
Spray reagent II: 0.5% aqueous potassium permanganate.
Procedure a: The chromatogram is sprayed with I and dried at room temperature.
Note: White, yellow or violet spots are seen in daylight; dark spots on a fluorescent background in long-wave UV light.
Procedure b: Spraying with II is then carried out.
Note: Yellow or white spots are yielded.

M. Frahm, A. Gottesleben u. K. Soehring: Pharm. Acta Helv. **38**, 785 (1963).

58. α-Cyclodextrin[1]: for straight-chain lipids.
Spray reagent: 30% ethanolic solution of α-cyclodextrin.
Treatment after spraying: The layer is dried at room temperature and the plate placed in a vessel containing iodine vapour.

[1] Preparation: K. Freudenberg et al.: Liebig's Ann. Chem. **558**, 1 (1947). D. French et al.: J. Am. Chem. Soc. **71**, 353 (1949).

D. C. Malins, and H. K. Mangold: J. Am. Oil Chemists' Soc. **37**, 576 (1960).
H. K. Mangold, J. L. Gellerman, and H. Schlenk: Federation Proc. **17**, 269 (1958).
H. K. Mangold, B. G. Lamp, and H. Schlenk: J. Am. Chem. Soc. **77**, 6070 (1955).

59. 3,5-Diaminobenzoic acid-phosphoric acid (quinaldine reaction): for 2-deoxysugars.

Spray reagent: 1 g 3,5-diaminobenzoic acid dihydrochloride is dissolved in 25 ml 80% phosphoric acid and diluted with 60 ml water.

Treatment after spraying: Heated 15 min at 100° C. The spots fluoresce green-yellow in long-wave UV light. Amounts of over 2 µg can be detected as brown spots in daylight.

M. Pesez: Bull. soc. chim. biol. **32**, 701 (1950).

60. o-Dianisidine: for aldehydes and ketones.

Spray reagent: Saturated solution of o-dianisidine in acetic acid.
Note: Colour differentiation is good.
o-Dianisidine may be replaced by 2,7-diaminofluorene in some cases.

R. Wasicky, u. O. Frehden: Mikrochim. Acta **1**, 55 (1937).

61. Diazotisation and Coupling with α-Naphthol: for aromatic primary amines and sulphonamides.

Spray reagent I: Freshly prepared 1% sodium nitrite solution in N hydrochloric acid.
Spray reagent II: Freshly prepared 0.2% α-naphthol solution in N potassium hydroxide.
Procedure: The layer is sprayed with I, left 1 min and then sprayed with II. The chromatogram is dried at 60° C.
Note: A 0.4% solution of N-(1-naphthyl)-ethylenediammonium dichloride in methanol may be used as coupling component instead of α-naphthol.

A. C. Bratton, and E. K. Marshall Jr., J. Biol. Chem. **128**, 537 (1939).
A. Wankmüller: Naturwissenschaften **39**, 302 (1952).
G. Wagner: Arch. Pharm. **285**, 409 (1952).
T. Bićan-Fišter, and V. Kajganović: J. Chromatog. **11**, 492 (1963).

62. 2,6-Dibromoquinonechloroimide (Gibbs reagent): for phenols.

Spray reagent: Freshly prepared 0.4% methanolic solution of 2,6-dibromoquinonechloroimide.
Treatment after spraying: Sprayed with 10% aqueous sodium carbonate solution or placed in a chamber containing 25% ammonium hydroxide.

E. Nürnberg: Dtsch. Apotheker-Ztg. **101**, 268 (1961).

63. 2′,7′-Dichlorofluorescein: fluorescence indicator for saturated and unsaturated lipids.

Spray reagent A: 0.2% ethanolic solution of 2′,7′-dichlorofluorescein.
Spray reagent B: 0.01% ethanolic solution of 2′,7′-dichlorofluorescein (for vitamin E).

Treatment after spraying: It is sometimes advisable to hold the plate in a current of warm air or steam after having dried, or to spray it with water. It is inspected in long-wave UV light.

D. C. MALINS, and H. K. MANGOLD: J. Am. Oil Chemists' Soc. **37**, 576 (1960).
P. J. DUNPHY, K. J. WHITTLE, and J. F. PENNOCK: Chem. & Ind. (London) **1965**, 1217.

64. 2,6-Dichlorophenol-indophenol-silver nitrate: for alkali chlorides.

Spray reagent: A solution is made of 0.2 g 2,6-dichlorophenol-indophenol (sodium salt) in 100 ml ethanol. 3 g silver nitrate are added and the mixture thoroughly shaken and then filtered. Prepared freshly each time.

T. BARNABAS, M. G. BADVE u. J. BARNABAS: Naturwissenschaften **41**, 478 (1954).

65. 2,6-Dichlorophenol-indophenol, sodium salt: for organic acids and keto acids.

Spray reagent: 0.1% ethanolic solution of the sodium salt of 2,6-dichlorophenol-indophenol.

Treatment after spraying: The acids appear as red spots on a blue background after briefly heating.

C. PASSERA, A. PEDROTTI, and G. FERRARI: J. Chromatog. **14**, 289 (1964).

66. 2,6-Dichloroquinonechloroimide: for antioxidants, adrenaline and its derivatives, cyanamide and derivatives.

Spray reagent: 0.1—1% solution of 2,6-dichloroquinonechloroimide in absolute ethanol; the solution is stable for about 3 weeks in the refrigerator (cannot be used for urea).

Note: The spot colours appear clearly after about 15 min. Some antioxidants show characteristic colour changes after spraying with a 2% borax solution in 40% ethanol.

A. SEHER: Fette u. Seifen, Anstrichmittel **61**, 345 (1959).
R. F. v. d. HEIDE u. O. WOUTERS: Z. Lebensm.-Untersuch. u. -Forsch. **115**, 129 (1962).
R. SEGURA-CARDONA u. K. SOEHRING: Med. Exp. **10**, 251 (1964).

67. Dicobalt octacarbonyl: for acetylene compounds.

Spray reagent I: 0.5 g dicobalt octacarbonyl is dissolved in 100 ml petroleum ether (BP 120—135° C).

Spray reagent II: 1 N hydrochloric acid.

Procedure: The layer is sprayed with I, left 10 min and sprayed with II; it is then allowed to dry and is peeled off using the Neatan® procedure (p. 128). Excess reagent is washed out and the chromatogram exposed to bromine vapour; yellow zones appear.

K. E. SCHULTE, F. AHRENS u. E. SPRENGER: Pharm. Ztg. **108**, 1165 (1963).

68. Diethylamine-cupric sulphate: for thiobarbiturates.

Spray reagent: 0.5 g cupric sulphate is dissolved in 100 ml methanol and 3 ml diethylamine are added to the solution.

Note: The reagent can be kept several days and must be shaken before use. Thiobarbiturates yield green spots.

W. DIETZ, u. K. SOEHRING: Arch. Pharm. **290**, 80 (1957).

69. Dimedone-phosphoric acid: for keto-sugars.
Spray reagent: 0.3 g dimedone (5,5-dimethylcyclohexane-1,3-dione) is dissolved in 90 ml ethanol and 10 ml 85% phosphoric acid are added.
Treatment after spraying: Heated 15—20 min at 110° C. Yellow spots on a white background are seen in daylight; blue fluorescent spots in long-wave UV light.

70. 4-Dimethylaminobenzaldehyde-acetic acid-phosphoric acid (EP-reagent): for proazulenes and azulenes.
Spray reagent: 0.25 g 4-dimethylaminobenzaldehyde is dissolved in a mixture of 50 g acetic acid, 5 g 85% phosphoric acid and 20 ml water (can be kept for months in a container of dark glass).
Note: Azulene hydrocarbons yield intense blue spots at room temperature. Proazulenes appear as blue spots only after 10 min heating at 80° C. The colours fade later and turn to green and yellow shades; the intense blue can be regenerated by exposure to steam over a water bath.

E. STAHL: Dtsch. Apotheker-Ztg. **93**, 197 (1953).
H. KAISER u. G. HASENMAYER: Arch. Pharm. **287**, 503 (1954).

71. 4-Dimethylaminobenzaldehyde-acetylacetone (Morgan-Elson reagent): for amino-sugars.
Spray reagent I: 0.5 ml of a mixture of 5 ml 50% aqueous potassium hydroxide and 20 ml ethanol with 10 ml of a solution of 0.5 ml acetylacetone in 50 ml n-butanol, both solutions freshly prepared, are mixed just before use.
Spray reagent II: 1 g 4-dimethylaminobenzaldehyde is dissolved in 30 ml ethanol and 30 ml 36% hydrochloric acid are added. The mixture may be diluted with 180 ml n-butanol as required.
Procedure: The layer is sprayed with I, heated 5 min at 105° C, sprayed with II and then dried 5 min at 90° C. Red spots are yielded.

L. A. ELSON, and W. T. J. MORGAN: Biochem. J. **27**, 1824 (1933).
R. BELCHER, A. J. MUTTEN, and C. M. SABROOK: Analyst **79**, 201 (1954).

72. 4-Dimethylaminobenzaldehyde-hydrochloric acid (Ehrlich reagent): for amines.
Spray solution A: A solution is made of 1 g 4-dimethylaminobenzaldehyde in a mixture of 25 ml 36% hydrochloric acid and 75 ml methanol.
Treatment after spraying: The plate has to be warmed in some cases.
Spray solution B: 1 g 4-dimethylaminobenzaldehyde is dissolved in 96% ethanol and the volume made up to 100 ml.
Treatment after spraying: The chromatogram sprayed with B is placed for 3—5 min in a vessel saturated with hydrogen chloride vapour, or is sprayed with 25% hydro-

chloric acid. The plate may sometimes need warming.
R. A. HEACOCK, and M. E. MAHON: J. Chromatog. **17**, 338 (1965).

73. 4-Dimethylaminobenzaldehyde-hydrochloric acid (van Urk reagent): for indole derivatives (modified according to Stahl).

Spray reagent: 1 g 4-dimethylaminobenzaldehyde is dissolved in 50 ml 36% hydrochloric acid and 50 ml ethanol are added.

Note: The plate must be warmed to about 50° C before spraying in order to evaporate any volatile, basic component of the developing solvent.

Procedure: The layer is liberally sprayed until it appears transparent; it is then exposed to vapours of aqua regia. Various colours (observed in daylight) are yielded.

E. STAHL u. H. KALDEWEY: Hoppe-Seylers Z. physiol. Chem. **323**, 182 (1961).

74. 4-Dimethylaminobenzaldehyde-sulphuric acid: for ergot alkaloids.

Spray reagent: 125 mg 4-dimethylaminobenzaldehyde are dissolved in a cooled mixture of 65 ml conc. sulphuric acid and 35 ml water; 0.05 ml 5% ferric chloride solution in water is then added. The solution can be kept about a week.

M. ZINSER u. C. BAUMGÄRTEL: Arch. Pharm. **297**, 158 (1964).

75. 5-(4'-Dimethylaminobenzylidene)-rhodanine: for Ag^+, Cu^{2+} and Hg^{2+} ions.

Spray reagent: 1 g of the reagent is dissolved in ethanol and the solution diluted to 100 ml with this solvent.

Treatment after spraying: The layer is sprayed with 25% ammonium hydroxide or placed in a chamber containing this reagent; pink to violet spots are yielded.

F. W. H. M. MERKUS: Pharm. Weekblad **98**, 955 (1963).

76. 4-Dimethylaminocinnamaldehyde: for indoles.

Stock solution: 2 g 4-dimethylaminocinnamaldehyde are dissolved in a mixture of 100 ml 6N hydrochloric acid and 100 ml ethanol and the solution stored in a refrigerator.

Spray reagent: The stock solution is diluted with 4 times its volume of ethanol.

Treatment after spraying: Heated 5 min at 105° C. Vapours of aqua regia blown over the layer intensify the spot colours.

Note: The reagent is unsuitable when ammonia-containing developing solvents are used since the background becomes coloured; this can be mitigated by briefly heating (10 min at 105° C) before spraying.

J. HARLEY-MASON, and A. A. P. G. ARCHER: Biochem. J. **69**, 60 (1958).

77. N,N-Dimethyl-p-phenylenediammonium dichloride: for bromine-containing hypnotics and chlorine-containing compounds (insecticides).

Spray reagent: 0.5 g N,N-dimethyl-p-phenylenediammonium dichloride is dissolved in 100 ml sodium alkoxide solution (1 g sodium in 100 ml ethanol).

Procedure: The chromatogram is moistened with a water spray immediately after spraying with the reagent and irradiated 1 min with unfiltered UV light. This liberates free halogen which oxidises the reagent to Wurster's red.

J. Bäumler u. S. Rippstein: Helv. Chim. Acta **44**, 1162 (1961).

78. N,N-Dimethyl-p-phenylenediammonium dichloride: for peroxides.
Spray reagent: 1.5 g N,N-dimethyl-p-phenylenediammonium dichloride is dissolved in a mixture of 128 ml methanol, 25 ml water and 1 ml acetic acid. Peroxides appear as purple spots.

E. Knappe u. D. Peteri: Z. anal. Chem. **190**, 386 (1962).

79. m-Dinitrobenzene: for 17-ketosteroids.
Solution a: 2% ethanolic solution of m-dinitrobenzene.
Solution b: 2.5 N methanolic potassium hydroxide.
Spray reagent: Equal volumes of a and b are mixed.
Treatment after spraying: Heated 1—2 min at 80° C, yielding violet spots.

T. Feher: Mikrochim. Acta **1965**, 105.
B. P. Lisboa: J. Chromatog. **16**, 136 (1964).
R. Neher: Steroid Chromatography, Elsevier 1964, Amsterdam, London, New York.

80. 3,5-Dinitrobenzoic acid (Kedde reagent): for cardiac glycosides.
Spray reagent A: 1 g 3,5-dinitrobenzoic acid is dissolved in a mixture of 50 ml methanol and 50 ml 2N potassium hydroxide.
Spray reagent B I: 2% methanolic solution of 3,5-dinitrobenzoic acid.
Spray reagent B II: 5.7 g potassium hydroxide are dissolved in methanol and the volume made up to 100 ml with it.
Procedure: The layer is first lightly sprayed with I and then with excess II. Blue-violet spots appear.

R. Tschesche, G. Grimmer u. F. Seehofer: Chem. Ber. **86**, 1235 (1953).
M. L. Lewbart, W. Wehrli u. T. Reichstein: Helv. Chim. Acta **46**, 505 (1963).

81. 2,4-Dinitrofluorobenzene: for amino acids.
Spray reagent I: Buffer solution, prepared by dissolving 8.4 g sodium bicarbonate in 80 ml water, adding 2.5 ml N sodium hydroxide and making up to 100 ml with water.
Spray reagent II: 10% solution of 2,4-dinitrofluorobenzene in methanol.
Procedure: The chromatogram is sprayed with I and then with II. A 5 mm margin is scraped from both sides of the plate. Two polyethylene strips of suitable breadth are laid on the margins so that a second glass plate can be placed on the layer. After heating 1 h at 40° C in the dark, the carrier plate is cooled and laid in an ether bath for 10 min. The layer is then dried briefly and the spots outlined.

G. Pataki: J. Chromatog. **16**, 541 (1964).

82. 2,4-Dinitrophenylhydrazine: for free aldehyde and keto groups and for ketoses.

Spray reagent A: 0.4% solution of 2,4-dinitrophenylhydrazine in 2N hydrochloric acid.

Spray reagent B: 10 ml 36% hydrochloric acid are added to a solution of 1 g 2,4-dinitrophenylhydrazine in 1000 ml ethanol.

Treatment after spraying: The 2,4-DNPs may be differentiated by subsequent spraying with a 0.2% solution of potassium ferricyanide in 2N hydrochloric acid.

Note: DNPs of saturated ketones give a blue colour immediately; those of saturated aldehydes react more slowly and turn olive green. The colours of the DNPs of unsaturated carbonyl compounds change only slowly or not at all.

A. MEHLITZ, K. GIERSCHNER u. T. MINAS: Chemiker-Ztg. **87**, 573 (1963).

83. Diphenylamine: for glycolipids.

Spray reagent: 20 ml of 10% ethanolic diphenylamine, 100 ml 36% hydrochloric acid and 80 ml acetic acid are mixed.

Treatment after spraying: Heated 5—10 min at 105° C, yielding blue-grey spots.

H. JATZKEWITZ: Hoppe-Seylers Z. physiol. Chem. **320**, 251 (1960).

84. Diphenylamine-palladium chloride: for nitrosamines.

Spray reagent: 1.5% ethanolic diphenylamine solution and a solution of 0.1 g palladium chloride in 100 ml 0.2% sodium chloride, are mixed in the volume ratio of 5:1.

Treatment after spraying: The substances appear as violet spots after exposure to short-wave UV light.

R. PREUSSMANN, D. DAIBER, and H. HENGY: Nature **201**, 502 (1964).
R. PREUSSMANN, G. NEURATH, G. WULF-LORENTZEN, D. DAIBER u. H. HENGY: Z. anal. Chem. **202**, 187 (1964).

85. Diphenylamine-zinc chloride: for chlorinated insecticides (DDT, CPCA, chlor-DDT, captan, methoxychlor, toxaphene).

Spray reagent: 0.5 g diphenylamine and 0.5 g zinc chloride are dissolved in 100 ml acetone.

Treatment after spraying: Heated 5 min at 200° C, giving a colour reaction.

D. KATH: J. Chromatog. **15**, 269 (1964).

86. Diphenylboric acid, β-aminoethyl ester (Neu's reagent for natural products): for α- and γ-pyrones (hydroxyflavonols).

Spray reagent: 1% methanolic solution of the β-aminoethyl ester of diphenylboric acid.

Procedure: The layer is sprayed with about 10 ml reagent and the fluorescence colours observed in long-wave UV light.

R. NEU: Naturwissenschaften **44**, 181 (1957).
E. STAHL u. P. J. SCHORN: Hoppe-Seylers Z. physiol. Chem. **325**, 263 (1961).

87. Diphenylcarbazide: for Ag^+, Pb^{2+}, Hg^{2+}, Cu^{2+}, Sn^{2+}, Mn^{2+}, Zn^{2+} and Ca^{2+} ions.

Spray reagent I: 1—2% ethanolic diphenylcarbazide solution.

Spray reagent II: 25% ammonium hydroxide solution (or an atmosphere of ammonia into which the plate can be introduced).
Note: Mercuric acetate adducts are best detected by heating briefly at 80° C, causing the zones to turn blue-violet.

F. W. H. M. MERKUS: Pharm. Weekblad **98**, 947 (1963).

88. **Diphenylpicrylhydrazyl:** for essential oils.
Spray reagent: 0.06 g diphenylpicrylhydrazyl is dissolved in 100 ml chloroform.
Treatment after spraying: Heated 5—10 min at 110° C; yellow zones on a violet background are formed.

G. BERGSTRÖM, and C. LAGERCRANTZ: Acta Chem. Scand. **18**, 560 (1964).

89. **2,5-Diphenyl-3-(4-styrylphenyl)-tetrazolium chloride** (TPTZ): for reducing steroids (corticosteroids).
Solution a: Freshly prepared 1% TPTZ solution in methanol.
Solution b: 3% aqueous sodium hydroxide.
Spray reagent: Equal volumes of a and b are mixed before use.

P. J. STEVENS: J. Chromatog. **14**, 269 (1964).

90. **Dipicrylamine:** for choline (non-specific).
Spray reagent: 0.2 g dipicrylamine is dissolved in a mixture of 50 ml acetone and 50 ml water.
Note: Choline and its derivatives appear as red spots on a yellow background.

K. P. ANGUSTINSSON, and M. GRAHN: Acta Chem. Scand. **7**, 906 (1953).

91. **α,α′-Dipyridyl-ferric chloride:** for phenols, vitamin E and other compounds with reducing properties.
Solution a: 0.5% ethanolic solution of ferric chloride (keep in the dark).
Solution b: 0.5% ethanolic solution of α,α'-dipyridyl.
Spray reagent: Equal volumes of a and b are mixed before use.

G. M. BARTON: J. Chromatog. **20**, 189 (1965).
R. STROHECKER u. H. M. HENNING: Vitaminbestimmungen, p. 311. Verlag Chemie, Weinheim, W. Germany 1963.

92. **Dithiooxamide:** for Pb^{2+}, Co^{2+}, Cu^{2+}, Mn^{2+}, Ni^{2+}, Hg^{2+} and Bi^{3+} ions.
Spray reagent I: 0.5% ethanolic solution of dithiooxamide.
Spray reagent II: 25% ammonium hydroxide solution.
Procedure: The chromatogram is sprayed with I, dried briefly and sprayed with II or placed in a chamber containing 25% ammonium hydroxide.

F. W. H. M. MERKUS: Pharm. Weekblad **98**, 955 (1963).
J. A. LEWIS, and J. M. GRIFFITHS: Analyst **76**, 388 (1951).

93. **Dithizone:** for ions of heavy metals.
Spray reagent I: 0.05% solution of dithizone in carbon tetrachloride.
Spray reagent II: 25% ammonium hydroxide solution or an atmosphere of ammonia into which the plate can be introduced.

T. BARNABAS u. J. BARNABAS: Naturwissenschaften **44**, 61 (1957).
F. W. H. M. MERKUS: Pharm. Weekblad **98**, 955 (1963).

94. Dragendorff's reagent: for polyethylene glycols, their ethers and esters.

Solution a: 1.7 g basic bismuth nitrate is dissolved in 20 ml acetic acid; 80 ml water, a solution of 40 g potassium iodide in 100 ml water and 200 ml acetic acid are added and the mixture made up to 1000 ml with water.

Solution b: 20% barium chloride solution in water.

Spray reagent: a and b are mixed in the volume ratio 2:1 before use.

K. THOMA, R. ROMBACH u. E. ULLMANN: Sci. Pharm. **32**, 216 (1964).

95. Dragendorff's reagent according to Bregoff-Delwiche: for quaternary nitrogen compounds.

Stock solution: 8 g basic bismuth nitrate are dissolved in 20 to 23 ml 25% nitric acid and the solution added slowly with stirring to a slurry of 20 g potassium iodide with 1 ml 6N hydrochloric acid and 5 ml water. Water is added to the dark precipitate until an orange-red solution is formed. The volume of the solution should be 95 ml. Any insoluble residue is filtered off and the solution made up to 100 ml with water. This product can be kept several weeks in dark glass in the refrigerator.

Spray reagent: The following are mixed in the order given:
20 ml water
6 ml 6N hydrochloric acid
2 ml stock solution
6 ml 6N sodium hydroxide.
A few drops of 6N hydrochloric acid should be added if all the bismuth hydroxide does not dissolve on shaking.

Note: The spray reagent can be kept about 10 days in the refrigerator.

H. M. BREGOFF, E. ROBERTS, and C. C. DELWICHE: J. Biol. Chem. **205**, 565 (1953).

96. Dragendorff's reagent according to Munier: for alkaloids and other nitrogen-containing compounds.

Solution a: 1.7 g basic bismuth nitrate and 20 g tartaric acid are dissolved in 80 ml water.

Solution b: 16 g potassium iodide are dissolved in 40 ml water.

Stock solution: A 1:1 (v/v) mixture of a and b is prepared. This may be kept several months in a refrigerator.

Spray reagent: 5 ml of the stock solution are added to a solution of 10 g tartaric acid in 50 ml water.

Note: The stock solution itself is used for detecting vitamin B_1.

R. MUNIER: Bull. soc. chim. biol. **35**, 1225 (1953).

97. Dragendorff's reagent according to Munier and Macheboeuf: for alkaloids and other nitrogen-containing compounds.

Solution a: 0.85 g basic bismuth nitrate is dissolved in a mixture of 10 ml acetic acid and 40 ml water.

Solution b: A solution is made of 8 g potassium iodide in 20 ml water.
Stock solution: Equal volumes of a and b are mixed (can be stored for a long time in dark glass vessels).
Spray reagent: 1 ml stock solution is mixed with 2 ml acetic acid and 10 ml water before use.

R. Munier et M. Macheboeuf: Bull. soc. chim. biol. **33**, 846 (1951).
H. Jatzkewitz: Hoppe-Seylers Z. physiol. Chem. **292**, 99 (1953).

98. **Dragendorff's reagent** according to Thies and Reuther, modified by Vágujfalvi: for alkaloids and other nitrogen-containing compounds.
    Stock solution: A mixture of 25 ml acetic acid, 2.6 g basic bismuth carbonate and 7 g sodium iodide is boiled for a few minutes. The copious precipitate of sodium acetate is filtered through a sintered glass filter after about 12 h. 20 ml of the clear red-brown filtrate are mixed with 80 ml ethyl acetate and 0.5 ml water is added. This solution must be stored in dark glass bottles.
    Spray reagent: A mixture is made of 10 ml stock solution, 100 ml acetic acid and 240 ml ethyl acetate. Alkaloids and a number of other compounds, some containing no nitrogen, appear as orange coloured spots after spraying with 5—10 ml reagent.
    Treatment after spraying: The sensitivity of detection can be increased appreciably by subsequent spraying with 0.05 to 0.1 N sulphuric acid. The optimum acid concentration and amount to be sprayed are determined in a preliminary trial. The spots are bright red to orange-red on a grey background.

H. Thies u. F. W. Reuther: Naturwissenschaften **41**, 230 (1954).
D. Vágujfalvi: Planta Med. **8**, 34 (1960).
E. Tyihák: J. Chromatog. **14**, 125 (1964).

99. **Ethylenediamine:** for catechol amines.
    Spray reagent: Ethylenediamine is mixed with an equal volume of water or dilute sodium hydroxide.
    Treatment after spraying: Heated 20 min at 50—60° C. The chromatogram is inspected in short- or long-wave UV light.

R. Segura-Cardona, and K. Soehring: Med. Exp. **10**, 251 (1964).

100. **Fast Blue B Salt** (diazonium reagent): for phenols and amines which can couple.
    Spray reagent I: A freshly prepared 0.5% aqueous solution of Fast Blue B salt.
    Spray reagent II: 0.1 N sodium hydroxide.
    Procedure: The layer is sprayed with I, then II.

H. Jatzkewitz u. U. Lenz: Hoppe-Seylers Z. physiol. Chem. **305**, 53 (1956).

101. **Ferric ammonium sulphate:** for *Vinca* alkaloids.
    Spray reagent: 1 g ferric ammonium sulphate is dissolved in 100 ml phosphoric acid (75 or 85%). The reagent is sprayed on to the heated (100° C) plate.

I. M. Jakovljevic, L. D. Seay, and R. W. Shaffer: J. Pharm. Sci. **53**, 553 (1964).

102. Ferric chloride: for phenols and hydroxamic acids.
Spray reagent: 1—5% solution of ferric chloride in 0.5N hydrochloric acid.
Note: Hydroxamic acids yield red spots, phenols blue or greenish.
K. Fink u. R. M. Fink: Proc. Soc. Exptl. Biol. Med. **70**, 654 (1949).

103. Ferric chloride-iodine: for xanthine derivatives.
Spray reagent: 5 g ferric chloride and 2 g iodine are dissolved in a mixture of 50 ml acetone and 50 ml 20% aqueous tartaric acid solution.
J. Zarnak u. S. Pfeifer: Pharmazie **19**, 216 (1964).

104. Ferric chloride-perchloric acid: for indoles (Salkowski reaction).
Spray reagent: 1 ml 0.5M aqueous ferric chloride solution is mixed with 50 ml 35% perchloric acid.
Treatment after spraying: Warmed for 5 min at 60° C. The spot colours are intensified by brief exposure to the vapours of aqua regia.
S. A. Gordon, and R. P. Weber: Plant Physiol. **26**, 192 (1951).

105. Ferric chloride-perchloric acid: for phenothiazines.
Spray reagent: A mixture is made of 5 ml 5% aqueous ferric chloride, 45 ml 20% perchloric acid and 50 ml 50% nitric acid. Coloured spots are formed.
A. Noirfalise et M. H. Grosjean: J. Chromatog. **16**, 236 (1964).

106. Ferric chloride-sulphosalicylic acid: for thiophosphate esters.
Spray reagent I: 0.1% solution of ferric chloride in 80% ethanol.
Spray reagent II: 1% solution of sulphosalicylic acid in 80%ethanol.
Procedure: The layer is exposed for about 10 min to a bromine atmosphere and then sprayed with I. After air-drying for 15 min, the layer is sprayed with II, yielding white spots on a violet background.
M. Salamé: J. Chromatog. **16**, 476 (1964).

107. Ferric chloride-sulphuric acid: for bile acids.
Spray reagent: 2 g ferric chloride are dissolved in 83 ml anhydrous n-butanol and 15 ml conc. sulphuric added.
Treatment after spraying: The plates are air-dried 15 min and then heated at 110° C for 25—30 min (for combined bile acids) or 45—50 min (for free bile acids): colours are yielded.
W. L. Anthony, and W. T. Beher: J. Chromatog. **13**, 567 (1964).

108. Ferric chloride-sulphuric acid: for indoles (Salkowski reaction).
Spray reagent: A mixture of 3 ml 1.5M aqueous ferric chloride, 100 ml water and 60 ml conc. sulphuric acid is prepared.
Treatment after spraying: Heated 5 min at 60° C. The spot colours are enhanced by blowing vapours of aqua regia over the layer.
P. E. Pilet: Rev. gén. bot. **64**, 1 (1957).

109. **Ferricyanide (potassium):** for adrenaline and derivatives.
Spray reagent: 0.6 g potassium ferricyanide is dissolved in 100 ml 0.5% sodium hydroxide solution. Red spots are yielded.

A. H. Beckett, M. A. Beavan, and A. E. Robinson: J. Pharm. Pharmacol. **12**, 203 T (1960).

110. **Ferricyanide (potassium):** for vitamin B_1 (thiochrome reaction).
Solution a: 1% aqueous potassium ferricyanide solution.
Solution b: 15% sodium hydroxide.
Spray reagent: 1.5 ml a is diluted with 20 ml water and 10 ml b are added. The dried chromatogram is inspected in long-wave UV light.

D. Siliprandi, and N. Siliprandi: Biochim. et Biophys. Acta **14**, 52 (1954)

111. **Ferricyanide (potassium)-ferric chloride:** for compounds with reducing properties, phenols, amines, thiosulphates and isothiocyanates.
Solution a: 1% aqueous potassium ferricyanide solution.
Solution b: 2% aqueous ferric chloride.
Spray reagent: Equal amounts of a and b are mixed just before use.
Treatment after spraying: The colours are intensified by subsequent spraying with 2 N hydrochloric acid.

G. M. Barton, R. S. Evans, and J. A. F. Gardner: Nature **170**, 249 (1952)
M. Gillio-Tos, S. A. Previtera, and A. Vimercati: J. Chromatog. **13**, 571 (1964).
H. Wagner, L. Hörhammer u. H. Nufer: Arzneimittel-Forsch. **15**, 453 (1965).

112. **Ferricyanide-ferrocyanide (both potassium salts):** for morphine.
Spray reagent: 57 mg potassium ferricyanide and 7.8 mg potassium ferrocyanide are dissolved in distilled water and the volume made up to 100 ml.

H. J. Kupferberg, A. Burghalter, and E. L. Way: J. Chromatog. **16**, 558 (1964).

113. **Ferrocyanide (potassium):** for ferric ions, Fe^{3+}.
Spray reagent: A freshly prepared 2% aqueous solution of potassium ferrocyanide.

F. H. Burstall, G. R. Davies, R. P. Linstead, and R. A. Wells: J. Chem. Soc. **1950**, 516.

114. **Ferrocyanide(potassium)-hydrogen peroxide:** for barbiturates.
Spray reagent I: 0.1 g potassium ferrocyanide is dissolved in 100 ml water containing 0.5 ml conc. (36%) hydrochloric acid. 5 g ammonium chloride are added to 10 ml of this solution and the whole made up to 100 ml with water.
Spray reagent II: 30% hydrogen peroxide solution.
Spray reagent III: 10% potassium carbonate solution.
Procedure: The layer is sprayed with I, dried at 100° C and, after cooling, sprayed with II. It is then heated 30 min at 150° C and finally sprayed with III which enhances the yellow and red spots.

This reaction may be carried out after the test with mercurous nitrate (No. 157).

H. WEICHSEL: Mikrochim. Ichnoanal. Acta **1965**, 325.

115. Ferrous thiocyanate: for peroxides.
Solution a: 4% aqueous ferrous sulphate.
Solution b: 1.3% solution of ammonium thiocyanate in acetone.
Spray reagent: 10 ml of a and 15 ml of b are mixed before spraying.
Note: Rapid formation of red-brown zones shows the presence of peroxidic compounds.

E. STAHL: Chemiker-Ztg. **82**, 323 (1958).
E. KNAPPE u. D. PETERI: Z. anal. Chem. **190**, 386 (1962).

116. Fluorescein: for lipids.
Spray reagent: 0.01% ethanolic solution of fluorescein.
Treatment after spraying: The layer is dried in a warm air current and then treated with steam or lightly sprayed with water.

117. Fluorescein-ammonia: for purines, pyrimidines and barbiturates.
Spray reagent: 0.005% solution of fluorescein in 0.5N ammonium hydroxide. The chromatogram is inspected in long- and short-wave UV light.

T. WIELAND u. L. BAUER: Angew. Chem. **63**, 511 (1951).

118. Fluorescein-bromine: for unsaturated compounds.
Spray reagent: 0.1 g fluorescein is dissolved in 100 ml ethanol.
Bromine solution: 5% in carbon tetrachloride.
Procedure: The plate is sprayed with the fluorescein solution and then placed in a chamber containing the bromine solution. Fluorescein is thereby converted to eosin which does not fluoresce in long-wave UV light. Unsaturated substances on the layer which add on bromine thus prevent eosin formation and the fluorescence remains.
Larger substance amounts yield yellow spots on a pink background.

F. RUNGE, A. JUMAR u. F. KOEHLER: J. prakt. Chem. **21**, 39 (1963).

Variation:
    The slurry is prepared with a 0.04% aqueous solution of sodium fluorescein instead of with water.
Procedure: Bromine vapour is blown over the chromatogram after development.

E. STAHL: Chemiker-Ztg. **82**, 323 (1958).

119. Fluorescein-hydrogen peroxide: for bromine-containing hypnotics.
Spray reagent I: 0.1% fluorescein solution in 50% ethanol.
Spray reagent II: 30% hydrogen peroxide-acetic acid (1 + 1).
Procedure: The layer is sprayed with I, then II and finally heated 20 min at 90° C.
Note: Bromine liberated by oxidation converts fluorescein to eosin.

H. WEICHSEL: Mikrochim. Ichnoanal. Acta **1965**, 325.

120. Fluorescein-rhodamine B-sodium carbonate: for chlorinated hydrocarbons and heterocyclic compounds.

Spray reagent I:	0.5% ethanolic solution of rhodamine B.
Spray reagent II:	10% aqueous sodium carbonate solution.
Procedure:	Layers impregnated with the sodium derivative of fluorescein are used. After development, these are sprayed with I, dried and liberally sprayed with II. The chromatograms are inspected in daylight and long-wave UV light.

121. Fluorescence indicators and materials as reagents for detection in general.

A. Spray reagents:
1. 0.2% ethanolic solution of 2',7'-dichlorofluorescein (Reagent No. 63).
2. 0.01% ethanolic fluorescein solution (Reagent No. 116).
3. 0.02% solution of methylumbelliferone in ethanol-water, (Reagent No. 163).
4. 0.1% ethanolic solution of morin.
5. 0.05% ethanolic solution of rhodamine B, Reagent No. 220.

B. Additives to Adsorbents:

6. 0.04% aqueous solution of the sodium derivative of fluorescein for preparing suspensions with adsorbents.
7. Leuchtstoff ZS-Super (Firm 118) added in 1% amount to the adsorbent.
8. Ultraphor WT, high concentration (Firm 16), added in 0.02% amount to the adsorbent.
9. Zinc silicate, fluorescent material (P 1, Type 118-2-7). General Electric, Cleveland/Ohio, USA, added in 0.8% amount to the adsorbent.

122. Folin-Ciocalteu reagent: for phenols.

Stock solution:	A solution is prepared of 10 g sodium tungstate and 2.5 g sodium molydate in 70 ml water; 5 ml 85% phosphoric acid and then 10 ml 36% hydrochloric acid added to it. The mixture is refluxed for 10 h, after which 15 g lithium sulphate, 5 ml water and 1 drop bromine are added. The resulting solution is refluxed 15 more minutes, cooled and made up to 100 ml with water. It should have no green tint.
Spray reagent I:	20% aqueous sodium carbonate.
Spray reagent II:	The stock solution is diluted with three times its volume of water before use.
Procedure:	Spraying is carried out first with I, then with II after briefly drying.

R. W. KEITH, D. LE TURNEAU, and D. MAHLUM: J. Chromatog. 1, 534 (1958).

123. Formaldehyde-hydrochloric acid (Procházka reagent): for indoles and their derivatives.

Spray reagent:	A mixture of 10 ml ca. 35% formaldehyde solution, 10 ml 25% hydrochloric acid and 20 ml ethanol is freshly made.
Treatment after spraying:	Heated 5 min at 100° C. The fluorescence colours (yellow-orange-greenish) in long-wave UV light may be enhanced by exposure to the vapours of aqua regia.

Z. Procházka: Chem. Listy **47**, 1643 (1953).
E. Stahl u. H. Kaldewey: Hoppe-Seylers Z. physiol. Chem. **323**, 182 (1961).

124. Formaldehyde-phosphoric acid: for steroid alkaloids, steroid sapogenins and phenothiazine derivatives.

Spray reagent: 0.03 g paraformaldehyde is dissolved in 100 ml 85% phosphoric acid by shaking at room temperature. The reagent may be kept several weeks.

K. Schreiber, O. Aurich u. G. Osske: J. Chromatog. **12**, 63 (1963).
E. G. C. Clarke: Nature **181**, 1152 (1958).

125. Formaldehyde-sulphuric acid: for polynuclear aromatic compounds.

Spray reagent: 0.2 ml 37% formaldehyde solution is dissolved in 10 ml conc. sulphuric acid.

Procedure: The layer is sprayed as soon as it has been removed from the development chamber. The different aromatic compounds yield different colours.

N. Kucharczyk, J. Fohl, and J. Vymetal: J. Chromatog. **11**, 55 (1963).

126. Furfuraldehyde-sulphuric acid: for carbamate esters, e. g., meprobamate.

Spray reagent I: 1% solution of furfuraldehyde in acetone.
Spray reagent II: 10% solution of sulphuric acid in acetone.
Procedure: The layer is sprayed with I, then with II.

A. Heyndrickx, M. Schauvliege, et A. Blommel: J. pharm. Belg. **20**, 117 (1965).
I. Sunshine: Am. J. Clin. Pathol. **40**, 576 (1963).

127. Glucose-aniline (Schweppe reagent): for acids.

Solution a: 10% aqueous glucose.
Solution b: 10% ethanolic aniline.
Spray reagent: 20 ml of each of a and b are mixed and diluted to 100 ml with n-butanol.
Treatment after spraying: Heated 5—10 min at 125° C, giving deep brown spots on a white background.

H. Schweppe: Dissertation, Münster 1954.

128. Glucose-phosphoric acid: for aromatic amines.

Spray reagent: 2 g glucose are dissolved in a mixture of 10 ml 85% phosphoric acid and 40 ml water; 30 ml ethanol and 30 ml n-butanol are added to this solution.
Treatment after spraying: Heated about 10 min at 115° C.

F. Micheel u. H. Schweppe: Mikrochim. Acta **1954**, 53.

129. Glyoxal-bis(2-hydroxyanil) (GBHA): for cations.

Spray reagent: 1 g reagent and 3 g potassium hydroxide are dissolved in methanol and the volume made up to 100 ml with this solvent.
Procedure: The chromatogram is dried, treated with the spray reagent and then dried again with a stream of air at 50° C. Coloured reaction products are yielded.

H. G. Möller u. N. Zeller: J. Chromatog. **14**, 560 (1964).

130. **Hydrazinium sulphate:** for piperonal, vanillin and ethylvanillin.
Spray reagent: 90 ml saturated, aqueous hydrazinium sulphate solution are mixed with 10 ml 4N hydrochloric acid.
Treatment after spraying: The moist chromatogram is inspected in long-wave UV light, both before and after exposure to ammonia vapour.
K. G. BERGNER u. H. SPERLICH: Dtsch. Lebensm.-Rundschau **47**, 134 (1951).

131. **Hydrochloric acid:** for glycals.
Spray reagent: 36% hydrochloric acid and ethanol are mixed in the volume ratio of 1:4.
Procedure: Glycals appear as pink spots on heating to 90° C.
Note: May be used also as a general spray reagent.
J. T. EDWARD, and D. M. WALDRON: J. Chem. Soc. **1952**, 3631.

132. **Hydrogen peroxide:** for aromatic acids.
Spray reagent: 0.3% aqueous hydrogen peroxide solution.
Treatment after spraying: The chromatogram is irradiated with long-wave UV light until the blue fluorescence of the spots has attained its maximum.
D. W. GRANT: J. Chromatog. **10**, 511 (1963).

133. **4-Hydroxybenzaldehyde-sulphuric acid** (Komarowsky reagent): for sapogenins and corticosteroids (3-ketosteroids, unsubstituted in the 2-position).
Solution I: 50% sulphuric acid.
Solution II: 2% methanolic solution of 4-hydroxybenzaldehyde.
Spray reagent: 5 ml I and 50 ml II are mixed just before use.
Treatment after spraying: Heated 3—4 min at 105° C or 10 min at 60° C, yielding yellow to pink spots.
P. J. STEVENS: J. Chromatog. **14**, 269 (1964).

134. **Hydroxylamine-ferric chloride:** for lactones, esters, amides and anhydrides of carboxylic acids.
Solution a: 20 g hydroxylammonium chloride are dissolved in 50 ml water and the solution made up to 200 ml with ethanol; it is stored in a cool place.
Solution b: 50 g potassium hydroxide are dissolved in a minimum of water and the solution diluted to 500 ml with ethanol.
Spray solution I: a and b are mixed in the proportion 1:2 and the precipitated potassium chloride filtered off. The resultant solution must be kept in a refrigerator and is stable for about 2 weeks.
Spray solution II: 10 g finely powdered ferric chloride ($FeCl_3 \cdot 6H_2O$) are dissolved in 20 ml 36% hydrochloric acid; this is shaken with 200 ml diethyl ether until a homogeneous solution is obtained. This reagent II can be kept for a longer time in a well-closed container.
Procedure: The chromatogram is sprayed with I, dried briefly at room temperature and then sprayed with II.
V. P. WHITTAKER, and S. WIJESUNDERA: Biochem. J. **51**, 348 (1952).

Spray Reagents

135. 8-Hydroxyquinoline: for Ba^{2+}, Sr^{2+} and Ca^{2+} ions.
Spray reagent: 0.5 g 8-hydroxyquinoline is dissolved in a mixture of 60 ml ethanol and 40 ml water.
Treatment after spraying: The chromatogram is sprayed with 25% ammonium hydroxide or placed in a tank containing this solution; the layer is observed in long-wave UV light.
W. A. REEVES, and T. B. CRUMLER: Anal. Chem. **23**, 1576 (1952).
T. V. ARDEN et al.: Nature **162**, 691 (1948).

136. 8-Hydroxyquinoline-hypobromite (Sakaguchi reagent): for arginine and other guanidine derivatives and for galegine.
Spray reagent I: 0.1% solution of 8-hydroxyquinoline in acetone.
Spray reagent II: 0.2 ml bromine is dissolved in 100 ml 0.5 N sodium hydroxide.
Procedure: The chromatogram is sprayed first with I, dried and then sprayed with II. Orange to red spots appear.
J. B. JEPSON, and J. SMITH: Nature **172**, 1100 (1953); **177**, 84 (1956).
J. KOLOUŠEK, M. KUTÁČEK, and J. BÍLEK: Českoslov. farm. **4**, 188 (1955).

137. 8-Hydroxyquinoline-kojic acid: for Al^{3+}, Mg^{2+}, Ca^{2+}, Sr^{2+} and Ba^{2+}.
Spray reagent I: 2.5 g 8-hydroxyquinoline and 0.5 g kojic acid are dissolved in 500 ml 90% ethanol.
Spray reagent II: 25% ammonium hydroxide solution.
The spots fluoresce in long-wave UV-light.
F. H. POLLARD, J. F. W. McOMIE, and I. I. M. ELBEIH: J. Chem. Soc. **1951**, 466.

138. Indanedione: for carotenoid aldehydes.
Spray reagent: 0.5 g 2-diphenylacetyl-indane-1,3-dione-1-hydrazone is dissolved in 20 ml water, the solution filtered after briefly warming and finally 0.3 ml 36% hydrochloric acid is added.
Treatment after spraying: The layer is dried in a current of cold air.
H. THOMMEN u. O. WISS: Z. Ernährungswiss. 1963, Suppl. 3, p. 18.

139. Iodide (potassium)-starch: for peroxides.
Spray reagent I: 10 ml 4% aqueous potassium iodide solution are mixed with 40 ml acetic acid and a pinch of zinc dust is added.
Spray reagent II: Freshly prepared 1% starch solution.
Procedure: The zinc dust is filtered off and the layer sprayed with I. After an interval of 5 min, II is sprayed on freely until the layer appears transparent. Peroxides are recognised through the blue colour from liberated iodine.
E. STAHL: Chemiker-Ztg. **82**, 323 (1958).

140. Iodide (potassium)-hydrogen sulphide: for heavy metal ions.
Spray reagent: 2% aqueous potassium iodide solution.
Procedure: The plate is dried after spraying and placed in a chamber containing 25% ammonium hydroxide for some minutes. It is then transferred to a second chamber into which hydrogen sulphide

is led from a Kipp apparatus. Caution in using the poisonous and explosive hydrogen sulphide (fume cupboard)!

H. Seiler u. M. Seiler: Helv. Chim. Acta **43**, 1939 (1960).

141. Iodine: general reagent for detection.

The chromatogram is introduced into a closed vessel on the floor of which some crystals of iodine have been placed. Iodine vapour is more quickly generated through gently warming the vessel. Many organic compounds yield brown spots.

Modification:

The plate is placed in a dense atmosphere of iodine vapour for 5 min or sprayed with an iodine solution (e. g., 0.5% in chloroform). Excess iodine evaporates on standing in the air. The spots turn blue on spraying with a starch solution (1% in water). The background also turns blue if there is too much iodine still on the layer (test on a corner or part of the covered layer).

G. C. Barrett: Nature **194**, 1171 (1962).
A. Bettschart u. H. Flück: Pharm. Acta Helv. **31**, 260 (1956).
G. Brante: Nature **163**, 651 (1949).
R. Munier et M. Macheboeuf: Bull. soc. chim. biol. **31**, 1144 (1949).
R. Munier: Bull. soc. chim. France **19**, 852 (1952).

142. Iodine azide: for sulphur-containing amino acids, for sulphides and penicillin.

Iodine azide solution:
Spray reagent: A solution of 3 g sodium azide in 100 ml 0.1N iodine is freshly prepared (solid iodine azide is explosive).

Iodine azide-starch reagent:
Spray reagent I: A solution of 1 g sodium azide in 100 ml 0.005N iodine is freshly prepared.
Spray reagent II: 1% starch solution in water.
Procedure: The chromatogram is sprayed with I, then II.

E. Chargaff, C. Levine, and C. Green: J. Biol. Chem. **175**, 67 (1948).
W. Awe, I. Reinecke u. J. Thum: Naturwissenschaften **41**, 528 (1954).

143. Iodine-iodide (potassium), acid: for alkaloids.
Spray reagent: 1 g iodine and 10 g potassium iodide are dissolved in 50 ml water by warming and 2 ml acetic acid added: the solution is made up to 100 ml with water.

F. Šantavý: unpublished.

144. Iodine-iodide (potassium), neutral: for organic compounds.
Spray reagent: 0.2 g iodine and 0.4 g potassium iodide are dissolved in 100 ml water.

A. Zaffaroni, R. B. Burton, and H. Kentmann: Science **111**, 6 (1950).
A. Bettschart u. H. Flück: Pharm. Acta Helv. **31**, 260 (1956).
J. Büchi u. H. Schumacher: Pharm. Acta Helv. **32**, 194 (1957).

145. Iodine-sulphuric acid: for organic nitrogen compounds, polyethylene glycols and their derivatives.
Spray reagent: A mixture of equal volumes of 0.1N iodine solution and 10% sulphuric acid.

H. Feltkamp u. F. Koch: J. Chromatog. **15**, 314 (1964).

146. Iodoplatinate(potassium): for alkaloids.
Spray reagent: 5 ml 5% hexachloroplatinic (IV) acid and 45 ml 10% aqueous potassium iodide solution are mixed and diluted to 100 ml with water.
The mixture is freshly prepared before use.
J. SMITH: Chromatographic and Electrophoretic Techniques, Vol. I, p. 396, Interscience, New York 1960.

147. Iodoplatinate(potassium): for alkaloids and other organic nitrogen compounds.
Spray reagent: 3 ml 10% hexachloroplatinic(IV) acid solution are mixed with 97 ml water and 100 ml 6% potassium iodide solution in water are added; the reagent is freshly prepared before use.
R. MUNIER: Bull. soc. chim. France **19**, 852 (1952).
R. HILZ, F. F. CASTANO, and G. A. LIGHTBOURN: J. Lab. Clin. Med. **54**, 634 (1959).

148. Isatin-sulphuric acid: for thiophene derivatives.
Spray reagent: 0.4 g isatin is dissolved in 100 ml conc. sulphuric acid.
Treatment after spraying: Heating to 120° C is occasionally needed; spots of various colours are yielded.
R. F. CURTIS, and G. T. PHILLIPS: J. Chromatog. **9**, 366 (1962).

149. Isatin-zinc acetate: for amino acids and some peptides.
Spray reagent: 1 g isatin and 1.5 g zinc acetate are dissolved in 100 ml 95% isopropanol by warming to 80° C; 1 ml acetic acid is added after cooling. The reagent can be stored in the refrigerator.
Treatment after spraying: Heated 30 min at 80—85° C or, better, the chromatogram is examined after standing 20 h at room temperature.
J. BARROLIER, J. HEILMAN u. E. WATZKE: Hoppe-Seylers Z. physiol. Chem. **304**, 21 (1956).

150. Isonicotinic acid hydrazide (INH): for Δ^4-3-ketosteroids.
Spray reagent: 1 g isonicotinic acid hydrazide and 1 ml acetic acid are mixed and diluted to 100 ml with ethanol.
Procedure: The layer is dried at room temperature after spraying. The spots fluoresce yellow in the long-wave UV lamp.
B. P. LISBOA: Acta Endocrinol. **43**, 47 (1963).
B. P. LISBOA: J. Chromatog. **16**, 136 (1964).

151. Lead acetate, basic: for flavonoids.
Spray reagent: 25% aqueous solution of basic lead acetate. The spots fluoresce in long-wave UV light.
L. HÖRHAMMER, H. WAGNER u. K. HEIN: J. Chromatog. **13**, 235 (1964).
R. NEU u. P. HAGEDORN: Naturwissenschaften **40**, 411 (1953).

152. Lead tetraacetate: for 1,2-diol groups.
Spray reagent: 1% solution of lead tetraacetate in benzene.
Treatment after spraying: Heated 5 min at 110° C, giving white spots on a brown background.
J. WRIGHT: Chem. & Ind. (London) **1963**, 1125.

153. Lead tetraacetate-rosaniline: for 1,2-diol groups.

Spray reagent I: 3 g red lead ($PbO_2 \cdot 2PbO$) are shaken from time to time with 100 ml acetic acid until completely dissolved.

Spray reagent II: 0.05 g rosaniline base is dissolved in acetic acid-acetone (10 + 90); a 0.1% methanolic fuchsine solution may be used also.

Procedure: The layer is sprayed with I, left 4—5 min and then with II.

K. SAMPSON, F. SCHILD, and R. J. WICKER: Chem. & Ind. (London) **1961**, 82.
K. G. BERGNER u. H. SPERLICH: Z. Lebensm.-Untersuch. u. -Forsch. **97**, 253 (1953).

154. Leuco-methylene blue: for ubi-, plasto- and tocopheryl quinones.

Spray reagent: A suspension of 0.25 g zinc dust in 1 ml acetic acid is added to 5 ml of a 0.02% solution of methylene blue in acetone.

T. W. GOODWIN: Lab. Practice **1964**, 295.

155. Magnesium acetate: for anthraquinone glycosides and their aglucones.

Spray reagent: 0.5% methanolic solution of magnesium acetate.
Procedure: Heated 5 min at 90° C; the spots are orange to violet.

S. SHIBITA, M. TAKIDO, and O. TANAKA: J. Am. Chem. Soc. **72**, 2789 (1950).

156. Mercuric diphenylcarbazone: for barbiturates.

A. Solution a: 2% ethanolic mercuric chloride.
Solution b: 0.2% ethanolic diphenylcarbazone.
Spray reagent: Equal volumes of a and b are mixed before use, yielding pink spots on a violet background.

E. K. J. CHRISTENSEN, T. VOS, and T. HUIZINGA: Pharm. Weekblad **100**, 517 (1965).

B. Spray reagent I: 0.1% ethanolic diphenylcarbazone.
Spray reagent II: 0.33% mercuric nitrate solution in 0.05N nitric acid.
Procedure: The layer is sprayed with I until it is faintly pink and then sprayed with II.
Note: Pink spots on a violet background are obtained; the latter is bleached by sunlight or UV light and the spots turn violet.

J. LEHMANN, and V. KARAMUSTAFAUGLU: Scand. J. Clin. & Lab. Invest. **14**, 554 (1962).

C. Spray reagent I: Mercuric sulphate solution: 5 g HgO are suspended in 100 ml water and 20 ml conc. sulphuric acid added with stirring. The volume is made up to 250 ml with water after cooling.
Spray reagent II: 0.01% solution of diphenylcarbazone in chloroform.
Procedure: The chromatogram is sprayed with I, dried and then sprayed with II.

I. SUNSHINE, E. ROSE, and J. LE BEAU: Clin. Chem. **9**, 312 (1963).

157. Mercurous nitrate: for barbiturates.
Spray reagent: 1% aqueous mercurous nitrate solution.
J. BÄUMLER: Mitt. Gebiete Lebensm. u. Hyg. **48**, 135 (1957).
R. DEININGER: Arzneimittel-Forsch. **5**, 472 (1955).

158. Metaperiodate(sodium)-benzidine: for substances with 1,2-diol groups, e. g., sugars, polyalcohols.
Spray reagent I: 0.1% aqueous solution of sodium metaperiodate.
Spray reagent II: 1.8 g benzidine is dissolved in 50 ml ethanol and 50 ml water, 20 ml acetone and 10 ml 0.2 N hydrochloric acid are added.
Procedure: The layer is sprayed with I, left 5 min and sprayed with II; white spots on a blue background are yielded.
J. A. CIFONELLI, and F. SMITH: Anal. Chem. **26**, 1132 (1954).

159. Metaperiodate(sodium)-benzidine-silver nitrate: for substances with the 1,2-diol group, e. g., sugars, polyalcohols.
Spray reagent I: 0.1% aqueous sodium metaperiodate solution.
Spray reagent II: 2.8 g benzidine are dissolved in a mixture of 80 ml ethanol, 70 ml water, 30 ml acetone and 1.5 ml N hydrochloric acid.
Spray reagent III: 1 ml saturated, aqueous silver nitrate solution is added with stirring to 20 ml acetone and water then added dropwise until the precipitated silver nitrate just dissolves.
Procedure: After having sprayed with I, the layer is air-dried, sprayed with II and placed for 5 min in a chamber containing 25% ammonium hydroxide. The chromatogram is sprayed finally with III, causing the white spots to assume a dark hue.
D. WALDI: J. Chromatog. **18**, 417 (1965).

160. Metaperiodate(sodium)-Nessler's reagent: for hydroxyamino acids (serine, threonine).
Spray reagent I: 1% aqueous sodium metaperiodate solution.
Spray reagent II: Nessler's reagent: 10 g mercuric iodide are rubbed into a thin paste with a little water and 5 g potassium iodide are added. A solution of 20 g sodium hydroxide in 80 ml water is added to this mixture. It is made up to 100 ml with water as soon as the paste has gone into solution. The cloudy solution is allowed to stand several days and then decanted from the precipitate which has settled out.
Procedure: The chromatogram is sprayed with I, dried at room temperature and subsequently sprayed with II.
R. CONSDEN, A. H. GORDON, and A. J. P. MARTIN: Biochem. J. **40**, 33 (1946).

161. Metaperiodate(sodium)-4-nitroaniline: for deoxy-sugars.
Spray reagent I: An aqueous, saturated solution of sodium metaperiodate is diluted with twice its volume of water.
Spray reagent II: 1% 4-nitroaniline in ethanol and 36% hydrochloric acid are mixed in the ratio 4:1 (by volume).

Procedure: Following spraying with I, the layer is left 10 min and then sprayed with II.
Note: Deoxy-sugars and glycals yield yellow spots which fluoresce strongly in long-wave UV light. The colour changes to green on spraying with 5% methanolic sodium hydroxide.

J. T. EDWARD, and D. M. WALDRON: J. Chem. Soc. **1952**, 3631.

162. Methylene blue: for sulphate esters of steroids.
Spray reagent: 0.025 g methylene blue is dissolved in 100 ml 0.05 N sulphuric acid. The solution is diluted with its own volume of acetone before use.
Note: The sulphate esters give spots of various colours on a blue background. The coloured complexes formed migrate on development with chloroform and leave white spots on the blue background where they were present before this development.

O. CRÉPY, O. JUDAS, and B. LACHESE: J. Chromatog. **16**, 340 (1964).

163. 4-Methylumbelliferone: for nitrogen-containing heterocyclic compounds (fluorescence indicator).
Spray reagent: 0.02 g 4-methylumbelliferone is dissolved in 35 ml ethanol and the solution made up to 100 ml with water.
Treatment after spraying: The chromatogram is introduced into a tank containing 25% ammonium hydroxide solution and subsequently inspected in long-wave UV light.

I. M. HAIS u. K. MACEK: Handbuch der Papierchromatographie I, p. 759. G. Fischer, Jena, 1958.

164. Methyl yellow-UV light: for chlorine-containing insecticides.
Spray reagent: 0.1 g methyl yellow (N,N-dimethyl-4-phenylazoaniline) is dissolved in 70 ml ethanol, 25 ml water are added and the volume is brought to 100 ml with ethanol.
Procedure: The chromatogram is air-dried after spraying and then exposed for 5 min to a UV lamp without filter. Red spots on a yellow background are yielded.

L. F. KRZEMINSKY, and W. A. LANDMANN: J. Chromatog. **10**, 515 (1963).

165. Millon's reagent: for phenols, phenol ethers and their glycosides.
Spray reagent: 5 g mercury are dissolved in 10 g fuming nitric acid (d = 1.40) and 10 ml water added. Yellow to orange spots on a white background are formed.
Treatment after spraying: Colour changes are often brought about by heating at 100—110° C.

E. STAHL u. P. J. SCHORN: Hoppe-Seylers Z. physiol. Chem. **325**, 263 (1961).

166. Molybdate(ammonium)-perchloric acid (Hanes reagent): for phosphate esters (sugar phosphates).
Spray reagent: 0.5 g ammonium molybdate is dissolved in 5 ml water and 1.5 ml 25% hydrochloric acid and 2.5 ml 70% perchloric acid are added. The solution, after having cooled to room temperature, is made

	up to 50 ml with acetone. This reagent should be left for a day before use and can be kept about 3 weeks.
Treatment after spraying:	The layer is exposed for 2 min to an infra red lamp placed 30 cm away and then for 7 min to long-wave UV light or heated 5—10 min at 110° C.

C. S. HANES, and F. A. ISHERWOOD: Nature **164**, 1107 (1949).
T. H. BEVAN, G. I. GREGORY, T. MALKIN, and A. G. POOLE: J. Chem. Soc. **1951**, 841.
S. BURROWS, F. S. M. GRYLLS, and J. S. HARRISON: Nature **170**, 800 (1952).
C. W. STANLEY: J. Chromatog. **16**, 467 (1964).

167. Molybdate(ammonium)-stannous chloride: for phosphoric acids.
Spray reagent I: 1% aqueous ammonium molybdate solution.
Spray reagent II: 1% solution of stannous chloride in 10% hydrochloric acid.
Procedure: The layer is sprayed with I, dried and then sprayed with II: heating 3—5 min at 105° C may sometimes be necessary.

H. SEILER: Helv. Chim. Acta **44**, 1753 (1961).

168. Molybdophosphoric acid: for reducing compounds, lipids, sterols and steroids.
A. Spray reagent: 5% ethanolic solution of molybdophosphoric acid.
Treatment after spraying: Heated at 120° C until the best spot formation is attained.
B. Spray reagent: 10% ethanolic solution of molybdophosphoric acid.
Treatment after spraying: Heated at 120° C until optimum spot formation is attained.
Note: The background can be rendered colourless by placing the plate in a tank containing 25% ammonium hydroxide solution.
C. Spray reagent: 20% solution of molybdophosphoric acid in ethanol or methyl cellosolve.
Antioxidants appear as blue spots after 1—2 min.

D. KRITCHEVSKY, and M. C. KIRK: Arch. Biochem. Biophys. **35**, 346 (1952).
A. SEHER: Fette u. Seifen, Anstrichmittel **61**, 345 (1959).

169. Morin: for Al^{3+} ions.
Spray reagent: 1% morin solution in acetic acid. Yields intensive light green fluorescence in long-wave UV light.

T. V. TORIBARA, and R. E. SHERMAN: Anal. Chem. **25**, 1594 (1953).

170. α-Naphthol-sulphuric acid: for sugars.
Spray reagent: A mixture is made of 10.5 ml 15% ethanolic α-naphthol, 6.5 ml conc. sulphuric acid, 40.5 ml ethanol and 4 ml water.
Treatment after spraying: Heated 3—6 min at 100° C.

H. JACIN, and A. R. MISHKIN: J. Chromatog. **18**, 170 (1965).

171. 1,2-Naphthoquinone-4-sulphonic acid (sodium salt): for aromatic amines.
Spray reagent: 5 ml acetic acid is added to s solution of 0.5 g sodium 1,2-naphthoquinone-4-sulphonate in 95 ml water. Any insoluble residue is filtered off.

Note: The colours are observed after 30 min reaction time.

R. B. SMYTH, and G. G. MCKEOWN: J. Chromatog. **16**, 454 (1964).

172. 1,2-Naphthoquinone-4-sulphonic acid (sodium salt) (Folin reagent): for amino acids.

Spray reagent: A solution of 0.02 g sodium 1,2-naphthoquinone-4-sulphonate in 100 ml 5% sodium carbonate is freshly prepared.

Procedure: The chromatogram is dried at room temperature after spraying; there is no further treatment. Various colours are yielded by the different amino acids.

D. MÜTING: Naturwissenschaften **39**, 303 (1952).

173. 1,2-naphthoquinone-4-sulphonic acid-perchloric acid: for sterols.

Spray reagent: 0.1 g 1,2-naphthoquinone-4-sulphonic acid is dissolved in 100 ml of a mixture of 20 ml ethanol, 10 ml 60% perchloric acid, 1 ml 40% formaldehyde solution and 9 ml water.

Procedure: The plate is heated to 70—80° C and the colour development observed. The spots turn first pink and then change to blue on prolonged heating.

E. RICHTER: J. Chromatog. **18**, 164 (1965).
C. W. M. ADAMS: Nature **192**, 331 (1961).

174. Naphthoresorcinol(1,3-dihydroxynaphthalene)-phosphoric acid: for sugars.

Spray reagent: 100 ml 0.2% ethanolic naphthoresorcinol solution are mixed with 10 ml 85% phosphoric acid.

Treatment after spraying: Heated 5—10 min at 100—105° C.

175. Naphthoresorcinol-sulphuric acid: for sugars.

Solution a: 0.2 g naphthoresorcinol is dissolved in 100 ml ethanol.
Solution b: 20% sulphuric acid.
Spray reagent: Equal volumes of a and b are mixed before use.
Treatment after spraying: Heated 5—10 min at 100—105° C.

176. Naphthoresorcinol-trichloroacetic acid: for sugars and uronic acids.

Solution a: 0.2% ethanolic naphthoresorcinol solution.
Solution b: 20% trichloracetic acid in water.
Spray reagent: Equal volumes of a and b are mixed as required.
Treatment after spraying: Heated 5—10 min in a drying oven at 100—105° C (for ketoses) or 10—15 min on the water bath in a moist atmosphere at 70—80° C (uronic acids).

Note: Any collidine or pyridine present interferes with the colour reaction. Resorcinol, orcinol, phloroglucinol or α-naphthol may be used instead of the naphthoresorcinol. The trichloroacetic acid solution may be replaced by a tenth of its volume of 85% phosphoric acid.

S. M. PARTRIDGE: Biochem. J. **42**, 238 (1948).

177. α-Naphthylamine: for 3,5-dinitrobenzoate esters and dinitrobenzamides.

Spray reagent I: 0.5% ethanolic solution of α-naphthylamine.
Spray reagent II: 10% methanolic potassium hydroxide solution.

Procedure: The chromatogram is sprayed with I, then with II; red-brown spots appear.

R. G. RICE, G. J. KELLER, and J. G. KIRCHNER: Anal. Chem. **23**, 194 (1951).

178. Ninhydrin: for amino acids, amines and aminosugars.

A. Spray reagent: 0.3 g ninhydrin is dissolved in 100 ml n-butanol and 3 ml acetic acid added.

B. Spray reagent: 0.2 g ninhydrin is dissolved in 100 ml ethanol.

Treatment after spraying: Heated at 110° C until the best colour development is reached. Heating at 160° C is recommended for pantothenic acid.

R. A. FAHMY, A. NIEDERWIESER, G. PATAKI u. M. BRENNER: Helv. Chim. Acta **44**, 2022 (1961).

A. R. PATTON, and P. CHISM: Anal. Chem. **23**, 1683 (1951).

Stabilisation of the ninhydrin-spots:

Spray reagent: A mixture is made of 1 ml saturated aqueous cupric nitrate, 0.2 ml 10% nitric acid and 100 ml 96% ethanol.

Procedure: The ninhydrin spots are sprayed with this reagent and the plate is then placed in a chamber containing 25% ammonium hydroxide. The red copper complex obtained in this way is stable only in the absence of free hydrogen ions and powerful complexing agents.

E. KAWERAU, and T. WIELAND: Nature **168**, 77 (1951)

179. Ninhydrin-cupric nitrate: for amino acids (polychromatic detection).

Solution I: 10 ml acetic acid and 2 ml collidine are added to a solution of 0.1 g ninhydrin in 50 ml absolute ethanol.

Solution II: 0.5 g cupric nitrate is dissolved in 50 ml absolute ethanol.

Spray reagent: Solutions I and II are mixed in the proportion 50:3 before use.

Treatment after spraying: The sprayed plate is held over a hot plate until colur development is just beginning. The gradual intensification of colour can be seen in transmitted light. Some amino acids appear first almost as points of colour; these are quickly marked with a sharp pencil. In this way it is often possible to detect individual components in spots which later merge into each other. Many amino acids show characteristic colours. They differ amongst themselves also in the speed with which they form the coloured products.

M. BRENNER u. A. NIEDERWIESER: Experientia **16**, 378 (1960).

180. 4-Nitroaniline, diazotised: for phenols, phenol carboxylic acids, amines and heterocyclic compounds which serve as coupling components.

Spray reagent: 10 ml 0.1% aqueous 4-nitroaniline solution are mixed with 10 ml 0.2% aqueous sodium nitrite solution and 20 ml 10% potassium carbonate solution in water are added. Coloured products are formed.

A. STURM u. H. W. SCHEJA: J. Chromatog. **16**, 194 (1964).

181. 4-Nitroaniline, diazotised (acid): for plasticisers.

Spray reagent I:	0.5 N alcoholic potassium hydroxide.
Spray reagent II:	0.8 g 4-nitroaniline is dissolved in 250 ml water, 20 ml 25% hydrochloric acid are added and the solution is diazotised with 5% aqueous sodium nitrite solution until colourless.
Procedure:	The layer is sprayed with I, dried for 15 min at 60° C and then sprayed with II; yellow to orange coloured spots are yielded.

J. W. Copius-Peereboom: J. Chromatog. **4**, 323 (1960).
D. Braun: Chimia (Switz.) **19**, 77 (1965).

182. 4-Nitroaniline, diazotised (buffered): for phenols.

Spray reagent:	5 ml 0.5% 4-nitroaniline solution in 2N hydrochloric acid are mixed under cooling with 0.5 ml 5% aqueous sodium nitrite solution and 15 ml 20% aqueous sodium acetate solution are added.

H. G. Bray, W. V. Thorpe, and K. White: Biochem. J. **46**, 271 (1950).
T. Swain: Biochem. J. **53**, 200 (1953).
C. F. van Sumere, G. Wolf, H. Teuchy, and J. Kint: J. Chromatog. **20**, 48 (1965).

183. 4-Nitrophenyldiazonium fluoborate: for phenols and amines, capable of coupling.

Spray reagent I:	Freshly prepared 1% solution of 4-nitrophenyldiazonium fluoborate in acetone.
Spray reagent II:	0.1 N methanolic potassium hydroxide.
Procedure:	The chromatogram is sprayed first with I, then with II.
Preparation of the reagent:	14 g 4-nitroaniline is dissolved by warming with 30 ml 36% hydrochloric acid and 30 ml water. After having cooled this solution to 5° C, a solution of 8 g sodium nitrite in 20 ml water is added and then 60 ml 40% fluoboric acid. The yellow precipitate formed is filtered off, washed successively with fluoboric acid, ethanol and ether and dried in vacuo in a desiccator.

J. H. Freeman: Anal. Chem. **24**, 955 (1952).
H. Seeboth, and H. Görsch: Chem. Techn. **15**, 294 (1963).

184. Nitroprusside (sodium): for compounds with the —SH group (cysteine), the —S—S— group (cystine) and for arginine.

Spray reagent I:	1.5 g sodium nitroprusside is dissolved in 5 ml 2N hydrochloric acid, 95 ml methanol and 10 ml 25% ammonium hydroxide solution are added and the solution filtered.
Note:	Thiols (SH-group) are visible as red spots. Arginine turns orange and later grey-blue.
Spray reagent II:	A solution of 2 g sodium cyanide in 5 ml water is diluted to 100 ml with methanol.
Note:	Disulphides (—S—S— group) appear as red spots on a yellow background when the layer from I is sprayed with II. Caution when using this highly toxic cyanide-containing reagent.

Variant for —S—S— *bridges:*

Spray reagent I: 5 g sodium cyanide and 5 g sodium carbonate are dissolved in 25 % ethanol and the volume made up to 100 ml with this aqueous solvent.
Spray reagent II: 2 g sodium nitroprusside are dissolved in 100 ml 75% ethanol.
Procedure: The layer is sprayed with I, allowed to dry in the air and then sprayed with II. Precautions must be taken against cyanide poisoning when spraying I.

G. TOENNIES, and J. J. KOLB: Anal. Chem. **23**, 823 (1951).

Variation for thiolactones:
Spray reagent I: N sodium hydroxide.
Spray reagent II: 2 g sodium nitroprusside are dissolved in 100 ml 75% ethanol.
Procedure: The layer is sprayed with I, allowed to dry in the air and then sprayed with II.

F. KORTE u. J. VOGEL: J. Chromatog. **9**, 381 (1962).

185. Nitroprusside(sodium)-acetaldehyde: for secondary aliphatic and alicyclic amines.
Solution a: 5 g sodium nitroprusside are dissolved in 100 ml of a 10% aqueous acetaldehyde solution.
Solution b: 2% sodium carbonate solution in water.
Spray reagent: Equal volumes of a and b are mixed before use.

F. FEIGL: Spot Tests in Organic Analysis, 7th Edition, p. 251, Elsevier Pub. Co., 1966.
K. MACEK, J. HACAPERKOVÁ u. B. KAKÁČ: Pharmazie **11**, 533 (1956).
E. STEIN v. KAMIENSKI: Planta **50**, 291 (1957).

186. Nitroprusside(sodium)-ammonia: for hemlock alkaloids.
Spray reagent I: 1% aqueous sodium nitroprusside solution.
Spray reagent II: 10% ammonium hydroxide.
Procedure: The chromatogram is sprayed with I, then II.
Note: γ-Coniceine turns red.

F. MOLL: Arch. Pharm. **296**, 205 (1963).

187. Nitroprusside(sodium)-ferricyanide (FCNP reagent): for aliphatic nitrogen compounds, e. g., cyanamide, guanidine, urea and thiourea and their derivatives; for creatine and creatinine.
Spray reagent: 10% aqueous sodium hydroxide, 10% aqueous sodium nitroprusside, 10% aqueous potassium ferricyanide and water are mixed in the ratio 1:1:1:3. The mixture is allowed to stand at least 20 min at room temperature before use. It can be stored several weeks in a refrigerator. It is mixed with an equal volume of acetone before use.

J. ROCHE et al.: Biochim. et Biophys. Acta **14**, 71 (1954).
L. FISHBEIN, and M. A. CAVANAUGH: J. Chromatog. **20**, 283 (1965).
L. FISHBEIN: Rec. trav. chim. **84**, 465 (1965).

188. Nitroprusside(sodium)-hydrogen peroxide: for guanidine, urea, thiourea and their derivatives, for creatine and creatinine.
Spray reagent: A mixture is made of 2 ml 5% sodium nitroprusside, 1 ml 10% sodium hydroxide and 5 ml

3% hydrogen peroxide (all aqueous solutions) and diluted with 15 ml water. It can be stored several days in a refrigerator.

E. HOFMANN u. A. WÜNSCH: Naturwissenschaften **45**, 338 (1958).

189. Nitroprusside(sodium)-hydroxylamine (Grote reagent): for thiourea derivatives.

Spray reagent: 0.5 g hydroxylammonium chloride and 1 g sodium bicarbonate are added to a solution of 0.5 g sodium nitroprusside in 10 ml water. 2 drops of bromine are added when gas evolution ceases and the total volume is made up to 25 ml with water. It is stable for about 2 weeks.

I. W. GROTE: J. Biol. Chem. **93**, 25 (1931).

190. Nitroprusside(sodium)-metaperiodate(sodium): for deoxy-sugars.

Spray reagent I: 2.5% solution of sodium metaperiodate in water.
Spray reagent II: A mixture of 7% aqueous sodium nitroprusside, water and saturated ethanolic piperazine in the volume proportion 1:3:20 is made.
Procedure: The layer is sprayed with I, dried 10 min at room temperature and sprayed with II. The blue spots attain maximum intensity in 5—10 min.

J. T. EDWARD, and D. M. WALDRON: J. Chem. Soc. **1952**, 3631.

191. Nitroprusside(sodium)-sodium hydroxide (Legal test): for methyl ketones and activated methylene groups.

Spray reagent: 1 g sodium nitroprusside is dissolved in 100 ml of a mixture of 2N sodium hydroxide and ethanol (1 + 1). Red to violet spots are yielded.

F. FEIGL: Spot Tests in Organic Analysis, 7th Edition, p. 208, Elsevier Pub. Co., 1966.

192. Orcinol-ferric chloride-sulphuric acid: for sugars.

Solution a: 1 g ferric chloride is dissolved in 10% sulphuric acid and the solution diluted to 100 ml with this medium.
Solution b: 6% ethanolic orcinol solution.
Spray reagent: 10 ml a and 1 ml b are mixed before use.
Treatment after spraying: Heated 10—15 min at 100° C.

193. Palladium(II) chloride: for thiophosphate esters and other sulphur compounds (e. g., phenothiazines).

Spray reagent: 0.5 g palladium(II) chloride is dissolved in 100 ml water containing a few drops 25% hydrochloric acid.

J. BÄUMLER u. S. RIPPSTEIN: Helv. Chim. Acta **44**, 1162 (1961).

194. Paraformaldehyde-phosphoric acid: for *Solanum* steroid alkaloids and steroid sapogenins.

Spray reagent: 0.03 g paraformaldehyde is shaken at room temperature with 100 ml 85% phosphoric acid until dissolved. The reagent can be kept for a few weeks.

K. SCHREIBER, O. AURICH u. G. OSSKE: J. Chromatog. **12**, 63 (1963).

195. Perchloric acid: for steroids and bile acids.
A. Spray reagent (for steroids): 20% aqueous perchloric acid.
B. Spray reagent (for bile acids): 60% aqueous perchloric acid.
Treatment after spraying: The plate is heated about 10 min at 150° C until maximum colour of the spots is reached. The fluorescence in long-wave UV light is also observed.
H. METZ: Naturwissenschaften 48, 569 (1961).
S. HARA, and M. TAKEUCHI: J. Chromatog. 11, 565 (1963).

196. Perchloric acid-ferric chloride: for indole derivatives.
Spray reagent: 100 ml 5% aqueous perchloric acid are mixed with 2 ml 0.05M ferric chloride solution.
Note: It does not react with isatin and other oxindole derivatives.
T. A. BENNET-CLARK, M. S. TAMBIAH, and N. P. KEFFORD: Nature 169, 452 (1951).

197. Permanganate (potassium), alkaline: for reducing compounds and aromatic polycarboxylic acids.
Solution a: 1% aqueous potassium permanganate.
Solution b: 5% aqueous sodium carbonate.
Spray reagent: Equal volumes of a and b are mixed.
O. B. MAXIMOV, and L. S. PANTHINKHINA: J. Chromatog. 20, 150 (1965).
I. M. HAIS u. K. MACEK: Papierchromatographie I, p. 735. G. Fischer, Jena 1958.

198. Permanganate (potassium), alkaline: for sugars and polyalcohols.
Spray reagent: 0.5 g potassium permanganate is dissolved in 100 ml N sodium hydroxide.
Treatment after spraying: The plate is heated at 100° C.
G. W. HAY, B. A. LEWIS, and F. SMITH: J. Chromatog. 11, 479 (1963).

199. Permanganate (potassium), neutral: for easily oxidised substances.
Spray reagent: 0.05% aqueous potassium permanganate.

200. Permanganate(potassium)-sulphuric acid (universal reagent).
Spray reagent: 0.5 g potassium permanganate is dissolved in 15 ml conc. sulphuric acid; danger of explosion of manganese heptoxide!
H. ERTEL u. L. HORNER: J. Chromatog. 7, 268 (1962).

201. Phenol-sulphuric acid: for sugars.
Spray reagent: 3 g phenol and 5 ml conc. sulphuric acid are dissolved in 95 ml ethanol.
Treatment after spraying: Heated 10—15 min at 100° C, yielding brown spots.

202. p-Phenylenediamine-phthalic acid: for 3-ketosteroid conjugates.
Spray reagent: 0.9 g p-phenylenediamine and 1.6 g phthalic acid are dissolved in n-butanol (saturated with water) and made up to 100 ml with this solvent.
Treatment after spraying: Heated at 100—110° C, giving yellow to orange spots.
B. P. LISBOA: Acta Endocrinol. 43, 47 (1963).
B. P. LISBOA: J. Chromatog. 16, 136 (1964).

203. o-Phenylenediamine-sulphuric acid: for dehydroascorbic acid.
Spray reagent: 0.1 g o-phenylenediamine is dissolved in a mixture of 50 ml 0.1 N sulphuric acid and 50 ml ethanol.

S. OGAWA: J. Pharm. Soc. Japan **73**, 59 (1953).

204. o-Phenylenediamine-trichloroacetic acid: for α-ketoacids.
Spray reagent: 0.05 g o-phenylenediamine is dissolved in 100 ml 10% aqueous trichloroacetic acid.
Procedure: Heated at 100° C in a drying oven for at most 2 min; spots are yielded which fluoresce green in long-wave UV light.

T. WIELAND u. F. FISCHER: Naturwissenschaften **36**, 219 (1949).
O. WISS: Hoppe-Seylers Z. physiol. Chem. **293**, 106 (1953).

205. Phenylhydrazine: for dehydroascorbic acid.
Spray reagent: 0.3 g phenylhydrazine hydrochloride and 0.45 g sodium acetate are dissolved in 10 ml water.

206. Phosphomolybdic acid see Molybdophosphoric acid.

207. Phosphotungstic acid see Tungstophosphoric acid.

208. Phosphoric acid: for sterols and steroids.
A. Spray reagent: 85% phosphoric acid and water are mixed 1:1 (volume).
B. Spray reagent: 15 ml 85% phosphoric acid are diluted to 100 ml with methanol.
Procedure: The layer is sprayed thoroughly until transparent and then heated 15—30 min at 120° C. The individual sterols or steroids require varying periods of heating for attainment of maximum colour intensity or fluorescence.
Note: All compounds of this class fluoresce in long-wave UV light. Larger amounts of substance yield spots which are visible in daylight.

R. NEHER u. A. WETTSTEIN: Helv. Chim. Acta **34**, 2278 (1951).

209. Phosphoric acid-bromine: for digitalis glycosides.
Spray reagent I: 10% aqueous phosphoric acid.
Spray reagent II: A mixture is made of 2 ml saturated aqueous potassium bromide, 2 ml saturated aqueous potassium bromate and 2 ml 25% hydrochloric acid.
Procedure: The layer is sprayed with I and the plate then heated 12 min at 120° C; digitalis glycosides of the series B, D and E yield spots which fluoresce blue in long-wave UV light.
The plate is heated again at 120° C and then lightly sprayed with II; the A and C glycosides then fluoresce in UV light with orange and with grey-green to grey-blue colours respectively.

L. FAUCONNET et M. WALDESBÜHL: Pharm. Acta Helv. **38**, 423 (1963).

210. Picric acid-alkali (Jaffe reagent): for creatinine, glycocyamidine.
Spray reagent I: 1% picric acid solution in ethanol.
Spray reagent II: 5% ethanolic potassium hydroxide solution.

Procedure: The chromatogram is sprayed with I, dried and then sprayed with II, yielding orange spots.

R. WILLIAMS: Biochem. Inst. Stud. IV, University of Texas, Publ., Austin/Texas No. 5 109, 205 (1951).

211. Picric acid-perchloric acid: for Δ^5-3β-hydroxysteroids.
Spray reagent: 0.1 g picric acid is dissolved in a mixture of 36 ml acetic acid and 6 ml 70% perchloric acid.
Treatment after spraying: Heated 3—5 min at 70—80° C, giving yellow-red colours.

W. R. EBERLEIN: J. Clin. Endocrinol. **25**, 288 (1965).

212. Pinacryptol yellow: for alkyl- and arylsulphonic acids.
Spray reagent: 0.05—0.1% aqueous solution of pinacryptol yellow. The spots fluoresce yellow-orange in the light of a long-wave UV lamp.

J. BORECKÝ: J. Chromatog. **2**, 612 (1959).

213. Potassium hydroxide, methanolic: for coumarins, anthraquinone glycosides and their aglucones.
Spray reagent: 5% methanolic potassium hydroxide solution. The dried chromatogram is inspected in daylight and long-wave UV light.

Z. LEDINOVÁ, and I. M. HAIS: Českoslov. Farm. **9**, 401 (1960).
L. HÖRHAMMER, H. WAGNER, and G. BITTNER: Arzneimittel-Forsch. **13**, 537 (1963).

214. 1-(2-Pyridylazo)-2-naphthol(PAN): for Cd^{2+}, Co^{2+}, Cu^{2+}, Mn^{2+}, Pb^{2+}, Ni^{2+}, Zn^{2+} and UO_2^{2+} ions.
Spray reagent: 0.25% ethanolic solution of PAN.
Treatment after spraying: The plate is placed in a chamber containing 25% ammonium hydroxide solution.

H. SEILER u. M. SEILER: Helv. Chim. Acta **44**, 939 (1961).
F. W. H. M. MERKUS: Pharm. Weekblad **98**, 947 (1963).

215. 1-(2-Pyridylazo)-2-naphthol(PAN)-cobalt(II) nitrate: for glucuronides of steroids.
Spray reagent I: 0.4% ethanolic PAN solution, diluted with four times its volume of methylene dichloride before use.
Spray reagent II: Solution a: 0.8% aqueous cobalt(II)nitrate.
Solution b: 2M acetate buffer, pH 4.6 (free of iron).
8 ml a and 4 ml b are mixed and diluted to 100 ml with water.
Procedure: The layer is sprayed with I until it is evenly yellow; it is then dried and sprayed with II. The glucuronides appear as rapidly fading violet spots, the colour of which turns greenish on drying.

O. CRÉPY, O. JUDAS, and B. LACHESE: J. Chromatog. **16**, 340 (1964).

216. Quercetin: for cations of the copper and ammonium sulphide groups, Al^{3+}, Mg^{2+}, UO_2^{2+} and WO_4^{2-} ions.
Spray reagent: 0.2% solution of quercetin in ethanol.
Treatment after spraying: The layer is sprayed with 25% ammonium hydroxide solution or the plate placed in a tank

containing this solution. Spots fluorescing in long-wave UV light are obtained.

A. WEISS u. S. FALLAB: Helv. Chim. Acta **37**, 1253 (1954).
E. PFEIL, A. FRIEDRICH u. T. WACHSMANN: Z. anal. Chem. **158**, 429 (1957).

217. p-Quinone: for ethanolamine.
Spray reagent: 0.5 g p-benzoquinone is dissolved in a mixture of 10 ml pyridine and 40 ml n-butanol. Ethanolamine yields red spots immediately after spraying; choline does not react.

218. Resorcinol-zinc chloride-sulphuric acid: for plasticisers (suitable especially for phthalate esters).
Spray reagent I: Some zinc chloride is added to a 20% solution of resorcinol in ethanol.
Spray reagent II: 4 N sulphuric acid.
Spray reagent III: 40% aqueous potassium hydroxide.
Procedure: The plate is sprayed with I, heated 10 min at 150° C, sprayed with II, heated 20 min at 120° C and finally sprayed with III. Orange-red spots on a yellow background are yielded.

J. W. COPIUS-PEEREBOOM: J. Chromatog. **4**, 323 (1960).
D. BRAUN: Chimia (Switz.) **19**, 77 (1965).

219. Resorcyl aldehyde-sulphuric acid: for 16-dehydrosteroids.
Solution a: 0.5% solution of resorcyl aldehyde (2,4-dihydroxybenzaldehyde) in acetic acid.
Solution b: 5% solution of conc. sulphuric acid in acetic acid.
Spray reagent: Equal amounts of a and b are mixed just before use.
Treatment after spraying: Heated at 100—110° C until the colour intensity of the spots has reached a maximum.

D. B. GOWER: J. Chromatog. **14**, 424 (1964).

220. Rhodamine B: general spray reagent.
A. Spray reagent: 0.025—0.05% ethanolic rhodamine B solution.
B. Spray reagent: 0.25% ethanolic rhodamine B solution. The chromatogram is inspected in long-wave UV radiation.

H. P. KAUFMANN u. J. BUDWIG: Fette u. Seifen, Anstrichmittel **53**, 390 (1951).

221. Rhodamine 6 G: for lipids.
Spray reagent: 1 mg rhodamine 6 G is dissolved in 100 ml acetone. The chromatogram is observed in long-wave UV light.

R. F. WITTER, G. V. MARINETTI, and A. MORRISON: Arch. Biochem. Biophys. **68**, 15 (1957).

222. Rhodanine: for carotenoid aldehydes.
Spray reagent I: 1—5% ethanolic solution of rhodanine.
Spray reagent II: 25% ammonium hydroxide or 27% sodium hydroxide solution.
Procedure: The layer is sprayed with I and then II and dried.

A. WINTERSTEIN u. B. HEGEDÜS: Chimia (Switz.) **14**, 18 (1960).

Spray Reagents

223. Rhodizonate (sodium): for Ba^{2+} and Sr^{2+} ions.
Spray reagent I: 1% aqueous solution of sodium rhodizonate.
Spray reagent II: 25% ammonium hydroxide solution.
Procedure: The layer is sprayed with I and then II.
T. V. ARDEN, F. H. BURSTALL, G. R. DAVIES, J. A. LEWIS, and R. P. LINSTEAD: Nature **162**, 691 (1948).

224. Silver nitrate: for phenols.
Spray reagent: 1 ml of saturated, aqueous silver nitrate is added with stirring to 20 ml acetone and the product treated dropwise with water until the precipitated silver nitrate has just dissolved. Light pink to deep green spots are yielded.
W. J. BURKE, A. D. POTTER, and R. M. PARKHURST: Anal. Chem. **32**, 727 (1960).

225. Silver nitrate-ammonium hydroxide (Tollens or Zaffaroni reagent) for reducing substances.
Solution a: 0.1 N silver nitrate.
Solution b: 5 N ammonium hydroxide.
Spray reagent: a and b are mixed in the proportion 1:5 as required. Longer standing may lead to formation of explosive silver azide.
Treatment after spraying: Heated 5—10 min at 105° C until the dark spots have become most intense.
E. C. BATE-SMITH, and R. G. WESTALL: Biochim. et Biophys. Acta **4**, 427 (1950).

226. Silver nitrate-ammonium hydroxide-chloride (sodium): for thioacids.
Spray reagent I: As needed, 50 ml 0.1 N silver nitrate solution are mixed with 50 ml 10% ammonium hydroxide. The reagent yields explosive silver azide on standing for a longer time.
Spray reagent II: 10% aqueous sodium chloride solution.
Procedure: The layer is sprayed with I, dried and then sprayed with II. It is then exposed to daylight until the yellow-brown spots have attained maximum colour intensity.

227. Silver nitrate-ammonium hydroxide-fluorescein: for halide ions.
Spray reagent I: 1 g silver nitrate is dissolved in 100 ml 0.5N ammonium hydroxide.
Spray reagent II: 0.1 g fluorescein is dissolved in 100 ml ethanol.
Procedure: The chromatogram is sprayed with I, briefly dried and sprayed with II.
H. SEILER u. T. KAFFENBERGER: Helv. Chim. Acta **44**, 1282 (1961).

228. Silver nitrate-ammonium hydroxide-methoxide (sodium): for sugars.
Solution a: 0.3% methanolic silver nitrate.
Solution b: Methanol, saturated with ammonia gas.
Solution c: Sodium methoxide solution, from dissolving 7 g sodium in 100 ml methanol.
Spray reagent: 20 ml a, 4 ml b and 8 ml c are mixed before use.
Treatment after spraying: Heated 10 min at 110° C.

229. Silver nitrate-dichromate (potassium): for barbiturates.

Spray reagent I:	25 ml saturated, aqueous silver nitrate are added to a mixture of 50 ml acetone and 2 ml water.
Spray reagent II:	0.3% potassium dichromate in water.
Spray reagent III:	2% sodium hydroxide in methanol.
Procedure:	The layer is liberally sprayed with I and air-dried. It is then sprayed with II, air-dried, re-sprayed with II and re-dried, after which III is sprayed on.

H. WEIDMANN: Dissertation, Berlin 1961.

230. Silver nitrate-fluorescein: for alkyl- and arylsulphonic acids.

Solution a:	10% aqueous silver nitrate.
Solution b:	0.2 g sodium-fluorescein is dissolved in 100 ml absolute ethanol.
Spray reagent:	This is made before use from 10 ml a and 50 ml b; yellow spots on a salmon-pink background are obtained.

F. H. POLLARD, G. NICKLESS, and K. W. C. BURTON: J. Chromatog. 8, 507 (1962).

C. M. COYNE, and G. A. MAW: J. Chromatog. 14, 552 (1964).

231. Silver nitrate-formaldehyde: for chlorinated insecticides (e. g., dieldrin, aldrin and lindane).

Spray reagent I:	0.05 N ethanolic silver nitrate.
Spray reagent II:	35% formaldehyde solution.
Spray reagent III:	2 N methanolic potassium hydroxide.
Spray reagent IV:	Freshly prepared mixture of equal volumes of 30% hydrogen peroxide and 65% nitric acid.
Procedure:	The layer is sprayed with I and then II, air drying for 30 min after each. It is then sprayed with III, dried 30 min at 130° C and finally sprayed with IV. The chromatogram is left 12 h in the dark and then exposed to sunlight, yielding dark grey spots on a light grey background.

L. C. MITCHELL: J. Assoc. Off. Agr. Chemists 35, 920 (1952).

232. Silver nitrate-hydrogen peroxide: for chlorinated hydrocarbons.

Spray reagent:	0.1 g silver nitrate is dissolved in 1 ml water and 10 ml phenyl cellosolve (2-phenoxyethanol) added. This solution is made up to 200 ml with acetone and 1 drop 30% hydrogen peroxide added.
Treatment after spraying:	The layer is irradiated with unfiltered UV light. Alumina layers must be exposed about 50 min and silica gel layers up to 15 min if long-wave UV light is used. Dark spots are formed.

M. F. KOVACS: J. Assoc. Off. Agr. Chemists 46, 884 (1963).

233. Silver nitrate-permanganate (potassium): for compounds with reducing properties.

Solution a:	0.1 N silver nitrate, 2 N ammonium hydroxide and 2 N sodium hydroxide are mixed 1:1:2 before use.
Solution b:	0.5 g potassium permanganate and 1 g sodium carbonate are dissolved in 100 ml water.

Spray reagent:	A mixture of equal volumes of a and b. The reagent is unstable and must be freshly made each time.
Note:	Reducing substances appear as light yellow spots on a green-blue background, immediately after spraying.

J. KELLEN: Chem. Listy **51**, 973 (1957).

234. Silver nitrate-sodium hydroxide: for sugars and polyalcohols.

Spray reagent I:	1 ml saturated aqueous silver nitrate solution is diluted to 200 ml with acetone and 5—10 ml water then added until the precipitate has dissolved.
Spray reagent II:	0.5 N aqueous-methanolic sodium hydroxide (20 g sodium hydroxide are dissolved in a minimum of water and the solution diluted to 1 l with methanol.
Procedure:	Spraying is carried out with I, then II and the plate finally heated 1—2 min at 100° C.

235. Sodium hydroxide: for Δ^4-3-ketosteroids.

Spray reagent:	10% sodium hydroxide in methanol-water (60+40).
Procedure:	Heated 10 min at 80° C, whereby the ketosteroids fluoresce yellow in long-wave UV light.

I. E. BUSH: Biochem. J. **50**, 370 (1951).

236. Stannic chloride: for triterpenes, sterols and steroids, phenols and polyphenols.

Spray reagent:	10 ml stannic chloride are added to 160 ml of a mixture of equal volumes of chloroform and acetic acid.
Treatment after spraying:	The layer is heated 5—10 min at 100° C and subsequently inspected in daylight and in long-wave UV light.

J. J. SCHEIDEGGER u. E. CHERBULIEZ: Helv. Chim. Acta **38**, 547 (1955).

237. Stannous chloride-iodide (potassium): for gold ions.

Spray reagent:	5.6 g stannous chloride are dissolved in 10 ml of 36% hydrochloric acid. The solution is diluted to 100 ml with water and 0.2 g potassium iodide added. Black spots are yielded.

F. H. BURSTALL, G. R. DAVIES, R. P. LINSTEAD, and R. A. WELLS: J. Chem. Soc. **1950**, 516.

238. Sulphanilic acid, diazotised (Pauly reagent): for phenols, amines and heterocyclic compounds which can couple.

Spray reagent:	4.5 g sulphanilic acid are dissolved in 45 ml 12 N hydrochloric acid by warming and the solution diluted to 500 ml with water. 10 ml of this diluted solution are cooled in ice and 10 ml of a cold 4.5% aqueous sodium nitrite solution added. The resulting reagent is maintained 15 min at 0° C (at which temperature it is stable for 1—3 days) and an equal volume of 10% aqueous sodium carbonate is added just before use as a spray reagent.

H. JATZKEWITZ: Hoppe-Seylers Z. physiol. Chem. **292**, 99 (1953)
M. R. GRIMMETT, and E. L. RICHARDS: J. Chromatog. **20**, 171 (1965).

239. Sulphanilic acid-α-naphthylamine: for nitrosamines.

Solution a:	1% solution of sulphanilic acid in 30% acetic acid.
Solution b:	0.1% solution of α-napththylamine in 30% acetic acid.
Spray reagent:	Equal volumes of a and b are mixed before use.
Procedure:	The layer is irradiated with short-wave UV light for about 3 min and then sprayed with the reagent.
Note:	Aliphatic nitrosamines yield red-violet spots; aromatic, green to blue.

R. Preussmann, D. Daiber, and H. Hengy: Nature **201**, 502 (1964).
R. Preussmann, G. Neurath, G. Wulf-Lorentzen, D. Daiber u. H. Hengy: Z. anal. Chem. **202**, 187 (1964).

240. Sulphanilamide, diazotised (Pauly reagent according to Kutáček): for phenols, amines and heterocyclic compounds which can couple.

Spray reagent I:	3 g sulphanilamide are dissolved in a mixture of 200 ml water, 6 ml 36% hydrochloric acid and 14 ml n-butanol; 0.3 g sodium nitrite is added to 20 ml of this solution before spraying.
Spray reagent II:	10% aqueous sodium carbonate solution.
Procedure:	The layer is sprayed with I, left 5—10 min and then sprayed with II.

I. M. Hais u. K. Macek: Handbuch der Papierchromatographie I, p. 743; G. Fischer, Jena, 1958.

241. Sulphuric acid reagents (reagents for general detection, in particular for sterols, steroids, bile acids and gibberellins.
Spray reagents:

A. Equal volumes of conc. sulphuric acid and methanol are cautiously mixed (cooling).
B. 5% solution of conc. sulphuric acid in ethanol.
C. 15% solution of conc. sulphuric acid in n-butanol.
D. 5% solution of conc. sulphuric acid in acetic anhydride.
E. Equal volumes of conc. sulphuric acid and acetic acid.

Procedure:	The chromatogram is sprayed with one of these reagents, allowed to dry for 15 min in the air and then heated at 110° C until the colour or fluorescence developed has reached its maximum.
Note:	Cholesterol and vitamin A and their esters and also many isoprenoid lipids are easily detected through the characteristic colours they yield on heating after having been sprayed with 50% sulphuric acid (reagent A): cholesterol and esters turn red, then red-violet and finally brown; vitamin A and esters first colour blue. Most compounds may be subsequently charred, yielding black spots. Oxidation to CO_2 may occur on layers impregnated with silver nitrate when heated, following spraying with sulphuric acid.

D. F. Jones, J. McMillan, and M. Radley: Phytochemistry **2**, 307 (1964), (gibberellins).
W. L. Anthony, and W. T. Beher: J. Chromatog. **13**, 570 (1964).
H. Jatzkewitz u. E. Mehl: Hoppe-Seylers Z. physiol. Chem. **320**, 251 (1960).
H. Metz: Naturwissenschaften **48**, 569 (1961).

242. Sulphuric acid-hypochlorite: for digitalis glycosides.
Spray reagent: A mixture is made of 10 ml 2N sulphuric acid and 3 ml sodium hypochlorite solution (10% active chlorine).
Treatment after spraying: Heated 10—15 min at 125° C.
Note: Fluorescence of various colours is shown in long-wave UV light by the digitalis glycosides of series A to E.

L. FAUCONNET et R. FAZAN: Bull. soc. vaud. sci. nat. **66**, 307 (1956).
L. FAUCONNET et M. WALDESBÜHL: Pharm. Acta Helv. **38**, 423 (1963).

243. Tetracyanoethylene: for aromatic hydrocarbons, phenols and heterocyclic compounds.
Spray reagent: 10% tetracyanoethylene solution in benzene.
Procedure: The layers are sprayed as soon as they are removed from the development chamber.
Note: Aromatic hydrocarbons show differentiating colours, sometimes evanescent.
JANÁK recommends heating at 100° C.

P. V. PEURIFOY, S. C. SLAYMAKER, and M. NAGER: Anal. Chem. **31**, 1740 (1959).
J. JANÁK: J. Chromatog. **15**, 15 (1964).
N. KUCHARCZYK, J. FOHL, and J. VYMĚTAL: J. Chromatog. **11**, 55 (1963).

244. 2,4,2′,4′-Tetranitrobiphenyl: for cardiac glycosides.
Spray reagent I: Saturated solution of 2,4,2′,4′-tetranitrobiphenyl in benzene.
Spray reagent II: 10% potassium hydroxide in methanol-water (1+1).
Procedure: The chromatogram is sprayed with I, dried at room temperature and then sprayed with II. Blue spots are obtained.

J. BINKERT, E. ANGLIKER u. A. v. WARTBURG: Helv. Chim. Acta **45**, 2122 (1962).

245. Tetraphenylborate (sodium) (Kalignost®): for alkaloids.
Spray reagent I: 1% solution of sodium tetraphenylborate in butanone, saturated with water.
Spray reagent II: 0.015% methanolic solution of fisetin or quercetin.
Procedure: After spraying with I, the layer is air dried, then sprayed with II and again air-dried. Orange to red spots are yielded which fluoresce in long-wave UV light.

R. NEU: J. Chromatog. **11**, 364 (1963).

246. Tetraphenylborate(sodium)-rhodamine B: for K^+ ions.
Spray reagent I: 0.1 N sodium hydroxide.
Spray reagent II: 1% ethanolic solution of sodium tetraphenylborate.
Spray reagent III: 0.5% ethanolic solution of rhodamine B.
Procedure: The layer is sprayed with I, allowed to dry and then sprayed with II and finally III. The spots show intense dark blue fluorescence in long-wave UV light. Larger amounts of K^+ appear in daylight as light red spots on a dark red background.

247. Tetrazolium, blue: for corticosteroids and other compounds with reducing properties.
Solution I: 0.5% blue tetrazolium in methanol.
Solution II: 6 N sodium hydroxide in water or water-methanol.
Spray reagent: Equal volumes of I and II are mixed before use. Violet spots are obtained at room temperature or on warming slightly.

O. ADAMEC: Steroids **1**, 495 (1963).
T. FEHER: Mikrochim. Acta **1965**, 105.
U. FREIMUTH, B. ZAWTA, u. M. BÜCHNER: Acta Biol. et Med. Ger. **13**, 624 (1964).
O. NISHIKAZE, R. ABRAHAM and H.-J. STAUDINGER: J. Biochem. (Tokio) **54**, 427 (1963).
I. E. BUSH, and M. WILLOUGHBY: Biochem. J. **67**, 689 (1957).

248. Thiobarbituric acid: for sorbic acid.
Spray reagent: Saturated, aqueous thiobarbituric acid. Red spots are yielded.

J. W. COPIUS-PEEREBOOM, and H. W. BEEKES: J. Chromatog. **14**, 417 (1964)

249. Thiocyanate(ammonium)-ferrous sulphate: for peroxides.
Spray reagent I: 0.4 g ammonium thiocyanate is dissolved in 30 ml acetone.
Spray reagent II: 1.2 g ferrous sulphate is dissolved in 30 ml water.
Procedure: The chromatogram is sprayed with I, dried briefly and then sprayed with II.

M. H. ABRAHAM, A. G. DAVIES, D. R. LLEWELLYN, and E. M. THAIN: Anal. Chim. Acta **17**, 499 (1957).

250. Thymol-sulphuric acid: for sugars.
Spray reagent: 5 ml of conc. sulphuric acid are cautiously added to a solution of 0.5 g thymol in 95 ml ethanol.
Treatment after spraying: Heated 15—20 min at 120° C, yielding pink spots

251. p-Toluenesulphonic acid: for steroids, flavonoids and catechins.
Spray reagent: 20% solution of p-toluenesulphonic acid in chloroform.
Treatment after spraying: Heated a few minutes at 100° C, giving spots which fluoresce in long-wave UV light.

D. G. ROUX: Nature **180**, 973 (1957).

252. Trichloroacetic acid: for steroids, digitalis glycosides and *Veratrum* alkaloids.
A. Spray reagent: 25% solution of trichloroacetic acid in chloroform.
B. Spray reagent (for vitamin D):
                                  1% solution of the acid in chloroform.
C. Spray reagent (for digitalis glycosides):
                                  3.3 g acid are dissolved in 10 ml chloroform and 1—2 drops 30% hydrogen peroxide added.
Treatment after spraying: Heated about 5—10 min at 120° C. The spots are inspected in daylight and in long-wave UV light.

B. J. ALDRICH, M. L. FRITH, and S. E. WRIGHT: J. Pharm. Pharmacol. **8**, 1042 (1956).
H. J. ZEITLER: J. Chromatog. **18**, 180 (1963).
H. SILBERMANN, and R. H. THORP: J. Pharm. Pharmacol. **6**, 546 (1954).

253. Trifluoroacetic acid: for steroids.
Spray reagent: 1% solution of trifluoroacetic acid in chloroform.
Treatment after spraying: Heated 5 min at 120° C.

254. 2,4,6-Trinitrobenzoic acid: for cardiac glycosides.
Spray reagent I: 0.1% solution of 2,4,6-trinitrobenzoic acid in water-dimethylformamide.
Spray reagent II: 5% sodium carbonate in water.
Spray reagent III: 5% aqueous sodium dihydrogen phosphate.
Procedure: The layer is sprayed with I, then II, heated 4—5 min at 90—100° C, cooled and finally sprayed with III. The cardiac glycosides give orange-red spots.

T. MOMOSE, T. MATSUKUMA, and Y. OHKURA: J. Pharm. Soc. Japan **84**, 783 (1964).

255. 2,3,5-Triphenyl-H-tetrazolium chloride (TTC): for reducing sugars, corticosteroids and other reducing compounds.
Solution a: 4% methanolic TTC solution.
Solution b: N sodium hydroxide.
Spray reagent: Equal volumes of a and b are mixed before use.
Treatment after spraying: Heated 5—10 min at 100° C, yielding red spots (blue tetrazolium is more sensitive).

F. G. FISCHER u. H. DÖRFEL: Hoppe-Seylers Z. physiol. Chem. **297**, 164 (1954).
H. METZ: Naturwissenschaften **48**, 569 (1961).

256. Tungstophosphoric acid: for reducing compounds, lipids, sterols and steroids.
Spray reagent: 20% ethanolic solution of tungstophosphoric acid.
Treatment after spraying: Heated at 120° C until optimum coloured spot formation is reached.

H. P. MARTIN: Biochim. et Biophys. Acta **25**, 408 (1957).

257. Urea-hydrochloric acid: for sugars.
Spray reagent: 5 g urea are dissolved in 20 ml 2N hydrochloric acid and 100 ml ethanol added.
Treatment after spraying: The layer is heated at 100° C until the spots show optimum colour formation. Ketoses and oligosaccharides containing ketose units turn blue.

R. DEDONDER: Bull. soc. chim. biol. **34**, 44 (1952).

258. Uvitex CF conc. (CIBA): for steroids.
Spray reagent: 0.05% solution of Uvitex CF conc. in acetone-water (85 + 15).
Note: Oestrogens and α,β-unsaturated ketosteroids quench fluorescence yielded by short-wave UV light; oestrogens, only that emitted in long-wave UV radiation.

R. NEHER u. E. v. ARX: unpublished.

259. Uvitex SWN conc. (CIBA): for steroids.
Spray reagent: 0.05% solution of Uvitex SWN conc. in acetone-water (90 + 10).
Note: Oestrogens quench fluorescence yielded by exposure to long-wave UV light; $\alpha\beta$-unsaturated ketones that caused by short-wave UV radiation.

R. NEHER, and E. v. ARX: unpublished communication.

260. Vanillin-phosphoric acid: for steroids.
Spray reagent: 1 g vanillin is dissolved in 100 ml 50% aqueous phosphoric acid.
Treatment after spraying: Heated 10—20 min at 120° C.
H. METZ: Naturwissenschaften 48, 569 (1961).

261. Vanillin-potassium hydroxide: for amino acids (ornithine, lysine, proline) and amines.
Spray reagent I: 2% solution of vanillin in propanol.
Spray reagent II: 1% ethanolic potassium hydroxide.
Procedure: The layer is sprayed with I and heated 10 min at 110° C; ornithine then fluoresces intensively green-yellow in long-wave UV light and lysine only weakly green-yellow. Spraying is then carried out with II, which is followed by similar heating; ornithine then turns salmon coloured, subsequently fading whereas proline, hydroxyproline, pipecolic acid and sarcosine turn red after a few hours; glycine yields greenish brown spots and the other amino acids turn to a feeble brown.
G. CURZON, and J. GILTROW: Nature 172, 356 (1953).

262. Vanillin-sulphuric acid: for higher alcohols, phenols, steroids and essential oils.
A. Spray reagent: 1 g vanillin is dissolved in 100 ml conc. sulphuric acid.
Treatment after spraying: Heating at 120° C is carried out until the spots attain maximum colour intensity.
E. TYIHÁK, D. VÁGUJFALVI, and P. L. HÁGONY: J. Chromatog. 11, 45 (1963).
A. L. LE ROSEN, R. T. MORAVEK, and J. K. CARLTON: Anal. Chem. 24, 1335 (1952).
B. Spray reagent: 0.5 g vanillin is dissolved in 100 ml sulphuric acid-ethanol (40 + 10).
Treatment after spraying: Heated at 120° C until maximum spot colour intensity is reached.
J. S. MATTHEWS: Biochim. et Biophys. Acta 69, 163 (1963).

263. Violuric acid: for alkali and alkaline earth metal ions.
Spray reagent: 1.5% aqueous violuric acid solution; it must not be heated above 60° C to effect solution.
Treatment after spraying: Heated 20 min at 100° C.
H. ERLENMEYER, H. v. HAHN u. E. SORKIN: Helv. Chim. Acta 34, 1419 (1951).

264. Xanthydrol: for tryptophan and other indole derivatives.
Spray reagent: 10 ml 36% hydrochloric acid are added to a solution of 0.1 g xanthydrol in 90 ml ethanol; the solution must be freshly prepared before use.
Treatment after spraying: Heated at 110° C until the spots acquire optimum colour intensity.
S. R. DICKMANN, and A. L. CROCKETT: J. Biol. Chem. 220, 957 (1956).

265. Zinc chloride: for steroid-sapogenins and steroids.
Spray reagent: 30 g zinc chloride are dissolved in methanol, the solution made up to 100 ml with this solvent and filtered.

Treatment after spraying: The plate is heated 1 h at 105° C and the layer then immediately covered with a glass plate for protection against the influence of moisture. The spots fluoresce in long-wave UV light.

P. J. STEVENS: J. Chromatog. 14, 269 (1964).

266. Zirconium-alizarin lake-hydrochloric acid: for F⁻ ions.
Spray reagent: 0.05 g zirconyl chloride ($ZrOCl_2 \cdot 8H_2O$) and 0.05 g sodium alizarinsulphonate are dissolved in 100 ml 2 N hydrochloric acid.

H. SEILER u. T. KAFFENBERGER: Helv. Chim. Acta 44, 1282 (1961).

II. Compounds or Compound Classes and Reagents for their Detection

Compound or Class	Spray Reagent Nos.
Acetylene compounds	67
Acids, aromatic	132
—, organic	25, 26, 27, 44, 65, 127, 132
Adrenaline and derivatives	66, 109
Alcohols, higher	262
Aldehydes	60, 82
Alkali chlorides	64
Alkaloids	36, 37, 53, 96, 97, 98, 143, 146, 147, 245
—, ergot	74
—, hemlock	186
—, *veratrum*	252
—, *vinca*	34, 101
Allyl isothiocyanate	111
Aluminium ions	169
Amides	41, 134
Amines	53, 72, 111, 178, 261
—, aliphatic	185, 187
—, aromatic	61, 100, 128, 171, 180, 183, 238, 240
Amino acids	81, 149, 172, 178, 179, 261
— —, sulphur-containing	142, 184
Anhydrides	134
Anions, inorganic	10, 27, 64, 227, 266
Anthraquinone glycosides and their aglucones	155, 213
Antioxidants	66, 168
Azulenes	70
Barbiturates	51, 52, 57, 114, 117, 156, 157, 229
Bases	25, 44
Bile acids	17, 107, 195, 241
Caffeine	38, 42, 43
Carbamate esters	126
Cardiac glycosides	39, 80, 209, 242, 244, 252, 254
Carotenoids	15

Compound or Class	Spray Reagent Nos.
Carotenoid aldehydes	138, 222
Catechins	251
Catechol amines	66, 99
Cations, inorganic	2, 75, 87, 92, 93, 129, 135, 137, 140, 214, 216, 223, 263
Chlorinated hydrocarbons (see insecticides)	120, 232
Choline	90
Coumarins	56, 213
Creatine	187, 188
Creatinine	187, 188, 210
Dehydroascorbic acid	203, 205
Detection tests, general	42, 46, 141, 144, 200, 220, 241
Dicarboxylic acids	28
Digitalis glycosides	39, 80, 209, 242, 252
3,5-Dinitrobenzoate esters	177
1,2-Diols	152, 153, 158, 159, 198
Disulphides	184
Essential oils	88, 262
Esters	134
Ferric ions	113
Flavonoids	3, 14, 56, 151, 251
Gibberellins	241
Glycals	131
Glycolipids	83
Gold ions	237
Guanidine and derivatives	136, 187, 188
Halide ions	27, 227, 266
Heterocyclic compounds	120, 163, 180, 238, 240, 243
Hydrocarbons, chlorinated (see insecticides)	120, 232
—, polynuclear	125, 243
Hydrastine	47
Hydroxamic acids	102
Hydroxyamino acids	160
Hypnotics, bromine-containing	77, 119
Indole and derivatives	41, 49, 73, 76, 104, 108, 123, 196, 264,
Insecticides	24, 77, 85, 106, 120, 164, 193, 231, 232
—, chlorine-containing	24, 77, 85, 120, 164, 231, 232
—, thiophosphate esters	24, 106, 193
Iodine-containing compounds	37
Keto acids	65, 204
Ketones	60, 82, 191
Ketoses	13, 69, 82

Compound or Class	Spray Reagent Nos.
Lactones	134
Lipids	31, 58, 63, 116, 168, 221, 256
Lipoids	31, 83
Mercaptans	184
Morphine	112
Narcotine	47
Nitrogen compounds, organic	96, 97, 98, 145, 147
— —, quaternary	95
Nitrosamines	84, 239
Oximes	54
Penicillins	142
Peroxides	78, 115, 139, 249
Persulphates	19
Phenols	4, 20, 29, 62, 91, 100, 102, 111, 122, 165, 180, 182, 183, 224, 236, 238, 240, 243, 262
Phenothiazines	105, 124, 193
Phosphate esters	50, 166
Phosphoric acids	167
Phthalate esters	218
Plasticisers	181, 218
Plastoquinones	154
Polyalcohols	33, 158, 159, 198, 234
Polycarboxylic acids, aromatic	197
Polyethylene glycols and derivatives	94, 145
Polyphenols	236
Polyphenyls	35
Potassium ions	246
Proazulenes	70
Purines	38, 42, 43, 117
Pyridine compounds	23, 40
Pyridine monocarboxylic acids	55
Pyrimidines	117
α- and γ-Pyrones	86
Reducing compounds	91, 111, 168, 197, 199, 225, 233, 247, 255, 256
Resins	18
Sapogenins	133, 265
Sorbic acid	248
Steroids	1, 11, 16, 45, 168, 195, 208, 211, 219, 236, 241, 251, 252, 253, 255, 256, 258, 259, 260, 262, 265
—, cortico-	89, 133, 247, 255
—, keto-	79, 150, 202, 235
Steroid alkaloids	36, 124, 194
— glucuronides	215

Compound or Class	Spray Reagent Nos.
Steroid glycosides	15
— sapogenins	15, 18, 36, 48, 124, 194, 265
— sulphates	162
Sterols	1, 22, 45, 168, 173, 208, 236, 241, 256
Sugars	5, 9, 11, 21, 32, 158, 159, 170, 174, 175, 176, 192, 198, 201, 228, 234, 250, 257
—, amino-	71, 178
—, 2-deoxy-	59, 161, 190
—, reducing	6, 8, 10, 12, 225, 247, 255
Sulphides	142
Sulphonamides	41, 61
Sulphonic acids, alkyl and aryl	212, 230
Terpenes	11, 15, 18
—, di-	16
—, tri-	1, 45, 236
Tetracyclines	7
Thioacids	226
Thiobarbiturates	57, 68
Thiols	184
Thiophene derivatives	148
Thiophosphate esters	106, 193
Thiourea and derivatives	187, 188, 189
Ubiquinones	154
Unsaturated compounds	118
Urea and derivatives	187, 188
Uronic acids	176
Vanillin	130
Vitamins	15, 18, 42, 91, 110
Vulcanisation accelerators	30
Xanthine derivatives	103

III. Names and Abbreviations of Reagents

Name	Rgt. No.	Name	Rgt. No.
Benedict reagent	56	Legal test	191
Blue tetrazolium reagent	247	Liebermann-Burchard reagent	1
Carr-Price reagent	15	Millon reagent	165
Diazonium-reagent	100	Morgan-Elson reagent	71
Dragendorff reagent	94—98	Neu reagent for natural products	86
Ehrlich reagent	72		
Emerson reagent	4	PAN-reagent	215
EP-reagent	70	Pauly reagent	238
FCNP-reagent	187	Pauly reagent according to Kutáček	240
Folin reagent	172		
Folin-Ciocalteu reagent	122	Procházka reagent	123
GBHA-reagent	129	Sakaguchi reagent	136
Gibbs reagent	62	Salkowski reagent	104, 108
Greig-Leaback reagent	43	Schweppe reagent	127
Grote reagent	189	Sonnenschein reagent	37
Hanes reagent	166	Tollens reagent	225
INH-reagent	150	TPTZ-reagent	89
Jaffé reagent	210	TTC-reagent	255
Kedde reagent	80	van Urk reagent	73
König reagent	23	Zaffaroni reagent	225
Komarowsky reagent	133	Zwikker reagent	51

Conversion table for *Rf* into *Rm* and vice versa

h*Rf*/	0	1	2	3	4	5	6	7	8	9		
00	∞	3,000	2,698	2,522	2,396	2,299	2,219	2,152	2,093	2,042	*—1,996*	*99*
01	1,996	1,954	1,916	1,881	1,848	1,817	1,789	1,762	1,737	1,713	*—1,690*	*98*
02	1,690	1,669	1,648	1,628	1,609	1,591	1,574	1,557	1,540	1,525	*—1,510*	*97*
03	1,510	1,495	1,481	1,467	1,453	1,440	1,428	1,415	1,403	1,392	*—1,380*	*96*
04	1,380	1,369	1,358	1,347	1,337	1,327	1,317	1,307	1,297	1,288	*—1,279*	*95*
05	1,279	1,270	1,261	1,252	1,243	1,235	1,227	1,219	1,211	1,203	*—1,195*	*94*
06	1,195	1,187	1,180	1,172	1,165	1,158	1,151	1,144	1,137	1,130	*—1,123*	*93*
07	1,123	1,117	1,110	1,104	1,097	1,091	1,085	1,079	1,073	1,067	*—1,061*	*92*
08	1,061	1,055	1,049	1,043	1,038	1,032	1,026	1,021	1,015	1,010	*—1,005*	*91*
09	1,005	1,000	0,994	0,989	0,984	0,979	0,974	0,969	0,964	0,959	*—0,954*	*90*
10	0,954	0,949	0,945	0,940	0,935	0,931	0,926	0,922	0,917	0,913	*—0,908*	*89*
11	0,908	0,904	0,899	0,895	0,891	0,886	0,882	0,878	0,874	0,869	*—0,865*	*88*
12	0,865	0,861	0,857	0,853	0,849	0,845	0,841	0,837	0,833	0,829	*—0,826*	*87*
13	0,826	0,822	0,818	0,814	0,810	0,807	0,803	0,799	0,796	0,792	*—0,788*	*86*
14	0,788	0,785	0,781	0,778	0,774	0,770	0,767	0,764	0,760	0,757	*—0,753*	*85*
15	0,753	0,750	0,747	0,743	0,740	0,736	0,733	0,730	0,727	0,723	*—0,720*	*84*
16	0,720	0,717	0,714	0,711	0,707	0,704	0,701	0,698	0,695	0,692	*—0.689*	*83*
17	0,689	0,685	0,682	0,679	0,676	0,673	0,670	0,667	0,665	0,662	*—0,659*	*82*
18	0,659	0,656	0,653	0,650	0,647	0,644	0,641	0,638	0,635	0,633	*—0,630*	*81*
19	0,630	0,627	0,624	0,621	0,619	0,616	0,613	0,610	0,607	0,605	*—0,602*	*80*
20	0,602	0,599	0,597	0,594	0,591	0,589	0,586	0,583	0,580	0,578	*—0,575*	*79*
21	0,575	0,572	0,570	0,567	0,565	0,562	0,560	0,557	0,555	0,552	*—0,550*	*78*
22	0,550	0,547	0,545	0,542	0,540	0,537	0,535	0,532	0,530	0,527	*—0,525*	*77*
23	0,525	0,523	0,520	0,518	0,515	0,513	0,511	0,508	0,506	0,503	*—0,501*	*76*
24	0,501	0,499	0,496	0,494	0,491	0,489	0,487	0,484	0,482	0,479	*—0,477*	*75*
25	0,477	0,475	0,472	0,470	0,468	0,465	0,463	0,461	0,459	0,456	*—0,454*	*74*
26	0,454	0,452	0,450	0,447	0,445	0,443	0,441	0,439	0,436	0,434	*—0,432*	*73*
27	0,432	0,430	0,428	0,425	0,423	0,421	0,419	0,417	0,414	0,412	*—0,410*	*72*
28	0,410	0,408	0,406	0,404	0,402	0,399	0,397	0,395	0,393	0,391	*—0,389*	*71*
29	0,389	0,387	0,385	0,383	0,381	0,378	0,376	0,374	0,372	0,370	*—0,368*	*70*
30	0,368	0,366	0,364	0,362	0,360	0,357	0,355	0,353	0,351	0,349	*—0,347*	*69*
31	0,347	0,345	0,343	0,341	0,339	0,337	0,335	0,333	0,331	0,329	*—0,327*	*68*
32	0,327	0,325	0,323	0,321	0,319	0,317	0,316	0,314	0,312	0,310	*—0,308*	*67*
33	0,308	0,306	0,304	0,302	0,300	0,298	0,296	0,294	0,292	0,290	*—0,288*	*66*
34	0,288	0,286	0,284	0,282	0,280	0,278	0,277	0,275	0,273	0,271	*—0,269*	*65*
35	0,269	0,267	0,265	0,263	0,261	0,259	0,258	0,256	0,254	0,252	*—0,250*	*64*
36	0,250	0,248	0,246	0,244	0,242	0,240	0,239	0,237	0,235	0,233	*—0,231*	*63*
37	0,231	0,229	0,227	0,225	0,224	0,222	0,220	0,218	0,217	0,215	*—0,213*	*62*
38	0,213	0,211	0,209	0,207	0,205	0,203	0,202	0,200	0,198	0,196	*—0,194*	*61*
39	0,194	0,192	0,190	0,189	0,187	0,185	0,183	0,181	0,180	0,178	*—0,176*	*60*
		9	8	7	6	5	4	3	2	1	0	h*Rf*/

Continuation

hRf	0	1	2	3	4	5	6	7	8	9		
40	0,176	0,174	0,172	0,170	0,169	0,167	0,165	0,163	0,162	0,160	—*0,158*	*59*
41	0,158	0,156	0,154	0,153	0,151	0,149	0,147	0,145	0,144	0,142	—*0,140*	*58*
42	0,140	0,138	0,136	0,135	0,133	0,131	0,129	0,127	0,126	0,124	—*0,122*	*57*
43	0,122	0,120	0,119	0,117	0,115	0,113	0,112	0,110	0,108	0,107	—*0,105*	*56*
44	0,105	0,103	0,101	0,100	0,098	0,096	0,094	0,092	0,090	0,089	—*0,087*	*55*
45	0,087	0,085	0,084	0,082	0,080	0,078	0,077	0,075	0,073	0,072	—*0,070*	*54*
46	0,070	0,068	0,066	0,065	0,063	0,661	0,059	0,057	0,056	0,054	—*0,052*	*53*
47	0,052	0,050	0,049	0,047	0,045	0,043	0,042	0,040	0,038	0,037	—*0,035*	*52*
48	0,035	0,033	0,031	0,030	0,028	0,026	0,024	0,022	0,020	0,019	—*0,017*	*51*
49	0,017	0,015	0,014	0,012	0,010	0,008	0,007	0,005	0,003	0,002	±*0,000*	*50*
	9	*8*	*7*	*6*	*5*	*4*	*3*	*2*	*1*		*0*	hRf

N. B. hRf-values from 0—50 : Rm positive
hRf-values from 50—100 (*italics*): Rm negative

Examples:

hRf	4,0	34,6	51,0[a]	65,4[a]
Rm	1,380	0,277	—0,017	—0,277[b]

[a] Use the column of hRf-values in *italics*.
[b] Rm-values which have been found with the help of the column in *italics*, are negative.

Terms Frequently Used in Thin-Layer Chromatography*

HELMUT K. MANGOLD and M. BRENNER

The terms are classified alphabetically and the German and French translations are given.

English	German	French
Acidic	sauer	acide
adhesive tape	Klebeband	bande adhésive
adjustable	verstellbar	réglable
adsorb (to)	adsorbieren	adsorber
adsorbent	Adsorptionsmittel	adsorbant; agent d'adsorption
adsorption	Adsorption	adsorption
aerosol package	Aerosolpackung	cartouche d'aerosol
air-dried	luftgetrocknet	séché à l'air
air dry (to)	lufttrocknen	sécher à l'air
aligning tray	Arbeitsschablone	gabarit
alkaline, basic	alkalisch, basisch	alcalin, basique
alumina	Aluminiumoxid	alumine
application	Anwendung, Auftragen	utilisation, application
application box	Auftragskammer	chambre d'application
applicator, spotter	Auftragegerät	dispositif d'application
applicator, multiple spot	Auftragegerät nach Morgan	dispositif d'application d'après Morgan
apply (to)	anwenden, auftragen	utiliser, appliquer, déposer
aqueous	wäßrig	aqueux
artifact (artefact)	Artefakt, Kunstprodukt	artefact
ascending	aufsteigend	ascendant
atomiser, sprayer, spray-gun, vaporiser	Sprüher	vaporisateur
Band, start band	Band, Startband	bande; zone
band pipette	Breitbandpipette	pipette pour dépot linéaire
basic, alkaline	basisch, alkalisch	basique, alcalin
binder	Bindemittel	liant
BN-chamber	BN-Trennkammer	cuve de séparation BN
buffer	Puffer	tampon
Calcinated calcium sulphate, gypsum, plaster of Paris	Gips (G)	sulfate de calcium
chamber, tank	Kammer, Trog	cuve
chamber saturation (CS)	Kammersättigung (KS)	saturation de la cuve de développement
chromatography on discontinuous layers	Mehrschicht-Chromatographie	chromatographie en couches discontinues
circular technique	Zirkulartechnik	technique circulaire
coat (to)	beschichten	recouvrir d'une couche

* See E. STAHL: J. Chromatog. 33, 273 (1968).

English	German	French
coloured	farbig	coloré
colourless	farblos	incoloré
colour reagent	Farbreagens	réactif colorant
column chromatography	Kolonnen- oder Säulen-Chromatographie	chromatographie sur colonne
compound	Verbindung (chemische)	composé chimique
continuous development	Durchlauftechnik	technique du développement au continu
coupling (of techniques)	Kopplungsverfahren	couplage des techniques
cover plate	Deckplatte	plaque couvrante
cryobox (jacketed chamber for TLC at low temperatures)	Kryobox (Trennkammer zur DC bei tiefen Temperaturen)	Kyrobox (chambre froide, pour chromatographie sur couche mince à basse temperature)
Daylight	Tageslicht	lumière du jour
demixing	Entmischen	demélange (démixtion)
descending	absteigend	descendant
desiccator	Exsiccator	dessicateur
detection	Nachweis	detection, dosage qualitatif, révélation
develop (to)	entwickeln, chromatographieren	réveler, développer, chromatographier
developer (photographic)	Entwickler	révélateur
developing solvent	Laufmittel	solvant de développement
development	Entwicklung	technique de développement
development time, duration	Laufzeit	durée du développement
deviation	Abweichung (Fehlerbreite)	écart
differentiate (to), distinguish (to)	unterscheiden, differenzieren	distinguer, différencier
dilute (to)	verdünnen	diluer
dip, immerse (to)	eintauchen	immerger
dipping technique	Eintauchen	immersion
directions, procedure	Vorschrift (Arbeits-)	mode opératoire
direction of development	Laufrichtung	direction de développement
dissolve (a substance in a solvent)	auflösen (eine Substanz in einem Lösungsmittel)	dissoudre
distinguish, differentiate (to)	unterscheiden, differenzieren	différencier, distinguer
divider box (gradient-TLC)	Teiler (Gradient-DC)	séparateur (gradient-CCM)
dry (to)	trocknen	sécher
drying rack	Trockengestell	séchoir
duration (time) of run or of development	Laufzeit, Entwicklungszeit, -dauer	duree du développement
dyes, test	Testfarben	couleurs témoins
Edge, border	Rand	bord
edge effect	Randphänomen	phénomène de bord
electrophoresis	Elektrophorese	électrophorèse
eluate	Eluat	éluat

English	German	French
eluent	Elutionsmittel	éluant, l'agent d'elution
elute (to)	eluieren	éluer
equilibration	Äquilibrierung	saturation de la cuve de développement
error	Fehler	erreur
estimation, determination	Bestimmung	dosage quantitatif
evaluation	Auswertung	évaluation quantitative
even, uniform	einheitlich, gleichmäßig	régulier, homogène
exchanger, ion	Austauscher, Ionen-	échangeur d'ions
Film, X-ray	Röntgenfilm	film pour rayons X
finger print	Fingerprint	empreinte
fixer	Fixierbad	bain de fixation
fluorescence indicator	Fluorescenzindikator	indicateur de fluorescence
fractionate (to) (separation into groups)	fraktionieren	fractionner
front, solvent	Front, Fließmittel-	front du solvant
Gas chromatography	Gaschromatographie, GC	chromatographie en phase gazeuse, CG
gelling agent	Geliermittel, gelbildende Substanz	agent gélifiant
GM (gradient-mixer)-spreader or -applicator	GM-Streicher	étaleur à gradients
gradient development, gradient elution	Gradient-Entwicklung	développement avec gradient
gradient layer	Gradient-Schicht	couche à gradient
grain size, particle size	Korngröße	grosseur des grains ou des particles
guide bar, guide strip	Führungsleiste, -schiene	guide
guide values	Richtwerte, Relativwerte	valeurs prévisibles, valeurs relatives
gypsum (G)	Gips (G)	sulfate de calcium
Hopper, divider	Teiler	
horizontal development	horizontale Entwicklung	développement horizontal
Identification	Identifizierung	identification
immerse, dip (to)	eintauchen	immerger
impregnation	Imprägnierung	imprégnation
indicate, render visible, visualise, detect (to)	nachweisen, sichtbar machen	révéler
indicator	Indicator	indicateur
inorganic	anorganisch	minéral
ion exchanger	Ionenaustauscher	échangeur d'ions
isolate (to)	isolieren, (wieder) gewinnen	isoler, récupérer
Jar, tank, chamber, vessel	Kammer, Tank, Gefäß zum Entwickeln	cuve
Kieselguhr	Kieselgur	kieselguhr

Terms Frequently Used in Thin-Layer Chromatography

English	German	French
Labelling (radioactive)	Markierung (mit einem Radioisotop)	marquage avec un isotope
labelling and measuring template	Auswerteschablone	gabarit d'évaluation
layer	Schicht	couche
length of run	Laufstrecke	parcours
locate, detect, indicate, visualise (to)	sichtbar machen, lokalisieren, nachweisen	détecter, révéler
loose layer	lose (aufgestreute) Schicht	couche pulvérante
low temperature chromatography	Tieftemperatur-Chromatographie	chromatographie à basse température
Map, pattern	chromatographisches Muster	modèle chromatographique
mark (to)	beschriften, markieren	marquer
migrate (to)	wandern	migrer
migration rate, speed	Wanderungsgeschwindigkeit	vitesse de migration, de déplacement
mix (to)	mischen	mélanger
mixed layer	Misch-Schicht	couche mélangée
moisture chamber	Klimakammer	pièce climatisée
mortar (and pestle)	Mörser und Pistill	mortier et pilon
multiple development	Mehrfachentwicklung	développement multiple
multiple spot applicator	Auftragegerät nach Morgan	dispositif d'application d'après Morgan
multipurpose template	Mehrzweckschablone	
Neutral	neutral	neutre
non-polar	unpolar, nicht polar	non-polaire
One-dimensional	eindimensional	uni-dimensional
optical density	optische Dichte	densité optique
overlap (to)	überlappen, überschneiden	chevaucher
Particle size, grain size	Korngröße	grosseur des grains ou des particles
partition	Verteilung	partage
path of migration	Trennstrecke	parcours de séparation, ligne de séparation
pestle and mortar	Mörser und Pistill	mortier et pilon
photometry	Photometrie	photométrie
plaster of Paris	Gips (G)	sulfate de calcium
plate	(Träger)platte	plaque (support)
point of application (of sample)	Startpunkt	point de départ
polar	polar	polaire
powder	Pulver	poudre
precoated plates	Fertigplatten	plaques preparées à l'avance
preparative TLC	präparative DC	chromatographie préparative sur couche mince
profile, chromatographic	chromatographisches Muster	modèle chromatographique

English	German	French
procedure	Vorschrift, Verfahren	mode opératoire
proportion, ratio	Verhältnis	rapport
Quantitative determination, estimation, evaluation	quantitative Bestimmung, Auswertung	évaluation ou dosage quantitative (f)
quenching (effect)	Löschung, Löscheffekt	effet d'étanchement
Radial chromatography	Zirkularchromatographie; Zirkulartechnik	technique circulaire
rate, speed of migration	Laufgeschwindigkeit	vitesse de migration
ratio, proportion	Verhältnis	rapport
reaction chromatography	Reaktions-Chromatographie	chromatographie réactionelle
reference material, mixture, substance; standard, standard blend, test mixture	Vergleichssubstanz; Referenzgemisch, Testgemisch, Bezugssubstanz(en)	solution étalon, témoin
reflection	Reflektion	réflexion
remission	Remission	rémission
resolution, sharpness of separation	Trennschärfe, Auflösung, Trenngüte	précision de séparation
reveal, visualise, locate, indicate (to)	sichtbar machen, nachweisen	révéler
reversed phase technique	Phasenumkehrtechnik	téchnique de réversion
Rf-value	Rf-Wert	valeur Rf
rinse, wash (to)	spülen, abwaschen	rinser, laver
Sample	Probe, Analysengemisch	échantillon, prise d'essai, mélange
saturate (to)	sättigen	saturer
scanner	„Scanner"	"Scanner"
S-chamber	S-Trennkammer	cuve de séparation S
scrape off, scratch off (to)	abschaben, abkratzen	détacher
separate (to)	trennen, auflösen	séparer
separation-reaction-separation (SRS)-technique	Trennung-Reaktion-Trennung (TRT)-Technik	technique séparation-réaction-séparation (SRS)
sharpness of separation, resolution	Trennschärfe, Auflösung, Trenngüte	précision de séparation
sheet	Folie	feuille
silica gel	Kieselgel	gel de silice
slurry, suspension	Streichmasse	produit à étaler (appliquer)
sorbent	Sorptionsmittel	adsorbant, l'agent d'adsorption
solvent (mixture)	Fließmittel	solvant
solvent front	Fließmittelfront	front du solvant
speed, rate of migration	Laufgeschwindigkeit	vitesse de migration
spot, zone	Fleck, Zone	tache
spray (to)	sprühen	vaporiser, pulvériser
sprayer, spray gun	Sprüher	vaporisateur
spray reagent	Sprühreagens	reactif à vaporiser
spreader, applicator (TLC-)	Streicher, Streichgerät (DC-)	dispositif d'étalement, dispositif d'application

English	German	French
standard, reference, test substance(s), mixture or solution	Vergleichs-, Referenz-, Test-, Bezugs-, Leit-Substanz(en) oder Gemisch	étalon. témoin
start(ing) point	Startpunkt	point de départ
stepwise development	Stufentechnik	technique par étapes
storage rack, drying rack	Trockengestell	séchoir
streak	Strich	ligne
strip	Streifen	bande
subfractionation	Subfraktionierung	sous-fractionnement
suction apparatus	Saugvorrichtung, Saugapparat	dispositif dé succion, d'aspication
syringe	Spritze	seringue
Tail formation, tailing	Schwanzbildung	formation de trainée, d'une queue de comète
tank, chamber, jar, vessel	(Trenn)kammer, Trog	cuve de développement
template	Schablone	gabarit
test dyes	Testfarben	couleurs témoins
thickness	Dicke	épaisseur
trace	Spur	trace
transfer technique	Transfer-Technik, Überführungs-Technik	technique de transfert
transmission	Durchlässigkeit, Transmission	transmission
trough	Trog	cuve
two-dimensional	zweidimensional	bi-dimensional
Uniform, even	einheitlich, gleichmäßig	homogène, régulier
"Vacuum-cleaner", vacuum zone-collector	„Staubsauger", Absaugevorrichtung	dispositif d'aspiration par le vide, "L'aspirateur"
visualisation, location, detection	Sichtbarmachung, Nachweis	révélation
volatile	flüchtig	volatil
Wash, rinse	waschen, spülen	laver, rinser
wedged-tip techîque	Keilstreifen-Technik	technique des bandes en cône
wick (for CS)	Papierstreifen zur Kammersättigung (KS)	bande de papier servant à la saturation de la cuve
X-ray film	Röntgenfilm	film pour rayons X
Yield	Ausbeute	rendement
Zone, spot	Zone, Fleck	zone, bande

List of Manufacturers and Suppliers*

1. Agfa-Gevaert AG, 5090 Leverkusen, W. Germany (X-ray film, salt for developing and fixing).
2. Aimer Products, Ltd., 56—58 Rochester Place, Camden Town, London N. W. 2, England (apparatus for TLC with loose layers).
3. Alupharm Chemicals, 616 Commercial Pl., P. O. Box 30628, New Orleans, Louisiana 70130, USA (US representatives for Firm 153).
4. American Instrument Company, Silver Spring, Maryland, USA.
5. American Viscose Co., Marcus Hook, Pennsylvania, USA (Avirin).
6. Aminco-Bowman, see 4 (Spectrofluorometers).
7. Analabs Analytical Engineering Laboratories, Inc., P. O. Box 5215, Hamden 18, Conn., USA (US representatives of Firm 106).
8. Analtech. Inc., 100 So. Justinson St., Wilmington, Delaware 19801, USA (Pre-coated plates and equipment).
9. Ansco, Vestal Parkway East, Binghampton, N. Y., USA (X-ray film and salts for developing and fixing).
10. Apeco American Photocopy-Equipment Comp., 2100 West Dempster St., Evanston, Illinois, USA (Apparatus for electrophotography.
11. Applied Science Laboratories, Inc., P. O. Box 440, State College, Pennsylvania, USA (Adsorbents for TLC, spreaders and accessories).
12. Atago Optics Co., Tokio, Japan (Densitometers).
13. Atomic Accessories, Inc., Subsidiary of Baird-Atomic, Inc., 811 W. Merrick Road, Valley Stream, New York, USA (TLC-scanners).
14. Bälz, W. & Sohn KG, 7100 Heilbronn a. N., Postfach 125, W. Germany (UV-lamps).
15. Baird and Tatlock (London) Ltd., Chadwell Heath, Essex, England (Automatic spreader, TLC-accessories).
15a. Baird-Atomic, Inc., 33, University Road, Cambridge, Mass. 02138, USA (Spectrofluorometers).
16. Badische Anilin & Sodafabrik (BASF), 6700 Ludwigshafen/Rhein, W. Germany (US representative is Firm 36) (Ultramid, ultraphor, polyethyleneimines and dyes).
17. Bausch & Lomb, Rochester 2, New York 14602, USA (Spectrophotometers).
18. Becco Chemical Division, Food Machinery and Chemical Corp., Buffalo 7, New York, USA (Peracetic acid).

---

* The list contains only those firms mentioned in the various chapters; it does not claim to be complete.

19. Beckmann GmbH., 8000 München 45, Frankfurter Ring 115, W. Germany.
20. Becton, Dickinson & Co., East Rutherford, N. J., USA (Self-filling and -adjusting pipettes).
21. Bender & Hobein, GmbH, Laborbedarf, 7500 Karlsruhe, Technische Hochschule, Kaiserstr., and 8000 München 15, Lindwurmstr., W. Germany ("Karlsruhe Apparatus").
22. Berthold, Prof., Laboratorium, 7547 Wildbad/Schwarzwald, Postfach 160, W. Germany (TLC-Scanners).
23. Bio-Rad Laboratories, 32nd & Griffin Avenue, Richmond, California 95804, USA (Adsorbents and exchangers).
24. Bioresearch, Schwartz, Inc., Orangeburg, New York 10962, USA (Radioactive ink, biochemicals).
25. Black Light Eastern Corp., Bayside, New York, USA (UV-lamps and filters).
26. Boehringer, C. F. & Soehne, GmbH, 6800 Mannheim, W. Germany (Purines, pyrimidines, nucleosides and nucleotides).
27. Braun, B., 3508 Melsungen, W. Germany (Continuous infusion apparatus).
28. Brinkmann Instruments, Inc., Cantiague Road, Westbury, New York 11590, USA (US-representatives of Firms 44, 83 and 88).
28a. Buma, S. A., Basel, Switzerland (Special graph paper).
29. Burroughs Wellcome & Co., London, England (Syringes, e. g. the Agla model).
30. Calbiochem, 6000 Lucerne, Löwengraben 14, Switzerland (Adsorbents).
31. California Corporation for Biochemical Research (Calbiochem), 3625 Medford St., Los Angeles 63, California, USA (Biochemicals).
32. Camlab (Glass) Ltd., Milton Road, Cambridge, England (English representatives of Firms 44, 83 and 153).
33. Camag AG, Muttenz, B. L., Homburger Str. 24, Switzerland (Representing Firm 136) (TLC-Spreaders, accessories and adsorbents).
34. Cellpack AG, 5610 Wohlen, AG, Switzerland (Special "UVEX" sheets).
35. Chemetron Milano, Via Gustavo, Modena 24, Italy (Automatic spreader and TLC-accessories).
36. Chemirad Corporation, P. O. Box 187, East Brunswick, N. J., USA (US-representative of Firm 16).
37. Cheng Chin Trading Co., Ltd., Importers & Exporters, No. 75, Section 1, Hankow Street, Taipei, Taiwan (Formosa) (Polyamide sheets coated on both sides).

38. Ciba AG, Basel, Switzerland (Optical bleach "Uvitex" and chemicals).
39. Colab Laboratories, Inc., Chicago Heights, Illinois, USA (US representative of Firm 129).
40. Columbia Chemical Division, Pittsburgh Plate Glass Co., Barberton, Ohio, USA (Silene EF adsorbent).
41. Columbia Organic Chemicals Co. Columbia, South Carolina, USA (Fluoroparaffins).
42. Conradi, Zürich, Switzerland (Planimeters).
43. Darco Dept., Atlas Powder Co., 60 East 42nd St. New York, N. Y., USA (Vegetable charcoal for TLC).
44. Desaga, C., GmbH, 6900 Heidelberg, Maaßstr. 26–28, W. Germany (TLC-Apparatus; Electrophoresis apparatus, scanners, UV lamps and TLC accessories).
45. Despatch Oven Co., 619 S. E. 8th St., Minneapolis, Minnesota, USA (Drying ovens).
46. Eugene Dietzgen Co., 407 10th St., N. W., Washington, USA (Photographic printing paper).
47. Distillation Products Industries, Division of Eastman Kodak Co., Rochester, New York 14603, USA (Coated TLC-sheets).
48. Dow Chemical Co., Midland, Michigan, USA (Ion exchangers).
49. Dow Corning Corp., Michigan, USA (Silicones).
50. Diazo Corp., Pasadena, California, USA (Photographic printing paper).
51. Drummond Scientific Company, 500 Parkway, Broomall, Pennsylvania 19008, USA (Self-filling micropipettes).
52. Eastman Kodak Co., Rochester 3, New York, USA (X-ray film, salts for developing and fixing).
52a. Engelhard Hanovia Lamps, Bath Road, Slough, Bucks., England (UV-light sources).
53. Eppendorf Gerätebau, 2000 Hamburg-Wellingsbüttel, Postfach 11130, W. Germany (Special microsyringe pipettes and spectrophotometers).
54. Erba, C., S. p. A., 24 Via Imbonati, Milano, Italy (Apparatus for automatic coating, TLC-accessories, adsorbents).
55. Farbenfabriken Bayer AG, Werk Dormagen, 4047 Dormagen, W. Germany (Nylon and perlon powders).
56. Farbwerke Hoechst AG, 6230 Frankfurt-Hoechst a. M., W. Germany (Polyethylene powder, Mowolith etc.).
57. Fabriek van chemische Producten, Vondelingenplaat N. V. (Netherlands) (Sheets transparent to UV).
58. Fisher Scientific Co., 633 Greenwich St., New York 14, N. Y. USA (Laboratory apparatus, chemicals).

59. Floridin Co., P. O. Box 989, Tallahassee, Florida, USA (Florisil).
60. Fluka AG., Buchs SG, Switzerland (Adsorbents for TLC; pure chemicals).
61. Frieseke & Hoepfner GmbH 8520 Erlangen-Bruck, W. Germany (TLC-scanners).
62. Gallard-Schlesinger, Chemical Manufacturing Corp., 584 Mineola Avenue, Carle Place, Long Island, N. Y. 11514, USA (US representatives of Firm 127).
63. Gelman Instrument Co., 600 South Wagner Road, P. O. Box 1448, Ann Arbor, Michigan 48106, USA (Glass fibre sheets for TLC).
64. General Electric, X-ray Dept., Milwaukee 1, Wisconsin 53201, USA ("Supermix" developer for X-ray film).
65. Gesellschaft für Teerverwertung mbH, 4100 Duisburg-Meiderich, W. Germany, now Rütgerswerke und Teerverwertungs AG (Standard substances; aromatic compounds).
66. Grumbacher, M., Inc., New York, N. Y. USA.
67. Haack, P., Wien IX/68, Garnisonsgasse 3, Austria (Glass apparatus for microanalysis).
68. Haltermann, J., 2000 Hamburg, Ferdinandstr., W. Germany (pure hydrocarbons).
69. Hamilton Co., Inc., P. O. Box 307, Whittier, California 90608, USA; European agents: Micromesure N. V., P. O. Box 205, Den Haag, Netherlands (see also Firm 122) (Microsyringes).
70. Heraeus, W. C., GmbH, 6450 Hanau, W. Germany (Drying ovens, IR-radiators).
71. Hercules Powder Co., 910 Market St., Wilmington 99, Delaware, USA (Plastics, etc.).
72. Hopkin & Williams, Ltd., Freshwater Road, Chadwell Heath, Essex, England (Adsorbents).
73. Hormuth, L., Inh. W. E. Vetter, 6908 Wiesloch, W. Germany (Equipment for layer spraying and accessories; UV lamps (representatives of Firm 140)).
74. Hupe, K.-P., Dr.-Ing., 7500 Karlsruhe-Rüppurr. Lange Str. 25, W. Germany (Regulating system for TLC-GC coupling).
75. Ilford-Ciba-Ilford, 6078 Neu-Isenburg, Frankfurt/Main, W. Germany (Film material and accessories).
76. International Chem. and Nuclear Corp., 13332 E. Amar Rd., City of Industry, California, USA (US representatives of Firm 60).
77. Johns Manville Products Corp., Celite Division, 22 E. Fortieth St., New York 16, N. Y., USA (Celites and kieselguhr preparations).
78. Joyce, Loebl and Co., Gateshead on Tyne, England (Representatives of Firm 90) ("Chromoscan" densitometer for evaluating TL-chromatograms).

78a. Kensington Scientific Corp., 1165 67th St., Oakland, California 94608, USA (Apparatus for coating, adsorbents).
79. Kissel u. Wolf GmbH, 6908 Wiesloch, W. Germany (Adhesives).
80. Koch-Light Laboratories, Ltd., Colnbrook, Bucks., England (Adsorbents and biochemicals).
81. Kodak-Pathé, Vincennes/Paris, France (Coated TLC-sheets).
82. Kopp Laboratory Supplies Inc., 70–73, 35th Rd., Jackson Heights 72, New York, USA (Glass apparatus for TLC).
83. Machery, Nagel & Co., 5160 Düren, W. Germany (MN-adsorbents, ready-coated sheets for TLC).
84. MacKay, A. D., Inc., 198 Broadway, New York, N. Y., USA (Fluorescence indicators).
85. Mallinckrodt Chemical Works, 2nd & Mallinckrodt Sts., St. Louis 7, Minnesota 63160, USA (Adsorbents for TLC).
86. Mann Research Laboratories, Inc., 136 Liberty St., New York 6, N. Y., USA (Adsorbents, pre-coated plates, spray reagents, biochemicals).
87. Markgraf, K., 1000 Berlin, Germany (Apparatus for TL-electrophoresis).
88. Merck, E., AG., 6100 Darmstadt, W. Germany (Adsorbents according to STAHL for TLC, pre-coated plates, spray reagents and solvents).
88a. Microchem. Specialties Co., 1825 Eastshore Highway, Berkeley 10, California 94710, USA.
89. Mikro-Technik, 8760 Miltenberg (Main), W. Germany (Cellulose exchangers).
90. National Instrument Laboratories, Inc., 12300 Parklawn Drive, Rockville, Maryland 20852, USA (US representatives of Firm 78).
91. National Lead Co., New York, N. Y., USA (Bentonites).
92. New England Nuclear Corp., 575 Albany St., Boston 18, Mass. 02118, USA (Radioactively labelled compounds).
93. Nuchar Industrial Chemical Sales, 230 Park Avenue, New York, N. Y., USA (Charcoal).
94. Nuclear Chicago Corp., 333 East Howard Avenue, Des Plaines, Illinois 60018, USA (Scanners for radioactive evaluation of TL-chromatograms).
95. Oak Ridge Institute of Nuclear Studies, Oak Ridge, Tennessee 37831, USA.
96. Osram GmbH., 8000 München 2, Dachauer Str. 112, W. Germany (UV lamps).
97. Ozalid Reproduction Products Division, General Aniline and Film Corp., 140 West 51st St., New York, N. Y., USA (Blueprint paper).

98. Pabst Laboratories, Division of Pabst Brewing Co., 1035 West McKinley Avenue, Milwaukee 5, Wisconsin, USA (Purines, pyrimidines, nucleosides, nucleotides).
99. Packard Instrument Co., Inc., P. O. Box 428, La Grange, Illinois, USA (TLC-scanners and chemicals for scintillation analysis).
100. Peninsular Chem. Research of Gainesville, P. O. Box 14318, Florida 32603, USA (supply products of Firm 41).
101. Perutz GmbH., 8000 München 25, Kistlerhofstr. 75, W. Germany (X-ray film, salts for developing and fixing).
102. Pharmacia, Uppsala, Sweden (Sephadex).
103. Philips N.V., Eindhoven, Netherlands and 2000 Hamburg 1, Mönckebergstr. 7, W. Germany (UV lamps).
104. Photovolt Corp., 1115 Broadway, New York 10, N. Y. 10010, USA (Photovolt densitometers for TLC).
105. Pierce Chemical Co., P. O. Box 117, Rockford, Illinois 61105, USA (Biochemicals).
106. Pleuger, G., SA, 511 Turnhoutsebaan, Wijnegem, Belgium (Apparatus for TLC).
107. Du Pont, E. I. de Nemours & Co. (Inc.), Photo Products Division, Wilmington 98, Delaware 19898, USA (Fluorescence indicators, X-ray film and accessories).
108. Du Pont, Towanda, USA (Chemicals).
109. Precision Scientific Co., Chicago, Illinois 60647, USA (Apparatus for centrifugal-TLC).
110. Quarzlampen Gesellschaft mbH, 6450 Hanau, W. Germany (UV lamps)
111. Quickfit Laborglas GmbH, 6200 Wiesbaden-Schierstein, Schloßbergstr. 11, W. Germany (Apparatus for TLC).
112. Radiochemical Centre, Amersham, Bucks., England (Radioactive compounds).
113. Rank-Xerox Corp., Rochester, USA (Photostat materials).
114. Reanal, Budapest, Hungary (Adsorbents).
115. Reeve, H., Angel & Co., Ltd., 9, Bridewell Place, London E.C. 4, England (Chromedia-adsorbents, Whatman paper).
116. Renker, Belipa GmbH, 5160 Düren, W. Germany (Photographic printing paper).
117. Research Specialties, Co., 200 S. Garrard Boulevard, Richmond, California, USA (Apparatus for TLC; spray reagents).
118. Riedel-de Haën AG, 3016 Seelze, Hannover, W. Germany (Adsorbents for TLC; fluorescence indicators).
119. Roth, Carl, OHG, 7500 Karlsruhe, Schoemperlenstr. 3–5, W. Germany (Representatives of Firms 33 and 60) (Biochemicals).

120. Sas, T. J., and Son Ltd., Victoria House, Vernon Place, London W. C. 1, England (TLC-spreaders).
121. Schleicher, C., u. Schüll, 3354 Dassel, Kreis Einbeck, W. Germany; in USA, 543 Washington St., Keene, New Hampshire (Cellulose powder, for TLC, etc.).
122. Schmidt, G., Labor-Geräte, 2000 Hamburg-Sasel, Saselbergweg 50, W. Germany (Micropipettes, etc., from Firm 69).
123. Schott u. Gen., Jenaer Glaswerk, 6500 Mainz, W. Germany (Glass apparatus, filters).
123a. Schuchardt, Dr. Theodor, GmbH, 8000 München 13, Ainmillerstr. 25 (Pure chemicals and biochemicals).
124. Schwarz Bioresearch Inc., Orangeburg, New York 10962, USA (Biochemicals, especially purine and pyrimidine derivatives; see also Firm 24).
125. Schwartz, A., Kunststoffwerk, 5000 Köln-Untereschbach, W. Germany.
126. Schwinherr, Alfred, 7070 Schwäbisch Gmünd, W. Germany (Multiple purpose motors).
127. Serva Entwicklungslabor, 6900 Heidelberg, Römerstr. 118, W. Germany (Spreading bar for TLC; adsorbents for TLC; reagents).
128. Severočeské chemické zavody, Lovosice n. p. (Factory "Rudnik"), Lovosice, Czechoslovakia (Silicones).
129. Shandon Scientific Co., 65 Pound Lane, London N. W. 10, England (Apparatus for TLC).
130. Sigma Chemical Co., 3500 DeKalb St., St. Louis, Minnesota 63118, USA (Biochemicals).
131. Smith Corona SCN Corp., 410 Park Avenue, New York 22, N. Y., USA (Electrofax procedure).
131a. Soc. d'Applications Industr. de la Physique, 38, Rue Gabriel Crie, Malakoff-Seine, France (TLC-Scanners).
132. Sprenger, A., KG, 3424 St. Andreasberg, W. Germany (Hot air driers).
133. Supelco, Inc., Bellefonte, Pennsylvania, USA (Lipids).
134. Sylvania Electric Products Inc., 60 Boston St., Salem, Mass., USA (UV fluorescent tubes).
135. Telefunken AG, 7900 Ulm, Postfach 627, W. Germany (TLC scanners).
136. Thomas, A. H., Co., 3rd and Vine St., Philadelphia 5, Pennsylvania, USA (TLC spreaders).
137. Touzart et Matignon, 3, Rue Amyot, Paris 5, France (TLC apparatus).
138. Toyo Rayon Co., Nakano-shima, Kita-ku, Osaka, Japan (Polyamides).

139. Turner, G. K., Associates, 2524 Pulgas Avenue, Palo Alto, California 94303, USA (Fluorometers).
140. Ultra-Violet Products, Inc., San Gabriel, California 91778, USA (UV lamps).
141. Umbin-Chemie Gesellschaft, 8436 Velburg, W. Germany (Undine-Sol Absorber-Lack [Varnish opaque to UV light]).
142. Union Carbide Corp., 30 East 42nd St., New York 17, N. Y., USA (Plastics).
143. Union Carbide Plastic and Co., Clifton, New Jersey, USA.
144. University of Minnesota, The Hormel Institute, Austin, Minnesota, USA (Lipids).
145. Vauka Lichttechnik und Elektronik GmbH, 3000 Hannover, Postfach 906, W. Germany.
146. VEB Farbenfabriken, Wolfen, E. Germany (Wofatits and adsorbents).
147. VEB-Leuchtstoffwerk, Bad Liebenstein, E. Germany (Fluorescent materials).
148. Wacker-Chemie, 8000 München, W. Germany (Silicones).
149. Wako Pure Chemicals Co., Tokio, Japan (Silica gel).
150. Waverly Chemical Co., Inc., Mamaroneck, New York, USA (Magnesol).
151. Westvaco Chemical Division, Food Machinery and Chemical Corp., 161 42nd St., New York 17, N. Y., USA (Magnesol).
152. Wilkens Instr. and Research, Inc., now Varian Aerograph, 2700 Mitchell Drive, Walnut Creek, California, USA (Microswitches, GC-TLC).
153. Woelm, M., 3440 Eschwege, W. Germany (Adsorbents for TLC).
154. Worthington Biochemical Corp., Freehold, New Jersey 07728, USA (Enzymes).
155. Zeiss, C., 7082 Oberkochen, W. Germany (Chromatogram-Spectrophotometer for TLC, according to STAHL).
156. Zeiss, VEB, now: Optische Werke VEB, Jena, E. Germany (ERI apparatus for light absorption measurement and recording).

Author Index

The numbers in square brackets are the numbers of the references in the bibliography. The page numbers in *italics* refer to the bibliography.

Abbott, D.C., H. Egan, E.W. Hammond, and J. Thomson [4] 32, *180*; [2] *654*
— — and J. Thomson [5] 94, *180*; [1] 645, *654*
— and J. Thomson [1, 2, 3] 32, 93, 94, *180*
Abdurakhimova, N., P. Kh. Yuldashev, and S. Yu Yunusov [1] 450, *462*
Aberhart, D.J., S. Chen, P. de Mayo, and J.B. Stothers [1] 718, *724*
Abou-Chaar, Ch. I. [1 b] 436, *462*
Abraham, E.P., see Lockhart, I.M. [162] 758, *784*
Abraham, M.H., A.G. Davies, D.R. Llewellyn, and E.M. Thain 902
Abraham, R., see Nishikaze, O. [133] 324, 340, *360*, 902
Abramson, D., and M. Blecher [6] 147, 154, *180*; [1] 389, 415, *415*
Abson, D., see Rosenthal, A. [91] 831, *837*
Acara, M., see Chmielewicz, Z.F. [10] 793, 794, *804*
Achaya, K.T., see Roo, M.K.G. [111] 284, 287, *310*
Achenbach, H., and K. Biemann [1 c] 447, *462*
— and H. Griesebach [1] *577*
Achenbach, M., and H. Griesebach [7] 178, *180*
Acher, R., and Ch. Crocker [77] 749, 751, *782*
— C. Fromageot, and M. Jutisz [64] 749, 750, *782*
Achmatowitcz jr., O., see Tsuda, Y. [247] 460, *469*
Achrem, A.A., and A.I. Kuznetsova [8, 9] 5, 42, 43, *180*
Acker, L., H. Grewe, and H.O. Beutler [3] 631, *654*
Ackermann, H. see Woggon, H. [104] 640, *656*

Ackermann, M., and M. Mühlemann [2] 712, *724*
Adachi, S. [9] 808, 811, 813, 814, 833, *835*, 861
Adam, E., u. C.L. Lapière [1] 547, *562*
Adam, G., and K. Schreiber [10] 148, *180*; [1] 322, 349, *357*; [2] 459, *462*
— see Schreiber, K. [207b] 459, *468*
Adamec, O. [3] 324, 340, *357*, 902
— J. Matis, and M. Galvanek [11] 148, *180*; [2] 324, 328, 340, *357*; [1] 601, *607*
— see Matis, J. [119] 316, 335, *360*
— see Stárka, L. [160] 338, *361*; [157] 602, *611*
Adams, C.W.M. 888
Adamski, R., J. Lutomski, and J. Wisiewski *470*
Adank, K., and W. Hammerschmidt [2] 509, *562*
Adhikari, V. [1] 210, 234, 237, *250*
— see Stahl, E. [257] 234, *257*
Aelion, R., A. Loebel, and F. Eirich [12] 9, *180*
Agranoff, B.W., see Suomi, W.D. [160] 599, *611*
Ågren, G., see Verdier, C.-H. de [18] 736, *781*
Aguiar, A.J., see Szulczewski, D.H. [162] 560, *566*
Aguilera, A., see Matthews, J.S. [437] 146, 147, 151, 153, *191*; [122] 322, 336, 339, 341, *360*
Agurell, S. [3] 451, *462*
— and E. Ramstad [4] 450, *462*
— and A.J. Ullstrup [5] *462*
— see Lundström, J. *470*
— see Ramstad, E. [99] 472, *493*
Ahmad, A.K.S., see Kaufmann, H.P. [91] 390, 415, *418*
Ahrens jr., E.H. 6
— see Grundy, S.M. [255] 178, *187*; [63] 604, *609*

Ahrens jr., E. H., see Miettinen, T. A. [444] 178, *192*; [88] 284, *310*; [106] 603, 604, *610*
— see Stoffel, W. [195] 372, *420*
Ahrens, F., see Schulte, K. E. [230] 235, 240, *256*, 867
Ahrland, S., I. Grenthe, and B. Norén [43] 739, *782*
Aiken, W. H. [14] 33, *180*
Airo, E., see Salo, T. [595] 37, *195*; [82] 648, *656*
Akazawa, T., J. Uritani, and Y. Akazawa [2] 213, *250*
Akazawa, Y., see Akazawa, T. [2] 213, *250*
Akita, E. [2] *577*
Alam, M. Z., see Shellard, E. J. [631] 91, 93, *196*, *471*
Alaupovic, P., see Crider, Q. [29] 376, *416*
Albanese, A. A. [1] 259, *308*
Albers, R. J., see Gritter, R. J. [69] 322, *359*
Albers, R. W., see Fahn, S. [22] 795, *804*
Albrecht-Recht, F., and J. A. Owen [15] 141, *180*
Albro, Ph. W., and Ch. K. Huston [2] 379, 384, 406, *415*
Alderhout, J. J. H., G. K. Koch, and A. H. W. Aten jr. [16] 176, *180*
Aldrich, B. J., M. L. Frith, and S. E. Wright 902
Alexander, G. B., W. M. Heston, and P. K. Iler [17] 13, *180*
Alexander, L. R., and E. R. Stanley [3] 541, 542, 545, *562*
Alexander, M., see Hesse, G. [288] 31, *187*
Alha, A. R., and R. Lindfors [4] 542, *562*
Alm, A., see Hansson, J. [10] 500, *505*
Alm, R. S., R. J. P. Williams, and A. Tiselius [18] *180*
Almy, E. F., see Paulson, C. [16] 736, *781*
Aloupovic, P., see Skinner, W. A. [120] 284, 287, *310*
Althoff, J., see Mohr, U. [19] 497, *505*
Ambrogi, V., see Fumagalli, U. [65b] 441, *464*
Amenta, J. S. [19] 147, *180*; [3] 414, *416*

Amin, G., see Zöllner, N. [786] 139, *200*; [189] 333, *361*; [200] 594, 595, *612*
Anacker, W. F., and V. Stoy [20] 30, *180*
Anand, D. R. [4a] 562, *562*
Anderer, A., see Schramm, G. [189] 774, 777, *785*
Anderer, F. A., see Stepanov, V. [24] 737, 752, *781*
Andersen, B., see Boll, P. M. [30] 458, *463*
Anderson, L. E., see Wolfrom, M. L. [767] 29, *200*; [56] 822, 828, *836*
Anderson, N. G., J. G. Green, M. L. Barber, and F. C. Ladd [1] 795, *804*
Anderson, R. C., see Emmerson, J. L. [49] 528, 529, *563*
— see Pollard, F. H. [57] 623, *629*
Anderson, T. T. [3] 572, 573, *577*
Andreas, H., see Jantzen, E. [68] 402, 403, *417*
Andrews, P. [21] *180*; [2] 591, *607*
Anet, E. F., and L. J. Anet [67] 826, 827, *836*
Anet, L. J., see Anet. E. F. [67] 826, 827, *836*
Anfinsen, C. B., see Katz, A. M. [122] 752, *783*
Angelico, R., G. Cavina, A. D'Antona, and G. Giocoli [190] 340, 341, *361*; [3] 599, *607*
Angelis, G. de, see Dati, T. [146] 73, *184*
Angliker, E., see Binkert, J. [20] 324, *357*, 901
Anguera, P., and L. Codern [22] 5, *180*
Angustinsson, K. P., and M. Grahn 872
Anim, G., see Zöllner, N. [223] 415, *421*
Anker, L., and D. Sonanini [4] 381, 411, 412, 413, *416*
— see Sonanini, D. [229] 337, *362*; [42] 573, *578*
Annuziata, A., see Scheig, R. L. [86] 792, 795, 796, *806*
Antener, I. 308
Anthony, W. L., and W. T. Beher [5] 323, 325, 352, 354, *357*, 858, 875, 900
Antonaccio, L. D., N. A. Pereira, B. Gilbert, H. Vorbrueggen, H. Budzikiewicz, J. M. Wilson, L. J. Durham, and C. Djerassi [6] 447, *462*

Antonaccio, L. D., see Djerassi, C. [46, 47] 447, *463*
Apgar, J., see Holley, R. W. [33] *805*
Applegarth, D. A., G. G. S. Dutton, and Y. Tanaka [69] 827, 828, *836*
Applewhite, T. H., M. J. Diamond, and L. A. Goldblatt [23] 54, *180*
Archer, A. A. P. G., see Harley-Mason, J. [30] *490*, 869
Arden, T. V., et al. 881
Arden, T. V., F. H. Burstall, G. R. Davies, J. A. Lewis, and R. P. Linstead 897
Arends, A., see Mandema, E. [103] 605, *609*
Arens, A., see Wallenfels, K. [159] 758, *784*
Arizan, S., see Cionga, E. *470*
Armah, J., see Shellard, E. J. [631] 91, 93, *196*
Arndt, C., see Bohlmann, F. [15] 236, *251*
Arnim, K. v., see Grassmann, W. [65] 749, *782*
Arnold, V. [3a] 599, *607*
Arvidson, G. A. E. [5] 398, 401, *416*
Arx, E. v., and R. Neher [24, 25] 36, 84, 88, *180*; [202] 740 743 745, *785*
— see Neher, R. [131] 317, 320, 322, 337, 338, 356, *360*, 903
Asahina, Y. [3] 687, *724*
d'Asaro, B. S., see Sheppard, H. [153] 540, *566*
Ash, L., see Li, C. H. [160] 758, 759 *784*
Ashurst, P. R., and D. R. J. Laws [4] 714, *724*
Ashworth, M. R. F., and G. Bohnstedt 862
— see Bohnstedt, G. [20a] 543, 549, *562*
Asmus, E., and G. Schulze [1] 501, *505*
Aspinall, D., see Paleg, L. G. [91a] 488, 489, *492*
Atal, C. K., see Bhatnagar, J. K. [68] 56, *181*
— see Schwarting, A. E. [211] 434, *468*
— see Sharma, R. K. [214b] 435, *468*
Aten jr., A. H. W., see Alderhout, J. J. H. [16] 176, *180*
Atkinson, R. E., and R. F. Curtis [3] 235, 237, 239, *250*
— — and G. T. Phillips [4] 235, 239, *250*

Attaway, J. A. [5] 214, *250*
— L. J. Barabas, and R. W. Wolford [6] 210, *250*
— A. P. Pieringer, and L. J. Barabas [6a] 210, 237, *250*
— R. W. Wolford, and G. J. Edwards [26] 139, *180*; [7] 214, 215, *250*
Audrin, P., F. C. Foussard, C. Bourgoin, L. Jung, and P. Morand [7] 341, *357*
— see Jung, L. [95] 351, *359*
Audus, L. J. see Burnett, D. [5] 473, *490*
Aue, W. A., see Hromatka, O. [38] 650, *655*; [34] 662, *685*
Aurell, B., see Raymond, S. [568] 112, *195*
Aurenge, J., M. Degeorges, and J. Normand [27, 28] 35, 138, *180*; [1] 848, *853*
Aurich, O., see Schreiber, K. [229] 241, 243, *256*; [148] 323, 324, 347, 349, *361*; [206, 207] 459, 460, *468*, 861, 879, 892
Austin, J. H. [4] 598, *607*
Auterhoff, H., and K. Kalpathy [7] 456, *462*
— and G. Theilacker [5] 711, *724*
— see Frauendorf, H. [66] 248, 250, *252*
Authaler, A., see Mohr, U. [19] 497, *505*
Avigan, J., D. S. Goodman, and D. Steinberg [6] 316, 323, 330, 332, *357*
— see Horlick, L. [88] 334, *359*
Awapara, J. [19] 736, 737, *781*
Awe, W., I. Reinecke, and J. Thum [84] 750, *782*, 882
— and W. Schulze [6] 508, 509, **514**, 518, *562*
— and H.-G. Tracht [5] 524, 526, *562*
— and W. Winkler [8, 9] 441, *462*
— see Winkler, W. [265] 440, 441, *469*
Axen, U. see Tschesche, R. [287] 243, *257*
Ayaz, A. A., see Hashmi, M. H. [278] 74, *187*; [10] 845, *853*
Ayer, W. A., A. N. Hogg, and A. C. Soper [10] 461, *462*
Aylward, F., and P. D. S. Wood [5] 597, *607*
Azarnoff, D. L., and D. R. Tucker [8] 323, 326, 331, *357*

Bache, C. A. [4] 647, *654*
Bachmann, O., see Drawert, F. [171] 164, *184*; [33] 585, *608*
Bacon, J. S. D., and J. Edelmann 859
Bacon, M. F. [30] 64, *181*; [2] 269, *308*
Badger, G. M., J. K. Donnelly, and T. M. Spotswood [31] 37, *181*; [1] 667, *684*
Badings, H. T. [32, 33] 31, 70, 71, *181*; [8] 222, *250*
— and J. G. Wassink [34] 31, *181*; [9] 219, 222, *250*
Badve, M. G., see Barnabas, T. 867
Bächle, O., see Konrad, E. [369] 8, *189*
Baehler, B. [35, 36] 82, *181*; [11] 457, *462*; [7, 8] 537, 549, 550, *562*
— see Cherbuliez, E. [4, 186] 732, 774, 775, 777, *781, 785*
Bähr, P., see Hänsel, R. [76] 698, *726*
Bäumler, J. [12] 509, *562*, 885
— and S. Rippstein [15] 433, 437, 439, 449, 454, *462*; [10 11] 508, 509, 514, 528, 534, 535, 536, 538, *562*; [6] 639, 642, *654*; [3] 846, *853*, 870, 892
Baev, A. A., see Venkstern, T. V. [98] 792, *806*
Bailey, K., and F. R. Bettelheim [171] 760, *784*
Bailey, R. W. [9] 526, 542, *562*
— and E. J. Bourne 856
Baitsholts, A. D., see Lestienne, A. [398] 52, 59, *190*
— see Przybylowicz, E. P. [546] 59, *194*
Baker, B. R., and D. H. Buss [87] 831, *837*
Baker, G. L., see Vroman, H. E. [209] 378, *421*
Baker, H., O. Frank, S. Feingold, H. Ziffer, R. A. Gellene, C. M. Leevy, and H. Sobodka [3] 303, 304, *308*
Baker, R. G., see Evans, J. R. [190] 179, *185*
Bakhsh, M. K., see Ikram, M. [93] 425, 433, 448, *465*
Bakinovskiĭ, L. V., see Khorlin, A. Y. [127] 243, 245, *253*
Balestra, G., see Fumagalli, U. [65b] 441, *464*
Balle, G., see Tschesche, R. [170a] 347, *361*; [70] 828, 829, 834, *836*
Ballin, G. [1] 473, 478, *489*
Balogh, B. [9] 314, *357*

Banasik, O. J., see Walsh, D. E. [213] 412, 414, *421*
Bancher, E., and H. Scherz [2] *684*
— — and K. Kaindl [38] 154, *181*; [47] 819, 820, 826, 827, 834, *836*
— — and V. Prey [37] 53, *181*; [5] 651, *654*; [203] 734, *785*
— see Prey, V. [539] 32, *194*; [17] 811, 812, 817, 820, *835*
— see Scherz, H. [604] 104, *195*
Bandi, Z., see Cubero, J. M. [31] *416*
— see Mangold, H. K. [124] 397, 401, 403, 406, 411, 412, *419*
Bandtlow, G., see Habermann, E. [258] 148, *187*; [53] 415, *417*; [67] 596, *609*
Banerjee, S. K., see Pfeifer, S. [167] 441, *467*
Bang, H. O. [39] 147, 154, *181*; [10] 322, 340, *357*; [6] 601, *607*
Barabas, L. J., see Attaway, J. A. [6, 6a] 210. 237, *250*
Barath, Z., see Betina, V. [5] 569, 576, *577*
Barber, M., see Sprecher, H. W. [192] 380, *420*
Barber, M. L., see Anderson, N. G. [1] 795, *804*
Barbier, M. [6] 697, *724*
— H. Jäger, H. Tobias, and E. Wyss [11] 323, 335, 355, *357*
Barclay, M., see Skipski, V. P. [637] 147, 154, *196*; [187, 188, 189, 190, 190a] 389, 391, 415, *420*; [153] 599, *611*
Barkemeyer, H., see Korte, F. [139] *254*
Barnabas, J., see Barnabas, T. 867, 872
Barnabas, T., M. G. Badve, and J. Barnabas 867
— and J. Barnabas 872
Barnes jr. A. J., see Svoboda, G. H. [235] 449, *469*
Barrell, B. G., see Sanger, F. [84] 802, *806*
Barret, J. F., and A. J. Ryan [1] 624, *628*
Barrett, C. B., M. S. J. Dallas, and F. B. Padley [40, 41] 144, *181*
Barrett, G. C. [12] 322, *357*, 882
Barrollier, J. [42] 128, 154, *181*; [50] 747, 771, *782*
— J. Heilman, and E. Watzke [66] 749, *782*, 883
Barry, R. D. [7] 687, *724*

Barry, V. C., and P. W. D. Mitchell [62] 825, *836*
Bartlett, M. F., B. F. Lambert, H. M. Werblood, and W. I. Taylor [12] 449, *462*
Barton, G. M. 872
— R. S. Evans, and J. A. F. Gardner 876
Batchelder, W. see Richardson, G. S. [573] 157, 159, *195*
Bate-Smith, E. C. [9] 687, *724*
— and R. G. Westall [8] 693, *724*, 897
Batkiewicz, E., see Borkowski, B. [32] 450, *463*
Battaile, J., and W. D. Loomis [43] 178, *181*
Battersby, A. R., and T. H. Brown [13] 444, *462*
— G. W. Evans, R. O. Martin, M. E. Warren jr., and H. Rapoport [13b] 444, *462*
— and M. Gregory [14] 448, *462*
Battista, O. A. [44] 34, *181*
Baucher, E. see Scherz, H. [139] 580, *610*
Baudler, M., and M. Mengel [45,46] 35, *181*; [2] 848, *853*
— and F. Stuhlmann [47] 35, *181*
Bauer, L., see Wieland, Th. [106] 751, *783*, 877
Bauer, R. D., and K. D. Martin [48] 47, *181*; [2] 802, *804*
Bauer, Š., see Mokrý, J. [136] 449, *466*
Baufeld, H., see Köhler, P. [96] 590, *609*
Baumann, H. [49] 13, *181*
Baumann, J., and G. Forthmann [50] 27, *181*
Baumann, W. J., and H. K. Mangold [51] 59, *181*; [8] *416*
— H. H. O. Schmid, and H. K. Mangold [10] 412, *416*
— — H. W. Ulshöfer, and H. K. Mangold [9] 381, 385, 414, *416*
— see Mangold, H. K. [425] 176, 177, *191*
— see Schmid, H. H. O. [182, 184] 373, 374, 383, 397, 401, 402, *420*
Baumgärtel, Ch., see Zinser, M. [268] 451, 453, *470*; [178] 562, *566*, 869
Baumgartner, R., and K. Leupin [10] *724*
Baun, R. M. de, see Kudzin, S. F. [115] 751, *783*

Baur, E. W. [52] 111, 112, *181*
Baur, H., see Edlbacher, S. [93] 750, *783*
Bayer, E. [53] 115, 181
— see Häfelinger, G. [24] 623, *628*
Bayer, I. [16] 438, 440, *462*
Bayer, J. [13] 529, *562*
Bayzer, H. [1a, 1b] 498, *505*
Beal, J., see Fong, H. H. S. [62b] 444, *464*
Beanlands, D. S., see Evans, J. R. [190] 179, *185*
Beau, J. le, see Sunshine, I. [160] 534, 535, 536, *566*
Beaugiraud, S., see Guilloux, E. [19] 812, 817, *835*
Beaven, M. A., see Beckett, A. H. [16] 524, 542, *562*, 876
Bebenburg, W. v., see Wolfrom, M. L. [96] 831, *837*
Beck, J. [4] 577, *577*
Becker, E., and M. Eder [7] 649, *654*
Becker, E. D., see Jones, W. A. [115] 711, *727*
Beckett, A. H., M. A. Beaven, and A. E. Robinson [16] 524, 542, *562*, 876
— and N. H. Choulis [2] 499, *505*; [15] 522, 541, 542, *562*
Beckett, A. M., E. J. Shellard, J. D. Phillipson, and Calvin M. Lee [18c] 454, *462*
— — and A. N. Tackie [17, 18, 18b] 437, 454, *462*
Bedo, E., see Rotbacher, H. [215a] 238, *256*
Beekes, H. W., see Copius-Peereboom, J. W. [19, 20] 269, 277, 278, *308*; [40, 43] 317, 322, 323, 326, 329, 331, 332, 333, 334, *358*; [90] *630*, 860
Beffa, F., P. Lienhard, E. Steiner, and G. Schetty [2] 623, *628*
Beglinger, U., see Brenner, M. [149] 755, *784*
Beggs, B. H., see Spencer, R. D. [799] 148, 150, *200*
Beher, W. T., see Anthony, W. L. [5] 323, 325, 352, 354, *357*, 858, 875, 900
— see Semenuk, G. [627a, 798] 139, *196*, *200*; [226] 354, *362*
Behrendt, H., see Poethke, W. [172, 173] 707, *728*
Beijeveld, W. M. [13] 323, 324, *357*
Beisenherz, G., see Koss, F. W. [376, 377] 154, 178, 179, *190*

Bekersky, I. [54] 57, *181*
Bekesy, N. v. [55] 98, *181*
Béla, D. [19] 437, 440, *462*
Belcher, R., A. J. Mutten, and C. M. Sabrook 868
Beldowicz, M., see Grynberg, H. [256] 139, *187*
Belič, I., and J. Bergant-Dolar [11] 691, *724*
Bell, C. E. [56] 98, *181*
Bell, N. H., see Pasalis, J. [102] 277, 278, *310*
Bello, D. di, and A. Signor [204] 778, *785*
Benassi, C. A., F. M. Veronese, and E. Gini [2] 474, *489*; [7] 584, *607*
Benesi, H. A., and A. C. Jones [57] 12, *181*
Benigni, J., see Daly, J. W. [4a] 499, *505*
Benjamin, J., see Brown, T. L. [82] 830, *836*
Benjaminov, B. S., see Lavie, D. [147] 243, 246, *254*
Benk, E. [4] 269, *308*
Bennett, R. B., see Wolfrom, M. L. [102] 832, 834, *837*
Bennett, R. D. [15] 316, 317, 325, 347 *357*
— and E. Heftmann [58] 69, 99, 100, 178, *181*; [14, 16] 325, 330, 332, 335, 347, *357*
Bennet-Clark, T. A., M. S. Tambiah, and N. P. Kefford 893
Benraad, T. J., and P. W. C. Kloppenborg [17] 341, *357* [191] 340, 341, *362*; [8] 601, *607*
Bentley, J. A. [3] 473, 474, *489*
Bentley, K. W. *461*
Bentley-Mowatt, J. A. [3a] 489, *489*
Berbalk, H., see Prey, V. [205] 211, *255*; [74] 650, 654, *656*; [70] *686*; [24, 50, 80] 814, 817, 820, 828, *835*, *836*
Berbéč, H., see Opiénska-Blauth, J. [89] 486, *492*
Berg, A., and J. Lam [3] 667, *684*
Bergant-Dolar, J., see Belič, I. [11] 691, *724*
Bergelson, L. D., E. V. Dyatlovitskaya, and V. V. Voronkova [11] 398, *416*
— V. A. Vaver, N. V. Prokazova, A. W. Ushakov, and G. A. Popkova [12] 379, 381, 385, *416*

Bergelson, L. D., see Dyatlovitskaya, E. V. [20] 662, *684*
Berger, A., see Prey, V. [205] 211, *255*
Berger, J. A., G. Meyniel, P. Blanquet, and J. Petit [61] 94, *181*
— — and J. Petit [59, 62, 63] 45, 46, *181*; [4] 842, 845, *853*
— — — and P. Blanquet [60] 46, *181*; [205] 780, *785*
Bergmann, E. D., R. Ikan, and S. Harel [18] 334, *357*
— see Ikan, R. [309] 37, *188*, 858; [102, 103] 242, *253*; [91] 323, 334, *359*
Bergmann, F., see Schraudolf, H. [108] 472, *493*
Bergner, K. G., and H. Sperlich 880, 884
Bergquist, P. L. [3] 801, 804, *804*
Bergström, G., and C. Lagercrantz [11] 212, *250*
Bergthaller, P., see Pailer, M. [162] 718, *728*
Bernasconi, R., St. Gill, and E. Steinegger [19b] 436, *462*
— see Steinegger, G. [231] 436, *469*
Bernauer, K. [20] 447, *462*
Bernauer, W. [64] 154, *181*; [19] 340, *357*; [9] 601, *607*
Bernfeld, P. [12] 687, *724*
Bernsohn, I., see Remenchik, A. P. [80] 792, 802, *806*
Beroza, M. [8] 647, *654*; [13] 711, *725*, 864
— and T. P. McGovern [65] 149, *181*
— see Jones, W. A. [115] 711, *727*
Beroza, M., see Walker, K. C. [83] 613, 618, 620, *630*; [102] 640, 642, 643, *656*; 859
Berry, J. S., see Kirby-Berry, H. [83] 750, 751, *782*
Bertenrath, T., see Hager, A. [59] *309*
Bertetti, J. 863
Berthold, K., and Hj. Staudinger [10] 601, *607*
Bessler, O., see Luckner, M. [125, 126] 432, 456 466; [137, 138, 139, 140] 717, 719, *727*
Best, P., see Kohlschütter, H. W. [360] 13, 20, *189*
Betina, V., and Z. Barath [5] 569, 576, *577*
Bettelheim, F. R., see Bailey, K. [171] 760, *784*

Betts,T.J. [12] 237, 239, *251*
Bettschart,A., and H.Flück 882
Beugelmans,R., see Potier,P. [179] 450, *467*
Beutler,H.O., see Acker,L. [3] 631, *654*
Bevan,T.H., G.I.Gregory, T.Malkin, and A.G.Poole 887
Bevenne,A., see Schwimmer,S. 856
Beyer,K.H. [21] 438, *462*; [14] *562*
Beyrich,T. [17] *562*; [14, 15] 692, 695, *725*
— and R.Pohloudek-Fabini [13] 239, *251*
Bhagavan,H.N., see Henninger,M.D. [65] 289, 290, *309*
Bhalerao,V.R., see Ramamurthy,M.K. [59] 613, *629*
Bhalla,A.K., see Rao, K.R.K. [71] 671, *686*
Bhandari,P.R. [67] 146, *181*; [16] 701, *725*
— B.Lerch, and G.Wohlleben [66] 53, 128, *181*
Bhatnagar,A.K., and S.Bhattacharji [21b] 444, *462*
Bhatnagar,J.K., K.K.Kapur, and C.K.Atal [68] 56, *181*
Bhatnagar,S.S., see Menrad,E.L. [165] 243, *254*
Bhattacharji,S., see Bhatnagar,A.K. [21b] 444, *462*
Bhattacharya,K.R., J.Datta, and D.K.Roy [74] 749, *782*
Bhramaramba,A., and G.S.Sidhu [14] 210, 237, *251*
Bhusman,B., see Swaleh,M. [277] 238, *257*
Bianco,S.lo, see Boissonnas,R.A. [25] 737, *781*
Bicán-Fister,T. [19] 558, *562*
— and V.Kajganović [69] 148, 149, 154, *181*; [18] 544, *562*, 866
Bichsel,K., see Flück,H. [64] 239, *252*
Bickel,H. [6] 570, 572, *577*; [11] 580, *607*
Bickoff,E.M., see Lymann,R.L. [141] 698, 699, *727*
Biebricher, Chr., see Seiler,H. [29] 846, *853*
Biedebach,F., see Kaiser,H. [98] 433, *465*
Biekert,E., see Butenandt,A. [112] 43, *183*

Bieleski,R.L. [4] 803, *804*
Biemann,K., see Achenbach,H. [1c] 447, *462*
— see Renner,U. [184] 447, *467*
Bienenfeld,W., see Reindel,F. [37] 738, *781*
Bieri,J.G., and E.L.Prival [5] 266, 284, 290, *308*
— see Plack,P.A. [107] 284, *310*
Biernoth,G., see Tschesche,R. [722] 27, 101, 146, *198*; [289] 243, *257*; [244] 422, *469*
Bigwood,E.J., see Drèze,A. [33] 737, *781*
Bílek,J., see Kaloušek,J. 881
Billek,G. [19] 687, 688, *725*
— and H.Kindl [17] 692, 693, *725*
— and W.Ziegler [18] 719, *725*
Billeter,M., and C.Martius [6] 290, *308*
Bine jr.,R., see Friedman,M. [48] 606, *608*
Binkert,J., E.Angliker, and A.v.Wartburg [20] 324, *357*, 901
Bird jr.,H.L., H.F.Brickley, J.P.Comer, P.E.Hartsaw, and M.L.Johnson [70] 103, 147, 149, 152, 153, *182*; [21] 341, *357*
Birkofer,L., and Ch.Kaiser [20, 21] 690, 691, 700, 704, *725*
— — H.-A.Meyer-Stoll, and F.Suppan [71] 73, *182*; [23] 813, *835*
Biserte,G., P.Boulanger, and P.Paysant [38] 738, *781*
— and M.Dautrevaux [148] 753, *784*
— J.W.Holleman, J.Holleman-Dehove, and P.Sautière [156] 756, 757, 758, *784*
— and R.Osteux [45] 743, 760, 763, 766, *782*
Bishop,C.T., M.B.Perry, F.Blank, and F.P.Cooper [72] 828, *836*
— see Tate,M.E. [54, 84] 822, 823, 825, 830, 831, *836*
Bister,F., see Westphal,O. [748] 42, *199*
Bittner,G., see Hörhammer,L. [294] 43, *188*; [104, 105, 106, 108] 691, 703, 706, 707, *727*, 895
Bláha,K., J.Hrbek jr., J.Kovář, L.Pijewska, and F.Santavy [22] 443, *462*
Blaine 35

Blakmore, R. C., K. Bowden, J. L. Broadbent, and A. C. Drysdale [22] 714, *725*
Blanchin, L., see Prettre, M. [538] 25, *194*
Blandenet, G., and J. P. Robin [72] *182*
Blank, F., see Bishop, C. T. [72] 828, *836*
Blank, M. L., L. J. Nutter, and O. S. Privett [14] 404, 406, 409, *416*
— J. A. Schmit, and O. S. Privett [73] 139, 140, 158, *182*; [13] 414, *416*
— see Privett, O. S. [540, 541, 542, 543, 544] 134, 139, 140, *194*; [165, 166, 168] 377, 408, 409, 412, 414, *420*; [128] 593, *610*
Blanquet, P., see Berger, J.-A. [60, 61] 46, 94, *181*; [205] 780, *785*
Blaschke, G., see Franck, B. [63, 64] 444, 460, *464*
Blaschke, N., see Flück, H. [62] 433, *464*
Blattná, J., see Davídek, J. [25] 263, 264, 265, 266, *308*
Blažek, J., V. Špinková, and Z. Stejskal [20] 508, *562*
Blecher, M., see Abramson, D. [6] 147, 154, *180*; [1] 389, 415, *415*
Bleecken, S., G. Kaufmann, and K. Kummer [74] 162, *182*
Bletzinger, J. C. [75] 33, *182*
Blinn, R. C. [9] 640, *654*
Bloch, K., see Erwin, J. [188] 177, *185*
— see Hulanicka, D. [307] *188*
— see Meyer, F. [442] 177, *191*
Block, R. J., see Winegard, H. M. [82] 749, 750, *782*
Blombäck, B., see Sjöquist, J. [132] 753, 774, *784*
Blomme, A., see Heyndrickx, A. M. [78] 520, 521, *564*, 879
Blomster, R. N., A. E. Schwarting, and J. M. Bobbit [23] *462*
— see Farnsworth, N. R. [51b, 53, 54] 449, *463*
— see Roper, E. C. [189b] 424, *467*
Bloor, W. R. [15] 370, *416*
Blumberg, J., see Schmid, E. [104] 486, 493
Blumenfeld, O. O., M. A. Paz, P. M. Gallop, and S. Seifter [68] 827, 828, *836*

Blunden, G., and R. Hardman [22] 347, *357*
Boak, W. K., see Connors, W. M. [134] 52, 141, *183*
Bobbitt, J. M. [76] 5, 54, *182*; [23] 311, *357*
— R. Ebermann, and M. Schubert [24] 460, *462*
— see Blomster, R. N. [23] *462*
— see Khanna, K. L. [102] 434, *465*
— see Leary, J. D. [119, 120] 431, 434, *465*
— see Rother, A. [190] 431, *467*
— see Schwarting, A. E. [211] 434, *468*
Bodo, G., see Tuppy, H. [161] 758, 759, *784*
Boeck, A. de, see Drèze, A. [32] 737, *781*
Böhm, H. [25b] 437, *462*
Boehm, H.-P., and G. Kämpf [77] 12, *182*
— and M. Schneider [78] 13, *182*
Böhme, H., and L. Kreutzig [24, 25, 26] 690, 706, 707, *725*
Böhmert, H., see Janiak, B. [114] 706, 707, *727*
Böhni, E., see Vogler, K. [146] 753, *784*
Boer, J. H. de [82] 24, 25, *182*
— J. M. H. Fortuin, B. C. Lippens, and W. H. Meijs [83] 26, *182*
— — and J. J. Steggerda [84] 24, *182*
— G. M. M. Houben, B. C. Lippens, W. H. Meijs, and W. K. A. Walrave [85] *182*
— J. J. Steggerda, and P. Zwietering [86] 25, *182*
— — J. M. H. Fortuin, and P. Zwietering [87] 25, *182*
Boernig, H., and C. Reinicke [5] 801, *804*
Böss, J., see Brieskorn, C. H. [98] *182*
Boggust, W. A., see Fearon, W. R. [101] 750, *783*
Bognár, R., and S. Makleit [25] 459, *462*
Bogoslovsky, N. A., L. O. Shnaidman, and E. N. Kuznetova [7] 277, *308*
Bogs, U., and G. Zessin [23] 706, *725*
Bohlmann, F., C. Arndt, K.-M. Kleine, and H. Bornowski [15] 236, *251*

Bohlmann, F., K.-M. Kleine, and H. Bornowski [16] 236, *251*
— E. Winterfeldt, and U. Friese [26] 436, *462*
— G. Winterfeldt, B. Janiak, D. Schumann, and H. Laurent [27] 436, *462*
Bohner, L. S. de, E. F. Soto, and T. de Cohan [79] 147, 154, *182*
Bohnstedt, G., and M. R. F. Ashworth [20a] 543, 549, *562*
— see Ashworth, M. R. F. 862
Bohrmann, H., see Stahl, E. [258] 215, *257*
Boissier, J. R., A. Bouquet, G. Combes, C. Dumont, and M. Debray [28] 444, *462*
Boissonnas, R. A., and S. lo Bianco [25] 737, *781*
— see Guttmann, St. [135, 136] 753, *784*
— see Huguenin, R. L. [137] 753, *784*
Boit, H.-G. 421, *461*
Boiteau, P., see Rahandraha, Th. [555] 30, *194*; [206, 207] 246, *255*
Bolden, A. H., see Eisenberg jr. F. [180] 179, *185*
Bolgar, M., see Rosenberg, J. [584] 160, *195*
Boll, P. M. [80] 56, *182*; [24] 317, 349, *357*; [29] 458, *463*
— and B. Andersen [30] 458, *463*
Bolliger, H.-R. [81] 146, *182*; [8] 259, 263, 264, 266, 267, 268, 273, 274, 276, 277, 278, 279, 280, 281, 284, 285, 287, 289, 290, 292, 293, 294, 295, 297, 298, 299, 300, 303, 305, 307, *308*, 334
— and A. König [10] 264, 266, 275, 276, 279, 280, 281, 282, *308*; [21] 558, *562*
— — and U. Schwieter [9] 266, 267, 268, *308*
— see Kofler, M. [79] 284, *309*
— see Pelick, N. [511] 1, *193*
— see Stahl, E. [663] 30, *197*; [125] 267, 268, 289, *311*
Bonati, A. [17] 245, *251*
Bondopadhyaya, C., see Kaufmann, H. P. [88] 398, 401, *418*
Bondy, P. K. see Hollingsworth, D. R. [214] 780, *785*
Bonner, W. A. [17] 407, *416*

Bonsen, P. P. M., G. H. de Haas, and L. L. M. van Deenen [16] 390, *416*
Booth, A. N., see Lymann, R. L. [141] 698, 699, *727*
Booth, D. A., H. Goodwin, and J. N. Cumings [12] 598, *607*
Booth, J., and E. Boyland [3] 500, *505*
Borden, W. T., see Desmond, C. T. [33] 388, *416*
Bordet, C., and G. Michel [18] 221, *251*
Borecký, J. 895
Boretti, G., see Erspamer, V. [111] 751, *783*
Borgström, B., see Dahlqvist, A. [23] 277, *308*
— see Hofmann, A. F. [61] 385, *417*
Borio, B. L., and A. Moreira [30b] 432, *463*
Borja, C. R., see Vahounty, G. V. [728] 179, *199*
Borke, M. L., and E. R. Kirsch [31] 437, *463*
Borkowski, B., E. Batkiewicz, and K. Drost [32] 450, *463*
— and B. Pasich [19] 245, *251*
Bornfleth, H., see Reisch, J. [134, 135, 136, 137] 533, 542, 543, 544, 550, 551, 558, *565*
Bornowski, H., see Bohlmann, F. [15, 16] 236, *251*
Borri, P., see Hooghwinkel, G. J. M. [64] 392, *417*
Bosly, J., see Stainier, C. [156] 537, *566*
Boucke, G. [88] 162, *182*
Boulanger, P., see Biserte, G. [38] 738, *781*
Bouquet, A., see Boissier, J. R. [28] 444, *462*
Bourgoin, C., see Audrin, P. [7] 341, *357*
— see Jung, L. [95] 351, *359*
Bourne, E. J., see Bailey, R. W. 856
Bouveng, H. O., I. Bremner, and B. Lindberg [73] 828, 830, *836*
Bové, J. M. [89] 159, *182*; [6] *804*
Bowden, K., see Blakmore, R. C. [22] 714, *725*
Bowman, R. R., see Duggan, D. E. [32] 303, 304, *308*
Bowyer, D. E., W. M. F. Leat, A. N. Howard, and G. A. Gresham [13] 592, *607*
Boyland, E., see Booth, J. [3] 500, *505*

Bradford, R. H., see Crider, Q. [29] 376, *416*
Bradley, D. C. [90] 9, *182*
Brady, L. R., see Montfort, M. L. [147] 718, *728*
— see Sullivan, G. [37] 498, *506*
Braeckmann, P., R. van Severen, and L. de Jaeger-van Moeseke [33, 34] 456, *463*
Braekkan, O. R. [11] 274, *308*
Braekman, J. C., M. Kaisin, J. Pecher, and R. Martin *470*
Braithwaite, C. H., see Johns, T. [40] 638, *655*
Braithwaite, D. P. [10] 642, *654*
Brand, J. M. [90a, 790] 146, 150, *182*, *200*
Brandner, G., and A. J. Virtanen [206] 740, *785*
— see Grisebach, H. [252] 179, *186*
Brante, G. [91] 147, *182*, 882
Brantner, A., see Végh, A. [252] 433, *469*
Bratton, A. C., and E. K. Marshall jr. 866
Brauer, G. [92] 9, *182*
Braun, D. [20] 221, 229, *251*; [4, 5] 660, 661, 662, *684*, 890, 896
— and H. Geenen [11] 650, *654*; [6] *684*
— and G. Vorendohre [21] 211, *251*; [7, 8] 657, 663, *684*
Braun, W., see Kortüm, G. [373] 143, *190*
Braunitzer, G. [172] 760, 761, *784*
Braunsteiner, W., see Prey, V. [36] 817, *835*
Bravo, R. O., and F. A. Herández [26] 541, *562*
Bray, H. G., W. V. Thorpe, and K. White [90] 750, 751, *783*, 890
Breccia, A., and F. Spalletti [93] 179, *182*
Breckenridge, W. C., see Kuksis, A. [109] 398, 414, *418*
Bredenberg, J. B., and R. Gmelin [21a] 246, *251*
Bregoff, H. M., E. Roberts, and C. C. Delwiche 873
Breidenbach, R. W., see Smith, L. W. [639] 40, *196*; [122] 269, *311*
Breinlich, J. [22, 23] 547, 551, 560, *562*

Bremner, I., see Bouveng, H. O. [73] 828, 830, *836*
Brenner, G. S., D. F. Hinkley, L. M. Perkins, and S. Weber [104] 832, 834, *837*
Brenner, M., and A. Niederwieser [95, 96] 75, 154, *182*; [24] 535, *562*; [7, 13] 733, 734, 740, 741, 742, 744, 755, 764, 765, 771, *781*, 889
— — and G. Pataki [94, 97] 132, 137, 138, *182*; [8] 733, 755, 756, 766, 777, *781*
— — — and A. R. Fahmy [12] 742, 744, *781*
— — — and R. Weber [97a] *182*
— and G. Pataki [140] 753, *784*
— J. P. Zimmermann, J. Wehrmüller, P. Quitt, A. Hartmann, W. Schneider, and U. Beglinger [149] 754, *784*
— see Fahmy, A. R. [9, 211] 733, 740, 743, 748, *781*, *785*, 889
— see Niederwieser, A. [476, 477] 76, *192*
— see Walz, D. [739] 131, 132, *199*; [180] 583, *611*; [231] 758, 771, *786*
Bressler, R., see Wittels, B. [220] 389, *421*
Brewer, H. W., see Djerassi, C. [45] 447, *463*
Brewington, R., see Schwartz, D. P. [39] 495, *506*
Brian, P. W., H. G. Hemming, and D. Lowe [4] 488, 489, *489*
Bricas, E., and Cl. Fromageot [120] 752, *783*
Brickley, H. F., see Bird, H. L. jr. [70] 103, 147, 149, 152, 153, *182*; [21] 341, *357*
Brieskorn, C. H., and J. Böss [98] *182*
— and S. Dalferth [22, 23] 239, *251*
— and A. Fuchs [24, 24a] 240, *251*
— and H. Klinger [25] 243, *251*
— — and W. Polonius [26] 245, *251*
— and W. Polonius [27] 242, 243, *251*
Brimacombe, J. S., M. Stacey, and L. C. N. Tucker [76] 828, *836*
Brintzinger, H., and B. Troemer [99] 8, *182*
Britt jr., R. D., see Keirs, R. J. [343] 141, *189*
Broadbent, J. H., J. A. Cornelius, and G. Shone [27] 696, *725*

Broadbent, J. L., see Blakmore, R. C. [22] 714, *725*
Brochmann-Hanssen, E., and T. Furuya [34b] 441, *463*; [25] 529, *562*
— and B. Nielsen [34c] *463*
— see Mary, N. Y. [434] 146, 151, *191*; [131] 440, *466*
Brockmann 203, 459, 613, 680
Brockmann, H., and H. Brockmann jr. [100] 26, 43, *182*
— and T. Waehneldt [20] 812, *835*
Brockmann jr., H., see Brockmann, H. [100] 26, 43, *182*
Broda, E., and T. Schönfeld [101] 155, 167, 170, *183*
Brodasky, T. F. [102] 32, *183*; [7] 574, *577*
Brodie, B., see Duggan, D. E. [32] 303, 304, *308*
Brodsky, A. L., see McCarthy, J. L. [123] 315, 340, 341, 346, *360*
Brody, S., see Korzun, B. P. [374, 375] 75, 98, *190*; [101, 102] 341, 349, 351, *359*; [109] 494, *465*; [89, 90] 541, 559, *564*
Brody, S. J., see Gluck, L. [58] 599, *608*
Broeck, J. van den [103] 27, *183*
Brooks, R. V. [14] 602, *607*
Brooksbank, B. W. L., and D. B. Gower [25] 339, *358*
Broom, A. D., L. M. Townsend, J. W. Jones, and R. K. Robins [7] 794, *804*
Brossi, A. [8] 574, *577*
Brossmer, R., see Weicker, H. [27] 814, *835*
Brotz, M., see Norton, W. T. [153] 392, 394, *419*
Brown, B. B., see Foppiano, R. [197] 148, 153, 154, *185*
Brown, J. C. [3, 4] 620, 621, 622, 623, *628*
Brown, J. F., D. Clark, and W. W. Elliott [104] 24, *183*
Brown, J. L., and J. M. Johnston [106] 163, 164, *183*
Brown, L., P. Holliday, and J. F. Trotter [105] 34, *183*
Brown, T. H., see Battersby, A. R. [13] 444, *462*
Brown, T. L., and J. Benjamin [82] 830, *836*
Brownlee, G. G., see Sanger, F. [84] 802, *806*

Bruchfield, H. P., and A. Harzell [12] 647, *654*
Brud, W., and W. Daniewski [28] 214, *251*; [27] 551, 554, *562*
Brud, W. S. [107] 30, *183*
Brühl, P., see Oertel, G. W. [113] 600, *610*
Bruinvels, J. [26] 341, *358*; [15] 601, *607*
Brunauer, St., P. H. Emmett, and E. Teller [108] *183*
Bruns, F. H., see Reinauer, H. [79] *806*
Brunschede, H., R. Hoffbauer, and H. W. Goedde [16] 584, 585, *607*
— see Goedde, H. W. [59] 584, *608*
Bryant, F., and B. T. Overell 859
Bryant, L. H. [109] 29, *183*
Bryson, I. L., and T. I. Mitchell 856
Brzuszkiewicz, H., see Opienska-Blauth, J. [492] 5, *193*; [90] 486, *492*; [115, 116] 582, 584, *610*
Buchan, J. L., and R. J. Savage 856
Buchner, H. [31] 737, *781*
Buchtela, K., and M. Lesigang [110] 31, *183*
Buchwald, H. D., see Mosher, H. S. [139b] 461, *466*
Budvari, R., see Végh, A. [252] 433, *469*
Budvari, Z., see Szász, G. [238] 423, *469*
Budwig, J., see Kaufmann, H. P. 896
Budzikiewicz, H., see Antonaccio, L. D. [6] 447, *462*
— see Djerassi, C. [45, 46] 447, *463*
Büch, H., see Rüdiger, W. [139] 526, *565*
Büchi, J., and J. A. Fresen [27a] 551, *562*
— and H. Schumacher 882
— and A. Zimmermann [34d] 433, *463*
Büchner, M., see Freimuth, U. [57] 316, 323, 324, 335, *358*, 902
Bürger, K. [18] 387, *416*; [9, 10] 664, 665, 674, *684*
Bürgi, W., J. P. Colombo, and R. Richterich [19] 583, *608*
Bürgin, see Mühlemann 526
Bürki, W. 308
Buettner, see Seydel, J. H. [151] 560, *566*

Buhrow, I., see Schlossberger, H. G. [103] 477, *493*; [141] 584, *610*
Bujard, El. [207] 740, *785*
— and J. Mauron [17] 586, *607*
Bukhari, M. A., A. B. Foster, and J. M. Webber [88] 831, *837*
Bukovac, M. I., see Wittwer, S. H. [133a] 489, *493*
Bulenkow, T. J. [28] 508, *562*
Bunt, J. S. [12] 269, *308*
Bunyan, P. J. [13] 640, 642, *654*
Burger, E. [29] 540, *562*
Burger, H. G., J. R. Kent, and A. E. Kellie [27] 339, *358*; [18] 602, *607*
Buriánek, J., and J. Cífka [111] 179, *183*
Burke, W. J., A. D. Potter, and R. M. Parkhurst 897
Burkhalter, A., see Kupferberg, H. J. [116] 438, *465*; [91] 529, 532, *564*, 876
Burlingame, A. L., see Renner, U. [184] 447, *467*
Burnett, D., L. J. Audus, and H. D. Zinsmeister [5] 473, *490*
Burrows, S., F. S. M. Grylls, and J. S. Harrison 887
Burstall, F. H., G. R. Davies, R. P. Linstead, and R. A. Wells 876, 899
— see Arden, T. V. 897
Burton, K. W. C., see Pollard, F. H. 854, 898
Burton, R. B., A. Zaffaroni, and E. H. Kentmann [86] 750, *783*
— see Zaffaroni, A. 882
Bush, I. E. [28] 314, 315, 316, 325, 326, 328, 336, *358*, 899
— and M. Willoughby 902
Buslanova, M. M., and W. F. Stepanovskaya [28a] 221, *251*
Buss, D. H., L. Hough, and A. C. Richardson [43] 818, *835*
— see Baker, B. R. [87] 831, *837*
Buswell, K. M., and W. E. Link [19] 387, 415, *416*
Butenandt, A. [6] 472, *490*
— E. Biekert, H. Kübler, and B. Linzen [112] 43, *183*
Bygdeman, M., and B. Samuelsson [20] 399, *416*; [20] 597, *608*
Byrne, G. A. [29] 210, 220, 223, *251*

Cabezas, J., see Faillard, H. [42] 597, *608*; [37] 817, *835*
Caggiano, E., and G. B. Marini-Bettolo [35] 448, *463*
Cain, L., see Kirby-Berry, H. [83] 750, 751, *782*
Caldin, D. J. Mc. [46] 749, *782*
Calvert, R. [113] 27, *183*
Cama, H. R., see John, K. V. [72] 273, 274, 275, *309*
— see Lakshmanan, M. R. [81] 274, *310*
Cammarato, L. V., see Farnsworth, N. R. [51b] 449, *463*
Campaigne, E., and M. Georgiadis [11] 680, *684*
Canić, V. D., and S. M. Petrović [115] 39, *183*; [5] 842, *853*
Cannon, C. G. [116] 41, *183*
Canonne, P., see Ritter, F. J. [75] 667, *686*
Cantoni, G. L., and D. R. Davis [8] 789, 791, *804*
Canuti, A., and B. Luboz Magrassi [5] 624, *628*
Capella, P., see Fedeli, E. [61] 219, 221, *252*
Cardinale, G., see Freudenberg, K. [53] 693, *725*
Carelli, V., A. M. Liquori, and A. Mele [117] 42, *183*
Cargill, D. I. [29] 326, 327, 330, 331, 337, *358*
— see Stansfield, D. A. [156] 340, *361*
Carlton, J. K., see Le Rosen, A. L. 904
Carman, P. C. [118] 18, *183*
Carnegie, P. R., and G. Pacheco [119] *183*
Carr, S. J., see Lees, M. [110] 370, *418*
Carreau, J.-P., and J. Raulin [120] 29, *183*
Carreras Matas, L. [30] 317, 346, *358*
Carsten, M. E. [28] 727, *781*
Carter, C. E., see Volkin, E. [100] 788, *806*
Carter, H. E., and H. S. Hendricksen [21] 412, *416*
— P. Johnson, D. W. Teets, and R. K. Yu [22] 381, 385, *416*
Carver, M. J., see Copenhaver, J. H. [31] 695, *725*
Casani, G. [9] 576, *577*

Cassidy, W., and A.J.Fisher [14] 632, *654*
Cassil, C.C. [15] 638, 654
Castagnou, M.R., and S. Larcebau [35b] 460, *463*
Castano, F.F., see Hilz, R. 883
Castrén, E. [13] 266, 280, 284, *308*
Catalin, J. see Vignoli, L. [253b] 440, *469*
Catch, J.R. [121] 167, *183*
Cava, M.P., see Fong, H.H.S. [62b] 444, *464*
— see Stuart, K.L. *461*
Cavanaugh, M.A., see Fishbein, L. 891
Cavina, G., and G. Moretti [29a] 556, *562*
— and C. Vicari [122] 148, 151, *183*; [31] 340, 341, *358*
— see Angelico, R. [190] 340, 341, *361*; [3] 599, *607*
Cecy, C., see Moreira, E.A. [171] 248, *254*
Ceder, O. [10] *577*
Cee, A., see Gasparič, J. [21] 622, *628*
Čekan, Z., see Heřmánek, S. [79] 323, 329, 335, 347, *359*; [87] 422, 439, 448, 451, 454, 464; [27] 613, *628*
— see Trojánek, J. [243] 449, *469*
Černý, J. [124] 56, *183*
Černý, V., J. Joska, and L. Lábler [123] 103, 149, *183*; [32] 315, 322, 330, 335, 347, *358*
— see Lábler, L. [105] 328, 349, *359*; [117] 459, *465*
Cerri, O., and G. Maffi [125] 5, *183*; [36] 457, *463*; [30] 549, *562*; [21] 606, *608*
Chakravarty, M.L., see Wolfrom, M.L. [42] 818, *835*
Chalmers, A.H., C.C.J. Culvenor, and L.W. Smith [36b] 435, *463*
Chalvardjian, A. [24, 25] 372, 398, *416*
— L.J. Morris, and R.T. Holman [23] 387, *416*
Chamberlain, J., A. Hughes, A.W. Rogers, and G.H. Thomas [126] 159, *183*
— and G.H. Thomas [33] 314, 328, *358*
Chandra, G., J. Clark et al. [30] 237, *251*
Chanez, M., see Rahandraha, Th. [555] 30, *194*; [206, 207] 246, *255*

Chang, E. [34] 320, 337, 338, *358*
Chang, T.-Ch. L., and C.C. Sweeley [26] 383, *416*
Chang Shen, N.-H., F.E. Francis, and R.A. Kinsella [35] 338, 340, *358*
Chapel, C.M.C., see Nadal, N.G.M. [176] 217, *255*
Charezinski, M., see Opiénska-Blauth, J. [89] 486, *492*; [53] 748, *782*
Chargaff, E., and J.N. Davidson [9] 789, 791, 795, *804*
— C. Levine, and C. Green [85] 750, *783*, 882
— see Tamm, C. [95, 96] 790, 793, *806*
— see Vischer, E. [99] 792, *806*
Chatten, L.G., see Morrison, J.C. [459] 135, 137, 138, *192*; [104, 105] 518, 537, *564*
Chattopadhyay, D.P., and E.H. Mosbach [192] 334, *362*
Chattoraj, S.C., see Guerra-Garcia, R. [64] 602, *609*
— see Wotiz, H.H. [184] 351, *361*
Chavré, V.J., see West, C.D. [188] 587, *611*
Chen, Ch.-L., see Freudenberg, K. [53] 693, *725*
Chen jr., P.S. [14] 278, *308*; [193] 322, 334, *362*
— A.R. Terepka, K. Lane, and A. Marsh [15] 275, 276, 280, *308*
Chen, S., see Aberhart, D.J. ̇[1] 718, *724*
Cherbuliez, E., Br. Baehler, M.C. Lebeau, and A.R. Sussmann [186] 774, *785*
— — and J. Rabinowitz [4] 732, 775, 777, *781*
— A.R. Sussmann, and J. Rabinowitz [187] 774, *785*
— see Scheidegger, J.J. 899
Chetaille, M., see Cotte, I. [9] 473, 483, *490*
Chevet, A. [127] 12, *183*
Chiang, S.P., and J.S. Schweppe [36] 338, *358*
Chierici, L., and M. Perani [4] *505*
Chihara, G., see Hara, S. [274] 139, *187*; [76] 354, *359*
Chih Tung, Y., and K. Tsung Wang [193a] 316, 350, *362*
Chino, H., and L.I. Gilbert [27] *416*
Chism, P., see Patton, A.R. [54] 749, *782*, 889

Chmielewicz, Z. F., and M. Acara [10] 793, 794, *804*

Chochin, J., and J. W. Daly [33, 34, 35] 509, 514, 518, 527, 528, 529, 533, 534, 535, 536, 538, 539, *563*

Cholnoky, L. v., see Zechmeister, L. [781] 1, *200*

Chopard-dit-Jean, L., see Planta, C. v. [108] 273, 274, *310*

Choulis, N. H. [31] 524, *562*
— see Beckett, A. H. [2] 499, *505*; [15] 522, 541, 542, *562*

Christ, B., see Müller, K. H. [149] 707, *728*

Christensen, E. K. J., Th. Vos, and T. Huizinga [32] 536, 537, 538, *562*, 884

Christian, J. C., S. Jakovcic, and D. Yi-Young Hsia [23] 596, *608*

Christian, J. E., see Seno, S. [150] 561, *565*

Christiansen, I., see Cotte, I. [9] 473, 483, *490*

Chu, F., see Stoffel, W. [195] 372, *420*

Chuchra, U., see Koss, F. W. [376, 377] 154, 178, 179, *190*

Chung, D., see Levy, A. L. [157] 757, 758, *784*

Čičiro, V. E. [37] 441, *463*

Cieślak, J., J. Kuduk, and F. Rulko [38] 431, *463*

Cifferi, O. [11] *577*

Cifka, J., see Buriánek, J. [111] 179, *183*

Cifonelli, J. A., and F. Smith 885

Ciglar, J., J. Kolšek and M. Perpar [6] 618, *628*

Cima, L., L. Levorato, and R. Mantovan [17] 277, 280, *308*
— and R. Mantovan [128] 146, 151, 155, *183*; [16] 302, 303, *308*

Ciocalteu, V., see Folin, O. [116] 751, *783*

Cionga, E., E. Nichiforesco, V. Mascov, N. Uricaru, and S. Arizan *470*

Claisse, J., L. Crombie, and R. Peace [21] 812, *835*

Clark, D., see Brown, J. F. [104] 24, *183*

Clark, H. M., see Overman, R. T. [498] *193*

Clark, J., see Chandra, G. [30] 237, *251*

Clarke, E. G. C. 879

Clarkson, T. W. [96] 750, *783*

Claude, J. R. [37] 316, 325, 332, *358*

Claussen, U., and F. Korte [28] 715, *725*

Clements, R., see Patterson, S. J. [224] 780, *786*

Clerc-Bory, M., H. Pachéco, and Ch. Mentzer [112] 751, *783*

Clesceri, N. L., and G. F. Lee [6] 848, *853*

Clifford, C. J., J. V. Wilkinson, and J. S. Wragg [37a] 337, 341, *358*

Close, R. [39] 738, *781*

Clotten, A., see Clotten, R. [129] 105, 106, *183*

Clotten, R., and A. Clotten [129] 105, 106, *183*

Cobb, W. Y. [31] 222, *251*
— L. M. Libbey, and E. A. Day [32] 221, *251*

Cochin, J., and J. W. Daly [39] 433, 440, 441, *463*

Cochran, C. N., see Russell, A. S. [589] *195*

Codding, D. W., see Privett, O. S. [544] 134, *194*

Codern, L., see Anguera, P. [22] 5, *180*

Coffey, R. G., and R. W. Newburgh [130] 36, 46, *183*; [11] 793, 795, *804*

Coffin, D. E., see McKinley, W. P. [59] 638, *655*

Coggins jr., C. W., see Khalifah, R. A. [52, 53] 489, *491*
— see Lewis, L. N. [70] 487, *492*

Cohan, T. de, see Bohner, L. S. de [79] 147, 154, *182*

Cohn, E. J., and J. T. Edsall [1] 730, *781*

Cohn, G. L., and E. Pancake [38] 320, *358*

Cohn, W. E. [12] 789, 794, 795, 801, *804*
— see Davidson, J. N. [16] 789, *804*

Coleman, M. H. [131] 99, *183*

Coleman, T. J., and D. V. Parke [33] 245, 246, *251*

Coli, F., see Luisi, M. [217] 340, 341, *362*

Collet, G. [132] 30, *183*; [7] 488, *490*
— J. Dubuochet, and P. E. Pilet [8] 478, 483, 488, *490*

Collins, R. P., and K. Kalnins [34] 221, *251*
— see Gaines, H. D. [69] 220, *252*

Collins, W. P., and J. F. Sommerville [39] 340, *358*

Collombel, C., see Cotte, J. [26] 814, *835*
Colman, B., and W. Vishniac [133] 40, *183*; [18] 269, *308*
Colobert, L., see Creach, O. [28] 379, *416*
Colombo, J. P., see Bürgi, W. [19] 583, *608*
Colowick, S. P., and N. O. Kaplam [13] 789, 791, 795, *804*
Combes, G., see Boissier, J. R. [28] 444, *462*
Comer, J. P., and P. E. Hartsaw [194] 340, 341, *362*; [36] 561, *563*
— see Bird jr., H. L. [70] 103, 147, 149, 152, 153, *182*
Comin, J. [12] *577*
Common, R. H., see Hertelendy, F. [80] 350, *359*
Cone, N. J., R. Miller, and N. Neuss [40] 449, 450, *463*
Connolly, J. P., P. J. Flanagan, R. Ó. Dorchaí, and J. B. Thomson [135] 99, *183*
Connors, W. M., and W. K. Boak [134] 52, 141, *183*
Conrad, K., see Köhler, K. H. [58] 475, *491*
Consden, R. [69] 749, 750, *782*
— A. H. Gordon, and A. J. P. Martin [136, 137] 3, 4, 105, *183*; [26, 47, 99] 737, 746, 747, 749, 750, *781, 782, 783*
Contessa, A. R., see Fassina, G. [60] 240, *252*
Cook, E. R., and M. Luscombe [34] 738, *781*
Cook, J. W. [16] 642, *654*
Cook, S., see Stanley, W. L. [268] 210, *257*
Coomes, T. J., P. C. Crowther, B. J. Francis, and G. Shone [29] 696, *725*
Cooper, F. P., see Bishop, C. T. [72] 828, *836*
Cooper, S. F. [30] 719, *725*
Copenhaver, J. H., and M. J. Carver [31] 695, *725*
Copet, A., see Geiss, F. [225] 150, *186*
Copius-Peereboom, J. W. [41, 42] 316, 317, 331, *358*; [17, 18, 19, 20, 21, 22] 631, 634, 635, 637, 638, 649, *654, 655*; [12] 659, *684*, 859, 890, 896
— and H. W. Beekes [19, 20] 269, 277, 278, *308*

Cornelius, J. A., see Broadbent, J. H. [27] 696, *725*
Corner, I. J., see Harborne, J. B. [84] 693, *726*
Cornwell, D. C., see Holla, K. S. [62] 374, *417*
Corona, G. L., and M. Raiteri [195] 345, *362*
Corral, R. A., see Djerassi, C. [45] 447, *463*
Corsano, S., and L. Pinizzi [35] 245, *251*
— and G. Spano [36] 245, *251*
Cotsis, T. P., and J. C. Garey [7] 624, *628*
— and J. C. Garey [86] *630*
Cotte, I., M. Chetaille, F. Poulet, and I. Christiansen [9] 473, 483, *490*
Cotte, J., M. Mathieu, and C. Collombel [26] 814, *835*
Cotthem, B. van, see Denöel, A. [43] 460, 461, *463*
Couchman, F. M. [37] *251*
Courtial, W., see Lück, E. [58] 637, *655*
Covello, M., and O. Schettino [21, 22] 302, 303, *308*
Coyne, C. M., and G. A. Maw 898
Craig, J. C., N. Y. Mary, N. L. Goldman, and L. Wolf [40c] 430, *463*
Cramer, F., and E. Ehrhardt [208] 778, *785*
— see Randerath, K. [62] 801, *805*
Cramer, F. J., see Reitsema, R. H. [210] 234, *255*
Cramp, W. A., see Wolfrom, M. L. [94] 831, *837*
Crane, F. L., see Dilley, R. A. [29] 284, *308*
— see Henninger, M. D. [64, 65] 289, 290, 292, *309*
Crane, J. C. [10] 473, *490*
Craven 291
Crawford, M. A., see Hansen, I. L. [28] 477, *490*
Crawhall, J. C., E. Saunders, and C. J. Thompson [22] 583, *608*
Creach, O., B. Entressangles, and L. Colobert [28] 379, *416*
Creech, B. G., see Horning, E. C. [89] 316, *359*
Crépy, O., O. Judas, and B. Lachese [138] 32, *183*; [44] 324, 357, *358*, 886, 895

Cress, E. A., see Snyder, F. [191] 384, *420*
Criddle, W. J., G. J. Moody, and J. D. R. Thomas [139] 111, *184*; [8, 9] 626, *628*
Crider, Q., P. Alaupovic, J. Hillsberry, C. Yen, and R. H. Bradford [29] 376, *416*
Crisan, C., see Rotbacher, H. [215a] 238, *256*
Crocker, Ch., see Acher, R. [77] 749, 751, *782*
Crockett, A. L., see Dickmann, S. R. 904
Crombie, L. [32] 687, *725*
— see Claisse, J. [21] 812, *835*
Cross, A. D., see Preininger, V. [179b] 444, *467*
Crowe, M. O'L. [140] 2, *184*
Crowther, P. C., see Coomes, T. J. [29] 696, *725*
Crumler, T. B., see Reeves, W. A. 881
Crump, G. B. [37] 540, *563*; [13, 14] 677, *684*
Csallany, A. S., and H. H. Draper [141] 160, *184*
Cuba, P., see Vácha, P. [250] 454, 455, 456, *469*
Cubero, J. M., Z. Bandi, and H. K. Mangold [31] *416*
— and H. K. Mangold [30] 397, 402, *416*
— see Schmid, H. H. O. [182] 397, 401, 402, *420*
Cudzinowski, M., see Ikan, R. [211] 332, *362*
Culvenor, C. C. J., see Chalmers, A. H. [36b] 435, *463*
Cumings, J. N., see Booth, D. A. [12] 598, *607*
— see Müldner, H. G. [146] 389, 392, 393, *419*; [110] 598, *610*
— see Wherrett, J. R. [218] 392, *421*; [190] *611*
Cummings, W. W., see Dugger, D. L. [173] 13, *184*
Currie, A. L., see Timell, T. E. 856
Curtis, R. F., and G. T. Phillips [38] *251*; [15] 681, 682, *684*, 883
— see Atkinson, R. E. [3, 4] 235, 237, 239, *250*
Curtius, H. Ch. [24] 603, *608*
— see Tancredi, F. [165] 584, *611*
Curzon, G., and J. Giltrow 904

Czarnocka, A., see Vaedtke, J. [174] 316, 335, *361*
Czeglédi-Jankó, G. [25] 597, *608*

Dahler, R. P. [26] *608*
Dahlke, E., see Graf, E. [66] 719, *726*
Dahlqvist, A., D. L. Thomson, K. Ekbohm, and B. Borgström [23] 277, *308*
Dahn, H., and H. Fuchs [143] 97, *184*; [45] 314, *358*
Daiber, D., see Preussmann, R. [32, 33] 497, *506*, 871, 900
Dain, J. A., H. Weicker, G. Schmidt, and S. J. Thannhauser [32] 392, 393, *416*
Dalen, E. van, see Vries, G. de 855
Dalferth, S. [39] 239, *251*
— see Brieskorn, C. H. [22, 23] 239, *251*
Dalgliesh, C. E. [98, 104] 750, 751, *783*
Dallas, M. S. J., see Barrett, C. B. [40, 41] 144, *181*; [10] 248, 249, *250*; [6, 7] 396, 398, 399, 414, *416*
Daly, J. W., J. Benigni, R. Minnis, Y. Kanaoka, and B. Witkop [4a] 499, 505
— see Cochin, J. [39] 433, 440, 441, *463*; [33, 34, 35] 509, 514, 518, 527, 528, 529, 533, 534, 535, 536, 538, 539, *563*
Dam, M. J. D. van, G. J. de Kleuver, and J. G. de Heus [175] 317, 329, *361*
— and S. P. J. Maas [144] 56, *184*
Dam-Karrer 291
Damm, K., see Noll, W. [483] 20, *192*
Damratoski, D., see Farnsworth, N. R. [51b] 449, *463*
Dancis, J., J. Hutzler and M. Levitz [23] 652, *655*
Danckwortt, P. W., and J. Eisenbrand [145] 77, *184*
Dang hahn Khoi [41] 460, *463*
Daniewski, W., see Brud, W. [28] 214, *251*; [27] 551, 554, *562*
D'Antona, A., see Angelico, R. [190] 340, 341, *361*; [3] 599, *607*
Darey, F. R., see Terner, Ch. [712] 177, *198*
Das, B., see Kaufmann, H. P. [81, 85, 86] 381, 401, 411, 412, *418*
Das, V. S. R., I. V. S. Rao, and K. U. K. Murthy [11] 490

Dass, R., see Verma, M. R. [81] 612, *630*
Dastoor, N., and H. Schmid [42] 447, *463*
Dastoor, N. J., see Menrad, E. L. [165] 243, *254*
Dati, T., G. de Angelis, P. Ippoliti, and C. Luly [146] 73, *184*
Datta, J., see Bhattacharya, K. R. [74] 749, *782*
Daum, M., see Kohlschütter, H. W. [367] *189*
Dautrevaux, M., see Biserte, G. [148] 753, *784*
Dauvillier, P. [147] 98, *184*
David, S., and H. Hirshfeld [24] 295, *308*
Davídek, J. [148, 149] 40, 44, *184*; [25] 634, *655*
— and J. Blattná [25] 263, 264, 265, 266, *308*
— and G. Janíček [11] 613, 616, *628*
— and J. Pokorný [24] 634, *655*
— — and G. Janíček [10] 613, *628*
Davidoff, F., and E. D. Korn [150] 177, *184*
Davidow, B. [26] 638, *655*
Davidson, I. W. F., and W. G. Drew [14] 792, 801, *804*
Davidson, J. N., and W. E. Cohn [16] 789, *804*
— and R. M. S. S ellie [15] 802, 804, *804*
— see Chargaff, E. [9] 789, 791, 795, *804*
Davies, A. G., see Abraham, M. H. 902
Davies, B. H. [151] 70, *184*; [26] 267, *308*
— D. Jones, and T. W. Goodwin [27] 267, *308*
— see Mercer, E. I. [85] 267, *310*
Davies, G. R., see Arden, T. V. 897
— see Burstall, F. H. 876, 899
Davis, D. R., see Cantoni, G. L. [8] 789, 791, *804*
Davis, R. B., see Macmahon, J. M. 858
Davoll, H., R. A. Turner, J. G. Pierce, and V. du Vigneaud [158] 758, *784*
Day, B. N., see Neill, J. D. [471] 118, *192*
Day, E. A., see Cobb, W. Y. [32] 221, *251*
— see Libbey, L. M. [149] 219, 221, 223, *254*; [56] 631, *655*

Day, K. C., see Murray, T. K. [91] 278, *310*
Dean, F. M. [33] 687, *725*
Deatherage, F. E., see Paulson, C. [16] 736, *781*
Debackere, M., see Eberhardt, H. [44] 519, *563*
Deboer 24
Debray, M., M. Plat, and J. LeMen 470
— see Boissier, J. R. [28] 444, *462*
Decker, K., and R. Sammeck [42b] 430, *463*
Dedonder, R. 903
Deenen, L. L. M. van, see Bonsen, P. P. M. [16] 390, *416*
— see Haverkate, F. [57] 383, 391, 398, 401, *417*
Deferrari, J. O., R. M. de Lederkremer, B. Matsuhiro, and J. F. Sproviero [53] 822, 823, 834, *836*
Deffner, G., see Grassmann, W. [246] 42, *186*
Degeorges, M., see Aurenge, J. [27, 28] 35, 138, *180*
Deicke, F., see Kaufmann, H. P. [96] 317, 323, 331, 332, 333, *359*; [79] 412, *418*; [87] 594, *609*
Deininger, R. 885
Delihas, N., see Tometsko, A. M. [97] 804, *806*
De Luca, H. F., see Norman, A. W. [93, 94, 95] 277, 279, 280, *310*
Delwiche, C. C., see Bregoff, H. M. 873
Demole, E. [152, 153] 5, 66, *184*; [40, 40a, 41, 42] 211, 216, 226, 241, *251*; [28] 268, *308*
— and E. Lederer [43] 241, *251*
— B. Willhalm, and M. Stoll [44] 216, *251*
Dénes, V. I., G. Giurdaru, and M. Farcasan [16] 680, *684*
Dengler, B., see Wagner, H. [733] 146, 151, *199*; [140, 141] 266, 288, 290, 292, *311*
Denis, W., see Folin, O. [117] 751, *783*
Dennis, D. T., see Graebe, J. E. [243] 178, *186*
Dennis, F. G., and J. P. Nitsch [11a] 489, *490*
Denöel, A., and B. van Cotthem [43] 460, 461, *463*
Dent, C. E., W. Stepka, and F. C. Steward [35] 738, *781*

Denti, E., and M. P. Luboz [46] *251*
Deshusses, J., and A. Gabbai [45] 217, *251*
Desimio, M. [5] 499, *505*
Desmond, C. T., and W. T. Borden [33] 388, *416*
Desnuelle, P., and C. Fabre [164] 759, *784*
Determann, H. [155] *184*
— and W. Michel [156] *184*
— see Wieland, Th. [755, 756] 37, 47, *199*; [191] 591, *611*; [104] 795, 796, *806*
Deters, R. [157] 146, *184*; [17] 676, *684*
Detter, F., J. Dietrich, and V. Klingmüller [27] 602, *608*
Detty, W. E., see Gage, T. B. [58] *726*
Dev, S., see Gupta, A. S. [83] 211, *252*
Devia, J. E., see Pérez-Medina, L. A. [164] 459, *467*
Dewey, L., see Elliot, W. H. [50] 527, 529, *563*
Deyl, Z., see Kutáček, M. [64] 479, 487, *491*
— see Rosmus, J. [585] 75, *195*; [214] 221, *256*
Deyrup, J. A., H. Schmid, and P. Karrer [44] 447, *463*
Dhont, J. H. [158] *184*
— and G. J. C. Dijkman [47] 217, 220, 223, *251*
— and C. de Rooy [48, 49] 220, 223, 228, 231, 235, *251*; [38] 540, *563*
Dhopeshwarkar, G. A., and J. F. Mead [159, 160] 177, 178, *184*
Diamantstein, T., and H. Ehrhart [12] 474, 483, *490*; [28] *608*
— and K. Lörcher [45a] 349, *358*
Diamond, M. J., see Applewhite, T. H. [23] 54, *180*
Dibbern, H. W., and H. Rochelmeyer [13] 485, *490*
Dicarlo, F. J., J. M. Hartigan jr., and G. E. Phillips [161] *184*; [18] 671, 673, *684*
Dickmann, S. R., and A. L. Crockett 904
Diczfalusy, E., see Lisboa, B. P. [112, 113] 323, 324, 325, 326, 349, 350, *360*
Diehl, H. W., see Fletcher jr., H. G. [99] 832, 834, *837*
Dietrich, C. P., S. M. C. Dietrich, and H. G. Pontis [162] 47, 149, 154, *184*; [18] 801, *804*; [11] 809, 821, 834, *835*

Dietrich, H., see Schwyzer, R. [139b] *784*
Dietrich, J., see Detter, F. [27] 602, *608*
Dietrich, S. M. C., see Dietrich, C. P. [162] 47, 149, 154, *184*; [18] 801, *804*; [11] 809, 821, 834, *835*
Dietz, W., and K. Soehring 868
Dijkman, G. J. C., see Dhont, J. H. [47] 217, 220, 223, *251*
Dillaha, J., see Landmann, W. A. [190] 774, 777, *785*
Dillard, M., see Hollingsworth, D. R. [214] 780, *785*
Dilley, R. A. [30] 266, 284, 290, 291, *308*
— and F. L. Crane [29] 284, *308*
Dimillier, I., and R. G. Trout [29] 585, *608*
Dimler, R. J., see Jeanes, A. [326] 86, *188*
Dische, Z., and K. Schwarz [19] 789, 802, *804*
Dittmann, J. [30, 31, 31a] 583, 590, *608*; [209] 740, *785*
Dittmer, J. C., see Wells, M. A. [217] 370, *421*; [185] *611*
Dittrich, S. [164] 148, *184*
Ditullio, N. W., C. S. Jacobs jr., and W. L. Holmes [196] 332, *362*
Djerassi, C., H. W. Brewer, H. Budzikiewicz, O. O. Orazi, and R. A. Corral [45] 447, *463*
— and R. McCrindle [50] 243, *251*
— Y. Nakagawa, J. M. Wilson, H. Budzikiewicz, B. Gilbert, and L. D. Antonaccio [46] 447, *463*
— R. J. Owellen, J. M. Ferreira, and L. D. Antonaccio [47] 447, *463*
— K. Undheim, R. C. Sheppard, W. G. Terry, and B. Sjöberg [184] *785*
— see Antonaccio, L. D. [6] 447, *462*
— see Ferreira, J. M. [57] 447, *464*
— see Joule, J. A. [96] 460, *465*
— see Vorbrueggen, H. [254] 460, *469*
— see Walser, A. [257b] 447, *469*
Dobas, J. [12] 620, *628*
Dobiásová, M. [35] *416*; [32] 599, *608*
Dobici, F., and G. Grassini [165] 30, 111, 112, *184*
Doctor, P. B., see Holley, R. W. [33] *805*
Dodgson, K. S., see Wusteman, F. S. [185] 356, *361*; [91] 679, 680, *686*

Doenecke, P., see Forth, W. [201] 354, 362; [46] 603, *608*
Döpke, W. [48] 441, 443, 444, 446, *463*
Dörfel, H., see Fischer, F. G. 903
Doerk, E., see Neurath, G. [20] 495, *505*
Doizaki, W. M., and L. Zieve [166] 148, *184*
Domagk, G. E., see Harbers, E. [28] 791, *805*
Domaglina, E., and J. Ochynska *470*
Donaldson, E. M., see Quesenberry, R. O. [223] 314, 317, *362*
Donaldson, K. O., V. J. Tulane, and L. M. Marshall [167] 90, *184*
Donnelly, J. K., see Badger, G. M. [31] 37, *181*; [1] 667, *684*
Donner, R., and K. Lohs 864
Doorenbos, H. J., R. F. Rekker, J. Gootjes, J. R. A. Simoons, and W. Th. Nauta [39] 518, *563*
Dorchaí, R. Ó., see Connolly, J. P. [135] 99, *183*
Dorfman, L., see Korzun, B. P. [374] 98, *190*; [102] 349, *359*; [109] 494, *465*
Dorfman, R. I., see Futterweit, W. [60] 340, *358*; [54] 602, *608*
— see Siegel, E. T. [634] 179, *196*; [151] 351, *361*
Dorfner, K. [168, 169] 44, 48, *184*
Dorofeenko, G. N., see Zhdanov, Yu. A. [30] 815, *835*
Dorp, D. A. van, see Riezebos, G. [175] 408, *420*
Dose, K., and G. Krause [170] 111, *184*
Doss, M., and K. Oette [36] 398, *416*
— see Klenk, E. [97] 394, *418*
Douglas, B., J. L. Kirkpatrick, R. F. Raffauf, O. Ribeiro, and A. J. Weisbach [49] 460, *463*
Douglas, C. D., see Gage, T. G. 856
Drake, L. C., see Ritter, H. L. [577] 28, *195*
Drake, M. P., see Landmann, W. A. [190] 774, 777, *785*
Draper, H. H., see Csallany, A. S. [141] 160, *184*
Draus, F. J., see Farnsworth, N. R. [53] 449, *463*
— see Roper, E. C. [189b] 424, *467*
Drawert, F. [34] 699, *725*
— O. Bachmann, and K.-H. Reuther [171] 164, *184*; [33] 585, *608*

Drawert, F., W. Heimann, and A. Ziegler [31] 307, *308*
Dressen, F.-P. [51] 240, *251*
Dressler, A. [40] 540, *563*
Drew, W. G., see Davidson, I. W. F. [14] 792, 801, *804*
Drèze, A., and A. de Boeck [32] 737, *781*
— S. Moore, and E. J. Bigwood [33] 737, *781*
Dreyer, W. J., see Katz, A. M. [122] 752, *783*
Drosdowsky, M., see Futterweit, W. [54] 602, *608*
Drost, K., see Borkowski, B. [32] 450, *463*
Druding, L. F. [172] *184*; [13] 620, *628*; [7] 842, *853*
Dryon, L. [41] 529, *563*
Drysdale, A. C., see Blakmore, R. C. [22] 714, *725*
Dubouchet, J., see Collet, G. [8] 478, 483, 488, *490*
Dúbravková, L., see Mokrý, J. [133] 449, *466*
Dünger, M., see Neurath, G. [57] 680, 681, *685*
Dugas, H., R. A. Ellison, Z. Valenta, K. Wiesner, and C. M. Wong [49b] 461, *463*
Duggan, D. E., R. R. Bowman, B. Brodie, and S. Udenfriend [32] 303, 304, *308*
Dugger, D. L., J. H. Stanton, B. N. Irby, B. L. McConnell, W. W. Cummings, and R. W. Maatman [173] 13, *184*
Dumazert, C., C. Ghiglione, and T. Pugnet [46] 316, *358*; [27] 650, *655*; [55] 822, 824, *836*
Dumont, C., see Boissier, J. R. [28] 444, *462*
Dunagin, P. E., E. H. Meadows, and J. A. Olson [33] 274, *308*
Duncan, G. R. [174] 56, *184*; [47] 314, 342, *358*
Duncan, G. W., see Neill, J. D. [471] 118, *192*
Dunlop, W. J., see Winkler, B. C. 471
Dunn, F., and P. Robson [37] 414, *416*
Dunphy, P. J., K. J. Whittle, and J. F. Pennock [34] 286, *308*; [38] 378, 412, *416*, 867
— — — and R. A. Morton [35] 284, *308*

Dunphy, P. J., see Pennock, J. F. [105] 283, 285, 286, *310*
— see Whittle, K. J. [146] 286, *311*
Dupaková, D., see Kraus, L. [129] 701, *727*
Duphorn, J., see Tschesche, R. [288, 290] 242, 243, 244, 245, *257*
Durant, J. A. [118] 751, *783*
Durham, L. J., see Antonaccio, L. D. [6] 447, *462*
Dusenberry, J. E., see Gröger, D. [84] 451, *464*
Dušinský, G., and M. Tyllová [52] 214, 237, *251*
Dutrieux, Fr., see Stainier, C. [156] 537, *566*
Dutta, N. L., A. C. Ghosh, P. M. Nair, and K. Venkataraman [35] 709, *725*
Dutton, G. G. S., K. B. Gibney, P. E. Reid, and K. N. Slessor [14] 810, 831, 832, 833, 834, *835*
— and K. N. Slessor [74] 828, *836*
— see Applegarth, D. A. [69] 827, 828, *836*
Dutz, H., see Lehmann, H. [393] 12, *190*
Duuren, A. J. van [53] 245, *251*
Duval, C., see Servigne, Y. 858
Duvivier, J. [197] 355, *362*
Dyatlovitskaya, E. V., V. V. Voronkova, and L. D. Bergelsson [20] 662, *684*
— see Bergelson, L. D. [11] 398, *416*
Dyer, T. A. [175] 46, *184*; [20] 795, *804*
Dyer, W. G., J. P. Gold, N. A. Maistrellis, C. T. Peng, and P. Ofner [48] 338, *358*; [34] 602, *608*

Eastoe, J. E., see Gordon, A. H. [240] 90, *186*
Eberhagen, D. [36, 37] 599, *608*
— and N. Zöllner [35] *608*
— see Zöllner, N. [225] 369, 374, 386, 398, *421*; [202] 592, 593, 594, *612*
Eberhardt, H., and M. Debackere [44] 519, *563*
— O. W. Lerbs, and K. J. Freundt [42, 46] 509, *563*
— and D. Norden [45] 527, 528, *563*
Eberhardt, W., K. J. Freundt, and J. W. Langbein [43] 533, 534, 536, 539, *563*
Eberlein, W. R. [198] 324, *362*, 895

Ebermann, R., see Bobbit, J. M. [24] 460, *462*
Ebing, W., see Henkel, H. G. [37] 648, *655*, 858
Eckenroth, H. [177] 41, *184*
Edelmann, J., see Bacon, J. S. D. 859
Eder, F., H. Schoch, and R. Müller [28] 638, 645, *655*
Eder, M., see Becker, E. [7] 649, *654*
Edlbacher, S. [94] 750, *783*
— H. Baur, H. R. Staehelin, and A. Zeller [93] 750, *783*
Edman, P. [155, 178, 179, 183] 756, 772, 773, 774, *784*, *785*
— and J. Sjöquist [192] 774, *785*
Edsall, J. T., see Cohn, E. J. [1] 730, *781*
Edward, J. T., and D. M. Waldron [102] 750, *783*, 880, 886, 892
Edwards, G. J., see Attaway, J. A. [26] 139, *180*; [7] 214, 215, *250*
Egan, H., see Abbott, D. C. [4, 5] 32, 94, *180*; [1,2] 645, *654*
Egels, W., see Schratz, E. [616] 86, *196*
Egge, H., see Kuhn, R. [106] 389, 394, *418*
Egger, H., and K. Schlögl [21] 665, *685*
Egger, K. [36, 37, 38, 39, 40] 266, 269, 271, 272, 273, 291, *308*, *309*; [36, 37, 38, 39, 40] 687, 692, 693, 700, 701, 703, *725*
— and M. Keil [41, 42] 700, 701, 703, 704, *725*
— and H. Kleinig [41] 265, 286, 289, 291, *309*
— and H. Voigt [42] 265, 270, 271, 272, *309*
— see Reznik, H. [177] 705, *728*, 865
Eggers, J. [178] 131, 133, *185*; [22] 676, *685*
Eggstein, M., and F. H. Kreutz [38] 593, *608*
Ehrhardt, E., see Cramer, F. [208] 778, *785*
Ehrhart, H., see Diamantstein, T. [12] 474, 483, *490*; [28] *608*
Ehrlich, E. N. [199] 340, *362*; [39] 601, *608*
Eich, E., and H. Rochelmeyer [14] 477, 481, 486, *490*
Eichenberger, W., and E. C. Grob [43, 44] 267, *309*
— see Grob, E. C. [56] 267, *309*

Eiden, F., and H.-D. Stachel [47] 508, 514, *563*
Eirich, F., see Aelion, R. [12] 9, *180*
Eisenberg jr., F. [179] 130, *185*
— and A. H. Bolden [180] 179, *185*
Eisenbrand, J., see Danckwortt, P. W. [145] 77, *184*
Eistert, B., and A. Langbein [23] 684, *685*
Ekbohm, K., see Dahlqvist, A. [23] 277, *308*
Ekholm, J., see Hanahan, D. J. [55] 374, *417*
Ekman, B. [113] 751, *783*
El-Basyomi, S. Z., and G. H. N. Towers [43] *725*
Elbeih, I. I. M., see Pollard, F. H. 881
Elbert, W. C. [181] 29, *185*
— see Sawicki, E. [598, 599, 600] 37, 141, 145, 155, *195*; [77, 78] 667, 669, *686*
El-Dakhakhny, M. [54] 238, *252*
El-Deeb, S. R. [55] 212, *252*
— M. S. Karawya, and S. K. Wahba [56] 237, *252*
Elgamal, M. H. A., and M. B. E. Fayez [57] 244, 247, *252*
El-Hamidi, A., and G. Richter [58] 237, *252*
— see Hörhammer, L. [90] 238, *252*
Elliot, K., and L. A. Telesz [14] 616, *628*
Elliot, W. H., N. Nomof, K. Parker, L. Dewey, and E. L. Way [50] 527, 529, *563*
Elliott, W. W., see Brown, J. F. [104] 24, *183*
Ellison, R. A., see Dugas, H. [49b] 461, *463*
El-Moghazy, A. M., see Khafagy, S. [101] 434, 435, *465*
Elmquist, A., see Lindner, E. B. [128] 752, *783*
El-Olemy, M. M., A. E. Schwarting, and W. J. Kelleher [150b] 431, *466*
El Sissi, H., see Grassmann, W. [249] 42, *186*
Elson, G. W., D. F. Jones, J. MacMillan, and P. J. Suter [15] 479, 487, 489, *490*
Elson, L. A., and W. T. J. Morgan 868
Elvehjem 369
Emich, F. [182] 63, *185*

Emmerson, J. L., and R. C. Anderson [49] 528, 529, *563*
Emmett, P. H., see Brunauer, St. [108] *183*
Endres, H. [183, 184] 43, 44, *185*; [44] 691, 699, 703, *725*
— W. Grassmann, and M. Oppelt [185] 43, *185*
— and H. Hörmann [186] 43, *185* [48] 541, *563*; [45] 691, 699, 700, 703, *725*
— see Grassmann, W. [248, 249] 42, 43, *186*
— see Grau, W. [80a] 231, *252*; [71] 541, *563*; [68] 699, 703, *726*
— see Hörhammer, L. [92] 691, 699, *726*
— see Stadler, P. [658] 44, *197*
Eneroth, P. [49] 352, 353, *358*
Eng, L. F., Y. L. Lee, R. B. Hayman, and B. Gerstl [187] 147, *185*; [39] 379, *417*
Engel, L. L., see Richardson, G. S. [573] 157, 159, *195*
Engelhardt, M., see Scheiffarth, F. [138] 601, *610*
Engelhorn, R., see Koss, F. W. [376] 154, 179, *190*
Entenman, C. [40] 369, *417*
— see Skidmore, W. D. [186] 389, 391, *420*
Entressangles, B., see Creach, O. [28] 379, *416*
Ercoli, A., R. Vitali and R. Gardi [50] 316, 341, *358*
Erde, D., see Gröger, D. [81] 451, *464*
Erhart, L., see Rey, E. [72] 673, 674, *686*
Erlenmeyer, H., H. v. Hahn, and E. Sorkin 904
— see Pfrunder, B. [528] 111, *194*
— see Seiler, H. [27, 29] 846, 850, *853*
Erspamer, V., and G. Boretti [111] 751, *783*
Ertel, H., and L. Horner [15] 810, *835*, 893
Erwin, J., and K. Bloch [188] 177, *185*
— see Hulanicka, D. [307] *188*
Erxleben, H., see Kögl, F. [56] 472, *491*
Es, T. van, see Whistler, R. L. [90] 831, *837*
Esser, K. [210] 747, *785*; [38] 817, *835*
Euler, H. v., and L. Hahn [21] 789, *804*

Euler, H. v., H. Hasselquist, and I. Limnell [40, 41] 585, 592, *608*
Euler, K. L., see Farnsworth, N. R. [52] 423, *463*
Evans, G. W., see Battersby, A. R. [13b] 444, *462*
Evans, J. R., R. W. Gunton, R. G. Baker, D. S. Beanlands, and J. C. Spears [190] 179, *185*
Evans, R. S., see Barton, G. M. 876
Evans, W. C., and W. J. Griffin [50] 435, *463*
Evelin, S. R., see Roux, D. G. [184] 693, *728*
Eyrich, W., see Hörhammer, L. [103] 695, 696, *727*

Fabian, J., see Mayer, R. [54] 682, *685*
Fabre, C., see Desnuelle, P. [164] 759, *784*
Fahmy, A. R. [175] 762, *784*
— A. Niederwieser, G. Pataki, and M. Brenner [9] 733, 740, 748, *781*
— see Brenner, M. [12] 742, 744, *781*
— see Walz, D. [739] 131, 132, *199*; [180] 583, *611*; [231] 758, 771, *786*
Fahn, S., R. W. Albers, and G. J. Koval [22] 795, *804*
Faillard, H., and J. Cabezas [42] 597, *608*; [37] 817, *835*
Fales, H. M., see Laiho, S. M. [117b] 446, *465*
Fallab, S., see Weiss, A. 896
Falzi, G., see Marozzi, E. [100] 509, *564*
Farcasan, M., see Dénes, V. I. [16] 680, *684*
Farell, G., see McIsaac, W. M. [78] 486, *492*
Farkas, L., see Hörhammer, L. [109, 110] 691, 707, *727*
Farmilo, C. G., see Genest, K. [68] 454, *464*
— see Smith, M. D. [241] 237, *256*
Farnsworth, N. R. [51] 423, 449, *463*
— R. N. Blomster, D. Damratoski, W. A. Meer, and L. V. Cammarato [51b] 449, *463*
— and K. L. Euler [52] 423, *463*
— H. H. S. Fong, R. N. Blomster, and F. J. Draus [53] 449, *463*
— and I. M. Hilinski [53b] 449, *463*
— W. D. Loub, and R. N. Blomster [54] 449, *463*

Farnsworth, N. R., see Roper, E. C. [189b] 424, *467*
Farrow, J., see Holley, R. W. [33] *805*
Fasella, P., A. Giartosio, and C. Turano [43] 591, *608*
Fass, W. E., see Reitsema, R. H. [210] 234, *255*
Fassina, G. [59] 240, *252*
— A. R. Contessa, and C. E. Tóth [60] 240, *252*
Fateh-Moghdam, A., see Knedel, M. [353] 90, *189*
Faubert Maunder, M. J. de [29] 639, *655*
Fauconnet, L., and R. Fazan [51] 323, *358*, 901
— and M. Waldesbühl [52] 323, 344, 358, 894, 901
Faugeras, G., R. Paris, and M. H. Meyruey [55, 56] 436, *464*
Fauss, R., see Noll, W. [483] 20, *192*
Favorskaya, J. A., see Shevchenko, Z. A. [239] 222, *256*
Fawkes, J., see Fishbein, L. [63] 230, *252*
Fayez, M. B. E., see Elgamal, M. H. A. [57] 244, 247, *252*
Fazan, R., see Fauconnet, L. [51] 323, *358*, 901
Fearon, W. R., and W. A. Boggust [101] 750, *783*
Feast, A. A. J., B. Lindberg and O. Theander [48] 820, *836*
Fechtig, B., see Urech, J. [311] 238, *258*
Fedeli, E., P. Capella, and L. Tadini [61] 219, 221, *252*
Fehér, T. [53, 54] 323, 324, 326, 328, 335, 338, 340, *358*, 870, 902
Fehlhaber, H.-W., see Tschesche, R. *471*
Feigl, F. 891, 892
Feingold, S., see Baker, H. [3] 303, 304, *308*
Fejér-Kossey, O. *470*
Fels, G. I., M. Kaufmann, and A. G. Karczmar [51] 508, *563*
Felt, V., see Vacíková, A. [203] *420*
Feltkamp, H. [191] 53, *185*; [6] 500, *505*
— and F. Koch 882
Ferrari, G., see Passera, C. [68] 650, *656*, 867

Ferrari, L., see Gerali, G. [63] 335, 341, *358*
Ferreira, J. M., B. Gilbert, R. J. Owellen, and C. Djerassi [57] 447, *464*
— see Djerassi, C. [47] 447, *463*
Ferrier, R. J. [103] 832, 834, *837*
Ferris, J. P. [58] 460, *464*
Fiebig, E. C., see McCoy, R. N. [411] 154, *191*
Fieser, L. F., and M. Fieser [55] 313, *358*
Fieser, M., see Fieser, L. F. [55] 313, *358*
Fike, W. W., and I. Sunshine [52] 514, 518, *563*
— see Sunshine, I. [159] 551, 554, *566*
Fikenscher, L. H., and M. R. Gibson [46] 714, *725*
— and R. Hegnauer [47, 48] 692, 714, *725*
Filippes, F. M. de [17] 804, *804*
Finch, N., and W. I. Taylor [59, 60, 61] 454, *464*
Fink, K., and R. M. Fink 875
Fink, R. M., see Fink, K. 875
Finkelstein, J. D., see Schachter, D. [117] 277, 278, 280, *310*
Finkelstein, M., see Ladany, S. [100] 602, *609*
Finston, H. L., and J. Miskel [192] 155, *185*
Fiori, A., and M. Marigo [62] 230, *252*; [53] 520, *563*
Firestone, D. [41] 381, *417*
Fischbach, H., and J. Levin [13] *577*
Fischer, A. [108] 751, *783*
Fischer, F., and H. Koch [30] 650, 655; [24] 680, *685*
— see Wieland, T. 894
Fischer, F. G., and H. Dörfel 903
Fischer, H. G., see Mosher, H. S. [139b] 461, *466*
Fischer, L. J., and S. Riegelman [192a, 791] 141, *185, 200*
Fischer, P. [200] 340, *362*
Fischer, R., and W. Klingelhöller [54] 563; [31] 639, 655
— and H. Lautner [14] *577*
— and H. Weixlbaumer [61b] 460, 461, *464*
Fischer, W., see Weidemann, G. [2] 807, 808, 811, 814, *834*

Fishbein, L. 891
— and M. A. Cavanaugh 891
— and J. Fawkes [63] 230, *252*
Fisher, A. J., see Cassidy, W. [14] 632, *654*
Fisher, N. E., and A. Y. Mottlau [193] 16, *185*
Fishman, W. H., F. Harris, and S. Green [56] 356, *358*; [44] 603, *608*
Fishwick, B., see Taylor, A. [98] 643, *656*
Fiskari, K., see Salo, T. [79] 639, 644, *656*
Fitelson, J., see Kohan, S. [136] 217, *254*
Flanagan, P. J., see Connolly, J. P. [135] 99, *183*
Flechter, R. A., and D. J. Osborne [15a] *490*
Fleischer, S., and G. Rouser [45] 266, *309*
Fletcher jr., H. G., and H. W. Diehl [99] 832, 834, *837*
— see Harrison, R. [45] 819, *836*
Fletcher, R. F., see Gloster, J. [47] 385, 414, 415, *417*
Flett, M. St. C. [194] 41, *185*
Flodin, P. [21] 737, 752, *781*
— see Porath, J. [126] 752, *783*
Floss, H. G., see Gröger, D. [82] 457, *464*
— see Mothes, K. [140] 454, *466*
— see Weygand, F. [238] 694, 695, *730*
Flowers, H. M., and D. Shapiro [46] 819, *836*
Flück, H., and K. Bichsel [64] 239, *252*
— and N. Blaschke [62] 433, *464*
— and C. Windeck-Lutz [65] 248, *252*
— see Bettschart, A. 882
— see Jaspersen-Schib, R. [108] 237, 238, 239, *253*
— see Oswald, N. [493, 494] 135, 136, 138, *193*; [152, 153] 425, 432, 434, 456, *466*
— see Rüttimann, O. [186] 710, *728*
Focke, J., G. Sembdner, and K. Schreiber [16] 487, 489, *490*
Földesi, D., see Tyihák, E. [301] 239, *258*
Foell, Th., see Smith, L. L. [153] 315, 317, 323, 335, 347, *361*
Förster, T. [195] 77, *185*
Fogg, A. G., and R. Wood [196] 31, *185*

Fohl, J., see Kucharczyk, N. [16] 504, 505; [46, 47] 667, 668, 669, *685*, 879, 901
Fokkens, J., and J. Polderman [55] 523, 560, *563*
Folch, J., M. Lees, and G. H. Sloane Stanley [42] 369, 370, *417*
— see Lees, M. [110] 370, *418*
Folin, O., and V. Ciocalteu [116] 751, *783*
— and W. Denis [117] 751, *783*
Folkers, K., see Masiti, D. [86] 290, *310*
— see Wagner, F. A. [142] 259, *311*
Fong, H. H. S., J. Beal, and M. P. Cava [62b] 444, *464*
— see Farnsworth, N. R. [53] 449, *463*
Fontell, K., see Morris, L. J. [133, 134, 135] 377, 386, 387, *419*
Foppiano, R., and B. B. Brown [197] 148, 153, 154, *185*
Foreman, E. M., see Patton, A. R. [88] 750, 751, *783*
Forsén, S., B. Lindberg, and B. Silvander [86] 831, *837*
Forth, W., P. Doenecke and H. Glasner [201] 354, *362*; [46] 603, *608*
— W. Rummel, and H. Glasner [45] 603, *608*
— see Glasner, H. [57] 603, *608*
Forthmann, G., see Baumann, J. [50] 27, *181*
Fortuin, J. M. H., see Boer, J. H. de [83, 84, 87] 24, 25, 26, *182*
Fosdick, L. S., see Piez, K. A. [29] 737, *781*
Foster, A. B. [198] 108, *185*
— M. H. Randall, and J. M. Webber [85] 831, *836*
— see Bukhari, M. A. [88] 831, *837*
Foussard, F. C., see Audrin, P. [7] 341, *357*
Foussard, J. C., see Jung, L. [95] 351, *359*
Fraenkel-Conrat, H. [185] 774, *785*
— and J. J. Harris [180] 772, *785*
Fragner, J. [46] 259, *309*
Frahm, M., A. Gottesleben, and K. Soehring [56] 533, 534, 535, 536, 538, 540, *563*, 865
Fraiture, W. H. de, see Mandema, E. [103] 605, *609*
Fram, D. H., and J. P. Green [47] 591, *608*

Franc, J., and M. Hájková [15] 626, *628*; [49] 709, *725*
Francis, B. J., see Coomes, T. J. [29] 696, *725*
Francis, F. E., see Chang Shen, N.-H. [35] 338, 340, *358*; [152] 602, *611*
Franck, B., and G. Blaschke [63, 64] 444, 460, *464*
Franck, H., see Weitz, E. [746] 9, 10, *199*
Frank, H., and H. Petersen [89] 750, 751, *783*
Frank, O., see Baker, H. [3] 303, 304, *308*
Franks, N. E., see Wolfrom, M. L. [52] 822, *836*
Franzke, Cl., and A. Jantz [199] 30, *185*
Fraser, D. R., and E. Kodicek [47] 277, 278, 280, *309*
Frauendorf, H., and H. Auterhoff [66] 248, 250, *252*
Fray, G., and J. Frey [200] *185*
Freed, S., see Gebicki, J. M. [24] 793, 801, *805*
Freeman, C. P., and D. West [43] 377, 414, *417*
Freeman, J. H. 890
Frehden, O., see Wasicky, R. [323] 225, *258*, 866
Frei, R. W., and M. M. Frodyma [48] 299, *309*
— and H. Zeitlin [201] 143, *185*
— see Frodyma, M. M. [204, 205] 144, *185*
Freimer, E. H. [44] 379, *417*
Freimuth, U., B. Zawta, and M. Büchner [57] 316, 323, 324, 335, *358*, 902
— see Ludwig, E. [83, 84] 263, 292, 294, *310*
Frencel, I. M. [64b] *464*
French, D. et al. 865
Frenzel, J., see Hänsel, R. [75] 697, *726*
Fresen, J. A. [57, 57a] 526, *563*
— see Büchi, J. [27a] 551, *562*
Freudenberg, K. [55] 687, *725*
— and K. Weinges [50, 51, 52, 56] 687, 688, 696, 712, *725*
— — Ch.-L. Chen, and G. Cardinale [53] 693, *725*

Freudenberg, K., and G. S. Sidhu [54] 711, *725*
Freudenberg, K. et al. 865
Freundt, K. J., see Eberhardt, H. [42, 43, 46] 509, 533, 534, 536, 539, *563*
Frey, A. J., see Hofmann, A. [90] 454, *465*
Frey, J., see Fray, G. [200] *185*
Freytag, W., see Tschesche, R. [168] 342, *361*
Freyvogel, T. A., see Honegger, C. G. [71] 598, *609*
Fricke, R. [202] 24, *185*
— and G. F. Hüttig [203] 8, *185*
Fried, M., see Schlögl, K. [84] 665, 666, *686*
Friedman, M., S. St. George, and R. Bine jr. [48] 606, *608*
Friedman, O. M., see Mahapatra, G. N. [417] 47, *191*; [44] 794, *805*
Friedrich, A., see Pfeil, E. 896
Friedrich, H., and R. Rangoonwala [57] 718, *725*
Fries, J., see Paris, R. [161 b] 422, *466*
Friese, U., see Bohlmann, F. [26] 436, *462*
Frith, M. L., see Aldrich, B. J. 902
Frodyma, M. M., and R. W. Frei [204] *185*
— — and D. J. Williams [205] 144, *185*
— see Frei, R. W. [48] 299, *309*
Frömming, K.-H. [67] 235, *252*
Frohne, D. [65] 433, *464*
Fromageot, C., see Acher, R. [64] 749, 750, *782*
— see Bricas, E. [120] 752, *783*
Frosch, B. [206] 149, *185*; [59, 202, 203] 354, *358, 362*; [52, 53] 603, *608*
— and H. Wagener [207, 208] 149, *185*; [58] 353, 354, *358*; [49, 50, 51] 603, *608*
— see Wagener, H. [179] 354, *361*; [172] 596, *611*
Frydrych, R. [209] 9, 10, *185*
Fuchs, A., see Brieskorn, C. H. [24, 24a] 240, *251*
Fuchs, H., see Dahn, H. [143] 97, *184*; [45] 314, *358*
Fuchs, J., see Kaldewey, H. [48] 478, *491*
— see Stahl, E. [259] 233, *257*
Fürst, A., see Meier, W. [145] 696, *728*
Fürst, W. [49, 50] 277, 280, *309*

Fugitt, C. H., see Steinhardt, J. [696] 42, *198*
Fuhrman, F. A., see Mosher, H. S. [139 b] 461, *466*
Fujii, J., see Wernze, H. [187] 585, *611*
Fujii, S., and M. Kamikura [16, 17] 613, 617, *628*; [87] *630*
Fukinbara, T., see Kagawa, T. [43] 479, 487, *491*
Fulco, A. J., and J. F. Mead [210] 177, *185*
Fumagalli, U., V. Ambrogi and G. Balestra [65 b] 441, *464*
Funck, F. W., see Scheiffarth, F. [138] 601, *610*
Furter, M. [67 a] 207, *252*
Furuya, T., see Brochmann-Hanssen, E. [34 b] 441, *463*; [25] 529, *562*
Futterweit, W., N. L. McNiven, and R. I. Dorfman [60] 340, *358*
— — R. Guerra-Garcia, N. Gibree, M. Drosdowsky, G. L. Siegel, L. J. Soffer, J. M. Rosenthal, and R. I. Dorfman [54] 602, *608*
Fuwa, T., T. Kido, and H. Tanaka [58, 59] 543, 551, *563*

Gabbai, A., see Deshusses, J. [45] 217, *251*
Gabel, E., K. H. Müller, and J. Schoknecht [68] 230, 233, 234, 235, 239, *252*, 856
Gabrilove, L. J., see Guerra-Garcia, R. [64] 602, *609*
Gadeva, V., see Popov, A. 860
Gänshirt, H. [211, 214, 215, 216, 218, 219] 125, 128, 131, 134, 146, 147, 148, 151, 153, *185*; [68a] 224, *252*; [66] 457, *464*; [63, 64, 65, 66] 518, 520, 524, 525, 526, 527, 537, 538, 539, 541, 542, 544, 545, 551, 554, 556, 557, 558, 559, 560, *563*
— F. W. Koss, and K. Morianz [212] 147, 149, 150, 154, *185*; [62] 323, 352, 353, 354, *358*
— and A. Malzacher [51] 292, 293, 294, 298, 299, 305, *309*; [67] 457, *464*; [62, 67] 557, 558, 560, *563*, 862
— and K. Morianz [213] 134, 146, 148, 150, 152, 153, *185*; [32] 636, *655*
— and J. Polderman [217] 135, 138, 146, 147, 149, 151, 153, 154, *185*; [61] 341, *358*; [61] 561, *563*

Gage, T. B., Q. L. Morris, W. E. Detty, and S. H. Wender [58] *726*
Gage, T. G., C. D. Douglas, and S. H. Wender 856
Gagliardi, E., and W. Likussar [8] 842, *853*
— and G. Pokorny [9] 847, *853*
Gaines, H. D., and R. P. Collins [69] 220, *252*
Gaither, R. A., see Mills, P. A. [62] 644, *656*
Gajdoš, M. [60] 543, *563*
Gajewska, A., see Vaedtke, J. [173, 174] 316, 317, 335, 337, *361*
Galletti, F. [204] 311, *362*
Galli, C., see Rouser, G. [177] 389, 391, 393, *420*
Gallop, P. M., see Blumenfeld, O. O. [68] 827, 828, *836*
Galvanek, M., see Adamec, O. [11] 148, *180*; [2] 324, 328, 340, *357*; [1] 601, *607*
— see Matis, J. [119] 316, 335, *360*
Gambassi, G., see Luisi, M. [217] 340, 341, *362*; [101] 600, *609*
Gamp, A., P. Studer, H. Linde, and K. Meyer [220] 53, *186*
Gamson, B. W., see Hull, W. Q [308] 28, *188*
Garbutt, J. L. [12] 809, 815, 816, 834, *835*
Gardi, R., see Ercoli, A. [50] 316, 341, *358*
Gardner, J. A. F., see Barton, G. M. 876
Garel, J.-P. [221, 222] 5, 88, *186*; [70] 211, 227, 230, 231, *252*
Garey, J. C., see Cotsis, T. P. [7] 624, *628*; [86] *630*
Garg, H. G., see Wolfrom, M. L. [95, 97] 831, *837*
Garrett, E. R., T. Suzuki, and D. J. Weber [23] 793, 794, *805*
Gasparič, J. [88] *630*
— and A. Cee [21] 622, *628*
— and I. Gemzová-Táborská [20] 616, *628*
— and M. Matrka [18, 22] 612, 618, *628*
— and I. Táborská [19] 616, *628*
— see Gemzová, L. [89] *630*
Gatsonis, Ch. D., see Gensler, W. J. [62] 710, *726*
Gautheret, R. J. [17] 472, 473, *490*

Gaver, R. C., and C. C. Sweeley [45] 394, *417*
Gay, R., see Maugras, M. [53] *685*
Gebicki, J. M., and S. Freed [24] 793, 801, *805*
Gebistorf, J., see Steinegger, E. [695] 146, 148, *198*; [208, 209] 694, 710, 717, *729*
Gee, M. [223] 145, 146, *186*; [46] 387, 414, *417*; [33] 649, *655*; [35, 81] 817, 828, 829, *835*, *836*
Geenen, H., see Braun, D. [11] 650, *654*; [6] *684*,
Gehl, A., see Rink, M. [137b] 549, *565*; [81] 801, *806*
Geiss, F., A. Klose, and A. Copet [225] 150, *186*
— and H. Schlitt [224] 66, *186*; [26] *685*, 861
— — and A. Klose [226, 227] 66, 67, 70, 71, 102, *186*
— — F. J. Ritter, and M. Weimar [25] 666, *685*
— see Ritter, F. J. [579] 32, *195*; [75] 667, *686*
Geissmann, T. A. [59, 60] 687, 691, 705, *726*
Geldmacher-Mallinckrodt, M., and L. Lautenbach [69] 540, *563*
Gellene, R. A., see Baker, H. [3] 303, 304, *308*
Gellerman, J. L., see Mangold, H. K. 866
Gelotte, B. J. [127] 752, *783*
Gemzová, L., and J. Gasparič [89] *630*
Gemzová-Táborská, I., see Gasparič, J. [20] 616, *628*
Gendi, S. E., W. Kisser, and G. Machata [70] 518, *563*
Genest, Chr., and D. M. Smith [61] 696, *726*
Genest, K. [67b] 451, *464*, *470*
— and C. G. Farmilo [68] 454, *464*
Gensler, W. J., and Ch. D. Gatsonis [62] 710, *726*
Gentili, B., see Stanley, W. L. [689] 72, *198*
George, S. St., see Friedman, M. [48] 606, *608*
Georgiadis, M., see Campaigne, E. [11] 680, *684*
Georgias, L., see Klenk, E. [98] 394, *418*

Georgopoulus, D., see Wieland, Th. [757] 76, *199*
Gerali, G., G. Lugaro, and L. Ferrari [63] 335, 341, *358*
Gerber, N. N. [25] *805*
Gerdes, H., and W. Staib [228, 229] 155, *186*; [205, 206] 340, 341, *362*; [55, 56] 601, *608*
Gerngross, O., K. Voss, and H. Herfeld [114] 751, *783*
Gerritsma, K. W., and M. C. B. v. Rheede van Oudtshoorn [63] 707, *726*
— see Rheede van Oudtshoorn, M. C. B. v. [179, 181] 706, *728*
Gerstl, B., see Eng, L. F. [187] 147, *185*; [39] 379, *417*
Gertig, H. [69, 70] 441, 443, *464*
Gesser, H. D., see Kramer, J. K. G. [380] *190*
Getrost, H., see Kohlschütter, H. W. [363] 23, *189*
Getz, H. R., and D. D. Lawson [230] 131, *186*
Ghiglione, C., see Dumazert, C. [46] 316, *358*; [27] 650, *655*; [55] 822, 824, *836*
Ghosal, C. R. [64] 695, *726*
Ghosh, A. C., see Dutta, N. L. [35] 709, *725*
Giacobazzi, C., and G. Gibertini [71] 242, *252*
Giacomo, A. di, see Rispoli, G. [211] 237, *256*; [114] 269, *310*
Giacopelle, D. [70b] 422, *464*
Giartosio, A., see Fasella, P. [43] 591, *608*
Gibertini, G., see Giacobazzi, C. [71] 242, *252*
Gibney, K. B., see Dutton, G. G. S. [14] 810, 831, 832, 833, 834, *835*
Gibree, N., see Futterweit, W. [54] 602, *608*
Gibson, M. R., see Fikenscher, L. H. [46] 714, *725*
— see Sullivan, G. [232] 435, *469*
Giddings, J. C., and R. A. Keller [114] 137, *183*
Gidez, C. J., see Korey, S. R. [104] 393, 394, *418*
Gielen, W., see Klenk, E. [95, 96] 389, 393, 394, *418*; [94] 598, *609*
Gierschner, K., see Mehlitz, A. [162, 163] 220, 223, 224, 225, 229, *254*, 871

Gigg, R., and C. D. Warren [100, 101] 832, 834, *837*
Gilbert, B., see Antonaccio, L. D. [6] 447, *462*
— see Djerassi, C. [46] 447, *463*
— see Ferreira, J. M. [57] 447, *464*
Gilbert, L. I., see Chino, H. [27] *416*
Gildemeister, E., and Fr. Hoffmann [72] 207, *252*
Giles, C. H., T. J. Rose, and D. G. M. Vallance [231] 41, *186*
Gill, S. [71, 72] 436, *464*
— and E. Steinegger [73, 74, 75] 436, *464*
— see Bernasconi, R. [19b] 436, *462*
Gilles, K. A., see Walsh, D. E. [213] 412, 414, *421*
Gillio-Tos, M., S. A. Previtera, and A. Vimercati [7] 500, 501, *505*; [23] 626, *628*, 876
Giltrow, J., see Curzon, G. 904
Gini, E., see Benassi, C. A. [2] 474, *489*; [7] 584, *607*
Ginsberg, H., W. Hüttig, and G. Strunk-Lichtenberg [232] 23, 24, *186*
Giocoli, G., see Angelico, R. [190] 340, 341, *361*; [3] 599, *607*
Giri, K. V., and A. Nagabhushanam [71] 749, *782*
Giurdaru, G., see Dénes, V. I. [16] 680, *684*
Glasner, H., and W. Forth [57] 603, *608*
— see Forth, W. [201] 354, *362*; [45, 46] 603, *608*
Glasstone, S. [233] *186*
Glaudemans, C. P. J., see Timell, T. E. 856
Glemser, O., and G. Rieck [234, 235] 12, 24, 26, 31, *186*
— — and H. Lackner [236] 31, *186*
Glombitzka, K.-W. [18, 19] 477, 481, 483, 486, *490*
— and T. Hartmann [20] 481, *490*
Gloor, U., see Thommen, H. [135, 137] 267, *311*
Gloster, J., and R. F. Fletcher [47] 385, 414, 415, *417*
Gluck, L., M. V. Kulovich, and S. J. Brody [58] 599, *608*
Glukhoded, I. S., see Kochetkov, N. K. [99, 100, 101] 392, *418*

Gmelin, R. [73, 74] 243, 246, *252*; [21] 472, 476, 479, *490*
— and A. I. Virtanen [22, 23] 472, 473, 474, 479, 486, *490*
— see Bredenberg, J. B. [21a] 246, *251*
Gnehm, R., H. U. Reich, and P. Guyer [237a, 792] 138, *186*, *200*; [70a] 554, *563*
Godon, M. [75] 237, *252*
— see Paris, R. [189] 237, 238, 239, *255*
Goebell, H., and M. Klingenberg [34] 654, *655*
Goedde, H. W., and H. Brunschede [59] 584, *608*
— see Brunschede, H. [16] 584, 585, *607*
Göldel, L., W. Zimmermann, and D. Lommer [76] 245, *252*; [65] 317, 337, *359*
Göndös, Gy., B. Matkovics, and Ö. Kovacs [66] 328, 341, 349, *359*
Görgényi, A., see Vecsei, P. [176] 324, *361*
Görlich, B. [67] 344, *359*
Görsch, H., see Seeboth, H. [623] 146, 154, *196*; [236] 230, *256*; [147] 541, 565, *890*
Göthe, P. O., see Marcuse, R. [154] 219, 221, *254*; [125] 411, 412, *419*
Gohlke, H., see Schmandke, H. [610] *196*
Golab, T., and D. S. Layne [64] 341, *358*
Gold, E. M., see Yawata, M. [186] 320, 337, *361*
Gold, J. P., see Dyer, W. G. [48] 338, *358*
Goldacre, P. L., see Kefford, N. P. [50] 472, *491*
Goldblatt, L. A., see Applewhite, T. H. [23] 54, *180*
Golder, H., see Köhler, M. [357] 37, *189*; [42] 667, *685*
Goldman, N. L., see Craig, J. C. [40c] 430, *463*
Goldrick, B., and J. Hirsch [238] 101, 162, *186*
Goller, R., see Prey, V. [36] 817, *835*
Gonatas, J., see Korey, S. R. [104] 393, 394, *418*
Goodman, D. S. [60] 597, *608*

Goodman, D. S., H. S. Huang, and T. Shiratori [52] 274, *309*
— and T. Shiratori [207] 333, *362*
— see Avigan, J. [6] 316, 323, 330, 332, *357*
— see Huang, H. S. [68] 275, *309*
Goodwin, H., see Booth, D. A. [12] 598, *607*
Goodwin, T. W. [53, 54] 259, 268, 291, *309*; [65] *726*, 884
— and R. J. H. Williams [55] 267, *309*
— see Davies, B. H. [27] 267, *308*
— see Holdgate, D. P. [31] 792, *805*
— see Johnson, D. B. [73] 295, 296, *309*
— see Mercer, E. I. [85] 267, *310*
— see Williams, B. L. [147] 267, *311*
— see Yusef, H. M. [153] 290, *311*
Gootjes, J., see Doorenbos, H. J. [39] 518, *563*
Goppelsroeder, F. [241] 4, *186*
Gordis, E. [48] 372, 398, *417*
Gordon, A. H., and J. E. Eastoe [240] 90, *186*
— see Consden, R. [136, 137] 3, 4, 105, *183*; [26, 47, 99] 737, 746, 747, 749, 750, *781*, *782*, *783*, 885
Gordon, H. T. [239, 239a, 793] 70, 130, 141, *186*, *200*
Gordon, S. A., and R. P. Weber [24] *490*, 875
Gottesleben, A., see Frahm, M. [56] 533, 534, 535, 536, 538, 540, *563*, 865
Goubeau, J., and R. Warncke [242] 8, *186*
Gouezo, F., see Vignoli, L. [253b] 440, *469*
Gould, J. P., see Dyer, W. G. [34] 602, *608*
Goutarel, R. *461*
— see Paris, M. [156] 435, *466*
— see Parello, J. [159] 460, *466*
Govindachari, T. R., K. Nagarajan, and H. Schmid [76] 447, *464*
— B. R. Pai, S. Rajappa, N. Viswanathan, W. G. Kump, K. Nagarajan, and H. Schmid [77] 447, *464*
— see Guggisberg, A. [86] 447, *464*
Gower, D. B. [68] 323, 325, 339, *359*, 896
— see Brooksbank, B. W. [25] 339, *358*

Goyan, J. E., see McLaughlin, J. L. [415] 153, *191*; [118] 451, 453, *465*
Gracza, L. and A. Zarándi [77] 237, *252*
Gracza, P., see Végh, A. [252] 433, *469*
Gräb, R. [78] 212, *252*
Graebe, J. E., D. T. Dennis, Ch. D. Upper, and Ch. A. West [243] 178, *186*
Gräfe, G. [68] 542, *563*
Graf, E., E. Dahlke, and H. W. Voigtländer [66] 719, *726*
— and W. Hoppe [79, 80] 226, 229, 245, *252*
— see Hörhammer, L. [113] 706, *727*
Graf, L., see Rapport, M. M. [172] 389, *420*
Grahn, M., see Angustinsson, K. P. 872
Gran, W., and H. Endres [80a] 231, *252*
Grant, D. W. 880
Grant, H. L. [67] 714, *726*
Granzer, E. [49] 394, *417*
Grasmeier, H., see Hörhammer, L. [94] 691, 699, *726*
Grasselli, J. G., see Snavely, M. S. [640] 150, 154, *196*
Grasshof, H. [244, 245] 29, *186*; [8] 495, *505*; [29] 815, *835*
Grassini, G., see Dobici, F. [165] 30, 111, 112, *184*
Grassmann, W., and K. v. Arnim [65] 749, *782*
— and G. Deffner [246] 42, *186*
— — — and H. El Sissi [249] 42, *186*
— — W. Pauckner, and H. Mathes [248] 43, *186*
— H. Hörmann, and A. Hartl [247] 42, *186*
— see Endres, H. [185] 43, *185*
Grau, R., and A. Schweiger [35] 649, *655*
Grau, W., and H. Endres [71] 541, *563*; [68] 699, 703, *726*
Graune, F. J., see Karting, Th. [118] 245, *253*
Gray, W. R., and B. S. Hartley [212] 778, *785*
Green, C., see Chargaff, E. [85] 750, *783*, 882
Green, J. G., see Anderson, N. G. [1] 795, *804*
Green, J. P., see Fram, D. H. [47] 591, *608*

Green, K., and B. Samuelsson [50] 399, *417*; [61] 597, *608*
Green, S., see Fishman, W. H. [56] 356, *358*; [44] 603, *608*
Greenberg, D. M., see Rehbinder, D. [569] 176, *195*
Greenway, R. M., P. W. Kent, and M. W. Whitehouse [18] 811, *835*
Gref, C.-G., and J. J. Saukkonen [250] 69, *186*
Gregory, G. I., see Bevan, T. H. 887
Gregory, M., see Battersby, A. R. [14] 448, *462*
Greig, C. G., and D. H. Leaback 863
Grenthe, I., see Ahrland, S. [43] 739, *782*
Gresham, G. A., see Bowyer, D. E. [13] 592, *607*
Grewe, H., see Acker, L. [3] 631, *654*
Gries, G., K. H. Pfeffer, and E. J. Zappi [62] 586, *608*
Griesebach, H., see Achenbach, H. [1] *577*
Griessbach, R. [251] 8, *186*
Griffin, W. J., see Evans, W. C. [50] 435, *463*
Griffith, C. M., J. C. MacWilliam, and T. Reynolds [24a] 488, 489, *490*
Griffith, M., see Richardson, G. S. [573] 157, 159, *195*
Griffiths, J. M., see Lewis, J. A. 872
Grimmelikhuysen, J. C., see Riezebos, G. [175] 408, *420*
Grimmer, G., see Tschesche, R. 870
Grimmett, M. R., and E. L. Richards 899
Grineva, N. I., see Lomakina, T. S. [43] 794, 801, *805*
Griot, R., see Hofmann, A. [90] 454, *465*
Grippo, P., M. Iaccarino, M. Rossi, and E. Scarano [26] 792, 793, 795, 796, *805*
Grisebach, H. [70] *726*
— and G. Brandner [252] 179, *186*
— and W. D. Ollis [69] 687, *726*
— and K.-O. Vollmer [253, 254] 178, *186*; [71] 692, *726*
— see Achenbach, M. [7] 178, *180*
Gritter, R. J., and R. J. Albers [69] 322, *359*
Grob, E. C., W. Eichenberger, and R. P. Pflugshaupt [56] 267, *309*

Grob. E. C., see Eichenberger, W. [43, 44] 267, *309*
Gröger, D. [78, 79, 80] 451, 454, *464*
— and D. Erde [81] 451, *464*
— and S. Johne [81 b] 460, *464*
— K. Mothes, H. Simon, H. G. Floss, and R. Weygand [82] 457, *464*
— and K. Stolle [82 b] 450, *464*
— and V. E. Tyler [83] 451, *464*
— — and J. E. Dusenberry [84] 451, *464*
— see Mothes, K. [140] 454, *466*
— see Tyler, V. E. [248] 447, *469*; [131] *493*
Groot, K., see Oertel, G. W. [490] 46, 47, *193*; [135, 220] 316, 340, 356, 357, *360, 362*
Grosjean, M. H., see Noirfalise, A. [112] 508, *565*, 875
Gross, D., and H. R. Schütte [85] 435, *464*
— see Mothes, K. *461*
Gross, R., see Sembdner, G. [111] 476, 479, 487, *493*
Grote, I. W. 892
Grothe, J. W. [194] 777, *785*
Grünberger, I., see Pailer, M. [188] 219, 220, *255*
Grundschober, F., and V. Prey [79] 828, *836*
Grundy, S. M., E. H. Ahrens jr., and T. A. Miettinen [255] 178, *187*; [63] 604, *609*
— see Miettinen, T. A. [444] 178, *192*; [88] 284, *310*; [106] 603, 604, *610*
Grylls, F. S. M., see Burrows, S. 887
Grynberg, H., and M. Beldowicz [256] 139, *187*
Gstirner, F. [81] 207, 246, 247, *252*; [57] 259, *309*
Guenther, E. [82] 207, *252*
Günther, H., and A. Schweiger [257] 35, *187*; [39] 817, 818, 834, *835*
— see Schweiger, A. [90] 804, *806*
Günther, H. H., and H.-G. Mautner [58] 298, 299, *309*
Guerra-Garcia, R., S. C. Chattoraj, L. J. Gabrilove, and H. H. Wotiz [64] 602, *609*
— see Futterweit, W. [54] 602, *608*
Guggisberg, A., T. R. Govindachari, K. Nagarajan, and M. Schmid [86] 447, *464*

Gugler, E., see Käser, H. [86] 587, *609*
Guillot, J., see Vignoli, L. [253 b] 440, *469*
Guilloux, E., and S. Beaugiraud [19] 812, 817, *835*
Guimarães, C. V., see Zelnik, R. [188] 345, *361*
Gundlach, G., see Turba, F. [170] 760, *784*
Gunstone, F. D., and F. B. Padley [51] 399, *417*
Gunton, R. W., see Evans, J. R. [190] 179, *185*
Gupta, A. K., sen., see Kaufmann, H. P. [121, 122, 123, 124] 240, 241, *253*
Gupta, A. S., and S. Dev [83] 211, *252*
Gupta, D., see Shellard, E. J. *471*
Guskova, L. I., see Lomakina, T. S. [43] 794, 801, *805*
Gut, M., see Noller, C. R. [185 a] 246, *255*
Guthrie, F. E., see Hodgson, E. [88 d] 430, 431, *465*, 859
Gutmann, H. 308
Guttmann, St. [143] 753, *784*
— and R. A. Boissonnas [135] 753, *784*
— J. Pless, and R. A. Boissonnas [136] 753, *784*
Gutzwiller, J., R. Mauli, H. P. Sigg, and Ch. Tamm [84] 226, *252*
Guyer, P., see Gnehm, R. [237 a, 792] 138, *186, 200*; [70 a] 554, *563*
Gyanchandani, N. [85] 212, 214, *252*

Haagen-Smit, A. G. [86] 207, *252*
Haahti, E., and T. Nikkari [65] 599, *609*
— T. Nikkari and K. Juva [70] 316, 322, 333, *359*; [52] 376, 398, *417*; [66] 599, *609*
— see Horning, E. C. [89] 316, *359*
Haas, G. H. de, see Bonsen, P. P. M. [16] 390, *416*
Haas, H. J., and A. Seeliger [64] 825, 826, 827, 828, *836*
Haas, K. H. 123
Haber 24
Habermann, E. [213] 777, *785*
— G. Bandtlow, and B. Krusche [258] 148, *187*; [53] 415, *417*; [67] 596, *609*
Hacaperková, J., see Macek, K. 891
Häfelinger, G., and E. Bayer [24] 623, *628*

Haefelfinger, P. 470
— B. Schmidli, and H. Ritter [259] 145, 151, *187*; [72] 558, *563*
Haenni, E. O., J. W. Howard, and F. L. Joe [36] 639, *655*
— see Morris, W. W. [66] 638, *656*
Hänsel, R. [74] 693, 705, *726*
— L. Langhammer, J. Frenzel, and G. Ranft [75] 697, *726*
— G. Ranft and P. Bähr, [76] 698, *726*
— and H. Rimpler [78] 704, *726*
— H. Schulz, and Ch. Leuckert [77] 711, *726*
Härri, E. [15] 574, *577*
Häusser, H. [260] 57, *187*; [26] 615, *629*
Haeussler, H. [261] 91, *187*
Hagedorn, P., see Neu, R. 857, 883
Hagen, U., see Keck, K. [342] 111, 112, 114, *189*
— see Keck, K. [37] 794, 803, *805*
Hagen-Smit, A. J., see Kögl, F. [56] 472, *491*
Hager, A., and T. Bertenrath [59] *309*
Hagerman, D. D., and J. M. Spencer [71] 319, *359*
Hágony, P. L., see Tyihák, E. [308, 309] 226, *258*, 904
Hagopian, L. M., see Reissell, P. K. [130] 591, 592, *610*
Hahn, H. v., see Erlenmeyer, H. 904
Hahn, L., see Euler, H. v. [21] 789, *804*
Haigh, W. G., and D. J. Hanahan [262] 159, 177, *187*
Hais, I. M., and K. Macek [263] 3, *187*; [9] *505*; [72] 691, 693, 705, *726*; [2] 732, 736, 737, 760, 763, *781*, 886, 893, 900
— see Ledinová, Z. 895
— see Macek, K. [413] 5, *191*
Hájková, M., see Franc, J. [15] 626, *628*; [49] 709, *725*
Hall, A. [71a] 341, *359*
Hall, N. F., see Meinhard, J. E. [439] 3, 73, *191*; [16] 837, *853*
Hall, R. J. [264] 98, *187*
Hall, S. W., see Morris, L. J. [141] 381, *419*
Hall, W. B., see Hilton, J. [291] 133, *188*
Halmekoski, J. [265] 31, *187*; [74] 522, 523, 524, *563*; [73] 698, *726*
— and H. Hannikainen [73] 540, *563*

Halpaap, H. [266, 267] 27, 87, 98, 100, *187*; [60] 262, *309*; [72] 314, 336, *359*
Halsall, T. G., see Jones, E. R. H. [109] 207, *253*
Haluk, J. P., see Urion, E. [221] 704, *729*
Hamilton, J. G. [73] 352, 353, 354, *359*
Hamman, B. L., and M. M. Martin [74] 338, *359*
Hammerschmidt, W., see Adank, K. [2] 509, *562*
Hammond, E. W., see Abbott, D. C. [4] 32, *180*; [2] *654*
Hammond, H. T., see Koehler, W. R. [102] 412, *418*
Hammonds, T. W., and G. Shone [54] 411, *417*
Haná, K., see Janák, J. [321] 118, *188*
Hanafi, S., see List, P. H. [123] 433, *465*; [133] 700, *727*
Hanahan, D. J., J. Ekholm, and C. M. Jackson [55] 374, *417*
— see Haigh, W. G. [262] 159, 177, *187*
Handa, K. L., see Vashist, V. N. [314, 315] 221, 236, *258*
Handa, N., and F. Smith [83] 830, *836*
Handschuh, D., see Stepanov, V. [24] 737, 752, *781*
Hanes, C. S., and F. A. Isherwood 887
Hanessian, S., and T. H. Haskell [106] 833, *837*
Hanewald, K. H., see Mulder, F. J. 280
Hanke, P., see Weill, C. E. [31] 816, 834, *835*
— see Wheeler, M. [33] 817, *835*
Hannig 746
Hannig, K. [268] 105, 106, *187*
Hannikainen, H., see Halmekoski, J. [73] 540, *563*
Hansbury, E., J. Langham, and D. G. Ott [269] 131, *187*
— and D. G. Ott [27] 793, *805*
— — and J. D. Perrings [270] 53, *187*
Hansen, I. A. [271] 177, *187*
Hansen, I. L., and M. A. Crawford [28] 477, *490*
Hansen, J., see Kauffmann, Th. [15] 501, *505*
Hanson, R. W., see Ziporin, Z. Z. [111] 801, *807*
Hansson, E., P. Hoffmann, and L. Kristerson [272] 179, *187*
Hansson, J. [27] 670, 671, *685*

Author Index

Hansson, J., and A. Alm [10] 500, *505*
Hanus, H., see Wassermann, L. [49] 820, *836*
Hara, S. [273] *187*
— and M. Takeuchi [75] 323, 352, 353, *359*, 893
— — M. Taschibana and G. Chihara [274] 139, *187*; [76] 354, *359*
— and H. Tanaka [75] 558, *564*
— — and M. Takeuchi [275] 5, *187*; [77] 354, *359*
— see Takeda, K. [166] 347, *361*, 855, 857, 863
Harada, H., and A. Lang [29] 489, *490*
Harbers, E., G. E. Domagk, and W. Müller [28] 791, *805*
Harborne, J. B. [79, 80, 81, 82, 83, 85, 86] 687, 693, 705, *726*
— and I. J. Corner [84] 693, *726*
Hardman, R., see Blunden, G. [22] 347, *357*
Hardmeier, E., and J. Schmidlein-Mèszáros [76] 532, *564*
Hardy, T. L., D. O. Holland, and J. H. C. Nayler [52] 747, 749, *782*
Harel, S., see Bergmann, E. D. [18] 334, *357*
— see Ikan, R. [102] 242, *253*; [91] 323, 334, *359*
Harlan jr., W. R., and S. J. Wakil [276] 177, *187*
Harley-Mason, J., and A. A. P. G. Archer [30] *490*, 869
Harris, F., see Fishman, W. H. [56] 356, *358*; [44] 603, *608*
Harris, G., and I. C. MacWilliam 859
Harris, J. F., see McKibbins, S. W. [414] 99, *191*
Harris, J. J., and P. Roos [188] 774, *785*
— see Fraenkel-Conrat, H. [180] 772, *785*
Harris, M., see Steinhardt, J. [696] 42, *198*
Harris, P., and F. W. Lindley [25] 616, *629*
Harrison, I. T., see Tursch, B. [299] 242, 243, *258*
Harrison, J. S., see Burrows, S. 887
Harrison, R., and H. G. Flechter jr. [45] 819, *836*
Harthon, J. G. L. [28] 669, 670, *685*

Hartigan jr., J. M., see Dicarlo, F. J. [161] *184*; [18] 671, 673, *684*
Hartikainen, I., see Schneckenburger, J. [204] 433, *468*
Hartl, A., see Grassmann, W. [247] 42, *186*
Hartley, B. S., see Gray, W. R. [212] 778, *785*
Hartmann, A., see Brenner, M. [149] 755, *784*
Hartmann, T., see Glombitzka, K.-W. [20] 481, *490*
Hartsaw, P. E., see Bird jr., H. L. [70] 103, 147, 149, 152, 153, *182*; [21] 341, *357*
— see Comer, J. P. [194] 340, 341, *362*; [36] 561, *563*
Hartwell, J. L., and A. W. Schrecker [87] 687, 710, *726*
Haruta, F., see Morita, K. [457] 57, *192*
Harzell, A., see Bruchfield, H. P. [12] 647, *654*
Hasenmayer, G., see Kaiser, H. 868
Hashim, S. A., see Krell, K. [383] 154, *190*; [105] 414, *418*; [98] 593, *609*
Hashimoto, A., K. Shiro, and K. Mukai [56] 406, *417*
Hashimoto, H., see Kuroiwa, Y. [145a] 247, *254*; [132] 714, *727*
Hashimoto, Y. [277] 5, *187*
Hashizume, T., and Y. Sasaki [29] 795, *805*
Hashmi, M. H., M. A. Shahid, and A. A. Ayaz [278] 74, *187*; [10] 845, *853*
Haskell, T. H., see Hanessian, S. [106] 833, *837*
Hasselquist, H., and M. Jaarma [61] 305, 306, *309*
— see Euler, H. v. [40, 41] 585, 592, *608*
Hatch, F. T., see Reissell, P. K. [130] 519, 592, *610*
Hatfield, D., see Zimmerman, M. [110] 794, 801, *806*
Hathway, D. E. [88] 693, *726*
— and J. W. T. Seakins [279] 43, *187*
Hatt, J. L., see Roche, J. N. [75] 749, *782*
Haub, H.-G., and H. Kämmerer [30] 658, *685*
Hauffe, K. [280] 133, *187*
Hauptmann, S., and J. Winter [76a] 524, *564*

Hausmann, W. [147] 753, *784*
Hausmann, W. K., see Lefemine, D. V. [391] 97, *190*
Haverkate, F., and L. L. M. van Deenen [57] 383, 391, 398, 401, *417*
Havinga, E., see Sanders, G. M. [116] 277, *310*
Haworth, R. D. [281] 43, *187*; [89] 709, *726*
Hawrylyshyn, M., B. Z. Senkowski, and E. G. Wollish [30] 794, *805*
— see Wollish, E. G. [774] 54, *200*; [175] 544, *566*
Hawthorne, B. E., N. Tuna, H. K. Mangold, and W. O. Lundberg [58] *417*
Hay, G. W., B. A. Lewis, and F. Smith [16] 810, 812, 819, 820, 830, 831, 834, *835*, 893
Hayaishi, O., see Olson, J. A. [98] 275, *310*
Hayashi, F., and L. Rappaport [31] 489, *490*
Hayashi, M., and T. Kamikubo [62] 302, 303, *309*
Hayman, R. B., see Eng, L. F. [187] 147, *185*; [39] 379, *417*
Haynes, C. G., see Rettie, G. H. [61] 613, 616, 618, 620, *629*
Heacock, R. A., and M. E. Mahon [32, 33, 34, 35] 477, 481, 486, *490*, 869
Head, C., see Levin, E. [399] 179, *190*
— see Levin, E. [112] 414, *418*
Heaysman, L. T., and E. R. Sawyer [282] 102, 135, *187*; [63] 278, 279, 280, *309*
Hecht, F., see Lesigang, M. [397] 31, *190*
— see Markl, P. [14, 15] 844, *853*
Hedin, P. A., see Thompson, A. C. [713 a, 800] 145, *198*, *200*
Hefendehl, F. W. [176] 771, *784*
Heftmann, E. [283] 90, *187*
— see Bennett, R. D. [58] 69, 99, 100, 178, *181*; [14, 16] 325, 330, 332, 335, 347, *357*
— see Ruddat, M. [587] *195*
Hegedüs, B., see Winterstein, A. [763] 135, *199*; [149, 150] 267, 272, 274, *311*, 896
Hegnauer, R., see Fikenscher, L. H. [47, 48] 692, 714, *725*
Heidbrink, W. [284] 81, 140, 147, *187*
Heide, R. ter [99] 631, *656*
— see Klouwen, M. H. [131, 132] 217, 223, 230, 231, *254*

Heide, R. F. v. d. [31, 32] 664, 665, *685*
— and O. Wouters [33] 673, *685*, 867
Heilman, J., see Barrollier, J. [66] 749, *782*, 883
Heimann, W., see Drawert, F. [31] 307, *308*
Hein, K., see Hörhammer, L. [111] 697, 727, 857, 883
Heiser, H. W., see Newsome, J. W. [474] 23, 24, *192*
Helbing, A. R. [87, 88] 245, *252*
Helmcke, J.-G. [285] 27, *187*
— and W. Krieger [286] 27, *187*
Hemming, F. W., see Pennock, J. F. [104] 283, 284, 285, 286, 287, *310*
Hemming, H. G., see Brian, P. W. [4] 488, 489, *489*
Hemming, R., see Schantz, M. v. [223] 211, *256*
Henckel, E., see Tschesche, R. [292] *257*
Hendricksen, H. S., see Carter, H. E. [21] 412, *416*
Heng, J., see Hörhammer, L. [93] 237, 238, *253*
Hengy, H., see Preussmann, R. [32, 33] 497, *506*, 871, *900*
Henke, S., see Jerchel, D. [328] 176, *188*
Henkel, H. G., and W. Ebing [37] 648, *655*, 858
Henneberg, D., see Klein, E. [129] 212, *253*
Henneberg, M., see Rusiecki, W. [588] 5, *195*; [193] 448, *467*; [138] 509, *565*
Henning, H., see Strohecker jr., R. [702] 146, 151, 155, *198*; [130] 259, 263, 264, 266, 274, 276, 280, 284, 295, 296, 297, 299, 300, 301, 305, 306, 307, *311*, 872
Henning, N., see Schmid, E. [142] 589, *610*
Henninger, M. D., H. N. Bhagavan, and F. L. Crane [65] 289, 290, *309*
— and F. L. Crane [64] 289, 290, 292, *309*
Henrichs, J. [77] 534, 536, *564*
Herández, F. A., see Bravo, R. O. [26] 541, *562*
Herberhold, C., and O. A. Neumüller [68] 586, *609*

Herbst, R., see Kartnig, Th. [118] 245, 253
Herfeld, H., see Gerngross, O. [114] 751, *783*
Heřmánek, S., V. Schwarz, and Z. Čekan [79] 323, 329, 335, 347, *359*; [87] 422, 439, 448, 451, 454, *464*; [27] 613, *629*
Herold, M., see Urx, M. 856
Herout, V., see Křepinský, J. [142] 239, *254*
— see Novotný, L. [158] 717, *728*
Herrmann, S., see Rink, M. [574, 575] 37, *195*; [131, 132] 580, 581, *610*; [75] 651, 652, *656*; [65] 825, 826, 827, 834, *836*
Herrscher, R. F., see McCarthy, J. L. [123] 315, 340, 341, 346, *360*
Hertelendy, F. [81] 351, *359*
— and R. H. Common [80] 350, *359*
Herz, W., and S. Inayama [89] 216, *252*
Herzmann, H., see Rabitzsch, G. [551] 177, *194*
Herzog, G., see Kortüm, G. [373] 143, *190*
Hesse, G. [287] 60, *187*
— and M. Alexander [288] 31, *187*
Hesse, M. *461*
— W. v. Philipsborn, D. Schumann, G. Spiteller, M. Spiteller-Friedmann, W. I. Taylor, H. Schmid, and P. Karrer [88] 447, *464*
Heston, W. M., see Alexander, G. B. [17] 13, *180*
Heus, J. G. de, see Dam, M. J. D. van [175] 317, 329, *361*
Heusser, D. [289] 153, 154, 155, *187*; [208] 345, *362*; [77a] 558, *564*; [90] 718, *726*
— and E. Jackwerth [290] 154, 155, *187*; [88b, 88c] 440, 441, *465*
Heuvel, W. J. A., see Horning, E. C. [89] 316, *359*
Heymanns, C., see Potter, W. P. de [132] 520, 524, *565*
Heyndrickx, A. M., Schauvliege, and A. Blomme [78] 520, 521, *564*, 879
Higaki, M., M. Takahashi, T. Suzuki, and Y. Sahashi [66] 277, *309*
Hilditch, T. P., and P. N. Williams [59] 367, *417*

Hilinski, I. M., see Farnsworth, N. R. [53b] 449, *463*
Hille, E., see Steuerle, H. [697] 43, *198*
Hiller, K., B. Linzer, and S. Pfeifer [89a] 246, 247, *252*
Hillsberry, J., see Crider, Q. [29] 376, *416*
Hilton, J., and W. B. Hall [291] 133, *188*
Hilz, R., F. F. Castano, and G. A. Lightbourn 883
Hindorf, H., see Schütte, H. R. [210] 436, *468*
Hinekl, R. D., see Philips, M. A. [73] 632, *656*
Hinkley, D. F., see Brenner, G. S. [104] 832, 834, *837*
Hirs, C. H. W. [44] 742, 759, 774, *782*
Hirsch, J., see Goldrick, B. [238] 101, 162, *186*
Hirshfeld, H., see David, S. [24] 295, *308*
Hjertén, S., see Tiselius, A. [714] 30, 90, *198*
Hoch, W., see Paulus, W. [124] 508, 509, 514, *565*
Hodes, M. E., see Tamm, C. [95] 790, *806*
Hodgson, E., E. Smith, and F. E. Guthrie [88d] 430, 431, *465*, 859
Hodosan, F., and A. Pop-Gocan [82] *359*
Höft, E., see Rieche, A. [74] 678, *686*
Högl, O., see Türler, M. [87] 664, *686*
Hörhammer jr., H. P., see Hörhammer, L. [108] 706, *727*
Hörhammer, L. [292] 44, *188*; [98, 107] 697, 703, *727*
— G. Bittner, and H. P. Hörhammer jr. [108] 706, *727*
— A. El-Hamidi, and G. Richter [90] 238, *252*
— L. Farkas, H. Wagner, and S. Imre [109] 707, *727*
— — — and E. Müller [110] 691, 707, *727*
— and H. Wagner [293] 43, *188*; [93, 96, 97] 691, 693, 697, 699, 700, *726, 727*
— — and G. Bittner [294] 43, *188*; [104, 105, 106] 691, 703, 706, 707, *727*, 895
— — and W. Eyrich [103] 695, 696, *727*

Hörhammer, L., H. Wagner, and E. Graf [113] 706, *727*
— — and H. Grasmeier [94] 691, 699, *726*
— — and K. Hein [111] 697, *727*, 857, 883
— — J. Isquierdo, and H. Endres [92] 691, 699, *726*
— — and F. Kilger [215] 740, *785*
— — and H. König [100] 719, *727*
— — and B. Lay [91] 247, *252*; [95, 101] 716, 717, *726, 727*
— — and W. Leeb [91] 699, *726*
— — and H. Reinhardt [112] 717, *727*
— — and G. Richter [92] 238, 239, *252*
— — — H. W. König, and J. Heng [93] 237, 238, *253*
— — and B. Salfner [99] *727*
— — and H. Schilcher [94] 238, *253*
— — and M. Seitz [102] 697, *727*
— see Wagner, H. [320] 235, *258*; [140] 290, 292, *311*; [210] 389, 390, 391, 393, *421*, 876
Hörmann, H., and H. v. Portatius [295] 53, *188*
— see Endres, H. [186] 43, *185*; [48] 541, *563*; [45] 691, 699, 700, 703, *725*
— see Grassmann, W. [247] 42, *186*
Hof, L., see Klenk, E. [98] 394, *418*
Hoffbauer, R., see Brunschede, H. [16] 584, 585, *607*
Hoffmann, Fr., see Gildemeister, E. [72] 207, *252*
Hoffmann, H., and K. Roller [79] 526, 527, *564*
Hoffmann, J., see Luckner, M. [125, 126] 432, 456, *466*
— see Poethke, W. [173] 432, *467*
Hoffmann, P., see Hansson, E. [272] 179, *187*
Hofmann, A. [89] 451, *461, 465*
— H. Ott, R. Griot, P. A. Stadler and A. J. Frey [90] 454, *465*
— and H. Tscherter [91] 451, *465*
— see Stadler, P. A. [224] 454, *468*
Hofmann, A. F. [296, 297, 298] 30, 53, *188*; [83, 84, 85, 86] 311, 316, 317, 352, 353, *359*; [60] 376, 385, 386, *417*
— and B. Borgström [61] 385, *417*
Hofmann, E., and A. Wünsch 892

Hofmann, G., see Kohlschütter, H. W. [364] 23, *189*
Hofmann, H. [95] 247, *253*
Hofstetter, J., see Schneider, H. [68] 626, *629*
Hofstetter, R., see Sonanini, D. [229] 337, *362*
Hogg, A. N., see Ayer, W. A. [10] 461, *462*
Hohita, T., see Kazuno, T. [100] 355, *359*
Hohmann, T., and H. Rochelmeyer [92] 451, *465*
Hohndorf, E., see Neidlein, R. [107b] 547, *564*
Holczabek, W. [69] 599, *609*
Holdgate, D. P., and T. W. Goodwin [31] 792, *805*
Holla, K. S., and D. C. Cornwell [62] 374, *417*
Holland, D. O., see Hardy, T. L. [52] 747, 749, *782*
Holleman, J. W., see Biserte, G. [156] 756, 757, 758, *784*
Holleman-Dehove, J., see Biserte, G. [156] 756, 757, 758, *784*
Holley, R. W., J. Apgar, P. B. Doctor, J. Farrow, M. A. Marini, and S. H. Merrill [33] *805*
— and S. H. Merrill [32] 801, 802, *805*
Holliday, P., see Brown, L. [105] 34, *183*
Hollingsworth, D. R., M. Dillard and P. K. Bondy [214] 780, *785*
Holman, R. T., W. O. Lundberg, and T. Malkin [63] 367, *417*
— see Chalvardjian, A. [23] 387, *416*
— see Morris, L. J. [133, 134, 135] 377, 386, 387, *419*
— see Sprecher, H. W. [192] 380, *420*
— see Vioque, E. [731] 147, *199*; [316] 237, *258*; [204, 205] 386, 414, *421*
— see Williams, J. A. [192] 597, *611*
Holmes, H. L., see Manske, R. H. F. *461*
Holmes, W. L., see Ditullio, N. W. [196] 332, *362*
Holton, S., see Wood, P. D. S. [193] 596, *611*
Holtzem, H., see Steiner, M. [271] 207, *257*
Holubek, J., O. Štrouf et al. *461*
— see Trojánek, J. [243] 449, *469*

Honegger, C. G. [299, 300, 301] 88, 93, 98, 109, 111, 112, 114, *188*; [87, 209] 314, 316, 320, 333, *359*, *362*; [11] 503, *505*; [70] 598, *609*; [123] 746, 752, *783*
- and T. A. Freyvogel [71] 598, *609*
- see Niederwieser, A. [478] 91, *192*

Honerlagen, H., see Müller, K. H. [465] 94, *192*; [143] 455, *466*

Hood, L. V. S., see Sherma, J. 858

Hooghwinkel, G. J. M., P. Borri, and J. C. Riemersma [64] 392, *417*

Hoopen, H. J. G. ten [302] 31, *188*; [96] 222, *253*

Hoppe, E., see Schmid, E. [144] 509, 519, *565*

Hoppe, W., see Graf, E. [79, 80] 226, 229, 245, *252*
- see Reindel, F. [570] 89, *195*; [73] 749, 755, *782*, 863

Horáková, E., see Kulenda, Z. [213] 340, *362*; [99] 601, *609*

Hori, M., see Kuwada, S. [80] 297, *310*

Horlick, L., and J. Avigan [88] 334, *359*

Horner, L., see Ertel, H. [15] 810, *835*, 893

Horning, E. C., A. Karmen, and C. C. Sweeley [65] 395, *417*
- T. Luukkainen, E. Haahti, B. G. Creech and W. J. A. van den Heuvel [89] 316, *359*

Horning, M. G. [66] 369, *417*

Hornung, W. [304] 130, *188*

Horrocks, L. A. [67, 67a] 388, 391, 397, *417*; [72] 598, *609*

Horský, O., see Urx, M. 856

Horstmann, C., see Schreiber, K. [207] 459, *468*

Horton, D. [57] 822, *836*
- see Wolfrom, M. L. [42, 77, 78, 93, 94, 95, 97] 818, 828, 831, *835*, *836*, *837*

Horvath, C. [304a, 794] 147, *188*, *200*

Hosogai, Y., see Kawashiro, I. [44a] 645, *655*

Houben, G. M. M., see Boer, J. H. de [85] *182*

Hough, L., see Buss, D. H. [43] 818, *835*

Houle, C. R., see Malins, D. C. [115, 116] 377, 383, *418*

Houle, C. R., see Mangold, H. K. [425] 176, 177, *191*
- see Wekell, J. C. [216] 388, *421*

Howard, A. N., see Bowyer, D. E. [13] 592, *607*

Howard, J. W., see Haenni, E. O. [36] 639, *655*

Hradil, M., see Šaršúnová, M. [201] 448, *468*

Hranisavljević-Jakovljević, M., I. Pejković-Tadić, and A. Stojilyković [29] 680, 681, *685*; [11, 12] 844, *853*, 865
- see Pejković-Tadić, I. [192] 219, *255*; [62] 680, 681, *685*

Hrbek jr., J., see Bláha, K. [22] 443, *462*
- see Potěšilová, H. [178] 426, 427, 429, *467*

Hromatka, O., and W. A. Aue [38] 650, *655*; [34] 662, *685*

Hruban, L., see Vácha, P. [250] 454, 455, 456, *469*

Hsiu, H.-Ch., see Huang, J.-T. 470

Huang, H. S., and D. S. Goodman [68] 275, *309*
- see Goodman, D. S. [52] 274, *309*

Huang, J.-T., H.-Ch. Hsiu, and K.-T. Wang 470

Huber, I., see Koss, F. W. [377] 178, *190*

Hückel, W., and H. Waiblinger [35] 680, *685*

Hündorf, H., see Zöllner, N. [203] 593, *612*

Hüttenrauch, R., L. Klotz, and W. Müller [305] 46, *188*; [67] 292, 293, 294, *309*
- and J. Schulze [306] 76, *188*; [16] *577*

Hüttig, G. F., see Fricke, R. [203] 8, *185*

Hüttig, W., see Ginsberg, H. [232] 23, 24, *186*

Hughes, A., see Chamberlain, J. [126] 159, *183*

Huguenin, R. L., and R. A. Boissonnas [137] 753, *784*

Huhnstock, K., and H. Weicker [73] 592, *609*

Huizinga, T., see Christensen, E. K. J. [32] 536, 537, 538, *562*, 884

Hulanicka, D., J. Erwin, and K. Bloch [307] *188*

Hull, W. Q., H. Keel, J. Kenney, and B. W. Gamson [308] 28, *188*
Hulyalkar, R. K., J. K. N. Jones, and M. B. Perry [75] 828, *836*
Huneck, S. [97, 98, 99] 242, 243, *253*
Hussey, H., see Speake, T. [220] 430, *468*
Huston, Ch. K., see Albro, Ph. W. [2] 379, 384, 406, *415*
Hutson, D. H., see Wolfrom, M. L. [93] 831, *837*
Hutzler, J., see Dancis, J. [23] 652, *655*
Hynie, I., see König, J. [88] *564*
Hyyrylainen, M. [100] 240, *253*

Iaccarino, M., see Grippo, P. [26] 792, 793, 795, 796, *805*
Ibayashi, H., M. Nakamura, S. Murakawa, T. Uchikawa, T. Tanioka, and K. Nakao [90] 339, *359*; [75] 602, *609*
Ichimura, Y., see Katsui, G. [74] 273, 284, 287, *309*
Ikan, R. [101] 242, 243, *253*
— and M. Cudzinowski [211] 332, *362*
— S. Harel, J. Kashman, and E. D. Bergmann [91] 323, 334, *359*
— J. Kashman, and E. D. Bergmann [103] 242, *253*, *858*
— — S. Harel, and E. D. Bergmann [102] 242, *253*
— I. Kirson, and E. D. Bergmann [309] 37, *188*
— see Bergmann, E. D. [18] 334, *357*
Ikeda, R. M., W. L. Stanley, S. H. Vannier, and L. A. Rolle [104] *253*
— see Stanley, W. L. [268] 210, *257*
Ikekawa, N., T. Kagawa, and Y. Sumiki [35a] 479, *491*
Ikekawa, T. [17, 18] 572, 574, 575, 576, 577, *577*
Ikram, M., and M. K. Bakhsh [93] 425, 433, 448, *465*
— G. A. Miana, and M. Islam [94] 438, 439, 448, *465*
Ikuta, M., see Takeda, K. [281] 213, *257*
Iler, R. K., see Alexander, G. B. [17, 310] 8, 11, 13, 27, *180*, *188*
Iliea, M., see Popova, Y. [109] 302, *310*
Imaichi, K., see Wood, P. [194] 593, *611*

Imelik, B., see Prettre, M. [538] 25, *194*
Imre, M., see Tyihák, E. [302] 237, *258*
Imre, S., see Hörhammer, L. [109] 707, *727*
Inayama, S., see Herz, W. [89] 216, *252*
Inazu, K. [69] 295, 296, *309*
Inglis, G. R. [60] 824, *836*
Inhoffen, see Lettré [108] 311, 313, *360*
Inhoffen, H. H., K.-H. Nordsiek, and H. Schäfer [36] *491*
Inscoe, M. N. [311] 133, *188*
Inubushi, Y., see Tsuda, Y. [297] 242, 243, *258*
Ippoliti, P., see Dati, T. [146] 73, *184*
Irby, B. N., see Dugger, D. L. [173] 13, *184*
Irrevere-Sullivan 291
Isherwood, F. A., see Hanes, C. S. 887
Ishii, S., and B. Witkop [19] *577*
Ishikawa, M., see Takeda, K. [280] 213, *257*
Ishikawa, S., and G. Katsui [70] 259, 263, 264, 265, 266, 277, 292, 293, 294, 297, 303, 304, 305, 307, *309*
— see Katsui, G. [75] 263, 284, *309*
Islam, M., see Ikram, M. [94] 438, 439, 448, *465*
Isler, O., see Mayer, H. [87] 284, *310*
— see Planta, C. v. [108] 273, *310*
— see Rüegg, R. [115] 289, 290, *310*
Ismailov, N. A. 1
— and M. S. Shraiber [313] 1, 2, 73, *188*
Isquierdo, J., see Hörhammer, L. [92] 691, 699, *726*
Ito, M. [105] 226, *253*
Ivars, L., see Schantz, M. V. [193] 714, *729*
Iwata, T., and K. Yamasaki [212] 354, *362*

Jaarma, M., see Hasselquist, H. [61] 305, 306, *309*
Jacin, H., and A. R. Mishkin [6] 808, 811, 814, *835*, 887
Jackson, C. M., see Hanahan, D. J. [55] 374, *417*
Jackson, D. S., see Myhill, D. [218] 740, *785*
Jackson, R. [314a, 788] *188*, *200*
Jackwerth, E., see Heusser, D. [290] 154, 155, *187*; [88b, 88c] 440, 441, *465*

Jacob,J. [106] 243, *253*
Jacobs jr.,C.C., see Ditullio,N.W. [196] 332, *362*
Jacobsohn,G.M. [316, 317] 142, *188*; [92, 93] 350, *359*; [74] 603, *609*
Jacobson,K.B. [34, 35] 793, *805*
Jäger,H., see Barbier,M. [11] 323, 335, 355, *357*
Jaeger-van Moeseke,L.de, see Braeckmann,P. [33, 34] 456, *463*
Jänchen,D. [314] 70, *188*; [37] 476, *491*
Jaenicke,W., and B.Lorenz [315] 133, *188*
Jakovcic,S., see Christian,J.C. [23] 596, *608*
Jakovljevic,I.M., L.D.Seay, and R.W. Shaffer [95] 449, *465*, 861, 874
James,A.T., and L.J.Morris [317a] 49, *188*; [94] 316, *359*
— see Nichols,B.W. [152] 375, *419*
— see Scott,R.P.W. [619] 94, *196*
James,C.N. [795] 150, *200*
Jamieson,G.R. [28] 618, 620, 624, *629*
Janák,J. [318, 319, 320] 89, 103, 117, 149, *188*, 901
— I.Klimeš, and K.Haná [321] 118, *188*
— see Ruseva-Atanasová,N. [178] 398, *420*
Janecke,H., and L.Maass-Goebels [322] 135, *188*; [71] 276, 278, 279, *309*
Janeway,C.M., see Randerath,K. [76] 801, *806*
— see Zamecnik,P.C. [108] 801, *806*
Janiak,B., and H.Böhmert [114] 706, 707, *727*
— see Bohlmann,F. [27] 436, *462*
Janíček,G., see Davídek,J. [10, 11] 613, 616, *628*
Janish,M.A.M., see Papariello,G. [21] 497, *505*
Janistyn,H. [107] 238, *253*
Janot,M.M., see Potier,P. [179] 450, *467*
Janousek,J. [29] 616, *629*
Janssen,E.G. [210] 345, *362*
Jantz,A., see Franzke,Cl. [199] 30, *185*
Jantzen,E., and H.Andreas [68] 402, 403, *417*

Jaquard,S., see Mouton,M. [464] 30, *192*; [175] 255
— see Rahandraha,Th. [555] 30, *194*; [207] 246, *255*
Jarczynski,R., and F.Kiermeier [39] 636, *655*
Jaspersen-Schib,H.P., see Jaspersen-Schib,R. [323] 54, *188*
Jaspersen-Schib,R., and H.Flück [108] 237, 238, 239, *253*
— and H.P.Jaspersen-Schib [323] 54, *188*
Jatzkewitz,H. [324] 148, *188*; [69, 71, 72] 375, 389, 393, 415, *417*; [78] 598, *609*, 871, 874, 899
— and U.Lenz 874
— and E.Mehl [70] 391, *417*; [76] 598, *609*, 860, 900
— H.Pilz, and K.Sandhoff [73] 389, 415, *417*; [79] 598, *609*
— and K.Sandhoff [77] 598, *609*
— see Pilz,H. [164] 393, 415, *420*
Jayme,G., and H.Knolle [325] 34, *188*
Jeanes,A., C.S.Wise, and R.J.Dimler [326] 86, *188*
Jenkin,H.M. [74] 390, *417*
Jennings,W.G., see Wrolstad,R.E. [331] 210, *258*
Jensen,A. [327] 5, *188*
Jensen,E.H., and D.J.Lamb [80] 560, *564*
Jensen,J. [80] 604, *609*
Jensen,R.G., see Komarek,R.J. [103] 414, *418*
Jentsch,J. 730
Jepson,J.B. [81] 582, *609*
— and I.Smith [76] 749, 750, *782*, 881
Jerchel,D., S.Henke, and Kl.Thomas [328] 176, *188*
— and R.Müller [107] 751, *783*, 864
— see Koss,F.W. [378, 379] 52, 173, 179, *190*; [103] 317, 352, *359*
Joe,F.L., see Haenni,E.O. [36] 639, *655*
Jörg,J., see Wessely,F. [90] 677, *686*
Johannesen,B., and A.Sandel [329] 149, *188*
Johanson,R. 857
Johansson,B.G., and L.Rymo [330, 331] 111, 112, *189*; [82, 83] 591, *609*
John,K.V., M.R.Lakshmanan, F.B.Jungalwala, and H.R.Cama [72] 273, 274, 275, *309*

Johne, S., see Gröger, D. [81b] 460, *464*
Johns, T., and C. H. Braithwaite [40] 638, *655*
Johns-Manville [332] 27, 28, *189*
Johnson, C. D., and L. A. Telesz [30] 616, *629*
Johnson, D. B., and T. W. Goodwin [73] 295, 296, *309*
Johnson, G. A., and R. H. McCluer [84] 598, *609*
Johnson, H., see Sawicki, E. [601] 52, 141, *195*
Johnson, M. L., see Bird jr., H. L. [70] 103, 147, 149, 152, 153, *182*; [21] 341, *357*
Johnson, P., see Carter, H. E. [22] 381, 385, *416*
Johnston, J. M., see Brown, J. L. [106] 163, 164, *183*
Jommi, G., P. Manitto, and M. A. Silanos [81] *564*
Jonas, J. [41] 634, *655*
Jones, A. C., see Benesi, H. A. [57] 12, *181*
Jones, C. R. [332a, 789] 132, *189*, *200*
Jones, D., see Davies, B. H. [27] 267, *308*
Jones, D. F. [38] 489, *491*
— J. MacMillan, and M. Radley [39] 479, 486, 487, 489, *491*, 900
— see Elson, G. W. [15] 479, 487, 489, *490*
Jones, E. R. H., and T. G. Halsall [109] 207, *253*
Jones, G. S., see Lau, H. L. [107] 340, *360*
Jones, J. B., see Vlasinich, V. [233] 323, *362*
Jones, J. K. N., see Hulyalkar, R. K. [75] 828, *836*
— see Szarek, W. A. [41] 818, *835*
Jones, J. W., see Broom, A. D. [7] 794, *804*
Jones, K. C. [39a] 489, *491*
Jones, L. L., see Schmid, H. H. O. [185] 377, 383, 384, *420*
Jones, L. R., and J. A. Riddick [42] 639, *655*
Jones, R. L., and J. D. J. Phillips [40] 475, *491*
Jones, W. A., M. Beroza, and E. D. Becker [115] 711, *727*

Jong, K., de, K. Mostert, and D. Sloot [110] 222, *253*
Jórgensen, C., and H. Kofod [116] 692, 693, *727*
Jørgensen, P. F. [61] 825, *836*
Jork, H. [333, 333a, 796] 70, 138, 143, *189*, *200*; [111, 112, 113, 114] 209, 227, 234, 237, 239, *253*; [41, 42] 485, *491*; [117] 711, *727*
— see Stahl, E. [260, 261, 262] 214, 226, 234, 237, 239, *257*; [229c] 438, *468*
Josefsson, L. [36] 792, 793, 794, 795, 801, *805*
Joshi, R. K., see Pijewska, L. *470*
Joska, J., see Černý, V. [123] 103, 149, *183*; [32] 315, 322, 330, 335, 347, *358*
Jost, K., J. Rudinger, and F. Šorm [42] 740, *782*
Joule, J. A., and C. Djerassi [96] 460, *465*
Joux, J. L. [43] 636, *655*
Judas, O., see Crépy, O. [138] 32, *183*; [44] 324, 357, *358*, 886, 895
Jumar, A., see Runge, F. [76] 683, *686*, 877
Jung, L., Ch. Bourgoin, J. C. Foussard, P. Audrin, and P. Morand [95] 351, *359*
— see Audrin, P. [7] 341, *357*
Jungalwala, F. B., see John, K. V. [72] 273, 274, 275, *309*
— see Lakshmanan, M. R. [81] 274, *310*
Jungbeck, J. [31] 616, 620, 622, *629*
Jurriens, G., and A. C. J. Kroesen [75] 398, *417*
— B. de Vries, and L. Schouten [76] 414, *417*
— see Koch, G. K. [355] 173, 175, *189*
— see Meijboom, P. W. [164] 221, *254*
— see Vries, B. de [208] 396, *421*
Jutisz, M., see Acher, R. [64] 749, 750, *782*
— see Llosa, P. de la [216] 740, *785*
Juva, K., see Haahti, E. [70] 316, 322, 333, *359*; [52] 376, 398, *417*; [66] 599, *609*
Juvonen, S. [115, 116] 211, 239, *253*
— see Schantz, M. v. [223] 211, *256*

Kabat, E. A., see Marcus, D. M. [40] 818, *835*

Kaćl, K., see König, J. [88] *564*
Kämmerer, H., see Haub, H.-G. [30] 658, *685*
Kämpf, G. 8
— and H.W. Kohlschütter [334] 16, 17, 19, *189*
— see Boehm, H.-P. [77] 12, *182*
— see Kohlschütter, H.W. [361] *189*
Kaempgen, D., see Ronkainen, P. [212] 220, *256*
Käser, H., and E. Gugler [86] 587, *609*
— and G. Masera [85] 579, *609*
Kaffenberger, T., see Seiler, H. [26] 847, *853*, 860, 897, 905
Kafka, F., see Procházka, V. [180b] 451, *467*
Kagawa, T., T. Fukinbara, and Y. Sumiki [43] 479, 487, *491*
— see Ikekawa, N. [35a] 479, *491*
Kahlke, W., see Klenk, E. [95] 594, *609*
Kaindl, K., see Bancher, E. [38] 154, *181*; [47] 819, 820, 826, 827, 834, *836*
— see Scherz, H. [604] 104, *195*
Kainer, F. [335] 27, *189*
Kaiser, Ch., see Birkofer, L. [71] 73, *182*; [20, 21] 690, 691, 700, 704, *725*; [23] 813, *835*
Kaiser, F., and A. Popelak [97] 448, *465*
Kaiser, H., F. Biedebach, and C. Manns [98] 433, *465*
— and G. Hasenmayer 868
Kaiser, R. [336, 337] 115, 117, *189*; [12, 13] 503, *505*
Kaiser, W., see Libbert, E. [71] 488, *492*
Kaisin, M. see Braekman, J.C. *470*
Kajganović, V., see Bićan-Fišter, T. [69] 148, 149, 154, *181*; [18] 544, *562*, 866
Kakáč, B., see Macek, K. 891
Kaldewey, H. [44, 45, 46] 473, 474, 476, 477, 479, 484, 488, *491*
— and E. Stahl [47] 483, 488, *491*
— — and J. Fuchs [48] 478, *491*
— see Stahl, E. [676] *197*; [120] 473, 476, 477, 478, 483, *493*, 869, 879
Kaley, G., see Samuel, P. [144] 341, *360*; [136] 603, *610*
Kalnins, K., see Collins, R.P. [34] 221, *251*
Kalnitzky, G., see Macmahon, J.M. 858

Kalpathy, K., see Auterhoff, H. [7] 456, *462*
Kaltenbach, U. [14] 495, *505*
— see Stahl, E. [670] 87, *197*; [154] 323, 344, *361*; [204] 694, 710, *729*; [1] 807, 808, 810, 812, 813, 833, *834*, 857
Kamikubo, T., see Hayashi, M. [62] 302, 303, *309*
Kamikura, M., see Fujii, S. [87] *630*; [16, 17] 613, 617, *628*
Kammereck, R., see Mangold, H.K. [426, 427] 140, 159, 162, 169, 179, *191*; [119, 121] 377, 378, 387, 388, 403, 404, 405, 406, 411, 412, 414, *418*; [51] 674, *685*
— see Tuna, N. [725] 171, *198*
Kamp, W., W.J.M. Onderberg, and W. A. van Seters [99] 422, 438, *465*
Kanaoka, Y., see Daly, J.W. [4a] 499, *505*
Kapadia, G.J., and G.S. Rao [20] 573, *577*
Kaplan, N.O., see Colowick, S.P. [13] 789, 791, 795, *804*
Kappeler, H., and R. Schwyzer [142] 754, *784*
— see Schwyzer, R. [139a] 753, *784*
Kapur, K.K., see Bhatnagar, J.K. [68] 56, *181*
Karácsony, E.M., and B. Szarvady [100] 451, *465*
— see Wolf, L. [266] 451, 453, *469*
Karamustafavglu, V., see Lehmann, J. [93] 534, 536, *564*, 884
Karawya, M.S., and S.K. Wahba [116a, 116b] 237, 239, *253*
— see El-Deeb, S.R. [56] 237, *252*
Karczmar, A.G., see Fels, G.I. [51] 508, *563*
Karlson, P., R. Maurer, and M. Wenzel [338] 149, *189*
Karmen, A., see Horning, E.C. [65] 395, *417*
Karpitschka, N. [82] 542, 544, *564*
Karrer, P., see Deyrup, J..A. [44] 447, *463*
— see Hesse, M. [88] 447, *464*
— see Nagarajan, K. [144] 447, *466*
— see Weissmann, Ch. [259, 260, 261] 447, *469*

Karrer, W. [117] 207, *253*; [118] 687, *727*
Kartnig, Th. [119] 240, *253*
— F. J. Graune, and R. Herbst [118] 245, *253*
Kashman, J., see Ikan, R. [102, 103] 242, *253*; [91] 323, 334, *359*, 858
Kassalitzky, H., see Ullmann, E. [249] 449, *469*
Katague, D. B., and E. R. Kirch [120] 215, 237, *253*
Kath, D. 871
Katsui, G. [76, 77] 259, 265, 273, 274, 287, 292, 304, 306, 307, *309*
— Y. Ichimura, and Y. Nishimoto [74] 273, 284, 287, *309*
— S. Ishikawa, M. Shimizu and Y. Nishimoto [75] 263, 284, *309*
— see Ishikawa, S. [70] 259, 263, 264, 265, 266, 277, 292, 293, 294, 297, 303, 304, 305, 307, *309*
Katz, A. M., W. J. Dreyer, and C. B. Anfinsen [122] 752, *783*
Katz, D., and I. Lempert [44] 639, *655*
Katzenmayer, W., see Kohlschütter, H. W. [365] *189*
Kauffmann, Th., J. Hansen, and R. Wirthwein [15] 501, *505*
Kaufmann 49
Kaufmann, G., see Bleecken, S. [74] 162, *182*
Kaufmann, H., W. Wehrli, and T. Reichstein [95a] 342, *359*
Kaufmann, H. P. [77] 367, 368, *417*
— and J. Budwig 896
— and B. Das [85, 86] 381, 411, 412, *418*
— and A. K. sen Gupta [121, 122, 123, 124] 240, 241, *253*, 858
— and T. H. Khoe [340] 30, 52, *189*; [84] 412, *418*
— and Y. S. Ko [82] 386, 411, 412, *418*
— and Z. Makus [339] 88, *189*; [78] 375, 377, 378, 386, 410, 411, 412, *418*
— — and B. Das [81] 401, 411, 412, *418*
— — and F. Deicke [96] 317, 323, 331, 332, 333, *359*; [79] 412, *418*; [87] 594, *609*
— — and T. H. Khoe [97] 326, 331, *359*; [80, 83] 411, 412, 413, *418*
— and K. D. Mukherjee [341] 140, *189*

Kaufmann, H. P., S. S. Radwan, and A. K. S. Ahmad [91] 390, 415, *418*
— and C. V. Viswanathan [87] 386, *418*; [88, 89] 597, 599, *609*
— and H. Wessels [89, 90] 401, 412, 413, 414, *418*
— — and C. Bondopadhyaya [88] 398, 401, *418*
Kaufmann, M., see Fels, G. I. [51] 508, *563*
Kaul, J. L. 435
— see Pijewska, L. *470*
Kausz, M., see Prey, V. [74] 650, 654, *656*; [70] *686*; [24, 50, 80] 814, 817, 820, 828, *835*, *836*
Kawaguchi, K., see Tsuda, Y. [297] 242, 243, *258*
Kawano, N., H. Miura, and H. Kikuchi [119] 690, *727*
Kawasaki, T., and K. Miyahara [125] 245, *253*
— and K. Miyahara [98] 346, 349, *359*
Kawashiro, I., and Y. Hosogai [44a] 645, *655*
Kawerau, E., and Th. Wieland [49] 747, *782*, 889
Kay, H. L., and F. L. Warren [99] 356, *359*
Kazuno, T., and T. Hohita [100] 355, *359*
Kean, E. L. [90] 598, *609*
Keck, K., and U. Hagen [342] 111, 112, 114, *189*; [37] 794, 803, *805*
Keel, H., see Hull, W. Q. [308] 28, *188*
Keeler, C. E., see Mellinger, T. J. [101] 508, 509, 514, *564*
Keenan, R. W., see Redman, C. M. [173] 390, *420*
Keeney, M., see Schwartz, D. P. [232] 222, *256*
Kefford, N. P. [49] 472, *491*
— and P. L. Goldacre [50] 472, *491*
— see Bennet-Clark, T. A. 893
Keil, M., see Egger, K. [41, 42] 700, 701, 703, 704, *725*
Keirs, R. J., R. D. Britt jr., and W. E. Wentworth [343] 141, *189*
Keith, R. W., D. le Turneau and D. Mahlum 878
Kelemen, J., and G. Pataki [344] 125, *189*
Kelleher, J., and J. G. Rollason [83] 536, *564*

Kelleher, W. J., see El-Olemy, M. M. [150b] 431, *466*
Kellen, J. 899
Keller, G. J., see Kirchner, J. G. [347, 348, 349] 3, 146, 149, *189*; [104] 437, *465*
— see Rice, R. G. 889
Keller, M., and G. Pataki [91, 92] 583, 586, *609*
— see Pataki, G. [504] 127, *193*; [117, 119, 120, 122] 579, 583, 585, 588, *610*
Keller, R. A., see Giddings, J. C. [114] 137, *183*
Keller-Schierlein, W., and G. Roncari [21] 572, *577*
Kellie, A. E., see Burger, H. G. [27] 339, *358*; [18] 602, *607*
Kelly, W., see Morris, L. J. [145] 399, *419*
Kemény, V., see Vecsei, P. [176] 324, *361*
Kende, H., and A. Lang [51] 475, 489, *491*
— see Ninnemann, H. [83] 487, 489, *492*
Kenney, J., see Hull, W. Q. [308] 28, *188*
Kent, J. R., see Burger, H. G. [27] 339, *358*: [18] 602, *607*
Kent, P. W., see Greenway, R. M. [18] 811, *835*
Kentmann, E. H., see Burton, R. B. [86] 750, *783*
Kentmann, H., see Zaffaroni, A. 882
Kern, W., K. J. Rauterkus, and W. Weber [36] 658, *685*
Kerr, J. D., see Pennock, J. F. [104] 283, 284, 285, 286, 287, *310*
Kesselring, K., and G. Siebert [38] 793, 794, 801, *805*
Kessler, W. V., see Seno, S. [150] 561, *565*
Keulemans, A. I. M. [345] 115, *189*
Keymer, R., see Paulus, W. [123, 124] 508, 509, 514, 539, 540, *565*
Khafagy, S., A. M. El-Moghazy, and F. Sandberg [101] 434, 435, *465*
Khajuria, G. S., see Sharma, R. K. [214b] 435, *468*
Khalifah, R. A., L. N. Lewis, and C. W. Coggins jr. [52] 489, *491*
— — — and P. C. Raddick [53] *491*

Khalifah, R. A., see Lewis, L. N. [70] 487, *492*
Khanna, K. L., A. E. Schwarting, A. Rother, and J. M. Bobbitt [102] 434, *465*
— see Leary, J. D. [120] 434, *465*
— see Schwarting, A. E. [211] 434, *468*
Khare, M. P., see Srivastava, R. M. [221, 222] 460, *468*
Kheifits, L. A., G. J. Moldovanskaya, and L. M. Shulov [126] 230, *253*
Khin, L., see Szász, G. [238] 423, *469*
Kho, B. T., and S. Klein [84] 543, 544, *564*
— see Klein, S. [86] 542, 544, *564*
Khoe, T. H., see Kaufmann, H. P. [340] 30, 52, *189*; [97] 326, 331, *359*; [80, 83, 84] 411, 412, 413, *418*
Khorana, H. G., see Lohrmann. R. [42] 794, *805*
Khorlin, A. Y., L. V. Bakinovskiĭ, V. E. Vas'kovskiĭ, A. G. Venyaminova, and Y. S. Ovodov [127] 243, 245, *253*
— Y. S. Ovodov, and N. K. Kochetkov [128] 245, *253*
— see Kochetkov, N. K. [135] 243, 245, *254*
Kiburis, J., see Miras, C. [146] 716, *728*
Kido, T., see Fuwa, T. [58, 59] 543, 551, *563*
Kiel, E. G., and G. H. A. Kuypers [32] 622, *629*
Kiermeier, F., see Jarczynski, R. [39] 636, *655*
Kiermayer, O., see Linser, H. [72] 487, *492*
Kikuchi, H., see Kawano, N. [119] 690, *727*
Kilger, F., see Hörhammer, L. [215] 740, *785*
Kim, M. J., see Sgoutas, D. S. [629] *196*
Kimble, H., see Snyder, F. [647] 149, 165, 166, 176, *197*
Kindl, H. [120] 719, *727*
— see Billek, G. [17] 692, 693, *725*
Kingdon, F., and R. E. Schranz [346] 131, *189*
Kinoshita, S. [92] 387, *418*
— and M. Oyama [93] 387, 414, *418*
Kinsell, L., see Wood, P. [194] 593, *612*
Kinsella, R. A., see Chang Shen, N.-H. [35] 338, 340, *358*; [152] 602, *611*

Kint, J., see Sumere, C. F. van [158] 540, *566*, 890
Kinze, W. [103] 423, *465*
— see Poethke, W. [535, 536] 139, 149, 151, 153, *194*; [174, 175, 176, 177] 425, 438, 440, *467*; [131] 543, *565*
Kirby-Berry, H., H. E. Sutton, L. Cain, and J. S. Berry [83] 750, 751, *782*
Kirch, E. R., see Katague, D. B. [120] 215, 237, *253*
Kirchner, J. G. [350] 31, *189*
— and G. J. Keller [347] 3, *189*
— — and J. M. Miller [348] 3, 146, *189*
— — and R. G. Rice [349] 3, 146, 149, *189*
— J. M. Miller, and G. J. Keller [104] 437, *465*
— see Miller, J. M. [446, 447, 448, 449] 3, 54, 97, *192*; [168] 206, 210, 216, *254*; [126] 325, *360*
— see Rice, R. G. 889
Kirchner, M. A., and M. B. Lipsett [93] 602, *609*
Kirk, M. C., see Kritchevsky, D. 887
Kirk, R. E., and D. F. Othmer [6] 732, *781*
Kirkpatrick, J. L., see Douglas, B. [49] 460, *463*
Kirsch, E. R., see Borke, M. L. [31] 437, *463*
Kirsch, K., see Zöllner, N. [204] 594, 595, 596, *612*
Kirson, I., see Ikan, R. [309] 37, *188*
Kishimoto, I., and N. S. Radin [94] 387, *418*
Kisser, W., see Gendi, S. E. [70] 518, *563*
— see Machata, G. [98] 536, 537, *564*
Kitamori, N., see Mima, H. [63] 649, *656*
Klaus, R. [351] 138, 141, *189*
Klavehn, M., and H. Rochelmeyer [352] 135, *189*; [105] 451, *465*
— — and J. Seyfried [106] 451, 453, *465*; [85] 562, *564*
Klein, E., W. Rojahn, and D. Henneberg [129] 212, *253*
Klein, S., and B. T. Kho [86] 542, 544, *564*
— see Kho, B. T. [84] 543, 544, *564*
— see Koller, D. [59] 473, *491*

Kleine, K.-M., see Bohlmann, F. [15, 16] 236, *251*
Kleinig, H., see Egger, K. [41] 265 286, 289, 291, *309*
Klement, R., and A. Wild [37] 678, *685*
Klenk, E., and M. Doss [97] 394, *418*
— and W. Gielen [95, 96] 389, 393, 394, *418*; [94] 598, *609*
— and W. Kahlke [95] 594, *609*
— W. Kunau, L. Hof, and L. Georgias [98] 394, *418*
Kleuver, G. J. de, see Dam, M. J. D. van [175] 317, 329, *361*
Klimeš, I., see Janák, J. [321] 118, *188*
Klingelhöller, W., see Fischer, R. [54] *563*; [31] 639, *655*
Klingenberg, M., see Goebell, H. [34] 654, *655*
Klinger, H., see Brieskorn, C. H. [25, 26] 243, 245, *251*
Klingmüller, L. [130] 244, 247, *253*
Klingmüller, V., see Detter, F. [27] 602, *608*
Klöcking, H.-P. [86a] 537, *564*
Kloppenborg, P. W. C., see Benraad, T. J. [17] 341, *357*; [191] 340, 341, *362*; [8] 601, *607*
Klose, A., see Geiss, F. [225, 226, 227] 66, 67, 70, 71, 102, 150, *186*
Klotz, L., see Hüttenrauch, R. [305] 46, *188*; [67] 292, 293, 294, *309*;
Klouwen, M. H., and R. ter Heide [131] 230, 231, *254*
— — and J. G. J. Kok [132] 217, 223, *254*
— and H. Weiffenbach [39] 801, *805*
Klügel, G., see Neidlein, R. [109] 550, 551, *564*
Knabe, J., and J. Kubitz [107] 444, *465*
— and N. Ruppenthal [108] 444, *465*
Knapp, R. [54, 55] 472, 473, 475, 488 *491*
Knappe, E., and D. Peteri [133] 214, *254*; [45, 46, 48, 49] 650, 651, 652, *655*; [38, 39] 664, *685*, 860, 870, 877
— — and I. Rohdewald [47] 650, *655*; [40, 41] 662, 664, *685*, 861
— and J. Rohdewald [133a] 233, *254*; [87] 540, *564*; [41a] 663, *685*; [121] 697, *727*
Knauff, H. G., H. Selmair, and H. Zickgraf [40] 738, *782*

Knedel, M., and A. Fateh-Moghdam [353] 90, *189*
Knight, C. A., see Rushizky, G. W. [82] 801, *806*
Knight, C. S. [354] 128, *189*
Knobloch, E. [78] 259, *309*
Knolle, H., see Jayme, G. [325] 34, *188*
Knowles, J., see Wood, P. [194] 593, *611*
Knütter, S., and R. Pohloudek-Fabini [134] 235, 236, *254*
Ko, Y. S., see Kaufmann, H. P. [82] 386, 411, 412, *418*
Koch, F., see Feltkamp, H. 882
Koch, G. K., and G. Jurriens [355] 173, 175, *189*
— see Alderhout, J. J. H. [16] 176, *180*
Koch, H., see Fischer, F. [30] 650, *655*; [24] 680, *685*
Koch, H. J., see Rosenthal, A. [92] 831, *837*
Kochetkov, N. K., A. Y. Khorlin and V. E. Vas'kovskii [135] 243, 245, *254*
— I. G. Zhukova, and I. S. Glukhoded [99, 100, 101] 392, *418*
— see Khorlin, A. Y. [128] 245, *253*
Kodak-Pathé [356] 59, *189*
Kodicek, E., and K. K. Reddi 859
— see Fraser, D. R. [47] 277, 278, 280, *309*
— see Murray, T. K. [91] 278, *310*
Kögl, F., A. J. Hagen-Smit and H. Erxleben [56] 472, *491*
Köhler, D., and A. Lang [57] 488, *491*
Koehler, F., see Runge, F. [76] 683, *686*, 877
Köhler, K. H., and K. Conrad [58] 475, *491*
Köhler, M., H. Golder, and R. Schiesser [357] 37, *189*; [42] 667, *685*
Köhler, P., and H. Baufeld [96] 590, *609*
Koehler, W. R., J. L. Solan, and H. T. Hammond [102] 412, *418*
König, A., see Bolliger, H. R. [9, 10] 264, 266, 267, 268, 275, 276, 279, 280, 281, 282, *308*; [21] 558, *562*
König, H., see Hörhammer, L. [100] 719, *727*
— see Schildknecht, E. [87] 648, *656*
König, H. W., see Hörhammer, L. [93] 237, 238, *253*
König, J., I. Hynie, and K. Kaćl [88] *564*

Kofler, L. [358] 81, *189*
Kofler, M., P. F. Sommer, H. R. Bolliger, B. Schmidli, and M. Vecchi [79] 284, *309*
Kofod, H., see Jórgensen, C. [116] 692, 693, *727*
Kofoed, J., see Korczak-Fabierkiewicz, C. [88 a] 508, *564*
Kofrányi, E. [68] 749, *782*
Kohan, S., and J. Fitelson [136] 217, *254*
Kohen, F., B. K. Patnaik, and R. Stevenson [137] 243, *254*
Kohlhepp, E. [359] *189*
Kohlschütter, H. W. [362] 8, 23, *189*
— P. Best, and G. Wirzing [360] 13, 20, *189*
— and M. Daum [367] *189*
— H. Getrost, and S. Miedtank [363] 23, *189*
— and G. Hofmann [364] 23, *189*
— and G. Kämpf [361] *189*
— and W. Katzenmayer [365] *189*
— A. Risch, K. Unger, and K. Vogel [366] 23, *189*
— and K. Unger 7
— see Kämpf, G. [334] 16, 17, 19, *189*
Kok, J. G. J., see Klouwen, M. H. [132] 217, 223, *254*
Kokoti-Kotakis, E. [368] 5, *189*
— see Synodinos, E. [278] 217, *257*; [93] *630*
Kolb, J. J., see Toennies, G. [55] 749, *782*, 891
Koller, A., and H. Neukom [32] 816, *835*
Koller, D., A. M. Mayer, A. Poljakoff-Mayber, and S. Klein [59] 473, *491*
Kolman, Z., see Michalec, C. [105] 596, *609*
Koloušek, J., M. Kutáček, and J. Bílek 881
Kolšek, J. [33, 34] 616, 620, *629*
— see Ciglar, J. [6] 618, *628*
Komarek, R. J., R. G. Jensen, and B. W. Pickett [103] 414, *418*
Kometani, K., see Tschesche, R. [245] 422, 432, *469*
Komori, T., see Tsukamoto, T. [48] 747, 749, *782*
Kompiš, I., see Mokrý, J. [134, 135, 136] 449, *466*

Kondo, S. [22] *577*
Konitzer, K., see Voigt, S. [171] 586, *611*
Konopka, W., see Reinhard, E. [100] 489, *493*
Konrad, E., O. Bächle, and R. Signer [369] 8, *189*
Koo, W. Y., see Lou, V. [74] *492*
Korczak-Fabierkiewicz, C., J. Koford, and G. H. W. Lucas [88a] 508, *564*
Korenman, S. G., H. Wilson, and M. B. Lipsett [97] 602, *609*
Korey, S. R., C. J. Gidez, A. Stein, J. Gonatas, and K. Suzuki [104] 393, 394, *418*
Korn, E. D., see Davidoff, F. [150] 177, *184*
Korn, O., and H. Woggon [43] 665, *685*
Kornhauser, A., S. Logar, and M. Perpar [122] 718, *727*
— and M. Perpar [108b] 454, *465*
Koronelly, T. V., see Kost, A. N. [60] 477, *491*
Korte, F. [138] *254*
— H. Barkemeyer, and J. Korte [139] *254*
— and H. Sieper [370] 154, *189*; [123, 124, 125] 715, 716, *727*
— — and S. Tira [126] 715, 716, *727*
— and J. Vogel [140] 215, *254*; [44] 683, *685*, 891
— see Claussen, U. [28] 715, *725*
— see Longo, R. [134] 707, *727*
— see Sieper, H. [189] 701, *728*
Korte, J., see Korte, F. [139] *254*
Kortüm, G., W. Braun, and G. Herzog [373] 143, *190*
— and J. Vogel [371, 372] 143, *190*
Korzun, B. P., and S. Brody [375] 75, *190*; [101] 341, 351, *359*; [89] 557, *564*
— — and F. Tishler [90] 541, *564*
— L. Dorfman, and S. M. Brody [374] 98, *190*; [102] 349, *359*; [109] 454, *465*
Koss, F. W., G. Beisenherz, U. Chuchra, and I. Huber [377] 178, *190*
— — R. Engelhorn, and U. Chuchra [376] 154, 179, *190*
— and D. Jerchel [378, 379] 52, 173, 179, *190*; [103] 317, 352, *359*

Koss, F. W., see Gänshirt, H. [212] 147, 149, 150, 154, *185*; [62] 323, 352, 353, 354, *358*
Kossmann, K., see Loth, H. [136] 709, *727*
Kost, A. N., T. V. Koronelly, and R. S. Sagitullin [60] 477, *491*
Katakis, G., see Synodinos, E. [278] 217, *257*; [93] *630*
Kottke, B. A., J. Wollenweber, and C. A. Owen [379a, 797] 150, *190*, 200
Kováč, J [50] 655
Kovacs, M. F. [51, 52] 640, 642, 643, 644, *655*, 898
Kovacs, Ö., see Göndös, Gy. [66] 328, 341, 349, *359*
Koval, G. J., see Fahn, S. [22] 795, *804*
Kovář, J., see Bláha, K. [22] 443, *462*
Kovařík, L., see Urx, M. 856
Kowarski, S., see Schachter, D. [117] 277, 278, 280, *310*
Kowitz, F., see Tschesche, R. [245] 422, 432, *469*
Kozuka, H. [110] 441, *465*
Kraczkowski, H., see Opienska-Blauth, [492] 5, *193*; [90] 486, *492*; [115, 116] 582, 584, *610*
Krämer, H., see Schildknecht, H. [194] 719, *729*
Kramer, J. K. G., E. O. Schiller, H. D. Gesser, and A. D. Robinson [380] *190*
Kratzl, K. [382] 178, *190*
— and G. E. Miksche [141] 230, *254*; [128] 712, *727*
— and G. Puschmann [381] 178, 179, *190*; [127] 698, *727*
Kraus, L. [130] 701, *727*
— and D. Dupaková [129] 701, *727*
Krause, G., see Dose, K. [170] 111, *184*
Krause, N., see Schön, H. [146] 599, *610*
Krause, U., see Meckel, L. [161] 238, 250, *254*; [42] 620, 622, 623, *629*
Kraut, H., see Willstätter, R. [760] 8, *199*
Krautheim, J., see Schmid, E. [104] 486, *493*; [143a] 590, 591, *610*
Krell, K., and S. A. Hashim [383] 154, *190*; [105] 414, *418*; [98] 593, *609*
Křepinský, J., M. Romańuk, V. Herout, and F. Šorm [142] 239, *254*

Kreutz, F. H., see Eggstein, M. [38] 593, *608*
Kreutzig, L., see Böhme, H. [24, 25, 26] 690, 706, 707, *725*
Krieger, H. [45] 677, *685*
Krieger, W., see Helmcke, J.-G. [286] 27, *187*
Kringstad, K. [22] 812, 813, *835*
Kristerson, L., see Hansson, E. [272] 179, *187*
Kritchevsky, D., and M. C. Kirk 887
— D. S. Martak, and G. H. Rothblatt [104] 323, 354, *359*
— see Paoletti, R. [158] 367, *419*
Kritchevsky, G., see Murphy, M. T. J. [147] 380, *419*
— see Rouser, G. [177] 389, 391, 393, *420*
Kröller, E. [53, 54] 649, *655*
Kroesen, A. C. J., see Jurriens, G. [75] 398, *417*
Krüll, H., see Neidlein, R. [108] 548, *564*
Krusche, B., see Habermann, E. [258] 148, *187*; [53] 415, *417*; [67] 596, *609*
Krzeminsky, L. F., and W. A. Landmann 886
Kubeczka, K.-H. [143] 220, *254*
Kubitz, J., see Knabe, J. [107] 444, *465*
Kučera, J. [55] 650, *655*
Kuch, H., see Schlossberger, H. G. [103] 477, *493*; [141] 584, *610*
Kucharczyk, N., and J. Fohl [47] 667, 669, *685*
— — and J. Vymétal [16] 504, *505*; [46] 667, 668, *685*, 879, 901
Kuduk, J., see Cieślak, J. [38] 431, *463*
Kudzin, S. F., R. M. de Baun, and F. F. Nord [115] 751, *783*
Kübler, H., see Butenandt, A. [112] 43, *183*
Kühn, G., see Müller, K. H. [149] 707, *728*
Kühn, L., and S. Pfeifer *461*
— see Pfeifer, S. *461*
Kuhn, A. [35] 627, 628, *629*
Kuhn, D., see Weicker, H. [183] 580, *611*
Kuhn, H., see Pailer, M. [188] 219, 220, *255*

Kuhn, H. J. [111] 426, *465*
— see Neumüller, O. A. [149] 426, *466*
Kuhn, M., and A. v. Wartburg [131] 710, *727*
— see Renz, J. [176] 710, *728*
— see Wartburg, A. v. [230] 710, *730*
Kuhn, R. 4
— and I. Löw [144] 243, *254*
— and H. Wiegandt [107] 389, 392, 394, *418*
— — and H. Egge [106] 389, 394, *418*
Kukkonen, E., see Schantz, M. V. [193] 714, *729*
Kuksis, A., and W. C. Breckenridge [109] 398, 414, *418*
— and J. Ludwig [108] 398, 414, *418*
Kulenda, Z., and E. Horáková [213] 340, *362*; [99] 601, *609*
Kulovich, M. V., see Gluck, L. [58] 599, *608*
Kumari, G. L., see Nigam, S. S. [181] 237, 238, 239, *255*
Kummer, K., see Bleecken, S. [74] 162, *182*
Kummerow, F. A., see Sgoutas, D. S. [628, 629] 176, 177, *196*
Kump, C., J. Seibl, and H. Schmid [114, 115, 115b] 447, *465*
Kump, W. G., and H. Schmid [112, 113] 447, *465*
— see Govindachari, T. R. [77] 447, *464*
Kump, W. H., see Pailer, M. [154] 430, *466*
Kunau, W., see Klenk, E. [98] 394, *418*
Kunovits, G. [145] 239, *254*
Kuntz, E. [48] 658, *685*
Kunz, A., see Pataki, G. [53] 794, *805*
Kupfer, W. [49] 657, *685*
Kupferberg, H. J., A. Burkhalter, and E. L. Way [116] 438, *465*; [91] 529, 532, *564*, 876
Kurek, E., see Oehlschläger, H. [116] 509, *565*
Kuroiwa, Y., and H. Hashimoto [145a] 247, *254*; [132] 714, *727*
Kuschke, H. J., see Schmid, E. [143] 590, *610*
Kuster, W., see Schetty, G. [64] 623, *629*
Kutáček, M. [61, 62] 472, 479, *491*
— and Ž. Procházka [63] 472, 474, 479, 486, *491*

Kutáček, M., J. Rosmus, and Z. Deyl [64] 479, 487, *491*
— see Kloušek, J. 881
Kuwada, S., and M. Hori [80] 297, *310*
Kuypers, G. H. A., see Kiel, E. G. [32] 622, *629*
Kuznetsova, A. I., see Achrem, A. A. [8, 9] 5, 42, 43, *180*; [4] 335, *357*
Kuznetsova, E. N., see Bogoslovsky, N. A. [7] 277, *308*
Kuzuya, T., E. Samols, and R. H. Williams [384] 179, *190*
Kyburz, E. [23] 574, *577*

Lábler, L. [385] 97, *190*; [106] *360*
— and V. Černy [105] 328, 349, *359*; [117] 459, *465*
— and Vl. Schwarz [386] 5, *190*
— see Černý, V [123] 103, 149, *183*; [32] 315, 322, 330, 335, 347, *358*
Lachese, B., see Crépy, O. [138] 32, *183* [44] 324, 357, *358*, 886, 895
Ladany, S., and M. Finkelstein [100] 602, *609*
Ladd, F. C., see Anderson, N. G. [1] 795, *804*
Lagercrantz, C., see Bersgtröm, G. [11] 212, *250*
Lagoni, H., and A. Wortmann [36] 612, *629*
Laiho, S. M., and H. M. Fales [117b] 446, *465*
Laitinen, L., see Schantz, M. V. [193] 714, *729*
Lakshmanan, M. R., F. B. Jungalwala, and H. R. Cama [81] 274, *310*
— see John, K. V. [72] 273, 274, 275, *309*
Lam, J., see Berg, A. [3] 667, *684*
Lamb, D. J., see Jensen, E. H. [80] 560, *564*
Lambert, B. F., see Bartlett, M. F. [12] 449, *462*
Lambertsen, G. [82] 284, 285, *310*
Lambie, D. A. [387] 155, 164, 170, *190*
Lamp, B. G., see Mangold, H. K. 866
Lampert, F., see Tschesche, R. [286, 293] 242, 243, 244, 247, *257*
Land, E. H. [388] 131, *190*
Landmann, W. A., M. P. Drake, and J. Dillaha [190] 774, 777, *785*
— see Krzeminsky, L. F. 886
Lane, E. S. [17] 496, *505*, 864

Lane, K., see Chen jr., P. S. [15] 275, 276, 280, *308*
Lang, A. [65] 473, *491*
— see Harada, H. [29] 489, *490*
— see Kende, H. [51] 475, 489, *491*
— see Köhler, D. [57] 488, *491*
— see Ninnemann, H. [83] 487, 489, *492*
— see Ruddat, M. [587] *195*
— see Skene, K. G. M. [118a] 489, *493*
Lang, D., see Wagener, H. [172] 596, *611*
Langbein, A., see Eistert, B. [23] 684, *685*
Langbein, J. W., see Eberhardt, W. [43] 533, 534, 536, 539, *563*
Lange, N. A. [92] 523, *564*
Langemann, A., see Weber, S. H. [171] 541, *566*
Langham, J., see Hansbury, E. [269] 131, *187*
Langhammer, L., see Hänsel, R. [75] 697, *726*
Lanz, P., see Vogler, K. [138, 146] 753, *784*
Lapière, C. L., see Adam, E. [1] 547, *562*
Lapina, T. G. [146] 230, *254*
Larcebau, S., see Castagnou, M. R. [35b] 460, *463*
Laroche, C., see Ruiz, S. L. [62] 613, 622, 624, *629*
Larose, J. A. G., see Sims, R. P. A. [636] 147, 148, *196*
Larsen, P. [66, 67] 473, 474, 487, *491*
Latinák, J. [50] 675, *685*
Lau, H. L., and G. S. Jones [107] 340, *360*
Laudi, B., see Schmid, E. [143a] 590, 591, *610*
Laurent, H., see Bohlmann, F. [27] 436, *462*
Lautenbach, L., see Geldmacher-Mallinckrodt, M. [69] 540, *563*
Lautner, H., see Fischer, R. [14] *577*
Lavie, D., and B. S. Benjaminov [147] 243, 246, *254*
Laws, D. R. J., see Ashurst, P. R. [4] 714, *724*
Lawson, D. D., see Getz, H. R. [230] *186*

Lawson, W. B., see Zachau, H. G. [107] 801, 802, *806*
Lay, B., see Hörhammer, L. [91] 247, *252*; [95, 101] 716, 717, *726, 727*
Layne, D. S., see Golab, T. [64] 341, *358*
Lea, C. H. [148] 254. *254*
Leaback, D. H., see Greig, C. G. 863
Leary, J. D., J. M. Bobbitt, A. Rother, and A. E. Schwarting [119] 431, *465*
— K. L. Khanna, A. E. Schwarting, and J. M. Bobbitt [120] 434, *465*
— see Schwarting, A. E. [211] 434, *468*
Leat, W. M. F., see Bowyer, D. E. [13] 592, *607*
Le Beau, J., see Sunshine, I. 884
Lebeau, M. C., see Cherbuliez, E. [186] 774, *785*
Lebert, U., see Neidlein, R. [109] 550, 551, *564*
Lecumberry, C., see Nadal, N. G. M. [176] 217, *255*
Lederer, E., see Demole, E. [43] 241, *251*
Lederkremer, R. M., de, see Deferrari, J. O. [53] 822, 823, 834, *836*
— see Wolfrom, M. L. [767] 29, *200*; [4, 56, 58] 807, 809, 814, 815, 818, 820, 821, 822, 828, 829, 834, *835, 836*
Ledinová, Z., and I. M. Hais 895
Lee, C. M., see Beckett, A. M. [18c] 454, *462*
Lee, G. F., see Clesceri, N. L. [6] 848, *853*
Lee, Y. L., see Eng, L. F. [187] 147, *185*; [39] 379, *417*
Leeb, W., see Hörhammer, L. [91] 699, *726*
Lees, M., J. Folch, G. H. Sloane Stanley, and S. J. Carr [110] 370, *418*
— see Folch, J. [42] 369, 370, *417*
Lees, T. M., M. J. Lynch, and F. R. Mosher [389] 75, *190*
— and P. J. deMuria [390] 57, *190*
Leete, G. [121] 456, *465*
Leevy, C. M., see Baker, H. [3] 303, 304, *308*
Lefar, M. S., and C. E. Weill [89] 831, *837*
Lefemine, D. V., and W. K. Hausmann [391] 97, *190*
Legler, G. [148a] 248, *254*
— and R. Tschesche [68] *491*

Legler, G., see Tschesche, R. [246] 441, *469*; [218, 219] 691, 695, 696, *729*
Lehmann, G. [392] 53, *190*
— and P. Martinod *470*
Lehmann, H., and H. Dutz [393] 12, *190*
Lehmann, J., and V. Karamustafavglu [93] 534, 536, *564*, 884
Lehner, H., and J. Schmutz [122] 447, *465*
Lehnert, L., see Stahl, E. [663] 30, *197*; [125] 267, 268, 289, 301, *311*
Lemaitre, H., see Vancraenenbroeck, R. [222] 704, *729*
LeMen, J., see Debray, M. *470*
— see Potier, P. [179] 450, *467*
Lemmich, J., see Nielsen, B. E. [155] 695, *728*
Lempert, I., see Katz, D. [44] 639, *655*
Lenk, H. P. [394] 86, *190*
Lennart-Harthon, J. G. [395] 94, *190*
Lenz, U., see Jatzkewitz, H. 874
Leopold, A. C., see Sen, S. P. [117] *493*
Lepage, M. [111] 389, 391, *418*
— see Negishi, T. [148] 379, *419*
— see Paquin, R. [136] 349, *360*; [158] 459, *466*
Lerbs, O. W., see Eberhardt, H. [42, 46] 509, *563*
Lerch, B., see Bhandari, P. R. [66] 53, 128, *181*
— see Stegemann, H. [692] 112, 113, 114, *198*
Lergier, W., see Studer, R. O. [139] 753, *784*
— see Vogler, K. [134, 138, 146] 753, *784*
Le Rosen, A. L., R. T. Moravek, and J. K. Carlton 904
Lesigang, M. [396] 31, *190*; [13] 843, *853*
— and F. Hecht [397] 31, *190*
— see Buchtela, K. [110] 31, *183*
Lestienne, A., E. P. Przybylowicz, W. J. Staudenmayer, E. S. Perry, A. D. Baitsholts, and T. N. Tischer [398] 52, 59, *190*
Letham, D. S. [68a] 475, *491*; [40] 794, *805*
— and C. O. Miller [69] 475, *492*
Lettré, Inhoffen, and Tschesche [108] 311, 313, *360*

Leuckert, Ch., see Hänsel, R. [77] 711, 726
Leupin, K., see Baumgartner, R. [10] 724
Levene, P. A., and H. Mandel [41] 790, 805
Levi, L., see Nigam, I. C. [480] 116, 192; [182, 182a, 183, 185] 212, 213, 217, 225, 238, 255
Levin, E., and C. Head [399] 179, 190; [112] 414, 418
Levin, J., see Fischenbach, H. [13] 577
Levin, Ö., see Tiselius, A. [714] 30, 90, 198
Levine, C., see Chargaff, E. [85] 750, 783, 882
Levitz, M., see Dancis, J. [23] 652, 655
Levorato, L., see Cima, L. [17] 277, 280, 308
Levy, A. L. [173] 760, 761, 784
— and D. Chung [157] 757, 758, 784
— and C. H. Li [163] 759, 784
Levy, G. B. [400] 141, 190
Lévy, J., see Quirin, M. [181b] 448, 467
Lewbart, M. L., W. Wehrli, and T. Reichstein [108a] 324, 342, 360, 870
Lewin, J. C. [401] 27, 190
Lewis, B. A., see Hay, G. W. [16] 810, 812, 819, 820, 830, 831, 834, 835, 893
Lewis, J. A., and J. M. Griffiths 872
— see Arden, T. V 897
Lewis, L. N., R. A. Khalifah, and C. W. Coggins jr. [70] 487, 492
— see Khalifah, R. A. [52, 53] 489, 491
Li, C. H., and L. Ash [160] 758, 759, 784
— see Levy, A. L. [163] 759, 784
Libbert, E., S. Wichner, U. Schiewer, H. Risch, and W. Kaiser [71] 488, 492
— see Schiewer, U. [102] 493
Libbey, L. M., and E. A. Day [149] 219, 221, 223, 254; [56] 631, 655
— see Cobb, W. Y. [32] 221, 251
Libiseller, R., see Pailer, M. [155] 431, 466
Libosvar, J. [24] 577
Lichtenberger, W. [402] 128, 190; [37] 614, 617, 619, 622, 629; [177] 771, 785
Lichti, H., see Wartburg, A. v. [230] 710, 730
Liebisch, M.-W., see Mothes, K. 461

Lie Kian Bo, and J. F. Nyc [403] 53, 190
Lienhard, P., see Beffa, F. [2] 623, 628
Lightbourn, G. A., see Hilz, R. 883
Likussar, W., see Gagliardi, E. [8] 842, 853
Lim, G. T., see Oelschläger, H. [489] 146, 151, 155, 193
Limnell, I., see Euler, H., v. [40, 41] 585, 592, 608
Lindberg, B., see Bouveng, H. O. [73] 828, 830, 836
— see Feast, A. A. J. [48] 820, 836
— see Forsén, S. [86] 831, 837
Linde, H. [150] 245, 254
— see Gamp, A. [220] 53, 186
Lindfors, R. [94] 520, 539, 564
— and A. Ruohonen [95] 538, 564
— see Alha, A. R. [4] 542, 562
Lindgren, I., see Schantz, M. V. [193] 714, 729
Lindlar, F., and H. Wagener [113] 384, 418
Lindley, F. W., see Harris, P. [25] 616, 629
Lindner, E. B., A. Elmquist, and J. Porath [128] 752, 783
Link, E., see Schlemmer, F. [607] 145, 195; [203] 448, 468
Link, W. E., see Buswell, K. M. [19] 387, 415, 416
Linow, F., H. Ruttloff, and K. Täufel [57] 649, 655
Linser, H., and O. Kiermayer [72] 487, 492
— H. Mayr, and F. Maschek [110] 751, 783
Linskens, H. F. [73] 473, 492; [17] 736, 737, 781
Linstead, R. P., see Arden, T. V. 897
— see Burstall, F. H. 876, 899
Linzen, B., see Butenandt, A. [112] 43, 183
Linzer, B., see Hiller, K. [89a] 246, 247, 252
Lippens, B. C., see Boer, J. H. de [83, 85] 26, 182
Lipsett, M. B., see Kirchner, M. A. [93] 602, 609
— see Korenman, S. G. [97] 602, 609
Lipshitz, R., see Tamm, C. [96] 793, 806

Liquori, A.M., see Carelli, V. [117] 42, *183*
Lisboa, B.P. [109, 110, 111, 214, 215, 216] 311, 312, 322, 323, 324, 325, 328, 335, 336, 337, *360, 362,* 857, 870, 883, 893
— and E. Diczfalusy [112, 113] 323, 324, 325, 326, 349, 350, *360*
Lissitzky, S., see Roche, J. [581] 155, *195*
List, P.H., and S. Hanafi [133] 700, *727*
— — and E. Stein [123] 433, *465*
Littler, J.S., and J.G. Sayce [151] 221, *254*
Liukonnen, A. [124] 449, *466*
Livingston, A.L., see Lymann, R.L. [141] 698, 699, *727*
Llewellyn, D.R., see Abraham, M.H. 902
Llosa, P. de la, C. Tertrin, and M. Jutisz [216] 740, *785*
Lloyd, A.G., see Wusteman, F.S. [185] 356, *361*; [91] 679, 680, *686*
Lobinger, K., see Willstätter, R. [760] 8, *199*
Locke, D.C., and C.E. Meloan [404] *190*
Lockhart, I.M., and E.P. Abraham [162] 758, *784*
Loebel, A., see Aelion, R. [12] 9, *180*
Lörcher, K., see Diamantstein, T. [45a] 349, *358*
Loev, B., and K.M. Snader [405] 94, 98, *190*; [114] 315, *360*
Löw, I., see Kuhn, R. [144] 243, *254*
Logar, S., J. Perkavec, and M. Perpar [38] 618, 619, 620, *629*
— see Kornhauser, A. [122] 718, *727*
Lohrmann, R., and H.G. Khorana [42] 794, *805*
Lohs, K., see Donner, R. 864
Lomakina, T.S., L.I. Guskova, and N.I. Grineva [43] 794, 801, *805*
Lommer, D., see Göldel, L. [76] 245, *252*; [65] 317, 337, *359*
Londong, W., see Zöllner, N. [204] 594, 595, 596, *612*
Longo, R., and G. Meinardi [135] 707, *727*
— — and F. Korte [134] 707, *727*
— see Sieper, H. [189] 707, *728*
Lontie, R., see Vancraenenbroeck, R. [222, 223] 704, *729*

Loomis, W.D., see Battaile, J. [43] 178, *181*
Lorenz, B., see Jaenicke, W. [315] 133, *188*
Loth, H., G. Schenck, and K. Kossmann [136] 709, *727*
Lou, V., W.Y. Koo, and E. Ramstad [74] *492*
Loub, W.D., see Farnsworth, N.R. [54] 449, *463*
Lowe, D., see Brian, P.W. [4] 488, 489, *489*
Luboz, M.P., see Denti, E. [46] *251*
Luboz Magrassi, B., see Canuti, A. [5] 624, *628*
Lucas, G.H.W., see Korczak-Fabierkiewicz, C. [88a] 508, *564*
Luckner, M., O. Bessler, and R. Luckner [139] 717, *727*
— — and P. Schröder [137, 138, 140] 717, 719, *727*
— K. Winkler, O. Bessler, J. Hoffmann, and W. Poethke [125] 432, *466*
— — — P. Schröder, J. Hoffmann, and W. Poethke [126] 456, *466*
Luckner, R., see Luckner, M. [139] 717, *727*
Ludwig, E. [406] 146, *190*
— and U. Freimuth [83, 84] 263, 292, 294, *310*
Ludwig, J., see Kuksis, A. [108] 398, 414, *418*
Lüben, G., see Wieland, Th. [755] 37, *199*; [104] 795, 796, *806*
Lück, E., and W. Courtial [58] 637, *655*
Lüderitz, O., see Westphal, O. [748] 42, *199*
Lüdy-Tenger, F. [407] 56, 63, *190*; [96] 558, *564*
Lüthi, U., and P.G. Waser [408] 159, 160, *190*
Lugaro, G., see Gerali, G. [63] 335, 341, *358*
Luisi, M., G. Gambassi, V. Marescotti, C. Savi, and F. Polvani [101] 600, *609*
— C. Savi, F. Coli, F. Panicucci, V. Marescotti, and G. Gambassi [217] 340, 341, *362*
— — and V. Marescotti [115] 350, *360*
Luly, C., see Dati, T. [146] 73, *184*

Lund, J., see Norman, A. W. [95] 277, 279, 280, *310*
Lundberg, W. O., see Hawthorne, B. E. [58] *417*
— see Holman, R. T. [63] 367, *417*
— see Mahadevan, V. [416] 176, *191*; [116] 334, *360*
— see Privett, O. S. [541] 139, 140, *194*; [166] *420*
Lundström, J., and S. Agurell *470*
Lundt, I., C. Pedersen, and B. Tronier [98] 831, *837*
Lunenfeld, B., see Sulimovici, S. [230] 340, *362*
Luscombe, M., see Cook, E. R. [34] 738, *781*
Lutomski, J., see Adamski, R. *470*
Luukkainen, T., see Horning, E. C. [89] 316, *359*
Lymann, R. L., A. L. Livingston, E. M. Bickoff, and A. N. Booth [141] 698, 699, *727*
Lynch, M. J., see Lees, T. M. [389] 75, *190*
Lyons, D., see Pollard, F. H. 854
Lytle, R. I., see Moffat, E. D. [3] 732, 747, 749, *781*

Maas, H. [409] 131, *191*
Maas, S. P. J., see Dam, M. J. D. van [144] 56, *184*
Maass-Goebels, L., see Janecke, H. [322] 135, *188*; [71] 276, 278, 279, *309*
Maatman, R. W. [410] 13, *191*
— see Dugger, D. L. [173] 13, *184*
Macek, K., J. Hacaperková, and B. Kakáč 891
— and I. M. Hais [413] 5, *191*
— J. Večerková, and J. Stanislavová [97] 507, *564*
— see Hais, I. M. [263] 3, *187*; [9] *505*; [72] 691, 693, 705, *726*; [2] 732, 736, 737, 760, 763, *781*
— see Hais, I. M. 886, 893, 900
Machata, G. [127] 437, 439, *466*; [102] 606, *609*; [39] 615, *629*
— and W. Kisser [98] 536, 537, *564*; [70] 518, *563*
Macheboeuf, M., see Munier, R. 874, 882
Machovičová, F., and V. Parrák [128] 444, *466*
— see Parrák, V. [162] 432, *467*
MacMahon, J. M., R. B. Davis, and G. Kalnitzky 858

MacMillan, J., J. C. Seaton, and P. J. Suter [75, 76] 475, *492*
— and P. J. Suter [77] 479, 486, 487, *492*
— see Elson, G. W. [15] 479, 487, 489, *490*
— see Jones, D. F. [39] 479, 486, 487, 489, *491*
MacWilliam, J. C., see Griffith, C. M. [24a] 488, 489, *490*
— see Harris, G. 859
Mäkinen, R., see Salo, T. [81, 83] 634, 635, 638, *656*
Märki, F., and B. Witkop [130] 460, *466*
Maffi, G., see Cerri, O. [125] 5, *183*; [36] 457, *463*; [30] 549, *562*; [21] 606, *608*
Mahadevan, V., and W. O. Lundberg [416] 176, *191*; [116] 334, *360*
Mahapatra, G. N., and O. M. Friedman [417] 47, *191*; [44] 794, *805*
Mahlum, D., see Keith, R. W. 878
Mahon, M. E., see Heacock, R. A. [32, 33, 34, 35] 477, 481, 486, *490*, 869
Maier, B., see Schütte, H. R. [210b] 461, *468*
Maier, R., and H. K. Mangold [418, 419] 5, 174, 179, *191*
— see Morris, L. J. [144] 380, *419*
— see Sprecher, H. W. [192] 380, *420*
Mainil, J., see Pais, M. [156] 435, *466*
Maistrellis, N. A., see Dyer, W. G. [48] 338, *358*; [34] 602, *608*
Makleit, S., see Bognár, R. [25] 459, *462*
Makus, Z., see Kaufmann, H. P. [339] 88, *189*; [96, 97] 317, 323, 326, 331, 332, 333, *359*; [78, 79, 80, 81, 83] 375, 377, 378, 386, 401, 410, 411, 412, 413, *418*; [87] 594, *609*
Malíková, J., see Stárka, L. [158] 323, 340, *361*
— see Vacíková, A. [203] *420*
Malins, D. C., and H. K. Mangold [420, 421] 5, 94, *191*; [114] 375, 377, 410, 411, 412, *418*, 866, 867
— J. C. Wekell, and C. R. Houle [115, 116] 377, 383, *418*
— see Mangold, H. K. [426] 159, 162, 169, 179, *191*; [118] 376, 377, 378, 381, 382, *418*
— see Wekell, J. C. [216] 388, *421*

Malkin, T., see Bevan, T. H. 887
— see Holman, R. T. [63] 367, *417*
Malzacher, A., see Gänshirt, H. [51] 292, 293, 294, 298, 299, 305, *309*; [67] 457, *464*, 862
Malzacher, F., see Gänshirt, H. G. [62, 67] 557, 558, 560, *563*
Man, J. M. de [117] 334, *360*
Mandel, H., see Levene, P. A. [41] 790, *805*
Mandel, P., see Rebel, G. [112] 290, 291, 292, *310*
Mandema, E., W. H. de Fraiture, H. O. Nieweg, and A. Arends [103] 605, *609*
Manegold, E. [422] 14, *191*
Mangold, H. K. [423, 424] 168, 173, 179, *191*, 312; [117, 120, 122] 375, 377, 378, 412, *418*, *419*
— and Z. Bandi [124] 397, 401, 403, 406, 411, 412, *419*
— W. J. Baumann, and C. R. Houle [425] 176, 177, *191*
— J. L. Gellerman, and H. Schenk 866
— and R. Kammereck [427] 140, *191*; [119, 121] 377, 378, 387, 388, 403, 404, 405, 406, 411, 412, 414, *418*; [51] 674, *685*
— — and D. C. Malins [426] 159, 162, 169, 179, *191*
— B. G. Lamp, and H. Schlenk 866
— and D. C. Malins [118] 376, 377, 378, 381, 382, *418*
— H. H. O. Schmid, and E. Stahl [428] 5, *191*
— and H. W. Ulshöfer [123] 399, *419*
— see Baumann, W. J. [51] 59, *181*; [9, 10] 381, 385, 412, 414, *416*; [8] *416*
— see Cubero, J. M. [30] 397, 402, *416*
— see Hawthorne, B. E. [58] *417*
— see Maier, R. [418, 419] 5, 174, 179, *191*
— see Malins, D. C. [420, 421] 5, 94, *191*; [114] 375, 377, 410, 411, 412, *418*, 866, 867
— see Morris, L. J. [144] 380, *419*
— see Pelick, N. [511] 1, *93*
— see Schmid, H. H. O. [181, 182, 183, 184, 185] 373, 374, 377, 383, 384, 385, 386, 397, 401, 402, *420*
— see Stahl, E. [683] *197*

Mangold, H. K., see Tuna, N. [724, 725] 169, 170, 171, 179, *198*; [202] 368, 373, 377, 383, 385, 386, *420*
Manitto, P., see Jommi, G. [81] *564*
Mann, I., see Pfeifer, S. *461*
Mann, J., see Marrinan, H. J. [431] 34, *191*
Manns, C., see Kaiser, H. [98] 433, *465*
Manske, R. H. F. 421
— and H. L. Holmes *461*
Mantovan, R., see Cima, L. [128] 146, 151, 155, *183*; [16, 17] 277, 280, 302, 303, *308*
Manzetti, A. R., and T. Reichstein [118] 317, 342, *360*
Marcucci, F., and E. Mussini [429] 58, *191*
Marcus, D. M., E. A. Kabat, and G. Schiffman [40] 818, *835*
Marcuse, R. [152, 153] 217, 219, 221, *254*
— U. Mobech-Hanssen, and P. O. Göthe [154] 219, 221, *254*; [125] 411, 412, *419*
Marderosian, A. D., and H. W. Youngken jr. [128b] 451, *466*
Marescotti, V., see Luisi, M. [115, 217] 340, 341, 350, *360*, *362*; [101] 600, *609*
Margasinski, Z. et al. [99] 509, *564*
Mariani, A., and O. Mariani-Marelli [129] 437, *466*
Mariani-Marelli, O., see Mariani, A. [129] 437, *466*
Marigo, M., see Fiori, A. [62] 230, *252*; [53] 520, *563*
Marinetti, G. V. [126] 378, 389, *419*
— see Witter, R. F. 896
Marini, M. A., see Holley, R. W. [33] *805*
Marini-Bettòlo, G. B. [430] 5, *191*, 859
— see Caggiano, E. [35] 448, *463*
Marion, L., see Tsuda, Y. [247] 460, *469*
Markham, K. R. [144] 698, *727*
Markham, R., and J. D. Smith [45] 792, 793, 796, 803, *805*
Markl, P., and F. Hecht [14, 15] 844, *853*
Marmur, J. [46a] 788, *805*
Marozzi, E., and G. Falzi [100] 509, *564*
Marquez, A. D., see Novotný, L. [158] 717, *728*

Marrinan, H. J., and J. Mann   [431] 34, *191*
Marsh, A., see Chen jr., P. S.   [15] 275, 276, 280, *308*
Marshak, A., and H. J. Vogel   [46] 790, 792, *805*
Marshall jr., E. K., see Bratton, A. C. 866
Marshall, L. M., see Donaldson, K. O. [167] 90, *184*
Marshall, M. O., see Morris, L. J.   [145] 399, *419*
Martak, D. S., see Kritchewsky, D.   [104] 323, 354, *359*
Martelli, A., see Nano, G. M.   [177, 177a] 226, 233, *255*
Marten, G.   [432] 111, *191*
Martin, A. J. P., and R. L. M. Synge [433] 3, *191*
— see Consden, R.   [136, 137] 3, 4, 105, *183*; [26, 47, 99] 737, 746, 747, 749, 750, *781, 782, 783*, 885
Martin, H. P.   903
Martin, K. D., see Bauer, R. D.   [48] 47, *181*; [2] 802, *804*
Martin, M. M., see Hamman, B. L.   [74] 338, *359*
Martin, R., see Braekman, J. C.   *470*
Martin, R. O., see Battersby, A. R. [13b] 444, *462*
Martinod, P., see Lehmann, G.   *470*
Martius, C., see Billeter, M.   [6] 290, *308*
Mary, N. Y., and E. Brochmann-Hanssen [434] 146, 151, *191*; [131] 440, *466*
— see Craig, J. C.   [40c] 430, *463*
Maschek, F., see Linser, H.   [110] 751, *783*
Mascov, V., see Cionga, E.   *470*
Masera, G., see Käser, H.   [85] 579, *609*
Masiti, D., H. W. Moore, and K. Folkers [86] 290, *310*
Masoro, E. J., L. B. Rowell, and R. M. McDonald   [104] 599, *609*
Massaglia, A., and U. Rosa   [435] 177, *191*; [217] 779, *785*
Massart, L., see Onckelen, H. A. van [88] 475, *492*
— see Petridis, C.   [93] 489, *492*
Masse, J.   [25] *577*
— and R. Paris   [155] 212, 236, 237, 238, 248, 250, *254*
Massie, H. R., and B. H. Zimm   [47] 788, 794, *805*

Máthé, I., and E. Tyihák   [156] 238, *254*; [132] 426, *466*
— see Tyihák, E.   [305, 306] 212, 238, *258*
Mathes, H., see Grassmann, W.   [248] 43, *186*
Mathieu, M., see Cotte, J.   [26] 814, *835*
Mathis, C.   [436] 94, *191*; [157] 206, 211, 212, 238, *254*
— and G. Ourisson   [158, 159] 206, 210, 212, 238, *254*
Matis, J., O. Adamec, and M. Galvánek [119] 316, 335, *360*
— see Adamec, O.   [11] 148, *180*; [2] 324, 328, 340, *357*; [1] 601, *607*
Matkovics, B., see Göndös, Gy.   [66] 328, 341, 349, *359*
Matrka, M., see Gasparič, J.   [18, 22] 612, 618, *628*
Matschke, E., see Poethke, W.   [172] 707, *728*
Matsuda, K., see Takitani, S.   [706, 707] 58, 76, *198*
Matsuhiro, B., see Deferrari, J. O.   [53] 822, 823, 834, *836*
Matsukuma, T., see Momose, T.   [219] 324, 345, *362*, 903
Matsumoto, N.   [120] 347, *360*
— see Takeda, K.   [166] 347, *361*, 855, 857, 863
Matsushita, H., Y. Suzuki, and H. Sakabe   [52] 667, *685*
Matthews, J. S.   [121] 323, *360*, 904
— A. L. Pereda, and A. Aguilera [437] 146, 147, 151, 153, *191*; [122] 322, 336, 339, 341, *360*
Matthias, W.   [438] 89, *191*
Mattocks, R. R.   *470*
Maturová, M., D. Pavlásková, and F. Šantavý   [132b] 444, *466*
Maugras, M., M. Ch. Robin, and R. Gay [53] *685*
Mauli, R., see Gutzwiller, J.   [84] 226, *252*
Maurer, R., see Karlson, P.   [338] 149, *189*
Mauron, J., see Bujard, El.   [17] 586, *607*
Mautner, H. G., see Günther, H. H.   [58] 298, 299, *309*
Maw, G. A., see Coyne, C. M.   898
Maximov, O. B., and L. S. Panthinkhina 893

Mayer, A. M., see Koller, D. [59] 473, *491*
Mayer, H., P. Schudel, R. Rüegg, and O. Isler [87] 284, *310*
Mayer, R., P. Rosmus, and J. Fabian [54] 682, *685*
Mayo, P. de [160] 207, *254*
— see Aberhart, D. J. [1] 718, *724*
Mayr, H., see Linser, H. [110] 751, *783*
McCarthy, J. L., A. L. Brodsky, J. A. Mitchell, and R. F. Herrscher [123] 315, 340, 341, 346, *360*
McCarthy, T. J., and C. H. Price [142] 706, *727*
— and M. C. B. van Rheede van Oudtshoorn [143] 706, *727*
McCloskey, P., see Speake, T. [220] 430, *468*
McCluer, R. H., see Johnson, G. A. [84] 598, *609*
— see Penick, R. J. [163] 389, 391, 415, *419*; [126] 598, *610*
McConnell, B. L., see Dugger, D. L. [173] 13, *184*
McCormick, D. B., see Sander, E. G. [83] *806*
McCoy, R. N., and E. C. Fiebig [411] 154, *191*
McCrindle, R., see Djerassi, C. [50] 243, *251*
McCully, K. A., see McKinley, W. P. [59] 638, *655*
McDonald, R. M., see Masoro, E. J. [104] 599, *609*
McDonald, R. S. [412] 12, *191*
McGovern, T. P., see Beroza, M. [65] 149, *181*
McInnes, A. G., see Towers, G. H. N. [217] 718, *729*
McIsaac, W. M., G. Farell, R. G. Taborsky, and A. N. Taylor [78] 486, *492*
McKeown, G. G., see Smyth, R. B. 888
McKernan, W., see Smith, R. H. [41] *578*
McKibbins, S. W., J. F. Harris, and J. F. Saeman [414] 99, *191*
McKillican, M. E., and R. P. A. Sims [127, 128] 389, 391, *419*
— see Negishi, T. [148] 379, *419*
McKinley, W. P., D. E. Coffin, and K. A. McCully [59] 638, *655*
McLaughlin, J. L., J. E. Goyan, and A. G. Paul [415] 153, *191*; [118] 451, 453, *465*

McMillan, J., see Jones, D. F. 900
McNeil, C. [40] 620, *629*
McNiven, N. L., see Futterweit, W. [60] 340, *358*; [54] 602, *608*
McOmie, J. F. W., see Pollard, F. H. 881
McSweeney, G. P. [151a] 225, 227, 241, *254*
McWain, P., see Wolfrom, M. L. [96] 831, *837*
Mead, J. F., see Dhopeshwarkar, G. A. [159, 160] 177, 178, *184*; [34] 368, 383, 385, 398, *416*
— see Fulco, A. J. [210] 177, *185*
— see Nevenzel, J. C. [149] 384, *419*
Meadows, E. H., see Dunagin, P. E. [33] 274, *308*
Meckel, L., [41] 622, *629*
— H. Milster, and U. Krause [161] 238, 250, *254*; [42] 620, 622, 623, *629*
Meer, W. A., see Farnsworth, N. R. [51b] 449, *463*
Mees, G., see Noirfalise, A. *470*
Mehl, E., see Jatzkewitz, H. [70] 391, *417*; [76] 598, *609*, 860, 900
Mehlitz, A., K. Gierschner, and T. Minas [162, 163] 220, 223, 224, 225, 229, *254*, 871
Meier, W., and A. Fürst [145] 696, *728*
Meijboom, P. W., and G. Jurriens [164] 221, *254*
Meijs, W. H., see Boer, J. H. de [83, 85] 26, *182*
Meinardi, G., see Longo, R. [134, 135] 707, *727*
Meinhard, J. E., and N. F. Hall [439] 3, 73, *191*; [16] 837, *853*
Meisler, M. H., see Penick, R. J. [163] 389, 391, 415, *419*; [126] 598, *610*
Mele, A., see Carelli, V. [117] 42, *183*
Melera, A., see Noller, C. R. [185a] 246, *255*
— see Parello, J. [159] 460, *466*
Mellinger, T. J., and C. E. Keeler [101] 508, 509, 514, *564*
Meloan, C. E., see Locke, D. C. [404] *190*
Men, J. le, see Quirin, M. [181b] 448, *467*
Mengel, M., see Baudler, M. [45, 46] 35, *181*; [2] 848, *853*
Menrad, E. L., J. M. Müller, A. F. Thomas, S. S. Bhatnagar, and N. J. Dastoor [165] 243, *254*

Mentzer, Ch., see Clerc-Bory, M. [112] 751, *783*
Mercer, E. I., B. H. Davies, and T. W. Goodwin [85] 267, *310*
Merck 441
Merkus, F. W. H. M. [17] 846, *853*, 869, 872, 895
Merrill, S. H., see Holley, R. W. [32, 33] 801, 802, *805*
Messerschmidt, W. [166, 167] 239, *254*
Měšťan, J., see Michaleč, C. [125] 317, 322, 332, *360*
Metcalfe, L. D., see Pelka, J. R. [512] 138, *193*; [162] 387, *419*; [25] 472, 497, *506*
Metche, M., see Urion, E. [221] 704, *729*
Metz, H. [440] 104, *191*; [124] 323, 341, *360*, 893, 900, 903, 904
Metzenberg, R. L., and H. K. Mitchell [100] 750, *783*
Metzger 273
Metzger, J., see Vernin, G. [47] 502, *506*
Metzler, D., see Smith, E. C. [121] 297, *311*
Metzner, H., and H. Volcsik [441] 64, *191*
Meyer, F., and K. Bloch [442] 177, *191*
Meyer, G. M., see Ritter, F. J. [578, 579] 99, 101, 149, *195*
Meyer, H. [443] 32, *191*; [60, 61] 631, 632, 633, 634, *655*
Meyer, K., see Gamp, A. [220] 53, *186*
Meyer, R., see Pietsch, H. P. [91] *630*
Meyer-Dulheuer, K.-H., and R. Ritter [102] 542, 543, *564*
Meyer-Stoll, H. A., see Birkofer, L. [71] 73, *182*; [23] 813, *835*
Meyl, M., see Neidlein, R. [108] 548, *564*
Meyniel, G., see Berger, J.-A. [59, 60, 61, 62, 63] 45, 46, 94, *181*; [205] 780, *785*; [4] 842, 845, *853*
Meyruey, M. H., see Faugeras, G. [55, 56] 436, *464*
Meythaler jr., Chr., see Schmid, E. [144] 509, 519, *565*
Miana, G. A., see Ikram, M. [94] 438, 439, 448, *465*
Michaels, G., see Wood, P. [194] 593, *611*

Michaleč, C. [129, 130, 131] 412, 414, *419*, 855
— and Z. Kolman [105] 596, *609*
— M. Šulc, and J. Měšťan [125] 317, 322, 332, *360*
Micheel, F., and H. Schweppe [43] 616, *629*, 879
Michel, G., see Bordet, C. [18] 221, *251*
Michel, K. H., see Sandberg, F. [199] 446, *468*
Michel, R., see Roche, J. [581] 155, *195*
Michel, W., see Determann, H. [156] *184*
Michelson, A. M. [48] 796, *805*
Middlebrook, W. R. [166] 759, *784*
Miedtank, S., see Kohlschütter, H. W. [363] 23, *189*
Miettinen, T. A., E. H. Ahrens jr., and S. M. Grundy [444] 178, *192*; [88] 284, *310*; [106] 603, 604, *610*
— see Grundy, S. M. [255] 178, *187*; [63] 604, *609*
Miksche, G. E., see Kratzl, K. [141] 230, *254*; [128] 712, *727*
Milborrow, B. V. [445] 105, *192*; [218] 322, *362*
Miller, C. O. [79] 473, 475, *492*
— see Letham, D. S. [69] 475, *492*
Miller J. M., and J. G. Kirchner [446, 447, 448, 449] 3, 54, 97, *192*; [168] 206, 210, 216, *254*; [126] 325, *360*
— see Kirchner, J. G. [348] 3, 146, *189*; [104] 437, *465*
Miller, R., see Cone, N. J. [40] 449, 450, *463*
Millett, M. A., W. E. Moore, and J. F. Saeman [451] 149, 151, 152, 153, *192*
— see Seikel, M. K. [625] 100, 101, *196*
Millon, E. [119] 751, *783*
Mills, G. L. [169] 760, *784*
Mills, J. S., and A. E. Werner [169] 248, *254*
Mills, P. A., J. H. Onley, and R. A. Gaither [62] 644, *656*
Milster, H., see Meckel, L. [161] 238, 250, *254*; [42] 620, 622, 623, *629*
Mima, H. [450] *192*
— and N. Kitamori [63] 649, *656*
Minas, T., see Mehlitz, A. [162, 163] 220, 223, 224, 225, 229, *254*, 871
Minato, H., see Takeda, K. [280] 213, *257*

Minnis, R., see Daly, J. W. [4a] 499, 505
Miras, C., S. Simons, and J. Kiburis [146] 716, 728
Mishkin, A. R., see Jacin, H. [6] 808, 811, 814, 835, 887
Miskel, J., see Finston, H. L. [192] 155, 185
Mistryukov, E. A. [452, 453] 75, 192; [18] 500, 505; [26] 577
Mitchell, H. K., see Metzenberg, R. L. [100] 750, 783
Mitchell, J. A., see McCarthy, J. L. [123] 315, 340, 341, 346, 360
Mitchell, L. C. [64] 638, 656; [44] 616, 629, 898
Mitchell, P. W. D., see Barry, V. C. [62] 825, 836
Mitchell, T. I., see Bryson, I. L. 856
Mitchell, W. D., see Truswell, A. S. [231] 332, 362
Mitsuhashi, H., U. Nagai, T. Muramatsu, and H. Tashiro [170] 215, 254
Mitsuhashi, M., and H. Shibaoka [80] 487, 492
Mittelstadt, K. A., see Whitaker, D. R. [750] 90, 199
Miura, H., see Kawano, N. [119] 690, 727
Miyahara, K., see Kawasaki, T. [125] 245, 253; [98] 346, 349, 359
Miyake, H., see Nakayama, F. [128] 354, 360
Miyawaki, M., see Takeda, K. [280, 281] 213, 257
Mizugaki, M., see Okui, S. [156] 387, 419
Moats, W. A. [65] 639, 656
Mobech-Hanssen, U., see Marcuse, R. [154] 219, 221, 254; [125] 411, 412, 419
Möller, H. G., and N. Zeller 879
Moffat, E. D., and R. I. Lytle [3] 732, 747, 749, 781
Moghissi, A. [454, 455] 32, 112, 179, 192; [18, 19] 847, 853
Mohar, A., see Schlögl, K. [79, 80, 81, 82] 665, 686
Mohr, U., A. Authaler, and J. Althoff [19] 497, 505
Mohrmann, H. L., see Schultz, O. E. [231] 235, 256

Mokrý, J., L. Dubravková, and P. Šefčovič [133] 449, 466
— and I. Kompiš [134, 135] 449, 466
— — P. Šefčovič, and S. Bauer [136] 449, 466
Moldovanskaya, G. J., see Kheifits, L. A. [126] 230, 253
Moll, F. [137] 431, 466, 891
Molnar, G., see Tyihák, E. [303, 304] 226, 258
Moloster, Z. [45] 617, 629
Momose, T., T. Matsukuma, and Y. Ohkura [219] 324, 345, 362, 903
Montag, A. [89] 269, 310; [46] 612, 629
Montfort, M. L., V. E. Tyler jr., and L. R. Brady [147] 718, 728
Moody, G. J., see Criddle, W. J. [139] 111, 184; [8, 9] 626, 628
Moore, H. W., see Masiti, D. [86] 290, 310
Moore, S., D. H. Spackman, and W. H. Stein [14] 735, 781
— see Drèze, A. [33] 737, 781
— see Stein, W. H. [30] 737, 781
— see Smyth, D. G. 774
Moore, W. E., see Millett, M. A. [451] 149, 151, 152, 153, 192
Morand, P., see Audrin, P. [7] 341, 357
— see Jung, L. [95] 351, 359
Moravek, R. T., see Le Rosen, A. L. 904
Moreira, A., see Borio, B. L. [30b] 432, 463
Moreira, E. A. [138] 449, 466; [103] 543, 564
— and C. Cecy [171] 248, 254
Moretti, G., see Cavina, G. [29a] 556, 562
Morgan, M. E. [456] 64, 192
— and R. L. Pereira [172] 220, 254
Morgan, W. T. J., see Elson, L. A. 868
Morianz, K., see Gänshirt, H. [212, 213] 134, 146, 147, 148, 149, 150, 185; [62] 323, 352, 353, 354, 358; [32] 636, 655
Morimoto, H., and H. Oshio [139] 466
Morin, R. J. [132] 415, 419
Morita, K., and F. Haruta [457] 57, 192
Moritz, O. [173, 174] 207, 255
Morris, C. J. O. R. [458] 192; [108] 591, 610

Morris, L.J. [127] 316, 322, 333, *360*; [136, 137, 138, 139, 143] 396, 397, 398, 401, *419*; [107] 594, *610*
— R.T. Holman, and K. Fontell [133, 134, 135] 377, 386, 387, *419*
— and S.W. Hall [141] 381, *419*
— R. Maier, and H.K. Mangold [144] 380, *419*
— M.O. Marshall, and W. Kelly [145] 399, *419*
— and D.M. Wharry [140, 142] 386, 387, 398, *419*
— see Chalvardjian, A. [23] 387, *416*
— see James, A.T. [317a] 49, *188*; [94] 316, *359*
— see Nichols, B.W. [152] 375, *419*
— see Williams, J.A. [192] 597, *611*
Morris, Q.L., see Gage, T.B. [58] *726*
Morris, W.W., and E.O. Haenni [66] 638, *656*
Morrison, A., see Witter, R.F. 896
Morrison, J.C., and L.G. Chatten [459] 135, 137, 138, *192*; [104, 105] 518, 537, *564*
Morton, M.J., and W.I. Rogers [49] 802, *805*
Morton, R.A. [90] 259, 284, 288, *310*
— see Dunphy, P.J. [35] 284, *308*
Mosbach, E.H., see Chattopadhyay, D.P. [192] 334, *362*
Mosbach, K. [148] 687, *728*
Moser, A.B., see Moser, H.W. [109] 594, *610*
Moser, H.W., A.B. Moser, and J.C. Orr [109] 594, *610*
Moses, A.J. [460] 170, *192*
Mosher, F.R., see Lees, T.M. [389] 75, *190*
Mosher, H.S., F.A. Fuhrman, H.D. Buchwald, and H.G. Fischer [139b] 461, *466*
Mosser, D.G., see Tuna, N. [724] 169, 170, 179, *198*
Mostert, K., see Jong, K. de [110] 222, *253*
Mothes, K. [81, 82] 471, 472, 473, *492*
— D. Gross, M.-W. Liebisch, and H.-R. Schütte 461
— and H.-B. Schröter 461
— K. Winkler, D. Gröger, H.-G. Floss, U. Mothes, and B. Weygand [140] 454, *466*
— see Gröger, D. [82] 457, *464*

Mothes, K., see Neubauer, D. [147] 437, *466*
Mothes, U., see Mothes, K. [140] 454, *466*
— see Weygand, F. [238] 694, 695, *730*
Mottier, M. [461, 462] 5, *192*; [47] 612, *629*; [11] 733, *781*
— and W. Potterat [463] 5, 149, *192*
— and M. Potterat [48] *629*
Mottlau, A.Y., see Fisher, N.E. [193] 16, *185*
Mouhgrabi, A. [109a] 586, *610*
Mouton, M., S. Jaquard, and M. Sagot-Masson [464] 30, *192*; [175] *255*
Moye, H.A., see Winefordner, J.D. [761] 155, *199*
Moza, B.K., and J. Trojánek [141, 142] 449, *466*
Mühlemann, and Bürgin 526
Mühlemann, H., see Sonanini, D. [229] 337, *362*
Mühlemann, M., see Ackermann, M. [2] 712, *724*
Müldner, H.G., J.R. Wherrett, and J.N. Cumings [146] 389, 392, 393, *419*
Müller, E., see Hörhammer, L. [110] 691, 707, *727*
Müller, J.M., see Menard, E.L. [165] 243, *254*
— see Thomas, A.F. [285] 243, *257*
Müller, K.H., B. Christ, and G. Kühn [149] 707, *728*
— and H. Honerlagen [465] 94, *192*; [143] 455, *466*
— see Gabel, E. [68] 230, 233, 234, 235, 239, *252*, 856
Müller, R., see Eder, F. [28] 638, 645, *655*
— see Jerchel, D. [107] 751, *783*, 864
Müller, W., see Harbers, E. [28] 791, *805*
— see Hüttenrauch, R. [305] 46, *188*; [67] 292, 293, 294, *309*
Müllhofer, G., see Simon, H. [635] 167, *196*
Müting, D. [70] 749, *782*, 888
Mukai, K., see Hashimoto, A. [56] 406, *417*
Mukherjee, K.D., see Kaufmann, H.P. [341] 140, *189*
Mulder, F.J., and K.H. Hanewald 280

Muldner, H. G., J. R. Wherett, and J. N. Cumings [110] 598, *610*
Mulé, S. J. [106] 527, 528, 529, 535, *564*
Mulryan, H. [466] 27, *192*
Munier, R. 873, 882, 883
— and M. Macheboeuf 874, 882
Munk, V. J., see Pásková, J. 860
Munson, P. L., see Peng, T. C. [125] 598, 602, *610*
— see Tai Chan, P. [232] 338, *362*
Munter, F., see Waldi, D. [257] 422, 424, 426, 433, 437, 439, 440, 448, 451, 454, *469*
Murakawa, S., see Ibayashi, H. [90] 339, *359*; [75] 602, *609*
Muramatsu, T., see Mitsuhashi, H. [170] 215, *254*
Murin, P. J. de, see Lees, T. M. [390] 57, *190*
Murphy, M. T. J., B. Nagy, G. Rouser, and G. Kritchevsky [147] 380, *419*
Murray, A. III., and D. L. Williams [467] 167, *192*
Murray, T. K., K. C. Day, and E. Kodicek [91] 278, *310*
— see Varma, T. N. R. [138] 273, 274, *311*
Murthy, K. U. K., see Das, V. S. R. [11] *490*
Mussini, E., see Marcucci, F. [429] 58, *191*
Mutschler, E., and H. Rochelmeyer [10] 733, 740, *781*
— see Röder, K. *471*
— see Schunack, W. [145a] 549, 550, *565*; [201] 718, *729*
— see Teichert, K. [241, 242] 422, 432, 433, 437, 439, 448, 451, 453, 455, 457, *469*; [45] 494, 495, *506*; [163], 522, 549, *566*; [213] *729*
Mutten, A. J., see Belcher, R. 868
Muus, L. T. [468] 33, *192*
Myhill, D., and D. S. Jackson [218] 740, *785*

Nadal, N. G. M., C. M. C. Chapel, and C. Lecumberry [176] 217, *255*
Nager, M., see Peurifoy, P. V. 901
Naff, A. S., see Naff, M. B. [469] 53, *192*; [49] 618, 620, *629*
Naff, M. B., and A. S. Naff [469] 53, *192*; [49] 618, 620, *629*

Nagabhushanam, A., see Giri, K. V. [71] 749, *782*
Nagai, U., see Mitsuhashi, H. [170] 215, *254*
Nagarajan, K., Ch. Weissmann, H. Schmid and P. Karrer [144] 447, *466*
— see Govindachari, T. R. [76, 77] 447, *464*
— see Guggisberg, A. [86] 447, *464*
Nagy, B., see Murphy, M. T. J. [147] 380, *419*
Nair, P. M., see Dutta, N. L. [35] 709, *725*
Nakagawa, Y., see Djerassi, C. [46] 447, *463*
Nakamura, M., see Ibayashi, H. [90] 339, *359*; [75] 602, *609*
Nakao, K., see Ibayashi, H. [90] 339, *359*; [75] 602, *609*
Nakayama, F., M. Oishi, N. Sakaguchi, and H. Miyake [128] 354, *360*
Nakon, R. S., see Schwane, R. A. [618] 160, *196*
Nano, G. M. [470] 145, 151, *192*
— and A. Martelli [177, 177a] 226, 233, *255*
— and P. Sancin [178, 179] 220, 222, *255*
— — and G. Tappi [107] 509, *564*
Natori, S., F. Sato, and S. Udagawa [150] *728*
Nauta, W. Th., see Doorenbos, H. J. [39] 518, *563*
Nayar, M. N. S. [50] 801, *805*
Nayler, J. H. C., see Hardy, T. L. [52] 747, 749, *782*
Nealey, R. H. [55] 678, *685*
Nefedov, V. D., see Vobecky, M. [88] 666, *686*
Negishi, T., M. E. McKillican, and M. Lepage [148] 379, *419*
Negwer, M. [107a] 507, 508, *564*
Neher, R. [129, 129a, 130] 312, 314, 315, 316, 317, 318, 319, 320, 322, 323, 324, 325, 326, 327, 328, 336, 337, 340, 341, 342, 346, 350, 355, *360*; [50] 613, *629*, 870
— and E. v. Arx [131] 317, 320, 322, 337, 338, 356, *360*, 903
— and A. Wettstein 894
— see Arx, E. v. [24, 25] 36, 84, 88, *180*; [202] 740, 743, 745, *785*

Neidlein, R., E. Hohndorf, and J. D. Rosenblath [107b] 547, *564*
— G. Klügel, and U. Lebert [109] 550, 551, *564*
— H. Krüll, and M. Meyl [108] 548, *564*
Neill, J. D., B. N. Day, and G. W. Duncan [471] 118, *192*
Neish, A. C., see Towers, G. H. N. [217] 718, *729*
Nesić, S., see Pejković-Tadić, I. [192] 219, *255*
Neu, R. [145] 424, *466*; [151, 152, 153, 154] 691, 705, *728*, 871, 901
— and P. Hagedorn 857, 883
Neubauer, D. [146] 437, *466*
— and K. Mothes [147] 437, *466*
Neubert, G. [56] 664, 665, *685*
Neuhard, J., E. Randerath and K. Randerath [472] *192*; [51] 795, 796, 797, *805*
Neukom, H., see Koller, A. [32] 816, *835*
Neumann, D., and H.-B. Schröter [148] 433, *466*
Neumüller, O. A., H. J. Kuhn, G. O. Schenck, and F. Šantavý [149] 426, *466*
— see Herberhold, C. [68] 586, *609*
Neurath, G., and E. Doerk [20] 495, *505*
— B. Pirmann, and M. Dünger [57] 680, 681, *685*
— see Preussmann, R. [33] 497, *506*, 871, 900
Neuss, N., see Cone, N. J. [40] 449, 450, *463*
Nevenzel, J. C., W. Rodegker, and J. F. Mead [149)] 384, *419*
Newburgh, R. W., see Coffey, R. G. [130] 36, 46, *183*; [11] 793, 795, *804*
Newman, A. A. [473] 130, *192*
Newsome, J. W., H. W. Heiser, A. S. Russell, and H. C. Stumpf [474] 23, 24, *192*
— see Stumpf, H. C. [704] 24, *198*
Nichaman, M. Z., C. C. Sweeley, N. M. Oldham, and R. E. Olson [475] 146, 147, *192*
Nichiforesco, E., see Cionga, E. *470*
Nicholas, H. J. [180] 240, *255*
Nicholls, P. B., and L. G. Paleg [82a] 488, *492*
— see Paleg, L. G. [91a] 488, 489, *492*

Nichols, B. W. [150, 151] 369, 389, 391, *419*
— L. J. Morris and A. T. James [152] 375, *419*
Nickell, E. C., see Privett, O. S. [544] 134, *194*; [167, 168, 169, 170] 398, 399, 406, 407, 408, 411, 414, *420*
Nickless, G., see Pollard, F. H. [57] 623, 629, 898
Nicolaus, B. J. R. [92] 304, *310*; [27, 28] 569, 573, 574, 577, *578*; [5] 732, *781*
Niederwieser, A. [479] 91, *192*
— and M. Brenner [476, 477] 76, *192*
— and C. G. Honegger [478] 91, *192*
— and G. Pataki [150] 756, *784*
— see Brenner, M. [94, 95, 96, 97, 97a] 75, 132, 137, 138, 154, *182*; [24] 535, *562*; [7, 8, 12, 13] 733, 734, 740, 741, 742, 744, 755, 756, 764, 765, 766, 767, 771, 777, *781*, 889
— see Fahmy, A. R. [9, 211] 733, 740, 743, 748, *781*, *785*, 889
— see Walz, D. [739] 131, 132, *199*; [180] 583, *611*; [231] 758, 771, *786*
Nielsen, B., see Brochmann-Hanssen, E. [34c] *463*
Nielsen, B. E., and J. Lemmich [155] 695, *728*
Nienstedt, W. [132] 316, 317, 335, *360*
Nieweg, H. O., see Mandema, E. [103] 605, *609*
Nieze, L. H., see Smith jr., C. R. [190] 719, *728*
Nigam, I. C., and L. Levi [182, 182a] 212, 213, *255*
— M. Sahasrabudhe, and L. Levi [480] 116, *192*; [185] 212, *255*
— see Nigam, M. C. [183] 217, 225, 238, *255*
Nigam, M. C., I. C. Nigam, and L. Levi [183] 217, 225, 238, *255*
— and R. M. Purohit [184] 237, *255*
Nigam, S. S., and G. L. Kumari [181] 237, 238, 239, *255*
Nikitin, N. I. [481] 34, *192*
Nikkari, T., see Haahti, E. [70] 316, 322, 333, *359*; [52] 376, 398, *417*; [65, 66] 599, *609*
Nilsson, M. [156] 691, *728*
Ninnemann, H., J. A. D. Zeevaart, H. Kende, and A. Lang [83] 487, 489, *492*

Nishikaze, O., R. Abraham, and Hj. Staudinger [133] 324, 340, *360*, 902
— and Hj. Staudinger [482] 146, 147, *192*; [134] 341, *360*; [111] 601, *610*
Nishimoto, Y., and S. Toyoshima [110] 522, 524, *565*
— see Katsui, G. [74, 75] 263, 273, 284, 287, *309*
Nitsch, J. P. [84, 85] 475, *492*
— see Dennis, F. G. [11a] 489, *490*
Niwaguchi, T., see Yamamura, J. [106] 642, *656*
Nobile, S., see Vuilleumier, J. P. [139] 305, *311*
Noirfalise, A. [111] 508, 509, 519, 520, *565*
— and M. H. Grosjean [112] 508, *565*, 875
— and G. Mees *470*
Noll, W., K. Damm, and R. Fauss [483] 20, *192*
Noller, C. R., A. Melera, and M. Gut [185a] 246, *255*
Nomof, N., see Elliot, W. H. [50] 527, 529, *563*
Nord, F. F., see Kudzin, S. F. [115] 751, *783*
Nordby, G. L., see Wallach, D. F. H. [738] 90, *199*
Norden, D., see Eberhardt, H. [45] 527, 528, *563*
Nordsiek, K.-H., see Inhoffen, H. H. [36] *491*
Norén, B., see Ahrland, S. [43] 739, *782*
Norin, T., and L. Westfelt [186] 248, 249, *255*
Norman, A. W., and H. F. DeLuca [93, 94] 277, 280, *310*
— J. Lund, and H. F. DeLuca [95] 277, 279, 280, *310*
Normand, J., see Aurenge, J. [27, 28] 35, 138, *180*; [1] 848, *853*
Norton, W. T., and M. Brotz [153] 392, 394, *419*
Novat, N. [86] *492*
Novotný, L., and F. Šorm [157] 717, *728*
— J. Toman, F. Starý, A. D. Marquez, V. Herout, and F. Šorm [158] 717, *728*
Noworytko, J., and M. Sarnecka-Keller [63] 749, *782*

Nowotny, E., and Hj. Staudinger [112] 601, *610*
Nüesch, J., see Urech, J. [311] 238, *258*
Nürnberg, E. [485] 5, *193*; [96, 97] 300, 301, *310*; [150] 441, *466*; [113, 114] 558, 560, 562, *565*; [41] 740, *782*, 862, 866
Nufer, H., see Wagner, H. [320] 235, *258*, 876
Nuno, M. [159] 695, *728*
Nussbaumer, P. A. [486] 146, 149, 151, *193*; [115] 555, *565*; [29, 30, 31, 32, 33, 35] 575, *578*
— and M. Schorderet [34] 574, *578*
Nutter, L. J., see Blank, M. L. [14] 404, 406, 409, *416*
Nybom, N. [487, 488] 56, 69, 100, *193*; [160, 161] 699, *728*
Nyc, J. F. see Lie Rian Bo [403] 53, *190*

Obreiter, J. B., and B. B. Stowe [87] 476, 478, *492*
Obruba, K. [154] 387, *419*; [58] 675, *685*
Ochynska, J., see Domaglina, E. *470*
O'Connor, R. [40b] 457, *463*
Oehlschläger, H., J. Volke, and E. Kurek [116] 509, *565*
— — and G. T. Lim [489] 146, 151, 155, *193*
Oertel, G. W., and P. Brühl [113] 600, *610*
— and K. Groot [220] 340, *362*
— M. C. Tornero, and K. Groot [490] 46, 47, *193*; [135] 316, 356, 357, *360*
— see Weinand, K. [184] 602, *611*
Oesch, M., see Sahli, M. [594] 149, *195*; [198] 451, *468*; [140] 534, 535, 536, 537, 538, 539, *565*
Oette, K. [155] 381, *419*
— see Doss, M. [36] 398, *416*
Ofner, P., see Dyer, W. G. [48] 338, *358*; [34] 602, *608*
— see Peng, T. C. [125] 598, 602, *610*
— see Tai Chan, P. [232] 338, *362*
Ogawa, S. 894
Ogihara, Y., see Shibata, S. [188] 719, *728*
Ognyanov, I. [59] 667, 668, *685*
Ohkura, Y., see Momose, T. [219] 324, 345, *362*, 903

Oishi, M., see Nakayama, F. [128] 354, 360
Oki, H., see Yatabe, M. [176] 551, 566
Okuda, T. [36] 578
Okui, S., M. Uchiyama, and M. Mizugaki [156] 387, 419
Oldham, N. M., see Nichaman, M. Z. [475] 146, 147, 192
Oliver 444
Ollis, W. D., see Grisebach, H. [69] 687, 726
Olson, J. A., and O. Hayaishi [98] 275, 310
— see Dunagin, P. E. [33] 274, 308
Olson, R. E., see Nichaman, M. Z. [475] 146, 147, 192
Onckelen, H. A. van, R. Verbeek, and L. Massart [88] 475, 492
Onderberg, W. J. M., see Kamp, W. [99] 422, 438, 465
Onley, J. H. [67] 644, 656
— see Mills, P. A. [62] 644, 656
Onoe, K. [491] 59, 193; [187] 219, 255
Ono, T. [99] 303, 310
Opienska-Blauth, J. [114] 582, 585, 610
— M. Charezinski, and H. Berběč [89] 486, 492
— H. Kraczkowski, and H. Brzuszkiewicz [492] 5, 193; [115] 582, 610
— — — and Z. Zagórski [90] 486, 492; [116] 584, 610
— M. Sanecka, and M. Charezinski [53] 748, 782
Oppelt, M., see Endres, H. [185] 43, 185
— see Grassmann, W. [249] 42, 186
Orazi, O. O., see Djerassi, C. [45] 447, 463
Oreskes, I., see Saifer, A. [62] 749, 782
Orr, J. C., see Moser, H. W. [109] 594, 610
Orth, H., see Preuss, R. [204] 243, 255
Osborne, D. J., see Fletcher, R. A. [15a] 490
Ościk, J., see Waksmundzki, A. [48] 500, 506
Oshio, H., see Morimoto, H. [139] 466
Osisiogu, I. V., see Shellard, E. J. [152] 534, 535, 536, 566, 864
Osske, G., see Schreiber, K. [229] 241, 243, 256; [148] 323, 324, 347, 349, 361; [206] 459, 468, 861, 879, 892

Osteux, R., see Biserte, G. [45] 743, 760, 763, 766, 782
Oswald, N. [151] 425, 432, 434, 456, 466
— and H. Flück [493, 494] 135, 136, 138, 193; [152, 153] 425, 432, 434, 456, 466
Otani, S. [100] 305, 310
Othmer, D. F., see Kirk, R. E. [6] 732, 781
Ott, D. G., see Hansbury, E. [269, 270] 53, 131, 187; [27] 793, 805
Ott, H., see Hofmann, A. [90] 454, 465
Ottenstein, D. M. [495] 28, 193
Otto, G. [496] 42, 193
Ottoviano, G., see Steinegger, G. [231] 436, 469
Ourisson, G., see Mathis, C. [158, 159] 206, 210, 212, 238, 254
— see Ponsinet, G. [202] 242, 255
Overath, P., and P. K. Stumpf [497] 177, 193
Overbeek, J. van [90a] 472, 492
Overell, B. T., see Bryant, F. 859
Overman, R. T., and H. M. Clark [498] 193
Ovodov, Y. S., see Khorlin, A. Y. [127, 128] 243, 245, 253
Owellen, R. J., see Djerassi, C. [47] 447, 463
— see Ferreira, J. M. [57] 447, 464
Owen, C. A., see Kottke, B. A. [379a, 797] 150, 190, 200
Owen, J. A., see Albrecht-Recht, F. [15] 141, 180
Owens, K. [157] 392, 419
Owoc, M., see Purzycki, J. [92] 630
Oyama, M., see Kinoshita, S. [93] 387, 414, 418
Ozarowski, A., see Żurkowska, J. [787] 29, 146, 200; [234] 345, 362

Pacheco, G., see Carnegie, P. R. [119] 183
Pachéco, H. [105] 751, 783
— see Clerc-Bory, M. [112] 751, 783
Padley, F. B., see Barrett, C. B. [40, 41] 144, 181; [10] 248, 249, 250; [6, 7] 396, 398, 399, 414, 416
— see Gunstone, F. D. [51] 399, 417
Pagnucco, R., see Wolfrom, M. L. [96] 831, 837

Pai, B.R., see Govindachari, T.R. [77] 447, *464*
Pailer, M., P. Bergthaller, and G. Schaden [162] 718, *728*
— H. Kuhn, and I. Grünberger [188] 219, 220, *255*
— and W.H. Kump [154] 430, *466*
— and R. Libiseller [155] 431, *466*
Pais, M., J. Mainil, and R. Goutarel [156] 435, *466*
Paleg, L.G. [91] 472, 488, *492*
— D. Aspinall, and P.B. Nicholls [91a] 488, 489, *492*
— see Nicholls, P.B. [82a] 488, *492*
Palvy, I., see Tyihák, E. [220] 718, *729*
Panalaks, T., see Varma, T.N.R. [138] 273, 274, *311*
Pancake, E., see Cohn, G.L. [38] 320, *358*
Panicucci, F., see Luisi, M. [217] 340, 341, *362*
Panteleeva, N.S. [52] 795, 801, *805*
Panthinkhina, L.S., see Maximov, O.B. 893
Paoletti, R., and D. Kritchevsky [158] 367, *419*
Papariello, G., and M.A.M. Janish [21] 497, *505*
Papée, D., and R. Tertian [499, 500] 23, 24, *193*
Papp, E., and Z. Szabo [157] 430, *466*
Paquin, R., and M. Lepage [136] 349, *360*; [158] 459, *466*
Parekh, C.K., and R.H. Wasserman [501] 177, *193*; [101] 277, 280, *310*
Parello, J., A. Melera, and R. Goutarel [159] 460, *466*
Parihar, D.B. [36] 738, *781*
— S.P. Sharma, and K.C. Tewari [21a] 500, *505*
Paris, M., see Paris, R. [161, 161b] 422, *466*; [164] 697, 699, *728*
Paris, R. [160] 422, *466*; [163] 697, *728*
— and M. Godon [189] 237, 238, 239, *255*
— and M. Paris [161] 422, 466; [164] 697, 699, *728*
— R. Rousselet, M. Paris, and J. Fries [161b] 422, *466*
— and M. Šaršúnova *470*
— see Faugeras, G. [55, 56] 436, *464*
— see Masse, J. [155] 212, 236, 237, 238, 248, 250, *254*

Parke, D.V., see Coleman, T.J. [33] 245, 246, *251*
Parker, F., and N.F. Peterson [159] 389, 415, *419*
Parker, K., see Elliot, W.H. [50] 527, 529, *563*
Parkhurst, R.M., see Burke, W.J. 897
— see Skinner, W.A. [119, 120] 284, 287, *310*
Parks, O.W. [190] *255*
— see Schwartz, D.P. [232, 233] 222, *256*; [39] 495, *506*
Parrák, V., E. Raděj ová, and F. Machovičová [162] 432, *467*
— see Machovičová, F. [128] 444, *466*
Parrish, J.R. [22] 502, *506*
Parsons, J.G., and S. Patton [159a] 391, 414, *419*
Partridge, S.M. 857, 888
Parups, E.V., see Sirois, J.C. [118] 489, *493*
Pasalis, J., and N.H. Bell [102] 277, 278, *310*
Pascaud, M. [221] 333, *362*
Pascher 746
Pasich, B. [191] 245, *255*
— see Borkowski, B. [19] 245, *251*
Pásková, J., and V.J. Munk 860
Passera, C., A. Pedrotti, and G. Ferrari [68] 650, *656*, 867
Pastor, J., and R. Raimondi [117, 118] 543, *565*
Pastuska, G. [120] 540, *565*; [165] 698, 699, *728*; [8] 808, 810, 811, 813, 814, *835*
— and H.-J. Petrowitz [191a] 230, *255*; [24] 500, *506*; [121] 540, 541, *565*; [60, 61] 676, *685*
— and H. Trinks [502, 503] 111, 112, *193*; [23] 502, *506*; [119] 540, *565*; [51] 621, *629*; [166] 699, *728*; [195] 746, *785*
— see Petrowitz, H.J. [30] 504, 505, *506*; [71] 650, *656*; [63, 69] 684, *685*, *686*
Pataki, G. [505, 506] 53, 138, *193*; [103] 299, *310*; [118, 121, 123] 582, 588, *610* [193, 219, 220, 221, 222, 223] 743, 751, 771, 777, 778, *785*, *786*, 863, 870
— and M. Keller [504] 127, *193*; [117, 119, 120, 122] 579, 583, 585, 588, *610*

Pataki, G., and A. Kunz   [53] 794, *805*
— see Brenner, M.   [94, 97, 97a] 132, 137, 138, *182*; [8, 12, 140] 733, 742, 744, 753, 755, 756, 766, 777, *781*, *784*
— see Fahmy, A. R   [9, 211] 733, 740, 743, 748, *781*, *785*, 889
— see Kelemen, J.   [344] 125, *189*
— see Keller, M.   [91, 92] 583, 586, *609*
— see Niederwieser, A.   [150] 756, *784*
— see Walz, D.   [739] 131, 132, *199*; [180] 583, *611*; [231] 758, 771, *786*
Patel, M. B., see Puisieux, F.   [181] 448, *467*
Patin, D. L., see Wolfrom, M. L.   [4] 807, 809, 814, 815, 818, 820, 821, 829, 834, *835*
Patnaik, B. K., see Kohen, F.   [137] 243, *254*
Patt, P.   [507] 145, *193*; [167] 718, *728*
Patterson, S. J., and R. Clements   [224] 780, *786*
Patton, A. R., and P. Chism   [54] 749, *782*, 889
— and E. M. Foreman   [88] 750, 751, *783*
Patton, S., see Parsons, J. G.   [159a] 391, 414, *419*
Pauckner, W., see Grassmann, W.   [248] 43, *186*
Paul, A. G., see McLaughlin, J. L.   [415] 153, *191*; [118] 451, 453, *465*
Pauling, L.   [508] *193*
Paulose, M. M.   [160] *419*
Paulson, C., F. E. Deatherage, and E. F. Almy   [16] 736, *781*
Paulus, W.   [122] 534, 538, *565*
— W. Hoch, and R. Keymer   [124] 508, 509, 514, *565*
— and R. Keymer   [123] 539, 540, *565*
— et al.   [125] *565*
Pauly, H.   [91, 92] 750, *783*
Pav   447
Pavlásková, D., see Maturová, M.   [132b] 444, *466*
Pavlíček, M., see Rosmus, J.   [585] 75, *195*
Payne, S. N.   [509] 144, *193*; [161] 415, *419*; [124] *610*
Paysant, P., see Biserte, G.   [38] 738, *781*
Paz, M. A., see Blumenfeld, O. O.   [68] 827, 828, *836*

Peace, R., see Claisse, J.   [21] 812, *835*
Pecher, J., see Braekman, J. C.   *470*
Pechtold, F.   [126, 127, 128] 526, 527, 560, *565*
Pedersen, C., see Lundt, I.   [98] 831, *837*
Pedrotti, A., see Passera, C.   [68] 650, *656*, 867
Peereboom, J. W. C.   [510] 27, *193*; [52] 613, *629*
— and H. W. Beekes   [90] *630*
Peifer, J. J.   [105] 833, *837*
Peirce, F. T.   [533] *194*
Pejković-Tadić, I., and M. Hranisavljević-Jakovljević   [62] 680, 681, *685*
— — and S. Nesić   [192] 219, *255*
— see Hranisavljević-Jakovljević, M.   [29] 680, 681, *685*, 865
Pekkarinen, A.   [222] 340, *362*
Pelick, N., H. R. Bolliger, and H. K. Mangold   [511] 1, *193*
Pelka, J. R., and L. D. Metcalfe   [512] 138, *193*; [162] 387, *419*; [25] 472, 497, *506*
Pelousek, H., see Schlögl, K.   [79, 80, 82] 665, *686*
Peng, C. T.   [513] 177, *193*
— R. Vena, P. Ofner, and P. L. Munson   [125] 598, 602, *610*
— see Dyer, W. G.   [48] 338, *358*; [34] 602, *608*
Penick, R. J., M. H. Meisler, and R. H. McCluer   [163] 389, 391, 415, *419*; [126] 598, *610*
Penna-Herreros, A.   [163] 441, *467*
Pennington, G. W., and D. Smith   [129] 537, *565*
Pennock, J. F., and P. J. Dunphy   [105] 283, 285, 286, *310*
— F. W. Hemming, and J. D. Kerr   [104] 283, 284, 285, 286, 287, *310*
— see Dunphy, P. J.   [34, 35] 284, 286, *308*; [38] 378, 412, *416*, 867
— see Whittle, K. J.   [146] 286, *311*
Pentillä, A., and J. Sundman   [168] 714, *728*
Perani, M., see Chierici, L.   [4] *505*
Pereda, A. L., see Matthews, J. S.   [437] 146, 147, 151, 153, *191*; [122] 322, 336, 339, 341, *360*
Pereira, N. A., see Antonaccio, L. D.   [6] 447, *462*

Pereira,R.L., see Morgan,M.E. [172] 220, *254*
Pérez-Medina,L.A., E.Travecedo, and J.E.Devia [164] 459, *467*
Peri,J.B. [514] 26, *193*
Perkavec,J., and M.Perpar [53, 54, 55] 620, 623, 626, 627, *629*
— see Logar,S. [38] 618, 619, 620, *629*
Perkins,L.M., see Brenner,G.S. [104] 832, 834, *837*
Perley,J.E., and B.B.Stowe [92] *492*
Pernoux,E., see Teichner,S. [710] 28, *198*
Perpar,M., see Ciglar,J. [6] 618, *628*
— see Kornhauser,A. [108b] 454, *465*; [122] 718, *727*
— see Logar,S. [38] 618, 619, 620, *629*
— see Perkavec,J. [53, 54, 55] 620, 623, 626, 627, *629*
Perrings,J.D., see Hansbury,E. [270] 53, *187*
Perry,E.S., see Lestienne,A. [398] 52, 59, *190*
— see Przybylowicz,E.P. [546] 59, *194*
Perry,M.B., see Bishop,C.T. [72] 828, *836*
— see Hulyalkar,R.K. [75] 828, *836*
Pertsev,I.M., and G.P.Pivnenko [193, 194] 237, 238, *255*
Pesch,L.A., see Scheig,R.L. [86] 792, 795, 796, *806*
Peschke,W. 857
Pesez,M. 866
Pesnelle,P., P.Teisseire, and M.Wichtl [195] 212, 228, *255*
Peteri,D., see Knappe,E. [133] 214, *254*; [45, 46, 47, 48, 49] 650, 651, 652, *655*; [38, 39, 40, 41] 662, 664, *685*, 860, 861, 870, 877
Peterlik,M., see Schlögl,K. [83] 666, *686*
Peters,K., see Steinegger,E. [210] 719, *729*
Petersen,H., see Frank,H. [89] 750, 751, *783*
Peterson,E.A., and H.A.Sober [516, 517] 37, 90, *193*; [54] *805*
Peterson,N.F., see Parker,F. [159] 389, 415, *419*
Peterson,R.F., see Skipski,V.P. [637] 147, 154, *196*; [187, 188, 189] 389, 391, 415, *420*; [153] 599, *611*

Petit,J., see Berger,J.-A. [59, 60, 61, 62, 63] 45, 46, 94, *181*; [205] 780, *785*; [4] 842, 845, *853*
Petitjean,M., see Prettre,M. [538] 25, *194*
Petridis,C., R.Verbeek, and L.Massart [93] 489, *492*
Petrović,S.M., see Canić,V.D. [115] 39, *183*; [5] 842, *853*
Petrowitz,H.J. [519, 520] 128, 137, *193*; [196, 197] 217, 226, *255*; [26, 27, 28, 29] 501, 503, 504, 505, *506*; [130] 541, *565*; [69, 70] 642, 648, 651, *656*; [64, 65, 66, 67, 68] 663, 667, 668, 676, *685, 686*
— and G.Pastuska [71] 650, *656*; [63] *685*
— — and S.Wagner [30] 504, 505, *506*; [69] 684, *686*
— see Pastuska,G., [191a] 230, *255*; [24] 500, *506*; [121] 540, 541, *565*; [60, 61] 676, *685*
Petschik,H., and E.Steger [72] 640, *656*
Pettersson,G. [169, 170] 694, 697, *728*
Peurifoy,P.V., S.C.Slaymaker, and M.Nager 901
Peyron,L. [521, 522, 523, 524] 73, *193*; [198, 199] 233, 238, *255*
Pfaff,J.D., and E.Sawicki [525] 141, *194*
— see Sawicki,E. [599, 600] 37, 145, 155, *195*; [77, 78] 667, 669, *686*
Pfannmüller,H., see Zahn,H. [174] 762, *784*
Pfeffer,K.H., see Gries,G. [62] 586, *608*
Pfeifer,J.J. [526] 53, 139, *194*; [137] 317, 334, *360*
Pfeifer,S. [165, 166] 441, *467, 470*
— and S.K.Banerjee [167] 441, *467*
— I.Mann, and L.Kühn 461
— see Hiller,K. [89a] 246, 247, *252*
— see Kühn,L. 461
— see Zarnack,J. [267] 422, 441, 457, *469*; [177] 524, 525, 542, 549, *566*; [109] *806*, 875
Pfeiffer,P. [527] 41, *194*
Pfeifle,J., see Stahl,E. [92] 646, 647, *656*; [228] 779, 780, *786*
Pfeil,E., A.Friedrich, and T.Wachsmann 896
Pflederer,G., see Wieland,Th. [753] 754] 109, 111, 112, 113, *199*

Pflugshaupt, R. P., see Grob, E. C. [56] 267, *309*
Pfrunder, B., R. Zurflüh, H. Seiler, and H. Erlenmeyer [528] 111, *194*
Philips, M. A., and R. D. Hinkel [73] 632, *656*
Philipsborn, W. v., see Hesse, M. [88] 447, *464*
— see Pinar, M. [168] 447, *467*
Phillips, B. M., and N. Robinson [127] 597, *610*
Phillips, G. E., see Dicarlo, F. J. [161] *184*; [18] 671, 673, *684*
Phillips, G. T., see Atkinson, R. E. [4] 235, 239, *250*
— see Curtis, R. F. [38] *251*; [15] 681, 682, *684*, 883
Phillips, J. D. J., see Jones, R. L. [40] 475, *491*
Phillipson, J. D., and E. J. Shellard *470*
— see Shellard, E. J. [18c, 215, 216, 217, 218] 454, *462, 468, 471*
Phinney, B. O., and C. A. West [94] 488, *492*
Pickett, B. W., see Komarek, R. J. [103] 414, *418*
Pierce, C. [532] 17, 33, *194*
Pierce, J. G., see Davoll, H. [158] 758, *784*
Pieringer, A. P., see Attaway, J. A. [6a] 210, 237, *250*
Pies, H., see Strohecker, jr. R. [701] 154, *198*; [129] 305, 306, *311*
Pietsch, H. P., and R. Meyer [91] *630*
Piez, K. A., E. B. Tooper, and L. S. Fosdick [29] 737, *781*
Pifferi, P. G. [7] 808, 810, 811, 813, 814, 833, *835*
Piironen, E., and A. I. Virtanen [106] 306, *310*; [95] *492*
Pijewska, L., J. L. Kaul, R. K. Joshi, and F. Šantavý *470*
— see Bláha, K. [22] 443, *462*
Pilet, P. E. [96] 492, 875
— see Collet, G. [8] 478, 483, 488, *490*
Pilz, H., and H. Jatzkewitz [164] 393, 415, *420*
— see Jatzkewitz, H. [73] 389, 415, *417*; [79] 598, *609*
Pinar, M., W. v. Philipsborn, W. Vetter, and H. Schmid [168] 447, *467*
— and H. Schmid [169] 447, *467*

Pinder, A. R. *461*
Pinizzi, L., see Corsano, S. [35] 245, *251*
Pinxteren, J. A. C., and M. E. van Verloop [170, 171] 438, 440, *467*
Pirmann, B., see Neurath, G. [57] 680, 681, *685*
Pisters, H. [200] 255
Pitra, J., and J. Štěrba [172] *467*
— see Procházka, V. [180b] 451, *467*
— see Reichelt, J. [140] 320, 323, 345, *360*
Pivnenko, G. P., see Pertsev, I. M. [193, 194] 237, 238, *255*
Plaa, G. L., see Whelan, F. J. [189] 605, *611*
Plack, P. A., and J. G. Bieri [107] 284, *310*
Planta, C. v., U. Schwieter, L. Choparddit-Jean, R. Rüegg, and O. Isler [108] 273, 274, *310*
Plat, M., see Debray, M. *470*
Pless, J., see Guttmann, St. [136] 753, *784*
Plummer, A. J., see Sheppard, H. [153] 540, *566*
Pöhm, M. [97] 485, *492*; [15] 736, *781*
Poel, G. H., van der [534] 31, *194*; [201] 255
Poethke, W., and H. Behrendt [173] 707, *728*
— — and E. Matschke [172] 707, *728*
— and J. Hoffmann [173] 432, *674*
— and W. Kinze [535, 536] 139, 149, 151, 153, *194*; [174, 175, 176, 177] 425, 438, 440, *467*; [131] 543, *565*
— see Luckner, M. [125, 126] 432, 456, *466*
Pötter, H., and R. Voigt *470*
Pohl, P., see Wagner, H. [734] 65, *199*
— see Wagner, H. [211, 212] 404, 406, *421*
Pohloudek-Fabini, R., see Beyrich, T. [13] 239, *251*
— see Knütter, S. [134] 235, 236, *254*
Poisson, J., see Puisieux, F. [181] 448, *467*
Pokorny, G., see Gagliardi, E. [9] 847, *853*
Pokorný, J., see Davídek, J. [10] 613, *628*; [24] 634, *655*
Polderman, J., see Fokkens, J. [55] 523, 560, *563*

Poldermann, J., see Gänshirt, H. [217] 135, 138, 146, 147, 149, 151, 153, 154, *185*; [61] 341, *358*; [61] 561, *563*

Poljakoff-Mayber, A., see Koller, D. [59] 473, *491*

Pollard, F. H., K. W. C. Burton, and D. Lyons 854
— J. F. W. McOmie, and I. I. M. Elbeih 881
— G. Nickless, and K. W. C. Burton 898
— G. Nickless, T. J. Samuelson, and R. G. Anderson [57] 623, *629*

Polonius, W., see Brieskorn, C. H. [26, 27] 242, 243, 245, *251*

Polvani, F., see Luisi, M. [101] 600, *609*

Ponsinet, G., and G. Ourisson [202] 242, *255*

Pontis, H. G., see Dietrich, C. P. [162] 47, 149, 154, *184*; [18] 801, *804*; [11] 809, 821, 834, *835*

Pont Lezia, R. F. [97a] 489, *492*

Poole, A. G., see Bevan, T. H. 887

Popelak, A., see Kaiser, F. [97] 448, *465*

Pop-Gocan, A., see Hodosan, F. [82] *359*

Popkova, G. A., see Bergelson, L. D. [12] 379, 381, 385, *416*

Popov, A., and V. Gadeva 860

Popov, K., see Popova, Y. [109] 302, *310*

Popova, Y., K. Popov, and M. Iliea [109] 302, *310*

Porath, J. [22, 23] 737, 752, *781*
— and P. Flodin [126] 752, *783*
— see Lindner, E. B. [128] 752, *783*

Porges, E., see Porgesová, L. [537] 81, *194*

Porgesová, L., and E. Porges [537] 81, *194*

Portatius, H. v., see Hörmann, H. [295] 53, *188*

Porter, R. [165] 759, *784*
— and F. Sanger [154] 756, 758, *784*

Porter-Silber 340

Portwich, F., see Seydel, J. H. [151] 560, *566*

Potěšilová, H., J. Hrbek jr., and F. Šantavý [178] 426, 427, 429, *467*
— see Vrublovský, P. [255] 441, 443, *469*

Potier, P., R. Beugelmans, J. Le Men, and M. M. Janot [179] 450, *467*

Potter 369

Potter, A. D., see Burke, W. J. 897

Potter, W. P. de, R. F. Vochten, and A. F. de Schaepdryver [31] 499, *506*
— — — and C. Heymanns [132] 520, 524, *565*

Potterat, M., see Mottier, M. [463] 5, 149, *192*; [48] *629*

Poulet, F., see Cotte, I. [9] 473, 483, *490*

Preininger, V., A. D. Cross, and F. Šantavý [179b] 444, *467*
— and P. Vrublovský [180] 440, 441, *467*
— see Vácha, P. [250] 454, 455, 456, *469*

Prétôt, M. 308

Prettre, M., B. Imelik, L. Blanchin, and M. Petitjean [538] 25, *194*

Preuss, K., see Riemenschneider, R. [72] 749, *782*

Preuss, R. [203] 237, *255*
— and H. Orth [204] 243, *255*

Preussmann, R., D. Daiber, and H. Hengy [32] 497, *506*, 871, 900
— G. Neurath, G. Wulf-Lorentzen, D. Daiber, and H. Hengy [33] 497, *506*, 871, 900

Previtera, S. A., see Gillio-Tos, M. [7] 500, 501, *505*; [23] 626, *628*, 876

Prey, V., H. Berbalk, and M. Kausz [74] 650, 654, *656*; [70] *686*; [24, 50, 80] 814, 817, 820, 828, *835*, *836*
— A. Berger, and H. Berbalk [205] 211, *255*
— W. Braunsteiner, R. Goller, and F. Stressler-Buchwein [36] 817, *835*
— H. Scherz, and E. Bancher [539] 32, *194*; [17] 811, 812, 817, 820, *835*
— see Bancher, E. [37] 53, *181*; [5] 651, *654*; [203] 734, *785*
— see Grundschober, F. [79] 828, *836*

Pribilla, O. [133] 509, *565*

Price, C. H., see McCarthy, T. J. [142] 706, *727*

Pridham, J. B. [171] 691, *728*

Prins, D. A., see Renner, U. [184] 447, *467*

Prival, E. L., see Bieri, J. G. [5] 266, 284, 290, *308*

Privett, O. S. 6

Privett,O.S., and M.L.Blank [540, 542, 543] 139, 140, *194*; [165, 168] 377, 408, 409, 412, 441, *420*; [128] 593, *610*
— — D.W.Codding, and E.C.Nickell [544] 134, *194*
— — and W.O.Lundberg [541] 139, 140, *194*; [166] *420*
— and E.C.Nickell [167, 168, 170] 398, 399, 406, 407, 408, 411, 414, *420*
— — and O.Romanus [169] 408, *420*
— see Blank,M.L. [73] 139, 140, 158, *182*; [13, 14] 404, 406, 409, 414, *416*
Procházka,V., F.Kafka, M.Prucha, and J.Pitra [180b] 451, *467*
Procházka,Ž. [545] 103, *194* [98] *493*; [109] 751, *783*, 879
— see Kutáček,M. [63] 472, 474, 479, 486, *491*
Prokazova,N.V., see Bergelson,L.D. [12] 379, 381, 385, *416*
Prucha,M., see Procházka,V. [180b] 451, *467*
Przybylowicz,E.P., W.J.Staudenmayer E.S.Perry, A.D.Baitsholts, and T.N.Tischer [546] 59, *194*
— see Lestienne,A. [398] 52, 59, *190*
Pufahl,R., and K.Schreiber [180c] 460, *467*
— see Schreiber,K. [207] 460, *468*
Pugnet,T., see Dumazert,C. [46] 316, *358*; [27] 650, *655*; [55] 822, 824, *836*
Puisieux,F., M.B.Patel, J.M.Rowson, and J.Poisson [181] 448, *467*
Purdy,S.J., and E.V.Truter [547, 548, 549] 137, 138, *194*; [171] 411, 415, *420*
Purohit,R.M., see Nigam,M.C. [184] 237, *255*
Purzycki,J., A.Szwark, and M.Owoc [92] *630*
Puschmann,G., see Kratzl,K. [381] 178, 179, *190*; [127] 698, *727*

Qedan,S., see Schratz,E. [228] 211, 217, 220, 230, 239, *256*
Quesenberry,R.O., E.M.Donaldson, and F.Ungar [223] 314, 317, *362*
— and F.Ungar [138] 320, 335, 340, 341, *360*
Quirin,M., J.Lévy, and J.le Men [181b] 448, *467*
Quitt,P., see Brenner,M. [149] 755, *784*

Raban,P. [58] 622, *629*
Rabenort,B. [550] 52, 149, *194*
Rabinowitz,J., see Cherbuliez,E. [4, 187] 732, 774, 775, 777, *781*, *785*
Rabitzsch,G., and H.Herzmann [551] 177, *194*
Radějová,E., see Parrák,V. [162] 432, *467*
Radin,N.S. [552] 130, *194*
— see Kishimoto,I. [94] 387, *418*
Radley,M. [98a] 489, *493*
— see Jones,D.F. [39] 479, 486, 487, 489, *491*, 900
Radlick,P.C., see Khalifah,R.A. [53] *491*
Radwan,S.S., see Kaufmann,H.P. [91] 390, 415, *418*
Raffauf,R.F. [182] 423, *467*
— see Douglas,B. [49] 460, *463*
Ragazzi,E., and G.Veronese [25] 814, *835*
Rahandraha,Th. [553, 554] 30, *194*
— M.Chanez, and P.Boiteau [206] 246, *255*
— — — and S.Jaquard [555] 30, *194*; [207] 246, *255*
Raimondi,R., see Pastor,J. [117, 118] 543, *565*
Raiteri,M., see Corona,G.L. [195] 345, *392*
Rajappa,S., see Govindachari,T.R. [77] 447, *464*
Ralston,H., see Thomas,A.E. III [201] 408, *420*
Ramamurthy,M.K., and V.R.Bhalerao [59] 613, *629*
Ramaut,J.L. [182b] 437, *467*; [174] 695, *728*
Ramsey,H.A. [556] 40, 111, 112, *194*
Ramstad,E., and S.Agurell [99] 472, *493*
— see Agurell,S. [4] 450, *462*
— see Lou,V. [74] *492*
Randall,M.H., see Foster,A.B. [85] 831, *836*
Randerath 39
Randerath,E., and K.Randerath [562, 566] 46, 104, *194*, *195*; [55, 56, 57] 793, 795, 796, 797, 800, 801, *805*
— see Neuhard,J. [472] *192*
— see Randerath,K. [51, 70, 72, 73, 75] 795, 796, 797, 799, 800, 801, *805*, *806*

Randerath, K. [557, 558, 559, 560, 561, 564, 565]; 5, 46, 47, 104, *194*; [110] 259, *310*; [139] 312, *360*; [58, 59, 60, 63, 64, 65, 66, 67, 69, 71, 74, 77] 792, 793, 794, 795, 796, 798, 799, 800, *805*, *806*
- and F. Cramer [62] 801, *805*
- C. M. Janeway, M. L. Stephenson, and P. C. Zamecnik [76] 801, *806*
- and E. Randerath [70, 72, 73, 75] 795, 796, 799, 800, 801, *806*
- and H. Struck [61] 795, 798, *805*
- and G. Weimann [563] 96, *194*; [68] 802, *806*
- see Neuhard, J. [472] *192*; [51] 795, 796, 797, *805*
- see Randerath, E. [562, 566] 46, 104, *194*, *195*; [55, 56, 57] 793, 795, 796, 797, 800, 801, *805*
- see Weimann, G. [745] 46, *199*; [103] 795, 801, *806*
- see Zamecnik, P. C. [108] 801, *806*
Ranft, G., see Hansel, R. [75, 76] 697, 698, *726*
Rangoonwala, R., see Friedrich, H. [57] 718, *725*
Rao, G. S., see Kapadia, G. J. [20] 573, *577*
Rao, I. V. S., see Das, V. S. R. [11] *490*
Rao, K. R. K., A. K. Bhalla, and K. Sinha [71] 671, *686*
Rao, M. K. G., S. V. Rao, and K. T. Achaya [111] 284, 287, *310*
Rao, S. V., see Rao, M. K. G. [111] 284, 287, *310*
Rapoport, H., see Battersby, A. R. [13b] 444, *462*
Rappaport, L., see Hayashi, F. [31] 489, *490*
Rapport, M. M., L. Graf, and H. Schneider [172] 389, *420*
Ratapongs, C. [78] 793, *806*
Ratshinsky, and Shraiber [567] 1, *195*
Ratshinsky, V. V., see Shostenko, U. V. [633] 1, *196*
Raulin, J., see Carreau, J.-P. [120] 29, *183*
Rauterkus, K. J., see Kern, W. [36] 658, *685*
Rawlings, F. I. G., and A. E. Werner [208] 248, *255*
Raymond, S., and B. Aurell [568] 112, *195*

Rebel, G., and P. Mandel [112] 290, 291, 292, *310*
Reddi, K. K., see Kodicek, E. 859
Redfield, R. R. [27] 737, *781*
Redgwell, R. J., see Turner, N. A. [230] 734, *786*
Redman, C. M., and R. W. Keenan [173] 390, *420*
Reeves, W. A., and T. B. Crumler 881
Rehbinder, D., and D. M. Greenberg [569] 176, *195*
Reich, H. U., see Gnehm, R. [237a, 792] 138, *186*, *200*
- see Gnehm, R. [70a] 554, *563*
Reichelt, J., and J. Pitra [140] 320, 323, 345, *360*
Reichstein, T. 345
- see Kaufmann, H. [95a] 342, *359*
- see Lewbart, M. L. [108a] 324, 342, *360*, 870
- see Manzetti, A. R. [118] 317, 342, *360*
Reid, P. E., see Dutton, G. G. S. [14] 810, 831, 832, 833, 834, *835*
Reif, J. [60] 623, *629*
Reinauer, H., and F. H. Bruns [79] *806*
Reindel, F., and W. Bienenfeld [37] 738, *781*
- and W. Hoppe [570] 89, *195*; [73] 749, 755, *782*, 863
Reinecke, I., see Awe, W. [84] 750, *782*, 882
Reinhard, E., W. Konopka, and R. Sacher [100] 489, *493*
Reinhardt, H., see Hörhammer, L. [112] 717, *727*
Reinicke, C., see Boernig, H. [5] 801, *804*
Reio, L. [175] *728*
Reisch, J., H. Bornfleth, and J. Rheinbay [134, 135, 136] 533, 542, 543, 544, 558, *565*
- - and G. L. Tittel [137] 550, 551, *565*
Reisert, P. M., and D. Schumacher [141] 338, *360*; [129] 602, *610*
Reissell, P. K., L. M. Hagopian, and F. T. Hatch [130] 591, 592, *610*
Reitsema, R. H. [571, 572] 3, *195*; [209] 241, *255*
- F. J. Cramer, and W. E. Fass [210] 234, *255*

Rekker, R. F., see Doorenbos, H. J. [39] 518, *563*
Remenchik, A. P., and I. Bernsohn [80] 792, 802, *806*
Renault, J. L. [183] 459, *467*
Renkonen, O. [174] 392, 401, 409, *420*
Renner, U. [183b] 448, *467*
— D. A. Prins, A. L. Burlingame, and K. Biemann [184] 447, *467*
Rentschler, H., see Tanner, H. [212] 699, *729*
Renz, J., M. Kuhn, and A. v. Wartburg [176] 710, *728*
Rettie, G. H., and C. G. Haynes [61] 613, 617, 618, 620, *629*
Reuter, G. [185] 423, *467*
Reuther 434
Reuther, F. W., see Thies, H. 526, 874
Reuther, K.-H., see Drawert, F. [171] 164, *184*; [33] 585, *608*
Rey, E. [72a] 673, 674, *686*
— and L. Erhart [72] 673, 674, *686*
Reynolds, T., see Griffith, C. M. [24a] 488, 489, *490*
Reznik, H., and K. Egger [177] 705, *728*, 865
Rheede van Oudtshoorn, M. C. B. v. [178, 180, 182] 706, *728*
— and K. W. Gerritsma [179, 181] 706, *728*
— see Gerritsma, K. W. [63] 707, *726*
— see McCarthy, T. J. [143] 706, *727*
Rheinbay, J., see Reisch, J. [134, 135, 136] 533, 542, 543, 544, 558, *565*
Ribeiro, O., see Douglas, B. [49] 460, *463*
Ribéreau-Gayon, P. [183] 704, *728*
Rice, R. G., G. J. Keller, and J. G. Kirchner 889
— see Kirchner, J. G. [349] 3, 146, 149, *189*
Richards, E. L., see Grimmett, M. R. 899
Richardson, A. C., see Buss, D. H. [43] 818, *835*
Richardson, G. S., I. Weliky, W. Batchelder, M. Griffith, and L. L. Engel [573] 157, 159, *195*
Richert, K. H., see Tschesche, R. [171] 312, 315, 327, 328, 329, 341, *361*
Richter, E. [142] 323, 334, *360*, 888
Richter, G., see El-Hamidi, A. [58] 237, *252*

Richter, G., see Hörhammer, L. [90, 92, 93] 237, 238, 239, *252*
— see Shalaby, A. F. [238] 236, *256*
Richterich, R., see Bürgi, W. [19] 583, *608*
Riddick, J. A., see Jones, L. R. [42] 639, *655*
Rieche, A., E. Höft, and H. Schultze [74] 678, *686*
— and M. Schulz [73] 678, *686*
Rieck, G., see Glemser, O. [234, 235, 236] 12, 24, 26, 31, *186*
Riedlová, J., see Stárka, L. [159, 160] 338, 340, *361*; [156, 157] 602, *611*
Riegelman, S., see Fischer, L. J. [192a, 791] 141, *185, 200*
Riemersma, J. C., see Hooghwinkel, G. J. M. [64] 392, *417*
Riemschneider, R., and K. Preuss [72] 749, *782*
Riess, J. [224] 356, *362*
Riezebos, G., J. C. Grimmelikhuysen, and D. A. van Dorp [175] 408, *420*
Riley, J. P., and T. R. S. Wilson [113] 269, *310*
Rimpler, H., see Hänsel, R. [78] 704, *726*
Rindt, W., see Weinand, K. [184] 602, *611*
Riniker, R. [141] 754, *784*
Rink, M., and A. Gehl [137b] 549, *565*; [81] 801, *806*
— and S. Herrmann [574, 575] 37, *195*; [131, 132] 580, 581, *610*; [75] 651, 652, *656*; [65] 825, 826, 827, 834, *836*
Rippel-Baldes, A., see Wallhäuser, K. H. [43] 569, *578*
Ripperger, H., see Schreiber, K. [208] 459, *468*
Rippstein, S., see Bäumler, J. [15] 433, 437, 439, 449, 454, *462*; [10, 11] 508, 509, 514, 528, 534, 535, 536, 538, *562*; [6] 639, 642, *654*; [3] 846, *853*, 870, 892
Risch, A., see Kohlschütter, H. W. [366] 23, *189*
Risch, H., see Libbert, E. [71] 488, *492*
Rispoli, G., and A. di Giacomo [114] 269, *310*
— — and M. E. Tracuzzi [211] 237, *256*

Ristić, S., and A. Thomas [186, 187] 460, *467*
Ritschard, W. J. [576] 76, 88, *195*
Ritschel-Beurlin [137a] 543, *565*
Ritter, F. J., P. Canonne, and F. Geiss [75] 667, *686*
— and G. M. Meyer [578] 99, 101, 149, *195*
— — and F. Geiss [579] 32, *195*
— see Geiss, F. [25] 666, *685*
Ritter, H., see Haefelfinger, P. [259] 145, 151, *187*; [72] 558, *563*
Ritter, H. L., and L. C. Drake [577] 28, *195*
Ritter, R., see Meyer-Dulheuer, K.-H. [102] 542, 543, *564*
Rivlin, R. S., and H. Wilson [580] 170, *195*; [143] 341, *360*
Roberts, E., see Bregoff, H. M. 873
Robin, J. P., see Blandenet, G. [72] *182*
Robin, M. Ch., see Maugras, M. [53] *685*
Robins, R. K., see Broom, A. D. [7] 794, *804*
Robinson, A. D., see Kramer, J. K. G. [380] *190*
Robinson, A. E., see Beckett, A. H. [16] 524, 542, *562*, 876
Robinson, N., see Phillips, B. M. [127] 597, *610*
Robles, M. A., and R. Wientjes [188] 423, 435, 441, *467*
Robson, P., see Dunn, F. [37] 414, *416*
Roche, J., S. Lissitzky, and R. Michel [581] 155, *195*
— et al. 891
Roche, J. N., van Thoai, and J. L. Hatt [75] 749, *782*
Rochelmeyer, H. [189] 423, 451, *467*
— see Dibbern, H. W. [13] 485, *490*
— see Eich, E. [14] 477, 481, 486, *490*
— see Hohmann, T. [92] 5 l, *465*
— see Klavehn, M. [352] 135, *189*; [105, 106] 451, 453, *465*; [85] 562, *564*
— see Mutschler, E. [10] 733, 740, *781*
— see Röder, K. *471*
— see Schunack, W. [209b] 461, *468*; [145a] 549, 550, *565*; [201] 718, *729*
— see Teichert, K. [241, 242] 422, 432, 433, 437, 439, 448, 451, 453, 455, 457, *469*; [45] 494, 495, *506*; [163] 522, 549, *566*; [213] 729

Rodegker, W., see Nevenzel, J. C. [149] 384, *419*
Röder, K., E. Mutschler, and H. Rochelmayer *471*
Römpp, H. [582] 12, *195*
Rönsch, H., and K. Schreiber *471*
Rössel, T. [583] 35, *195*; [20] 849, 850, *853*
Rogers, A. W., see Chamberlain, J. [126] 159, *183*
Rogers, W. I., see Morton, M. J. [49] 802, *805*
Rogirst, A., see Vancraenenbroeck, R. [222] 704, *729*
Rohdewald, J., see Knappe, E. [133a] 233, *254*; [87] 540, *564*; [47] 650, *655*; [40, 41, 41 a] 662, 663, 664, *685*; [121] 697, *727*, 861
Rohrbaugh, L. M., see Winkler, B. C. *471*
Rojahn, W., see Klein, E. [129] 212, *253*
Rokkones, T. [133, 134] 582, *610*
Rollason, J. G., see Kelleher, J. [83] 536, *564*
Rolle, L. A., see Ikeda, R. M. [104] *253*
Roller, K., see Hoffmann, H. [79] 526, 527, *564*
Rollins, C. B., and R. D. Wood [176] 378, *420*
— see Zabin, B. A. [780] 32, 47, *200*
Románuk, M., see Křepinský, J. [142] 239, *254*
Romanus, O., see Privett, O. S. [169] 408, *420*
Rombach, R., see Thoma, K. [95, 96] 662, *686*, 873
Roncari, G., see Keller-Schierlein, W. [21] 572, *577*
Ronkainen, P. [76] 651, 652, *656*
— D. Kaempgen, and H. Suomalainen [212] 220, *256*
— T. Salo, and H. Suomalainen [213] 220, *256*
— see Suomalainen, H. [276] 221, *257*
Roos, P., see Harris, J. J. [188] 774, *785*
Rooy, C. de, see Dhont, J. H. [48, 49] 220, 223, 228, 231, 235, *251*; [38] 540, *563*
Roper, E. C., R. N. Blomster, N. R. Farnsworth, and F. J. Draus [189b] 424, *467*

Rosa, U., see Massaglia, A. [435] 177, 191; [217] 779, 785
Rose, E., see Sunshine, I. [160] 534, 535, 536, 566, 884
Rose, F. A., see Wusteman, F. S. [185] 356, 361; [91] 679, 680, 686
Rose, T. J., see Giles, C. H. [231] 41, 186
Rosebeek, S. [67] 749, 782
Rosenberg, J., and M. Bolgar [584] 160, 195
Rosenblath, J. D., see Neidlein, R. [107b] 547, 564
Rosenthal, A., and D. Abson [91] 831, 837
— and H. J. Koch [92] 831, 837
Rosenthal, J. M., see Futterweit, W. [54] 602, 608
Rosmus, J., and Z. Deyl [214] 221, 256
— M. Pavliček, and Z. Deyl [585] 75, 195
— see Kutáček, M. [64] 479, 487, 491
Rosmus, P., see Mayer, R. [54] 682, 685
Rossi, M., see Grippo. P. [26] 792, 793, 795, 796, 805
Rotbacher, H., C. Crisan, and E. Bedo [215a] 238, 256
Roth, M., see Wagner-Jauregg, Th. [319] 245, 258
Rothblatt, G. H., see Kritchewsky, D. [104] 323, 354, 359
Rothenheimer, C. [215] 248, 256
Rother, A., J. M. Bobbitt, and A. E. Schwarting [190] 431, 467
— see Khanna, K. L. [102] 434, 465
— see Leary, J. D. [119] 431, 465
— see Schwarting, A. E. [211] 434, 468
Rothweiler, W., see Seiler, H. [25] 842, 853
Roucayrol, J. C., and P. Taillandier [586] 162, 195
Rouser, G., C. Galli, and G. Kritchevsky [177] 389, 391, 393, 420
— see Fleischer, S. [45] 266, 309
— see Murphy, M. T. J. [147] 380, 419
Rousselet, R., see Paris, R. [161b] 422, 466
Roux, D. G. 902
— and S. R. Evelin [184] 693, 728
Rovigati da Silva Jardim, I. [191] 441, 467
Rowe, J. W. [216] 256

Rowe, J. W. see Seikel, M. K. [237] 226, 256
— see Zinkel, D. F. [334] 249, 258
Rowell, L. B., see Masoro, E. J. [104] 599, 609
Rowell, R. M., see Whistler, R. L. [90] 831, 837
Rowson, J. M., see Puisieux, F. [181] 448, 467
Roy, D. K., see Bhattacharya, K. R. [74] 749, 782
Różyzo, J., see Waksmundzki, A. [48] 500, 506
Rubenstein, D., see Smith, L. W. [639] 40, 196; [122] 269, 311
Rucker, W., see Scherz, H. [139] 580, 610
Ruddat, M., E. Heftmann, and A. Lang [587] 195
Rudinger, J., see Jost, K. [42] 740, 782
Rudloff, E. v., and A. Sato [185] 712, 728
Rücker, G. [217] 235, 256
Rüdiger, W., and H. Büch [139] 526, 565
Rüegg, R., and O. Isler [115] 289, 290, 310
— see Mayer, H. [87] 284, 310
— see Planta, C. v. [108] 273, 274, 310
— see Winterstein, A. [764] 178, 199; [148] 267, 271, 274, 311
Rüegger, A., and D. Stauffacher [192] 456, 457, 467
Rüttimann, O., and H. Flück [186] 710, 728
Ruffini, G. [218] 218, 221, 223, 256
Ruggli, and Ziegler 658
Ruhemann, S. [56, 57, 58, 59, 60] 747, 749, 782
Ruitz 447
Ruiz, S. L., and C. Laroche [62] 613, 622, 624, 629
Rulko, F., see Cieślak, J. [38] 431, 463
Rummel, W., see Forth, W. [45] 603, 608
Runge, F., A. Jumar, and F. Koehler [76] 683, 686, 877
Ruohonen, A., see Lindfors, R. [95] 538, 564
Ruppenthal, N., see Knabe, J. [108] 444, 465
Ruseva-Atanasová, N., and J. Janák [178] 398, 420

Rushizky, G. W., and C. A. Knight [82] 801, *806*
Rusiecki, W., and M. Henneberg [588] 5, *195*; [193] 448, *467*; [138] 509, *565*
Russel, J. H. [590] 5, *195*; [194] 460, *467*
Russell, A. S., and C. N. Cochran [589] *195*
— see Newsome, J. W. [474] 23, 24, *192*
— see Stumpf, H. C. [704] 24, *198*
Rustici, L., see Tubaro, E. [723] *198*
Rutkowska, U., and K. Wojsa [195] 449, *468*
Ruttloff, H., see Linow, F. [57] 649, *655*
Ruzicka, L. [219] 206, *256*
Ryan, A. J., see Barrett, J. F. [1] 624, *628*
Rybicka, S. M. [591] 90, *195*
Rybicka, W. S. [179] 377, *420*
Rymo, L., see Johansson, B. G [330, 331] 111, 112, *189*; [82, 83] 591, *609*

Saalfeld, H. [592] 23, 24, 25, *195*
Sabrook, C. M., see Belcher, R. 868
Saccardi, A., see Sundt, E. [273] 215, 217, *257*
Sacher, R., see Reinhard, E. [100] 489, *493*
Sachs, B. A., and L. Wolfman [180] 385, 386, *420*; [135] 592, *610*
— see Wolfman, L. [183] 317, 331, *361*
Sachs, L., and Z. Szereday [593] 70, *195*
— see Szereday, Z. [164] 602, *611*
Sachs, R. M. [101] 472, *493*
Saeman, J. F., see McKibbins, S. W. [414] 99, *191*
— see Millett, M. A. [451] 149, 151, 152, 153, *192*
— see Seikel, M. K. [625] 100, 101, *196*
Sagitullin, R. S., see Kost, A. N. [60] 477, *491*
Sagot-Masson, M., see Mouton, M. [464] 30, *192*; [175] *255*
Sahashi, Y., see Higaki, M. [66] 277, *309*
Sahasrabudhe, M. R. [77] 635, *656*
— see Nigam, I. C. [480] 116, *192*; [185] 212, *255*
Sahli, M. [196, 197] 449, *468*
Sahli, M., and M. Oesch [594] 149, *195*; [198] 451, *468*; [140] 534, 535, 536, 537, 538, 539, *565*
Saifer, A., and I. Oreskes [62] 749, *782*
Saito, A., see Yamada, M. [152] 300, 301, *311*
Sakabe, H., see Matsushita, H. [52] 667, *685*
Sakaguchi, N., see Nakayama, F. [128] 354, *360*
Sakaguchi, S. [78, 79, 80] 749, *782*
Salamé, M. [78] 639, 640, 642, *656*, 875
Salfner, B., see Hörhammer, L. [99] *727*
Salminen, K., see Salo, T. [595, 596] 37, *195*; [63] 626, *629*; [79, 80, 82, 83] 634, 635, 639, 644, 648, *656*
Salo, T., E. Airo, and K. Salminen [595] 37, *195*; [82] 648, *656*
— and R. Mäkinen [81] 638, *656*
— — and K. Salminen [83] 634, 635, *656*
— and M. Salminen [596] 37, *195*; [63] 266, *629*; [79] *656*
— — and K. Fiskari [79, 80] 639, 644, *656*
— see Ronkainen, P. [213] 220, *256*
Sammeck, R., see Decker, K. [42b] 430, *463*
Samols, E., see Kuzuya, T. [384] 179, *190*
Sampson, K., F. Schild, and R. J. Wicker 884
Samuel, P., M. Urivetzky, and G. Kaley [144] 341, *360*; [136] 603, *610*
Samuelson, T. J., see Pollard, F. H. [57] 623, *629*
Samuelsson, B., see Bygdeman, M. [20] 399, *416*; [20] 597, *608*
— see Green, K. [50] 399, *417*; [61] 597, *608*
Sancin, P., see Nano, G. M. [178, 179] 220, 222, *255*; [107] 509, *564*
Sandberg, F., see Khafagy, S. [101] 434, 453, *465*
— and K. H. Michel [199] 446, *468*
Sandel, A., see Johannesen, B. [329] 149, *188*
Sander, E. G., D. B. McCormick, and L. D. Wright [83] *806*
Sander, H. [145] 347, *360*
Sanders, G. M., and E. Havinga [116] 277, *310*

Sanders, J., see Skipski, V. P. [188] 389, 391, *420*
Sandhoff, K., see Jatzkewitz, H. [73] 389, 415, *417*; [77, 79] 598, *609*
Sanecka, M., see Opienska-Blauth, J. [53] 748, *782*
Sanger, F. [130, 151, 152, 153] 752, 756, *783*, *784*
— G. G. Brownlee, and B. G. Barrell [84] 802, *806*
— and H. Tuppy [95] 750, 756, *783*
— see Porter, R. [154] 756, 758, *784*
Sankoff, I., and T. L. Sourkes [597] 146, *195*; [137] 589, *610*
Sano, T., see Tsuda, Y. [297] 242, 243, *258*
Šantavý, F. [200] 435, *468*, 882
— see Bláha, K. [22] 443, *462*
— see Maturová, M. [132b] 444, *466*
— see Neumüller, O. A. [149] 426, *466*
— see Potěšilová, H. [178] 426, 427, 429, *467*
— see Preininger, V. [179b] 444, *467*
— see Vácha, P. [250] 454, 455, 456, *469*
— see Vrublovský, P. [255] 441, 443, *469*
— see Pijewska, L. *470*
Sanz, M. C. 598
Sargeant, K. [187] 696, *728*
Sárkány-Kiss, J., see Tyihák, E. [305, 306] 212, 238, *258*
Sarnecka-Keller, M., see Noworytko, J. [63] 749, *782*
Šaršúnová, M. [141] 551, 554, *565*
— and V. Schwarz [201b, 201c] 422, 424, 457, *468*; [142] 536, *565*
— see Paris, R. *470*
— J. Tölgyessy, and M. Hradil [201] 448, *468*
— see Schwarz, V. [212] 422, 432, 437, 448, 451, 457, *468*
Sasaki, Y., see Hashizume, T. [29] 795, *805*
Sato, A., see Rudloff, E. v. [185] 712, *728*
Sato, F., see Natori, S. [150] *728*
Sattler, L., and F. W. Zerban 856
Saukkonen, J. J. [85] 794, 795, 801, *806*
— see Gref, C.-G. [250] 69, *186*
Saunders, E., see Crawhall, J. C. [22] 583, *608*

Sautière, P., see Biserte, G. [156] 756, 757, 758, *784*
Savage, R. J., see Buchan, J. L. 856
Savi, C., see Luisi, M. [115, 217] 340, 341, 350, *360*, *362*; [101] 600, *609*
Savidge, R. A., and J. S. Wragg [143] 526, *565*
Sawicki, E., T. W. Stanley, and W. C. Elbert [598] 141, *195*
— — — and J. D. Pfaff [599] 37, 145, 155, *195*; [77] 669, *686*
— — and H. Johnson [601] 52, 141, *195*
— — J. D. Pfaff, and W. C. Elbert [600] 37, *195*; [78] 667, 669, *686*
— see Pfaff, J. D. [525] 141, *194*
— see Stanley, Th. W. [690] 141, *198*
Sawyer, E. R., see Heaysman, L. T. [282] 102, 135, *187*; [63] 278, 279, 280, *309*
Sayce, I. G., see Littler, J. S. [151] 221, *254*
Scanes, F. S., and B. T. Tozer [167] 759, *784*
Scarano, E., see Grippo, P. [26] 792, 793, 795, 796, *805*
Schacht, U., see Tschesche, R. [218, 219] 691, 695, 696, *729*
Schachter, D., J. D. Finkelstein, and S. Kowarski [117] 277, 278, 280, *310*
Schachter, M. [121] 752, *783*
Schaden, G., see Pailer, M. [162] 718, *728*
Schäfer, F. [220] 238, *256*
Schäfer, H., see Inhoffen, H. H. [36] *491*
Schäfer, L. 23
Schaepdryver, A. F. de, see Potter, W. P. de [31] 499, *506*; [132] 520, 524, *565*
Schantz, M. v. [602] 100, *195*; [221, 222, 223a] 236, 237, 238, 239, *256*; [202] 422, 441, *468*; [191, 192] 713, 714, *729*
— L. Ivars, I. Lindgren, L. Laitinen, E. Kukkonen, H. Walenius, and C. J. Widén [193] 714, *729*
— S. Juvonen, and R. Hemming [223] 211, *256*
Scharoun, J. E., see Thomas, A. E. III [201] 408, *420*
Schauvliege, M., see Heyndrickx, A. [78] 520, 521, *564*, 879

Scheidegger, J. J., and E. Cherbuliez 899
Scheiffarth, F., and L. Zicha [146] 341, 361
— — F. W. Funck, and M. Engelhardt [138] 601, *610*
Scheig, R. L., A. Annuziata, and L. A. Pesch [86] 792, 795, 796, *806*
Scheit, K. H. [87] 792, 795, 802, *806*
Scheja, H. W., see Sturm, A. 889
Schellenberg, P. [144] 753, 756, *784*
Schenck, G., see Loth, H. [136] 709, *727*
Schenck, G. O., see Neumüller, O. A. [149] 426, *466*
Scherf, H., see Zenk, M. H. [135, 136] 472, *493*
Scherz, H., E. Bancher, and K. Kaindl [604] 104, *195*
— W. Rucker, and E. Baucher [139] 580, *610*
— see Bancher, E. [37, 38] 53, 154, *181*; [5] 651, *654*; [2] *684*; [203] 734, 785; [47] 819, 820, 826, 827, 834, *836*
— see Prey, V. [539] 32, *194*; [17] 811, 812, 817, 820, *835*
Schettino, O., see Covello, M. [21, 22] 302, 303, *308*
Schetty, G. [65, 66, 67] 623, *629*
— and W. Kuster [64] 623, *629*
— see Beffa, F. [2] 623, *628*
Scheu, D., see Stahl, E. [263, 264] 224, 226, 236, 237, *257*
Scheuer, P. J., see Werny, F. [262] 457, *469*
Schiesser, R., see Köhler, M. [357] 37, *189*; [42] 667, *685*
Schiewer, U., and E. Libbert [102] *493*
— see Libbert, E. [71] 488, *492*
Schiffman, G., see Marcus, D. M. [40] 818, *835*
Schilcher, H. [605] 104, 149, *195*; [224, 224a, 224b] 210, 212, 239, *256*
— see Hörhammer, L. [94] 238, *253*
Schild, F., see Sampson, K. 884
Schildknecht, E., and H. König [87] 648, *656*
Schildknecht, H., and H. Krämer [194] 719, *729*
— and O. Volkert [606] 102, *195*
Schilke, J. F., see Stowe, B. B. [124] 473, *493*

Schiller, E. O., see Kramer, J. K. G. [380] *190*
Schink, W., and H. Struck [147] 341, *361*
Schlemmer, F., and E. Link [607] 145, *195*; [203] 448, *468*
Schlenk, H., see Mangold, H. K. 866
Schlierf, G., and P. Wood [608] 135, 137, *196*; [140] 592, 593, *610*
Schlitt, H., see Geiss, F. [224, 226, 227] 66, 67, 70, 71, 102, *186*; [25, 26] 666, *685*, 861
Schlögl, K. [88] 654, *656*
— and M. Fried [84] 665, 666, *686*
— and A. Mohar [81] 665, *686*
— — and H. Pelousek [79, 82] 665, *686*
— H. Pelousek, and A. Mohar [80] 665, *686*
— and M. Peterlik [83] 666, *686*
— see Egger, H. [21] 665, *685*
Schlossberger, H. G., H. Kuch, and I. Buhrow [103] 477, *493*; [141] 584, *610*
Schmall, M., see Wollish, E. G. [774] 54, *200*; [175] 544, *566*
Schmandke, H. [609] *196*
— and H. Gohlke [610] *196*
Schmeiser, K. [611] 157, 170, *196*
Schmialek, P. [612] 86, *196*; [225, 226] 226, *256*
Schmid, E., and N. Henning [142] 589, *610*
— E. Hoppe, Chr. Meythaler jr., and L. Zicha [144] 509, 519, *565*
— and H. J. Kuschke [143] 590, *610*
— B. Laudi, J. Krautheim, and N. A. Tautz [143a] 590, 591, *610*
— L. Zicha, J. Krautheim, and J. Blumberg [104] 486, *493*
— see Tautz, N. A. [166] 589, *611*
Schmid, H. [195] 687, 706, *729*
— see Dastoor, N. [42] 447, *463*
— see Deyrup, J. A. [44] 447, *463*
— see Govindachari, T. R. [76, 77] 447, *464*
— see Hesse, M. [88] 447, *464*
— see Kump, C. [114, 115, 115b] 447, *465*
— see Kump, W. G. [112, 113] 447, *465*
— see Nagarajan, K. [144] 447, *466*
— see Pinar, M. [168, 169] 447, *467*

Schmid, H. see Schumann, D. [209] 447, 468
— see Weissmann, Ch. [259, 260, 261] 447, 469
Schmid, H. H. O., W. J. Baumann, J. M. Cubero, and H. K. Mangold [182] 397, 401, 402, 420
— — and H. K. Mangold [184] 373, 374, 383, 420
— L. L. Jones, and H. K. Mangold [185] 377, 383, 384, 420
— and H. K. Mangold [181, 183] 373, 377, 383, 384, 385, 386, 420
— see Baumann, W. J. [9, 10] 381, 385, 412, 414, 416
— see Mangold, K. H. [428] 5, 191
— see Stahl, E. [683] 197
Schmid, M., see Guggisberg, A. [86] 447, 464
Schmidlein-Mèszáros, J., see Hardmeier, E. [76] 532, 564
Schmidli, B. 308
— see Haefelfinger, P. [259] 145, 151, 187; [72] 558, 563
— see Kofler, M. [79] 284, 309
Schmidt, G., and S. J. Thannhauser [88] 790, 806
— see Dain, J. A. [32] 392, 393, 416
Schmidt, O. Th. [196, 197] 687, 688, 729
— and W. Schönleben [613] 43, 196; [198] 691, 729
Schmit, J. A., see Blank, M. L. [73] 139, 140, 158, 182; [13] 414, 416
Schmitt, G., see Stahl, E. [265] 237, 257
Schmittmann, B. [227] 243, 256
Schmutz, J., see Lehner, H. [122] 447, 465
Schnackerz, K., see Waldi, D. [257] 422, 424, 426, 433, 437, 439, 440, 448, 451, 454, 469
Schneckenburger, J., and I. Hartikainen [204] 433, 468
— see Schultz, O. E. [208b] 441, 468
Schneider, C., see Schneider, G. [145] 586, 610; [229] 779, 786
Schneider, E. [89] 634, 656
Schneider, G. [205] 457, 468
— and C. Schneider [145] 586, 610; [229] 779, 786
— G. Sembdner, and K. Schreiber [105, 106] 472, 481, 493

Schneider, G. see Sembdner, G. [112, 113, 114] 472, 475, 486, 489, 493
Schneider, H., and J. Hofstetter [68] 626, 629
— see Rapport, M. M. [172] 389, 420
Schneider, H. P. G., and Z. Szereday [[225] 340, 362; 144] 601, 610
Schneider, J. W., see Schramm, G. [189] 774, 777, 785
Schneider, M., see Boehm, H.-P. [78] 13, 182
Schneider, W., see Brenner, M. [149] 755, 784
Schoch, H., see Eder, F. [28] 638, 645, 655
Schön, H., and N. Krause [146] 599, 610
Schönfeld, T., see Broda, E. [101] 155, 167, 170, 183
Schönleben, W., see Schmidt, O. Th. [613] 43, 196; [198] 691, 729
Schönthal, H., see Weicker, H. [182] 597, 611
Schoknecht, J., see Gabel, E. [68] 230, 233, 234, 235, 239, 252, 856
Scholderer, D. [147] 596, 610
Schorderet, M., see Nussbaumer, P. A. [34] 574, 578
Schorn, H., and C. Winkler [148] 587, 610
Schorn, P.-J. [614, 615, 615a] 49, 91, 93, 127, 152, 196; [38] 500, 504, 506; [199, 200] 710, 712, 713, 714, 729
— and E. Stahl [69] 615, 619, 629
— see Stahl, E. [229b] 434, 468; [203, 205, 206, 207] 694, 695, 696, 697, 698, 699, 707, 713, 714, 719, 722, 729, 871, 886
— see Weigert, E. [324] 244, 245, 246, 258; [233] 715, 730
Schouten, L., see Jurriens, G. [76] 414, 417
Schramm, G., J. W. Schneider, and A. Anderer [189] 774, 777, 785
Schranz, R. E., see Kingdon, F. [346] 131, 189
Schratz, E., and W. Egels [616] 86, 196
— and S. Qedan [228] 211, 217, 220, 230, 239, 256
Schraudolf, H. [107] 472, 474, 493
— and F. Bergmann [108] 472, 493
Schrecker, A. W., see Hartwell, J. L. [87] 687, 710, 726

Schreiber, K., O. Aurich, and G. Osske [229] 241, 243, *256*; [148] 323, 324, 347, 349, *361*; [206] 459, *468*, 861, 879, 892
— — and K. Pufahl [207] 460, *468*
— C. Horstmann, and G. Adam [207b] 459, *468*
— and H. Ripperger [208] 459, *468*
— see Adam, G. [10] 148, *180*; [1] 322, 349, *357*; [2] 459, *462*
— see Focke, J. [16] 487, 489, *490*
— see Pufahl, K. [180c] 460, *467*
— see Ronsch, H. *471*
— see Schneider, G. [105, 106] 472, 481, *493*
— see Sembdner, G. [111, 112, 113, 114, 115, 116] 472, 475, 476, 479, 486, 487, 489, *493*
Schröder, P., see Luckner, M. [126] 456, *466*; [137, 138, 140] 717, 719, *727*
Schröter, H.-B., see Mothes, K. *461*
— see Neumann, D. [148] 433, *466*
Schubert, M., see Bobbit, J.M. [24] 460, *462*
Schuchard, M., see Weitz, E. [746] 9, 10, *199*
Schudel, P., see Mayer, H. [87] 284, *310*
Schueler, W., see Stroh, H.H. [66] 826, 827, *836*
Schüppel, R., and Kl. Soehring [146] 526, *565*
Schütte, H. R., and H. Hindorf [210] 436, *468*
— and B. Maier [210b] 461, *468*
— see Gross, D. [85] 435, *464*
— see Mothes, K. *461*
Schütte, J., see Vidic, E. [253] 441, *469*
Schütz, G. P., see Vries, G. de 855
Schulte, K. E., F. Ahrens, and E. Sprenger [230] 235, 240, *256*, 867
Schultz, O. E., and H. L. Mohrmann [231] 235, *256*
— and J. Schneckenburger [208b] 441, *468*
— and D. Strauss 861
— and F. Zymalkowski *461*
Schultze, H., see Rieche, A. [74] 678, *686*
Schulz, H., see Hänsel, R. [77] 711, *726*
Schulz, M., see Rieche, A. [73] 678, *686*
Schulze, G., see Asmus, E. [1] 501, *505*

Schulze, J., see Hüttenrauch, R. [306] 76, *188*; [16] 577
Schulze, P.-E., and M. Wenzel [617] 160, 162, 170, 177, *196*
Schulze, W. [145] 508, 509, 514, 518, *565*
— see Awe, W. [6] 508, 509, 514, 518, *562*
Schumacher, D., see Reisert, P.M. [141] 338, 360; [129] 602, *610*
Schumacher, H., see Büchi, J. 882
Schumann, D., and H. Schmid [209] 447, *468*
— see Bohlmann, F. [27] 436, *462*
— see Hesse, M. [88] 447, *464*
Schunack, W., E. Mutschler and H. Rochelmeyer [145a] 549, 550, *565*; [201] 718, *729*
— and H. Rochelmeyer [209b] 461, *468*
Schunk, R., see Volk, O.H. [317] 239, *258*
Schwander, H., see Signer, R. [92] 788, *806*
Schwane, R. A., and R. S. Nakon [618] 160, *196*
Schwarting, A. E., J. M. Bobbitt, A. Rother, C. K. Atal, K. L. Khanna, J. D. Leary, and W. G. Walter [211] 434, *468*
— see Blomster, R.N. [23] *462*
— see El-Olemy, M.M. [150b] 431, *466*
— see Khanna, K.L. [102] 434, *465*
— see Leary, J.D. [119, 120] 431, 434, *465*
— see Rother, A. [190] 431, *467*
Schwartz, A.N., A.W.G. Yee, and B.A. Zabin [89] *806*
Schwartz, D.P. [97] 750, *783*
— R. Brewington, and O.W. Parks [39] 495, *506*
— M. Keeney, and O.W. Parks [232] 222, *256*
— and O.W. Parks [233] 222, *256*
Schwarz, H., see Tschesche, R. [169] 347, *361*
Schwarz, K., see Dische, Z. [19] 789, 802, *804*
Schwarz, V. [149] 316, 335, 352, *361*
— and M. Šaršúnová [212] 422, 432, 437, 448, 451, 457, *468*

Schwarz, V. and K. Sykora [150] 317, 335, 355, *361*
— see Heřmánek, S. [79] 323, 329, 335, 347, *359*; [87] 422, 439, 448, 451, 454, *464*; [27] 613, *629*
— see Lábler, L. [386] 5, *190*
— see Šaršúnová, M. [201b, 201c] 422, 424, 457, *468*; [142] 536, *565*
Schweiger, A. [90, 91] 649, 651, *656*; [3] 807, 811, 814, 815, 833, *835*
— and H. Günther [90] 804, *806*
— see Grau, R. [35] 649, *655*
— see Günther, H. [257] 35, *187*; [39] 817, 818, 834, *835*
Schweiger, M., see Zachau, H. G. [107] 801, 802, *806*
Schweppe, H. [70, 71] 612, 616, 617, 618, 623, *629*, 879
— see Micheel, F. [43] 616, *629*, 879
Schweppe, J. S., see Chiang, S. P. [36] 338, *358*
Schwieter, U., see Bolliger, H. R. [9] 266, 267, 268, *308*
— see Planta, C. v. [108] 273, 274, *310*
Schwimmer, S., and A. Bevenne 856
Schwyzer, R., and H. Dietrich [139b] 784
— and H. Kappeler [139a] 753, *784*
— see Kappeler, H. [142] 754, *784*
Scott, R. P. W., and A. T. James [619] 94, *196*
Scott, T. A., see Speake, T. [220] 430, *468*
Seakins, J. W. T., see Hathway, D. E. [279] 43, *187*
Searle, N. E. [109] 473, *493*
Sears jr., G. W. [620] 15, *196*
Sease, J. W. [621] 146, *196*
Seaton, J. C., see MacMillan, J. [75, 76] 475, *492*
Seay, L. D., see Jakovljevic, I. M. [95] 449, *465*, 861, 874
Šebánek, J. [109a] 489, *493*
Seeboth, H. [622] 32, *196*; [234, 235] 230, *256*
— and H. Görsch [623] 146, 154, *196*; [236] 230, *256*; [147] 541, *565*, 890
Seehofer, F., see Tschesche, R. 870
Seeliger, A., see Haas, H. J. [64] 825, 826, 827, 828, *836*
Šefčovič, P., see Mokrý, J. [133, 136] 449, *466*

Segura-Cardona, R., and K. Soehring [34] 499, *506*; [148] 524, *565*; [149] 589, *611*, 867, 874
Seher, A. [603, 624] 131, 136, *195*, *196*; [118] 277, 284, 285, 287, *310*; [84, 85, 86] 632, 633, 649, *656*, 867, 887
Seibl, J., see Kump, C. [114, 115, 115b] 447, *465*
Seifert, H. [151] 598, *611*
— and G. Uhlenbruck [150] 598, *611*
Seifter, S., see Blumenfeld, O. O. [68] 827, 828, *836*
Seikel 705
Seikel, M. K., M. A. Millett, and J. F. Saeman [625] 100, 101, *196*
— and J. W. Rowe [237] 226, *256*
Seiler, H. [21, 24, 30, 31] 838, 841, 848, 851, *853*, *854*, 887
— Chr. Biebricher, and H. Erlenmeyer [29] 846, *853*
— and H. Erlenmeyer [27] 850, *853*
— and T. Kaffenberger [26] 847, *853*, 860, 897, 905
— and W. Rothweiler [25] 842, *853*
— and M. Seiler [627] 179, *196*; [22, 23, 28] 839, 840, 844, 850, *853*, 882, 895
— see Pfrunder, B. [528] 111, *194*
Seiler, M., see Seiler, H. [627] 179, *196*
— see Seiler, H. [22, 23, 28] 839, 840, 844, 850, *853*, 882, 895
Seiler, N., G. Werner, and M. Wiechmann [626] 141, *196*; [110] 485, *493*
— and M. Wiechmann [35, 36] 495, 496, 499, *506*; [149] *565*; [225] 778, *786*
Seitz, M., see Hörhammer, L. [102] 697, *727*
Selmair, H., see Knauff, H. G. [40] 738, *782*
Sembdner, G., R. Gross, and K. Schreiber [111] 476, 479, 487, *493*
— G. Schneider, and K. Schreiber [112] 489, *493*
— — J. Weiland, and K. Schreiber [113, 114] 472, 475, 486, 489, *493*
— and K. Schreiber [115, 116] 475, 479, 486, 489, *493*
— see Focke, J. [16] 487, 489, *490*
— see Schneider, G. [105, 106] 472, 481, *493*

Semenuk, G., and W. T. Beher [627a, 798] 139, *196*, *200*; [226] 354, *362*
Sen, S. P., and A. C. Leopold [117] *493*
Senanayake, U. M., and R. O. B. Wijesekera *471*
SenGupta, A. K., see Kaufmann H. P. 858
— see Tschesche, R. [291] 243, *257*
Senkowski, B. Z., see Hawrylyshyn, M. [30] 794, *805*
Senn, G., see Tanner, H. [212] 699, *729*
Seno, S., W. V. Kessler, and J. E. Christian [150] 561, *565*
Sensi, P. [37, 38] *578*
Servigne, Y., and C. Duval 858
Seters, W. A. van, see Kamp, W. [99] 422, 438, *465*
Severen, R. van [213] 456, *468*
— see Braeckmann, P. [33, 34] 456, *463*
Severin, M. [237a] 217, *256*
Seydel, J. H., Buettner, and F. Portwich [151] 560, *566*
Seyfried, J., see Klavehn, M. [106] 451, 453, *465*; [85] 562, *564*
Sgoutas, D. S., M. J. Kim, and F. A. Kummerow [629] *196*
— and F. A. Kummerow [628] 176, 177, *196*
Shaffer, R. W., see Jakovljevic, I. M. [95] 449, *465*, 861, 874
Shahid, M. A., see Hashmi, M. H. [278] 74, *187*; [10] 845, *853*
Shalaby, A. F., and G. Richter [238] 236, *256*
— and E. Steinegger [214] 436, *468*
Shamma, M., and W. A. Slusarchyk *461*
Shapiro, D., see Flowers, H. M. [46] 819, *836*
Shapiro, H. S., see Tamm, C. [96] 793, *806*
Sharma, A., see Williams, J. A. [192] 597, *611*
Sharma, R. K., G. S. Khajuria, and C. K. Atal [214b] 435, *468*
Sharma, S. P., see Parihar, D. B. [21a] 500, *505*
Sharon, N., see Zehavi, U. [44] 818, *835*
Shasha, B., and R. L. Whistler [91] 793, *806*; [13] 809, 815, 820, 834, *835*

Shearer, C. M., see Szulczewski, D. H. [162] 560, *566*
Shelesnyak, M. C., see Sulimovici, S. [230] 340, *362*
Shellard, E. J. [630] 5, *196*
— and M. Z. Alam *471*
— — and J. Armah [631] 91, 93, *196*
— and I. V. Osisiogu [152] 534, 535, 536, *566*, 864
— and J. D. Phillipson [215, 216, 217, 218] 454, *468*
— — and D. Gupta *471*
— see Beckett, A. M. [17, 18, 18b, 18c] 437, 454, *462*
— see Phillipson, J. D. *470*
Shen, N.-H. C., F. E. Francis, and R. A. Kinsella [152] 602, *611*
Sheppard, H., B. S. d'Asaro, and A. J. Plummer [153] 540, *566*
— and W. H. Tsien [632] 159, *196*
Sheppard, R. C., see Djerassi, C. [184] 785
Sherma, J. [32] 845, *854*
— and L. V. S. Hood 858
Sherman, R. E., see Toribara, T. V. 887
Shevchenko, Z. A., and J. A. Favorskaya [239] 222, *256*
Shibaoka, H., see Mitsuhashi, M. [80] 487, *492*
Shibata, S., and Y. Ogihara [188] 719, *728*
— M. Takido, and O. Tanaka 884
Shimizu, M., see Katsui, G. [75] 263, 284, *309*
Shiratori, T., see Goodman, D. S. [52] 274, *309*; [207] 333, *362*
Shiro, K., see Hashimoto, A. [56] 406, *417*
Shnaidman, L. O., see Bogoslovsky, N. A. [7] 277, *308*
Shone, G., see Broadbent, J. H. [27] 696, *725*
— see Coomes, T. J. [29] 696, *725*
— see Hammonds, T. W. [54] 411, *417*
Shostenko, U. V., and V. V. Ratshinsky [633] 1, *196*
Shraiber, M. S. 1
— see Ismailov, N. A. [313] 1, 2, 73, *188*
— see Ratshinsky [567] 1, *195*
Shulov, L. M., see Kheifits, L. A. [126] 230, *253*

Sidhu, G. S., see Bhramaramba, A. [14] 210, 237, *251*
— see Freudenberg, K. [54] 711, *725*
— see Swaleh, M. [277] 238, *257*
Siebert, G., see Kesselring, K. [38] 793, 794, 801, *805*
Siegel, E. T., and R. I. Dorfman [634] 179, *196*; [151] 351, *361*
Siegel, G. L., see Futterweit, W. [54] 602, *608*
Sieper, H., R. Longo, and F. Korte [189] 707, *728*
— see Korte, F. [370] 154, *189*; [123, 124, 125, 126] 715, 716, *727*
Sigg, H. P. [39, 40] 573, *578*
— see Gutzwiller, J. [84] 226, *252*
Signer, R., and H. Schwander [92] 788, *806*
— see Konrad, E. [369] 8, *189*
Signor, A., see Bello, D. di [204] 778, *785*
Silanos, M. A., see Jommi, G. [81] *564*
Silbermann, H., and R. H. Thorp 902
Siliprandi, D., and N. Siliprandi 876
Siliprandi, N., see Siliprandi, D. 876
Silvander, B., see Forsén, S. [86] 831, *837*
Simon, H., and G. Müllhofer [635] 167, *196*
— see Gröger, D. [82] 457, *464*
— see Weygand, F. [238] 694, 695, *730*
Simons, S., see Miras, C. [146] 716, *728*
Simonsen, J. [240] 207, *256*
Simoons, J. R. A., see Doorenbos, H. J. [39] 518, *563*
Sims, R. P. A., and J. A. G. Larose [636] 147, 148, *196*
— see McKillican, M. E. [127, 128] 389, 391, *419*
Sinha, K., see Rao, K. R. K. [71] 671, *686*
Sinotova, E. N., see Vobecky, M. [88] 666, *686*
Sirois, J. C., and E. V. Parups [118] 489, *493*
Sjöberg, B., see Djerassi, C. [184] *785*
Sjöholm, I. [152] 323, 345, *361*; [226] 740, *786*
Sjöquist, J. [133, 181, 182, 191] 753, 772, 773, 774, 777, *784*, *785*
— B. Blombäck, and P. Wallén [132] 753, 774, *784*

Sjöquist, J. see Edman, P. [192] 774, *785*
Sjövall, J. [227] *362*
Skenne, K. G. M., and A. Lang [118a] 489, *493*
Skidmore, W. D., and C. Entenman [186] 389, 391, *420*
Skinner, W. A., and R. M. Parkhurst [119] 284, *310*
— — and P. Aloupovic [120] 284, 287, *310*; [187, 189] 389, 415, *420*
Skipski, V. P., R. F. Peterson, and M. Barclay [637] 147, 154, *196*; [153] 599, *611*
— — J. Sanders, and M. Barclay [188] 389, 391, *420*
— A. F. Smolowe, and M. Barclay [190a] *420*
— — R. C. Sullivan, and M. Barclay [190] 377, *420*
Slavík, J., see Slavíková, L. [219] 441, 443, *468*
Slavíková, L., and J. Slavík [219] 441, 443, *468*
Slaymaker, S. C., see Peurifoy, P. V. 901
Slessor, K. N., see Dutton, G. G. S. [14, 74] 810, 828, 831, 832, 833, 834, *835*, *836*
Sloane Stanley, G. H., see Folch, J. [42] 369, 370, *417*
— see Lees, M. [110] 370, *418*
Sloot, D., see Jong, K. de [110] 222, *253*
Slusarchyk, W. A., see Shamma, M. *461*
Smellie, R. M. S., see Davidson, J. N. [15] 802, 804, *804*
Smith jr., C. R., L. H. Nieze, H. F. Zobel, and I. A. Wolff [190] 719, *728*
Smith, D., see Pennington, G. W. [129] 537, *565*
Smith, D. M., see Genest, Chr. [61] 696, *726*
Smith, E., see Hodgson, E. [88d] 430, 431, *465*, 859
Smith, E. C., and D. Metzler [121] 297, *311*
Smith, F., see Cifonelli, J. A. 885
— see Handa, N. [83] 830, *836*
— see Hay, G. W. [16] 810, 812, 819, 820, 830, 831, 834, *835*, 893
Smith, G. A. L., and P. J. Sullivan [638] 145, 151, *196*; [242] 231, *256*; [154] 540, *566*; [78a] 664, *686*

Smith, I. [103] 751, *783*
— see Jepson, J. R. [76] 749, 750, *782*
Smith, J. 883
— see Jepson, J. B. 881
Smith, J. D., see Markham, R. [45] 792, 793, 796, 803, *805*
Smith, L. L., and Th. Foell [153] 315, 317, 323, 335, 347, *361*
Smith, L. W., R. W. Breidenbach, and D. Rubenstein [639] 40, *196*; [122] 269, *311*
— see Chalmers, A. H. [36 b] 435, *463*
Smith, M. D., and C. G. Farmilo [241] 237, *256*
Smith, R. H., and W. McKernan [41] *578*
Smith, W. K., see Speake, T. [220] 430, *468*
Smolowe, A. F., see Skipski, V. P. [190, 190 a] 377, *420*
Smyth, D. G., W. H. Stein, and S. Moore 774
Smyth, R. B., and G. G. McKeown 888
Snader, K. M., see Loev, B. [405] 94, 98, *190*; [114] 315, *360*
Snatzke, G., see Tschesche, R. [286, 288, 289, 290, 292, 293] 242, 243, 244, 245, 247, *257*; [168, 169, 170] 334, 342, 347, *361*; [245] 422, 432, *469*
Snavely, M. S., and J. G. Grasselli [640] 150, 154, *196*
Snell, C. T., see Snell, F. D. [155] 524, *566*
Snell, F. D., and C. T. Snell [155] 524, *566*
Snyder, F. [641, 642, 644, 645, 646] 52, 157, 163, 164, 165, 166, 179, *196*, *197*
— E. A. Cress, and N. Stephens [191] 384, *420*
— and H. Kimble [647] 149, 165, 166, 176, *197*
— and F. Stephens [648, 649] 148, 163, 164, *197*
— see Wood, R. [222] 374, *421*
Snyder, L. R. [643] 90, *196*
— and H. D. Warren [650] 90, *197*
Sober, H. A., see Peterson, E. A. [516, 517,] 37, 90, *193*; [54] *805*
Sobodka, H., see Baker, H. [3] 303, 304, *308*
Söding, H. [119] 472, 474, 487, *493*

Soehring, K., see Dietz, W. 868
— see Frahm, M. [56] 533, 534, 535, 536, 538, 540, *563*, 865
— see Schüppel, R. [146] 526, *565*
— see Segura-Cardona, R. [34] 499, *506*; [148] 524, *565*; [149] 589, *611*, 867, 874
Soffer, L. J., see Futterweit, W. [54] 602, *608*
Solan, J. L., see Koehler, W. R. [102] 412, *418*
Solle, M., see Voigt, S. [171] 586, *611*
Sommer, P. F. [123] 289, 291, *311*
— see Kofler, M. [79] 284, *309*
Sommerville, J. F., see Collins, W. P. [39] 340, *358*
Sonanini, D. [228] 344, *362*
— and L. Anker [42] 573, *578*
— R. Hofstetter, L. Anker, and H. Mühlemann [229] 337, *362*
— see Anker, L. [4] 381, 411, 412, 413, *416*
Soper, A. C., see Ayer, W. A. [10] 461, *462*
Sorkin, E., see Erlenmeyer, H. 904
Šorm, F. [243] 240, *256*
— see Jost, K. [42] 740, *782*
— see Křepinský, J. [142] 239, *254*
— see Novotný, L. [157, 158] 717, *728*
— see Wollrab, V. [330] 239, 240, *258*
Soto, E. F., see Bohner, L. S. de [79] 147, 154, *182*
Sourkes, T. L., see Sankoff, I. [597] 146, *195*; [137] 589, *610*
Spackman, D. H., see Moore, S. [14] 735, *781*
Spalletti, F., see Breccia, A. [93] 179, *182*
Spano, G., see Corsano, S. [36] 245, *251*
Speake, R. N. [119a] 487, *493*
Speake, T., P. McCloskey, W. K. Smith, T. A. Scott, and H. Hussey [220] 430, *468*
Spears, J. C., see Evans, J. R. [190] 179, *185*
Specht, W., see Stier, A. [78] 618, *630*
Spencer, J. M., see Hagerman, D. D. [71] 319, *359*
Spencer, R. D., and B. H. Beggs [799] 148, 150, *200*
Sperlich, H., see Bergner, K. G. 880, 884

Spikner, J. E., and J. C. Towne [651] 103, 149, 155, *197*

Špinková, V., see Blažek, J. [20] 508, *562*

Spiteller, G., see Hesse, M. [88] 447, *464*

Spiteller-Friedmann, M., see Hesse, M. [88] 447, *464*

Spotswood, T. M., see Badger, G. M. [31] 37, *181*; [1] 667, *684*

Spranger, D., see Woggon, H. [104] 640, *656*

Sprecher, E. [244] 237, *256*

Sprecher, H. W., R. Maier, M. Barber, and R. T. Holman [192] 380, *420*

Sprenger, E., see Schulte, K. E. [230] 235, 240, *256*, 867

Sprenger, H.-E. [652, 653, 654, 655] 78, 130, 133, *197*

Spritz, N. [153a] 354, *361*

Sproviero, J. F., see Deferrari, J. O. [53] 822, 823, 834, *836*

Squibb, R. L. [656] 52, 139, *197*; [154, 155] 586, *611*

Šrámek, J. [72, 73, 74] 612, 616, 622, 623, *630*

Srepel, B. [657] 5, *197*

Srivastava, R. M., and M. P. Khare [221, 222] 460, *468*

Stacey, M., see Brimacombe, J. S. [76] 828, *836*

Stachel, H.-D., see Eiden, F. [47] 508, 514, *563*

Stadler, P., and H. Endres [658] 44, *197*

Stadler, P. A. [223] 454, *468*
— and A. Hofmann [224] 454, *468*
— see Hofmann, A. [90] 454, *465*

Staehelin, H. R., see Edlbacher, S. [93] 750, *783*

Staehelin, M. [93] 801, *806*

Stafford, H. A. [202] *729*

Stahl, E. [659, 660, 661, 662, 664, 665, 666, 667, 668, 671, 672, 673, 674, 675, 677, 678, 680, 681, 682, 684, 685, 686] 1, 4, 5, 7, 48, 55, 58, 59, 66, 69, 73, 74, 76, 84, 85, 87, 88, 89, 91, 93, 94, 95, 96, 97, 98, 99, 100, 101, 128, 134, 179, *197*, *198*; [245, 246, 247, 248, 249, 250, 251, 252, 253, 254, 255, 256] 208, 210, 211, 212, 215, 225, 233, 234, 236, 238, 239, 240, 245, 246, 248, *256*, *257*; [124, 126, 127] 262, 267, 273, *311*; [193, 194] 375, 376, *420*; [225, 226, 227, 228, 229] 422, 423, 437, 441, 444, 445, 450, *468*; [40, 41, 42] 500, 503, *506*, 604; [75, 76, 77, 77a] 612, 621, *630*; [93, 94] 645, 646, 647, 649, *656*; [196, 197, 198, 199, 200, 227] 730, 733, 734, *785*, *786*, 810, 857, 858, 868, 877, 881

Stahl, E. and V. Adhikari [257] 234, *257*
— and H. Bohrmann [258] 215, *257*
— H. R. Bolliger, and L. Lehnert [663] 30, *197*; [125] 267, 268, 289, 301, *311*
— and J. Fuchs [259] 233, *257*
— and H. Jork [260, 261, 262] 214, 226, 234, 237, 239, *257*; [229c] 438, *468*
— and H. Kaldewey [676] *197*; [120] 473, 476, 477, 478, 483, *493*, 869, 879
— and U. Kaltenbach [670] 87, *197*; [154] 323, 344, *361*; [204] 694, 710, *729*; [1] 807, 808, 810, 812, 813, 833, *834*, 857
— H. K. Mangold, and H. H. O. Schmid [683] *197*
— and J. Pfeifle [92] 646, 647, *656*; [228] 779, 780, *786*
— and D. Scheu [263, 264] 224, 226, 236, 237, *257*
— and G. Schmitt [265] 237, *257*
— and P. J. Schorn [229b] 434, *468*; [203, 205, 206, 207] 694, 695, 696, 697, 698, 699, 707, 713, 714, 719, 722, *729*, 871, 886
— and L. Trennheuser [669] 118, *197*; [266] 209, *257*
— and H. Vollmann [679] 86, 91, *197*; [267] 228, *257*
— see Kaldewey, H. [47, 48] 478, 483, 488, *491*
— see Mangold, K. H. [428] 5, *191*
— see Schorn, P. J. [69] 615, 619, *629*

Staib, W., see Gerdes, H. [228, 229] 155, *186*; [205, 206] 340, 341, *362*; [55, 56] 601, *608*

Stainier, C., J. Bosly, Fr. Dutrieux, and R. Stainier [156] 537, *566*

Stainier, R., see Stainier, C. [156] 537, *566*

Stammbach, K. [687] 74, *198*

Stanislavová, J., see Macek, K. [97] 507, *564*

Stanley, C. W.   [95] 640, 642, *656*, 887
Stanley, E. R., see Alexander, L. R.   [3] 541, 542, 545, *562*
Stanley, T. W., and E. Sawicki   [690] 141, *198*
— see Sawicki, E.   [598, 599, 600, 601] 37, 52, 141, 145, 155, *195*; [77, 78] 667, 669, *686*
Stanley, W. L., R. M. Ikeda, and S. Cook   [268] 210, *257*
— and S. H. Vannier   [688] 72, *198*
— — and B. Gentili   [689] 72, *198*
— see Ikeda, R. M.   [104] *253*
— see Vannier, S. H.   [729] 155, *199*
Stansfield, D. A.   [691] *198*; [155] 315, 322, *361*
— and D. I. Cargill   [156] 340, *361*
Stanton, J. H., see Dugger, D. L.   [173] 13, *184*
Stárka, L.   [157] 328, 335, 349, 350, *361*
— and J. Malíková   [158] 323, 340, *361*
— and J. Riedlová   [159] 340, *361*; [156] *611*
— J. Šulcová, J. Riedlová, and O. Adamec   [160] 338, *361*; [157] 602, *611*
Starý, F., see Novotný, L.   [158] 717, *728*
Staudenmayer, W. J., see Lestienne, A.   [398] 52, 59, *190*
— see Przybylowicz, E. P.   [546] 59, *194*
Staudinger, Hj., see Berthold, K.   [10] 601, *607*
— see Nishikaze, O.   [482] 146, 147, *192*; [133, 134] 324, 340, 341, *360*; [111] 601, *610*, 902
— see Nowotny, E.   [112] 601, *610*
Stauffacher, D., and H. Tscherter   [230] 451, *468*
— see Rüegger, A.   [192] 456, 457, *467*
Steele, J. A.   [230 b] 441, *468*; [157] 528, 529, *566*
Stegemann, H., and B. Lerch   [692] 112, 113, 114, *198*
Steger, E., see Petschik, H.   [72] 640, *656*
Steggerda, J. J., see Boer, J. H. de   [48, 86, 87] 24, 25, *182*
Stegmann, H., see Weicker, H.   [183] 580, *611*
Steidle, W.   [693] 146, *198*; [161, 162] 344, *361*

Stein, A., see Korey, S. R.   [104] 393, 394, *418*
Stein, E., see List, P. H.   [123] 433, *465*
Stein, G.   [269, 270] 226, *257*
Stein, O., see Stein, Y.   [694] 177, *198*
Stein, W. H., and S. Moore   [30] 737, *781*
— see Moore, S.   [14] 735, *781*
— see Smyth, D. G.   774
Stein, Y., and O. Stein   [694] 177, *198*
Stein von Kamienski, E.   [43] 494, *506*, 891
Steinberg, D., see Avigan, J.   [6] 316, 323, 330, 332, *357*
Steinegger, E., and J. Gebistorf   [695] 146, 148, *198*; [208, 209] 694, 710, 717, *729*
— and K. Peters   [210] 719, *729*
— and J. H. van der Walt   [163] 345, *361*
— see Bernasconi, R.   [19 b] 436, *462*
— see Gill, S.   [73, 74, 75] 436, *464*
— see Shalaby, A. F.   [214] 436, *468*
Steinegger, G., R. Bernasconi, and G. Ottoviano   [231] 436, *469*
Steiner, E., see Beffa, F.   [2] 623, *628*
Steiner, M., and H. Holtzem   [271] 207, *257*
Steinhardt, J., C. H. Fugitt, and M. Harris   [696] 42, *198*
Stejskal, Z., see Blažek, J.   [20] 508, *562*
Stepanov, V., D. Handschuh, and F. A. Anderer   [24] 737, 752, *781*
Stepanovskaya, W. F., see Buslanova, M. M.   [28 a] 221, *251*
Stephens, F., see Snyder, F.   [648, 649] 148, 163, 164, *197*
Stephens, N., see Snyder, F.   [191] 384, *420*
Stephenson, M. L., see Randerath, K.   [76] 801, *806*
— see Zamecnik, P. C.   [108] 801, *806*
Stepka, W., see Dent, C. E.   [35] 738, *781*
Šterba, J., see Pitra, J.   [172] *467*
Steuerle, H., and E. Hille   [697] 43, *198*
Stevens, P. J.   [164] 323, *361*, 872, 880, 905
Stevenson, R., see Kohen, F.   [137] 243, *254*
Steward, F. C., see Dent, C. E.   [35] 738, *781*

Stickland, R. G. [698] 91, *198*; [94] 795, *806*
Stier, A., and W. Specht [78] 618, *630*,
Stiller, R. L., see Weiss, B. [215] 394, *421*
Stock, E. [272] 248, *257*
Stocker, H. R. [699] 64, *198*; [211] 714, *729*
Stoddart, J. L. [121, 122] 489, *493*
Stöber, W. [700] 13, *198*
Stoffel, W., F. Chu, and E. H. Ahrens jr. [195] 372, *420*
Stojilykovic, A., see Hranisavljević-Jakovljević, M. [29] 680, 681, *685*, 865
Stoll, M., see Demole, E. [44] 216, *251*
— see Sundt, E. [274] 237, *257*
Stolle, K., see Gröger, D. [82b] 450, *464*
Storherr, R. W. [96] 640, *656*
Stothers, J. B., see Aberhart, D. J. [1] 718, *724*
Stowe, B. B. [123] 472, *493*
— and J. F. Schilke [124] 473, *493*
— and K. V. Thimann [125] 473, *493*
— and T. Yamaki [126, 127] 472, *493*
— see Obreiter, J. B. [87] 476, 478, *492*
— see Perley, J. E. [92] *492*
Stowe, H. D. [128] 284, 285, *311*
Stoy, V., see Anacker, W. F. [20] 30, *180*
Strauss, D., see Schultz, O. E. 861
Streibl, M., see Wollrab, V. [330] 239, 240, *258*
Stressler-Buchwein, F., see Prey, V. [36] 817, *835*
Striegler, H., see Tschesche, R. [294] 245, *258*
Stroh, H. H., and W. Schueler [66] 826, 827, *836*
Strohecker, R. [97] 636, *656*
Strohecker jr., R. [131] 262, 264, 283, 287, 300, 305, *311*
— and H. Henning [702] 146, 151, 155, *198*; [130] 259, 263, 264, 266, 274, 276, 280, 284, 295, 296, 297, 299, 300, 301, 305, 306, 307, *311*, 872
— and H. Pies [701] 154, *198*; [129] 305, 306, *311*
Štrouf, O., see Holubek, J. *461*
— see Trojánek, J. [243] 449, *469*
Struck, H. [703] 146, *198*; [165] 350, *361*; [158] 602, *611*

Struck, H. see Randerath, K. [61] 795, 798, *805*
— see Schink, W. [147] 341, *361*
Strunk-Lichtenberg, G., see Ginsberg, H. [232] 23, 24, *186*
Stuart, K. L., and M. P. Cava *461*
Studer, A., see Winterstein, A. [764] 178, *199*; [148] 267, 271, 274, *311*
Studer, P., see Gamp, A. [220] 53, *186*
Studer, R. O., K. Vogler, and W. Lergier [139] 753, *784*
— see Vogler, K. [134, 138, 146] 753, *784*
Stüttgen, G., and K. H. Vogelberg [159] 599, *611*
Stuhlmann, F., see Baudler, M. [47] 35, *181*
Stumpf, H. C., A. S. Russell, J. W. Newsome, and J. W. Tucker [704] 24, *198*
— see Newsome, J. W. [474] 23, 24, *192*
Stumpf, P. K., see Overath, P. [497] 177, *193*
Sturm, A., and H. W. Scheja 889
Sugita, J., and Y. Tsujino [157a] 508, *566*
Šulč, M., see Michaleć, Č. [125] 317, 322, 332, *360*
Šulcová, J., see Stárka, L. [160] 338, *361*; [157] 602, *611*
Sulimovici, S., B. Lunenfeld, and M. C. Shelesnyak [230] 340, *362*
Sullivan, G., and L. R. Brady [37] 498, *506*
— and M. R. Gibson [232] 435, *469*
Sullivan, P. J., see Smith, G. A. L. [638] 145, 151, *196*; [242] 231, *256*; [154] 540, *566*; [78a] 664, *686*
Sullivan, R. C., see Skipski, V. P. [190] 377, *420*
Sumére, C. F. van [224] 698, *729*
— G. Wolf, H. Teuchy, and J. Kint [158] 540, *566*, 890
Sumiki, Y., see Ikekawa, N. [35a] 479, *491*
— see Kagawa, T. [43] 479, 487, *491*
Sundaresan, P. R., see Yagishita, K. [151] 274, *311*
Sundman, J., see Pentillä, A. [168] 714, *728*
Sundt, E., and A. Saccardi [273] 215, 217, *257*

Sundt, E., B. Willhalm, and M. Stoll [274] 237, *257*
Sunshine, I. [161] 521, *566*
— and W. W. Fike [159] 551, 554, *566*
— E. Rose, and J. le Beau [160] 534, 535, 536, *566*, 884
— see Fike, W. W. [52] 514, 518, *563*
Sunshine, L. 879
Suomalainen, H. [275] 220, *257*
— and P. Ronkainen [276] 221, *257*
— see Ronkainen, P. [212, 213] 220, *256*
Suomi, W. D., and B. W. Agranoff [160] 599, *611*
Suppan, F., see Birkofer, L. [71] 73, *182*
— see Birkofer, L. [23] 813, *835*
Sussmann, A. R., see Cherbuliez, E. [186, 187] 774, *785*
Suszko-Purzycka, A., and W. Trzebny [233, 233b] 455, 456, *469*
Suter, P. J., see Elson, G. W. [15] 479, 487, 489, *490*
— see MacMillan, J. [75, 76, 77] 475, 479, 486, 487, *492*
Sutter 57
Sutton, H. E., see Kirby-Berry, H. [83] 750, 751, *782*
Suzuki, K. [196, 197] 394, 415, *420*
— see Korey, S. R. [104] 393, 394, *418*
Suzuki, T., see Garrett, E. R. [23] 793, 794, *805*
— see Higaki, M. [66] 277, *309*
Suzuki, Y., see Matsushita, H. [52] 667, *685*
Svennerholm, E. [163] 598, *611*
— and L. Svennerholm [198] 389, 394, *420*; [161, 162] 597, *611*
Svennerholm, L. [199, 200] 370, 389, 391, 393, *420*
— see Svennerholm, E. [198] 389, 394, *420*; [161, 162] 597, *611*
Svoboda, G. H. [234] 450, *469*
— and A. J. Barnes jr. [235] 449, *469*
Swain, T. 890
Swaleh, M., B. Bhusman, and G. S. Sidhu [277] 238, *257*
Sweeley, C. C., see Chang, T.-Ch. L. [26] 383, *416*
— see Gaver, R. C. [45] 394, *417*
— see Horning, E. C. [65] 395, *417*
— see Nichaman, M. Z. [475] 146, 147, *192*

Sykora, K., see Schwarz, V. [150] 317, 335, 355, *361*
Synge, R. L. M., see Martin, A. J. P. [433] 3, *191*
Synodinos, E. [79] 624, *630*
— E. Kokoti-Kotakis, and G. Kotakis [278] 217, *257*
— G. Kotakis, and E. Kokoti-Kokatis [93] *630*
Syper, L. [236] 456, *469*
Szabo, Z. [237] 450, *469*
— see Papp, E. [157] 430, *466*
Szarek, W. A., and J. K. N. Jones [41] 818, *835*
Szarvady, B., see Karácsony, E. M. [100] 451, *465*
— see Wolf, L. [266] 451, 453, *469*
Szász, G., L. Khin, and Z. Budvári [238] 423, *469*
— see Szasz, Z. M. [279] 239, *257*
— see Végh, A. [252] 433, *469*
Szasz, Z. M., and G. Szasz [279] 239, *257*
Székely, G. [705] 128, 130, *198*
Szendey, G. L. [161a] 549, *566*
Szereday, Z., and L. Sachs [164] 602, *611*
— see Sachs, L. [593] 70, *195*
— see Schneider, H. P. G. [225] 340, *362*; [144] 601, *610*
Szulczewski, D. H., C. M. Shearer, and A. J. Aguiar [162] 560, *566*
Szwark, A., see Purzycki, J. [92] *630*

Taber, W. A. [239, 240] 451, *469*
Táborská, I., see Gasparič, J. [19] 616, *628*
Taborsky, R. G., see McIsaac, W. M. [78] 486, *492*
Tachibana, M., see Hara, S. [76] 354, *359*
Tackie, A. N., see Beckett, A. M. [17, 18, 18b] 437, 454, *462*
Tada, M., see Zachau, H. G. [107] 801, 802, *806*
Tadini, L., see Fedeli, E. [61] 219, 221, *252*
Täufel, K., see Linow, F. [57] 649, *655*
Tai Chan, P., R. Venna, P. Ofner, and P. L. Munson [232] 338, *362*
Taillandier, P. see Roucayrol, J. C. [586] 162, *195*
Takacs, F. [85] 677, *686*

Takacs, F., see Zbiral, E. [94] 678, *686*
Takahashi, M., see Higaki, M. [66] 277, *309*
Takeda, K., S. Hara, A. Wada, and N. Matsumoto [166] 347, *361*, 855, 857, 863
— M. Ikuta, and M. Miyawaki [281] 213, *257*
— H. Minato, M. Ishikawa, and M. Miyawaki [280] 213, *257*
Takeuchi, M. [167] 349, *361*
— see Hara, S. [274, 275] 5, 139, *187*; [75, 76, 77] 323, 352, 353, 354, *359*, 893
Takido, M., see Shibita, S. 884
Takitani, S., and K. Matsuda [706, 707] 58, 76, *198*
Tambiah, M. S., see Bennet-Clark, T. A. 893
Tamioka, T., see Ibayashi, H. [75] 602, *609*
Tamm, C., M. E. Hodes, and E. Chargaff [95] 790, *806*
— H. S. Shapiro, R. Lipshitz, and E. Chargaff [96] 793, *806*
— see Gutzwiller, J. [84] 226, *252*
Tamm, J., see Voigt, K. D. [178] 339, *361*; [170] 602, *611*
Tamura, Z. [708] 65, *198*
Tanaka, H., see Fuwa, T. [58, 59] 543, 551, *563*
— see Hara, S. [275] 5, *187*; [77] 354, *359*; [75] 558, *564*
Tanaka, O., see Shibita, S. 884
Tanaka, Y., see Applegarth, D. A. [69] 827, 828, *836*
Tancredi, F., and H. C. Curtius [165] 584, *611*
Tanioka, T., see Ibayashi, H. [90] 339, *359*
Tanker, M. [282] 230, *257*
Tanner, H., H. Rentschler, and G. Senn [212] 699, *729*
Tappi, G., see Nano, G. M. [107] 509, *564*
Taschibana, M., see Hara, S. [274] 139, *187*
Tashiro, H., see Mitsuhashi, H. [170] 215, *254*
Tatar, J. [283] 237, *257*
Tate, M. E., and C. T. Bishop [54, 84] 822, 823, 825, 830, 831, *836*

Tautz, N. A., G. Voltmer, and E. Schmid [166] 589, *611*
— see Schmid, E. [143a] 590, 591, *610*
Taylor, A., and B. Fishwick [98] 643, *656*
Taylor, A. N., see McIsaac, W. M. [78] 486, *492*
Taylor, A. O., see Wong, E. [239] 692, 693, *730*
Taylor, E. H. [709] 53, *198*; [44] 498, *506*
Taylor, W. I., see Bartlett, M. F. [12] 449, *462*
— see Finch, N. [59, 60, 61] 454, *464*
— see Hesse, M. [88] 447, *464*
Teets, D. W., see Carter, H. E. [22] 381, 385, *416*
Teichert, K., E. Mutschler, and H. Rochelmeyer [241, 242] 422, 432, 433, 437, 439, 448, 451, 453, 455, 457, *469*; [45] 494, 495, *506*; [163] 522, 549, *566*; [213] 729
Teichner, S., and E. Pernoux [710] 28, *198*
Teijgeler, C. A. [711] 5, *198*
Teisseire, P., see Pesnelle, P. [195] 212, 228, *255*
Telesz, L. A., see Elliot, K. [14] 616, *628*
— see Johnson, C. D. [30] 616, *629*
Teller, E., see Brunauer, St. [108] *183*
Tenhunen, R. [167] 604, *611*
Terepka, A. R., see Chen jr., P. S. [15] 275, 276, 280, *308*
Terner, Ch., and E. R. Darey [712] 177, *198*
Terry, W. G., see Djerassi, C. [184] *785*
Tertian, R., see Papée, D [499, 500] 23, 24, *193*
Tertrin, C., see Llosa, P. de la [216] 740, *785*
Tétenyi, P., and D. Vágujfalvi [284] 238, *257*
Teuber, M. [214] 714, *729*
Teuchy, H., see Sumere, C. F. van [158] 540, *566*, 890
Teuscher, E. [128, 129] 472, *493*
— see Teuscher, G. [130] *493*
Teuscher, G., and E. Teuscher [130] *493*
Tewari, K. C., see Parihar, D. B. [21a] 500, *505*
Thain, E. M., see Abraham, M. H. 902

Thannhauser, S. J., see Dain, J. A. [32] 392, 393, *416*
— see Schmidt, G. [88] 790, *806*
Theander, O., see Feast, A. A. J. [48] 820, *836*
Theidel, H. [86] 675, *686*
Theilacker, G., see Auterhoff, H. [5] 711, *724*
Thiele, O. W., see Wober, W. [221] 389, *421*
Thieme, H. [215] 687, *729*
Thies, H. 434
— and F. W. Reuther 526, 874
Thimann, K. V., see Stowe, B. B. [125] 473, *493*
Thoai van, see Roche, J. N. [75] 749, *782*
Tholey, G., and B. Wurtz [713] 146, *198*
Thoma, F. [164, 164a] 526, 560, *566*
Thoma, K., R. Rombach, and E. Ullmann [95, 96] 662, *686*, 873
Thomas, A., see Ristić, S [186, 187] 460, *467*
Thomas, A. E. III., J. E. Scharoun, and H. Ralston [201] 408, *420*
Thomas, A. F., and J. M. Müller [285] 243, *257*
— see Menrad, E. L. [165] 243, *254*
Thomas, G. H., see Chamberlain, J. [126] 159, *183*; [33] 314, 328, *358*
Thomas, J. D. R., see Criddle, W. J. [139] 111, *184*; [8, 9] 626, *628*
Thomas, Kl., see Jerchel, D. [328] 176, *188*
Thommen, H. [132] 267, 308, *311*
— and U. Gloor [137] 267, *311*
— — and O. Wiss [135] 267, *311*
— and H. Wackernagel [134, 136] 267, *311*
— and O. Wiss [133] 267, 272, *311*, 881
Thompson, A. [168] 759, *784*
Thompson, A. C., and P. A. Hedin [713a, 800] 145, *198*, *200*
Thompson, C. J., see Crawhall, J. C. [22] 583, *608*
Thomson, D. L., see Dahlqvist, A. [23] 277, *308*
Thomson, J., see Abbott, D. C. [1, 2, 3, 4, 5] 32, 93, 94, *180*; [1, 2] 645, *654*
Thomson, J. B., see Connolly, J. P. [135] 99, *183*

Thomson, R. H. [216] 687, 689, *729*
Thornburg, W. [100] 638, *656*
Thorp, R. H., see Silbermann, H. 902
Thorpe, W. V., see Bray, H. G. [90] 750, 751, *783*, 890
Threlfall, D. R., see Yusef, H. M. [153] 290, *311*
Thum, J., see Awe, W. [84] 750, *782*, 882
Thyiák, E. [46] 498, *506*
Timell, T. E., C. P. J. Glaudemans, and A. L. Currie 856
Tira, S., see Korte, F. [126] 715, 716, *727*
Tischer, T. N., see Lestienne, A. [398] 52, 59, *190*
— see Przybylowicz, E. P. [546] 59, *194*
Tiselius, A. [20] 737, *781*
— S. Hjertén, and Ö. Levin [714] 30, 90, *198*
— see Alm, R. S. [18] *180*
Tishler, F., see Korzun, B. P. [90] 541, *564*
Tittel, G. L., see Reisch, J. [137] 550, 551, *565*
Tobias, H., see Barbier, M. [11] 323, 335, 355, *357*
Tölgyessy, J., see Šaršúnová, M. [201] 448, *468*
Toennies, G., and J. J. Kolb [55] 749, *782*, 891
— see Winegard, H. M. [82] 749, 750, *782*
Toman, J. ,see Novotný, L. [158] 717, *728*
Tometsko, A. M., and N. Delihas [97] 804, *806*
Tomko, J., and A. Vassová [242b] 461, *469*
Tooper, E. B., see Piez, K. A. [29] 737, *781*
Topham, J. C., and J. W. Westrop [80] 613, *630*
Tore, J. P. [518] 30, *193*; [28, 63] 815, 825, 827, *835*, 836
Toribara, T. V., and R. E. Sherman 887
Toribio, F., see Weinges, K. [236] 696, 712, *730*
Torkar, K., u. Mitarb. [715, 716, 717] 23, 24, *198*
Tornero, M. C., see Oertel, G. W. [490] 46, 47, *193*; [135] 316, 356, 357, *360*

Tóth,C.E., see Fassina,G. [60] 240, 252
Tove,B.S. [718] 164, *198*
Towers,G.H.N., A.G.McInnes, and A.C. Neish [217] 718, *729*
— see El-Basyomi,S.Z. [43] *725*
Towne,J.C., see Spikner,J.E. [651] 103, 149, 155, *197*
Townsend,L.M., see Broom,A.D. [7] 794, *804*
Toyada,H. [87] 750, *783*
Toyoshima,S., see Nishimoto,Y. [110] 522, 524, *565*
Tozer,B.T., see Scanes,F.S. [167] 759, *784*
Tracht,H.-G., see Awe,W. [5] 524, 526, *562*
Tracuzzi,M.E., see Rispoli,G. [211] 237, *256*
Trappe 202
Travecedo,E., see Pérez-Medina,L.A. [164] 459, *467*
Treiber,E. [719] 33, *198*
Trennheuser,L., see Stahl,E. [669] 118, *197*; [266] 209, *257*
Trinks,H., see Pastuska,G. [502, 503] 111, 112, *193*; [23] 502, *506*; [119] 540, *565*; [51] 621, *629*; [166] 699, *728*; [195] 746, *785*
Troemer,B., see Brintzinger,H. [99] 8, *182*
Trojánek,J., O.Štrouf, J.Holubek, and Z.Čekan [243] 449, *469*
— see Moza,B.K. [141, 142] 449, *466*
Tronier,B., see Lundt,I. [98] 831, *837*
Trotter,I.F., see Brown,L. [105] 34, *183*
Trout,R.G., see Dimillier,I. [29] 585, *608*
Truswell,A.S., and W.D.Mitchell [231] 332, *362*
Truter,E.V. [720, 721] 5, 69, 137, *198*
— see Purdy,S.J. [547, 548, 549] 137, 138, *194*; [171] 411, 415, *420*
Trzebny,W., see Suszko-Purzycka,A. [233, 233b] 455, 456, *469*
Tscherter,H., see Hofmann,A. [91] 451, *465*
— see Stauffacher,D. [230] 451, *468*
Tschesche,R. 855, 863
— and U.Axen [287] 243, *257*
— and G.Balle [70] 828, 829, 834, *836*

Tschesche,R., G.Biernoth, and G.Snatzke [289] 243, *257*
— — and G.Wulff [722] 27, 101, 146, *198*; [244] 422, *469*
— J.Duphorn, and G.Snatzke [288, 290] 242, 243, 244, 245, *257*
— W.Freytag, and G.Snatzke [168] 342, *361*
— G.Grimmer, and F.Seehofer 870
— E.Henckel, and G.Snatzke [292] *257*
— K.Kometani, F.Kowitz, and G.Snatzke [245] 422, 432, *469*
— F.Lampert, and G.Snatzke [286, 293] 242, 243, 244, 247, *257*
— U.Schacht, and G.Legler [218, 219] 691, 695, 696, *729*
— H.Schwarz, and G.Snatzke [169] 347, *361*
— and A.K.SenGupta [291] 243, *257*
— and G.Snatzke [170] 334, *361*
— and H.Striegler [294] 245, *258*
— R.Welters, and H.-W.Fehlhaber *471*
— P.Welzel, and G.Legler [246] 441, *469*
— and G.Wulff [295] 245, *258*; [71] 829, 834, *836*, 863
— — and G.Balle [170a] 347, *361*
— — and K.H.Richert [171] 312, 315, 327, 328, 329, 341, *361*
— and N.Ziegler [296] 243, *258*
— see Legler,G. [68] *491*
— see Lettré [108] 311, 313, *360*
Tsien,W.H., see Sheppard,H. [632] 159, *196*
Tsuda,Y., O.Achmatowitcz jr., and L.Marion [247] 460, *469*
— T.Sano, K.Kawaguchi, and Y.Inubushi [297] 242, 243, *258*
Tsujino,Y., see Sugita,J. [157a] 508, *566*
Tsukamoto,T., and T.Komori [48] 747, 749, *782*
TsungWang,K., see ChihTung,Y. [193a] 316, 350, *362*
Tubaro,E., and L.Rustici [723] *198*
Tucker,D.R., see Azarnoff,D.L. [8] 323, 326, 331, *357*
Tucker,J.W., see Stumpf,H.C. [704] 24, *198*
Tucker,L.C.N., see Brimacombe,J.S. [76] 828, *836*

Tucker, T.C., see Vomhof, D.W. [10] 809, 814, 815, 833, 834, *835*
Tudball, N., see Wusteman, F.S. [185] 356, *361*; [91] 679, 680, *686*
Türler, M., and O. Högl [87] 664, *686*
Tulane, V.J., see Donaldson, K.O. [167] 90, *184*
Tuna, N., R. Kammereck, and H.K. Mangold [725] 171, *198*
— and H.K. Mangold [202] 368, 373, 377, 383, 385, 386, *420*
— — and D.G. Mosser [724] 169, 170, 179, *198*
— see Hawthorne, B.E. [58] *417*
Turano, C., see Fasella, P. [43] 591, *608*
Turba, F., and G. Gundlach [170] 760, *784*
Turneau, D. le, see Keith, R.W. 878
Turner, N.A., and R.J. Redgwell [230] 734, *786*
Turner, R.A., see Davoll, H. [158] 758, *784*
Tursch, B., and E. Tursch [298], 241, *258*
— — and I.T. Harrison [299] 242, 243, *258*
Tursch, E., see Tursch, B. [298, 299] 241, 242, 243, *258*
Tuppy, H. [81] 749, *782*
— and G. Bodo [161] 758, 759, *784*
— see Sanger, F. [95] 750, 756, *783*
Tyihák, E. [300] 226, *258*; [46] 498, *506*, 874
— and D. Földesi [301] 239, *258*
— and M. Imre [302] 237, *258*
— and G. Molnar [303, 304] 226, *258*
— and I. Palvy [220] 718, *729*
— J. Sárkány-Kiss, and I. Máthé [305, 306] 238, *258*
— and D. Vágujfalvi [307] 226, *258*
— — and P.L. Hágony [308, 309] 226, *258*, 904
— see Máthé, I. [156] 238, *254*; [132] 426, *466*
— see Vágujfalvi, D. [312] 229, 240, *258*
Tyler, V.E., and D. Gröger [248] 447, *469*; [131] *493*
— see Gröger, D. [83, 84] 451, *464*
Tyler jr., V.E., see Montfort, M.L. [147] 718, *728*
Tyllová, M., see Dušinský, G. [52] 214, 237, *251*

Uchikawa, T., see Ibayashi, H. [90] 339, *359*; [75] 602, *609*
Uchiyama, M., see Okui, S. [156] 387, *419*
Udagawa, S., see Natori, S. [150] *728*
Udenfriend, S. [726] 77, *199*
— see Duggan, D.E. [32] 303, 304, *308*
Uhlenbruck, G., see Seifert, H. [150] 598, *611*
Uhlmann, H.-J. [166] 533, 534, 535, 536, 539, *566*
Ullmann, E., and H. Kassalitzky [249] 449, *469*
— see Thoma, K. [95, 96] 662, *686*, 873
Ullstrup, A.J., see Agurell, S. [5] *462*
Ulshöfer, H.W., see Baumann, W.J. [9] 381, 385, 414, *416*
— see Mangold, H.K. [123] 399, *419*
Undheim, K., see Djerassi, C. [184] *785*
Ungar, F., see Quesenberry, R.O. [138, 223] 314, 317, 320, 335, 340, 341, *360, 362*
Unger, K., see Kohlschütter, H.W. [366] 7, 23, *189*
Upper, Ch.D., see Graebe, J.E. [243] 178, *186*
Urbach, G. [310] 219, 222, 223, *258*
Urech, J., B. Fechtig, J. Nüesch, and E. Vischer [311] 238, *258*
Uricaru, N., see Cionga, E. *470*
Urion, E., M. Metche, and J.P. Haluk [221] 704, *729*
Uritani, J., see Akazawa, T. [2] 213, *250*
Urivetzky, M., see Samuel, P. [144] 341, *360*; [136] 603, *610*
Urx, M., J. Vondráčková, L. Kovařík, O. Horský, and M. Herold 856
Ushakov, A.W., see Bergelson, L.D. [12] 379, 381, 385, *416*
Usui, T. [172] 323, 326, 352, 353, 354, *361*

Vácha, P., P. Čuba, Vl. Preininger, L. Hruban, and F. Šantavý [250] 454, 455, 456, *469*
Vacíková, A., V. Felt, and J. Maliková [203] *420*
Vaedtke, J., and A. Gajewska [173] 317, 337, *361*
— — and A. Czarnocka [174] 316, 335, *361*

Vágujfalvi, D. [251, 251b] 424, 433, 434, *469*, 874
— and E. Tyihák [312] 229, 240, *258*
— see Tétenyi, P. [284] 238, *257*
— see Tyihák, E. [307, 308, 309] 226, *258*, 904
Vahounty, G. V., C. R. Borja, and S. Weersing [728] 179, *199*
Valenta, Z., see Dugas, H. [49b] 461, *463*
Valentin, H. [313] 248, *258*
Vallance, D. G. M., see Giles, C. H. [231] 41, *186*
Vanclef, A., see Vancraenenbroeck, R. [223] 704, *729*
Vancraenenbroeck, R., A. Rogirst, H. Lemaitre, and R. Lontie [222] 704, *729*
— A. Vanclef, and R. Lontie [223] 704, *729*
Vannier, S. H., and W. L. Stanley [729] 155, *199*
— see Ikeda, R. M. [104] *253*
— see Stanley, W. L. [688, 689] 72, *198*
Varma, T. N. R., T. Panalaks, and T. K. Murray [138] 273, 274, *311*
Vashist, V. N., and K. L. Handa [314, 315] 221, 236, *258*
Vas'kovskiĭ, V. E., see Khorlin, A. Y. [127] 243, 245, *253*
— see Kochetkov, N. K. [135] 243, 245, *254*
Vassová, A., see Tomko, J. [242b] 461, *469*
Vaver, V. A., see Bergelson, L. D. [12] 379, 381, 385, *416*
Vecchi, M., see Kofler, M. [79] 284, *309*
Večerkova, J., see Macek, K. [97] 507, *564*
Vecsei, P., V. Kemény, and A. Görgényi [176] 324, *361*
Végh, A., R. Budvári, G. Szász, A. Brantner, and P. Gracza [252] 433, *469*
Vegis, A. [132] 473, *493*
Venkataraman, K. [225] *729*
— see Dutta, N. L. [35] 709, *725*
Venkstern, T. V., and A. A. Baev [98] 792, *806*
Venna, R., see Peng, T. C. [125] 598, 602, *610*
— see Tai Chan, P. [232] 338, *362*

Venturini, D., see Verderio, E. [315a] 237, *258*
Venyaminova, A. G., see Khorlin, A. Y. [127] 243, 245, *253*
Verbeek, R., see Onckelen, H. A. van [88] 475, *492*
— see Petridis, C. [93] 489, *492*
Vercellotti, J. R., see Wolfrom, M. L. [77, 78] 828, *836*
Vercruysse, A. [167] 540, *566*
Verderio, E., and D. Venturini [315a] 237, *258*
Verdier, C.-H. de, and G. Ågren [18] 736, *781*
Veres, E. 308
Verloop, M. E. van, see Pinxteren, J. A. C. [170, 171] 438, 440, *467*
Verma, M. R., and R. Dass [81] 612, *680*
Vermeulen, A., and J. C. M. Verplancke [730] 179, *199*; [177] 339, *361*; [168] 602, *611*
Vernin, G., and J. Metzger [47] 502, *506*
Veronese, F. M., see Benassi, C. A. [2] 474, *489*; [7] 584, *607*
Veronese, G., see Ragazzi, E. [25] 814, *835*
Verplancke, J. C. M., see Vermeulen, A. [730] 179, *199*; [177] 339, *361*; [*168*] 602, *611*
Vetter, W., see Pinar, M. [168] 447, *467*
Vicari, C., see Cavina, G. [122] 148, 151, *183*; [31] 340, 341, *358*
Vidic, E. [168] 528, *566*
— and J. Schütte [253] 441, *469*
Vigneaud, V. du, see Davoll, H. [158] 758, *784*
Vignoli, L., J. Guillot, F. Gouezo, and J. Catalin [253b] 440, *469*
Vikrot, O. [169] 596, *611*
Vimercati, A., see Gillio-Tos, M. [7] 500, 501, *505*; [23] 626, *628*, 876
Vioque, A., see Vioque, E. [206] 414, *421*
Vioque, E., and R. T. Holman [731] 147, *199*; [316] 237, *258*; [204, 205] 386, 414, *421*
— and A. Vioque [206] 414, *421*
Virtanen, A. I., see Brandner, G. [206] 740, *785*

Virtanen, A. I., see Gmelin, R. [22, 23] 472, 473, 474, 479, 486, *490*
— see Piironen, E. [106] 306, *310*; [95] *492*
Vischer, E., and E. Chargaff [99] 792, *806*
— see Urech, J. [311] 238, *258*
Vishniac, W., see Colman, B. [133] 40, *183*; [18] 269, *308*
Viswanathan, C. V., see Kaufmann, H. P. [87] 386, *418*; [88, 89] 597, 599, *609*
Viswanathan, N., see Govindachari, T. R. [77] 447, *464*
Vitali, R., see Ercoli, A. [50] 316, 341, *358*
Vlasinich, V., and J. B. Jones [233] 323, *362*
Vobecky, M., V. D. Nefedov, and E. N. Sinotova [88] 666, *686*
Vochten, R. F., see Potter, W. P. de [31] 499, *506*; [132] 520, 524, *565*
Völksen, W. [226] 718, *729*
Vogel, G. [316a] 245, *258*
Vogel, H. J., see Marshak, A. [46] 790, 792, *805*
Vogel, J., see Korte, F. [140] 215, *254*; [44] 683, *685*, 891
— see Kortüm, G. [371, 372] 143, *190*
Vogel, K., see Kohlschütter, H. W. [366] 23, *189*
Vogelberg, K. H., see Stüttgen, G. [159] 599, *611*
Vogler, K. [145] 753, 755, *784*
— R. O. Studer, P. Lanz, W. Lergier, and E. Böhni [146] 753, *784*
— — and W. Lergier [134] 753, *784*
— — — and P. Lanz [138] 753, *784*
— see Studer, R. O. [139] 753, *784*
Voigt, H., see Egger, K. [42] 265, 270, 271, 272, *309*
Voigt, K. D., U. Volkwein, and J. Tamm [178] 339, *361*; [170] 602, *611*
Voigt, R., see Pötter, H. *470*
Voigt, S., M. Solle, and K. Konitzer [171] 586, *611*
Voigtländer, H. W., see Graf, E. [66] 719, *726*
Volcsik, H., see Metzner, H. [441] 64, *191*
Volk, O. H., and R. Schunk [317] 239, *258*

Volke, J., see Oelschläger, H. [489] 146, 151, 155, *193*; [116] 509, *565*
Volkert, O., see Schildknecht, H. [606] 162, *195*
Volkin, E., and C. E. Carter [100] 788, *806*
Volkwein, U., see Voigt, K. D. [178] 339, *361*; [170] 602, *611*
Vollmann, H., see Stahl, E. [679] 86, 91, *197*; [267] 228, *257*
Vollmer, K.-O., see Grisebach, H. [253, 254] 178, *186*; [71] 692, *726*
Voltmer, G., see Tautz, N. A. [166] 589, *611*
Vomhof, D. W., and T. C. Tucker [10] 809, 814, 815, 833, 834, *835*
Vondráčková, J., see Urx, M. 856
Vorbrüggen, H. [318] 242, 243, *258*
— and C. Djerassi [254] 460, *469*
— see Antonaccio, L. D. [6] 447, *462*
Vorendohre, G., see Braun, D. [21] 211, *251*; [7, 8] 657, 663, *684*
Voronkova, V. V., see Bergelson, L. D. [11] 398, *416*
— see Dyatlovitskaya, E. V. [20] 662, *684*
Vos, Th., see Christensen, E. K. J. [32] 536, 537, 538, *562*, 884
Voss, K., see Gerngross, O. [114] 751, *783*
Vries, B. de 316; [207] 396, *421*
— and G. Jurriens [208] 396, *421*
— see Jurriens, G. [76] 414, *417*
Vries, G. de, G. P. Schütz, and E. van Dalen 855
Vroman, H. E., and G. L. Baker [209] 378, *421*
Vrublovský, P., H. Potěšilová, and F. Šantavý [255] 441, 443, *469*
— see Preininger, Vl. [180] 440, 441, *467*
Vuilleumier, J. P., and S. Nobile [139] 305, *311*
Vymetal, J., see Kucharczyk, N. [16] 504, *505*; [46] 667, 668, *685*, 879, 901

Wachsmann, T., see Pfeil, E. 896
Wachtmeister, C. A. [228] 695, *729*
Wackernagel, H., see Thommen, H. [134, 136] 267, *311*
Wada, A., see Takeda, K. [166] 347, *361*, 855, 857, 863

Waehneldt,T., see Brockmann,H. [20] 812, *835*
Wagener,H., and B.Frosch [179] 354, *361*
— D.Lang, and B.Frosch [172] 596, *611*
— see Frosch,B. [207, 208] 149, *185*; [58] 353, 354, *358*; [49, 50, 51] 603, *608*
— see Lindlar,F. [113] 384, *418*
Wagner,A. [173, 175] 596, 598, *611*
— and H.Weicker [174] 599, *611*
— see Weicker,H. [182] 597, *611*
Wagner,F.A., and K.Folkers [142] 259, *311*
Wagner,G. 866
— and J.Wandel [168a] 543, *566*
Wagner,H. [732] 179, *199*
— and B.Dengler [733] 146, 151, *199*; [141] 266, 288, 290, 292, *311*
— L.Hörhammer, and B.Dengler [140] 290, 292, *311*
— — and H.Nufer [320] 235, *258*, 876
— — and P.Wolff [210] 389, 390, 391, 393, *421*
— and P.Pohl [734] 65, *199*; [211, 212] 404, 406, *421*
— see Hörhammer,L. [293, 294] 43, *188*; [91, 92, 93, 94] 237, 238, 239, 247, *252*; [91, 92, 93, 94, 95, 96, 97, 99, 100, 101, 102, 103, 104, 105, 106, 109, 110, 111, 112, 113] 691, 693, 695, 696, 697, 699, 700, 703, 706, 707, 716, 717, 719, *726, 727*; [215] 740, *785*, 857, 883, 895
Wagner,S., see Petrowitz,H.J. [30] 504, 505, *506*; [69] 684, *686*
Wagner-Jauregg,Th., and M.Roth [319] 245, *258*
WagnerRomero,F. [176] 591, *611*
Wahba,S.K., see El-Deeb,S.R. [56] 237, *252*
— see Karawya,M.S. [116a, 116b] 237, 239, *253*
Waiblinger,H., see Hückel,W. [35] 680, *685*
Waisvisz,J.M. [45] *578*
Wakhloo,J.L. [133] 473, 485, *493*
Wakil,S.J., see Harlan,W.R.jr. [276] 177, *187*
Waksman 566

Waksmundzki,A., J.Różyzo, and J.Ościk [48] 500, *506*
Waldesbühl,M., see Fauconnet,L. [52] 323, 344, *358*, 894, 901
Waldi,D. [735, 736, 737] 32, 45, 80 135, *199*; [143] 290, 295, 296, 297, 301, *311*; [180, 181, 182] 312, 323, 332, 336, 340, 342, 345, *361*; [256] 441, *461*, 469; [169, 170] 524, 536, *566*; [177, 178, 179] 589, 600, 601, *611*; [82] 612, 614, 615, 618, 619, 620, 621, 624, *630*; [101] 643, 644, 650, 652, *656*; [89] 661, 662, *686*; [229] 718, *729*; [5] 808, 812, 813, 819, 820, 833, *835*, 855, 862, 885
— K.Schnackerz, and F.Munter [257] 422, 424, 426, 433, 437, 439, 440, 448, 451, 454, *469*
Waldron,D.M., see Edward,J.T. [102] 750, *783*, 880, 886, 892
Walenius,H., see Schantz,M.V. [193] 714, *729*
Walker,K.C., and M.Beroza [83] 613, 618, 620, *630*; [102] 640, 642, 643, *656*, 859
Wallach,D.F.H., and G.L.Nordby [738] 90, *199*
Wallach,O. [321] 207, *258*
Wallen,P., see Sjöquist,J. [132] 753, 774, *784*
Wallenfels,K., and A.Arens [159] 758, *784*
Wallhäuser 547
Wallhäuser,K.H. [44] 577, *578*
— and A.Rippel-Baldes [43] 569, *578*
Walrave,W.K.A., see Boer,J.H.de [85] *182*
Walser,A., and C.Djerassi [257b] 447, *469*
Walsh,D.E., O.J.Banasik, and K.A. Gilles [213] 412, 414, *421*
Walsh,M.P. [144] 299, *311*
Walt,J.H.van der, see Steinegger,E. [163] 345, *361*
Walter,W.G., see Schwarting,A.E. [211] 434, *468*
Walz,D., A.R.Fahmy, G.Pataki, A. Niederwieser, and M.Brenner [739] 131, 132, *199*; [180] 583, *611*; [231] 758, 771, *786*
Wandel,J. [170a] 543, 545, *566*
— see Wagner,G. [168a] 543, *566*
Wang,K.-T., see Huang,J.-T. *470*

Wankmüller, A. 866
Waring, P. P., and Z. Z. Ziporin [740] 35, *199*; [101] 801, *806*; [51] 821, 822, 834, *836*
Warncke, R., see Goubeau, J. [242] 8, *186*
Warren, B. [741], 93, *199*
Warren, C. D., see Gigg, R. [100, 101] 832, 834, *837*
Warren, F. L., see Kay, H. L. [99] 356, *359*
Warren, H. D., see Snyder, L. R. [650] 90, *197*
Warren jr. M. E., see Battersby, A. R. [13 b] 444, *462*
Wartburg, A. v., M. Kuhn, and H. Lichti [230] 710, *730*
— see Binkert, J. [20] 324, *357*, 901,
— see Kuhn, M. [131] 710, *727*
— see Renz, J. [176] 710, *728*
Waser, P. G., see Lüthi, U. [408] 159, 160, *190*
Wasicky, A. [258] 422, *469*
Wasicky, R. [742, 743, 744] 53, 56, *199*; [322] 239, *258*; [231] 715, *730*
— and O. Frehden [323] 225, *258*, 866
Wasserman, R. H., see Parekh, C. K. [501] 177, *193*; [101] 277, 280, *310*
Wassermann, L., and H. Hanus [49] 820, *836*
Wassink, J. G., see Badings, H. T. [34] 31, *181*; [9] 219, 222, *250*
Watkin, J. E. [232] 691, *730*
Watzke, E., see Barrollier, J. [66] 749, *782*, 883
Way, E. L., see Elliot, W. H. [50] 527, *563*
— see Kupferberg, H. J. [116] 438, *465*; [91] 529, 532, *564*, 876
Webber, J. M., see Bukhari, M. A. [88] 831, *837*
— see Foster, A. B. [85] 831, *836*
Weber, C. J. [102] 789, *806*
Weber, D. J., see Garrett, E. R. [23] 793, 794, *805*
Weber, F., and O. Wiss [145] 284, 285, 287, *311*
Weber, R. [201] *785*
— see Brenner, M. [97a] *182*
Weber, R. P., see Gordon, S. A. [24] *490*, 875
Weber, S., see Brenner, G. S. [104] 832, 834, *837*

Weber, S. H., and A. Langemann [171] 541, *566*
Weber, W., see Kern, W. [36] 658, *685*
Weersing, S., see Vahounty, G. V. [728] 179, *199*
Wehrli, A. [172] 542, 543, 545, *566*
Wehrli, W., see Kaufmann, H. [95a] 342, *359*
— see Lewbart, M. L. [108a] 324, 342, *360*, 870
Wehrmüller, J., see Brenner, M. [149] 755, *784*
Weichsel, H. [173] 537, *566*, 877
Weicker, H. [214] 375, *421*; [181] 592, *611*
— and R. Brossmer [27] 814, *835*
— D. Kuhn, and H. Stegmann [183] 580, *611*
— A. Wagner, and H. Schönthal [182] 597, *611*
— see Dain, J. A. [32] 392, 393, *416*
— see Huhnstock, K. [73] 592, *609*
— see Wagner, A. [174] 599, *611*
Weidemann, G., and W. Fischer [2] 807, 808, 811, 814, *834*
Weidmann, H. [174] 534, 536, 537, 538, *566*, 864, 898
Weiffenbach, H., see Klouwen, H. M. [39] 801, *805*
Weigert, E., and P. J. Schorn [324] 244, 245, 246, *258*; [233] 715, *730*
Weiland, J., see Sembdner, G. [113, 114] 472, 475, 486, 489, *493*
Weill, C. E., and P. Hanke [31] 816, 834, *835*
— see Lefar, M. S. [89] 831, *837*
— see Wheeler, M. [33] 817, *835*
Weimann, G., and K. Randerath [745] 46, *199*; [103] 795, 801, *806*
— see Randerath, K. [563] 96, *194*; [68] 802, *806*
Weimar, M., see Geiss, F. [25] 666, *685*,
Weinand, K., W. Rindt, and G. W. Oertel [184] 602, *611*
Weinges, K. [234, 235] 712, *730*
— and F. Toribio [236] 696, 712, *730*
— see Freudenberg, K. [50, 51, 52, 53, 56] 687, 688, 693, 696, 712, *725*
Weisbach, A. J., see Douglas, B. [49] 460, *463*
Weiser, H. 308
Weiss, A., and S. Fallab 896

Weiss, B., and R. L. Stiller [215] 394, *421*
Weissmann, Ch., H. Schmid, and P. Karrer [259, 260, 261] 447, *469*
— see Nagarajan, K. [144] 447, *466*
Weitz, E., H. Franck, and M. Schuchard [746] 9, 10, *199*
Weixlbaumer, H., see Fischer, R. [61b] 460, 461, *464*
Wekell, J. C., C. R. Houle, and D. C. Malins [216] 388, *421*
— see Malins, D. C. [115, 116] 377, 383, *418*
Weliky, I., see Richardson, G. S. [573] 157, 159, *195*
Wellendorf, M. [325] 237, 238, 239, *258*
Wells, M. A., and J. C. Dittmer [217] 370, *421*; [185] *611*
Wells, R. A., see Burstall, F. H. 876, 899
Welters, R., see Tschesche, R. *471*
Welzel, P., see Tschesche, R. [246] 441, *469*
Wender, S. H., see Gage, T. B. [58] *726*, 856
— see Winkler, B. C. *471*
Wendt, H., see Weygand, F. [237] 695, *730*
Wentworth, W. E., see Keirs, R. J. [343] 141, *189*
Wenzel, M., see Karlson, P. [338] 149, *189*
— see Schulze, P.-E. [617] 160, 162, 170, 177, *196*
Werblood, H. M., see Bartlett, M. F. [12] 449, *462*
Werner, A. E., see Mills, J. S. [169] 248, *254*
— see Rawlings, F. I. G. [208] 248, *255*
Werner, G., see Seiler, N. [626] 141, *196*; [110] 485, *493*
Werny, F., and P. J. Scheuer [262] 457, *469*
Wernze, H. [747] 35, *199*; [186] 605, *611*
— and J. Fujii [187] 585, *611*
Wessels, H., see Kaufmann, H. P. [88, 89, 90] 398, 401, 412, 413, 414, *418*
Wessely, F., E. Zbiral, and J. Jörg [90] 677, *686*
— see Zbiral, E. [94] 678, *686*
West, Ch. A., see Graebe, J. E. [243] 178, *186*

West, Ch. A., see Phinney, B. O. [94] 488, *492*
West, C. D., V. J. Chavré, and M. Wolfe [188] 587, *611*
West, D., see Freeman, C. P. [43] 377, 414, *.417*
Westall, R. G., see Bate-Smith, E. C. [8] 693, *724*, 897
Westfelt, L. [326] 211, *258*
— see Norin, T. [186] 248, 249, *255*
Westphal, O., O. Lüderitz, and F. Bister [748] 42, *199*
Westrop, J. W., see Topham, J. C. [80] 614, *630*
Wettstein, A., see Neher, R. 894
Weygand, B., see Mothes, K. [140] 454, *466*
Weygand, F., H. Simon, H. G. Floss, and U. Mothes [238] 694, 695, *730*
— and H. Wendt [237] 695, *730*
Weygand, R., see Gröger, D. [82] 457, *464*
Weyl, W. A. [749] 11, *199*
Wharry, D. M., see Morris, L. J. [140, 142] 386, 387, 398, *419*
Wheeler, M., P. Hanke, and C. E. Weill [33] 817, *835*
Whelan, F. J., and G. L. Plaa [189] 605, *611*
Wherrett, J. R., and J. N. Cumings [218] 392, *421*; [190] *611*
— see Müldner, H. G. [146] 389, 392, 393, *419*; [110] 598, *610*
Whistler, R. L., T. van Es, and R. M. Rowell [90] 831, *837*
— see Shasha, B. [91] 793, *806*; [13] 809, 815, 820, 834, *835*
Whitaker, D. R., and K. A. Mittelstadt [750] 90, *199*
White, H. B. jr. [219] 404, *421*
White, K., see Bray, H. G. [90] 750, 751, *783*, 890
Whitehead, J. K. [751] 176, *199*
Whitehouse, M. W., see Greenway, R. M. [18] 811, *835*
Whittaker, V. P., and S. Wijesundera 880
Whittle, K. J., P. J. Dunphy, and J. F. Pennock [146] 286, *311*
— see Dunphy, P. J. [34, 35] 284, 286, *308*; [38] 378, 412, *416*, 867
Wichner, S., see Libbert, E. [71] 488, *492*

Wichtl, M. [327] 228, *258*
— see Penelle, P. [195] 212, 228, *255*
Wickberg, B. [59] 822, *836*
Wicker, R. J., see Sampson, K. 884
Widén, C. J., see Schantz, M. V. [193] 714, *729*
Wiechmann, J., see Seiler, N. [225] 778, *786*
Wiechmann, M., see Seiler, N. [626] 141, *196*; [110] 485, *493*; [35, 36] 495, 496, 499, *506*; [149] *565*
Wiedenhof, N. [752] 32, *199*; [34] 817, *835*
Wiegandt, H., see Kuhn, R. [106, 107] 389, 392, 394, *418*
Wieland, Th., and L. Bauer [106] 751, *783*, 877
— and H. Determann [756] 47, 90, *199*; [191] 591, *611*
— and F. Fischer 894
— and D. Georgopoulus [757] 76, 90, *199*
— G. Lüben, and H. Determann [755] 37, *199*; [104] 795, 796, *806*
— and G. Pfleiderer [753, 754] 109, 111, 112, 113, *199*
— see Kawerau, E. [49] 747, *782*, 889
Wieme, R. J. [758] 109, *199*
Wientjes, R., see Robles, M. A. [188] 423, 435, 441, *467*
Wiesner, K., see Dugas, H. [49b] 461, *463*
Wiggins, L. F., and J. H. Williams [61] 749, *782*
Wijesekera, R. O. B., see Senanayake, U. M. 471
Wijesundera, S., see Whittaker, V. P. 880
Wild, A., see Klement, R. [37] 678, *685*
Wilkinson, J. V., see Clifford, C. J. [37a] 337, 341, *358*
Willhalm, B., see Demole, E. [44] 216, *251*
— see Sundt, E. [274] 237, *257*
Williams, B. L., and T. W. Goodwin [147] 267, *311*
Williams, D. J., see Frodyma, M. M. [205] 144, *185*
Williams, D. L., see Murray, A. III. [467] 167, *192*
Williams, E. I. [103] 649, *656*

Williams, J. A., A. Sharma, L. J. Morris, and R. T. Holman [192] 597, *611*
Williams, J. H., see Wiggins, L. F. [61] 749, *782*
Williams, P. N., see Hilditch, T. P. [59] 367, *417*
Williams, R. 895
Williams, R. H., see Kuzuya, T. [384] 179, *190*
Williams, R. J. H., see Goodwin, T. W. [55] 267, *309*
Williams, R. J. P., see Alm, R. S. [18] *180*
Williams, T. I. [759] 2, *199*
Willoughby, M., see Bush, I. E. 902
Willstätter, R., H. Kraut, and K. Lobinger [760] 8, *199*
Wilson, H., see Korenman, S. G. [97] 602, *609*
— see Rivlin, R. S. [580] 170, *195*; [143] 341, *360*
Wilson, J. M., see Antonaccio, L. D. [6] 447, *462*
— see Djerassi, C. [46] 447, *463*
Wilson, T. R. S., see Riley, J. P. [113] 269, *310*
Windeck-Lutz, C., see Flück, H. [65] 248, *252*
Winefordner, J. D., and H. A. Moye [761] 155, *199*
Winegard, H. M., G. Toennies, and R. J. Block [82] 749, 750, *782*
Winkler, B. C., W. J. Dunlop, L. M. Rohrbaugh, and S. H. Wender 471
Winkler, C., see Schorn, H. [148] 587, *611*
Winkler, H. [49] 496, *506*
Winkler, K., see Luckner, M. [125, 126] 432, 456, *466*
— see Mothes, K. [140] 454, *466*
Winkler, W. [328] 246, *258*; [263, 264] 447, *469*
— and W. Awe [265] 440, 441, *469*
— see Awe, W. [8, 9] 441, *462*
Winter, J., see Hauptmann, S. [76a] 524, *564*
Winterfeldt, E., see Bohlmann, F. [26] 436, *462*
Winterfeldt, G., see Bohlmann, F. [27] 436, *462*
Winterstein, A. [762] 178, *199*
— and B. Hegedüs [763] 135, *199*; [149, 150] 267, 272, 274, *311*, 896

Winterstein, A., A. Studer, and R. Rüegg [764] 178, *199*; [148] 267, 271, 274, *311*
Wirthwein, R., see Kauffmann, Th. [15] 501, *505*
Wirzing, G. [765] 12, *199*
— see Kohlschütter, H. W. [360] 13, 20, *189*
Wise, C. S., see Jeanes, A. [326] 86, *188*
Wisiewski, J., see Adamski, R. *470*
Wiss, O. 894
— see Thommen, H. [133, 135] 267, 272, *311*, 881
— see Weber, F. [145] 284, 285, 287, *311*
Witkop, B., see Daly, J. W. [4a] 499, *505*
— see Ishii, S. [19] *577*
— see Märki, F. [130] 460, *466*
Wittels, B., and R. Bressler [220] 389, *421*
Witter, R. F., G. V. Marinetti, and A. Morrison 896
Wittwer, S. H., and M. I. Bukovac [133a] 489, *493*
Wober, W., and O. W. Thiele [221] 389, *421*
Woggon, H., D. Spranger, and H. Ackermann [104] 640, *656*
— see Korn, O. [43] 665, *685*
Wohlleben, G. 761
— see Bhandari, P. R. [66] 53, 128, *181*
Wohnlich, J. J. [766] 52, *199*
Woiwod, A. [51] 747, 749, *782*
Wojsa, K., see Rutkowska, U. [195] 449, *468*
Wolf, G., see Sumere, C. F. van [158] 540, *566*, 890
— see Yagishita, K. [151] 274, *311*
Wolf, L., B. Szarvady, and E. M. Karácsony [266] 451, 453, *469*
— see Craig, J. C. [40c] 430, *463*
Wolfe, M., see West, C. D. [188] 587, *611*
Wolff, I. A., see Smith jr., C. R. [190] 719, *728*
Wolff, P., see Wagner, H. [210] 389, 390, 391, 393, *421*
Wolfman, L., and B. A. Sachs [183] 317, 331, *361*

Wolfman, L., see Sachs, B. A. [180] 385, 386, *420*; [135] 592, *610*
Wolford, R. W., see Attaway, J. A. [26] 139, *180*; [6, 7] 210, 214, 215, *250*
Wolfram, G. [195] 594, 595, *612*
— see Zöllner, N. [784, 785, 786] 139, 140, *200*; [189] 333, *361*; [223, 224] 386, 415, *421*; [198, 199, 200, 201, 203, 204] 592, 593, 594, 595, 596, *612*
Wolfrom, M. L., W. v. Bebenburg, R. Pagnucco, and P. McWain [96] 831, *837*
— and R. B. Bennett [102] 832, 834, *837*
— P. Chakravarty, and D. Horton [42] 818, *835*
— W. A. Cramp, and D. Horton [94] 831, *837*
— and N. E. Franks [52] 822, *836*
— H. G. Garg, and D. Horton [95, 97] 831, *837*
— D. Horton, and D. H. Hutson [93] 831, *837*
— and R. M. de Lederkremer [58] 822, *836*
— — and L. E. Anderson [767] 29, *200*; [56] 822, 828, *836*
— D. L. Patin, and R. M. de Lederkremer [4] 807, 809, 814, 815, 818, 820, 821, 829, 834, *835*
— J. R. Vercellotti, and D. Horton [77, 78] 828, *836*
Wollenweber, J., see Kottke, B. A. [379a, 797] 150, *190*, *200*
Wollenweber, P. [768, 769, 770, 771, 772, 773] 33, 36, 37, *200*; [84] 616, 624, 625, *630*; [232] 740, *786*
Wollish, E. G. [775] 5, *200*
— M. Schmall, and M. Hawrylyshyn [774] 54, *200*; [175] 544, *566*
— see Hawrylyshyn, M. [30] 794, *805*
Wollrab, V. [329] 238, 239, 240, *258*
— M. Streibl, and F. Šorm [330] 239, 240, *258*
Wong, C. M., see Dugas, H. [49b] 461, *463*
Wong, E., and A. O. Taylor [239] 692, 693, *730*
Wood, P., K. Imaichi, J. Knowles, G. Michaelis, and L. Kinsell [194] 593, *612*
— see Schlierf, G. [608] 135, 137, *196*; [140] 592, 593, *610*

Wood, P. D. S., and S. Holton [193] 596, *612*
— see Aylward, F. [5] 597, *607*
Wood, R., and F Snyder [222] 374, *421*
— see Fogg, A. G. [196] 31, *185*
Wood, R. D., see Rollins, C. B. [176] 378, *420*
Wortmann, A., see Lagoni, H. [36] 612, *629*
Wotiz, H. H. [196] 603, *612*
— and S. C. Chattoraj [184] 351, *361*
— see Guerra-Garcia, R. [64] 602, *609*
Wouters, O., see Heide, R. F. v. d. [33] 673, *685*, 867
Wragg, J. S., see Clifford, C. J. [37a] 337, 341, *358*
— see Savidge, R. A. [143] 526, *565*
Wren, J. J. [776] 90, *200*
Wright, J. [105] 650, *656*, 883
Wright, L. D., see Sander, E. G. [83] *806*
Wright, S. E., see Aldrich, B. J. 902
Wrolstad, R. E., and W. G. Jennings [331] 210, *258*
Wünsch, A., see Hofmann, E. 892
Wulf-Lorentzen, G., see Preussmann, R. [33] 497, *506*, 871, 900
Wulff, G., see Tschesche, R. [722] 27, 101, 146, *198*; [295] 245, *258*; [170a, 171] 312, 315, 327, 328, 329, 341, 347, *361*; [244] 422, *469*; [71] 829, 834, *836*, 863
Wunderly, C. [777] 105, *200*
Wurtz, B., see Tholey, G. [713] 146, *198*
Wusteman, F. S., K. S. Dodgson, A. G. Lloyd, F. A. Rose, and N. Tudball [185] 356, *361*; [91] 679, 680, *686*
Wyatt, G. R. [105] 790, 792, *806*
Wyss, E., see Barbier, M. [11] 323, 335, 355, *357*

Yagishita, K., P. R. Sundaresan, and G. Wolf [151] 274, *311*
Yamada, M., and A. Saito [152] 300, 301, *311*
Yamaguchi, M. [778] 5, *200*
Yamaki, T., see Stowe, B. B. [126, 127] 472, *493*
Yamamura, J., and T. Niwaguchi [106] 642, *656*

Yamasaki, K., see Iwata, T. [212] 354, *362*
Yanari, S. [125] 752, *783*
Yasuda, St. K. [50] 497, *506*; [92, 93] 670, *686*
Yatabe, M., and H. Oki [176] 551, *566*
Yawata, M., and E. M. Gold [186] 320, 337, *361*
Yee, A. W. G., see Schwartz, A. N. [89] *806*
Yen, C., see Crider, Q. [29] 376, *416*
Yi-Young Hsia, D., see Christian, J. C. [23] 596, *608*
Yoshitake, D. [240] 719, *730*
Young, G. J. [779] 12, *200*
Youngken jr., H. W., see Marderosian, A. D. [128b] 451, *466*
Yu, R. K., see Carter, H. E. [22] 381, 385, *416*
Yuldashev, P. Kh., see Abdurakhimova, N. [1] 456, *462*
Yusef, H. M., D. R. Threlfall, and T. W. Goodwin [153] 290, *311*
Yu Yunusov, S., see Abdurakhimova, N. [1] 450, *462*

Zabin, B. A., and Ch. B. Rollins [780] 32, 47, *200*
— see Schwartz, A. N. [84] *806*
Zachau, H. G. [106] 801, *806*
— M. Tada, W. B. Lawson, and M. Schweiger [107] 801, 802, *806*
Zaffaroni, A., R. B. Burton, and H. Kentmann 882
— see Burton, R. B. [86] 750, *783*
Zagórski, Z., see Opiénska-Blauth, J. [90] 486, *492*; [116] 584, *610*
Zahn, H. [85] 616, *630*
— and H. Pfannmüller [174] 762, *784*
Zamecnik, P. C., M. L. Stephenson, C. M. Janeway, and K. Randerath [108] 801, *806*
— see Randerath, K. [76] 801, *806*
Zamojski, A., and F. Zamojska [332] 220, *258*
Zamojska, F., see Zamojski, A. [332] 220, *258*
Zappi, E. J., see Gries, G. [62] 586, *608*
Zarándi, A., see Gracza, L. [77] 237, *252*
Zarnack, J., and S. Pfeifer [267] 422, 441, 457, *469*; [177] 524, 525, 542, 549, *566*; [109] *806*, 875

Zawta, B., see Freimuth, U. [57] 316, 323, 324, 335, *358*, 902
Zbiral, E. [51] 502, *506*
— F. Takacs, and F. Wessely [94] 678, *686*
— see Wessely, F. [90] 677, *686*
Zechmeister, L. [333] 235, *258*
— and L. v. Cholnoky [781] 1, *200*
Zeevaart, J. A. D., see Ninnemann, H. [83] 487, 489, *492*
Zehavi, U., and N. Sharon [44] 818, *835*
Zeitler, H. J. [267c] 461, *470*, 902
Zeitlin, H., see Frei, R. W. [201] 143, *185*
Zeitmann, B. B. [782] 130, *200*
Zelenskaya, S. V., see Zhdanov, Yu. A. [30] 815, *835*
Zeller, A., see Edlbacher, S. [93] 750, *783*
Zeller, N., see Möller, H. G. 879
Zelnik, R., and L. M. Ziti [187] 345, *361*
— — and C. V. Guimarães [188] 345, *361*
Zenda, H. [267b] 460, *469*
Zenk, M. H. [134] 472, 488, *493*
— and H. Scherf [135, 136] 472, *493*
Zerban, F. W., see Sattler, L. 856
Zessin, G., see Bogs, U. [23] 706, *725*
Zhdanov, Y. A. et al. [783] 30, *200*
— G. N. Dorofeenko, and S. V. Zelenskaya [30] 815, *835*
Zhukova, I. G., see Kochetkov, N. K. [99, 100, 101] 392, *418*
Zicha, L., see Scheiffarth, F. [146] 341, *361*; [138] 601, *610*
— see Schmid, E. [104] 486, *493*; [144] 509, 519, *565*
Zickgraf, H., see Knauff, H. G. [40] 738, *782*
Ziegler, see Ruggli 658
Ziegler, A., see Drawert, F. [31] 307, *308*
Ziegler, N., see Tschesche, R. [296] 243, *258*
Ziegler, W., see Billek, G. [18] 719, *725*
Zieve, L., see Doizaki, W. M. [166] 148, *184*
Ziffer, H., see Baker, H. [3] 303, 304, *308*

Zimm, B. H., see Massie, H. R. [47] 788, 794, *805*
Zimmerman, M., and D. Hatfield [110] 794, 801, *806*
Zimmermann, A., see Büchi, J. [34d] 433, *463*
Zimmermann, J. P., see Brenner, M. [149] 755, *784*
Zimmermann, W., see Göldel, L. [76] 245, *252*; [65] 317, 337, *359*
Zinkel, D. F., and J. W. Rowe [334] 249, *258*
Zinser, M., and Ch. Baumgärtel [268] 451, 453, *470*; [178] 562, *566*, 869
Zinsmeister, H. D., see Burnett, D. [5] 473, *490*
Ziporin, Z. Z., and R. W. Hanson [111] 801, *807*
— see Waring, P. P. [740] 35, *199*; [101] 801, *806*; [51] 821, 822, 834, *836*
Ziti, L. M., see Zelnik, R. [187, 188] 345, *361*
Zobel, H. F., see Smith jr., C. R. [190] 719, *728*
Zöllner, N., 312, 333, 340; [197] 599, *612*
— and D. Eberhagen [225] 369, 374, 386, 398, *421*; [202] 592, 593, 594, *612*
— H. Hündorf, and G. Wolfram [203] 593, *612*
— and G. Wolfram [784, 785] 140, *200*; [224] 386, *421*; [198, 199, 201] 592, 593, 594, 595, *612*
— — and G. Amin [786] 139, *200*; [189] 333, *361*; [223] 415, *421*; [200] 594, 595, *612*
— — W. Londong, and K. Kirsch [204] 594, 595, 596, *612*
— see Eberhagen, D. [35] *608*
Zubay, G. [112] 788, *807*
Zuber, H. [131] 753, *784*
Zurflüh, R., see Pfrunder, B. [528] 111, *194*
Žurkowska, J., and A. Ozarowski [787] 29, 146, *200*; [234] 345, *362*
Zwietering, P., see Boer, J. H. de [86, 87] 25, *182*
Zymalkowski, F., see Schultz, O. E. *461*

Subject Index

Acetates, sugar 822—825
Acetone 220
—, detection in urine 581
Acetylation 175, 205, 325
Acetylcarbromal 538, 539
Acetylsalicylic acid 524—527, 556, 558
Acid dyes 620, 621
Acids 48, 620—621
—, amino 33, 111, 113, 139, 144, 154, 581—592, 730—786
—, cutin 37
—, detection 654
—, dicarboxylic 51, 651—653
—, fatty 50, 139, 168, 174, 386—388, 406, 412
—, heteropoly 31
—, organic 650—654
—, phosphoric 35
Acoxyacetaldehydes 408
Acridine and derivatives 500, 503, 618, 619
Acrylic acid derivatives 657
Actinomycins 576
Activation analysis 172
Activity grades 203
Acylation 315
Adenine 457, 549, 791, 793, 794, 800
— nucleotides 47
Adenosine 791, 794, 800
— phosphates 111, 788, 791, 798—800
Adrenaline 31, 499, 520—524, 589
Adsorbents, miscellaneous inorganic 30—32
Adsorption TLC, fundamental principles 201—203
AE-cellulose 36, 38
Aerosol spray vessels 58
Aflatoxins 695
Agar culture 568—570
Alcohols 200, 412, 650, 660—663
—, fatty 173
—, sugar 820
Aldaric acids 819
Aldehydes 200, 373, 384, 399, 412

Aldehydic cores 408, 409, 414
Aldonic acids 819
Aldosterone 146, 155, 336, 341, 601
Aldrin 642—644, 676
Alkali group 842
Alkalies 46
Alkaline earths 46
Alkaloids 48, 51, 421—471
—, adsorbent layers 421—423
—, Amaryllidaceae 445—446
—, aporphine 442
—, aspidospermine 447
—, benzofluorene 458, 461
—, benzophenanthridine 443
—, benzylisoquinoline 442
—, bis(benzylisoquinoline) 444
—, Catharanthus 449—450
—, cinchona 454—456
—, colchicine 425—430
—, diterpene 460
—, eburnamine 447
—, ellipticine 447
—, elypticin 460
—, fluorocurine 447
—, Funtumia 459
—, furanoquinoline 456—457
—, Heimia 460
—, hemlock 431
—, Holarrhena 349, 459
—, indole 446—454
—, ipecacuanha 444—445
—, isopavine 443
—, Kurchi 349
—, Lobelia 432
—, lycopersicum 458
—, mavacurine 447
—, narceine 442
— of unknown constitution 443, 460
—, opium 139, 146, 436—444
—, oxindole 454
—, Papaveraceae 436—444
—, pavine 443
—, phthalide-isoquinoline 442—444
—, piperidine 430—432

Alkaloids, proaporphine 442
—, protoberberine 442
—, protopine 443
—, purine 457
—, pyridine 430—432
—, pyrrolidine 430—432
—, pyrrolizidine 435
—, quantitative determination by TLC 425
—, quinazoline 460
—, quinolizidine 435—436
—, Rauwolfia 145
—, rotundin 460
—, secale 66, 451
—, separation scheme 424
—, Solanum 458
—, solvents 421—423
—, sparteine 436
—, steroid 348, 460
—, sterol 457
—, strychnine 447—448
—, tetrahydroisoquinoline 442, 444, 460
—, tropane 432—435
—, tylocrebin 460
—, Vinca 449—450
—, visualisation 423—424
Alkenyl ethers 384
— glycerol ethers 363
Alkoxylipids 373, 414
Alkoxyphenols 680
Alkyl diglycerides 381—385, 400
— ethers 372, 373, 374, 399, 400
— gallates 633—635
— glycerol ethers 363
Alkylphenols 680
Allobarbital 534, 537
Aloe species 706—709
Alumina 23—27
Amaryllidaceae alkaloids 445—446
Amides 387, 400
Amines 48, 111, 124, 387, 400
—, thin-layer electrophoresis 502
—, aliphatic 440, 494—497
—, —, detection of 496
— and metabolites 589
—, aromatic 124, 500—502
—, 2,4-dinitrophenyl 51
—, electrophoresis 112, 502
—, heterocyclic 500—502
—, tertiary 138
Amino acids 33, 111, 113, 139, 144, 154, 581—592, 730—786

Amino acids and derivatives 730—786
— —, detction 746—751
— —, DNP derivatives 583, 584
— —, extraction from biological material 738
— — in biological material 736
— — in blood and organs 585
— — in other body fluids 586
— — in urine 582, 763
— —, limits of detection 748
— —, preparation of layers for TLC 732—734
— —, preparation of solution for investigation 734
— —, recovery 738
— —, solubilities 730
— —, solvents for TLC 739—746
— —, their derivatives and metabolites 581—592
— —, thin-layer electrophoresis 746
— —, TLC-technique 734
Aminoalcohols 498
Aminobenzoate esters 552—553
Aminobenzoic acids 500, 501, 626
ε-Aminocaproic acid 587
β-Aminoisobutyric acid 584, 607
6-Aminopenicillanic acid 567, 575
Aminophenols 501, 560, 626, 676
p-Aminophenol derivatives 524—527
Aminopyrine 151, 524—527, 558
4-Aminosalicylic acid 560
Aminosteroids 328, 348
Amino sugars 817—819
Ammonium carbonate group 841
— sulphide group 840
Amphetamine 519, 520
β-Amyrene 241
Amyrins 242
Anabasine 155, 431
Anaesthetics, local 551—554
—, —, detection 554
Analeptics 518, 520
Analgesics 147, 524—532
— with narcotic activity 527—532
Androgens 334—341
Androsterone 339
Anethole 231, 232, 233
Anhydro-vitamin A 274, 275
Animal lipids 369, 370
— —, non-polar 370
— —, polar 370, 391
Anions 111, 179, 847—851
Anisaldehyde 218

Subject Index

Anisidines 501, 626
Anthocyanidins 690, 699
Anthocyanins 693, 694, 699
Anthracene 668, 669
Anthraquinone 688
— derivatives 694, 706—709
— —, visualisation 709
— drugs 706—709
— dyes 37, 612, 616, 617, 620, 622
Anti-allergics 507—517
Antibiotics 82, 566—577
—, classification according to biosynthetic structural units 567
—, detection 568
—, miscellaneous 576
Anticoagulants(4-hydroxycoumarin group) 532
Antidepressives 518—521
Antidiabetic agents, oral 550
Antihistamines 137, 507—518
Antioxidants 32, 37, 631—636, 673, 674
—, detection 632, 635
—, hRf-values 635
—, quantitative determination 635—636
Antipyretics 524—532
Antirheumatic agents 524—532
Apigenin 689, 697
Apiole 224, 230, 231, 232
Apo-β-carotenals 267—269
Aporphine alkaloids 441, 442
Appetite depressants 518—520
Application, band 64
— methods 63—65
—, spot 63
—, streak 64
Aprobarbital 534, 536
Argentation chromatography 394, 396—400
— —, solvents 399, 400
Aristolochia acids 145, 718
Aromatic compounds, polynuclear 667—669
Artificial sweeteners 648
Asarinin 711
Asarone 230
Ascaridole 212—214
Ascending development (chambers for) 68
Ascorbic acid 292—294, 304—306
Ascorbyl palmitate 634—636
Aspidospermine alkaloids 447
Astaxanthin 267,

Astragalin 690, 701, 705
Aurones 688, 697
Autoliner 99
Automatic coating 58
— development 76
— spreaders 85
Autoradiography 157—160, 168, 169, 173, 178
Auxins, adsorbent layers 476
—, biological detection 487, 488
—, diffusible 474, 476—479
—, free 474, 476—479
—, solvents 477
—, two-dimensional separation 481
—, visualisation 485, 486
Avicel 34—36, 422, 809
A-vitamin group 273—275
Aza compounds 669
Azobenzene derivatives 615
Azo dyes 145, 151, 612—618, 620—622
Azulenes 96, 234

Bactericidal substances 540—547
Bacteriostatic substances 540—547
Balsams 138, 247—250
—, detection 250
Band application 64
Barbital 534, 536, 537
Barbiturates 533—537
—, bromine-containing 537
—, detection 536
—, N_1-methylated 536
—, quantitative evaluation 537
Basic dyes 618—620
— equipment 83—85
— — for TLC 83—85
— kit 83—85
Bayerite 24, 25
Benegride 519, 520, 538
Benzene hexachloride 643, 644, 676
Benzoates, sugar 822—825
Benzocaine 551, 552
Benzofluorene alkaloids 458, 461
Benzoic acid 81, 637, 704
Benzophenanthridine alkaloids 443
Benzophenones 51
Benzo-(α)-pyrene 155, 669
Benzoquinone 688, 697
Benzyl alcohol 227, 229
— mandelate 147, 151, 558
Benzylisoquinoline alkaloids 442
Betaines, visualisation 424
BET-method 14—16, 18, 26

Betulinic acid 244, 247
BHA 632—636
γ-BHC 137, 643, 644, 676
BHT 632—636
Biacetyl 222
Bile acids 139, 154, 178, 351—354, 603
— —, detection 323
— alcohols 334, 355
— pigments 604
Bilirubin 604
Bioautograph 568
Biological detection, antibiotics 568—570
— —, auxins 487
— —, gibberellins 488
— visualisation 82
Bio-Rad (ion exchangers) 32, 45—47
Biotests for antibiotics 568—570
— for auxins 487
— for gibberellins 488
Biotin 293, 294, 306
Biphenyl 638, 667, 668
Bisabolene oxides 212, 213
Bis(benzylisoquinoline)alkaloids 444
Bitter principles 82, 241, 246, 247, 714, 715, 718
— woods 715
Bleaches, optical 675
Blue-print paper (for copies of TLCs) 130
BN-chamber 73, 75, 76
Body substances, endogenous 579—604
— —, exogenous 604—607
Böhmite 24—26
Borate, impregnation with 31, 523, 662, 813
Borneol 227, 228
Brain lipids 598
— sphingolipids 598
Brighteners, optical 675
Brominating solvents 412, 413
Bromination 315, 331, 412, 413, 637
Bromisoval 538, 539
Bromoanilines 501, 626
Bromophenols 677
Bromosulphalein test 605
Bromoureides 538
Brucine 448
Bufadienolides 341—346
Bulnesol 212, 228
Bush system 317, 320
Butabarbital 534, 536
Butalbital 534, 536

Butobarbital 534, 536
Butylhydroxyanisoles 632—636
Butylhydroxytoluene 632—636
B-vitamin-group 292—304
B_1-vitamin-group 295, 296
B_2-vitamin group 296, 297
B_6-vitamin group 300, 301
B_{12}-vitamin group 301—303

Cab · O · Sil 163—165
Cadmium (II) 173, 839, 840, 844, 845
Cafestol 240, 241
Caffeic acid 689, 698, 700
Caffeine 81, 82, 151, 457, 548—550, 556, 558
Campesterol 329, 333
Camphor 219, 224, 519, 520
Cannabinols 154, 716
Canthaxanthin 267—270
Capsanthin 268, 270, 271
Capsorubin 268, 270, 271
Carbamate esters 514, 520, 521
Carbazole 500, 503
Carbobenzoxy compounds 778
Carbohydrates 32, 141, 807—837
Carbohydrate chemistry, monitoring of reactions by TLC 831—833
Carbonisation (of spots) 138—140
Carbonyl compounds 217—225
— —, detection 224
Carboxylic acids, aromatic 178
Carbromal 538, 539
Cardenolides 341—346
—, detection 323, 324
Cardiac glycosides 177, 341—346
— —, aglycones 341—345
— —, detection 323
Cardiolipin 390, 391
Carnitine 307
Carotenes 68, 264, 268, 270
Carotenoid aldehydes 178, 267, 271, 272
— esters 412
Carotenoids 265, 266—273
Carotol 96, 225, 227, 228
Carvacrol 230, 232
Carvone 217, 219, 225
Caryophyllene 96, 210
Catechins 689, 693
Catechol 230, 232, 541, 689
— (derivatives) 31
— amines 499, 524
Catharanthus alkaloids 449, 450

Cations 39, 47, 111, 179, 839—847
—, separation by circular TLC 845
—, — of 839—846
—, — on layers of ion exchangers 845
Cedrol 227, 228
Celite for TLC 28
— — of sugars 899
Celliton dyes 616
Cellulose, acetylated (as layer) 37
— (and derivatives) (as layers) 32, 33—36
—, AE- 36, 38
—, CM- 36, 38, 39, 46
—, DEAE- 36, 38, 39, 46
—, ECTEOLA 36, 38, 39, 47
— ion-exchanger 37, 39
—, P- 36, 38, 46
—, PEI- 36, 38, 39, 46
—, Poly-P-, 36, 38, 39
—, PP- 46
—, QA- 47
—, SE- 46
Centrifugal technique 73, 75
Cerebrosides 390—393, 414
CFCM 206
Chalkones 697
Chamazulene 210, 212
Chamber, BN- 73, 75, 76
— saturation (CS) 66
Chambers, for ascending development 68
—, GS-moisture 71
—, moisture 71, 72
—, rectangular 68
—, S- 69
—, sandwich- 70
—, setting up 68
Chanoclavine 451, 452
Charring (of spots) 138—140
Chemical races 234, 236
Chenodeoxycholic acid 351, 352
Chimyl alcohol 177
— —, labelled 176
Chloramphenicol 576
Chlorinated hydrocarbons 642—645
Chloroanilines 501, 626
Chlorobenzene (impregnation agent) 51
Chlorophenols 680
Chlorophylls 266—273
Chloroplast pigments 40, 50
Chlorothiazides 547, 548
Cholestanols 329—334

Cholesterol 137, 277, 318, 329—334, 375, 385, 400, 415, 597
—, detection 323
Cholesteryl acetate 318
— esters 50, 139, 177, 178, 398, 400, 415
— —, detection 323
— — in plasma 593—595
— — in serum 593—595
— —, labelled 176
— —, quantitative determination 594
— oleate 375, 385
— sulphate 356, 595
Cholic acid 351, 353, 603
Cholines 307, 498
Chromatobars 97
Chromato-charger 99
Chromatograms, documentation 125—133
Chromatographic patterns 402
Chromatography, continuous 69
—, disc 73
—, radial 73
—, surface 3
Chromatograms, ring 2
Chromatostack-technique 69
Chromatostrip technique 3, 97
Chromato-Vue 79
Chrysene 668, 669
Cinchona alkaloids 454—456
— —, visualisation 424
Cinchonidine 454—456
Cinchonine 454—456
1,8-Cineole 212, 213
Cinnamaldehyde 217, 218
Cinnamic acid 81
— — glycosides 693, 704
Circular TLC 73, 422, 845
Citral 218, 220
Citronellal 218, 220
Citronellol 229
Civetone 216
Clinical diagnosis, TLC in 578—612
CM-cellulose 36, 38, 39, 46
— -Sephadex 47
Coating, automatic 58
Cobalamin 148, 151
Cobinamide 302
Cocaine 553
Cocoa butter 401, 413
Codeine 437—441, 529, 530
Colchicine alkaloids 425—430
Colouring materials, synthetic 612

65*

Colour reagents 148, 854—909
Combination (of procedures) 204
Condensed phosphates 848
Conessine 348, 349
Continuous chromatography 69, 75, 86, 100
— development 69, 75, 86, 100
Copper (II) 173, 839, 840, 844—846
— group, separation 839
Coproporphyrins 604
Copying of thin-layer chromatograms, graphical 128, 129
Corrinoids 301—303
Corticosteroids 147, 154, 334—341, 601
Corticosterone 318, 336, 601
Cortisol 155, 318, 336, 341, 601
Cortisone 318, 336, 340
Coumarin glycosides 699
Coumarins 81, 115, 215—217, 688, 695, 696
Counting tubes 160—166
Coupling, azo 205, 229, 626, 665, 705
—, GC-TLC 115—125, 205
— procedures, chromatographic 114—125, 204, 205
Creatinine 588
Critical pairs 7, 277, 301, 326, 329, 332, 347, 541
Cryobox 68, 95
Cryptoxanthin 268—271
Cucurbitacins 243, 246
Cumic alcohol 227, 228
Cutin acids 37
C-vitamin group 304—306
Cyanocobalamin 301—303
Cyclamate 648
Cyclobarbital 534, 536, 537
Cyclohexylamines 500
Cyclopal 534, 536
Cytidine 791, 794
— phosphates 791, 798, 799
Cytokinins 475
Cytosine 791, 794

DANS-amides 495
— -amino acids 778
Daucol 213, 225, 228
Daucus oils 210, 234
DDT 137, 643, 644, 676
DEAE-cellulose 36, 38, 39, 46
— -Sephadex (ion exchangers) 44, 45, 47
Decolin (impregnation agent) 50

Dehydroascorbic acid 304—306
Dehydrocholesterols 275, 277, 278, 330, 332, 333
11-Dehydrocorticosterone 336
Dehydroepiandrosterone 338, 339, 602
Dehydroergosterol 333, 334
16-Dehydroprogesterone 337
— -Dehydrosteroids 339
Delfter-system 99
Densitometres 140, 141
Deoxycholic acid 351, 352, 353, 603
Deoxyribonucleic acids 111, 786—792, 801
— —, cleavage 790
— —, distinction from ribonucleic acids 789
Deoxyribonucleoside triphosphate 179
Deoxyribonucleotides 47
Deoxy-sugars 817—818
Derivatives, preparation for TLC 205, 325
Descending development (equipment for) 72
Desmesterol 329—333
Detection, apparatus for 77—81
—, biological 487, 488, 568—570
—, reagents for 855—909
Detergents 674, 675
Determination, quantitative 135—155
Development, automatic 76
— in preparative TLC 100
—, iterating 76
—, polyzonal 76
— procedures, special 86—96
— —, standard 68—77
— types 100
Dextran gels 33, 40
Diagnostics, clinical 578—612
Dialkyl hydrazines 680
Diazepines 508—514
Diazogram procedure 78
Diazo-type paper (for copies of TLCs) 129
Dicarboxylic acids 51, 651—653
Dieldrin 642—644
Digitalis glycosides 154, 342—345
— —, detection 323
Diglycerides 50, 173, 400, 408, 414
Dihydroxy acids 174
Dihydroxyanthraquinones 706
Dimethylacetals 372, 373, 383, 384, 399, 400

Dimethylaminonaphthalene-
   sulphonamides   495, 778
Dimethyl sulphoxide (impregnation
   agent)   51
Dinitrobenzamides   495, 497
Dinitrobenzoates   228, 299, 540,
   660—662
—, visualisation   229
2,4-Dinitrophenylamines   51, 496
Dinitrophenylamino acids   756—772
Dinitrophenylation of amino acids   757
— of peptides   758
— of polypeptides   759
Dinitrophenylhydrazones   50, 51, 145,
   151, 154, 218—225
2,4-Dinitrophenylhydrazides   145
— -Dinitrophenylosazones   50, 827,
   828
Dinitropyridyl amino acids   777
Diolein   375
Diol lipids   363, 379, 381, 385, 414
Diols   650, 662—663
Diosgenin   347, 458
Diphenylpicrylhydrazyl   208, 497
Dipicrylamine   670, 671
Direct dyes   621, 622
Disc chromatography   73
D-isovitamins   278
Disperse dyes   616—617
Disulphides   683
Diterpene alkaloids   460
Diterpenes   240
Dithizonates   844
Diuretics   547, 548
DNP-amino acids   111, 583, 584,
   756—772
— — —, documentation   771
— — —, hR_f-values   765
— — —, solvents and separation ef-
   ficiency   760—771
Documentation (TLC)   125—133
— of DNP-amino acids   771
— with light-sensitive paper   129—133
Dowex (ion exchanges)   45, 46
D-previtamins   276—279
Drug combination preparations
   554—558
— extracts   716, 717
— preparations, analysis   554—558
Drugs, characterisation   720—724
—, stability tests   558—562
Dry-column chromatography   98
Drying (of plates)   60, 61, 85

Dulcin   648
D-vitamin group   275—283
Dye intermediates   626—628
Dyes   111, 612—630
—, acid   620, 621
—, anthraquinone   612, 616, 617, 620,
   622
—, azo   145, 151, 612—618, 620—622
—, basic   618—620
—, Celliton   616
—, direct   621, 622
—, disperse   616, 617
—, fluorescent   619
— for microscopy   619
—, indicator   621
— in motor fuels   615
— in polystyrene   616
—, metal complex   623
—, Oracet   613
—, Palatine Fast   620, 621
—, reactive   622, 623
—, solvent   612—616
—, Sudan   65, 613, 614
—, triarylmethane   617—620
—, Waxoline   613

Eburnamine alkaloids   447
Ecdysone   177
Echinenone   267—270
ECTEOLA-cellulose   36, 38, 39, 47
Edge effects   66
Electrophoresis, thin-layer   105—114
Electrophotography (for copies of TLCs)
   133
Electrovisualisation procedure   78
Elemol   230
Elimicin   231, 232, 234
Ellipticine alkaloids   447
Eluents (for spots)   149—151
Eluotropic series   202
Elution technique   101, 102, 148
Elymoclavine   451, 452
Elypticin alkaloids   460
Emetine alkaloids   444
Emodi-podophyllin   88, 146
Emodin   81, 708
Emulsifiers   649
Endogeneous body substances   579—604
Endrin   642—644
Enteromycin   575
Eosin B   144
Ephedrine   499, 520—524
Epiquinine   454—456

Epoxidation 205, 326
Epoxides 212—214, 659
Epoxy-acids 386, 387, 399
— —, triglycerides 380
Epoxydihydrocaryophyllene 96, 213, 234
Equilibration (of the layer) 66
— reversal 71
Equipment, basic 4, 83—85
— for descending development 72
Equol 351
Ergometrine 66, 451, 452
Ergometrinine 66, 451, 452
Ergosterol 275, 277, 278
Ergot alkaloids 66, 450—454, 562
— alkaloids, visualisation 424
— pigments 718
Ergotamine 66, 451, 452
Eschscholtzxanthin 270, 271
Essential oils, separation by TLC 236—239
Esterases 111
Esters 138, 200
—, cholesteryl 50
—, xanthophyll 50
Estolide-triglycerides 380, 381
Ethanolamine-cephalin 390, 391
Ethenzamide 524—527
Etherification 325
Ethinamate 538, 538
Ethylvanillin 217
Eucalyptol 212, 213
Eudesmols 226
Eugenol methyl ether 230—234
Evaluation, automatic 140
—, quantitative 135—155
E-vitamin group 283—288
Exogenous (body) substances 604—607
Explosives 669—673

Farnesols 226, 227, 229
Fast bases 626, 627
— salts 626, 627
Fat-soluble vitamins 263—291, 558
— — —, detection 265
— — —, determination 265
— — —, separation 263—265
Fatty acid derivatives, technically important 387
— acids 50, 139, 168, 174, 386—388, 406, 412
— —, labelled 176
— —, long-chain 177

Fatty acid, methyl esters 177, 178, **412**
— —, unsaturated 177
— alcohols 173
Fenchone 219, 220, 224
Ferrocenes 665
—, visualisation 666
Ferulic acid 81, 689, 698, 700
Filicin, crude 712—714
Filix-phloroglucides 712
— -phloroglucinol butanones 712
Films on TLC 5, 6
Filters for UV lamps 78
Finger print technique 114, 250, 649, 706
Fixed spreader (Kirchner-type) 54
Flavaspidic acid 713, 723
Flavone aglycones 689, 692—694
— glycosides 690, 692, 693, 699, 701—705
Flavonoid acetates 696
— aglycones 697, 701
— methyl ethers 696
Flavonoids 693
Flavonol glycosides 700, 702
Florisil 29
Fluoranthene 668, 669
Fluorescein 619, 620
Fluorescein, tetrachloro- 146
Fluorescence, quantitative determination 141
—, tritium-induced 159
Fluorescent dyes 619
— indicators 27, 77, 146
— layers 146
— spots, evaluation 141, 142
Fluorocurine alkaloids 447
Fluorograms 159, 160
Fluotest 79
Folic acid group 293, 294, 303—304
Food colorants, synthetic 624—626
Foodstuffs and their additives 630—656
Formamide (impregnation agent) 51
Frangula species 707—709
Fructose 579—581, 812—815
—, determination in human ejaculate 579
—, visualisation 810
Fucoxanthin 270, 271
Fume cupboards for spraying 79—81
Function tests 605
Fungicides 638
Funtumia alkaloids 459

Furanoquinoline alkaloids 456
Furfuraldehyde 218, 220, 225
Furfuryl alcohol 227, 229
Furocoumarins 695

Galactose 580, 581, 813—815
Gallates 633—635
Gangliosides 392—394, 415, 598
Gas chromatography 114—125, 204, 205
GC-TLC coupling 114—125, 204, 205
Gel filtration 40, 41, 737
Gentisic acid 689, 698
Geraniol 96, 227—229, 234
— epoxide 212
Gestagens 334—341
Gibberellins 178, 475, 479—481
—, biological detection 488, 489
—, visualisation 486, 487
Glass powder for TLC 30, 246
Glucobrassicin 478, 479
Glucose 146, 179, 580, 581, 810—816
—, visualisation 810
Glucuronic acid 604, 704, 820
Glucuronides 356, 604, 690, 704, 820
Glucuronosides 46, 47
Gluthethimide 538—540
Glycerides 144, 363—367, 375, 376, 380—386, 399—401, 408—415
— in plasma 593
Glycerol 619, 620, 820
— alkyl ethers 374, 383, 385, 399, 400
— ethers 177
— —, labelled 176
Glycine 583, 585, 730—786
Glycolipids 148, 381, 388—394, 415, 597
Glycols 650, 662, 663
Glycoproteins 580
Glycosides 87
—, cardiac 177, 341—346
—, Scilla 344
Glycyrrhetic acid see Glycyrrhetinic acid
Glycyrrhetinic acid 245, 723, 724
Glycyrrhizic acid 724, 734
Glyoxal 220, 222
Gradient-elution 90
— layers 91, 422
— techniques 89—94
Grain size 6
Graphical copying 128
Griseofulvin 141

Groups, functional, influence 201—203, 231, 326—328, 693, 694
GS-moisture chamber 71
Guaiazulene 65, 212
Guaiol 212, 227—229
Guanine 457, 549, 791, 794
Guanosine 791, 794
— phosphates 791, 798, 799

Haemolysing substances 82
Halide anions 847, 848
Halogens 46
Handling (of plates) 60
Hashish 715, 716
HCH 643, 644, 676
Heating after TLC 81
Heavy metals 46, 173, 837—847
Heimia alkaloids 460
Hemlock alkaloids 431
Heptachlor 643, 644
— epoxide 643, 644
Herbicides 647, 648
— (chlorine-containing) 32
Hetarines 501
Heterocyclic compounds (antibiotics) 575
Heteropoly acids as adsorbent 31
Hexachlorocyclohexane 643, 644, 676
Hexadienolides 146
Hexapropymate 521, 539
Hexobarbital 535—537
Hexogen 670, 671
Histamine 494, 499
HMX 670, 671
Holarrhena alkaloids 349, 459
Homovanillic acid (HVA) 146, 589—591
Hop bitter principles 247, 714
Horizontal development (devices for) 73, 75
hRf-values, definition 127
hR_{St} values, definition 127
Humulene 210
Humulones 714
Hydantoins 537, 538
Hydrargillite 24—26
Hydrazones, sugar 825—828
Hydrocarbons 60, 396, 400, 406
—, chlorinated 642—645
—, monoterpene 210—212
—, polynuclear 145, 666—669
—, sesquiterpene 210—212
Hydrochlorothiazides 547, 548

Hydrocortisone 601
Hydrogenation 205, 326, 652, 653
Hydroquinone 232, 541, 689, 704
— glucosides 719
Hydroquinones 290, 291, 718
Hydroxocobalamin 301—303
Hydroxy-acids 50, 386, 387, 399, 400, 412
p-Hydroxybenzoates 152, 636, 637
Hydroxybenzoic acids 689, 692—694, 696, 698, 699, 705
2-Hydroxybenzophenones 664
Hydroxycorticosteroids 148, 341, 600, 601
Hydroxycortisol 601
Hydroxycoumarins 689, 694, 695
Hydroxyethyltheophylline 549, 550
Hydroxy-fatty acids 50, 386
5-Hydroxyindoleacetic acid (5-HIA) 589—591
Hydroxyindoles 481
Hydroxylapatite 30
Hydroxyoestrenes 146, 151, 558—561
Hydroxypropyltheobromine 549, 550
Hydroxyquinones 694, 697
Hydroxyskatoles 481
Hydroxysteroids 338, 341, 600, 601
—, detection 324
Hydroxysterols 331, 334
Hyocholic acid 353
Hypnotics 533—540
Hypoxanthine 457, 549

Immersion procedures (layer preparation) 53
Impregnation 48—52
— agents 50, 51
Indicator analysis 171
— dyes 621
Indole-3-acetic acid 472, 478, 479
— alkaloids 446
— —, visualisation 424
— derivatives, abbreviations Table 89, p. 472
— —, identification 483—489
— — in urine 474, 476—479
— —, multiple development 481—483
— —, "simple" 471—493
— —, stepwise development 481—483
— —, two-dimensional development 481—483
— —, visualisation 483—489
Indolyl alkylamines 446

Inhibitors 673, 674
Inorganic ions, quantitative determination 851—853
— —, separation scheme 838
— —, thin-layer electrophoresis 112
— —, TLC 837—853
Inositol 179, 307
Insecticides 82, 146, 638—647
Insulin 177
Intermediates in organic synthesis 677—684
Iodates 30, 111, 847
Iodine-containing compounds 779—780
— — —, detection 780
—, location with 147, 322, 882
Iodoamino acids 586, 779, 780
Ion exchangers in TLC 44—48
Ionones 217, 220
Ipecacuanha alkaloids 444, 445
Ipomeamarone 213
Ipomeanine 213
Isoasarone 224, 231, 232
Isocaryophyllane 210
Isoeugenol methyl ether 231—233
Isoflavones 179, 688
Isoniazide 521
Isopavine alkaloids 443
Isophytol 241
Isoquercitrin 690, 701
Isoquinoline 503
Isohamnetin 689, 697, 701
Isothiocyanates 235, 240
Isotope dilution method 171
— techniques 155—179
— derivative method 172
Iterating development 76

Junenol 225, 227

Kaempferol 689, 701, 705
Kahweol 240, 241
Karlsruhe apparatus 207
Keto acids 37, 386, 412
— —, dinitrophenylhydrazones 652, 653
— -fatty acids 50
Ketone bodies in urine 581
Ketones 51, 200, 217—225, 412
17-Ketosteroid conjugates 357
— Ketosteroids 335, 336, 338, 339, 600, 602
Ketosteroids, detection 323, 324
Kieselguhr 27—29
Kirchner-type spreader 54

Kit, basic 4, 83—85
Kurchi alkaloids 349
K-vitamin group 264, 288—292

Labelled substances, preparation 167, 175
Laboratory equipment for TLC 83—85
Lactams 215, 683
Lactones 50, 214—217, 412, 683
Lactose 580, 812—815
—, visualisation 810
Lanatosides 146
Lanosterol 333, 334
Layer preparation, immersion procedures 53
— —, pouring procedures 53
— —, spraying procedures 57
— —, spreading procedures 54
Laxatives 551, 560
Lead (II) 94, 839, 840, 844, 845
Lecithins 390, 391, 398, 401, 406, 409, 415
Legally binding drug characterisation 720—724
Lichen constituents 694
Lignan glycosides 710, 711
Lignans 87, 693, 694, 709—712
—, visualisation 710
Lignin 178
Limonene epoxide 212, 213
— hydroperoxide 213
— peroxides 214
Linalool 227, 229
— epoxide 212
Lindane 643, 644
Linderene derivatives 213
Lindestrene 213
Lipid analysis, older methods 366
— separation, newer procedures 367, 368
Lipids 147, 363—421, 592—599
—, aldehydes from 373
—, alkaline hydrolysis 371
—, animal 369, 370
—, brain 598
—, degradation 370
—, detection 323
—, extraction 368—370
— in blood 592
— in faeces 597
— in secreted and excreted material 597
— in serum 401, 592

Lipids in tissue 598, 599
—, methanolysis 372
—, neutral 375—388, 414, 593
—, neutral, detection 378, 379
—, — in human tissue 385
—, — in microorganisms 379
—, polar 388—394
—, polar animal 370, 391
—, polar in human tissue 392—394
—, —, in microorganisms 390
—, polar plant 369, 390
—, quantitative evaluation 139, 144, 414, 415
—, reduction 373
—, TLC of 374—415
—, unsaturated 384—414
Lipoic acid 307
Lipoquinones 265, 291
Liquid paraffin, impregnation with 411
Lithocholic acid 351, 353
Liquorice root 723
Lobelia alkaloids 432
Local anaesthetics 551—554
Lunipyrethrins 645, 646
Lupulones 714
Lutein 268—271
— epoxide 269, 271
— esters 270, 271
Lycopene 268, 270
Lycopersicum alkaloids 458
Lysergic acid 452, 454
Lysolecithin 390, 391

Macrocyclic peptides (antibiotics) 576
Macrolides 178
— (antibiotics) 572
Malachite green 618
Male fern rhizome 722
Maltol 229
Maltose 146, 810—816
—, visualisation 810
Mannitol 820
— as layer 40
Marihuana 715, 716
Mavacurine alkaloids 447
Menadione 288, 289, 688
Menthofuran 178, 225, 226
Menthols 225, 226
Menthone 178, 219, 220, 224, 225
Meprobamate 521, 539
6-Mercaptopurine 549
Mercuric acetate adducts 211, 212, 395, 402—406, 657

Mercuric acetate adducts, visualisation 404
Mercury(I) 94
— (II) 173, 839, 840, 844—846
Metal complex dyes 623
Methyl chaulmoograte 397, 403
— elaidate 177
— esters 372, 373, 383, 384, 398, 399
— glycosides 820
— linoleate 403, 408, 410
— oleate 375, 385, 397, 403, 408, 410
— palmitate 177
— stearate 403, 410
Methylene blue 618
Methanolysis of lipids 372
Methaqualone 539, 540
Methoxyphenylpropane derivatives 229—235
Methylchavicol 230, 232, 233
Methylcyclohexanones 221
Methylglyoxal 222
Methylhistamine 591
Methylionones 217, 220
Methylnaphthalenes 668
Methylphenobarbital 152, 535—537
Methylprylone 538—540
Mineral oils 676, 677
— — (impregnation agents) 50, 219
Mixed layers 32
Mixtures, standard 7
—, test 7
Moisture chambers 71, 72
Molybdate, impregnation with 31, 49, 523, 662
Monoglycerides 173, 385, 400, 408, 414, 593
Monohydroxy acids 174
Monoolein 375, 386
Monosaccharides 154
Monoterpene alcohols 225—229
— hydrocarbons 210—212
Morphine 152, 154, 437—441, 529, 530
Motor fuels, dyes in 615
Movable spreader (Stahl-type) 55
Mucic acid 152
— — derivatives 151
Multidimensional technique 88
Multiple development 86, 100
Muscone 216
Myoporone 213
Myrcene oxide 213
Myricetin 689, 701
Myristicin 224, 230—232

Naphthalene 668
Naphthols 111
Narceine alkaloids 437—441
Narcotine 437—442
NDGA 632—636
Neatan, new 128
Necine alkaloids 435
Neomycin (sulphate) 32, 153, 154
Neoxanthin 270, 271
Nerol 227—229
— epoxide 212
Nerolidol 227—229
Neutral fat in plasma 593
— lipids 375—388
— — in human tissue 385
— — in microorganisms 379
— plasmalogens 381—386
Nicotinamide 293, 294, 299, 300
Nicotine 155, 430—432
— acid 151, 293, 294, 299, 300, 562
Niemann-Pick disease 393, 598
Nikethamide 519, 520
Ninhydrin reaction 747, 889
— —, polychromatic 747, 889
Nitrate esters 671—673
Nitriles 387, 400
Nitrosamines 670
Nitroanilines 501, 626
Nitroazobenzamides 495
Nitrodiphenylamines 497, 670
Nitrofurans (chemotherapeutic agents) 547
Nitromethane (impregnation agent) 51
Nitrophenols 680
Nitrosamines 497, 680
Nitrosodiphenylamines 670
Nitrosomorphine 155
Nitrotoluenes 670—672
Noradrenaline 499, 520—524, 589
Nordihydroguaiaretic acid 632—636
Normal saturation (NS) 66
Nornicotine 155, 430—432
Noscapine 529, 531
Novalgin 524, 525
Novolaks 658
Nucleic acid derivatives 33, 36
— — — (antibiotics) 575
— — —, UV-spectra 791
— acids 786—804
— —, electrophoresis 802—804
— —, hydrolysis 789—791
— —, isolation 788
Nucleobases 46, 792, 794, 802

Nucleosides 46, 111, 792—794, 802, 803
Nucleotides 46, 47, 111, 794—804
Nucleotide-coenzymes 46, 794—801
Nujol 50, 411

Octogen 670, 671
Oestradiols 350, 351, 602, 603
Oestrenols 146, 151, 558—561
Oestriols 350, 351, 602, 603
Oestrogens 349—351, 600, 602, 603
Oestrone 318, 350, 602, 603
Oils, mineral 676
—, — (impregnation agents) 50
—, sulphur-containing 235
Oleanolic acid 244, 247
Oleic acid 375, 385, 386, 411
Oligonucleotides 96, 801, 802
—, thin-layer electrophoresis 804
Oligosaccharides, TLC 816, 817
Olive oil 413
Opium alkaloids 139, 146, 422, 437—444
Optical bleaches 675
— brighteners 675
Oracet dyes 613
Organic phosphorus compounds 148, 639—642, 794—804, 821, 822
— pigments 617
Organo-metallic compounds 664—666
— -tin compounds 664, 665
Orotic acid 307
Osazones 221
—, 2,4-dinitrophenyl 50, 827, 828
—, sugar 825—828
Oxidation 205, 326
Oxides, terpene 212—214
Oxidising, solvents 411
Oximes 219, 225, 328, 341, 349, 680, 681
Oxindole alkaloids 454
Ozonides 395, 406—409
—, visualisation 407
Ozonolysis, reductive 408, 409, 414

Packing density 14
Pairs, critical 7, 277, 301, 326, 332, 347, 541
Palatine fast dyes 620, 621
Palmitic acid 177, 411
Palmityl alcohol 177
Palm oil 401
Panthenol 298

Pantothenic acid group 293, 294, 297—299
Papaveraceae alkaloids 436—444
Papaverine 437—442, 529, 531, 557
Papaverrubins 441, 443
—, visualisation 424
Paracetamol 524—527
Paraffin, liquid (impregnation agent) 50, 411
Paramorphan 529, 530
Pavine alkaloids 443
P-cellulose 36, 38, 46
PEI-cellulose 36, 38, 39, 46
Peltatins 87, 710, 711
Penicillin preparations 555
Penicillins 146, 151, 575
Pentaerythritol nitrates 154, 673
Pentetrazole 519, 520
Pentobarbital 534, 536
Peptides 111, 751—756
—, antibiotic 576
—, detection 755, 756
—, hydrolysis 735, 736
Periodates 30, 111
Peroxides 678
— (polymerisation initiators) 664
—, terpene 212—214
Peru balsam 248, 250
Pesticides 638—648
Pethidine 529, 531
Pharmaceutical combination preparations 554—558
— products, synthetic 506—566
—, stability tests 558—562
Pharmacopoeias, TLC in 6, 720
Phase reversal 200
Phenacetin 151, 524—527, 556, 558
Phenanthrene 668, 669
Phenazone 524—527
Phenobarbital 152, 534, 536, 557
Phenol aldehydes 146, 540
— derivatives in plants 687—706
— — —, concentration from plant material 690, 691
— — —, medium polar 689
— — —, slightly polar 688
— — —, strongly polar 690
— — —, TLC on cellulose 692, 693
— — —, — on ion exchangers 704
— — —, — on polyacrylonitrile 704
— — —, — on polyamide and other polymers 699—704
— — —, — on silica gel 693—699

Phenol derivatives in plants, visualisation 705
Phenols 32, 48, 111, 146, 154, 663, 664
—, azo coupling products 663
—, detection 541
—, mononuclear 540
— of pharmaceutical interest 540, 541
—, polynuclear 540, 541
Phenolcarboxylic acids 111, 540, 689, 692—694, 696, 698, 699
— —, visualisation 705
Phenolic steroids 349—351
Phenothiazines 508—514
2-Phenoxyethanol (impregnation agent) 51
Phenylalanine 582, 607, 744, 745
Phenylalkylamines 460, 499
Phenylalkylamines, visualisation 424
Phenylbutazone 524—527
Phenylenediamines 501, 626
Phenylhydrazones, 2,4-dinitro 50, 51, 145, 151, 154, 218—225
Phenylosazones 580, 825—828
Phenylpropane derivatives 207
Phenylthiohydantoins 772—777
—, detection 777
—, solvents and separation efficiency 774—777
Phloroglucinol derivatives 712—715
— —, visualisation 713
— drugs 712—714
Phloroglucinols 541, 689, 722
Phosphate esters 639—642, 659, 660, 678, 794—804
— —, detection 642
Phosphates 32, 35, 847, 848
—, condensed 848—850
— for TLC (adsorbents) 30
—, inorganic 847, 848
—, sugar 821
Phosphatide plasmalogens 383, 388
Phospholipids 176—178, 388—394, 415
— in plasma 596
— in secreted and excreted material 597
— in serum 596
—, labelled 176
Phosphoric acids 35
Phosphorimetry 141
Phosphorus compounds, organic 148, 639—642, 794—804, 821, 822
Photography, chemicals for 676—677
— (for copies of TLCs) 131

Phthalate esters 659, 660
Phthalide-isoquinoline alkaloids 442, 443
Phthalides 207, 215, 217
Phytol 227, 229, 241
Picric acid 670, 671
Picrosalvin 240
Pigments, organic 617
α-Pinene epoxide 213
Piperidine alkaloids 430—432
Piperitone 219, 224, 225
— oxide 212
Plant growth regulators 471—493
— lipids, polar 369, 390
— waxes 240
Plasmalogens 373, 388, 392
—, neutral 381—386
Plasma, neutral fat in 593
Plasticisers 659, 660
—, visualisation 660
Plastics 657—664
Plastoquinones 288, 291
Plates, ready-coated 5
Podophyllum lignans 710
Polar lipids 388—394
— — in human tissue 392
— — in microorganisms 390
Polyacetylenes (antibiotics) 571
— in essential oils 235, 236, 240
Polyalcohols 650, 662, 663
—, detection 662, 663
—, nitrate esters 671—673
Polyamide layers 259, 699
Polyamides 30, 41—44
Polyamines 502
Polycyclic compounds 37, 133, 141, 145, 668, 669
Polyenes (antibiotics) 571
Polyethylene glycols (impregnation agents) 51
Polyglycerols 649
Polymers 657, 658
Polynuclear aromatic compounds 667—669
Polyolefineic acids 147, 398
Poly-P-cellulose 36, 38, 39
Polyphenols 689, 692, 697
Polyphenyl ethers 678
Polyphenyls 32, 37, 666
Polystyrene, dyes in 616
Polyterpenes 247
Polythionates 850
Polyurethanes 658

Polyzonal development 76
Pore diameter (or radius) 7, 14, 21
— radius, average 17—19, 26, 29
— — distribution 14, 21, 22
— —, effective 17
— volume 7, 14, 16
— —, specific 15, 16, 19, 21, 22
Porphyrins and metabolites 604
Positive copies (of TLCs) 130
Pouring procedures (layer preparation) 53
PP-Cellulose 46
Prednisolone 601
Pregnanediols 154, 340, 600, 602
Pregnanetriols 340, 602
—, detection 323
Preparation box 68
Preparative TLC 97—102
Preservation of thin-layer chromatograms 127, 128
Preservatives 636—638
Previtamin D 276—279
Proaporphine alkaloids 442
Procaine 551, 552
Progesterone 151, 318, 335, 337, 340, 555, 557, 600
Prostaglandins 387, 399, 597
Proteins, hydrolysis 735, 736
Protoberberine alkaloids 442
Protocatechuic acid 689, 698
Protopine alkaloids 443
Protoporphyrins 604
PTH-amino acids 772—777
— — —, detection 777
— — —, solvents and separation efficiency 774—777
Pulchellins 216
Pulegone 178, 220
Purine alkaloids 457
— derivatives 548—550
Purines 46
—, detection 549
— from nucleic acids 792—794
—, visualisation 424
Pyrazolines 684
Pyrazolones 524—527, 560
Pyrene 668, 669
Pyrethrin peroxides 645, 647
— synergists 645, 711
Pyrethrins 645—647
—, detection 646
Pyridine alkaloids 430—432
— derivatives 504

Pyridine nucleotides 796
Pyridines 684
Pyridoxal and derivatives 300, 301
Pyridoxine (salts) 293, 300, 301
Pyrimidines 46
— from nucleic acids 792—794
Pyrogallol 541, 689
Pyrones 686—706
Pyrrolidine alkaloids 430—432
Pyrrolizidine alkaloids 435

QA-cellulose 47
Quantitative evaluation after extraction from the adsorbent layer 145—155
— — by fluorescence 141
— — by reflectance measurements 142
— — of fluorescing spots 141, 142
— — of light-absorbing spots 138-141
— — of TLCs 133—155
— — —, by spot area measurement 135
— — —, visual comparison 135
— — on the thin layers 135—144
Quaternary ammonium bases 387
— — salts 498
Quercetin 689, 701, 705
Quinazoline alkaloids 460
Quinidine 454—456
Quinine 454—456
Quinolines 684
Quinones 288—292, 688, 693, 697, 703, 719
Quinolizidine alkaloids 435, 436
Quinoxaline derivarives 155

Radial chromatography 73
Radioactive compounds, commercially available 167, 179
— —, preparation 167, 175, 176
Radioisotopes 156
— in analysis 170—175
Radionuclides 847
Radix Liquiritiae 723
Raffinose 146, 813
Rauwolfia alkaloids 448, 449
— —, visualisation 424
RDX 670, 671
Reactions at start point 206
Reactive dyes 622, 623
Ready-coated plates and sheets 59
Rectangular chambers 68
Reduction at the start 205, 326

Reductive ozonolysis 408, 409, 414
Reference chromatograms 145
Reflectance spectra 142
Remission spectra 142
Resin acids 249
Resins 70, 138, 247—250
—, detection 250
Resorcinol 232, 541, 704
Retinal (vitamin A_1-aldehyde) 267, 273, 274
Reversed phase partition chromatography 409—414
R_f-R_m-conversion table 180, 910
— -Values, definition 127
Rhamnetin 697, 701
Rhamnose 581, 812—815
—, visualisation 810
Rhamnus species 706—709
Rheum species 707—709
Rhizoma Filicis 722
Riboflavin 293, 296, 297
Ribonucleic acids 786—792, 801
— —, cleavage 790
— —, distinction from deoxyribonucleic acids 789
— —, transfer 801
Ribonucleoside triphosphate 179
Ribose 580, 581, 812
—, visualisation 810
Ring chromatograms 2
Rose oxides 213
Rotenones 645, 719
Rotundin alkaloids 460
Rutin 293, 294, 307, 690, 701

Sabouraud medium 568—570
Saccharin 648
Saccharinic acids 819
Safrole 230—232
Salicylamide 524—527, 556, 558
Salicylic acid 81, 524, 526, 556, 704
Sandwich chambers 68—70, 72
Santonin 215
Sapogenins 243, 347, 458
—, detection 323
Saponins 241, 245—247, 346
—, detection 246, 323, 347
Saturated fatty acids, esters 403, 405, 406, 410
— lecithins 409
Saturation of the chamber (CS) 66, 67
Scanners 160—162
S-chambers 68—70, 72

Scilla glycosides 344
Scintillation counters 160—166
— solutions 162, 163
Scrapers, zonal 166
Scraping off spots (zones) 148, 165
Secale alkaloids 66, 451
SE-cellulose 46
Secobarbital 534, 536
Senna species 706—709
Separation chambers 65
Sephadex 40, 41
Sequence analysis of proteins 751—756
—, chromatographic, steroids 326, 353, 354
SE-Sephadex (ion exchangers) 44, 47
Serum lipids 401, 592
— proteins 111, 591
Sesame oil 351, 555, 557
Sesamin 647, 711
Sesamolin 711
Sesquiterpene alcohols 225—229
— epoxides 212—214
— hydrocarbons 210—212
— lactones 216
— oxides 212—214
— peroxides 212—214
Siam benzoin 248, 250
Silene EF 29
Silica gel 7—23
— —, cavity system 13—22
— —, formation 8
— — for TLC 22
— —, framework 8—13
— —, nomenclature 11
— —, reproducibility of preparations 21
Silicates for TLC 29, 30
Silicones, impregnation with 50, 411
Silver (I) 94, 844, 845
— nitrate impregnation 52, 177, 211, 248, 259, 269, 277, 289, 291, 347, 349, 394, 396—402
Sintering 18, 19, 22, 25
β-Sitosterol 330—334
Solanum alkaloids 348, 458
— glycosides 348
Solvent dyes 612—616
Solvents for extraction 149—151
— for TLC 202
Sorbic acid 637, 638
— —, bromination in TLC 637
Sparteine alkaloids 436
Sphingolipids 393, 415

Sphingolipids, brain 598
Sphingomyelins 390—393, 414, 596
— in plasma 596
— in serum 596
Spot application 63
Spots, scraping off 148, 165
Spotting methods 63—65
Spray-gun 58, 80
— reagents names and abbreviations 909
— —, preparation and application 855—905
Sprayers 79, 80
—, Aerosol 58
Spraying procedures (layer preparation) 57
Spreaders, automatic 85
—, fixed 54
—, movable 55
Spreading procedures (layer preparation) 54
— rods 56
— troughs 56
Squalane (impregnation agent) 50
Squalene 241, 597
SRS-technique 88, 206
Stability tests on pharmaceutical preparations 558—562
Stahl-type spreader 55
Standard conditions (in TLC) 85
— mixtures 7
Starch (as layer) 33, 39
Stearoptenes 240
Stepwise development 87
Steroid alkaloids 348, 460
— carboxylic acids 351—355
— conjugates 351—355
— glucuronides 355—357
— glucuronides, detection 324
— oximes 349
— phosphates 356
— sulphates 46, 47, 356
— sulphates, detection 324
Steroids 32, 142, 146, 147, 170, 177, 178, 311—362, 599—604
—, adsorbents for TLC 316, 317
—, chromatographic behaviour 326—329
—, — methods 313—316
—, derivative formation 325, 326
—, detection 322—325
—, labelled 176
—, nomenclature 312

Steroids, partition TLC 328
—, reducing, detection 324
—, solvents for TLC 317—320
—, structure and chromatographic behaviour 326—329
—, TLC-conditions 316—326
—, — -procedures 320—322
—, urinary 335, 338—341, 351, 599—503
C_{18}-Steroids 602
C_{18}-C_{22}-Steroids, neutral 334—341
C_{19}-Steroids 602
C_{21}-Steroids 600
Sterol alkaloids 457—461
Sterols 177, 178, 329—334
— in faeces 603
Steryl esters 412
Stigmasterol 329—331, 333, 334
Stilbenes 693
Storage (of plates) 60, 61, 85
Streak application 64
Streptomycin 574
Strophanthus glycosides 342, 345
Strychnine 448
— alkaloids 447, 448
Sublimation 81
Sucrose 40, 146, 580, 810—817
— as layer 33, 40
— esters 146
—, visualisation 810
Sudan dyes 65, 613, 614
Sugar acetates 51, 822—825
— alcohols 820
— benzoates 822—825
— derivatives, miscellaneous 830, 831
—, detection in urine and faeces 579
— hydrazones 825—828
— methyl ethers 828—830
— — —, quantitative analysis 829
— osazones 825—828
— phosphates 47, 821
Sugars and derivatives 32, 33, 807—837
— in urine 579, 580, 826
—, their derivatives and metabolites 579—581
—, thin layers for 807—810
—, visualisation 810, 811
Sulphate esters 679
Sulphates 32
—, inorganic 850
Sulpholipids 363, 388—394, 415
Sulphonamides 32, 154, 541—547

Sulphonamides, detection 543
Sulphur compounds, heterocyclic 681—684
— -containing anions 850
— — essential oils 235
Sumatra benzoin 248, 250
Sunflower oil 413
Suppositories, analysis 557
Surface area, specific 7, 14—16, 19, 21, 22, 35
— chromatography 3
Sweetening materials 37, 648
Swelling agents 649
Sympathomimetics, adrenaline-type 520—524
Synergists 645—647
Syringic acid 689, 698

Tar bases 503
— dyes 111
Tay-Sachs disease 393, 598
Temperature-TLC 94
Terminology in TLC 912
Terpene alcohols 216, 225—229
— aldehydes 217—225
— —, detection 224
— derivatives 206—258
— epoxides 212—214
— esters 214—217
— —, detection 216
— hydrocarbons 96, 210—212, 241
— —, visualisation 211
— ketones 217
— —, detection 224
— lactones 214—217
— —, detection 216
— oxides 212—214
— peroxides 212—214
Terpineol 225, 227, 229
Test mixtures 7, 86
— organisms for antibiotic detection 569
Testosterone 318, 339, 602
Tetrachlorofluorescein 46
Tetracyclines 573
Tetradecane (impregnating agent) 50, 411
Tetrahydrocorticosterone 340
Tetrahydroisoquinoline alkaloids 442, 444, 460
Tetrahydro-S 340
Tetramethoxyallylbenzene 230—232
Thalidomide 539, 540

Thebaine 437—441, 529, 530
Theobromine 82, 457, 548—550
Theophylline 82, 457, 548—550, 557
Theoretical fundamentals of TLC 179
Thiamine and derivatives 293, 295, 296
Thin-layer electrophoresis 105—114
— — —, apparatus 109, 110
— — —, experimental conditions 111—114
— — —, two-dimensional 113
Thiobarbituric acids, detection 537
Thiochrome 296
Thiolactones 215, 683
Thiophene derivatives 236, 681—682
Thiophosphate esters 639—642
Thujyl alcohols 227, 228
Thymol 230, 232
Tobacco alkaloids 430, 431
Tocopherols 264, 283—288, 290
—, detection 287
—, quantitative determination 287
Tocopheronolactone 155
α-Tocophenyl-acetate 264, 285, 287
Tocopheryl esters 264, 285
Tocopherylquinones 285, 286, 288—291
Tocotrienols 283—288
Tolbutamides 550, 551
Toluidines 501, 626
Tomatidine 348, 349, 458
Toxicological analysis 507, 605
Toxic metals 846
Transfer ribonucleic acids 801
— techniques 102—105, 205
Triarylmethane dyes 617—620
Triglycerides 50, 177, 178, 363, 367, 375, 380—384, 398—401, 408, 411—414, 597
— in urine 597
Trilinolein 410
Trimethoxyallylbenzene 231, 232
Trimyristin 410
Triolein 169, 170, 375, 385, 410, 412
Tripalmitin 169, 410, 412, 415
Triterpene acids 243—245
— glycosides 245—247
Triterpenes 241—243
—, detection 247
Tropane alkaloids 432—435
Tropolone alkaloids 425—430
— —, visualisation 426
Tropones 678
Tryptophan and metabolites 583—585

Tungstate, impregnation with 31, 49, 523, 662
Two-dimensional separation 88
Tylocrebin alkaloids 460
Tyrosine 586, 744, 779
—, detection 751

Ubiquinones 50, 146, 151, 288, 291
Umbelliferae drugs 695
Umbelliferone 81, 688
Unsaturated fatty acids, esters 403, 405, 406, 408, 410
— lecithins 409, 414
Undecane (impregnating agent) 50, 411
Uracil 791, 794
Urethanes 658
Uric acid 457, 549
Uridine 791, 794
— phosphates 791, 798, 799
Urinary amino acids 582, 763
— steroids 335, 338—341, 351, 599—603
— sugars 579, 580, 826
Urine metabolites 485, 486, 578—612
— metabolites, visualisation 485, 486
Uronic acids 819, 820
Ursolic acid 244, 245, 247
Uvanalys 79
UVIS 79
UV-lamps 77—79

Vacuum extractors 101—103, 148, 149
Vanillic acid (VA) 589—591, 689, 698
Vanillin 217, 218
— mandelic acid (VMA) 589—591
Vegetable oils, fats (impregnation agents) 50
Veratraldehyde 218
Vinca alkaloids 449, 450
Vinyl compounds 657
Violaxanthin 268—271
Visualisation 77—83
— in preparative TLC 100
Vitamin A-acetate 261, 264, 274
— -acid 273—275
— A_2-acid 274
— A-alcohol 264, 265, 273, 274
— A-aldehyde 267, 273—275
— A_1-aldehyde (Retinal) 267, 273, 274
— A-esters 261, 264, 274, 400
— A_2-esters 274
— A group 273—275, 400
— B_6 factors 146, 155

Vitamin B group 46, 292—304
— B_1-Group 293, 295, 296
— B_2 group 293, 296, 297
— B_6 group 293, 300, 301
— B_{12} group 293, 301—303
— Bp 307
— B_t 307
— C 154, 292—294, 304—306
— D 135, 265
— D_2 264, 265
— D_3 177, 264
— D-esters 277
— D group 275—283
— —, quantitative determination 280—283
— E-group 283—288
— H 306
— K group 264, 288—292
— P 307
— PP 299
Vitamins 259—311
—, fat-soluble 263—266, 558
—, water-soluble 292—307, 558

Water-soluble vitamins 292—307, 558
— — —, detection 294
— — —, separation 292—294
Wax esters 381, 398, 399, 412
Waxes 240, 377, 381
Waxoline dyes 613
Wedged-tip technique 89, 331
Wheat coleoptile test 30, 487
Wilzbach procedure 167, 177
Wofatit CP 300 45, 46
Wood protective agents 675, 676

Xanthine 457, 549
Xanthones 698
Xanthophyll esters 50
Xerogels 10, 11, 15, 21
Xylenols 124, 233
Xylose 581, 812—816
—, visualisation 810

Zaffaroni system 317, 320
Zeaxanthin 268—271
Zirconium compound (ion exchangers) 32
Zonal scrapers 166
Zone collection 101, 102, 104, 148—151
— extraction 101
Zones, scraping of 148, 165

Druck: Brühlsche Universitätsdruckerei Gießen

QD271
.S7613
1969

Stahl

Thin-layer chromatography